ENCYCLOPEDIA OF

ENERGY TECHNOLOGY AND THE ENVIRONMENT

VOLUME 4

WILEY ENCYCLOPEDIA SERIES
IN ENVIRONMENTAL SCIENCE

ATTILIO BISIO AND SHARON BOOTS
ENCYCLOPEDIA OF ENERGY TECHNOLOGY AND THE ENVIRONMENT

ROBERT MEYERS
ENCYCLOPEDIA OF ENVIRONMENTAL ANALYSIS AND REMEDIATION

EDITORIAL BOARD

ENCYCLOPEDIA OF ENERGY TECHNOLOGY AND THE ENVIRONMENT

VOLUME 4

Attilio Bisio
Sharon Boots
Editors

A Wiley-Interscience Publication
John Wiley & Sons, Inc.
New York / Chichester / Brisbane / Toronto / Singapore

This text is printed on acid-free paper.

Library of Congress Cataloging in Publication Data:

Encyclopedia of energy technology and the environment / Attilio Bisio, Sharon Boots, editors.
p. cm.
Includes bibliographical references (p.) and index.
ISBN 0-471-54458-2
1. Power resources—Handbooks, manuals, etc. 2. Environmental protection—Handbooks, manuals, etc. I. Bisio, Attilio. II. Boots, Sharon.
TJ163.235.E53 1995
333.79'03—dc20 94-44119

Printed in the United States of America

10 9 8 7 6 5 4 3 2 1

ENCYCLOPEDIA OF

ENERGY TECHNOLOGY AND THE ENVIRONMENT

VOLUME 4

R

RADIATION HAZARDS, HEALTH PHYSICS

Diane E. Vance
William D. Ehmann
University of Kentucky
Lexington, Kentucky

Health physics is the area of environmental/occupational health that is concerned with the effects of radiation on biological systems, the process of monitoring exposure to radiation, and the means for protecting people who must work with, or in close proximity to, radiation and radioactive materials. The topics included here give an overview of health physics concepts that all users of radioactivity, and that even the general public, should be familiar with. This includes an explanation of the ways in which radiation affects living organisms, and the units used to measure these effects. Common sources of radiation exposure are discussed, and an attempt is made to give the reader some perspective on the real risks associated with various levels of radiation exposure. Finally, a summary of the recommended limits on radiation exposure for both workers and the public is given. More technical aspects of health physics, such as detailed calculations of dose and dose rate, the procedures for obtaining licensing for a laboratory, decontamination activities, personnel or facility monitoring, shielding considerations, and the details of regulations governing radiation exposure are not covered in this article.

By its nature, health physics is an interdisciplinary field. The health physicist must be familiar with the processes of radioactive decay, and understand how radiation interacts with matter, especially in biological systems. He/she must be able to use a wide variety of nuclear instrumentation. A knowledge of radiation dosimetry, the intricacies of dose rate calculations, and the governmental regulations dealing with radiation protection is also essential.

See also Nuclear power: decomissioning power plants; Managing nuclear materials; Safety of aging power plants; Ozone, stratospheric.

RADIATION QUANTITIES AND UNITS

In health physics work, or in related areas where the effects of radiation on a sample or subject are of importance, simple knowledge of the activity level of a radioactive source does not convey sufficient information to assess health risk. For radioactive materials, the activity is defined as the number of decay events that occur in a given period of time. However, this gross activity level does not provide direct information about the amount of radiation actually received by the absorbing medium or about the potential for adverse effects in that medium. Therefore, different units that can describe these effects of radiation are needed.

Exposure

There is a distinction between radiation exposure and radiation dose. Exposure (X) is the term used to refer to the total electrical charge (the ionization) produced in a given mass (or volume) of air:

$$X = \frac{dQ}{dm}$$

where Q is the charge and m the mass.

The unit of radiation exposure is the röntgen (R). The röntgen was originally defined as that quantity of x or γ radiation that would produce 1 esu of electrical charge of either sign in 0.001293 g of air (a volume of 1.00 cm^3, at STP). This can be shown to be approximately equivalent to 1.61×10^{12} ion pairs per gram of dry air, or the release of about 84 ergs of energy per gram of dry air at STP. The equivalent radiation exposure in water or tissue is about 93 ergs per gram of water. The röntgen is now exactly defined as 2.58×10^{-4} coulombs/kg. The number of röntgens produced by a radioactive source is easily measured using air ionization chambers. However, because the röntgen is a unit of radiation exposure, not dose, it does not provide exact information about the amount of radiation that is actually absorbed by a medium other than air or about the effects of the radiation on the medium.

Absorbed Dose

The absorbed dose refers to the quantity of energy imparted by ionizing radiation to a given mass of matter. The unit that was traditionally used to measure absorbed dose is the rad. One rad is defined as that dose of any type of radiation that will deposit 0.01 J of energy in one kilogram of absorbing material:

$$1 \text{ rad} = \frac{10^{-2}\text{ J}}{\text{kg}}$$

The SI unit of absorbed dose is the gray (Gy), defined as that amount of radiation that will deposit one joule of energy in one kilogram of absorbing material:

$$1 \text{ Gy} = \frac{1\text{ J}}{\text{kg}}$$

Hence, the relationship between gray and rad is that 1 Gy equals 100 rad. Although the gray is the recommended unit, the traditional units used in health physics are still used in every day work in some nuclear facilities. It is important to remember that the gray and the rad are units of absorbed dose, in contrast to the röntgen, which is a measure of exposure or field strength of the radiation. There is no simple quantitative relationship between the röntgen and the rad or gray.

Dose Equivalent

While the rad is more useful than the röntgen in assessing the effects of radiation, it still does not give all the information needed. This is because the physiological effects resulting from radiation absorption will vary greatly depending on which type of radiation deposits the energy. The severity of the molecular damage caused by radiation is directly related to the degree of ionization produced by the radiation in the tissue. Specific energy loss, or stopping power, is one way to measure the energy lost by radiation as it passes through an absorbing material. A related quantity used in health physics is the linear energy transfer (LET), defined as the rate at which energy is transferred to a given region of matter:

$$LET = \frac{dE_{abs}}{dx}$$

The units most often used by health physicists to measure LET are keV/μm. The higher the LET of the radiation, the greater the damage it can do in biological tissue.

The specific ionization and thus the LET values of the three main types of radiation are quite different. Alpha particles show the largest specific ionization, β particles have a lower specific ionization than alphas, and gamma radiation has a still lower specific ionization compared to particles and radiations of similar energy. Thus, over a given range, α particles will produce much more biological damage than the other two types of radiation. Therefore, α particles are the most biologically hazardous if they are ingested and incorporated into internal tissue. However, because α particles are not very penetrating, it is easy to shield against them. Alpha particles do not even penetrate the skin, so they are not very hazardous with respect to external exposure. At the other extreme, γ rays do the least biological damage over a given range, but are much more penetrating than α or β particles, and thus require much more protective shielding. The other types of radiation that must be considered besides α, β, and γ radiations include energetic charged particles from cosmic rays and particle accelerators, and both thermal and fast neutrons. Positively charged particles of all masses interact with matter like α particles. They are not very penetrating, but have high LET values, so they are very damaging if acting on internal organs regulating physiological functions. Neutrons are not charged, so they have high penetrating power. They do not cause ionization directly by interaction with atomic electron clouds, but they will induce nuclear reactions or undergo elastic scattering processes which will produce charged particles. Therefore, neutrons are also very biologically hazardous.

The point of the above discussion of LET values and biological effects of radiation is that identical doses of the various types of radiation may not have identical biological effects. Therefore, simple knowledge of the absorbed dose will be of limited usefulness when trying to assess the potential for adverse biological effects due to different radiations. To compensate for this, other quantities which assign numerical values to the relative amounts of damage done by radiation have been defined. One of these is called relative biological effectiveness (RBE), and is defined as the ratio of the amount of 200-keV x rays or γ rays needed to produce a specific biological effect to the amount of any other radiation needed to produce the same effect. The RBE is a very specific quantity used in precise radiobiological studies and so has limited usefulness in assessing general biological effects. Therefore, another factor, called the quality factor (Q), has been defined. Q is related to the LET, as shown in Table 1. Note that the higher the LET, the higher is the value of Q. Table 2 gives the value of Q for several common radiations.

Table 1. Relationships between LET and Quality Factor[a]

Linear Energy Transfer in Water, keV/μm	Q
<10	1
10 to 100	0.32(LET) − 2.2[b]
>100	$300(\text{LET})^{1/2}$[b]

[a] Reference 1.
[b] Calculations using Q should be rounded to the nearest whole number.

The concepts of quality factor and dose can be used to define the rem, which is the traditional unit of dose equivalent (H). This is the unit used by health physicists in discussing radiation effects. Dose equivalent is the absorbed dose multiplied by the quality factor. For the rem:

$$H(\text{rem}) = \text{rads} \times Q$$

The newer SI unit of dose equivalent is the Sievert (Sv), defined as:

$$H(\text{Sievert}) = \text{Grays} \times Q$$

The relationship between the Sievert and the rem is that 1 Sv = 100 rem.

To illustrate the relationships just discussed, consider the value of the dose equivalent for a dose of 1.0×10^{-4} Gy of γ, α, and thermal neutron radiation. The quality factors for these radiations are 1, 20, and 5, respectively. A simple calculation shows that 1.0×10^{-4} Gy of gamma radiation would produce a dose equivalent of 1.0×10^{-4} Sv, while identical exposures to α particles and thermal neutrons would produce dose equivalents of 2.0×10^{-3} and 5.0×10^{-4} Sv, respectively. Thus, the same amount of absorbed dose does not necessarily result in the same

Table 2. Values of Q for Some Common Types of Radiation[a]

Radiation	Q
X and γ rays	1
Beta particles	1
Neutrons	
<10 keV	5
10 keV to 100 keV	10
>100 kev to 2 MeV	20
>2 MeV to 20 MeV	10
>20 MeV	5
Protons <2 MeV	2
Alpha particles, Fission fragments	20

[a] Reference 2.

amount of biological damage. In the example, the α radiation would produce more harmful effects than the γ or neutron radiation, even though the dose of all three is the same.

There are two other terms relating to dose equivalent that may be encountered in the literature. One is the effective dose equivalent. It refers to the dose equivalent weighted for the differing susceptibilities of specific types of tissue to radiation. The second term comes into play because doses are often expressed in terms of an entire population, rather than just one person. The effective dose equivalent to a whole group of people is called the collective effective dose equivalent.

Simple Calculations of Dose and Exposure

Exact calculations of exposure and dose can be quite complicated, and will not be discussed here. However, there are some simple rules of thumb that can be used to get rough estimates of exposure or dose. In some cases, these estimates give results that are quite close to those that would be obtained using more exact methods. In other cases, they will not, so the results must be used with caution.

The exposure rate from γ rays is directly related to the energy and number of γ rays emitted per decay, and inversely related to the square of the distance between the source and the object. An equation that gives the exposure rate for γ rays is

$$\textit{Exposure rate,}\ (\text{mR/h}) = \frac{6\,AEn}{d^2}$$

where A = activity (mCi), E = energy of the γ ray (MeV), n = number of γ rays of energy E emitted per decay, and d = distance to the source (feet). For γ rays, the dose rate is approximately equal to the exposure rate, because $Q = 1$ for γ rays.

Beta particles will experience significant attenuation as they traverse the air between the source and an object. For low energy β particles, like the 0.0186 MeV β from tritium, dose rates are small even a few centimeters away from the source. For higher energy β particles, an estimate of the upper limit of dose can be made using this equation

$$\textit{Dose rate}\ (\text{mrad/h}) = \frac{338{,}000A}{d^2}$$

where A = activity (mCi) and d = distance to the source (centimeters).

BIOLOGICAL EFFECTS OF RADIATION

The dangers of radiation to human health were recognized during the early years of research into the phenomenon, but little was done to recommend safe handling practices. Many researchers suffered radiation burns and eventually died from radiation-induced cancers. Eve Curie writes in her biography of her mother about a gathering of friends held to celebrate the granting of Mme. Curie's doctorate. A ZnS-coated tube containing a strong radium solution was held up for all to admire. The solution glowed beautifully in the dark, but in the light from the radium solution, the radiation burns on Pierre Curie's hands could be seen. Pierre Curie died prematurely in a street accident, but Marie Curie's death in 1934 from a bone marrow disease called aplastic anemia may well have been due to her long years of exposure to radiation.

Radiation interacts with biological material in the same ways it interacts with other types of matter. Ionization and atomic excitation are the most important effects. In living organisms, where proper physiological functioning is often acutely sensitive to correct chemical structure, the alterations induced in molecules by radiation almost always are deleterious. Radiation effects on biological systems can be put into two broad classes: stochastic and nonstochastic, or deterministic.

Stochastic Effects in Biological Systems

Stochastic effects are those whose probability is a function of the radiation dose, but whose severity is not related to the dose. The word stochastic refers to events that occur by chance, that is, according to the laws of probability. The appearance of a stochastic effect is, therefore, something of a hit-or-miss affair. The relationship between smoking and lung cancer is an example of a stochastic effect. Some persons who have never smoked at all will develop lung cancer, while other people who have been heavy smokers for many years never develop it. The two most important stochastic effects of radiation exposure are cancer and genetic defects in offspring.

Stochastic effects are generally related to long-term, low level exposures to radiation, and are usually the outcome of radiation damage done initially to only a few cells. It is extremely difficult to determine accurately the effects of low levels of radiation (0.1 to 0.2 Sv, or less) because there are no immediately observable effects. Because of the probabilistic nature of stochastic effects, and the fact that the effect may appear decades after the initiating event, it is virtually impossible to link a specific radiation exposure to a given stochastic effect in a single person.

Figure 1 illustrates some of the characteristics of stochastic effects. Below about 0.1 to 0.2 Sv, there are no immediately observable responses to radiation exposure. To predict effects at low levels, the approach that is taken is to extrapolate the known effects of radiation at higher doses down to lower doses. Most health physicists assume that there is a linear relationship between dose and effect at these lower levels, rather than an increased risk (shown by line 2) or a decreased risk (shown by line 3), and that there is no threshold. This is referred to as the linear dose-effect extrapolation with no threshold hypothesis. The implication of this assumption can be illustrated with a simple example. If one person who is exposed to 3 Sv of radiation dies of cancer, the hypothesis above would lead to the conclusion that a dose of 0.01 Sv given to 300 people would cause one of those people to die of cancer. Figure 1 also shows that there is an increasing likelihood of a stochastic effect with increasing dose. However, the severity of response is independent of the dose.

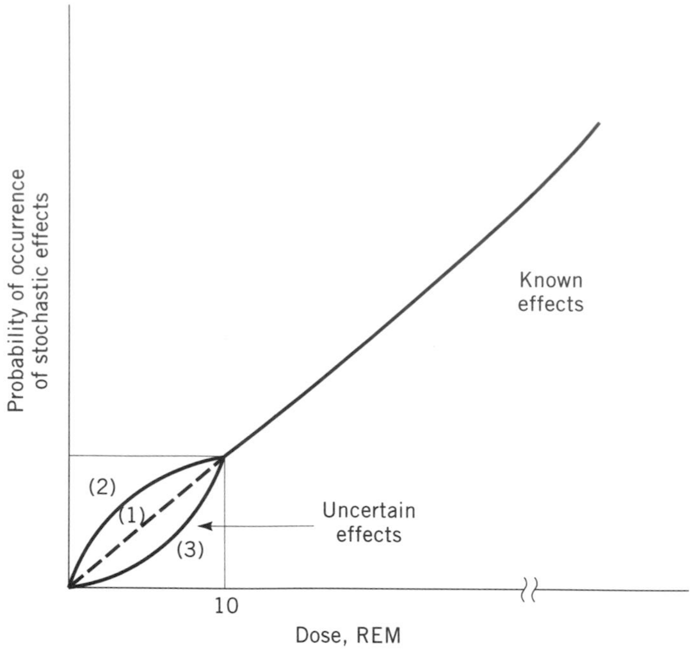

Figure 1. Dose-response curve for stochastic effects. The curve shows no threshold level, or minimum radiation dose at which no damage will ever occur. Zero radiation can not be obtained. The area outside the box in the lower left corner represents the levels at which radiation effects are directly observable. The area inside the box is the area of difficulty.

Whenever the stochastic effects of radiation are discussed, large populations of subjects must be considered, because prediction of effects in a single individual is nearly impossible. Estimates of the occurrence of radiation-induced stochastic effects in humans have been made by studying groups of people who have been exposed to larger than normal doses of radiation. These groups include the survivors of the atomic bombs dropped on Japan at the end of World War II, survivors of certain nuclear weapons tests, workers in plants where radium-containing paints were used for luminous dials, and persons exposed to large doses of radiation for therapeutic reasons. The Japanese group, in particular, has been regularly studied by several groups, among them the National Academy of Science Committee on the Biological Effects of Ionizing Radiation (BEIR), which published its most recent study (BEIR V) in 1990 (3).

The initial effects of radiation occur at the molecular level. The ionization and excitation of the molecules in an organism may alter the chemical nature of the molecules. The biological consequences of these ionizations depend on the identity and number of molecules that are affected. Radiation damage to protein molecules, which are the most important structural and physiological molecules in a living system, could cause immediate adverse effects in the organism's structure and/or function. It is more likely, however, that the change is one which would not be immediately observable, but will lead to the development of some abnormality in the future. A discussion of the chemical mechanisms by which radiation damages matter is beyond the scope of this article, but one important mechanism worth mentioning here is the production of free radicals. These are highly reactive species that undergo many reactions with biomolecules. The production of free radicals usually results from radiation interactions with the most abundant molecules in a cell, which are those of water. The radiation-induced dissociation of water can lead to the subsequent formation of hydrogen peroxide, a strong oxidizing agent that can interact adversely with many other molecules.

The observable stochastic effects, as mentioned above, are cancer and hereditary effects. Cancer is a somatic effect, meaning that it is a disorder affecting the nonreproductive cells of the organism. Somatic effects will affect only one organism, not its offspring. Leukemia is the most likely type of cancer to be induced, because of the greater susceptibility of blood system components to radiation. Cancers of the breast, thyroid, and lung are also common among radiation-induced cancers. Other types of cancer are less likely to result from radiation exposure. The National Council on Radiation Protection and Measurements (NCRP) has estimated that the value of a lifetime risk of a fatal cancer for low dose rate exposure would range from 3.3 to 5 per Sv for a population of all ages, and 2.7 to 4 per Sv for a working population (1). There is no significant difference between these two ranges.

The molecules that transmit hereditary information can also be damaged by radiation. If damage is done to the genetic material of the organism, mutations may result that would cause alterations that are passed on to the next generation. These are called genetic or hereditary effects. Extensive studies of the Japanese atomic bomb survivors have not shown the presence of significant genetic effects in humans (4). Therefore, most predictions of the chances of genetic effects are based on animal experiments.

Deterministic Effects

Deterministic effects are defined by the NCRP as those effects which increase in severity with increasing dose rate above a certain threshold level. Deterministic effects are those that do not occur by chance, but rather can be directly related to a particular causative agent. For example, alcohol consumption in humans results in deterministic effects. Drinking a small amount of alcohol will not produce any observable effects. However, after a certain level of alcohol is consumed, individuals will exhibit certain systemic effects, which will worsen in direct proportion with the amount of alcohol consumed. A blood-alcohol level of 0.4% is almost certainly fatal. This example illustrates two significant ways in which deterministic effects differ from stochastic effects. One is the existence of a threshold level, below which no observable changes occur and above which noticeable changes are sure to occur. The second is that there is a direct relationship between the size of the dose and the severity of the effects after the threshold level is reached. Figure 2 shows a dose-response curve for a deterministic effect, which illustrates the presence of a threshold level, and the linear relationship between dose and effect. A useful index which may be obtained from a dose-response curve is the dose to which 50% of the population responds with a specific biological effect. If the response is death, this dose is called the le-

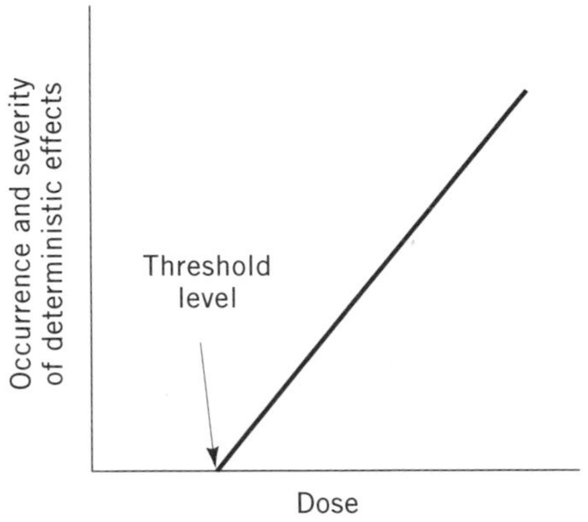

Figure 2. Dose-response curve for deterministic effects. Note presence of a threshold level.

thal dose for 50% of the population, or the LD_{50}. The time needed for 50% of the population to die is usually specified also. For example, an LD_{50}/30-day dose would be one which would result in death in 50% of the population within 30 days.

The deterministic effects of radiation in humans are generally related to large radiation doses received over a short period of time. These effects are usually seen on whole organs or body systems, rather than in only a few cells. In contrast to stochastic effects, deterministic effects occur promptly after an acute, high radiation exposure, and a definite link between the cause and the effects can be established. The severity of the effects depends on the total dose, the time over which it was received, the body part irradiated, whether the dose was external or internal, and the age of the person. Young children are much more susceptible to the harmful effects of radiation than older adults are. The most important deterministic effects include skin changes, alterations in the blood, gastrointestinal problems, and central nervous system changes.

The first observable changes that take place in a person who has received a dose of radiation are blood alterations. The threshold for these effects for β or γ radiation ($Q = 1$) is around 250–500 mGy. With a dose of this size, the number of white blood cells is seen to decrease within a few hours after exposure. Red blood cells and platelets decrease at a slower rate, perhaps days or weeks depending on the dose. Recovery to normal conditions can take weeks or months. The rate and degree of cell loss, and the time needed for recovery, are directly related to the dose received. Doses in this range also induce temporary sterility in both men and women, but fatalities would not be expected to occur.

A dose of 2 Gy will result in damage to the bone marrow, the part of the body responsible for production of blood cells. The bone marrow damage is not complete and is reversible. A dose of this magnitude will also result in gastrointestinal symptoms, including nausea and vomiting, and general malaise and fatigue. Loss of hair is almost always seen. Some deaths may occur.

Doses of 4–6 Gy result in complete, although reversible, loss of bone marrow function. More severe levels of the blood cell and gastrointestinal symptoms listed above also occur, but with medical treatment survival is still possible. The LD_{50}/30-day for man is around the 4–5 Gy level.

Above 7 Gy the bone marrow is irreversibly damaged, and survival is very unlikely. At 10 Gy the desquamation (sloughing off) of the intestinal epithelia leads to severe diarrhea, nausea, and vomiting soon after exposure. Death is highly probable within 1–2 weeks. At 20 Gy and higher, the central nervous system is damaged. Unconsciousness follows within minutes after exposure, and death occurs within hours or a few days.

SOURCES OF RADIATION EXPOSURE

The radiation to which we are exposed in everyday life comes from both natural and synthetic sources. Figure 3 summarizes the most important radiation sources as percentages of the total dose received. It is readily apparent from this Figure that the most plentiful source of radiation to which most people are exposed is from natural background, which includes cosmic radiation and the natural radioactive materials in the earth, atmosphere, and our own bodies. The average annual effective dose equivalent due to natural radiation is about 2.4 mSv per year, and the range for natural exposure is 1.5–6.0 mSv per year.

Natural Sources of Radiation

Natural sources of radiation include cosmic rays, naturally radioactive materials in the surrounding earth environment, and naturally radioactive elements in our bodies.

Cosmic rays originate in outer space and consist mainly of high energy protons. When these particles encounter earth's atmosphere, they induce nuclear reactions that produce showers of other subatomic particles, especially electrons, muons, neutrinos, neutrons, and photons. At sea level, a person will receive an annual dose of around 0.36 Sv from cosmic radiation. The cosmic ray dose increases at higher elevations and at higher latitudes. For example, a person flying in a jet at an altitude

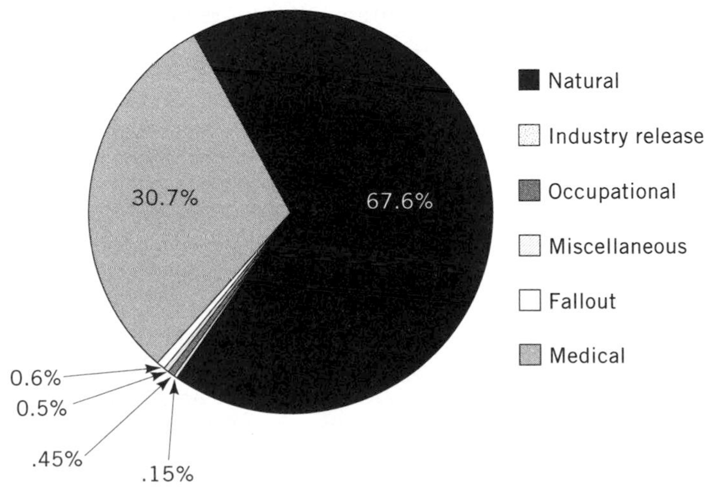

Figure 3. Sources of radiation exposure (percentage of total yearly dose).

of 6 miles (10 km) receives an additional 5 μSv every hour. Similarly, people who live in cities at high altitudes will receive a cosmic ray dose of more than 0.36 Sv per year.

The earth contains many radioactive elements, with the largest contributors to dose being ^{40}K, ^{87}Rb, ^{232}Th, ^{235}U, and ^{238}U, and the radioactive daughters in the U and Th natural decay chains. The dose received averages from 0.3 to 0.7 mSv annually, but varies greatly depending on location. There are several hot spots that have been identified in Brazil, India, Iran, and some other countries. Many of these are areas where thorium-rich sands are high in abundance. Annual doses in these areas can be up to 250 mSv per year (near the city of Pocos de Caldas, in Brazil). No significant adverse effects have been observed in persons living in or near these high radiation background areas.

People are also constantly irradiated by the radioactive atoms within their own bodies and in the food and water they ingest. Potassium is an element which is essential to life, and the ^{40}K isotope is radioactive, with a very long half-life. In a person of 70 kg mass, the ^{40}K contributes an activity of around 10^5 disintegrations per minute, resulting in an annual dose of about 0.18 mSv. There are also significant amounts of ^{14}C and ^{3}H present in the body, but these contribute little to the dose because they emit low energy β radiation. The air we breathe contains radioactive gases, especially the inert gas radon. The food and water we ingest contains ^{40}K, ^{226}Ra, and a variety of radioactive U and Th decay products. All of these sources contribute an additional annual dose of about 0.40 mSv.

One of the most significant sources of natural radiation is radon gas. The problem of indoor radon gas contamination was first noted in 1984, when it was discovered that one of the employees of a nuclear power plant set off radiation detection alarms when he came into the plant. Subsequent investigations led to the finding that the worker's home had high levels (10^5 becquerels/m^3) of the radioactive gas radon. (A becquerel (Bq) is a unit of radioactivity equal to one disintegration per second). For comparison, the average indoor Rn level is only about 50 Bq/m^3 (1.4 picocuries/L), and the average outdoor level is 5–10 Bq/m^3.

Radon gas (^{222}Rn) is a natural decay product of ^{238}U. Its decay chain is:

$$^{222}\mathrm{Rn} \xrightarrow[3.82\,\mathrm{d}]{\alpha} {}^{218}\mathrm{Po} \xrightarrow[3.11\,\mathrm{m}]{\alpha} {}^{214}\mathrm{Bi} \xrightarrow[19.8\,\mathrm{m}]{\beta} {}^{214}\mathrm{Po} \xrightarrow[164\,\mu\mathrm{s}]{\alpha} {}^{210}\mathrm{Pb}$$

$$\xrightarrow[22.3\,\mathrm{y}]{\beta} {}^{210}\mathrm{Bi} \xrightarrow[5.01\,\mathrm{d}]{\beta} {}^{210}\mathrm{Po} \xrightarrow[138.4\,\mathrm{d}]{\alpha} {}^{206}\mathrm{Pb}$$

Because Rn is gaseous, it is inhaled with the air we breathe. If the Rn is trapped in the lungs, it emits α radiation and gives rise to the long series of radioactive decay products listed above. The natural dose equivalent received from radon gas is now believed to be greater than that received from all other sources, and this is why radon has become a great concern.

The amount of Rn that is found in any individual home is a function of the amount of Rn generated in the soil underneath the home, the rate at which the Rn enters, and the rate at which air flows through the house (the ventilation rate). In most cases, if excessive Rn levels are found, there are steps that can be taken to prevent the Rn gas from entering the house, or to increase the ventilation rate through it, so that amounts are lowered to acceptable levels.

Synthetic Sources of Radiation

The remainder of the radiation people receive comes from synthetic sources. As shown in Figure 3, this amounts to only about 1/3 of the amount received from natural sources (around 0.8 mGy/year). By far the largest amount of radiation exposure from synthetic sources is from medical and radiopharmaceutical procedures. A small number of these procedures will result in extremely high doses, in cases where radiation is being used therapeutically to treat disease. However, the exposure that most people receive comes from x-ray procedures, both medical and dental. The dose received from an x-ray is variable, depending on the part of the body examined and the type of x-ray performed. A typical chest x-ray would give a dose of around 0.5 mSv.

The dose received from all other synthetic sources is much smaller than that from natural sources or from medical procedures. The dose due to fallout from nuclear weapons testing has decreased significantly in the years following the atmospheric nuclear test ban treaty. The main nuclides of interest from fallout are ^{14}C, ^{95}Zr, ^{90}Sr, and ^{137}Cs. Presently, fallout contributes less than 0.01 mSv to the annual dose. The routine operation of nuclear power plants contributes very little to the annual dose. The only radionuclide that is normally emitted from a nuclear power plant is the gaseous fission product ^{85}Kr. The dose to persons in the vicinity of a plant is, at most, a few percent of the dose received from natural sources. Probably, most people fear the potentially high doses that could be received in an accident more than they fear the low level dose due to routine operation. However, even in the 1979 Three Mile Island accident, which was the worst in U.S. history, the maximum dose to the surrounding population was only about 0.1 mSv, due mainly to the release of ^{133}Xe. The accident that occurred at Chernobyl in the Soviet Union in 1986 was the most severe commercial nuclear power accident that has ever occurred. Thirty-one people died within a short time from a combination of explosion, burns, and radiation exposure as a result of this accident. Surrounding areas, exposed to doses of a few hundredths of a Sv, had to be permanently evacuated.

Radiation is a natural phenomenon to which living things have been exposed since the beginnings of life. The bulk of the dose (>95%) that is received now by people is either from natural sources that cannot be avoided or comes from voluntary medical procedures. In everyday life, the public is not exposed to untoward amounts of radiation from any synthetic sources.

RADIATION PROTECTION AND CONTROL

As the understanding of the effects of radiation on organisms improved, so did efforts to control unnecessary exposure. The process of setting limits on the amounts of radiation to which people may be exposed at work or in the normal environment is fraught with difficulties and is still an inexact science. The reasons for this include the difficulty of accurately assessing the effects of low level radiation and of defining what is meant by an acceptable risk.

Risk

Risk can be defined as the chance that a given activity will produce illness, injury, or death in an individual. All life activities involve a certain degree of risk. People accept these risks because they judge that the benefits to be gained from the activity outweigh the risk involved. One difficulty in developing radiation protection standards lies in deciding in the first place what constitutes an acceptable risk. This judgement is highly variable from one person or group to another. The risks encountered in a sport like sky-diving are unacceptably high to some people, but obviously are not too high for others. The risks associated with consumption of alcohol and smoking cigarettes are well documented, but people continue to smoke and drink because, for them, the benefits outweigh the risks. Generally, risks are more acceptable to people when they feel they have some measure of control over them, and when they have the option of whether or not to take the risk. Besides these factors, the better understood a risk is, the more likely it is to be acceptable. It may be that the risks associated with radiation are not well accepted because they are not understood, and because people feel that the risks are forced on them. Some life-shortening risks are shown in Table 3 (5, 6).

There is no question that excessive radiation exposure is detrimental to living organisms, but the actual risks from the radiation we receive are much less than the level of risk that is perceived by the general public. Exposure to radiation, at the levels commonly encountered in everyday life or even occupationally, poses a very small risk in reality, but it is perceived as a large risk. This is illustrated by a 1979 study in which three groups of Americans were asked to rate the risk of 30 activities. The rankings of the first 20 of these activities are shown in Figure 4. Nuclear power was rated as a high risk by all three groups, but in actuality, it results in far fewer deaths than activities such as smoking that were rated low by the three groups.

In setting radiation protection standards, a commonly used measure of acceptable risk is that the activity will cause only 1 death in 100,000 people per year in the general population. For industrial or work settings, the level of acceptable risk is 1 death in 10,000 workers per year. Activities that result in a fatality rate less than this are generally considered safe.

Table 3. Estimated Loss of Life Expectancy from Health Risks and Occupational Hazards[a]

Risks	Average Days of Life Expectancy Lost
Health Risks	
Smoking 20 cigarettes a day	2370
Overweight by 20%	985
Accidents of all types	435
Alcohol consumption, U.S. average	130
Drowning	41
Natural background radiation	8
Catastrophies, ie, earthquakes, etc	3.5
Industrial Occupational Hazards	
Mining and quarrying	328
Construction	302
Agriculture	277
Transportation and utilities	164
Service industries	47
All industry, average[b]	74
Radiation Related	
Medical diagnostic x-rays, average	6
0.01 Sv occupational single radiation dose	1
0.01 Sv/year dose for 30 years	30

[a] References 5 and 6.

[b] The nuclear industry high dose occupational average is 0.0065 Sv/year. The maximum population dose associated with the Three Mile Island accident is approximately 0.001 Sv.

Regulatory Bodies and Objectives of Radiation Standards

The first real step in radiation protection came in the late 1920s when the International Conference on Radiology began work aimed at the definition of methods used for measuring radiation. Some years later, the International Commission on Radiological Protection (ICRP) grew out of this. Its purpose is to provide recommendations for safe levels of radiation exposure. In the United States, a similar group called the National Council on Radiation Protection and Measurements (NCRP) was formed. These two agencies, along with the International Atomic Energy Agency (IAEA) and the International Radiation Protection Association (IRPA), are the professional organizations that provide recommendations on radiation protection to regulatory agencies and promote communication among members of the international radiation protection community. These groups do not write the regulations specifying legal standards for radiation exposure. That task falls to governmental regulatory organizations like the United States Nuclear Regulatory Commission (NRC) or the Environmental Protection Agency (EPA). The NRC and EPA would use the technical information provided by the NCRP and ICRP to formulate their regulations.

The objectives of radiation protection are twofold: (*1*) to prevent the occurrence of clinically significant radiation-induced deterministic effects by adhering to dose limits that are below threshold levels, and (*2*) to limit the risk of stochastic effects to an acceptable level in relation to societal needs, values, benefits gained, and economic factors (1). Figure 2 illustrated the fact that deterministic effects have a threshold level. If the threshold level is known, then limiting dose to an amount below threshold should protect people against the occurrence of deterministic effects. The radiation doses required to induce observable deterministic effects are really quite large, in relation to exposures that ordinarily occur in daily life, and such doses occur very infrequently. There is greater need

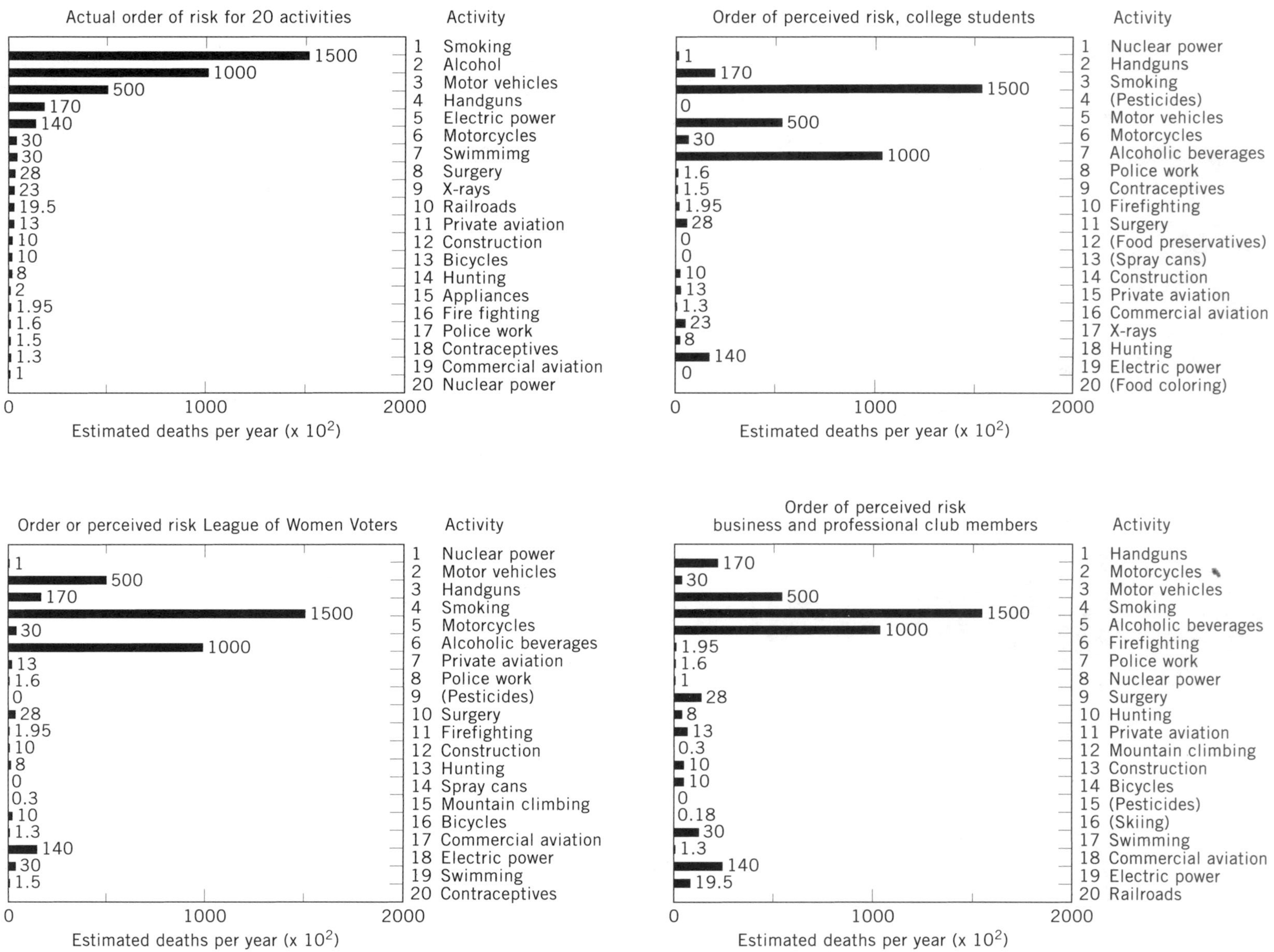

Figure 4. Perceived risk versus actual risk for three groups of the American population. (**a**) Actual order of risk for twenty activities; (**b**) order of perceived risk, college students; (**c**) order of perceived risk, League of Women Voters; (**d**) order of perceived risk, business and professional club members (7).

for radiation protection standards that deal with routine exposures, rather than with the accidental high level doses that occur only rarely. So, most regulatory requirements are aimed at reducing stochastic effects to an acceptable limit. Remember that the stochastic effects, because of their probabilistic nature, can never be completely eliminated. A general guiding principle is that of ALARA: keep all exposure As Low As Reasonably Achievable.

The quantity that is really of interest in radiation protection studies is the risk to the individual, over a lifetime, of experiencing a severe negative effect (eg, cancer) from radiation exposure. It is very difficult to state this risk in readily measurable terms. Because the risk is assumed to be related to the dose of radiation received, radiation protection standards are stated in terms of allowable doses for a given period of time. The most recent recommendations from the NCRP regarding exposure for both occupationally exposed workers and the general public are cited in Table 4.

RADIATION SURVEYS AND MONITORING

An essential part of a radiation protection program is the process of monitoring personnel, the workplace, and the environment for the presence of radiation. All facilities that handle radioactive materials must have a well-defined radiation monitoring program to ensure that regulatory limits on personnel exposure are not exceeded. The extent of this program will vary with the type of facility and the kinds of radioactive materials being handled. This article will describe some of the points to be considered in radiation monitoring and some of the instruments used.

Area Surveys

An area survey refers to the overall process of establishing the conditions for operation of a facility where radiation levels may be above natural background levels (>0.1 μcurie) (9). The international symbol that indicates the presence of ionizing radiation is shown in Figure 5.

There are five phases involved in an area survey. First, an investigation is performed to obtain information about the workplace itself, such as the types of radiation expected to be present, the type of work to be performed, and the geometrical configuration of the workplace. An inspection is then done by a person trained in radiation protection to verify this information. Next, measurements of the radiation fields and any surface or airborne contamination in the facility are taken. These measurements are used to make recommendations for operation of the facility. In all radiation protection work, extensive record keeping is a necessity. The frequency and extent of an area survey will be dependent on the individual facility.

Personnel Monitoring

An important facet of radiation protection programs is monitoring the amount of radiation to which personnel are exposed in their working environment. People may be

Table 4. Recommended Exposure Limits[a]

Exposures	Limits
Occupational	
Effective dose limits	
Annual	50 mSv
Cumulative	10 mSv × age
Equivalent dose annual limits for tissues and organs	
Lens of eye	150 mSv
Skin, hands, feet	500 mSv
Public	
Effective dose limit, continuous or frequent exposure	1 mSv
Effective dose limit, infrequent exposure	5 mSv
Equivalent dose limits for tissues and organs	
Lens of eye	15 mSv
Skin, hands, feet	50 mSv
Education and Training, Annual	
Effective dose limit	1 mSv
Equivalent dose limit for tissues and organs	
Lens of eye	15 mSv
Skin, hands, feet	50 mSv
Embryo-fetus, Monthly	
Equivalent dose limit	0.5 mSv
Individual	
Negligible dose, annual	0.01 mSv

[a] Reference 8.

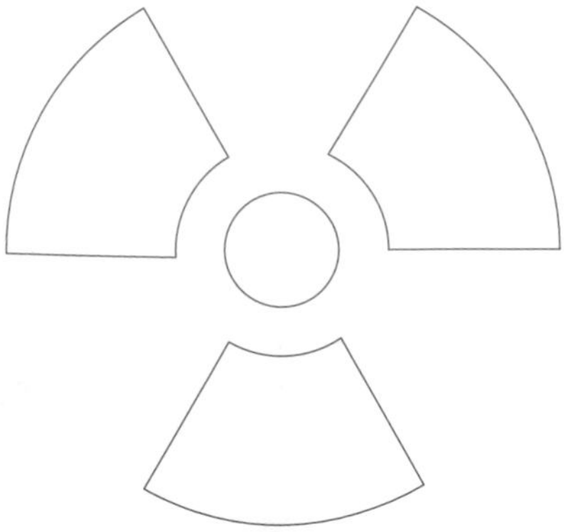

Figure 5. International symbol for the presence of ionizing radiation. Colors are black or magneta on yellow background.

exposed to radiation either externally or internally, if radioactive materials have been ingested, inhaled, or absorbed through the skin. Therefore, both external and internal personnel monitoring is routinely done for workers in installations or laboratories where radiation exposure is possible.

Information regarding the extent of external exposure of personnel is most often obtained through the use of a personnel dosimeter. This is a device that the person wears at all times while in the facility. The choice of dosimeter is based on a knowledge of the types and levels of radioactivity to which the worker is exposed. Ionization chambers, in the form of the pocket dosimeter, can be used to monitor exposure to x-rays, γ rays, and electrons. They are an appropriate choice in situations where monitoring needs to be done for a short period of time. For example, visitors to a nuclear facility are often required to wear a pocket dosimeter during their stay. For longer term measurements, the thermoluminescent detector, or TLD badge, is the most widely used personnel dosimeter. These badges can be made with a variety of thermoluminescent crystals and filters that can respond to different types and energies of radiation. The time over which these badges are worn may range from a week to a few months, depending on expected exposure. The badges are then collected and measured by radiation protection personnel.

Internal exposure monitoring is routinely done only in facilities where unsealed radioactive materials are present. Both *in vitro* and *in vivo* methods are used. Bioassay is the analysis of body tissues for radioactivity after the tissues have been removed from the body (*in vitro* analysis). The body fluid most used for internal exposure monitoring is urine. The urine is analyzed for specific nuclides, usually α emitters such as uranium and plutonium, and β emitters. Other tissues used for internal monitoring include blood, feces, sputum, nasal smears, and breath analysis for gaseous radionuclides.

Internal monitoring is also done using *in vivo* methods, in which instruments are configured to count the whole body or specific parts of the body. Whole body counters typically use NaI(Tl) scintillation detectors or large-volume liquid or plastic scintillators to measure γ-emit-

Table 5. Instruments for Measuring Activity[a]

Instrument	Counting Range	Comments
	Beta-gamma Surface Monitors	
Geiger-Müller portable count rate meter	0–100,000 counts/min	Surfaces, hands, clothing
Geiger-Müller hand and shoe monitor	From $1\frac{1}{2}$ to 2 times background	Rapid contamination monitoring
Geiger-Müller floor or doorway monitors	From $1\frac{1}{2}$ to 2 times background	Convenient, rapid monitoring
	Alpha Surface Monitors	
Proportional counter, portable	0–100,000 disintegrations/min over 100 cm^2 area	Surfaces, hands, clothing
Scintillation counter, portable	0–100,000 disintegrations/min over 100 cm^2 area	Surfaces, hands, clothing
	Personnel Monitors	
Proportional counter, hand and shoe monitor	0–2,000 counts/min over 300 cm^2 area	Rapid monitoring
Scintillation counter, hand and shoe monitor	0–4,000 counts/min over 300 cm^2 area	Rapid monitoring
	Air Monitors	
Filter paper, high volume	40 ft^3/min	For quick grab samples; use intermittently; count separately
Filter paper	0.1 to 10 ft^3/min	For continuous monitoring; count separately
Impinger	20 to 40 ft^3/min	For alpha contamination; count separately

[a] Reference 9.

ting nuclides. These detector systems are expensive and often quite specialized, so whole body counting is not done as routinely as *in vitro* measurements.

It is essential for radiation protection workers to keep accurate records of personnel exposure, and to maintain these records for many years.

Instrumentation

Radiation detection instruments can be divided into two types. The first type is used for survey work and measures radiation fields or exposures. This type of detector provides readouts in units of exposure or dose, such as mR/h or μSv/h. The other type of radiation detection instruments measure the activity of a radioactive source and read out in activity units, such as counts per minute or counts per second.

Survey instruments include ionization chambers, Geiger-Müller counters, scintillation instruments, and thermoluminescent dosimeters. Of these, the ionization chamber is the most versatile and widely used. Most gamma and x-ray survey exposure measurements are made with the ionization chamber. They can also be configured to measure β particles and neutrons. Ionization chambers, in the form of pocket dosimeters, are also used for personnel monitoring. Geiger counters are used primarily for x- and γ-ray fields. Scintillation instruments such as sodium iodide detectors and organic scintillators are used for x- and γ-ray fields, β measurements, and neutron determinations. Thermoluminescent materials can be designed to respond to γ and x rays, low energy electrons, and neutrons.

Measurements of activity in radiation protection work are usually made to detect the presence of contamination (that is, radioactive materials that are located anywhere besides their appropriate place). Activity measurements are made of personnel, work areas, air and liquids. The instruments used for this purpose should be reliable and sensitive, because they are intended to detect small amounts of contamination. Table 5 summarizes the various types of instruments used for activity measurements.

BIBLIOGRAPHY

1. National Council on Radiation Protection and Measurements, *Limitation of Exposure to Ionizing Radiation: Recommendations of the National Council on Radiation Protection and Measurements,* Report No. 116, NCRP Publications, Bethesda, Md., 1993, p. 18.
2. *Ibid.,* p. 20.
3. National Academy of Sciences/National Research Council, *Health Effects of Exposure of Low Levels of Ionizing Radiation,* Report of the Committee on the Biological Effects of Ionizing Radiations, BEIR V, National Academy Press, Washington, D.C., 1990.
4. J. V. Neel and W. J. Shull, eds., National Academy of Sciences/National Research Council, *The Children of Atomic Bomb Survivors, A Genetic Study,* National Academy Press, Washington, D.C., 1991.
5. B. L. Cohen and I. S. Lee, *Health Physics* **36,** 707 (1979).
6. World Health Organization, *Health Implications of Nuclear Power Production,* WHO, Geneva, 1975.
7. W. K. Sinclair, *Radiology* **138**(1), 149 (1988).
8. Ref. 1, p. 56.
9. National Council on Radiation Protection and Measurements, *Instrumentation and Monitoring Methods for Radiation Protection,* NCRP Report No. 57, NCRP Publications, Bethesda, Md., 1978.

General References

H. Cember, *Introduction to Health Physics,* Pergamon Press, New York, 1988.

W. D. Ehmann and D. E. Vance, *Radiochemistry and Nuclear Methods of Analysis,* John Wiley & Sons, Inc., New York, 1991.

J. R. Greening, *Fundamentals of Radiation Dosimetry,* 2nd ed., Medical Physics Handbooks 15, Adam Hilger Ltd., Boston, 1985.

F. F. Hahn and co-workers, *Health Physics* **55**(2), 303–313 (1988).

N. Howard and S. Atilla, *Dun's Rev.* **114,** 48 (1979).

International Atomic Energy Agency, *Basic Principles for Occupational Radiation Monitoring,* IAEA, Vienna, Austria, 1987.

International Atomic Energy Agency, *Radiation Protection Glossary: Selected Basic Terms Used in IAEA Publications,* IAEA, Vienna, Austria, 1986.

International Commission on Radiological Protection, *1990 Recommendations of the International Commission on Radiological Protection,* ICRP Publication 60, Annals of the ICRP, Pergamon Press, New York, 1991.

G. D. Kerr, *Health Physics* **55**(2), 241–249 (1988).

D. C. Kocher, K. F. Eckerman, and R. W. Leggett, *Health Physics* **55**(2), 339–347 (1988).

A. D. Martin and S. A. Harbison, *An Introduction to Radiation Protection,* Chapman and Hall, New York, 1986.

G. Medvedev, *The Truth About Chernobyl,* Basic Books, New York, 1991.

F. A. Mettler and W. K. Sinclair, *Health Physics* **58**(3), 241–250 (1990).

K. L. Miller, *CRC Handbook of Management of Radiation Protection Programs,* CRC Press, Boca Raton, Fl., 1992.

A. Nero, *Physics Today,* 32–39 (Apr. 1989).

National Council on Radiation Protection and Measurements, *Radiation Exposure of the U.S. Population from Consumer Products and Miscellaneous Sources,* Report No. 95, NCRP Publications, Bethesda, Md., 1987.

National Council on Radiation Protection and Measurements, *Exposure of the U.S. Population from Diagnostic Medical Radiation,* Report No. 100, NCRP Publications, Bethesda, Md., 1989.

National Council on Radiation Protection and Measurements, *Exposure of the U.S. Population from Occupational Radiation,* Report No. 101, NCRP Publications, Bethesda, Md., 1989.

National Council on Radiation Protection and Measurements, *Evaluation of Risk Estimates for Radiation Protection Purposes,* NCRP Report No. 115, NCRP, Bethesda, Md., 1994.

G. Paic, *Ionizing Radiation: Protection and Dosimetry,* CRC Press, Boca Raton, Fl., 1988.

J. Shapiro, *Radiation Protection,* 2nd ed., Harvard University Press, Cambridge, Mass., 1981.

J. Shapiro, *Radiation Protection: A Guide for Scientists and Physicians,* Harvard University Press, Cambridge, Mass., 1981.

W. K. Sinclair, *Health Physics* **55**(2), 149–157 (1988).

J. N. Stannard, *Radioactivity and Health: A History,* Office of Scientific and Technical Information, National Technical Information Service, Springfield, Va., 1988.

J. E. Turner, *Atoms, Radiation, and Radiation Protection,* Pergamon Press, New York, 1986.

J. E. Turner, *Problems and Solutions in Radiation Protection,* Pergamon Press, New York, 1988.

United Nations Environment Programme, *Radiation Doses, Effects, Risks,* United Nations Publication, Nairobi, Kenya, 1985.

United Nations Scientific Committee on the Effects of Atomic Radiation, *Sources, Effects, and Risks of Ionizing Radiation,* United Nations Publication, New York, 1988.

H. N. Wagner and L. E. Ketchum, *Living with Radiation, the Risk, the Promise,* The Johns Hopkins University Press, Baltimore, Md., 1989.

RADIATION MONITORING

DIANE E. VANCE
WILLIAM D. EHMANN
University of Kentucky
Lexington. Kentucky

RADIATION DETECTION

Living organisms do not possess any natural sensors that can directly detect radiation. Therefore, instruments that respond to the passage of radiation with some observable signal must be used to detect it. No single type of detection scheme could be equally useful for all types of radiation, because of differences in the ways these types interact with matter and because the levels of radioactivity and the energies of the radiations vary drastically. Measurable activity levels can range from a few decays per day up to 10^{13} decays per second or more. The energies of common radiations can be as low as a small fraction of an electron volt (a slow or thermal neutron) or as high as several gigaelectron volts (cosmic rays). Therefore, a variety of detection schemes appropriate for different types, energies, and activity levels have been developed.

Detector systems can be divided into two major groups: electronic, and all others. Electronic detectors, which rely on the electrical signals generated either directly or indirectly when radiation passes through matter, are the most widely used. The electronic detectors include gas-filled chambers, scintillation detectors, and solid-state semiconductor detectors. Other types of detectors include photographic plates, chemical reactions, calorimetric detectors, cloud and bubble chambers, and thermoluminescence detectors (TLD). In environmental work, the types of radiation most often detected include alpha, beta, and gamma radiation. This chapter will concentrate on methods used for these three types. Methods of neutron detection are also briefly described. It is assumed that the reader is familiar with these modes of radioactive decay.

See also NUCLEAR POWER: DECOMISSIONING POWER PLANTS; MANAGING NUCLEAR MATERIALS; SAFETY OF AGING POWER PLANTS; OZONE, STRATOSPHERIC.

RADIATION INTERACTIONS WITH MATTER

In order to be detected, radiation must interact with matter in some way. There are five common modes of interaction. *Ionization* refers to the process by which incoming radiation removes an electron from an atom, resulting in the formation of an ion pair consisting of the electron and the remaining positive ion. *Kinetic energy transfer* can occur when the radiation has sufficient energy not only to ionize atoms, but to give them additional kinetic energy. *Molecular* and *atomic excitation* can occur when the radiation excites electrons to higher energy levels without causing ionization, or when it causes rotational, translational, or vibrational excitations in molecules. *Nuclear reactions* occur when incoming radiation interacts with the nuclei of atoms, rather than with the orbital electrons. Finally, *radiative processes* can occur, in which the interaction of radiation with certain materials results in the emission of electromagnetic radiation. Of these five modes, the ionization process forms the basis for most of the detection systems to be discussed here.

The detection system used for a given type of radiation reflects the characteristic ways in which that radiation interacts with matter. Alpha particles, which have a high mass and high charge, interact primarily with orbital electrons through ionization processes. They undergo a large number of small interactions with surrounding atoms and create large numbers of ion pairs. They move in straight-line paths through matter, and travel short distances. For example, an alpha particle with an energy of 3 MeV will travel only about 1.6 cm (.63 in) in air. Alpha particles are easily stopped even by low-density materials. Beta particles also induce ionization, but to a lesser degree than alpha particles. They travel through matter in an irregular path, and have longer ranges. A beta particle with an energy of only 1 MeV will have a range in air of about three m (10 ft). Radiative processes are important for high energy beta particles. Gamma rays induce some ionization in matter, but at a density less than for beta particles of corresponding energy. More commonly, gamma rays undergo only one or a few direct interactions with orbital electrons and the original energy of the γ ray is dissipated through secondary electron interactions. Gamma rays have long ranges and are absorbed most efficiently in high-atomic-number, high-density materials.

DEFINITIONS OF DETECTOR OPERATING CHARACTERISTICS

There are three parameters that will be mentioned often in this article in characterizing the various types of detectors. These are the efficiency, the resolution, and the dead time (resolving time) of the detector.

The efficiency (ε) of a detector refers to the number of radiations actually detected out of the number that are emitted by the source. There are two ways to define efficiency, one based on the number of radiations actually emitted by the source, and one based on the number of radiations that actually strike the detector. The former is called the absolute efficiency, and the latter the intrinsic efficiency.

$$\varepsilon_{\text{abs}} = \frac{\text{Number pulses recorded}}{\text{Number radiations emitted by source}}$$

$$\varepsilon_{\text{intr}} = \frac{\text{Number pulses recorded}}{\text{Number radiations striking detector}}$$

It is desirable to have as high an efficiency as possible. For nonpenetrating radiations, such as alpha particles, 100% efficiency may be achieved relatively easily. However, for the more penetrating uncharged radiations, such as gamma rays, efficiencies will typically be much lower.

Energy resolution (R) refers to the ability of the detector to discriminate between two radiations of different energies. Not all detectors are capable of giving energy information, so this parameter is not relevant for those detectors. Resolution is defined with reference to a plot of the number of radiations detected versus the radiation energy, as shown in Figure 1.

$$R = \frac{\Delta E}{E_0}$$

where R is the resolution, E_0 is the centroid of the peak, and ΔE refers to the width of the peak halfway between the baseline and the top of the peak. This is also called the full-width half-maximum, or FWHM. The resolution is often expressed as a percentage. The smaller the value, the better the detector will be at separating two radiations of similar energy. Resolution is never perfect because of electronic noise and the statistical nature of the interactions of radiation with matter. Resolution varies greatly for different types of detectors.

Detector dead time (τ) refers to the amount of time needed before a detector can recover from one incoming radiation and respond to the next. The dead time is affected by the nature of the detector and by the associated electronics. Very fast detector systems may be able to respond to a second event within nanoseconds.

GAS-FILLED DETECTORS

All gas-filled detectors consist of a volume of gas surrounded by a housing that may be either sealed or designed to permit a continuous flow of the counting gas. A voltage is applied across electrodes within the gas volume, creating an electric field. As radiation passes through the gas-filled chamber, it ionizes the gas to form ion pairs consisting of an electron and a positive ion. The ions are attracted to the electrodes, and this generates the electrical signal that indicates the passage of radiation. The electric signal that is measured may be either a current, a voltage pulse, or the total accumulated charge. The choice will depend on the way the instrument is configured. The energy lost by the incoming radiation for the formation of one ion pair (called the W-value) in a gas is approximately 30-35 eV. The actual value varies by a few eV depending on the type of gas as well as the identity and energy of the incoming radiation.

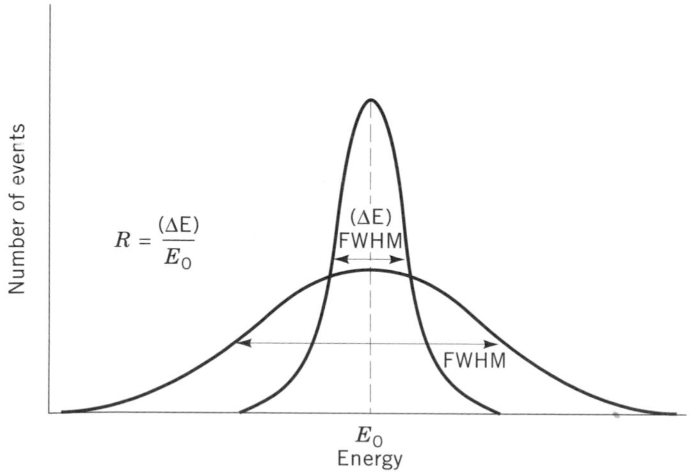

Figure 1. Calculation of the energy resolution of a pulse-type detector.

Three kinds of gas-filled detectors will be discussed: ionization chambers, proportional counters, and Geiger-Müller counters. These differ primarily in the strength of the electric field applied across the electrodes. These variations in voltage lead to differences in the ionization processes taking place in the chamber. A graph illustrating the relationship of the output signal of a gas-filled counter to the applied voltage is shown in Figure 2. Curves for sources undergoing decay via single α- or β-particle transitions are illustrated. In region I, called the recombination region, the voltage is too low for all the ion pairs to be collected. Many of the ions simply recombine to form neutral atoms or gas molecules again. This region is not useful for detection. In region II, the voltage is high enough so that virtually all the ion pairs that are formed by a single ionization event are collected. However, the ions are not accelerated enough to cause secondary ionization events. Therefore, the amplitude of the signal is related to the energy of the incoming radiation and will remain constant even with some minor variations in detector voltage. This is the region in which the ionization chamber is operated. At higher voltages, in region III, the electrons formed in the original ionization event are accelerated toward the electrodes with enough kinetic energy to induce limited secondary ionization events as they migrate. The amplitude of the signal is not constant with varying detector voltage, but is still linearly related to the energy of the incoming radiation. This is the region of operation for the proportional counter. At still higher voltages, in region IV, the linear relationship between the energy of the incoming radiation and the output signal begins to deteriorate. This region of limited proportionality is not used for detection. In region V, the voltage has been raised so high that a single ionizing event results in the electrical discharge of the entire tube. The Geiger-Müller (G-M) counter is operated in this region. Because only a single amplitude signal is produced by any radiation interacting with the detector, G-M counters do not permit particle energy discrimination. At very high voltages (region VI), continuous discharge occurs in the gas. This region is not useful for radiation detection.

Ionization Chambers

The ionization chamber is one of the oldest and simplest radiation detection devices. A simplified diagram of an ionization chamber is shown in Figure 3. The chamber, typically a few centimeters in diameter, is filled with a gas at pressures ranging from 0.1 to 10 atmospheres. An electric field is applied across the electrodes. As radiation passes through the gas, ion pairs form. If no electric field were present, these pairs would simply recombine. How-

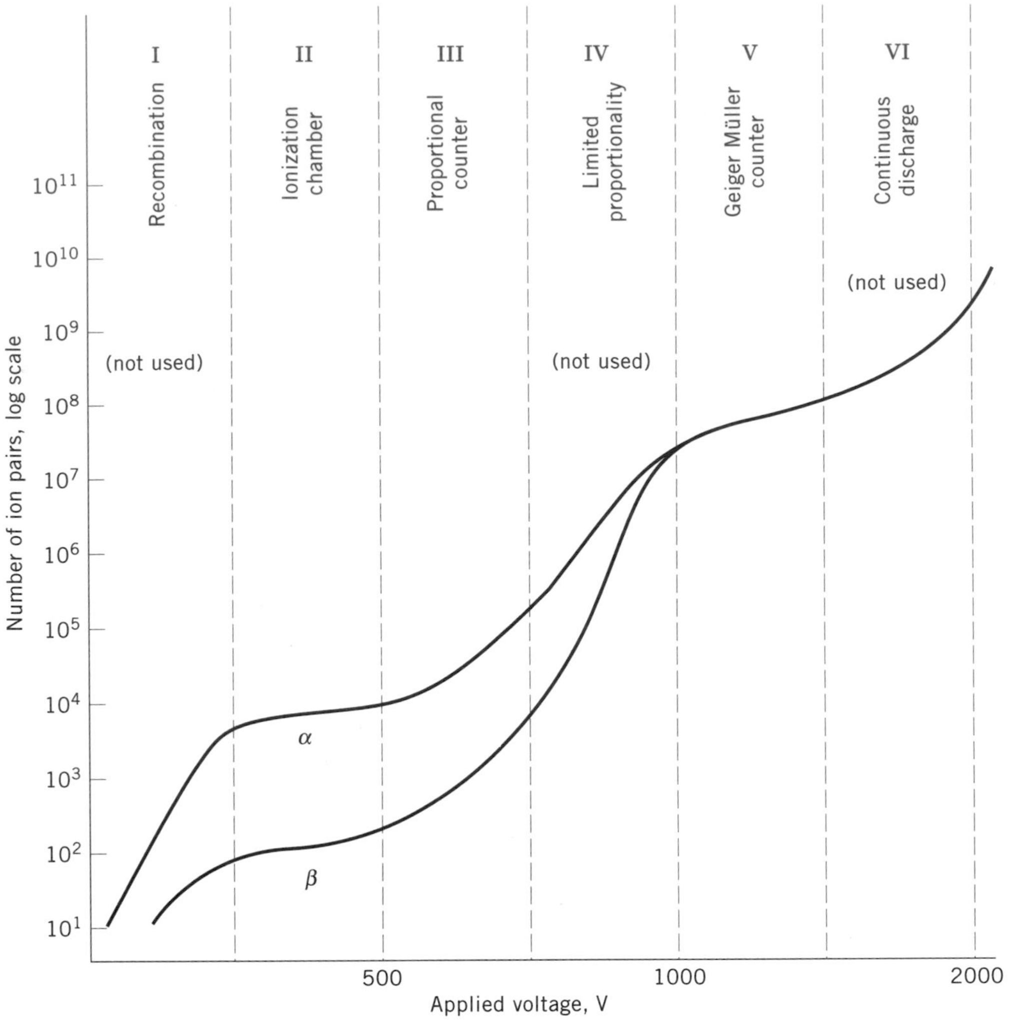

Figure 2. Operational regions for gas-filled detectors.

ever, if a large enough electric field is applied across the electrical plates, the ions will drift toward the electrodes. The electrical signal measured can be either a current, a voltage pulse, or the accumulation of the total amount of charge, depending on the type of electrical circuit used.

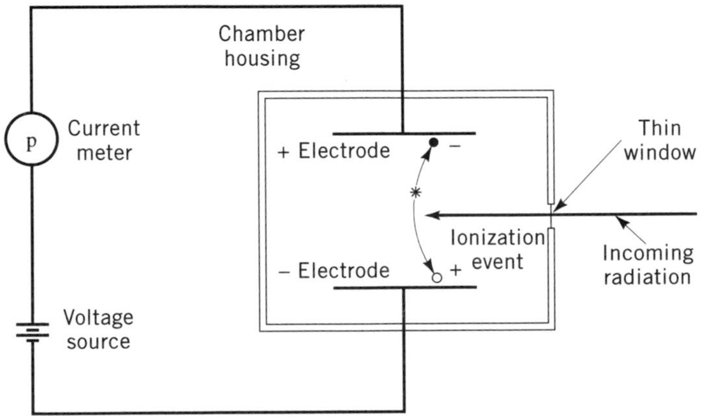

Figure 3. Simplified diagram of a dc ionization chamber. The radioactive sample (gas or solid) can also be placed inside the chamber.

Current Mode

The most common way of operating an ionization chamber is in the current mode, that is, the electrical signal from the chamber is in the form of a current measurement. When operated in this mode, the chamber is referred to as a dc ion chamber, or a mean-level ion chamber. The current is due to the movement of the ion pairs in response to the applied electric field, and the magnitude of the current is related to the rate of formation of ion pairs. Figure 4 shows a plot of the current from an ion chamber versus the applied voltage. At low voltages, only a portion of the ion pairs are collected (see recombination region in Figure 2). Eventually a voltage is reached at which virtually all the ion pairs are being collected, and the resulting current is stable for a constant activity source. This is called the saturation region, and represents the proper operating voltage for the dc ion chamber.

The output signal from an ionization chamber relies on the collection of both the positive and negative ions. The negative charges are the free electrons or any negative ion formed with the electrons. Because only ions produced by primary ionization events are collected, almost any kind of gas can be used as the fill gas in an ionization

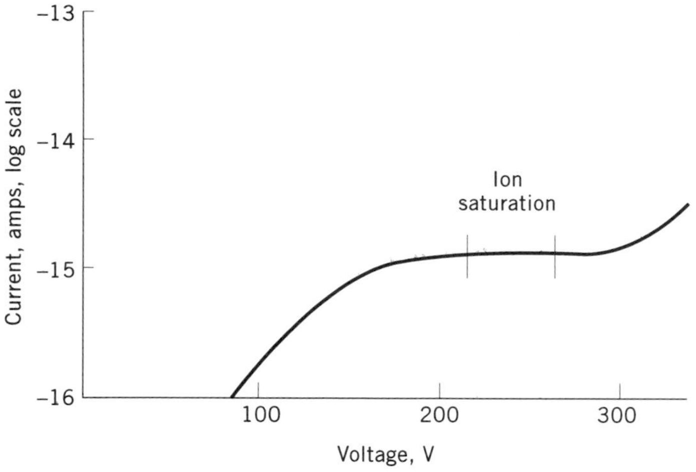

Figure 4. Illustration of the saturation region for a dc ionization chamber. The counter is normally operated at a bias voltage of one-third to one-half way up this region of constant response.

chamber. Even air is often used, in which case small gas leaks in the chamber are not troublesome. Applied voltages up to a few hundred volts are commonly used. These low voltages are easily supplied by batteries for portable radiation monitors. The currents generated in an ion chamber are quite small. For example, an alpha source with an activity of 10^3 disintegrations per second that is emitting 5 MeV alpha particles would produce a current of only around 10^{-11} amps. Therefore, electronic devices such as dc electrometers or vibrating reed electrometers are often used to amplify the signal.

Pulse Mode

The operation of an ionization chamber in pulse mode is based on the measurement of a voltage pulse produced by individual interactions of charged particles with the chamber gas. (This is in contrast to the dc ion chamber, where current flow produced by the cumulative interactions of many radiations entering the chamber is measured). An ion chamber configured to operate in pulse mode is diagrammed in Figure 5. The measured signal is the voltage across the resistor in the circuit. If no radiation enters the chamber, all of the applied voltage appears across the electrodes in the ionization chamber, and the voltage across the resistor is zero. However, when radiation enters the chamber, the ion pairs that are formed collect at the electrodes and reduce the voltage across the electrodes. This results in the appearance of a voltage difference across the resistor that is equal to the voltage drop in the chamber. This voltage difference is the measured quantity. As with the dc ion chamber, the size of the voltage change is quite small, so amplification of some type must be done.

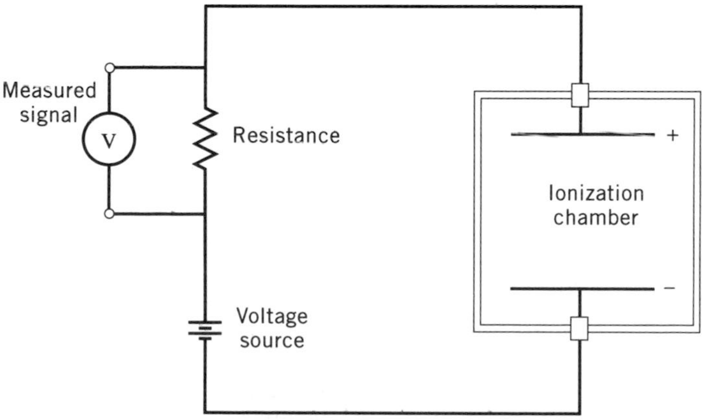

Figure 5. Diagram of an ionization chamber system configured to operate in the pulse mode.

Pulse mode ionization chambers may be operated in one of two ways: collection of both the positive ions and the electrons, or collection of electrons only. In the first case, the time needed to collect the positive ions is long (on the millisecond time scale) due to the lower mobility of the heavy positive ions. Therefore, when positive-and-negative ion collection is used, only low activities can be measured without experiencing pileup of signals. The advantage of this mode is that the amplitude of the output pulse is strictly proportional to the energy of the ionizing radiation, regardless of the location in the chamber of the ionizing event. Collection of only the electrons results in a faster response time, so higher activity levels can be handled. However, the amplitude of the resulting pulse now is no longer precisely related to the incident particle energy, and is sensitive to the position of the interaction in the chamber. These problems may be overcome through use of more complex designs of ionization chambers.

Electrostatic or Charge Integration Ionization Chambers

This is the oldest type of ionization chamber. These devices are similar to electroscopes in principle. Pairs of lightweight metal foils or fibers (alternatively, a foil or fiber suspended from a rigid conductor) are given a static electric charge, resulting in their physical separation because of the repulsion of like charges. The passage of radiation creates ion pairs that migrate to the foils and gradually discharge them. This causes the foils to return to their original positions. The deflection of the foils over time is measured and related to the amount of radiation that has passed through the chamber. The "pocket dosimeter" (or "pen dosimeter") that is often used in health physics applications for measurement of gamma doses is an example of the electrostatic chamber.

Ionization chambers have many applications in radiochemistry. Their advantages include comparative simplicity in construction and a relatively low cost. The system can operate with almost any kind of gas in the chamber, making it very convenient for measuring samples of radioactive gases. Electrostatic chambers can be configured to handle both high- and low-activity levels, and can give information about the energy of the incoming radiations. Because the electronics are simple and stable, they do not require frequent recalibration. They have good detection efficiencies for α and β particles, which are less penetrating, but poor efficiencies for more penetrating radiation, such as gamma rays. A disadvantage of ionization chambers is that the output signal is quite small, so external amplification is necessary.

Direct current (dc) and charge integration chambers are used often as survey instruments or dosimeters in health physics. Because of their high efficiency for α particles and ability to give energy information, they are quite useful for measuring the precise energies of alpha particles and other heavy charged particles, such as fission fragments. The dc ion chamber is especially useful for measuring α (Rn measurements) and weak β (^{14}C) activities of gases. For example, ^{14}C dating measurements can be done inexpensively with a simple dc ionization chamber by incorporating the ^{14}C into CO_2 gas, and running the gas through the chamber. In this case, the counting efficiency for ^{14}C is nearly 100%.

Proportional Counters

A proportional counter is a gas-filled detector that is operated at a higher voltage than the ionization chamber (see region III in Figure 2). As a consequence of the higher electric field, the ion pairs created by the incident radiation are accelerated to a greater velocity as they drift toward the electrodes. The increased amount of kinetic energy means that the collisions between the electrons and other gas molecules along the path are energetic enough to induce secondary ionization, with the release of more free electrons. Therefore, an internal multiplication of the original signal occurs. This multiplication process, called the Townsend avalanche or Townsend cascade, is illustrated in Figure 6. If conditions are proper, the amplification can be kept linearly proportional to the original number of ion pairs produced by the incident radiation. Therefore, energy information is retained.

A diagram showing two common configurations for proportional counters is given in Figure 7. Part a shows a sealed, cylindrical tube counter. The tube itself serves as the cathode, and a fine wire passing through the center of the tube acts as the anode. The sample must be placed outside the tube, so a thin window is required to allow passage of the incident radiation. A different configuration, which allows for continuous gas flow and internal sample placement, is shown in Figure 7b. This arrangement is better for radiations that have little penetrating power (soft radiations) and would be partly absorbed in passage through a detector window. Proportional counters are nearly always operated in the pulse mode.

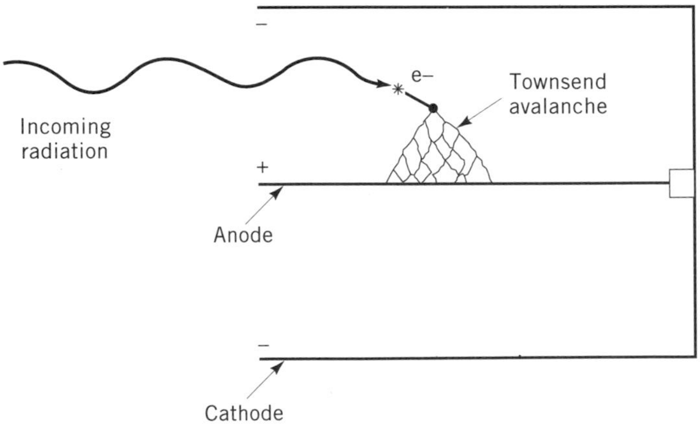

Figure 6. Illustration of the Townsend avalanche in a gas-filled radiation detector (adapted from Ref. 1).

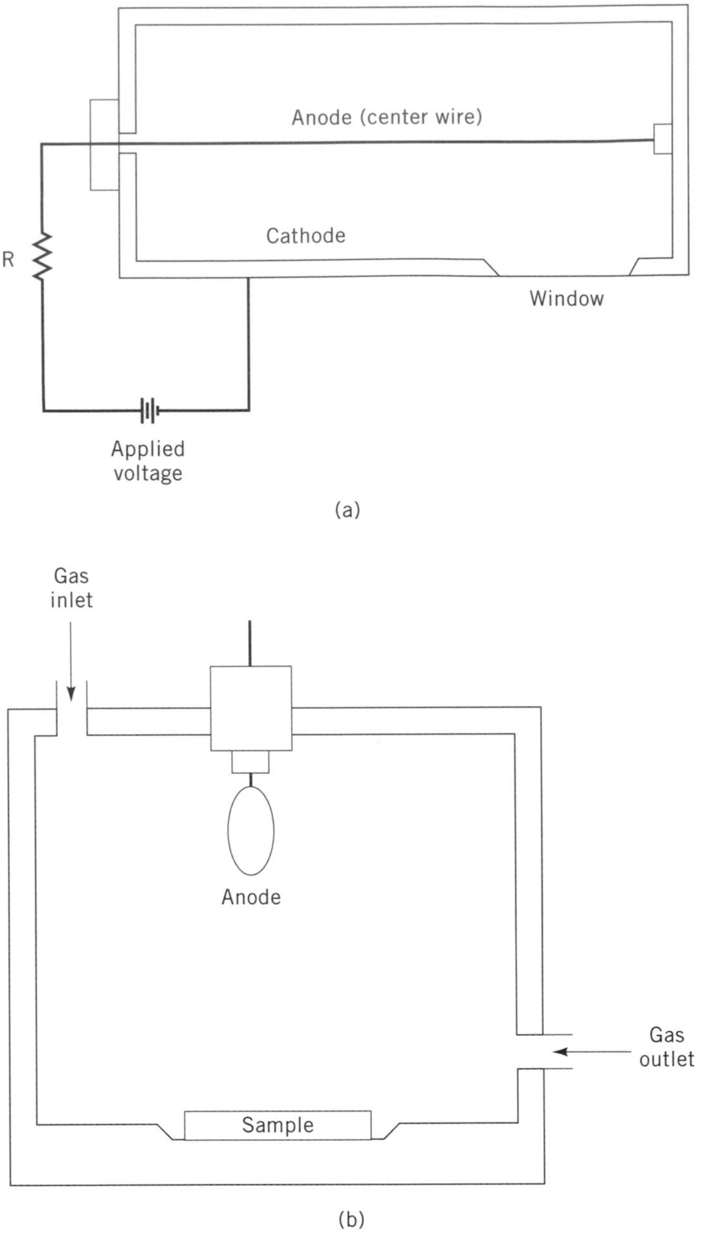

Figure 7. Typical proportional counter configurations: (a) a sealed cylindrical detector; (b) a gas-flow system with a loop electrode.

While nearly any gas will work in an ionization chamber, this is not true for proportional counters. In proportional counters, the signal results from the collection of free electrons, and the formation of negative ions is undesirable. Therefore, the fill gas should not contain any components with a strong electron-attracting tendency that would form anions. The gases that best fill this requirement are the noble gases, mainly argon, helium, and xenon. A second requirement for optimum operation of the proportional counter is that secondary Townsend avalanches not occur with any great frequency. These secondary avalanches can be induced by ultraviolet photons produced by atomic de-excitation processes in the filler gas. To prevent this, another component called a quenching agent is added to the filler gas. This component is a molecular species that will receive energy from the excited noble gas ions and then de-excite mainly by dissociation or other nonradiative modes. With proper selection of filler gas and quenching agent, the Townsend avalanche in the proportional counter remains localized and does not spread throughout the tube. A commonly used proportional counter gas mixture is called P-10 gas. It consists of 90% Ar, with 10% CH_4 acting as the quenching agent.

Proportional counters have the advantage of producing a large output signal, so less external amplification is required than for ionization chambers. The internal amplification factor may be as much as 10^3 to 10^4. This simplifies circuitry and eliminates some sources of electronic noise. Because only the voltage pulse resulting from electron collection is utilized, and not that of slow-moving positive ions, proportional counters have a fast response time. Dead times as low as a few tenths of a microsecond can be obtained in some applications. They also have a very high efficiency for particles with short ranges, such as heavy charged particles or low-energy β particles. However, they have poorer energy resolution than ionization chambers because of the voltage dependence of the internal secondary ionization process. Proportional counters are used for the detection of low energy x rays, neutrons, and mixed α-β sources. They are rarely used for gamma detection, because of the low efficiency for high-energy, penetrating radiations.

Proportional counters are the most commonly used gas-filled detectors for determinations of environmental radionuclides. Proportional counters are used to determine gross alpha/beta activity, alpha emitters such as ^{226}Ra, ^{232}Th, U, the Pu isotopes, ^{237}Np, and ^{244}Cm, and beta emitters such as ^{89}Sr and ^{90}Sr. The ability of proportional counters to distinguish the alpha and beta activities in a mixed source is very useful in this regard. This can be accomplished because of differences in the interactions of α and β particles in the proportional counter. For a given voltage, the internal counter multiplication factor is constant. However, the output pulse will be directly related to the number of ion pairs formed in the primary interaction of the incident radiation. Alpha particles will produce a larger output pulse than β particles of the same energy because they produce higher specific ionization (more ion pairs/mm of path), and because they are completely stopped within the detector volume. Beta particles of moderate energy are more penetrating and may leave the active detector volume before losing all their energy. With a counting system employing a lower-level electronic discriminator (LLD), pulses due to α particles can be detected at a lower applied voltage than can β particles of comparable energies.

The counting characteristics of a proportional counter exposed to a mixed α and β source are illustrated in Figure 8. It is important to note that this figure plots counting rate versus voltage, not number of ion pairs versus voltage as in Figure 2. For very low detector voltages, as in the lowest voltage setting of this figure, only high-energy background radiations or sporadic electrical noise would generate voltage pulses sufficient to exceed the LLD setting. This region is not used. In the α plateau region, the voltage is sufficient so that α particles produce pulses large enough to pass the discriminator, but beta particles cannot. Therefore, with a detector voltage set in this region, only α particles are detected. The third region represents a detector voltage region where all α's are counted, but only some of the β particles. The number of β particles counted will depend strongly on the detector voltage, because electrons emitted in β decay have a distribution of energies up to a maximum, rather than a single energy. This region is not useful for counting. In the $\alpha + \beta$ plateau region, the voltage is high enough so that all α and nearly all the β particles produce pulses large enough to exceed the LLD and be counted.

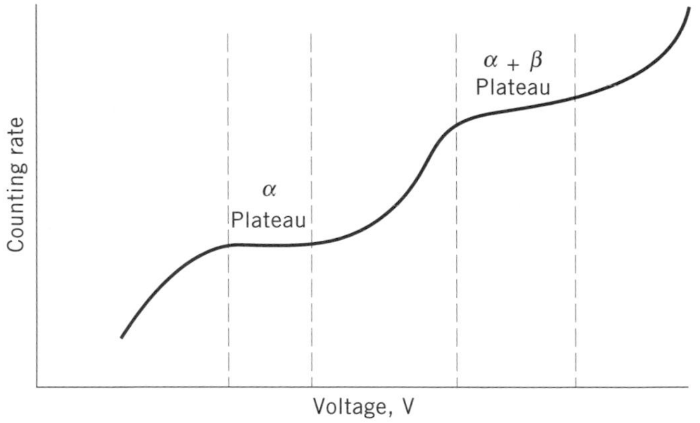

Figure 8. A plot illustrating the two plateaus obtained in a counting rate versus bias voltage plot for a proportional counter and a source that is emitting both α and β particles.

The β plateau in the operating region of the proportional counter is not flat as it was for a pulse ion chamber or the proportional counter α plateau. This is because β particles are not monoenergetic, and higher detector voltages will cause lower-energy β particles in the spectrum to pass the LLD. Beta particles with energies that are very low with respect to the most probable energy in the β spectrum are also low in abundance. Hence, small changes in detector voltage will have only a small effect on the counting rate in the β plateau region. The β plateau is reached after pulses from β particles with energies somewhat below the most probable energy in the spectrum pass the LLD.

Counting efficiencies for α and β particles may vary. This is because some β particles, which are more penetrating than α particles, may escape from the active volume of the counter while depositing an insufficient amount of energy to be recorded. The difference in count rate between region IV and region II in Figure 8 can give a measure of the β activity of the mixed sample.

Geiger-Müller Counters (G-M Counters)

The third type of gas-filled detector is the Geiger-Müller counter, often referred to simply as a Geiger counter. These detectors are the most commonly used counting devices when gross activity level alone is of primary interest. In the G-M counter system, the voltage applied to the chamber is higher than for either ionization chambers or proportional counters. As a result, the electrons formed when radiation passes through the filler gas are accelerated strongly toward the anode and attain high kinetic energies. The electrons will induce secondary ionization, similar to what took place in proportional counters. However, in the G-M tube, the greater energy and larger number of collisions result in the excitation of many gas molecules, some of which will de-excite by emission of energetic photons. These photons can themselves trigger other cascades elsewhere in the tube. The result is that a propagated Townsend avalanche spreads around the entire anode of the detector. Figure 9 illustrates this phenomenon.

When a detector is operating in the G-M voltage region, the interaction of any radiation, regardless of type or energy, leads to the discharge of the entire tube. The discharge eventually stops due to the development of a positive ion sheath around the anode. The sheath forms because the more massive positive ions cannot migrate as rapidly as the electrons. In contrast to the other gas-filled counters, the G-M counting rate reaches a single plateau with increasing voltage, regardless of particle type or energy. This is similar to the α plateau in pulse ionization chambers and proportional counters. If the voltage applied to the detector is increased significantly, the insulating capacity of the fill gas is exceeded and the detector enters the region of continuous discharge. In this region the detector is not directly responsive to sample activity. The operating voltage for all pulse gas-filled counter systems is set at about one-third to one-half of the way through the plateau in the counting rate versus applied detector voltage curve.

The Geiger-Müller counter is similar to the proportional counter in construction and in requirements for fill gas. Sealed tubes are also readily available. Because the output signal from the G-M tube relies on electron production, the fill gas must not contain any high concentrations of strongly electron-attracting species. The noble gases, especially helium and argon, are most used.

The problem of repetitive pulses occurring in the tube due to secondary electron release by the positive ions as they reach the cathode is more severe for the G-M tube than for proportional counters. Quenching agents are essential for suppression of these repetitive pulses. In gas-flow detectors, these agents are usually organic molecules, such as ethyl alcohol, isobutane, or ethyl formate. In sealed tubes, halogen gases, mostly small amounts of Br_2 or Cl_2, are used. The excess energy released in the neutralization of the positive ions at the cathode is dissipated by dissociation of the quench gas molecules or excitation of vibrational or rotational modes of freedom in these molecules. Thus, subsequent emission of secondary electrons at the cathode is suppressed. Dissociation results in a gradual depletion of the organic quench-gas molecules in a sealed tube over time. Because halogens can recombine after dissociation, they retain their usefulness as quenching agents much longer and are preferred for use in sealed G-M tubes. Q-gas is a gas mixture commonly used for gas-flow G-M tubes. It contains 97% He and 3% of an organic quenching agent such as isobutane.

The advantages of G-M counters include simplicity of design, durability, insensitivity to small changes in applied voltage (in the plateau region), and low cost. Because the entire tube discharges in response to the interaction of radiation, the signal size is large enough so little external amplification is needed. This simplifies design and construction. However, their disadvantages include the lack of ability to provide information about the energy or identity of incoming radiation, and a relatively long dead time (100–500 μs) required to clear the many positive ions formed. The latter limits their usefulness at high counting rates unless corrections are made. Geiger-Müller tubes are frequently used in survey meters in both the laboratory and under field conditions, because of their low cost and sensitivity to most common types of radiation.

SCINTILLATION DETECTORS

Scintillation detectors, like ionization chambers, have a long history in the study of radioactivity. By the early 1900s, workers had observed that certain materials would emit a flash of light when α particles were allowed to impinge on them. (The word *scintillation* comes from the Greek word for spark). Ernest Rutherford used a scintillation detector, in the form of a screen coated with zinc sulfide (ZnS), to detect the α particles used in the experiments that established the nuclear model of the atom. In those days, a microscope was used to watch for and count the light flashes. Radiation detection by this means was extremely tedious, and it provided the incentive for Rutherford's co-worker Geiger to begin working on an electronic method for counting α radiations. Scintillation de-

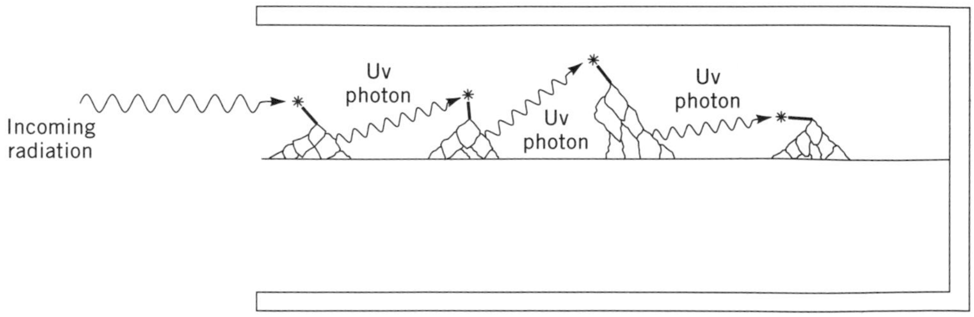

Figure 9. The propagated Townsend avalanche in a G-M detector (adapted from Ref. 1).

tectors operate on a much more sophisticated basis now than the visual counting of light flashes. However, the basis for scintillation detection is still the radiation-induced emission of light by the detecting material.

Although the details of the process will differ among the various types of scintillation detector systems, there are some similarities in their basic operating principles. The incoming radiation interacts with the scintillating material in the usual ways, producing ionization and atomic or molecular excitation. However, scintillation detectors are not based on direct measurement of electrical signals from the ion pairs that are produced in the radiation-absorbing medium. Rather, the excited atoms or molecules undergo de-excitation by emission of a photon of light. This light passes through the scintillating material, then through a "light pipe" to a photomultiplier tube, which converts the light to an electrical signal. The signal is then processed electronically for display or storage.

From this simple description, some of the properties desirable in a good scintillating material may be deduced. First, it must have the appropriate molecular or crystalline structure to react to the passage of radiation with the emission of light, and it should be able to do this efficiently. The light production process is called luminescence. The light must pass through the material to be detected, so the scintillation material must be optically transparent to the wavelengths of light generated. Long delays between the time of radiation interaction and the actual emission of light (phosphorescence) are undesirable for high counting rates, so the light emission should take place quickly (fluorescence) after interaction. Solid scintillating materials should have the appropriate physical properties to allow them to be formulated into crystals and easily machined to enable coupling with other system components. Ideally, the scintillator should be chemically stable and not hygroscopic, so that handling is easier. However, the crystals that have the best overall properties as scintillation detectors are hygroscopic. They must be maintained in air-tight containers.

Scintillation materials fall into two major groups: inorganic and organic.

Inorganic Scintillation Detectors

The most widely used inorganic scintillation detectors are the alkali halide crystals, especially sodium iodide (NaI). These crystals can function as radiation detectors because of electron transitions that take place between energy levels in the crystal in response to the passage of radiation.

In crystalline solids, the individual atomic energy levels form groups of very closely spaced orbitals called bands. In the crystal, the electrons can occupy two possible energy levels, as shown in Figure 10. The valence band is a lower-energy band containing the electrons bound to the crystal lattice. These electrons are not free to move through the crystal. If only these bands are occupied, the material will be an electrical insulator. The conduction band is at a higher-energy level and contains electrons that are free to move within the crystal. If this level is populated with electrons, the material is called an electrical conductor. Between these two energy levels is the band gap, which contains forbidden energy levels. Absorption of energy (from incoming radiation, for example) can excite an electron into the conduction band. As the excited electrons return to the valence band, they lose energy by emitting a photon of light.

The scintillation process described above is not very efficient for a crystal of pure NaI for two reasons. One is that the band gap is so large that electron transitions are not highly probable. Second, even if electron transitions

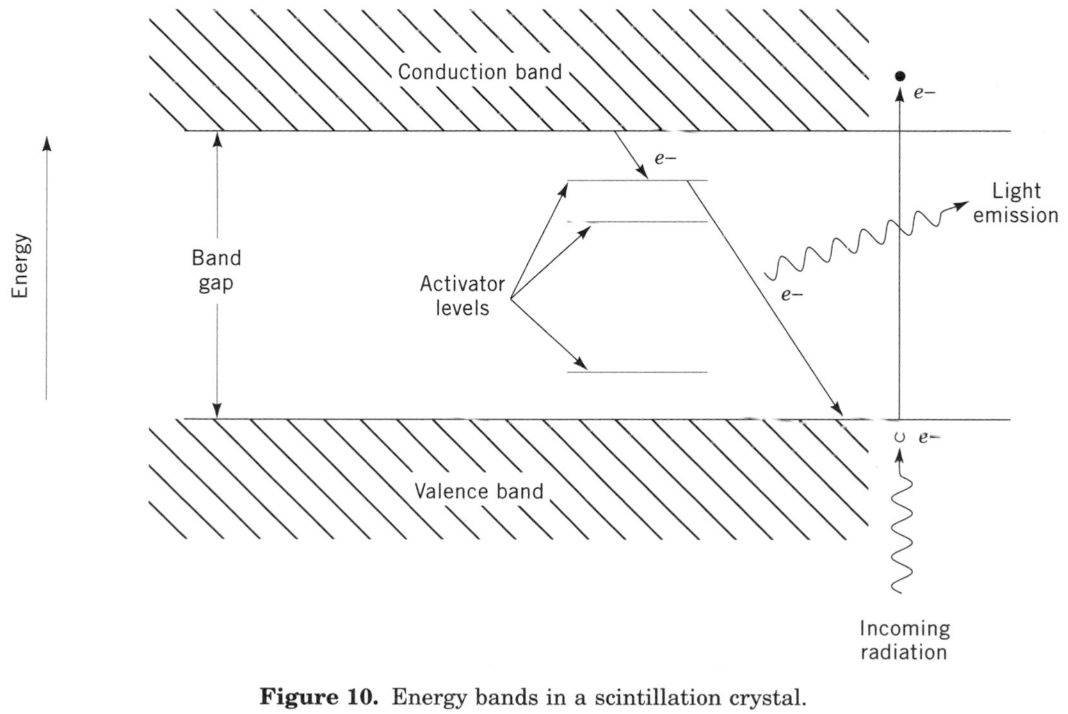

Figure 10. Energy bands in a scintillation crystal.

did occur, the emitted photons would be of such a high energy that they would not be near the visible part of the spectrum where amplification via a photomultiplier tube is most efficient. To overcome these problems, an impurity called an activator is added to the crystal. This creates new energy levels within the band gap. Electron transitions to these new levels (activator levels) do not require as much energy, so they are more likely to occur. In addition, when de-excitation takes place, the emitted light has a lower energy, and is more likely to be in the near-ultraviolet or visible region of the electromagnetic spectrum. For NaI scintillation crystals, thallium (Tl) is a good activator. This system is commonly designated as an NaI(Tl) scintillation crystal.

The properties of the NaI(Tl) scintillation detectors make them best suited for the detection of γ rays, x rays, and high-energy electrons. These detectors are very widely used, and are likely to be found in almost any laboratory doing radioactivity measurements. The characteristics that make them useful for γ-ray spectroscopy include good efficiency, high light output, and a linear energy response. The good efficiency for γ rays arises from the fact that the crystals are moderately high density solids (in contrast to the gas-filled counters) and contain a high-Z element, iodine, which enhances the photoelectric process interactions of γ rays. The efficiency advantage is further enhanced by the large size to which NaI(Tl) crystals can be grown. Cylindrical sizes of 12.7 cm in diameter by 12.7 cm in length are common. These are usually referred to in the U.S. literature as 5-by-5's, in reference to their dimensions in inches. The NaI crystals can also be formed into unusual shapes for special applications. One shape that is very often used is a well crystal. This is a cylindrical crystal into which a cylindrical well has been drilled. The sample is put into the well for counting. For radiations whose energy is not too large, high counting efficiencies can be achieved for this configuration. The light yield from the NaI(Tl) detector is better than any of the other solid scintillators, and the relationship between the energy of the incoming radiation and the light output is linear.

The biggest disadvantage of NaI(Tl) crystals for gamma-detection is their poor energy resolution. The best resolution currently available would be about 7.5%, as measured by a 3×3 NaI(Tl) for the 662-keV gamma ray emitted by ^{137}Cs. The reasons for the poor resolution are intrinsic to the scintillation process itself, so improving technology will not be able to improve the resolution. The time needed for the scintillation to decay is around 230 ns. This is a long time, in comparison with other scintillation materials. Therefore, NaI(Tl) detectors are not ideal for high-count rate applications or for experiments requiring very fast timing responses. NaI crystals are very hygroscopic and so must always be covered with sealed metal containers.

The simplified gamma spectrum sketched in Figure 11 illustrates the response of a NaI(Tl) detector to a source consisting of ^{137}Cs and ^{60}Co. A spectrum obtained with a semiconductor detector is also shown. The latter will be discussed shortly. The broad peaks in the NaI(Tl) spectrum reflect the poor energy resolution of this detector, while the greater areas under the peaks illustrate the superior efficiency of the NaI(Tl) detector over a semiconductor detector of similar size. As will be noted later, high-efficiency semiconductor detectors have recently become available, but they are much more expensive than equivalent-efficiency NaI(Tl) detectors.

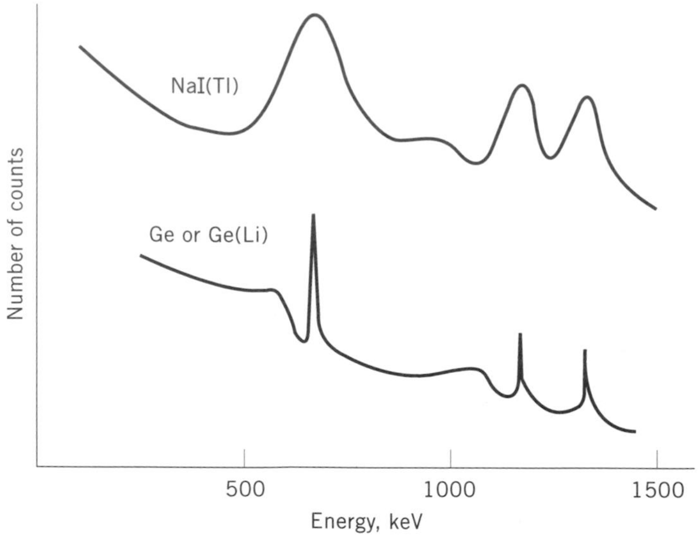

Figure 11. Comparison of γ-ray spectra for a NaI(Tl) scintillation detector and a semiconductor detector using a mixed ^{60}Co and ^{137}Cs source.

Other alkali halides used as detectors include CsI(Tl), LiI(Eu), and CsF. CsI(Tl) has the highest gamma-ray efficiency (photons produced per MeV of energy deposited) of any scintillation crystal. This is due in part to the high-Z element composition of the crystal. It also has the unusual property of having a variable pulse decay time for different radiations. For example, it would be easy to distinguish positive charged-particle interactions from electron interactions by looking at the pulse shape. The LiI(Eu) crystal is used largely for neutron detection, because of the high probability of interactions of neutrons with Li. CsF has the advantage of a very fast decay time (a few ns). None of these detectors is as widely used as the NaI(Tl) detectors.

There are a few other inorganic materials that are used as scintillation detectors. These include glasses that contain B and Li (used for neutron detection), ZnS-coated screens (used for heavy charged-particle detection), CaF_2 (useful under hostile environmental conditions), and some noble gases. A recently developed inorganic scintillation detector that is gaining in popularity is bismuth germanate ($Bi_4Ge_3O_{12}$), abbreviated as BGO. Because of the high-Z elements it contains, its efficiency for gamma rays is very high. Its light output is lower than NaI(Tl), however, and it is more expensive.

Organic Scintillators

Most of the organic molecules that are used as scintillation detectors have a pi-electron system. For these molecules, light emission occurs as a result of electron transitions within the individual molecules, rather than between the energy levels that exist due to the crystal lattice. Because crystal structure is not important, organic scintillators may be either liquid or solid. They are used primarily for β- or α-counting rather than for γ counting, because most organic molecules are composed of only low-Z elements.

Liquid Scintillation Counting (LSC). Liquid-scintillation detection systems are used extensively in environmental studies to obtain quantitative data for α-, β-, and γ-emitting nuclides. The capability of liquid scintillation methods to determine the weak beta emitters ^{3}H and ^{14}C has also made this detection method indispensable for biological, medical, biochemical, and organic chemistry studies.

The performance of liquid scintillation counting requires that the sample be mixed with a scintillating solute and a solvent. This mixture is called a cocktail. The cocktail is placed into a small vial, about 6 cm (2.36 in) tall and 3 cm (1.18 in) wide, made of glass or translucent plastic. The radiation emitted by the sample interacts with the solvent molecules first, because there are many more of them than there are of the solute. The solvent transfers the absorbed energy to the scintillating solute, which then de-excites by emission of a photon. The emitted photon must be at a wavelength that the photomultiplier tube can effectively respond to. If it is not, another component, called a wavelength shifter or secondary scintillator, can be added to the liquid scintillation cocktail. The wavelength shifter will efficiently absorb the higher energy light emitted by the primary scintillator and re-emit it at a lower energy that is more suitable for the phototube. The emitted photons strike the photocathode and cause the emission of photoelectrons. These electrons are multiplied in the photomultiplier tube and a signal is recorded. The magnitude of the pulse is proportional to the energy deposited in the scintillator, but the amplitude and shape of the pulse are related to the type of radiation that deposited the energy in the cocktail.

Toluene and *p*-xylene are the solvents that have been used traditionally for liquid scintillation counting. Three common scintillating solutes are 2,5-diphenyloxazol (PPO), *p*-terphenyl, and tetraphenylbutadiene, with PPO the most widely used. A substance that has been frequently used as a wavelength shifter is POPOP (1,4-bis-[2-(5-phenyloxazolyl)]-benzene). A related compound, *p*-bis-(o-methylstyryl)-benzene (bis-MSB) is even more efficient. Aqueous samples will not readily dissolve in organic solvents, so sometimes still another component (an emulsifier) must be added to the cocktail to aid in the dissolution process. Three emulsifiers in common use are Triton N57, Hyamine 166, and dodecylbenzene sodium sulfonate.

There have been many advances in scintillation cocktails in recent years due to analytical requirements for increased sensitivity and compatibility with samples, and due to regulatory restrictions on the use and disposal of chemical reagents. Manufacturers now market a variety of premade cocktails that contain the reagents appropriate for various applications. These cocktails are effective scintillators, easy and safe to handle, and at the same time environmentally more friendly than traditional cocktail ingredients.

Environmental applications of liquid scintillation have been mainly for beta emitters such as tritium, ^{32}P, ^{90}Y, ^{63}Ni, and ^{99}Tc. Liquid scintillation has been less used for α-emitters because of poorer efficiency of light production and interference from β- and γ-produced scintillations. Recently, however, the application of liquid scintillation to α determinations has increased.

Solid Organic Scintillators. The solid organic scintillators are less widely used for environmental applications than either the inorganic scintillation detectors or liquid scintillation counting. Stilbene and anthracene are examples of crystalline organic scintillators. Anthracene has good light output for an organic scintillator, and stilbene, like CsI(Tl), exhibits different decay times for different types of radiation. It can, therefore, be used to identify the type of particle inducing the scintillations. Another solid organic material that has been used as a detector is a polymethyl methacrylate plastic that has had an organic scintillator dissolved in it before polymerization. Plastic has the obvious advantage of being able to be made into many different shapes. None of the solid organic scintillants is as frequently used as the solid inorganic scintillants or the liquid scintillants.

The Photomultiplier Tube

The radiation-induced photon emitted by a scintillation detector is only the first step in the detection process. These very weak light signals must be converted into a useful electric signal. The photomultiplier tube (PMT) accomplishes this task.

Figure 12 shows the basic structure of a PMT coupled to a scintillation detector. The PMT is housed in an opaque container, because external light must be completely excluded for the PMT to respond only to the scintillation photons. The scintillation process occurs in the detector and the photon passes first through a light pipe that serves as an optical coupler between the detector and the PMT. At the end of the PMT tube is a layer of a photosensitive material called the photocathode. The photon from the radiation interaction strikes this photocathode and causes the emission of electrons referred to as photoelectrons. The photoelectrons are attracted toward the anode. Before reaching the anode, however, they encounter a series of dynodes whose function is to increase the number of electrons that ultimately reach the anode. When the photoelectrons strike the first dynode, they induce the emission of more electrons. These electrons are accelerated to the next dynode, where each causes still more electrons to be emitted. This electron multiplication process continues for several (8–12) dynodes, with the end result being a large shower of electrons collected at the anode for each single photoelectron originally emitted from the photocathode. The gain (number of electrons out per photoelectron in) can range from 10^{7} to 10^{10} for a typical PMT. The pulse produced, however, is still proportional to the number of original photoelectrons, which in turn is controlled by the intensity of the light produced in the interaction. The light intensity, in turn, is proportional to the energy of the radiation that caused the scintillation. Thus, energy information can be obtained. Photomultiplier tubes are used with all of the scintillation detectors discussed here.

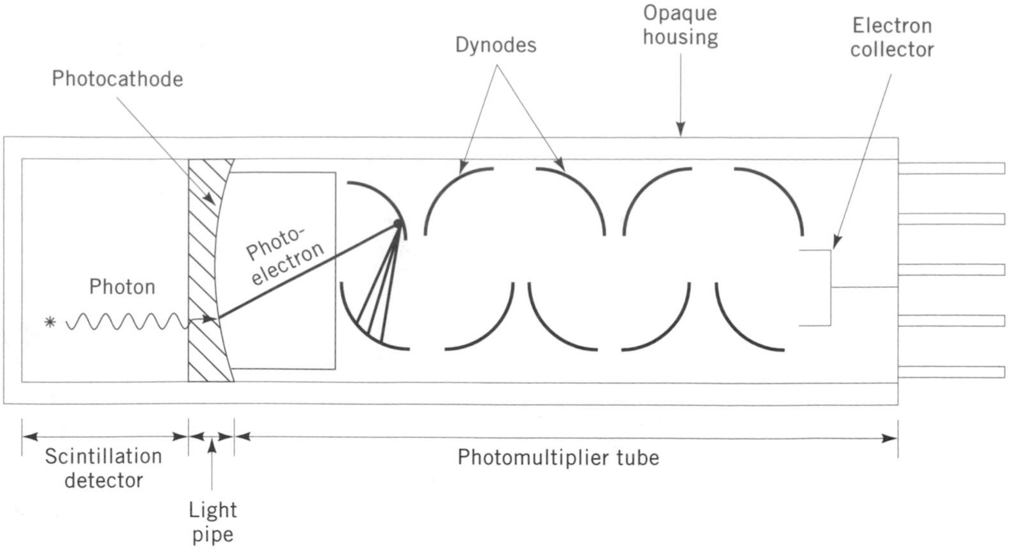

Figure 12. Schematic diagram of a solid-crystal scintillator coupled through a light pipe to a PMT.

There are sources of noise in circuits using a PMT that cause a background counting rate not related to sample activity. It is possible for photoelectrons to be emitted spontaneously from the photocathode due to thermal energy, in a process called thermionic emission. Because each background signal from this source is the result of the thermionic emission of a single electron, the amplitude of the signal produced is very small, even after electron multiplication. A discriminator may be used to eliminate this low-amplitude noise. Reductions in thermionic emission may be achieved by cooling the PMT and by varying the photosensitive material used on the photocathode. An electronic solution to the problem is to use a coincidence unit. In this arrangement, two PMTs are arranged at 180° to each other around the sample. A true scintillation event will trigger pulses in both PMTs that will arrive simultaneously at the coincidence unit and will be recorded as a true event. Thermionic emission in only one PMT is not likely to be matched by a similar event in the other, so this false signal will be excluded by the coincidence circuit, which records only coincident events. Another source of PMT noise is the production of photons within the tube, due to the naturally occurring ^{40}K or U and Th natural decay series radionuclides in the glass envelope or other construction materials of the PMT. These radiations could produce electrons at the photocathode that would be multiplied together with true photoelectrons related to the sample activity. This source of electrical noise is referred to as the dark current. These are also low-intensity events. Light leaks in the housing will, of course, also result in noise. However, this source of noise is easily eliminated.

SOLID STATE SEMICONDUCTOR DETECTORS

Gamma-ray spectrometry was advanced greatly by the development of solid-state semiconductor detectors in the 1960s and 1970s. The newest semiconductor detectors offer a combination of good efficiency and high-energy resolution for gamma rays that is not matched by the gas-filled or scintillation detectors.

Theory of Semiconductor Detectors

Semiconductor detectors are usually made of germanium (Ge) or silicon (Si) crystals. A band structure exists in the semiconductor crystal, similar to that described earlier for the solid scintillator (see Fig. 13). However, the band gap is much smaller for semiconductors than for the NaI(Tl) detector. For Si, the gap is about 1.1 eV, and for Ge, about 0.66 eV. The passage of radiation may inject enough energy into the system to raise an electron from the valence band to the conduction band, thus creating an electron-hole pair, analogous to the ion pair created in the gas-

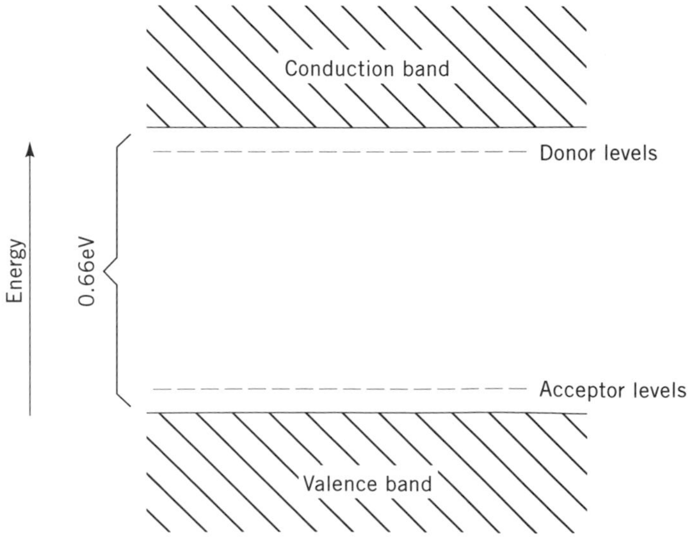

Figure 13. Donor and acceptor energy levels in the band gap of a semiconductor crystal.

filled counters. The hole (which is really the absence of an electron) and the electron can both migrate through the crystal in response to an electric field and produce an electrical signal that marks the passage of the radiation.

Completely pure semiconductor crystals composed *only* of atoms of Si or Ge can be described in theory but cannot be made in practice. They are called intrinsic semiconductors. Real-world semiconductor materials contain various impurities that have a great effect on their electrical properties. Sometimes the impurities are added purposely, in a controlled manner, to effect some desirable change in the electrical characteristics of the semiconductor. Therefore, when discussing the use of semiconductor materials for radiation detection, the nature of the impurities and the effects they have must be understood.

The type of impurities that are important in semiconductor crystals have either one more or one less valence electron than the four normally possessed by neutral Ge or Si atoms. As an example, consider the addition of P to a Ge crystal, as shown in Figure 14. Because P has one more valence electron than Ge, there will be an excess of electrons present. A semiconductor of this type is called an n-type semiconductor. The presence of the P also creates a new energy level in the band gap, called a donor level, that is very close to the conduction band (see Fig. 13). This means that electrons are promoted very easily into the conduction band, and the electrical properties of the semiconductor are controlled by these excess electrons in the conductor band. The electrons in an n-type semiconductor are referred to as the majority carriers. On the other hand, if one of the Ge atoms is replaced with a B atom, an excess of holes in the semiconductor will result. This forms a p-type semiconductor, in which the electrical properties are dominated by the presence of excess holes. The presence of the B creates an acceptor energy level in the band gap that is very close to the valence band (Figure 13). Electrons from the valence band may have enough thermal energy to be promoted to the acceptor level, leaving behind the excess holes. Thus, the presence of the impurities increases the number of charge carriers, either electrons or holes, that are available.

To use a semiconductor as a radiation detector, the first approach might be to apply an electrical field to a block of n- or p-type semiconductor material and collect the charge carriers created when ionizing radiation passes through the crystal. However, this is not a practical way to use a semiconductor for radiation detection. The designs that have been found to be most useful for radiation detectors are based on the effects that occur when an n-type and a p-type semiconductor are in contact, forming a p-n junction.

A diagram of a p-n junction is shown in Figure 15. If the two types can be brought into intimate contact, movement of carriers in the immediate vicinity of the p-n junction will occur. Electrons will migrate toward the p zone, and holes toward the n zone. This will create an area where there are neither excess holes nor excess electrons, and thus a depletion of charge carriers. This is called a depletion region, and it strongly resists the passage of electric current. If radiation strikes the depletion region, charge carriers are created and are attracted toward the n or p side. Therefore, a depletion zone should be able to serve as a useful radiation detector.

A p-n junction created this way, without application of any external voltage, does not work very well as a detector. The depletion zone is very small, so most types of

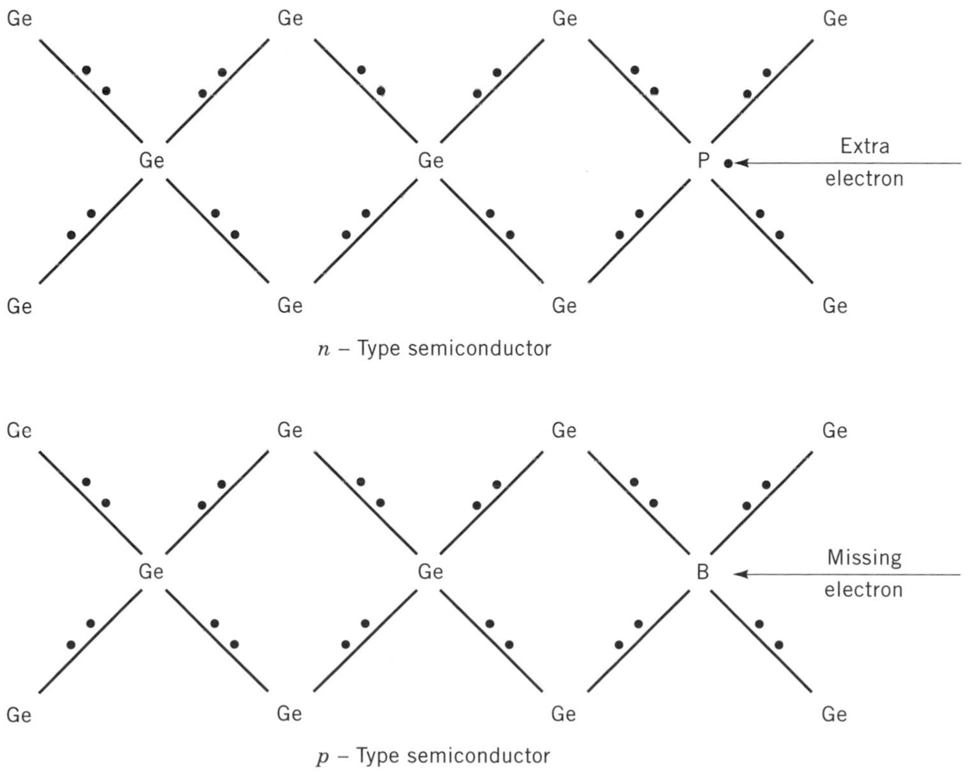

Figure 14. Semiconductors with n-type and p-type impurities.

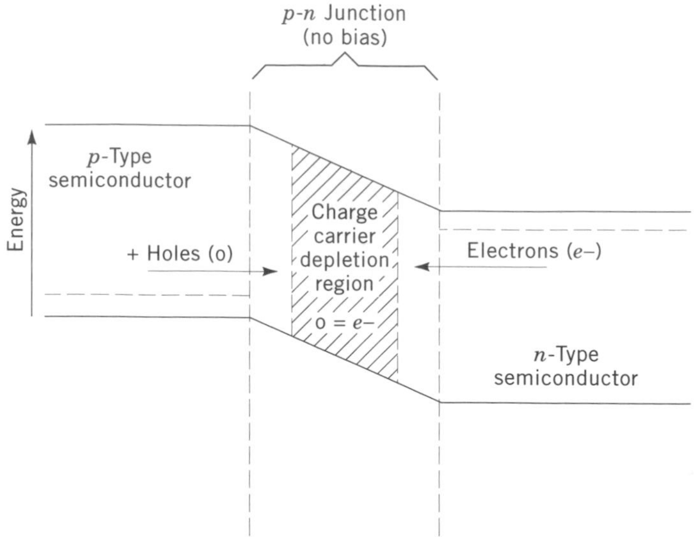

Figure 15. Generation of a charge-carrier depletion region in a semiconductor p-n junction with no applied bias.

radiation would pass right through it undetected. In addition, the difference in electrical potential that is naturally created between the p and n sides is not large enough to collect all the charge carriers that form in the depletion zone. Therefore, a more useful detector results if a reverse bias is applied to the p-n junction, as diagrammed in Figure 16. Application of a positive voltage to the n side results in the movement of the majority carriers (electrons) away from the depletion zone, while the negative voltage applied to the p side causes the holes to move away from the depletion zone. Thus, the application of a reverse bias intensifies the situation that developed spontaneously in

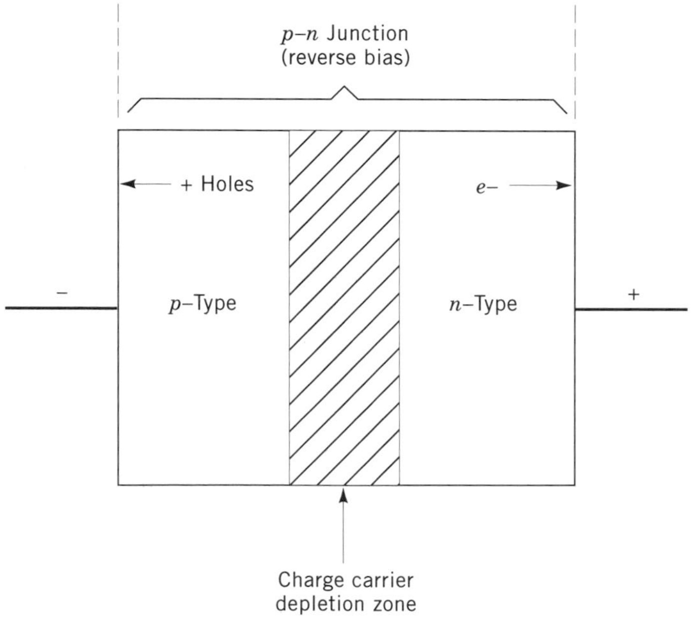

Figure 16. Generation of a large charge-carrier depletion zone in a semiconductor p-n junction by applying reverse bias.

the p-n junction. As the electrons and holes both move farther away from the junction, the size of the depletion region increases, and its resistance to the passage of electricity increases even more. All of the semiconductor detectors are based on the properties of these reverse-biased p-n junctions. It is not practical to create a p-n junction by physically joining the two types of semiconductor, because void spaces left between the p and n types are too large. Therefore, the p-n junction must be created within a single crystal. There are a variety of semiconductor detectors that differ primarily in the way in which the p-n junction is formed.

Surface Barrier Detectors

A diagram of a surface barrier detector is shown in Figure 17. The base material for these detectors is n-type silicon. The top surface is chemically oxidized to produce a p-type region on the surface. A thin layer of gold is then applied to provide an electrical contact. Application of tens of volts creates a small depletion zone with a maximum size of 2 or 3 mm (.079 or .118 in). There is only a very thin region above the depletion zone that is not responsive to radiation, so these detectors have a small dead zone.

Because of their small size and dead zone, the surface barrier detectors are best used for measuring α particles or other heavy charged particles with short ranges. They are not very expensive and have good energy resolution. However, they are not very durable and are not applicable to detecting other types of radiation.

Ion-implanted Detectors

A newer method for the fabrication of charged-particle detectors is the ion implantation-silicon dioxide passivation technology. In this method, the dopants are implanted into the surface of the detector by exposing it to a beam of particles from an accelerator. Ion beams of fixed energy and composition will have well-defined ranges in the semiconductor. Annealing follows the ion implantation step. The advantages of detectors made in this way are that they are more rugged and stable than the surface barrier detectors, can be touched and cleaned without destroying the contact surface, and have very thin entrance windows.

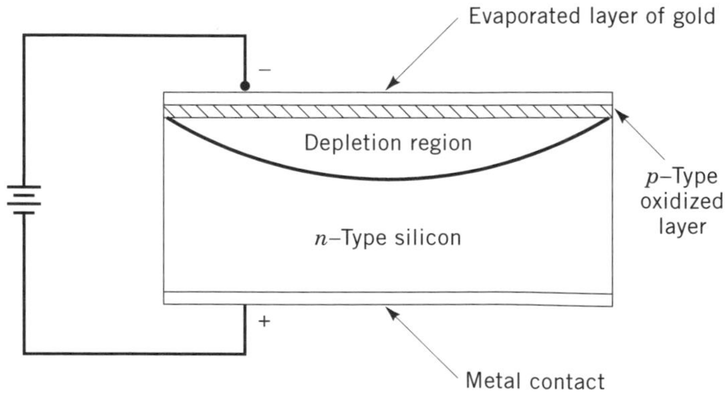

Figure 17. Schematic diagram of a surface barrier semiconductor charged-particle detector.

Lithium-drifted Semiconductor Detectors

It is not possible to use a simple reverse-bias to create depletion zones that are large enough to detect γ rays efficiently. The applied voltage can only be increased to a certain value before the system breaks down. Therefore, a different approach must be used. One of these is referred to as ion drifting. The principle is that ions opposite in charge to the majority carriers of the semiconductor are made to enter the semiconductor. The numbers of these ions must be sufficient to just equal the numbers of majority carriers. This process is called compensation. Compensated semiconductor material is similar to intrinsic material in that there is no excess of either carrier. Therefore, the compensated region, like a depletion zone, can act as a radiation detector.

When Ge and Si are manufactured, both tend to end up as p-type materials. To create a compensated region in a piece of Ge or Si, atoms that will donate electrons and act as n-type material must be added. Lithium is the atom used for this purpose because of its small size and high mobility.

To form the compensation region, lithium metal is coated onto one end of a piece of p-type Ge or Si and allowed to diffuse into the crystal at elevated temperatures over a period of days or weeks. The electrons donated by the lithium will exactly compensate for the acceptor impurities already present in the Ge or Si, and a depletion zone some 10 to 15 mm (.39 to .59 in) long can be created. Although it is not strictly correct to do so, the depletion zone is sometimes called the intrinsic region. This process is illustrated in Figure 18. Once the lithium-drifting process is completed, the crystal is cooled to prevent the lithium from drifting farther into or back out of the crystal. The lithium-drifted germanium detector is written as Ge(Li), and pronounced like the word *jelly*. Similarly, lithium-drifted silicon detectors are written Si(Li), and are pronounced like the word *silly*.

The Ge(Li) and Si(Li) detectors are quite different in their applications. Germanium is more desirable than Si for γ-ray spectrometry applications, because its higher Z value makes γ-ray interactions more efficient. Silicon, on the other hand, is more efficient for detection of lower-energy γ rays, x rays, and α and β particles. The difference in the band gap for Ge and Si results in differences in the way the crystals must be used and stored. The smaller band gap in Ge results in more spontaneous electron excitations into the conduction band, and Ge has more noise problems than does Si. This, coupled with the higher mobility of Li in Ge than in Si, means that a Ge(Li) detector must be kept cooled to liquid nitrogen temperatures (77 K) at all times. Even one warming cycle is usually enough to destroy the lithium drifting and ruin a Ge(Li) detector. The Si(Li) detectors, on the other hand, are cooled to liquid nitrogen temperatures only when they are actually in use. They may be stored at room temperature. Because of the need for liquid nitrogen cooling during operation, both Si(Li) and Ge(Li) detectors are attached to large cryostats (see Figure 19), which makes them bulky and difficult to maneuver in some instances.

The semiconductor γ-ray detectors are usually characterized by their resolution, efficiency, and peak-to-Compton ratios. The resolution is given as the value of the FWHM (in keV) for a given gamma peak. Resolutions down to $\approx$1.5 keV at the 1332 keV ^{60}Co peak can be achieved with the best detectors. The efficiency of a semiconductor detector is usually given as a relative efficiency, that is, the efficiency compared to a standard detector. The standard is taken to be a 7.6-cm-diameter by 7.6-cm-long (3 in $\times$ 3 in) NaI(Tl) detector with a point source of activity at a distance of 25 cm (9.8 in) from the detector. For Ge(Li) detectors, relative efficiencies up to $\approx$15% with resolutions better than 2.0 keV at ^{60}Co are common. The peak-to-Compton ratio refers to the ratio of the number of counts in the 1332 keV ^{60}Co peak compared to the number of counts in a selected region of the Compton continuum. Values up to 50 are common for this parameter, with the highest values being preferred.

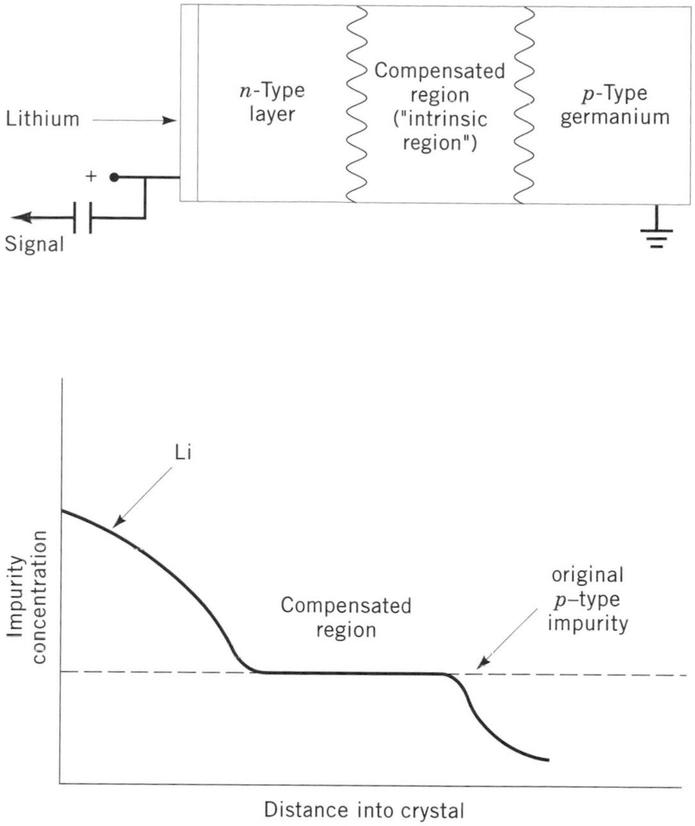

Figure 18. Illustration of lithium drifting to form a charge-carrier compensated region in a germanium semiconductor crystal. The lithium concentration as a function of distance is also shown. Note that the lithium concentration in the compensated region is constant.

The advantages of using drifted semiconductor detectors include their good energy resolution and their reasonable efficiency for γ-ray or high-energy particulate radiation. Lithium-drifted silicon detectors can be made with thin windows for α detection. All semiconductor detectors have low dead times and are relatively insensitive to magnetic fields. Their disadvantages include the necessity for liquid nitrogen cooling, the small size of the output signal (which entails the use of more electronic components), and their greater cost compared to solid scintillation detectors for γ-ray spectrometry.

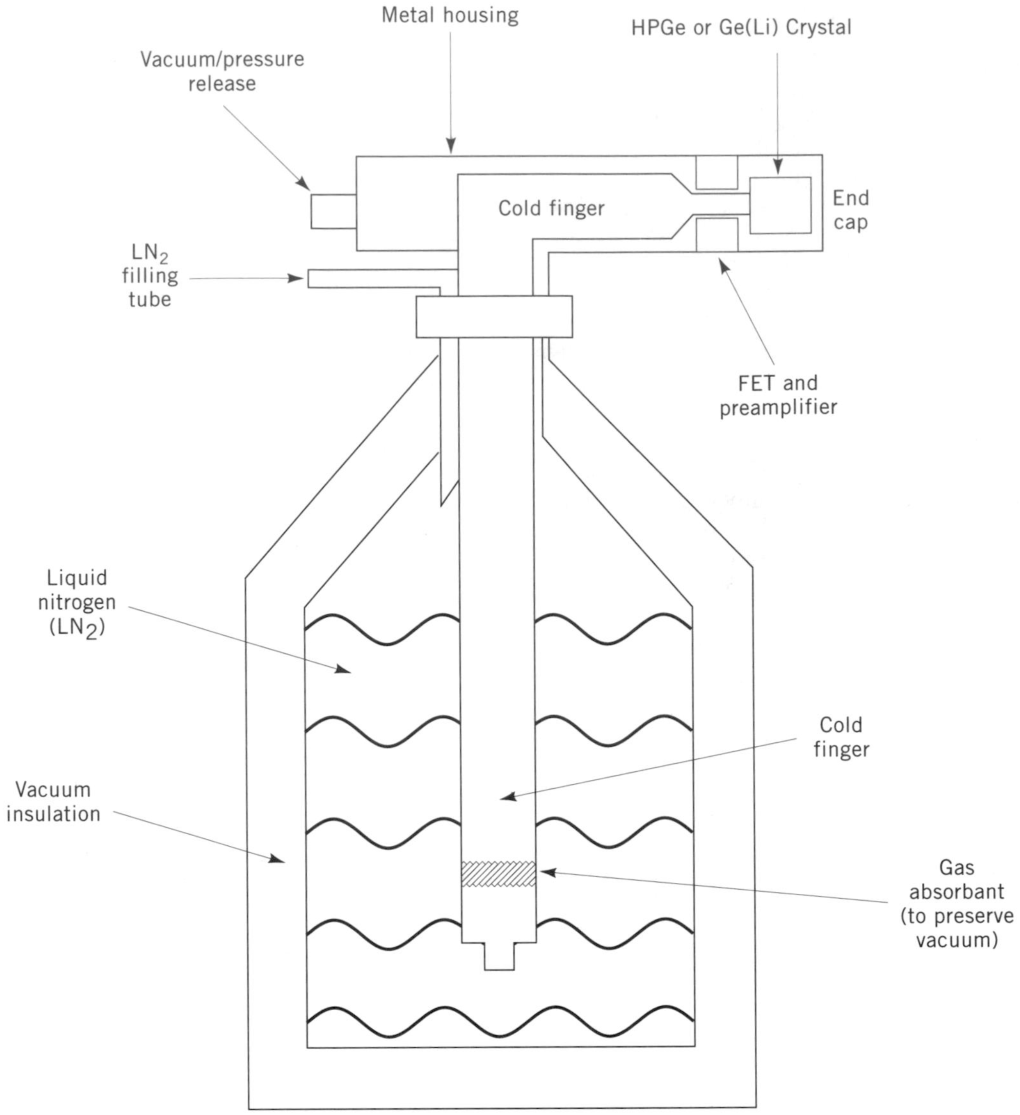

Figure 19. Schematic diagram of a HPGe or Ge(Li) detector system with liquid nitrogen Dewar.

Intrinsic Germanium Detectors

In recent years it has become possible to produce Ge in sufficient purity (less than 10^{10} impurity atoms per cm^3) so that it approaches the properties of the theoretical true intrinsic (pure) semiconductor material. It can therefore be used as a detector without any lithium drifting. These Ge detectors without any Li are called high-purity germanium (HPGe) or intrinsic germanium (correctly, near-intrinsic) detectors. Their great advantage is that they can be stored at room temperature, because there is no Li to drift out of the crystal. They still must be cooled to liquid nitrogen temperatures for operation because the problem of thermal excitation of electrons still exists. Intrinsic Ge detectors are often better than Ge(Li) detectors in their efficiency and resolution characteristics. Some detectors have been reported to have relative efficiencies greater than 100%, but costs can approach $1,000 for each 1% relative efficiency achieved. Other than being able to be stored at room temperature when not in use, intrinsic Ge detectors share most of the advantages and disadvantages of Ge(Li) detectors. They have largely replaced the Ge(Li) detector in the current market.

Gamma spectrometry is used frequently in environmental analysis to perform gamma scans. Samples are counted and data collected over the energy range of 0.1 to 2.0 MeV. The purpose of these scans is to determine the identities and activity levels of any gamma emitters in the sample. Often, sample preparation is minimal. Gamma-ray spectrometry is also used in neutron activation analysis. This technique offers the possibility for simultaneous determination of 30 to 40 elements in a single sample with minimal sample preparation.

OTHER COMPONENTS OF ELECTRONIC DETECTION SYSTEMS

Electronic detector systems based on voltage pulse measurement require many other components besides the detector. A block diagram of some of these is shown in Figure 20. The upper dashed box after the amplifier shows the components for a single-channel analyzer, while the bottom dashed box is for a multichannel analyzer. The brief discussion to follow is by no means an exhaustive list of electronic components that are used in radiation detection systems.

All of the detectors discussed above require the application of a voltage, often called the detector bias. Therefore, a high-voltage supply, or detector bias supply, is an essential component of the counting system. The high-voltage supply should be able to deliver a range of stable dc voltages to meet the requirements of different detectors.

For detectors that innately generate small amplitude pulses, a preamplifier is essential. This device is located very close to the detector to avoid interferences and loss of signal. It serves to maximize the signal-to-noise ratio from the detector and to provide preliminary amplification and pulse-shaping of the small detector signal. The amplifier accepts the pulse from the preamplifier and provides further amplification. It also further shapes the pulse for improved processing of the signal by the remaining components of the signal-processing and data-storage system. The height (or amplitude) of the pulse that comes from the amplifier is related to the energy of the radiation producing the pulse.

If radiations of more than one energy are striking the detector, pulses of different amplitudes will be produced. If radiation of only one particular energy is needed, the pulses corresponding to that energy can be selected with a discriminator. This is an electronic device that can be set to accept pulses above, or between, certain preset amplitudes. A lower-level discriminator (LLD) sets the lower-amplitude limit, while an upper-level discriminator (ULD) sets the upper limit. This is illustrated in Figure 21. The figure shows 5 pulses of differing amplitudes. Pulses 2 and 5 are the only ones that will be accepted by the system, because they fall within the limits set by the LLD and ULD. Pulses 1 and 4 are too low in amplitude, and pulse 3 is too large in amplitude to be accepted. A simple discrimination device that allows both the LLD

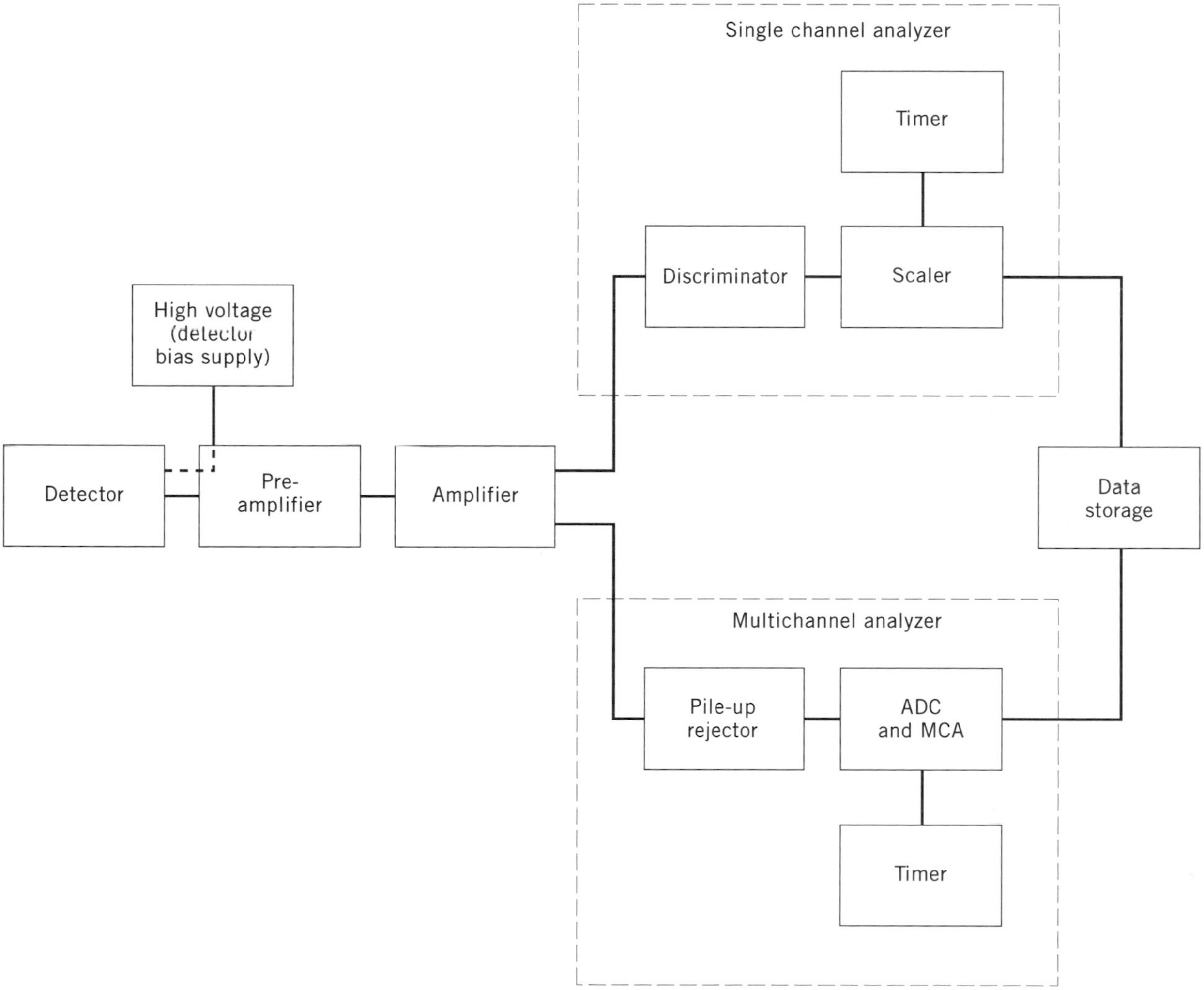

Figure 20. Schematic diagram of an electronic counting system configured for pulse analysis. The amplifier output may be routed to a single-channel analyzer (SCA) and scaler to count a single energy radiation (*top*), or to an analog-to-digital converter (ADC) and a multichannel analyzer/computer (MCA) to obtain an energy spectrum (*bottom*).

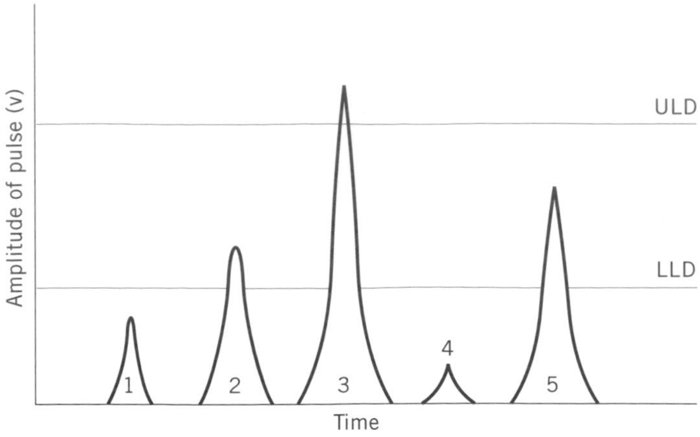

Figure 21. Utilization of a single-channel analyzer (SCA) for energy discrimination.

and the ULD to be preset is called a single-channel analyzer (SCA). The SCA can be connected to a scaler (a device that counts signals received, also called simply a counter), so that the number of pulses that pass through the SCA may be recorded.

In many spectrometry applications, it is desirable to collect data from all the radiations of differing energies that are emitted by one source. It is then desirable to sort the pulses according to amplitude (and thus according to energy) as they come from the detector, and to record the number of each different amplitude pulse that is received. This task is accomplished by an analog-to-digital converter (ADC). The ADC takes analog signals from the amplifier and converts them into digital data for storage in a multichannel analyzer (MCA). Each incoming pulse is sorted into an appropriate channel (a storage location in a computer memory), which corresponds to a specific incremental range of incident particle energies, and the number of events in each channel is tallied. The information processed by an MCA can be displayed on a computer terminal. The horizontal axis of the display gives the channel number, or equivalent energy, of the radiation, and the vertical axis is the number of pulses or "counts" received for each energy increment. Modern spectroscopy systems are computer controlled and the spectra from the MCA are recorded in the computer for future processing with appropriate software. MCA boards are now available that plug directly into microcomputers (PCs) and convert them into efficient and relatively inexpensive nuclear spectrometry systems.

All of the components of the detector system are linked to one another by appropriate cables. Cables used for high dc voltages, low dc voltages, and signals differ from one another and must be properly selected and matched with the circuits they connect. The detector can be placed at some distance from most of the remainder of the counting system, if required by the experiment. The pre-amplifier, however, is normally attached directly to the detector housing for reasons stated earlier. The high-voltage supply, the amplifier, and the SCA or ADC are separate modules that fit into special housings. The most popular type of module housing also provides the low-voltage power for these components and is called a NIM bin. NIM (Nuclear Instrument Module) and CAMAC (Computer Automated Measurement and Control) are international systems of nuclear electronics that allow for maximum compatibility of components. This is achieved through common power supply requirements, connector configurations, pulse shape requirements, and physical dimensions of modules. The CAMAC system is oriented toward large mainframe computer operations and is much more costly than the more widely used NIM systems. Multichannel analyzers can be stand-alone minicomputer-drive units with a variety of sophisticated data-storage capabilities, or simple boards.

NONELECTRONIC DETECTOR SYSTEMS

The bulk of the work done in environmental radiochemistry applications is done with one of the electronic systems described above. The nonelectronic detection schemes are rarely used in environmental work, so they are discussed here very briefly.

Photographic Plates

Photographic plates were the first type of radiation detector, used by Becquerel in 1895. Incoming radiation acts on the photographic emulsion just as light does and sensitizes the silver halide grains. Development of the emulsion produces dark areas where radiation interactions have occurred.

In autoradiography, a thin section of a material labeled with radionuclides is placed on a photographic emulsion. The darkened areas that appear after development of the film will produce an image of the object. Autoradiography has been widely used in biological studies to show how a radiotracer is distributed in an organism. Film badges used in health protection to monitor radiation exposure of personnel are another way photographic detectors are used. These badges can be quite versatile in measuring different radiations by varying types of filters placed on either side of the film.

Photographic emulsions can also be useful to detect single nuclear particles. For this purpose, a special nuclear emulsion is used. It is thicker than the usual photographic emulsions and contains larger amounts of silver halides. As a particle passes through the emulsion, it will leave a track of activated silver halide grains that can be made visible as silver deposits after photographic development. The density (darkness) of the track gives information about the charge on the particle, and the length of the track can give particle energy information. If the plate has been placed in a magnetic or electrical field during data accumulation, the direction of deflection can provide the sign of the charge on the particle. This method is still used for cosmic ray studies.

Chemical Detectors

Chemical detectors are usually used as "disaster dosimeters," where extremely high levels (kCi or more) of γ-ray emitters are present. These detectors rely on the production of free radicals by interactions of radiations with a chemical system. The free radicals induce chemical reactions that may be monitored using conventional chemical techniques. An example is the ferrous sulfate dosimeter. The reactions are

$$\gamma + H_2O \rightarrow OH^{\cdot} + H^{\cdot}$$

$$H^{\cdot} + H^{\cdot} \rightarrow H_2\,(g)$$

$$OH^{\cdot} + Fe^{2+} \rightarrow Fe^{3+} + OH^{-}$$

$$Fe^{3+} + SCN^{-} \rightarrow Fe(SCN)^{2+}.$$

The $Fe(SCN)^{2+}$ is a red complex that may be detected spectrophotometrically. The system may be calibrated using known radiation doses. To enhance stability, the solutions should be kept in oxygen-free closed containers.

Calorimetric Detectors

This is a detector system used for very high levels (kCi) of α or β emitters, or radionuclides undergoing spontaneous fission. The source is placed into a sensitive calorimeter in which all of the energy of the radiations will be completely dissipated. The energy deposited by the radiation eventually shows up as an increase in the temperature of the system. The change in temperature with time is used to indicate the activity.

Cloud and Bubble Chambers

The cloud chamber was first used by C.T.R. Wilson in the early 1900s to detect nuclear particles. A volatile material such as ethanol is added to the chamber, which is then cooled with dry ice. This results in the formation of a supersaturated vapor. As radiation passes through the chamber, ion pairs form in the gas. The supersaturated vapor condenses out as droplets along the path of ion formation, thus marking the particle track. The process is similar to the formation of rain droplets on dust particles in the atmosphere. Photographic and visual observations can be made of the track. As with the photographic plate, track density is related to the particle charge, and path length to particle energy. Cloud chambers can be easily set up for classroom demonstrations, but they are no longer used for research, having been supplanted by the bubble chamber.

The operating principle of the bubble chamber is similar to that of the cloud chamber, except that the bubble chamber contains a superheated liquid (ie, one that has been heated slightly past its normal boiling point, but is not boiling). The passage of radiation through the liquid forms ion pairs, and bubbles form along the path of the ions to mark the track. The process is similar to the effect obtained by dropping a grain of salt into pure water heated slowly to just above its boiling point. Both cloud and bubble chambers are used primarily by nuclear physicists interested in cosmic-ray studies or subatomic particles. They do not have analytical applications.

Thermoluminescence Dosimeters (TLD)

The exposure of certain inorganic crystals to radiation results in excitation of electrons from the valence to the conduction band. This process is familiar from the discussion of scintillation detectors above. Crystals used in thermoluminescence dosimetry, however, have the property of "trapping" the electrons, so light emission does not occur immediately. If the crystals are heated at some time after their radiation exposure, sufficient energy is provided to release the electrons from the traps, and de-excitation occurs with emission of photons. The intensity of the light emitted is proportional to the dose of radiation received by the crystal.

The type of material chosen for a TLD depends on what kind of radiation is to be detected and the time over which detection is to occur. TLD crystals can respond to x rays, γ rays, β particles, and protons over a wide range of exposure intensities. Lithium fluoride crystals are a favorite choice. They are reasonably sensitive, yet also stable enough so that they do not reemit energy too easily. Its effective atomic number is also close to that of biological tissue (8.1 versus 7.4). A $CaSO_4$:Mn crystal (where the Mn is an activating impurity) is very radiation sensitive, but does not hold the trapped electrons very well, so fading is a problem for this material. On the other hand, CaF_2:Mn does not fade, but is also not very sensitive to radiation. Selection of the appropriate crystal is dictated by the needs of the situation. TLDs are used for personnel monitoring devices in nuclear facilities.

NEUTRON DETECTION

Neutrons are not often determined in routine environmental analysis, but must be monitored for personnel who work with radioactive materials and for some of the processes that occur in nuclear facilities. Neutrons will not be detected efficiently by any of the basic detectors discussed earlier because neutrons interact with matter via scattering and nuclear reactions, rather than by atomic ionization or excitation. However, the products of neutron-induced reactions are often charged particles or γ rays, which can be detected with the instruments just discussed. Therefore, modifications of the detector systems are needed for neutron detection.

Thermal neutrons are low-energy neutrons. These neutrons undergo nuclear reactions that form charged particles. Detection of these charged particles provides the basis for the detection of slow neutrons.

The thermal neutron interaction most often used for its detection is shown by the following nuclear reaction: $^{10}B + n \text{ ---> } \alpha + {}^{7}Li$. (The shorthand way to write a nuclear reaction like this is $^{10}B(n,\alpha)^{7}Li$). It is actually the α particles formed in the reaction that are detected. Ionization chambers and proportional counters may be adapted for neutron detection by using $^{10}BF_3$ as the gas in the chamber, or by lining the chamber with a material containing ^{10}B. The use of boron-containing gas is preferred, because it is difficult to make the solid lining material thin enough for the α particles to escape after they are produced. The $^{3}He(n,p)^{3}H$ reaction can also be used in ionization chambers and proportional counters. ^{3}He is used as the gas, and the proton is the particle detected. ^{3}He gas is expensive, so this approach is not as widely used as the ^{10}B reaction. Another possible reaction appropriate for ionization chambers and proportional counters is the neutron-induced fission of ^{235}U. The chamber can be lined with ^{235}U (or another fissile material), and the fission products detected.

Another useful reaction with a high cross section for thermal neutron interaction is the $^{6}Li(n,\alpha)^{3}H$ reaction. This is not used in ionization chambers or proportional counters because no suitable gas exists, but it is possible to use this reaction conveniently with scintillation detectors. Lithium iodide crystals, analogous to NaI crystals, can serve as both the neutron detector and the scintillating medium. This method gives good neutron-detecting efficiency because the detecting medium is a moderately high-density solid, rather than a gas. Lithium compounds may also be dispersed in other scintillating media, such as ZnS(Ag). Light produced by the secondary α particles interacting with the ZnS scintillator is measured by use of a PMT.

The radiative capture reactions (in which a nucleus captures a neutron and emits a γ ray) are rarely used for simple neutron detection. The $^{113}Cd(n,\gamma)^{114}Cd$ and $^{115}In(n,\gamma)^{116}In$ reactions both have quite high cross sections (i.e., high probabilities of occurring), and can be used to determine neutron fluxes from a nuclear reactor or other neutron sources.

Neutrons having energies in the MeV range are known as fast neutrons. Fast neutrons, like slow neutrons, undergo nuclear reactions that result in the production of charged particles. However, for fast neutrons, scattering reactions may also be an important means of interaction, and can be used for detection purposes.

The detection schemes described above for slow neutrons can all be used for fast neutron detection if a suitable way to moderate, or slow down, the fast neutrons can be found. The thermalization process involves the collisions of the fast neutrons with moderating material, and this requires time. Thus, this method is not appropriate where fast response is needed, nor is it useful when information regarding neutron energy is important.

Scintillation detection is often used to detect the scattered protons produced by fast neutron collisions with hydrogen atoms. Many of the scintillation materials are organic, and therefore contain a great deal of hydrogen. Inorganic scintillants like ZnS can act as neutron detectors when they are embedded in a hydrogenous medium, such as a plastic.

All these neutron detectors simply detect the presence of neutrons, but do not give much energy information beyond "thermal" and "fast" classifications. Determination of neutron energies requires more sophisticated techniques.

BIBLIOGRAPHY

Reading List

G. R. Choppin and J. Rydberg, *Nuclear Chemistry,* Pergamon Press, Oxford, 1980.

G. T. Cook, "Recent advances and environmental applications in liquid scintillation spectrometry," *Anal. Proc.* (London) **29**(1), 4-6, (1992).

P. N. Cooper, *Introduction to Nuclear Radiation Detectors,* Cambridge University Press, Cambridge, UK, 1986.

K. Debertin and R. G. Helmer, *Gamma- and X-Ray Spectrometry with Semiconductor Detectors,* North-Holland, Amsterdam, 1988.

S. DeFilippis, "Activity Analysis in liquid scintillation counting: A technical comparative summary—Part I," *Radioact. Radiochem.* **1**(4), 22, 24–35 (1990).

I. G. Draganic, Z. D. Draganic, and J. P. Adolff, *Radiation and Radioactivity on Earth and Beyond,* CRC Press, Boca Raton, Florida, 1990.

A. Dyer, *Liquid Scintillation Counting Practice,* Heyden, London, 1980.

W. D. Ehmann and D. E. Vance, *Radiochemistry and Nuclear Methods of Analysis,* John Wiley and Sons, New York, 1991.

G. G. Eichholz and J. W. Poston, *Principles of Nuclear Radiation Detection,* Ann Arbor Science, Ann Arbor, Michigan, 1980.

M. Eisenbud, *Environmental Radioactivity: From Natural, Industrial, and Military Sources,* Academic Press, Orlando, 1987.

G. Espinosa, "Past, Present, and Future of Materials, Methodology, and Instrumentation in Particle Tracks in Solids," *Nucl. Tracks Radiat. Meas.* **19**(1–4), 835–842, (1991).

R. C. Fernow, *Introduction to Experimental Particle Physics,* Cambridge University Press, Cambridge, UK, 1986.

G. Friedlander, J. W. Kennedy, E. S. Macias, and J. M. Miller, *Nuclear and Radiochemistry,* John Wiley and Sons, New York, 1981.

G. Hall, "Modern Charged Particle Detectors," *Contemp. Phys.* **33**(1), 1–14, (1992).

Y. S. Horowitz, ed., *Thermoluminescence and Thermoluminescent Dosimetry,* CRC Press, Boca Raton, Florida, 1984.

International Atomic Energy Agency, *Methods of Low-Level Counting and Spectrometry,* STI/PUB/592, IAEA, Vienna, Austria, 1981.

M. J. Kessler, ed., *Liquid Scintillation Analysis,* Packard Instrument Company, Publication No. 169-3052, 1989.

G. F. Knoll, "Some Recent Developments in Charged-particle and Gamma-ray Detectors," *Nucl. Instrum. Methods Phys Res.* **B24/25,** 1021–1027, (1987).

G. F. Knoll, *Radiation Detection and Measurement,* (2nd ed.), John Wiley & Sons, New York, 1989.

J. Krugers, ed., *Instrumentation in Applied Nuclear Chemistry,* Plenum Press, New York, 1973.

W. B. Mann, R. L. Ayres, and S. B. Garfinkel, *Radioactivity and its Measurements: SI Units,* Pergamon Press, New York, 1980.

W. B. Mann, A. Rytz, and A. Spernol, *Radioactivity Measurements: Principles and Practice,* Pergamon Press, Oxford, 1988.

G. D. Miller, *Radioactivity and Radiation Detection,* Gordon & Breach Science, New York, 1972.

National Council on Radiation Protection and Measurements, *A Handbook of Radioactivity Measurements,* Report No. 58, NCRP Publications, Bethesda, Maryland, 1985.

C. Passo and M. Kessler, *The Essentials of Alpha / Beta Discrimination*, Packard Instrument Company, 1992.

H. Ross, J. E. Noakes, and J. D. Spaulding, *Liquid Scintillation Counting and Organic Scintillators,* Lewis Publishers, Inc., Chelsea, Michigan, 1991.

G. Shani, "Instrumentation, Particle Accelerators, and Particle and Radiation Detection," *Elem. Anal. Part. Accel.* 1–74, 1992.

J. Tolgyessy and E. H. Klehr, *Nuclear Environmental Chemical Analysis,* John Wiley and Sons, New York, 1987.

C. H. Wang, D. L. Willis, and W. D. Loveland, *Radiotracer Methodology in the Biological, Environmental, and Phyical Sciences,* Prentice-Hall, Inc., Englewood Cliffs, New Jersey, 1975.

RAILROADS

CONAN FURBER
Hallowell, Maine

The railroad industry in the United States has always been a dynamic and evolving industry. However, the last decade has seen the industry undergo greater changes than at any other time in its history. As a result of mergers, regulatory reform, and technical innovations the size of the industry has diminished. Yet despite the reduction in size, the railroad industry is annually transporting more revenue ton-miles of freight than at any previous time in history. Also, as a result of the inherent advantage of the steel wheel on steel rail, the railroads have been able to move this tremendous amount of freight using less energy and with less environmental impact per ton-mile of freight than any other mode of transportation.

U.S. railroads are the most productive in the world, and according to U.S. Labor Statistics they are continuing to improve productivity (1). From 1982 through 1992, productivity as measured by revenue ton-miles of freight moved per employee hour steadily increased from 863 to 12,176 (1). While there are many factors affecting productivity, some of the increase can be attributed to labor force reductions, rationalization of fixed plant, and increased equipment utilization brought about through mergers and regulatory reform. As shown in Table 1, from 1981 to 1991 for the Class I railroads (defined as having in excess of $251.4 million dollars in annual revenues) the miles of track, number of freight cars, and number of employees have decreased while at the same time revenue ton-miles of freight carried has steadily increased. It should be noted that in 1989, for the first time in history, the railroad industry carried in excess of one trillion revenue ton-miles of freight. Since 1989, revenue ton-miles of freight have not only exceeded the one trillion mark each year but have continued to increase.

In 1978 there were 54 Class I railroads operating in the United States (2). Through mergers huge super railroads were formed reducing the number of Class I railroads to twelve in 1992. However, although the number of Class I railroads has decreased, the total number of railroads in operation today has actually increased. Regional and local railroads have been incorporated to provide a service in areas no longer served by the primary railroads. Regional railroads are defined as line-haul railroads operating at least 350 miles of track and/or earning revenues between $40 million and the $251.4 million threshold for Class I railroads (1). Local railroads are line-hauls that do not meet the regional criteria, plus switching and terminal railroads. As shown in Table 2, in 1992 there were 33 regional railroads and 464 local railroads. The regional and local railroads account for 26% of the mileage operated by railroads, 11% of railroad employees, and receive 9% of the revenues (1).

Table 1. Class I Railroads[a]

Units	1981	1992
Ton-miles, billions	910	1,067
Miles of track, thousands	268	191
Freight cars, thousands	1,111	605
Employees, thousands	436	197

[a] Reference 1.

Table 2. Railroad Statistics, 1992[a]

Railroad	Number	Miles operated	Year-end employees	Freight revenue, $
Class I	12	126,237	194,120	27,507,607
Regional	33	20,697	11,600	1,514,195
Local	464	22,730	12,885	1,242,463

[a] Reference 1.

See also ENERGY CONSUMPTION IN THE UNITED STATES; DIESEL FUEL; MIDDLE DISTILLATE; PETROLEUM PRODUCTS AND USES; AIR POLLUTION.

ENERGY

No other mode of transportation is as energy efficient as the freight railroads; energy efficiency continues to improve each year assisted by new technological improvements. In 1980, the freight railroads consumed over 3.9 billion gallons of diesel fuel in moving 918 billion revenue ton-miles of freight for an average rate of consumption of 235 revenue ton-miles per gallon of fuel consumed (1). In 1992, the railroads consumed a little over 3.0 billion gallons of diesel fuel in moving 1.07 trillion revenue ton-miles of freight at an average rate of 355 ton-miles of freight per gallon of fuel (1). To move this freight, the Class I freight railroads had 18,004 diesel electric locomotives in service at the end of 1992 (1). The new generation of locomotives allows the railroads to move more freight with fewer locomotives; for example, in 1983, the railroads had a fleet of 25,448 diesel electric locomotives in service and moved less freight than in 1992 (1). The freight railroads rely exclusively on diesel electric locomotives for moving their trains. Amtrak, the National Passenger Service Corp., uses electric locomotives within portions of the Northeast corridor and diesel–electric locomotives elsewhere.

Locomotives

A typical diesel electric locomotive contains a medium speed diesel engine (approximately 900 to 1100 RPM) that is used to power either a generator or an alternator, which in turn provides the electricity for the AC or DC electric traction motors. The electric traction motors are mounted on each axle of the locomotive and provide the tractive effort necessary to move the locomotive and train. Power to rail is controlled by means of sophisticated interactive wheel-slip control systems in order to be able to obtain maximum utilization of available adhesion. Current state-of-the-art locomotives with dc propulsion can achieve all-weather adhesion levels of 24–28%, and ac propulsion systems can achieve 32–35% adhesion levels (3).

Over the years, the medium speed diesel engines have

been available in six, eight, twelve, sixteen, eighteen, and twenty cylinder models. The majority of road locomotives, at this time, are sixteen cylinder models. Typical engine output powers range from 2,000 hp for road switcher units to 4,300 hp for mainline freight service. Current engine series operate with an overall thermal efficiency of 34–36% (3). Power levels are constrained by adhesion control systems and a railroad imposed requirement that the maximum axle load for a six-axle locomotive, including fuel, not exceed 65,000 lbs per axle (3). The use of ac traction motors is expected to allow engine size to increase, reaching power ratings of 6,000 hp/unit in the near future.

When dieselization of the railroads began in earnest in the late 1940s there were five principal U.S. builders of locomotives: the ElectroMotive Division of General Motors (EMD), General Electric (GE), Fairbanks Morse, American Locomotive Co. (ALCO), and Baldwin (4). By the early 1960s, only EMD and GE remained. Essentially, all diesel electric road locomotives in use by the large railroads were manufactured by either EMD or GE. There are some firms which remanufacture locomotives and these units may have, for example, engines built by Caterpillar, Cummins, or Detroit Diesel.

EMD Diesel Engines

EMD has manufactured over 35,000 locomotives (4). These locomotives contain a two-stroke medium speed diesel engine with either eight, twelve, sixteen, or twenty cylinders arranged in a 45° vee. In the early 1930s, EMD produced the 201A engines, the first in a series of four engines. The 201A was followed by the 567, 645, and 710 series. Although the 201A engine is no longer used in railroad services, engines from the other series are all in active service. The number in the model designation denotes the cubic inch displacement per cylinder for that engine. A letter suffix refers to the crankcase design, and if followed by the numeral 3 indicates that the engine is turbocharged. Engine ratings have increased over the years with the original sixteen cylinder model 567 Roots blown engine being rated at 1,350 brake horsepower (bhp) at 800 rpm (4). The 645F3 was rated at 3,800 bhp at 950 rpm and the latest model, the 710 series, has the 710G3 rated at 3,950 bhp (4). The 710 series engine has a bore of 9.06 in. and a stroke of 11.00; full speed operation is at 900 rpm (4). Newer models approaching 6,000 bhp are expected to be ready for marketing in the near future.

GE Diesel Engines

General Electric produced its first diesel electric locomotive in 1918, and over the years has used a variety of engines supplied by Ingorsol-Rand, Cooper-Bessemer, Cummins, and Caterpillar (4). Since the 1950s GE has primarily used the Cooper-Bessemer FDL engine design, purchased and manufactured under GE's own trademark. The Model FDL engine is four-stroke and has a bore of 9.0 in. and a stroke of 10.5 in. for a displacement per cylinder of 668 cubic inches (4). The FDL engine is available in eight, twelve, and sixteen cylinders aligned in a 45° vee and operates at 1,050 RPM at full load.

FUELS

Prior to the oil shortages of 1973, the railroads primarily purchased diesel fuel meeting the ASTM specification for DF-2 diesel fuel oil. Faced with the potential for continuing fuel shortages and steadily increasing fuel oil prices, the railroad industry directed the Association of American Railroads (AAR) to establish an alternative fuels program (5). The program had two objectives: to identify fuels suitable for use as extenders or replacements for No. 2 fuel oil and to identify fuels having a potentially lower cost than No. 2 fuel oil. The AAR established a full-size medium speed diesel engine test facility consisting of two twelve cylinder locomotives engines (one GE 7FDL and one EMD 645E3) at Southwest Research Institute (SwRI). Emergency fuels tested included: alcohols, vegetable oils, shale oils, liquefied coals (SRC II), water emulsions, coal slurries, and off-specification petroleum products. Potential cost-reducing fuels included off-specification middle distillates and heavy oil blends. Results of the engine tests showed that the medium speed diesel engine could satisfactorily use a much broader range of fuel than that specified by the ASTM DF-2 specification. Based on the results of the AAR Alternative Fuels Program many railroads reduced their fuel specifications to be able to utilize the lower cost fuels that did not meet the ASTM DF-2 specification.

Emergency Fuels

Alcohols, sunflower oil, and synthetic fuels were evaluated through the AAR Alternative Fuels Program (5). Methanol and ethanol were tested neat (100%), in stabilized emulsions, and in unstablized emulsions in the full-size medium speed engines. With a cetane rating of zero, the alcohols did not provide satisfactory combustion in the unmodified diesel engines (5,6). By emulsifying up to about 15% alcohol in No. 2 diesel fuel, the resulting fuel could be satisfactorily used in the engines (5,6).

Sunflower oil both neat and blended 50% by volume with No. 2 diesel fuel was tested (5,6). Engine performance characteristics comparable to baseline diesel fuel were obtained using fully refined sunflower oil both when blended with standard diesel fuel and neat.

Two synthetic fuels were also evaluated. The Paraho DFM shale oil was an excellent grade of fuel, meeting No. 2 diesel fuel specification; it provided engine performance test results equivalent to standard diesel fuel (5,6). The SRC II fuel was derived from coal and had a very low cetane value. However, the fuel did not provide satisfactory engine test results in the unmodified engines, and required complicated handling and safety measures.

Lower Cost Fuels

While medium speed diesel engines can satisfactorily use a blend of No. 2 and No. 6 residual fuel, other considerations beyond engine performance may reduce the potential cost savings to a point at which use of the fuel is not warranted. These considerations arise because of problems involving fuel handling and distribution, cold weather filtration plugging, excess wear from heavy metals, and manifold fires caused by accumulations of unburned fuel (5).

Middle distillate fuels not meeting the ASTM specification for No. 2 diesel fuel offer the greatest potential for reducing fuel costs (5,6). Medium speed diesel engines are quite fuel tolerant, and middle distillate fuel specification can be broadened to take advantage of lower cost nonspecification fuels available from petroleum refiners. Therefore, the railroad industry adopted a middle distillate fuel specification that reduces the minimum cetane requirement from 40 to 32 (5).

The AAR Alternative Fuels Program gradually evolved into a comprehensive energy research program. Included within the overall Energy Program were components addressing such diverse areas as locomotive and railcar aerodynamics, track lubrication, comprehensive computer simulation of energy consumed in the operation of trains, and the efficiency of all components onboard a locomotive. Results obtained through this program has assisted the railroad industry in further reducing energy consumption and increasing productivity.

ENVIRONMENTAL CONSIDERATIONS

Environmental regulation will have a profound impact on locomotive engine design, cost, maintenance, and performance, throughout the 1990s. The Federal Clean Air Act requires the Environmental Protection Agency (EPA) to regulate exhaust emissions from all new engines and to set standards at reasonable levels, given existing technologies and the likely ability of manufacturers to adapt those technologies to locomotive engines at a reasonable cost (3). In meeting the proposed emission regulations, industry can reduce locomotive exhaust emissions by: introducing changes in technology, changing operational procedures, and/or using a combination of technological changes combined with changes in operational procedures. Technological changes can be achieved by either modifying the combustion process, or by modifying the post-combustion process, or by a combination of both. Alternative fuels (other than standard diesel fuel) and new prime mover technologies may hold potential for meeting the emission regulations.

Estimates of locomotive emissions prepared for the California Air Resources Board (CARB) indicate that locomotive emissions are relatively low when compared to emissions from all other sources (7). As shown in Table 3, locomotives contribute only about 0.12% of hydrocarbons (HC), 0.10% of carbon monoxide (CO), 3.38% of nitrogen oxide (NO_x), 1.55% of sulfur oxide (SO_x) and 0.06% of particulate matter (PM) to the total emissions estimated for the six California basins. The majority of the remaining emissions result from the use of highway vehicles and stationary sources. When compared to other sources, the relatively low emissions levels resulting from the use of locomotives would appear to indicate that dramatic improvements in air quality might be achieved by simply shifting freight and passenger movements to rail.

Table 3. California Locomotive Emissions[a]

Sources	HC	CO	NO_x	SO_x	PM
Stationary	1,862	2,087	804	183	3,711
On-road	1,375	9,943	1,678	111	152
Other mobile	250	1,552	2,934	480	58
Total	*3,487*	*13,582*	*5,416*	*764*	*3,921*
Trains	4	13	99	7	2
Trains, % of total	0.12	0.10	3.38	1.55	0.06

[a] Reference 7.

Table 4. Comparative Diesel Engine Emissions[a]

	g/bhp-hr			
Engine	HC	CO	NO_x	PM
EPA heavy-duty, 1991	1.3	15.5	5.0	0.25
Truck standard, 1994	1.3	15.5	5.0	0.10
1998	1.3	15.5	4.5	0.10
EMD 12-645E3B	0.29	0.81	10.46	0.15
GE 12-7FDL	0.42	1.98	11.21	0.19

[a] Reference 7.

Whereas emission standards for trucks have been established, standards governing new and existing locomotive engines are not presently available. To obtain an idea of the magnitude of the problem, Table 4 provides a comparison of existing truck emission standards and typical locomotive emission levels obtained from tests conducted by the AAR.

Conventional Diesel Engine

From the comparative engine emissions contained in Table 4, it is apparent that the existing medium speed diesel engine used in locomotives will meet the standards for HC and CO. Reductions in NO_x and PM will probably be required. Engine changes currently under consideration to achieve the required reductions include: changes in the combustion bowl, increased charge-air cooling, increased fuel injection pressures, incorporation of electronic fuel injection, and methods for reducing lubrication oil consumption.

Post-combustion treatments are also being considered as means for reducing exhaust emissions. Whereas combustion modification processes attempt to prevent the formation of NO_x, aftertreatment processes are used to reduce NO_x to its basic components of nitrogen and oxygen before releasing it to the environment. A variety of methods that have been successfully used to reduce NO_x emissions from boilers, furnaces, and stationary diesel engines are being investigated to determine their adaptability for use in mobile source applications. Size constraints, variations in exhaust temperatures, and control systems are the principal problems confronting the transfer of technology from stationary source to mobile source applications. The primary aftertreatment methods currently being investigated include the use of chemicals either metered directly into the exhaust stream or as fuel additives; dissociation of No_x through the use of electronically produced plasma; and selective catalytic reduction (SCR).

Cost of Reducing Locomotive Emissions

California is in the process of promulgating regulations limiting the emissions from locomotives within their

State. Studies undertaken for the State provide an indication of the costs that will have to be borne by the railroads in order to achieve the required reductions in emissions and continue to conduct business. The method proposed for achieving the reductions includes the use of LNG as fuel with and without SCR aftertreatment, the use of SCR aftertreatment on existing diesel engines, and electrification. The effectiveness of each of the proposals vary ranging from achieving a 70% reduction in No_x emissions by switching to LNG to a maximum of a 98% reduction if the railroads are electrified. However, as can be seen from Table 5, there is a considerable cost differential depending on the method selected for implementation. Shifting to LNG as a fuel carries a penalty of $22,000,000 a year and goes up to $67,000,000 if SCR is required (7). Electriciation of the railroads would cost an estimated $524,000,000 per year (7).

The annual cost for emission reduction in California alone could cost the railroads that operate there between $22,000,000 and $95,000,000 annually to reduce their percent of the total No_x produced from 3.38% to 0.33%. Much larger reductions in total No_x emissions could be achieved by simply diverting traffic from other mobile sources to rail.

ALTERNATIVE PRIME MOVER TECHNOLOGIES

Some industry experts predict that, even with the most rigorous development program possible, petroleum fueled diesel engines will fall short of meeting the emissions standards likely to be required by the year 2010 (3). Candidate technologies, likely to meet the more stringent standards, have been identified as natural gas diesel engines, electrification, multiple small diesel engines (with and without distributed power), gas turbines, and fuel cells. Nuclear powered locomotives are not considered to be feasible alternatives.

Natural Gas

Natural gas as an alternative fuel for diesel powered locomotives is currently receiving considerable attention by the railroad industry. Two retrofit units, employing dual fuel pilot injection, are in service on the Burlington Northern (BN) Railroad, and four newly manufactured units (two by EMD and two by GE) have been ordered by the Union Pacific (UP) Railroad (3).

For use in these locomotives the liquid methane is obtained by refining compressed natural gas, and is carried in cryogenic fuel tenders. There are three combustion technologies under development or consideration: dual fuel pilot injection, direct gas injection, and direct gas liquid injection. Dual fueling requires the injection of a small amount of diesel fuel to initiate the combustion process. Direct gas injection technology uses natural gas as the only fuel and relies on either spark or glow plug ignition to achieve combustion. Direct liquid gas injection is expected to produce the cleanest burning engines; spark or glow plug assistance may be required to control the timing of the combustion process.

Table 5. Cost Estimates[a]

Method	Reduction, %	Cost, $/yr
LNG dual fuel	70	22,000,000
LNG + SCR	97	67,000,000
Diesel + SCR	90	95,000,000
Electrification	98	524,000,000

[a] Reference 7.

Electrification

Electrification of railroads is a mature and established technology. In electrified territories the locomotive draws current for propulsion from an overhead catenary and returns it to the running rails and local ground. In North America, the use of direct current (dc) has been limited to rapid transit and commuter rail operations. Catenary voltages for dc systems are generally limited to 750–1500 volts whereas AMTRAK operates its passenger system using a 25 kilovolt, 60 hertz single-phase ac supply system (3). The primary obstacle to the widespread use of electrification for freight railroads is the high first cost for construction of the system, including the cost for modifying overhead clearances and the additional costs resulting from the added complexity of signaling equipment and track circuits. When the higher costs and lack of operational flexibility are considered, it is unlikely that additional miles of railroad track will be electrified in the foreseeable future.

Multiple Small Diesel Engines

Small engines, because of shorter life cycles, lower first cost per unit, and the large number of engines produced, are much more technologically advanced than are medium speed diesel engine. There are two ways that the railroad industry could use this advanced technology: use multiple small engines onboard a locomotive instead of a single large engine, and use distributed power (3). Under this concept, small diesel engines complete with traction motors (minilocomotives) would be distributed throughout the train. The use of distributed power could result in lower in-train forces and better train control.

Gas Turbines

Gas turbine generator sets for stand-by power are commonplace. While gas turbines have been used both in freight and passenger service, when compared to other prime movers they have the disadvantage of being not very efficient at partial load. As the railroads operate a considerable portion of the time at partial load, gas turbines have not been able to compete with diesel engines (3). Because gas turbines produce significantly lower levels of emissions than diesel engines they are being considered for future specialized locomotives applications.

Fuel Cells

Although some experts predict that the fuel cell will eventually replace the internal combustion engine, fuel cell technology is still in its infancy. Because the locomotive already uses electric traction motors for propulsion, a switch to fuel cells as a replacement for the diesel engine would be less of a problem than for trucks. However, at the present state of fuel cell development it is estimated that this will not be a fully cost competitive technology until the year 2020 (3).

SUMMARY

Over the last two decades the freight railroads in the United States have, as a result of mergers, regulatory reform, and technical innovations, brought about changes that enable them to achieve unprecedented levels of productivity. Today the railroads are transporting more freight while using less energy and having less environmental impact than at any other time in history. Unfortunately, in meeting the proposed new locomotive emission standards, railroad energy consumption may increase and productivity may decrease. This is ironic when considering that far greater gains in reducing air pollution could be achieved by simply shifting freight and passengers to rail.

BIBLIOGRAPHY

1. Railroad Facts, Association of American Railroads, Washington D.C., 1992.
2. Railroad Facts, Association of American Railroads, Washington D.C., 1982.
3. G. R. Cataldi, and co-workers, *Locomotive 2000*. Association of American Railroads, Washington D.C. July 1993, R-840.
4. C. P. Furber, *Railroad Medium-speed Diesel Engine Research*, Association of American Railroads, Washington D.C., July 1987.
5. C. P. Furber, *Association of American Railroads Alternative Fuels Program*, SAE, Warrendale, Pa., May 1985, no. 851222.
6. Q. A. Baker and co-workers, *Alternative Fuels for Medium-speed Diesel Engines (AFFMSDE) Project, First Research Phase Final Report Off-Specification Diesel Fuels, Simulated Coal-derived Fuel and Methanol*, Southwest Research Institute, San Antonio, Tx., June 1981.
7. C. Weaver and co-workers, *Controlling Locomotive Emissions in California*, Engine, Fuel, and Emissions Engineering, Inc., Sacramento, Calif., Oct. 1993.

RECYCLING

Thomas A. Polk
Annandale, Virginia

Recycling is a method of materials management in which discarded materials are separated from waste, processed to acceptable standards, and re-enter the economy as usable product. Recycling of items which would otherwise become waste occurs in three distinct settings: municipal solid waste (MSW), sewage treatment or biosolids management, and industrial and mining waste (1). For each area of waste to be managed, recycling takes a particular form. MSW recycling includes the collection and separation of various waste items by manual and automated systems, processing of the materials to industry specifications, and use of these materials, known as secondary materials or commodities, in a manufacturing facility or as an end product. Biosolids management usually occurs at sewage treatment or in-vessel composting plants, where a humus-like soil amendment product (compost) is produced. Industrial and mining wastes are usually managed on-site through a variety of treatment methods which may include recovery of materials for recycling. Recycling of MSW is the focus herein.

Whereas all waste materials are theoretically capable of being recycled, they may instead undergo a number of other processes due to the potentially dangerous nature of the waste itself, ie, infectious medical waste from health care sources and regulated hazardous waste.

The waste management hierarchy, or integrated waste management (2), underlies many recycling efforts and aims to handle wastes most cost effectively and with the least impact on the environment by combining four approaches in descending order of preference. Waste prevention attempts to reduce the entire cost of waste management by reducing the amount of material originally discarded. A subcategory of waste prevention is reuse, in which a product is either cleaned and reused as is, such as a refillable bottle, or remanufactured, using new or reconditioned parts and resold for the same use. Recycling is another approach, followed by waste incineration for energy recovery, also known as resource recovery. At the bottom of the hierarchy is landfilling, which acknowledges that there are waste materials that cannot be safely or efficiently recycled or incinerated, and that both recycling and incineration will have some residue which must be disposed.

In 1992 the United States generated over 291 million tons of MSW, of which 17% was recycled (3). Materials currently recycled from MSW include paper, metal, glass, plastics, scrap tires, wood, yard trimmings, textiles, and construction aggregates. See also Environmental economics; Waste management planning; Waste to energy; Fuels from waste.

The MSW Recycling Cycle

Recycling is a system in which a number of interrelated processes occur in sequence (4). Each recyclable material has its own physical and chemical attributes which give it a unique set of processes, however, the concept of a recycling cycle or loop system can be applied to most materials with slight modifications. The "three chasing arrows" symbol created by the American Paper Institute (American Forest Products Association ca 1993) demonstrates the processes of recycling, as seen in Figure 1. This logo entered the public domain and is used widely in commerce and by recycling practitioners (5).

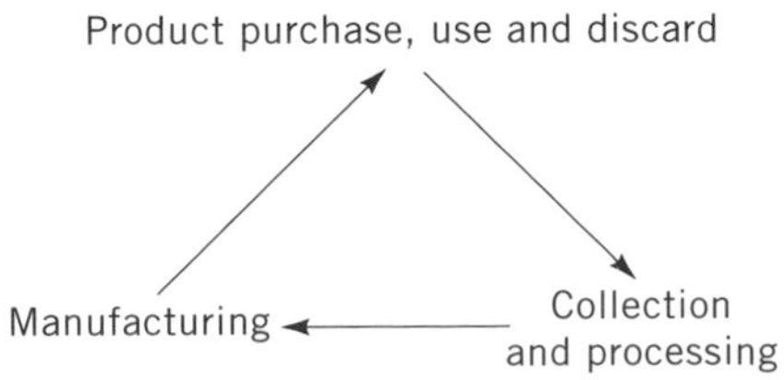

Figure 1. The recycling cycle.

At the top of the cycle, the product enters the economy and is purchased by a consumer; consumers may be individuals, governments, businesses, or other manufacturers. Once purchased, the consumer makes use of the product and discards it in one of several ways to facilitate its further handling. Depending on the characteristics of the consumer and the product, a source-separated collection system may be used. In such a system, one or more recyclable materials are stored separately from nonrecyclables, and may be collected by special recycling vehicles. Alternatively, a recyclable may be mixed or commingled with other recyclables, or all materials, ie, recyclables and trash, may be mixed and collected in one vehicle. Another collection approach requires that the generator of source-separated material transport it to a location where it is accepted for processing. The key element is that the discard is treated as a potential raw material input and is kept free of contaminants, so that the maximum value may be obtained.

Material Sorting and Processing. Once removed from its point of generation the recyclable material must be processed in some way to add value and conform to the standards of a manufacturer or other customer. For source-separated materials little work may be involved to attain industry quality specifications. For items which are commingled, either with other recyclables or with waste, the items of value must be culled from the rest of the mix, and this occurs usually through a combination of manual, mechanical, and electronic means. For some materials the processing stage may require several facilities, each of which performs a specific value-added service to the material. Typically, after each processing phase, the resulting material is densified to reduce freight costs before being transported to the next stage of processing or to its final user. The final stage is also the first stage, ie, once the recyclable item has been transformed from a discarded product into a commodity, it may be used as a product in its own right or as a raw material in the manufacture of a good which is in turn solid in the marketplace and potentially the cycle begins again.

Historical Background

Prior to the large-scale manufacturing and consumption of consumer goods, the opportunities to develop the complex network of producers, consumers, and materials handlers were limited; in an agricultural society food scraps, animal manure, and other wastes were commonly recovered and used on the farm. With mass production factories, the quantities and types of material available expanded dramatically, and recycling, as a means to reduce manufacturing costs, quickly emerged in the paper and metals industries.

At this point almost all materials were recycled, ie, waste generated in the manufacturing process itself was commonly returned to the hydrapulper (paper) or steel mill, foundry, or smelter (metals). This material, known as mill broke in the paper industry and home scrap in the ferrous and nonferrous industry, is often generically termed pre-consumer or in-plant scrap, and is the source of unresolved debates about the definition of a recycled product. Instead of being sold, used, and discarded, pre-consumer scrap went from manufacturer, was collected, and processed back into a raw material for production, frequently at the same location. In any event, use of pre-consumer scrap demonstrates an important economic principle in recycling, namely that if an item has value, it will not become waste. Because this scrap was homogeneous, free of contaminants, and its use entailed little if any additional costs, it was efficient materials handling to recycle it.

The Scrap Dealer Infrastructure. Shortly before the turn of the twentieth century a new source of recyclables emerged, and with it a new industry: the scrap dealer. Scrap dealers, or junkmen, responded to a need from manufacturers looking to lower raw material costs and find out who could use more recyclables than were generated by in-plant scrap. In response, scrap dealers offered a price for scrap paper, metal, and rags and induced a supply from households and businesses. The scrap market was driven by the supply, demand, and prices in the virgin materials markets and was entirely a function of market forces. During World War II, junkmen rose to a higher profile as recycling became a national security endeavor rather than strictly economic. As raw materials were consumed in greater quantities for the war effort, a large number of materials were recycled.

The 1960s and 1970s: Ecological Concerns Spur Recycling. After World War II, the scrap dealers and private recycling industries continued to thrive, but faded out of public consciousness. With the growing awareness of air and water pollution which emerged in the 1960s, the idea that recycling should be undertaken for the benefit of the environment took hold. This phenomenon culminated in the first Earth Day in April 1970. In addition to community clean-up programs, educational efforts, and protests against industrial polluters, many nonprofit groups and student organizations set up collection centers for newspapers, bottles, and cans. Some of these programs continued far beyond Earth Day but it was not until the mid-1980s that a more specific environmental concern arose, and with it another change in the nation's understanding of recycling.

The 1980s and 1990s: Government Mandated Recycling. Once the domain of private scrap dealers and manufacturers, recycling became the concern of local and state governments as more stringent landfill requirements saw the number of operating facilities decline from 8000 in 1988 to 5386 in 1992; an estimated 1000 landfills were scheduled to close in 1993 (6). Coupled with this was the increasingly difficult process of siting new facilities, known as the "Not In My Back Yard" (NIMBY) syndrome. The landfill crisis did not result from a lack of space, but rather the realities of operating and siting MSW disposal locations. The most visible sign of this was the Mobro garbage barge which in the summer of 1987 carried a load of New York garbage down to the Caribbean Sea and back up the East Coast attempting to find a site to accept its cargo (7). Garbage became a public policy issue. Governments, usually municipalities with MSW management responsibilities, and states, with larger land use and planning concerns, enacted laws requiring collection of recyclables, termed post-consumer materials, usually through curbside source-separation programs (8). The direct involvement of the government set in motion a new dynamic in recycling, one which mandated a supply of materials, but not the demand for these new quantities. These market imbalances continue to affect the recycling of some post-consumer materials into the mid-1990's.

Paper and Paperboard Recycling

In 1992, 83.68 million tons of paper and paperboard entered the MSW stream, composed of packaging materials (boxes, bags, and wrappings), printing and writing papers, newspapers, tissues and towels, and miscellaneous paper and nonpaper items. Of these products, 32.21 million tons (38.5%) was recycled (9,10). This material is termed wastepaper or paper stock and is divided into two general groups known as high grades and low or bulk grades. High grades refer to the quality of the paper fibers present, and low/bulk grades to either the fiber quality or the fact that these grades are used in large quantities (11). Within the high grades, statistics are kept for two sub-groups, pulp substitutes and deinking papers. Under each of these headings there are numerous individual grades. For the bulk grades, there are three sub-groups: newspaper, corrugated containers, and mixed papers, and again these are further divided into industry-specified grades. All told, 46 general grades and 33 speciality grades are in use in the United States (12).

Pulp Substitutes. In 1991, 3.41 million tons of pulp substitutes were recycled (13). Pulp substitutes are the most valued category of paperstock, and are composed of paper that contains little or no contaminants, specifically inks or glues. These materials are literally able to substitute for virgin wood pulp and are typically generated as in-plant scrap by manufacturers (also called converters) of products containing paper or board. Trim from the paper used to make envelopes is one example; likewise, makers of bleached paperboard products also generate scrap which becomes a pulp substitute. These items are segregated, compressed, and tied with wire into bales of between 227–450 kg (500 and 1000 lbs). Bales are collected by a paper stock dealer and delivered to a sorting facility, where the bales may be loaded into tractor trailers or railway box cars and shipping as is, or the bales may be broken open and any materials that may contaminate the specific grade are removed. Once sorted, this paper is compressed into bales having a density of 680–900 kg (1500–2000 lbs). The bales are then loaded into trucks or railway cars for shipment. Pulp substitutes are usually the highest priced wastepaper grades because they require the least preparation within the papermaking process, giving them a premium over paperstock from which contaminants must be removed. Pulp substitutes are used primarily in the production of printing and writing papers and packaging products, though they may be used in tissue products (14).

High Grade Deinking Papers. In 1991, 3.24 million tons of deinking papers were recycled (15). While high grade deinking papers are considered valuable because they have few nonink contaminants and contain relatively strong paper fibers, inks are usually present and must be removed for use in high value end products. Generators of these types of paperstock include printing plants and office buildings. Printing facilities separate materials much as converters do, but deinking papers from offices require different separation methods. Since in a given office location, paper is generated in a number of locations, paper stock dealers usually opt for a source-separation program, where employees separate the high grade papers from other materials (food waste, low grade paper, etc), and place them in separate containers. Office employees or building maintenance personnel then combine these containers into large bins which may hold up to 230 kg (500 lbs). When these bins are filled, an empty one is exchanged and the full ones transported to the paper sorting facility. The paper is fed onto a series of conveyor belts where contaminants are removed, or specific paper stock grades are selected out.

In other office paper recycling programs there is minimal or no source separation; all waste is removed together and sorted at the paper stock dealer's location. This collection and processing strategy offers the dealer more flexibility in what types of grades are sorted, but a higher level of contaminants must be removed. Primary uses of deinking grades are in the manufacture of tissues and towels and a small amount of printing and writing papers (16).

Newspapers. In 1992, 7.14 million tons of newspapers were recycled (17). Known as ONP (Old NewsPapers), this bulk grade is probably the one most closely identified with recycling in the public mind. Since the 1960s ONP has been collected by nonprofit groups raising funds for charitable causes while forming the basis of the collection system used by paper stock dealers and mills. With the growth of government source-separation programs, this supply source has all but disappeared. Typically ONP is the first item a community recycling program designates as a recyclable material, with the paper being tied in bundles, or placed in paper bags or recycling bins. Collection may occur at the curbside or residents may deliver the material to a drop-off site. Processing then follows the same pattern as for the high grades, with contaminants such as plastic bags and cardboard being removed. To be used at a paper mill to make new newsprint, ONP must be cleaned and all ink removed by means of mechanical and chemical processes. Historically, in the United States, a system known as the washing deinking system has been used, which could only operate successfully on ONP; small amounts of other papers would significantly reduce the quality of the end product. In the late 1980s an additional process was installed in North American mills called the floatation deinking system, requiring the presence of clay to aid in the removal of ink; in these systems, a combination of ONP and lightweight coated papers (such as magazines and catalogs) are needed (18). ONP is also used in conjunction with other paperstock grades in the production of paperboard, and is found in a number of nonpaper products including animal bedding, gypsum drywall, cellulose insulation, hydromulch for erosion control, and molded pulp products such as egg cartons. For these applications no deinking is required (19).

Corrugated Containers. In 1992, 15.37 million tons of corrugated containers were recycled (20). Generators of this material, referred to as OCC (old corrugated containers), are found predominantly in the commercial rather than residential sector, with retail outlets and distribution centers as the primary sources. These generators usually perform some source separation and if the quantities produced are large enough, may use balers on-site to reduce storage space and make transportation to the paper stock sorting facility more economical. Contaminants that must be removed to meet the requirements of the OCC grade are primarily nonpaper items, noncorrugated paper, and corrugated paper with a waxy coating. OCC is used extensively in the production of the middle layer of a corrugated box, ie, the corrugating medium, and may form part of the inner layer known as linerboard (21).

Mixed Papers. In 1992, 4.02 million tons of mixed papers were recycled (16). As its name suggests, this category of bulk grades is not identified by any one type of paper, and hence the quality of the paper fibers is less homogeneous. Mixed paper lowers the sorting costs and produces the lowest revenue per ton of any grade. Generators of mixed paper include offices, other commercial establishments, and households, with collection efforts focused on the commercial sector where the better quality fibers are usually present. Mixed paper is usually combined with OCC or news in the manufacture of paperboard, though it may also be used with ONP for nonpaper products (22).

Issues in Paper Recycling. Paper recycling faces a continuing period of changes in a number of areas. For all paper industry sectors, especially the printing and writing segment, the debate over what constitutes a recycled-content product (and with it the specific definitions of pre- and post-consumer) continues to be controversial. The absence of agreed standards and definitions slow the growth of markets for paper stock. For other grades, such as ONP, supply–demand imbalances caused by the increase in curbside collection programs continue to lessen as more of planned deinking capacity becomes operational in the mid-1990s. If municipalities initiated collection of mixed paper from residential sources, the potential supply could dwarf the oversupply problem already encountered with ONP (23).

Ferrous Metal Recycling

Recycling of ferrous metal is, along with paper, one of the oldest and most well-established recycling industries. In 1992, 91.60 million tons of steel products, from automobiles to tin cans, were generated with an estimated 56.03 million tons recovered for recycling, giving an industry recycling rate of 61% (24). Ferrous metal is collected and processed differently, depending on the type of product and its source. These can be divided into two major groups: scrap steel and steel cans. However, once processed, both groups find markets in the steel industry, either in electric arc minimills or integrated steel mills.

Scrap Steel. Scrap steel is broadly defined to include any durable product composed mainly of a steel product. It includes most vehicles, appliances, industrial machinery, and steel used in building construction. Scrap steel may also be recovered, using magnets, from resource recovery incinerator ash. In addition, it is generated by steel fabricators as home scrap. Most steel scrap is brought by the generator or independent collectors to processing facilities where a number of metals may be accepted, and where a range of technology is used to convert the scrap to a marketable commodity. Typically, a load of incoming material is weighed, dumped, and inspected for hazardous items that the facility does not accept. Automobiles, for example, must have the fuel tank and battery removed, and appliances with capacitors containing polychlorinated biphenols (PCBs) must have these items removed prior to acceptance at a processing facility. If no prohibited materials are present, the items are placed in a conveyor where they are fed to a hammermill or other type of shredder. This equipment pulverizes the product and the nonferrous items drop through rotating or vibrating screens. This leaves the scrap steel in the processing system where it is usually shredded again to reduce volume before being loaded into trucks, barges, or gondola rail cars for shipment.

Steel Cans and De-tinning. Popularly known as "tin" cans, steel cans and bimetal cans (with tops of aluminum and sides of steel) are predominantly made of steel with a thin tinplate layer on the interior to protect the product. Historically, steel cans were not recovered by scrap metal processors due to smaller volume (1.3% of MSW), and lack of an economic collection system. It was not until the public sector recycling collection programs emerged that steel cans could be collected and processed more efficiently (25). Steel cans are collected from curbside source-separation programs or by means of drop-off sites, and are delivered to a recycling facility that typically handles commingled materials (metals, glass, paper, etc, depending on recycling requirements in that area). These plants are called material recovery facilities (MRFs) and do not accept the bulkier scrap steel items. At the MRF, loads are dumped on a tipping floor where manual sorting of nonrecyclable items takes place before the commingled materials are fed onto an inclined conveyor belt. Additional sorting stations may be placed prior to the first mechanical separation system. This is usually an electromagnet in the form of a rotating belt that is suspended above the conveyor. As the materials move under it, the ferrous items are lifted up and deposited to the side into a collection container or onto another conveyor. Final processing is some form of densification using a baler or briquetter (26).

For both home scrap and cans with tin plating, another processing step is used. Known as de-tinning, the plated items are placed in an electrically charged chemical bath which causes the tin to separate from the steel, producing a high quality tin, which can be sold to instrument makers and others, and a purer grade of steel for that industry. Increasingly, however, the economics of de-tinning are reducing its role. Can manufacturers have reduced the amount of tin contained in steel cans, so there is less for a de-tinner to recover and sell, and steel mills have begun to accept cans which have not been de-tinned since the lower level of tin in the cans is not considered a contaminant (27).

Issues in Ferrous Recycling. The major issue in scrap steel recycling is the nature and extent of the regulations governing processing facilities. The U.S. EPA has been reviewing the treatment of residue from scrap steel processing, which has been found to contain PCBs and other hazardous materials. The processors are also subject to liabilities if their facilities are designated as hazardous waste sites. In addition, the Clean Air Act of 1990 prohibits the venting of chlorofluorocarbons (CFCs) from air conditioners and other appliances when they are collected for recycling or disposal; this will lead to changes in the way scrap metal processors operate and the creation of a new infrastructure to comply with the law (28). The recovery of steel cans has been increasing, and the industry trade association is seeking to widen the types of cans collected by encouraging government recycling agencies to include aerosol and paint cans in their programs (29).

Nonferrous Metals: Aluminum

In 1967 the aluminum industry initiated the first can collection center. In 1989, 4.42 million tons of aluminum products were manufactured in the United States, and 2.39 million tons of home scrap and aluminum beverage containers (UBCs) were recovered (30). UBCs represent 1.6% of MSW, and 984,000 thousand tons were recovered in 1991; the recovery of UBCs was 62.4% in 1992 (31). These high recovery figures and early start in collection are due to the much lower energy costs associated with scrap aluminum versus virgin ore; use of recycled materials reduces energy costs by 95% over virgin materials (32). The high intrinsic value of aluminum scrap is worth the cost to industry to set up voluntary collection centers and pay for cans.

Processing Infrastructure. For large, bulky aluminum items, eg, aluminum siding, window frames, etc, collection and processing occurs in the same system as with ferrous scrap. UBC collection may take place through the "cash-for-cans" programs operated by the aluminum industry, via curbside recycling/MRF processing, or by using the beverage distributors in the 10 states which have deposit systems. Under such systems, consumers pay a deposit at the point of purchase and claim the deposit by returning the beverage container to the retailer, who in turn directs it to the distributor who sells it to the aluminum manufacturer.

Other Nonferrous Metals

In addition to aluminum, most other nonferrous metals are recycled to the degree to which it is profitable. Metals such as gold, silver, copper, nickel, cadmium, zinc, brass, and bronze are generated in small quantities and recovered at high rates (33). Lead, found in vehicle batteries, is also recovered, both for its value as a commodity but also because of its potential disposal cost as a hazardous item; mercury from household batteries is recovered for similar reasons.

Glass Recycling

Glass, primarily in the form of bottles and jars, accounted for 12 million tons (9%) of MSW in 1990, and 2.60 million tons were recycled. A small amount of noncontainer glass, eg, windows, mirrors, light bulbs, and ceramics, are also present in the waste stream (10% of the total, or 1.2 million tons). These types of glass contain additives that are incompatible with bottle making and hence are not recovered in significant quantities (34).

Collection and Sorting of Container Glass. Unlike paper and metal, very little in-plant scrap ever leaves the manufacturing facility because the container produced is not further converted into another product. Glass manufacturers use in-plant scrap because it reduces emissions, energy costs, and furnace maintenance costs (35). As a result, glass collection has been the province of voluntary drop-off programs, beverage container deposit systems, and municipal curbside programs. A distinguishing feature of glass collection and handling is the industry requirement for color separation. Historically, container glass has been produced in three colors: flint (clear), amber (brown), and emerald (green), and while flint may be used in the production of either of the other colors, and amber may be used in small quantities in the making of emerald, no other combinations are possible, given the aesthetic demands of the food and beverage industry.

Once delivered to a MRF, glass is usually the second material after steel to be sorted from the incoming commingled stream. All glass sorting is done by hand, with different workers each removing whole bottles/jars of one of the three colors. Inevitably through the collection, transportation, and initial unloading, bottles break and glass fragments cannot be picked by hand therefore becoming a residue item. For the sorted bottles, MRFs typically crush them to reduce volume. Before the product, known as cullet, can be melted down and used to make another container, it must undergo additional cleaning and crushing. This process, called beneficiation, may take place at the MRF, at a separate third-party facility, or at the glass manufacturer's location.

Markets for Glass. There are 75 container glass manufacturing plants in the United States, and they typically use 30% cullet in their operations, though the industry has a target of 50% cullet use by 1995 (36). Other uses, particularly for mixed color and noncontainer cullet, are being found in the construction industry. Use of cullet in asphalt paving (glassphalt) is gaining acceptance, and use of mixed glass to aid drainage or as a clean fill is also an alternative use for mixed glass (37).

Issues in Glass Recycling. Changes in glass recovery seem unlikely. The container glass industry has been losing market share to both paper and plastics due to consumer preference for lighter weight and safer packaging alternatives. In addition, the industry was the subject of several leveraged buy-outs in the 1980s and has fewer financial resources with which to invest in improved technology or additional marketing efforts. One innovation which may improve the efficiencies in glass recycling is a system for covering flint bottles with a thin plastic coating containing pigments. The coating is burned off when the cullet is melted, and this practically eliminates the need for color sorting and allows hundreds of "designer" color glass bottles (38).

Plastics Recycling

Production and use of plastics is a new phenomenon compared to paper, metal, and glass, because plastic containers were not marketed until the 1960s. While the industry's existence is recent, the amount of plastics in the waste stream is growing. Generation of plastic waste has risen substantially, from 2.7% of MSW in 1970, to 7.3% in 1986, and is projected to increase to 9.2% in 2000 (39). Several characteristics of plastics are relevant to understanding plastics recycling. First, the development of the infrastructure and technology to achieve the recycling loop is in its early stages. For example, it has taken 16 years for the plastic bottle to reach a 17.7% recycling rate, while glass containers have been produced for much longer and have a recycling rate of 33% (40).

A second factor which affects plastic recycling is the large and growing number of plastic resins, each of which has a specific chemical formula the characteristics of which are more difficult to detect than the difference between flint and amber glass, or between ferrous and nonferrous metals. The Society of Plastics Industries (SPI), an industry group, has developed a voluntary coding system for the most common resins. The resins, codes, and common products are polyethylene terephthalate (PETE, #1; soda bottles); high density polyethylene (HDPE, #2; milk and detergent bottles); polyvinyl chloride (V, #3; mineral water bottles); low density polyethylene (LDPE, #4; plastic dry cleaner and grocery bags); polystyrene and expanded polystyrene (PS, #5; frozen yogurt containers, meat trays, and fast-food containers); polypropylene (PP, #6; margarine tubs, ketchup bottles, automotive battery casings); and other, classified as #7 (41).

Collection and Processing. Collection of plastics for recycling follows many of the same means as those of other food and beverage containers: drop-off sites, curbside source separation, and bottle deposit systems. In addition, individual trade groups representing resin producers, container manufacturers, and retailers have established collection efforts for specific resins. Large grocery store chains, for example, have collection containers for used grocery bags, and concessionaires at U.S. National Parks provide recycling receptacles for polystyrene, UBCs, and glass containers (42). Curbside collection and transport to MRFs has proven to be a very costly area for efficient plastics recycling, owing to high volume of plastic items; with low weights, this means that processing throughput is low and costs are high. In addition, once at the MRF, hand-sorting of plastics is slow and inefficient, because container makers shift from one resin to another for the same product. Containers themselves look very much the same, with PETE and PVC being particularly difficult to distinguish visually. Automated sorting equipment that can sort PETE and HDPE by color and separate out PVC is in operation, but so far the equipment is costly and slow (43). Once separated by resin and color, most MFRs bale whole containers and sell them to plastic reclaimers, who wash, grind, and pelletize the resins so that they can then be sold to users of plastic resins. Unfortunately, by this stage, the costs of collection, initial processing, transportation, and additional cleaning and processing make it difficult for the recycled resin pellet to compete against a virgin pellet.

Recycled Plastic Products. Once reclaimed to a pellet which meets the same quality specifications as a virgin resin, there are no technical reasons why recycled plastics cannot be used in the manufacture of the same products; the plastic bottle industry has a goal of using 25% recycled content. Concerns regarding the presence of recycled resins in packages where there is contact with food or liquids for human consumption and the difficulties in achieving lighter colors from previously colored resins are two factors limiting the increase of the "bottle to bottle" recycling loop. As a result recycled resins are often used in products where color is not a factor: audio and video cassette cases, padding and insulation for clothing, PVC pipes and drainage channels, polypropylene battery cases, etc (44).

In an effort to lower processing costs, a number of companies have developed extrusion technologies that use a mixture of resins. This mixture varies by manufacturer, but usually includes some PETE and HDPE, with the resulting product a substitute for lumber products. Known generically as plastic lumber, it can be extruded into dimensional sizes which may be used to fashion benches, boardwalks, and picnic tables, or formed into marine pilings, car stops, and road signs. While promising, the plastic lumber industry suffers from a lack of capital and difficulty in penetrating large markets (45).

Issues in Plastics Recycling. The infrastructure for the full recycling cycle in plastics is not fully developed, and it is difficult to say whether recycling will be a cost-effective waste management strategy into the year 2000. Collection and processing costs must be reduced to enable recycled resins to compete with their virgin counterparts, which will take financial resources and technical developments, particularly in automated sorting. In addition, the resin-producing industry plans to expand virgin production capability which may retard the demand for recycled resins. These factors may lead to increased use of resource recovery incineration or landfilling for plastics (46).

Scrap Tire Recycling

It is estimated that 200–250 million car and truck tires are generated annually in the United States, representing 1.5% of MSW, with an additional three billion tires in unregulated stockpiles (47). In 1992, approximately 10 million tires were recycled in paving and civil engineering applications with an additional 58 million scrap tires used for energy recovery (48). The disposal of tires is of concern because there are aesthetic, public health, and safety problems which arise when tires are improperly disposed. Illegally dumped tires are a litter problem, may act as breeding habitats for disease-bearing insects, and create fires having noxious fumes that are difficult to extinguish (49).

Collection and Processing. As with many of the larger items in MSW, tire collection and processing have historically been the domain of the private sector. Retail tire dealers, service stations, and automobile sales operations typically remove tires from their customers' vehicles and replace them with new ones; the discarded tires are then collected by specialized haulers called tire jockeys. Government programs concerning scrap tires usually involve removal of tires that have been improperly disposed, with tire jockeys or waste haulers performing these services under contract. Once collected, tires are manually sorted into two groups: those that are capable of being retreaded, and those which are not. Tires for retreading are transported to retreading facilities. All other scrap tires are shredded into pieces between 5 and 20 cm (2 and 8 inches) long. Shredded tires may then be further processed to fine grains of crumb rubber.

Uses for Scrap Tires. Retreading is the preferred use for truck tires in which the carcass, or tire casing, has retained its strength; retreading of passenger car tires, except for taxi fleets, is limited. This is due to the lower cost of new passenger tires versus retreaded tires. Tire shreds may be used as fuel in cement kilns and in power generating stations, in construction for fill material and in composting as a bulking agent. Markets for crumb rubber include asphalt rubber and consumer products (50). The use of asphalt rubber is projected to increase from five million tires per year in 1992 to 80 million tires in 1997. This is due to the implementation of the national Intermodal Surface Transportation Act (ISTEA), which requires contractors on federally funded highway projects to use increasing amounts of asphalt rubber, reaching a cap of 20% of asphalt use by 1996 (51).

Wood Recycling

It is difficult to estimate the amount of wood that is discarded because wood waste is usually captured by construction and demolition facilities (52) rather than household or commercial waste haulers. Waste wood is termed urban or dry wood to distinguish it from green or dry wood generated by wood products manufacturing. Sawmill and lumber waste, for example, has a moisture content of approximately 30%; urban wood, such as packing crates and pallets, typically has 10% moisture. Urban wood is generated by construction and demolition activities, packaging, and trimmings from the furniture industry. Most urban wood is collected commingled with other materials and is processed at facilities known as construction and demolition transfer (C&D) stations. At a C&D station, wood items are separated by hand or, for larger pieces, by use of a grappling device. The wood is then sorted into two categories, treated and untreated. Treated wood is covered in some way with materials that are contaminants for recycling, eg, paint and creosote. Untreated wood is ground into chips of 5–10 cm (2–4 inches) or may be ground as fine as sawdust, depending on its intended market. Metal scrap and other nonwood items are removed using magnets and screens.

Uses and Issues of Urban Wood Recycling. A major use for urban wood is in the generation of power in cogeneration or biomass facilities, with over 94 million tons used in 1991 (53). Other uses include mulch, compost bulking agent, and landfill cover. Two to five times more wood is available for recycling than is currently reaching these uses due to the costs of processing and the low prices of both disposal and the virgin materials with which urban wood competes (54). The processing industry also lacks standard specifications and there are concerns among potential users of the material that it may contain hazardous substances, such as lead dust from paint.

Composting

Composting is a specific type of recycling that involves the biological decomposition of organic materials through bacterial and fungal activity. Typical compost materials are yard trimmings and food waste; in 1992, these accounted for 33% of MSW, and 2981 composting facilities were in operation (55). Compostable materials are usually source separated and may be composted on-site or transported to another location. Composting of unsorted MSW also occurs. Households generate small quantities of yard trimmings and food, and if space permits, "backyard composting" is effective and eliminates transportation costs. Larger generators, such as parks and highway agencies, food service establishments and landscapers, may also have space available to compost on-site, but many do not. In these situations, materials are transported to one of two types of facilities; a windrow composting site or an in-vessel plant.

Windrow Composting. In a windrow compost operation, materials are unloaded at an open-air site of between 2 and 40 acres and are formed into windrows, long piles of material arranged in parallel rows. Windrows range from 1.8–3 m (6–10 ft) in height and are roughly triangular when viewed from the end. To ensure that the decomposition process takes place at a uniform rate and temperature (approximately 60°C), windrows are watered regularly and turned once or twice per day by front-end loading tractors or specialized vehicles. If the material to be composted has a high water content or has been compacted during collection, bulking agents may be added. A bulking agent, such as wood or tire chips, allows air to circulate and assists bacterial activity. Depending on climate, windrows may produce finished compost in nine to twelve months. Composted material looks like a dark brown humus or grainy soil. Prior to use, compost is usually cured to remove contaminants and bulking agents, or to produce a product with specific characteristics (56).

In-Vessel Composting. In contrast to windrow composting, in-vessel composting takes place within an enclosed space and produces a final product in time periods of four to ten weeks. Though there are a number of different technologies in use, most use the same principles to accelerate the decomposition process. These include continuous turning of the material, regulated temperatures, protection from the elements and mechanical air circulation systems.

Compost Products and Issues. Both methods of composting produce a soil amendment suitable for fertilizing crops and gardens or providing erosion control. Compost may be used by its original generator, given away to citizens, or sold in competition with other fertilizers. While the mechanics of composting are well understood, land use regulations, citizen concerns over odor, education on backyard composting, and the need for additional marketing to potential users are several issues that presently limit the amount of composting.

Textile Recycling

Textiles make up 4.9% of MSW, totaling 7.6 million tons in 1990, and include clothing, bedding materials, leather goods, and shoes (57). Historically, textiles were recovered for papermaking beginning in the eighteenth century, with the rag dealer obtaining supplies and selling them to mills. Like the paper and metal dealers, textile dealers form the basis of the processing infrastructure (58), while nonprofit organizations are responsible for the bulk of the collections.

Collection and Processing of Textiles. Nonprofit organizations such as Goodwill Industries, the Saint Vincent de Paul Society, and the Salvation Army obtain donations from the public at drop-off sites or provide collection from the household. Of the donated textile material, approximately 25% is selected for free distribution or sale in the stores operated by these groups. The remaining material is sold as mixed rags or institutional rags to the private sector textile dealers. Mixed rags are hand sorted by type of garment, size, color, fabric, and condition, into as many as 150 grades, and then compressed into bales weighing between 150–300 pounds (59).

Markets for Textiles and Issues for Textile Recycling. The graded textiles are sold to three types of markets. The majority are sold as used clothing and are produced primarily for export. Approximately 20% are used in the production of industrial wiping cloths, while another 26% are fashioned into fiber for reprocessing, a shredded mixture of fabric colors and materials which are used in automobile upholstery and carpet padding. Also in this category is a small amount of cotton, which continues to be used by the paper industry in the manufacture of rag bond papers. Conditions appear good for increased amounts of textile recycling. Municipal governments also have shown interest in collecting textiles not donated to nonprofit institutions (60).

Construction Aggregates

Construction aggregates consist of rock, stone, masonry, concrete, and brick and are generated in the construction of buildings and roadways, or in their demolition or renovation. Typically, construction aggregates are generated with scrap wood and metal and are processed at C&D stations. Processing techniques are designed to reduce volume and form a product of uniform size.

Uses of aggregates are found primarily in the construction industry. Crushed rock and stone is used as a clean fill material, as part of hot mix asphalt, as sub-roadbed material, and as landfill cover. Asphalt millings from street repaving projects can also be crushed and returned to asphalt forming recycled asphalt pavement (RAP). Market prices for these products are low. If processing costs are not met by product revenue, the only cost-effective processing is simple crushing for volume reduction and disposal in a landfill (52).

Policy Issues and Future Developments

The challenges facing recycling are many, and for recycling to play the role envisioned in the waste management hierarchy, developments will need to take place through a variety of means and by a number of participants. Some issues are specific to individual recyclable materials, others affect all recycling sectors, while still others are caused by changes in MSW systems (waste retention, incineration, landfilling) which, by the nature of the rated approach, affect each of the other methods. Following the flow of activities which constitute recycling, the major areas of concern are public education, materials handling technology, market development, product design, and solid waste regulations.

Public Education. Curbside recycling for households and commercial recycling programs for business represent a change in behavior in the way in which refuse is conceived and handled. The overriding need will continue to be that of providing information and motivation to habituate such behavior among individuals and institutions. Only when the consumer of products is aware of their potential value as recyclable commodities will the extra effort and costs of recycling be undertaken (61). Private industry, schools, government at all levels, and civic and environmental organizations must all communicate to the public and the business community the value of recycled materials; the costs of alternative MSW methods, and the appropriate measures to take to ensure the maximum recovery of materials that are collected.

Technologies for Collection, Processing, and Manufacturing. Advances in technology needs to continue, particularly for the materials that have technical bottlenecks preventing efficient recycling, such as collection and processing problems with plastic resins. Lowering overall sorting costs at MRFs is another issue, affecting all materials, in which technological changes allowing for increased automation, higher throughput, and lower operating costs further the economic value of recovered materials (62). Also of importance for technical research and development is the area of difficult-to-handle materials (batteries, PCB- and CFC-containing appliances), and research on materials for which few end uses exist (scrap tires, some plastic resins).

Market Development. As government-mandated collection programs expand, the flow of recyclable materials being processed for manufacturing or other uses has been greater than the ability of industry to use these materials; partly as a result, demand is slack and scrap prices are low. Government, environmental organizations, and the private sector are all involved in attempts to bring about changes in the supply–demand equation. Purchasing policies favoring products with recycled content is one way enacted by local, state, and the federal government to increase demand (63).

Legislative means of spurring demand go beyond buy-recycled programs. Discussion on the reauthorization of the Resource Conservation and Recovery Act (RCRA) in 1991 highlighted several possible courses of action. Voluntary commitments by industry to expand capacity for use of recyclables have been effective at the state level (64) and it has been suggested that a national agreement would provide the best impetus for additional investments. It has been argued that the federal government should establish minimum content or utilization rates for each recyclable. An addition to this approach is to provide for recycling credits which could be traded among companies that exceeded the required levels and those for whom the cost of changing production systems would be unsustainable. Other approaches to develop markets include use of advanced disposal fees or taxes on disposal of items; disposal bans; expansion of deposit systems to include more beverage containers, batteries, and tires; and investment tax credits. In addition, use of existing economic development tools, ie, tax credits, assistance in siting, use of tax-exempt bond financing, with a focus on recycling processors or manufacturers has also been suggested (65).

Environmental groups have identified more wide-ranging changes which may improve the markets for recyclables. Removal of tax advantages accruing to the users of virgin timber and the low royalties paid for hard rock mining are two areas where existing natural resource policies pose barriers to use of recycled commodities (66). Programs that shift the MSW function from government to the manufacturers of products, termed producer responsibility programs, are in place in Germany and France, and have been advocated for the United States (67). In such programs, product makers assume responsibility for managing their goods after they have been used by consumers, and in Germany this has taken the form of the Green Dot program where packaging manufacturers employ a private recycling consortium to handle product waste (68).

Product Design. Another policy area concerns the design and initial manufacture of products. The aim of legislative and voluntary initiatives here is to reduce the overall volume waste generated, while at the same time eliminating as many toxic elements and product design features which render the spent product expensive, hazardous, or difficult to recycle. By focusing at the beginning of the production–use–recycling circle, costs for the ensuing collection, processing, and disposal of residue can all be lowered and the entire system made more functional. Automobile makers, for example, are changing design and production to make car disassembly and recycling easier (69).

Solid Waste Regulation. A cluster of existing MSW rules, regulations, and practices are in the process of changing and regardless of how these developments emerge, recycling will be affected for better or worse. Flow control, the principal that government has the right to direct waste to specific incinerators or landfills, has been challenged as infringing commerce. Many public sector recycling programs direct the recyclables in a similar way; if they were unable to do so, it could add uncertainty and costs. If interstate transportation of waste were to be restricted or made more costly, as some state governments have argued, the relative cost of recycling could fall, improving its economics. In a similar way, if municipal waste incinerators are required to dispose of incinerator ash as a regulated hazardous waste and not use existing MSW landfills, costs of this waste management method would increase, again making recycling relatively less expensive.

BIBLIOGRAPHY

1. U.S. Congress, Office of Technology Assessment, *Facing America's Trash: What Next for Municipal Solid Waste*, U.S. GPO, Washington, D.C., 1989, pp. 74–75.
2. World Resources Institute, *1993 Information Please Environmental Almanac*, Houghton Mifflin, New York, 1993, pp. 52–53.
3. R. Steuteville and N. Goldstein, *Biocycle* **34**(5), 46 (May 1993).
4. Ref. 1, p. 135.
5. C. Thompson, *Recycled Papers: The Essential Guide*, MIT Press, Cambridge, Mass., 1992, p. 75.
6. Ref. 3, p. 47.
7. S. J. Reaven, "Long Island's Garbage Barge: What Really Happened – and Why," *National Recycling Coalition 1987, Recycling Congress,* Austin, Tex.
8. Ref. 3, pp. 43–46; Ref. 2, p. 59.
9. J. Powell, *Resource Recycl.* **10**(4), 41–41 (Apr. 1993).
10. *Recovered Paper Statistical Highlights 1991*, American Forest Products Association (AFPA), Washington, D.C., 1992, pp. 1–20.
11. W. E. Franklin, M. A. Franklin, and R. G. Hunt, *Waste Paper: The Future of a Resource*, Franklin Association Ltd., Prairie Village, Kans., 1982, p. 238.
12. *Scrap Specifications Circular 1993*, Institute of Scrap Recycling Industries (ISRI), Washington, D.C., 1993, pp. 28–32.
13. Ref. 10, p. 14.
14. *1990 Annual Statistical Summary of Waste Paper Utilization*, American Forest Products Association, Washington, D.C., 1991, p. 39.
15. Ref. 13, p. 14.
16. Ref. 14, p. 40.
17. Ref. 9, p. 43.
18. New York State Newspaper Recycling Task Force, *Final Report*, N. Y. S. Dept. of Economic Development, Albany, New York, 1989, pp. 43–46.
19. E. Hanig, *Markets for New York City's Old Newspapers: Current and Potential Uses*, N. Y. C. Dept. of Sanitation, New York, 1989, pp. 2–51.
20. Ref. 9, p. 42.
21. R. W. Beck, *Market Study of Old Corrugated Cardboard*, R. W. Beck and Associates, Waltham, Mass., 1990, pp. 2-1–2-3.
22. *Household Mixed Waste Paper Recycling: Paper Mill Survey Results*, Sound Resource Management Group, Inc., Seattle, Wash., 1990, pp. 11–17.
23. *Developing Markets for Recycling Multiple Grades of Residential Paper*, Environmental Defense Fund, New York, 1992, pp. i–ix.
24. Ref. 9, p. 40.
25. C. Miller, *Waste Age*, 69 (Jan. 1993).
26. G. Crawford, *Resource Recycl.* **12**(2), 27–34 (Feb. 1993).
27. Ref. 25, p. 70.
28. J. T. Aquino, *Waste Age*, 39–42 (Feb. 1993).
29. Ref. 26, p. 31.
30. *Facts: 1989 Yearbook*, Institute of Scrap Recycling Industries, Washington, D.C., 1990, p. 2.
31. C. Miller, *Waste Age*, 79 (Oct. 1992).
32. *Fourth Report to Congress: Resource Recovery and Waste Reduction*, U.S. Environmental Protection Agency, Washington, D.C., 1977, p. 39.
33. Ref. 30, pp. 5–7; 12–17.
34. *Solid Waste Manage. Newsl.* **7**(4), 1–3 (Apr. 1993).
35. *Waste Glass Use and Demand in the New York Region*, Resource Management Associates, Napa, Calif., 1990, pp. 3–6.
36. Ref. 34, p. 2.
37. J. Slovin, *World Wastes*, 14–16 (June 1993).
38. N. Y. S. Dept. of Economic Development, Office of Recycling Market Development, *Status of the Markets Update* 1:2, pp. 3–4 (February, 1992).
39. Ref. 1, p. 80.
40. Ref. 9, pp. 43–45.
41. *A Survey of Mixed Plastics Recycling Programs*, Resource Management Associates, Napa, Calif., 1990, pp. II-6–II-7.
42. *Resource Recycl.* **12**(7), 16 (July 1993).
43. G. Brewer, *Resource Recycl.* **9**(6), 42–46 (June 1990); P. Dinger, *Biocycle* **32**(3), 80–82 (Mar. 1992).

44. *An Overview of Markets for Used Plastics Collected in the New York City Area*, Resource Integration Systems, Hartford Conn., 1989, pp. 15–29.
45. *Ibid.*, pp. 32–33; Ref. 41, pp. II-8–II-9.
46. K. Meade, *Waste Age*, 39–48 (Aug. 1992).
47. *Report on Scrap Tire Management Recycling and Disposal in New York City*, Recycling Research Institute, Suffield, Conn., 1990, pp. 1–8.
48. M. Phillips, *Tire Rev.*, 29 (Mar. 1993).
49. M. Marriot, "In Growing Piles, Dead Tires Haunt New York," *New York Times*, May 17, 1993, pp. A1–B6.
50. N. Getz and M. F. Teachey, *Waste Age*, 81–88 (Oct. 1992).
51. Ref. 48, p. 40.
52. B. Lee, *Solid Waste Power*, 45–49 (Oct. 1991).
53. *1993 State Biomass Statistical Directory*, National Wood Energy Association, Washington, D.C., 1993.
54. *Preliminary Recycling Plan Fiscal Year 1991 Appendix III: Market Development*, N. Y. C. Dept. of Sanitation, New York, 1990, p. 47.
55. Ref. 3, p. 48.
56. Ref. 1, pp. 184–187.
57. T. Polk, *Resource Recycl.* **11**(2), 56 (Feb. 1992).
58. Ref. 5, pp. 22–23.
59. E. Stubin, *A Survey and Market Research Study of the Post-Consumer Textile Product Waste Recycling Industry*, Trans-Americas Clothing Corp., Brooklyn, New York, 1991, pp. 3–14.
60. *Ibid.*, p. 15.
61. J. Holusha, "Who Foots the Bill For Recycling?" *New York Times*, Apr. 25, 1993, pp. E4–5.
62. J. Glenn, *Biocycle* **33**(8), 34–39 (Aug. 1992); C. Miller, *Waste Age*, 26–33 (Oct. 1992).
63. *Progress in Implementing the Federal Program to Buy Products Containing Recovered Materials*, U.S. General Accounting Office, Washington, D.C., Apr. 1992, pp. 1–10; N. VandenBerg, "Expanding and Protecting the Buy-Recycled Agenda," presentation at the *U.S. Conference of Mayors*, Apr. 1, 1993, pp. 1–8.
64. New York State Newspaper Recycling Task Force, *Progress Report*, N. Y. S. Dept. of Economic Development, Albany, New York, 1992, pp. 8–9.
65. A. D. Reamer, *Commentary*, 20–29 (Winter 1991).
66. Ref. 1, pp. 197–201.
67. A. Hershkowitz, *The Atlantic Monthly*, 108–109 (June 1993).
68. F. Cairncross, *Harvard Business Rev.*, 34–45 (Mar./Apr. 1992); J. E. McCarthy, *Recycling and Reducing Packaging Waste: How the United States Compares to Other Countries*, Congressional Research Service, The Library of Congress, Washington, D.C., 1991, pp. 80–90.
69. E. Culp, *Mod. Plast.*, 68–70 (Oct. 1992).

RECYCLING—HISTORY IN THE UNITED STATES

NEIL N. SELDMAN
Institute for Local Self-Reliance
Washington D.C.

Before the turn of the century, the nature of local regional economies made recycling and composting relatively easy: milk and juice bottles were returned to local vendors to be washed and refilled; organic food and plant wastes were used on nearby gardens or farms; quilts, rags, and even paper were created from discarded textiles. Entire cities such as New York and Philadelphia developed productive exchanges, delivering organic wastes to rural areas for use as fertilizer, and food and fuel were returned to the cities (1–5). The old economy had a two-way system, or a closed loop.

In the late 20th century, however, economic and production trends undermined this cycle, creating a one-way flow of materials, from producer to consumer, ending at the garbage dump. Consumption has soared, and producers are separated, sometimes by thousands of miles, from the consumer. Disposable products have become the norm, replacing once durable goods such as cameras, staplers, and diapers. Packaging, which once consisted only of paper, now often includes a combination of plastics, paper, and even metal, and plastic packages often combine several different resins. Efficiency took a back seat to convenience (6). "Our economy became extremely wasteful with respect to the way we use natural resources," one technological analyst concluded.

> Our automobiles convert only fifteen to twenty percent of fuels' energy content into motion of the car; our power plants are only thirty percent efficient. Cities discard approximately 250 million tons of materials in a waste stream richer in minerals than many industrial mining operations; we produce roughly 350 million tons of agricultural wastes annually. Eighty percent of our products are used once and thrown away. This accounts for well over a billion tons of discarded materials, an inviting target for a war on waste (7).

In the 1960s, several publications awakened citizens and officials to the dark side of the emerging consumer society. Rachel Carson's *The Silent Spring* delivered a profoundly disturbing account of hazardous materials' threat to environmental security and public health. John Kenneth Galbraith's *The Affluent Society* provided equally troublesome insights into the consumer economy, and Vance Packard's *The Waste Makers* illustrated the prevalence and problems of superfluous production. The message of these publications was clear: every product consumed eventually becomes a product discarded, and methods of disposal have a permanent impact on the environment, economy, and the well-being of people.

In the 1970s, grassroots environmental activists and leading Wall Street and business analysts predicted the trend toward recycling. Cliff and Mary Humphrey made recycling the centerpiece of their teach-in style "Survival Walks" in 1968. Local recycling centers were established in each town they visited. Both symbolic and practical, the centers allowed citizens to participate directly and immediately in environmental action, while laying the ground for the development of a recycling infrastructure (8–10). Cliff Humphrey explained that more materials could have been economically recovered from commercial generators, but grassroots strategists knew they had to start mobilizing voters to change rules and policies, hence the early focus on household recycling. A 1975 Dun & Bradstreet confidential report to investors advised that by the year 2000 waste reclamation would become the second largest industry in the world, second only to agriculture (11), and the spring 1978 issue of *California Management Review* introduced the concept of reverse channels of distribution, referring to the phenomenon of consumers not only purchasing products from manufacturers but also returning materials to the manufacturers (12).

In the 1980s, citizens began to take action against old leaking landfills, the construction of new landfills, and the profligate generation of waste. As garbage became one of the central problems affecting both metropolitan and rural areas, there was an unprecedented rise in the cost of waste disposal. Communities and businesses turned to the two major alternatives to land filling: incineration and materials recovery (recycling and composting). In the early 1980s, it appeared that waste incineration would become the disposal method of choice; however, since 1985 there has been a steady decline in the number of plants ordered.

By the end of the decade, no less than 42 states had mandated minimum recovery goals for residential, commercial, and government consumers. After 1990, materials recovery was recognized as an economic development tool, prompting another round of federal, state, and local government laws and policies to stimulate investment in and the expansion of secondary materials use. Four trade journals regularly report on state and local legislative and policy developments: *Recycling Today, Biocycle, Resource Recycling,* and *Recycling Times.* Materials recovery and economic development became the primary goal of solid waste management planning, and coalitions of recycling advocates, business leaders, elected and appointed officials, and community development activists joined in coalition to redefine acceptable solid waste management procedures that incorporated environmental safety and economic growth criteria.

See also ENVIRONMENTAL ECONOMICS; WASTE MANAGEMENT PLANNING, WASTE TO ENERGY; FUELS FROM WASTE.

REVIEW OF SOLID WASTE MANAGEMENT IN THE UNITED STATES, 1945–1990

Rarely if ever found in solid waste management plans of the 1970s, recycling has become the law of the land. Materials recovery plans are now required in every state in the union as part of their overall solid waste management program. In November 1989, citizen organizations shocked the waste incineration industry by convincing the U.S. Environmental Protection Agency (EPA) to reissue New Source Performance Standards (NSPS) for solid waste incinerators. The NSPS required that communities achieve a minimum 25% materials recovery rate to obtain a permit to construct a new incineration facility.

Recycling activists had hoped for a 50% recycling minimum requirement, but were satisfied that, for the first time, EPA policy established pollution prevention as official solid waste policy. EPA staff had begun planning the extension of the concept to landfill permitting when suddenly, in December 1990, the Competitiveness Council (established by the Office of Management and Budget) rejected the policy as an anticompetitive practice. The issue has been rejoined in subsequent Congressional debate on the reauthorization of the Resource Conservation and Recovery Act (RCRA).

Also, the Pollution Prevention and Incineration Alternatives Act of 1993 would place a moratorium on the construction of new solid waste incinerators until 1997. The bill also places limits on the permitting authorities of state and local governments as well as the EPA. After 1996, incineration applicants must divert 65% of newspapers, 80% of metals, 50% of plastic containers, 90% of yard waste, and 10% of food waste by the year 2000 to obtain permit approval. The bill also requires the applicant to provide $50,000 in monitoring grants to concerned local citizens every 6 months until the plant is sited.

The principal argument was the Best Available Control Technology (BACT) provision of the 1969 National Environmental Protection Act (NEPA), which stated that commercially available, economically competitive technologies known to reduce pollution must be incorporated into new programs (13–15).

To best understand the current options before solid waste management decision makers, it is important to review the history of landfilling, incineration, and recycling in the United States. After two decades of pent-up consumer demand (the Great Depression of the 1930s, when people had little money to spend, and the World War II era, when there were few products to be purchased), Americans went on a spending spree. Television and other forms of mass advertising fueled the fury of consumption.

Recent research shows that the expanded waste stream resulted primarily from new packaging and the transition from once durable or reusable goods (razors, beverage containers, food utensils) to disposable or nondurable goods, often discarded within minutes of purchase (see Table 1). These new products and packaging

Table 1. Products Discarded into the Municipal Solid Waste Stream[a,b]

	1970		1988	
Products	Tons[c]	Percent	Tons[c]	Percent
Durable goods	13.7	12.4	22.6	13.9
Nondurable goods	32.2	20.9	45.7	28.1
Containers and packaging	39.5	35.7	51.5	31.6
Other wastes	34.3	31.0	43.1	26.4
Total	*110.6*	*100.0*	*162.4*	*100.0*

[a] Data from Ref. 16.
[b] Municipal solid waste is defined here to include residential, commercial, and institutional solid waste.
[c] Millions of tons.

transformed the waste stream from one that was readily handled by open-bodied trucks to a waste stream that required large 9-t capacity compactor trucks to compress the garbage, thereby increasing the volume being hauled to landfills. These new vehicles brought their loads to transfer stations, where the waste, compressed yet again by stationary hydraulic components, was deposited into 27-t tractor trailers for transport to ever more distant landfills.

The amount of solid waste in the United States quadrupled between 1960 and 1980. Per capita, Americans generate more than 1.8 kg solid waste daily, more than any other country in the industrialized world. (The Japanese and Danes generate just over 0.45 kg per capita per day.) In addition to the sharp increase in solid waste generation, suitable landfill sites were becoming scarce. The U.S. population explosion immediately following World War II took families from cities to the suburbs. In practical solid waste management terms, this move drastically decreased available landfill sites for urban centers, as increasingly powerful suburban jurisdictions rejected the landfill applications of their urban neighbors.

By the mid-1960s, the federal government began to react to the nation's growing solid waste management crisis. New laws permitted the Department of Health, Education, and Welfare to implement research and technical assistance programs on behalf of local governments, which, up to that time, had sole responsibility for solid waste management decisions. Spurred by a spontaneous national environmental happening, Earth Day 1970, the environmental movement took on new dimensions. By 1972 the EPA had been established and had assumed expanded research, development, and implementation programs; and the 1976 passage of the Resource Conservation and Recovery Act (RCRA) bolstered the EPA's authority (17). The first result was increased pressure to clean up and close landfills. The discovery that existing landfills were leaking dangerous materials into soil and groundwater quickly raised the consciousness of citizens and planners. In the late 1980s, it was revealed that more than 50% of the country's Superfund sites were former solid waste landfills (18). A movement began to close thousands of unpermitted and poorly maintained (yet permitted) landfills and halt the development of new landfills. Between 1960 and 1990, the number of landfills in the country declined from an estimated 19,000 to 6,000 sites. As a result of RCRA Subtitle D requirements, only 1,600 landfills may be operating by 1995. Thus, waste generation increased as disposal capacity decreased.

Industry, too, became aware of the problems. Recognizing that the lack of convenient disposal methods could limit sales of traditional products and draw increasingly negative attention to the array of disposable products and throw-away packages (19), leading product and packaging manufacturing companies formed Keep America Beautiful (KAB). While KAB's programs initially focused on litter, by 1969 it had created a spin-off group, the National Center for Solid Waste Disposal, a nonprofit industry research organization. The Washington, D.C.–based agency was soon renamed the National Center for Resource Recovery, but its resource recovery was limited to incineration technology research and development. The energy crisis of the 1970s painted energy recovery through solid waste incineration as the ideal solution to two national concerns (20).

Initially, the EPA offered only a lukewarm reception to the waste-to-energy approach, which depended on large capital outlays and as yet unproven technologies. Early EPA studies underscored the environmental superiority of recycling and waste reduction (Fig. 1). For example, for every ton of waste *not produced,* 20×10^9 J could be saved, while incineration recovered only 10×10^9 J per ton (17,21).

By the late 1970s, however, EPA was convinced of the viability of the technology. The agency also recognized the appeal that waste-to-energy facilities held for local officials, for whom one large plant could not only replace distant landfills but would allow municipalities to continue collecting waste in traditional collection, transfer, and disposal cycles. In 1979, EPA and the U.S. Department of Energy (DOE) formally adopted waste-to-energy as the disposal technology of choice (22). These agencies teamed with industry to promote waste incineration through a comprehensive set of commercialization programs and regulatory adjustments, including grants, below market rate loans, loan guarantees, arbitrage and municipal bonding rules, price supports, energy entitlements, guaranteed resale of electricity (PURPA rates), and the reclassification of ash as a nonhazardous material. While EPA formed technical assistance teams to work with cities, the DOE created the Office of Commercialization of Municipal Waste to Energy. DOE's announced goal was to build between 200 and 250 large (907 t per day minimum) waste-to-energy plants between 1980 and 1992, rendering a 108×10^6 t/yr capacity, at a cost estimated at $20 to $40 billion (23). Many plants were built to accommodate more waste than was available, resulting in financial strain on government sponsors of the projects, who assumed the risk of guaranteeing minimum tons of waste for the plants (24,25).

The American public said no. Beginning in the late 1970s and maturing by the mid-1980s, a citizen-based movement effectively stopped this EPA/DOE plan. Since 1985, more than 150 waste-to-energy plants have been canceled (26); less than 30 large facilities have been built, 6 of which are scheduled for completion in 1994 (27,28). In 1987, for the first time, more plant capacity was canceled (32,347 t/day) than ordered (18,675 t/day). Between 1985 and 1988, 28 plants were canceled in California, and another 16 New Jersey plants were canceled in 1989. The declining number of plant orders from 1985 to 1990 is listed below.

Year	Number of Plants Ordered
1985	42
1986	25
1987	25
1988	22
1989	12
1990	10

Figure 2 shows the declining demand for incineration capacity from 1980 to 1993. One industry analyst commented that the waste incineration industry seemed to be

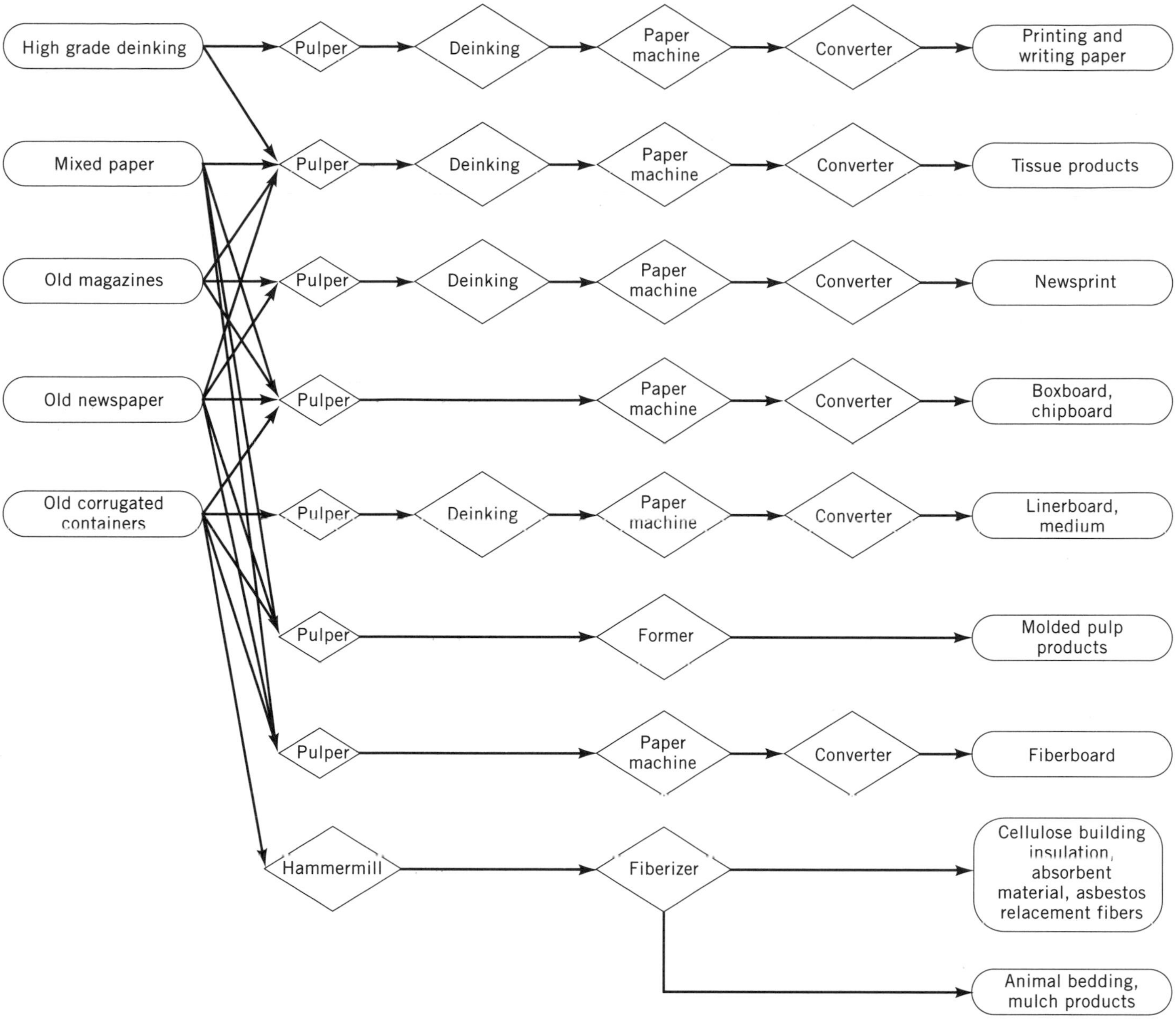

Figure 1. Paper remanufacturing flow diagram. Courtesy of Institute for Local Self-Reliance.

mimicking the precipitous decline of the nuclear energy industry (29,30).

The movement to stop waste-to-energy plants was based on opposition to the facilities' pollution emissions, the environmental destruction caused when virgin materials are mined and processed to replace incinerated materials, and the high capital and operating costs of incinerators. While environmental and health concerns mobilized the citizen opposition to incinerators, it was the unprecedented rise in disposal costs that forced officials to reconsider the technology.

THE RECYCLING AND NIMBY MOVEMENTS

The recycling movement was not a reaction to the disposal crisis; there was no disposal crisis in the late 1960s when the movement emerged. Rather, it was a call to action; within 6 months of Earth Day 1970, 2000 drop-off recycling centers were opened. EPA's first recycling technician documented these centers and the citizens who supported them, activists who, the technican concluded, did not want to change solid waste disposal practices, but rather "wanted to change history" (31).

These drop-off centers were typically started in backyards and assisted by local paper and metals brokers and tolerated by dubious solid waste officials. By 1974, however, only a handful had survived the deep recession that cut brokers' prices for materials. Those that did survive made the transition from volunteer projects to bonafide recycling enterprises, and many of the country's recycling technicians are veterans of these formative years. Among the early cadres of recyclers were Rick Anthony (director of solid waste, San Diego County), Peter Grogan (partner, A. W. Beck and Assoc.), Gary Peterson (Waste Management, Inc.), Steve Apotheker and Jerry Powell (*Resource*

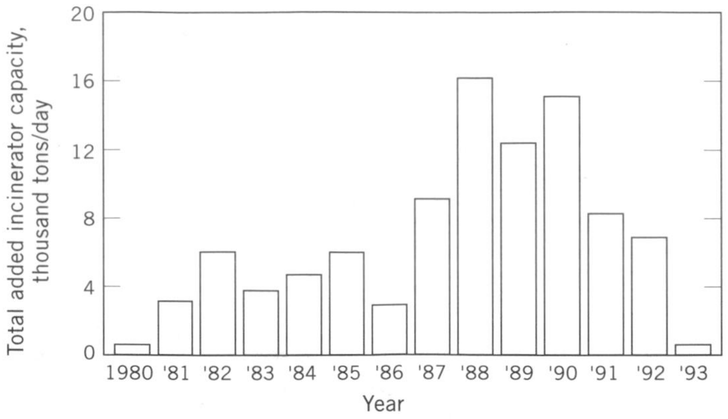

Figure 2. Demand for incinerators is falling. Figures are for energy-producing incinerators, by far the majority of all facilities. Plants under construction will add capacity of 8800 t/day by 1995. Courtesy of *BioCycle Magazine,* Integrated Waster Services. Assoc.

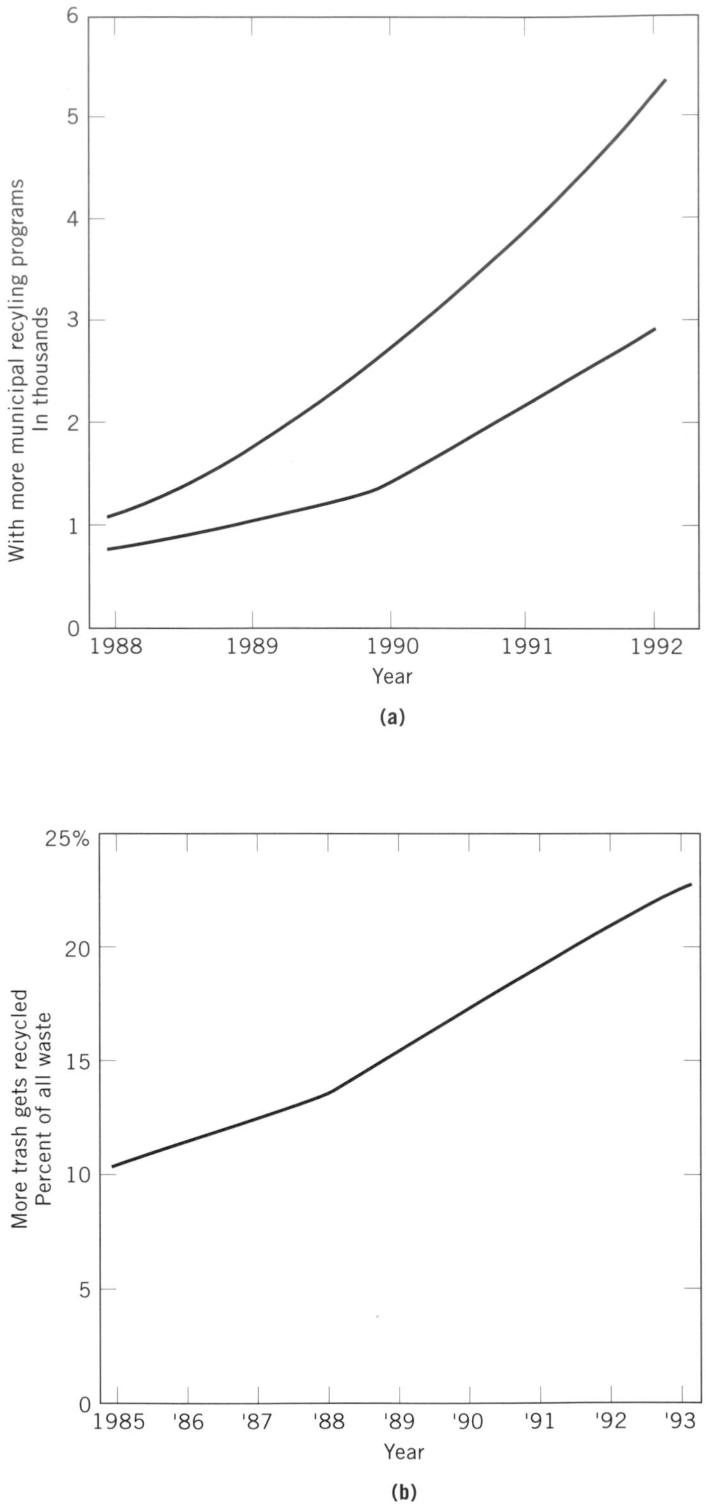

Figure 3. **(a)** With more municipal recycling programs. **(b)** More trash gets recycled. 1993 is an estimate. Courtesy of *BioCycle Magazine,* Integrated Waste services Assoc.

Recycling Magazine), Bernie Meyerson (Secondary Resources, Inc.), Margaret Gainer (Gainer and Assoc.), Mike Anderson (Sonoma County Community Recycling Center), Paul Relis (California State Solid Waste Management Board), Hal Conklin (City Council, Santa Barabara), Arman Stepanian, Hoosier Environmental Council, Indianapolis, and Dan Knapp (Urban Ore, Berkeley).

As the economy rebounded in the late 1970s and 1980s, recycling programs spread rapidly. By the end of the 1980s, there were more than 5000 municipal recycling programs nationwide, a dramatic increase from the 10 in existence in 1975. (Figure 3 depicts the rapid increase in programs and percent of materials recycled.)

While this resurgence was spurred by new alliances among recyclers, brokers, and end-use industries, state recycling associations were the crucial catalysts for the movement, linking industries with state and local officials. The 1978 passage of California Litter Tax Law, which created a surcharge fund for investment in recycling equipment and training, was due in large part to the lobbying efforts of the California Resource Recovery Association (CRRA). Then only 3 yr old, the CRRA was the nation's first state recycling organization; Washington and Oregon soon followed suit. The pressure created by similar state organizations led to a series of new funding mechanisms for recycling investments (32,33). The state associations sponsored hundreds of conferences that included workshops on equipment, public policy, bookkeeping lobbying, contract negotiations, and environmental education (34–40). In 1980, a coalition of 400 recycling organizations held the First National Recycling Congress and consequently formed the National Recycling Coalition, which now boasts more than 5000 members. That same year, 120 recycling practitioners prepared the National Recycling Research Agenda, which outlined a national investment program for solid waste management (41).

As the incineration industry advanced, so did citizen action in opposition to planned facilities. Incineration firms, consultants, Wall Street underwriters, and local officials dubbed the citizen activists "NIMBY's" (not in my back yard). This became a catch phrase used to describe what proincineration forces described as nay-sayers with a know-nothing, knee-jerk reaction to any effort to resolve the nation's solid waste issues. Even the director of the New Jersey Department of Environmental Protection referred to these citizen activists as "environmental terrorists" bent on crippling the progress of technology.

Claims and counterclaims were raised in every corner of the country, as large urban centers faced simultaneous disposal capacity shortage. Jurisdictions in the rural Midwest as well as urban centers on the East and West coasts announced plans to use rail and barge systems to carry solid waste to New Mexico, Texas, Montana, and West Virginia.

In reality, the NIMBY movement initiated a scientific review of environmental and health standards that proved both accurate and reliable (42). As the recycling movement gained momentum via the hundreds of conferences and workshops conducted by state recycling associations, the NIMBY movement sponsored thousands of citizen meetings at which the dangers of incineration were detailed in both technical and lay terms. For example, Kenneth Lasser, of San Diego, translated the projected emissions from a proposed plant with state-of-the-art pollution control technologies in terms of their automobile emission equivalents. Lasser explained that sulfur dioxide emissions from the plant would be equivalent to an additional 187,000 cars traveling through San Diego neighborhoods; oxides of nitrogen emissions would equal an additional 134,000 cars (43). Epidemiologists, engineers, and doctors affiliated with both grassroots groups and national organizations disseminated scientific research on ash, dioxin, heavy metals, acid gasses, and mercury (43–48). Information and organizing strategies were relayed across the country by local and regional groups. These groups included Sierra Club affiliates, Kentuckians for the Commonwealth, Hoosier Environmental Council, New York Public Interest Research Group, Waste Not, New Jersey Environmental Federation, Grassroots Environmental Organization, Northern California Resource Recovery Association, Garbage Reincarnation and the Institute for Local Self Reliance. National networks focused on grassroots agendas, and by the late 1980s, traditional environmental groups had joined the anti-incineration efforts. These networks included Citizens Clearinghouse for Hazardous Waste, Center for the Biology of Natural Systems, Institute for Local Self-Reliance, Clean Water Action, Environmental Action, the American Public Health Association Natural Resources Defense Fund, Environmental Defense Fund, and the Sierra Club Legal Defense Fund.

Initially, the national environmental organization supported incineration of solid waste as an improvement over open dump burning, which was prevalent throughout the country despite the RCRA requirement to halt such operations. Controlled burning in modern facilities would vastly improve air and ground water quality. By 1985, the national environmental movements joined the grassroots efforts against incineration.

The flood of scientific data threw into question the assumptions put forward in the *Environmental Impact Statement,* drafted by incinerator facility developers. These discrepancies, coupled with the revolving door between government regulators and industry and the rough-shod methods of handling dissent at public hearings, led to the creation of *ad hoc* political coalitions that took local officials by storm. Wheat, of Gainesville, Florida, was told to go back to her kitchen and leave decision making to elected officials. Indeed, Wheat did retreat to the kitchen, where she started her successful campaign for county commissioner. The Duarte, California, City Council supported a mass burn plant until 300 citizens showed up at a council meeting. Despite its prompt reversal, the entire council was defeated in the next election by a slate of anti-incineration candidates. In Lowell, Massachusetts, when the city council voted 8–0 in favor of a mass incineration plant, citizens announced that not only would they never vote for council members but that their children would never vote for them. Within 2 months, an 8–0 vote was recorded against the plant. As stated by the outgoing mayor of Lincoln, Nebraska, "garbage is one way to unseat an incumbent official."

A series of shock waves in the late 1980s underscored the urgency and intensity of these solid waste debates. The slow but gradual increase in landfill costs was abruptly changed. A 1986 court ruling allowed a New Jersey landfill receiving 907 t of waste from Philadelphia to deny further shipments. Overnight, Philadelphia had to identify alternate capacity in Baltimore and Harrisburg, increasing its disposal costs to more than $100/t. On Long Island, state-mandated landfill closings forced jurisdictions to transport wastes to Ohio and West Virginia, where tip fees were low; however, the transportation costs drove overall costs to more than $150/t.

Local problems became national headlines; efforts to accommodate changes in solid waste disposal regulations resulted in the sagas of the Islip, Long Island, "garbage barge" and Philadelphia's "ash ship." For several months in 1989, Americans watched these vessels, laden with wastes formerly disposed of locally, traverse the East Coast, the Caribbean, and the Atlantic Ocean in a vain search for a disposal site. There could have been no better forum to bring the solid waste debate before the nation. Ultimately, the Islip ship docked in Brooklyn, where its 2722 t of waste were incinerated, and the ash residue was deposited in Islip's landfill. In 1991, it was revealed that the Philadelphia ship was lost at sea in the Atlantic Ocean.

By the close of the 1980s, combined efforts at the local and national levels resulted in an entirely new regulatory approach to solid waste incineration. New facilities were required to have bag houses for particulate emissions, acid gas scrubbers, oxides of nitrogen and mercury controls, and special monofills for ash disposal (See Table 2). As states added new pollution control requirements and retrofits for new or existing incinerators, the cost of incineration increased dramatically, paralleling the steady rise in landfill costs (see Tables 3 and 4). In October 1993, Florida announced requirements for further mercury emissions control, including the need for activated carbon injection as well as a source reduction program for wastes with mercury content. Compliance is required by September 1995. A task force of the New Jersey Department of Environmental Protection and Energy in the same month recommended even more strident mercury emissions standards (51). These and other requirements are being applied to existing plants as well as new facilities. One industry analyst calculated the retrofit costs for the 2722 t/day Pinnellas County, Florida, incinerator at between $100 and $200 million (52,53).

These increases forced officials to reconsider disposal options. What was once a closed decision-making process slowly opened to citizen groups, small businesses, and en-

Table 2. Types of Air Pollution Control Equipment (APC): 1982–1993[a]

APC Equipment	Year 1982	1984	1986	1988	1991	1993
Electrostatic precipitators	59.0%[b]	52.7%	46.2%	41.3%	36.5%	40.4%
	(49)[c]	(59)	(79)	(83)	(72)	(69)
Bag house or fabric filter	16.9%	16.1%	34.5%	44.3%	53.3%	50.3%
	(14)	(18)	(59)	(89)	(105)	(86)
Dry scrubbers	6.0%	8.0%	35.1%	43.3%	51.8%	50.9%
	(5)	(9)	(60)	(87)	(102)	(87)
Wet scrubbers	7.2%	4.5%	5.3%	3.0%	4.6%	6.4%
	(6)	(5)	(9)	(6)	(6)	(11)
Afterburn or 2-chamber system	19.3%	14.3%	15.2%	12.4%	10.7%	9.4%
	(16)	(16)	(26)	(25)	(21)	(16)
NO_x control	0.0%	0.9%	1.8%	2.5%	14.7%	17.5%
	(0)	(1)	(3)	(5)	(29)	(30)
Mercury control	0.0%	0.0%	0.0%	0.0%	0.0%	6.4%
	(0)	(0)	(0)	(0)	(0)	(11)
Other technology	1.2%	0.9%	4.1%	2.5%	2.5%	1.8%
	(1)	(1)	(7)	(5)	(5)	(3)
Nothing used	4.8%	6.3%	4.7%	3.0%	4.1%	2.9%
	(4)	(7)	(8)	(6)	(8)	(5)
Total percent	*114.5%*	*103.6%*	*146.8%*	*152.2%*	*178.2%*	*186.0%*
Total of facilities	*88*	*113*	*166*	*201*	*202*	*171*

[a] Data from Ref. 50.
[b] Percentages were derived by dividing the number of responses by number of projects reporting air pollution control data for each year. Up to three types of pollution control equipment were recorded for each plant.
[c] Numbers in parentheses are number of facilities.

Table 3. Average Tipping Fees by Region of Year[a]

Region	Year 1982	1984	1986	1988	1991	1993
Northeast	$11.82	$15.62	$23.32	$41.88	$62.34	$66.67
	(22)[b]	(29)	(53)	(63)	(62)	(57)
South	$7.43	$13.42	$16.70	$25.28	$37.24	$45.41
	(14)	(25)	(33)	(48)	(48)	(44)
North central	$8.82	$13.65	$17.5	$29.08	$46.33	$53.46
	(7)	(12)	(16)	(33)	(36)	(31)
West	$10.99	$11.71	$16.18	$31.61	$45.03	$55.61
	(4)	(12)	(8)	(17)	(16)	(15)
All regions	*$9.77*	*$14.00*	*$19.62*	*$33.23*	*$49.64*	*$56.39*
Total number	*42*	*78*	*110*	*161*	*162*	*147*

[a] Data from Ref. 50.
[b] Numbers in parentheses are total number of facilities.

Table 4. Tip Fee Data Set ($/t)

Region	Number of Sites Surveyed	Year 1985	1986	1987	1988	1990	1992
Northeast	16	12.66	17.11	52.41	61.11	64.76	65.83
Mid-Atlantic	39	16.99	22.08	26.32	33.84	40.75	47.94
South	42	3.24	5.76	13.13	16.46	16.92	22.48
Midwest	99	7.23	11.75	16.42	17.70	23.15	27.10
West Central	19	5.36	6.21	7.23	8.50	11.06	12.62
South Central	35	7.24	7.61	10.17	11.28	12.50	12.53
West	32	10.96	11.10	13.92	19.45	25.63	27.92
National		*8.57*	*11.81*	*19.40*	*22.74*	*26.56*	*31.20*

vironmental organizations, which brought environmental and economic factors to the surface. Finally, citizens secured the opportunity to alter drastically the solid waste disposal agendas of their communities and the nation. In 1982, a coalition of Berkeley citizens, environmental activists, and business leaders defeated plans for an incinerator that would have burned 50% of the city's waste stream. The fight centered on the issue of whether nonprofit and for profit organizations could recycle this portion of the waste stream more economically. Today, six organizations are recycling 46% (78,019 t) of Berkeley's waste stream, creating 94 jobs and returning $335,000 to the local tax base (54). In 1985, Philadelphia adopted an anti-incineration solid waste management plan, followed in 1987 by a mandatory recycling law that established a fifty percent recycling goal. Since its passage, thirty-six processing and manufacturing plants have located in the city, creating 700 jobs (55,56).

An Austin, Texas, incinerator already under construction was canceled in 1987 when the business and real estate communities joined local environmental groups opposed to the plant. The city council chose to lose the $23 million already invested in the project, based on projected savings of $100 million over 20 yr through the implementation of recycling and composting programs. The city's recycling program is currently overseen by a citizens' advisory committee (57–59).

In 1989, citizens of King County, Washington, insisted on a total review of all solid waste options. The Department of Public Works commissioned the study, which determined that the best alternative for long-term disposal capacity was an aggressive recycling and composting program. The county subsequently established a 65% recycling goal (60). In 1992, Rhode Island passed a moratorium on incineration and landfilling, establishing a 70% recycling goal. One study revealed that incineration technology actually required more landfill capacity than traditional waste disposal, due to the necessity of landfilling incinceration ash and bypass waste that does not fit in the boiler and given the down time, during which unprocessed waste must be landfilled (60).

In 1994 new developments posed yet higher future costs for incineration technology. In May, the U.S. Supreme Court ruled that ash must be tested and, if found to be hazardous, disposed of in special landfills. In Sept, the U.S. EPA acknowledged that dioxin is a major threat to humans and that 99% of known dioxin emissions come from waste incinerators.

RECYCLING AT THE TAKE-OFF STAGE

The publicity surrounding the nation's inability to identify sites for its garbage literally brought the sight of the country's waste into our living rooms. A variety of factors (media attention; newly compiled scientific data that undermined previous assumptions regarding the health, environmental, and economic impacts of incineration; spiraling disposal costs; and the formation of powerful coalitions of recyclers, NIMBYs, and reprocessing and remanufacturing firms) changed the climate of opinion and set the stage for the introduction of new solid waste management rules, policies, and procedures. No longer could decision makers focus on constructing large landfills, transfer stations, or incinerators; planners and politicians were forced to broaden their scope. Recycling required new contracts, new processing infrastructures, new markets for materials, and new public awareness and education programs. From 1988 to 1990, thousands of local laws and regulations were effected, and tens of millions of dollars were allocated for tax credits, grants, pooled bonds for small manufacturers, surcharges, and system bonds for cities in need of investment in small facilities.

The result was an unprecedented development of recycling programs. By 1989, 17 cities and rural areas recycled more than 25% of their waste stream. By 1990, 15 jurisdictions had achieved or exceeded 40% recycling levels, with some approaching 60% (see Table 5). These dramatic results forced the EPA to question its 1988 national goal of 25% (61,63–65). Rural jurisdictions joined the recycling effort sweeping the nation, establishing cooperative marketing arrangements that allowed small communities with no proximate markets to take advantage of national and international markets. At this writing, 61 such interlocal agreements have been created (66).

Soon recycling became almost too successful, a consequence of dealing solely with the collection, or supply-side, of the materials equation. By 1989, the nation was suffering a newsprint glut. Recycling programs had overextended the capacity of existing domestic and foreign used paper mills. With brokers charging $5 to $25 per ton to remove collected newsprint, the issue was seen as a vital flaw in the recycling phenomenon. *Waste Age, Recycling Today, Resource Recycling Magazine,* and *Recycling Times* all reported on this. Nonetheless, it was cheaper to pay brokers' fees than to absorb the high costs of disposal. The backlash proved short lived. By 1991, no less than 11 mills were under construction in the United States and overseas. In addition, new uses for old newspaper often brought higher prices than ever paid by the mills for the materials.

As cities and states began investigating the demand-side of the equation, a flood of new legislation was passed. Maryland's procurement initiatives, including the purchase and storage of large quantities of recycled paper products for resale to local governments, became a model for other states and cities. Procurement preferences for recycled-content products (allowing differentials of 5–10% over the price of products made exclusively from virgin materials) and minimum-content laws (mandating that products purchased by state agencies have a minimum recycled content) were enacted in six states. "Procurement preferences apply to government spending. Under Senate Bill 337, Pennsylvania offers companies tax credits equal to 50 percent of the difference paid to purchase products made with recycled rather than virgin materials."

Representatives of California's Integrated Solid Waste Management Board affirmed that "the market development impact of minimum content laws is impressive." Consumption of newsprint increased from 569,722 t in 1991 to 1,678,320 t in 1992 (67). Minimum-content laws have since been applied to newsprint, plastic and glass containers, film plastic, telephone books, and fiberglass insulation. President Clinton's October 1993 executive order establishing a minimum content requirement for all federal government paper purchases is expected to in-

Table 5. Record-Setting Recycling and Composting Programs[a,b]

Community	Population	Year Data Collected	Residential Recovery Rate	Commercial Recovery Rate	MSW Recovery Rate[c]	Total Recovery Rate
Lincoln Pk, N.J.	10,980	1990	49%	70%	62%	NA
Berlin Township, N.J.	5,630	1990	56%	61%	57%	NA
Bowdoinham, Maine	2,190	FY90[e]	NA[f]	NA	54%	53%
West Linn, Oreg.	16,560	1990	NA	NA	50%	46%
Longmeadow, Mass.	16,310	1989	45%	53%	49%	NA
Perkasie, Pa.	7,880	1990	52%	NA	NA	NA
Haddonfield, N.J.	12,150	1989	51%	24%	49%	NA
Rodman, N.Y.	850	1989	NA	NA	43%	NA
Upper Township, N.J.	10,860	1990	50%[g]	34%[h]	NA	43%
Wellesley, Mass.	26,590	1989	NA	NA	41%	NA
Seattle, Wash.	516,260	1990	45%	40%	40%	NA
Hamburg, N.Y.	11,000	1989	39%	18%	40%	NA
Wilton, Wis.	470	1989	40%	38%	40%	NA
Fennimore, Wis.	2,380	1990	51%	25%	38%	NA
Woodbury, N.J.	10,450	1989	49%	11%	32%	NA
Newark, N.J.	275,220	1989	10%[g]	46%[h]	NA	30%
LaCrescent, Minn.	4,300	1990	41%	9%	29%	41%
Monroe, Wis.	10,220	1989	32%	27%	28%	50%
Peterborough, N.H.	5,240	1990	42%	4%	19%	18%

[a] Data from Refs. 61 and 62.
[b] All recovery rates represent proportions by weight and include material recycled and composted.
[c] Municipal solid waste recovery rate may take into account tonnages that cannot be broken down into commercial and residential, such as bottle bill tonnages and landscapers' waste.
[d] Total waste is the sum of MSW and C&D debris.
[e] *Fy,* fiscal year.
[f] *NA,* not available.
[g] Publically collected waste.
[h] Privately collected waste.

crease demand for recycled newsprint by 272,160 t annually.

The surge of minimum-content laws and increased overall demand for paper has prompted the construction of six new paper mills for the Mid-Atlantic region. Most will produce "market pulp" for sale to virgin paper mills that must add recycled pulp to their feedstock to meet content requirements. Far from the glut of 1989, The American Paper Institute reports that there may be a shortage of recycled paper by 1995 (Table 6).

The 1991 Intermodal Surface Transportation and Efficiency Act (ISTEA) requires that states use asphalt with ground rubber (from discarded tires) in at least 5% of their highway projects beginning in 1994, with the percentage increasing in 5% increments each year through 1997.

Eight directories of recycled product manufacturers are currently available (69,70), listing the hundreds of new recycled content products available to state and local procurement officials, from roadbed and surface materials to pens, paper products, construction materials, compost, antifreeze, lubricating oil, oil filters, floppy disks, plastic film, carpeting, furniture, clothing, and computers. The nation's product and packaging manufacturers rushed to accommodate the new consumer activism.

- Proctor & Gamble introduced plastic packaging made from recycled resins, and initiated a source reduction program by replacing large laundry detergent containers with small packages of concentrated detergents (71–75).
- In 1990, Coca-Cola and Pepsi-Cola announced that they would use recycled plastics in the liner of their PET soda bottles. By 1993, Pepsi-Cola had announced its intention to manufacture containers made from 100% terepthalic acid derived from recycled PET bottles. Both leading beverage producers have also switched to reusable rigid plastic containers for shipments to retailers.
- In 1989, the Aseptic Packaging Council was formed by polycoated drink container manufacturers in reaction to criticism from recyclers. Industry investment has resulted in new techniques for recovering and processing packages. By 1992, recovered containers were valued at $150/t, and pulp derived from processing at $500/t (76). In 1992, the Buy Recycled Business Alliance, made up of large product and packaging companies, spent $4 billion on the purchase of recycled materials.
- Automakers BMW, Volvo, and Volkswagen introduced new "design for disassembly," which includes snap-on panels to replace welded panels, making the parts easier to reuse and recycle.
- Weyerhauser Corp., citing feedback from a "stakeholders meeting," announced in September 1993 that it will double its use of recycled feedstocks within 1 year.

The advent of green consumer activism and plentiful feedstock stimulated the American entrepreneur as well. New commercially viable technologies and production

Table 6. Minimum-Content Goals for Selected States[a]

Product	California[b]	Connecticut	District of Columbia[c]	Illinois	Oregon	Maryland	Missouri	Wisconsin
Newsprint	10% by 1/1/94 12% by 1/1/91 20% by 1/1/00	11% by 1/1/92 20% by 1/1/94 31% by 1/1/96 50% by 1/1/99	12% by 1/1/92 20% by 1/1/94 40% by 1/1/98	28% by 1/1/94 35% by 1/1/95 45% by 1/1/97	7.5% by 1/1/95	12% by 1/1/92 20% by 1/1/94 30% 1/1/96 40% by 1/1/98	10% by 1/1/93 20% by 1/1/94 40% by 1/1/96 50% by 1/1/00	10% by 1/1/92 25% by 1/1/94 45% by 1/1/01
Fiberglass	10% by 1/1/92 30% by 1/1/95							
Glass containers	15% by 1/1/92 35% by 1/1/96 55% by 1/1/02 65% by 1/1/05				35% by 1/1/95 50% by 1/1/00			
Plastic trash bags	10% by 1/1/93 (for bags ≥1.0 mil thick) 30% by 1/1/95 (for bags ≥0.75 mil thick)							
Plastic containers	25% by 1/1/95							10% by 1/1/95
Phone directories		10% by 1/1/95 20% by 1/1/97 30% by 1/1/99 40% by 1/1/01			25% by 1/1/95	12% by 1/1/94 20% by 1/1/96 40% by 1/1/00		
High grade paper			50% by 1/1/94[d]					
Tissue products			5–40% by 1/1/94[d]					
Unbleached packaging			5–35% by 1/1/94[d]					

[a] Data from Institute for Local Self-Reliance and Ref. 68.
[b] California has established the following percentage goals for "recycled-content newsprint": 25% by 1994, 40% by 1998, and 50% by 2000. The percentages listed in the table reflect 40% of these goals, as California's law defines recycled-content newspaper as containing not less than 40% postconsumer waste paper.
[c] The District of Columbia has also established goals for paperboard.
[d] Designated percentages are based on recommendations of the EPA pursuant to Section 6002 of the Resource Conservation and Recovery Act.

methods that use recycled materials as a feedstock continue to come on-line. Coon Manufacturing Co. (Spickard, Missouri) pioneered in the use of recycled plastics to produce high value end products. While plastic lumber products sell for roughly $0.50 per pound of recycled plastic content, Coon's propane gas tanks, construction siding, helicopter seats, outdoor toilets, and commercial garbage bins sell for $4 per pound of recycled plastic content.

The Ceres Co. (St. Paul, Minnesota) has developed a new yard debris processing system, producing mulch and compost from yard waste collected from jurisdictions that have banned the materials from their landfills. Ceres now supplies 80% of the nursery industry in the Twin Cities metropolitan area.

Three companies have developed aluminum dross and salt cake recycling processes that recover valuable industrial chemicals from the residues of secondary aluminum smelting plants. Champion Paper Co. collects recycled paper from four cities in Texas, processes and remanufactures the material, and returns finished paper products to the cities.

Since its creation in 1992, the American Appliance Recycling Association (AARA) has extended its reuse and recycling program to dozens of cities. When the steel industry announced the creation of the Steel Recycling Institute that same year, it introduced a 21-city demonstration project to increase rapidly steel recycling levels (77). The advent of minimills, with 90,720 t/yr capacities, compared with the 907,200 t/yr capacity of traditional steel mills, have bolstered the scrap steel market. Minimills, which enjoy a 40% share of the U.S. steel market, are currently using 100$ scrap as feedstock to electric arc furnaces. As a result, scrap steel investors are at their lowest level since the Korean War (1950–1953).

Cumulatively, industry measures have sharply increased the use of recycled materials in common consumer goods. From 1988 to 1991, the use of recycled glass increased from 21% to 31%; recycled plastics, from 5% to 14%; and paper, from 24% to 36% (78–81).

Increased collection and processing efficiency and increased industrial demand for recycled materials has lowered the solid waste management costs of cities that have invested in comprehensive programs (82,93). A city of 1 million people, whose disposal costs equal the national average of $26/t, can save its government, businesses, and households $7 million annually through recycling and reduction programs. Communities with $65/t disposal costs can save $30 million per year (84,85). The city of Los Angeles, through information dissemination and technical assistance to the private sector, will save businesses $300 million (86).

CREATING WEALTH FROM WITHIN

The new solid waste management agenda, prompted by environmental, recycling, and NIMBY organizations, has increased disposal costs in the United States over the last 25 yr. Activists argue that previous accounting systems allowed companies and cities to forego the real price of disposal, providing false messages to the economy, which in turn promoted the wasteful use of resources:

> This enduring legacy, is the recognition that there is a distinction between price and cost. Price is what the individual pays, cost is what the community as a whole pays. A free market system functions effectively only when it relies on accurate price signals. . . . The prices we pay for goods and services are woefully out of line with the full cost of extracting raw materials, processing, manufacturing and send-

ing them to as final products to the ultimate consumer, and disposing of the product and all the wastes generated along the way (6).

Local governments and businesses spent an estimated $30 billion for solid waste collection and disposal services in 1990, up from an estimated $2 billion in 1970. Wall Street's Morgan Stanley estimates that the national garbage bill may reach $45 billion by 1995 and $75 billion by the year 2000 (87). Redirecting funds from expenditures for disposal to investments in processing and manufacturing industries has become the focus of national attention, creating opportunities for processors and entrepreneurs to recycle and remanufacture products at costs lower than disposal (88,89).

Waste-based local economic development expands the community's employment base, particularly for the disadvantaged, and allows communities to use local resources to create and retain local wealth. Each processing step employs more people, uses more utilities, requires more administrative and maintenance services, and requires the purchase of more supplies (90–95). Table 7 provides an estimate of the full potential of waste-based economic development based on actual commercial operations.

Table 7. Potential Scrap-Based Manufacturing for a City of 1 Million[a]

Products	Product Postconsumer Content	Number of Plants	Local Postconsumer Scrap Used (TPY)	Jobs	Annual Gross Revenue	Value-added to 1 Local Scrap
Asphalt and concrete						
Pothole patch and paving[b]	25%	1	40,000	20	$4,000,000	$65
Aggregate	100%	1	20,000	5	$160,000	$10
Subtotal		2	60,000	25	$4,160,000	$45
Glass						
Tile	30%	1	3,600	75	$40,000,000	$1,700
Art glass	90%	1	500	5	$300,000	$500
Containers	52%	1	38,000	250	$37,000,000	$200
Glasphalt	10%	1	4,300	5	$1,500,000	$5
Subtotal		4	46,400	335	$78,800,000	$300
Metals						
Aluminum products[b]	60%	1	6,000	50	$150,000,000	$1,800
Aluminum dross	10%	1	18,000	20	$2,500,000	$250
Detinned steel	100%	1	9,000	10	$1,900,000	$170
Steel products[b]	100%	1	10,000	50	$150,000,000	$1,200
Subtotal		4	43,000	130	$304,400,000	$860
Paper						
Tissue	100%	2	41,000	250	$30,000,000	$640
Newsprint[b]	100%	1	37,000	220	$57,000,000	$480
Printing and Writing	50%	1	24,000	120	$50,000,000	$470
paperboard	100%	1	79,000	180	$21,000,000	$220
Cellulose building insulation	82%	1	7,000	10	$2,000,000	$180
Subtotal		6	188,000	780	$160,000,000	$390
Plastics						
PS packaging	25%	1	500	150	$1,000,000	$1,500
Dimensional lumbar	100%	1	3,400	50	$3,000,000	$1,000
LDPE bags	30%	1	2,000	50	$4,000,000	$1,000
PVC products	10%	1	25	10	$250,000	$1,000
PET packaging	100%	1	960	25	$1,800,000	$950
HDPE sheeting	90%	1	960	20	$1,500,000	$900
Subtotal		6	7,845	305	$11,550,000	$1,015
Rubber						
Flooring	85%	1	1,700	25	$1,200,000	$600
Wood						
Hardboard[b]	45%	1	20,000	100	$12,000,000	$150
Pallets	100%	1	6,000	15	$600,000	$100
Mulch	100%	2	14,000	20	$1,200,000	$76
Stable bedding	100%	1	4,500	5	$200,000	$44
Subtotal		5	44,500	140	$14,000,000	$110
Other						
Auto batteries[b]	99%	1	6,500	120	$210,000,000	$1,000
Textile products	100%	1	10,000	120	$3,000,000	$200
Subtotal		2	16,500	240	$213,000,000	$515
Total		*30*	*407,945*	*1,980*	*$787,110,000*	*$365*

[a] Data from the Institute for Local Self-Reliance.
[b] Additional scrap is required from the surrounding region to supply this minimum-size manufacturing plant.

In addition, up to $760 million in value-added can be generated in the local economy each year. Loose paper is sold for $30; baled the paper triples in value; pulped, the paper is worth $400/t; and when converted into writing paper, the value doubles once again. Similar increases are enjoyed by other materials: the plastic used to manufacture soda bottles is worth $120/t baled, $500/t when granulated, and up to $8000/t when remanufactured. Recycled glass cullet, worth $30/t when color sorted, is worth $780/t when processed into industrial abrasives.

Over the past several years, state and local governments have embarked on programs and policies aimed at achieving the full potential of scrap-based manufacturing in their jurisdictions. In 1990, 8951 citizens were employed by New Jersey end use manufacturers with sales of $902 million. An additional 225 workers were employed in processing enterprises. In Massachusetts, recycling and manufacturing industry contributed $743 million in sales to the state economy, employing just under 10,000 workers. According to the Department of Environmental Protection, industries in the 1980s sited in states with cheaper labor and raw materials; however, today recycling mills are the only industrial plants in the state garnering new investments. 1990 Department of Labor statistics showed that recycling accounted for 1194 jobs in rural Maine compared with 1070 jobs in radio and television communications, and 1030 jobs in the boat building and repair industry. Most of the recycling jobs paid above minimum wages, with annual average salaries of $18,476 (96–98).

In 1985, Newark, New Jersey, became the first city in the United States to merge its enterprise development and scrap-based manufacturing programs. The concept has been replicated in cities such as Denver, Colorado; Maywood Village, Illinois; and Evansville, Indiana, which have created industrial parks reserved exclusively for scrap-based manufacturers.

New Jersey, Michigan, Pennsylvania, and New York have each committed at least $10 million in grants and loans to industries that use hard to recycle materials (eg, old tires, yard debris, and plastics). Washington State, constitutionally prohibited from awarding grants to private firms, has created the Clean Washington Center, which provides research, technology evaluation and technical assistance to the private sector.

To date, California has progressed the furthest in establishing an entire government infrastructure to promote and develop recycling and scrap-based manufacturing. In 1992, the state created 12 Recycling Market Development Zones (RMD2), which provide technical assistance, tax incentives, and employment tax credits to manufacturers. In the program's first year, Sacramento County has attracted seven firms, creating 400 jobs. After extending its RMDZ to include any industrially zoned area within the county, Ventura County attracted three new firms. The Integrated Solid Waste Management Board, Department of Conservation, California Coastal Commission, and Department of Energy have all developed programs; and California cities, counties, and state agencies have formed working partnerships with federal agencies, including the Department of Commerce, Environmental Protection Agency, National Aeronautics and Space Agency, and National Oceanographic and Atmospheric Administration, to support waste-based economic growth and job creation. Private corporations and foundations in the state established the Materials for the Future Foundation, which provides technical assistance and investment dollars to nonprofit, community-based enterprises.

The Clinton Administration mandate to create "jobs, jobs, and more jobs" has mobilized federal agency support for economic growth through recycling and manufacturing. The Department of Commerce has channeled $6.6 million to the Clean Washington Center and the National Recycling Coalition to support technology research, evaluation, and direct technical assistance throughout the country. (Funding is derived from military conversion programs as well as traditional sources.) The EPA has also initiated economic growth partnerships programs with Los Angeles, Philadelphia, Baltimore, Washington, D.C., and Richmond (Virginia) and the Council of New England Governors (CONEG). In 1994, the EPA will provide up to $500,000 in investments to expand state and local government projects through is Urban Initiatives Program. (Funding for the Recycling Technical Assistance Project is derived from the departments of Energy and Housing and Urban Development and the Small Business Administration.)

The most profound impact on local scrap based manufacturing could derive from the administration's Enterprise/Empowerment Zone program, scheduled for implementation in December 1994. This economic development package will provide $30 to $40 billion in investment capital for local investment (99). Cities have already begun incorporating scrap-based manufacturing into their Empowerment Zone Program proposals and integrating Community Reinvestment Act resources and community bank loan fund programs into their development packages (100–102).

In addition, the McArthur Foundation recently announced a new Program Related Investment (PRI) initiative for scrap-based manufacturing projects. PRI programs, initiated by the Fund for New Jersey and Moriah Fund in 1992, provided low or no capital investment loans for nonprofit organizations. The Alliance for Sustainable Materials Economy, a coalition of 300 government, community, and private sector agencies, has called for a recycling investment tax credit similar to the federal housing investment tax credit. Several states already have tax credits for enterprise investment in recycling-related equipment. A national tax credit for firms that use recycled materials in manufacturing has been proposed (103). Florida has implemented an advanced disposal fee on products made from hard to recycle materials (104).

Traditionally, federal programs have favored virgin material use by implementing policies that provide economic subsidies to virgin material users (eg, energy subsidies, mining regulations, and tax code allowances), increase the cost of using secondary materials (eg, environmental rules and discriminatory transportation pricing), and impose market and procurement barriers (105,106). However, the Clinton Administration's proposal to increase mineral extraction fees on federal lands indicates a shift in policy. According to the Mineral Policy

Center, mining operations on federal lands currently produce \$3 to \$4 billion worth of minerals each year with no payment to the federal government (107–109). The changing federal climate was also illustrated by the Bureau of Mines' recent announcement of its plans to focus on scrap utilization rather than extraction.

There are ample opportunities for each of these federal, state, local, and private sector initiatives in the following areas.

Construction and Demolition Debris

The estimated 54 million t of construction and demolition (C&D) wastes discarded in the United States each year represents a huge potential for job creation for both skilled and unskilled labor forces. Data from projects in the U.S. Northwest show that manual housing demolition is cost-competitive with mechanical demolition. The former generates 400 labor h (for a 390 m^2 house), while the latter generates just under 100 labor h. The Metropolitan Portland (Oregon) planning staff estimated that manual demolition could extract almost 1 million board feet of reusable lumber from the waste stream, creating 50 new jobs. Furthermore, the value of the material extracted from hand demolition was \$4/ft^2 roughly 8 to 10% of the construction cost. The cost of building a new house can be reduced by \$1000 by cutting waste by 80%.

The recovery and reuse of C&D and other household items (eg, appliances and plumbing fixtures) offers further job creation and training opportunities. Urban Ore (Berkeley, California) is a commodity reuse organization that handles 2540 t of material annually. The organization employs 20 workers, most of whom were trained on the job. Knapp, of Urban Ore, estimates that 5% (by weight) of the nation's waste stream can be recovered through commodity reuse enterprises. The Loading Dock, a nonprofit Baltimore-based organization, employs 12 workers to gather, sort, grade, and price commodities donated by local firms and/or delivered by individuals. The Loading Dock handles 10,886–12,700 t of material annually. Recovered materials are then sold exclusively to the city's low-income housing residents, which further reduces the cost of rehabilitating and furnishing living units for residents of low income housing.

These companies have begun networking with each other, allowing, for instance, The Loading Dock to forward shipments of excess doors to a similar organization in Philadelphia, which in turn sends excess appliances to Baltimore.

Agricultural Waste

The United States generates 318 million t of agricultural waste annually. These materials can replace up to 318 million t of petrochemicals currently used annually by U.S. industry (110–112). Companies now seeking supplies of cellulosic waste for use in chemical and fuel production are targeting agricultural wastes, yard debris, and sewage sludge from urban centers that have banned these materials from landfills and incinerators. Such biomass projects are planned for the Philadelphia Navy Yard and Maywood Village, Illinois.

Aluminum Dross and Salt Cake

Three U.S. companies have developed the ability to recover aluminum, salts, exothermics, and nonmetallic products (oxides and other nonreactive compounds) from the 90.72 million kg dumped by aluminum smelters each month in the Midwest, Southeast, and southern California. Cleveland-based Aluminum Waste Technology, Inc., which employs 50 workers, processes aluminum dross and salt cake for \$0.03/lb tip fee, diverting billions of pounds of this chemically active waste from the nation's landfills. This process also releases the aluminum industry, currently responsible for the dumps, from their liability. National replication of this industry could create 2000 industrial jobs.

Recycling has provided citizens, regulators, and entrepreneurs with a gateway to new thinking, programs, and actions. As the country becomes recycling literate, understanding how and why to recycle, waste handling costs are reduced and the raw materials generated serve as the foundation of economic growth. As *The Wall Street Journal* recently observed, economies and firms that adjust to the new realities of waste recycling will be the most competitive in the years to come.

Managers of universities, hospitals, hotels, airlines, and office and apartment buildings are all learning the impact of reduced costs and positive publicity derived from recycling. In 1992, the by-product and waste search service helped redirect 27,216 t of commodities and materials from landfills to industry (114–117).

OUTSTANDING ISSUES

Economic, political, and environmental factors combined to effect a reversal in solid waste management policies, shifting the focus from disposal to recovery. Prorecycling government regulations and programs, municipal institutions, and private sector investments have become a permanent part of the national landscape. Recycling caught the imagination of the public; common sense, practical application, and cost-effectiveness reinforced the concept and stimulated recycling activity. Recycling curricula introduced into all education levels will guarantee recycling's role in the solid waste management and economic development policies of future generations, providing desperately needed opportunities for economic growth and revitalization in the national economy's neediest sectors.

It is still important, however, to fulfill the promise of a closed-loop waste and production economy. Continued citizen efforts and strategic planning, such as those listed below, are critical to this process.

Least Cost Planning for Disposal Facilities

The need for disposal facilities of any kind must be measured against the costs of *not* producing or recycling those materials scheduled for disposal. The city of Toronto embarked on a special vermicomposting program for apartment house residents. Worm bins were distributed at no cost, based on the calculation that siting future landfills would be more expensive. King County and Seattle, Washington, and a dozen California cities have provided tens of thousands of backyard composting bins and established technical assistance programs for households in an effort to avoid collection and processing costs.

The Environmental Policy Institute (SPI) of Georgia has used least cost planning to argue against the theory of Subtitle D landfill requirements. These federal regulations, effected in October 1993, require landfill operators to monitor entombed waste materials for 30 yr. Rather than commit local resources to managing sites, EPI believes that investment should focus on the frontend, by diverting materials from landfills and developing value-added operations. Only dense, inert residues would be landfilled (118). San Diego County, California, anticipates a 90% (by volume) reduction in its waste stream by 1995 by combining mandatory source reduction, source separation composting, mixed-waste composting programs, and energy recovery from biomass.

Strategic Planning for Market Development

The goals of a market development program are to attract firms that maximize the value of materials and firms or agencies that guarantee a use for waste materials, even a low value use, during periods of low demand. For instance, the manufacturing specialty products made from individual polymers would be the optimal use of recycled plastics; manufacture of plastic lumber made from mixed plastics grades would be a back-up market. Composting would be a back-up use for recycled low grade or soiled papers; roadbed and surface repair would be back-up end uses for construction and demolition aggregate, glass, plastics and old tires. Each region must develop a unique mix of firms and facilities that guarantees a high value end use and low value but mass extras.

Focusing Economic Growth Where It Is Needed Most

Federal, state, and local programs have given priority to linking the expanding recycling infrastructure with the revitalization of the country's neediest economic sectors. The most effective means of achieving this end are through direct contracting and encouragement of joint venture enterprises, approaches that are easily integrated into plans for enterprise development zones.

When Chattanooga, Tennessee, awarded its recyclable processing contract to the Orange Grove Center, a non-profit organization that employs more than 800 differently abled workers, it prompted three other jurisdictions to enter similar contracts with the center. Pepsi-Cola intervened on behalf of the project, securing an exclusive contract to deliver glass (all colors) to a bottle manufacture at optimal prices. This, in turn, allowed the center to attract glass and other materials from other regional recycling programs. Within 1 yr of winning the contract, the Orange Grove Center attracted $6 million in grants to build a modern recycling facility and on-site housing for its employees (119).

At the urging of citizen groups, the Los Angeles Public Works Department altered its Request for Proposals process for recycling services. As a result of the new guidelines, which favor companies that form joint ventures with community development corporations, four such ventures have been awarded city contracts. Browning Ferris Industries illustrated its commitment to the joint venture initiative by providing a $1.5 million grant to a local minority business development fund.

In Easton, Pennsylvania, the city entered a joint venture with Webform Industries, which has granted the city stock equity in exchange for start-up capital. Webform can regain the shares through stock purchase or by providing other services to the city.

Transportation of Solid Waste

In 1975, the U.S. Supreme Court refused to allow New Jersey to ban out-of-state waste from its landfills. By 1985, New Jersey landfills were full, and the state now exports about 50% of its waste. In the early 1990s, states in the Midwest and Southeast sought to prevent the importation of East Coast wastes into their jurisdictions; however, courts consistently labeled this discrimination against out-of-area waste as a violation of the Constitution's Commerce Clause.

Barring an act of Congress, there seem to be only two ways for a community to reserve its landfills for local use. Wisconsin has declared that by 1996 all recyclable items will be banned from all disposal sites in the state. Thus any out-of-area waste containing recyclable materials can be rejected. This approach does not discriminate exclusively against out-of-area waste but merely requires it to meet the same standards as in-state waste. Alternatively, state and federal courts have consistently upheld communities' rights to ban waste from publicly owned landfills designed and financed for local use only (120–122).

Control of the Supply of Waste and Materials

"The industrial history of the past three decades shows an increasing differentiation between solid waste and recycling, as the garbage industry failed to implement the public will and the recycling industry filled that service void" (123,124). As illustrated in Rancho Mirage, California, flow control ordinances threaten the recycling industry's access to materials and commodities supply. Waste Management, Inc., when awarded that city's residential recycling collection contract, demanded that all county recycling activities be halted. The action affected nonprofit recycling centers, brokers, and processors, some in business for three generations.

Unless the right to recycle is upheld in this and other flow control cases, the opportunity to develop a diverse, efficient, complex commercial network that generates local jobs, industries, and wealth through the collection, processing, and remanufacturing of recyclable materials will be lost.

BIBLIOGRAPHY

1. M. Melosi, *Garbage in the Cities 1880–1980,* Texas A&M Press, 1981.
2. M. Melosi, *Recovering the Past: A Handbook of Community Recycling Programs, 1890–1945,* Public Works Historical Society, Chicago, 1979.

3. T. Ferrand, *Washington Post,* Dec. 15, 1979.
4. C. Glaab and T. Brown, *A History of Urban America,* Maxmillian Publishing Co., New York, 1976.
5. A. D'Arnay, *Salvage Markets,* U.S. Environmental Protection Agency, Washington, D.C., 1972.
6. D. Morris, *Economic Development Review* (Summer 1992).
7. D. Morris, *Democracy and Autonomy,* paper presented at The Other Economic Summit, Houston, July 8, 1990.
8. C. Humphrey in R. Nader, ed., *Ecotactics,* Public Citizen, Washington, D.C., 1972.
9. M. Humphrey and C. Humphrey in G. De Bell, ed., *The Environmental Handbook,* Ballantine Books, New York, 1970.
10. T. Radke, *Clear Creek Magazine* (Aug. 1971).
11. National Credit Office, *Reclamation 1975–2000: A Key to Economic Survival,* Dun & Bradstreet, New York, 1974.
12. P. Gunter and J. Starling, *California Management Review* (Spring 1978).
13. B. Platt and co-workers, *Union County New Jersey: The Pitfalls and Promise of Waste Incineration,* Institute for Local Self-Reliance, Washington, D.C., 1988.
14. *Environmental Reporter,* Washington, D.C., 1990.
15. S. Darcy, *World Wastes* (Oct. 1993).
16. U.S. Environmental Protection Agency, *Characterization of Municipal Solid Waste in the United States: 1990 Update,* EPA, Washington, D.C., June 1990.
17. U.S. Environmental Protection Agency, *Report to Congress,* EPA, Washington, D.C., 1972–1975.
18. *RACHEL Newsletter.*
19. N. Seldman and J. Huls, *Environment* (Nov. 1981).
20. M. Wentworth, *Resource Recovery: Truth and Consequences,* Environmental Action, Washington, D.C., 1976.
21. U.S. Environmental Protection Agency, *Engergy Conservation Through Solid Waste Management,* EPA, Washington, D.C., 1974.
22. U.S. Environmental Protection Agency and U.S. Department of Energy, *Memorandum of Understanding,* May 29, 1979.
23. H. Yakowitz, *Recent Federal Actions Which Will Affect Prices and Demand for Discarded Fibers,* National Bureau of Standards, Washington, D.C., 1980.
24. J. Bailey, *Wall Street Journal,* Aug. 16, 1993.
25. J. Mathews, *Washington Post,* Oct 30, 1993.
26. E. Connett, *Waste Not,* Canton, N.Y.
27. B. Gershman and Bratton, *District of Columbia Comprehensive Solid Waste Management Plan,* unpublished manuscript, Washington, D.C., Nov. 1993.
28. M. A. Charles, *Waste Age* (Nov. 1993).
29. *Resource Recovery Industry Annual Report,* Kidder Peabody, New York, 1988.
30. B. Richards, *Wall Street Journal,* June 16, 1988.
31. P. Hanson, *Recycling Programs in the U.S.,* EPA, Washington, D.C., 1972.
32. C. Leach, *Financing Mechanisms to Promote Recycling at the State and Local Level,* Institute for Local Self-Reliance, Washington, D.C., 1974.
33. State Capacity Task Force, *Alternative Financing Mechanisms for Environmental Programs,* EPA, Washington, D.C., 1992.
34. *Sorting It Out: Recycling Options in California,* Office of Appropriate Technology, Sacramento, Calif., 1976.
35. N. Seldman, *Economic Development Potential of Recycling,* Economic Development Administration, Washington, D.C., 1978.
36. Management Assistance Program, Recycling Consultants in California, *California's Model Recycling Programs,* California Resource Recovery Association, Santa Barbara, 1981.
37. *Reduce, Reuse, Recycle,* Oregon Environmental Education Association, Portland, 1991.
38. *Garbage Reincarnation,* Sonoma County Community Recycling Center, Santa Rosa, Calif., 1979.
39. U.S. Environmental Protection Agency, *Recycling Equipment Guide,* EPA, Washington, D.C., 1979.
40. M. Applehof, *Worms Eat My Garbage,* Flower Press, Kalamazoo, Mich., 1982.
41. N. Seldman, ed., *National Recycling Research Agenda,* National Science Foundation, Washington, D.C., 1980.
42. P. Montague, *Rachel's Hazardous Waste Newsletter* (140) (Aug. 1, 1989).
43. K. Lasser in N. Seldman, ed., *Environmental Impacts of Waste Incineration,* Institute for Local Self-Reliance, Washington, D.C., 1985.
44. P. Klein, *Processing of Fly Ash,* Citizens Against the Burn Plant, Coxsackie, N.Y., 1986.
45. G. Harhay, *Testimony before the Massachusetts Department of Environmental Quality,* Oct. 1986.
46. *The Rush to Burn,* New York Public Interest Group, N.Y., 1986.
47. P. Connett, *Waste Management as if the Future Mattered,* St. Lawrence University, Canton, N.Y., 1988.
48. Clean Water Action, *Mercury Rising from Incineration to the Food Chain: The Growing Threat of Mercury,* Washington, D.C., 1990.
49. Berkeley Recycling Group, *The Burn Papers,* Berkeley, Calif., 1982.
50. Resource Recovery Yearbook, 1982–1993, Governmental Advisory Associates, Inc.
51. *Resource Recovery Report,* Washington, D.C., Oct. 1993.
52. J. Bailey, *Wall Street Journal,* Aug. 11, 1993.
53. *Waste to Energy Fever Burns Local Tax Payers,* Waste Not, Canton, N.Y., Sept. 1993.
54. *Urban Ore, Reuse, Recycle, Refuse and the Local Economy,* Center for Neighborhood Technology, Chicago, 1993.
55. City Council, Rules Committee, *Solid Waste Plan for Philadelphia,* Jan. 1985.
56. N. Weisberg, Office of Recycling, Philadelphia, Sept. 1993, personal communication.
57. G. Vittori, *Economic Analysis: Mass Burn vs. Composting/Recycling,* Center for the Maximum Potential Building Systems, Austin, Tex., 1988.
58. B. McCann, *Austin American Statesman,* June 24, 1988.
59. B. McCann, *Austin American Statesman,* July 15, 1988.
60. Parametrix, Inc., *Solid Waste Options for King County,* Department of Public Works, 1989.
61. B. Platt and co-workers, *Beyond 40 Percent: Record-Setting Recycling and Composting Programs,* Island Press, 1990.
62. B. Platt and co-workers, *In-Depth Studies of Recycling and Composting Programs: Designs, Costs, and Results,* Institute for Self-Reliance, Washington, D.C., 1992.
63. U.S. Environmental Protection Agency, *Agenda for Action,* EPA, Washington, D.C., 1988.

64. B. Platt and co-workers, *Beyond 25 Percent Recycling,* Institute for Local Self-Reliance, Washington, D.C., 1989.
65. B. Platt and co-workers, *Waste Prevention, Recycling, and Composting Options: Lessons from Thirty U.S. Communities,* unpublished manuscript, EPA, Washington, D.C.
66. L. Schoenrich, *Cooperation Benefits Recyclers in the US and Canada,* Cooperative Connections, Omaha, Fall 1993.
67. *Buy Recycled,* National Environmental Law Center, Portland, Oreg., 1993.
68. M. Reisch, *Encouraging Recycling: State Minimum Content Laws,* Congressional Research Service, Washington, D.C., Dec. 24, 1991.
69. *Recycled Products Business Letter,* Herndon, Va., Aug. 1993.
70. *King County (Washington) Model Recycled Product Procurement Policy,* King Country Purchasing Agency, Apr. 1993.
71. *Cost-Benefit Analysis of Six Market Development Programs,* Integrated Solid Waste Management Board, Sacramento, Calif., 1993.
72. *Source Reduction Planning Checklist,* Inform, N.Y., 1993.
73. *Making Less Garbage: A Planning Guide for Communities,* Inform, N.Y., 1992.
74. H. J. H. Whiffen and co-workers, *Waste Prevention Saves Energy,* Florida Cooperative Extension Service, Gainesville, 1993.
75. M. Hammer, *Source Reduction: The #1 Waste Management Alternative,* Florida Cooperative Extension Service, Gainesville, 1993.
76. P. Hood, *Waste Age* (Apr. 1993).
77. *A Comprehensive Guide to Steel Recycling through Curbside Collection,* Steel Can Recycling Institute, Washington, D.C., 1993.
78. *1992 Recycling Almanac: Statistics and Projections,* Recycling Media Group, Washington, D.C., 1993.
79. J. McCarthy, *Recycling and Reducing Packaging Waste: How the United States Compares to Other Countries,* Congressional Research Service, Washington, D.C., 1991.
80. D. Morris and co-workers, *Beating the Btu Tax: The Six Percent Solution.*
81. D. Morris and co-workers, *Saving Btus Through Recycling: How Manufacturers Can Offset the Proposed Energy Tax.*
82. B. Platt and J. Zachary, *Co-Collection of Recyclables and Mixed Waste: Problems and Opportunities,* Institute for Local Self-Reliance, Washington, D.C., 1992.
83. S. Frysinger in *Getting the Most from Our Materials: Strategies from the New Jersey Conference,* Institute for Local Self-Reliance, Washington, D.C., 1992.
84. B. Platt and D. Morris, *The Economic Benefits of Recycling,* Institute for Local Self-Reliance, Washington, D.C., 1993.
85. S. Bogart and J. Morris, *The Economics of Recycling and Recycled Materials,* Clean Washington Center, Olympia, 1993.
86. J. Edwards, speech presented to the Local Government Commission, Santa Barbara, Calif., Sept. 10, 1993.
87. A. Hershkowitz in *Alliance for a Sustainable Materials Economy, Materials Policy and Recycling Infrastructure,* Chicago, 1993.
88. I. Ahmed, *Alcohol Fuels From Whey: Novel Commercial Uses for a Whey Product.*
89. *Recycling Today Magazine* (Nov. 1992).
90. Self-Reliance, Inc. Waste Utilization Study for Newark, New Jersey, Department of Engineering, Newark, 1985.
91. C. Rennie, *Salvaging the Future: Waste Based Production,* Institute for Local Self-Reliance, Washington, D.C., 1988.
92. D. Dillaway, *Capturing the Local Economic Benefit of Recycling: A Strategy Manual for Local Government,* Local Government Commission, Sacramento, Calif., 1992.
93. C. Hildebran, *Engines of Recycling: Strategies for Using Recycling to Drive Economic Development,* Californians Against Waste Foundation, Sacramento, Calif., 1992.
94. Gainer and Assoc., *Recycling Entrepreneurship: Creating Local Markets for Recycled Materials,* Arcata Community Recycling Center, Arcata, Calif., 1991.
95. M. Lewis, *Recycling Economic Development: Case Studies of Twenty-five Select Manufacturers,* unpublished manuscript, EPA, Washington, D.C.
96. D. Morris and coworkers, *Getting the Most from Our Materials: Making New Jersey the State of the Art,* Institute for Local Self-Reliance, Washington, D.C., 1991.
97. R. Ingerthron, *Massachusetts Recycling Coalition Newsletter* (Fall 1993).
98. Land and Water Assoc., *Recycling and the Maine Economy,* Maine Waste Management Agency, Portland, Oreg., 1993.
99. Department of Public Administration, *Empowerment Zones and Enterprise Communities: Moving from Legislative Concept to Successful Implementation,* George Washington University, Washington, D.C., 1993.
100. J. Vitarello, *Testimony before the House Small Business Committee,* June 7, 1993.
101. B. Stein, *Size Efficiency and Community Enterprise,* Center for Community Economic Development, Cambridge, Mass., 1974.
102. J. Blair, *Economic Concentration,* Harcourt Brace Jovanovich, New York, 1972.
103. W. Kovacs, *The Coming Era of Conservation and Industrial Utilization of Recyclable Materials,* Resource Recovery Report, Washington, D.C., 1988.
104. C. Hill, *1993 Solid Waste Management Act: How Solid Waste Programs in Florida Are Affected by Recent Legislation,* Deerfield Beach, Fla., 1993.
105. D. Rogich, *Introduction and Overview.*
106. R. Hurdelbrink, *Implications of the Materials Shifts,* The New Materials Society, Bureau of Mines, Washington, D.C., 1991.
107. W. Nixon, *E Magazine* (Sept. 1993).
108. U.S. Senate, *Tax Expenditures: Compendium of Background Material on Individual Provisions,* Congressional Research Service, Washington, D.C., 1992.
109. J. Powell, *Resource Recycling* (June 1992).
110. U.S. Department of Agriculture, *Situation and Outlook Report,* Economic Research Service, Washington, D.C., June 1993.
111. D. Morris and I. Ahmed, *The Carbohydrate Economy: Making Chemical and Industrial Materials from Plant Matter,* Institute for Local Self-Reliance, Washington, D.C., 1992.
112. D. Morris, *Rural Development, Birefineries, and the Carbohydrate Economy,* Institute for Local Self-Reliance, Washington, D.C., Sept. 1993.
113. *Wall Street Journal,* Oct 15, 1993.
114. R. Buck, *Seattle Times,* Dec. 19, 1993.
115. M. Dore and P. Nahmias, *New Jersey Law Journal* (Nov. 29, 1993).
116. D. Biddle, *Howard Business Review* (Nov. 1993).
117. C. Steinberg, *U.S.A. Today,* Jan. 6, 1994.

118. W. Sheehan and L. Fowler, *Economics of Small Landfills: An Analysis of Operational and Proposed Subtitle D Landfills in Georgia Taking Less than Two Hundred Tons of Waste per Day,* Georgia Environmental Policy Institute, 1993.
119. M. Sampson, *The Elwyn Report,* EPA Region III, Philadelphia, 1993.
120. W. Kovacs and M Pellegrini, *Flow Control: The Continuing Conflict Between Free Competition and Monopoly Public Service,* Resource Recovery Report, Washington, D.C., 1992.
121. Facts to Act On, *Solid Waste and Interstate Commerce,* Institute for Local Self-Reliance, Washington, D.C., Dec. 1991.
122. Facts to Act On, *Trashing Transport: Strategies to Ban Imported Garbage,* Institute for Local Self-Reliance, Washington, D.C., Feb. 1991.
123. *Brief and Amicae Curiae,* Waste Management of the Desert, Inc.
124. *City of Rancho Mirage v Palm Springs Recycling Center,* Supreme Court of the State of California, Sept. 9, 1993.

RECYCLING, TERTIARY—PLASTICS

RAYMOND J. EHRIG
Aristech Chemical Corp.
Monroeville, Pennsylvania

BACKGROUND

Until recently, plastics recycling activities focused on primary and secondary recycling (1). Improvements in secondary recycling continue; however, recycling innovations will be made mostly through new developments in tertiary technology. To classify plastics recycling by technologies, primary indicates simple regrinding and re-extrusion of production scrap back into manufacturing and fabrication processes; secondary employs physical methods to separate and purify postconsumer (and postindustrial) waste plastics, after which the recycled resins are used to fabricate products having less demanding performance requirements; tertiary involves chemical and thermal processes; and quatenary is burning with or without energy recovery.

The life cycle of plastics involves four phase: resin formation, part fabrication, product service, and disposal (2). In secondary recycling, plastics are refabricated into a new product, thus prolonging their lives. The secondary recycling cycle can be repeated only a finite number of times, however, because plastics deteriorate as they are reheated and reworked; some disposal is inevitable. To meet new-product performance requirements, the recycled resin must be upgraded by the addition of virgin resin. At some point, part of the recycled material must be discarded (ie, incinerated, landfilled, or treated via tertiary technology).

In tertiary recycling, either chemical or thermal technologies are employed to decompose plastics wastes to chemicals and fuels. Chemical processes decompose plastics by the action of chemicals on plastics wastes at increased temperatures and/or pressures, usually in the presence of a catalyst; thermal processes decompose plastics by heating them to high temperatures in an oxygen-free or oxygen-deficient atmosphere (plastic does not burn in pyrolysis).

Whereas primary and secondary recycling provide longevity to plastics, tertiary recycling is regenerative (see Fig. 1) (2). For example, established chemical processes decompose waste plastics under known and controlled conditions to yield chemical compounds that become intermediates, used in the formation of the same or entirely new products. Monomers and low molecular-weight polymers or oligomers formed upon depolymerization are starting materials for the regeneration of new resins, usable for fabrication into the same or new plastic products. Similarly, the simple chemical compounds derived from pyrolysis can be used either as chemical feedstocks or as industrial and residential fuels.

Chemical treatment that reduces polymers to monomers or low molecular-weight components is applicable only to certain polymers in the wastestream (3). These polymers, formed by step-growth reactions which are reversible, have specific functional groups in the backbone. The functional groups contain weak chemical bonds susceptible to dissociation by certain chemical agents. These agents may be water (usually in the form of steam), alcohols, amines, ammonia, and acids. The depolymerization

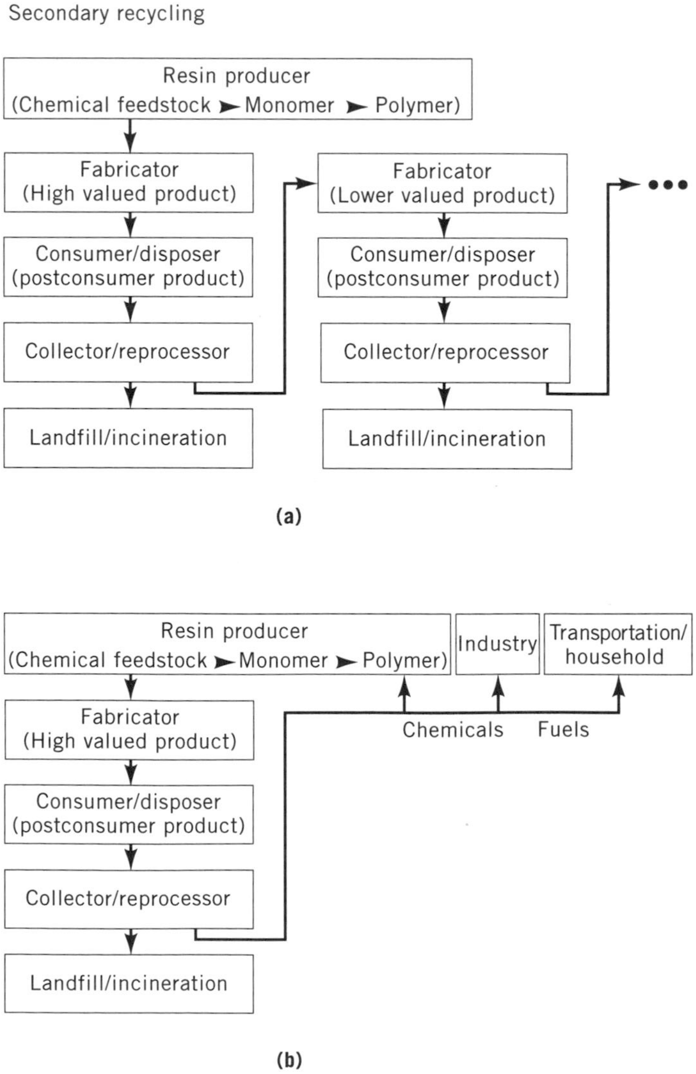

Figure 1. Secondary (**a**) versus tertiary (**b**) recycling.

reactions effected by the specific chemicals indicated are known respectively as hydrolysis, alcoholysis (methanolysis, glycolysis), aminolysis, ammonolyses, and acidolysis. Solvolysis is the general term that describes all these reactions.

The total quantity of thermoplastic and thermoset step-growth polymers sold in the United States in 1993, potentially available for chemical decomposition, is approximately 6.9×10^6 t or about 22% of all polymers (see Table 1 (4). The most common of these polymers are the polyurethanes (PUR), phenolics, and the thermoplastic polyesters with polyethylene terephthalate (PET) at 1.24×10^6 t. Excluding the thermoset PUR, the remaining thermosets constitute 10.5% of the total. Recycling of the phenolic, urea, epoxy, and unsaturated polyester (UPE) thermosets by solvolysis appears impractical in the foreseeable future, which leaves approximately 3.6×10^6 t of step-growth polymers, or 12% of the total available. These polymers are concentrated in specific wastestreams and provide good feedstock for chemical conversion processes.

Tertiary recycling also involves the decomposition of polymers by heating to high temperatures to give monomers, chemicals, fuels as liquids and gases, and solid residues. Thermal decomposition includes pyrolysis and the various refinery processes such as coking, catalytic cracking, hydrocracking, and gasification (5). Within these processes, decomposition is accomplished under a variety of conditions, eg, at increased temperatures and pressures, under vacuum, with or without catalysts, in an oxygen-free or oxygen-limited atmosphere, and in the presence of steam, hydrogen, or other gases. Depending on the specific reaction conditions; residence time; purity, particle size, and type of waste feedstream; and product treatment, varying amounts of solids, liquids, and gases can be produced. For example, pyrolysis processes, usually operated in the absence of oxygen, without pressure, and at temperatures ranging from 400 to 900°C, generally yield 40–60% gases, 20–40% oils, and 6–30% solids. Hydrocracking processes (6), usually conducted at the lower temperature zones but at increased pressures, give higher percentages of oil, lower gases, and decreased solids.

Thermolysis has greater diversity and fewer limitations than solvolysis, thus increasing potential to treat mixed waste streams successfully (7). Provided the waste materials contain high hydrocarbon contents, thermolytic processes can be applied to municipal solid wastes, automotive-shredder residues, mixed textile wastes, medical wastes, and other nuisance plastics to recover valuable products. However, the applicability of a particular process can be limited by the existence of certain contaminants in the waste-streams, and some pretreatment may be necessary.

Table 1. Step-Growth Polymers for Chemical Decomposition

	Polymers	1993 Sales, %
Thermoplastics	Polyesters	4.4
	Polyamides	1.1
	Polycarbonates	0.9
	Polyacetals	0.2
Thermosets	Polyurethanes	5.1
	Phenolics	4.5
	Ureas, melamines	2.9
	Unsaturated polyesters	1.8
	Epoxies, Alkyds	1.3
Total		*22.2*

See also Fuels from waste; Fuels, synthetic—gaseous fuels; Fuels, synthetic—liquid fuels; Pyrolysis; Waste-to-energy.

CURRENT TRENDS

Tertiary Versus Secondary Plastics Recycling

As the mid-1990s approach, many of the legislative bodies of the main industrial nations have realized that the postconsumer plastics recycling initiatives previously established could not be accomplished through present-day, available commercial technologies. One extreme example is the ordinance established in Germany in 1990 and 1991 which required that by 1995, 80% of plastics packaging waste had to be collected and 80% of the collected waste had to be recycled. This translates to a recycle goal of 64%, a significant increase from the rate of 9% achieved in 1993 (8). Because of insufficient commercial recycling capacity, particularly for mixed plastics, incineration with energy recovery is now permitted on an interim selective basis. German authorities have established that by the end of 1995, the processing technologies of tertiary recycling should provide the necessary capacity, particularly for the mixed plastics now being collected.

Although the above example is specific to Germany, it reflects the interest and emphasis placed on tertiary recycling by many of the main companies involved in polymer recycling worldwide (9). Activities involve more trials with mixed and/or partially fractionated municipal solid wastes, automotive-shredder residues, packagings, and carpet—textile wastes in thermolytic units to generate chemical and petrochemical feedstock. Increased emphasis on using mixed plastic waste is also a current trend in chemical decomposition. For example, developments are being pursued to process postconsumer wastes containing higher levels of contamination, mixtures of known composition, and mixed durables. Other waste plastics not previously handled, eg, polycarbonates and acetals, are also receiving attention.

Trends in secondary plastics recycling somewhat parallel tertiary recycling. For example, there has been an increase in developing the technology of plastic lumber and in combining mixed wastes, selected fractions of mixed wastes, and blends of other materials for producing lumber products. The Plastics Lumber Trade Association was formed in 1993 to codevelop (with the American Society for Testing and Materials) test methods and product standards for the industry (10). In addition, chemical companies have also become involved; Dow Chemical is part of a venture with Advanced Environmental Recycling Technologies, Inc., and Mobil Chemical has purchased Timbrex, a manufacturer of plastic lumber from wood and plastic wastes (11).

Efforts in other areas are directed toward more efficient and innovative techniques for separating mixed plastic containers and for handling waste film. Improving the separation and cleaning process, overall product quality, and product performance are continuing trends in secondary recycling. Studies on compatibilization, chemical modification, fillers, and reinforcements are directed toward product upgrading.

Chemical Decomposition

As of the early 1990s, few companies practice chemical decomposition on a commercial scale using postconsumer plastics waste to generate feedstock for polymer production. However, companies do use the technology internally for recycling plant feedstreams containing semifinished or other clean industrial plant scrap and off-grade finished product. This is usually plant practice for mature polymerization production facilities if the process is so designed. In cases where recycled materials can be adequately controlled in the production facility and the economics are favorable, some companies have initiated closed-loop systems. Here, product scrap generated at customers' fabrication shops is returned to the resin producers, and recycled back to the polymerization facility as monomers or polymer intermediates through chemical decomposition (12).

Hydrolysis—PET. Hydrolysis is a reverse reaction with water to depolymerize a polyester to acid and alcohol. If the polyester is PET, hydrolysis of the polymer yields the starting diacid, terephthalic acid, and starting dialcohol, ethylene glycol. The hydrolysis of waste PET can be conducted in an acid, base, or neutral medium under mild conditions of temperature and pressure (12). The process has yet to be commercialized due to the inherent difficulty of isolating and purifying the terephthalic acid (TPA) from the other reaction products.

The estimated economics of the hydrolysis of recycled PET, although not yet commercialized, are comparable to that of glycolysis, which has been commercialized and which will be discussed later. Clean, recycled, bottle-PET resin—at prices ranging from $0.61/kg to $0.66/kg—is required. Operating costs are similar to those for glycolysis, whereas capital costs may be higher. The economic advantage of hydrolysis over methanolysis and glycolysis is that the production of terephthalic acid has greater potential as starting material for the regeneration of PET than does dimethyl terephthalate (DMT) or the polyols (2).

Hydrolysis—PUR. Hydrolysis of PUR to polyols and diamines has been achieved in solution with superheated steam in stirred-tank vessels, fluidized beds, and tubular reactors. When the hydrolysis is conducted in a twin-screw extruder, the polyol product can be used directly in the preparation of new polyurethanes. The diamines must be isolated and purified or treated as a hazardous waste, however, which is costly and complex; the glycolysis approach (discussed later), in which diamines are not formed, is favored (13).

Hydrolysis—Nylon. Most nylon-6 producers recover caprolactam monomer internally by hydrolyzing plant waste and scrap. For example, BASF recovers caprolactam from plant liquid and solid-waste streams at its nylon plants in North Carolina and Ontario (14). Allied Signal has been recovering caprolactam internally for many years by phosphoric acid steam hydrolysis of nylon-6 scrap (15,16). More recently, the company has initiated a closed-loop system with customer fabricators to return scrap. DuPont has a number of early patents on depolymerization of nylon; the processes described are possibly used to recover monomer from internal scrap to increase plant yield (17).

Currently, there is no commercial plant depolymerizing postconsumer nylon waste. DuPont has started up a 230-t/year capacity pilot facility in Gloscow, Delaware, to depolymerize polyamides by "new" proprietary processes (18). The recycling technologies apparently have the capabilities to depolymerize postconsumer nylon fibers, moldings, and composite nylon structures to component monomers, which can then be repolymerized into various polyamides to serve current nylon markets (19).

BASF has also developed processes for the recovery of monomers from postconsumer polyamide wastes (20). For nylon-6, 6, an alkaline hydrolysis is followed by electrolysis to recover both hexamethylene diamine and salt-free adipic acid. The method can be used to recover the monomers from all types of nylon-6,6 wastes, including reinforced and colored automotive parts. A pilot stage is now in progress. BASF has also developed a process for the recovery of caprolactam from postconsumer nylon-6—polypropylene carpet wastes. In a recently patented process (21), waste carpet materials are continuously depolymerized at the melting temperature of nylon-6 in the presence of phosphoric acid and superheated steam. The monomer is distilled and subjected to an oxidation to remove other components and to obtain pure caprolactam.

Methanolysis—PET. Both Eastman Chemical and Hoechst Celanese depolymerize postconsumer and postindustrial PET commerically by reacting with methanol to form DMT and ethylene glycol (EG). By the end of 1991, the two companies received letters from the Food and Drug Administration (FDA) indicating there was no objection to their respective processes for the regeneration of DMT from PET and for the reuse of the monomer in producing PET for food packaging applications (22).

In the methanolysis processes described to date, waste—scrapped PET is dissolved in a mixture of EG and oligomers, and reacted with excess liquid methanol or, alternatively, added directly in the melt phase to an excess of methanol in the liquid state, in the presence of a catalyst at increased pressures. The resulting depolymerizate contains DMT, other esters and glycols derived from variants of PET production, and various contaminants found in postconsumer materials. Proprietary purification methods used possibly include filtration, washing, centrifuging, and distillation (23).

The latest economic evaluation of the methanolysis process estimates a cost of $1.30–1.50/kg of product (DMT and EG) derived from waste PET-bottle flake, compared to petroleum-derived virgin monomers at $0.77/kg. This is based on a 25×10^3 t depolymerization plant costing $25–30 million, and waste PET flake priced at $0.55–0.77/kg. For PET containing 25% recycling content, the increased cost is $0.13–0.18/kg or slightly less than $0.01 per 2-liter bottle.

Eastman and DuPont have developed vapor-phase

methanolysis, and both have had patents issued in this area (22, 24). In these processes, DMT and EG may be recovered in the methanol vapor stream from the reactors and separated by distillation. Any catalyst employed in the preparation of the original PET and other nonvolatile waste impurities, such as paper, polyolefins, and metal, remain in the bottoms of the reactors. The processes may be bulk, semicontinuous, or continuous. DuPont is operating this process at a 7,500-t/year semicommercial facility in Old Hickory, Tennessee, and is projecting to operate at a commercial level sometime in 1994 (17). The versatility of the process was demonstrated by depolymerizing reinforced automotive PET and repolymerizing the recovered monomers into quality-grade polymer used in fiber applications (25).

Alcoholysis—PET. The methanolysis reaction, also known as transesterification, is an example of a general class of reactions known as alcoholysis. Any number of alcohols can be used to obtain the corresponding ester. Synergistics Industries of Canada has extended the solvolysis of PET by methanolysis to higher alcohols. In one of a series of reactions with C_6–C_{20} linear and branched alcohols, scrap PET reacted with 2-ethylhexanol yielded the product dioctyl terephthalate (DOTP). The product is a cost-effective plasticizer for PVC and a potential replacement for the current commercial plasticizer, dioctyl phthalate (DOP). The process is patented, and a license for the technology is offered by Synergistics (26,27).

Glycolysis—PET. The only commercial U.S. plant producing PET from oligomers by the glycolysis of postconsumer PET is operated by Shell Chemical Co. (formerly Goodyear Tire and Rubber Co.) in Point Pleasant, West Virginia. In the process, clean postconsumer flake is depo lymerized during esterification in the presence of TPA and EG at a temperature greater than 260°C and at a positive pressure (28). Shell's purification of the reaction mixture is proprietary and may include a special separation device for removing any contaminated materials from the oligomers (29). In 1991, the FDA issued a "letter of no objection" on the process and for the use of the PET produced by the process in food packaging.

Companies worldwide are producing PET from a combination of virgin starting material and oligomers obtained by glycolyzing internally generated plastic scrap. Several companies also use some postconsumer plastics waste. One company in Stuttgart, Germany, is said to be using 50% postconsumer-derived oligomers in the preparation of film-grade PET. Also, the DuPont Dacron operations in Kingston, North Carolina, claim to have recycled over 4.95×10^5 t of PET polymers by glycolysis over the last 10 years (some of which was postconsumer PET waste) (30).

A comparison of methanolysis and glycolysis of PET is given in Table 2. Methanolysis does not require a high quality feed, but the process is somewhat energy-intensive. Polymer purification by crystallization and distillation is simple, controllable, and results in a product of good quality with high recycle potential. The overall cost is estimated to be higher than for glycolysis, although no economic data have been published. For the glycolysis reaction, a high quality feed is required; the process requires less energy than methanolysis. The filtration required during product purification is more complex than distillation. The product can be slightly colored, and the potential for continued recycling is probably less than for the PET methanolysis product.

Table 2. Comparison of PET Depolymerization Processes[a]

	Methanolysis	Glycolysis
Feed quality required	Low	High
Estimated energy required	Moderate	Low
Degree of ease of purification	Low	Moderate
Overall product quality	High	Moderate
End-use versatility	High	Moderate
Overall cost	High	Moderate

[a] Ref. 31.

There are also a number of U.S. companies glycolyzing postconsumer and postindustrial PET scrap to oligomers and using these polyols internally to prepare UPE, PUR, and polyisocyanurates (PIUR) (32). These commercial operations glycolyzed approximately 25×10^3 t of PET in 1993. For these processes, the feedstock purity requirement is less than for the methanolysis reaction, and purification of the polyols is also less than that required for bottle-grade product. The polyols are used at various concentrations in the preparation of UPE, PUR, and PIUR. No detrimental effects on polyester process or product performance characteristics are noted at low concentrations of PET-derived polyols. Product differences become noticeable at concentrations above 50% because of compositional differences between PET-derived and virgin-derived polyols. Similar effects on PUR and PIUR are noted but at lower concentrations of polyols than for UPE.

The economics are favorable for glycolyzing waste PET and using the polyols for the preparation of UPE, PUR, and PIUR. For example, UPE prepared with PET-derived polyols has an $0.18–0.24/kg cost advantage over conventionally prepared UPE, estimated using a waste—PET price of $0.77/kg (2).

Glycolysis—PUR. Only one commercial installation glycolyzes waste PUR to polyols. Pebra of Germany is operating a small plant using the technology developed at the Technical University, Aalen, Germany. The plant recovers polyols from automotive seats for reuse in rigid-foam production.

BASF has established a pilot unit at its plant in Schwarzheide, Germany, to develop and pilot a process for glycolyzing PUR foam automotive seats to polyols (33). A consortium of companies including Dow, ICI, and Bayer plans to research new automotive applications for the recovered polyols (34).

Dow Chemical, Europe, had developed a new two-step process of reacting postconsumer and postindustrial waste PUR foams to given polyols free of amines (35). The process, developed at the Netherlands Research Laboratory and now in the pilot stage, is applicable to flexible foams, Reaction Injection Moldings (RIM), and semirigid elastomers. The process is a combination of glycolysis and alkoxylation that negates the need to separate the amine and glycolate mixtures. Foams containing 20–40 wt% recycled-derived polyols were made equivalent to reference foams by varying the index. At 12% recycled polyol, the resulting foams have properties suitable for a variety of sound- and energy-absorbing applications.

Rapra Technology Ltd. has developed a cavity-transfer mixer to increase the efficiency of PUR recycling (36). A continuous glycolysis of microcellular PUR was conducted, and the product was reformulated to a cast elastomer without any intermediate extraction or distillation.

An interesting application with possible economic advantages was reported by INOAC Corp. of Japan (37). Waste rigid PUR and PIUR foams were decomposed with different glycol—ethanolamine systems. The decomposition products, with separation or purification, acted as hardeners and/or toughening agents for general-purpose and structural epoxy adhesives.

Three other activities have recently been reported. First, Reinforced Reaction Injection Molding (RRIM) foam production scrap and sorted washed automotive waste was glycolyzed, and the resulting polyol at 10%, mixed with 10% regrind, was used to form an amine—cross-linked RRIM part. With the exception of the elongation at break (117 vs 146%), the properties exceeded those of another part prepared from virgin glycol—cross-linked RIM system. Second, microcellular PUR scrap from the footwear industry was glycolyzed with 1,4-butanediol and triethanolamine. The processibility of a new microcellular formation with 10% incorporation of the recovered polyol was similar to the original system. The final product showed a slight improvement in mechanical properties. And third, polyols recovered from flexible foam automotive seats were used to prepare Structural Reaction Injection Molding (SRIM), hollow-structured foam, and elastomers. Product properties depended on the specific glycol used to obtain the recovered polyol. Up to 50% recovered polyol gave a satisfactory product, although some property differences were noted (38).

Potential. Although the total quantity of step-growth polymers available for monomer recovery is less than 25% of all polymers produced in the United States, enough volume exists for significant future growth in solvolysis. The step-growth polymers are the highest valued polymers, PET has the highest recycling rate of all polymers and is expected to increase, and markets for high quality recycled polymers will continue to grow.

The activities by chemical companies on the solvolysis of PET, PUR, and nylons are indicative of a good future. In addition, there is potential growth for solvolysis of other step-growth polymers. For example, General Electric has a patented process for the recovery of bisphenol A from polycarbonate scrap (39). The first efficient process for the chemical decomposition of acetal has been developed by Hoechst of Germany. The acid-catalyzed depolymerization is said to be less costly than PET and nylon chemical decomposition processes. In small-scale tests, the recovered product obtained from contaminated scrap has been used to prepare virgin acetal (40). DuPont also claims to have proprietary technology for chemically reprocessing acetals and is planning a facility to do this (41). Possibilities also exist for the solvolysis of polyureas, other polyesters such as polybutylene terephthalate, and alkyds.

Overall, even though few commercial solvolytic processes using postconsumer plastics waste exist today, the potential for chemical recycling remains positive for the future. However, this growth will be affected by factors such as landfill costs, mandated recycling legislation, availability of adequate feedstock, collection infrastructure, improved separation techniques, market demand for high valued products, and improved technologies and economics of the solvolytic process.

Thermal Decomposition

Whereas solvolysis is limited to step-growth polymers, thermolysis is not specifically limited by polymer type. However, addition polymers and polyolefins in particular are preferred, as thermal decomposition of these polymers yields maximum amounts of desired products, ie, oils for fuel and chemicals. Exceptions are selective-addition polymers that decompose mechanistically by unzipping of the polymer chains to give monomers in high yields. This is differentiated from other polymers that produce liquid or gaseous precursor products when pyrolyzed. Polymer to monomer is often referred to as thermal depolymerization, whereas polymer to liquids or gases is known as pyrolytic liquefaction and gasification, respectively. There are, of course, process and product variations that prevent these from being considered rigid classifications.

Thermal Depolymerization. Few of the common plastics depolymerize upon heating to monomers in high yield. Polymethyl methacrylate (PMMA), polytetrafluoroethylene, and nylon-6 will depolymerize, but only the depolymerization of PMMA has been practiced commercially in the United States (42). Although there may still be a unit in operation in Florida, other commercial units are operating in India and Europe (3). The depolymerization of nylon-6 to monomer from waste nylon—polypropylene carpeting by a fast pyrolysis technique was patented by Midwest Research Institute in 1993 (43). The technique involves selecting and controlling a temperature profile to cause pyrolysis of the polyamide to caprolactam before other components are pyrolyzed. Specific carrier gas, catalyst, and support are used to optimize the monomer yield to 85%. Conditions are also given for isolating individual monomers from the PET, PUR, and polyphenylene oxide—polystyrene present in postconsumer and automotive waste and engineering resin scrap.

Liquefaction. The liquefaction of polymers by pyrolysis constitutes the largest segment of thermal decomposition technology under study today. A variety of processes are used, and conditions can be selected to meet the decomposition requirements of most polymers. Individual or commingled polymer wastes can be treated.

Currently, the commercial status of liquefaction of plastics waste is similar to that of solvolysis; there are more commercial solvolytic process units in operation than thermolytic units. Many demonstration plants, large enough to be considered commercial units, have and continue to run thermolytic processes today. A demonstration "commercial" fluidized bed pyrolysis unit of 5,000-t/year capacity operated successfully in Germany for many years but eventually closed down because of poor economics (7).

The only process practiced on a significant commercial

scale today is a combination thermolytic—catalytic cracking liquefaction system. Fuji Recycle Industry (44) is operating a 5,000-t/year capacity commercial plant in Aioi, Japan, using industrial plastics waste; 400-t/year capacity demonstration unit in Okegawa treats mixed postconsumer plastics waste. In the Aioi plant, mixed polyolefin—polystyrene waste is thermally cracked at 400°C and catalytically cracked at lower temperatures using a Mobil Zeolite ZSM-5 catalyst to 80% yield of a high grade liquid fraction containing 50–60% gasoline, 20–30% kerosene, and 10–30% light oil. A gas (15%), used as a fuel, and 5% residue are also formed. The plant is estimated to cost approximately $4.2 million with annual employee cost of $1 million for a three-shift operation. Power is the largest single utility cost for the unit, but the cost can be minimized by using gas and heavy diesel fuel oil produced from the unit. The process is claimed to be commercially profitable at the 5,000-t/year level, if the cost of waste plastics is calculated as zero and if the tipping fees are not counted as income. With a tipping fee of $35/t, a 3,000-t/year plant is estimated to have favorable economics.

The demonstration plant at Okegawa is treating postconsumer plastics waste with a maximum of 20% polyvinyl chloride (PVC) and producing a gasoline—kerosene mixture containing less than 1 ppm chlorides. The clorides are removed at three stages in the process: at an extruder used to homogenize the waste mix, from a mixing vessel, and finally by passing through a bed of calcium oxide before the catalyst bed. The resulting product mix is 65% liquid, 15% gas, and 20% residue.

Wayne Technology Corp. has a demonstration pyrolysis unit in Macedon, New York, which employs a rotary kiln, or converter, operating at a positive pressure and at a temperature of about 980°C (45). The feed for this unit can consist of mixed plastics, tires, and cardboard; the resulting products are three grades of oil, gas used as a fuel for distillation, and char. Compared to a fluidized bed system, a rotary kiln system usually requires a longer residence time and gives greater amounts of residue. The economics for this system have been estimated for a 45-t/day unit operating for 270 days/year, with an output of 119 barrels of oil per day, selling at $10 per barrel, and with assumed tipping fees of $66/t. Assuming further a three-year payback and an interest rate of 9% on the capital cost of the converter at $2.5 million, a possible pretax profit of $250 million per year is calculated.

Nikon Rikazaku Ltd, Japan, has developed a catalytic process for the recovery of oil from mixed plastic waste and has an 8-t/day facility in operation at Matsue Shemane Prefecture (46). The technique is said to have marked advantages over other pyrolytic—catalytic processes, as low operating temperatures are used. Also, the process has the capability to treat wastes containing 20% PVC and up to 50% nonplastic material.

BASF is building a 15,000-t/year pilot plant near Ludwigshafen, Germany, to convert mixed postconsumer plastics into special α-olefins by a thermal noncatalytic patented process (47). Postconsumer waste containing up to 2% PVC can be treated. If pilot-plant trials are successful, the company plans to build a commercial-scale plant of 1–3 $\times$ 10^5t/year capacity. Estimated cost of the pilot plant is $18–25 million (48).

E.T.P. Technologies and the University of Laval, Quebec, Canada, have completed the first-phase development of a vacuum–pyrolysis process for the decomposition of automotive-shredder residue to oils, metals, inorganics pyrolytic water, and gases (49). A 4-t/hr demonstration unit at an estimated cost of $5 million is planned. A potential operating profit of $27/t for a 24-hr/day, 330-day operation is estimated, assuming various selling prices for the products.

Mazda of Japan claims to have developed a special metallic salt containing catalyst, capable of catalyzing the decomposition of all types of mixed plastics wastes (including PVC) to gasoline, kerosene, and light oils (50). Toshiba Corp. has reported a similar process using an alkaline wash solution during heating to decompose chloride-based polymers to gasolines and kerosenes.

Recently, oil companies have been investigating plastics waste as another potential source of feedstock for the refineries. The companies have made significant progress, and developments have moved rapidly, partly because thermal decomposition processes have been known and practiced in refineries for decades. Ideally, plastics waste would be fed directly to the refinery cracking units, but because of limitations at the refineries for homogeneous liquid feedstock and large volume throughputs, some pretreatment of the material is necessary. Companies have already moved in this direction. For example, in pilot tests conducted by BP Chemicals (9), a fluidized bed pyrolysis unit was used to thermally crack mixed plastics waste containing 2% PVC to determine its use as a refinery feedstock. The resulting liquid product, containing $<$ 5 ppm chloride, was within refinery—petrochemical feedstock specification and provided excellent yields of ethylene and propylene as a mixed feed with refinery naptha. Petrofina, DSM, Elf-Atochem, and Enichem have joined with BP in a consortium to further the commercialization of the process. The first phase is a 100-kg/hour pilot plant now in the design stage by BP (51).

Veba Oel of Germany has developed a complete system for the preparation of refinery-grade synthetic cruide oil from mixed plastics waste (52). The process consists of thermally decomposing the plastics waste premixed with crude oil, hydrogenating in the liquid stage, followed by a second hydrotreating in the gas phase over a fixed-catalyst bed. The liquid product obtained in 90% yield is a high quality synthetic crude oil suitable for direct feeing to refineries. Over 60 t of plastics waste, including PVC from the "Duales system," has been successfully converted. Technically, the process is viable, as evidenced by the success already achieved. The plastics waste need not be separated, PVC and inert solid materials (glass, dirt, pigments) can be tolerated, the conversion of feed to product is energy-efficient, and product quality is high. No economics have been published, but equipment and process costs may be high. Equipment for the four major stages (blending, depolymerizing, liquid and gaseous hydrogenation) and a number of other steps (separation, purification, etc.) is required. Pressures of 150–250 bars are used; hydrogenation catalyst life is unknown. Also, the treatment facility will have to be closely linked to a refinery to use the crude oil and to dispense the syncrude product.

REW-Entersorgung of Essen, Germany, has developed a process for the hydrogenation of mixed plastics waste to syncrude (53). During a successful pilot test, specification syncrude was produced for the Wesseling refinery.

A hydrocracking process is also being investigated by researchers from the University of Utah and the Institute of Mining and Minerals Research, University of Kentucky, part of the Consortium for Fossil Fuel Liquefaction Science (54, 55). Waste plastics are converted almost completely to liquid products, with approximately 90% yield of quality oil when treated at 350–450°C and 80–105 bar hydrogen in the presence of zeolite-based catalysts. When coliquified with coal with tetraline as a hydrogen feedstock, a zeolite catalyst, and heated to 400–500°C at 55 bar for 1 h, oil yields are less than with plastics alone, but the total liquid and oil yields from the coal are increased. Metals from the plastics are scavenged by the insoluble fraction of the coal, leading to a lighter, cleaner oil.

The American Plastics Council is funding a pyrolysis program to convert waste polyolefins initially, and later, mixed plastics waste to oils using a kiln reactor (14). The program includes upgrading a tire pyrolysis plant in Chehalis, Washington, and sponsoring research efforts at the site. The primary objective is to demonstrate technology feasibility and economic viability (56). A test pilot program on the pyrolysis of polyethylene, polypropylene, and polystyrene was completed, and 22.7 kl of the liquid product was successfully processed as feedstock to a commercial coking unit by Lyondell Petrochemical Co. (57).

Gasification. Pyrolytic gasification is the technology of thermally decomposing materials to predominately gaseous products. Harsh conditions of high temperatures with increased pressures and optional catalysts are used. The technology includes a number of processes recently developed for treating waste materials, as well as the refining operation of steam cracking and partial oxidation or thermoxidation for the preparation of hydrogen and carbon monoxide (syngas).

A new gasification system using a sprouted bed reactor (SBR) for the treatment of waste plastics, including automotive-shredder residue (ASR), has been developed by the Energy and Environmental Research Corp. with funding from the E.P.A. and the Gas Research Institute (58). A pilot unit of 18-t/day capacity consists of an extrusion feed system, the SBR, a secondary hot cyclone, and units for heat recovery, steam superheating, fines recycle, and scrubber water reinjection. Preliminary tests were conducted on unsegregated ASR at a feed rate of 635 kg/hour in a reducing atmosphere, at reactor temperatures of 540–870°C. The gaseous products obtained directly from the reactor contained a high level of hydrocarbons, together with CO, CO_2, and H_2. Follow-up work is targeted to generate a clean syngas product. The estimated cost for a 227-t/day plant is $5 million; at a tipping fee of $100/t, a two-year simple payback on the investment is calculated.

General Motors has been granted three patients (59–61) for a pyrolysis apparatus and process for treating waste thermoplastics and thermosets, ASR, paint sludge, tires, and so forth. Temperatures range from 590 to 870° C. Inorganic material, such as $CaCO_3$ from sheet-molded composite thermosets, is recovered and reused as a filler or as a flux to remove impurities during an iron-making process. The organic components, recovered as gases and oils, are used to fuel the iron making or to produce other chemical products.

Thermoset of Locarno, Switzerland, has developed a gasification process for treating unsegregated commingled household wastes at 2,000°C in an oxygen atmosphere (9). For each metric ton of waste, approximately 66% syngas is generated together with mineral slag for concrete, industrial-grade salts, metal alloys, and treatable water. Virtually no offgas, ash, or dust is formed. A pilot plant in Italy is treating 4.1 t/hour of municipal waste, and a 20-t/hour plant is slated for startup in 1995 by a German company.

Shell International Chemical Co. is developing a gasification system with a pretreatment of dehydrochlorination and thermal cracking (52, 62). The process is claimed to be capable of handling different types of plastic wastes and providing a variety of end-products. For example, if the pretreated output contains inorganic materials, it can be treated by a gasifier for syngas production; if free of inorganics, the product can be used as feedstock for the most demanding refinery processes, including steam cracking. The process is known as the "SPROUT" technology, ie, the Shell Process for Recovery of Used Thermoplastics (79).

Texaco has had a commercial gasification process in operation for over 40 years, converting various carbonaceous feeds to predominantly hydrogen and carbon monoxide (63). The feedstock can be in any solid, liquid, or gaseous form, including coal, petroleum coke, other petroleum byproducts, and natural gas. Operating conditions of the process are severe; pressures of 20–80 bars and temperatures between 1,200 and 1,500°C are used, depending on the specific feedstock. Oxygen input is tightly controlled to maintain an oxygen-deficient atmosphere. The product syngas and/or separated components are used as feedstock for the preparation of ammonia, methanol, acetic acid, methyl-*t*-butyl ether, plastics, fertilizers, and other valued products.

Pilot-scale tests conducted by Texaco confirmed that the gasification technology is applicable for mixed plastic wastes (64). Both postconsumer and postindustrial wastes containing materials such as PVC and nylon were used; one waste product contained 30% PVC. Conversions of the plastics waste to syngas were in excess of 99%. Computer simulation of the operating performance of the process on a postconsumer packaging waste mix from the "Duales system" verified that this material is an excellent gasifier feed and that the gasification gives a simulated high quality product. An economic evaluation by computer modeling was favorable for the system using this feed.

The Texaco gasification process can apparently convert mixed unsorted plastic wastes of various origins to a suitable product. Not enough data have been published to evaluate the overall economics, however. The usual dependencies exist, ie, tipping fees, product revenues, and capital debt repayments, and product revenue may be limited by the potential future growth of syngas (no significant new future uses of syngas are projected). Another cost factor may be oxygen purchases. Finally, for capital debt repayment, Texaco estimates that the gasification unit will cost $10–15 million.

Another near-commercialization system is Molten

Metal Technology's catalytic extraction processing (65–67). The system uses a molten-metal medium in reactors originally designed for steel making; batch demonstration and commercial tests have been conducted on a variety of waste materials. Operating temperatures are in the range of 1,370 to 2,200°C, and residence time is about two seconds. All organic material is converted to gases, primarily carbon monoxide and hydrogen (68).

This technology has the potential to treat plastics waste not easily handled by other processes, such as ASR, medical wastes, wire and cable, residential plastics wastes, and PVC.

Other Activities. Battelle has a patent application on a gasification unit that pyrolyzes commingled wastestreams containing polyolefins, polystyrene, and PVC into a gaseous mix of 40% ethylene, 27% methane, 17% hydrogen, and other fractions (69).

A process for the recovery of hydrocholoric acid from the pyrolyzate of PVC has been developed by three engineering firms in Germany (70). PVC producers in that country are jointly planning to develop and pilot the process further.

A partial oxidation degradative extrusion process is under study at the Institute of Kunststoffverabeitung, Aachen, Germany (71). Co- and counter-rotating twin-screw extruders operating at 300–400° are used to decompose mixtures of polyolefins and unsorted plastics wastes containing PVC. Volatiles from PVC are removed upstream in the extruder, and oxygen is added downstream. Products are of low viscosity and should be suitable as a feedstock for refinery liquefaction or gasification.

Other current research activities are concerned with pyrolyzing polyurethanes (72), upgrading the quality of coke and graphite products of postconsumer plastic waste pyrolyzates, and converting chars from specific polymer pyrolyzates to value-added activated carbons (73,74).

Potential. The broad scope of active development programs in thermolysis is indicative of the positive potential for this technology. The capability to process mixed plastics waste untreatable by other means and to generate valued chemical and fuel products provides the impetus for future growth of the technology. The different processes available provide opportunities for recycling various plastics wastes, from simple depolymerization of individual polymers to complex decompositions of mixed materials. The greatest potential for future growth exists with the processes being developed to treat mixed materials, such as automotive-shredder residue, wire and cable, medical waste, postconsumer plastics waste tailings, and other nuisance plastics. Refinery processes also have potential, but these will be limited to treating postconsumer plastics waste and other wastes containing relatively high levels of hydrocarbons and low levels of contaminants. It is unlikely that wastes such as automotive-shredder residue, wire and cable, and medical plastics will ever be used as direct refinery feedstock; facilities to pretreat these materials for refinery operations will have to be installed.

The future growth of thermal decomposition may also be limited by other factors. Only a few of these are technically related (eg, feed-system designs, mechanical seals, material preparation, and product purification) and should have only limited effects. Plant size, site location, adequate raw material supply, raw material, and capital costs will have greater effects. However, the deciding factors that will determine if thermolysis becomes a major technology in the recycling of plastics waste will be regulatory legislation, public perception, and economics.

Energy and Environmental Aspects

As indicated earlier, most solvolytic process reactions in operation today are run in units integral with polymerization production units of chemical companies. A few are stand-alone units; none are grassroots plants constructed to treat postconsumer plastics waste. Eastman Chemical is considering such a unit for Columbia, South Carolina, as the current methanolysis repolymerization unit is operating at an aging facility in Rochester, New York (75).

Solvolysis units have become part of the polymerization facilities through plant optimization programs. Recycle streams are part of normal plant practices to obtain greater efficiencies and better economics of plant operations; recycle streams often include internal wastes or unreacted materials that are recovered, treated, and returned to the polymerization unit. Thus, scrap or waste plastics are treated solvolytically within the plants, and the resulting monomers or low molecular-weight components are used as feedstock for polymer preparation. The energy involved in these solvolytic processes depends on the particular reaction. For example, as indicated in Table 2, the PET methanolysis reaction is estimated to have a moderate energy requirement whereas glycolysis is considered low. The environmental considerations are similar, ie, dependent on the particular process involved. However, as these systems are plant-integrated, there are no significant negative or positive energy or environmental effects to be considered.

As in solvolysis, the energy and environmental implications of pyrolysis depend on the specifics of the process being used and the feedstream, eg, municipal solid waste, mixed-plastics waste, or singular-plastic waste. In general, pyrolytic processes are energy-intensive, but more energy is recovered through pyrolysis than via combustion, and useful materials are recovered as well. The gaseous fractions produced during pyrolysis are sufficient and high in energy content to sustain the processes. In a U.S. Bureau of Mines study (76), raw municipal solid waste containing 1.5% plastics, and processed waste containing mainly plastics waste, were pyrolyzed. The gases produced at lower temperatures had a higher heat value, whereas the high gas yields produced at increased temperatures resulted in a high total energy per ton of refuse. The pyrolysis requires approximately 2.11×10^9 J/t (2×10^6 Btu/t) of refuse at 900°C but more than enough gas with a calorific value of 8.43×10^9 J (8×10^6 Btu) is produced to sustain the process.

Pyrolyses are nonpolluting, as the systems are operated in closed units (77). Gases are used within the system or sold; liquids are condensed, further treated, and used directly as fuel or feedstock. Some of the solids formed are being utilized as fillers. However, solid residues do present some possible environmental complications. Currently, a number of research programs are directed toward beneficial use of such material. Overall, the environmental implications for pyrolyses are less than for incineration, but are not zero in any case.

BIBLIOGRAPHY

1. R. J. Ehrig, ed., *Plastics Recycling: Products and Processes*, Hanser Publishers, New York, 1992.
2. R. J. Ehrig, "Development and Commercialization of Tertiary Plastics Recycling," *Spectrum*, Decision Resources, Inc., July 31, 1992.
3. J. C. Randall and co-workers, "Chemical Recycling," *Modern Plastics*, 37–38 (mid-Nov. 1993).
4. *Modern Plastics*, 73 (Jan. 1994).
5. W. W. Irion, O. S. Neuwirth, "Oil, Oil Refining" *Ullmann's Encyclopedia of Industrial Chemistry* Vol. A18 5th ed. VCH Publishers, New York, 1991.
6. V. Hoffman and M. Gebauer, "Raw Materials Recycling: An Approach to the Reuse of Scrap Plastics," *Kunststoffe* **83,** 8–11 (1993).
7. W. Kaminsky, "Possibilities and Limits of Pyrolysis," *Makromol. Chem. Macromol. Symp.* **57,** 145–160 (May 1992).
8. J. Chowdhury, "Chementator," *Chemical Engineering*, 27 (Oct. 1992).
9. P. Mapleston, "Chemical Recycling Scales Up," *Modern Plastics*, 58–61 (Nov. 1993).
10. *Discussions of the American Society for Testing and Materials*, Pittsburgh, Penn., Nov. 1993.
11. J. Powell, "Thermal Plastics Processing: Is It Recycling?" *Resource Recycling, Recovered Plastics Supplement*, 52–55 (May 1993).
12. *Plastics Recycling Technology Course*, Plastics Institute of America, Washington, D.C., June 1993.
13. Ref. 1, pp. 233–259.
14. *Chemical and Engineering News*, 15 (Dec. 29, 1993).
15. U.S. Pat. 4,582,642 (Apr. 15, 1986), L. Crescentini and co-workers (to Allied Chemical Corp.).
16. U.S. Pat. 4,764,607 (Aug. 16, 1988), L. J. Balint and J. Greenberg (to Allied Chemical Corp.).
17. B. E. Gracon, "Monomers From Nylon Waste," *Conference Proceedings of the Plastics Institute of America*, June 1993, pp. 219–245.
18. *Plastics Business News*, 2 (Oct. 19, 1992).
19. *Plastics World*, 3 (Nov. 1992).
20. *European Chemical News*, 23 (June 6, 1993).
21. U.S. Pat. 5,169,870 (Dec. 8, 1992), T. F. Corbin and co-workers (to BASF Corp.).
22. M. M. Nir and co-workers, "Update on Plastics and the Environment: Progress and Trends," *Plastics Engineering*, 75–93 (Mar. 1993).
23. D. D. Cornell, "Depolymerization of PET for Food Packaging," *Conference Proceedings of the Society of Plastics Engineers*, June 1993, pp. 107–110.
24. U.S. Pat. 5,051,528 (Sept. 24, 1991), A. A. Naujokas and K. M. Ryan (to Eastman Kodak Co.).
25. *Advanced Composites* **8**(5), 5 (Sept.–Oct. 1993).
26. L. A. Dupont and V. P. Gupta, "Terephthalate Polyester Recycling: A Unique Method Yielding Value Added Products," *Conference Proceedings of the Society of Plastics Engineers*, May 1991, pp. 2139–2141.
27. L. A. Dupont and V. P. Gupta, "Degradative Transesterification of Terephthalate Polyesters to Obtain DOTP Plasticizer for Flexible PVC," *Davos Recycle '92*, Apr. 1992, pp. 3-1–3-15.
28. R. M. Oblath and E. N. Nowak, "A Protocol to Verify Contaminant Removal from Post Consumer PET—Goodyear's Experience," *Conference Proceedings of the Plastics Institute of America*, May 1992, pp. 191–204.
29. U.S. Pat. 5,223,544 (June 29, 1993), E. J. Burkett and R. S. Jenks (to Shell Oil Co.).
30. D. Gentis, "Glycolytic Recycle of PET," *Makromol Chem. Macromol. Symp.* **57,** 185–190 (May 1992).
31. R. J. Ehrig, *Tertiary Plastics Recycling: An Update*, Canadian Plastics Institute, Sept. 1993.
32. R. J. Ehric, *Proceedings of the "Tertiary Recycling / Polymer Decomposition," of the American Chemical Society*, Pittsburgh, Penn., Oct. 1993.
33. *European Plastics News*, 30 (Apr. 1993).
34. L. White, "PUR Projects Take Off," *Urethanes Technology*, (Feb/March 1993) p 28.
35. Ref 17, pp 247–276.
36. *Plastics Industry News* **39** (7), 108 (July 31, 1993).
37. Y. Imai and co-workers, "General Purpose Adhesives Prepared from Chemically Decomposed Waste Rigid Polyurethane Foams," *Proceedings of the Polyurethanes World Conference*, Oct. 1993, pp. 97–103.
38. *Ibid.*, 108–119, 218–223.
39. U.S. Pat. 4,885,407 (Dec. 5, 1989), D. W. Fox and E. N. Peters (to General Electric Co.).
40. J. R. Best, Acetal Recycling *Plastics Business News*, 3 (Dec. 6, 1993).
41. M. Berins, *Plastics Focus* (Oct. 19, 1992).
42. Ref. 1, pp. 171–185.
43. U.S. Pat. 5,216,149 (June 1, 1993), R. J. Evans and L. Chum (to Midwest Research Inst.).
44. F. N. Fagan, "Liquefaction of Plastics," *Conference Proceedings of the Plastics Institute of America*, May 1993, pp. 170–190.
45. G. W. Gunderson, "Plastics Recycling Through Pyrolysis: A Project Report," *Conference Proceedings of the Plastics Institute of America*, May 1992, pp. 159–169.
46. *Modern Plastics International* **23** (3), 22 (Mar. 1993).
47. Anon., BASF Pilots New Technology, *Chemical Week*, 19 (Sept. 29, 1993).
48. J. R. Best, "Chemical Recycling," *Plastics Business News* (Oct. 11, 1993).
49. C. Roy and co-workers, "Vacuum Pyrolysis of Automobile Shredder Residue," *Abstracts of the 43rd Canadian Chemical Engineering Conference*, Oct. 1993, p. 177.
50. T. Furukawa, "Recyclable Composite Unveiled," *American Metal Market*, 10 (June 10, 1992).
51. M. Heathcote, "Back to Basics for Polymers Recycling," *European Chemical News*, 27–28 (Feb. 15, 1993).
52. R. Holighaus, "From Plastic Wastes Back to Oil," *Conference Proceedings of the Plastics Institute of America*, May 1993, pp. 201–218.
53. *Chemical Week*, 22 (Aug. 19, 1992).
54. *Progress Report*, Consortium for Fossil Fuel Liquefaction Science, Dec. 1993.
55. J. R. Best, "Issues & Your Business," *Plastics Business News* (Aug. 30, 1993).
56. *Advanced Recycling Technology*, American Plastics Council, 1993.
57. *Plastic Focus* (Dec. 6, 1993).
58. R. Koppang and co-workers, "Pilot Plant Demonstration of

ASR Gasification," *Conference Proceedings of the 9th Annual ASM / ESD*, Nov. 1993, pp. 713–726.
59. U.S. Pat. 5,129,995 (July 14, 1992), K. B. Agarwal (to General Motors Corp.).
60. U.S. Pat. 5,198,018 (Mar. 30, 1993), K. B. Agarwal (to General Motors Corp.).
61. U.S. Pat. 5,244,490 (Sept. 14, 1993), K. B. Agarwal (to General Motors Corp.).
62. D. C. Stannard, "Feedstock and Energy Recovery," *Conference Proceedings 'DAVOS Recycle' 92*, Apr. 1992, pp. 1–14.
63. R. Perry. "Process Profiles," *Chemical Engineering*, 23 (Feb. 1993).
64. J. Sandersen, "Gasification of Mixed Plastics Waste," *Conference Proceedings of the Plastics Institute of America*, May 1993.
65. "Business Technology," *The New York Times* (June 16, 1993).
66. "Hot Solution," *The Economist* (July 10–16, 1993).
67. *Technical Brochures*, Molten Metal Technology, 1993.
68. A. Pierce, "Gasification of Heterogeneous Polymers Via Catalytic Extraction Processing," *Conference Proceedings of the 9th Annual ASM / ESD, Nov. 1993, pp. 727–731.*
69. M. A. Paisley, "Ethylene Recovery From Waste Plastics," *The Battelle Perspective* (1991).
70. *European Chemical News*, 57 (Sept. 27, 1993).
71. W. Michaeli and V. Lackner, "Chemical Recycling and Pretreating of Plastics Waste Into Industrial Gases and Oils by Degradative Extrusion," *Conference Proceedings of the Society of Plastics Engineers*, May 1994.
72. G. J. Wasilczyk and T. J. Cerobona, "Developing a Viable Polyurethane Waste Management Plan: A PURRC Update," *Proceedings of the Polyurethanes World Conference—1993*, Oct. 1993, pp. 196–204.
73. A. A. Merchant and M. A. Petrich, "Pyrolysis of Scrap Tires and Conversion of Chars to Activated Carbon," *AIChE Journal* (Jan. 25, 1993).
74. M. A. Petrich, "Production of Valuable Carbons by Pyrolysis of Waste Plastics and Rubber," *The Environmental Technology Exposition*, Chicago, Apr. 1991.
75. D. Loepp, "If demand comes, Eastman may build plant," *Plastics News*, 4 (Oct. 4, 1993).
76. *Investigation 7428 Report*, U.S. Bureau of Mines, 1970.
77. J. Leidner, *Plastics Waste,* Marcel Dekker, Inc., New York, 1981.

RECYCLING AND REUSE—ENERGY SAVINGS

SCOTT WHITTIER CHAPLIN
Rocky Mountain Institute
Snowmass, Colorado

Concern over energy use in recent years has led to the promotion of more efficient cars, light bulbs, showerheads, and many other technological and resource management innovations, including recycling. This chapter reviews some of the issues surrounding the potential energy savings from recycling and reuse and focuses on the most common materials found in municipal solid waste (MSW), or what is commonly referred to as "garbage and trash." These materials include paper and paperboard products, glass containers, steel and aluminum cans, and plastics.

The ultimate question for many is not simply finding out which package or product saves the most energy if recycled, but rather, which product or delivery system has the lowest overall environmental impact. In some situations, product reuse or material substitution may be more desirable than recycling. From an environmental perspective, energy savings are only one of several important considerations. In many cases there is a trade off between saving energy and reducing water and air pollution as well as solid waste and hazardous waste production. Analyzing these considerations for different materials using techniques such as life-cycle assessments has been controversial at best, and often misleading. The actual impacts to the environment, including energy savings, will vary greatly depending on diverse factors such as population density, available fuel sources, transportation infrastructure, and other variables. Therefore, generalizations regarding energy savings from recycling should not be used as the sole criteria for determining what options have the lowest environmental impacts.

Experience in industry has confirmed that significant energy savings are achievable. However, for certain materials such as plastics, the energy requirements to collect and process them can be greater than the energy savings that result from reduced manufacturing needs. The potential savings for the recycling of all materials are somewhat hard to state as improved technologies and collection techniques continually increase the efficiency with which materials can be reused and/or recycled. In addition, increased efficiency in the use of various materials, from reducing package weight and materials substitution, have lowered total energy requirements.

While the energy savings available from recycling are frequently touted, such praise often confuses measured savings at individual production facilities with the technical potential for savings, and with the system-wide savings that can be achieved. The often quoted 90 to 95% energy savings from using recycled aluminum, for instance, compares production using 100% scrap with production using only primary resources. In reality, however, there is a metal loss of approximately 10% in the recycling loop due to losses during the shredding and casting processes. This loss must be replaced by energy-intensive primary aluminum, which lowers the actual energy savings to approximately 70% (1).

Analysts are increasingly moving toward comparisons of the overall energy impact caused by a given package or product's *function*, and not simply the energy requirements per pound or the percentage of energy saved through recycling. This broader level of discussion is perhaps best illustrated by the debate in the beverage packaging industry over the relative merits of various packaging materials. The debate involves comparisons of energy impact per serving or gallon of product delivered to a customer using a variety of containers which are recycled and/or reused. Reuse has shown to have a much greater potential to save energy than recycling, even at high recycling rates. The latter portion of this article examines this function-based, or "end-use" approach.

See also ENERGY EFFICIENCY; ENVIRONMENTAL ECONOMICS; WASTE-TO-ENERGY; LIFE CYCLE ANALYSIS.

DEFINITIONS AND TERMINOLOGY

There is no widely accepted definition for the term "recycling." Different political regions have adopted definitions to suit their particular legislative needs, but these definitions are often not compatible. Incineration to recover energy, for example, is considered a form of recycling in some statutes, while in others it is not. From an environmental standpoint, the **recycling** of a material displaces the need for the extraction of **virgin,** or **primary,** resources. The recycling of old newspapers into new newspapers, for instance, displaces the need for harvesting trees. In practice, however, any secondary use of a material is often considered recycling, even if the material in the new product is of a lower quality and no significant amount of primary material extraction is avoided. A prime example of this is the remanufacturing of plastics from old milk jugs and soda bottles into plastic lumber and flower pots. The same level of petroleum products must be extracted to make new jugs and bottles, and markets for the new products must be found.

In the smelting or milling of paper, steel, and other materials there is often a significant amount of scrap that accumulates in the initial production process, which is referred to as **industrial** or **home scrap.** After initial formation, most materials are sent to a separate facility to be cut, shaped, rolled or finished. In this second stage, more scrap is generated, which is referred to as **prompt industrial scrap.** These two types of scrap are very clean and are usually reincorporated into furnaces or pulp mills in most industries. With the exception of the paper industry, use of such scrap is generally not considered recycling. In most areas, the criteria for recycling is the use of **post-consumer,** or **old scrap,** which includes products and packaging that have been sold to consumers and then collected for the purpose of remanufacturing into new products. In the past, most post-consumer scrap has been landfilled. Not all post-consumer scrap can be recycled. Failure to differentiate among these three types of scrap and the recyclability of certain types of scrap has often led to confusion among consumers over the labels "recycled" and "recyclable." **"Recycled"** generally refers to products or packaging that were manufactured using post-consumer scrap. **"Recyclable,"** on the other hand, simply means that it is technically possible to use a scrap material to create new products. In contrast to "recycled," the term **reusable** refers to products or packaging that can be utilized several times for the same function, without requiring remanufacturing, such as **refillable** boxes and bottles.

PAPER AND PAPERBOARD PRODUCT RECYCLING

Estimates of energy savings that can be obtained by recycling paper and paperboard products vary widely from negative to 74% (2). In general, most studies have concluded that the production of paper products (like newsprint, office paper, and tissue paper) using recycled fiber will lead to energy savings, while the production of paperboard or kraft products (such as liner board, construction board, and box board) will require additional energy inputs compared to production from primary materials. The one exception to this generalization is the production of corrugated cardboard from recycled fibers, which can lead to marginal energy savings. A study by the Environmental Protection Agency has cited energy savings of 57% for tissue paper, 22% for newsprint, 36% for office paper, 3% for corrugated cardboard. Also, increased energy requirements of 150% are estimated for linear board, 40% for box board, and 2% for construction board (3). These figures, however, only represent the increase or decrease of purchased fuels such as oil, gas, and electricity, and don't take into account the use of by-product scrap wood-fuel, which is the main fuel used in manufacturing kraft paper products from primary materials. In fact, the recycling of any paper product will save on *total* energy consumption, but the recycling of kraft papers, substitutes the use of fossil fuels and electricity for energy derived from the burning of bark, wood scraps, and unsalable residues.

Improvements in paper making technologies as well as the relocation of paper mills will increase the energy savings available from recycling. In the past, most mills have been large and located in forest areas, near raw wood supplies, cheap energy, and far from cities that might complain about adverse air and water quality impacts. Today, many new smaller mills, sometimes referred to as "minimills," are located near large cities where they are closer both to a supply of scrap paper and markets for products. Such scrap is generally cheap and purchased on a contractual basis. Although energy rates can be higher near cities, total energy consumption will be lower. Of greater concern to recyclers is the cost of fiber, which is generally the single largest operating cost (4). It is the reduced need for energy, not fiber costs, however, that can make some types of recycled paper production more economical. In the newsprint industry, for example, the raw material costs to produce 100% recycled newsprint are higher than the raw material costs of producing news print from primary pulp; however, the energy costs are over 40% lower. In all, the operating costs for producing newsprint with recycled fibers are roughly 20% lower (5).

GLASS CONTAINER RECYCLING

As with paper recycling, estimates for energy savings that can be had by recycling glass vary widely from 4% to 32% (6). Much of the energy used in glass container production stems from bottle formation processes and does not decrease if recycled glass, called cullet, is used. In glass furnaces, however, cullet requires only 1/2 to 2/3 of the energy to reach melting point compared to that required for the endothermic chemical reactions needed to transform primary raw materials—mainly silica sand, feldspar, limestone, and soda ash—into melted glass. The use of cullet also reduces the energy needed to mine and transport the primary materials, although these savings may be offset by the energy requirements of collecting, transporting, and beneficiating (removing contaminants) processes (7). Unlike the recycling of other materials, such as aluminum and paper, there is no material loss during the melting and remanufacturing of glass (although there are some losses during beneficiation). While glass furnaces can and have used up to 100% cullet, most manufacturers prefer to use levels between 30% to 70% in order to control product quality and aid in the removal of bubbles. At this level, energy savings are estimated to be 2–2.5% for each 10% addition of cullet (8). In addition to saving operational energy, the use of cullet prolongs furnace life and thus lowers the embodied energy consumption per container from the construction of the production facility. Energy savings from recycling glass can easily be lost, however, due to increased energy needs for transportation if the cullet is shipped more than a few hundred miles.

Glass offers several other advantages over other materials used for similar purposes. Raw materials for the production of glass are relatively abundant and cheap. The manufacturing of glass products, using primary or recycled materials, is relatively benign compared to the production of similar products made of aluminum or plastic, especially when the analysis examines both normal impacts of mining and production as well as the common accidents involving toxic chemicals for each industry (9). Improvements in glass container manufacturing technologies as well as improved recycling collection and beneficiation technologies will further add to the potential savings available from recycling glass.

ALUMINUM CAN RECYCLING

As mentioned earlier, the often quoted figure of 90% to 95% energy savings from using recycled aluminum is misleading because it compares production using 100% scrap with production using primary resources. Due to a metal loss of approximately 10% in the recycling loop during the shredding and casting processes, actual system-wide energy savings are probably closer to 70%. Although well over 50% of aluminum cans are recycled in the United States, the amount of aluminum in the remaining cans now dumped in landfills and along roadsides in the U.S. is greater than the total use of aluminum by all but seven nations and represents a tremendous energy loss (10).

Improvements to recycling systems, melting technologies, and aluminum can production may lead to additional energy savings; however, the use of aluminum for disposable products is likely to come under increased criticism from resource efficiency advocates. In addition, the cost of aluminum production would increase in the absence of energy subsidies for the industry. Throughout the world aluminum manufacturers pay only a fraction of the electric rates that their neighbors do. Electric rates paid by many U.S. aluminum manufacturers are approximately half of the national average rate (11). Similarly, in France, aluminum smelters are charged only 1.6 cents per kWh while other industries in that country are charged 6 cents per kWh. Also, resource efficiency advocates will likely push for changes in how aluminum is used. Substituting aluminum for other metals in automobiles, for example, will decrease vehicle weight and improve fuel efficiency. Such substitutions may offer an energy savings potential far beyond the savings available from recycling aluminum cans, even if recycling rates were to increase dramatically (12). Moreover, old aluminum automobile parts can also be recycled which will lead to significant energy savings.

PLASTIC PACKAGE RECYCLING

Plastics present a dilemma for many environmentalists due to a lack of energy savings and primary material displacement from recycling, and to the various hazardous substances used in plastics production. For some types of plastics these include benzene, para-xylene, styrene, as well as various heavy metals. The use of such chemicals has increased the need for hazardous waste disposal facilities and has led to significant numbers of hazardous waste spills. The recycling of plastic generally does not displace the need for primary materials because many post-consumer plastics cannot be remanufactured into the same product. While new products have been developed to utilize these materials, primary materials are still required to make the original products.

Although different plastic resins have very different chemical compositions, the question of energy savings from recycling is the same for most types of plastics. Over half of the energy embodied in many plastics is in the form of petroleum-based feedstocks. This embodied energy can be recovered to a greater or lesser degree depending on the level of contamination in the scrap. Remelting scrap plastics requires relatively little energy—1,000 Btu/lb compared to over 30,000 Btu/lb for primary production (including feedstock) (13). At this stage of the process energy savings are over 80% (14). The stages of plastics recycling that require significant energy inputs are the collection, separation, and beneficiation processes. Energy demand for these stages outweighs the savings incurred during remanufacturing. The bulkiness of scrap plastics before they are shredded or pelletized severely reduces the efficiency with which they can be collected. Once collected, plastics must be separated by resin type, which is performed mostly by hand. Contamination of scrap of one resin type by small amounts of another resin type greatly reduces its value and usefulness. A small amount of scrap polyvinyl chloride (PVC) in a load of polyethylene terephthalate (PET), for example, can cause an entire shipment to be rejected. The contamination problem is compounded because some containers are made of layers of PVC and PET which cannot be separated or recycled easily. The energy needed to collect scrap, even if it is rejected, must be included when calculating energy savings.

Another important factor is the suitability of plastic scrap for various products. While plastic lumber, flower pots, and other similar products can be made with 100% scrap plastic, the market for these products is limited and does not displace the need for the primary materials. Also, many products are limited with regard to the amount of scrap they can contain. Scrap plastics, from industrial, prompt industrial, and post-consumer sources is generally only usable for middle layers, especially in food product containers. Primary materials must be used to sandwich the scrap plastic. These problems are perhaps best illustrated in the use of PET, which is most commonly used for soft drink bottles. If 50% of all PET in bottles was recycled this process would only replace 4% of primary PET use. Recycling of high-density polyethylene (HDPE), commonly used for milk jugs, faces similar restrictions, with the highest recycled content use still less than 50% (15). True recycling occurs when post-consumer scrap material is used to manufacture products of the same or similar value, such as new bottles from old bottles. Such an accomplishment will not be easy for the plastics industry, whose current goal of 25% recycled content by 1995 is considered overly ambitious by many in the field.

Unlike technologies for the production of other materials, such as paper, glass, and steel, the technologies for producing plastics are relatively new and efficient. Increases in the energy savings from recycling plastics, if any, will come from improved collection and processing technologies for post-consumer scrap.

STEEL CAN RECYCLING

Commonly cited figures for energy savings available from recycling steel range from 47 to 74% (16). This range assumes 100% scrap use, which although possible, is not common for most products. Most steel products contain either 25% recycled content or 100% recycled content depending on how they are produced.

Use of 100% scrap is common in "minimills" which use electric arc furnaces (EAF) to produce low grade steel products such as reinforcing bar (rebar) to medium structural steel (17). Steel cans, however, are made with a higher grade steel that is produced using exothermic reactions in basic oxygen process (BOP) furnaces. This process relies primarily on the combustion of oxygen and excess carbon for heat production instead of fossil fuels or electricity. The use of scrap in such furnaces is usually limited for technical reasons to 25% to 30%. After sheets are produced from these furnaces, the fabrication of cans (cutting, shaping, and finishing) accounts for over 50% of the energy input to new cans. The energy needed for can fabrication is the same regardless of whether or not recycled materials are used and thus the potential energy savings from using scrap steel to make cans is approximately 33% (18).

Improvements in EAF and other minimill technologies are permitting the production of higher grade steel products using 100% scrap. In general EAFs are more energy efficient than BOP furnaces and have been replacing them rapidly over the past 20 years. These mills have an additional advantage over large primary mills in that, due in part to their small size, they are more easily located closer to large cities with steady scrap supplies. In the near future it is likely that minimills will be able to produce a wide variety of high grade steel products, including cans, which would significantly raise the potential energy savings available from recycling steel.

INCINERATION FOR ENERGY RECOVERY

In some countries incineration for energy recovery is considered a form of recycling. Due to toxic ash problems, air pollution, and high financial requirements, it has been considered inferior to recycling in the United States. On a ton for ton basis, the cost of building incineration capacity is roughly 10 times that of building recycling capacity (19). Incinerators in the United States face some of the strictest environmental standards in the world, yet are still plagued by problems resulting from the emission of dioxin, mercury and other airborne toxins as well as toxic ash disposal problems. In addition to creating new environmental problems, incineration represents a loss of potential energy savings. A study conducted by Argonne National Laboratory has estimated that 0.6 quad (1 quad = 1 quadrillion Btu) could be recovered by burning packaging materials found in municipal solid waste, but 1.5 quad could be saved if these materials were recycled (20).

In addition, the construction of incinerators has a perverse effect on the viability of recycling. Many of the materials with the highest Btu values, such as newspaper, cardboard, and some plastics, are also the easiest to recycle. The key to successfully operating an incinerator is to maintain a high Btu level in the fuel source. Once large financial investments are made in incinerators, they must be operated as near to capacity as possible to be financially successful. Such operation often requires the implementation of "flow control" laws, that mandate that materials in the waste stream, including paper and plastic must be sent to the incinerator, even if markets for recycling are available. Moreover, financial resources that could have been used to construct recycling capacity are tied up in incineration.

END-USE ANALYSIS AND REUSE

Much of the interest in recycling has come about because of the concern over the "garbage crisis" and the need to reduce landfill inputs. This approach has led to a focus on what to do with materials after they become waste rather than how to prevent waste in the first place. However, such an emphasis may not lead to the most efficient use of resources in the long term, because the need to use a particular material is not questioned. An end-use analysis, on the other hand, would focus on the function of a product and ways to maximize total materials use efficiency. The main function, or end-use, of packaging, for

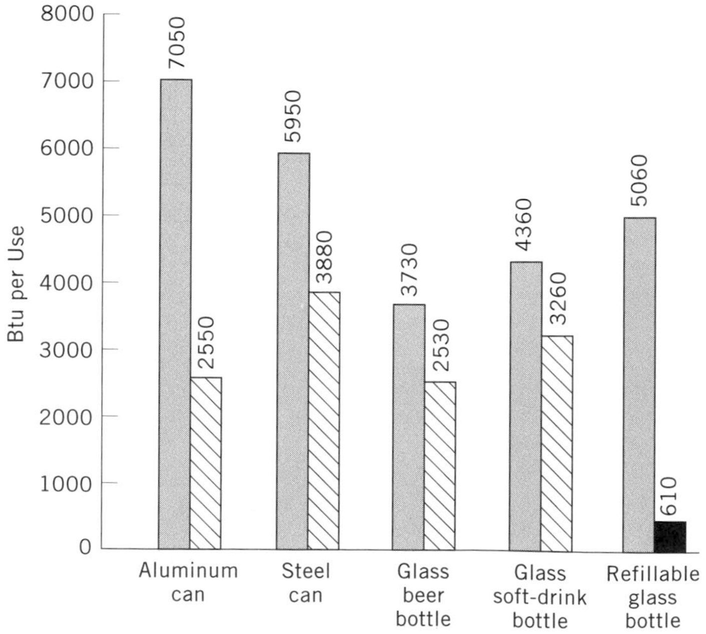

Figure 1. Energy consumption per use for 12-oz (.35 liter) beverage containers. ▢, Virgin Material, used once and discarded; ▨, Material recycled many times; ■, Container used ten times. Courtesy of Argonne National Laboratory.

instance, is to provide consumers with a convenient and attractive carrying device that will protect a product on its journey from the manufacturer to the consumer. The question then shifts from "how much energy is saved if a paper or plastic grocery bag is recycled?" to "what is the most energy efficient method of providing shoppers with a carrying device that is economical, marketable, and practical?" The results of an end-use analysis in this case might conclude that a reusable cloth bag provides the most energy efficient grocery carrying device. Conducting such analysis for other products could open the door to significant reductions in material flows through the economy while maintaining or improving quality of life (21). Energy savings from the reuse and/or elimination or substitution of a particular material may, for example, be far greater than those available from the highest achievable levels of recycling. What's more, most reusable products may be recycled when they become broken or worn.

A useful example of an end-use analysis comes from the beverage container industry. Rather than focusing solely on the percentage of recycled content, or energy savings from recycling any particular package type, com-

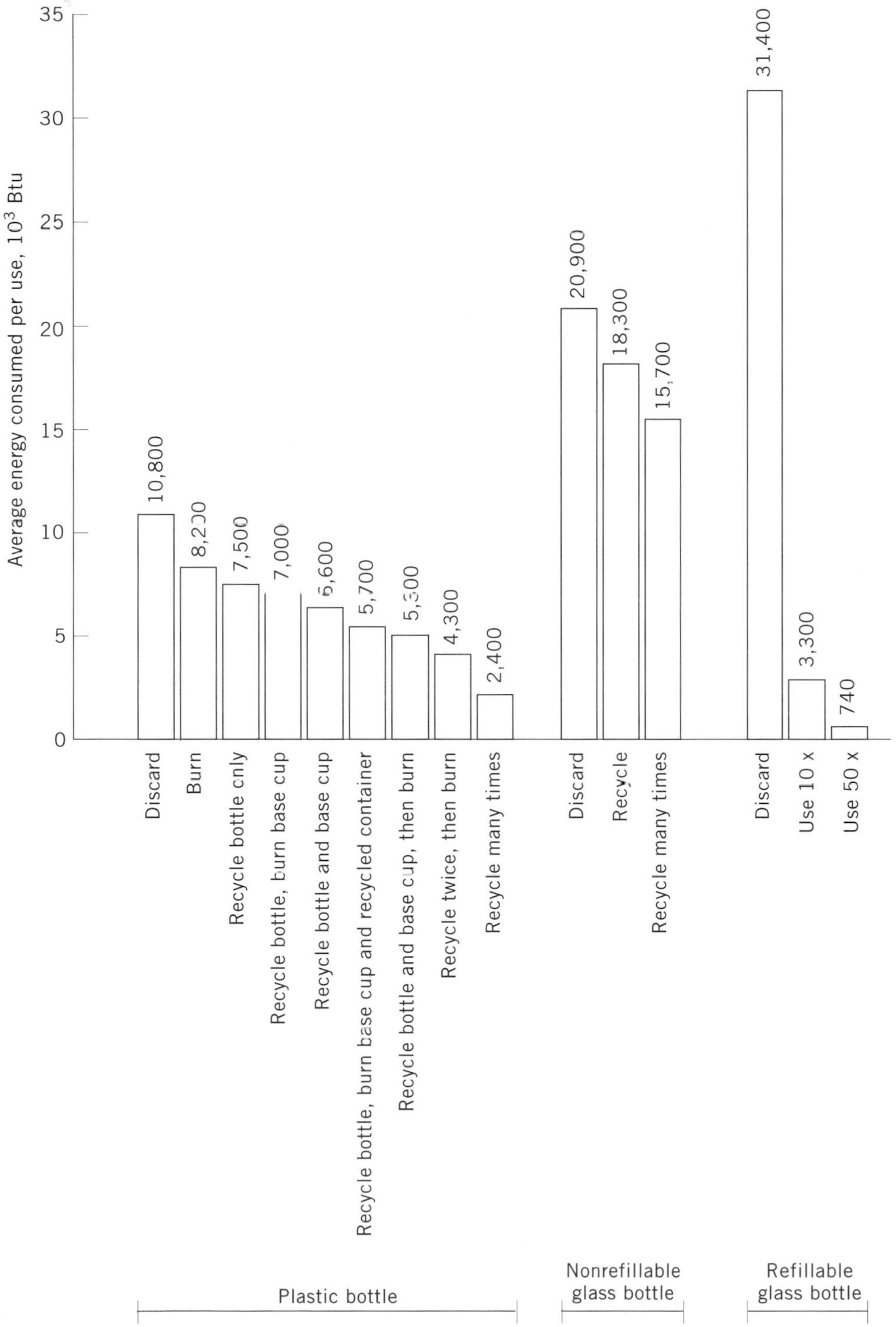

Figure 2. Energy consumption per use of 2-L bottles. Courtesy of Argonne National Laboratory.

parisons are being made based on the quantity of beverage delivered to consumers (the "end use"). In terms of delivering beverages, aluminum cans are approximately as efficient as single serving glass and plastic bottles, although they are less than half as efficient as refillable bottles (22). Increasing the use of refillables for beer and soft drinks in the United States to just 30% would save 40 trillion Btu annually, or 17% of the energy now used to deliver these beverages (23). Such information, along with information regarding other environmental, economic, and social impacts, allows planners to balance the various impacts and decide what option is best for their needs.

A 1981 study conducted by researchers at Argonne National Laboratory performed two end-use analyses of beverage packaging. First, they compared energy requirements for delivering 12 ounces of beverage in various containers using both primary materials and materials reused or recycled at achievable rates. The results are shown in Figure 1. Their second analysis compared energy requirements for delivering beverages in two liter bottles (see Figure 2). For both analyses, the use of refillable glass bottles proved to be the most energy-efficient beverage delivery container (24).

A more recent study compared the delivery of soft drinks in various containers, including 16 ounce PET bottles, 12 ounce aluminum cans, 16 ounce refillable glass bottles, and 16 ounce non-refillable glass bottles. Based on 1987 recycling rates, refillable glass bottles were found to consume less than half the energy per 1,000 gallons of beverage delivered compared to other single serving containers. Moreover, the use of refillable bottles was found to produce fewer other adverse environmental impacts such as atmospheric emissions, waterborne wastes, and solid wastes (25).

Several other studies conducted since 1974 have yielded similar results (26). The estimates for potential energy savings due to recycling and reuse in these studies have tended to decrease over time as all industries increased their operating efficiency. In the future, such technological advances in packaging as the refillable PET soft drink bottles used by Coke and Pepsi in Europe, or the refillable lexan milk jugs produced by General Electric, will add new dimensions to these types of analysis. The successful implementation of a refillable system will require some type of deposit system or other financial incentive to encourage consumers to return empties. Such systems have become common in most of Europe as well as parts of Canada, Mexico, and Asia (27).

CONCLUSION

Recycling is important because it will generally result in energy savings and a reduction in pollution and primary resource depletion that would have occurred in its absence. Reuse, for these same concerns, may be superior in many respects. By using an end-use analysis to focus on a product's function, rather than composition, decisionmakers are better able to address environmental concerns such as energy use.

BIBLIOGRAPHY

1. C. L. Kusik and C. B. Kenahan, *Energy Use Patterns for Metal Recycling,* Bureau of Mines Information Circular 8781, U.S. Department of Interior, Wash., D.C., 1978.
2. C. Pollock, *Worldwatch Paper 76-Mining Urban Wastes: The Potential for Recycling,* Worldwatch Institute, 1987, p. 22.
3. *Facing America's Trash: What Next for Municipal Solid Waste?,* Congress of the United States Office of Technology Assessment, Wash., D.C., 1989, pp. 143–144.
4. R. Montanari, "Small Mills Could Find a Niche As Communities Look for Markets," *Fiber Market News–Annual Paper Stock Issue, 1990,* pp. 29–35.
5. F. D. Iannazzi, "The Economics Are Right for U.S. Mills to Recycle Old Newspapers," *Resource Recycling,* 34–35 (July 1989).
6. C. Pollock, *Worldwatch Paper 76-Mining Urban Wastes: The Potential for Recycling,* Worldwatch Institute, 1987, p. 22, and S. Hassol and B. Richman, *Recycling,* Windstar Foundation, 1989, p. 48.
7. *Facing America's Trash: What Next for Municipal Solid Waste?,* Congress of the United States Office of Technology Assessment, Wash., D.C., 1989, pp. 151–152.
8. T. J. Roberts, *Glass Container Recycling Update with a Commitment to the Future,* Owens-Illinois, 1984.
9. J. Wolfe, "Aluminum and Glass," *The Southface Journal of Energy and Building Technology,* (1), 10–13 (1993) and H. S. Cole and K. A. Brown, *Advantage Glass,* Kenneth Brown and Associates, 1993.
10. J. E. Young, "Aluminum's Real Tab," *Worldwatch,* **5,** (2) p. 33 (Mar. 1992).
11. "Nonferrous Metals," *U.S. Industrial Outlook 1989,* U.S. Department of Commerce. pp. 18-1–18-2.
12. J. Wolfe, "Aluminum and Glass," *The Southface Journal of Energy and Building Technology,* (1), 10–13 (1993) and J. E. Young, "Aluminum's Real Tab," *Worldwatch,* **5** 26–33 (Mar. 1992).
13. L. L. Gaines, *Energy and Materials Use in the Production and Recycling of Consumer-Goods Packaging,* Argonne National Laboratory, 1981, p. 3.
14. C. Pollock, *Worldwatch Paper 76-Mining Urban Wastes: The Potential for Recycling,* Worldwatch Institute, 1987, pp. 30–31.
15. *Facing America's Trash: What Next for Municipal Solid Waste?,* Congress of the United States Office of Technology Assessment, Wash., D.C., 1989, p. 184.
16. C. Pollock, *Worldwatch Paper 76-Mining Urban Wastes: The Potential for Recycling,* Worldwatch Institute, 1987, p. 22.
17. J. P. Raymond, *Trends in the U.S. Ferrous Scrap Industry,* U.S. Department of Commerce, Wash., D.C., 1987, pp. 26–31.
18. L. L. Gaines, *Energy and Materials Use in the Production and Recycling of Consumer-Goods Packaging,* Argonne National Laboratory, 1981, p. 19.
19. J. E. Young, *Worldwatch Paper 101—Discarding the Throwaway Society,* Worldwatch Institute, 1991, p. 19.
20. L. L. Gaines, *Energy and Materials Use in the Production and Recycling of Consumer-Goods Packaging,* Argonne National Laboratory, 1981, p. 3.
21. B. Nordman, *The Case for Materials End-Use Analysis,* Energy and Environmental Division, Lawrence Berkeley Laboratory, 1992.

22. V. R. Sellers and J. D. Sellers, *Comparative Energy and Environmental Impacts For Soft Drink Delivery Systems,* Franklin Associates, Ltd., 1989, p. I–5.
23. G. Lomax, *Many Happy Returns: The Environmental and Economic Benefits of Refillable Beverage Packaging,* National Environmental Law Center, June, 1992; and *Testimony of Geoffrey Lomax, National Environmental Law Center, On the Energy Benefits of Increasing Beverage Container Reuse and Recycling as Outlined in S. 2335, "The National Beverage Container Reuse and Recycling Act of 1993," Before the Committee on Energy and Natural Resources, United States Senate,* Sept. 1992.
24. L. L. Gaines, *Energy and Materials Use in the Production and Recycling of Consumer-Goods Packaging,* Argonne National Laboratory, 1981; Subsequent research has shown that the energy consumption estimates used for steel cans in this figure were too high and that actual consumption is probably closer to that of non-refillable glass bottles.
25. V. R. Sellers and J. D. Sellers, *Comparative Energy and Environmental Impacts For Soft Drink Delivery Systems,* Franklin Associates, Ltd., 1989.
26. B. Nordman, *Regarding Potential Energy Impacts of S. 2335, "The National Beverage Container Reuse and Recycling Act of 1992,"* (written testimony for the hearing of the Energy and Natural Resources Committee, U.S. Senate, Sept. 17, 1992), Energy and Environment Division, Lawrence Berkeley Laboratory, 1992, pp. 2–5.
27. S. W. Chaplin, *Environmentally-Responsible Packaging Trends Around the World: Implications for the U.S.,* Rocky Mountain Institute, 1993; and, *Refillable Bottles: An Idea Whose Time Has Come, Again!,"* Container Recycling Institute, 1993.

REFORMULATED GASOLINE

UGO G. BOZZANO
UOP Corporation
Des Plaines, Illinois

HISTORICAL PERSPECTIVE

The Clean Air Act (CAA) Amendments of 1990 represent the greatest challenge to the U.S. refining industry since the mandated removal of lead from gasoline. As a result of these amendments, refinery changes will be necessary, and gasoline production costs will increase. The challenge for the refiner is to make investments that meet the immediate demands of product quality and offer long-term flexibility.

In the early 1970s, U.S. Legislation prompted the removal of lead additives from gasoline, and the industry responded. New gasolines were produced that efficiently replaced the lost octane by using new refining processes selected for flexibility and profitability. The environmental results were positive. Lead-free gasoline enabled automakers to install catalytic converters, and the combination of new fuel and new automotive hardware resulted in a 96% reduction in hydrocarbon and carbon monoxide (CO) emissions and a 76% reduction in nitrogen oxide (NO_x) emissions from automobile exhaust.

In the 1990s, new demands on gasoline refining have been legislated. Required changes in gasoline will reduce the remaining hydrocarbon emissions by no less than 15% starting in 1995, and further reductions will be mandated in the year 2000. Estimates from air-modeling projections of areas with the poorest air quality indicate that in 20 years, reducing hydrocarbon emissions will lower the ozone contribution from automobiles to only 5% to 9% of the total ground-level ozone (7).

Changes in gasoline production to meet the new specifications will be complex. In the context of any specific refinery, these production changes will be constrained by refining hardware, feedstock availability, and market factors.

The refining industry is aggressively pursuing the challenges that have been defined in the CAA Amendments. As in the case of lead phasedown, refiners must find long-term refinery-process alternatives to meet complex and difficult product specifications and still retain profitability. This article summarizes the issues facing the refining industry as defined in the CAA Amendments and suggests technical solutions to the challenges posed by the new legislation. The first part of this discussion addresses the broad requirements of the legislation as they affect gasoline production. The second part discusses the various ways for refiners to meet the new requirements. The third part addresses the cost of making reformulated gasoline.

See also CLEAN AIR ACT OF 1990, MOBILE SOURCES; AIR POLLUTION AUTOMOBILE; AIR POLLUTION: AUTOMOBILE, TOXIC EMISSIONS; AUTOMOBILE EMISSIONS, CONTROL; AUTOMOTIVE ENGINES; PETROLEUM REFINING.

THE 1990 CAA AMENDMENTS

Amendments to the CAA establish broad enforcement powers within the U.S. Environmental Protection Agency (EPA) that affect almost every aspect of American industry. Titles cover mobile sources, air toxics, permitting, stationary sources, and many other issues. In the section on mobile sources, the intent of Congress was to specifically legislate fuel changes that will result in reduced emissions and to legislate separately a reduction in automobile emissions that is independent of the fuel, An overview of the timing of critical events in gasoline regulation is shown in Figure 1.

Demands on Gasoline

The CAA Amendments define two categories of regulated gasoline: oxygenated gasoline (OxyFuel) and reformulated gasoline (RFG). RFG must be sold starting January 1, 1995 in the nine extreme severe ozone nonattainment areas and in opt-in areas. RFG must comply with the following specifications

- General Requirements:
  - $\leq$ 1.0 vol % benzene
  - $\geq$ 2.0 wt % oxygen
  - No heavy metals, unless waived
- Performance Standards: reduction of VOC, toxic, and NO_x emissions

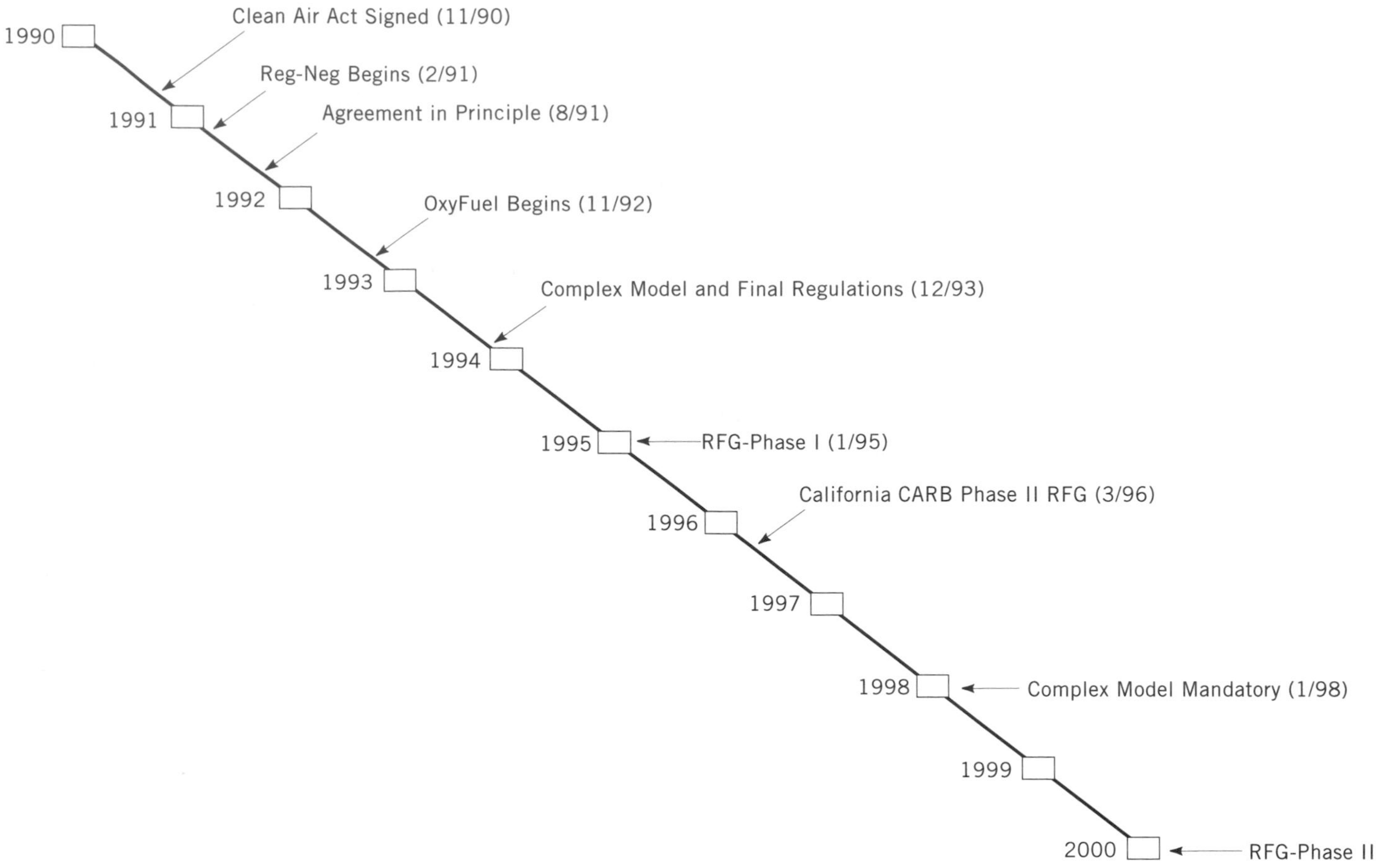

Figure 1. Gasoline reformulation time line.

OxyFuel, which is gasoline with an oxygen content of 2.7 wt %, is specified for CO nonattainment areas during the winter months when CO emissions are high. Beginning November 1, 1992, OxyFuel was supplied to 37 out of 41 cities that at the time of the legislation were in violation of federal CO air-quality standards.

Reformulated gasoline refers to a more extensive change in gasoline properties that reduces VOC (volatile organic compounds) emissions and toxic emissions in nine extreme and severe ozone nonattainment areas of the United States. A list of all the ozone nonattainment areas is shown in Table 1. A provision in the legislation allows other high-ozone areas the option to use RFG (to opt-in), but only after the EPA is petitioned by the state governor. At the end of 1993, 12 governors had petitioned the EPA to allow their high-ozone areas to opt into the RFG program.

The standards for RFG are complex, and the remainder of this section clarifies the definition as stated in the CAA Amendments. The legislation defines RFG in terms of general requirements and peformance standards. General requirements call for a minimum of 2.0 wt % oxygen, a maximum of 1.0 vol % benzene, and no heavy metals. The RFG must also meet emission-reduction standards (performance standards), which have been legislated in terms of ozone-forming VOC emissions and emissions of toxic compounds. Toxics are defined as benzene, 1,3-butadiene, formaldehyde, acetaldehyde, and polycyclic organic matter (POM). Exhaust NO_x emissions must not increase as a result of reformulation.

The standards specify measuring emissions from a baseline vehicle using a defined fuel (baseline gasoline). The baseline vehicle is a group of vehicles representative of the 1990 model year with regard to combustion and emission-control technology. Baseline gasoline is the average of all U.S. gasoline marketed in 1990 as determined by an industry survey. The baseline is different in the summer and winter (Table 2).

Emissions from the entire vehicle consist of the sum of exhaust, evaporative, running-loss, and refueling-loss emissions, all of which are based on mass measurement. Baseline emissions are the emissions from the baseline vehicle using the baseline gasoline. The emissions generated from a particular gasoline are compared with the emissions from the baseline gasoline to determine the emission potential of that gasoline compared with the baseline gasoline.

Performance Standards. Defined in the CAA Amendments as the reduction in the emission of ozone-forming VOC during the high-ozone season and the emission of toxics during the entire year. Both emissions are measured on a mass basis. For 1995, the emission reductions for per-liter compliance under the Simple Model are set at 15%. The specifications for VOC and NO_x reduction are met under the Simple Model by compliance with the regu-

Table 1. Ozone Nonattainment Areas[a]

Extreme Areas

Los Angeles, Calif.

Severe Areas

Baltimore, Md.
Chicago, Ill., Ind., Wisc.
Houston, Texas
Hartford, Conn.
Milwaukee, Wisc.
New York, New York, New Jersey, Conn.
Philadelphia, Penn. New Jersey, Del., Md.
San Diego, Calif.

Serious Areas

Atlanta, Ga.
Bakersfield, Calif.
Baton Rouge, La.
Beaumont, Texas
Boston, Mass., N.H.
El Paso, Texas
Fresno, Calif.
Huntington, W. Va., Ky., Ohio
Muskegon, Mich.
Parkersburg, W. Va., Ohio
Portsmouth, N.H., Maine
Providence, R.I.
Sacramento, Calif.
Sheboygan, Wisc.
Springfield, Mass.
Washington, D.C., Md., Va.

Moderate Areas

Atlantic City, N.J.
Bowling Green, Ky.
Charleston, W. Va.
Charlotte, N.C., S.C.
Cincinnati, Ohio, Ky., Ind.
Cleveland, Ohio
Dallas, Texas
Dayton-Springfield, Ohio
Detroit, Mich.
Grand Rapids, Mich.
Greensboro, N.C.
Jefferson Co., N.Y.
Kewaunee Co., Wisc.
Knox Co., Maine
Louisville, Ky., Ind.
Memphis, Tenn., Ark., Miss.
Miami, Fla.
Modesto, Calif.
Nashville, Tenn.
Pittsburgh, Pa.
Portland, Maine
Raleigh-Durham, N.C.
Reading, Pa.
Richmond, Va.
Salt Lake City, Utah
San Francisco–Oakland–San Jose, Calif.
Santa Barbara, Calif.
Smyth Co., Va.
St. Louis, Mo., Ill.
Toledo, Ohio
Visalia, Calif.
Worcester, Mass.

Marginal Areas

Albany, N.Y.
Allentown, Pa., N.J.
Altoona, Pa.
Birmingham, Ala.
Buffalo, N.Y.
Canton, Ohio
Columbus, Ohio
Erie, Pa.
Essex Co., N.Y.
Evansville, Ind., Ky.
Fayetteville, N.C.
Greenbrier Co., W. Va.
Greenville-Spartanburg, S.C.
Hancock Co., Maine
Harrisburg, Pa.
Indianapolis, Ind.
Johnson Co.-Kingsport-Bristol, Tenn.
Johnstown, Pa.
Kansas City, Mo., Kans.
Knoxville, Tenn.
Lake Charles, La
Lancaster, Pa.
Lewiston, Maine
Lexington, Ky.
Lincoln Co., Maine
Manchester, N.H.
Montgomery, Ala.
Norfolk, Va.
Owensboro, Ky.
Paducah, Ky.
Poughkeepsie, N.Y.
Scranton, Pa.
South Bend, Ind.
Stockton, Calif.
Sussex Co., Del.
Tampa, Fla.
Waldo Co., Maine
York, Pa.
Youngstown, Ohio; Sharon, Pa.

[a] EPA may redesignate areas to another classification if their design value is within 50% of the other classification.

Table 2. 1990 U.S. Industry Average Gasoline

Gasoline Parameters	Summer	Winter
RVP, kPa	59.99	79.29
Benzene, vol %	1.53	1.64
Total aromatics, vol %	32.0	26.4
Olefins, wt %	9.2	11.9
Oxygen, wt %	0.0	0.0
T50, °C	103.3	93.3
T90, °C	165.5	167.2
Sulfur, ppm	339	338

lations on oxygen, RVP, and benzene. Under the Complex Model, VOC reduction for per-liter compliance is a minimum of 15.6% during Phase I (1995 to 1999) and 25.9% under Phase II of the program (2000+). Toxic emission reduction is 15% under Phase I of the program and 20% under Phase II. The NO_x emission reduction from the 1990 baseline gasoline must be a minimum of 0% under Phase I and a minimum of 5.5% under Phase II.

In a broad sense, the concept of performance can be considered as the percent reduction of VOC, toxic, and NO_x emissions relative to emissions from baseline vehicles operating on baseline gasoline. Performance is attributed to the improved fuel properties of the reformulated fuel over the baseline gasoline.

The emission reduction required by the CAA Amendments will take place in two phases. Phase I begins in 1995. For Phase 1 reformulation, the law specifies a minimum of 15% reduction in VOC emissions from the baseline during the high-ozone season and a minimum of 15% reduction in toxics emissions from the baseline during the entire year. In addition to limits on VOC and toxics, the CAA Amendments state that NO_x emissions cannot increase.

For Phase II emission reduction, which starts in the year 2000, the law establishes that the reduction in VOC and air toxic emissions be increased from 15 to 25% compared with the baseline. If such a reduction proves technically unfeasible, emission reduction shall be at least 20%.

Another aspect of the CAA legislation covers conventional gasoline, which is the gasoline sold outside of areas where RFG is required. These regulations are designed to prevent refiners from meeting the RFG specifications by simply removing undesirable components from RFG and dumping them into the conventional gasoline. These so-called antidumping provisions of the CAA are designed to ensure that the conventional gasoline produced after 1995 is no more polluting than the gasoline produced in 1990.

The legislation only sets a framework for regulating gasoline. Congress did not attempt to work out the intricate details necessary to implement the law within the industry. Instead, the EPA was charged with writing the specific rules that define, clarify, and quantify the meaning of the 1990 CAA Amendments.

EPA Rule Making

Negotiated rule making, called Reg-Neg, was chosen by the EPA to allow the principals who are affected by the legislation to negotiate the details of the way the law is enforced in everyday business. As part of the Reg-Neg process, the EPA set up an umbrella committee composed of representatives from the automobile and oil industries, environmental groups, pipeline and distribution organizations, the oxygenate industry, state lobbyists, and representatives from the EPA. This diverse representation indicates the many interests affected by the legislation and the difficulty in drafting rules that represent fairly the interests of all concerned parties.

The CAA Amendments require a formal fuel certification procedure for demonstrating emission performance. Fuel certification can be achieved by actually testing the fuel in a fleet of vehicles representative of the baseline vehicle or by emission modeling. Emission modeling is the mathematical relationship between fuel composition and emission performance that is derived by extensive testing of a number of fuel compositions in automobiles representing the baseline vehicle. Emission modeling establishes a means of predicting the emission performance of a gasoline on the basis of composition and reduces the need to test each batch of gasoline in laborious fleet testing. The gasoline parameters used in modeling emissions are Reid vapor pressure (RVP); distillation characteristics (T50 and T90, which are the temperatures at which 50% and 90% of the gasoline is evaporated, or E200 and E300, the percent evaporated at 93.3 and 148.8°C, respectively); and benzene, total aromatics, sulfur, olefins, and oxygenate contents.

Agreement in principle between the various parties involved in Reg-Neg was established on August 16, 1991. In addition to fuel-certification features, the agreement in principle contains provisions regarding:

- *RFG enforcement for oxygen, benzene, toxics, and VOC:* The enforcement is divided into per-liter compliance and compliance on averaging. The provisions for averaging are more stringent than for per-liter compliance.
- *Antidumping:* Antidumping provisions are specific for gasoline sold in non-RFG markets. The intention of these rules is to ensure that components removed from RFG are not added (or dumped) into conventional gasoline.
- *Test Tolerances:* Tolerances are given for the various analytical tests.

The compromise solution was actively negotiated to closure by the EPA. However, significant negotiations with the Reg-Neg parties were required to gain agreement on the detailed rules drafted by the EPA. All existing data bases relating vehicle emission to gasoline quality were scrutinized. New experimental programs were statistically designed and carried out by diverse organizations to encompass the large majority of gasolines and vehicles presently in use in the United States. Principal contributors to the generation of new data were the EPA, other federal and state agencies, private companies, and the Auto/Oil Air Quality Improvement Research Program. The last group is a cooperative effort between the three largest American vehicle manufacturers, the American Petroleum Institute, and other oil company associations.

Data generation was followed by similar efforts in statistical analysis to merge data bases, validate data, and generate significant correlations. At the same time, such correlations had to be kept fairly simple so that gasoline quality can be monitored through the entire production and distribution chain: from refinery production of gasoline intermediates to blending them into finished gasoline, transporting the finished gasoline through the complex pipeline system in use in the United States, distributing the gasoline to gas stations, and delivering it to the fuel tank of the vehicle. The development of the mathematical model, which relates gasoline composition to emissions, took two and a half years. During that time, emission correlations went through seven iterations. The final model and the final regulations were issued by the EPA in December 1993 (2).

As part of Reg-Neg, the refining industry was given four years to work with the emission model and make the necessary changes in the refineries to comply with the required emission reduction. By August 1991, everyone realized that the final version of the emission model would clearly not be available in time to give the refiners a four-year lead time and still begin the RFG program on January 1, 1995. As a result, an interim emission model was developed in 1991. This interim model, called the Simple Model, was based on a fewer number of gasoline parameters and was to be used until the more complete Complex Model was made mandatory, four years after its development.

Simple Model

Under the Simple Model, the accepted RFG for the period January 1, 1995, to December 31, 1997, will be manufactured to the following specifications:

- Benzene: 1.0 vol % maximum.
- Oxygen: 2.0 wt % minimum.
- Summer RVP: 49.64 kPa (7.2 psi) in VOC Control Region 1 (approximately equivalent to ASTM volatility Class A and B areas) or 55.85 kPa (8.1 psi) in VOC Control Region 2 (ASTM volatility Class C areas) (Fig. 2).
- Toxic emission reduction: at least 15%.
- Sulfur, T90 or E300, and olefins: capped at the average values in each refiner's 1990 gasoline (T50 or E200 is not controlled under the Simple Model.)

The Simple Model equations for calculating emissions are shown later. The Simple Model can be used only in the ranges shown below

Fuel Parameter	Valid Range
RVP, kPa	45.51–62
Oxygen, wt %	0–3.5
Benzene, vol %	0–2.5
Aromatics, vol %	10–45

VOC Reduction. The two specific changes in gasoline composition required to reduce VOC emissions under the Simple Model are the reduction in RVP and the addition of oxygenates. Under the Simple Model, reducing RVP to 55.85 kPa (8.1 psi) in VOC Control Region 2 or 49.64 kPa (7.2 psi) in VOC Control Region 1 and adding a minimum of 2.0 wt % oxygen are sufficient to meet the VOC-reduction requirements of the CAA Amendments. Under the

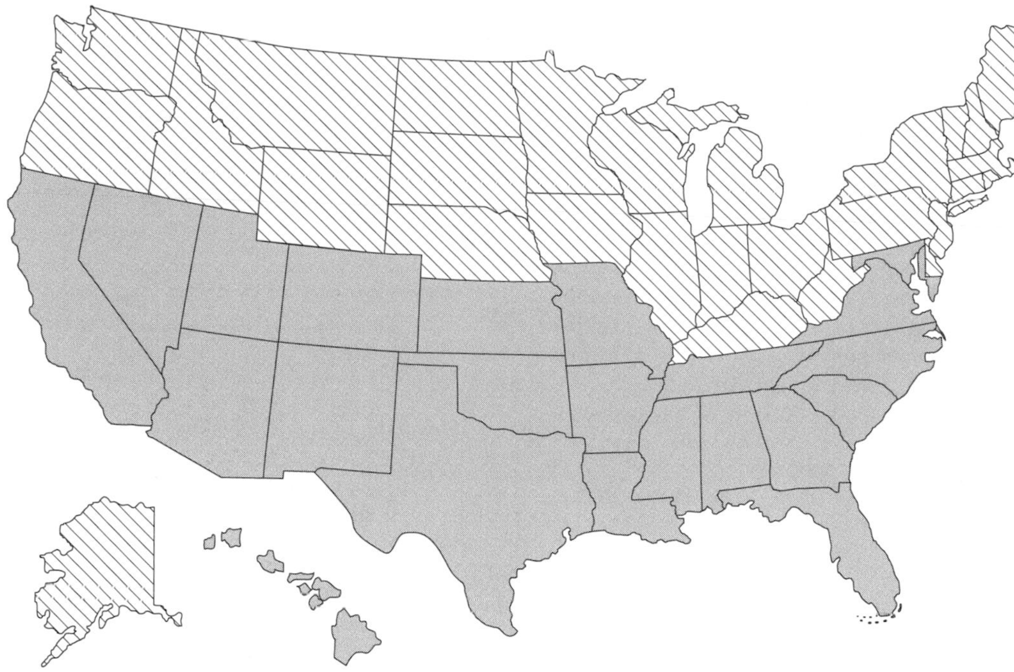

Figure 2. RVP control regions. M898A, Region 2 (Class C). M898D, Region 1 (Class B).

Simple Model, the T90 or E300 and the levels of sulfur and olefins in the refiner's RFG may not exceed the levels in the refiner's 1990 average gasoline.

NO_x Reduction. Under the Simple Model, NO_x emissions do not have to be calculated because all RFG is assumed to meet the specification of no increase in NO_x emissions. The EPA stipulates that adding any oxygenate up to the level of 2.7 wt % satisfies the requirements for no NO_x increase in Simple Model RFG during the summer ozone season (May 1 to September 15). In the winter months, oxygenates can be added up to 3.5 wt % oxygen in Simple Model RFG without increasing NO_x emissions.

Toxics Reduction. The EPA developed a specific set of equations to calculate toxic emissions under the Simple Model. The fuel properties used in these equations are benzene, total aromatics, RVP, and oxygen content. These equations calculate the concentrations of the following toxic compounds: benzene, formaldehyde, acetaldehyde, 1,3-butadiene, and POM.

An indication of the comparative effect of the fuel variables on toxics for RFG in VOC Control Region 2 is shown in Table 3. The first column represents summer baseline toxic emissions. At 48.4 mg/km (30.1 (mg/mi), benzene is 74% of the total toxics emitted. The next three columns of the table represent the cumulative effects of the changes required to meet the specifications under the Simple Model: RVP reduction from 59.99–55.85 kPa (8.7–8.1 psi), oxygenate addition to 2.0 wt %, benzene reduction from 1.53 to 1.0 vol %. The last column shows the estimated impact on aromatics content that resulted from adding oxygen to a typical gasoline pool: aromatics are reduced from 32 to 26.3 vol %. The impact of benzene reduction is clearly the greatest. In fact, in this example, which is for Volatility Control Region 2, reducing benzene is 17 times more effective in reducing toxics than is reducing aromatics. (Because of the effect of RVP on toxics, reducing benzene in VOC Control Region 1 is only 15 times more effective than reducing aromatics). In other words,

Table 3. Simple Model Toxic Emissions for VOC Control Region 2 RFG

	Baseline Gasoline, RVP = 59.99 kPa	Reduce RVP 0.6 to 2.0 wt %	Add Oxygen to 2.0 wt %	Reduce Benzene 0.53 to 1.0 vol %	Reduce Aromatic s to 26.3 vol %
Exhaust toxics, mg/km:					
Benzene	48.43	48.43	43.87	41.01	36.83
Acetaldehyde	6.37	6.37	6.10	6.10	6.10
Formaldehyde	8.98	8.98	10.67	10.67	10.67
1,3-Butadiene	3.97	3.97	3.60	3.60	3.60
Polycyclics	2.25	2.25	2.04	2.04	2.04
Subtotal	69.99	69.99	66.27	63.41	59.24
Nonexhaust benzene, mg/km	13.89	11.71	10.72	7.02	7.02
Total toxics, mg/km	83.88	81.71	76.99	70.41	66.24
Toxic reduction, %	Base	2.6	8.2	16.1	21.0

according to the Simple Model, reducing benzene an additional 0.1 vol % enables the refiner to increase aromatics by 1.5 to 1.7 vol % without increasing toxics.

Conventional Gasoline. Under the Simple Model, a refiner's conventional gasoline can have no greater exhaust emissions of benzene than the refiner's 1990 gasoline. Sulfur, olefins, and T90 (or E300) are capped at 1.25 times the levels in the refiner's 1990 gasoline.

Simple Model Calculations. The Simple Model establishes three different sets of equations to calculate toxics reduction (ToxRed) from the baseline gasoline for summer VOC Control Regions 1 and 2 and for winter.

The general form of the toxics reduction equation is

$$\text{ToxRed} = f(\text{ExhBen}, \text{EvpBen}, \text{RunBen}, \text{RefBen}, \text{Form}, \text{Acet}, \text{Buta}, \text{POM})$$

where

$$\text{ExhBen} = f(\text{Ben}, \text{Aro}, \text{Oxy})$$

$$\text{EvpBen} = f(\text{Ben}, \text{MTBE}, \text{RVP}, \text{RVP}^2)$$

$$\text{RunBen} = f(\text{Ben}, \text{MTBE}, \text{RVP}, \text{RVP}^2)$$

$$\text{RefBen} = f(\text{Ben}, \text{MTBE}, \text{RVP})$$

$$\text{Form} = f(\text{Oxy}, \text{MTBE}, \text{TAME}, \text{EtOH}, \text{ETBE}, \text{ETAE})$$

$$\text{Acet} = f(\text{Oxy}, \text{MTBE}, \text{TAME}, \text{EtOH}, \text{ETBE}, \text{ETAE})$$

$$\text{POM} = f(\text{Oxy})$$

Ben and Aro are the benzene and aromatics contents of the test fuel in vol-%. RVP is the Reid vapor pressure in Pa (psi). Some of the emission terms depend on the total oxygen content of the gasoline, and some terms depend on the type of oxygenate used. Oxy is the total oxygen content of the gasoline in wt %. The terms MTBE, TAME, EtOH, ETBE, and ETAE represent the wt % oxygen from MTBE, TAME, ethanol, ETBE, and ETAE. If oxygenates other than these are used in the test fuel, they should be treated as follows: higher alcohols as ethanol, methyl ethers as MTBE, and ethyl and higher ethers as ETBE.

ExhBen, EvpBen, RunBen, and RefBen are the exhaust, evaporative, running-loss, and refueling-loss emissions of benzene. Form, Acet, and POM are the emissions of formaldehyde, acetaldehyde, and POM. All emissions are in milligrams per km.

Complex Model

The Complex Model includes a complete range of fuel parameters: RVP; E200; E300; sulfur; olefins; benzene; aromatics; and oxygen content, which is differentiated by type of oxygenate. (In the Complex Model, T50 and T90 are not used because they do not blend volumetrically; instead, E200 and E300 are used.) Three separate sets of equations are used to calculate VOC, toxic, and NO_x emissions. The Complex Model can be used only over the range of gasoline composition shown in Table 4. This range is more restrictive for RFG than for conventional gasoline. Certification of gasoline that falls outside the valid range may require testing in automobiles. The use of this model becomes mandatory on January 1, 1998.

Table 4. Valid Range for Complex Model

Gasoline Parameters	RFG	Conventional Gasoline
Oxygen, wt %	0–3.7	0–3.7
Sulfur, ppm	0–500	0–1000
RVP, kPa	44.13–68.95	44.13–75.84
E200, vol %	30–70	30–70
E300, vol %	70–100	70–100
Aromatics, vol %	0–50	0–55
Olefins, vol %	0–25	0–30
Benzene, vol %	0–2.0	0–4.9

Under the Complex Model, the calculation of emission reduction for a given RFG is based on the deviations of the RFG properties from baseline gasoline. Each refinery will have a different set of deviations from the baseline gasoline and a different ability to change each of the critical fuel specifications. Thus, the problems facing each refiner will be different, and so each refiner's solution will be different.

One important difference between the Simple and Complex models is the difference in gasoline parameters used as a basis. The Simple Model merely limits a refinery's gasoline to its own 1990 average properties. The Complex Model is based on the 1990 U.S. industry-average (baseline) gasoline.

For example, refiners who process a high-sulfur crude may have a 1990 average sulfur content of 500 ppm in their gasoline pool instead of the 1990 industry baseline of 339 ppm. During the period when the Simple Model applies, the sulfur level in RFG is capped at the 500-ppm level, which cannot be exceeded. Reducing RVP and benzene and adding oxygenates are sufficient to certify this gasoline. However, after January 1998, when the Complex Model is used to certify emissions performance, this same gasoline may not pass certification because of its high sulfur level. Sulfur must then be reduced, or a compensating change in some other parameter must be made. In other words, the debit in performance resulting from the high sulfur must be either eliminated by lowering sulfur or offset by an improvement over baseline gasoline in some other fuel parameter.

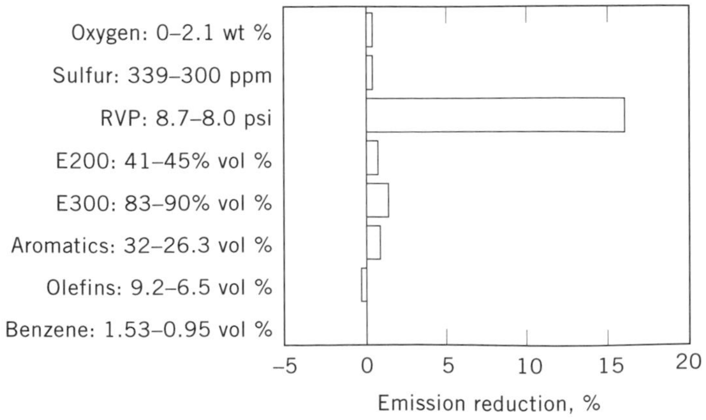

Figure 3. VOC reduction using complex model.

VOC Reduction. Under the Complex Model, VOC emissions must be reduced by 15.6% relative to the 1990 industry baseline. The impact on VOC emissions from changing one fuel parameter at a time from the 1990 industry baseline is shown in Figure 3. For example, VOC emissions are reduced by 0.3% by adding 2.1 wt % oxygen in the form of methyl tertiary butyl ether (MTBE) to 1990 baseline gasoline. Reducing RVP from the 1990 baseline to 55.16 kPa (8.0 psi) reduces VOC by 16.0%. Clearly, the greatest impact on VOC reduction is due to the reduction of RVP. None of the other fuel parameters has as great an impact.

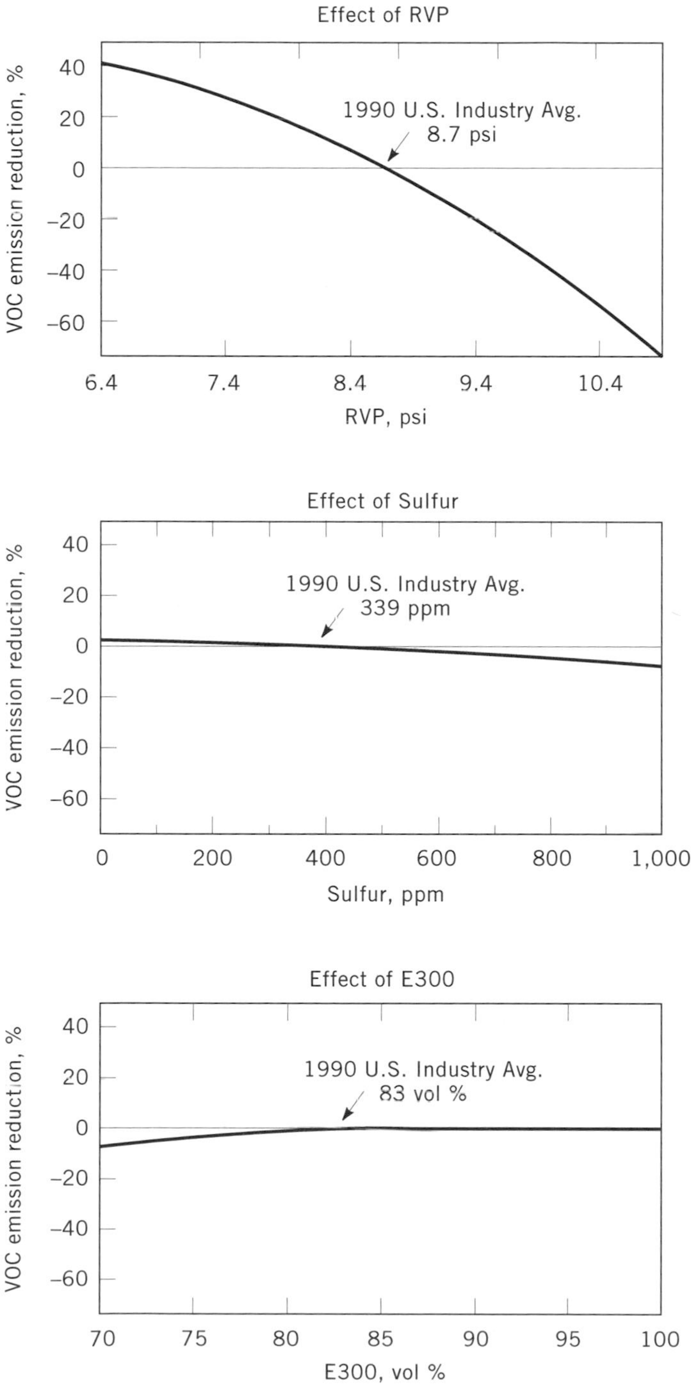

Figure 4. Impact on VOC emission reduction.

Because the Complex Model is nonlinear, the impact of each parameter on emissions can change depending on its value. Figure 4 shows that reducing RVP has a somewhat lower impact on reducing VOC emissions at either end of its valid range. Increasing E300 has the greatest effect on reducing VOC emissions at low E300, and reducing sulfur has about the same effect on VOC emissions over the entire sulfur range.

NO_x Reduction. Under the Complex Model, NO_x emissions are calculated using the same fuel parameters that are used to monitor VOC emissions. Under Phase I of the RFG program, NO_x emissions cannot increase relative to the 1990 industry baseline. Figure 5 shows the impact on NO_x emissions from changing each gasoline parameter individually from the 1990 industry baseline. The results show that the most influential gasoline parameters NO_x emissions are olefins, sulfur, and aromatics. Increasing E200 actually causes an increase in NO_x emissions. At concentrations above 12 vol %, olefins have a big impact on NO_x emissions (Fig. 6). Reducing sulfur below 300 ppm has a strong impact on NO_x emissions, and aromatics levels between 10 and 30 vol % also have an impact.

As discussed earlier, the impact of each parameter on emissions may depend on its value. As shown in Figure 6, the impact of olefins on NO_x emissions is clearly nonlinear. Olefins begin to play a significant role on NO_x emissions above a concentration of 12 vol % olefins. Reducing olefins below the baseline value of 9.2 vol % has little benefit. Sulfur also shows a nonlinear relationship with NO_x emissions; the impact of reducing sulfur increases at low concentrations. The effect of aromatics on NO_x emissions is also nonlinear, but the benefit from reducing aromatics is greatest between 30 and 15 vol %.

Toxics Reduction. The Complex Model calls for a 15% reduction in toxics emissions relative to the 1990 industry baseline. Figure 7 shows the impact of changing each gasoline parameter from the 1990 industry baseline. The results indicate that toxics are most affected by benzene and aromatics, and toxic emissions are affected somewhat less by RVP and sulfur. Olefins, E200, and E300 have a much smaller impact on toxics emissions. The effect on

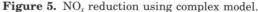

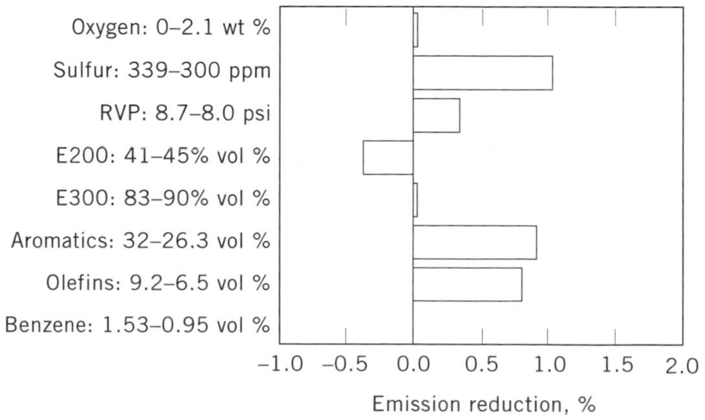

Figure 5. NO_x reduction using complex model.

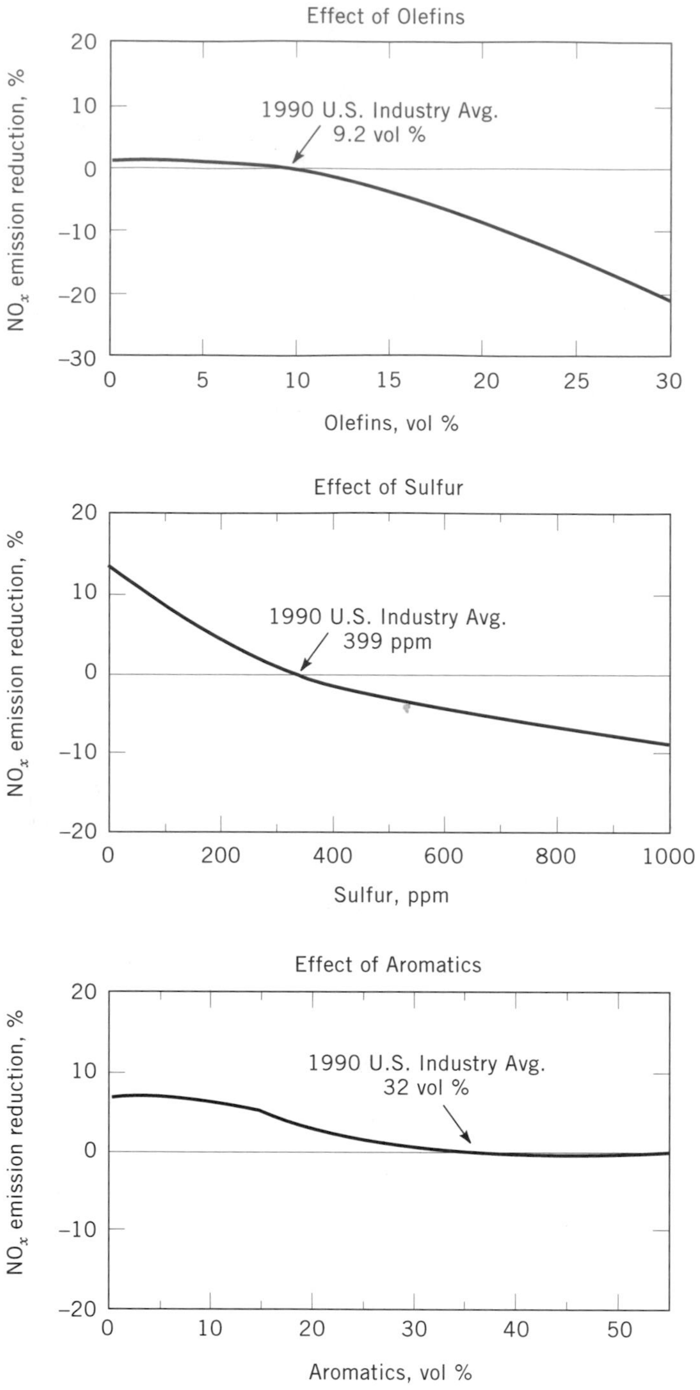

Figure 6. Impact on NO_x emission reduction.

toxic emissions from varying benzene, aromatics, or sulfur over their valid ranges can be seen in Figure 8.

Conventional Gasoline. With the advent of the Complex Model, all the gasoline parameters come into play when evaluating conventional gasoline, and individual gasoline properties are not capped. Under the Complex Model, the exhaust toxic emissions from conventional gasoline can be no greater than those from the refinery's 1990 gasoline.

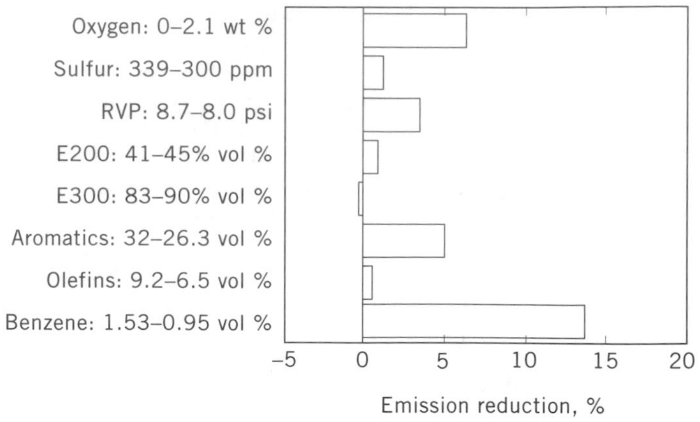

Figure 7. Toxics reduction using complex model.

Early Use of the Complex Model

During the time the Simple Model is in effect, a refiner may choose the Complex Model to certify its emission reductions. Using the Complex Model in the interim period from 1995 to 1997 gives the refiner flexibility in meeting emission reductions. However, if a refiner chooses to use the Complex Model during this time, the RFG must have no worse VOC, NO_x, or toxic-emission performance than would be predicted by the Complex Model for a gasoline meeting the specifications of the Simple Model fuel, that is, the gasoline must meet the minimum 2.0 wt % oxygen, maximum 1.0 vol % benzene, and maximum RVP of 55.85 kPa (8.1 psi) in VOC Control Region 2 or 49.64 kPa (7.2 psi) in VOC Control Region 1 and must have that refiner's 1990 levels of sulfur, olefins, and T90 (or E300). Finally, gasoline certified under early use of the Complex Model cannot be mixed with gasoline certified by the Simple Model.

Phase II Reformulation

Under the Phase II RFG requirements that go into effect in the year 2000, VOC emission reduction from the 1990 industry baseline increases from 15.6% to 25.9%, and toxic emission reduction increases from 15% to 20%. In addition, refiners will be required to reduce NO_x emissions by 5.5% relative to the 1990 industry baseline. The equations used to certify Phase II gasoline are somewhat different than those used for Phase I gasoline.

Renewable Oxygenate Requirement

In December 1993, the EPA proposed a rule specifying that 30% of the RFG oxygen requirements on a year-round basis must come from renewable sources (3). The primary renewable oxygenate will be ethanol or ethanol derivatives, such as ethyl *tertiary* butyl ether (ETBE) and ethyl *tertiary* amyl ether (ETAE). Because of concern over its high vapor pressure, neat ethanol is not classified as a renewable oxygenate if used in summertime RFG. As a result, ETBE and ETAE will be the main sources of renewable oxygenates for summertime RFG, and neat ethanol will be used primarily in wintertime RFG. Although

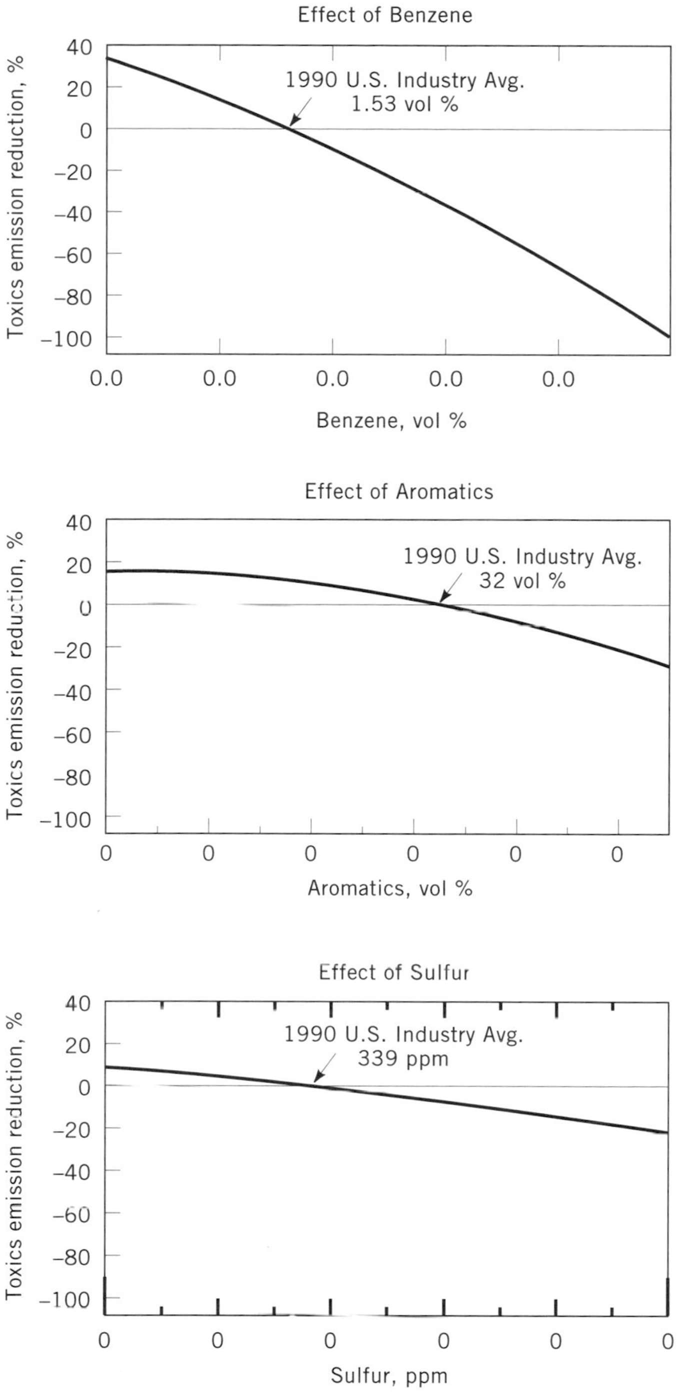

Figure 8. Impact on toxics emission reduction.

methanol can be made from renewable sources, it is not expected to be a significant source of renewable oxygenates for the foreseeable future.

As part of the renewable oxygenate mandate, refiners using more than the required level of 30% renewable oxygenates can sell the credits to other refiners. Because the renewable oxygenate requirement can be met on a year-round average, the refiner using an excess of renewable oxygenates in one season can bank the credits for use in another season. In this manner, the refiner could use ethanol to meet his entire oxygenate requirements for winter RFG and then use the credits to meet the renewable oxygen requirements for summer RFG, thereby avoiding physical blending of renewable oxygenates into summer RFG.

Complex Model Calculations. The Complex Model consists of different systems of equations to calculate VOC NO_x, and toxics emissions for gasoline designated for VOC Control Regions 1 and 2 in the summertime and wintertime. Different sets of equations have been established for Phase I (1995 to 1999) and (Phase II (2000 and beyond).

Weightings for normal and high emitters, the latter being defined as vehicles that have malfunctioning combustion and emission-control systems, are provided separately for VOC or toxics and NO_x and are different for Phases I and II. Baselines are provided for exhaust VOC, NO_x, and individual toxics, for nonexhaust VOC and benzene, and for total emissions.

VOC Emissions

Total VOC emissions (VOC) are the sum of exhaust (Exh) and nonexhaust (Nonexh) and are determined through different sets of equations for summer and winter in VOC Control Regions 1 and 2:

$$\mathrm{VOC} = \mathrm{VOCExh} + \mathrm{VOCNonexh}$$

The percent change in VOC exhaust emissions of the test fuel (TF) from the fuel baseline depends on the weightings for normal (N) and high (H) emitters and is an exponential function of the fuel properties:

$$\mathrm{VOCExh} = \mathrm{VOCExh}_{\mathrm{Baseline}} \mathrm{x} \; [1 + \%\mathrm{VOCChange(TF)}/100]$$

$$\%\mathrm{VOCChange(TF)} = \mathrm{Nx}\frac{\exp \nu_{\mathrm{N}}(\mathrm{TF})}{\exp \nu_{\mathrm{N}}(\mathrm{Base})} + \mathrm{Hx}\frac{\exp \nu_{\mathrm{H}}(\mathrm{TF})}{\exp \nu_{\mathrm{H}}(\mathrm{Base})}$$

$$\mathrm{V_i} = \mathrm{V_i}(\mathrm{Oxy, Sul, RVP, E200, E300, Aro, Ole, nonlinear\ terms})$$

Aro and Ole are aromatic and olefin contents of test fuel in vol %, Oxy and Sul are the oxygen in wt % and sulfur in wt-ppm, RVP is the Reid vapor pressure in psi, and E200 and E300 are the vol % evaporated at 93.3 and 148.9°C.

Nonexhaust VOC emissions are in turn the sum of diurnal (Diu), hot-soak (Hot), running (Run), and refueling (Ref) losses.

$$\mathrm{VOCNonexh} = \mathrm{VOCDiu} + \mathrm{VOCHot} + \mathrm{VOCRun} + \mathrm{VOCRef} = f(\mathrm{RVP}, \mathrm{RVP}^2)$$

Nonexhaust VOC emissions are set to zero under winter conditions.

NO_x Emissions. NO_x emissions are solely exhaust products. These emissions are determined through different sets of equations for the summer and winter seasons. The form of the equations is similar to the one for VOCExh, where v_i are replaced by n_i:

$$NOx = NOx_{Baseline} x$$
$$[1 + \%NOxChange(TF)/100]$$

$$\%NOxChange(TF) = Nx \frac{\exp n_N(TF)}{\exp n_N(Base)} + Hx \frac{\exp n_H(TF)}{\exp n_H(Base)}$$

$$n_i = n_i(Oxy, Sul, RVP, E200, E300, Aro, Ole, nonlinear\ terms)$$

Toxics Emissions. Toxics emissions (Tox) are the sum of exhaust (Exh) and nonexhaust (Nonexh) benzene and of formaldehyde, acetaldehyde, and polycyclic organic matter, the latter three being exclusively exhaust-type emissions:

$$Tox = ExhBen + NonexhBen + Form + Acet + Buta + POM$$

The form of the equations for all the exhaust terms is similar to the one for VOCExh, where v_i is replaced by similar b_i, f_i, a_i, bdi, and p_i equations for benzene, formaldehyde, acetaldehyde, butadiene, and polycyclic, respectively. As an example, the equations for benzene will be

$$ExhBen = ExhBen_{Baseline} x$$
$$[1 + \%BenChange(TF)/100]$$

$$\%BenChange(TF) = Nx \frac{\exp b_N(TF)}{\exp b_N(Base)} + Hx \frac{\exp b_H(TF)}{\exp b_H(Base)}$$

$$b_i = b_i(Oxy, Sul, Aro, E200, E300, Ole, Ben).$$

Nonexhaust benzene emissions are the sum of diurnal (Diu), hot-soak (Hot), running (Run), and refueling (Ref) losses. Nonexhaust benzene emissions are different for summer VOC Control Regions 1 and 2 and are set to zero under winter conditions.

$$NonexhBen = BenDiu + BenHot + BenRun + BenRef = f(Ben, MTBE, RVP, RVP^2)$$

IMPACT ON THE REFINERY

The new regulations are having a substantial impact on the petroleum refining industry. Significant refinery changes are required to meet the specification for RFG and OxyFuel because a large proportion of the U.S. gasoline production will be affected. During wintertime, the proportion of gasoline subject to the new legislation stemming from the 1990 CAA Amendments is greatest because both RFG and OxyFuel are required. As of 1993 the production of OxyFuel has been estimated to be approximately 12% of the annual U.S. gasoline pool (4). Estimates in 1993 of the quantity of RFG that will be required to satisfy the demands of the nine ozone nonattainment areas and the opt-in areas range from 30% to 37% of the annual U.S. gasoline production (3,5). Additional quantities of RFG will be needed to overcome distribution inefficiencies.

The impact of the legislation on each refinery will vary. First, the mix of OxyFuel, RFG, and conventional gasoline for each refiner will depend on what distribution network is available and what oxygenates are selected. Multirefinery companies may find new reasons to trade products between their refineries to meet the specifications. In addition, groups of refiners with synergistic capabilities may have new motivations for trading gasoline components or increasing leverage on issues of distribution, such as alcohol handling.

Reformulation has a major impact on the material balance of a refinery because it affects the quantity of gasoline produced from a fixed quantity of crude oil, the amount of hydrogen available to the refinery, and the amount of butanes allowed in the pool. A brief example shown in Table 5 illustrates the impact of reformulation on gasoline production in the interim period from 1995 to 1998. This example assumes a generic refinery processing a constant amount of Alaskan North Slope crude oil. Crude conversion in this refinery comes from fluid catalytic cracking (FCC) and coking units. Before reformulation, the benzene is 1.5 vol % and the RVP is 62.06 kPa (9.0 psi) (VOC Control Region 2 gasoline). As a result of reformulation, which includes the addition of purchased MTBE, the volume of gasoline produced increases by about 14%. The 14% increase is a result of two factors: the volume of oxygenates and the increased yield of reformate. The addition of MTBE supplies octane to the pool, and the result is a reduction in the reformer severity needed to meet the pool octane requirements. Reduced reformer severity increases liquid reformate yield but reduces the hydrogen that is available for hydrotreating and hydrocracking operations. In this example, the reformer severity was reduced by about seven research octane numbers (RONs), and hydrogen production decreased by more than 20%. Benzene reduction was achieved by saturating the light reformate. As a result of RVP reduction, about 10% of the butane was removed from the pool.

Many different solutions can be implemented to meet the requirements of the CAA Amendments. Selecting the best long-range solution requires a detailed analysis of each refinery and an evaluation of all production options

Table 5. Effect of Reformulation on an FCC-Coking Refinery Operation

	Before RFG	After RFG
Gasoline pool properties		
RVP, kPa	62.055	55.85
Benzene, vol %	1.5	1.0
Oxygen, wt %		2.0
Change in refinery operation		
Gasoline production, %	Base	+14
Pool butanes, %	Base	−10
Reformer severity, RON	Base	−7
Reformer hydrogen production, %	Base	−22

within a total refining strategy. The remainder of this article discusses the process options that address the following issues:

- Meeting the Simple Model requirements
  * RVP reduction
  * Benzene reduction
  * Oxygenate production in the refinery
  * Toxics reduction
- Meeting the Complex Model requirements
  * Olefins reduction
  * Sulfur reduction
  * E300 increase (T90 decrease)
  * E200 increase (T50 decrease)
  * Aromatics reduction
- Meeting hydrogen demand

MEETING THE SIMPLE MODEL REQUIREMENTS

Reducing the RVP to the appropriate level—55.85 kPa (8.1 psi) in VOC Control Region 2 or 49.64 kPa (7.2 psi) in VOC Control Region 1—and adding 2.0 wt % oxygen are sufficient to meet the targeted VOC reduction in the interim period from January 1, 1995, to December 31, 1997, when the Simple Model applies. In addition to reducing RVP and adding oxygenates, benzene must be reduced to less than 1.0 vol % and total toxics emissions must be reduced by at least 15% from the baseline. Toxic emission reduction during this period is calculated using the Simple Model equations.

RVP Reduction

The reduction in RVP is of critical importance to the refinery in meeting new gasoline emission regulations. Under the Simple Model, the 49.64-kPa (7.2-psi) RVP required in the VOC Control Region 1 summertime RFG is near that of a butane-free gasoline pool. Achieving the 55.85-kPa (8.1-psi) RVP required for summertime RFG in VOC Control Region 2 may also be difficult.

Butane is the first gasoline component to be targeted in reducing RVP. However, butane removal from the gasoline pool eliminates a versatile, high-octane component and forces the refiner to find another use for this material. The next components to be considered when reducing RVP are the C_5 hydrocarbons. These C_5 hydrocarbons can be the isopentanes and normal pentanes found in straight-run naphtha, isomerate, and light hydrocrackate; they can also be the various amylene isomers present in light FCC and coker gasolines.

The relative concentration of C_4 and C_5 hydrocarbons in the gasoline components depends on the refinery-process configuration and the performance of product stabilizers. Gasoline RVP can be affected in two ways: by removing butane and C_5s through fractionation or reaction, and by reducing butane and C_5 formation.

Butane Fractionation. Most refining units have undergone extensive revamps over their lifetimes. As a result, product stabilizers may be operating significantly above design and may not meet the typical design parameter of 1 mol % butane in the bottom fraction. Higher butane levels in the stabilized gasoline components make controlling the RVP of the finished gasoline more difficult. Some of the problems associated with operating beyond the design capacity of the column can be overcome with advanced fractionation and heat-exchange technology, which may substantially improve the refinery's ability to debutanize products and minimize the RVP in gasoline components.

Alternatives to Butane Fractionation. When butane control is difficult or costly, the removal of C_5 hydrocarbons may be preferred to meet the RVP specifications in the finished gasoline. The isopentanes and normal pentanes can be fractionated from the gasoline components and sold as petrochemical feeds. Reduced vapor pressure will also result from reacting the high-vapor-pressure amylenes into low-vapor-pressure gasoline components. The production of tertiary amyl methyl ether (TAME) from isoamylene reduces not only the requirement for purchased oxygenates but also reduces pool RVP as a result of removing isoamylene from the gasoline pool. The alkylation of C_5s in the TAME raffinate further removes amylenes from the pool and reduces RVP. The formation of butane and C_5s in the reformer can be reduced by lowering the operating pressure, which lowers the RVP of the reformate and thereby lowers the RVP of the gasoline pool. The cost and difficulty of RVP compliance will vary from refiner to refiner.

Benzene Reduction

The major source of benzene in the gasoline pool is catalytic reformate, which typically contributes 50% to 75% of the benzene in the gasoline pool; only minor contributions of benzene come from light hydrocrackate and light FCC naphtha (Fig. 9). Therefore, the best target for benzene reduction is reformate.

Two basic approaches can reduce net benzene production: minimize its formation by removing the precursors charged to the reformer, or fractionate a benzene-rich light reformate stream for subsequent benzene conversion

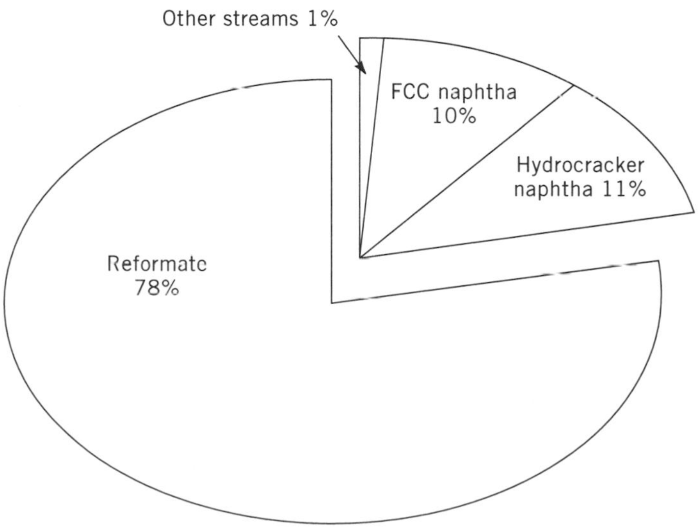

Figure 9. Benzene contribution in gasoline pool.

or extraction. Preventing the formation of benzene rather than converting it after it is formed makes good engineering sense. However, in certain cases, practical and economic constraints may limit the ability of refiners to implement this strategy.

Precursor Removal. The first approach in removing benzene consists of lifting the C_6 cyclics (methylcyclopentane, cyclohexane, and benzene) overhead in the naphtha splitter and including these materials in the feed to the isomerization unit, or if octane is not required, blending this C_5–C_6 stream directly into the gasoline pool (Figure 10). In this case, the naphtha splitter is designed to produce a bottom stream, containing 0.5% C_6 cyclics, that is fed to the reformer. Relatively sharp fractionation is required to minimize the C_7 carryover into the isomerization unit feed and yet achieve the desired C_6 cyclic recovery. Typically, a 40-tray tower operating at a reflux-to-feed ratio of 0.5 to 0.7 is required. The benzene in the splitter overhead can be saturated in a suitably designed (or revamped) isomerization unit.

Isomerization provides a net increase in the octane of the light straight-run (LSR) naphtha stream and at the same eliminates benzene. Alternatively, if octane is not required, the LSR naphtha stream can be processed in a simple benzene-saturation unit.

Light Reformate Processing. Light reformate processing is a more direct benzene control alternative than benzene precursor removal. When choosing between light reformate processing or precursor removal for benzene reduction, refiners must consider three variables: the level of benzene allowed in the regulated pool, the benzene-producing tendency of the reformer, and the overall refinery flow scheme and crude source. If reducing benzene below 1% is necessary, then precursor removal may not be adequate. If the reformer operating pressure is high, the production of benzene can occur even after precursor removal because the rate of dealkylation of heavier hydrocarbons to C_6 cyclics followed by conversion to benzene increases with pressure. Light reformate processing may also be necessary if the refinery flow scheme is heavily dependent

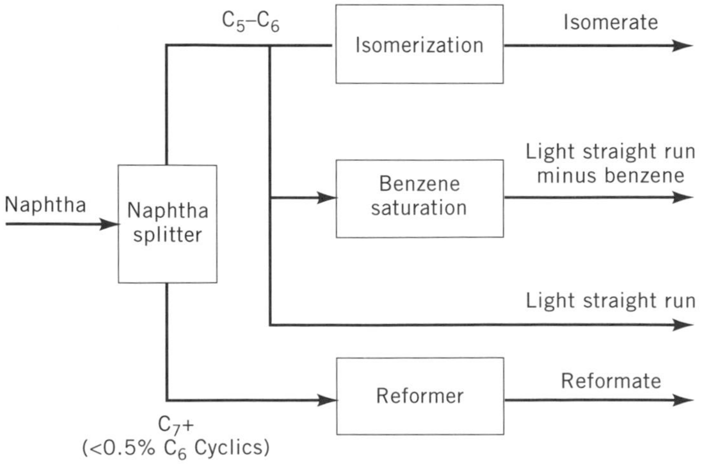

Figure 10. Prefractionation of C_6 cyclics.

on the reformer as a gasoline pool component, if the crude used is highly cyclic, or if the reformer feed comes from a number of other process units in which accurate control of benzene precursors is difficult. Four options for benzene removal from light reformate can be considered: saturation, saturation with isomerization, alkylation with propylene, or extraction (Fig. 11).

The saturation of benzene to cyclohexane can be accomplished in a simple stand-alone saturation unit or in a suitably designed isomerization unit. The benzene saturator selectively hydrogenates benzene to cyclohexane and minimizes side reactions and consequent hydrogen consumption. However, the conversion of benzene to C_6 naphthenes (cyclohexane and methyl cyclopentane) entails an octane loss of 30 RON and typically reduces the octane of the light reformate stream by five to six octane numbers.

The isomerization unit supplements benzene saturation with isomerization of the unconverted C_6 paraffins present in the light reformate. The resulting net gain in octane is approximately 5 RON in a once-through isomerization unit. Using isomerization to reduce benzene consumes slightiy more hydrogen relative to simple benzene saturation and produces a product with a slightly higher RVP.

A viable option for benzene removal from gasoline produced in an FCC refinery is to alkylate the benzene in the light reformate fraction with propylene. In addition to producing an octane comparable to that from isomerization, benzene alkylation is the only benzene conversion route that does not require hydrogen. The economics of benzene alkylation strategy depend on the value and availability of propylene.

Finally, the extraction option is a clear choice for refiners who are in the petrochemical market for benzene. The economics of installing new benzene extraction and fractionation facilities indicate a potentially attractive return based on the inside-battery-limits investment. However, the cost of offsites burdened by health, safety, and environmental regulations must be accurately assessed. In general, the extraction option appears to be limited to those refiners who are already in the merchant benzene business or who can enter the market through a close association with an existing benzene-processing facility.

Meeting the Oxygenate Requirements

The legislated oxygen requirements of OxyFuels and RFG create a significant oxygenate demand. Refiners are looking to outside merchant suppliers to provide MTBE, ethanol, and ETBE. A number of MTBE complexes based on butane dehydrogenation have been built or are currently under construction to provide MTBE to refiners. Grain-derived ethanol can also provide a portion of the total oxygenates required by the CAA Amendments. However, the availability of merchant oxygenates in the face of RFG demand raises a significant uncertainty for refiners. In addition, because of their high-octane content, the use of oxygenates in gasoline reduces the need for octane from the reformer and leads to lower hydrogen production. Finally, the addition of oxygenates to gasoline upsets the refinery product slate because gasoline production is in-

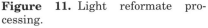

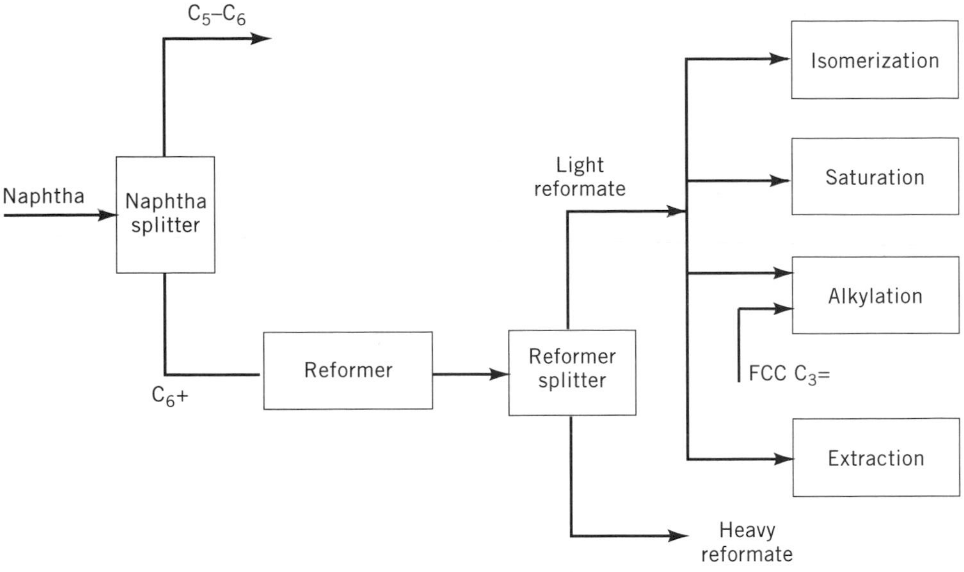

Figure 11. Light reformate processing.

creased by as much as 15%, depending on the type and source of oxygenate used.

Naturally, the refining industry would like to regain control of their entire gasoline pool and be independent of outside suppliers for a critical component of their primary product. Internal production of oxygenates is possible by using available FCC olefins, and many refineries have chosen to install MTBE units. The production of ETBE from C_4 olefins and TAME or ETAE from C_5 olefins is also under evaluation by the refining industry. Alcohols are being reconsidered for gasoline blending because of their high oxygen content; for example, isopropyl alcohol (IPA) can be produced from FCC propylene and contains 46% more oxygen than MTBE. Propylene from the FCC unit can also be converted to diisopropyl ether (DIPE), an attractive oxygenate for gasoline blending. Properties of the most common oxygenates are listed in Table 6.

Ethers from FCC and Coker Olefins. The refiner can make oxygenates from the C_3, C_4, and C_5 olefins available in the refinery is a number of ways. In addition, changes in the FCC unit can increase olefin production within the refinery. Butyl and amyl ethers (MTBE, ETBE, TAME, and ETAE) can be produced from refinery C_4 and C_5 olefins. Isopropyl alcohol and DIPE can be made from refinery C_3 olefins.

Table 6. Oxygenate Properties

Oxygenates	Octane (R + M)/2	Blending RVP, kPa[a]	Oxygen, wt %
Alcohols			
Methanol	116	344.75–413.7	49.9
Ethanol	113	117.22–151.69	34.7
Isopropyl Alcohol	108	68.95–103.42	26.0
Ethers			
MTBE	108	55.16–68.95	18.2
TAME	105	20.68–34.48	15.7
ETBE	110	20.68–34.48	15.7
DIPE	104	27.58–34.48	15.7

[a] To convert kPa to psi, multiply by 0.145.

In the past, MTBE was used only sparingly as an octane enhancer. However, as a result of the new emissions regulations, the role of MTBE is switching from an octane enhancer to an oxygen carrier, and plants to make MTBE from FCC isobutylene have been and are being built in many refineries. Because MTBE has an average octane-blending value of approximately 108 (R + M)/2 and a blending RVP of 55.16 to 68.95 kPa (8 to 10 psi), it is an extremely attractive oxygenate.

Also being considered by refiners is TAME, which is produced from FCC isoamylenes. An ideal gasoline-blending component, TAME has a blending octane number in the range of 105 (R + M)/2 and a blending RVP of only 20.69–34.48 kPa (3–5 psi). The formation of TAME from isoamylenes, which have high vapor pressures, reduces the gasoline RVP and adds oxygen to the pool. The production of TAME, which requires the installation of a depentanizer and an amylene-treating unit, may justify the subsequent alkylation of the unreacted amylenes. The alkylation of C_5s is attractive when alkylation capacity already exists or is made available by the depletion of C_4 olefins from the alkylation feed as a result of the production of MTBE. In addition, the alkylation of C_5s can be a convenient way to remove olefins from the gasoline pool and further reduce pool RVP. The C_5-derived alkylate has a blending octane of approximately 90 (R + M)/2 and an RVP of 20.69–41.37 kPa (3–6 psi), depending on the type of alkylation process used. Existing alkylation units can be revamped to process C_5s. However, when alkylating C_5 olefins, significant care in feedstock pretreatment must be exercised to minimize acid consumption and maintain alkylate quality.

Refiners are also looking at the production of ETBE and ETAE, which are ethanol homologues of MTBE and TAME. One advantage of ETBE is that it has a low blending RVP of 20.69 to 34.48 kPa (3–5 psi). Thus, ETBE provides an attractive route for adding ethanol to RFG and will probably be the major source of renewable oxygenates for summertime RFG. Another advantage is that using ETBE allows the gasolines to be transported through wet distribution systems. Because of its miscibility with water, ethanol must be blended into gasoline at the terminal.

IPA and DIPE from FCC Olefins. The total amount of MTBE and TAME that can be produced from FCC isobutylenes and tertiary amylenes in an average refinery is approximately 0.6 wt % oxygen in the overall gasoline pool. Propylene is the only remaining light olefin that is available for oxygenate production. The hydration of propylene to IPA results in a substantial boost in oxygenate production. A recycle IPA unit, which converts virtually all the FCC propylene to IPA, can add 1.3 to 2.0 wt % oxygen to the gasoline pool, depending on the FCC conversion level. Isopropyl alcohol has a blending octane number of 108 (R + M)/2, which is comparable to MTBE, and has a higher oxygen content (26.0 versus 18.2 wt % for MTBE).

Drawbacks to the use of IPA in gasoline are its higher blending RVP—68.95–103.43 kPa (10–15 psi) versus 55.16–68.95 kPa (8–10 psi) for MTBE at 2.7 wt % oxygen—and the handling problems typical of alcohols. Using alcohols in gasoline requires either handling gasoline in dry tankage and distribution systems or splash-blending the alcohol at the terminal.

Producing DIPE instead of IPA from the refinery propylene results in an ether that can be blended into gasoline at the refinery. This gasoline can then be transported in the same manner as gasoline containing MTBE or TAME. Diisopropyl ether can be wholly produced in the refinery from FCC or coker propylene without the need to import any alcohol. This ether has a blending octane of 104 (R + M)/2, an oxygen content of 15.7 wt %, and a low blending RVP of 27.58–34.48 kPa (4–5 psi). A DIPE unit that converts all the FCC propylene adds 0.6 to 1.1 wt % oxygen to the gasoline pool, depending on the FCC conversion level.

Maximum Oxygenates from Light Olefins. Making changes in the FCC unit may be necessary to increase light olefin yields if refinery oxygenate and alkylate production are to be increased. Using a different catalyst can improve light olefin yield. Other process changes to increase light olefin production include modifying FCC operating conditions and revamping FCC units to include new advanced engineering technology.

Low-hydrogen-transfer FCC catalysts can boost the production of C_3 and C_4 olefins by more than one-third. New FCC catalysts under development specifically target selective production of isobutylene and will further increase isobutylene production relative to current FCC operations. Alternatively, refiners may use zeolitic additives for incremental light olefin production at the expense of C_5+ gasoline yield.

Changes in FCC operating conditions to increase light olefin production include raising the FCC reactor temperature. Raising temperatures increases conversion and thus can increase the production of light olefins by more than 50%. Increasing olefin production by raising temperatures will partially come at the expense of gasoline yield. The change in yield pattern will have an economic impact that must be carefully evaluated.

A more selective way to increase light olefin yield is through increased conversion coupled with advanced reactor design. The combination of these two changes improves the yield of light olefins, and the loss of gasoline yield is smaller than if temperature is the only change.

Each option for increasing light olefin production must balance the increased yield against its cost. An extensive revamp of the reactor section may be required if the temperature is increased beyond the limits of the existing metallurgy. Increasing the yield of light products beyond the capacity of the gas-concentration and fractionation sections of the FCC unit requires significant revamps. The selection of the appropriate FCC operation from the many viable options depends on the FCC unit design and the relative value that light olefins and gasoline have for each refiner.

The use of C_3, C_4, and C_5 olefins to produce oxygenates comes at the expense of alkylate production and is a potential concern to refiners. As a result, refiners must look carefully at the balance between the alkylation, etherification, and IPA or DIPE units. Alkylate is an excellent RFG component because of its blending properties and its limited impact on gasoline emissions. Economic analysis shows that keeping the alkylation unit at full capacity is often the most desirable solution. The final choice on whether to maximize oxygenate production at the expense of alkylate is based on the relative value of isobutane, alkylate, and oxygenates.

Impact of Oxygenates on the Hydrogen Supply. Ethers and alcohols are high-octane components that supply approximately two additional octane numbers to the gasoline pool at the legislated 2 wt % oxygen level. This octane contribution has significant influence on the operation of the catalytic reformer, which is the main unit for controlling octane in the gasoline pool.

The catalytic reformer took a prominent role in the refinery when improved reformer technology was required to supply the octane lost by removing lead from gasoline. The additional hydrogen available from high-severity reforming was used by the refinery in such applications as FCC feed desulfurization and hydrocracking.

A substantial reduction in reformer severity (approximately six to eight octane numbers on average) may result from the mandated addition of oxygenates to the gasoline pool. The reduction in reforming severity increases reformate yields but decreases hydrogen production. At the same time that hydrogen production is decreasing, hydrogen consumption is increasing to meet new specifications on diesel sulfur introduced in the United States in October 1993.

Additional demands on the hydrogen supply may result from the refiner's decision to limit the sulfur in gasoline to meet the emissions specifications of the Complex Model. Therefore, the role of the reformer may change from an octane producer to a hydrogen producer. The issue of increasing hydrogen production is discussed in greater depth in the section "Meeting Hydrogen Demand."

Impact of Oxygenates on Refinery Product Slate. Another potential change in refinery operations results from the addition of oxygenates to the pool. The addition of purchased MTBE to meet the oxygenate requirements may increase the gasoline yield by 11% to 15% if the crude rate is held constant. If crude rate is reduced to compensate for this increase in gasoline production, the production of other refined products, especially middle distillates, decreases. To rebalance the product slate, refiners must consider adjustments in cutpoint, reforming severity levels, and feeds to heavy oil cracking. The impact of these changes on gasoline-stream properties must then be reexamined in the light of the RFG specifications.

Toxics Reduction

Gasoline reformulation causes a reduction in aromatics content as a result of benzene reduction and oxygenate addition. Although the CAA Amendments do not impose a direct limit on the aromatics content of RFG, fuel aromatics affect exhaust benzene, which affects toxic emissions. Applying the toxics equations of the EPA emission models to the relevant gasoline parameters determines whether the final level of aromatics is sufficiently low to meet the legislated toxics reduction. Gasoline at baseline levels of aromatics will not need any additional aromatics reduction once the RVP, benzene, and oxygen specifications are met. However, gasoline pools with high aromatics content may require additional process changes to meet the toxic emissions standards.

Reducing aromatics by decreasing reforming severity means removing components with octane numbers in the 105 to 115 RON range from the gasoline pool. In addition, reducing reformer severity decreases hydrogen production. Instead of directly reducing aromatics, the refiner should first look at the economics of lowering the benzene level below the legislated 1 vol %.

The additional benzene reduction compensates for a higher aromatics level in the gasoline. According to the Simple Model, the impact on toxic emissions from a 0.1 vol % reduction in benzene is comparable to a reduction in aromatics of 1.7 vol % in VOC Control Region 2, and 1.5 vol % in VOC Control Region 1. Under the Complex Model, the impact of a 0.1-vol % reduction in benzene is equivalent to reducing aromatics by 2.0 to 2.5 vol %. If a further reduction of aromatics is required, the octane lost from lowering reformer severity can be regained by increasing the octane of the C_5-C_6 LSR naphtha fraction by recycle isomerization or through the addition of more alkylate or oxygenates to the gasoline pool.

MEETING THE COMPLEX MODEL SPECIFICATIONS

From January 1998 onward, emissions from gasoline, and hence its composition, will be controlled by the Complex Model. The various parameters that are included in the Complex Model are RVP, benzene, oxygen, olefins, sulfur, T50 or E200, T90 or E300, and aromatics. The impact of RVP, benzene, and oxygen was discussed under the Simple Model; this section discusses the remaining parameters.

Olefin Reduction

Olefins are a major source of NO_x emissions at high concentration in gasoline. The major sources of olefins in the U.S. gasoline pool are FCC and coker naphthas. Minor sources of olefins are dimate and polymer gasoline.

The light olefins that constitute 40% to 60% of light FCC or light coker gasolines have octane numbers much higher than the corresponding aliphatics. Hydrotreating these light fractions will consume substantial amounts of hydrogen and produce a low-octane fraction that may require isomerization to regain the lost octane. A better option for reducing the olefin content is to distill the C_5 fractions from the FCC and coker naphthas and then either convert the amylenes to alkylate or etherify the tertiary amylenes to TAME and then alkylate the unreacted amylenes.

Sulfur Reduction

Sulfur in gasoline is widely recognized as having an adverse effect on exhaust emissions because sulfur causes the temporary deactivation of catalytic converters. The major contributors of sulfur to the pool are unhydrotreated LSR naphtha, thermal-cracked gasoline, and FCC gasoline (Fig. 12). The FCC gasoline in the pool can contribute more than 90% of the total sulfur, about half of which is in the last 10 vol % of the FCC gasoline (Fig. 13).

Normally, FCC naphtha is not hydrotreated because saturation of olefins may occur with consequent loss of octane and high consumption of hydrogen. The low level of olefins in the heavy fraction of the FCC naphtha makes it a suitable candidate for hydrodesulfurization. Treatment of this fraction results in minimal loss of octane and low hydrogen consumption and has a maximum impact on the pool sulfur. For example, the sulfur level of an average gasoline pool containing one-third FCC gasoline can be reduced by 100 to 200 ppm by hydrotreating the heavy tail of the FCC gasoline.

Another way to reduce the sulfur content of the gasoline pool is to increase the hydrotreating of the vacuum

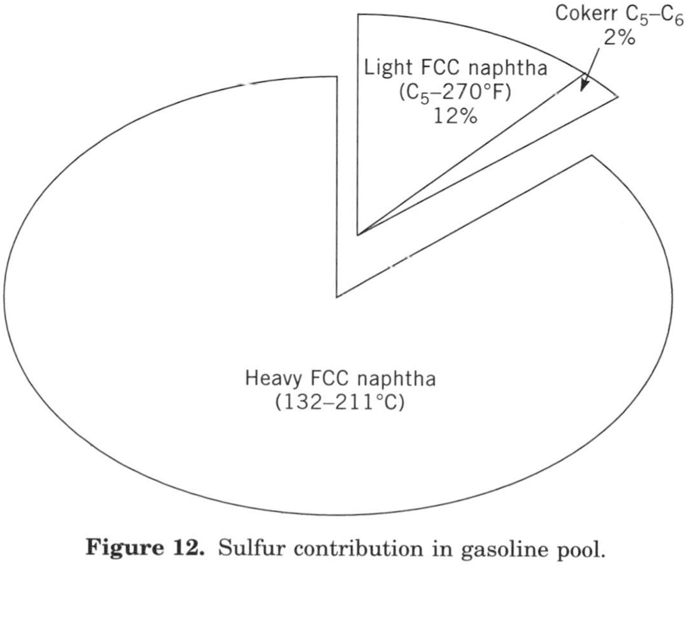

Figure 12. Sulfur contribution in gasoline pool.

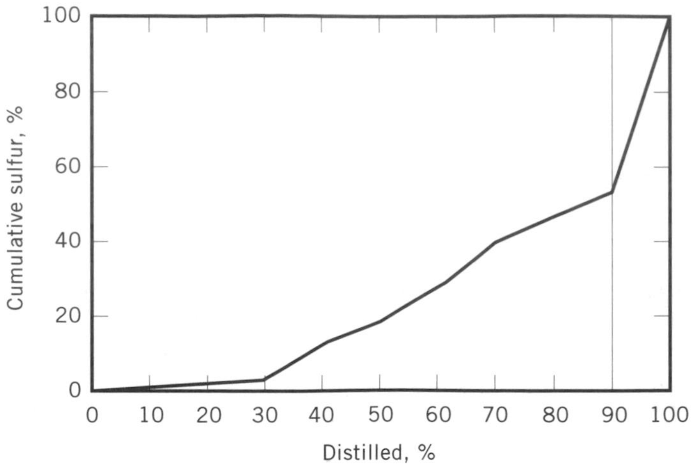

Figure 13. Sulfur distribution in FCC gasoline.

gas oil (VGO) feed to the FCC unit. Hydrotreating VGO has a number of benefits: overall conversion in the FCC unit increases; sulfur in all the products, not just gasoline, decreases; and overall emissions of SO_x decrease. However, VGO hydrotreating is expensive. The unit typically operates at a minimum of 6.206 MPa (900 psi); compared with 2.758–3.448 MPa (400–500 psi) for an FCC naphtha hydrotreater. In addition, because it must process all the feed to the FCC unit, the relative feed rate to the VGO hydrotreater is 16 times that of a hydrotreater processing the heavy 10% of the FCC naphtha. Hydrogen consumption is another concern, and it is about 21 times greater for hydrotreating VGO than for hydrotreating the heavy 10% of the FCC naphtha.

T90 Reduction (Increasing E300)

Refiners should consider two streams for T90 reduction: FCC gasoline and the naphtha feed to the reformer (Figure 14). Two options for treating these streams are to fractionate out the heavy portions and blend them into middle distillates, or to crack these portions into lighter components. No matter which option is chosen to reduce gasoline T90, the primary impact on the refinery from cracking the heavy gasoline faction is a further demand on the refinery's hydrogen balance. The effect of lower reformer hydrogen production resulting from the addition of oxygenates is now coupled with a significant new hydrogen demand: the hydrotreating and hydrocracking of heavy naphtha.

Demand for middle distillates is expected to increase, and demand for gasoline is expected to be fairly flat. If gasoline demand is constant, the gasoline volume increase from adding oxygenates to the pool is likely to cause an imbalance in the ratio of gasoline to middle distillates. To keep up with the distillate demand, the refiner may need to decrease the gasoline endpoint and shift the tail end of gasoline to distillates. The net effect will be a gradual reduction in gasoline T90 (increase in E300).

Heavy Gasoline to Middle Distillates. Reformate endpoint can be lowered by sending the heavy fraction of straight-run or hydrocracked naphtha to middle distillates. The portion of the straight-run naphtha that boils from 160°C to endpoint is certainly a good diesel component, and for most crudes, this naphtha is an acceptable jet fuel component. Hydrocracked naphtha in this same boiling range is an excellent component for either diesel or jet fuels.

The heavy FCC fraction constitutes a more difficult problem. This stream is highly aromatic and contains high levels of sulfur and nitrogen. The heavy FCC gasoline may be used as a cutter stock for fuel-oil blending up to the limit of flash point, or it can be blended into the distillate pool to the extent allowed by its poor cetane index and high sulfur content. Hydrotreating to improve the properties of this FCC fraction increases hydrogen consumption but enables this fraction to be used in higher-value distillate products.

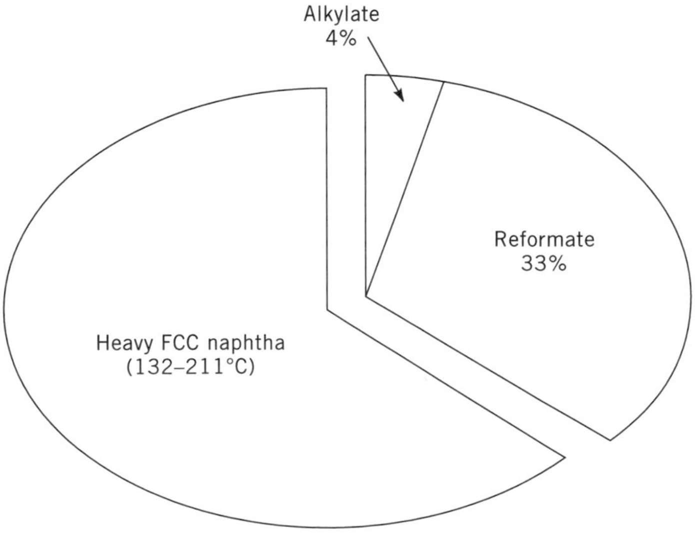

Figure 14. Contribution to Pool T90 over 165.5°C.

Cracking Heavy Gasoline. An alternative to blending or hydrotreating the heavy tail of the FCC gasoline is to crack it in a hydrocracker or in the FCC unit itself. The latter entails undercutting the heavy FCC gasoline, hydrotreating the wider boiling-range light cycle oil (LCO) to make it more crackable, and recycling a portion of this hydrotreated LCO back to the FCC unit. The hydrotreater required for this procedure is a medium-severity unit that saturates one ring of the double-aromatic-ring structure and thereby allows additional cracking in the FCC unit.

The heavy LSR naphtha or heavy fraction of hydrocracker naphtha can be cracked in an existing hydrocracker or within a highly selective cracking-reforming scheme. The latter route allows the selective production of isobutane, with yields of more than 30 vol % on the heavy naphtha charge, and the production of a high yield of low-T90 gasoline component. The efficiency of the hydrogen utilization, yield structure, and capital cost for these options must be evaluated on the basis of the refiner's individual needs and economics.

T50 Reduction (Increasing E200)

No effective, direct way has been found to control T50 or E200. The addition of oxygenates to the gasoline pool will reduce T50 (increase E200). However, reducing the gasoline RVP increases T50 (lowers E200). In practice, the addition of oxygenates is expected to be sufficient to control the middle portion of the distillation curve and thereby control T50.

Aromatics Reduction

The Complex Model includes gasoline aromatics as a variable. However, aromatics have little impact on NO_x or VOC emissions and a moderate impact on toxics. The primary source of aromatics in gasoline is reformate. Lowering reformer severity reduces aromatics in gasoline and also reduces hydrogen production. Restricting reformate in gasoline reduces an inexpensive source of high-octane, low-RVP, low-olefin blendstock. As mentioned previously, reducing benzene beyond the legislated level is probably more cost-effective than reducing aromatics and can be used to compensate for high aromatics levels in gasoline.

MEETING HYDROGEN DEMAND

Hydrogen management will be a critical issue for refiners as a result of the CAA Amendments. The problem stems from a reduction in hydrogen production and an increase in consumption. Hydrogen production from the reformer decreases primarily as a result of lower reformer severity. When oxygenates are added to the gasoline pool, less octane from the reformer is required, and reformer severity is lowered. Another reason for decreased hydrogen production is reduced reformer feed resulting from benzene precursor removal or reduction of naphtha endpoint. The other contribution to the hydrogen imbalance is the increased hydrogen consumption to meet benzene specifications, reduce FCC naphtha T90, and meet diesel-fuel requirements. The overall effect of lower hydrogen production coupled with increased hydrogen demand may mean a deficit of 1 to 2 MM SCFD of hydrogen per 10,000 BPSD of crude refining capacity. Solutions to the hydrogen imbalance depend on the configuration of the individual refinery.

The reformer continues to be the least expensive refinery source for hydrogen, and as a result of the CAA Amendments, the operating objective of the reformer may change from an octane producer to a hydrogen producer. The reduction in reformer severity reduces hydrogen production but also provides a way to get some hydrogen back. Lower reformer severity significantly reduces catalyst deactivation and thereby allows a decrease in operating pressure that was not practical at pre-RFG severities.

Reducing reactor pressure can significantly improve the refinery hydrogen balance. For example, lowering the reformer pressure from 2.069 MPa to 344.75 kPa (300 to 50 psi) increases hydrogen production by more than 400 SCFB, or 8 to 11 MM SCFD in a nominal 100,000 BPSD refinery. A reduced-pressure operation provides a higher reformate yield and lowers the benzene content of the reformate.

Other means of overcoming the hydrogen deficit are available. One of the most obvious is to build or expand a hydrogen plant. However, these plants can be expensive and can increase emissions from the refinery. Refiners must also look at nontraditional hydrogen-containing streams to supplement the hydrogen available from traditional sources. Hydrocracker and hydrotreater offgas and FCC dry gas may be considered as potential new sources of hydrogen. Applications of pressure-swing adsorption (PSA), cryogenics, and membrane technologies to recover hydrogen from these streams may be economically attractive in an RFG refinery. Other technologies, such as LPG dehydrogenation or aromatics production from LPG, are also excellent sources of hydrogen.

The preceding discussion is merely a summary of the various process options available to the refiner in meeting the requirements for making RFG. The various technical solutions are further summarized in Table 7.

COST PROJECTIONS FOR MAKING RFG

Assessing the cost of RFG is a rather complex task and cannot be done in isolation from the preexisting conditions of the refinery where RFG is to be produced. To produce RFG, each individual refinery will incur site-specific investment and manufacturing costs that depend on the supply of crude and intermediates, the process-flow scheme, the storage and delivery systems, and the product demand. Special consideration must also be given to the ratio of reformulated to conventional gasoline. In addition, costs for distribution, testing enforcement, and record keeping must be considered. Therefore, because of differences in assumptions, the costs generated by different organizations for making RFG cover a wide range.

The National Petroleum Council (NPC), a federal advisory committee to the Secretary of Energy, completed a two-and-a-half-year comprehensive study on the future of U.S. petroleum refining (6). The NPC study used a preliminary version (April 1992) of the Complex Model to set RFG properties. The study assumed that in 1995 only the nine extreme and severe ozone nonattainment areas of the country would use RFG. The study further assumed that by the year 2000 all opt-in areas and the entire Northeast Ozone Transport Corridor would use RFG and California would use California Resource Board (CARB) reformulated gasoline. The preliminary Complex Model used by NPC did not foresee any reduction in NO_x emissions for the year 2000, and the more severe VOC reduction of Phase II RFG was met primarily by a reduction in RVP. The present Phase II RFG regulations and Complex Model are likely to increase slightly the cost of reformulation.

Cost increases, above 1989 costs, were separately estimated for

- New stationary-source emission controls and health and safety regulations

Table 7. Technical Solutions for Meeting RFG Specifications

Issue	Approach	Considerations

RVP reduction	Debutanization
	Depentanization
	Amylene conversion:
	  TAME
	  Alkylate
	Low-pressure reforming
	Butane Isomerization
		Sulfur removal
		Diolefin removal

Benzene reduction	Precursor fractionation and processing
	Reformate processing:
	  Reformate splitting
	  Saturation
	  Isomerization
	  Alkylation
	  Extraction
		Hydrogen availability
		Gasoline pool octane requirements
		RVP of pool

Oxygenate production	MTBE, ETBE, TAME ETAE, IPA, DIPE from FCC olefins Olefin production from LPG Increase yield of FCC olefins	Sulfur removal Oxygenate removal Diolefin removal Butane isomerization
T90 reduction (increasing E300)	Heavy gasoline cracking and hydrocracking	Hydrogen availability
Sulfur reduction	FCC feed hydrodesulfurization Heavy FCC naphtha or LCO hydrodesulfurization	Hydrogen availability
T50 reduction (increasing E200)	Increase E300 to compensate for lower RVP Oxygenates	
Olefin reduction	Amylene conversion to TAME, ETAE, alkylate	Sulfur removal Diolefin removal
Aromatics reduction	C_5–C_6 isomerization Alkylate production Benzene removal	Hydrogen availability
Hydrogen management	Increase reformer hydrogen production H_2 purification and LPG recovery	

- Refining costs, including oxygenates
- Logistics and other costs
- Retail marketing regulations
- Lower fuel economy resulting from the low energy content of oxygenates.

Table 8 shows the average cost increase per gallon of RFG in 1995 (Phase I), 2000 (Phase II), and 2010. The cost increases are on an annual basis relative to 1989 gasoline production costs and are in 1990 U.S. dollars. The NPC study estimates that in the summertime, because of the additional VCO reduction requirement needed, the cost of RFG will be approximately 3.785 cents/L (1 cent/gal) higher than the total annual average cost increase to the consumer shown in Table 7.

The NPC study used regional aggregate refinery models, one for each Petroleum Administration for Defense District (PADD). Figure 15 shows the composition of the five PADDs. The cost of reformulation changes substantially in different PADDs: it is expected to be lower for PADDs III and V, which are characterized by larger, more complex high-conversion refineries; intermediate for PADDs II and III; and higher for the smaller and simpler refineries of PADD IV. Actual refineries are likely to incur different costs than the aggregate.

In its final rule document, the EPA states that the cost of producing RFG under Phase I is expected to be 11.36 to 18.93 cents/L (3 to 5 cents/gal) higher than the cost of conventional gasoline in 1995 (2). The cost of producing Phase II RFG during the VOC-control period, when the more stringent VOC and NO_x standards are in effect, is estimated by the EPA to be approximate 4.54 cents/L (1.2 cents/gal) higher than for Phase I RFG. The EPA expects no additional costs during the non-VOC-control season.

Clearly, making RFG will cost more than making conventional gasoline. The bulk of the cost comes from the requirement for adding oxygenates. However, other costs will result from reducing RVP and controlling some of the

Table 8. Annual Average Cost Increase of RFG over 1989 Gasoline

	Cost to Consumer, cents/L Gasoline		
Contributions to Cost	1995	2000	2010
Stationary source control:			
Health and safety	9.84	17.03	24.6
Refining costs	18.17	24.22	24.22
Logistics and other	3.78	3.78	3.78
Subtotal	*31.79*	*45.04*	*52.61*
Retail marketing	5.68	5.68	5.68
Lower fuel economy	9.46	9.46	9.46
Total added cost to consumer on annual basis	*46.93*	*60.18*	*67.75*

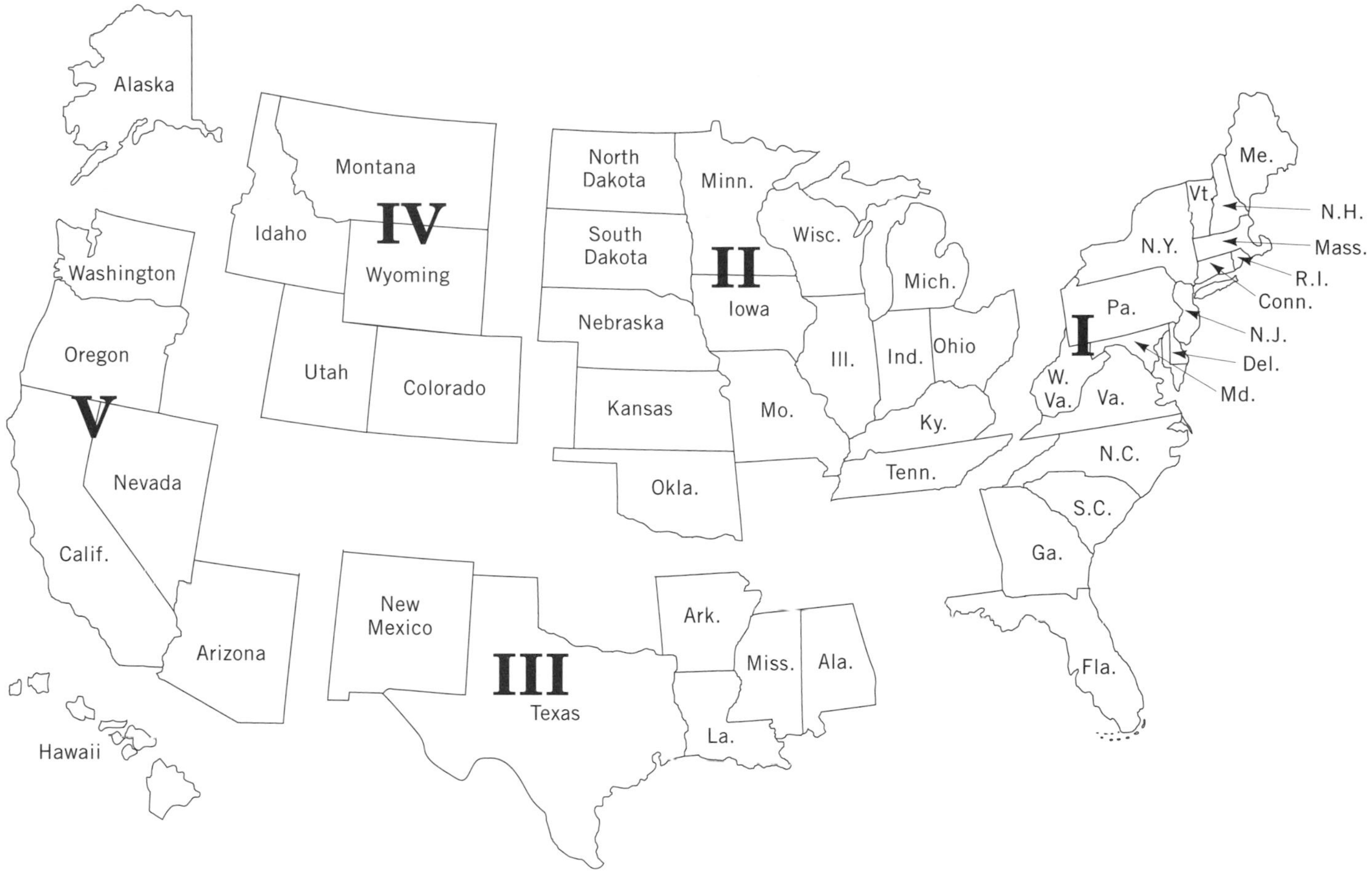

Figure 15. Petroleum Administration for Defense Districts (PADDs).

other gasoline pool properties. Finally, the logistical costs from analyzing, certifying, and distributing RFG will be higher than for conventional gasoline.

CONCLUSION

The CAA Amendments passed in 1990 legislate radical changes in gasoline quality. The wintertime oxygenated gasoline program was started in November 1992 in CO nonattainment areas, and RFG is mandated in the nine severe ozone nonattainment areas and the opt-in areas starting in 1995.

Detailed rules on how the CAA Amendments will be implemented were negotiated, and an agreement in principle was reached in August 1991 by the representatives of the oil industry, oxygenated fuel interests, state air-quality groups, environmental groups, and the EPA. The negotiations resulted in a simplified certification approach, the Simple Model, that is based on a limited number of fuel parameters. These parameters will be used to certify RFG from January 1, 1995, to December 31, 1997. In the meantime, a Complex Model, which goes into effect January 1, 1998, has been developed on the basis of a more complete range of fuel parameters. The final version of the Complex Model together with the final legislation was released on December 15, 1993.

The new regulations will greatly affect the refining industry in the years to come. To comply with the Simple Model during the period from 1995 to December 31, 1997, refiners will reformulate gasoline as follows:

- *Reduce RVP.* The reduction of RVP can be achieved by the removal of C_4 or C_5 hydrocarbons by fractionation or by the reaction of amylenes. Fractionation may be improved by retrofitting stabilizers with advanced distillation and heat-exchange technology. The reaction of amylenes to low-vapor-pressure components (TAME or alkylate) can be achieved with refinery process changes.
- *Reduce benzene.* Reformate benzene can be reduced by the removal of precursors or through reformate treatment. The benzene precursors can be fractionated from the reformer feed and processed in an isomerization unit. The reformate can be fractionated to concentrate benzene for subsequent processing through benzene saturation, isomerization, or alkylation. Benzene can also be removed from the reformate through extraction.
- *Add oxygenates.* Oxygenates are mandated by the CAA Amendments for use in OxyFuel and RFG. The resulting demand has caused refiners to look at the production of oxygenates in the refinery, especially

MTBE and ETBE from refinery isobutylene, TAME from tertiary amylenes, and IPA and DIPE from propylene. Changes in the FCC process can increase the yield of light olefins for further oxygenate production. However, the balance of the oxygenate requirement will have to be met by merchant supplies of MTBE or other oxygenates.

- *Control toxic emissions.* Calculated toxic emissions are primarily a function of fuel benzene and aromatics. The mandated benzene reduction will decrease toxic emissions. Aromatics will automatically decrease as a result of the addition of high-octane oxygenates combined with the reduction of reformer severity. Therefore, after these changes, further decrease in aromatics may not be necessary to achieve the required reduction in toxic emissions. If further toxics reduction is necessary, reducing benzene rather than aromatics is more cost-effective.

After January 1995, refiners will have to meet RFG specifications by applying the Complex Model. This model includes the additional fuel parameters for olefins, sulfur, E300, and E200 in calculating emissions.

- *Reduce olefins.* At high concentrations, olefins cause a significant increase in NO_x emissions. Olefin reduction can be achieved by converting the C_5 olefins to alkylate or TAME.
- *Reduce sulfur.* Because half of the FCC naphtha sulfur is concentrated in the last 10 % of the heavy FCC gasoline fraction, hydrotreating the tail end of the FCC gasoline is an effective way to reduce sulfur. Sulfur plays a significant role in NO_x toxics emissions.
- *Reduce T90 (increase E300).* Reducing gasoline T90 (increasing E300) has a small role in reducing emissions. The T90 may be lowered by fractionating the product stream to divert the gasoline tail end to distillates or to crack it to a lighter fraction. A number of options exist to integrate a cracking process into the refinery and thereby achieve the target of reducing T90 and producing additional isobutane.
- *Reduce T50 (increase E200).* Adding oxygenates to the gasoline pool will probably be sufficient to control T50. Decreasing T50 (increasing E200) causes a small increase in NO_x and a small decrease in toxics.

Gasoline reformulation coupled with the 1993 specifications for lower sulfur in highway diesel will result in a hydrogen shortfall. Thus, hydrogen management will become a strategic issue. Refiners may look to low-pressure reforming to partially reduce the hydrogen deficit. Management of nontraditional hydrogen sources may also help to offset the refinery hydrogen deficit.

The NPC has estimated that the cost of making reformulated gasoline will range from 30.28 to 53 cents/L (8 to 14 cents/gal) for Phase I and Phase II RFG. Another 15.14 cents/L (4 cents/gal) will be required to overcome the lower fuel economy from using RFG and for the higher costs associated with retail marketing of RFG.

To comply with the new gasoline specifications, refiners must find cost-effective process solutions that offer flexibility and meet the requirements of the CAA Amendments. Wisely choosing the process alternatives that meet product quality and environmental constraints is a significant challenge facing the U.S. refining industry in the coming years.

BIBLIOGRAPHY

1. Auto/Oil Air Quality Improvement Research Program, Technical Bulletin No. 3, "Air Quality Modeling Results for Reformulated Gasolines in Year 2005/2010," May 1991.
2. Environmental Protection Agency, 40 *CFR* Part 80, "Regulation of Fuels and Fuel Additives: Standards for Reformulated and Conventional Gasoline, Final Rule," *Federal Register,* **59,** 7716. (Feb. 16, 1994).
3. Environmental Protection Agency, 40 CFR Part 80, "Regulation of Fuels and Fuel Additives: Renewable Oxygenate Requirement for Reformulated Gasoline," *Federal Register,* **58**(246), Dec. 27, 1993.
4. "First Oxygenated Gasoline Season Shakes Out Differently Than Expected," *Oil & Gas Journal,* Oct. 25, 1993.
5. "State Interest in RFG Option Has Slowed Dramatically," *Octane Week,* (June 21, 1993).
6. *U.S. Petroleum Refining: Meeting Requirements for Cleaner Fuels and Refineries,* vol. 6, National Petroleum Councli, Aug. 1993.

REFRIGERANT ALTERNATIVES

Richard L. Powell
J. Hugo Steven
R&T Centre
Runcorn, Cheshire, United Kingdom

The collection of ice from ponds and its storage in specially constructed ice houses for subsequent summer use was common practice in the early 19th century, both in the United States and Europe. Indeed, Thoreau estimated that in the winter of 1847, 10,000 tons of ice were removed from Walden pond and stacked in huge heaps, 11 m (35 ft) high, insulated with hay and boards. However, losses due to melting were so great that only 25% was expected to reach the market, presumably to cool the summer drinks of the wealthier Bostonians (1). The concept of freezing water artificially had been attempted in the early years of the 19th century, but commercial-scale refrigeration did not become practical until the 1870s, when developments in engineering enabled reliable compressors and other high pressure components to be constructed (2).

Early refrigeration equipment was based on simple fluids already in substantial production by the chemical industry in the latter half of the 19th century, notably ammonia, sulfur dioxide, hydrocarbons, and methyl chloride. In the early 20th century, ice generated by central plants was distributed in urban areas to those private homes and small commercial businesses who had ice boxes to keep food cool. An enduring reminder of this earlier era is the continuing U.S. practice of quoting the cooling capacity of refrigeration and air-conditioning equipment in "tons," where 1 ton of refrigeration represents the enthalpy required to produce 1 ton of ice in 24 hours.

The introduction of electrical distribution systems in the 1880s prompted the development of small refrigeration systems suitable for homes and small commercial premises. Various fluids were used, including diethyl ether and ethane, in the late 19th and early 20th centuries, but sulfur dioxide emerged as the preferred refrigerant, with the introduction of the "Domelre" (Domestic Electric Refrigerator) in 1913. The demand for such devices, which obviated the need for frequent ice deliveries, increased rapidly in the United States, but by the mid-1920s, considerable concern was being raised in the press over accidents caused by exposure to catastrophic leaks of toxic sulfur dioxide. Frigidaire, realizing that the further development of its market crucially depended on the discovery of a safe refrigerant to replace sulfur dioxide, approached General Motors (GM) for help. GM turned over the task to one of their senior engineers, Thomas Midgley, who, in collaboration with chemists Albert Henne and Robert McNary, systematically scanned the Periodic Table for suitable alternatives. They identified as potential candidates the one and two carbon-containing chlorofluorocarbons (CFCs) and hydrochlorofluorocarbons (HCFCs), first reported by the Belgian chemist F. Swarts in the 1890s (3).

CFCs and HCFCs combined low toxicity, nonflammability, and good stability with acceptable thermodynamic properties. The first announcement of the CFC refrigerants was made at the American Chemical Society meeting in 1930; production began in 1931 by Kinetic Chemicals Inc, a joint GM—Du Pont company, only three years after the inception of the research program. The first CFC-containing unit was an ice cream freezer containing CFC 12, dichlorodifluoromethane. The first CFC-containing domestic unit, the "Meter-Miser," was introduced in 1933 and contained CFC 114, 1,2-dichloro-1,1,2,2-tetrafluoroethane.

The introduction of low toxicity CFCs and HCFCs revolutionized refrigeration and the nascent air-conditioning industry. Although competing initially with other refrigerants such as propane, ammonia, sulfur dioxide, and carbon dioxide, and with other technologies such as ammonia absorption, by the 1950s the CFC—HCFC mechanical vapor recompression refrigeration system dominated refrigeration and air-conditioning technology. Only in the food-processing industry did a nonfluorine-containing refrigerant, ammonia, retain its pre-eminence.

See also REFRIGERATION; GLOBAL HEALTH INDEX; COMMERCIAL AVAILABILITY OF ENERGY TECHNOLOGY; OZONE, STRATOSPHERIC; GLOBAL CHANGE.

THE CFC AND HCFC REFRIGERANTS

By the mid-1960s, a full range of CFC and HCFC fluids had been introduced, and worldwide production for all purposes, including aerosol propellants and the blowing of insulating foams, exceeded 1 million tons. Table 1 summarizes the range of CFCs—HCFCs and their main applications. To facilitate reference to the various compounds, a numbering system has been devised based on their chemical structures. This system is based on a three-digit code, where the first digit represents the number of carbon atoms in the compound minus one, the second digit represents the number of hydrogen atoms plus one, and the third digit indicates the number of fluorine atoms. If the first digit is zero, corresponding to a halogenated methane, then the digit is dropped so only two dig-

Table 1. CFC and HCFC Refrigerants

Compound	Code Number	Normal bp, °C	Refrigerant Applications	Other Applications
Trifluoromethane (CF_3H)	23	−81	Azeotrope with 13; bp, −88°C; in biomedical freezers	
Chlorotrifluoromethane (CF_3Cl)	13	−82	Azeotrope with 13; bp, −88°C; in biomedical freezers	
Bromotrifluoromethane (CF_3Br)	13*b*1	−57	Biomedical freezers	Fire extinguisher
Chlorodifluoromethane ($CHClF_2$)	22	−41	Room air conditioning, supermarket freezers, industrial refrigeration, Japanese domestic refrigerators	Tetrafluoroethene feed stock, polystyrene foam blowing (with chlorodifluoroethane)
Chloropentafluoroethane (CF_3CF_2Cl)	115	−38	Azeotrope with 22 (R502); bp, −46°C for supermarket freezers	
Dichlorodifluoromethane (CCl_2F_2)	12	−30	Domestic refrigeration, automobile air conditioning, supermarket fresh foods stores, industrial refrigeration, hot climate air conditioning	Polystyrene foam blowing, medical aerosols, carrier for epoxyethene, sterilant
1,1-difluoroethane (CHF_2CH_3)	152*a*	−24	Azeotrope with R12 (R500)	
1,2-dichloro-1,1,2,2,-tetrafluoroethane	114	3	Low pressure air conditioning (e.g., in nuclear submarines)	
(+ isomer CF_3CCl_2F)	(114*a*)			
Trichlorofluoromethane (CCl_3F)	11	24	Water chiller air-conditioning systems	Polyurethane foam blowing, medical aerosols
1,1,2-trichloro-2,2,1,-trifluoroethane (CCl_2FCClF_2)	113	48	Train switch gear coolant	Solvent, chlorotrifluoroethene feed stock

its appear in the code. If two carbon atoms are present, then two or more isomers are possible and are distinguished by adding a letter (*a*, *b*, or *c*), selected according to the differences in the formula weights of the two methyl groups that form the compound. The isomer in which these weights are closest has no letter; the next closest is assigned the letter *a*, the next *b*, and the next *c*; no further letters are needed. The system is best understood by applying these rules to some of the compounds listed in Table 1 and comparing the result with the number.

Most of the CFCs—HCFCs produced have been released to the atmosphere either during use, eg, aerosol propellants, or through leaks and disposal, as with refrigeration equipment. The environmental fate of these chemically stable fluids in the environment was unknown, so in 1972, the refrigerant manufacturing companies initiated and set up a joint organization, eventually known as the Fluorocarbon Program Panel, to assess the environmental impact of the CFCs and HCFCs. Independently, two years later, Molina and Rowland hypothesized that the unreactive CFC molecules might be capable of migrating unchanged to the upper atmosphere, the stratosphere (4). There, these molecules might be broken down by intense solar uv radiation into chlorine atoms, considered capable of catalyzing the decomposition of ozone (O_3) back to dioxygen (O_2), the most common form of oxygen in the Earth's atmosphere.

The stratospheric ozone layer not only protects the Earth's surface from intense solar uv radiation but is itself generated by the uv irradiation of dioxygen. The atmospheric chemistry is highly complex, but the essential steps are summarized below to demonstrate that chlorine atoms can catalyze the decomposition of many ozone molecules (5). Ultimately, chlorine is removed from the stratosphere by conversion to hydrogen chloride (HCl), which is transported to the lower atmosphere from where it is "rained out." The contribution to "acid rain" is, however, negligible compared to other tropospheric sources of strong mineral acids, especially sulfuric and nitric acids from fossil fuel combustion.

Ozone formation:

$$O_2 \rightarrow O + O$$
$$O_2 + O \rightarrow O_3$$

Chlorine atom formation:

$$CCl_2F_2 \rightarrow CClF_2 + Cl$$

Ozone destruction:

$$O_3 + Cl \rightarrow O_2 + ClO$$
$$ClO + O \rightarrow Cl + O_2$$

Net:

$$O_3 + O \rightarrow 2O_2$$

In contrast to CFCs, HCFCs are now considered to be decomposed mainly in the lower atmosphere by reaction with hydroxyl radicals ($HO^{\cdot}$) as indicated for HCFC 22 below.

$$CHClF_2 + HO^{\cdot} \rightarrow CClF_2 + H_2O$$

Until the late 1980s, techniques were inadequate to separate the putative CFC ozone loss from the large daily and seasonal variations resulting from weather, climate change, and fluctuations in solar uv. Instead, complex computer models of the atmosphere were constructed with experimentally determined rate constants for the chemical reactions considered important in the depletion of stratospheric ozone (5,6). By 1980, the general consensus was that even if any CFC-induced ozone depletion occurred, it would be sufficiently small so that it could be controlled, if necessary, by limiting CFC production capacity.

In 1984, renewed political pressures, coupled with the discovery of larger-than-expected ozone depletion over Antarctica during its spring (triggered by special meteorological conditions) (7), resulted in an international initiative to control the production of CFCs. This culminated in the Montreal Protocol of 1987, the first international treaty to protect the environment. In addition to provisions mandating reductions in the production of CFCs, the Protocol also called for the situation to be reviewed at regular intervals, to revise the controls on CFCs and other potentially ozone-depleting substances based on the most recent scientific data. Following this 1987 Protocol, the 1990 London Conference fixed a timetable for the complete phase-out of specified CFCs. The 1992 Copenhagen Conference moved forward the date of CFC phase-out and, for the first time, included a timetable for HCFC phase-out. However, the European Union has since introduced regulations to phase-out CFCs on an even shorter time scale than that required by the revised Protocol. Table 2 summarizes the current position.

REPLACEMENTS FOR CFCs—HCFCs

The chlorine-containing refrigerants owe their undoubted success to low toxicity and nonflammability, safety characteristics that are more valued now than when the CFCs were introduced in the 1930s. Clearly, their preferred replacements should be at least as good in this respect, if not better. The implication of atomic chlorine as the key factor in human-induced stratospheric ozone depletion immediately suggests that all chlorine compounds should be rejected (as well as bromine compounds, which potentially may be even worse). Only hydrofluorocarbons (HFCs), compounds containing carbon, hydrogen, and fluorine atoms, possess all the health and safety attributes of the CFCs—HCFCs, yet avoid their damaging ozone-depletion effects. Table 3 lists the main HFCs that have already been introduced or are under development as of early 1994, together with their actual or intended applications.

The HFC fluids have been selected due to their comparable physical properties to the CFCs—HCFCs that they are intended to replace, thus facilitating phase-out by minimizing the engineering modifications that the re-

Table 2. Regulatory Reductions in the Consumption of CFCs—HFCs[a]

	Copenhagen Amendments to Montreal Protocol			European Community's Regulation		
	Agreed and in Force			Enacted and in Force		Proposed
	CFCs[b]	Other CFCs[c]	HCFCs	CFCs[b]	Other CFCs[c]	HCFCs
Cap level[d]			3.1%			2.6%
Base year	1986	1989	1989	1986	1986	1989
Year:						
1993		−20%		−50	−50	
1994	−75	−75		−85	−85	
1995	−75	−75		−100	−100	Freeze and use controls
1996	−100	−100	Freeze			
2000						
2004			−35			−35
2007						−60
2008						
2010			−65			−80
2012						
2013						−95
2015			−90			−100
2020			−99.5			
2030			−100			

[a] Ref. 8.
[b] 11, 12, 113, 133*a*, 114, 114*a*, 115.
[c] 13, 112, 112*a*, 111, and the fully halogenated C_3 CFCs.
[d] The cap is defined as the percentage of the calculated level of chlorofluorocarbons consumed in the base year, plus the calculated level of hydrochlorofluorocarbons consumed in the same base year; it applies only to HCFCs.

frigeration and air-conditioning industries must make. Even so, an immense development effort is required to ensure that the equipment continues to operate as efficiently and as reliably as it did with the CFCs. For example, it has been necessary to develop new ester lubricants, compatible with HFCs, since the mineral and synthetic hydrocarbon oils used with CFCs—HCFCs lack adequate miscibility with the HFCs. Although HFC 134*a* is sufficiently close in properties to CFC 12 to be a direct substitute, other CFCs—HCFCs are being replaced by fluid blends. These blends contain up to three components in order to match the physical properties and performance

Table 3. Replacements for CFCs and HCFCs

Replacement	Code Number	Normal bp, °C	Refrigerant Applications
Hexafluoroethane (CF_3CF_3)	116	−86	Azeotrope with 23 for biomedical refrigeration down to −90°C
Difluoromethane (CH_2F_2)	32	−51	High capacity industrial refrigerator set refrigeration down to −50°C Component of blends to replace 502 and 22
Pentafluoroethane (CF_3CF_2H)	125	−48	Component of blends to replace 502 and 22
1,1,1-Trifluoroethane (CF_3CH_3)	143*a*	−48	Component of blends to replace 502
1,1,1,2-Tetrafluoroethane (CF_3CH_2F)	134*a*	−27	Domestic refrigeration, automobile air conditioning, supermarket fresh food stores, industrial refrigeration, hot climate air conditioning, component of blends to replace 502 and 22, insulating foam blowing
1,1-Difluoroethane (CHF_2CH_3)	152*a*	−24	Proposed for domestic refrigerators
1,1,1,2,3,3,3-Heptafluoropropane	227*ea*	−20	
1,1,1,3,3,3-Heptafluoropropane ($CF_3CH_2CF_3$)	236*fa*	−5	Proposed replacement for CFC 114*i* naval air conditioning
1,1,2,2,3-Pentafluoropropane ($CHF_2CF_2CH_2F$)	245*ca*	26	Proposed replacement for CFC 11 in water chiller air conditioners
1,1,1,3,3,3-Hexafluorobutane ($CF_3CH_2CH_2CF_3$)	356*ffa*	26	Foam blowing, replacement for CFC 11 in water chiller air conditioners
1,1,1,3,3-Pentafluoropropane	365*fcb*	40	Foam blowing agent

of the CFC—HCFC refrigerants, while ensuring nonflammability under all possible operating conditions, despite the presence of a mildly flammable component such as HFC 32 or HFC 143*a*. The compositions of these blends are currently being optimized so it is not yet possible to be definitive about the identity of the final products.

However, it is clear that both HCFC 22, the major air-conditioning unit fluid, and Azeotrope 502 (R502), the supermarket freezer refrigerant, are likely to be replaced by ternary mixtures of HFCs 32, 125, and 134*a*. CFC 114, a small tonnage refrigerant mainly used in military applications (such as the air conditioning of naval vessels including nuclear submarines), is likely to be replaced by HFC 236*fa* or 236*cb* for which the U.S. Environmental Protection Agency (EPA) is initiating toxicity trials in 1994. There is to date no proven, nonchlorine-containing replacement for CFC 11, which is used extensively in water chiller air-conditioning systems for large buildings, especially in the United States. HCFC 123 is technically suitable, and HFC 245*ca* has been mooted as a chlorine-free potential replacement, but has proven to be marginally flammable.

The HFCs clearly offer the quickest route for the elimination of the CFCs—HCFCs, thus enabling stratospheric ozone concentrations to recover as quickly as possible, while maintaining the high safety and performance characteristics demanded by the refrigeration and air-conditioning industries. Other options, however, have been advocated and may be used for specific applications. Hydrocarbon refrigerants, notably propane, *i*-butane, and their blends, have been promoted, especially in Germany, for domestic refrigerators (9). These fluids have been available and offered as refrigerants in competition with CFCs—HCFCs as early as the 1930s, but, despite their lower costs, were only recently accepted by the equipment manufacturers due to their flammability. In addition, the hydrocarbons are classed in the United States as volatile organic compounds (VOCs), as they react rapidly with oxygen in the presence of sunlight in the lower atmosphere, generating ozone which is a significant contributor to photochemical smog. In contrast, HFCs, which photolyze more slowly, do not cause the buildup of tropospheric ozone. Clearly, a balanced and integrated environmental assessment is needed of all CFC—HCFC alternatives, whether flammable or not.

To hasten CFC phase-out transitional fluids have been developed (see Table 4). HCFCs were originally considered to be potential long-term replacements for CFCs, and new products, such as HCFC 123 and HCFC 141*b*, entered into development. However, with increasing concern over the stratospheric ozone depletion, HCFCs are now considered transitional products that will facilitate CFC phase-out. The 1992 amendments to the Montreal Protocol already mandate phase-out of HCFCs to begin in 2004 and to be completed by 2030. HCFC 123, calculated to have the least ozone-depleting effect of any of the HCFCs being considered, will play an important role in replacing CFC 11 in chillers for which no adequate HFC alternative has yet been identified.

Perfluorocarbons (PFCs) have also been mooted and azeotropes of octafluoropropane (C_3F_8, R218), with HCFC 22 and/or propane, are currently being offered as replacements for Azeotrope 502. Although PFCs contain no chlorine or bromine and therefore cannot be ozone-depleting, concern has been expressed over their long atmospheric lifetimes and their high potential for global warming (*vide infra*).

A return to "natural fluids," i.e., those existing in the biosphere independently of human activity, has been proposed (10). However, the low vapor pressure of water limits its range of applicability. Ammonia has coexisted with the CFCs—HCFCs, but its toxicity is still likely to confine it to leak proof applications, to avoid endangering life or poisoning the immediate environment. Ammonia is also marginally flammable. Carbon dioxide has previously been used in refrigeration and could compete in some applications with HFCs, provided equipment can be built to contain the high pressures yet still be cost- and performance-competitive. Sulfur dioxide still has the inherent toxicological disadvantages, which originally prompted Midgley's development of the CFCs.

DIRECT GLOBAL WARMING

Calculations based on the solar energy flux suggest that the Earth's surface temperature would be on average 33°C lower than its actual value (11). Naturally occurring greenhouse gases in the atmosphere (carbon dioxide, water vapor, and methane, plus clouds), aerosols, and partic-

Table 4. Transitional Hydrochlorofluorocarbons

Compound	Code Number	bp, °C	Applications
1,1,-Dichloro-1-fluoroethane (CCl_2FCH_3)	141*b*	32	Polyurethane foam blowing, solvent
1,1-Dichloro-2,2,2,-trifluoroethane (CCl_2HCF_3)	123	26	Direct replacement for 11 in chillers
1-Chloro-difluoroethane ($CClF_2CH_3$)	142*b*	−9	Blended with 22 to replace 12 in polystyrene foam blowing
HCFC 22—HFC 125 blend		−46	Azeotrope 502 replacement
HCFC 22—perfluoropropane azeotrope		−46	Azeotrope 502 replacement
HCFC 22—perfluoropropane—propane near azeotrope			Blend with 22 152*a* for retrofitting into 12 containing units
1-Chloro-1,2,2,2-trifluoroethane (CF_3CHClF)	124	−12	

ulate matter absorb ir radiation and re-emit part of it back to the surface, thus maintaining a higher temperature. However, the emission of additional greenhouse gases, especially carbon dioxide from the combustion of fossil fuels, might increase the average temperature of the Earth's surface. This hypothesis notes that this temperature increase could melt the polar ice caps, causing a significant rise in sea level, the flooding of low-lying land, and significant changes in weather patterns worldwide. Despite the considerable controversy, there is currently no unequivocal scientific evidence for global warming. However, the governments of the developed nations have adopted policies to reduce emissions of greenhouse gases over the next decade in anticipation of a possible long-term problem. An international Convention on Climate Change is expected in early 1994.

A comprehensive and authoritative analysis of the potential impact of refrigeration and air conditioning on global warming has been provided by the Alternative Fluorocarbons Environmental Acceptability Study (AFEAS) and the U.S. Department of Energy (12). Figure 1 shows the estimated contribution of main greenhouse gases to global warming in the 1980s.

Although the CFCs were emitted in much smaller quantities than carbon dioxide, their greater global warming ability per molecule renders their contribution significant. The estimated global warming effect of a gas per unit mass is represented by its Global Warming Potential (GWP). The GWP measures the total energy reflected back to the Earth's surface integrated over a given time, the Integrated Time Horizon (ITH), following a hypothetical instantaneous release of a unit mass of the gas, compared to the warming effect expected from the release of the same mass of carbon dioxide. The calculation also requires data about the intensity of its ir absorption bands at those wave lengths where the atmosphere would otherwise be transparent to ir radiation. Normalizing the GWP scale to carbon dioxide, which by definition has a GWP of 1, allows the estimated global warming effects of other gases to be converted to their equivalent emissions of carbon dioxide, thus providing a common basis for comparison.

However, GWP values are markedly dependent on the ITH chosen, a consequence of the much lower rate at which carbon dioxide is purged from the atmosphere compared to the HFCs. The longer the ITH chosen, the smaller the GWP values for the HFCs become. Figure 2 clearly illustrates this important point by demonstrating the case for 100-year and 500-year ITHs. The former does not properly reflect the true situation because the effect of a large proportion of a short-lived HFC would be included, but a substantial proportion of the ultimate carbon dioxide contribution will be excluded. Scientifically, the 500-year values, listed in Table 5, provide the best of the currently available comparisons of the global warming effects of HFCs.

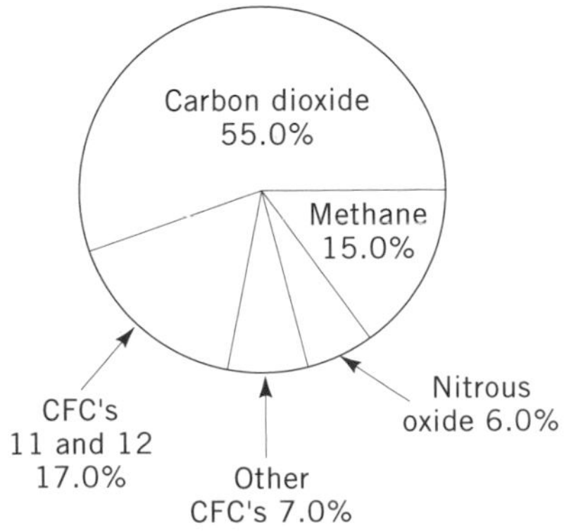

Figure 1. Estimates of the contribution of human-induced greenhouse gases to global warming, from 1980 to 1990.

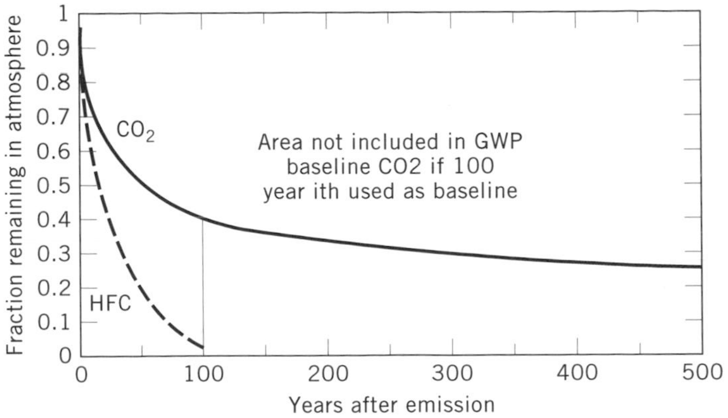

Figure 2. Global Warming Potential (GWP) equivalence of fluorocarbons based on the 500-year Integrated Time Horizon (ITH) for carbon dioxide: (**a**), HFC; (**b**), carbon dioxide; (**c**), area not included in GWP baseline carbon dioxide if 100-year ITH is used.

INDIRECT GLOBAL WARMING AND ENERGY EFFICIENCY

The operation of any power-consuming device, whose primary energy source is a fossil fuel, generates carbon dioxide and thus contributes to global warming. In the case of air-conditioning and refrigeration units, this is referred to as "indirect" global warming, to distinguish it from the "direct" effect resulting from the release of refrigerant to the atmosphere. Consequently, any refrigerant, ammonia, sulfur dioxide, water, or HFC can contribute indirectly to global warming, unless the equipment used is powered by electricity derived from a nuclear or renewable energy source.

The Total Equivalent Warming Impact (TEWI) concept has been introduced to combine both direct and indirect contributions in a single parameter. The TEWI expresses

Table 5. HFC Global Warming Potentials

Gas	Atmospheric Lifetime, Years	Global Warming Potential, 500-Year ITH[a]
CO_2	1	1
HFC 125	40.5	1,200
HFC 134*a*	15.6	400
HFC 143*a*	64.2	1,600
HFC 152*a*	1.8	49

[a] ITH = Integrated Time Horizon.

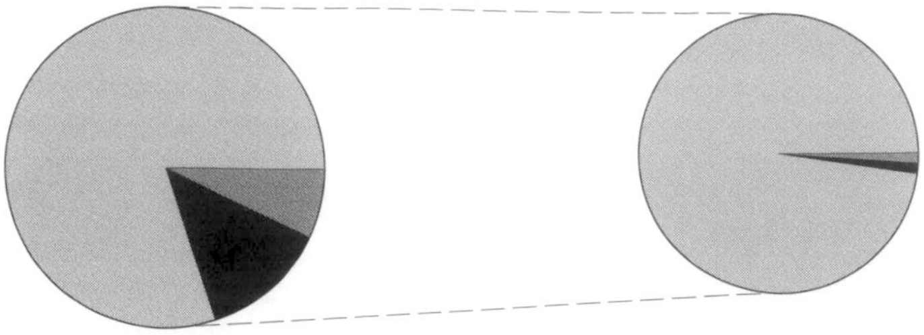

Figure 3. The Total Equivalent Warming Impact (TEWI) applied to domestic refrigeration. (**a**): □, carbon dioxide from energy generation; ■, CFC 11 in insulation; ■, CFC 12 in refrigerant. (**b**): □, carbon dioxide from energy generation; ■, HCFC or HFC in insulation; ■, HCFC or HFC refrigerant.

the direct (chemical emission) effect as a carbon dioxide equivalent using the GWP, and adds it to the indirect (energy-related) carbon dioxide emission. TEWI values are used to compare changes in estimated global warming when CFC—HCFCs are replaced in a specific application by HFCs or even another technology. Figure 3 illustrates the change in TEWI when HFCs replace CFCs in a domestic refrigerator, assuming that the energy efficiencies of the two units are identical. (Figure 3 also shows that the TEWI concept can be applied to the effects in changing the blowing agent in the insulating foam.) The relative importance of the indirect and direct contributions to global warming vary considerably between the various applications, direct being more significant for those uses such as automobile air conditioning and retail refrigeration, where refrigerant losses are higher (see Fig. 4).

The TEWI Concept Applied to Insulation

CFC 11 is used extensively to blow polyurethane insulation foams both for refrigerators and buildings. In the latter area, it competes with polystyrene foam, blown until recently by CFC 12 but currently by a mixture of HCFC 22 and HCFC 142*b*. A variety of options have been offered to replace CFC 11, notably HCFC 141*b* and HCFC 123. Although both compounds were originally viewed as long-term products, the inclusion of HCFCs in the Montreal Protocol means that at best they can only be considered transitional. The toxicity data on HCFC 123 also renders it unacceptable as a blowing agent, but HCFC 141*b* is likely to be used in Japan and the United States. Figure 5 shows the effects of replacing CFC 11 by CFC 141*b* in various applications.

In Europe, cyclopentane has been strongly promoted as an environmentally acceptable blowing agent, despite concerns about its flammability and the absence of any published toxicity data. HFC 356*ffa* and HFC 345*fcb* have also been mooted. Not surprisingly, it is difficult to identify a clear long-term solution for insulation. For refrigerators, vacuum panels may provide the solution, rather than a foam blown with a volatile fluid.

Energy Efficiency

The energy efficiency of an air-conditioning or refrigeration unit determines its contribution to indirect global warming, but its effect depends on a number of factors:

1. The fundamental thermodynamic properties of the fluid
2. The transport properties
3. The viscosity of the lubricant
4. The design of the compressor
5. The size of the heat exchangers
6. System design
7. The consumer's behavior

The fundamental thermodynamic properties of a refrigerant, which are generally presented in the form of a pressure—enthalpy diagram or tables, are often assumed to be pre-eminent in determining the performance of a re-

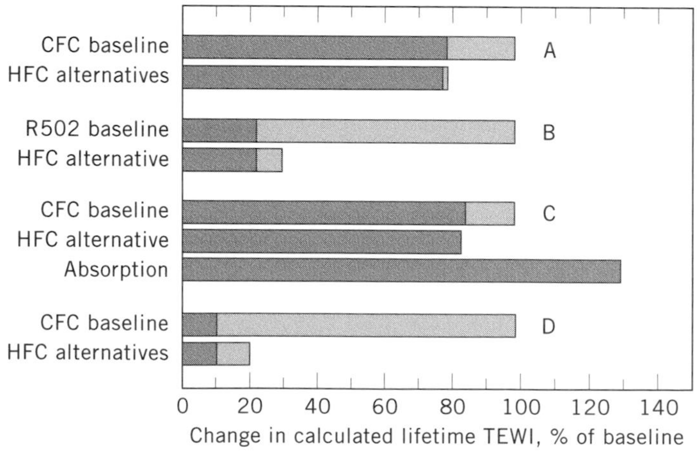

Figure 4. Changes in the Total Equivalent Warming Impact (TEWI) for selected refrigeration and air-conditioning applications, relative to baseline CFC technology: ■, direct global warming (chemical emissions); ■, indirect global warming. (**a**), refrigerators and freezers; (**b**), retail refrigeration; (**c**), chillers; (**d**), automobile air conditioning.

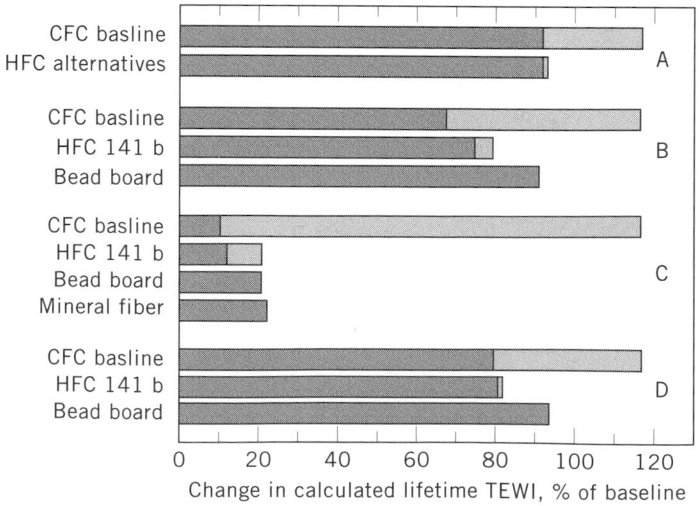

Figure 5. Changes in the Total Equivalent Warming Impact (TEWI) for selected insulation appliances, relative to baseline CFC technology: ■, direct global warming (chemical emissions); ■, indirect global warming (energy use). (**a**), refrigerators and freezers; (**b**), residential walls; (**c**), commercial-building low slope roofs and chillers; (**d**) commercial-building walls.

frigeration unit. In theory, any stable chemical compound could be considered for a cooling duty within its fluid range, but in practice, the need to design a cost-effective machine considerably limits the choices available to the design engineer. Rankine cycle efficiency improves as the operating temperature drops below the critical temperature. However, the lower the temperature, and hence the pressure, of a fluid, the larger the compressor needs to be to provide adequate cooling capacity. For systems operating over large temperature ranges, larger compressors are generally more expensive; the design engineer thus needs to balance high cycle efficiency against cost-effectiveness.

Although transport properties, which determine heat transfer in the condenser and evaporator of a cooling unit, are generally ignored in cycle calculations, the unit's practical performance can be better than simple estimates predict. For example, as a result of its lower molecular weight, HFC 134*a* has better heat-transfer characteristics than CFC 12, and in real systems is able to return at least as good of a performance despite an apparent poorer calculated cycle efficiency.

Higher viscosity oils are more effective in protecting compressors against wear but reduce energy efficiency through increased viscous drag. On the other hand, too low a viscosity not only increases wear but also allows high pressure vapor to escape past the piston or the rotor, thereby also reducing efficiency. For each compressor type, it is necessary to select a lubricant that optimizes energy efficiency. Compressor design itself not only strongly influences efficiency but can determine the selection of the fluid. For example, in an hermetic compressor found in refrigerators, air-conditioning units, and many commercial systems, the refrigerant provides both the cooling duty and removes heat from the electric motor. Analysis of the system shows that refrigerants with greater intrinsic heat capacity are superior for this application. In this respect, HFC 134*a* compares favorably with CFC 12. Centrifugal compressors are able to compress large volumes of vapor through relatively small pressure differences, enabling fluids to be used at temperatures well below their critical points, where their intrinsic thermodynamic energy efficiencies are high. For example, CFC 11 is used in high efficiency centrifugal water chillers to air-condition large buildings.

Increasing the surface areas of evaporator and condenser heat exchangers reduces temperature differentials and exploits liquid supercooling, thus boosting cycle efficiency. However, this increases the unit's retail cost, which must be balanced against the improved energy efficiency. More sophisticated system designs can be employed to enhance performance, for example, by using proportional speed control of compressors, which continuously matches capacity to the required duty. The technology for using the temperature glides produced in heat exchangers by wide boiling blends of refrigerants has existed for many years, but attempts to introduce this technology have proven unsuccessful. In part, this is due to the refrigeration and air-conditioning industries' concerns over composition shifts arising from differential leakage of one component relative to another. However, with increasing confidence that blends based on HFCs 32, 125, and 134*a* can successfully substitute for HCFC 22 and Azeotrope 502 (R502), wide boiling mixtures are becoming more generally accepted. Indeed, recent research suggests that if an air-conditioning unit is properly designed for wide boiling blends (eg, by ensuring that the vapor and liquid flow rates in the heat exchangers are approximately equal), then energy efficiency improvements of 20–30% can be realized over an equivalent commercial HCFC 22 unit (13).

"NOT IN KIND" (NIK) REPLACEMENTS FOR CFCs—HCFCs

The search for CFC—HCFC refrigerant replacements has been mostly centered on finding HFCs that have similar physical properties. The adoption of such an approach has been vital in order to meet the increasingly tight deadlines set by the Montreal Protocol for CFC phase-out while retaining the same energy efficiency achieved by existing equipment. However, a number of alternatives have been promoted which do not require HFCs. The wider use of ammonia, for example, as a replacement for Azeotrope 502 in supermarket refrigeration, is occurring despite its toxicity and flammability hazards.

Absorption cooling, which replaces the compressor, has been used for over 60 years in refrigeration and for over 40 years in large water chillers. Ammonia—water has been the only widely used refrigerant—absorbent pair for refrigeration, as the molecular interactions between the components of the pair must be strong. For large building air conditioning, absorption water chillers are based on a water—concentrated lithium bromide (LiBr) solution operating under subatmospheric conditions. Although ammonia—water refrigerators are still used in niche applications, such as hotel rooms or for camping (where their near silent operation or their ability to utilize a variety of energy sources are especially advantageous), they are unable to provide the full range of cooling required in modern domestic refrigerators and freezers. Despite their ability to be gas-powered, absorption refrigerators are not as efficient in their use of fossil fuel as conventional units and therefore contribute more to indirect global warming. There is, however, renewed interest in more sophisticated designs of water—LiBr chillers, based on more energy-efficient "double effect" cycles, and these will probably capture a portion of the market currently satisfied by the CFC 11 centrifugal chillers.

A near-market alternative for large building air conditioning is adsorption cooling, which uses the water vapor in the incoming airstream as the refrigerant. The system operates by adsorbing and desorbing the water from a solid inorganic material called a zeolite. Competitive energy efficiencies are claimed for adsorption systems, resulting from the use of slowly rotating heat exchangers, called "heat wheels," for highly efficient heat exchange. A further advantage is the ability to optimize the design of each unit specifically and independently for the humidity and cooling duties of the local climate where each is installed.

The demise of the CFCs has prompted the inception of programs to develop numerous nonhydrofluorocarbon technologies. These range from high pressure, helium-based Stirling cycle refrigeration, to low pressure butane systems, intended for automobile air conditioning. Perhaps the most intriguing development is the thermoacoustic refrigerator, which is based on the temperature difference across the ends of a set of plates in a resonating column of helium powered by a loud speaker diaphragm, a principle previously unapplied to cooling (14). In contrast to standard refrigeration designs, the device can also operate in zero gravity; a lightweight prototype was flown on a Shuttle mission to provide cooling duty for an experiment.

CONCLUSION

Refrigeration is fundamental to the well-being of modern society; it allows for the storage of food, blood, vaccines, and drugs at low temperatures. Refrigeration is also required for the production of many artifacts and services. CFCs and HCFCs, by making refrigeration universally available, have greatly contributed to the current standard of living. Although the problem of ozone depletion has made phase-out of CFCs and HCFCs inevitable, their replacement by more environmentally benign alternatives, with the minimum of disruption to the maintenance of existing and the introduction of new refrigeration equipment, is clearly essential. This is a role which the HFCs fulfill admirably. If a less-developed or even unproven technology had to be used with lower energy efficiencies, then indirect global warming attributable to refrigeration and air conditioning could have been substantially higher. Undoubtedly, as their performance is improved, some of the nonhydrofluorocarbon systems will also enter the market. As indicated above, methods for substantially improving the energy efficiencies of the latter are being developed, and air conditioning and refrigeration will be even more energy-efficient in the future, along with a reduction in carbon dioxide emissions.

BIBLIOGRAPHY

1. Henry David Thoreau, *Walden,* Signet Classics ed., The New American Library Inc., New York, 1960, p. 197.
2. B. A. Nagengast, *CFCs: Time of Transition,* ASHRAE, 1989.
3. F. Swarts, *Bull. Acad. Roy. Belg.* **24,** 309, 474 (1892).
4. M. J. Molina and F. S. Rowland, *Nature* **810,** 249 (1974).
5. United Kingdom Stratospheric Ozone Review Group, *Stratospheric Ozone 1988, Second Report,* Her Majesty's Stationary Office, London, 1988.
6. *Scientific Assessment of the Stratosphere 1992,* United Nations Environment Program—World Meteorological Organisation of Global Ozone Research and Monitoring Project, 1992.
7. J. C. Farman, B. G. Gardiner, and J. D. Shanklin, *Nature* **315,** 207 (1987).
8. United Kingdom Stratospheric Ozone Review Group, *Stratospheric Ozone 1993, Fifth Report,* Her Majesty's Stationery Office, London, 1993.
9. M. D. A. Meyer, "International Conference on Energy Efficiency in Refrigeration Global Warming Impact," *Proceedings of the Commission B1/2,* Ghent, 1993, p. 218.
10. G. Lorentzen, "International Conference on Energy Efficiency in Refrigeration Global Warming Impact," *Proceedings of the Commission B1/2,* Ghent, 1993, p. 55.
11. J. E. Lovelock, *Gaia: The Practical Science of Planetary Medicine,* Gaia Books Ltd., London, 1991, p. 134.
12. S. K. Fischer, P. J. Hughes, P. D. Fairchild, C. L. Kusik, J. K. Dieckmann, E. M. McMahon, and N. Hobday, *Energy and Global Warming Impacts of CFC Alternative Technologies,* sponsored by the U.S. Department of Energy (DOE) and the Alternative Fluorocarbons Environmental Acceptability Study (AFEAS), 1991.
13. J. Chen and G. G. Haselden, *Mixed Refrigerants for Air Conditioning and Heat Pumping,* paper presented at the Institute of Refrigeration, Dec. 3, 1993.
14. G. W. Swift, *J. Acoust. Soc. Am.* **84,** 1145 (1988).

General References

J. J. Sciarra, "Aerosols," in *Kirk-Othmer Encyclopedia of Chemical Technology,* 4th ed., Vol. 1, John Wiley & Sons, Inc., New York, 1991, p. 670.

K. W. Cooper, "Air Conditioning," in *Kirk-Othmer Encyclopedia of Chemical Technology,* 4th ed., Vol. 1, John Wiley & Sons, Inc., New York, 1991, p. 686.

K. W. Cooper and K. E. Hickman, "Refrigeration," in *Kirk-Othmer Encyclopedia of Chemical Technology,* 3rd ed., Vol. 20, John Wiley & Sons, Inc., New York, 1982, p. 78.

B. E. Smart, "Fluorinated Aliphatic Compounds," in *Kirk-Othmer Encyclopedia of Chemical Technology,* 3rd ed., Vol. 10, John Wiley & Sons, Inc., New York, 1980, p. 856.

REFRIGERATION

HARRY J. SAUER, JR.
University of Missouri, Rolla
Rolla, Missouri

According to *Webster's* (1), refrigerate means to "to make or keep cold or cool." In the *ASHRAE Terminology of Heating, Ventilation, Air Conditioning, and Refrigeration* (2), refrigeration is the process of extracting heat from a substance or space by any means; usually at a low temperature. Continuous refrigeration can be accomplished by several different processes. In the great majority of applications, and almost exclusively in the low horsepower range, the vapor-compression system is used for the refrigeration process and for the heating process of heat pumps. In large equipment centrifugal systems are used, which are basically adaptations of the compression cycle. However, absorption systems and steam-jet vacuum systems are also being successfully used in many cooling applications. For aircraft cooling, air-cycle refrigeration is often used instead of vapor compression. See also THERMODYNAMICS; COMMERCIAL AVAILABILITY OF ENERGY TECHNOLOGY; REFRIGERANT ALTERNATIVES; ENERGY EFFICIENCY—ELECTRIC UTILITIES.

Premechanical refrigeration for the production, utilization, and storing of cold extends back into history to the use of mountain snows, pond and lake ice, and the manufacture of ice by evaporative and radiative cooling of water on clear nights. Complementing the use of mountain snow and pond and lake ice in warm climates was the production of cold by producing a bath solution of "frigoric" mixtures such as saltpeter mixed with water and ice. Some mixtures, such as calcium chloride and snow, which made possible a temperature down to −32.8°C, were introduced for commercial use. Chemical mixture freezing machines were introduced in Britain for the production of low temperatures, but by the time these inventions were available for exploitation, mechanical ice-making processes made the chemical mixture methods for freezing noncompetitive except for some bath processes like ice-cream making, which used a mixture of common salt and ice.

In both Egypt and India, the evaporative cooling process supported by radiation to clear skies at night furnished ice for the royal tables as early as 500 B.C. About 375 B.C., Alexander the Great insisted that ice be stored along the route his armies would take to cool their wine. During the Dark Ages in Europe, refrigeration was practiced as sorcery, the general public regarding it as a "black art" until the early 1800s. During the Renaissance in Europe refrigeration made a comeback. By 1825 the concept of air cycle and vapor compression for refrigeration was written about in the technical literature although there were no working devices at the time. In 1834, Jacob Perkins received British patent no. 6662 for a closed-cycle ice machine using ether or other volatile fluids as refrigerants. Perkins's machine was never produced commercially. It was in 1844 that Dr. John Gorrie (1803–1855), director of the U.S. Marine Hospital at Apalachicola, Florida, described his new refrigeration machine. In 1851, he was granted U.S. Patent No. 8080. This was the first commercial machine in the world built and used for refrigeration and air-conditioning. Gorrie's machine received international recognition and acceptance. In France, Ferdinand Carré designed the first ammonia absorption unit in 1851. In 1853, Professor Alexander Twining of New Haven, Connecticut, employing the unsuccessful invention of Jacob Perkins, produced 1600 pounds of ice a day with a double-acting vacuum and compression pump using sulfuric ether as the refrigerant. In 1872, David Boyle invented the ammonia compression machine, producing ice with it the following year.

Until 1880, the story of the advance in mechanical refrigeration (worldwide) was primarily the progress being made in ice making and in meat and fish processing and preservation, with a few notable examples of beer making, oil dewaxing, and wine cooling in the United States, Europe, and Australia. Air cooling for comfort was obtained by ice and by chilling systems from either "lake" or "ice plant" ice.

The American Society of Refrigerating Engineers (ASRE) was organized in 1904. (In 1959 ASRE merged with the American Society of Air-Conditioning Engineers to form the American Society of Heating, Refrigerating, and Air-Conditioning Engineers [ASHRAE]). The new Society had 70 charter members and took pride in being the only engineering group in the world confining its activities to refrigeration. In 1905, the American Society of Mechanical Engineers (ASME) established 303,552,000 J (288,000 Btu) in 24 hours as the commercial ton of refrigeration within the United States. The same year the New York Stock Exchange was cooled by refrigeration.

In 1925 the first theater in New York City was air-conditioned using industrial refrigeration technology. It was shortly thereafter that a distinction was made between industrial refrigeration and refrigeration for air-conditioning. The needs of the air-conditioning refrigeration system were low first cost and ease of operation and installation. The industrial refrigeration system had to be reliable and cost-effective over the life of the plant. The two types of refrigeration systems were not compatible and developed separately. By 1925 the industrial refrigeration industry was booming when the General Electric Company decided to develop a household refrigerator to replace the ice box. The ice plant business was soon on the skids, making surplus equipment available for breweries, meat-packing plants, and refrigerated warehouses. By 1930 the bottom fell out of the stock market and a worldwide depression set in from which the refrigeration industry did not fully recover until World War II, when new food facilities were needed.

However, refrigeration research did go forward during the 1930s. Leading the procession was the development of the Freon® refrigerants by the General Motors team at its Frigidaire Division headed by Thomas Midgley, Jr. Commercial production of Freon 11 and Freon 12 was begun in 1931 by Kinetic Chemicals Company, a joint stock company formed by General Motors and E. I. Du Pont De Nemours and Co., Inc. Their safety and desirable properties made the small reciprocating compressor practical and economical for the commercial and residential markets. Soon manufacturers started to build room air conditioners using "Freon-12." Larger sizes of self-contained units suitable for stores and restaurants followed. Downing (3) and Nagengast (4) provide historical data on the development of these and other CFC (chlorofluorocarbon) refrigerants.

There were several other noteworthy achievements during this decade. In 1931, Servel introduced its first lithium bromide absorption machine for residences. The same year, Carrier marketed a line of steam ejector cooling units for railroad passenger cars. By 1940, air-conditioning was being introduced to an ever-increasing number of applications. Cooling of department stores, restaurants, hotels, and hospitals became accepted practice. For air travel, cabin conditioning became a must. Soon all passenger ships, buses, and railroad passenger cars found space cooling essential. Product technology then advanced by leaps and bounds. The air source heat pump and the large lithium bromide water chiller were important innovative breakthroughs right after World War II. Then in the 1950s came automobile air conditioners and the small outdoor-installed ammonia absorption chiller. By the 1960s, air-conditioning of almost all spaces of public entertainment, office buildings, school classrooms, and residences south of 40 degrees North latitude became prevalent. Today, hundreds of applications of refrigeration exist for conservation and preservation for museums, libraries, printing plants, textile processing, metal treatment, frozen foods, fruits, vegetables, and food of all kinds.

A concise history of refrigeration from 1748 to 1968 has been prepared by Woolrich (5).

VAPOR-COMPRESSION CYCLE AND COMPONENTS

Figure 1 illustrates the basic vapor-compression cycle and shows the relative location of the four basic components

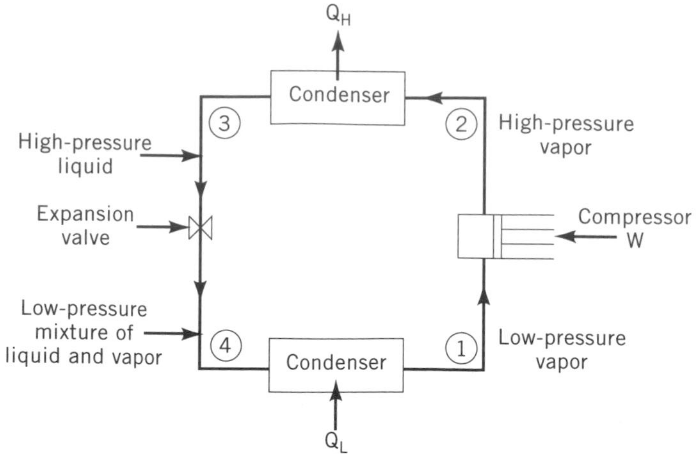

Figure 1. Schematic of the basic vapor-compression cooling cycle.

of the cycle: compressor, condenser, expansion valve, and evaporator.

A large number of working fluids (refrigerants) are used in vapor-compression refrigeration systems than in vapor-power cycles. Ammonia and sulfur dioxide were first used as vapor-compression refrigerants. Until recently, the main refrigerants were the halogenated hydrocarbons, also called chlorofluorocarbons (CFCs), which are marketed under trade names such as Freon® and Genetron.® The choice of a refrigerant for a particular application often depends on properties not directly related to its ability to remove heat. Such properties are flammability, toxicity, density, viscosity, and availability. As a rule, the selection of a refrigerant is a compromise between conflicting desirable properties. For example, the pressure in the evaporator should be as high as possible whereas, at the same time, a low condensing pressure is desirable.

The vapor-compression refrigeration cycle can be plotted on a (p,h) diagram, as shown in Figure 2. Subcooled liquid, at point *A*, decreases in pressure as it goes through the metering valve located at the point where the vertical liquid line meets the saturation curve. As it leaves the metering point, some of the liquid flashes into vapor and cools the liquid entering the evaporator at point *B*. Note that there is additional reduction in pressure from the metering point to point *B* but no change in enthalpy. As it passes from point *B* to *C*, the remaining liquid receives heat and changes from a liquid to a vapor, but it does not increase in pressure. However, enthalpy does increase. Superheating occurs between point *C*, where the vapor passes the saturation curve, and point *D*. As the vapor passes through the compressor, point *D* to *E*, its temperature and pressure increase markedly, as does the enthalpy, because of the energy input for compression. Line *E–F* indicates that the vapor must be desuperheated within the condenser before it attains a saturated condition and begins to condense. Line *F–G* represents the change from vapor to liquid within the condenser. Line *G–A* represents subcooling within the liquid line or the capillary tube, before the liquid flows through the metering device.

The pressure remains essentially constant but the temperature is increased beyond the saturation point, because of superheating, as the refrigerant passes through the evaporator and before it enters the compressor. The pressure likewise remains constant as the refrigerant enters the condenser as a vapor and leaves as a liquid. Although the temperature is constant through the condenser, it decreases as the liquid is subcooled before

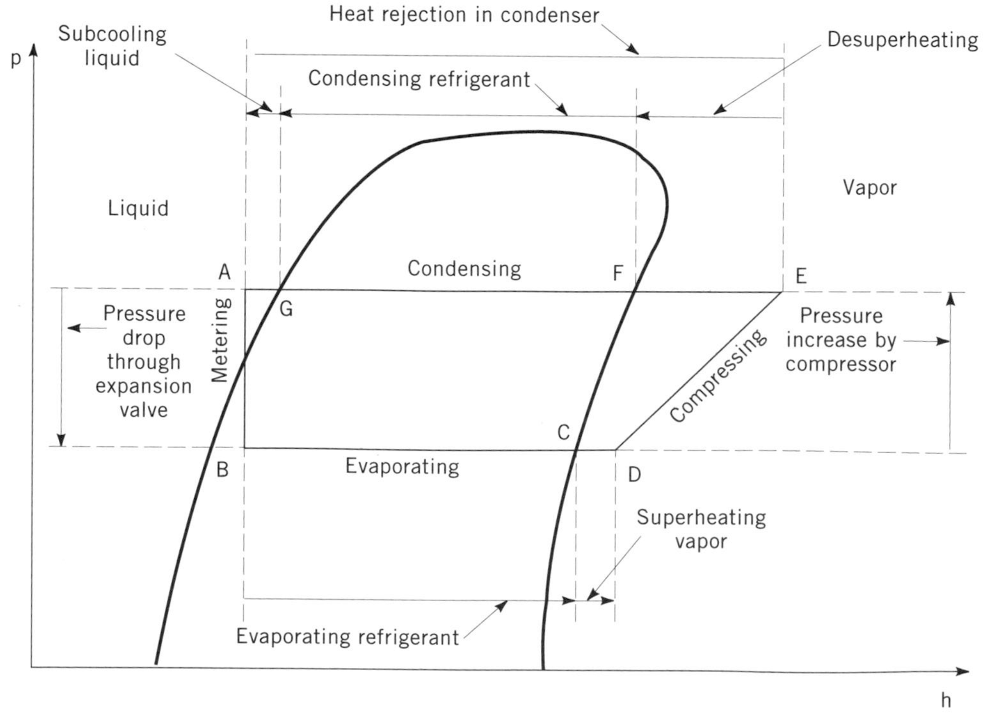

Figure 2. Typical pressure–enthalpy diagram for ideal refrigeration cycle.

entering the metering valve. The change in enthalpy as the refrigerant passes through the evaporator is almost all latent heat, because the temperature does not change appreciably. In this ideal cycle, the only pressure changes are the result of the compression and expansion process. The actual system experiences many pressure drops due to friction.

Applying the first law of thermodynamics to the system as a whole yields

$$Q_L + W = Q_H$$

It is apparent from this relation that every refrigeration cycle operates at all times as a heat pump. The household refrigerator absorbs a quantity of heat (Q_L) at a low temperature in the vicinity of the ice-making section (the evaporator) and rejects heat (Q_H) at a higher temperature to the air in the room where it is located. The rate of heat rejection (Q_H) is greater than the rate of absorption (Q_L) by the power input (W) to drive the compressor.

Historically, the phrase "tons of refrigeration" (that is, the number of tons of ice that melts in one day) was used to describe the size of a refrigeration system. This is, of course, an insufficient definition from an engineering standpoint. Today this phrase is still used but is explicitly defined as:

$$\begin{aligned} 1 \text{ ton of refrigeration} &= 12{,}648{,}000 \text{ J (12,000 Btu) of refrigeration/hr} \\ &= 210{,}800 \text{ J (200 Btu/min)} \\ &= 3.514 \text{ kW} \\ &= 4.712 \text{ hp} \end{aligned}$$

Several parameters are used in describing the energy performance of refrigeration systems, including the following:

Coefficient of Performance, Cooling (COP_c)—Ratio of the rate of heat removal from the conditioned space to the rate of energy input, in consistent units, for a complete heat pump or refrigerating plant, or some specific portion of that plant, under designated operating conditions.

Energy Efficiency Ratio, Cooling (EER_c)—Ratio of the rate of heat removal from the conditioned space, J/hr in (Btu/hr) to the rate of energy input, in watts, for a complete heat pump or refrigerating plant, or some specific portion of that plant, under designated operating conditions (same as COP_c except for units).

Seasonal Energy Efficiency Ratio, Cooling ($SEER_c$)—Ratio of the total heat removed, in J, during the normal usage period of cooling (not to exceed 12 months) to the total energy input, in watt-hours, during the same period.

Refrigerants

Ammonia and sulfur dioxide were among the first refrigerants in vapor-compression systems. Most of the early refrigerants were thermodynamically suitable. Some of them, such as methylchloride, sulfur dioxide, carbon dioxide, ammonia, and some hydrocarbons, became commercially important. However, all of these refrigerants suffered from one or more properties that restricted their general use. Some are toxic or flammable, and some are both. Others are lacking in chemical stability or operate at very high pressures. In the 1930s the fluorocarbon refrigerants were developed and introduced into the refrigeration and air-conditioning industry.

The refrigerant system of numbering and categorizing refrigerants has been greatly expanded since the 1930s by the American Society of Heating, Refrigerating, and Air-Conditioning Engineers and now includes inorganic, hydrocarbon, azeotropic, and mixed refrigerants. The *ASHRAE Handbook 1993 Fundamentals* (6) provides a description and explanation of the refrigerant numbers, as well as tabular and graphic data for thermodynamic and thermophysical properties of refrigerants.

Table 1 shows the theoretical calculated performance of a number of refrigerants for the U.S. standard cycle of −15°C evaporation and 30°C condensation. The table can be used to compare the properties of various refrigerants, but actual operating conditions will be somewhat different from the theoretical calculated data.

The choice of refrigerant for an application often depends on characteristics not directly related to its ability to remove heat. Such characteristics are flammability, toxicity, density, viscosity, condensing and evaporating pressure, and availability. As a rule, the choice of a refrigerant is a compromise between conflicting desirable properties. For example, the pressure in the evaporator should be as high as possible to avoid air leakage into the evaporator, but at the same time, a low condensing pressure is desirable so that dangerous pressure conditions do not prevail.

Refrigerant-12 is the refrigerant that has been used most widely with reciprocating compressors in domestic refrigerators and freezers, commercial and industrial air conditioners, and some automotive refrigeration systems. Refrigerant-22 competed with refrigerant-12 in smaller air-conditioning units and is competitive with ammonia in industrial low-temperature systems where toxicity is of concern. The greatest use of ammonia is for large industrial systems and low-temperature applications. Refrigerant-11 and refrigerant-113, which have much lower densities, have been the most widely used refrigerants in centrifugal compressor systems. The advantages of refrigerant-12 over refrigerant-22 were lower cost, head pressure, and compressor discharge temperature, and less tendency to leak. An advantage of refrigerant-22 is a lower volume flow per unit of refrigeration and hence a smaller compressor for a given refrigeration capacity. However, all this is changing.

In the search for alternatives to CFCs, many industry sectors have chosen R-134a as the preferred replacement for R-12. Considerable effort has been made to provide practical solutions to the challenges of new equipment with this new refrigerant and its associated lubricants. In addition, there is a considerable stock of existing equipment (valued at hundreds of billions of dollars) that must be maintained, despite vanishing supplies of CFCs. Increasing attention is being paid to the retrofit of existing

Table 1. Comparative Refrigerant Performance Per Ton of Cooling[a]

No.	Refrigerant Name	Evaporator Pressure, lb/in² gauge	Condensing Pressure lb/in² gauge	Compression Ratio	Net Refrigerating Effect, Btu/lb	Refrigerant Circulated, lb/min	Liquid Circulated, in³/min	Specific Volume of Suction Gas, ft³/lb	Compressor Displacement, ft³/min	Horse-Power, hp	Coefficient of Performance	Compressor Discharge Temperature, °F
170	Ethane	221.3	661.1	2.86	58.6	3.41	342.9	0.53	1.82	1.953	2.41	122
744A	Nitrous oxide	294.3	922.3	3.03	85.2	2.35	71.2	0.28	0.66	1.310	3.60	
744	Carbon dioxide	317.5	1031.0	3.15	55.5	3.62	167.1	0.27	0.96	1.840	2.56	151
13B1	Bromotrifluoromethane	63.2	247.1	3.36	29.3	6.86	123.8	0.38	2.63	1.030	4.25	124
1270	Propylene	37.0	167.0	3.51	173.0	1.1	61.5	2.61	3.03	1.046	4.51	108
290	Propane	27.2	140.5	3.70	121.0	1.65	94.0	2.48	4.09	1.030	4.58	97
502	22/115 azeotrope	36.0	175.1	3.75	45.7	4.38	99.4	0.82	3.61	1.079	4.37	99
22	Chlorodifluoromethane	28.2	158.2	4.03	70.0	2.86	67.4	1.24	3.55	1.011	4.66	128
115	Chloropentafluroethane	24.0	135.8	3.89	29.1	6.88	151	0.77	5.30	1.17	4.02	86
717	Ammonia	19.6	154.5	4.94	474.4	0.422	19.6	8.15	3.44	0.989	4.76	210
500	12/152a azeotrope	16.4	112.9	4.12	60.6	3.30	80.3	1.50	4.95	1.01	4.65	105
12	Dichlorodifluoromethane	11.8	93.3	4.08	50.0	4.00	85.6	1.46	5.83	1.002	4.70	101
40	Methyl chloride	6.5	80.0	4.48	150.2	1.33	40.9	4.47	5.95	0.962	4.90	172
600a	Isobutane	3.3[c]	44.8	4.54	111.5	1.79	91.0	6.41	11.50	1.083	4.36	80
764	Sulfur dioxide	5.9[c]	51.8	5.63	141.4	1.41	26.6	6.42	9.09	0.968	4.87	191
630	Methylamine	9.9[c]	46.8	6.13	304.0	0.66	28.2	15.54	10.23	0.978	4.81	
600	Butane	13.2[c]	26.9	5.07	128.6	1.56	75.9	9.98	15.52	0.953	4.95	88
114	Dichlorotetrafluoroethane	16.1[c]	22.0	5.42	43.1	4.64	89.2	4.34	20.14	1.049	4.49	86
21	Dichlorofluoromethane	19.2[c]	16.5	5.96	89.4	2.24	45.7	9.13	20.43	0.941	5.01	142
160	Ethyl chloride	20.5[c]	12.4	5.83	142.3	1.45	45.8	17.06	24.82	0.906	5.21	106
631	Ethylamine	23.1[c]	10.0	7.40	225.5	0.89	349.0	32.32	38.67	0.855	5.52	
11	Trichlorofluoromethane	23.9[c]	3.5	6.19	66.8	2.99	56.6	12.21	36.54	0.938	5.03	111
611	Methyl formate	26.3[c]	1.6[c]	7.74	189.2	1.06	29.9	48.25	51.00			
610	Ethyl ether	26.9[c]	4.9[c]	8.20	126.3	1.58	62.9	35.00	55.40	0.822	5.74	
30	Methylene chloride	27.6[c]	9.5[c]	8.60	134.6	1.49	30.9	49.90	74.30	0.963	4.90	205
113	Trichlorotrifluoroethane	27.9[c]	13.9[c]	8.02	53.7	3.73	66.5	27.38	102.03	0.973	4.84	86
1130	Dichloroethylene	28.3[c]	15.8[c]	8.42	114.3	1.75	38.3	63.60	111.20	0.973	4.83	
1120	Trichloroethylene	29.6[c]	26.2[c]	11.65	91.7	2.18	41.6	229.40	502.00	0.980	4.82	

[a] Reprinted with permission of the American Society of Heating, Refrigerating and Air-Conditioning Engineers.
[b] lb/in² gauge = 6.894 kPa gauge; 1 Btu/lb = 2326 J/kg; 1 lb/min = 0.00756 kg/s; 1 in³/min = 2.73×10^{-7} m³/s; 1 ft³/lb = 0.062 m³/kg; 1 ft³/min = 4.72×10^{-4} m³/s; 1 hp = 0.7457 kW; °C = (°F − 32)/1.8.
[c] Inches Hg vacuum.

equipment, with respect to both cost and performance. R-134a is an attractive alternative to R-12 for many reasons. It is a good performance match to R-12 at medium- and high-temperature refrigeration conditions, and it has zero ozone depletion potential (ODP) and low direct global warming potential (GWP). In addition, R-134a is nonflammable, it has extremely low toxicity and high thermal stability, and is now available in commercial quantities. R-12 systems are being successfully retrofitted with R-134a today. Retrofitting to R-134a has been possible largely because of the development of synthetic polyolester lubricants. These lubricants exhibit excellent physical and chemical stability characteristics under retrofit conditions. It is possible to predict and achieve excellent performance in retrofitted systems with every expectation that such retrofitted equipment will serve out its normal, useful life.

It has been estimated that approximately 58,000 to 65,000 centrifugal chillers employing R-11 as the refrigerant as currently in use in the United States. Accordingly, finding a suitable alternative for R-11 has presented the HVAC&R industry with a formidable challenge. R-123 (1,-dichloro-2,2,2-trifluorethane) is generally recognized as the leading alternative refrigerant for near-term use. It is being adopted by chiller manufacturers for use in new equipment and, to some extent, by users for retrofitting existing equipment.

Because of its chlorine content and consequent potential to deplete stratospheric ozone, however, R-123 will eventually be phased out of production. Therefore, it is viewed as a transitional alternative. In November 1992, parties to the Montreal Protocol agreed to a phaseout schedule for all HCFCs including R-123. This schedule calls for placing a production cap on all HCFCs beginning in 1996. Additional reductions from the 1996 cap are prescribed in a series of graduated steps, with total production phaseout to occur by the year 2030. This phaseout time frame should allow new equipment designed for use with R-123 to be placed into service within the near term to continue to be used over an equipment lifetime of approximately 30 years. Several years are normally required to evaluate thoroughly any new alternative refrigerant for toxicity, environmental acceptability, end-use applicability, and production viability and cost, and to bring the new chemical to market. Recognizing this long lead time and the need to ensure that acceptable alternatives are available in the far term to take the place of all ozone-depleting substances, the EPA's Air and Energy Engineering Research Laboratory (AAERL) and the Electric Power Research Institute (EPRI) initiated a project in 1988 to investigate an array of new potential alternative refrigerants.

This effort has revealed several promising new refrigerants including at least one that appears to be worthy of consideration as a far-term replacement for R-11 and R-123. The compound is HFC-245ca (1,1,2,2,3-pentafluoropropane). Refrigerant-245ca has an ozone depletion potential of zero. A measured rate of reaction with hydroxyl radicals indicates a relatively short atmospheric lifetime of 6.4 years of R-245ca. This, in turn, leads to an estimated global warming potential for the refrigerant only approximately one-third that of R-134a. Toxicity testing has been very limited, but no signs of acute inhalation toxicity were observed at a concentration level of 993 parts per million (ppm). With the long lead times (7 to 10 years) typically required for any new candidate chemical to be brought to market, it seems prudent to evaluate compounds such as R-245ca now in anticipation of the eventual phaseout of all ozone-depleting refrigerants.

A commonly used refrigerant for low-temperature applications, R-502 is a binary minimum boiling azeotrope composed of HCFC-22 (chlorodifluoromethane) and YEAR-115 (Chloropentafluoroethane). R-502 has a low boiling point, which makes it the refrigerant of choice in many applications such as supermarket frozen food cases and transport refrigeration.

Because of the large installed investment in low-temperature refrigeration systems operating with R-502, an alternative with similar properties is needed to permit continued use of this equipment. Two near-azeotropic mixtures of HFC-125 (pentafluoroethane), HC-290 (propane) and HCFC-22 have been developed with thermophysical properties and refrigeration performance parameters very similar to R-502. These alternative refrigerants for R-502 provide significant reductions in ozone depletion and global warming potentials, plus cooling performance that is essentially the same as R-502.

Compressors

The compressor, of which there are many types, is one of the four essential parts of the vapor-compression system. Reciprocating compressors are most commonly used for systems in the general range of 0.45 to 90.7 t (0.5 to 100 tons) and larger. They are used in unitary heat pumps, are in most cases either fully or accessibly hermetic, and should be designed for this purpose. The hermetic compressor has an integral motor with a separate control panel. The open compressor has no motor, base, or controls.

Another type of compressor that is commonly used is the positive-displacement compressor. One of the important thermodynamic considerations for this compressor is the effect of the clearance volume (that is, the volume occupied by the refrigerant within the compressor that is not displaced by the moving member). This effect is illustrated, in the case of the piston-type compressor, by considering the clearance volume between the piston and the cylinder head when the piston is in a top, dead-center position. The clearance gas remaining in this space after the compressed gas is discharged from the cylinder reexpands to a larger volume as the pressure falls to the inlet pressure (see Figure 3). As a consequence, the mass of refrigerant discharged from the compressor is less than the mass that would occupy the volume swept by the piston, measured at inlet pressure and temperature.
This effect is quantitatively expressed by the volumetric efficiency. η_v:

$$\eta_v = \frac{m_a}{m_i}$$

where

m_a = the actual mass of new gas entering the compressor per stroke

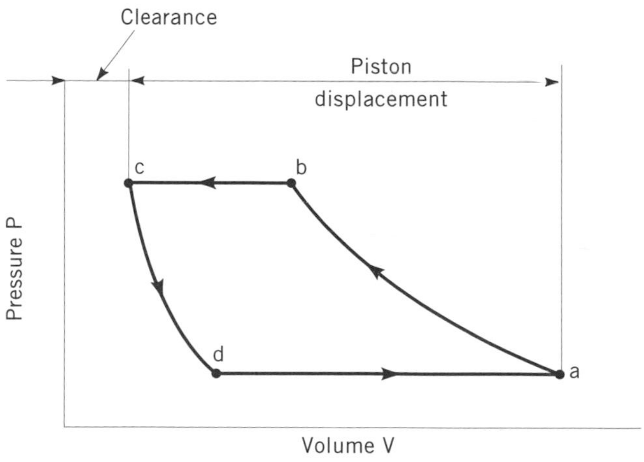

Figure 3. Cycle for idealized compressor.

m_i = the theoretical mass of gas represented by the displacement volume and determined at the pressure and temperature at the compressor inlet

If the effect of clearance alone is considered, the resulting expression may be termed clearance volumetric efficiency. The expression used for grouping into one constant all the factors affecting efficiency may be termed total volumetric efficiency. The clearance volumetric efficiency can be calculated with reasonable accuracy. For the simple cycle, the clearance volumetric efficiency becomes

$$\eta_v = \frac{V_b - V_a}{V_b - V_d}.$$

If $C = V_d/(V_b - V_d)$ = clearance ratio, this equation can be rearranged to

$$\eta_v = 1 + C - C\left(\frac{V_a}{V_d}\right).$$

The clearance volumetric efficiency is a measure of how well the piston displacement (size) of the compressor is used in moving refrigerant vapor through the cycle. The choice of refrigerant greatly affects V_a and thus the mass flow that a given compressor displacement can deliver, because one of the most significant differences among modern refrigerants is the specific volume V_a at a given evaporator temperature and pressure.

The total volumetric efficiency of a compressor is best obtained by actual laboratory measurements of the amount of refrigerant compressed and delivered to the condenser. The differences between actual and predicted volumetric efficiency, considering only clearance volume effects, is shown in Figure 4.

An example of a useful presentation of capacity data is given in Figure 5. This is a typical set of curves for a 4-cylinder, semihermetic compressor with a 60.3-mm ($2\frac{3}{8}$ in) bore, 94.4-mm ($1\frac{3}{5}$ in) stroke, and 1720 rpm, operating with Refrigerant-22. Figure 5 also shows a set of power curves for the same compressor. Figure 5 shows the heat rejection curves.

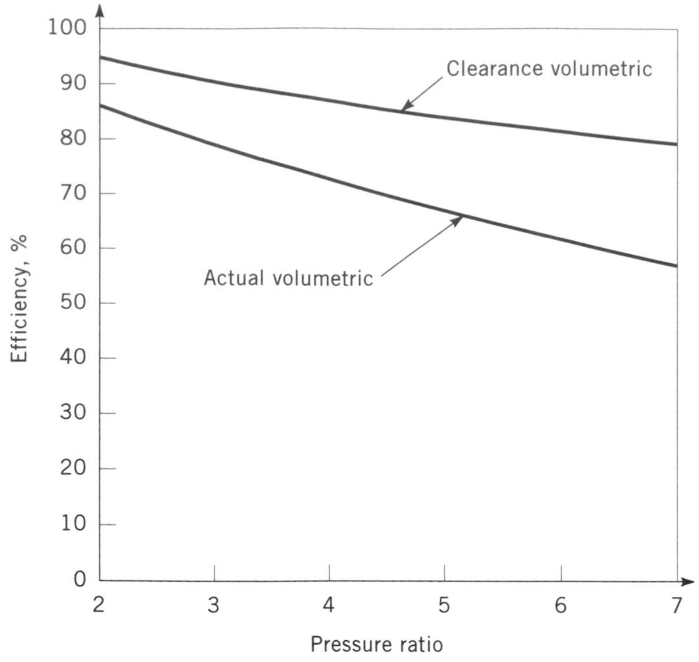

Figure 4. Volumetric efficiency as a function of pressure ratio.

Centrifugal compressors, or turbocompressors, are characterized by a continuous exchange of angular momentum between a rotating mechanical element and a steadily flowing fluid. Because their flows are continuous, turbomachines have greater volumetric capacities, size for size, than do positive displacement devices. For effective momentum exchange, their rotative speeds must be higher, but little vibration or wear results because of the steadiness of the motion and the absence of contacting parts.

Centrifugal compressors are used in a wide variety of refrigeration and air-conditioning applications. As many as eight or nine stages can be installed in a single casing.

Condensers

The condenser in a refrigerating system removes, from the compressed refrigerant gas, the energy gained during compression and the heat absorbed by the refrigerant in the evaporator. The refrigerant is thereby converted back into the liquid phase at the condenser pressure and is available for reexpansion into the evaporator. The common forms of condensers may be broadly classified on the basis of the cooling medium: (a) water-cooled, (b) air-

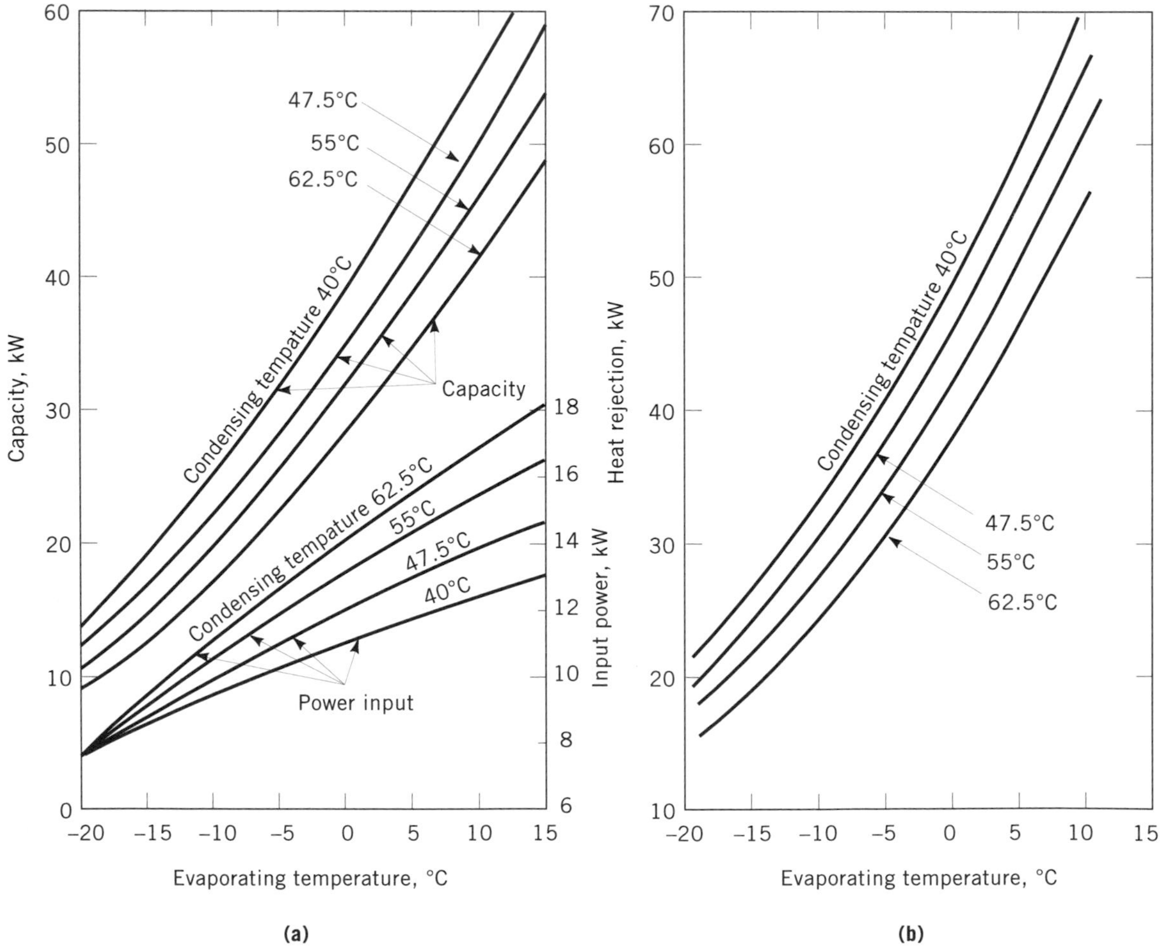

Figure 5. Typical capacity, power input, and heat rejection curves for hermetic reciprocating compressor. Refrigerant 22, 5.5°C liquid subcooling, 11°C superheated vapor, 29 r/s compressor speed. Courtesy of the American Society of Heating, Refrigerating, and Air-Conditioning Engineers.

cooled, and (c) evaporative (air- and water-cooled). Because it gives up heat to the air pushed across the condenser surface, refrigerant vapor is condensed inside the tubes of the air-cooled condenser. Water-cooled condensers have cooling water flowing inside the tubes; thus the refrigerant condenses inside the shell but outside the tubes. The selection of a condenser depends on the cooling load, refrigerant used, source and temperature of the available cooling fluid, amount of coolant that can be circulated, condenser location, required operating pressures, and maintenance considerations.

The heat-transfer process in an air-cooled condenser has three main phases: (a) desuperheating, (b) condensing, and (c) subcooling. Desuperheating, condensing, and subcooling zones will vary 5% to 10%, depending on the temperature of the entering gas and of the existing liquid. Condensing takes place in approximately 85% of the condenser area, at a substantially constant temperature. The indicated drop in condensing temperature is due to the friction loss through the condenser coil.

Coils in air-cooled condensers are commonly constructed of copper, aluminum, or steel tubes, ranging from 6.3 to 19 mm ($\frac{1}{4}$ to $\frac{3}{4}$ in.) in diameter. Copper, the most expensive material, is easy to use in manufacturing and requires no protection against corrosion. Aluminum requires exact manufacturing methods, and special protection must be provided if aluminum-to-copper joints are made. Steel tubing is also used, but weather protection must be provided.

Fins are used to improve the air-side heat transfer. Most fins are made of aluminum, but copper and steel are also used. The most common forms are plate fins making a coil bank, plate fins individually fastened to the tube, or a fin spirally wound onto the tube. The number of fins per meter varies from 160 to 1180 (4 to 30 per in.). The most common range at present is 315 to 700 fins per meter (8 to 18 per in.).

Evaporators

Direct expansion and shell-and-tube evaporators, or flooded coolers, are the types employed in most refrigeration systems. The direct expansion evaporator is basically a coil that contains refrigerant over which air is passed for cooling purposes.

In the flooded cooler, the refrigerant is vaporized on the outside of bar or augmented surface tubes that are submerged in evaporating liquid refrigerant within a closed shell. The cooled liquid flows through these tubes, which may be straight, U-shaped, or coiled. Space is usually provided above the tubes submerged in the boiling refrigerant for the separation of liquid droplets from the exiting vapor. The size and number of tubes determine the velocity of the fluid being cooled. The velocity should be held between the limits of 2 and 4 m/s, but it may vary above and below these limits when clean liquid, devoid of suspended abrasive or fouling substances, is used.

Expansion Devices

The control of refrigerant flow is an essential feature of any refrigeration system. The expansion device is a flow restriction—a small opening or a long length of small-bore tubing—so neither work nor any significant amount of heat transfer occurs. Hence, the enthalpy of the refrigerant does not change significantly during the expansion process. Every refrigerating unit requires a pressure-reducing expansion device to meter the flow or refrigerant to the low side in accordance with the demands placed on the system. With the advent of the hermetic compressor and halocarbon refrigerants, the capillary tube became practical and rapidly achieved popularity, especially with the smaller unitary hermetic equipment such as household refrigerators and freezers, dehumidifiers, and room air conditioners. Recently it has been used in larger units such as unitary air conditioners in sizes up to 365.2-kW (10-ton) capacity.

The capillary tube is a small-bore tube that acts as a flow restriction and reduces the pressure. The capillary operates on the principle that liquid passes through it much more readily than gas goes . It consists of a small-diameter line that when used to control the flow of refrigerant in a system, connects the outlet of the condenser to the inlet of the evaporator. It is sometimes soldered to the outer surface of the suction line for heat exchange.

The capillary tube is popular for smaller, unitary hermetic equipment, such as household refrigerators and freezers, dehumidifiers, and room air conditioners in sizes up to 35-kW (10-ton) capacity. If a system unbalance occurs so that some gas (uncondensed refrigerant) enters the capillary, this gas tends to reduce considerably the mass flow of refrigerant with little or no change in the system pressures. If the opposite type of imbalance occurs, liquid refrigerant backs up in the condenser. This condition tends to cause subcooling and increases the mass flow of refrigerant. Thus, a capillary properly sized for the application tends to compensate automatically for load and system variations and gives acceptable performance over a wide range of operating conditions.

The expansion devices normally used to control the refrigerant flow in larger refrigeration units and in heat pumps are thermostatic expansion valves. These valves are more complex and have better control capability for the evaporator. They are commonly used in domestic and commercial air-conditioning and refrigeration units because they automatically control the refrigerant superheat and the flow of refrigerant to the evaporator. Expansion valves are also used as metering devices in many air-conditioning systems, although they are normally not used with systems under 4 tons. The expansion valve controls the flow of refrigerant by means of a needle valve placed in the refrigerant line. Liquid refrigerant enters the valve from the high-pressure side of the system and passes through the needle valve to the low-pressure side. Here its pressure is reduced, causing a portion of it to vaporize immediately, cooling the balance of the refrigerant.

ABSORPTION REFRIGERATION

Absorption Cycles

Absorption-refrigeration cycles are heat-operated cycles in which a secondary fluid—the absorbent—is used to absorb the primary fluid—gaseous refrigerant—that has been vaporized in the evaporator. The basic absorption cycle is shown in Figure 6.

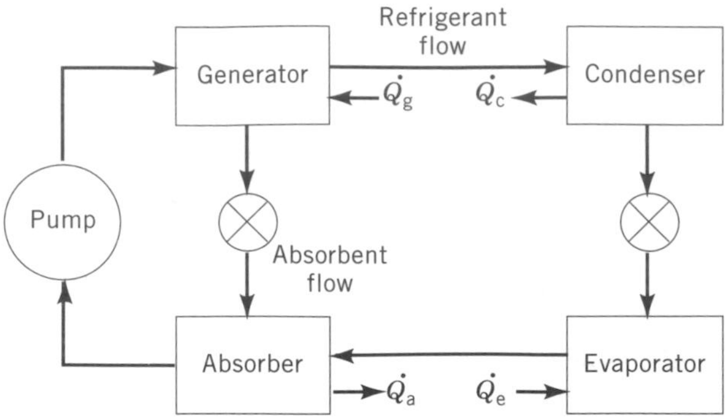

Figure 6. Basic absorption refrigeration cycle.

In the basic absorption cycle, low-pressure refrigerant vapor is converted to a liquid phase (solution) while still at low pressure. Conversion is made possible by the vapor being absorbed by a secondary fluid, the absorbent. Absorption proceeds because of the mixing tendency of miscible substances and, generally, because of an affinity between absorbent and refrigerant molecules. Thermal energy released during the absorption process must be disposed of to a sink. This energy originates in the heat of condensation, sensible heat, and normally, the heat of dilution.

The frigerant-absorbent solution is pressurized in the generator, where refrigerant and absorbent are separated (ie, regenerated) by distillation. A simple still is adequate for the separation when the pure absorbent material is nonvolatile, as in the water–lithium bromide system. However, fractional distillation equipment is required when the pure absorbent material is volatile, as in the ammonia-water system. If the refrigerant is not essentially free of absorbent, vaporization in the evaporator is hampered. The regenerated absorbent normally contains a substantial amount of refrigerant. If the absorbent material tends to become solid, as in the water–lithium-bromide system, enough refrigerant must be present to keep the pure absorbent material in a dissolved state at all times. Certain practical considerations—particularly, the avoidance of excessively high temperatures in the generator—make it generally desirable to leave a moderate amount of refrigerant in the regenerated absorbent. The high-temperature energy required for regeneration approximately equals the immediate-temperature energy released in absorption.

As shown in Figure 6, the refrigerant and absorbent have different circulating patterns. The refrigerant goes from generator to condenser, evaporator, absorber, and solution pump—and back to the generator. The absorbent may be thought of as a carrier fluid, that is, it carries spent refrigerant from the low-pressure side of the cycle to the high-pressure side.

The absorption cycle and the mechanical-compression cycle have the evaporation and the condensation of a refrigerant liquid in common; these processes occur at two pressure levels within the unit. The two cycles differ in that the absorption cycle uses a heat-operated generator to produce the pressure differential, whereas the mechanical-compression cycle uses a compressor. The absorption cycle substitutes physiochemical processes for the purely mechanical processes of the compression cycle. Both cycles require energy for operation: heat in the absorption cycle, mechanical energy in the compression cycle. The distinctive feature of the absorption system is that very little work input is required (low cost) because the pumping process involves a liquid. However, more equipment is involved (high cost) in an absorption system than in the vapor-compression cycle.

The principal source of inefficiency in the basic absorption-refrigeration cycle is sensible-heat effects. Conveying hot absorbent from the generator into the absorber wastes a considerable amount of thermal energy. A liquid-to-liquid heat exchanger, which transfers energy from this stream to the refrigerant-absorbent solution being pumped back to the generator, saves a major portion of the energy. There is further need of heat exchange in an ammonia-water machine. The absorbent reaches such a high temperature in the generator that it is necessary to transfer some heat by a coil in the analyzer or stripping section. Also, heat must be removed in a partial condenser that forms reflux for the system. Means of accomplishing these two purposes are shown in the flow diagram for an ammonia-water machine (Figure 7).

Of the many combinations of working fluids that have been tried, only the lithium-bromide–water and the ammonia-water cycles remain in common use in air-conditioning equipment. Ammonia-water absorption equipment has also been used in large-tonnage industrial applications requiring low temperatures for process work.

Lithium Bromide–Water Equipment

The process conditions in a lithium bromide absorption-refrigeration unit require that all operations be conducted in a vacuum 5 to 100 mm Hg. (0.1 to 2 lbf/in 2 abs). To minimize the penalties of vapor-flow pressure drops, operations at about the same pressure are conducted in the same vessel, the respective liquid pools being separated by a partition. Hence, the evaporator and the absorber are combined in one vessel and the regenerator and the condenser in the other. Mounting the regenerator-condenser vessel above the evaporator-absorber lets the condensed refrigerant (water) and the concentrated lithium-bromide solution flow by gravity. The diluted lithiumbromide solution is pumped from the absorber to the regenerator. It is a common industrial practice to provide for pumped recirculation of adsorbate or condensate over the various cooling coils to promote effective heat transfer.

In small commercial units, the concentrations of lithium-bromide solutions typically are 54% and 58.5% to ensure against crystallization of the salt, particularly on shutdown. High concentrations, normally about 60% and 64.5%, are used in large commercial units that operate at higher absorber temperatures and thus save on heat-exchanger costs. Control (mixing, temperature, flow velocity, and so on) and a shutdown dilution cycle are used with these large units to prevent crystallization.

Figure 8 is a typical diagram of machines that are available in the form of indirect-fired liquid chillers (two-shell) in capacities of 1976 to 5280 kW (50 to 1500 tons).

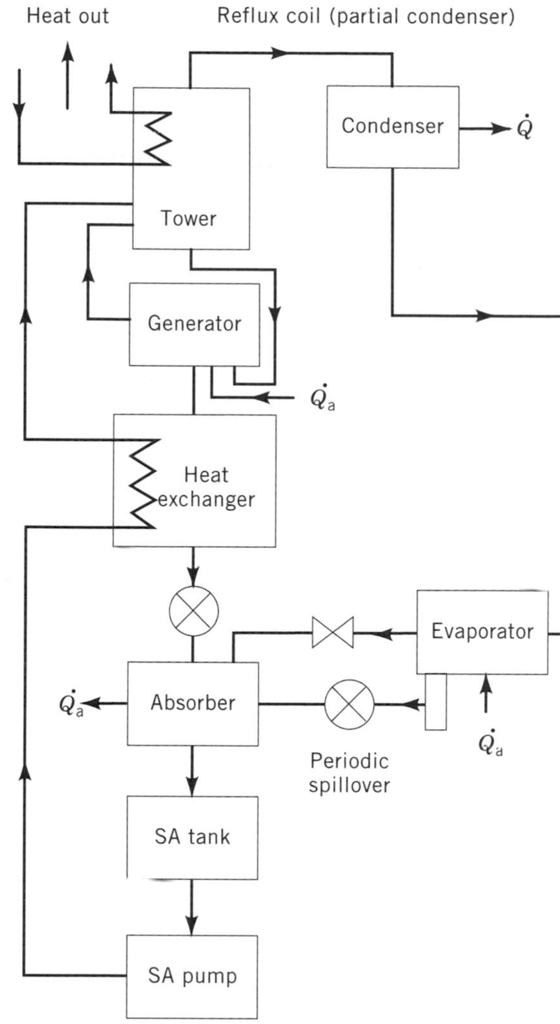

Figure 7. Ammonia absorption refrigeration cycle. Courtesy of the American Society of Heating, Refrigerating, and Air-Conditioning Engineers.

All lithium-bromide–water-cycle absorption machines meet load variations and maintain chilled water temperature control by varying the rate of reconcentration of the absorbent solution. At any given constant load, the chilled water temperature is maintained by a temperature difference between the refrigerant and the chilled water. The refrigerant temperature is maintained in turn by the absorber being supplied with a flow rate and a concentration of solution, and by the absorber's cooling water temperature. Load changes are reflected by corresponding changes in chilled water temperature. A load reduction, for example, results in less temperature difference being required in the evaporator and a reduced requirement for solution flow or concentration. The resulting chilled water temperature drop is met basically by adjusting the rate of reconcentration to match the reduced requirements of the absorber. The coefficient of performance (COP) of a lithium-bromide–water-cycle absorption machine operating at nominal conditions is typically in the range of 0.65 to 0.70.

Small-tonnage units are designed for residential or limited commercial use in the 10- to 90-kW (3- to 25-ton)

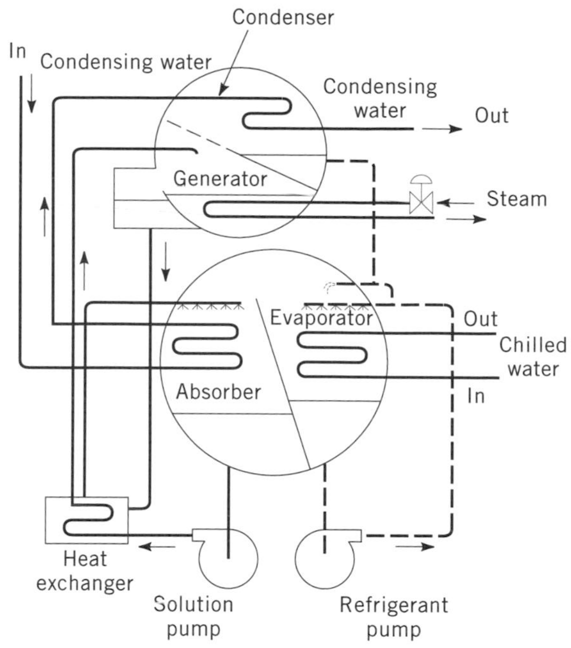

Figure 8. Diagram of two-shell lithium bromide cycle water chiller. Courtesy of the American Society of Heating, Refrigerating, and Air-Conditioning Engineers.

capacity range. They have been produced as indirect- and direct-fired liquid-chiller, chiller-heater, and air-conditioning forms. Currently available equipment uses one or more of the following unique capabilities of the water-cooled cycle:

1. The cycle can be efficiently heated by a flat-plate collector heat source.
2. Heating can be derived from the cooling cycle by stopping condenser-water flow.
3. Solution can be circulated thermally by vapor-lift action in a "pump" tube.

Small-tonnage units operate without mechanical pumps. Solution circulation between the absorber and the generator is by vapor-lift action in the pump tube. The absorber and evaporator tubes are wetted by a one-pass flow of liquid delivered through capillary drippers. Generators are of steel fire-tube construction, usually with atmospheric gas burners. Power gas burners and oil burners have also been used.

Aqua-Ammonia (Ammonia-Water) Equipment

Figure 9 is a diagram of a typical aqua-ammonia absorption system. The design of ammonia-water equipment varies from lithium-bromide–water equipment to accommodate three major differences:

1. Water (the absorbent) is also volatile, so the regeneration of weak absorbent to strong absorbent is a fractional distillation process.
2. Ammonia (the refrigerant) causes the cycle to operate at condenser pressures in the 2070 kPa (300

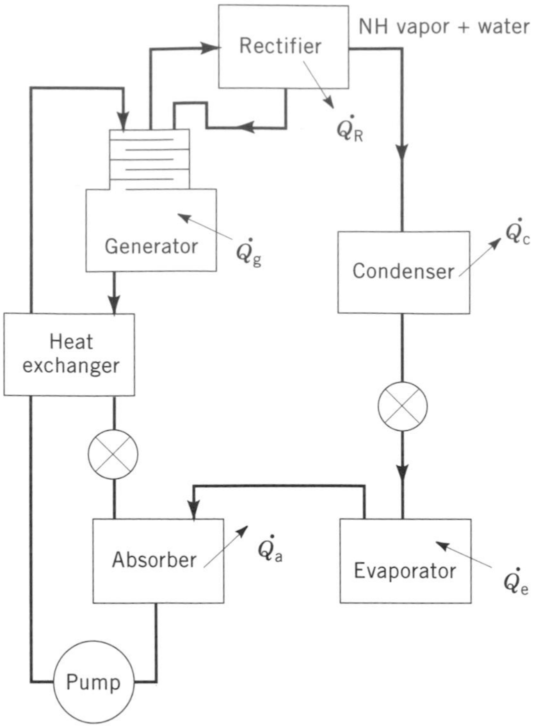

Figure 9. Aqua–ammonia absorption cycle.

psia) range and at evaporator pressures in the 480 kPa (70 psia) range, so that vessel sizes are held to a diameter of 152 mm (6 in), and solution pumps are positive-displacement types.

3. Air cooling requires condensation and absorption to take place inside the tubes so that the outside surface of the tubes can be finned for greater contact with the air.

As indicated, the ammonia-water cycle must be more complex than the water–lithium-bromide cycle to provide acceptable performance. More heat recovery means are required, and the simple still must be replaced by a complete distillation system.

Solar-Activated Absorption Refrigeration

When solar energy is to be used for cooling as well as heating, the absorption system shown in Figure 10 can be used. The collector and storage subsystem must be able to operate at temperatures approaching 93.3°C on hot summer days when the water from the cooling tower exceeds 26.7°C, but considerably lower operating water temperatures can be used when cooler water is available from the tower. The controls for the collection, cooling, and distribution subsystems are generally separated, with the circulating pump, p_1, operating in response to the collector thermostat, T_1. The heat-distribution (system responds to thermostat T_2, which is located within the air-conditioned space. When T_2 calls for heating, valves V_1 and V_2 are positioned to direct the water flow from the storage tank through the inactivated auxiliary heater to the fan coil in the air-distribution system. The fan, F_1, in this unit may also respond to the thermostat, or it may have its own control circuit so that it can bring in outdoor air during suitable temperature conditions.

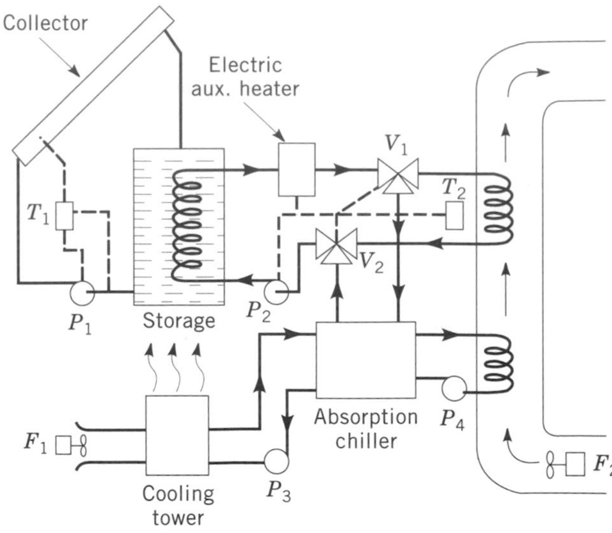

Figure 10. Space heating and cooling system using powered lithium bromide system. Courtesy of the American Society of Heating, Refrigerating, and Air-Conditioning Engineers.

When thermostat T_2 calls for cooling, the valves are switched to direct the hot water into the absorption unit's generator, the pumps P_3 and P_4 are activated to cause the cooling-tower water to flow through the absorber and condenser circuits and the chilled water to flow through the cooling coil in the air-distribution system. A large hot water storage tank enables the unit to continue operating when there is no sunshine available. It may also be desirable to include a chilled-water storage tank (not shown) so that the absorption unit can operate during the day whenever there is water available at a sufficiently high temperature to make the unit function properly. The coefficient of performance of a typical absorption unit of the lithium-bromide–water type may be as high as 0.75 under favorable conditions, but frequent cycling of the unit to meet a high-variable cooling load will result in significant loss in performance owing to the necessity of heating the unit to operating temperature after each shutdown.

Water-cooled condensers are required with the absorption cycles available today because the lithium-bromide–water cycle must maintain a relatively delicate balance among the temperatures of the three fluid circuits—cooling-tower water, chilled water, and activating water. The steam-operated absorption systems, from which today's solar cooling systems are derived, customarily operate at energizing temperatures of 100 to 115.6°C, but these temperatures cannot be obtained by most flat-plate collectors. The solar cooling units are designed to operate at considerably lower temperatures, but then unit ratings are also lower.

AIR-CYCLE REFRIGERATION

As the name implies, an air-cycle refrigeration unit uses air as the refrigerant. The most common cycle of this type is the reversed Brayton cycle, in which the air is compressed, cooled in a heat exchanger, and expanded through a turbine to a low temperature at which it is capable of having a cooling effect.

Air-cycle equipment is lightweight and small. The turbine may weigh only several kilograms and spin at speeds up to 100,000 rpm. The complete refrigeration unit is light and compact. However, one disadvantage of the air cycle is that it is not as efficient as the vapor-compression cycle.

Because of their light weight and compactness, air-cycle refrigeration systems are more commonly used in the air-conditioning of aircraft than in surface and stationary applications. However, the development of lightweight vapor-cycle refrigeration with high-speed compressors has somewhat decreased the use of air-cycle equipment in aircraft systems. Air-cycle systems are used in the Boeing 707, 727, 737, 757, and 767 and the McDonnell-Douglas DC-10. Vapor-cycle systems have been used in the Douglas DC-8, Convair 880, and some versions of the Boeing 707. Combined systems are sometimes used (eg, in the Lockheed Electra). Air-cycled systems for military and commercial aircraft involve various equipment arrangements. Figure 11 shows an elementary type of system for a jet fighter plane. All the air that circulates through the cabin is exhausted. Air-cycle air-conditioning has been found uneconomical in residential and commercial buildings because of the high power required.

In a refrigeration system using the open air-cycle principle, the refrigerant is air, which is used directly to cool the space requiring refrigeration. Unlike the refrigerant of a vapor-cycle system, which continuously changes phase from a liquid to a gas and back to a liquid again, the refrigerant of an air-cycle system remains in the gaseous phase throughout the cycle. Refrigeration in a simple, open air-cycle system is obtained by three basic steps: (a) compression, (b) heat transfer, and (c) expansion accompanied by work extraction. Air is first compressed in a compressor to a pressure higher than that of the space to be cooled. The heat of compression then is removed from the air in a heat exchanger by rejecting it to a suitable heat sink such as air. In the third step, air is cooled by extracting work from it as it expands through a turbine. Under humid, ambient conditions, condensation occurs in the expansion process of the air discharged from an open air-cycle refrigeration unit, and the air often contains water in the form of mist, fog, or even snow. It is important to install a water separator downstream of the air-cycle refrigeration unit for the removal of a sufficiently large fraction of the entrained moisture.

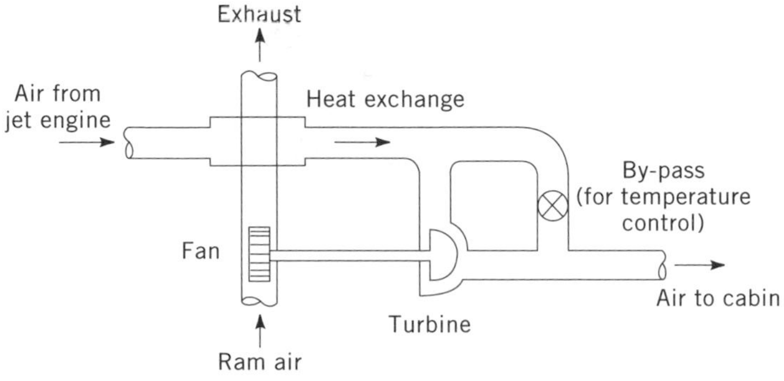

Figure 11. Open cycle cooling system for a jet fighter aircraft.

Aircraft cabins require cooling to offset heat gains from passengers, electrical and mechanical equipment, solar radiation, and heat transmission through the walls of the plane. In addition, the ram air temperature rise resulting from adiabatic stagnation of air taken into the plane and pressurization of the cabin further contributes to cooling needs. Air circulated through the refrigeration equipment is also circulated through the cabin. Advantages claimed for air-cycle refrigeration systems compared with vapor-compression systems in aircraft include (a) less weight per ton of refrigeration, (b) a refrigeration unit that can be easily removed and repaired, and (c) insignificance of refrigeration leakage.

The following are common sources of high-pressure air used in air-cycle refrigeration systems for aircraft:

1. Jet engine and prop-jet engine compressors
2. Auxiliary air compressors driven by the main engine, either through an air turbine or by means of a shaft
3. Auxiliary gas turbine compressor

The bleed air from some jet engines is sometimes contaminated with oil or toxic products formed by the decomposition of oil at high temperatures. If, for this reason, the turbojet engine compressor cannot be used as a source of high-pressure air, then one or more auxiliary air compressors are installed in the airplane. These air compressors can be equipped with air turbine drives and can use engine bleed air as a power source, or they can be shaft-driven from the main engine accessory gearbox. In some instances, a gas turbine compressor is used for supplying high-pressure air for the air-conditioning of airplanes on the ground. Some airplanes carry an auxiliary gas turbine on board and use it during flight for cabin pressurization and air-conditioning. Generally, to justify the weight penalty of an on-board auxiliary gas turbine, which may weigh 450 kg (1000 lb), the unit must be capable of supplying additional services, such as ground electrical power and engine starting.

The most commonly used air-cycle systems on aircraft are (a) basic (or simple) type, (b) bootstrap type, and (c) regenerative type (the basic and bootstrap types are used the most). There are also many other air-cycle systems that are variations or combinations of the three common systems.

The *basic* air-cycle refrigeration system consists of an air-to-air heat exchanger and a cooling turbine. High-pressure, high-temperature air is initially cooled in the heat exchanger and then cooled further in the cooling turbine by the process of expansion with work extraction. The heat sink for this system is ram air (the ambient air rammed into an airplane through a scoop, such as an engine scoop, as the airplane moves through the air). The work from the turbine can drive a fan, which pulls ambient air over the heat exchanger.

The term *bootstrap*, as used in air-cycle refrigeration

systems, indicates a system in which the pressure of the working fluid (high-pressure air) is raised to a higher level in the compressor section of the cooling turbine unit before the working fluid (air) expands in the turbine section. The bootstrap air-cycle refrigeration system consists of a primary heat exchanger, a secondary heat exchanger, and a cooling turbine. High-pressure air is first cooled in the primary heat exchanger. The air is then compressed to a higher pressure as the heat of compression is removed in the secondary heat exchanger, and the air is cooled further as it expands through the turbine section of the cooling turbine. Ram air is used as a heat sink in the primary and secondary heat exchangers. The bootstrap air-cycle refrigeration system is used most frequently in transport-type aircraft.

If the turbine-discharge temperature of the basic system is too high, the *regenerative* system may be necessary. In the regenerative cycle, some of the turbine-discharge air passes back to cool the air entering the turbine. Air can then enter the turbine at a lower temperature than would be possible by cooling with ambient air.

EJECTOR REFRIGERATION (FLASH COOLING)

Flash cooling is commercially important in obtaining chilled water and in the manufacture of dry ice. In this process, a flash chamber is maintained under an extremely low absolute pressure. Liquid admitted to the chamber is partially vaporized so that the remaining liquid is cooled to the saturation temperature corresponding to the chamber pressure. A relatively enormous volume for flash vapor must be handled, usually by a jet ejector.

A steam-jet compressor is commonly employed for producing chilled water. Steam-jet–water-vapor systems have few moving parts and need only low maintenance, use a cheap, nontoxic refrigerant (water vapor), and have minimum power requirements, but they also require relatively large quantities of motive steam and condensing water and are limited to flash-chamber temperatures higher than about 5°C. They can be more economical than mechanical compression systems for chilling water if low-cost steam or waste steam and sufficient condensing water are available. The principle of flash cooling is also used in the manufacture of dry ice, CO_2, with triple point of −56.5°C, and 517.8 kPa (75.1 psia) is used instead of water. Steam-jet refrigeration units were used during the early 1930s for air-conditioning large buildings. Today, the steam-jet unit retains some importance for industrial uses such as the chilling of water to moderate temperatures in process industries and, infrequently, for vacuum precooling of vegetables and concentrating fruit juices.

When water is the refrigerant in a steam-jet system, evaporation provides the refrigeration. As shown in Figure 12, the water boils in the evaporator. A sufficient quantity evaporates to cool the water that is returning from the refrigerant load. Low pressure must be maintained in the evaporator (eg, for water to evaporate at 5°C, the pressure must be 0.85 Kpa [0.25 in Hg absolute]). Some type of compressor must continuously remove the vapor from the evaporator, or refrigeration will cease. A steam jet will pump large volumes of vapor with no contamination by the steam, because water is the refrigerant. Motive steam expands through a converging-diverging nozzle and rushes out at supersonic speed. In the mixing section, the high-velocity steam entrains the slow-moving vapor from the flash chamber. The diffuser compresses the mixture to the condenser pressure. The condensate from the condenser supplies makeup water for the flash chamber and also is pumped back to the boiler. The condenser must be equipped with an air ejector to remove air that was originally in the system and that seeps in through any leaks. Steam-jet refrigeration systems have some disadvantages: (a) Steam-jet units cannot be used for refrigerating temperatures lower than 0°C; (b) about twice as much heat must be removed from the condenser of the steam-jet unit per ton of refrigeration as must be removed in the vapor-compression system.

Solar-Powered Jet Refrigeration

The coupling of a solar collector to a vapor-jet system, as shown in Figure 13, offers an alternative to more conventional solar-powered cooling systems. The solar energy vaporizes the power-cycle working fluid to saturated vapor in the boiler. Instead of the conventional turbine compres-

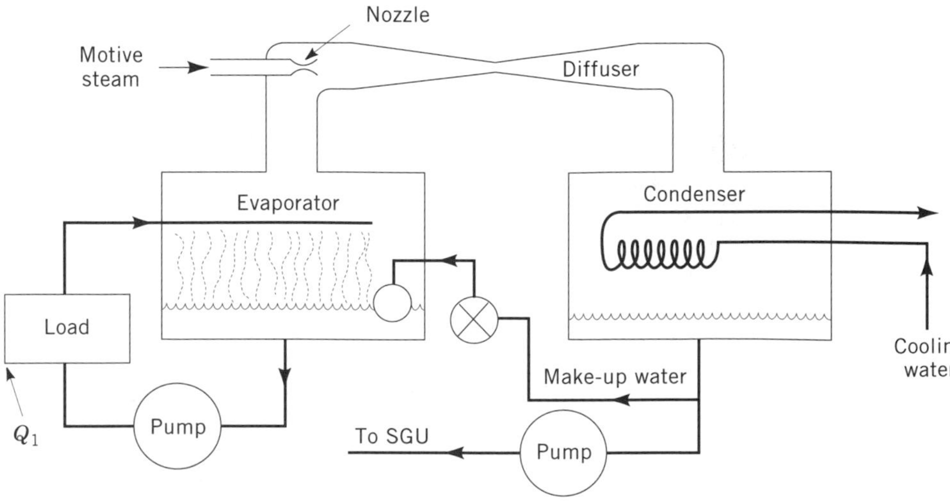

Figure 12. Steam-jet refrigeration system.

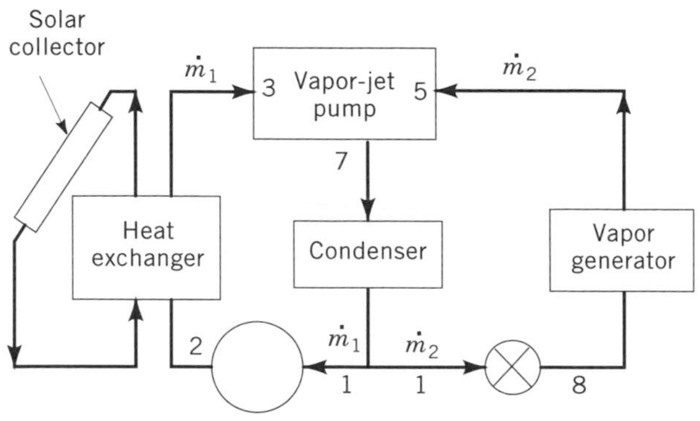

Figure 13. Solar powered ejector refrigerator.

sor, a vapor-jet pump supports all the enthalpy conversions. The balance of the system operates as a conventional single-fluid refrigerator. In an evaluation of one hypothetical system, with water as the working fluid, a coefficient of performance of 0.465 was obtained. This compares favorably with coefficients of 0.5 for a more complex solar-powered absorption refrigerator. Moreover, coefficients above 0.6 are predicted for a solar-powered vapor-jet refrigerator if the nozzle and diffuser efficiencies are raised to 0.9 each. In addition, if water is the working fluid, the coefficient of performance is double that of several organic fluids (such as butane).

VORTEX TUBE REFRIGERATION

Another type of air-cycle cooling is the Ranque-Hilsch vortex tube. Formerly a laboratory curiosity, the vortex tube has found recent applications in supplying small amounts of refrigeration, such as for small drinking-water coolers, spot cooling of a hard-to-reach critical component in an electronic control system, and air-conditioning of helmets and suits for workers in hot, humid, or toxic locations. This device, invented by Ranque in 1931 and improved by Hilsch in 1945, converts compressed air into hot and cold air. The vortex tube contains no moving parts; it is simply a straight length of tubing into which compressed air is admitted tangentially at the outer radius and so throttled that the central core of the resulting airstream can be separated from the peripheral flow. This is usually done on a counterflow arrangement, as shown in Figure 14. The central core of air is cold relative to the hot gases at

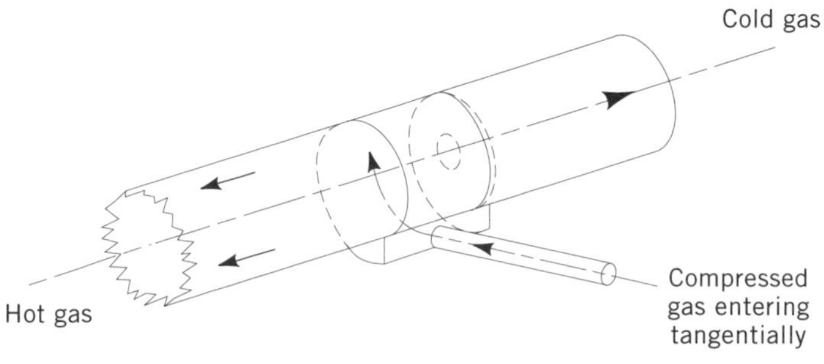

Figure 14. Vortex refrigeration tube.

the periphery. Cold-air temperatures of 38°C below nozzle temperatures are readily obtainable with moderate inlet pressures.

A gas at a moderately high pressure (eg, 5 atmospheres) enters a tube tangentially, expands in the nozzle to nearly atmospheric pressure, and thereby attains a very high velocity. This produces a rapid rotation of the gas in the tube near the nozzle. The gas is permitted to escape by two avenues, to the left through the unobstructed full diameter of the tube and to the right through a small central aperture. A throttling valve situated some distance (about 30 tube diameters) to the left of the vortex allows the operator to adjust the ratio of the amounts of gas that depart via the two exits. The gas that emerges through the central aperture is cold, whereas that which departs through the unobstructed part of the tube is warm. By proper adjustment of the flows in the two exits, a temperature drop of as much as 40°C can be obtained.

Vortex tube efficiency is low (about 5% of a comparable vapor-refrigeration cycle), but its simplicity, light weight, and reliability make it attractive for some applications. However, where large-scale cooling is required, it is doubtful that it will ever be competitive with other systems. Typical coefficients of performance range from 0.10 to 0.15. It appears that the lowest possible temperature drop that could be obtained with a device of this nature would be that defined by reversible adiabatic expansion from the initial state to the final state. About one-half of this temperature drop has been obtained in practice. Because of flow is highly turbulent and supersonic in the free vortex, it is doubtful that much improvement can be expected.

THERMOELECTRIC COOLING

In 1822 Seebeck observed that if a closed circuit were made of two dissimilar metals, an electrical current would flow in the circuit when the two junctions were maintained at different temperatures. In 1834, Peltier observed the inverse effect: if an electrical current flowed across the junction between two dissimilar materials, heat would be either absorbed or evolved.

For many years, practical application of these thermoelectric effects was almost exclusively restricted to thermocouples for temperature measurement, because metals exhibit a comparatively small Seebeck effect. However, the Seebeck effect in semiconductors can be considerably greater. Since the advent of the transistor and other semiconductors, materials have been developed in which thermoelectric effects are of sufficient magnitude that fabrication of useful devices has become a reality.

A thermoelectric system is composed of individual thermocouples, each with a relatively small capacity for cooling. Thermoelectric systems may be either air or water cooled and of any convenient size. The capacity of each system can be very accurately determined and thereby matched to the requirements, a distinct advantage when compared to the large incrementation of vapor-cycle equipment.

A basic thermoelectric element is shown in Figure 15. It consists of one leg made of a positive, or P-type, semi-

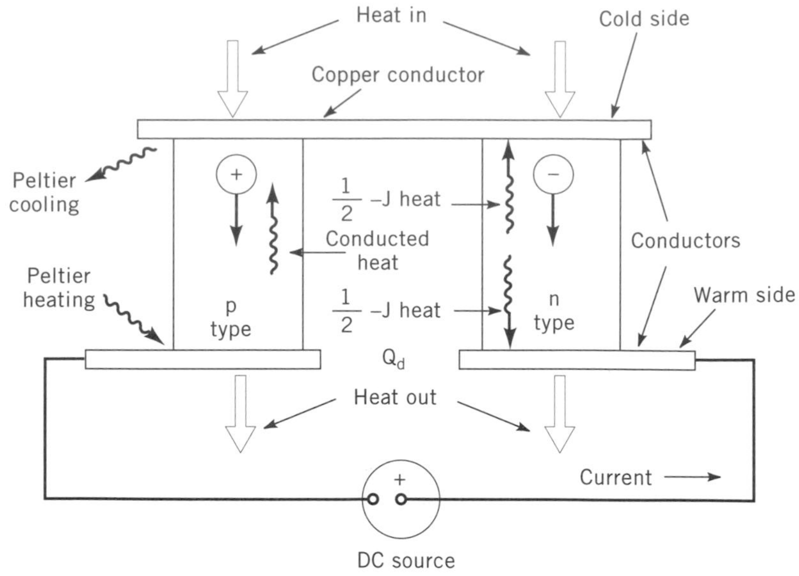

Figure 15. Schematic of thermoelectric cooling unit.

conductor and a second leg made of a negative, or N-type, semiconductor. The legs are mounted between conductors that are usually made of copper. If connected to a battery, as shown in the figure, the flow of electrons from the negative terminal of the battery through the two legs of the couple and back to the battery provides the power needed for the heat-pumping action. At each location where electrons leave a P material and then enter an N material, cooling will occur. The heat received from the surroundings by this cooled junction is pumped by the couple to the opposite conductors, causing them to be heated and to discharge the heat to their surroundings. Thus, the thermocouple performs the same functions as the evaporator, compressor, and condenser of the vapor-cycle refrigeration system.

Thermocouples are readily available assembled into groups called modules, as shown in Figure 16. The couples are electrically in series, arranged so that current flows in sequence, first through a P, then through an N material, and then to another P and another N material. Thus, when a voltage is applied one side of the module is cooled and the other side is heated. Physically, the two surfaces of the module are quite close together, usually between 0.16 and 0.64 cm ($\frac{1}{16}$ and $\frac{1}{4}$ in). Therefore, thermoelectric systems are designed as though the evaporator, compressor, and condenser were immediately adjacent to each other.

Thermoelectric cooling is different from conventional compression refrigeration in that there are no moving parts in the process of producing cold. However, because energy is being transferred as heat in the cooling process, heat must be removed from the hot side of the couples so that cold can be produced. For improved efficiency on the cold side, heat-transfer surfaces are frequently used. These heat-transfer surfaces take the form of fins and fans in gas systems, or with heat transfer to fluid systems using pumps for circulation. The heating and cooling functions of a thermoelectric system can be interchanged by reversing the polarity of the direct current applied to it. Because there are no moving parts, there is nothing to wear out and nothing to generate noise. There is no refrigerant to contain, so the problem of handling a two-phase changeover is simplified. The pressure-tight tubing is replaced by electrical wiring.

Because the cooling capacity of a single thermoelectric couple is small, it is practical to make systems of low refrigerating capacity. In the cooling of infrared detectors, small thermoelectric assemblies can cool to temperatures as low as 145 K, using cascading. Thermoelectric systems have built-in capacities up to 10 tons of air-conditioning by using many couples.

For some applications, the advantages of thermoelectric cooling outweigh its major disadvantages of high cost of couples and low coefficient of performance. The lack of moving parts and elimination of cooling liquids are appealing features. Another advantage is that a thermoelectric system will operate under zero gravity, or many times the force of earth gravity, and will operate in any orientation. This has been an important consideration in selecting thermoelectric cooling applications in the U.S. space program.

There are also applications in commercial and appliance refrigeration: a drinking-water cooler for use on diesel locomotives; a small, bottle-type water cooler for offices; and a small ice maker for hotel rooms. In the commercial area, thermoelectric cooling is best suited to small specialty devices such as cold junction references and humidity control for instruments.

Information similar to that required for design of a compression refrigerating system must also be established to design a thermoelectric system. Some factors to be considered are ambient temperature of air or water to

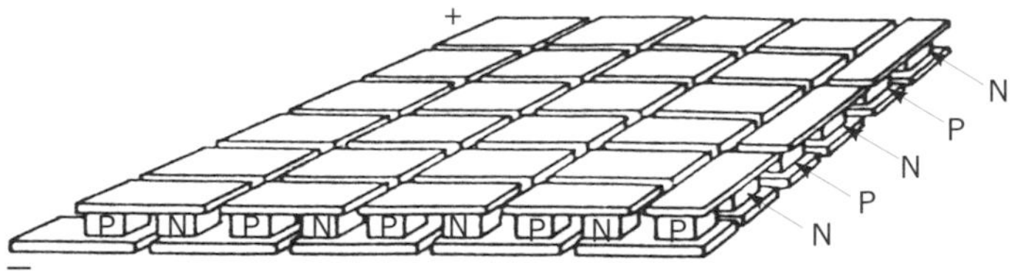

Figure 16. Multiple couple thermoelectric module.

remove heat from hot junctions, means of heat absorption at cold junctions, and power available.

CONCLUSION

For additional reading on refrigeration, there are a number of text and reference books available (7, 8, 9, 10). A wealth of information is available from the many publications of the professional society in the field, the American Society of Heating, Refrigerating, and Air-Conditioning Engineers (ASHRAE), and from the refrigeration trade association, the Air-Conditioning and Refrigeration Institute (ARI). ARI is a trade association of manufacturing companies that produce residential, commercial, and industrial air-conditioning and refrigeration equipment, as well as machinery, parts, accessories, and allied products for use with such equipment. ARI traces its history back to 1903 when it started as the Ice Machine Builders Association of the United States. ASHRAE and ARI cosponsor an annual International Air-Conditioning-Heating-Refrigeration Exposition, which normally draws from 30,000 to 40,000 people in the field.

In addition to suggested background material on the critical issue of CFCs and depletion of the ozone layer (12, 13), timely reporting of related activities is provided weekly by the *Air Conditioning, Heating and Refrigeration News* (14).

BIBLIOGRAPHY

1. *Webster's Ninth New Collegiate Dictionary*, Merriam-Webster Company, Springfield, Mass, 1987.
2. *ASHRAE Terminology of Heating, Ventilation, Air Conditioning, & Refrigeration*, 2nd Ed., American Society of Heating, Refrigerating, and Air-Conditioning Engineers, Atlanta, Ga., 1991.
3. R. Downing, "Development of Chlorofluorocarbon Refrigerants," *ASHRAE Transactions*, **90**(2), 1990.
4. B. Nagengast, "A Historical Look at CFC Refrigerants," *ASHRAE Journal*, Nov. 1988.
5. W. R. Woolrich, "The History of Refrigeration; 220 Years of Mechanical and Chemical Cold: 1748–1968," *ASHRAE Journal*, July 1969.
6. *ASHRAE Handbook 1993 Fundamentals*, American Society of Heating, Refrigerating and Air-Conditioning Engineers, Altanta, Ga., 1993.
7. R. B. Stewart, R. T. Jacobsen, and S. G. Penoncello, *ASHRAE Thermodynamic Properties of Refrigerants*, American Society of Heating, Refrigerating and Air-Conditioning Engineers, Atlanta, Ga., 1986.
8. W. F. Stoecker and J. D. Jones, *Refrigeration and Air-Conditioning*, McGraw-Hill, Inc. New York, 1982.
9. H. J. Sauer, Jr., and R. H. Howell, *Heat Pump Systems*, John Wiley & Sons, Inc., New York, 1983.
10. *Refrigeration and Air-Conditioning*, 2nd ed., Air-Conditioning and Refrigeration Institute, Prentice-Hall, Inc., Englewood Cliffs, NJ, 1985.
11. *ASHRAE Handbook 1990 Refrigeration*, American Society of Heating, Refrigerating and Air-Conditioning Engineers, Atlanta, Ga., 1990.
12. M. O. McLinden and D. A. Didion, "CFCs: Quest for Alternatives," *ASHRAE Journal*, **29**(12), Dec. 1987.
13. EPA, "Refrigeration," *Regulatory Impact Analysis: Protection of Stratospheric Ozone*, EPA, Office of Air and Radiation, Vol. III, Part 4A, August 1, 1988.
14. *The Air Conditioning, Heating and Refrigeration News*, Business News Publishing Co., Troy, Mich. (weekly).

RENEWABLE ENERGY TECHNOLOGIES

Renewable energy technologies are those that convert some natural phenomena, eg, falling water or natural products, into a useful energy form, most often electricity. In future years, the use of renewable energy technologies should expand as the cost of energy from conventional sources increases and as concerns over the environmental impact of the use of fossil fuels grows.

Renewable energy technologies today are a mix of mature technologies, developments, and advanced ideas, as shown in Table 1. In 1992, renewable energy technologies, principally hydropower and combustion of biomass, produced less than 8% of the energy used in the United

Table 1. Maturity of Renewable Energy Technologies

Proven Capability	Transition Phase	Future Potential
Hydropower	Wind	Advanced Turbines
Geothermal	Geothermal	Geothermal
Hydrothermal	Hydrothermal	Hot dry rock
		Geopressure
		Magma
Biomass	Biofuels	Biofuels
Direct combustion	Ethanol from corn	Methane
Gasification	Municipal wastes	
Passive Solar	Active Solar	Ocean Thermal
Buildings	Buildings	
	Process heat	
Photovoltaics	Solar Thermal	Solar Thermal
Small remote	Thermal/gas hybrid	Advanced electricity
Specialty products		High temp processes
	Photovoltaics	Photovoltaics
	Remote power	Utility power
	Diesel hybrids	

Table 2. Current Contributions of Renewable Energy Technologies to U.S. Energy Supply

Renewable Energy	% Contribution
Biomass	48.7
Hydropower	46.7
Geothermal	3.4
Solar heat	0.7
Wind	0.3
Solar electric	0.1

States; a breakdown of the relative contribution made by various renewable energy technologies is shown in Table 2. One may expect that the relative contribution will change significantly in the next decade. For example, the use of solar energy and photovoltaics, for example, should increase considerably.

Renewable energy technologies are considered in many of articles in the *Encyclopedia.* An overall perspective of the status of these technologies is provided in the article, COMMERCIAL AVAILABILITY OF ENERGY TECHNOLOGY. The relative economics of renewable energy technologies compared to more conventional technologies based on fossil fuel are critical to their future utilization. Different aspects of the economic questions are considered in ENERGY CONSUMPTION IN THE UNITED STATES; ENERGY TAXATION, BIOMASS; ENERGY TAXATION, SUBSIDIES FOR BIOMASS; and ENVIRONMENTAL ECONOMICS.

There are many technical aspects to the combustion and gasification of biomass or its use to produce liquid fuels. These aspects are covered in AGRICULTURE AND ENERGY; FOREST RESOURCES; FORESTRY, SUSTAINABLE, FUELS FROM BIOMASS; FUELS FROM WASTE; and WOOD FOR HEATING AND COOKING. There are many articles on specific technologies in the Encyclopedia: BUILDINGS, GEOTHERMAL ENERGY; OCEAN THERMAL ENERGY CONVERSION; OCEANOGRAPHY; RENEWABLE RESOURCES FROM THE SEA: SOLAR CELLS; SOLAR COOKING; SOLAR HEATING; SOLAR THERMAL ELECTRIC; TIDAL POWER; THERMAL ELECTRIC ENERGY CONVERSION; WAVE POWER; and WIND POWER.

Reading List

The Potential of Renewable Energy, an Inner Laboratory Analytical Paper, prepared for the Department of Energy, Office of Policy, Planning and Analysis, by INEL, LANL, ORNL, SNI, and SERI, SERI/TpP-260-3674, March 1990.

New Electric Power Technology, Problems and Prospects for the Nineties, U.S. Office of Technology Assessment, OTA-E-246, July 1985.

M. Mintzer, *A Matter of the Degrees, The Potential for Controlling Degree Sound Effects,* World Resources Institute, Washington, D.C., Research Report 5, April 1987.

RENEWABLE RESOURCES—FROM THE OCEAN

RICHARD J. SEYMOUR
Texas A&M University
College Station, Texas

The source of almost all renewable energy is the sun. Because the ocean covers the majority of the earth's surface, the oceans and the atmosphere above them collect the majority of the sun's radiation—about 80 trillion kW, or about one thousand times the amount of energy used by all of the earth's population at current rates. Using the simplest of analyses, this would indicate that a large fraction of human energy needs could be extracted from the ocean without seriously affecting its characteristics. The assumption here is that changing the heat balance in the ocean by a small fraction of one percent is unlikely to have serious environmental consequences. This idea, as yet not tested seriously even in computer simulations, was first put forward in the 1970s (1,2).

The sun's energy heats both the surface waters of the ocean and the air overlying them. Light energy is transmitted very poorly by seawater and most of it is absorbed in the upper few meters, limiting the direct effect of the sun's radiation. However, the heating of the air, which occurs unevenly over the earth's surface, results in atmospheric circulation, the formation of zones of relative high or low atmospheric pressure, and ultimately in winds blowing over the sea surface. By this roundabout scheme, some of the thermal energy from the sun is converted to kinetic energy in the winds.

These winds stress the surface of the ocean in three important ways. First, some of the wind energy is transferred in a complicated and poorly understood process into the generation of waves. Light winds make ripples. Strong winds blowing over large areas for long periods of time generate very large and energetic waves. Mechanical energy can be extracted from these waves. Secondly, the winds stir the upper layers of the ocean so that the temperature in the first 10 to 30 meters of depth is nearly constant. Below this mixed layer, the temperature gradually drops to near freezing at great depths. Thus, in places like the tropics where the mixed surface layer is quite warm, it overlays very cold water. This temperature difference can be used to drive a heat engine to extract energy. Finally, the wind stress causes mass transfer in the surface waters, resulting in the formation and maintenance of currents. These currents are modified by varying water depth, by the existence of land masses and the rotation of the earth, resulting in particular predictable patterns of strong currents, as shown schematically in Figure 1. These strongest currents, achieving modest speeds (up to about 3 knots) in the areas of concentration, represent significant levels of kinetic energy only because of the huge size of their cross-sectional areas.

Finally, the heat from the sun evaporates water from the ocean. This occurs most rapidly when the overlying air is warm and dry and most slowly when the air is cold and saturated. The rate is also influenced by cloud cover. The result is an uneven distribution of evaporation, onto which is superimposed an uneven distribution of precipitation in the form of snow or rain. As a consequence, small differences in the salinity of water masses can occur and the resulting density differences can drive currents. Although these currents are important to overall ocean circulation and mixing, they do not reach an intensity that makes them attractive as energy sources. However, one aspect of the solar distillation of seawater is of interest. A disproportionate amount of the evaporated water falls on land because of the chilling effect of the air pass-

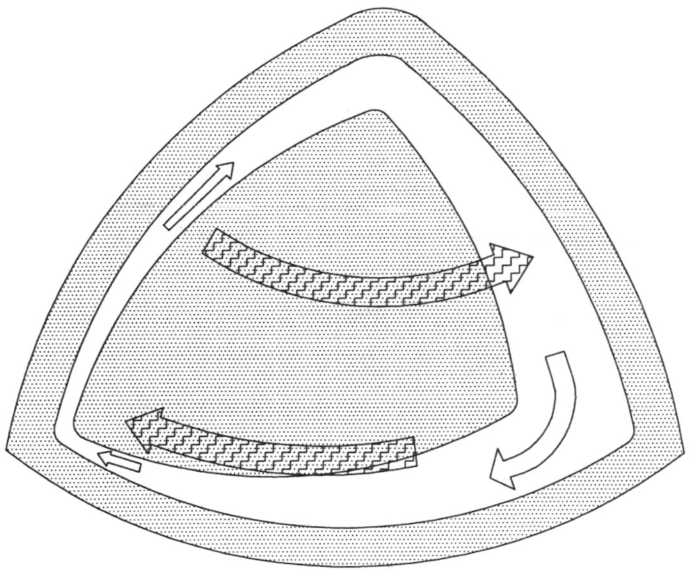

Figure 1. A schematic representation of wind-forced currents in a typical northern hemisphere ocean basin (dots). The equator is at the bottom. The roughly circular and clockwise current path is shown in white. Near the equator, the prevailing winds (lower wavy pattern arrow) are called trade winds and the blow from the east, forcing water to move to the west. At higher latitudes, the persistent winds blowing out of the west, called westerlies (upper wavy arrow), apply an opposing force to the water. In combination with the land masses (left and right boundaries of the ocean basin) this results in a rotary motion and a current near the edges of the basin. The effects of the earth's rotation and the interactions with the shoreline cause the current to be narrow and fast on the western side of the basin (lower left corner of the flow) and to be broad and slow on the eastern side of the basin (right-hand leg of current loop).

ing over high mountains. When this water falls, it remains relatively salt free. As it collects into rivers and flows into the ocean it has a very significant chemical energy relative to the receiving seawater. This osmotic energy is equivalent to the potential energy in water flowing over a dam that is 240 m high (1).

In the same sense that the sun's energy is considered inexhaustible within our time frame, so is the internal energy of the earth. This energy is manifested in the creation of new rock through volcanism and spreading centers driven by plate tectonics. This is in rough balance with the loss of terrestrial material through erosion by winds and rain, largely created from solar radiation collected over the oceans. The rain water in turn leaches out chemicals from these eroded materials and deposits some of them into the ocean. If these chemicals represent a fuel value, then they can be considered renewable in the solar framework. That is, they can be tapped in moderation without serious environmental disruption.

The final solar-driven energy source in the ocean is the creation of biomass. That is, the fixing of carbon dioxide by chlorophyll in marine plants. Because the vast majority of the plants in the ocean are microscopic, it is only in fertile coastal waters protected from extreme storms that large plants occur naturally. Some of these plants have extremely rapid growth rates—among the highest anywhere on the planet—and are harvested already for their food or chemical content. In China, over one million tons are harvested each year for food and in the United States the rate is about 150 thousand tons for chemicals and food additives.

It should be noted that there is another energy source in the ocean that depends only on the mass of the sun, not on its radiation, although the moon is more important to this process. The gravitational interaction among these three bodies gives rise to small stirring forces in the ocean at periods of about 12 and 24 hours that excite and maintain resonant long waves with comparable periods wherever the oceans geometry will reinforce these particular resonances. We call these long waves tides. The high tide is merely the passage of a wave crest while the low tide is a trough. Mechanical energy from these tides can be extracted at locations where there are extreme resonant conditions. It is interesting to note that the friction in this system, caused by stress on the bottom from tidal currents, is compensated by a slowing of the earth's rotation. Angular momentum is conserved by the moon moving further from the earth.

See also OCEANOGRAPHY; ELECTRIC POWER GENERATION; FUELS FROM BIOMASS.

Ocean Currents

Resource Availability. As shown in a schematic sense in Figure 1, there are only a few places in the world's oceans where strong, persistent, unidirectional surface currents can be found. The Gulf Stream off the east coast of Florida (see Fig. 2) is the strongest such flow and is relatively close both to shore and to potential energy markets. The Kiroshio Current in Japan offers a similar situation. Note that tidal currents, which are reversing—but not necessarily symmetrical—are not discussed in detail here, even though their maximum speeds may greatly exceed the average values of the unidirectional ocean currents discussed.

The quality of an ocean energy resource can be defined in an engineering or economic sense by its density, that is, the magnitude of this energy type found in a square (or cubic) meter of the ocean surface (volume). The concept here is very simple. Using the analogy of a hydroelectric dam, if the dam is very high, the trapped water has a high potential energy per unit of surface area. This means that the water will be moving faster through the turbines and a turbine of a given size will produce more power. This almost always means lower power costs, if only from reduced capital investment. Similar arguments apply to other forms of energy and extraction schemes. Ocean currents have very low energy densities. At a maximum speed of about 3 knots the energy density is equivalent to a hydroelectric dam less than one foot high. Obviously it would take a very large turbine to generate much power and only the very enormous size of the resource would lead anyone to consider tapping it.

Technology for Ocean Current Power Recovery

Ocean currents held considerable interest in the 1970s and a MacArthur Workshop was held on the subject in 1974 (3). As a result, venture capital was put into re-

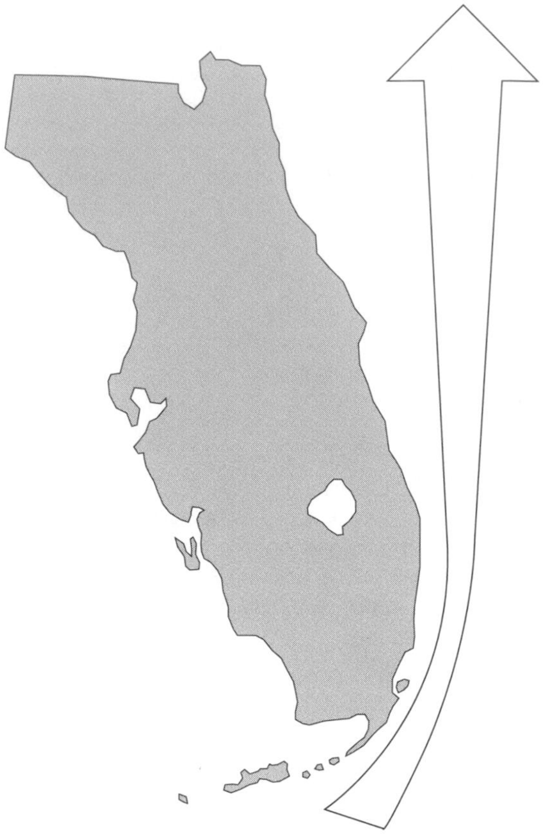

Figure 2. The Gulf Stream close to the state of Florida in southeastern United States. This corresponds to the lower left portion of the schematic current shown in Figure 1. The width of the arrow represents the relative width of the major part of the flow. The current achieves its greatest speed at the narrowest part of the flow, close to the city of Miami. Both the current intensity and its relative proximity to land and to large energy markets was responsible for strong interest in exploiting current energy during the 1970s and 1980s.

search and development. At that time, it was generally believed that the velocities could be increased substantially by using a converging section ahead of the turbine. To the surprise of the investigators who attempted to model this at smaller scale, the ducted turbine actually had lower flow speeds through it than a similar unit that was freely mounted in the stream. Rather than causing the water to accelerate, the free stream saw the converging section as an impediment and simply diverted around it. This meant that small, fast-flow turbine technology could not be applied.

In spite of this, conceptual design studies continued until 1981, culminating in a huge free stream counter-rotating turbine 91 m in diameter (the length of a football field, as illustrated in Fig. 3) that would be buoyant and moored in the Gulf Stream in large numbers (4). It was planned to moor the units below the keels of ships. Because the current intensity does not decrease substantially at these depths, there is little penalty for this decision. Each turbine was to have carried a rating of 7.5 MW. The counter rotating blades were to be 65 m in diameter with rigid rims that would drive generators within the buoyant shroud. A ducted shroud was shown in model tests to give increases in efficiency.

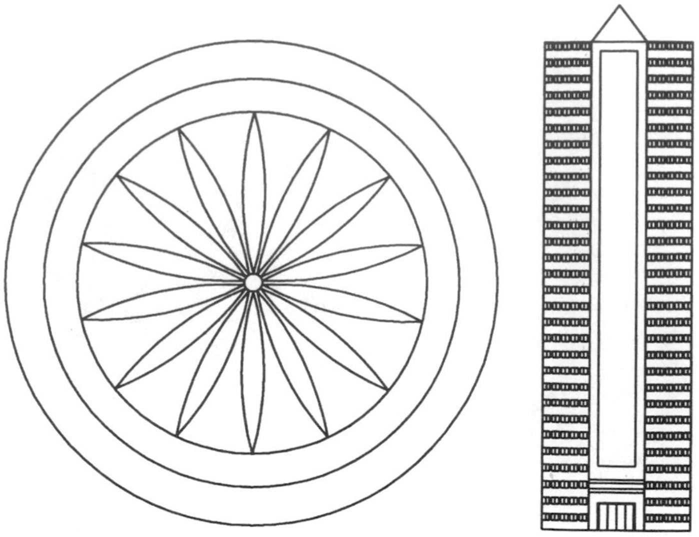

Figure 3. A frontal view of one of the large turbines proposed for mooring in the Gulf Stream. A 30-story building is shown next to the structure to indicate its huge size. The unit was intended to be buoyant and to be maintained below the keel depths of ships by massive mooring lines. There are two sets of counter-rotating blades with generators mounted in the shroud surrounding the blades.

There was no consideration given, as is often the case with designs for ocean equipment by engineers unfamiliar with the environment, to fatigue of the mooring lines and the electrical cables caused by vibrations induced from eddy shedding in these currents. The mooring loads would also have exceeded any known technology for anchoring these structures.

At a different scale, free stream ("run of the river") economical turbines have been developed subsequently for rapidly flowing rivers without impoundment dams (5). This technology could, at least theoretically, be scaled to ocean currents. However, the river units are firmly mounted to foundations in the bottom (river flows typically extend close to the bottom) and so they offer no anchoring solutions to the floating systems required by the ocean. There are presently no development programs anywhere in the world for harnessing large ocean currents.

Ocean Winds

Winds over the ocean are quite similar to those over land. That is, they are highly variable over time at any location and are stronger on average in some locations than in others. There has been some conjecture about offshore windmill farms (6) but, in general, the problems associated with maintenance and power transmission are greater at sea than on land, and research and development has been concentrated entirely on terrestrial sites.

Marine Biomass Conversion

The Resource. Ocean plants are quite different from terrestrial plants because over most of the ocean there is such a great distance between the soil, where gravity tends to concentrate the nutrients from dead plants, and the surface, where there is access to sunlight for carbon fixation. Most oceanic plants are small enough to remain suspended without root systems in near-surface waters and extract nutrients directly from compounds in solution in the surrounding waters. Large, firmly fixed plants are extremely rare. Among those rarities are the marine macroalga—communal plants that attach themselves to rocky substrate with holdfasts that look like roots, and which grow large upright structures that are analogous to the branches and leaves of terrestrial plants. The foliage, which absorbs nutrients directly, is denser than water and is held up by gas filled chambers—an example of the specialized functions of this communal organism. One species of these plants, the giant kelp (*Macrocystis pyrifera*), grows as much as 50 cm per day in certain seasons but requires moderately cool water temperatures, clear and nutrient-rich waters, and rocky bottoms for attachment. A typical *Macrocystis* plant is shown in Figure 4. It survives best when protected from severe storms as its dense canopies are ripped away by the force of large waves (7). Because of its high rate of growth, it has attracted the interest of several investigators (8,9) over the past several decades as a source of biomass which could potentially be processed into methane (natural gas). As noted above, *Macrocystis* is presently harvested along the California coast by mechanically cutting off the very tops of the plants near the surface, this crop is then processed for a variety of extracts, many of which are used as food and beverage additives. In especially clear and protected waters, as in the lee of offshore islands, *Macrocystis* can grow to heights of 25 m or more and then spread another 10 m on the surface.

Cultivating Marine Biomass. Natural sources of rapidly growing plants, such as *Macrocystis,* are extremely limited in size and could not possibly supply sufficient biomass to be a viable energy source. Research was therefore centered on finding the means to cultivate it over great expanses of the ocean and increasing its output by several orders of magnitude. This is roughly analogous to developing a commercial crop on terrestrial grasslands from a plant that had a specialty niche in the mountains. There appeared to be no economical means of dedicating the required millions of hectares of ocean surface close to shore, so all of the research programs envisioned deep water ranches in which the plants would grow on submerged floating frames. Several development programs were initiated in the early 1980s when there was a perceived shortage on natural gas. As new reserves were discovered, the shortage turned to a glut by the end of the 1980s and all effort ceased.

During the research conducted under these programs, it was discovered that surface waters offshore did not have sufficient nutrients to support the growth rates observed nearshore. Therefore, plans were made to pump up water from great depth to fertilize the surface waters. The amount of water required was determined from nutrient uptake experiments on kelp plants. However, it was found that as much as seven times the energy was required for pumping the deep water than could be extracted, under the most favorable assumptions, from the methane produced by the natural gas.

Experience nearshore (7) has shown that extensive beds of kelp plants can be nearly completely destroyed by the occasional severe storm that penetrates the areas of natural growth. Severe and continuing loss of plants by wave action was found throughout the few at-sea experiments that were undertaken. The broken kelp fronds are divided into two populations: those that retain enough gas chambers to float, and those that sink. In the very large masses of plants envisioned for these farms, even a modest amount of shedding during storms—much less their complete loss—would have resulted in two very significant conditions tending to degrade the environment. The sinking kelp could provide sufficient oxygen demand on the bottom water to cause them to become anoxic and to kill all of the bottom fauna, because of the slow rate of

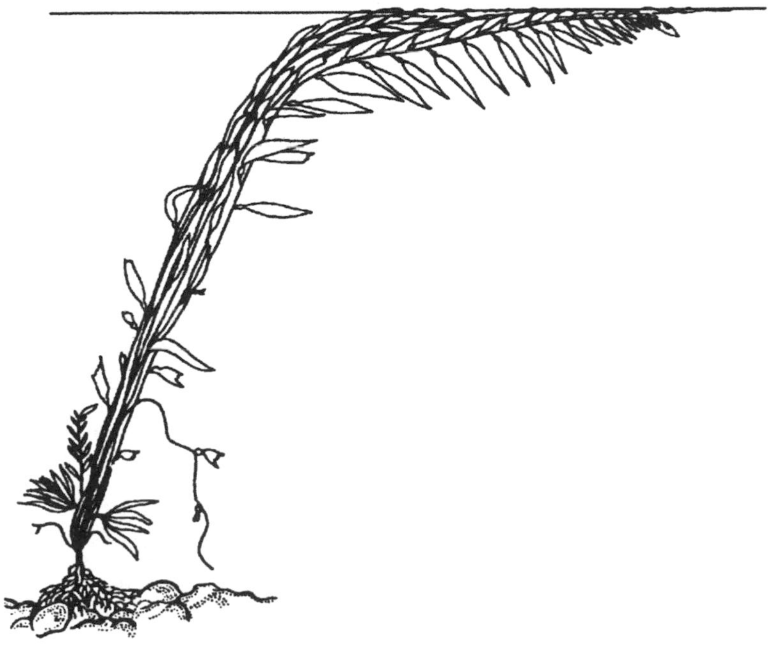

Figure 4. A giant kelp plant, *Macrocystis pyrifera*. This is one of the fastest growing plants on earth and the largest that grows in the ocean, reaching lengths of over 30 m. It was the focus of research into the feasibility of growing the plant in deep water (attached to submerged floating frames) to provide biomass for conversion into methane (natural gas).

oxygen replenishment at depth. Much of the floating material would come ashore on local beaches and would represent an impossible burden on local resources to remove and dispose of the bulky mass. In Southern California, the removal of naturally occurring kelp flotsam, a tiny fraction of what would occur with a huge offshore ranch, is already an expensive public service necessary to keep the beaches attractive for recreational use. In addition, environmental studies showed that the residue from the processing represented a huge disposal problem. It could not be dumped at sea because of the oxygen demand problems discussed above. Further, the air pollution associated with the very large number of round trips required for the harvesting vessels would seriously impact coastal air quality.

Chemical Energy from the Ocean

The huge volume of the ocean ensures that even trace constituents of seawater are available in massive quantities. With few exceptions, however, these chemical constituents are not conveniently clustered in ore deposits, as often occurs on land, but are more or less evenly dispersed throughout the ocean. Both magnesium and deuterium (an isotope of hydrogen, often called heavy water) have been successfully extracted from the ocean by processing enormous amounts of seawater.

All of the presently proposed methods for achieving electrical power generation through controlled fusion reactions have proposed using one or more of the elements deuterium (found at about 160 ppm in the ocean), lithium (at about 425 ppm, one of the more abundant elements of seawater), or tritium (another hydrogen isotope). Tritium can be bred from a lithium blanket around a reactor, so that the ocean could conceivably become a source from which fusion reactants are mined in the future. Lithium is undoubtedly involved in many biological processes in the ocean, given its relative availability, and any extraction process might locally reduce the concentration to a level which could affect marine life. The output of the processing plant would have to be treated as a pollutant in this case and care taken to dilute it through diffused discharge as is accomplished with treated sewage.

Energy from Salinity Differences

The Resource. The chemical energy resulting from the osmotic potential between seawater and fresh water is the equivalent of a height difference of 240 m, which is roughly equivalent to the total hydroelectin head of a heavily exploited river like the Columbia, the Northwestern part of the United States. This enormous energy source, the equivalent of a 240 m dam at the mouth of every river in the world that discharges into the ocean, has been recognized for decades (10,11). It probably represents the largest renewable resource in the ocean. However, there are many easily recognizable environmental problems associated with using this resource, most resulting from the intricate natural ecological balance that exists in bays and estuaries where fresh and saline waters mix.

Another approach might be to utilize the ocean as the low salinity side of the process and a very high salinity source to replace the ocean. Possible brine sources would include solar ponds, saline lakes, and water leached from salt domes. Calculations have indicated that the energy in the salt (relative to the ocean) in a salt dome could exceed that of the petroleum or natural gas that they often contain (12). These brine to seawater gradients can, of course, have a significantly greater energy density (potential) than the seawater to fresh system.

The salinity difference is a surprising energy source, largely because the mixing occurs without a significant temperature rise and is therefore difficult to discern. There are three major schemes for extracting this energy.

The first scheme utilizes osmotic pressure differences in the two fluids. Less saline water is capable of forcing fresh water into more saline water that is at a higher pressure if they are separated by a suitable osmotic membrane (11,13). It is clear that work can be done by this system since there is an increase in the volume of water on the saltier side with no external energy input. In practice, the system would employ a pump to pressurize the higher salinity side of the membrane and then the output would be run through a turbine (see Fig. 5). If the amount of energy added by the now-pressurized fresh water passing through the turbine exceeded the losses in the pump and the turbine, net power output is achieved. Note that the work put into pressurizing the saltier water, less losses, is recovered in the turbine. Very salty water is an advantage, in theory, but it causes corrosion of the membranes, and the high pressures necessary to achieve the economies are difficult for low cost membranes to support mechanically.

The second proposed method utilizes the vapor pressure differences between the two fluids through a reversal of the typical vapor pressurization scheme for the production of fresh water (14,15). The air and other noncondensibles are removed from adjacent chambers over fresh and saline water sources. If the water masses are initially at the same temperature, the vapor pressure over the

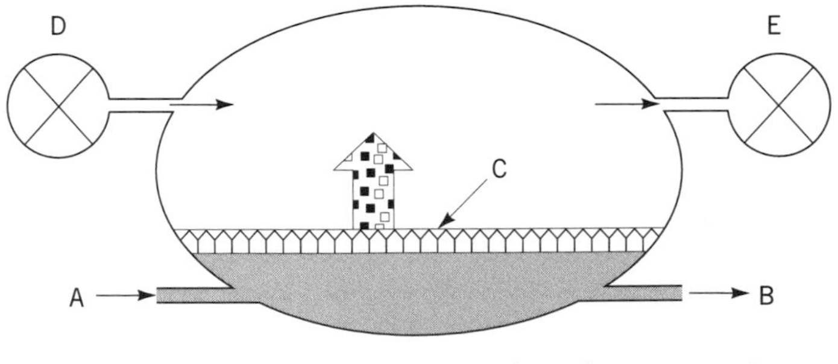

Figure 5. The schematic representation of a process for extracting energy from the osmotic differences between fresh and saline water. The lower salinity water is added at **A** and flushed at **B**. An osmotic membrane is shown at **C** which passes fresh water towards the more saline water in the upper chamber. This saline supply is delivered under pressure by a pump **D**. If this pressure is less than the osmotic pressure across the membrane caused by the salinity difference, fresh water will flow from the lower to the upper chamber. Energy is extracted through a water turbine **E**. Most of the energy put into the saline water pumping at **D** is recovered at the turbine and the fresh water flow (now pressurized without mechanical energy input) provides the positive output.

fresher source will be greater than over the saltier source so that evaporation will be greater over the fresher side. This will drive vapor through a turbine that connects with the saltier side (see Fig. 6). The process removes heat from the fresher side, reducing its vapor pressure and, thus, potentially stopping the process. This is overcome by a heat exchanger which transfers heat back to the fresher side. This method eliminates all membranes, but substitutes heat exchangers with their own cost and maintenance problems. The use of low pressure turbines and large heat exchangers provide this method with many technologies in common with the extraction of energy from thermal, rather than salinity, gradients in the ocean. A further modification of the vapor-pressure difference approach (16) uses the increase in temperature in the saline side, under vacuum, to power a stage of desalination. The "fresh" side could be seawater or brackish water and the saline side a concentrated brine. One source of such a brine is as a by product of salt evaporation ponds. The product here is not energy in an exportable form, but fresh water, with a large avoided energy cost compared to reverse osmosis, thermal desalination and other energy intensive schemes for desalination.

The third method, and the only one presently under serious development, is reverse electrodialysis (RED), which can be thought of as a direct current battery that produces electrical energy directly from salinity differences (17,18). The most effective method eliminates conventional electrode stacks and allows for compounding of voltages through the use of alternating membranes that pass either cations or anions (see Fig. 7). These membranes separate the fresh and salt water, which ultimately mix and are discharged as brine. Great progress has been achieved recently in overcoming the problems of membrane cost, swelling and leakage. This method is probably closer to commercial realization than any of the other alternative energy sources discussed above.

In all three of the basic salinity gradient approaches discussed above there are environmental effects which must be considered. If fresh water is used as the low salinity fluid, then it is degraded to a brackish brine or mixed with seawater in such a way that its inherent value as a source of irrigation or potable water is forever lost. Given the very high energy cost of producing fresh water from brackish water and the near-universal shortage of fresh water, it will seldom make sense economically—or environmentally—to destroy fresh water for energy generation. However, if seawater is used in conjunction with a higher salinity brine, the environmental degradation could be more readily handled because of the enormous capacity of the ocean for dilution. Local oversalinity at the point of discharge must be considered. However, the dilution—and therefore the disposal—of a highly saline source on land (such as the residue from solar ponds) may serve as an environmental enhancement. There are a number of sources of brackish water that could be effectively used as the low salinity side of a salinity difference process. Because a very small increase in

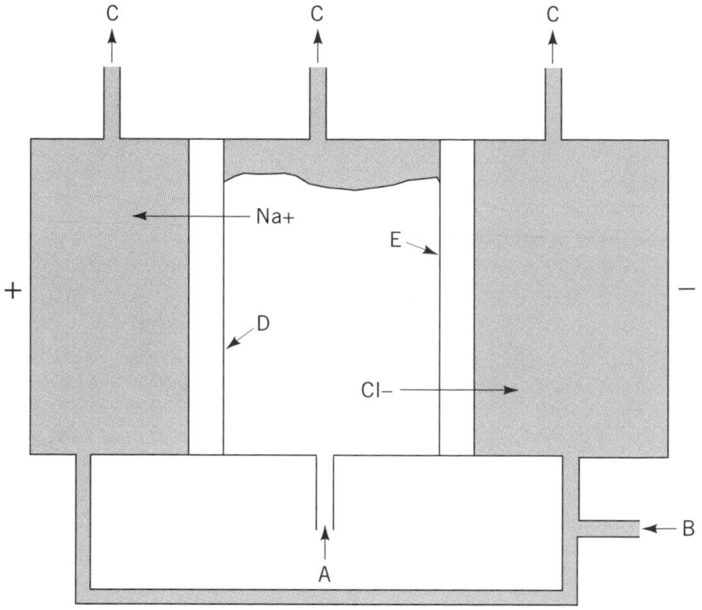

Figure 7. This is a schematic diagram of a single cell in a reverse dialysis process for generating direct current from the chemical energy associated with the differences in salinity between water sources. The saline water enters at **A** and the less saline at **B**. Brine is exhausted from all the cells at points **C**. A selective membrane at **D** passes only cations and the membrane at **E** passes only anions. Thus an electrical potential is developed that can be extracted through terminals at the ends. In practice, many such cells are stacked together in sequence so that larger voltages can be generated.

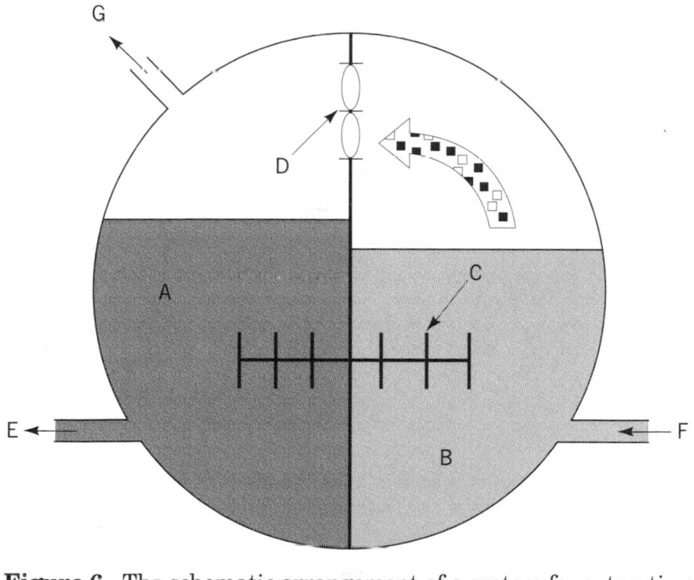

Figure 6. The schematic arrangement of a system for extracting mechanical energy from the difference in vapor pressure between fresh and saline waters. In this diagram, **A** represents the supply of saline water and **B** the fresh water source. The chambers above both water pools are initially pumped down with a vacuum pump at **G** to remove the air and other noncondensibles. The fresh water side has a higher vapor pressure (increased evaporation) so that there is a flow of vapor through the turbine, **D**. Heat is carried with the vapor which slows the process. Therefore, a heat exchanger, **C**, is supplied to restore the temperature of the fresh water side. Fresh water is supplied to the system at **F** and excess brine removed at **E**.

salinity compared to fresh water reduces the value of the resource significantly, it may be promising to utilize these slightly salty fluids for power generation or as a feed stock for salinity difference desalination.

BIBLIOGRAPHY

1. J. D. Isaacs and R. J. Seymour, "The Ocean as a Power Resource," *Int. J. Environmental Studies* **4,** 201–205 (1973).
2. J. D. Isaacs and W. R. Schmitt, "Ocean Energy: Forms and Prospects," *Science* **207**(4428), 265–273 (1980).
3. H. B. Stewart, Jr., *Proceedings of the MacArthur Workshop on Energy from the Florida Current*, Palm Beach Shores, Fla., Feb. 1974.
4. P. B. S. Lissaman "The Coriolis Program," *Oceanus, Ocean Energy* **22**(4), 23–28 (1979).
5. G. Miller, and co-workers, "Kinetic Hydro Energy Conversion System Study," *Proceedings of Waterpower '85*, American Society of Civil Engineers, Las Vegas, Nev., Sept. 25–27, 1985, pp. 1022–1033.
6. A. L. Burton and S. C. Roberts, "The Outline Design and Costing of 100 m Diameter Wind Turbines in an Offshore Area," A. Garrad, ed., *Proceedings of the 1985 Seventh British Wind Energy Association Conference*, Oxford, March 27–29, 1985, Mechanical Engineering Publications, Ltd., London, 1985, pp. 269–277.
7. R. J. Seymour and co-workers, "Storm Wave Induced Mortality of Giant Kelp, *Macrocystis pyrifera*, in Southern California," *Estuarine, Coastal and Shelf Science*, **28,** 277–292 (1989).
8. OTA, Fuel from Open Ocean Kelp Farms, Office of Technology Assessment, U.S. Government Printing Office, Washington, D.C., 1980.
9. L. B. McKay, ed., *Seaweed Raft and Farm Design in the United States and China*, New York Sea Grant Institute, Albany, N.Y., 1983.
10. G. L. Wick and W. R. Schmitt, "Prospects for Renewable Energy from the Sea," *MTS J.* **11** (5/6), 16–21 (1977).
11. G. L. Wick, "Power from Salinity Gradients," *Energy* **3,** 95–100 (1978).
12. G. L. Wick, 1981. "Salinity Energy," in G. L. Wick and W. R. Schmitt, eds., *Harvesting Ocean Energy,* UNESCO Press, Paris, 1978, pp. 111–131.
13. S. Loeb, F. Van Hessen, and D. Shahaf, "Production of Energy from Concentrated Brines by Pressure-retarded Osmosis: II. Experimental Results and Projected Energy Costs," *J. Membrane Science* **1,** 249–269 (1976).
14. M. S. Olsson, G. Wick, and J. D. Isaacs, "Salinity Gradient Power: Utilizing Vapor Pressure Differences," *Science* **206,** 452–454 (1979).
15. M. S. Olsson, "Salinity-Gradient Vapor-Pressure Power Conversion," *Energy* **7**(3), 237–246 (1982).
16. R. J. Seymour and P. Lowrey, "State of the Art in Other Ocean Energy Sources," in R. J. Seymour, ed., *OCEAN ENERGY RECOVERY: The State of the Art*, American Society of Civil Engineers, New York, 1992, Chapt. 10, pp. 1–25.
17. A. T. Emren, "Concentration Cell for Salinity Power Production, Economic Potential of the Concentration Cell," *Alternate Energy Sources 3, Proc. of the 3rd Miami Int. Conf. on Alternate Energy Sources*, **4,** 229–241 (Dec. 1980).
18. R. N. O'Brien, "Reverse electrodialysis—an unused power source?" *Chemistry in Britain* **22**(10), 927–929 (1986).

RISK ASSESSMENT

VINCENT COVELLO
Columbia University
School of Public Health
New York, New York

Risk assessment can be defined as the systematic use of data to identify and quantify risks. Risk assessment can be used to analyze a variety of risk situations, ranging from occupational risks, risks in the home, and recreational risks to the risks of transportation accidents and risks from the production of energy, chemicals, or consumer products.

There are three elements to a risk assessment: a description of the hazard (ie, the source of risk); an assessment of probability of exposure to the hazard; and an assessment of the probability of harm or adverse consequences from the exposure. A complete risk assessment describes the activity or substance that can cause harm; the events that create the possibility of harm; and a statistical estimate of the likelihood of exposure and that the harm will occur from the exposure.

Risk assessors commonly define risk as the possibility of suffering harm from exposure to a risk agent. Risk agents of concern include, most predominantly, radiation, chemical substances, heavy metals, and biological organisms. In most texts, the term "risk" does not mean the same as "hazard." Risk is created by a hazard. "Hazard" refers only to the source of a risk. The concept of exposure is critical to the distinction between hazard and risk. For a hazard to constitute a risk there must be an exposure.

See also LIFE CYCLE ANALYSIS; ENVIRONMENTAL ECONOMICS; GLOBAL HEALTH INDEX; WATER QUALITY ISSUES.

GOALS AND OBJECTIVES OF RISK ASSESSMENT

Techniques for analyzing and estimating risks have evolved from a variety of fields such as medicine, toxicology, industrial hygiene, occupational safety, engineering, environmental impact studies, and epidemiology. Progress has been rapid, spurred in part by the passage of numerous laws designed to protect public health and the environment. From these efforts has emerged a formal process called risk assessment. The power of the process derives primarily from the systematic use of data, assumptions, and expert judgments to exploit fully all available information to estimate the probability of adverse health, safety, and environmental consequences.

Risk assessments are employed for a variety of purposes. For example, they can be used to:

- identify and quantify the risks of activities or substances;
- compare and rank the risks of different activities or substances;
- compare the effectiveness of different strategies for managing or reducing risks;
- set priorities for action or scientific study.

Risk assessment was first introduced into regulatory decision-making in the early 1970s. It is now an established part of the regulatory process. As described below, regulatory agencies use risk assessment for a variety of purposes.

- Provides the logic by which different regulatory actions can reduce risk
- Presents a range of plausible risk estimates reflecting uncertainty in theory and data
- Reduces the range of uncertainty in regulatory decision-making
- Helps set regulatory priorities and standards
- Describes and quantifies levels of risk that remain after regulatory actions have been taken
- Provides an empirical foundation for balancing risks against benefits
- Identifies subpopulations that are especially sensitive or vulnerable
- Identifies critical areas where resolution of uncertainty can be most effective in reducing risk
- Reduces the guesswork in health, safety, and environmental rulemaking
- Aids decision-makers in identifying those areas where research is most needed

Regulatory applications of risk assessment include the evaluation of hazardous activities and the effects of exposure to risk agents in the air, water, land, and food.

The goal of risk assessment is to estimate the severity and likelihood (or probability) of harm or injury to human health or the environment occurring from exposure to a risk agent. Table 1 lists categories of risks identified by the U.S. Environmental Protection Agency.

Several types of analyses may be conducted in assessing a risk. These include:

1. **Hazard Identification.** Hazard identification provides data on the potential for a risk agent to cause adverse health or environmental effects.
2. **Source/Release Assessment.** Source/release assessment—often called probabilistic risk assessment—provides data on the amounts, frequencies, and locations of releases of risk agents from specific sources into specific environments.
3. **Exposure Assessment.** Exposure assessment provides data on populations or ecosystems that may be exposed to a risk agent, the concentrations of the risk agent, the duration, and other characteristics of exposure.
4. **Dose-Response Assessment.** Dose-response assessment provides data on the amounts of a risk agent that may reach exposed individuals or populations. It also attempts to estimate the percentage of the exposed populations that might experience harm or injury and the relevant characteristics of such populations.

Risk estimates—one of the main outcomes of the risk assessment process—are statistical descriptions of risks based on available information. Risk estimates cover a wide array of phenomena, ranging from the number of excess cancers expected to be caused by the use of a new pesticide, to the health or environmental consequences of low probability, to the high consequence events such as explosions at industrial facilities. In some cases, risk estimates inspire great confidence in their accuracy; in other cases, they are not much better than guesses. To help users understand the value of a risk estimate, risk analysts often provide statistical information on the strength of support for the estimate.

Table 1. Categories of Risk Identified by the Environmental Protection Agency

| Cancer Risks | Noncancer Health Risks | | Ecological Risks | Welfare Risks |
	Type of Risk	Example		
Lung	Cardiovascular	Increased heart attacks	*At ecosystem level:*	Material damages and soiling
Colon	Developmental	Birth defects		
Breast	Hematopoietic	Impaired heme synthesis	Direct physical destruction or major alteration	Reduced recreational opportunities
Pancreas			Changes in ecological community structure/ function	
Prostate	Immunological	Increased infections		Damage to natural resources (eg, crop, timber)
Stomach	Kidney	Dysfunction	Changes in species richness and diversity	
Leukemia	Liver	Hepatitis A	Threats to/loss of rare or endangered species	Damage to commercial and public property and groundwater supplies
Other cancers	Mutagenic	Hereditary disorders		
	Neurotoxic/Behavioral	Retardation		
	Reproductive	Increased spontaneous abortions	*At population level:*	Impairment of aesthetic values
	Respiratory	Emphysema	Changes in mortality or reproductive rates	
	Other	Gastrointestinal disease	Changes in growth rates of individuals	
			Physiological or behavioral abnormalities	
			Susceptibility to environmental stresses

RISK ASSESSMENT AND RISK MANAGEMENT

Much debate within the risk literature has centered on the distinction between risk assessment and risk management. The basic point of contention is the degree to which risk assessment and risk management are different and should be viewed independently.

As noted above, risk assessment involves the systematic analysis of risk-related information to identify and estimate the probability and magnitude of risks associated with a particular hazard. Decision-makers often find risk assessments most useful in situations where answers are not obvious and information is missing, ambiguous, or uncertain. Risk assessment techniques provide one means for organizing the relevant information and estimating the consequences of different decisions.

Risk management, in contrast, refers primarily to the integration of risk information with information about social, economic, and political values and various response options to determine what actions to take. These actions include, but are not limited to, reducing or eliminating the risk. Critical factors influencing the decision process are the resources and the technical capabilities available. Risk management also includes the design and implementation of policies and strategies that result from this decision process. It involves weighing the risks of alternatives and weighing tradeoffs between the health and environmental benefits of efforts to reduce risks and the costs to society of using its resources to obtain those benefits. Making such tradeoffs is usually keyed to a consideration of what constitutes an acceptable level of risk.

The risk-management process can be highly political. Balancing the costs of risk reduction against the benefits of the activity is a controversial and value-based process. More generally, the extent to which risk assessment and risk management can be distinct is still unresolved. The controversy revolves around the degree to which, in practice, the scientific risk assessment is, or should be, kept free from biases or values that typically are part of a management decision. For example, the practice of risk assessment within the federal government has been criticized from both ends of the environmental political spectrum. Proponents of less restrictive environmental regulation have charged that the techniques and assumptions used in risk assessment reflect an unjustified bias in favor of overly protective risk management values. Others have argued precisely the opposite, claiming that risk assessments used to support government decisions reflect techniques and assumptions that understate risks for the purpose of relieving regulatory burdens on industry.

A classic case in this regard was the controversy about Alar, a growth hormone used on apples, that erupted in 1989. In this and many similar cases, the public was exposed to inconsistent and contradictory information about risks. Experts from environmental groups said the risks were high. Experts from industry said the risks were negligible.

Aside from technical issues, an understanding of the differences between risk assessment and perception can help sort out these differences. When scientific experts talk about risk, they generally mean risk as estimated by the basic risk assessment process; that is, the likelihood or probability of specific adverse health or environmental effects. In contrast, there is a commonsense notion of risk, as people perceive it in their everyday lives, that is linked with numerous other social and psychological considerations.

Researchers from the fields of psychology, social psychology, and decision analysis have identified several qualities or dimensions of risk that influence risk perceptions. For example, some people judge riskiness solely on the basis of the likelihood of the risk actually occurring, while others are primarily concerned about effects or consequences; for example, who it affects, how widespread the effects may be, and how familiar and dreadful the effects are. Furthermore, perceptions of risks are influenced strongly by issues of choice and control. Risks often seem riskier to people if they have not voluntarily agreed to bear the risks and if they have no control over the source and management of the risks. In addition, people often incorporate in otheir perceptions of risks consideration of the benefits derived from accepting the risks. Trust, credibility, catastrophic potential, familiarity, uncertainty, dread, fairness, equity, and the distribution of risks and benefits are also critical factors in risk perception.

Based on these factors, a small actual risk may be perceived as a large risk, and a large actual risk may be perceived as a small risk. An understanding of risk perception is critical to understanding the dynamics of public debates about risk.

The Risk Assessment Process: Hazard Identification

Risk assessment begins with the seemingly straightforward question: Does a hazard exist?

The term hazard is not absolute. Its usage depends on the characteristics of the substance and the circumstances of use. For the purposes of this section, hazard refers only to the potential for a risk agent to cause adverse health or environmental effects.

For some chemical substances, the physical effects, such as corrosiveness, ignitability, explosiveness, and reactivity with other substances, can be tested by using relatively well-established laboratory techniques. Health effects, however, are often more subtle. Effects such as toxicity, carcinogenicity, mutagenicity, teratogenicity, and behavioral modifications require more sophisticated detection methods.

To identify health effects, risk analysts use four types of analytical techniques:

- Epidemiological studies
- In vivo animal bioassays
- Short-term in vitro cell and tissue culture tests
- Structure–activity relationship analyses

Epidemiological Studies. Epidemiological studies are concerned with the patterns of disease and injury in human populations and the factors that influence these patterns. In general, scientists view well-conducted, epidemiological studies as the most valuable information from which to draw inferences about human health risks. For example, a positive finding in a well-conducted epidemiological study is viewed by most risk analysts as strong evidence that a chemical poses a carcinogenic risk to humans.

Unlike the other analytical approaches described in this section, risk analysts can use epidemiological methods to study the direct effects of hazardous substances on human beings. Also, epidemiological studies help identify actual hazards to human health without prior knowledge of disease causation. They also can complement and validate information about hazards generated by animal laboratory studies.

Compared to other techniques used in risk assessment, epidemiology is relatively well suited to situations in which exposure to a risk agent is high (such as cigarette smoking) or in which health effects are unusual (such as rare forms of cancer). However, most risk agents currently subject to regulatory and societal concern do not fall into these two categories. As a result, epidemiological studies used in risk analysis have important limitations that constrain their usefulness. These limitations arise not from epidemiology per se, but rather from the nature of the specific risk analysis needs to which epidemiology is sometimes applied.

One limitation of epidemiological studies is that they have poor sensitivity and are generally unable to detect small increases in risk unless very large populations are studied. At low exposure levels, adverse effects may be very difficult to detect. For example, to identify any change in the number of genetic defects that could be caused by one additional x-ray (a 3-millirem dose) per person per year, a study would need to observe a population of 700 million people for three generations. A positive result cannot prove 100% certainty of harm, and, of course, a negative effect cannot prove certainty of zero risk. If a study of 1,000 people showed no excess cancer, it would be inappropriate to infer that the risk is nonexistent.

These points underscore the importance of the concept of statistical power in epidemiological studies. The concept of statistical power is used by epidemiologists to decide whether statistical techniques are sufficient to reveal the effect under investigation. It also provides information relevant to the decision of whether the investigation should proceed.

After an epidemiological study is completed, its statistical value is determined by its confidence limits. Its epidemiological significance is also determined by:

- The consistency of association
- The strength of association
- The temporal relationship of association
- The specificity of association
- The biological plausibility of association

In general, statistical significance and statistical power are not by themselves sufficient for results to be considered positive.

A second limitation of epidemiological studies is that they can be conducted only for hazards to which people already have been exposed. They are not useful for predicting effects of new substances or technologies.

A third limitation is the lack of experimental controls. For obvious moral and ethical reasons, as well as pragmatic ones, researchers can neither control nor account completely for the behavior of study subjects that may affect their health and therefore influence the study results.

A fourth limitation is that epidemiological studies may fail to account for the effects of multiple sources of exposure. If both exposed and unexposed groups in a study (for example, in a study of a chemical in the water supply) are exposed to a risk agent from another unidentified source or medium (such as air), the study results may not show an association although one is actually present.

Epidemiological studies are often described as being either descriptive or observational. Furthermore, different types of observational studies exist, including case-control studies and cohort studies. Although other types of epidemiological studies exist, such studies are infrequently undertaken. For example, experimental epidemiological studies are sometimes conducted, but they require the deliberate application or withholding of a possible disease factor and observing the appearance or lack of appearance of any effect. For obvious ethical reasons, such studies are rare.

Descriptive and observational epidemiological studies are described below.

Descriptive Epidemiology. Descriptive epidemiological studies examine the distribution and extent of disease in populations according to such basic characteristics as age, gender, and race. The primary goal of a descriptive epidemiological study is to generate clues about the causes of disease. Descriptive epidemiological studies often provide data for more detailed follow-up epidemiological and toxicological studies.

Descriptive epidemiological studies have produced important findings about patterns of disease in human populations and the factors that influence these patterns. Many of these findings come from international comparisons of disease patterns and comparisons of smaller geographic regions. For example, descriptive epidemiological studies first suggested that occupational conditions might be the cause of excess mortality rates from cancer of the mouth, colon, rectum, larynx, and bladder in industrialized parts of the northeastern United States.

Data for descriptive epidemiological studies come from several sources, including summaries of self-reported symptoms in exposed populations, case reports prepared by medical personnel, and studies that demonstrate a correlation between occurrences of health problems and the presence of a hazardous environmental or occupational condition.

Observational Epidemiology. Data for observational epidemiological studies come from observations of individuals or relatively small groups of people. These studies are analyzed using generally accepted statistical methods to determine if an association exists between a hazardous substance and a disease and, if so, the strength of the association. Often the hypothesis to be investigated is generated by a descriptive epidemiological study.

Observational epidemiological studies have also produced important findings about patterns of disease in human populations and the factors that influence these patterns. For example, observational epidemiological studies have provided data on elevated risk for lung cancer among smokers and shipbuilders who worked with asbestos. Two of the most common types of observational epidemiological studies are case-control studies and cohort studies.

Case-Control Studies. Case-control studies compare the exposure histories of two matched groups: a group of individuals who exhibit symptoms of a particular disease, called cases, and a group of otherwise similar individuals who do not exhibit disease symptoms, called controls. The statistical measure most often used in case-control studies is the odds ratio, which is an estimate of relative risk. The odds ratio represents the odds, or probability, of disease occurrence in exposed individuals compared with unexposed individuals.

A well conducted case-control study requires a control group that is as similar as possible to the case group in all other ways except the presence of disease symptoms. Eligibility criteria are used to eliminate confounding factors such as age, race, gender, and smoking. The relationship between the exposure and health effect can then be seen more clearly. Eligibility criteria are applied equally to both the case and control groups. In practice, however, assembling an appropriate control group is one of the most challenging tasks in a case-control epidemiological study. For example, the circumstances that lead to exposure in the case group (a person's occupation, for instance) is likely to be correlated with other risk factors, such as lifestyle and economic status. These factors may be difficult to match properly in a control group. An additional problem with case-control studies is that individuals in both groups may differ systematically in their recollection of past events.

Despite these problems, case-control studies are the most commonly practiced epidemiological studies. They have been used to establish the carcinogenicity of several chemicals. Case-control studies are less expensive and require fewer subjects than cohort studies (see below). In addition, a case-control study takes only two to three years to complete because the adverse health effect being studied is already manifested in the exposed group.

Cohort Studies. Cohort studies compare groups exposed to different levels of a hazardous substance. They permit a direct estimate of the risk of disease associated with a particular exposure. The usual statistical measure is the relative risk, which is the ratio of the disease rate in the exposed group to that in the unexposed group.

There are two kinds of cohort studies: prospective and retrospective. Prospective cohort studies measure exposures while they occur and monitor study subjects for adverse health effects for extended periods–sometimes as long as 20 or 30 years or more. Retrospective cohort studies, by contrast, more closely resemble case-control studies in that known exposures have occurred in the past and effects may already be evident. Retrospective studies also can be conducted in only about two years.

Prospective cohort studies have several advantages over case-control studies. Because exposure and life-style information is collected before adverse health effects develop, biases attributable to errors in recollection can be reduced. In addition, because the studies focus on groups receiving exposures to a common hazard, they allow researchers to determine associations with more than one health outcome. (In contrast, case-control studies select cases with the same health problem and compare past exposure patterns to look for associations.)

Cohort studies are expensive and complex to conduct, however. They require follow-up periods, which frequently cause subjects to drop out because of relocation or other lifestyle changes. For this reason, and also because the effects to be detected are unknown in advance, cohort studies require large numbers of exposed individuals to detect relatively rare effects.

Although epidemiological data can be useful in assessing human health risks, several questions regarding the validity and applicability of the studies must be resolved by the risk analyst. These questions, as first described by the National Research Council, include:

- What relative weights should be given to studies with differing results? For example, should positive results outweigh negative results if the studies that yield them are comparable? Should a study be weighted in accordance with its statistical power?
- What relative weights should be given to results of different types of epidemiological studies? For example, should the findings of a prospective study supersede those of a case-control study, or should those of a case-control study supersede those of a descriptive study?
- What statistical significance should be required for results to be considered positive?
- Does a study have special characteristics (such as the questionable appropriateness of the control group) that lead one to question the validity of its results?
- What is the significance of a positive finding in a study in which the route of exposure is different from that of a population at potential risk?
- Should evidence on different types of responses be weighted or combined (for example, data on different tumor sites and data on benign versus malignant tumors)?

Despite these uncertainties, epidemiological studies are increasingly being used to assess risks. One type of epidemiological study in particular, community health studies, have received increased attention in recent years. The discovery of hazardous waste in a community, for example, often sparks local concerns about the degree to which cancer cases, miscarriages, and other health problems in the community may have been caused by exposure to the waste through food, water, air, or soil. This concern may prompt officials, or citizens themselves, to conduct a community health study.

Community health studies are frequently used by researchers or public groups to draw attention to environmental health problems. In a typical community health study of chemical emissions, the rate of occurrence of disease for people living near the source of the emissions is compared to the average rate for that disease in other communities or in the United States as a whole. The investigators survey people who have lived near the source and collect information on the age, gender, and disease history of residents during a defined period of time. If the rate of occurrence of disease is greater than the average rate used for comparison, and if the difference is greater than would be expected owing to chance variation, then further, more detailed epidemiological investigation is typically undertaken.

Community residents often see a positive correlation in a community health study as evidence of a cause-and-effect relationship between a risk agent (eg, industrial chemical emissions) and adverse health effects (eg, miscarriages or diseases such as cancer). However, risk analysts view this type of study differently. A community health study is seen as useful primarily in developing hypotheses for further analytical studies. Being a type of correlational study, a community health study is usually not sufficient by itself to prove that the exposure of concern is associated with adverse health effects, unless the adverse health effects are unusually rare or specific to the risk agent of concern.

A significant difficulty with community health studies and other such correlational studies is that the exposure history of individuals with and without disease is unknown. Also, the occurrence of disease can be affected by such factors as smoking, diet, and genetic influences. These factors, which are difficult to account for even in analytical epidemiology studies, are not usually considered in correlational studies. Thus, except for rare health problems that have well-understood causes, observations from community health studies are often too difficult to isolate from multiple confounding causes (as in the case of smoking). Consequently, they cannot provide definitive evidence of a hazard. On the other hand, community health studies can be useful in identifying unusual patterns of disease occurrence and in indicating potential links between exposures and disease.

In Vivo Animal Bioassays. For most hazards, adequate epidemiological or human clinical data are not available. Many risk analyses prepared for regulatory risk decisions rely on information from laboratory studies on live animals. Most of the pathogenic studies of microorganisms have been done in animals–primarily guinea pigs, mice, and some primates. Studies of chemical carcinogenicity are most often conducted in rats and/or mice. Test procedures for animal bioassays fall into three general categories: acute exposure tests, subchronic exposure tests, and long-term chronic exposure tests (including carcinogenicity).

Acute Exposure Tests. Tests for acute (short-term) effects generally involve exposure of one or more species of test animals to chemical, radioactive, pathogenic, or infectious agents. The route of exposure typically depends on the exposure route or routes of concern. In many cases, however, risk assessments may use data from tests conducted for other purposes, using different exposure routes. In oral toxicity studies, researchers administer a single oral dose to each test animal and a placebo (or no dose) to a control group. When the concentrations of the test substance cannot be carefully controlled in test animals, food via a feeding tube is used in a technique called gavage. In dermal toxicity studies, skin exposures are generally continuous for 24 hours. In inhalation toxicity studies, there is typically up to eight hours of continuous exposure.

In a well-conducted study, three to six dose levels are tested, depending upon the route of exposure, using five to ten animals of each sex for each dose level. Exposed animals are observed for signs of disease or toxic effects, and the death of any animal is recorded. After a short time, typically 14 days, any surviving animals are sacrificed and all test animals and controls are examined for signs of disease or toxic effects.

For chemicals, the most common measure of acute toxicity is the "median lethal dose" (LD_{50}), the dose level that is lethal to 50% of the test animals exposed to that dose: the lower the LD_{50}, the more potent the risk agent.

In aquatic toxicity tests, an analogous measure–the "median lethal concentration," or LC_{50}–is used to refer to the concentration of the test substance in the water that results in 50% mortality in the test species. LD_{50} values can be used to compare the relative acute toxicity of different risk agents. For pathogenic bacteria or viruses, potency can be measured by the LD_{50} or by the infectious dose for 25 to 50% of the test subjects.

Growing concern about the use of animals in laboratory research, combined with the sometimes poor reproducibility of LD_{50} results between laboratories and the increasing development and sophistication of other tests (see below), have resulted in efforts to substitute other tests for LD_{50} animal studies as a measure of acute toxicity. Officials in Cambridge, Massachusetts, for example, banned the use of LD_{50} tests and Draize tests (which test for irritancy on the corneas of rabbits) within the city limits.

Subchronic Exposure Tests. Properly conducted subchronic tests for chemicals involve repeated exposure of two species of test animals for 5 to 90 days, by exposure routes corresponding to expected human exposures. Often, more than one exposure route is tested. At least three dose levels are used to help define the dose-response relationship and to identify what are called the no observed adverse effects level, or NOAEL, and the lowest observed adverse effects level, or LOAEL. As implied, the NOAEL is the dose at which no adverse effects are observed, and the LOAEL is the lowest dose at which adverse effects are observed. Another important level that may be estimated from subchronic tests is the maximum tolerated dose, or MTD. The MTD is the largest dose a test animal can receive for the majority of its lifetime without demonstrating adverse effects other than carcinogenicity.

Subchronic tests provide useful information about the dose-response relationship of a test substance, and they also help to identify pharmacokinetic and metabolic reactions and define appropriate dose levels for chronic toxicity tests (as described below).

Long-term Chronic Exposure Tests. In studies of chronic effects of chemicals, test animals receive daily doses of the test agent for approximately two years. In well-conducted experiments, at least three dose levels are used: the MTD, a fraction of the MTD, and unexposed control.

For studies of carcinogenicity, however, the use of the MTD is controversial. Doses used in bioassays are necessarily relatively high so as to increase the sensitivity of the experiments, whereas human exposures tend to be much lower (and, as discussed below, actual doses to human tissues may be lower still). Thus, extrapolation from higher test doses to lower environmental concentrations creates large uncertainties in the validity of such bioassays as a means of identifying potential chemical or radioactive hazards.

This issue is critical for risk assessment. It underlies the controversy that exists concerning the use of animal studies in risk assessment. The main point of contention is that testing protocols call for high doses to compensate for the insensitivity of the bioassays. In cancer testing, this method is done because cancer occurs at very low rates; for example, lung cancer affects fewer than eight in 10,000 people in the United States every year, an incidence rate of less than 0.08%. Such low incidence rates are virtually impossible to detect in animal studies.

In a very large animal study, popularly known as the Megamouse Study, 24,192 mice were subjected to seven different dose levels of a potent liver and bladder carcinogen. Even with that many test animals, the lowest incidence of observed tumors was about 1%. Practical considerations prohibit using such large numbers of animals in most studies; typical animal tests may use 50 or 60 animals of each species and sex at each dose level. Very high doses of substances may be necessary to ensure that chemicals that are carcinogens do not yield negative results.

The highest dose used in carcinogenicity tests is the maximum tolerated dose. The importance of the MTD in carcinogenicity testing is demonstrated by results of long-term animal tests conducted for the National Toxicology Program. Of 52 chemicals judged as carcinogens, two-thirds would not have been found carcinogenic if the high dose selected had been one-half the MTD actually used.

One problem with doses as high as the MTD is that an organism's normal mechanisms of self-defense may be overwhelmed, allowing cancer to be induced or promoted. Additionally, very high doses of a substance may have a qualitatively different impact on the organism than the same substance at a low dose: the distribution of a chemical risk agent in the body may be altered, or processes that detoxify or eliminate the risk agent may be affected. A carcinogenic response at high dose levels may not be indicative of effects at low exposure levels.

In response to this problem, many researchers suggest that greater attention should be given to the mechanisms that lead to cancer and how the mechanisms may change as the dose changes. Information on the distribution of the substance within the organism and metabolism of the substance also need to be incorporated in selecting the MTD. This information is equally if not more important in extrapolating results of tests on one species to other species. For example, exposure to gasoline vapors has been shown to induce tumors in rat kidneys by mechanisms apparently unique to that species. Most chemicals, even when administered at the MTD, do not induce cancer. Ideally, the doses used in animal bioassays will allow detection of a carcinogen without introducing distortions in the test results. However, proponents of the MTD, as currently defined, argue that if cancer is not observed in an experiment using a dose lower than the MTD, the possibility still remains that the substance is a carcinogen.

Because of the need to detect small effects, and because the doses in chronic studies are lower than in acute or subchronic studies, a larger number of test animals—usually about 50 of each sex for each dose level—must be used to detect statistically significant effects. Thus, two major drawbacks of chronic animal bioassays are the time they take to conduct and analyze (from two to three years) and their large cost owing to the number of animals required (several hundred, depending upon the number of species, sexes, and doses tested).

Test animals are observed during the experiment and sacrificed at the end of the exposure period for examination of organs and tissues. Abnormal behavior, physiological damage, or other signs of adverse effects (such as differences in organ weights) in test animals are recorded and compared with control animals. Significant adverse effects are inferred from several pieces of data, including data indicating a statistically significant increase in the incidence of the effect between dosed animals and controls at each dose level; an increase in the severity of the adverse effects with increases in dose; an accelerated emergence of adverse effects with increases in dose; and an increase in the numbers (types) of adverse effects with increases in dose.

As predictors of acute and chronic adverse effects in other species, and in humans particularly, in vivo animal bioassays are limited and controversial. For example, long-term animal bioassays may not be completely reliable predictors of carcinogenicity in humans. In evaluating over 700 chemicals, groups of chemicals, and industrial processes, the International Agency for Research on Cancer (IARC) has concluded that the available data provide causal evidence of human carcinogenicity for only 30 chemicals. Of these 30 human carcinogens, only 19 chemicals have exhibited significant evidence of carcinogenicity in animals; data on the remaining 11 were limited, inadequate, or unavailable to determine animal carcinogenicity. Many of these substances were identified in the 1970s or earlier as human carcinogens, and initial animal studies have indicated nearly complete correspondence between human and animal carcinogenicity.

A listing of chemical and physical agents for which there is sufficient evidence of carcinogenicity to humans is listed below.

4-aminobiphenyl
arsenic and certain arsenic compounds
asbestos
azathioprine
benzene
benzidene
betel quid
bis(chloromethyl) ether

chemotherapy for lymphomas (MOPP)
chlorambucil
chlornaphazine
chromium and certain chromium compounds
coal tars and coal-tar pitches
conjugated estrogens
cyclophosphamide
diethylstilbestrol
ionizing radiation
melphalan
methoxsalen with ultraviolet A therapy (PUVA)
mineral oils
mustard gas
myleran
2-naphythlamine
shale oils
soots
tobacco smoke and smokeless tobacco
treosulphan
ultraviolet radiation
vinyl chloride

However, agreement among agencies regarding such lists is seldom unanimous given that different organizations use different criteria for determining carcinogenicity. Part of the problem is that the mechanisms by which cancerous tumors form are still poorly understood. Currently accepted theories of carcinogenesis characterize the formation of tumors as a multistage process. Risk agents (for example, chemicals, viruses, and radioactive materials) may "cause" cancer by initiating the process by which tumors form ("initiators") or by promoting the unrestrained growth of cells once a cell has been transformed to a precancerous state ("promoters"). Risk agents that are both initiators and promoters are often called complete carcinogens. In addition, some carcinogens may not initiate or promote cancerous growth but may cause other metabolic changes that in turn lead to cancer.

For many carcinogenic agents, the exact mechanism by which they "cause" the cancer is not well understood. In the absence of scientific data on the mechanics by which an agent causes cancer, regulatory agencies typically infer that the agent has carcinogenic properties by weighing the evidence from both human and animal studies. Several different agencies and international institutions have developed various weight-of-evidence criteria for determining whether a risk agent should be considered carcinogenic. These criteria are shown in Table 2.

Not all chemicals that cause cancer in one animal species also cause cancer in other species. The National Cancer Institute and the National Toxicology Program tested 224 chemicals for carcinogenicity in rats and mice; of these 224 chemicals, 75 were carcinogenic in both rats and mice, 32 were carcinogenic only in rats, and 33 were carcinogenic only in mice.

As well as producing varying effects in different species, many carcinogens affect different organs in different species. Thus, from test results in one or two animal species, it is not possible to predict which organs will be affected or which species most closely represents the human in likely response to a substance.

There are several reasons for the difficulties in estimating human responses on the basis of in vivo animal bioassay results. Physiology and metabolic pathways that affect the response to a chemical differ considerably among species. For tests of chemicals or radiation, another limitation is the large differences between the doses used (per kilogram of body weight) in the bioassays and the doses to which humans are typically exposed.

Table 2. Weight-of-Evidence Criterion Used by EPA and IARC for Determining if a Risk Agent Should be Considered a Carcinogen

Category	Criterion
	U.S. Environmental Protection Agency
A	Human carcinogen, with sufficient evidence from epidemiological studies
B1	Probable human carcinogen, with limited evidence from epidemiological studies
B2	Probable human carcinogen, with sufficient evidence from animal studies and inadequate evidence or no data from epidemiological studies
C	Possible human carcinogen, with limited evidence from animal studies in the absence of human data
D	Not classifiable as to human carcinogenicity, owing to inadequate human and animal evidence
E	Evidence of noncarcinogenicity for humans, with no evidence of carcinogenicity in at least two adequate animal tests in different species, or in both adequate animal and epidemiological studies
	International Agency for Research on Cancer
1	Carcinogenic to humans, with sufficient epidemiological evidence
2A	Probably carcinogenic to humans, with (usually) at least limited human evidence
2B	Probably carcinogenic to humans, but having (usually) no human evidence
3	Sufficient evidence of carcinogenicity in experimental animals

The scientific validity of using animal bioassay data to infer adverse human health effects is generally accepted among scientists. However, the use of these data requires the use of assumptions or judgments on the part of the risk analyst. For example, using a positive result from an animal test to infer an adverse human health effect represents a more conservative (risk adverse) judgment, while using negative animal data to infer the absence of adverse human health effects represents a less conservative judgment.

In addition to these issues, the National Research Council has identified the following questions related to animal testing:

- What degree of confirmation of positive results should be necessary? Is a positive result from a single animal study sufficient, or should positive results from two or more animal studies be required? Should negative results be disregarded or given less weight?
- Should a study be weighted according to its quality and statistical power?
- How should evidence of different metabolic pathways or vastly different metabolic rates between animals and humans be factored into a risk assessment?
- How should the occurrence of rare tumors be treated? Should the appearance of rare tumors in a treated group be considered evidence of carcinogenicity even if the finding is not statistically significant?

- How should data for experimental animals be used when the exposure routes in experimental animals and humans are different?
- Should a dose-related increase in tumors be discounted when the tumors in question have high or extremely variable spontaneous rates?
- What statistical significance should be required for results to be considered positive?
- Does an experiment have special characteristics (for example, the presence of carcinogenic contaminants in the test substance) that lead one to question the validity of its results?
- How should findings of tissue damage or other toxic effects be used in the interpretation of tumor data? Should evidence that tumors may have resulted from these effects be taken to mean that they would not be expected to occur at lower doses?
- Should benign and malignant lesions be counted equally?
- Into what categories should tumors be grouped for statistical purposes?
- Should only increases in the numbers of tumors be considered, or should a decrease in the latent period for tumor occurrence also be used as evidence of carcinogenicity?

Despite these uncertainties and shortcomings of animal bioassays, they are widely viewed as the best available analogy for identifying adverse human health effects in the absence of good human epidemiological data. In addition, tests on animals are still widely used in estimating ecological effects.

Short-Term In Vitro Cell and Tissue Culture Tests. In the past decade, there has been an explosive growth in the quantity, variety, and quality of laboratory tests using in vitro cell and tissue cultures to study the effects of risk agents on biological organisms. The low cost and relative speed with which cell and tissue tests can be conducted have earned them a popular role in screening potentially hazardous substances and in providing additional sources of evidence for carcinogenicity.

This popularity has resulted primarily from (1) an increased demand for quicker and less expensive techniques to screen chemicals for potentially adverse effects on a routine basis; (2) rapid and revolutionary advances in biological techniques for manipulating genetic material, monitoring cellular behavior, and initiating tissue culture growth; and (3) increasing pressures from animal welfare activists and others to develop alternatives to live animal assays.

Because of the great number and variety of test systems and protocols used in in vitro tests, only representative details are presented here. A description of some general types of in vitro assays will follow.

Much of the development of in vitro techniques has occurred in the area of genetic toxicity tests for identifying chemicals that may be human mutagens and carcinogens. In vitro tests for genetic toxicity can be organized into the following categories: assays for genetic mutation; assays for chromosome effects; assays for disruption of DNA repair or DNA synthesis mechanisms; and assays for cellular transformation.

The predictive value of in vitro genetic toxicity tests for carcinogenicity in vivo is controversial. Most of the debate has centered around the predictive value of the Ames test, which is the most widely used mutagenicity assay. Researchers investigating the relationship between carcinogenicity and mutagenicity in the Ames test have reported correlations exceeding 90%. More recent evidence, which includes testing of a larger number of noncarcinogens, indicates the correlation is not as great as originally suggested. In examining the in vivo and in vitro test results of 73 chemicals studied in the National Toxicology Program, researchers have found that 83% of chemicals with positive Ames tests have also been rodent carcinogens. However, only 51% of the chemicals with negative Ames test results were noncarcinogens. Similarly, in an evaluation of results for 224 chemicals studied by the National Cancer Institute and the National Toxicology Program, 69% of Ames test mutagens were also animal carcinogens, while only 43% of nonmutagens were noncarcinogens.

Other short-term tests do not appear to predict carcinogenicity better than the Ames test. In two tests for chromosome effects, approximately 70% of chemicals with positive results were carcinogens, and 50% of chemicals with negative results were noncarcinogens. As might be expected, results of short-term tests on known human carcinogens show mixed results.

In general, the accuracy of in vitro tests for carcinogenicity is compromised by (1) the lack of direct correspondence between the mechanisms of genetic toxicity used in the assays and the current understanding of carcinogenesis; (2) the difficulty of extrapolating from simple cellular systems to complex, higher organisms; (3) the use of different protocols and different interpretive rules by different laboratories; (4) the need for externally constructed metabolic activation systems that differ from in vivo activation systems; (5) the limited number of risk agents tested (except for the Ames test) by the assays, making validation difficult; and (6) the limited number of noncarcinogens tested.

The usefulness of short-term test data in risk assessment is controversial. The National Research Council has identified the following questions as decision points in the use of short-term tests in carcinogenicity risk assessment:

- How much weight should be placed on the results of various short-term tests?
- What degree of confidence do short-term tests add to the results of animal bioassays in the evaluation of carcinogenic risks for humans?
- Should in vitro transformation tests be accorded more weight than bacterial mutagenicity tests in seeking evidence of a possible carcinogenic effect?
- What statistical significance should be required for results to be considered positive?
- How should different results of comparable tests be weighted? Should positive results be accorded greater weight than negative results?

These questions primarily relate to the use of in vitro tests for carcinogenicity. The use of in vitro tests for adverse effects other than carcinogenicity has been much less widely examined. In vitro toxicity tests have been used mainly to study mechanisms of toxicity, but they are increasingly being developed as complements or alternatives to existing in vivo animal bioassays for estimating toxic effects. Scientists conduct in vitro toxicity tests on cell and tissue types ranging from microorganisms to tissue cultures from human organs. The choice depends partly upon the specific site and nature of the toxic effects of concern. For example liver cells are used extensively, especially in hepatotoxicity tests, because the liver plays such an important role in the removal of toxic substances. Similarly, rat vaginal tissue has been used to test vaginal irritancy of contraceptives, and some tests with nerve tissue have been used to correlate neurotoxic effects with the presence of chemicals around the tissue. Measures of toxic damage in in vitro tests include changes in rates of cell reproduction, rates of synthesis of certain substances, changes in membrane permeability, and damage to some part of the cell structure.

As with toxicity tests, in vitro tests of viruses and pathogenic bacteria have been conducted on other microorganisms–especially those known to be present in the human respiratory, circulatory (blood), and digestive systems–as well as on cells and tissue cultures. These tests have generally been limited to infectious organisms that have known effects. As the use of techniques for genetic modification (for example, recombinant DNA) increases, in vitro tests for adverse effects of organisms may be developed further.

Structure–Activity Relationship Analysis. The simplest and most frequently employed first step in any analysis of a potentially hazardous risk agent or new technology is to compare it with similar agents or technologies for which the presence or absence of hazard is known. EPA's Office of Toxic Substances, for example, relies heavily on a technique known as structure–activity relationship (SAR) analysis for analyzing hazards associated with most of the new chemicals for which it receives premanufacture notices. EPA evaluates this information to determine whether more data (for example, from in vitro or in vivo tests) may be required.

SAR analysis involves comparing the molecular structure and chemical and physical properties of a chemical having unknown hazards with the molecular structures and physiochemical properties of other, similar chemicals having known toxic or carcinogenic effects. Knowledge of the relationship between a particular structural feature or physiochemical property and the physiological or molecular mechanisms by which the feature or property induces adverse effects, strengthens the usefulness of SAR analysis.

Analogous techniques are used to assess potential hazards of novel biological organisms created by genetic manipulation. For these organisms, data on unmodified but biologically similar species or subspecies are evaluated.

Frequently, as in the examples cited above, SAR analysis (or its biological analogue) is the only readily available source of information about the hazards of new technological developments. The predictive value of such data, however, is uncertain. Seemingly insignificant differences in chemical structures or physiochemical properties (or species classification) may in fact obscure significant differences in hazardous characteristics. For example, the chemical 2-acetylaminofluorene is a well-documented carcinogen, whereas its close chemical relative, 4-acetylaminofluorene is not carcinogenic.

SAR analysis is potentially the most widely applicable and usable technique for assessing hazards of chemicals because molecular structures are readily identifiable and, in theory, any adverse effect can be assessed. In practice, however, predicting an adverse effect from a substance may not always be possible because of the lack of data about the particular adverse effect in structurally comparable substances.

THE RISK ASSESSMENT PROCESS: SOURCE/RELEASE ASSESSMENT

Types of situations to which source/release assessment are applicable include the incidental or accidental release of toxic chemicals or other hazardous materials from a facility or a transportation vessel; storm water runoff in urban areas; accidental releases of radioactive material from a nuclear power plant; leakage from a lined hazardous-waste landfill, waste pond, or underground storage tank; and the incidental or accidental release of pathogenic microorganisms from a research facility or hospital.

Source/release assessment methods were refined in the mid-1970s and were aimed primarily at quantifying the sources and likelihood of environmental releases of radiation from nuclear power plants. The analytical processes developed at this time have been collectively termed "probabilistic risk assessment" or "probabilitistic safety assessment." A decade later, the explosion at the Union Carbide facility in Bhopal and a series of major industrial accidents in other parts of the world has led to broadened and intensified research on techniques of probabilistic risk assessment designed to quantify the sources and likelihood of environmental releases of potentially hazardous chemicals and other toxic materials from industrial facilities.

Most conventional techniques for source/release assessment focus on engineering applications, such as the failure of structural or mechanical equipment and accidents at chemical or nuclear power plants. For example, probabilistic risk assessment techniques have been developed to estimate the chances of a serious reactor accident. For chemical plants or industrial facilities, several engineering safety review techniques have been developed to provide systematic, but mostly qualitative, analyses about the potential sources and consequences of explosions, fires, toxic chemical releases, and other accidents. In recent years, however, analysts have attempted to incorporate quantitative estimation into these and other safety review methods. These traditional, qualitative methods thus provide an important source of information, which, when combined with the quantitative techniques described below, can yield numerical estimates of the types, amounts, timing, duration, and probabilities of releases.

Four kinds of techniques that can be used alone or, more typically, in combination, to assess sources and releases:

- Monitoring
- Performance testing and accident investigation
- Statistical methods
- Modeling

These techniques may be used to evaluate the likelihood and quantity of releases of an identified hazard or they may be used to identify the types, locations, and amounts of potentially hazardous releases from a runoff area, containment structure, or facility.

Monitoring. Monitoring generally involves a regular or ongoing program of sampling in an area near or around a risk source to detect and, if possible, quantify the amount of potentially harmful gases, vapors, effluents, particles, radioactive particles, or organisms escaping from the source (such as an industrial facility). Monitoring data can be used to estimate past and current emissions or releases; they can also be used as a basis for extrapolating potential or maximum emissions or releases if control techniques of known or estimated efficiencies are applied to the source. Monitoring programs are also essential for calibrating models of technological and natural systems.

Because monitoring focuses on past and current experience, however, it has some important limitations. Use of monitoring data to estimate future emissions or releases from a source requires assumptions that may add significant uncertainties to the projected estimates. Similarly, monitoring is not very helpful for rare events, such as explosions or disasters, because the number of observations is too small to support precise projections. Wide data variation may result from monitoring systems depending upon the location of the monitors in relation to the sources (for example, upwind, downwind, ground level, or elevated) and upon the existence of other, unknown sources in the vicinity of the monitors. Also, the sensitivity and specificity of monitoring technologies vary, depending upon the type of monitoring equipment used; the nature of the chemical, radioactive particles, or organisms being monitored; and the skill with which samples are treated and analyzed.

Finally, monitoring programs can be time-consuming and expensive, and even at that, not everything can be monitored.

Performance Testing and Accident Investigation. Performance testing involves assessing a system's behavior under controlled, and sometimes stressed, conditions. By observing and measuring important features of the system during a performance test, analysis may be able to predict the system's behavior under a variety of operating conditions (for mechanical systems) or environmental conditions (for living organisms or chemicals).

Accident investigation involves the interpretation of the causes and sequences of events after a disruption in a system occurs.

Performance testing and accident investigation provide important information that routine monitoring cannot: the behavior of systems under unusual conditions that may cause the release of a potential hazard. Simulating or predicting those unusual conditions may be difficult with some systems; however, accidents, by definition, are difficult to predict and prepare for. Accident investigation is further complicated by the frequent loss or destruction of important evidence owing to the effects of the accident itself, lack of experimental controls that would permit more conclusive judgments about causes, and delays between the accident and investigation necessitated by the need to manage any injuries and other consequences of the accident.

Statistical Methods. Statistical methods are used to analyze previously collected data on a risk source–either from monitoring programs or from accident records–to estimate the likelihood of a particular accidental release or hazardous event. In the simplest case, the release or event of concern may occur with such historical frequency that accurate records are available. Based upon these historical data (or a sample from them), simple statistical estimates of risk can be made. For example, if a certain type of valve has been in use in great enough numbers for a sufficient period of time, it may exhibit an average failure rate of one failure every seven years. This rate can then be used to predict the failure of similar valves.

In some cases, historical data are limited or of little use (because, for example, of the development of new technologies or changes in a process). Under such circumstances, the risk analyst may assume that events that cause releases or accidents are random events that with some assumptions, can be described as probability distributions, which assign probabilities to events of different frequencies and severities.

Statistical methods are limited by the data from which they are derived, however. The fewer data available from which to infer a probability distribution, the more assumptions or expert judgments must be used and the greater the uncertainty or potential for error. Even if some data exist, they may not be representative of the actual distribution of events (that is, they may be extreme values), and extrapolation from the data can bias the estimated frequencies or quantities of releases.

Modeling Risk assessments sometimes rely on formal models of source systems to estimate releases. Frequently, these models require extensive information about system processes, data from monitoring programs, historical event records, or assumptions about probability distributions to estimate key model parameters. Depending upon the characteristics of the risk source, the risk analyst may select any of several possible models for estimating releases. Many of the models used can be characterized by graphic representations, such as flow diagrams or trees.

Fault Trees. Fault trees describe in graphic form the specific chain of events or conditions required for an undesired event (for example, a release of a hazardous material) to occur. Fault trees start at the undesired event and work backwards, identifying the full range of possible causes of that event. Those causes, in turn, become events characterized in terms of their causative elements. The process continues until the most basic initiating events are identified. The rates of occurrence of these basic events may be known; if not, the analyst characterizes the likelihood of their occurrence using probability distributions derived from whatever data are available. Once rates or probability distributions are assigned to each basic initiating event, the analyst works back up the fault tree, calculating the probabilities of each successive event until a probability distribution is determined for the final, undesired event.

Event Trees. Event trees are similar to fault trees, except that they begin with some identified, undesired initiating event. From that initiating event, a sequence of corrective responses to that event is arrayed in steps, and the probabilities of successful and unsuccessful occurrence of the corrective responses are estimated from data or other component models. A typical event tree is shown in Figure 1.

The probabilities of each sequence of events are aggregated mathematically to provide estimates of the probabilities of various outcomes.

Difficulties with Fault and Event Trees. A major problem in designing fault or event trees is omitting or leaving this out, with the result that the true risk may not be accurately represented. For example, omissions may lead to an underestimation of the true risk.

One type of omission is the failure to consider the varied and imaginative ways in which people can thwart or undermine the system. For instance, facility operators may ignore or violate approved system operating procedures. This type of failure occurs in part because of a lack of understanding of factors that lead to operator errors. Another type of omission is the failure to consider unanticipated changes in the world outside of the technology under assessment. For instance, an engineer may design a facility using incorrect assumptions about the availability of backup electrical power, should the plant generator fail.

Omissions also occur because of overconfidence in current scientific and technical knowledge, that is, assuming that there are no new chemical, physical, or biological effects yet to be discovered. Another omission problem is the failure to see how the system functions as a whole and how different components interact (for example, a system may fail because a backup component has been removed for routine maintenance or was damaged during testing). Yet another type of omission consists of insensitivity to the possibility of "common mode failures," which occur if the same event can cause failure of both the system and its backup systems (for example, if pipes in several nuclear plants were made from the same batch of defective steel).

Other major difficulties with fault and event trees are the uncertainties arising from the estimates of the success or failure of mechanisms. For mechanical systems, data from performance testing or engineering analysis may provide limited confidence in probabilities of successful operation. Estimating the probabilities of successful human intervention in a potential release incident is even more difficult. Uncertainties in a model may also arise from oversimplification or incorrect specification of the sequence of events.

When faced with these types of uncertainty, fault and event tree analysts typically use conservative, ie, pessi-

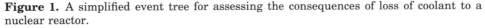

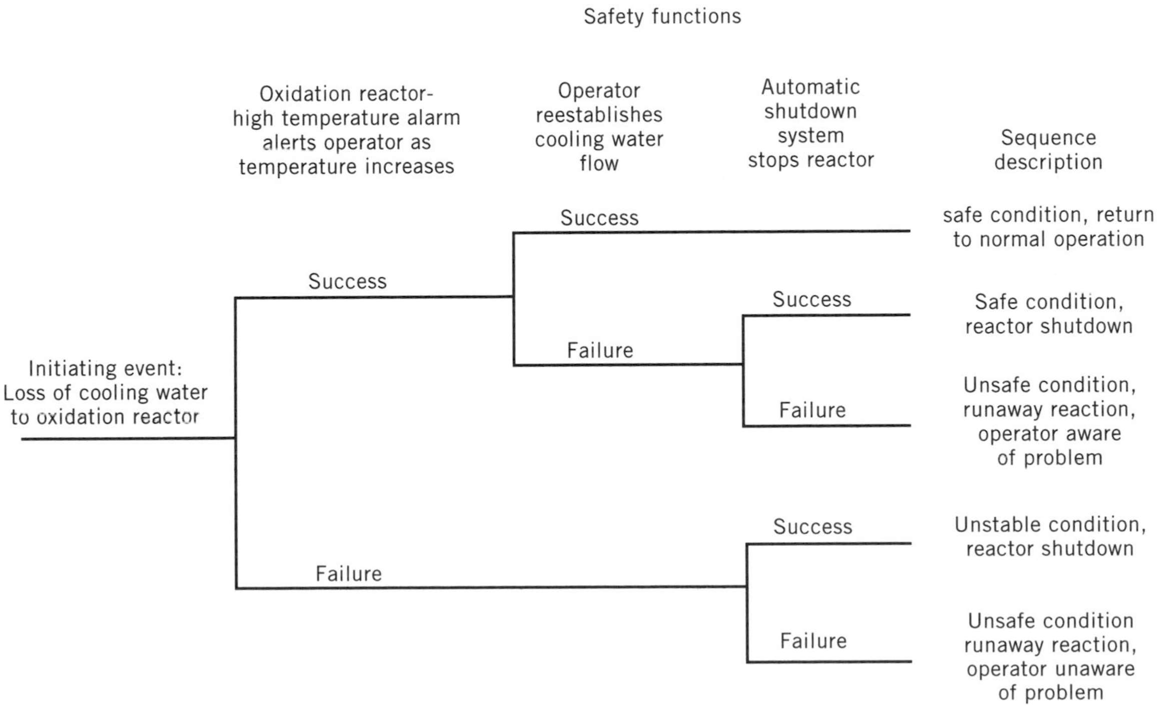

Figure 1. A simplified event tree for assessing the consequences of loss of coolant to a nuclear reactor.

mistic, assumptions. For example, given the difficulties and uncertainties in modeling human behavior, fault and event tree analysts generally do not assume that operators can intervene creatively to prevent an accident from progressing to a more serious stage.

Because of these uncertainties, limited information and value are gained from focusing too much attention on the accuracy of risk estimates derived from fault and event tree analyses. In many cases, the primary value of such analyses is in identifying and correcting specific weaknesses in the facility design and in assessing whether research efforts are aimed at the major contributors of risk.

THE RISK ASSESSMENT PROCESS: EXPOSURE ASSESSMENT

Exposure-assessment techniques estimate or directly measure the quantities or concentrations of risk agents (and any by-products or transformation products) received by individuals, populations, or ecosystems. Exposure assessments address the following multiple part question: What target organisms, species, or environments are, or may be, exposed to how much, or what substance(s) (or risk agents), in what way, for how long, and under what circumstances?

Exposure assessments, as defined above, typically estimate concentrations of a risk agent at a particular point of contact with the exposed organisms (that is, the skin or the surface of the respiratory or gastrointestinal tracts). Some exposure assessments rely on data from fluid or tissue samples–called biomonitoring–to identify exposed populations and to determine exposure levels. In risk assessments, estimates of exposure are necessary to estimate doses–concentrations or quantities of risk agents reaching tissues, organs, or cells within the exposed organisms where damage may occur.

Most exposure assessments are complicated by the fact that humans and other organisms may move from place to place and engage in a variety of activities that affect when and for how long they are exposed to risk agents. Some exposure assessments attempt to control for this variability by assuming or tracking—by using daily logs or surveys, for instance–activity patterns for exposed populations. The amount of time spent in various microenvironments, such as the home, car, or office, may be combined with data on environmental concentrations of risk agents in those microenvironments to estimate exposure.

Gathering the data necessary to account for such time-activity patterns is difficult and costly. Thus, many risk assessments use average exposures when better information is not available. This method is analytically equivalent to assuming that all members of an exposed population are exposed to the same amount or concentration of a risk agent, although this assumption is not actually made. In addition, constant or average exposure levels over time may be assumed. Whether these assumptions are reasonable depends upon the specific risk being assessed and the purpose of the risk assessment.

The remainder of this section describes how exposure assessments estimate the amount of a risk agent that may come in contact with an exposed population or environment. The specific methods or techniques used in exposure assessments vary depending upon the nature of the risk agent or agents, knowledge about the source or sources, and knowledge about the activity patterns of exposed populations. For environmentally mobile risk agents, such as chemicals, biological organisms, and radioactive materials, exposure assessments typically employ three kinds of approaches: analogies, monitoring, and modeling which are discussed below.

Analogies With Other Risk Agents. Information about the transport and fate of hazardous substances in the environment can, in some cases, be inferred by comparison with what is known about the transport and fate of similar hazardous substances. The use of analogies to predict the environmental behavior of risk agents is similar in concept to the use of structure-activity relationships to identify hazards. For example, some classes of chemicals, owing to their physiochemical characteristics, are known to disperse and react with other chemicals in relatively predictable ways under given environmental conditions. This disposition is also true of some species or genera of biological organisms, although the behavior of some non-native organisms may be more uncertain.

Analogies with other hazardous substances generally do not predict the environmental transport and fate of a risk agent as accurately as actual measurements of environmental concentrations. However, such analogies may identify general characteristics of the risk agent that support specific judgments or assumptions used in exposure models.

Exposure Monitoring The most accurate information about exposure comes from monitoring data. Moreover, monitoring data provides an accuracy benchmark for exposure models.

Monitoring techniques for exposure assessments are similar in most respects to those used for source/release assessment. For exposure assessment, however, monitoring data is typically collected in close proximity to the populations or environments of concern.

There are two basic kinds of exposure monitoring: personal monitoring and ambient (or site or location) monitoring.

Personal Monitoring. Personal monitoring involves using one or more techniques to measure the actual concentrations of hazardous substances to which individual humans are exposed, regardless of the location. Personal monitoring may include sampling of the air inhaled by people or sampling of the water consumed by them. In occupational settings where radiation is present, personal monitoring consists of routinely monitoring exposed individuals for radiation. For example, radiation workers may carry dosimeters, film badges, or other radiation sensitive devices to monitor their exposures. However, this system has possible drawbacks if procedures controlling the use of such monitoring devices are not stringent. For example, individuals may forget to wear their monitors, or individuals may deliberately or inadvertently be tampering with the monitor.

Personal monitoring may also involve, especially in occupational settings, the sampling of human body fluids (for example, blood, urine, or semen). This type of monitoring is often referred to as biological monitoring. (The term "biological monitoring," also called biomonitoring, is also used to refer to the regular sampling of animals, plants, or microorganisms in an ecosystem to determine the presence and accumulation of pollutants in an ecosystem, as well as their effects on ecosystem components.)

Ambient Monitoring. Ambient (or site or location) monitoring, in contrast to personal monitoring, involves collecting samples from the air, water, or soil at fixed locations and then analyzing the samples to determine environmental concentrations of hazardous substances at that location. Ambient monitoring is clearly a more practical technique than personal monitoring when large geographical areas or human populations are exposed. It is especially practical when emissions are routine and the data are used to construct regional indices of ambient air pollution.

Several major ambient monitoring activities are currently underway. For example, an extensive system of monitoring stations–known as the Houston Regional Monitoring Network–is operational and is measuring levels of airborne chemicals in the Houston area. One limitation of such systems is that individuals may move between locations covered by different monitoring stations or may live or work in areas between monitoring stations. To cover such contingencies, careful processing of data is required.

As noted in the earlier discussion of source/release assessment, there are potential sources of error in monitoring techniques including poorly designed samples. One important variable in sample design, for example, is the time period during which a sample reading is taken. Readings taken during different time periods may show different patterns of pollutant emissions. Pollutant emissions from a plant may be low for 8 hours and high for the next 16 hours. Similarly, monitoring results can be biased by the selection and location of monitoring sites. In some cases, extrapolations from monitoring point sites must also be corrected for weather or topographical features.

In addition to these problems, monitoring studies can be costly, time-consuming, and difficult to generalize to other environmental situations. Furthermore, monitoring data may not be representative of the full range of exposure situations since such data are collected under a defined set of conditions. For example, air monitoring systems are typically designed to measure outdoor concentrations of chemical pollutants; however, most people spend most of their time indoors.

Despite these limitations, the main advantage of monitoring techniques over other methods is that they provide actual measured data from locations that are close to the exposed populations and environments of concern. For the same reason, monitoring data are typically more accurate than model predictions (see below). Monitoring data can be erroneous, however, and need to be periodically checked and questioned. This problem is especially true if monitoring data differ substantially and inexplicably from expected values. In some cases the monitoring data will be found to be in error; however, in most cases, the model data are in error and the monitoring data can be used to improve the model's accuracy.

Exposure Modeling. When monitoring data are either inadequate or inappropriate for estimating exposures to a risk agent, risk analysts frequently use models to simulate the behavior of risk agents in the environment. In most of the literature on risk assessment, such models are generally known as exposure models. In many engineering and environmental impact studies, however, models of pollutant behavior and environmental quality are called natural systems models. Hundreds of such models exist, although many are highly specialized. For example, models exist that are specific to particular sources (eg, point sources such as factories), risk agents (such as a specific class of chemicals), and environmental settings (such as watersheds and lakes).

All environmental models are limited by two types of uncertainty: first, there is uncertainty in the data used in the models. Many existing databases can provide information about chemicals, radioactive materials, biological organisms, meteorological conditions, and geohydrological parameters for specific types of environments. Nevertheless, these data may not be precisely applicable to environments different from the ones used to create the databases. Site-specific data may be gathered to calibrate models and reduce data uncertainties, but this procedure is quite costly. In evaluating a risk assessment, therefore, it is important to consider the uncertainties introduced into the analysis by site-specific parameters that are estimated rather than measured directly.

The second type of uncertainty involves specification of the model. Because models are only mathematical representations of environmental conditions, they are necessarily imperfect predictors of the behavior of risk agents in the real environment. There is a complex relationship between the degree of simplification of a model and the amount of uncertainty in the model. Although complicated models may represent more characteristics of the real environment, their generally greater demand for data means that many parameters must be estimated. Also, uncertainty can multiply with the number of parameters estimated, sometimes creating greater aggregate uncertainty than is generated in a simpler, less data-intensive model.

The predictive capabilities of a model are ideally validated and improved through field testing. However, the data required for such testing are difficult to collect and interpret. Although some models have been validated in specific cases, procedures for validating models are not universally accepted. Moreover, validation of a model under one set of environmental conditions does not necessarily support its applicability to a different set of conditions.

Although a detailed discussion of exposure models is beyond the scope of this article, five general types of models are briefly described below: atmospheric models, surface-water models, groundwater and saturated-zone models, multimedia models, and food chain models. For historical reasons, these five types of models are primarily applicable to chemicals or to radioactive materials associated with dusts and other particles. Recently, however, new concerns have emerged about the hypothetical risks from genetically modified or nonnative microorganisms introduced into new environments. Accordingly, following the discussion of the five types, there is a brief description of issues related to modeling the environmental behavior of microorganisms.

Atmospheric Models. The atmospheric transport of a risk agent may be estimated by applying known laws of physics and chemistry to data from monitoring systems. However, in many cases, monitoring data are not available and the releases of risk agents occur with substantial variability. Attempts to predict and to regulate the effects of gaseous or particulate emissions have led to the development of many sophisticated models that simulate the transport and transformation of materials in the atmosphere; for instance, Gaussian plume models, long-range atmospheric transport models, and so-called puff models.

Gaussian plume models are widely used in estimating airborne concentrations of pollutants in the vicinity of a source. Downwind concentrations are calculated based on the height of release, emission rate from the source, wind speed, and atmospheric conditions. Such models assume that the pollutants released–referred to as a plume–will spread out both laterally and vertically, like a bell-shaped curve. Such shapes are described mathematically as normal, or Gaussian, statistical distributions. Some research indicates that Gaussian plume models predict annual average concentrations at a point of exposure within a factor of two to four for pollutants released continuously over flat terrain. Complex meteorology or uneven terrain increases the uncertainty associated with the predictions of these models.

Long-range atmospheric transport models attempt to predict pollutant concentrations over geographical regions ranging in size from a few states to entire continents. Trajectories that released pollutants might follow are computed based on historical wind data from weather stations within the region. Generally, these models are thought to predict annual average concentrations at a point of exposure within a factor of three to five.

Puff models are based on the same principles as the continuous-release models described above, but they are used to simulate the transport of emissions after episodic or short-duration releases, such as explosions or accidental releases.

Many atmospheric models also incorporate equations that account for the rates of transformation, degradation, or deposition–that is, through settling or precipitation–of a risk agent released in the atmosphere. These phenomena are especially important when the transformation products of a risk agent pose a significantly greater or lesser hazard than the risk agent emitted from the source.

Surface-Water Models. Risk assessments often incorporate surface-water models in exposure assessments when risk agents released into the environment may affect drinking water, food supplies, or recreational resources derived from streams, rivers, lakes, or other surface water bodies.

Many surface-water models divide surface-water bodies into compartments or boxes. The transport of most risk agents is traced by mathematical equations that assume conservation of energy and mass going into and out of each box. Refinements to the compartmental model concept attempt to account for complex flow conditions (eg, tidal forces), sedimentary deposition, biological degradation and uptake, and chemical transformation and decay, and the time-varying introduction of pollutants (or risk agents) into the model system.

Groundwater and Unsaturated-Zone Models. Risk agents released or deposited on or in the ground can move through soil and rocks into groundwater, thereby threatening drinking-water supplies and possibly discharging into surface-water bodies.

Models of the unsaturated soil zone estimate the vertical movement of contaminated water into the soil and between the ground surface and the saturated (groundwater) zone. Traditional soil models rely on mathematical equations to statistically describe the behavior of risk agents moving with moisture through the soil. Newer soil-modeling techniques, though, emphasize compartmental, mass balance approaches.

Although compartmental models may provide more sophisticated treatment of geochemical effects than traditional differential equation models, no single model yet exists that simulates all physical, chemical, and biological processes associated with the behavior of pollutants in soils.

The simplest groundwater transport models calculate the movement of risk agents as a function of the linear velocity of the groundwater (known as advection), and the spreading or mixing (dispersion) of the risk agent in the groundwater. More complex groundwater models account for differences in density of chemical risk agents, adsorption of risk agents on soil or rock particles, chemical and biological transformations, and flow through fractured geological media.

Multimedia Models. Although some air and water models account for deposition of pollutants into another medium (such as air, water, soil, or groundwater), most of the models in the categories described above are limited to the analysis of a single environmental medium or exposure pathway. However, the movement of chemicals among different media is a critical factor in exposure assessment. For example, one recent study estimated that 50 to 85% of all chemicals released to a single medium will eventually have at least 5% of their mass in another medium. Because many risk agents are released simultaneously into two or more media, risk analysts have devoted increasing attention to multimedia pollutant behavior.

Risk analysts have begun to develop and use multimedia models to account for transfers of chemicals among different media (known as partitioning) and exposures from multiple environmental pathways. Most of these models consist of linked, single-medium models which may simulate the physical and chemical processes that drive the transport of chemicals across the air/water, air/soil, and water/soil interfaces. However, the data requirements for such models are substantial and scientific understanding of intermedia transport processes is still embryonic.

An alternative approach to intermedia transport modeling involves the use of compartmental models that estimate the relative distribution of pollutants among environmental compartments–air, soil, water, sediment, and biota. These models are typically based on the concept of fugacity, which expresses the tendency of a chemical to escape from one chemical phase into another. For example, fish may have a higher capacity for retaining an organic chemical than does water, so the concentration of the chemical in fish will be higher than the concentration in water. The fugacities, however, may be equal: because of the higher capacity of the fish, there is no tendency for the chemical to escape from the fish to the surrounding water. Compartmental models based on fugacity are particularly helpful in identifying compartments of concern, but they are not designed to estimate absolute concentrations of chemicals in each compartment.

Food Chain Models. Some risk assessments incorporate models to estimate risks of exposure through the food chain. Food chain models differ somewhat from the multimedia models described above. Food chain models simulate the transport, transformation, and accumulation of risk agents in the environment as they are deposited (for example, in sediments) and ingested by different species representing the various trophic levels of the food chain.

As with other models, the data requirements for food chain models are substantial. Moreover, large uncertainties exist on such critical subjects as the uptake, chemical behavior, and retention of risk agents in plants and animals later consumed as food.

Modeling the Environmental Behavior of Microorganisms. Some of the features of the models described above are applicable to the behavior of microorganisms. For example, the transport of some microorganisms can be predicted by atmospheric, surfacewater, and groundwater models developed for chemical risk agents. Microorganisms can also be transported by animal vectors. Their survival and multiplication in new environments depend upon species interactions and environmental conditions that are generally absent from most existing exposure models–for example, availability of essential nutrients, competition with other species, and predator/prey relationships. In general, the prediction of the behavior of genetically modified organisms introduced into the environment currently depends largely on expert judgment. Expert judgments are based primarily on information about the unmodified organism, the traits added to the organism, and the behavior, if known, of unmodified strains of the organism in the environment.

The decision points encountered in exposure assessment vary widely since most approaches are medium- or route-specific. For example, many uncertainties arise in modeling the transport of a chemical through air, and other uncertainties arise for the same chemical when considering its movement through water. Different assumptions in these models, and in estimates of human exposure can produce widely varying risk estimates. The National Research council has identified the following questions as decisions points in exposure assessment:

- How should one extrapolate exposure measurements from a small segment of a population to the entire population?
- How should one predict dispersion of air pollutants into the atmosphere attributable to convection, wind currents, etc, or predict seepage rates of toxic chemicals into soils and groundwater?
- How should dietary habits and other variations in lifestyle, hobbies, and other human activity patterns be taken into account?
- Should point estimates or a distribution be used?
- How should differences in timing, duration, and age at first exposure be estimated?
- What is the proper unit of dose?
- How should one estimate the size and nature of the populations likely to be exposed?
- How should exposures of special risk groups, such as pregnant women and young children, be estimated?

THE RISK ASSESSMENT PROCESS: DOSE-RESPONSE ASSESSMENT

Dose-response assessment, the next step in the risk-assessment process, involves (1) determining the dose of the risk agent received by exposed populations, and (2) estimating the relationship between different doses and the magnitude of their adverse effects. In essence, dose-response assessment involves efforts to quantify or statistically describe the qualitative relationship identified in the hazard-identification stage between a risk agent and its adverse effects. A general dose-response relationship chart is shown in Figure 2.

The determination of "delivered dose," ie, the amount of a hazardous substance that is received by exposed populations, is sometimes discussed in the risk assessment

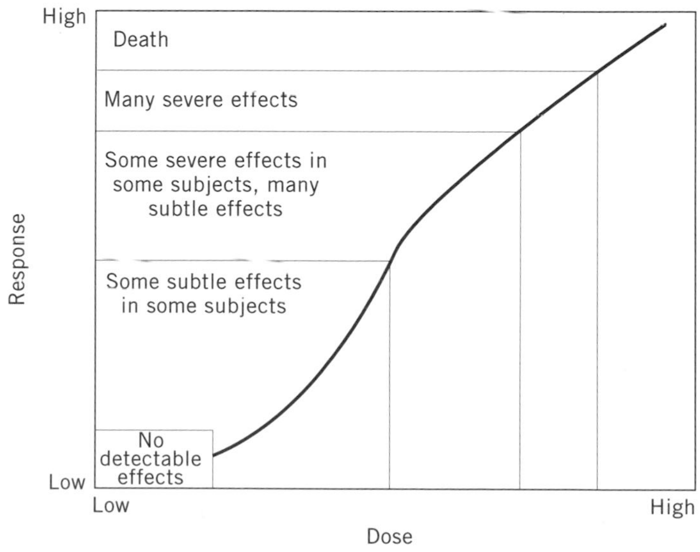

Figure 2. General dose–response relationship.

literature under the heading "exposure assessment" rather than under the heading "dose-response assessment." Many such inconsistencies can be found in the risk assessment literature and illustrate the newness of the field.

Determining Dose. Some risk assessments assume that the concentration of a risk agent that reaches the immediate vicinity or the body surfaces of the target population, represents the dose to which the target population is exposed. The difficulty with this simple assumption is that the physiological and metabolic systems of organisms (and the functional interrelationships of ecosystems) can act on risk agents and potentially increase, decrease, or modify the amounts of the risk agent received by relevant parts of the human body or other organism of concern. In response to this problem, increasing numbers of scientists have focused their attention on developing improved methods for estimating the amount of a risk agent that is actually taken into the human body and how the risk agent interacts with particular tissues or organs.

To distinguish between ambient exposures and actual doses, analysts use two measures of dose: one is the absorbed dose–the amount of a risk agent that is absorbed by:

(1) the lungs (for inhalation exposures),

(2) the gastrointestinal tract (for ingestion exposures), and

(3) the skin (for dermal exposures).

The other measure is the internal, or effective, dose–the amount of a risk agent reaching a tissue or an organ where it inflicts damage. These routes are described in Figure 3.

Determining the absorbed dose of a risk agent usually is performed by applying a number of standard adjustment factors to environmental concentrations of a risk agent. Such factors may include, where appropriate, the number of adults and children in the exposed population; the amounts of drinking water consumed by members of the population (for risk agents in drinking water); the amounts of air inhaled by exposed individuals, adjusted for age and level of physical activity (for inhaled risk agents); the size of the risk agent and site of absorption in the respiratory or intestinal tract; and the absorption rates of skin, lung tissue, bronchial tissue, and intestinal tissue for the risk agent. Generally, the data needed to determine the values of these adjustment factors are not available for the specific populations exposed to the risk agent. Standard values or statistical distributions are typically used, but these values may introduce some error into estimations of dose for actual exposed populations.

Determining the effective dose of a risk agent is more difficult than calculating the absorbed dose. For example, often the risk agent is a metabolite, ie, the risk agent becomes active only after it has been metabolized by the body. Moreover, the human body has complex mechanisms for responding to chemical or biological stimuli. Some risk agents can be metabolized and excreted, whereas others may accumulate in certain tissues or or-

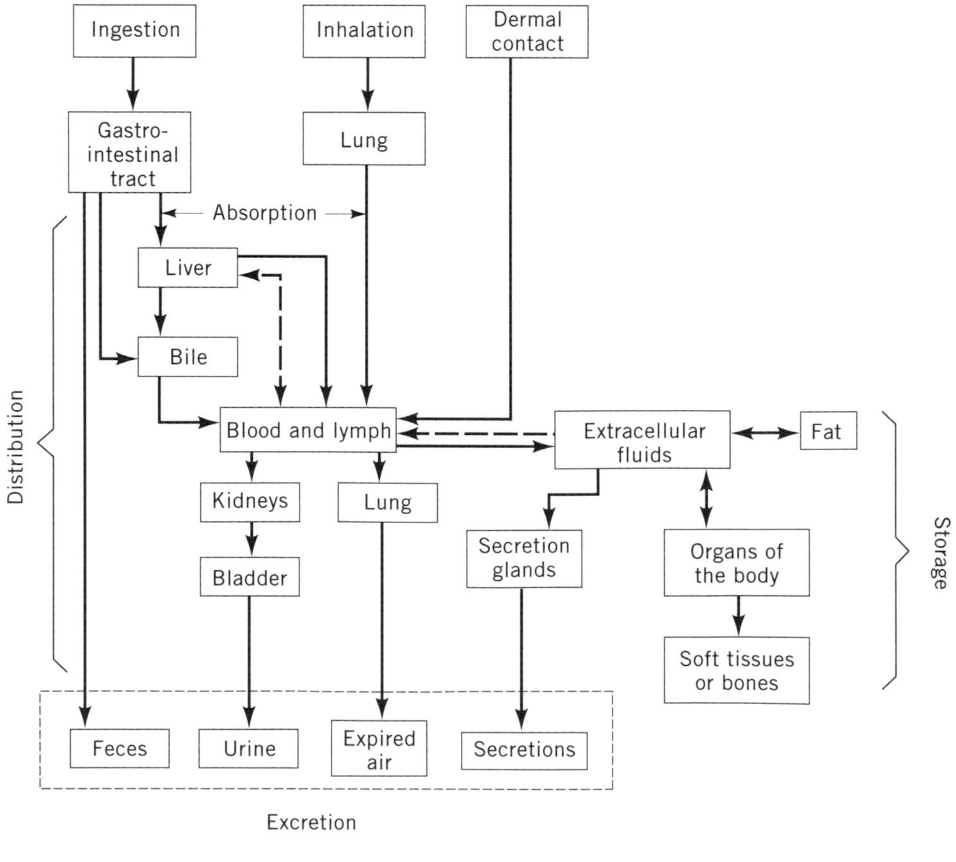

Figure 3. Key routes of chemical absorption, distribution, and excretion.

gans. Some risk assessments use models of chemical and metabolic activity to determine the fate of risk agents from the time they enter the bloodstream and/or fluids of an organism to the time they interact with tissues or organs. Such models are often called physiologically based pharmacokinetic models, or PB–PK models.

PB–PK models require enormous amounts of data on: anatomy, physiology, the partitioning of risk agents into specific tissues and fluids, rates of decay and metabolism, and biochemical interactions between risk agents and tissues. Such data may be unavailable or unconfirmed for the majority of risk situations, however. When such data are available, they are usually derived from experiments on laboratory animals (usually rats or mice) and so must be carefully weighed in light of the poorly understood biological differences between species and also between relatively homogeneous, inbred organisms (ie, laboratory strains) and more heterogeneous human populations (ie, exposed groups).

Given these difficulties and problems, the typical risk assessment equates effective dose with absorbed dose (or even environmental concentrations).

Dose-Response Estimation. To estimate the responses of populations exposed to a given dose (or distribution of doses), risk analysts conduct mathematical extrapolations. In rare instances, epidemiological data are available that demonstrate the magnitudes of adverse effects across a range of exposures (doses) that are comparable to the doses of concern. Most epidemiological data, though, are available only for doses significantly higher than the doses of concern. Analysts extrapolate the effects resulting from high doses of the risk agent to the lower doses to which the populations at risk are exposed. For most risk agents, adequate epidemiological data do not exist, and data from laboratory animal studies (similar to those described for hazard identification) are used. In such instances, analysts also extrapolate from laboratory species to humans or other species to account for differences in size, metabolism, and population heterogeneity.

As discussed above, animals used in toxicity studies receive high doses of risk agents to compensate for the low sensitivity of such studies (ie, their poor ability to indicate a positive result when a condition is present). Toxicologists generally make two assumptions about the effects of exposure to risk agents at low doses: the first is that thresholds exist in a population for most biological effects. That is, for noncarcinogenic, nongenetic, toxic effects, doses exist below which no adverse effects are observable in a population of exposed individuals. The second assumption is that no thresholds exist for genetic damage or carcinogenic effects. That is, for genetic damage, including carcinogenic effects, any level of exposure to the risk agent corresponds to some nonzero increase in the likelihood of inducing genotoxic or carcinogenic effects.

The first assumption is widely accepted by the scientific community and is supported by empirical evidence. The threshold for a chemical risk agent is often called the no observed effects level (NOEL) or the no observed adverse effects level (NOAEL). The difference between the NOEL and the NOAEL hinges upon the definition of adverse effects. As discussed above (see section on hazard identification), toxicologists usually estimate NOELs or NOAELs from toxicity studies.

The second assumption, regarding no thresholds, is more controversial. This assumption has been adopted by regulatory agencies as precautionary measure. However, there is a strong ongoing debate among scientists over whether chemical carcinogens in small doses can be detoxified, or whether exposure to minute amounts of a carcinogen lead to the development of cancer. If a chemical carcinogen can be tolerated in small amounts, then a threshold dose can be posited below which the organism is not affected by the chemical. If very small doses can cause an adverse reaction, then no threshold dose exists for that chemical.

Currently, it is not agreed whether thresholds exist for any carcinogens. Proponents of the no-threshold view argue that irreversible, self-replicating lesions may result from a mutation in a single cell. Those arguing against the no-threshold hypothesis cite the existence of metabolic detoxification, DNA repair, and other mechanisms that may act to overcome the effects of potential carcinogens at low doses. In addition, some carcinogens are suspected of promoting cancer by acting on transformed cells at the later stages of the carcinogenic process. Such carcinogens may not cause cancer by genetic mutation and may exhibit threshold effects. In any case, scientific evidence has not yet proven the existence or nonexistence of threshold doses for carcinogens.

Dose-Response Extrapolation. Extrapolations from high experimental doses to low doses require the use of mathematical models, which are often represented graphically as dose-response curves. Choosing an appropriate dose-response curve or extrapolation model for carcinogens is difficult because relatively little is known about the biological mechanisms involved in carcinogenesis, especially the latency period between exposure and effects. Different extrapolation models are based on different approaches to characterize carcinogenic activity. The following is a brief discussion of some of these models.

One category of models are termed mechanistic models. These models derive from particular theories about the biological steps involved in the formation of cancerous tumors. Mechanistic models include one-hit and multihit models, which are based on informed judgments about the number of "hits" (or interactions between a risk agent and a cell) required to make a cell become cancerous. Another mechanistic model, the multistage model, is the most frequently used extrapolation model. It derives from the theory that developing tumors go through several stages before they become clinically detectable.

Another category of models are termed threshold distribution models. These models are not based on mechanistic theories of carcinogenesis. Rather, they assume that different individuals have different tolerances for risk agents and that this variability in tolerance in an exposed population can be described as a probability distribution. Within this category, the probit, logit, and Weibull extrapolation models each adopt different probability distributions for the tolerance levels of individuals in the exposed population.

A third category of models are termed time-to-tumor models. These models attempt to describe the relationships among dose, tumor latency, and cancer risk. Several models of this type have been developed but none seems to have significantly improved the precision of estimates of risk.

In general, there is no scientific basis for deciding which of the many low-dose extrapolation models most accurately simulates the true dose-response relationship, and no single model has gained universal acceptance within the scientific community. In some cases, risk analysts estimate dose-response using several extrapolation models as an indication of the uncertainties in extrapolating to low doses.

For noncarcinogenic health effects, extrapolation from high to low doses involves identifying the NOEL or the NOAEL and describing the dose-response relationship that best fits the observed data from in vivo, in vitro, or epidemiological studies.

Species Extrapolation. Differences in size, metabolism, anatomy, physiology, and population heterogeneity make dose extrapolation among species a highly uncertain activity. Risk analysts frequently seek to adjust for size differences by standardizing the experimental dosage on the basis of body weight or body surface area. For example, dose may be measured in terms of milligrams per kilogram of body weight or milligrams per square meter of body surface area. There is no clear consensus as to which extrapolation principle is more appropriate; substantial variability results in risk estimates derived from the two methods.

Differences in genetic variability or heterogeneity between the populations of clinical and environmentally exposed species are difficult to adjust for on the basis of empirically derived principles. Laboratory species are generally inbred, and so they are thought to have a relatively homogeneous response to risk agents in comparison to highly variable exposed populations. For example, sensitive populations, such as elderly, young, or ill people, may be much more susceptible to harm than an average member of the population. Assuming that the exposed population will respond identically to groups of laboratory subjects introduces considerable variability and uncertainty into risk estimates.

Metabolic, physiological, and anatomical differences among species also complicate interspecies extrapolation. Where differences, eg, in enzyme activity, are known, such information may be useful in qualitative judgments about risk. Often, however, data on these differences are insufficient to make reliable quantitative adjustments to dose-response estimates.

Estimating the relationship between dose and response requires many assumptions and judgments on the part of the risk analyst. High- to low-dose extrapolation is often called for because environmental exposures are usually lower than occupational exposures for which epidemiological data may be available. The choice of one model over another will lead to a more or less conservative (eg, upper bound) estimate of risk. Questions identified as decision points by the National Research Council concerning the use of epidemiological data include:

- What dose-response models should be used to extrapolate from observed doses to relevant doses?
- Should dose-response relations be extrapolated according to best estimates or according to upper confidence limits?
- How should risk estimates be adjusted to account for a comparatively short follow-up period in an epidemiological study?
- For what range of health effects should responses be tabulated? For example, should risk estimates be made only for specific types of cancer that are unequivocally related to exposure, or should they be made for all types of cancers?
- How should exposures to other carcinogens, such as cigarette smoke, be taken into consideration?
- How should one deal with different temporal exposure patterns in the study population and in the population for which risk estimates are required? For example, should one assume that lifetime risk is only a function of total dose, irrespective of whether the dose was received in early childhood or in old age? Should recent doses be weighted less than earlier doses?
- How should physiological characteristics be factored into the dose-response relation? For example, is there something about the study group that distinguishes its response from that of the general population?

In addition to these sources of uncertainty, many scientific uncertainties confront the risk analyst in attempting to extrapolate results of animal tests to expected effects on humans. For example, the decision to use test results from the most sensitive species to estimate the human dose-response curve will produce a more conservative (eg, upper bound) estimate of the effect on humans.

The National Research Council has identified the following questions as decision points regarding animal-to-human extrapolation procedures:

- What mathematical models should be used to extrapolate from experimental doses to human exposures?
- Should dose-response relations be extrapolated according to best estimates or according to upper confidence limits? If the latter, what confidence limits should be used?
- What factor should be used for interspecies conversion of dose from animals to humans?
- How should information on comparative metabolic processes and rates in experimental animals and humans be used?
- If data are available on more than one nonhuman species or genetic strain, how should they be used? Should only data on the most sensitive species or strain be used to derive a dose-response function, or should data be combined? If data on different species and strains are to be combined, how should this be accomplished?

- How should data on different types of tumors in a single study be combined? Should the assessment be based on the tumor type that was affected the most (in some sense) by the exposure? Should data on all tumor types that exhibit a statistically significant dose-related increase be used? If so, how? What interpretation should be given to statistically significant decreases in tumor incidence at specific sites?

The importance of these questions is clearly illustrated by the controversies surrounding the chemical *dioxin*. Dioxins are a family of chemical compounds that have inspired considerable scientific and public controversy. Dioxins are formed as by-products from a number of processes, including the incineration of solid wastes, the making of 2,4,5-T (a herbicide used on crops and known as Agent Orange), the making of the wood preservative pentachlorophenol, the burning of wastes containing chlorine, and the bleaching of wood pulp to make paper. The controversy over the uses and health effects of dioxins illustrates the challenges that risk assessors face in interpreting the results of studies.

Most of what is known about dioxins is based upon studies of one particular dioxin (2,3,7,8 tetrachlorodibenzo-*p*-dioxin), or TCDD. As shown in animal tests, TCDD is one of the most acutely toxic man-made compounds known. However, the animal species used in these tests have differed widely in susceptibility. For example, guinea pigs have proved to be 5,000 times more susceptible than hamsters.

Humans have been exposed to elevated concentrations of dioxin through industrial accidents and occupational exposures. Studies of exposed populations after these incidents have documented an association between TCDD and acute adverse health effects, including liver disorders, loss of appetite and weight, loss of sex drive, nerve damage, severe fatigue, and–most commonly–a severe skin condition called chloracne. All of these effects, except chloracne, diminished after exposure stopped.

Most of the controversy over TCDD has related to the presence or absence of chronic, as opposed to acute, adverse health effects in humans. Epidemiological studies have yielded ambiguous results and these results have been used to support conflicting interpretations. The first known dioxin accident occurred in Nitro, West Virginia, in 1949, when a vat contaminated with TCDD exploded. Twenty years later, according to one study, the death and cancer rates of 122 workers were no higher than normal. The highly publicized explosion at Seveso, Italy, in 1976, resulted in TCDD contamination of a large residential area; although acute adverse health effects on the population were manifested, no long-term effects have yet been documented.

Factory workers, farmers, and Vietnam veterans exposed to TCDD have also been studied. According to some scientists, no long-term adverse human health effects have been clearly demonstrated. However, others have argued that the weight of evidence from the epidemiological studies supports the conclusion that TCDD has adverse effects on the human immune system and causes birth defects and damage to fetuses.

Several epidemiological studies have also produced conflicting results regarding the carcinogenic effects of TCDD. Some of this uncertainty may be the result of population study groups that were too small, doses that were too low, or exposures that were too short in duration to detect adverse health effects. Some epidemiological studies attempting to link TCDD to specific cancers have been complicated by the fact that the exposed groups were exposed to mixtures of chemicals.

In 1982, the National Toxicology Program rated the evidence from carcinogenicity assays in mice as "inadequate." However, other federal agencies, including the Environmental Protection Agency (EPA) and National Institute for Occupational Safety and Health (NIOSH), have used results from animal tests to make a weight-of-evidence determinations that TCDD is carcinogenic (NIOSH) or probably carcinogenic (EPA).

In vitro cell culture tests of TCDD have been less ambiguous. Researchers have demonstrated that TCDD is a potent promoter of cell transformation. Studies also suggest that it may initiate genetic damage as well. TCDD has also been shown in some studies to suppress cell-mediated immunity, suggesting that such immune effects could make exposed individuals more susceptible to cancer from other sources.

Given these uncertainties, it is clear that the debate and research on TCDD will continue. TCDD is extremely persistent in the environment, and it has been detected in many locations. Given its ubiquitous nature and the suggestive evidence of its health effects, concern about TCDD and other dioxins is unlikely to diminish until the substantial uncertainties are reduced.

THE RISK ASSESSMENT PROCESS: RISK CHARACTERIZATION

Risk characterization is the final step of the risk-assessment process. It is designed to integrate the results of source/release assessment, exposure assessment, and dose-response assessment to generate several types of estimates, including estimates of the types and magnitudes of adverse effects that the risk agent may cause and estimates of the probabilities that each effect will occur. Ideally, these estimates are accompanied by a description and discussion of uncertainties and analytic assumptions, since characterizing risk includes characterizing uncertainty and assumptions. Presented in written form, the results of the risk characterization step constitute what is termed the risk statement.

In practice, the ways in which risk assessments characterize risk vary widely, depending upon the nature and purposes of the assessment and the preferences of the risk analyst or decision-maker.

As described above, risks assessments typically focus on one or two adverse human health effects. Since these effects do not reflect the full range of adverse effects that the risk agent may cause, the risk statement ideally includes a discussion of adverse effects that were not assessed and why they were excluded from the assessment.

One of the most ambitious efforts to charactize enviromental risks to date was conducted in 1987 by the Environmental Protection Agency. The EPA evaluated the risks of 31 environmental problems and published the results in a report titled *Unfinished Business*. In the report, the EPA Cancer Risk Work Group presented its findings using a common summary format for each of the 31 problems. Each summary provided a brief problem description, identified the substances or risk agents addressed by the assessment, and discussed the risk-assessment methodology employed. Results of the assessment were presented in qualitative and quantitative terms. Where possible, uncertainties were explicitly identified.

In many of the calculations, the risk characterization methods used by the EPA produced an upper bound estimate of risk. One advantage of an upper bound estimate of risk is that the actual risk of harm or injury is unlikely to be greater than the estimate. By possibly overstating the actual risk, upper bound estimates of risk reflect a conservative approach to risk assessment and management. However, they do not necessarily represent the best estimate of risk based on the strongest supporting evidence.

Several measures of risk are used by researchers to display results. One of the most common risk measures is individual lifetime risk. This risk is the excess, or increase in probability that an individual will experience a specific adverse effect as a result of exposure to the risk agent of concern. Individual lifetime risks are often presented as very small probabilities–such as 1×10^{-6}, or a one-in-a-million chance that an individual will develop the adverse effect. Unless otherwise indicated, individual lifetime risk means the likelihood of experiencing an effect owing to a continuous lifetime exposure to the risk agent. Individual lifetime risk is calculated by multiplying the potency of a substance by the dose an individual receives. For substances with a threshold for exposure, the potency of the substance is zero for dose levels below the threshold. For substances, such as carcinogens, which are assumed to have no threshold, a linear dose-response relationship is assumed. The risk at a high dose level is divided by that dose to determine potency (also known as unit cancer risk). Potency is then multiplied by the low dose expected to calculate risk.

Population risk, also known as societal risk, is another frequently used measure of risk. Population risk may be expressed as the number of cases resulting from one year of exposure, or as the number of cases occurring in one year. Population risk measures presented in a risk assessment need to be explicitly defined because the risk will vary greatly if it is calculated for one year or for lifetime exposure. It is usually calculated as individual risk times the number of people exposed. This calculation assumes a linear dose-response relationship, which may or may not be appropriate.

Relative risk is a comparison of rates of disease occurrence between exposed and unexposed (or differently exposed) populations. Thus, relative risk is the risk in the exposed population compared to the risk in the unexposed population. If rates of disease incidence are not available for an appropriate unexposed group, then incidence rates for the general population can be used for comparison. The standardized mortality (or morbidity) ratio is the number of deaths or cases of disease observed in an exposed group, divided by the number that would be expected if the group experienced the same mortality or morbidity rate as the general population. "Standardized" means that factors, such as age and time period, have been taken into account.

Loss of life expectancy is another measure of risk. It is expressed as days or years of life expected to be lost owing to a particular exposure or activity. For example, smoking will shorten the average male smoker's life by 6.2 years. Any risk factor to which a person is exposed can affect that person's life expectancy; the decrease (or increase) in life expectancy can be calculated using tables of death rates. Approximate conversion factors, averaged over all ages and assuming a 72-year life span, are given in various texts.

The choice of risk measures, as noted above, is often a reflection of the way in which the analyst collects and organizes data. The choice of a particular measure is not value-neutral, however, and should be considered as only one possible way of characterizing risk.

The expressions of risk developed during the risk characterization stage require combining information from the previous steps. It is particularly important that uncertainties encountered in the previous steps are reflected in the final risk characterization.

The National Research Council has identified some of the key questions that serve as decision points facing the risk analyst at this final step:

- What are the statistical uncertainties in estimating the extent of health effects? How are these uncertainties to be computed and presented?
- What are the biological uncertainties in estimating the extent of health effects? What is their origin? How will they be estimated? What effect do they have on quantitative estimates? How will the uncertainties be described to agency decision-makers?
- Which dose-response assessments and exposure assessments should be used?
- Which population groups should be the primary targets for protection, and which provide the most meaningful expression of the health risk?

ECOLOGICAL RISK ASSESSMENT

In several respects, ecological risk assessment is more complex and more uncertain than assessment of human health risks. This section describes the applicability of the previously described techniques for estimation of ecological risks and discusses the specific assessment techniques.

Ecological Hazard Identification. Ecological risk assessment is primarily concerned with the adverse effects of risk agents on populations of particular animal, plant, or microbial species and on the structure and function of ecosystems. The methods for evaluation of population-level risks to a species are analogous to those used for human populations. Endpoints of concern include changes in mortality rates, reproductive rates, and growth rates of individuals, as well as physiological or behavioral abnormalities and susceptibility to environmental stresses.

Ecosystem-level risks include such endpoints of concern as changes in species diversity, species location, productivity and biomass accumulation, connectivity among living and nonliving elements of the ecosystem, resistance and resilience to disruptive events, species interactions, taxonomic variability, energy and nutrient cycling, and the composition of functional groups, such as decomposers.

Techniques for assessing the potential impacts of a risk agent fall into four general categories: single-species tests; multiple-species tests; microcosm tests; and field tests. Each of these categories is discussed below.

Single-Species Tests. Single-species tests are laboratory tests analogous to in vivo bioassays for acute and chronic toxicity in human beings. Testing protocols are relatively standardized and easily replicable. The choice of test species depends on the ecosystem exposed to the risk agent, judgments about the species potentially affected by the risk agent, and availability of and experience with the test species. Usually, several species are tested and employed as indicators of adverse effects on all species populations within a potentially affected ecosystem.

Single-species tests used for ecological risk assessments suffer from the same limitations described above for in vivo animal bioassays. In addition, the value of such tests as indicators for other species is limited by differences among species populations in an ecosystem. Relatively little scientific attention has been given to these differences. As a result, large data gaps exist. Another limitation of single-species tests is that they are generally insufficient or inappropriate as indicators of ecosystem-level effects.

Multiple-species Tests. Laboratory tests using more than one species are sometimes used to estimate the effects of a risk agent on populations of species that interact in the environment. Tests of competitive aquatic and terrestrial plants and predator/prey relationships have been developed to determine the effects of risk agents (mostly chemicals) on the survival of selected species. However, few tests exist for other species interactions, such as symbiosis or parasitism.

In comparison with single-species tests, multiple-species tests broaden the range of ecosystem phenomena that can be tested. As a result, they increase the usefulness of ecological testing. Nevertheless, they still lack significant predictive capability for ecosystem-level effects. The number and complexity of interspecies relationships in an ecosystem far exceed those of simple multiple-species tests.

Microcosm Tests. Microcosms are simplified living models of natural ecosystems housed in artificial containers and kept in a laboratory environment. Microcosms are designed to represent, in a simplified form, the ecosystem of concern. One advantage of microcosms over natural ecosystems is that test variables and relationships can be more easily manipulated and measured in a controlled manner. Because many species from the environment are often represented in a microcosm, some effects beyond the level of one or two species can be observed and measured. Attention to microcosm systems has increased in recent years as a result of advances in microbiology and interest in the impacts of microorganisms introduced into the environment.

Because microcosms are simplifications of ecosystems, and because key components of the natural environment are missing, important variables and interactions may be absent. The choice of microcosm components, therefore, can greatly influence the results of the tests conducted in the system.

Field Tests. Field tests in natural environments, rather than in a laboratory, are conducted on various scales. Control techniques range from greenhouse-like enclosures to biological controls to no containment at all. Field tests rely on biomonitoring and census techniques to identify the results of introducing a risk agent into the ecosystem. Field tests provide the maximum realism and insight into a risk agent's ecosystem-level effects because they are performed in the natural environment.

Field tests are sometimes difficult to replicate when ecological conditions vary geographically and over time. Natural variability and gaps in knowledge about ecosystem dynamics also make it difficult to arrive at unequivocal conclusions about the effects of a risk agent.

Ecological Exposure and Dose-Response Assessment. Many of the approaches to exposure (eg, monitoring) and dose-response assessment described in previous sections are directly applicable to ecological risk assessment. For example, ecological risk analysts use field monitoring to obtain baseline data before an environment is disturbed. Dose-response assessment is more difficult, however. Although dose-response curves may be derived for the effects of risk agents on ecological populations (for example, crops, trees, or species of fish), there are no dose response functions for ecosystem-level effects. Even at the population level, however, activity patterns of exposed animals and organisms and metabolic/physiological aspects of nonhuman species are generally more variable and less well documented than those of humans. This variability makes dose determination (especially absorbed and effective dose) and dose-response extrapolation more difficult for most ecological risk assessments.

THE RISK ASSESSMENT PROCESS: IDENTIFYING AND EVALUATING ERRORS AND UNCERTAINTIES

By definition, error and uncertainty are inherent in all assessments of risk. Frequently the source and magnitude of the errors or the uncertainties are not clearly or adequately described. One critical aspect of evaluating risk assessments or other risk information, therefore, is to identify and evaluate the analysis and expression of error and uncertainty. Users of risk assessment must determine: (1) which sources of error and uncertainties can be accepted in making a risk-management decision, and (2) when sources of error and uncertainties are so great that they severely limit the applicability of the risk assessment to the decision.

Error and uncertainty arise from three kinds of sources. First, they occur as a result of natural variability over time and space in the environment. For example, rainfall, wind velocity, temperature, and other environmental conditions vary from location to location and from time period to time period. These variations make statistical descriptions of the behavior of risk agents inherently uncertain. Only in rare instances are sufficient data available to accurately estimate and characterize the probabilities of the environmental conditions being in a particular state.

Second, error and uncertainty arise in the measurement or estimation of variables used to characterize the risk agent, the releases from risk sources, the environment, and exposed populations. For example, failure rates for safety systems may be inaccurately estimated or census data may be in error.

Third, error and uncertainty arise from models or test systems that do not accurately reflect the real environments or exposed populations of concern. Examples include uncertainties arising from untested assumptions in a dose-response model and the use of a groundwater model that fails to account for biodegradation of the risk agent.

Analysis of Uncertainty

Critical to the evaluation of a risk assessment is knowledge of how different sources of uncertainty contribute to the overall variability of the final risk estimates. The kinds of error and uncertainty described above affect all stages of the risk-assessment process: hazard identification, exposure assessments, source/release assessments, and dose-response assessments. Analyzing the sources and magnitudes of uncertainties can help focus debate and identify areas of missing scientific information.

Sensitivity analysis is a technique used to test the effects of the different component estimates and event probabilities on the final risk estimates in a risk assessment. The sensitivity of a risk estimate to a particular element of a risk assessment (such as the estimated velocity of groundwater flow) is the magnitude of the change in the final risk estimate that results when that element is replaced with a substitute value and the analysis is recalculated.

Monte Carlo simulation is an analysis of the variability in the risk estimates attributable to uncertainties in the variables used to determine the risk. This and other simulation techniques use computers to quickly generate a large number of recalculations of the equations in a risk assessment (or in a part of the assessment) using randomly chosen sets of values (from a known or assumed distribution of possible values) for the different estimated variables in an assessment.

Expressions of Uncertainty

Uncertainty in a risk assessment is typically expressed as probability distributions. Probability distributions are mathematical equations or graphical representations of the relationship between all possible values (or outcomes) a variable can have, and the likelihood (expressed as a number from zero to one) that the variable will have a particular value.

Probability distributions can be discrete or continuous. Discrete distributions can be represented as bar graphs that describe the probabilities of a specific, finite number of values a variable can have. Continuous distributions–sometimes called probability density functions (pdf)–are graphed as smooth curves and describe probabilities for variables that can have a continuous range and infinite number of possible values. The area under the smooth curve between two points represents the probability that the true value of the variable lies between those points.

A special type of probability distribution frequently used by analysts is the cumulative distribution. The cumulative distribution shows the probabilities of a variable being equal to or less than each value within the appropriate range for that variable.

Probability distributions can be generated from data (for example, from historical records or monitoring programs) or from knowledge of the underlying processes or systems. Probability distributions are extremely useful vehicles for conveying the uncertainty in a variable because several important features of the variable can be derived from the distribution. These features include measures of central tendency, such as the mean (or average), the median, and the mode; the variation around the central tendency (that is, variance or standard deviation); the range of possible values; and the asymmetry, or skewness, if any, in the distribution of probable values.

There is a practical problem with the use of probability distributions in risk assessments: the propagation of uncertainty. As the previous sections of this chapter indicate, uncertainty enters the analysis of risk all through the assessment process. For example, the uncertainty in a variable representing a failure rate for a particular piece of plant equipment will expand the uncertainty in later estimates of exposed populations, which in turn will contribute to the uncertainty in the dose estimate and, ultimately, to the total uncertainty in the final risk estimate (as presented in the risk statement). Uncertainties accumulate rapidly in a risk assessment. As noted above, the challenge for the user of risk assessment is to determine which sources of uncertainty can be legitimately dismissed.

Other, less complicated approaches to analyzing and expressing uncertainty have both the advantages and disadvantages of greater simplicity. The available information may be more understandable to risk managers and citizens, but it may also be less useful or even misleading.

Worst-case/best-case analysis requires propagating the lowest and highest extreme values through risk assessments to determine the upper and lower bounds of risk estimates. The upper bound, or highest risk, represents the worst possible case, while the lower bound, or lowest risk, represents the best or most optimistic case. Like the propagation of probability distributions, however, the range of final risk estimates obtained with this method of analysis is likely to be extremely large for most assessments. Moreover, extreme estimates of risk (upper or lower bounds) can be misleading in the absence of some indication of how probable the extreme estimates are.

Confidence intervals are a relatively common method of limiting the problems with extreme upper- and lower-bound estimates. Risk analysts may refer to the "95th percent confidence interval" to indicate that they are 95% sure that the true value lies within that interval. Such intervals are still likely to be large, however, and there is an implicit policy judgment in selecting 95% as a measure of confidence.

In many risk assessments, analysts simply acknowledge the existence of uncertainty but make no efforts to analyze or quantify it. The only number calculated is a point estimate of risk that represents the best estimate.

The best point estimate is usually the mean or median (based on sample data or expert judgment). If the probabilities of different values for a variable are known (or estimated), the mean of the variable may be calculated by taking the (weighted) average of all possible values multiplied by their probabilities of occurrence.

Point estimates are easy to understand–especially for people with little or no statistical training–but such simplicity can obscure important considerations in a policy or decision-making context. For example, point estimates of risk, including expected values, can mislead risk-management decision-makers by hiding the existence of uncertainty and variability.

CONCLUSIONS

Given these observations on risk assessment, several conclusions can be drawn. These conclusions can be organized around three topics: (1) standarization, (2) uncertainty characterization, and (3) value judgments.

Standardization

In recent years, a variety of efforts have been made to standardize risk assessment. For example, a major focus of the U.S. Environmental Protection Agency in recent years has been to develop and promulgate a set of standardized, formal guidelines describing the scientific basis for each of the analytic steps involved in risk assessment, and specifying how the analysis at each step is to be approached.

Some of the principal arguments for developing standardized guidelines include: achieving greater quality control; improving consistency among risk-management activities; increasing the predictability of the regulatory process; focusing attention and debate over the critical judgmental aspects of risk assessment; fostering greater public understanding; and promoting increased administrative efficiency. These are legitimate goals. However, the specification of risk assessment guidelines should be approached with caution.

Many of the difficulties associated with attempts to standardize risk assessment stem from (1) the different types of risks that need to be addressed; (2) the different objectives 98 served by risk assessment; (3) the different types and amounts of analytic resources available for an assessment; (4) the different types and quality of available data; and (5) the different contexts in which risk assessment is applied; which different risk assessment models and methods are appropriate. The descriptions and evaluations of risk assessment methods presented in this chapter illustrate the wide variety of analytic options available to the risk analyst. The capabilities of these methods depend heavily on available data and other resources. Methods that are useful in one situation may be counterproductive in another. As a result, standardized guidelines that require the use of particular methods or that establish the weight to be given to particular types of information may be inappropriate in many cases. More importantly, standardized guidelines may lead to significant evidence being disregarded, the generation of biased assessments, or assessments that fail to make use of all important and relevant information. Another concern is that the very act of formalizing and institutionalizing risk assessment guidelines may create the erroneous impression of a policy-free or value-free science.

The middle course that best serves all parties is a strategy that includes the development of flexible risk assessment guidelines. Flexible guidelines, designed to serve as benchmarks rather than as rules, are less likely to oversimplify the assessment process or stultify the development of scientific knowledge.

Uncertainties

A critical flaw in many risk assessments is the failure to describe and characterize adequately the uncertainties in the estimates of risk outcomes. Although methods have been developed to account for and characterize the effects produced by uncertainty, few risk assessments explicitly address the various types of uncertainty listed below.

- Statistical uncertainty
  Origin: randomness of nature as represented in the variability of available data.
- Parameter uncertainity
  Origin: imprecision in statistical estimates of model parameters made from limited or incomplete data.
- Judgmental uncertainty
  Origin: inability to specify precise values for model inputs for which statistical data are unavailable.
- Model uncertainty
  Origin: lack of understanding of physical processes or engineering behavior and failure of the models to capture the salient features of the risk.
- Completeness uncertainty
  Origin: omissions of important processes or events from the analysis.

Most assessments compute only a single value for the level of adverse health or environmental consequences. Occasionally, statistical uncertainties are estimated and, less frequently, parameter uncertainties. Only rarely are judgmental uncertainties quantified. Model uncertainty is almost never addressed, even qualitatively.

A risk assessment that reports only a single value for the level of adverse health or environmental consequences ignores the range of possibilities and provides decision-makers with an inaccurate picture of the risk being faced. To make effective, efficient, and equitable decisions, decision-makers need information not only about what adverse consequences can happen, including worst cases, but also about the relative likelihood of these consequences. Similarly, a risk assessment that omits important sources of uncertainty provides an incomplete picture of the risk. The preferences of decision-makers toward various policies and actions may be strongly influenced not only by the probable range of risk consequences, but also by the potential for consequences that might fall outside of the probable range, including the probability and severity of extreme possibilities. It is therefore critical that comprehensive measures of uncertainty (eg, probabilities, confidence intervals, or standard errors) be estimated and provided. Given the increased importance assigned to risk assessment in the regulatory decision-making process, risk assessors have a professional responsibility to present a clear and comprehensive statement of uncertainties that affect the scope and interpretation of their work.

Value Judgments

Many of the judgments made in conducting a risk assessment are, in effect, policy choices. Policy choices enter into risk assessment through a variety of routes, each associated with a different entry point.

One entry point for policy choices is through the decision of what problems to investigate and what level of resources (eg, financial and manpower resources) to allocate to the assessment. Another entry point is the choice of what adverse health or environmental consequences to assess. As noted earlier, for example, most risk assessments have focused on cancer. One consequence of the overriding emphasis on cancer is that fewer risk assessment studies have been conducted on other recognized adverse health effects such as cardiovascular, developmental, immunological mutagenic, neurotoxic/behavioral, reproductive, and respiratory effects. Even fewer risk assessment studies have focused on adverse effects on animals, plants, or ecosystems.

A second entry point for policy choices is through methodological choices and decisions that may have a subtle but significant impact on the results of an assessment. Examples include decisions and choices about what data to collect, about what data are relevant, about what methods to use, and about what simplifying assumptions to make in constructing workable models.

A final entry point for considerations is the selection of one risk assessment assumption over another at different stages in the risk assessment process. The lack of scientific consensus on many issues provides ample opportunity for policy choices to significantly affect risk estimates. Table 3 provides examples of the kinds of judgments that are necessary in the conduct of a risk assessment. At each point in the assessment process, assumptions must be made in the face of insufficient scientific knowledge. The scientific knowledge necessary to support one assumption over another is often extremely limited or poor in quality. Yet these assumptions have policy consequences. The policy consequences of selecting one assumption over another may be substantial, reflecting either a conservative or nonconservative bias.

One conclusion from these observations is the critical need for procedures that ensure the soundness and objectivity of risk assessments. These procedures include extensive peer review and sensitivity analyses that evaluate the implications of different assumptions. An even more important conclusion is that the current state of the art of risk assessment does not permit questions of science to be clearly separated from questions of policy. Despite the

Table 3. Examples of Scientific and Policy Judgments Necessary for Risk Assessment

Component of Risk Chain	Method Used	Example of Judgment Required
Releases	Monitoring	What method of interpolation or extrapolation should be used to deal with omissions or errors in the recording of data?
	Performance testing and accident investigation	How can isolated events be generalized to produce conclusions relevant to risk assessment?
	Statistical methods	How should surrogate databases be adjusted to account for known differences between the variable of concern and the surrogate variable for which data are available?
	Modeling	What assumptions should be made concerning the nature and likelihood of externally initiated events—for example, attempts by terrorist groups to produce radioactive releases from a nuclear reactor?
Exposures	Monitoring	How should exposure measurements representing a small segment of the population be extrapolated to the entire population?
	Modeling	How should dispersion of air pollutants into the atmosphere due to convection, wind currents, etc, be represented?
Consequences	Animal tests	How should evidence of different metabolic pathways or vastly different metabolic rates between animals and humans be factored into a risk assessment?
	Epidemiology	How should risk estimates be adjusted to account for a comparatively short follow-up period in an epidemiologic study?
	Modeling	What dose-response models should be used to extrapolate from experimental doses to relevant doses?
Risk estimation	Statistical analysis	How should statistical estimates based on "independence assumptions" be adjusted to account for known but analytically untractable dependencies?
	Judgmental methods	How should differences in expert opinions be aggregated?
	Modeling	How should probability distributions over health consequences be estimated and displayed?

value of institutionally and analytically separating risk assessment from risk management, policy choices enter into risk assessment at virtually every stage of the process. The goal of a risk assessment that is free, or nearly free, of policy considerations is an ideal that cannot be achieved.

BIBLIOGRAPHY

Reading List

B. Ames, "Identifying Environmental Chemicals Causing Mutations and Cancer," *Science* **204,** 587–593 (1979).

B. Ames, R. Magaw, and L. Gold, "Ranking Possible Carcinogenic Hazards," *Science* **236,** 271–279 (Apr. 17, 1987).

E. J. Calabrese, *Principles of Animal Extrapolation*, Wiley, New York, 1983.

Conservation Foundation, *Risk Assessment and Risk Control,* The Conservation Foundation, Washington, D.C., 1985.

V. T. Covello, and M. Merkhofer, *Risk Assessment Methods* Plenum Press, New York, 1993.

K. Crump, "Mechanisms Leading to Dose-Response Models," in P. Ricci ed, *Principles of Health Risk Assessment,* Prentice-Hall Inc., Englewood Cliffs, N.J., 1985, pp. 235–277.

U.S. Environmental Protection Agency, "Principles of Risk Assessment—A Nontechnical Review," *Workshop on Risk Assessment,* Easton, Maryland, March 17–18 1985, Washington, D.C., 1985a.

U.S. Environmental Protection Agency, "Status of Extrapolation Modeling Research Needed to Extrapolate from Animal Data to Human Risk, from High to Low Doses, and from Acute to Chronic Effects," Office of Research and Development, 1985.

U.S. Environmental Protection Agency, "Superfund Public Health Evaluation Manual," *EPA / 540 / 1-86 / 061,* Office of Emergency and Remedial Response, EPA, Washington, D.C., 1986a.

U.S. Environmental Protection Agency, "Guidelines for Carcinogen Risk Assessment," *Fed. Regist.* **51,** 34000 (Sept. 24, 1986).

U.S. Environmental Protection Agency, "Guidelines for Mutagenicity Risk Assessment," *Fed. Regist.* **51,** 34006 (Sept. 24, 1986).

U.S. Environmental Protection Agency, "Guidelines for Health Risk Assessment of Chemical Mixtures," **51,** 34014 (Sept. 24, 1986).

U.S. Environmental Protection Agency, "Guidelines for Health Risk Assessment of Suspect Developmental Toxicants," *Fed. Regist.* **51,** 34028 (Sept. 24, 1986).

U.S. Environmental Protection Agency, "New Chemical Review Process Manual," EPA 560/3-86-002, Office of Toxic Substances, Washington, D.C., 1986.

U.S. Environmental Protection Agency, "Unfinished Business: A Comparative Assessment of Environmental Problems: Vol. 1, Overview Report," Office of Policy Analysis, Office of Policy, Planning and Evaluation, Washington, D.C., Feb. 1987.

U.S. Environmental Protection Agency, "Unfinished Business: A Comparative Assessment of Environmental Problems: Appendix II, Non-Cancer Risk Work Group," Office of Policy Analysis, Office of Policy, Planning and Evaluation, Washington, D.C., Feb. 1987.

U.S. Environmental Protection Agency, "Unfinished Business: A Comparative Assessment of Environmental Problems: Appendix III, Comparative Ecological Risk, a Report of the Ecological Risk Work Group," Office of Policy Analysis, Office of Policy, Planning and Evaluation, Washington, D.C., Feb. 1987.

U.S. Environmental Protection Agency "Unfinished Business: A Comparative Assessment of Environmental Problems Appendix IV, An Assessment of Welfare Effects from Environmental Pollution, a Report of the Welfare Risk Work Group," Office of Policy Analysis, Office of Policy, Planning and Evaluation, Washington, D.C., Feb. 1987.

U.S. Environmental Protection Agency, "Statistical Methods for Evaluating Ground Water from Hazardous Waste Facilities," Office of Solid Waste, 1988.

U.S. Environmental Protection Agency, "Superfund Exposure Assessment Manual," *EPA / 5401-88 / 001,* Office of Emergency and Remedial Response, 1988.

U.S. Environmental Protection Agency, "Selection Criteria for Mathematical Models Used in Exposure Assessments: Groundwater Models," *EPA / 600 / 8-88 / 075,* Office of Health and Environmental Assessment, 1988.

U.S. Environmental Protection Agency, "Risk Assessment Guidance for Superfund: Vol. I—Human Health Evaluation Manual (Part A), Interim Final," *EPA / 540 / 1-89 / 002,* Office of Emergency and Remedial Response, Washington DC, Dec. 1989.

U.S. Environmental Protection Agency, "Ecological Assessment of Hazardous Waste Sites: A Field and Laboratory Reference," *EPA / 600 / 3-89 / 013,* W. W. Hicks, B. R. Parkhurst, and S. S. Baker, Jr., eds., EPA Environmental Research Laboratory, Corvallis Oreg., Mar. 1989.

U.S. Food and Drug Administration, *Toxicological Principles for the Safety Assessment of Direct Food Additivies and Color Additives Used in Food,* USFDA, Bureau of Foods, Washington, D.C., 1982.

J. R. Fiksel, and V. T. Covello, eds., *Biotechnology Risk Assessment,* Pergamon Press, Inc., Elmford, N.Y., 1986.

J. R. Fiksel, and V. T. Covello, eds. *Safety Assurance for Environmental Releases of Genetically Engineered Organisms,* Springer-Verlag, Berlin, 1989.

Health and Welfare Canada, *Carcinogen Evaluation,* Health & Welfare Canada, Ottawa, Ont. 1992a.

D. Hoel, R. Merrill, and F. Perera, eds., *Risk Quantification and Regulatory Policy,* 19 Banbury Report, Cold Spring Harbor Laboratory, Cold Spring Harbor, N.Y., 1985.

International Agency for Research on Cancer, *IARC Monographs on the Evaluation of the Carcinogenic Risk of Chemicals in Humans: Chemicals, Industrial Processes, and Industries Associated with Cancer in Humans* (IARC Monographs Vols. 1–29), Supp. No. 4, IARC, Lyon, France, 1982.

International Agency for Research on Cancer, *IARC Monographs on the Evaluation of the Carcinogenic Risk of Chemicals in Humans: Overall Evaluations of Carcinogenicity* (Updating IARC Monographs Vols. 142), Supp. No. 7, IARC, Lyon, France, 1987.

International Commission on Radiological Protection: "Recommendations of the International Commission on Radiological Protection" (ICRP Publication 26), *Annals of the ICRP* **1**(3), (1977).

D. Krewski, and D. Murdoch: "Quantitative Factors in Carcinogenic Risk Assessment," in R. W. Hart and F. D. Hoerger eds., *Carcinogen Risk Assessment: New Directions in the Qualitative and Quantitative Aspects,* 31 Banbury Report, Cold Spring Harbor Laboratory, Cold Spring Harbory, N.Y., 1988.

D. Krewski, D. Wigle, D. B. Clayson, and G. Howe: "Role of Epidemiology in Health Risk Assessment," in P. Band ed., *Occupational Cancer Epidemiology,* Recent Results in Cancer Research, Vol. 120, Springer–Verlag, Berlin, pp. 1–24.

L. B. Lave, and A. C. Upton, eds., *Toxic Chemicals, Health and the Environment,* Johns Hopkins University Press, Baltimore, Md., 1987.

W. W. Lowrance, *Of Acceptable Risk: Science and the Determination of Safety,* Kaufman, Los Altos, Calif., 1976.

National Research Council, *Risk Assessment in the Federal Government: Managing the Process,* NAS-NRC Committee on the Institutional Means for Assessment of Risks to Public Health, National Academy Press, Washington D.C., 1983.

U.S. Nuclear Regulatory Commission, "Reactor Safety Study: An Assessment of Accident Risks in U.S. Commercial Nuclear Power Plants," *NUREG-75/014* (WASH 1400), Washington, D.C., Oct. 1975.

U.S. Nuclear Regulatory Commission, "PRA Procedures Guide: A Guide to the Performance of Probabilistic Risk Assessments for Nuclear Power Plants," prepared under the auspices of the American Nuclear Society and the Institute of Electrical and Electronics Engineers, NUREG/CR-2300, NTIS, Jan. 1983.

U.S. Nuclear Regulatory Commission, "Probabilistic Risk Assessment (PRA): Status Report and Guidance for Regulatory Application, Appendix A," Draft Report, Division of Risk Analysis, Washington, D.C., Feb. 1984.

U.S. Office of Science and Technology Policy, "Chemical Carcinogens; A Review of the Science and Associated Principles," *Fed. Regist. 50*(50) (Mar. 14, 1985).

U.S. Office of Technology Assessment, U.S. Congress, "Assessment of Technologies for Determining Cancer Risks from the Environment," U.S. Government Printing Office, Washington, D.C., 1981.

U.S. Office of Technology Assessment, U.S. Congress, Biological Effects of Power Frequency Electric and Magnetic Fields, U.S. Government Printing Office, Washington, D.C., 1989.

C. N. Park, and R. D. Snee, "Quantitative Risk Assessment: State-of-the-Art for Carcinogenesis," *Am. Stat.* **34**(4), 427–441 (Nov. 1983).

P. F. Ricci, ed.: *Principles of Health Risk Assessment,* Prentice-Hall, Inc., Englewood Cliffs, N.J., 1985.

J. V. Rodricks, and R. G. Tardiff eds.: *Assessment and Management of Chemical Risks,* American Chemical Society, Washington, D.C., 1984.

(WHO) World Health Organization: *Guidelines on Studies in Environmental Epidemiology,* Environ. Health Crit. 27, Geneva, 1983.

World Bank: *Manual of Industrial Hazard Assessment,* World Bank, Washington, D.C., 1985.

RISK COMMUNICATION

VINCENT COVELLO
Columbia University
School of Public Health
New York, New York

Risk communication can be defined as the exchange of information among interested parties about the nature, magnitude, significance, or control of a risk. Interested parties include government agencies, corporations or industry groups, unions, the media, scientists, professional organizations, special interest groups, communities, and individual citizens.

Information about risks can be communicated through a variety of channels, ranging from media reports and warning labels or signs to public meetings or hearings involving representatives from government agencies, industry, the media, and the general public. These communication efforts can be a source of difficulty for risk communicators and for the intended receivers. Government officials, industry representatives, and scientists, for example, often argue that laypeople do not accurately perceive and evaluate risk information. Representatives of citizen groups and individual citizens, in turn, often argue that government officials, industry representatives, and scientists are uninterested in their concerns and unwilling to take actions to solve seemingly straightforward environmental problems. In this context, the media often serve as transmitters and translators of risk information. But the media have been criticized for exaggerating risks and for emphasizing drama over scientific facts.

In recent years, the literature on risk communication has grown rapidly with hundreds of articles and books being published on the topic. Most of these works focus on problems and difficulties in communicating information during crisis and noncrisis situations about risks of exposures to environmental risk agents – particularly the risks of exposures to chemicals, heavy metals, and radiation in the air, water, land, and food.

Why the interest in risk communication? One explanation is the increased number of hazard communication and environmental right-to-know laws relating to exposures to environmental risk agents. Another explanation stems from increased public fear and concern about exposures to environmental risk agents from past, present, and future industrial activities and the corresponding demand for risk information. A third explanation is the expansion of media interest coverage of environmental issues, which in turn reflects greater public interest in environmental issues. But a fourth underlies the first three: the loss of faith and trust in government and industry officials as environmental managers and as credible sources of risk information.

The purpose of this article is to review the literature on risk communication. The specific aim is to provide the reader with a general outline of the field and to relate this work to the practical needs of those with risk communication responsibilities.

See also GLOBAL HEALTH INDEX; COMMERCIAL AVAILABILITY OF ENERGY TECHNOLOGY; LIFE CYCLE ANALYSIS.

SCIENCE AND PERCEPTION

A significant part of the risk communication literature focuses on problems and difficulties in communicating risk information effectively. These problems and difficulties revolve around issues of science and perception and can be organized into four categories: (1) characteristics and limitations of scientific data about risks; (2) characteristics and limitations of government officials, industry officials, and other spokespersons in communicating information about risks; (3) characteristics and limitations of the media in reporting information about risks; and (4) characteristics and limitations of the public in evaluating and interpreting risk information. Each category is described below:

Characteristics and Limitations of Scientific Data About Risks

One source of difficulty in communicating information about risks is the uncertainty and complexity of data on health, safety, and environmental risks. Risk assessments, despite their strengths, seldom provide exact answers. Due to limitations in scientific understanding, data, models, and methods, the results of most risk assessments are at best approximations. Moreover, the resources needed to resolve these uncertainties are seldom adequate to the task.

These uncertainties invariably affect communications with the public in the hostile climate that surrounds many environmental issues. For example, uncertainties in environmental risk assessments often lead to radically different estimates of risk. An important factor underlying many debates about risks are the different assessments of risk produced by government agencies, industry, and public interest groups.

Given these uncertainties, one goal of risk communication is to provide detailed information on the assumptions underlying the calculation of risks. Since many disagreements revolve around these assumptions (eg, assumptions about doses and exposures), such disclosures are critical to public confidence and understanding.

A related flaw is the failure to describe and characterize uncertainties. Risk reports that contain only single risk values or point estimates of risk (eg, 4000 excess cancers over a lifetime) ignore the true range of risk possibilities. Moreover, they provide an inaccurate picture to the public of the ability of risk assessment to produce precise numbers. A more accurate picture is provided by describing best and worst cases based on optimistic or pessimistic (conservative) assumptions.

Characteristics and Limitations of Industry Officials, Government Officials, and Other Spokespersons in Communicating Information about Risks

A central question addressed by the literature on risk communication is why some individuals and organizations are trusted as sources of risk information and others are not. This question takes on special importance given that two of the most prominent sources of risk information–industry and government–often lack trust and credibility. In the United States, for example, overall public confidence and trust in government and industry as trusted sources of information has declined precipitously over the past two decades. The majority of people in the United States, for example, view industry and government as among the least trusted sources of information about the risks of environmental exposures to environmental risk agents. At the same time, the majority of people view industry and government as among the most knowledgeable sources of information about such risks.

Public distrust of government and industry is grounded in several beliefs: that they have been insensitive to public concerns and fears about environmental risks, unwilling to acknowledge problems, unwilling to share information, unwilling to allow meaningful public participation, and negligent in fulfilling their environmental responsibilities. Compounding the problem are beliefs that current environmental laws are too weak, that the environment is worse today than it was twenty years ago, and that government and industry have done a poor job protecting the environment.

Several factors compound these perceptions and problems. First, many officials have engaged in highly visible debates and disagreements about the reliability, validity, and meaning of the results of environmental health risk assessments. In many cases, equally prominent experts have taken diametrically opposed positions on the risks of nuclear power plants, hazardous waste sites, asbestos, electric and magnetic fields, lead, radon, PCBs, arsenic, and dioxin. While such debates may be constructive for the development of scientific knowledge, they often undermine public trust and confidence in industry and government.

Second, resources for risk assessment and management are seldom adequate to meet demands by citizens and public interest groups for definitive findings and rapid action. Explanations by officials that the generation of valid and reliable toxicological or epidemiological data is expensive and time consuming (or that risk assessment and management activities are constrained by resource, technical, statutory, legal, or other limitations) are seldom perceived to be satisfactory. Individuals facing what they believe is a new and significant risk are especially reluctant to accept such claims.

Third, coordination among responsible authorities is seldom adequate. In many debates about risks, for example, lack of coordination has severely undermined public faith and confidence. Compounding such problems is the lack of consistency in approaches to risk assessment and management by authorities at the local, state, regional, national, and international levels. For example, few requirements exist for regulatory agencies to develop coherent, coordinated, consistent, and interrelated plans, programs, and guidelines for managing risks. As a result, regulatory systems tend to be highly fragmented. This fragmentation often leads to jurisdictional conflicts about which agency and which level of government has the ultimate responsibility for assessing and managing a particular environmental activity or risk. Lack of coordination, different mandates, and confusion about responsibility and authority also lead, in many cases, to the production of multiple and competing estimates of risk. A commonly observed result of such confusion is the erosion of public trust, confidence, and acceptance.

Fourth, many officials in government and industry lack adequate training in communications and in the specific requirements of risk communication. For example, many officials use complex and difficult technical language and jargon in communicating information about risks and benefits to the media and the public. The use of technical language or jargon is not only difficult to comprehend but can also create a perception that the official or expert is being unresponsive, dishonest, or evasive.

Finally, government and industry officials have often been insensitive to the information needs of the public and the differences between expert and lay perceptions of risk. Officials often operate on the assumption that they and their audience share a common framework for evaluating and interpreting risk information. However, this assumption is often not the case. One of the most important findings to emerge from risk perception and communication studies is that people take into consideration a complex array of qualitative and quantitative factors in defining, evaluating, and acting on risk information.

One of the costs of this heritage of mistrust and loss of confidence is the public's reluctance to believe information from government and industry about the risks of exposure to chemicals, heavy metals, and radiation in the workplace and environment.

Programs and plans for overcoming distrust of government and industry require improvements in three areas: in risk assessment; in risk management; and in risk communication. In the risk communication area, better risk communication requires improvements in (1) the credibility of individual spokespeople, (2) the credibility of organizations with risk assessment and management responsibilities, and (3) the quantity and quality of third party support and endorsements.

Table 1. Developing Effective Risk Messages

JARGON	Avoid; define all terms for 12–14 year olds.
ORGANIZATIONAL IDENTITY	Use personal pronouns; avoid organization's name.
ATTACKS	Attack issues; avoid attacking those with higher credibility.
HUMOR	Avoid jokes, cartoons, sarcasm, etc.
RISK/BENEFIT COMPARISONS	Discuss in separate communications.
RISK COMPARISONS	Use for perspective, not acceptability.
NEGATIVE ALLEGATIONS	Do not repeat.
SPECULATION	Do not speculate about worst cases or "what if" situations.
MONEY	Avoid references to money spent on environment programs.
KEY MESSAGES	Stress performance, trend lines, and achievements.
TECHNICAL DETAILS AND DEBATES	Avoid providing too much.
QUANTITATIVE HEALTH RISK NUMBERS	Avoid risk statements stated as quantitative probabilities.
GUARANTEES	Talk about achievements, not about lack of guarantees.
ZERO RISK	Emphasize value of zero risk as a goal.
COMPARISONS WITH OTHERS	Do not compare your company to others.
LENGTH OF ANSWERS TO QUESTIONS	Limit to 2–4 minutes or less.
LENGTH OF PRESENTATIONS	Limit to 15–30 minutes or less.
VISUALS	Use as much as possible.
ABSTRACTIONS	Use examples, analogies, and stories to establish meaning.
NONVERBAL MESSAGES	Be sensitive to them.
TESTING	Test and practice all answers and presentations.

Table 2. Elements of Effective Responses to Difficult Environmental Questions from the Public and the Media

1. EXPRESS CARING/EMPATHY
   Characteristics:
   - Length: Males = Ideally 25% of length of complete answer
     Females = Ideally 10% of length of complete answer
   - Contains a personal story that illustrates caring/empathy
   - Includes a transition to conclusion
2. STATE THE CONCLUSION
   Characteristics:
   - Addresses underlying concern
   - Short 7 to 12 words
   - Positive
   - Set apart/framed
   - Stated with intensity/enthusiasm
3. STATE THE FIRST SUPPORTING FACT
   Characteristics:
   - Cites 3rd party support
   - Contains a personal story/narrative
4. STATE THE SECOND SUPPORTING FACT
   Characteristics:
   - Cites 3rd party support
   - Contains a personal story/narrative
5. REPEAT CONCLUSION
   (Same as Step 2 above)
6. FUTURE ACTION
   Characteristics:
   - Provides details – who, what, where, when, how – on how to obtain more information
   - Provides details – who, what, where, when, how – on commitment to continuous improvement in knowledge and/or management

Improving the credibility of individual spokespeople means that those in government and industry who interact with the public on risk issues must develop better risk communication skills. These include verbal risk communication skills (Tables 1 and 2) and nonverbal risk communication skills (Table 3).

For skill development to take place, three fundamental formulas/principles of risk communication need to be recognized: that perceptions are realities, that the goal of risk communication is to establish trust and credibility when trust and credibility are low, and that effective risk communication is a skill requiring knowledge, preparation, practice, and execution:

$$P = R$$

Perceptions (P) equal realities (R): what is perceived as real will be real in its consequences.

$$G = T + C$$

The primary goal (G) of risk communication is to establish high levels of trust (T) and credibility (C). The secondary goal is to convey facts and figures.

$$C = S$$

Effective risk communication (C) is a complex skill (S). As with any complex skill, it requires significant amounts of knowledge, training, and practice.

Table 3. Nonverbal Communications Checklist

Poor Eye Contact. Messages: dishonest, closed, unconcerned, nervous.
Excellent Eye Contact. Messages: honest, open, competent, sincere, dedicated, confident, knowledgeable, interested.
Sitting Back in Chair. Messages: uninterested, unenthusiastic, unconcerned, uncooperative.
Sitting Forward in Chair. Messages: interested, enthusiastic, concerned, cooperative.
Standing with Weight on Back Leg. Messages: uninterested, unenthusiastic, unconcerned, uncooperative.
Standing with Weight on Forward Leg. Messages: interested, enthusiastic, concerned, cooperative.
Arms Crossed on Chest. Messages: uninterested, unconcerned; defiant, not listening, arrogant, impatient, defensive, stubborn.
Infrequent Hand Gestures/Body Movements. Messages: dishonest, deceitful, nervous, lack of self-confidence.
Rocking Movements. Messages: nervous, lack of self confidence.
Frequent Hand-to-Face Contact. Messages: dishonest, deceitful, nervous.
Hidden Hands. Messages: deceptive, guilty, insincere.
Open Hands. Messages: open, sincere.
Buttoned Coat. Messages: closed, unfriendly, disagreeable.
Unbuttoned Coat. Messages: open, friendly, agreeable.
Coat On. Messages: closed, unfriendly, disagreeable, uncooperative.
Coat Removed. Messages: open, friendly, agreeable, cooperative.
Speaking from Behind Barriers. Messages: dishonest, deceitful, formality, unconcerned, uninterested, superior.
Speaking Outdoors. Messages: dedicated, hardworking, involved, concerned.
Speaking from an Elevated Position. Messages: superiority, dominance, judgmental.
Steepling/Joined Fingertips. Messages: self-confidence, self-assurance, self-importance, superiority.
Touching and/or Rubbing Nose. Messages: doubt, disagreement, nervous, deceitful.
Touching and/or Rubbing Eyes. Messages: doubt, disagreement, nervous, deceitful.
Hands on Hips. Messages: dedication, self confidence, cooperative, arrogant.
Pencil Chewing/Hand Pinching. Messages: lack of self confidence, doubt.
Jingling Money in Pockets. Messages: nervous, lack of self confidence, lack of self control, deceitful.
Drumming on Table, Tapping with Feet. Messages: nervous, hostile, impatience, anxiety, boredom.
Head in Hand. Messages: boredom.
Hand to Chest. Messages: openness, honesty, dedication, sincerity.
Clenched Hands. Messages: anger, hostility, determination, uncooperative.
Locked Ankles. Messages: apprehensive, deceitful, nervous.
Crossed Legs. Messages: defensive, closed.
Palm to Back of Neck. Messages: frustration, anger, irritation, hostility.
Straight Posture. Messages: self-confidence, self control, assertive, determination.
Slumping Posture. Messages: nervous, poor self control.
Glasses on. Messages: competent, hiding something, deceitful.
Glasses off. Messages: caring, concerned, confident.
Frequent Blinking. Messages: nervous, deceitful, inattentive.
Infrequent Blinking. Messages: self confidence, honest, listening.
Lowering Voice. Messages: self assurance, honesty, caring.
Raising Voice. Messages: nervous, deceitful.
Clearing Throat. Messages: nervous, deceitful.

Table 3. Continued

Tapping Pencil/Clicking Pen. Messages: nervous, lack of self confidence, lack of self control, deceitful.
Looking at Watch. Messages: disinterested, unconcerned.
Shrugging Shoulders. Messages: unconcerned, indifferent.
Tight Lipped. Messages: nervous, deceitful, angry, hostile.
Licking Lips. Messages: nervous, deceitful.
Indoor/Behind a Desk: Messages: bureaucratic, uncaring, removed, uninvolved.
Outdoors with Sleeves Rolled Up: Messages: Hard working, involved, caring, concerned.

For some audiences, such as the media, spokespersons must develop highly refined skills. But more sophisticated communication skills alone will not suffice. Improving the credibility of spokespeople also calls for improvements in actions by institutions. People judge others more on their actions than on their words; when their actions fail, they distrust their words. Because spokespeople speak for their institutions, their credibility depends on and will be enhanced only by improvements in the actions of the institutions they represent. Finally, gaining better credibility for spokespeople involves thinking broadly about risk communication and attending to the importance of developing ongoing partnerships with employees and the public. Environmental risk problems are generally long-term, complex problems that require active listening and two-way communication.

Improved credibility for organizations involves making improvements both in environmental performance and outreach efforts, based on a credo of ethical and responsible care. It also involves respect for the differing values and world views that employees and the public use when they evaluate risk information. Their evaluations are often broader than those of experts, and take into account a wide range of considerations, including information about fairness, benefits, control, voluntariness, alternatives, catastrophic potential, familiarity, process, and trust in institutions.

Improvements in the quantity and quality of third party support and endorsements include improvements in coordination and collaboration with organizations and individuals that are perceived by the public to be credible. Surveys indicate that organizations and individuals perceived to be relatively high to medium in credibility on environmental risk issues include health and safety professionals, educators, professional organizations, the media, nonmanagement employees, nonprofit voluntary health organizations, environmental groups, and local citizens who are respected, neutral, and informed on environmental issues:

Top Third

Local citizens who are perceived to be neutral, respected, and well-informed about the issue

Health and safety professionals

Educators (especially those from respected local schools)

Nonprofit voluntary health organizations

Nonmanagement employees
Professional societies

Middle Third
Media
Environmental groups

Bottom Third
Industry officials
Federal government officials
Environmental consultants from for-profit firms

Changes from Previous Years
Environmental groups, 10–15% loss of credibility
Media, 5–10% gain in credibility
Government and industry, 10% loss in credibility

Better coordination and collaboration with such groups can, in turn, result in alliances, joint communications, and support.

A common thread in all three strategies is the need to be proactive in establishing high levels of trust and credibility. Only when trust and credibility have been established can other goals, such as education and the sharing of information, follow.

Surveys, case studies, and experimental research all indicate that trust and credibility cannot be built quickly. Instead, they are the result of ongoing partnerships, actions, performance, and skill in communications. More specifically, trust and credibility are built on a foundation of perceived caring and empathy; perceived competence and expertise; perceived honesty and openness, and perceived dedication and commitment.

- Perceived caring/empathy (50%) eg, perceived sincerity, ability to listen, ability to see issues from the perspective of the other.
- Perceived competence/expertise (15–20%) eg, perceived intelligence, training, authoritativeness, experience, educational level, and professional attainment, knowledge, command of information.
- Perceived openness/honesty (15–20%) eg, perceived truthfulness, candidness, justness, objectivity, sincerity, disinterestedness.
- Perceived dedication/commitment (15–20%) eg, perceived altruism, diligence, self-identification, involvement, hard work.

These four factors form the criteria that people use to evaluate the credibility of an organization or individual. Risk communication is effective to the degree to which all actions and communications–verbal and nonverbal–convey caring and empathy, competence and expertise, honesty and openness, and dedication and commitment.

Of the four factors, caring and empathy appear to be the most important in low trust/high concern situations. For example, is the communicator perceived to be a good listener? Is the communicator perceived to be concerned first and foremost about the public's concerns, be they health, safety, environmental, economic, aesthetic, fairness, or process concerns? Is the communicator perceived to be empathetic, capable of seeing the issue from another persons point of view? Is the communicator perceived as following the golden rule? Often, people make their initial judgments relating to these questions within a relatively short period of time–often within the first 30 seconds. Once made, such judgments are extremely resistant to change.

Perceptions of trust and credibility also derive from perceptions of competence and expertise, honesty and openness, and dedication and commitment. Perceptions of competence and expertise are influenced largely by an organization's environmental record, and by the spokespersons education, track record, experience, knowledge, presentation skills, education, and professional recognition.

Perceptions of honesty and openness derive largely from actions, words, and nonverbal cues that convey truthfulness and candidness. In oral communications, nonverbal cues to honesty and openness, such as poor eye contact and physical barriers between the speaker and the audience, play a particularly important role.

Perceptions of dedication and commitment are influenced largely by perceptions of hard work and diligence in the pursuit of health, safety, and environmental goals. An audience will gauge a communicator's dedication and commitment by a number of verbal and nonverbal cues: is the communicator willing to be available to the community after hours? Does the communicator come to public meetings early and stay late? Failure to communicate verbal and nonverbal messages of dedication and commitment can diminish a speaker's credibility by as much as one-fifth.

Male/female differences on these dimensions of credibility are often pronounced. Women, for example, often receive relatively low initial ratings on competence and expertise. But because women in general receive substantially higher initial ratings than men on the other trust and credibility factors, a woman who is perceived as competent and expert outranks most men in credibility. Further, because men are generally perceived to be less honest, caring, and dedicated to others than women, the rare man who does rank high on all four trust and credibility factors outranks even the most competent woman in credibility.

At the institutional level, prospects for establishing high levels of public trust are modest at best–at least in the short run. They do, however, appear to be better locally than nationally or globally. For example, the thrust of many risk communication training programs sponsored by industry and government is to develop, at the local level, a track record of dealing openly, fairly, safely, and responsibly with neighboring communities.

To further address the trust issue, many government agencies and companies have initiated programs aimed at increasing public trust. These strategies have included enhanced community outreach and public dialogue programs combined with extensive field testing of risk communication messages. Several government agencies, including the Environmental Protection Agency (EPA) and various branches of the military, have moreover launched ambitious risk communication training programs for all staff with environmental communication responsibilities. Finally, many organizations have published codes of management practice for effective risk communication. In 1988, for example, the Environmental Protection Agency published a set of principles and guidelines for effective risk communication titled "The Seven Cardinal Rules of Risk Communication." In introducing these principles and rules, the EPA noted that the goal of risk communication is not to diffuse public concern or avoid action, but rather to produce an informed public that is involved, interested, reasonable, thoughtful, solution-oriented, and collaborative.

Characteristics and Limitations of the Media in Reporting Information About Risks

The mass media play a critical role in transmitting risk information. Given this importance, researchers have focused their attention on the role of the mass media and on characteristics and limitations of the media that contribute to problems in risk communication.

One of the main conclusions to emerge from risk communication research is that the media are biased toward stories that contain drama, conflict, expert disagreements, and uncertainties. The media are especially biased toward stories that contain dramatic or sensational material, such as a minor or major accident at a chemical manufacturing facility or a nuclear power plant. Much less attention is given to daily occurrences that kill or injure far more people each year but take only one life at a time. In reporting about risks, journalists often focus on the same concerns as the public, eg, potentially catastrophic effects, lack of familiarity and understanding, involuntariness, scientific uncertainty, risks to future generations, unclear benefits, inequitable distribution of risks and benefits, and potentially irreversible effects.

Media coverage of risks is frequently deficient in that many stories contain oversimplifications, distortions, and inaccuracies in reporting risk information. Media coverage is also deficient not only in what is contained in the story but in what is left out. For example, analyses of media reports on cancer risks show that these reports are often deficient in providing few statistics on general cancer rates for purposes of comparison; providing little information on common forms of cancer; not addressing known sources of public ignorance about cancer; and providing little information about detection, treatments, and other protective measures.

Many of these problems stem from characteristics of the media and the constraints under which reporters work. First, most reporters work under extremely tight deadlines that limit the amount of time for research and for the pursuit of valid and reliable information. Second, with few exceptions, reporters do not have adequate time or space to deal with the complexities and uncertainties surrounding many risk issues. Third, journalists achieve objectivity in a story by balancing opposing views. Truth in journalism is different from truth in science. In journalism, there are only different or conflicting views and claims to be covered as fairly as possible. Fourth, journalists are source dependent. Under the pressure of deadlines and other constraints, reporters tend to rely heavily on sources that are easily accessible and willing to speak out. Sources that are difficult to contact, hard to draw out, or reluctant to provide interesting and nonqualified statements are often left out. Finally, few reporters have the scientific background or expertise needed to evaluate the complex scientific data and disagreements that surround many debates about risks. Given these limitations, effectiveness in communicating with the media about risks depends in part on understanding the constraints and needs of the media and adapting one's behavior and information to meet these needs.

Characteristics and Limitations of the Public in Evaluating and Interpreting Risk Information

Much of the risk communication literature focuses on characteristics and limitations of the public in evaluating and interpreting risk information. These include: (1) inaccurate perceptions of levels of risk, (2) difficulties in understanding probabilistic information related to unfamiliar activities or technologies, (3) strong emotional responses to risk information, (4) desires and demands for scientific certainty, (5) strong beliefs and opinions that are resistant to change, (6) weak beliefs and opinions that are easily manipulated by the way information is presented, (7) ignoring or dismissing risk information because of its perceived lack of personal relevance, (8) perceiving accidents and mishaps as signals, (9) using health and environmental risks used as proxies or surrogates for other concerns, and (10) distinguishing perceptions of risk from judgments of risk acceptability. Each of these characteristics is described below.

Inaccurate Perceptions of Risk. People often overestimate some risks and underestimate others. For example, people tend to overestimate the risks of dramatic or sensational causes of death, such as accidents at manufacturing plants or waste disposal facilities, and underestimate the risks of undramatic causes, such as asthma, emphysema, and diabetes. This bias is caused in part by the tendency for risk judgments to be influenced by the memorability of past events and by the imaginability of future events. A recent disaster, intense media coverage, or a vivid film can heighten the perception of risk. Conversely, risks that are not memorable, obvious, palpable, tangible, or immediate tend to be underestimated.

Difficulties Understanding Probabilistic Information Related to Unfamiliar Activities or Technologies. A variety of cognitive biases and related factors hamper people's understanding of probabilities. These difficulties, in turn, hamper discussions of risk probabilities between experts and nonexperts. For example, risk experts are often confused by the public's rejection of the argument that a cancer risk from a new activity is acceptable if it is smaller than one in a million. This attitude is especially frustrating given that a one in a million risk is an extremely small number and that the background chance of cancer is approximately 1 in 4. In rejecting this argument, people respond with one or more objections. First, they personalize the risk (for example, What if that one is me or my child? Second, they raise questions of trust; for example,

Why should I believe you–didn't you and your colleagues do the calculations? Third, they raise concerns about cumulative risks; for example, I am already exposed to enough risks in life–why do I need one more? Fourth, they question whether the risks are worth the benefits; for example, Is the activity that generates the risk really worth losing one life? Finally, they raise ethical questions; for example, Who gave government and industry the right to play God and choose who will live or die? Exacerbating the problem is the difficulty people have understanding, appreciating and interpreting small probabilities–such as the difference between one chance in a hundred thousand and one chance in a million.

Given these difficulties, risk communication experts often avoid using probabilities in explaining risk decisions. This avoidance is especially true when there is little time available for discussion. Instead, the communicator should focus on other messages, including (1) the degree to which the activity meets health, safety, or environmental standards as set or reviewed by trusted and credible authorities; (2) the relationship between the risk of activity in question and other risks; and (3) the degree to which the risk estimate is based on worst case or pessimistic assumptions that are biased toward public health and safety.

These same problems hamper discussions between experts and nonexperts on low probability/high consequences events and "worst case scenarios." In many such cases, the imaginability of the worst case makes it difficult for people to distinguish between what is remotely possible and what is probable.

Strong Emotional Responses to Risk Information. Strong feelings of fear, hostility, anger, outrage, panic, and helplessness are often evoked by exposure to unwanted or dreaded risks. These emotions often make it difficult to engage in rational discourse about risk in public settings. Emotions tend to be most intense when people perceive the risk to be involuntary, unfair, not under their personal control, and low in benefits. More extreme emotional reactions often occur when the risk affects children, when the adverse consequences are particularly dreaded eg, cancer and birth defects, and when worst case scenarios are presented.

Desires and Demands for Scientific Certainty. People often display a marked aversion to uncertainty and use a variety of coping mechanisms to reduce the anxiety generated by uncertainty. This aversion often translates into a marked preference for statements of fact over statements of probability–the language of risk assessment. People often demand to be told exactly what will happen, not what might happen.

Strong Beliefs and Opinions That Are Resistant to Change. People tend to ignore evidence that contradicts their current beliefs. Strong beliefs about risks, once formed, change very slowly and are extraordinarily persistent in the face of contrary evidence. Initial beliefs about risks tend to structure the way that subsequent evidence is interpreted. New evidence, eg, data provided by a government or industry official, appears reliable and informative only if it is consistent with the initial belief; contrary evidence is dismissed as unreliable, erroneous, irrelevant, or unrepresentative.

Weak Beliefs and Opinions That Are Easily Manipulated by the Way Information Is Presented. When people lack strong prior beliefs or opinions, subtle changes in the way that risks are expressed can have a major impact. To test this hypothesis, one group of researchers asked two groups of physicians to choose between two therapies–surgery or radiation. Each group received the same information but with one major difference–probabilities were expressed either in terms of dying or in terms of surviving. Even though these two numbers are the same, the difference resulted in dramatic differences in the choice of therapy. Virtually the same results were observed for other test populations.

A variety of studies have demonstrated the powerful influence of such presentation or "framing effects." The experimental demonstration of these effects suggests that risk communicators can, under some circumstances, easily manipulate risk perceptions.

Ignoring or Dismissing Risk Information Because of its Perceived Lack of Personal Relevance. Most risk data relate to society as a whole. These data are usually of little interest or concern to individuals, who are more likely to be concerned about risks to themselves than about risks to society.

Perceivings Accidents and Mishaps as Signals. The significance of an accident is determined only in part by its health, safety, or environmental consequences, eg, the number of deaths or injuries that occur. Of equal if not greater importance is what the accident or mishap signifies or portends. A major accident with many deaths and injuries, for example, may have only minor social significance (beyond that to the victims' families and friends) if it occurs as part of a familiar and well-understood system (eg, a train wreck). However, a minor accident in an unfamiliar or poorly understood system, such as a leak at a radioactive waste disposal site, can have major social significance as a harbinger of future, possibly catastrophic events.

Using Health and Environmental Risks Used as Proxies or Surrogates for Other Concerns. The specific risks that people focus on reflect their beliefs about values, social institutions, nature, and moral behavior. Risks are exaggerated or minimized accordingly. Debates about risks often are a proxy or surrogate for debates about other, more general societal concerns. The debate about nuclear power, for example, has often been interpreted as less a debate about the specific risks of nuclear power than about other concerns, including the proliferation of nuclear weapons, the adverse effects of nuclear waste disposal, the value of large-scale technological progress and growth, and the centralization of political and economic power in the hands of a technological elite.

One conclusion that can be drawn from these observations is that risk is not an objective phenomenon perceived in the same way by all interested parties. Instead, it is a psychological and social construct with its roots deeply embedded in specific social contexts. A variety of scientific, psychological, social, and cultural factors determine which risks will ultimately be selected for societal attention and concern. Scientific evidence about the magnitude of possible adverse consequences is only one of these factors.

Distinguishing Between Perceptions of Risk and Judgments of Risk Acceptability. Even though the level of risk is related to risk acceptability, it is not a perfect correlation. Two factors affect the way people assess risk and evaluate acceptability; these factors modify the correlation.

First, the level of risk is only one among several variables that determines acceptability (Table 4). Some of the most important variables that matter to people in evaluating and interpreting risk information are described below:

1. Catastrophic potential, ie, people are more concerned about fatalities and injuries that are grouped in time and space (eg, fatalities and injuries resulting from a major accidental release of toxic chemicals or radiation) than about fatalities and injuries that are scattered or random in time and space (eg, automobile accidents);
2. Familiarity, ie, people are more concerned about risks that are unfamiliar (eg, leaks of chemical or radiation from waste disposal facilities) than about risks that are familiar (eg, household accidents);
3. Understanding, ie, people are more concerned about activities characterized by poorly understood exposure mechanisms or processes (eg, long term exposure to low doses of toxic chemicals or radiation) than about activities characterized by apparently well understood exposure mechanisms or processes (eg, pedestrian accidents or slipping on ice);
4. Uncertainty, ie, people are more concerned about risks that are scientifically unknown or uncertain (eg, risks from a radioactive waste facility designed to last 20,000 years) than about risks that are relatively known to science (eg, actuarial data on automobile accidents);
5. Controllability, ie, people are more concerned about risks that they perceive to be not under their personal control (eg, accidental releases of toxic chemicals or radiation from a waste disposal facility) than about risks that they perceive to be under their personal control (eg, driving an automobile or riding a bicycle);
6. Voluntariness, ie, people are more concerned about risks that they perceive to be involuntary (eg, exposure to chemicals or radiation from a waste or industrial facility) than about risks that they perceive to be voluntary (eg, smoking, sunbathing, or mountain climbing);
7. Effects on children, ie, people are more concerned about activities that put children specifically at risk (eg, milk contaminated with radiation or toxic chemicals; pregnant women exposed to radiation or toxic chemicals) than about activities that do not put children specifically at risk (eg, adult smoking);
8. Effects manifestation, ie, people are more concerned about risks that have delayed effects (eg, the development of cancer after exposure to low doses of chemicals or radiation) than about risks that have immediate effects (eg, poisonings).
9. Effects on future generations, ie, people are more concerned about activities that pose risks to future generations (eg, genetic effects due to exposure to toxic chemicals or radiation) than to risks that pose no special risks to future generations (eg, skiing accidents);

Table 4. Factors Important in Risk Perception and Evaluation

Factor	Conditions Associated with Increased Public Concern	Conditions Associated with Decreased Public Concern
Catastrophic potential	Fatalities and injuries grouped in time and space	Fatalities and injuries scattered and random
Familiarity	Unfamiliar	Familiar
Understanding	Mechanisms or process not understood	Mechanisms or process understood
Uncertainty	Risks scientifically unknown or uncertain	Risks known to science
Controllability (personal)	Uncontrollable	Controllable
Voluntariness of exposure	Involuntary	Voluntary
Effects on children	Children specifically at risk	Children not specifically at risk
Effects manifestation	Delayed effects	Immediate effects
Effects on future generations	Risk to future generations	No risk to future generations
Victim identity	Identifiable victims	Statistical victims
Dread	Effects dreaded	Effects not dreaded
Trust in institutions	Lack of trust in responsible institutions	Trust in responsible institutions
Media attention	Much media attention	Little media attention
Accident history	Major and sometimes minor accidents	No major or minor accidents
Equity	Inequitable distribution of risks and benefits	Equitable distribution of risks and benefits
Benefits	Unclear benefits	Clear benefits
Reversibility	Effects irreversible	Effects reversible
Personal stake	Individual personally at risk	Individual not personally at risk
Origin	Caused by human actions or failures	Caused by acts of nature or God

10. Victim identity, ie, people are more concerned about risks to identifiable victims (eg, a worker exposed to high levels of toxic chemicals or radiation) than about risks to statistical victims (eg, statistical profiles of automobile accident victims);
11. Dread, ie, people are more concerned about risks that are dreaded and evoke a response of fear, terror, or anxiety (eg, exposure to radiation or carcinogens) than to risks that are not especially dreaded and do not evoke a special response of fear, terror, or anxiety (eg, common colds and household accidents);
12. Trust in institutions, ie, people are more concerned about situations where the responsible risk management institution is perceived to lack trust and credibility (eg, lack of trust in certain government agencies for their perceived close ties to industry) than they are about situations where the responsible risk management institution is perceived to be trustworthy and credible (eg, trust in the management of recombinant DNA risks by universities and by the National Institutes of Health);
13. Media attention, ie, people are more concerned about risks that receive much media attention (eg, accidents, leaks, and other problems at waste disposal facilities) than about risks that receive little media attention (eg, on-the-job accidents);
14. Accident history, ie, people are more concerned about activities that have a history of major and sometimes minor accidents (eg, leaks at waste disposal facilities) than about activities that have little or no history of major or minor accidents (eg, recombinant DNA experimentation);
15. Equity and fairness, ie, people are more concerned about activities that are characterized by a perceived inequitable or unfair distribution of risks and benefits (eg, inequities related to the siting of waste disposal facilities) than about activities characterized by a perceived equitable or fair distribution or risks and benefits (eg, vaccination);
16. Benefits, ie, people are more concerned about hazardous activities that are perceived to have unclear, questionable, or diffused benefits (eg, waste disposal facilities) than about hazardous activities that are perceived to have clear benefits (automobile driving);
17. Reversibility, ie, people are more concerned about activities characterized by potentially irreversible adverse effects (eg, nuclear war) than about activities characterized by reversible adverse effects (eg, injuries from sports or household accidents);
18. Personal stake, ie, people are more concerned about activities that they believe place them (or their families) personally and directly at risk (eg, living near a waste disposal site) than about activities that do not place them (or their families) personally and directly at risk (eg, disposal of hazardous waste in remote sites or in other nations);
19. Nature of evidence, ie, people are more concerned about risks that are based on evidence from human studies (eg, risk assessments based on adequate epidemiological data) than about risks based on animal studies (eg, laboratory studies of the effects of radiation using animals); and
20. Human vs. natural origin, ie, people are more concerned about risks caused by human actions and failures (eg, accidents at waste disposal sites caused by negligence, inadequate safeguards, or operator error) than about risks caused by acts of nature or God (eg, exposure to geological radon or cosmic rays).

These factors explain, in large part, the aversion of parts of the public toward activities and technologies such as nuclear power. They also help to explain phenomena such as the "not in my backyard" (NIMBY) response to chemical, nuclear, and related facilities. For example, many residents in communities where unwanted industrial facilities exist or are planned believe that government and industry officials: (1) have excluded them from meaningful participation in the decision-making process; (2) have denied them the resources needed to evaluate or monitor independently the associated health, safety, or environmental risks; (3) have denied them the opportunity to give their "informed consent" to management decisions that affect their lives and property; (4) have imposed or want to impose upon them facilities that provide few local economic benefits; (5) have imposed or want to impose upon them facilities that entail high costs to the community (eg, adverse impacts on health, safety, wildlife, recreation, tourism, property values, traffic, noise, visual aesthetics, community image, and quality of life); (6) have imposed or want to impose on them facilities that provide most of the benefits to other parties or to society as a whole; and (7) have dismissed their opinions, fears, and concerns as irrational and irrelevant.

Critical to resolving NIMBY and related environmental controversies is recognition that a fairly distributed risk is more acceptable than an unfairly distributed one. A risk entailing significant benefits to the parties at risk is more acceptable than a risk with no such benefits. A risk for which there are no alternatives is more acceptable than a risk that could be eliminated by using an alternative technology. A risk that the parties at risk have control over is more acceptable than a risk that is beyond their control. A risk that the parties at risk assess and decide voluntarily to accept is more acceptable than a risk that is imposed on them. These statements are true in exactly the same sense in which it is true that a small risk is more acceptable than a large risk. Risk is multidimensional; size is only one of the relevant dimensions.

If the validity of these points is accepted, then a whole range of risk communication and management options present themselves. Because factors such as fairness, familiarity, and voluntariness are as relevant as size in judging the acceptability of a risk, efforts to make a risk fairer, more familiar, and more voluntary are as appropriate as efforts to make the risk smaller. Similarly, because control is important in determining the acceptability of a risk, efforts to share power, such as establishing and assisting community advisory committees or supporting third party research, audits, inspections, and monitoring, can be effective in making a risk more acceptable.

Second, deciding what level of risk ought to be acceptable is not a technical question but a value question. People vary in how they assess risk acceptability. They weigh the various factors according to their own values, sense of risk, and stake in the outcome. Because acceptability is a matter of values and opinions, and because values and opinions differ, debates about risk are often debates about values, accountability, and control.

RISK COMPARISONS

A significant part of the risk communication literature focuses on risk comparisons. Interest in risk comparisons derives in part from the perceived difficulties in communicating complex, quantitative risk information to laypersons and the need to put risk information in perspective. Several authors have argued that comparisons provide this perspective.

In a typical risk comparison, the risk in question is compared with the risks of other substances or activities. Because comparisons are perceived to be more intuitively meaningful than absolute probabilities, it is widely believed that they can be used effectively for communicating risk information. A basic assumption of the approach is that risk comparisons provide a conceptual yardstick for measuring the relative size of a risk, especially when the risk is new and unfamiliar.

Risk comparisons have several strengths that address important facets of this problem. They present issues in a mode that appears compatible with intuitive, natural thought processes, such as the use of analogies to improve understanding; they avoid the difficult and controversial task of converting diverse risks into a common unit, such as dollars per life lost or per day of pain and suffering; and they avoid direct numerical reference to small probabilities, which can be difficult to comprehend and evaluate in the abstract.

Many risk comparisons are advanced not only for gaining perspective and understanding but also for setting priorities and determining which risks are acceptable. More specifically, risk comparisons have been advocated as a means for determining which risks to ignore, which risks to be concerned about, and how much risk reduction to seek. A common argument in many risk comparisons, for example, is that risks that are small or comparable to already accepted risks should themselves be accepted.

The risk comparison literature contains two basic types of risk comparisons: (1) comparisons of the risks of diverse activities; and (2) comparisons of the risks of similar or related activities. Each type is described below.

Comparisons of the Risks of Diverse Substances, Activities, and Technologies

The basic strategy in this type of comparison is to compare–along a common scale or metric–the risk of a new or existing substance, activity, or technology to the risks of a diverse set of substances, activities, or technologies. For example, the health risks of a new pesticide might be compared to the risks of sunbathing, smoking, and driving. An underlying but untested assumption is that the health risks of the new or existing substance, activity or technology can be more easily appreciated by people if placed in comparative perspective.

Approaches. A variety of different scales have been used by researchers for comparing risks, including scales based on the annual probability of death, the risk per hour of exposure, and the overall loss in life expectancy. Data for constructing such scales are typically drawn from diverse sources, including public health and accident data collected by various government agencies.

One of the most commonly used scales for comparing risks is the annual death rate. Using such a measure, it can be shown that Americans on average face a 2 in 1000 risk of dying from cancer, a 5 in 10,000 risk of dying in an accident, a 2.5 in 10,000 risk of dying in a motor vehicle accident, a 1 in 10,000 risk of being murdered, and a 5 in 10 million risk of being killed by a lightning bolt. These risks can be contrasted with other risks. For example, the average smoker faces a 3 in 1000 risk of dying each year from smoking, the average mountain climber faces a 6 in 10,000 risk of being killed in a climbing accident, and the average hang glider faces a 4 in 10,000 annual risk of being killed in a hang gliding accident. On an annual basis, the risk of smoking is substantially greater than the risk of hang gliding, mountaineering, boxing, and working in a mine; somewhat greater than the risk of military service during the Vietnam era; and nearly as great as the risk of stunt flying.

One deficiency in these risk comparisons is their lack of sensitivity to age differences. For example, at age five the risk of dying from all causes is less than 1 in 1000; at age 40 it is about 2 in 1000; and at age 80 it is about 83 in 1000. Given the large effect that age can have on risk estimates, an alternative procedure that takes this factor into account is to calculate the expected loss in life expectancy due to various causes. Several authors have taken this approach and have shown that the risk of dying from cigarette smoking is twice as great as the risk of being a coal miner; and that the risk of dying in a motor vehicle accident is twice as great as the risk of dying in an accident at home.

Other formats for comparing risks have also been developed. For example, one format is to compare a set of activities with approximately equal risks, such as the risk of activities estimated to increase a person's chance of death (during any year) by one in a million. Using this measure, researchers have found that each of the following activities presents the same risk: smoking 1.4 cigarettes, riding ten miles by bicycle, eating 40 tablespoons of peanut butter, drinking 30 12-ounce cans of diet soda containing saccharin, and living within five miles of a nuclear reactor for 50 years.

Adopting a somewhat different approach, researchers have calculated the time needed to accumulate a one-in-a-million risk of death from a variety of activities. Using this measure, it can be shown that the risk of dying in a motor vehicle accident is approximately equivalent to the risk of dying from being on police duty; and that the risk of dying from employment in trade or manufacturing is approximately equivalent to the risk of dying from a fall.

Several of the best known risk comparisons are contained in the Reactor Safety Study published by the Nuclear Regulatory Commission in 1975. The study (also known as the Rasmussen Report, after its chairman, Norman Rasmussen) compared the risk of nuclear reactor accidents with the risk of a variety of other activities and events. Specifically, the study included a comparison of the risk of death from nuclear power plant accidents with the risk of death from (1) natural hazards–such as hurricanes, earthquakes, and meteorite impacts; (2) technological hazards, such as air crashes, fires, explosions, and dam failures. The analysis concluded that the probability of 100 or more people dying in a nuclear reactor accident (based on the assumption that 100 plants are operating) is (1) about the same as the probability of the same number of fatalities due to the impact of a meteor, and (2) substantially less than the probability of the same number of fatalities due to other natural and technological hazards.

Several reviewers have criticized the Reactor Safety Study for its use of risk comparisons. For example, one review of the study, sponsored by the Nuclear Regulatory Commission itself, noted that the comparisons contained in the Executive Summary of the study, ". . . which is by far the most widely read part of the report among the public and policy makers, does not sufficiently emphasize the uncertainties involved in the calculation of their probability. It has therefore lent itself to misuse in the discussion of reactor risk."

Another criticism of the study was that the risk estimates for nuclear power referred only to immediate fatalities. Nuclear power plant accidents, however, can also have a significant number of delayed fatalities, eg, increases in cancer rates 30–40 years after an accident. Although the Reactor Safety Study contained estimates of these cancer fatalities, they were not factored into the risk comparisons. In fairness, the tables and figures in the report do note that the comparisons were limited only to "early fatalities." Nonetheless, the tables and figures can be misleading. Whereas the estimates for the other activities represented an approximation of the total risk, the estimate for nuclear power plant accidents represented only a part of the total risk.

Several important deficiencies in the Reactor Safety Study were addressed by a comparative risk study conducted by researchers at Clark University. A central objective of this study was to characterize the total risk of a technology more completely. Based on the analysis, the study produced a scale that compared a variety of technological risks on the basis of twelve descriptors defined in terms of several social, physical, and biological dimensions: intentionality, spatial extent, concentration, persistence, recurrence time, population at risk, delay of consequences, annual mortality, maximum potentially killed, transgenerational effects, maximum potential nonhuman mortality, and experienced nonhuman mortality. In contrast to conventional, unidimensional risk comparisons, the descriptors were used by the researchers to generate a multidimensional risk profile for each of the technological risks.

Risk Comparisons and Risk Acceptability. Many of the risk comparisons described above have been advanced not only for gaining perspective and understanding but also for setting priorities and determining which risks are acceptable. More specifically, they have been advocated as a means for determining which risks to ignore, which risks to be concerned about, and how much risk reduction to seek.

Based on such arguments, researchers have constructed scales ranking risks from acceptable to unacceptable. In one such study, activities falling in the upper zone, representing risks of death per year of exposure of less than one in a million, were deemed acceptable. The basic argument was that the risks of these activities were insignificant: insignificance being defined as the level of risk that individuals routinely accept in their personal and daily activities. For example, since individuals routinely accept the risk of being struck by lightning, which poses a risk of death of 1 in a million per year of exposure, risks of this size can be regarded as acceptable. Following the same logic, it was argued that activities representing risks of death that are greater than one in a thousand per year of exposure can be regarded as unacceptable. Activities falling in the middle zone of the scale were identified as the most problematic: the acceptability of these could not be determined a priori. Instead, they must be closely scrutinized and subjected to analysis and societal debate.

Numerous authors have criticized studies that use this type of approach for determining which risks are acceptable. The basic criticism is that such efforts fail to recognize the importance and legitimacy of basing decisions about the acceptability of a risk on factors other than the size of the risk.

Comparisons of the Risks of Similar or Related Substances, Activities, or Technologies

Some researchers have adopted a narrower approach to risk comparison–limiting their comparisons to risks that are similar or closely related. Several of the most important studies are described below.

Foods, Food Products, and Food Additives. To gain perspective and improved understanding, a large number of studies have compared the risks posed by different foods, food products, and food additives. For example, a study conducted by the National Academy of Sciences in 1978 compared the risks of saccharin, where the primary concern was cancer, with the risks of sugar, where the primary concerns were heart disease and diabetes.

Another study compared the risks of processed meats treated with sodium nitrite with the risks of processed meats that are not. This type of comparison have been labelled a "risk-risk" comparison, where risks and benefits are associated with each alternative and where the decision problem is to select the best alternative. In the sodium nitrite case, for example, the alternatives are to permit or prohibit sodium nitrite in food. The choice of one alternative over the other depends in part on whether the benefits of adding sodium nitrite (ie, the decreased risk of botulism provided by adding sodium nitrites to food) exceed the risks of adding nitrite (ie, the increased risk of cancer). One criticism of such comparisons is that they can be misleading if only part of the total risk is represented. In the sodium nitrite case the comparison focuses only on health risks to consumers and fails to consider the possibility of additional risks to workers in the food industry.

Perhaps the best known comparative analysis of the risks of different foods and food products are the studies on food risks, diet, and cancer by Ames and his colleagues at the University of California, Berkeley. These studies have compared the cancer risks of foods that contain synthetic chemicals (eg, food additives and pesticide residues) with the risks of natural foods. An important conclusion is that synthetic chemicals represent only a very small fraction of the total carcinogens in foods. The basic argument underlying this conclusion is that natural foods are not benign. Large numbers of potent carcinogens (eg, aflatoxin in peanuts) and other toxins are present in foods that contain no synthetic chemicals. Many of these natural carcinogens are produced by plants as part of their natural defense processes. Analysis shows that human dietary intake of these natural carcinogens in food is likely to be at least 10,000 times greater than the intake of potentially carcinogenic synthetic chemicals in food (although partial protection against the effects of natural carcinogens is provided by the many natural anti-carcinogens that also appear in food).

Some of Ames's critics have argued that his risk estimates are inflated. The same critics have argued against an implicit, and sometimes explicit, risk comparision argument that natural carcinogens in foods deserve greater societal and regulatory attention and concern than synthetic chemicals.

Energy Production Technologies. In the last two decades, a large number of studies have attempted to compare the risks of alternative energy production technologies. Perhaps the best known comparison of risks from alternative energy production technologies was an analysis conducted by Inhaber in 1979 for the Atomic Energy Control Board of Canada. The study compared the total occupational and public health risks of different energy sources for the complete energy production cycle–from the extraction of raw materials to energy end-use. The study examined the risks of eleven methods of generating electricity–coal, oil, nuclear, natural gas, hydroelectricity, wind, methanol, solar space heating, solar thermal, solar photovoltaic, and ocean thermal. Two types of risk data were analyzed: (1) data on public health risks from industrial sources or pollutant effects, and (2) data on occupational risks derived from statistics on injury, death, and disease rates for workers. Total risk for the energy source was calculated by summing the risks for the seven components of complete energy production cycle: (a) materials acquisition and construction, (b) emissions from materials acquisition and energy production, (c) operation and maintenance, (d) energy back-up system, (e) energy storage system, (f) transportation, and (g) waste management.

The report concluded (a) that most of the risk from coal and oil energy sources is due to toxic air emissions arising from energy production, operation, and maintenance; (b) that most of the risk from natural gas and ocean thermal energy sources is due to materials acquisition; (c) that most of the risk from nuclear energy sources is due to materials acquisitions and waste disposal; and (d) that most of the risk from wind, solar thermal, and solar photovoltaic energy sources is due to the energy backup system required (assumed to be coal). Alternative sources were compared on the basis of the calculated number of man-days that would be lost per megawatt year of electricity produced.

The most controversial aspect of the report was the conclusion that nuclear power carries only slightly greater risk than natural gas and less risk than all other energy technologies considered. Inhaber reported, for example, that coal has a 50 fold larger mortality rate than nuclear power. The report also argued (a) that, contrary to popular opinion, nonconventional energy sources, such as solar power and wind, pose substantial risks; and (b) that the risks of nuclear power are significantly lower than those of nonconventional energy sources. The relatively high risk levels associated with nonconventional energy sources were traced by Inhaber, in part, to the large volume of construction materials required for these technologies and to the risks associated with energy back-up systems and energy storage systems.

Following publication of the report, its methodology was severely criticized. Critics claimed (a) that the study mixed risks of different types, (b) that it used risk estimators of dubious validity, (c) that it made questionable assumptions to cover data gaps, (d) that it failed to consider future technological developments, (e) that it made arithmetic errors, (f) that it double counted labor and back-up energy requirements, and (g) that it introduced arbitrary correction factors. Perhaps the most damaging criticism was that the study was inconsistent in applying its methodology to the various energy technologies. For example, while the study considered materials acquisition, component fabrication, and plant construction in the analysis of unconventional energy sources and of hydropower, critics have claimed that the study did not follow the same approach for coal, nuclear power, oil, and gas. Furthermore, the labor figures for coal, oil, gas, and nuclear power included only on-site construction, while those for the renewable energy sources included on-site construction, materials acquisition, component manufacture.

Despite these criticisms, Inhaber's study represented a landmark effort in the literature on risk comparisons. It made a significant conceptual contribution by attempting to compare, in a systematic and rigorous way, the risks of alternative technologies intended to serve the same purpose. Also important were Inhaber's observations (a) that risks occur at each stage in product development (eg, raw material extraction, manufacturing, use, and disposal), and (b) that risks from each stage need to be added together to obtain an accurate estimate of the total risk.

Cancer. A variety of studies have used risk comparisons to put cancer risks in perspective. In perhaps the best known such study, Sir Richard Doll and his colleagues analyzed data for a variety of causes of cancer, including industrial products, pollution, food additives, tobacco, alcohol, and diet. Results of the study provided an important comparative perspective on cancer risks. The study found, for example, that the combined effect of food additives, occupational exposures to toxic agents, air and water pollution, and industrial products account for only about 7% of U.S. cancer deaths. These results suggest that removing all pollutants and additives in the air, water, food, and workplace would result in only a small decrease in cancer mortality (although even this small percentage represents a substantial number of lives). By contrast, the combined effects of alcohol, diet, and smoking are related to 70% of U.S. cancer deaths. Consequently, even a modest change in personal habits would result in a significant decrease in cancer mortality.

Other authors have adopted a different approach for comparing the risks of cancer. For example, researchers have compared data on the annual risks of cancer from various common or everyday activities. Smoking clearly poses the largest risk of cancer, with an annual risk of 1.2 in 1000. Drinking one beer per day, receiving an average number of diagnostic X-rays, and background radiation at sea level all pose about the same annual risk of cancer–2 in 100,000.

Other Types of Comparisons

A large number of other studies have also compared the risks of similar or diverse activities and technologies. Because of their policy significance, two types of risk comparisons have received special attention in the risk communication literature: (a) comparisons of different occupations or industries; and (b) comparisons of different sources of radiation. In the occupational arena, one study found that the annual job-related risk of death experienced by timber workers, miners, and pilots–the most risky occupations–is several orders of magnitude greater than for school teachers, dentists, and librarians–among the least risky occupations. Occupational risks of death range from one chance in one thousand per year to one chance in one million per year. In the radiation field, one study compared different sources of radiation exposure (eg, natural, medical, and occupational) and found that the single largest source of radiation exposure is radon emanating from the ground or construction materials and accumulating in closed buildings. Such a finding is, of course, only as useful as it is accurate. Only a few years ago, diagnostic X-rays were identified as the largest source of radiation exposure. Radon was not even listed in most tables as a source of radiation exposure.

One of the most ambitious attempts to compare risks was a 1987 study by the Environmental Protection Agency (EPA) titled *Unfinished Business: A Comparative Assessment of Environmental Problems*. A critical part of the study is a comparative ranking of 31 health and environmental risks. The ranking was performed by EPA experts and covered nearly every environmental problem addressed by EPA programs, including air pollution, water pollution, and hazardous waste. Risks were ranked in four major categories: cancer effects; noncancer causing human health effects; ecological effects; and welfare effects (eg, damage to materials). Factors not included in ranking environmental problems were benefits to society; qualitative aspects, such as whether the risk is voluntary or equitable; and economic or technical controllability.

An important finding of the study was that the EPA's ranking of health and environmental risks did not correlate well with the public's ranking of health and environmental risks. For example, risks associated with hazardous waste sites were in the middle of the EPA ranking for cancer effects and at the lower end of the EPA ranking for noncancer causing health effects. By comparison, risks associated with hazardous waste sites were at the top of the public's rankings. Risks associated with radon were at the top of EPA's ranking, while the same risks were ranked at the bottom of the public's ranking. The report notes that the EPA risk rankings also do not correspond well with the agency's program priorities. Agency priorities appear to be more closely aligned with public opinion than with estimated risks.

Another type of comparison that attempts to put risk information in perspective are "concentration" comparisons. In such comparisons, the concentrations of a toxic substance in the environment is compared to other measurement units such as length or time (Table 5). For example, one part per billion of a toxic chemical in drinking water is equivalent to one inch in 16,000 miles and one second in 32 years. Although some insight is provided by such comparisons, the data often lack relevance and meaning unless coupled with data on the toxic potency of a particular chemical.

In addition to these comparisons, several studies have attempted to compare the costs of different risk reduction programs. For example, one study compared the cost per fatality averted implied by various activities aimed at reducing risks. Researchers have also compared the cost per year of life saved of different various investments, and the cost per life saved of various risk-reduction regulations.

One general conclusion from these studies is that different health, safety, and environmental programs and investments vary enormously in their costs per lives saved. Disproportionate amounts of money are spent to reduce risks in some areas, while other areas are relatively neglected. One study found, for instance, that the cost-per-fatality-averted for different risk reduction programs ranges between $10,000 and $1 billion; another study found that the cost-per-life saved for various health investments ranges between $540 and $6.6 million per year; and a third study showed that the cost-per life saved of various proposed, rejected, and final regulations ranges between $100,000 and $72 billion (for final regulations only, the range is $100,000 to $132 million, with an average and median of $23 million and $2 million respectively).

Limitations of Risk Comparisons

As indicated in the examples cited above, critics have noted significant limitations of the risk comparison approach. The most important are (1) failing to identify and emphasize uncertainties involved in the calculation of comparative risk estimates; (2) failing to consider the broad set of quantitative dimensions that define and measure risk; and (3) failing to consider the broad set of qualitative dimensions that underlie people's concerns about

Table 5. Concentration Comparisons Organized by Unit Categories

Unit	1 Part per Million	1 Part per Billion	1 Part per Trillion
Length	1 in./16 miles	1 in./16,000 miles	1 in./16,000,000 miles (a 6-in. leap on a journey to the sun)
Time	1 min./2 years	1 s/32 years	1 s/320 centuries (or 0.06 s since the birth of Jesus Christ)
Money	1 cent/$10,000	1 cent/$10,000,000	1 cent/$10,000,000,000
Weight	1 oz/31 tons	1 pinch salt/10 tons of potato chips	1 pinch salt/10,000 tons of potato chips
Volume	1 drop vermouth/80 "fifths" of gin	1 drop vermouth/500 barrels of gin	1 drop of vermouth in a pool of gin covering the area of a football field 43 ft deep
Area	1 square ft/23 acres	1 in./160-acre farm	1 sq ft/the state of Indiana; or 1 large grain of sand on the surface of Dayton Beach
Action	1 lob/1,200 tennis matches	1 lob/1,200,000 tennis matches	1 lob/1,200,000,000 tennis matches
Quality	1 bad apple/2,000 barrels	1 bad apple/2,000,000 barrels	1 bad apple/2,000,000,000 barrels

the acceptability of risks and technologies. Each limitation is described below.

Failing to identify and emphasize uncertainties involved in the calculation of comparative risk estimates.

A critical flaw in many risk comparisons is the failure to provide information on the assumptions underlying the calculation of comparative risk estimates. Since risk estimates are typically drawn from a variety of different data sources, tables of comparative risks may contain risk estimates based on one set of assumptions together with noncomparable estimates based on another set of assumptions. Similarly, tables of comparative risks may contain risk estimates based on actuarial statistics (eg, deaths from motor vehicle accidents) together with estimates based on controversial models, assumptions, and judgments (cancer deaths from chronic exposure to pesticides or air pollutants).

A related flaw in many risk comparisons is the failure to describe and characterize uncertainties. This flaw can seriously undermine the value and potential usefulness of risk comparisons for risk communication purposes. Tables of risks that report only single values for adverse health, safety, or environmental consequences, for example, ignore the range of possibilities and may provide an inaccurate picture of the risk problem to the public. Given the various biases, errors, and other sources of uncertainty that can undermine the validity and reliability of a risk assessment, it is critical that tables of comparative risks provide the fullest possible information on potential errors and inaccuracies in each computed risk value—including qualifiers, ranges of uncertainty, confidence intervals, and standard errors. To date, it is more the exception than the rule for results of comparative risk studies to be presented with full disclosure of the strengths and limitations of the assessment and with full disclosure of the degree to which assessment results are based on controversial data and judgments. Risk comparisons that do not include such information can produce a false sense of certainty.

Failing to consider the broad set of quantitative dimensions that define and measure risk.

Most lists of comparative risks are unidimensional. They present statistics for only one dimension of risk, such as expected annual mortality rates or reductions in life expectancy. The use of such narrow quantitative measures of risk can, however, obscure the importance of other significant quantitative dimensions, such as expected annual probability of injury or disability, spatial extent, concentration, persistence, recurrence, population at risk, delay, maximum expected fatalities, transgenerational effects, expected environmental damage (eg, ecological damage or adverse effects on endangered species), and maximum expected environmental damage.

Significant distortions and misunderstandings also result from comparative analyses that fail to provide the full range of relevant quantitative risk information. Consider some of the problems involved in comparing the risks of airplane travel to the risks of traveling by automobile or train. Using a measure of risk to an individual based on the number of deaths per hundred million passenger miles, traveling as an airplane passenger appears to pose slightly less risk to an individual (0.38 deaths per hundred million passenger miles) than being an automobile passenger (0.55 deaths per hundred million passenger miles); and slightly more risk than traveling as a train passenger (0.23 deaths per hundred million passenger miles). However, for airplane travel, the landing and take-off phase represents the period of highest risk; thus, it can be argued that a better estimate of individual risk is the number of passenger journeys rather than the number of miles traveled. Using this measure, traveling as an airplane passenger (1.8 deaths per million passenger journeys) poses slightly greater risk than traveling as an automobile passenger (0.027 deaths per million passenger journeys) or as a train passenger (0.59 deaths per million passenger journeys). As a result, if distance traveled is the selected measurement criterion, then airplane travel is marginally safer than automobile travel and marginally less safe than train travel; but if number of journeys is the selected measurement criterion, then airplane travel is marginally less safe than both automobile travel and train travel.

A related deficiency is the failure in most risk comparisons to estimate the total quantitative risk of technologies and activities included in the risk comparison. Technological activities encompass a variety of different components; stages of development (eg, extraction of raw materials, production, consumption and disposal); and relationships (direct and indirect) with other technological and societal activities. Detailed examination of the risks of these different components, stages of development, and relationships may significantly alter the overall ranking of a technology or activity. Consequently, any risk comparison that claims to be comprehensive must either present risk data for each of these aspects, or explicitly acknowledge those aspects of the analysis that have been excluded.

Even when the analyst provides data on the total quantitative risk of an activity or technology, the comparison can nonetheless be misleading if it fails to provide risk data for sensitive, suspectible, or high risk groups. These groups include children, pregnant women, the elderly, and individuals who are particularly vulnerable or susceptible because of illness or disease. Most lists of comparative risks present population averages. However, population averages often mask important subpopulation variations in susceptibility.

Important distinctions also can be masked in other ways. For example, it is not always clear from risk comparison tables what is included in the specific risk entries. For example, do deaths from smoking include cardiovascular disease and emphysema as well as lung cancers? Are the risk estimates based on the entire population or only the population that is exposed?

Even if the analyst carefully and accurately reports risk data, misunderstandings can develop if important situational qualifiers are left out. For example, the risk calculation for driving includes many different driving situations. Yet speeding home from a party just before dawn is two orders of magnitude more dangerous than driving to the supermarket. Similarly, the risk of being hit by lightning for people who remain on a golf course during a thunderstorm is much higher than the average risk for the U.S. population.

A related deficiency stems from the failure to recognize the importance of framing effects on risk comparisons. Different impressions are created by different presentation formats. Each format for presenting or expressing risk information, such as deaths per million people, deaths per unit of concentration, or deaths per activity, is likely to have a different impact on the audience. Context can be equally important. For example, an individual lifetime risk of one in a million in the U.S. is mathematically equivalent to approximately 0.008 deaths per day, 3 deaths per year, or 200 deaths over a 70-year lifetime. Many people will view the first two numbers as small and insignificant, whereas the latter two statistics are likely to be perceived as sufficiently large to warrant societal or regulatory attention.

A final deficiency is the failure in most risk comparisons to acknowledge deficiencies in the quality of the data. Most risk comparisons draw on diverse data sources that vary considerably in quality. Because of the high cost and difficulty of collecting original data, researchers seldom have access to data developed exclusively for the comparison. Instead, a variety of existing data sources are used, each varying in quality. As a result, comparative risks often contain data of high quality together with data of questionable scientific validity and reliability.

Failing to consider the broad set of qualitative dimensions that underlie people's concerns about the acceptability of risks and technologies.

Risk comparison is often advocated as a means for setting priorities or for determining which risks are acceptable. A common argument is that risks that are small, or that are comparable to risks that are already being accepted, should themselves be accepted. A number of critics have argued, however, that such claims cannot be defended. Although carefully prepared lists of comparative risk statistics can provide insight and perspective, they provide only a small part of the information needed for setting priorities or for determining which risks are acceptable.

Judgments of acceptability are related not only to annual mortality rates, the focus of most risk comparisons, but also to a multiplicity of qualitative dimensions or factors. These factors have been discussed earlier in the section on risk perceptions and include voluntariness, controllability, fairness, effects on children, familiarity, and benefits.

Because of the importance of these factors, comparisons showing that the risk of a new or existing activity or technology is higher (or lower) than the risks of other activities or technologies may have no effect on public perceptions and attitudes. For example, comparing the risk of living near a nuclear power or chemical manufacturing plant with the risk of driving *X* number of hours, eating *X* tablespoons of peanut butter, smoking *X* number of cigarettes a day, or sunbathing *X* number of hours may provide perspective but may also be highly inappropriate. Since such risks differ on a variety of qualitative dimensions (eg, perceived benefits, extent of personal control, voluntariness, catastrophic potential, familiarity, fairness, origin, and scientific uncertainty) it is likely that people will perceive the comparison to be meaningless.

For example, it is often tempting for government or industry to use the following argument in meetings with community groups or members of the public. The risk of *a* (eg, emissions from an incinerator or facility) is lower than the risk of *b* (driving to the meeting or smoking during breaks). Since you (the audience) find *b* acceptable, you are obliged to find *a* acceptable.

This argument has a basic flaw in logic; its use can severely damage trust and credibility. Some listeners will analyze the argument this way: "I do not have to accept the (small) added risk of living near an incinerator just because I accept the (perhaps larger, but voluntary and personally beneficial) risk of sunbathing, bicycling, smoking or driving my car. In deciding about the acceptability of risks, I consider many factors, only one of them being the size of the risk and I prefer to do my own evaluation. Your job is not to tell me about what I should accept but to tell me about the size of the risk and what you are doing about it."

The fundamental argument against the use of risk comparisons is that it is seldom relevant or appropriate to compare risks with different qualities for risk acceptability purposes, even if the comparison is technically accurate. Several reasons underlie the argument. First, there are important psychological and social differences among risks with different qualities. Risks that are involuntary and result from lifestyle choices, for example, are more likely to be accepted than involuntary and imposed risks.

Second, people recognize that risks are cumulative; that each additional risk adds to their overall risk burden. The fact that a person is exposed to risks resulting from voluntary lifestyle choices does not lessen the impact of risks that are perceived to be involuntary and imposed.

Finally, people perceive many types of risk in an absolute sense. An involuntary increased risk of cancer or birth defects is a physical and moral insult regardless of whether the increase is small or whether the increase is smaller than risks from other exposures.

Aggrevating the problem is the lack of attention given in most risk comparisons to how people actually make decisions about the acceptability and tolerability of a risk. Judgments about risks are seldom separated from judgments about the risk decision process. Public responses to risk are shaped both by the characteristics of the activity or technology and by the perceived adequacy of the decision-making process. Risk comparisons play only a limited role in such determinations.

Guidelines for Improving the Effectiveness of Risk Comparisons

Despite the limitations reviewed above, a well constructed and well documented risk comparison can be useful in communicating risk information. It can, for example, provide (1) a benchmark and yardstick against which the magnitude of new or unfamiliar risks can be calibrated and compared; (2) a means for determining and communicating the relative numerical significance and seriousness of a new or existing risk; and (3) a means for informing and educating people about the range and magnitude of risks to which they are exposed.

For a risk comparison to achieve these goals and purposes, however, the limitations and deficiencies of the approach must specifically be addressed. Results from recent analyses and case studies suggest that the following guidelines may be useful for this purpose.

These results suggest that risk comparisons, if provided, should have the following characteristics: (1) risks that are compared should be as similar as possible; (2) dissimilarities between the compared risks should be identified; (3) sources of data on risk levels should be credible and should be identified; (4) limitations of the comparision should be described; (5) all comparisons should have only one purpose–perspective; and (6) all comparisons should be pilot-tested with a surrogate for the intended audience.

In summary, the risk comparison approach can be a powerful tool in risk communication. However, the simplicity and intuitive appeal of the method is often deceptive. Many factors play a role in determining the legitimacy and effectiveness of a risk comparison. The success of the comparison will depend on the degree to which these factors have been adequately recognized, considered, and addressed.

CONCLUSIONS

Given the passage of increasing numbers of right-to-know laws, and given increasing demands by the public for environmental information, risk communication will be the focus of increasing attention in years to come. Officials in both the public and private sector will continually be asked to provide information about health, safety, and environmental risks in crisis and noncrisis situations. How they answer this challenge will have a profound affect on environmental initiatives and programs.

The findings reported in this chapter represent only a sampling of results from the emerging area of risk communication research. However, several general principles and guidelines for communicating information about risks can be extrapolated from this literature. Although many of these principles and guidelines may seem obvious, they are so often violated in practice that a useful question is why are they so frequently not followed.

Principle 1. Accept and Involve the Public as a Legitimate Partner

Discussion. Two basic tenets of risk communication in a democracy are generally understood and accepted. First, people and communities have a right to participate in decisions that affect their lives, their property, and the things they value. Second, the goal of risk communication should not be to diffuse public concerns or avoid action. The goal should be to produce an informed public that is involved, interested, reasonable, thoughtful, solution-oriented, and collaborative.

Guidelines. Demonstrate your respect for the public and your sincerity by involving the community early, before important decisions are made. Make it clear that you understand the appropriateness of basing decisions about risks on factors other than the magnitude of the risk. Involve all parties that have an interest or a stake in the particular risk in question.

Principle 2. Plan Carefully and Evaluate Performance

Discussion. Different goals, audiences, and media require different risk communication strategies. Risk communication will be successful only if carefully planned.

Guidelines. Begin with clear, explicit objectives–such as providing information to the public, motivating individuals to act, stimulating emergency response, or contributing to conflict resolution. Evaluate the information you have about risks and know its strengths and weaknesses. Classify the different subgroups among your audience. Aim your communications at specific subgroups in your audience. Recruit spokespersons who are good at presentation and interaction. Train your staff–including technical staff–in communication skills; reward outstanding performance. Whenever possible, pretest your messages. Carefully evaluate your efforts and learn from your mistakes.

Principle 3. Listen to Your Audience

Discussion. People in the community are often more concerned about issues such as trust, credibility, control, competence, voluntariness, fairness, caring, and compassion than about mortality statistics and the details of quantitative risk assessment. If you do not listen to people, you cannot expect them to listen to you. Communication is a two-way activity.

Guidelines. Do not make assumptions about what people know, think or want done about risks. Take the time to find out what people are thinking: use techniques such as interviews, focus groups, and surveys. Let all parties that have an interest or a stake in the issue be heard. Recognize people's emotions. Let people know that you understand what they said, addressing their concerns as well as yours. Recognize the "hidden agendas," symbolic meanings, and broader economic or political considerations that often underlie and complicate the task of risk communication.

Principle 4. Be Honest, Frank, and Open

Discussion. In communicating risk information, trust and credibility are your most precious assets. Trust and credibility are difficult to obtain. Once lost they are almost impossible to regain.

Guidelines. State your credentials; but do not ask or expect to be trusted by the public. If you do not know an answer or are uncertain, say so. Get back to people with answers. Admit mistakes. Disclose risk information as soon as possible (emphasizing any appropriate reservations about reliability). Do not minimize or exaggerate the level of risk. Speculate only with great caution. If in doubt, lean toward sharing more information, not less–or people may think you are hiding something. Discuss data uncertainties, strengths and weaknesses, including the ones identified by other credible sources. Identify worst-case estimates as such, and cite ranges of risk estimates when appropriate.

Principle 5. Coordinate and Collaborate with Other Credible Sources

Discussion. Allies can be effective in helping you communicate risk information. Few things make risk communication more difficult than conflicts or public disagreements with other credible sources.

Guidelines. Take time to coordinate all interorganizational and intraorganizational communications. Devote effort and resources to the slow, hard work of building bridges with other organizations. Use credible and authoritative intermediaries. Consult with others to determine who is best able to answer questions about risk. Try to issue communications jointly with other trustworthy sources such as credible university scientists, physicians, trusted local officials, and opinion leaders.

Principle 6. Meet the Needs of the Media

Discussion. The media are a prime transmitter of information on risks. They play a critical role in setting agendas and in determining outcomes. The media are generally more interested in politics than in risk; more interested in simplicity than in complexity; and more interested in danger than in safety.

Guidelines. Be open with and accessible to reporters. Respect their deadlines. Provide information tailored to the needs of each type of media, such as graphics and other visual aids for television. Prepare in advance and provide background material on complex risk issues. Follow-up on stories with praise or criticism, as warranted. Try to establish long-term relationships of trust with specific editors and reporters.

Principle 7. Speak Clearly and with Compassion

Discussion. Technical language and jargon are useful as professional shorthand. But they are barriers to successful communication with the public.

Guidelines. Use simple, nontechnical language. Be sensitive to local norms, such as speech and dress. Use vivid, concrete images that communicate on a personal level. Use examples and anecdotes that make technical risk data come alive. Avoid distant, abstract, unfeeling language about deaths, injuries and illnesses. Acknowledge and respond (both in words and with actions) to emotions that people express–anxiety, fear, anger, outrage, helplessness. Acknowledge and respond to the distinctions that the public views as important in evaluating risks. Use risk comparisons to help put risks in perspective; but avoid comparisons that ignore distinctions that people consider important. Always try to include a discussion of actions that are under way or can be taken. Tell people what you cannot do. Promise only what you can do, and be sure to do what you promise. Never let your efforts to inform people about risks prevent you from acknowledging–and saying–that any illness, injury or death is a tragedy.

Analyses of case studies suggest that these principles and guidelines can form the basic building blocks for effective risk communication. Each principle recognizes, in a different way, that effective risk communication is an interactive process based on mutual trust, cooperation, and respect among all parties. Each principle recognizes that effective risk communication is a complex art and skill that requires substantial knowledge, training, and practice. Each principle recognizes that there are no easy prescriptions for effective risk communication; that there are limits on what can be accomplished through risk communication alone–no matter how skilled, committed, and sincere an organization or person is. And each principle addresses, from a different perspective, the most important obstacles to effective risk communication: lack of trust and credibility.

BIBLIOGRAPHY

Reading List

M. Bean, "Tools for Environmental Professionals Involved in Risk Communication at Hazardous Waste Facilities Undergoing Siting, Permitting, or Remediation." Report No. 87-30.8. 1987, Air Pollution Control Association. Reston, Va.

B. Cohen, and I. Lee, "A Catalog of Risks," *Health Physics* **36,** 707–722 (1982).

J. Cohrssen, and V. Covello, *Risk Analysis*. White House Council on Environmental Quality, Washington, D.C., 1989.

B. Combs, and P. Slovic, "Newspaper Coverage of Causes of Death," *Journalism Quarterly* **6,** 837–843 (1979).

Conservation Foundation, *Risk Assessment and Risk Control*, Conservation Foundation, Washington, D.C., 1985.

V. T. Covello, "The Perception of Technological Risks: A Literature Review," *Technological Forecasting and Social Change* **23,** 285–297 (1983).

V. T. Covello, "Informing the Public About Health and Environmental Risks: Problems and Opportunities for Effective Risk Communication," in N. Lind, ed., *Risk Communication: A Symposium*, University of Waterloo, Waterloo, Ont., 1988.

V. Covello, "Issues and Problems in Using Risk Comparisons for Communicating Right-to-Know Information on Chemical Risks," *Environmental Science and Technology* **23**(12), 1444–1449 (1989).

V. T. Covello, "Risk Comparisons and Risk Communication," in R. Kasperson, and P. J. Stallen, eds., *Communicating Risks to the Public*, Kluwer Academic Publishers, Boston, Mass., 1991.

V. T. Covello, "Risk Communication, Trust, and Credibility," *Health and Environmental Digest,* **6**(1), (Apr. 1992).

V. Covello, and F. Allen, *Seven Cardinal Rules of Risk Communication*, U.S. Environmental Protection Agency, Office of Policy Analysis, Washington, D.C., 1988.

V. Covello, D. von Winterfeldt, and P. Slovic, "Communicating Risk Information to the Public," *Risk Abstracts*, 1986.

V. Covello, D. von Winterfeldt, and P. Slovic, "Communicating Risk Information to the Public," in J. C. Davies, V. Covello, and F. Allen, eds., *Risk Communication:* The Conservation Foundation, Washington, D.C., 1987.

V. T. Covello, P. Sandman, and P. Slovic, *Risk Communication, Risk Statistics, and Risk Comparisons*. Chemical Manufacturers Association, Washington, D.C., 1988.

V. Covello, D. McCallum, and M. Pavlova, eds., *Effective Risk Communication: The Role and Responsibility of Governmental and Non-Governmental Organizations*, Plenum Press, New York, 1989.

V. Covello, E. Donovan, and J. Slavick, *Community Outreach*, Chemical Manufacturers Association, Washington, D.C., 1991.

V. Covello, and M. Merkhofer, *Risk Assessment Methods*, Plenum Press, New York, 1992.

J. C. Davies, V. T. Covello, and F. W. Allen, eds., *Risk Communication*. The Conservation Foundation, Washington, D.C., 1987.

M. Douglas, and A. Wildavsky, *Risk and Culture*, University of California Press, Berkeley, Calif., 1982.

J. Fessenden-Raden, J. Fitchen, and J. Heath, "Risk Communication at the Local Level: A Complex Interactive Progress," *Science, Technology and Human Values* (Dec. 1987).

B. Fischhoff, "Protocols for Environmental Reporting: What to Ask the Experts," *The Journalist* 11–15 (Winter 1985).

B. Fischhoff, "Managing Risk Perception," *Issues in Science and Technology*, **2,** 83–96 (1985).

B. Fischhoff, S. Lichtenstein, P. Slovic, S. L. Derby, and R. L. Keeney, *Acceptable Risk*, Cambridge University Press, New York, 1981.

B. Fischhoff, "Treating the Public with Risk Communications: A Public Health Perceptive," *Science, Technology, and Human Values* **12**(3, 4), 13–19 (1987).

S. M. Friedman, "Blueprint for Breakdown: Three Mile Island and the Mass Media before the Accident," *Journal of Communications,* **31,** 85–96 and 116–128 (1981).

S. Hadden, *Citizen Right to Know: Communication and Public Policy*, Westview, Boulder, Colo., 1989.

B. Hance, C. Chess, and P. Sandman, *Improving Dialogue with Communities: A Risk Communication Manual for Government*, Office of Science and Research, New Jersey Department of Environmental Protection, Trenton, New Jersey, Dec., 1987.

B. Johnson, and V. Covello, eds., *The Social and Cultural Construction of Risk: Essays on Risk Selection and Perception*, Reidel, Boston, Mass., 1987.

R. Kasperson, "Six Propositions on Public Participation and Their Relevance to Risk Communication," *Risk Analysis* **6,** 275–282 (1986).

R. Kasperson, and J. Kasperson, "Determining the Acceptability of Risk: Ethical and Policy Issues," in J. Rogers and D. Bates, eds., *Risk: A Symposium*, The Royal Society of Canada, Ottawa, Ontario, 1983.

R. Kasperson, and P. J. Stallen, eds., *Communicating Risks to the Public*, Kluwer Academic Publishers, Boston, Mass., 1991.

R. Kates, C. Hohenemser, and R. Kasperson, *Perilous Progress: Managing the Hazards of Technology*, Westview, Boulder, Colo., 1985.

S. Klaidman, "Health Risk Reporting," Institute for Health Policy Analysis, Georgetown University, Washington, D.C., 1985.

S. Krimsky, and A. Plough, *Environmental Hazards: Communicating Risks as a Social Process*, Auburn House, Dover, Mass., 1988.

W. W. Lowrance, *Of Acceptable Risk: Science and the Determination of Safety*, Kaufman, Los Altos, Calif., 1976.

A. Mazur, "Media Coverage and Public Opinion on Scientific Controversies," *Journal of Communication* 106–115 (1981).

A. Mazur, "The Journalists and Technology: Reporting about Love Canal and Three Mile Island," *Minerva,* **22,** 45–66 (Spring 1984).

D. McCallum, S. Hammond, and V. Covello, *Public Knowledge and Perceptions of Risks in Six Communities: Analysis of a Baseline Survey*, Report No. EPA 230-01-90-074. Environmental Protection Agency, Washington, D.C., 1990.

M. G. Morgan, P. Slovic, I. Nair, D. Geisler, D. MacGregor, B. Fischhoff, D. Lincoln, and K. Florig, "Powerline Frequency and Magnetic Fields: A Pilot Study of Risk Perception," *Risk Analysis* **5,** 139–149 (1985).

National Research Council, *Risk Assessment in the Federal Government: Managing the Process*, National Academy Press, Washington, D.C., 1983.

National Research Council, *Improving Risk Communication*, National Academy Press, Washington, D.C., 1989.

D. Nelkin, *Science in the Streets*, Twentieth Century Fund, New York, 1984.

D. Nelkin, "Communicating Technological Risk: The Social Construction of Risk Perception," *American Review of Public Health*, **10,** 95–113 (1989).

President's Commission on the Accident at Three Mile Island, *Report of the Public's Right to Information Task Force*, U.S. Government Printing Office, Washington, D.C., 1979.

F. Press, "Science and Risk Communication," in J. C. Davies, V. T. Covello, and F. W. Allen, eds., *Risk Communication*, The Conservation Foundation, Washington, D.C., 1987, pp. 11–17.

O. Renn, and D. Levine, "Trust and Credibility in Risk Communication," in H. Jugermann, R. Kasperson, and P. Wiedermann, eds., *Risk Communication*. Kernforschungsanlage Julich GmbH, Julich, Germany, 1988, pp. 51–82.

E. Roth, G. Morgan, B. Fischhoff, L. Lave, and A. Bostrom, "What Do We Know About Making Risk Comparisons?" *Risk Analysis* **10,** 375–387 (1990).

N. Rothchild, "Coming to Grips with Risk," *Wall Street Journal* (May 13, 1979).

W. D. Ruckelshaus, "Science, Risk, and Public Policy," *Science*, **221,** 1026–1028 (1983).

W. D. Ruckelshaus, "Risk in a Free Society," *Risk Analysis*, 157–163 (Sept. 1984).

W. D. Ruckelshaus, "Communicating About Risk," in J. C. Davies, V. T. Covello, and F. W. Allen, eds., *Risk Communication*. The Conservation Foundation, Washington, D.C., 1987, pp. 3–9.

P. M. Sandman, "Getting to Maybe: Some Communications Aspects of Hazardous Waste Facility Sitting," *Seton Hall Legislative Journal* (Spring 1986).

P. M. Sandman, *Explaining Environmental Risk*. U.S. Environmental Protection Agency, Office of Toxic Substances, Washington, D.C., 1986.

P. Sandman, D. Sachsman, M. Greenberg, and M. Gotchfeld, *Environmental Risk and the Press*, Transaction Books, New Brunswick, New Jersey, 1987.

P. Sandman, D. Sachsman, and M. Greenberg, *Risk Communication for Environmental News Sources*, Industry/University Cooperative Center for Research in Hazardous and Toxic Substances, New Brunswick, New Jersey, 1987.

H. Sharlin, "EDB: A Case Study in the Communication of Health Risk," in B. Johnson and V. Covello, eds., *The Social and Cultural Construxction of Risk: Essays on Risk Selection and Perception*, Reidel, Boston, Mass., 1987.

P. Slovic, "Informing and Educating the Public About Risk," *Risk Analysis*, **4,** 403–415 (1986).

P. Slovic, "Perception of Risk," *Science* **236,** 280–285 (1987).

P. Slovic, B. Fischhoff, and S. Lichtenstein, "Facts Versus Fears: Understanding Perceived Risk," in D. Kahneman, P. Slovic, and A. Tversky, eds., *Judgment Under Uncertainty: Heuristics and Biases*, Cambridge University Press: Cambridge, UK, 1982.

P. Slovic, and B. Fischhoff, "How Safe is Safe Enough? Determinants of Perceived and Acceptable Risk," in L. Gould and C. Walker, eds., *Too Hot to Handle*, Yale University Press, New Haven, Conn., 1982.

P. Slovic, N. Krauss, and V. Covello, "What Should We Know about Making Risk Comparisons," *Risk Analysis*, **10,** 389–392 (1990).

V. K. Smith, W. Desvousges, A. Fisher, and R. Johnson, *Communicating Radon Risk Effectively: A Mid-Course Evaluation*, Report No. CR-811075. Environmental Protection Agency, Washington, D.C., 1987.

L. M. Thomas, "Why We Must Talk About Risk," in J. C. Davies, V. T. Covello, and F. W. Allen, eds., *Risk Communication*. The Conservation Foundation, Washington, D.C., 1987, pp. 19–25.

R. Wilson, "Analyzing the Daily Risks of Life," *Technology Review*, **81,** 40–46 (1979).

R. Wilson, "Commentary: Risks and Their Acceptability," *Science, Technology and Human Values* **9**(2), 11–22 (Spring 1984).

R. Wilson, and E. Crouch, "Risk Assessment and Comparisons: An Introduction," *Science* **236,** 267–270 (Apr. 17, 1987).

SAE VISCOSITY GRADES

RAMON ESPINO
Exxon Research and Engineering
Annandale, New Jersey

Running an engine efficiently and free of damage depends greatly on the viscosity of the lubricant being used. The lubricant must be able to flow at low temperatures to lubricate all parts and thus allow sufficient movement of the engine parts under cold start conditions. The lubricant must also be sufficiently viscous so as to protect all engine parts at high temperatures. Under normal driving conditions certain portions of the engine that need lubrication can reach temperatures over 150°C.

The Society of Automotive Engineers (SAE), with the cooperation of the American Society of Testing Materials (ASTM), developed a viscosity classification system that quantifies the requirements described in the introductory paragraph. Essentially the system defines winter grades by their viscosity at a temperature below 0°C and at 100°C, and for regular grades only by the viscosity of the oil at 100°C. These SAE grades, by the way, are used worldwide and accepted as standards by most countries and organizations dealing with lubrication systems (Table 1).

A very important advance in engine lubrication was the design of engine oils with the viscosity of a low winter grade, for example 5W at −25°C, and the viscosity of a higher grade, for example 30 at 100°C. These "multigrade" engine oils were designed by adding polymers that are soluble in the mineral base oils used in engine oils. These polymers are called viscosity index improvers (VI improvers) or viscosity modifiers (VM). The polymers significantly increase the viscosity of the base oil at high temperatures (100°C) but are relatively ineffective viscosity enhancers at low temperatures (below 0°C). The end result is that the multigrade engine oil has a viscosity of a 5W or 10W oil at low temperatures and the viscosity of a 30 or 40 oil at 100°C.

The original polymers used as VI improvers tended to shear under the high shear conditions encountered in the engines. During the last five decades significant improvements in the polymer structure have resulted in polymer families that are very shear stable under the very demanding conditions of today's engines. Moreover, the viscosity temperature behavior of the base stocks can be changed to provide many of the benefits of VI improvers. A widely used parameter of the viscosity–temperature behavior of base oils takes advantage of the essentially linear character of this function. A base oil vintage 1928 with the smallest slope was given a VI index of 100, and the base oil with the steepest slope a value of 0. Obviously what is desired is a base stock whose viscosity changes very little with temperature and thus has a high Viscosity Index (VI). Synthetic and semisynthetic oils have a VI higher than 100, with poly alpha olefins in the 140 range and polyol esters in the 170 range.

Solvent extracted mineral base oils can have their VI increased from their normal range of 90–100 VI by increasing the severity of the extraction and hydrofining steps. Severe hydrofining or hydroisomerization of heavy petroleum cuts or waxes can yield base oils with VI in the 140 range. As a result, the engine oils formulators have a number of options to make very cost-effective engine oils that can be used under a wide range of ambient conditions.

Most engine manufacturers (OEMs) recommended viscosity grades in their manuals for a range of ambient temperatures. Improvements in engine design and materials of construction has broadened the ambient temperature range that OEMs are willing to recommend in multigrade oils such as 5W-30 and 10W-30. Table 2 presents recommendations that are common to Japanese and North American OEMs.

It is important to keep in mind that these recommendations also take into account that in order to meet the fuel economy requirements in the United States, low viscosity oils must be used. From the exclusive view of lubrication, engine oils with viscosity grades 15W-40 and 15W-50 can be used all year around in North America and Europe, with the exception of the very northern regions of the continents where 10W-30 or 5W-30 oils are required in the winter. See also AUTOMOBILE ENGINES; LUBRICANTS.

Table 1. SAE Viscosity Grades

Grade	Max Viscosity, kPa's at °C	Viscosity at 100°C in mm^2/s	High Temp (150°C) High Shear, $(10^6 s^{-1})$, kPa's
0W	3250 at −30	3.8 min	2.4 min
5W	3500 at −25	3.8 min	2.9 min
10W	3500 at −20	4.3 min	2.9 min
15W	3500 at −15	5.6 min	3.7 min
20W	4500 at −10	5.6 min	3.7 min
25W	6000 at −5	9.3 min	3.7 min
20		5.6–9.3	
30		9.3–12.5	
40		12.5–16.3	
50		16.3–21.9	
60		21.9–26.1	

Table 2. Japanese and North American OEMs Recommendations

Ambient Temperature,°C	Viscosity Grade
−30°C to 10	5W–20
−30°C to 30	5W–30
−10°C to 35	10W–30
−10°C to 40	10W–40

BIBLIOGRAPHY

Reading List

Society of Automotive Engineers (SAE), J300, March 93.

SODIUM HEAT ENGINES

The Sodium Heat Engine (SHE), also known as an Akali Metal Thermoelectric Converter (AMTEC), is a device that converts heat directly to electricity. It does this by carrying out a thermodynamic conversion cycle essentially equivalent to the isothermal expansion of an ideal gas. Electrochemists describe this cycle as an electrochemical concentration cell which can be continuously regenerated (recharged) by application of heat.

See also ELECTRIC POWER GENERATION; THERMODYNAMICS.

ORIGINS

In 1962 J. T. Kummer and N. Weber, at the Ford Motor Company Scientific Laboratory, conceived the SHE/AMTEC. This invention followed on their earlier work on the sodium sulfur battery (1) which had evolved to utilize the unique ionic conductivity and electrolyte properties of the sodium beta and beta″-aluminas. Weber and Kummer demonstrated the principles of the SHE/AMTEC device and obtained a patent (2) on it in 1969. In a seminal paper published in 1974 (3), Weber described the basic principles of SHE/AMTEC operation in considerable detail. The operating cycle is quite simple. Weber initiated research on the new device and in 1974 T. K. Hunt and T. Cole joined in this work as collaborators. It was at this stage that the device became known as the Sodium Heat Engine to distinguish it, as a simple thermodynamic heat engine, from Seebeck type thermoelectric devices. Shortly afterwards the more descriptive name Alkali Metal Thermoelectric Converter (AMTEC) was coined.

SHE/AMTEC devices are of practical significance primarily because the crucial electrolyte role can be filled by beta″-alumina solid electrolyte (BASE) (4). BASE has remarkable electrical properties, acting as a solid electrolyte with an extraordinarily high conductivity (0.6 Ω^{-1} cm^{-1} at 800°C), for sodium ions and an electronic conductivity some five orders of magnitude lower. It is a transparent crystalline solid, with the approximate composition, $Na_{1.72}Al_{10.66}Li_{0.30}O_{17}$ and it is chemically stable in metallic sodium at temperatures up to 1100°C. The small lithium oxide component acts to stabilize the crystal structure. This stabilization is also achieved by some manufacturers with the addition of small amounts of magnesium oxide. Polycrystalline BASE in sintered shapes is available commercially as a ceramic in the form of tubes and plates (Ceramatec, Inc., Salt Lake City, Utah).

FEATURES

SHE cells are typically characterized by low voltage, high current output, by overall thermal to electric conversion efficiencies in the 15–30% range, and by relative insensitivity to the heat sink/exhaust temperature as long as that temperature is maintained below approximately 300°C. Because all of the mechanisms contributing to loss and inefficiency scale with the area of the cell electrodes in the same way that the power generated does, SHE/AMTEC efficiency is essentially independent of system size. Uncommon in energy conversion devices, this feature is responsible for the inherent modularity of the AMTEC concept. Large systems can thus be assembled from an appropriate number of smaller modules which can be efficient individually at their own scale. The relatively high exhaust temperature capability allows the "waste" heat to be made available at temperatures such that it can be used for secondary purposes, such as the raising of process steam for industry, or residential and commercial space conditioning. Because these systems operate with no moving parts, they are silent and lifetimes of many years are expected.

OPERATING PRINCIPLES

Figure 1 shows a schematic diagram of a simple SHE from which the electricity generating cycle can be described. A closed container partially filled with liquid sodium is divided into two regions by a pump and a BASE membrane which has on one surface an electronically conducting permeable electrode (PE, generally made of a refractory metal). In this diagram, the upper section of the device is maintained at temperature T_2 by the heat source, while the lower section is maintained at a lower temperature T_1 by a heat sink. The temperature differential imposed between the two regions leads to a difference in the sodium vapor pressure (activity) on the two sides of the BASE membrane.

During operation, sodium travels a closed cycle through the device. Starting in the high temperature

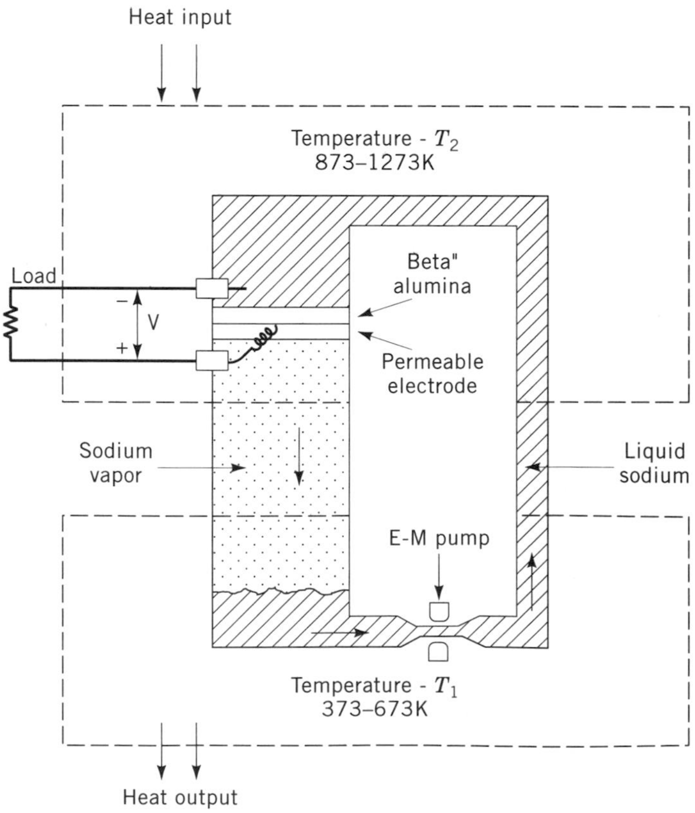

Figure 1. Schematic diagram of a simple SHE.

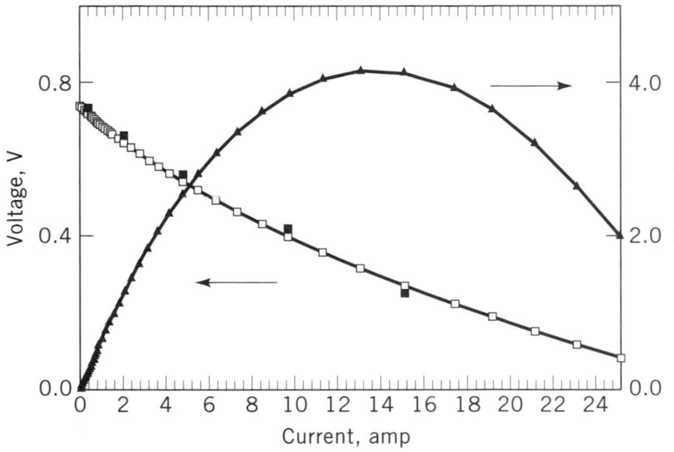

Figure 2. Typical current-voltage and power data for a compact AMTEC cell.

(high pressure) region, the external heat source raises the incoming sodium to a temperature T_2 generally in the range of 650–850°C. In this temperature range the vapor pressure of sodium varies from 7×10^3 Pa to 7.5×10^4 Pa. At the liquid sodium-BASE interface, the sodium ionizes and sodium ions enter and then migrate through the BASE membrane in response to the pressure differential (difference in the Gibbs free energy) across the BASE. Since beta″-alumina has a high sodium ion conductivity and negligible electronic conductivity, a mole of electrons must exit the high temperature zone for each mole of sodium ions entering the beta-alumina. After passing through the external load and delivering the useful work, the electrons are returned on the permeable electrode and recombined with sodium ions at the PE/BASE interface. The neutral sodium atoms then diffuse through the PE and evaporate at a pressure close to P_1 (the vapor pressure corresponding to the condenser temperature T_1) and temperature T_2, then pass through the vapor space to the condenser which is maintained at a temperature T_1 in the range of 120–500°C. At this lower temperature, the sodium vapor pressure is from 10^{-4} Pa to 500 Pa and the sodium condenses as a liquid. It is then returned to the high temperature zone either by an electromagnetic pump or by some form of wicking structure. The cycle is sustained by heat input to the high pressure side and removal of heat from the condenser surface upon which the condensing sodium vapor deposits its latent heat of vaporization. The sodium flow rate is quite small (a current of 1 ampere corresponds to a flow of 1 cm^3/hour) and the required sodium inventory can be as small as 40 cm^3/kWe.

It is remarkable that one of the simplest thermodynamic cycle calculations, for an ideal gas expanding at constant temperature, actually describes SHE operation fairly well. In an ideal gas, there are assumed to be no interactions between the molecules beyond simple collisions and the internal energy of the gas therefore depends only on temperature. Thus in an isothermal process in an ideal gas, no work is expended either in heating the gas (since the temperature is constant, the internal energy remains constant) or in separating the molecules during an expansion. The work performed in a quasireversible cycle (infinitesimal current output) is obtained by integrating the pressure over the volume change corresponding to the end point pressures:

$$Work = \int_{P_1}^{P_2} p\,dv; \quad pv = RT_2 \qquad (1)$$
$$= RT_2 \ln \frac{P(T_2)}{P(T_1)}$$

The unit of charge corresponding to a mole of sodium (or electrons) is the Faraday F = 96,520 coulombs and the work obtained by passing a mole of sodium through a SHE device is therefore just the voltage times the Faraday, F and the open circuit voltage is approximately,

$$Voltage = \frac{RT_2}{F} \ln \frac{P(T_2)}{P(T_1)} \qquad (2)$$

When current is drawn, allowance must be made for the internal resistive losses and for the increase in the pressure at the electrode–electrolyte interface that is necessary to drive the sodium flux through the electrode and across the vacuum space. At practical current densities and heat sink temperatures up to approximately 300°C, this latter pressure usually limits the overall expansion ratio and hence the output voltage. A more detailed analysis of AMTEC operation presented recently (5) has led to engineering models capable of accurate prediction of cell performance. Typical current-voltage and power data for a compact AMTEC cell are shown in Figure 2.

The efficiency of an energy conversion system is usually defined as the ratio of the output work to the input energy. Here the major, parasitic losses are thermal radiation from the hot electrode to the cooler condenser and conduction of heat along the electrical output leads and mechanical structure from the hot cell to the relatively cool world where the electrical output is used. The relationship is

$$Efficiency = \frac{Electrical\ output\ power - Pump\ power}{Output\ power + Latent\ heat + Reheat\ power + Radiation + Conduction}. \qquad (3)$$

SHE/AMTEC systems take advantage of several fortunate circumstances. (*1*) The radiation losses, which could be severe (at 800°C they could reach 7 watts/cm^2 compared to a SHE output of ~1 watt/cm^2), can be greatly reduced due to reflection of the infrared radiation from the sodium film collecting on the condenser surface. (*2*) In contrast to the situation for Seebeck(thermocouple) type generators, conduction losses can be greatly reduced by series connecting several SHE cells at the high temperature T_2, because both the positive (permeable metal) and negative (liquid sodium) electrodes operate nearly at the high temperature, T_2. (*3*) Because the sodium flow is relatively small, an electromagnetic pump or capillary structure uses less than 1/10th of 1% of the SHE output power. Further, because EM pumps are very low impedance devices, the power to run them can be derived conveniently by connecting them in series in the SHE electrical output line without causing important losses. (*4*) While a number of alkali metals can be incorporated into the beta″-alumina structure, the sodium form of beta″-alumina not only has the highest ionic conductivity, but is by far the easiest of the beta-aluminas to make. (*5*) Sodium, a by-product of commercial chlorine production, is abundant and inexpensive.

EARLY EXPERIMENTS

The earliest experiments were carried out by Weber in 1968 and later by his group at Ford from 1974–1981. The results verified the basic understanding of the conversion process and demonstrated fully recirculating operation, self-pumping with an electromagnetic pump and power levels up to 100 watts (6). An efficiency of 19% for a thermal conduction loss-compensated single cell device was also measured (7). During this period a cell was operated under load for over 14,000 hours with no loss of output.

In the early experimental devices sodium recirculation was accomplished using simple electromagnetic pumps as shown in Figure 1. It was recognized that simpler systems could be built if the sodium could be recirculated using a passive, capillary structure. A device based on wick recirculation was tested by Sherrit and Sayer in 1988 (8) but was limited to a peak temperature of about 600°C by the pressure capability of its wick structure. At this temperature with the electrodes available to them, the output power reached was only 0.25 watts for a 120 cm^2 electrode. A simple wick-fed cell design is shown schematically in Figure 3.

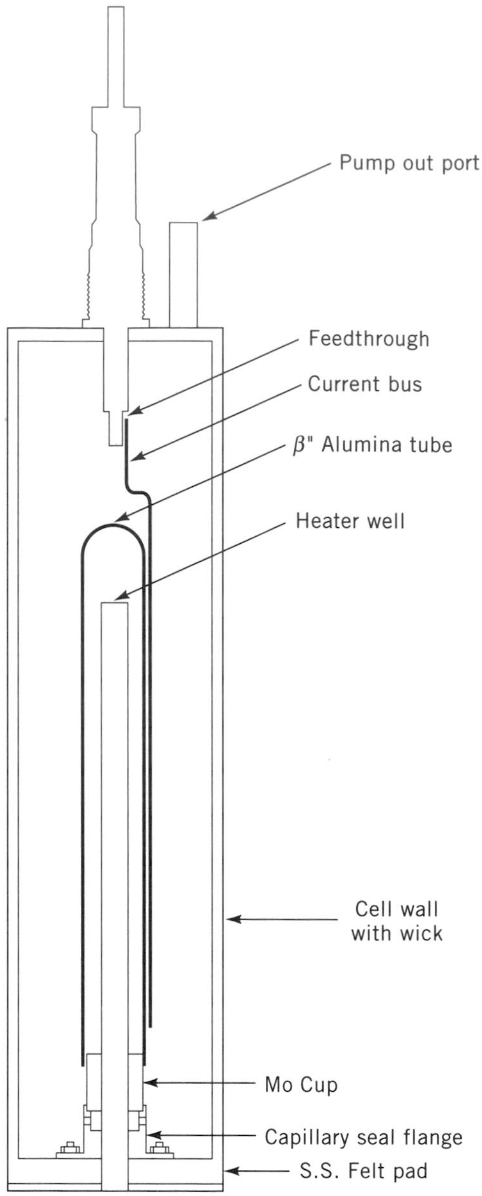

Figure 3. Schematic of a simple wick-fed cell design.

STATE OF ART

State of the art AMTEC designs and test systems use wicks to return the sodium to the hot zone and radiation shields to minimize the direct heat loss from the hot electrode to the cool condenser. Radiation transfer is the major source of inefficiency in these devices. Recent developments have produced wick structures capable of recirculating sodium to the hot zone at temperatures up to 750°C. At this temperature, conversion efficiencies up to 18% and power densities up to 170 watts/kg are achievable in high power devices assembled with such cells. Higher temperatures enabled by improved wicks will lead directly to even higher efficiencies and power densities. Cells of this remotely condensed, wick return type have operated under continuous load for over 8400 hours without degradation of output (9) and appear to be mechanically rugged. In tests, a compact cell design prepared for space power applications has withstood random acceleration loads up to 18 G (rms) and shock loads up to 3,000 G at 1000 Hz.

FUTURE DEVELOPMENTS

The direction of future AMTEC development appears clear. At the cell level, large numbers of BASE tubes, series connected within a single vacuum shell with the sodium delivered by vapor transport, will be used to form higher voltage units. An example of a multitube module is shown in Figure 4. With higher voltages, these modules will have reduced output lead size and shorter cell interconnects leading to reduced thermal conduction losses, higher efficiency, and lower mass. Use of high temperature vacuum multifoil insulation, BASE tubes with enhanced surface area, and optimized remote condensing will lead to system efficiencies above 25% with power densities above 250 watts/kg and 140 watts/liter. System exhaust temperatures will be made adjustable and regulated passively over the range from 150°C to 500°C to serve a variety of cogeneration applications. A variety of heat input methods will be employed. These will include the catalytic combustion of liquid fuels and natural gas, parabolic solar concentrators, and conventional burners for home furnace applications. AMTEC will become a candidate for use as a bottoming cycle for certain industrial processes whose exhaust temperatures are suitably high.

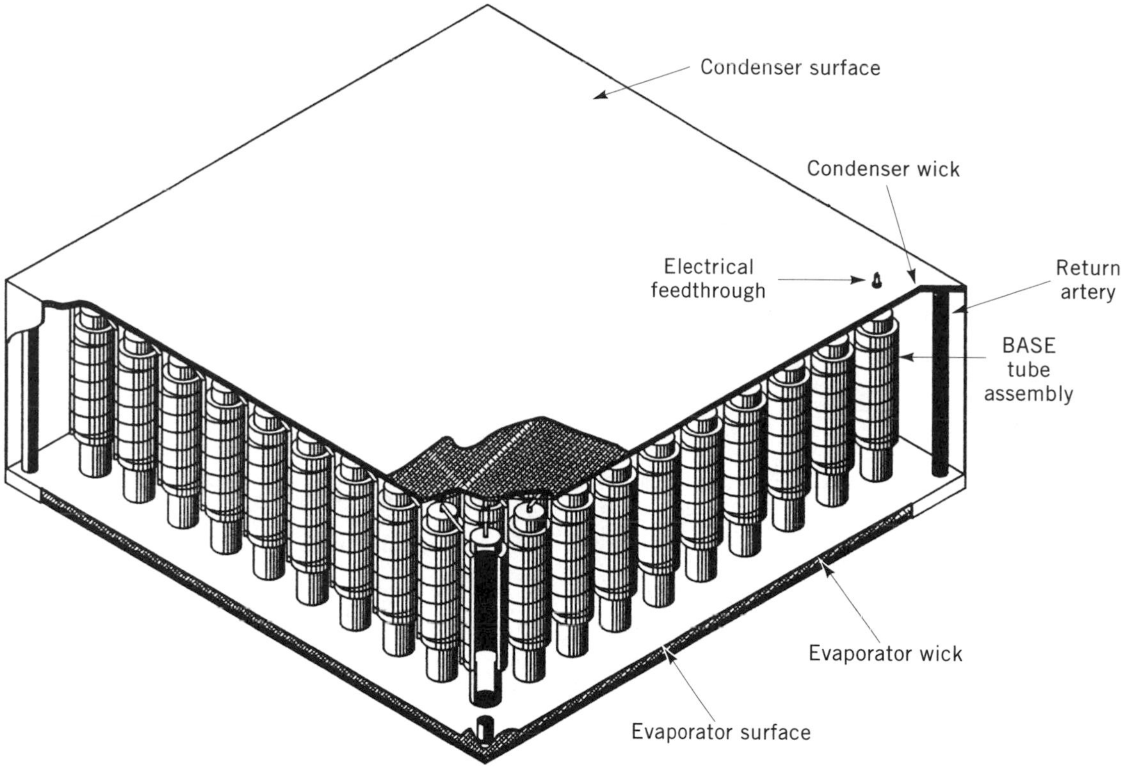

Figure 4. A multitube module.

AMTEC will become cost-competitive for many applications. BASE tube costs are approximately \$30/tube at the production levels of current manufacturers, bringing system costs to near \$1500/kWe. Reduction in BASE costs to \$5/tube as projected by all the major sodium sulfur battery developers will lower AMTEC system costs to ~\$700/kWe. Comparable costs for central station electric generating capacity are now on the order of \$1000/kWe. At the projected efficiencies, lifetimes and costs, SHE/AMTEC systems are expected to find widespread use for both terrestrial and space power applications.

BIBLIOGRAPHY

1. J. L. Sudworth and A. R. Tilley, eds., *The Sodium Sulfur Battery,* Chapman and Hall, New York, 1985.
2. U.S. Pat. 3,404,036 (July 29, 1969), J. T. Kummer and N. Weber (to Ford Motor Company).
3. N. Weber, *Energy Convers.* **14,** 1–8, (1974).
4. J. T. Kummer, *Progress in Solid State Chemistry,* Pergamon Press, 1972, Chapt. 5.
5. R. M. Williams and co-workers, *J. Electrochem. Soc.* **137,** 1709–1716 (1990); R. M. Williams and co-workers, *J. Electrochem. Soc.* **137,** 1716–1723 (1990).
6. T. K. Hunt, Neill Weber, and Terry Cole, "Research on the Sodium Heat Engine," *Proc. 13th Intersociety Energy Conversion Engineering Conference (IECEC),* San Diego, Calif., August 20–25, 1978, pp. 2011–2017.
7. T. K. Hunt, N. Weber, and T. Cole, *Solid State Ionics* **5,** 263–266, 1981.
8. S. Sherrit, M. Sayer, and B. Kindl, "Electrode Systems and Heat Transfer in Thermoelectric Generator Design," *Proc. 23rd IECEC,* 1988, vol. 1, pp. 241–247.
9. T. K. Hunt and co-workers, "Small Capillary Pumped AMTEC Systems," *Proceedings of the 10th Symposium on Space Nuclear Power and Propulsion,* Albuquerque, N. Mex., 1993, vol. 2, p. 891.

SOLAR CELLS

Dan Halacy
Lakewood, Colorado

PHOTOVOLTAIC CELLS

A History of Photovoltaics

In 1822, German scientist Thomas Johann Seebeck discovered the photovoltaic (PV) phenomenon but considered it simply a form of magnetism. Even when Jean Charles Peltier later reversed the Seebeck effect and made heat (or cold) from electricity, PVs remained a laboratory curiosity. Edmond Becquerel in 1839 was first to comprehend the physical phenomena we now make use of in transforming solar energy directly into electric power.

Some progress in PVs came in 1930, when Russia's Abram Joffe made advances with thermoelectricity to the point of coupling solar collectors to his generators to produce electricity from sunshine. A quarter of a century later, Bell Laboratories initiated the modern PV age with its 4% efficient single-crystal silicon solar cell. Bell unveiled the device to the scientific community in 1955 at the First World Symposium on Solar Energy in Phoenix, Arizona. The first use by a utility of photovoltaics was for powering rural telephone lines in Americus, Georgia, in 1955 (Fig. 1).

By June of 1957, a small array of these solar cells had been mounted on a U.S. Navy Aerobee-Hi rocket to permit telemetry of its flight. The U.S. Army's Vanguard TV-4 satellite was next, with a slightly larger array of Bell solar cells. Vanguard transmitted so successfully that for many years its signals were considered a nuisance to space communications (1).

Except for a few that used nuclear power, all spacecraft soon relied on PV cells for electricity, since there were no fuel-based competitors for these applications. Photovoltaic cells also receive more solar energy in the absence of an atmosphere. But the space PV market required only about 80 kW a year and only two solar cell manufacturers survived this period.

During the oil price rise in the early 1970s, the U.S. government began an accelerated program of renewable-energy development, including PVs. By 1975, government research and development (R&D) programs accelerated the development of PV cells economical enough for some terrestrial power needs. Japan, with its Sunshine Project, and several European countries also began serious PV research and development.

A Presidential Initiative in 1978 raised the U.S. PV R&D budget from $6 million to $30 million. Establishment of the Energy Research and Development Agency [ERDA, (which later became the U.S. Department of Energy) (DOE)] and the Solar Energy Research Institute [now the National Renewable Energy Laboratory (NREL)] further accelerated PV development among other technologies.

Figure 1. Photovoltaic cells power rural phone lines in Georgia. Courtesy of Bell Telephone Laboratories.

By 1992, after the U.S. government had invested more than $1.2 billion in PV development and Japanese and European countries had spent a comparable amount, production plants around the world were producing about 60 MW of PV modules a year. Most of these were used in remote, stand-alone applications. The U.S. private sector's investment has been more than double that of the government during the past 15 years.

See also RENEWABLE ENERGY TECHNOLOGIES; COMMERCIAL AVAILABILITY OF ENERGY TECHNOLOGY; BUILDING SYSTEMS.

Potential of PV Electricity

Annual mean daily intensity of sunlight reaching the Earth is about 5.3 kW·h/m^2, varying from 3.3 to 6.7 kW·/m^2 depending on geographical location. At noon on a clear day, about 1000 W/m^2 of sunlight reaches the Earth's surface. Production of 100 MW of PV electric power thus requires only about one square mile of land. On an annual basis, U.S. sunlight averages about 2.4 MW·h/m^2 on a tracking PV array. Almost 90% of the country is within 25% of this amount. This excellent sunlight resource has positive implications for the use of solar energy (see Fig. 2).

The United States presently uses about 2.4 trillion kW·h of electricity a year. Generating this amount with PV modules would require about 30,000 km^2 (allowing space between modules to prevent shadowing losses). This area is equivalent to only about 0.3% of the 9 million km^2 our country covers. Put another way, about 20% of the land area of Nevada would suffice. Twenty times that much land is presently set aside in the United States to prevent overproduction crops (2). The difference in solar radiation between the sunniest parts of our country and those with the least sunshine varies by a factor of about 2. Coincidentally, that closely matches the range in utility rates for conventionally produced electricity in the United States.

Photovoltaic plants should be sited in many high-insolation areas of the United States to be closer to electric power markets. Land costs do not adversely affect PV projects until they reach about $50,000 per acre. Distributed in this way they will have the potential of serving all U.S. needs economically.

Present Status

In the nearly four decades since Bell Laboratory's invention, millions of PV modules have been used in hundreds of different applications. These range from spacecraft to earth-bound calculators and computers, to residential electric power rooftop arrays, utility power plants, PV-powered vehicles, and even a few man-carrying aircraft. There are many thousands of PV arrays in operation, manufactured by commercial firms worldwide. Costs have dropped more than 20-fold since the 1950s, and the United States continues to expand its share of the growing PV market.

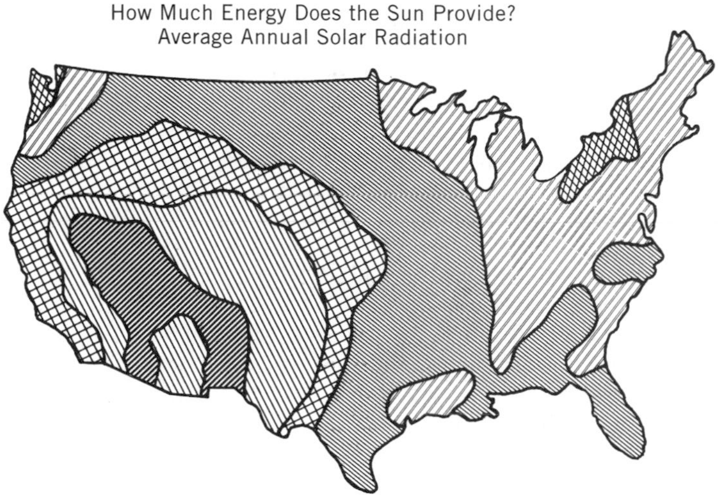

Figure 2. Average annual solar radiation. Courtesy of NREL.

Megawatt-hours per square meter

1.6–1.8	2.6–3.0
1.8–2.2	3.0–3.4
2.2–2.6	3.4–3.8

U.S. Government Involvement

The Department of Defense (DOV) is the largest consumer of electric power in the United States, and the DOE projects deployment of more than 100 MW of additional PV power for the DOD in the 1990s. The DOE is also working with the U.S. Agency for International Development to provide technical support to Mexico in implementing a rural electrification program, emphasizing renewable energy, including PVs. This effort will be expanded to other Central American, South American, and Caribbean countries, including Guatemala, Costa Rica, and the Dominican Republic. Through its Sandia National Laboratories, the DOE also provides design assistance for more than 100 PV projects in the United States, the Pacific Rim, Latin America, the Caribbean, and Africa (3).

The DOE 1991–1995 PV Plan. Aided by the spending of nearly \$4 billion by government and industry in the last two decades, PV module prices dropped to \$4.00 from \$4.50 per watt in 1990, resulting in a significant decrease in the costs of installing PV electric power (4). The price of PV electricity has dropped to about 30 cents/kW·h. U.S. companies continue to lead the world in the production and sale of PV cells and modules, with a 32% share of the global market. Sales of U.S.-manufactured PV modules totaled nearly 20 MW in 1992 (4). Increased federal R&D support for manufacturers could result in faster price reductions and greater market impact for PV power (Fig. 3).

With its 1991–1995 PV program plan, the DOE expects to further reduce PV electricity costs to about 12–15 cents/kW·h by the late 1990s and as low as 6 cents/kW·h in the early decades of the next century. This will match conventional utility power costs and provide the added benefits of a cleaner, safer environment (4). Unlike the previous 5-year plan, which emphasized PV materials and cells research, the 1991–1995 DOE plan aims primarily at accelerating broad commercial acceptance of PVs as a cost-competitive alternative source of electricity, rapidly transferring the technology to a growing PV industry serving electric utilities and other markets globally. The probably overoptimistic goals of the plan include the following:

- Increase of PV module efficiencies to 10–20% for flat-plate and 15–25% for concentrator technologies.
- Installation of 1000 MW of PV capacity in the United States by the year 2000.
- Installation of at least 500 additional megawatts of PV capacity internationally by the year 2000.

Utility PV Projects

The Electric Power Research Institute (EPRI) and many individual electric utility companies have been major factors in the rapid technical development and production of PV cells as well as for a broad range of commercial applications for them (5). Large corporate sponsors, including several major oil companies, have invested more than \$1 billion for PV research and development. In 1993 there were about 70 manufacturers of PV cells worldwide; 35 of them were in the United States.

There are presently more than 100 utility PV demonstration or research power plants in operation. More than 35 electric utilities in 20 states have cost-effective PV systems in service. Some of these have operated for more than a decade; Public Service Company of New Hampshire recently replaced modules powering a remote micro-

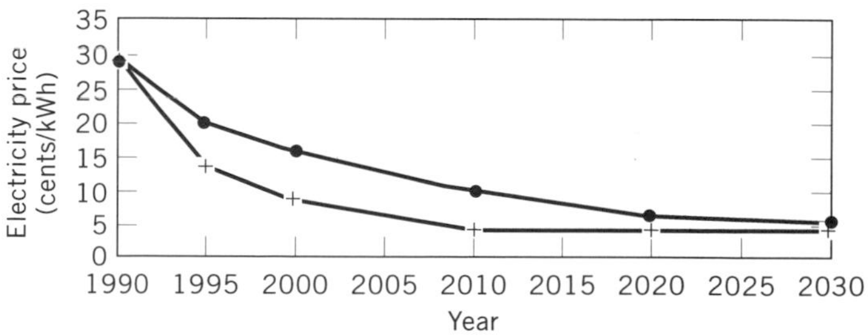

Figure 3. National energy strategy projected price of PVs. Courtesy of NREL. ——, Baseline budget; —+— R&D Market stimulus.

wave facility after 12 years of satisfactory operation in rigorous environmental conditions. It has been found that for remote sites requiring 300 kW·h or more per month PV arrays are cheaper than a 3000-m utility grid connection.

Many utilities are involved in the design and development of next-generation PV power plants, among them Arizona Public Service, Pacific Gas & Electric, Sacramento Municipal Utility District, Southern California Edison, and Idaho Power. Edison and GM Hughes Electronics are part of the PV industry's effort to reduce installed costs of PV balance of systems (Table 1).

The Coast Guard operates more than 10,000 PV-powered navigation aids throughout the United States. More than 500 PV-powered remote watering systems for farms and ranches are installed and operating. A Pacific Gas & Electric Company survey found about 3500 PV standalones in California by the end of 1989 and estimated them to be growing at about 30% a year. This suggests more than 8000 California PV installations in 1993. The telecommunications industry has been successfully installing PV systems around the world for more than a decade, many of them in harsh environments. Photovoltaic systems, being entirely electrical, actually function better in low temperatures.

Austin's 300-kW PV Plant. In 1986, the City of Austin's Electric Utility Department began planning for the additional 125 MW of capacity it will need by 2001. Photovoltaic power was chosen because of low maintenance costs, the coincidence of solar power production with utility peak demand periods, environmental considerations, and an awareness of high interest in renewable energy use among its 300,000 rate payers.

As a result, Austin constructed PV 300, a 300-kW PV array. Two roof-mounted 1-kW systems were also built so that the utility could evaluate small stand-alone residential applications as well as grid-connected users. Two years of testing proved good coincidence with electric power demand profile and excellent reliability and low maintenance costs for the PV arrays. However, the capital costs of fixed-array grid-connected PV systems remained high. Further studies were made, this time comparing the costs of future fixed-tilt and single-axis-tracking PV plants of 25, 125, and 250 MW output with costs of atmospheric fluidized-bed or combined-cycle gas turbine units. These studies showed that the two larger PV plants would result in capacity benefits from deferring fossil fuel plants at least a year. Fossil plants have high fuel costs and, in the event of penalties for SO_2 and other emissions, could be considerably more expensive than future, lower cost PV electric power.

Austin calculated the total environmental benefits for PVs at an installed cost of $175/kW in 2001 dollars. Projected break-even costs of single-axis tracking PV systems in that year ranged from $876 to $3885/kW, depending on system size, type of conventional plant deferred, environmental constraints, future fuel costs, and plant design but not including possible savings from the deferral of new T&D lines. Reduction of PV electricity cost to 10 cents/kW·h (equivalent to about $1700 per installed kilowatt), expected early in the next century, will open up large new markets, including peaking power production, PV/diesel/battery hybrid systems, and remote power plants. Including the undeveloped world as a market, these PV applications represent potential multibillion-dollar global sales. By 2010, PVs could be producing bulk power for all utilities situated in areas with sufficient sunlight (6).

Table 1. Representative Utility PV Systems

Utility	Number of Facilities
Pacific Gas & Electric, California	>1100
Florida Power Corporation	>90
The Salt River Project	>80
Arizona Public Service	11
Austin, Texas	3
U.S. Coast Guard	>10,000

From Ref. 5.

Photovoltaics for Utility-Scale Applications. The DOE, EPRI, California Energy Commission, Pacific Gas & Electric Company, and nine other U.S. utilities have set up a large test facility called Photovoltaics for Utility-Scale Applications (PVUSA). This facility initially installed and tested 20-kW PV arrays using various PV modules in Davis, California, and on the island of Maui. Later PVUSA began installing 200- and 400-kW arrays at the Davis site. Other significant PV demonstration projects include the Sacramento Municipal Utility District's 2 MW of PV modules and large PV arrays at Austin, Texas, Idaho Power, and Niagara Mohawk (6).

Residential PV Applications

Homes powered by PV electricity were developed early and have been built by owners, homebuilders, and utilities as well as through government programs (7). The first American PV residence was the Carlisle home, designed by Solar Design Associates of Lincoln, Massachusetts, and built in 1982 in Carlisle, Massachusetts (Fig. 4). The PV array on the roof produced 7.5 peak kilowatts on a sunny day and the owner sold excess electric power to the local utility.

Another early residential PV power installation, the 1980 John Long/Arizona Public Service Utility intertie model home in Glendale, Arizona, led to an all-PV neighborhood of 24 custom homes with PV roofs included in the purchase price and deeded by the buyers.

San Diego Gas & Electric Company (SDG&E) has been installing PV residential systems in the San Diego area since 1986. They consider rooftops as "free land" for power generation. Photovoltaics are of special importance in San Diego since the utility's summer-peaking load coincides with the season of maximum sunshine. The SDG&E's first PV project included 36 town homes in Laguna del Mar, with 1-kW, single-crystal silicon arrays installed by the developer and owned by the homebuyer. Surplus electric power was sold back to the utility at retail cost.

Typical peak output in summer was 700–800 W per system, benefiting peak load with the power produced di-

Figure 4. The Carlisle House. Courtesy of Steven Strong.

vided equally between the homeowner and the utility. There was no peaking power for SDG&E during winter months. The utility's initial concerns about harmonics and the possibility of "islanding" or "run-on" did not develop into a safety problem.

Although standard Public Utilities Regulatory and Policy Act (PURPA) interconnection agreements were available, SDG&E further clarified and simplified these to avoid possible income tax liabilities and additional insurance requirements for homeowners. It was also necessary to educate insurance brokers about the safety of PV systems.

The utility has had no serious problems with its residential PV systems. The SDG&E also considers PV systems benign with respect to power quality and safety concerns and sees bureaucratic barriers as all that remain to be eliminated for widespread residential PVs. The utility estimates that south-facing roofs of moderately priced and higher priced homes in San Diego represent about 2000 MW of potential electric power. Assuming a 1-kW PV array per residence, this translates to a potential of 2 million PV-powered homes.

Since the mid-1980s, The New England Electric System (NEES) has operated a residential PV project in two of its subsidiaries: the Massachusetts Electric Company (MECo) and the Narragansett Electric Company (NECo) in Rhode Island. Photovoltaic arrays 22.3 m^2 (240 ft^2) in area were installed on homes with southern exposure in Gardner, Massachusetts. All receive electric power from the same substation through one phase of a distribution feeder. Electric power is metered in either direction depending on whether the homes have sufficient power or not, with homeowners credited for any excess energy routed to other users. This project documented the performance, reliability, and cost-effectiveness of 30 residential PV systems on a single distribution feeder.

Most of the NEES homes are 102 m^2 (1100 ft^2) ranch-style, with an average south-facing roof pitch of 23°. Six had window-mounted air conditioners, and baseboard heat is used in 11 of them. None used central air conditioning. The 11% efficiency crystalline silicon cells were manufactured by Mobil Solar Energy Corporation and mounted in 1.2 × 1.8-m (4 × 6-ft) aluminum panels bolted to the roofs. Connected in series, the modules delivered 220 W per module at 250 V dc. Fully automatic inverters provide 240-V ac power. The PV arrays start automatically when there is sufficient sunlight, shut down in its absence, and automatically disconnect from the house service panel upon loss of utility power. The system is effective in providing summer peaking power, reaching maximum output shortly after the NEES reaches peak load.

The 30 PV-powered NEES homes (Fig. 5) can supply electricity as needed to another 25 homes without PV systems from noon until 2 p.m. on clear summer days, helping the utility with peaking power and reducing the homeowner's summer electricity bills by about 50%. Typical annual electric power production is 2200 kW·h, ranging from about 50 kW·h/month in midwinter to 270 kW·h/month in midsummer. The NEES also installed PV arrays on five institutional or commercial buildings and three institutional sites in Rhode Island. All of these buildings use considerably more electricity than do the residences.

In another Massachusetts PV project, the Boston Edison electric utility developed its own "Impact 2000" model PV home with a much larger 4-kW array on the roof.

Government Residential PV Projects. In the 1980s, the Department of Energy established two PV residential test facilities: (1) The Southwest Residential Experimental Station in Las Cruces, New Mexico, and (2) the Northeast Photovoltaic Residence Experimental Station and the Southeast Residential Experimental Station (SERES) at the Florida Solar Energy Center at Cape Canaveral, Florida.

Photovoltaic-powered Vehicles

A variety of demonstration vehicles have been powered with PV cells, including the SunRaycer built by Paul

Figure 5. The New England Electric System PV homes. Courtesy of NREL.

Macready, which set a world speed record of 49 miles an hour on an Arizona racetrack. Macready also built the world's first PV-powered airplane, Solar Challenger, which flew from France to England. This aircraft used gallium arsenide PV cells, designed for space power uses.

On June 20, 1993, 34 PV-powered cars built by university students left Arlington, Texas for the 7-day Sun Race to Minneapolis/St. Paul. Weighing from 181 to 454 kg (400–1000 lb) and equipped with 1000- to 1500-W electric motors, these solar racers cost as much as $200,000. Although the cars averaged more than 80 km/h (50 mph) [with top speeds of more than 145 km/h (90 mph) during the 160-km (100-mile) qualifying races], rainy weather kept average speeds between 40 and 64 km (25 and 40 miles) an hour for the first 4 days. The University of Michigan's entry finished first, followed by those of California State Polytechnic and California State (Fig. 6.) (8).

Commuter Solar-Electric Vehicles

More practical is the solar-powered electric commuter car. Charged by PV panels during the day and off-peak electricity at night, this type of vehicle has been suggested for solving transportation and environmental issues at the same time. Forward-looking Europeans have for several years commuted in electric cars, recharging them with PV panels at home and in some cases also at work. A few American drivers are beginning to follow suit.

In March of 1993 Southern California Edison and the South Coast Air Quality Management District opened Edison's first PV-powered Solar Chargeport. This facility charges as many as 12 "impact"-class electric vehicles at a time.

Edison has also built the world's largest fully transportable PV powerplant in the United States presently in-

Figure 6. University of Michigan's solar car wins Sunrayce 1993. Courtesy of NREL.

stalled at California State University at Fullerton's Desert Studies Center near Baker. Replacing diesel generators previously used at the center, the mobile solar powerplant consists of 10 kW of PV modules plus 5180 W·h of sealed batteries to provide electric power day and night for lighting and other electrical requirements for the Studies Center. Edison is seeking regulatory approval of a rate base so it can begin offering portable powerplants to other remote locations not served by utility power lines.

FOREIGN INVOLVEMENT IN PVs

Austria has prohibited nuclear power plants and uses a large amount of hydropower to produce electricity (7). However, 40% of the country's electricity is still produced by imported fossil fuels. A national PV program to introduce more PV utility power is coordinated by the Austrian Federal Electricity Board and a number of provincial and municipal utilities.

Germany encourages the independent production of electric power and requires utilities to pay competitive prices to producers using renewable energy. The country's Federal Ministry for Research and Technology is implementing a program to install 1500 grid-connected rooftop PV arrays producing from 1 to 5 kW. Qualifying homeowners pay between 25 and 50% of installed costs, depending on state subsidies.

Japan pioneered residential PVs with its Sanyo house many years ago, and its Central Research Institute for the Electric Power Industry (CRIEPI) believes that residential PVs could be cheaper than utility power by the year 2000. The CRIEPI also says that PV central utility power will not be cost-effective in Japan until well into the twenty-first century because very high land prices add about \$2000/kW to installed costs. In addition to being a large manufacturer of PV cells, Japan is very active in the development of PV electric utility projects as well as stand-alones, including the pioneering Sanyo Home.

Land costs are a problem for large utility PV plants in Japan, resulting in a sizable number of residential PV installations, although the reverse flow of electric power is prohibited by law. With a goal of 2% of all electric needs met with residential PV installations, the government is providing an attractive set of financial incentives, including tax credits and low-interest loans. Residential PV arrays are projected to be cheaper than grid power by the year 2000.

In Switzerland private PV systems are allowed grid connection by law, and utilities must pay nearly peak load prices (9–12 cents/kW·h) for such electricity fed into the grid. Ten large Swiss utilities participate in these programs (7).

In 1990, Austria, Germany, Italy, and Japan all had PV research budgets higher than those of the United States.

Third World Photovoltaic Projects

Brazil/NREL Rural Electrification Project. In July 1992, during the Global Summit held in Brazil, the DOE's NREL in Golden, Colorado, announced the Brazilian Rural Electrification Project just agreed upon with Brazil (9). This electrification project will provide PV arrays for 1000 remotely located homes in Pernambuco and another 1000 homes plus a model village in the state of Ceara. Including back-up batteries, the project will cost about \$1000 per home, instead of the \$9000 or so per home required to extend electric power lines from the existing grid.

On January 26, 1993, dedication services were held commemorating the wiring of the first of nearly 1000 rural homes to be supplied with U.S.-manufactured PV modules. Phase 1 will wire more than 400 homes in 14 villages in the Brazilian state of Ceara for electric lights, together with power for refrigeration and educational television in village schools. The PV-powered street lights will illuminate village common areas and rural clinics will get PV power for lighting and vaccine refrigeration. The project is part of the DOE's Solar 2000 strategy to develop new markets for renewable energy technologies.

The ultimate goal is to electrify 500,000 homes, schools, and clinics in Brazil, thus opening up a substantial market for the U.S. PV industry. The PV panels for this first phase are of crystalline silicon, manufactured by Siemens Solar Industries in Camarillo, California.

The PV electrification projects are proceeding in many other countries as well. By 1990, Colombia had 17,000 small PV stand-alones operating and South Africa had installed about 700 kW of PVs. Only about one-third of Indonesia's scattered island villages were electrified, but the country planned to power an additional 12,000 villages with small PV plants in 1993.

THE PV INDUSTRY

Pioneered in the United States, PV production long ago spread overseas. The United States, Japan, and Germany, Italy, and Spain in Europe produce most of the PV modules. Algeria, Brazil, India, and Venezuela manufacture commercial PV modules on a smaller scale. With lower production costs, these latter countries could compete in the rapidly growing global PV market (see Fig. 7).

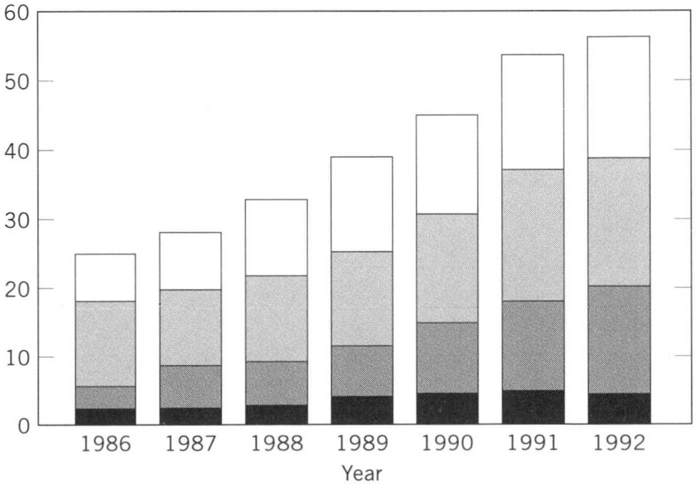

Figure 7. World PV module shipments, consumer and commercial MW. Courtesy of NREL. □, U.S.; ▩, Japan; ▩, Europe; ■, Rest of the world.

Table 2. Percentage of U.S. Sales of PV Products by Category (1990)

Calculators, etc.	22%
Commercial systems	21%
Grid systems	10%
Buoys, etc.	9%
Pumps	6%
Other	30%

Table 2 shows percentage of world shipments of PV modules approaching 60 MW. Some estimates suggest that by the year 2030, the United States may have more than 100,000 MW of installed PV capacity. This represents cumulative sales of PV equipment of more than $200 billion (3).

Manufacturing

In 1992 there were 32 PV module manufacturers in the United States, of a total of 69 in the world. In the United States annual manufacturing capacity reached 17 MW of a world production of 47 MW. The PV market consists principally of consumer products, remote stand-alone electric power systems, residential PVs, and utility applications.

DEPARTMENT OF ENERGY'S PV BONUS PROGRAM

In late 1992, the DOE began a new solar energy initiative called Building Opportunities in the U.S. for Photovoltaics (PV BONUS for short.) This program is a 5-year, $25-million cost-shared initiative to disseminate PV skills into schools of architecture and industry, toward the accelerated use of PV technology and development of products and markets for a variety of applications (Fig. 8).

Presently involved in the program are Advanced Photovoltaic Systems, Delmarva Power & Light, Energy Conversion Devices, FIR Solar Technology, and Solar Design Associates. Table 3 lists preliminary responses to the PV BONUS solicitation.

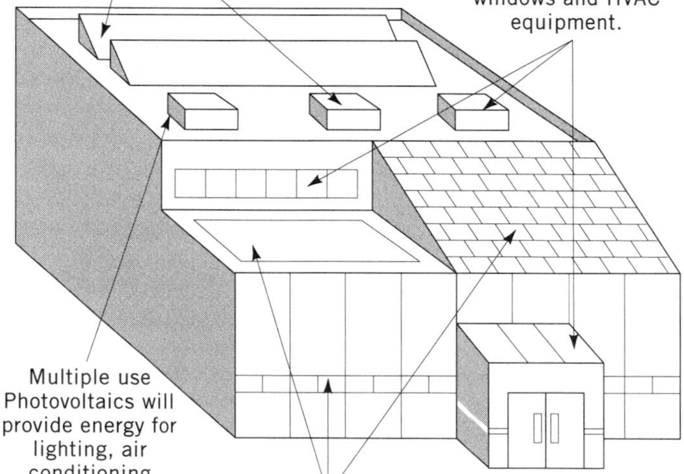
Demand-side management
Photovoltaics can reduce the peak load of a building to the electric utility.
Control
Photovoltaics can control active (electrochromic) windows and HVAC equipment.
Multiple use
Photovoltaics will provide energy for lighting, air conditioning and fans (HVAC) equipment), and/or storage.
Architecture
Photovoltaics can form part of the roof wall, or windows of a building.

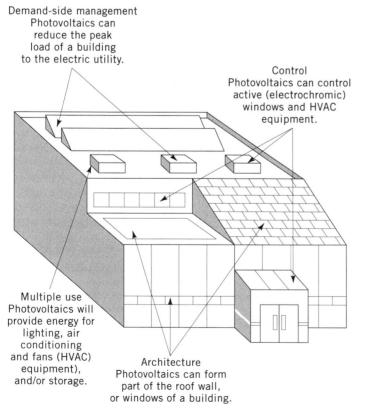

Figure 8. Potential applications in PV-BONUS solicitation.

PHOTOVOLTAIC CELL TECHNOLOGIES

Since the original Bell silicon solar cells of 1954, many different PV cell technologies using a broad variety of materials and fabrication methods have been developed. As noted above, most of today's solar cells are made of single-crystal silicon. Expensive crystalline PV cells still accounted for almost half the total, with low-cost amorphous silicon cells taking a smaller share of the market. As of 1993, the efficiency of these commercial modules range from 11 to 17%, with individual cells having attained a record 34%. But electricity from crystalline PV cells still costs three to seven times as much as conventional utility electricity, and PVs are generally economical only in remote areas or for special tasks. However, with industry investment exceeding $100 million annually, improvements in cell technology and reduced costs of production are expected to become cost competitive with fossil-fueled power plants early in the next century (10).

Crystalline Silicon

First used 40 years ago by Bell Laboratories, this is the most mature PV material. Since Bell's pioneering 4% cells

Table 3. Preliminary Responses to PV BONUS Solicitation

PV building products
Curtain walls
Blinds
Greenhouses
Dynamic facades
Roofing systems
Variable transmission windows
Smart windows
Roofs for manufactured homes
PV hybrid products
AC lighting
DC lighting
Air handling
Condenser fans
Heat pumps
Air cleaner
Vent fans
Applications
Gas stations
Fast food restaurants
Pay phones
EV charging stations
Residential/commercial development
Parking lots and walkways

in 1955, improved silicon cells have achieved efficiencies as high as 23%. The cost of the high-purity silicon used to make them has also been reduced. However, crystalline PV materials of several hundred micrometers resulted in very high manufacturing costs. Achievements over the last two decades include:

- Finding semiconductor materials that absorb sunlight in 1-μm layers.
- Depositing these films on readily available, low-cost substrates, including plastic, glass, or metal foil.
- Fabricating entire modules of interconnected solar cells during film deposition to reduce manufacturing costs.
- Achieving efficient conversion of solar energy to electricity.

Thin-Film Cells

Some researchers believe the future of commercial PV lies with *thin-film* amorphous solar cells. In 1993, the front-running PV thin-film materials appeared to be:

Thin-film crystalline silicon
Amorphous silicon (ASi)
Copper indium diselenide (CIS)
Cadmium telluride (CdTe)
Gallium arsenide (GaAs)

These cells yield efficiencies in the 10% plus range, although amorphous silicon degrades between 10 and 25% (to 7.5 to 9%) during the first month or so of operation before stabilizing (Fig. 9). In theory, large PV modules should be as efficient as small ones, but such is not yet the case. Researchers at the NREL believe that present manufacturing problems can be solved and lead to sizable 15% efficiency modules. However, only CIS is currently produced in power modules as large as 1.2 m^2 (4 ft^2).

Industry goals for 1995 include 10% efficiency and module cost of \$100/m^2, reducing the price of PV electricity to about half the present 35 cents/kW·h. Further reduction to between 10 and 15 cents is expected by the year 2000 (2).

Performance of Thin-Film PV Cells. ***Thin-Film Crystalline Silicon.*** Thin-film crystalline silicon cells are only 100 μm thick, rather than the 200–300 μm of conventional crystalline cells (2). These cells are fabricated by depositing crystalline silicon on low-cost substrates. The result is thinner cells at very low manufacturing costs and with efficiencies in the range of 15%. Conventional crystalline-silicon modules cost from \$300 to \$500 per square meter. Thin-film modules are expected to be from 2 to 10 times less expensive.

Amorphous Silicon (A-Si). Photovoltaic cells made from hydrogenated amorphous silicon are a leading alternative to crystalline silicon (2). First produced in 1974 with an efficiency of only 1%, in 1993 these single-junction devices yield efficiencies of about 12% but initially lose several percent before stabilizing. Multijunction A-Si cells have reached 13%, and very large modules exceeding 1 m^2 (10

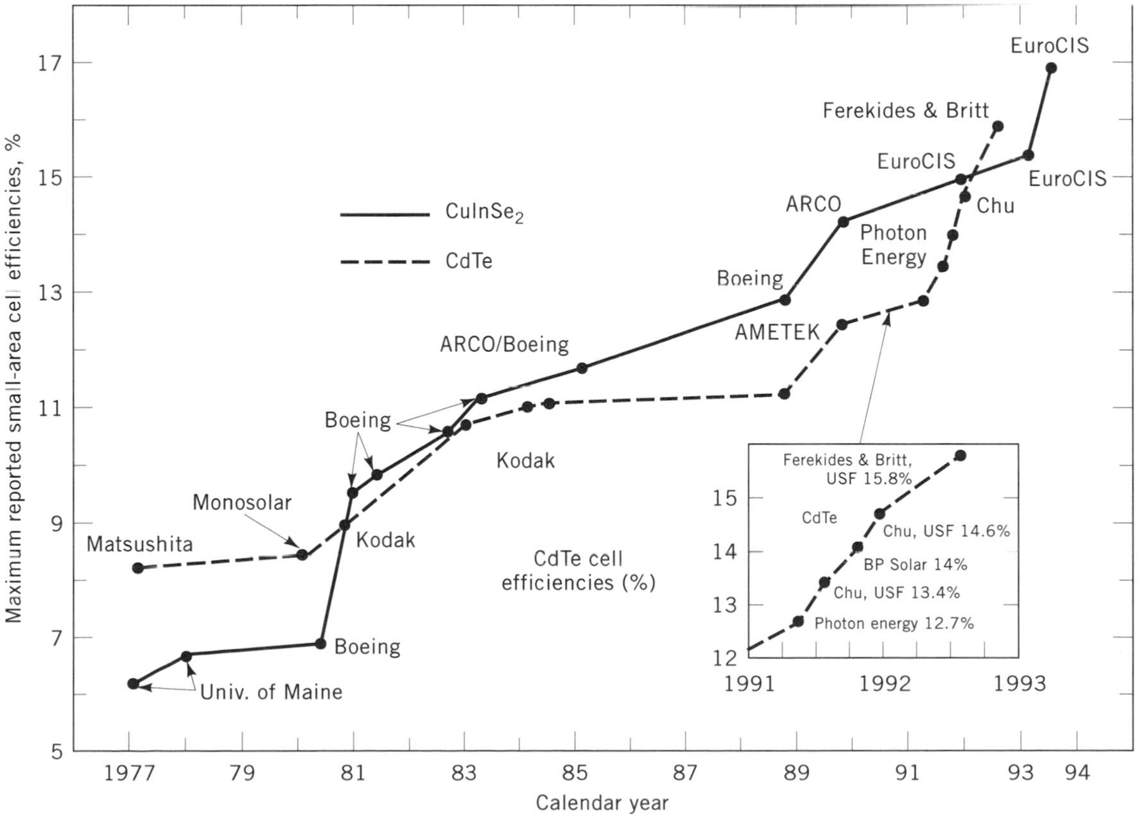

Figure 9. Polycrystalline thin-film efficiencies. Courtesy of NREL.

s ft^2) offer 4–5% efficiency. These lowest cost cells presently make up about 30% of the world market for PV devices.

Gallium Arsenide (GaAs). These cells have the highest efficiencies, near 25% at one-sun conditions and more than 30% under concentrated sunlight (2). The cells can also be made on germanium or silicon substrates at lower efficiencies and can be tailored to produce multijunction devices of very high efficiency. Individual multijunction GaAs cells have reached 31–34% efficiency under 100-sun concentration in the laboratory.

Copper Indium Diselenide (CIS). These polycrystalline thin films have yielded 14% cell efficiency and a submodule efficiency of 11% (2). Some combinations of CIS and amorphous silicon have reached 14.6%.

Cadmium Telluride (CdTe). These polycrystalline thin films have reached 12% cell efficiency and submodule efficiency of 7% (2). Low-cost manufacturing techniques including electrodeposition, screening, and screen printing make them very attractive, and several U.S. companies are developing optimum fabrication methods. Recent progress in laboratory thin-film cells has been extraordinary.

Cadmium Selenide (CdS). To investigate the scale-up of its new method of making CdS cells, the NREL fabricated a 6.6-cm^2 device (2). This 13.9% efficient cell is the most efficient thin-film cell of its size and indicates that the process has good potential for scale-up to larger areas.

Early in 1993, the NREL awarded contracts to six U.S. PV manufacturing companies for continued development of thin-film PV technologies. Five of the six manufacture amorphous silicon PV arrays, the lowest cost cell technology. British Petroleum Solar, Golden Photon, Matushita, and Solar Cells Incorporated are working on cadmium telluride thin films. Copper indium diselenide thin-film devices are being developed by Energy PV, International Solar Electric Technologies, Martin Marietta, Siemens Solar Industries, and Solarex. Meanwhile, a variety of thin-film modules in 1-kW sizes are being tested at the NREL and in 20–400-kW arrays at PVUSA in Davis, California.

Worldwide, about 30 PV cell manufacturers are producing thin-film cells, a number of them supported by industry and government alike. However, it is probable that not just thin-film technology but a number of other technologies will succeed in the various market areas.

State of the Art of Thin-Film PVs

As a result of 20 years of R&D at a cost of $1.2 billion by the DOE and more than $2 billion by private industry, the cost of PV cells has been reduced by a factor of more than 20. In 1993, costs of PV-generated electricity range from about 30 to 50 cents/kW·h depending on size and applications. New PV technologies have been developed, including thin-film technologies that yield lower costs while delivering high efficiencies.

Researchers continue to improve the efficiency, stability, and reliability of thin-film PVs: to reduce manufacturing costs, to develop safe handling methods for potentially hazardous materials, to assume a sufficient supply to exotic materials, with the help of university research centers (in particular the University of Delaware's Institute of Energy Conversion) to improve the theoretical understanding of PV phenomena; to demonstrate to the private sector the value and potential of PVs; and to encourage the expansion of markets for PVs by utilities and also rural electrification in developing countries.

One of the greatest benefits the broad use of PV power can bring is a cleaner and safer planet in the decades ahead (10).

PHOTOVOLTAIC MODULES AND ARRAYS

Flat-Plate Collectors

Flat-plate collectors accept "global" solar radiation: the combination of sunlight direct from the sun and that reflected and/or refracted by the atmosphere to strike the collector. They are basically more efficient since global sunlight can be 25–75% more intense than direct sunlight; they are also not limited to maximum-insolation sites (see Fig. 10).

Concentrating Collectors

These collectors accept only the direct rays of the sun unless expensive optical technologies are used. However, even without global sunlight, concentrating collector systems can be made to generate more electric power per cell area by using lenses or mirrors to focus very large amounts of sunlight onto small PV cells. Much research has gone into producing cost-effective PV concentration modules.

Concentration of solar energy requires constant tracking of the sun in two axes with complex and expensive equipment; such a trade-off may be cost-effective only in areas of very high direct insolation, such as the southwestern United States. Concentrating systems are being developed at three levels of concentration:

Low: less than 30-fold
Mid: 100- to 400-fold
High: above 400-fold

The EPRI PV Concentration Project

Beginning in the mid-1980s, the EPRI in Palo Alto, California, sponsored the development of high-concentration PV techniques that resulted in a then world record of more than 28% sunlight-to-electricity conversion efficiency (11). The EPRI and two PV component manufacturers are continuing this work toward the production of low-cost PV electricity using high-concentration techniques.

The EPRI's concentrating systems use Fresnel lenses, parabolic mirrors, or heliostats plus sensors and electric motors to precisely track the sun and concentrate its rays onto very small cell areas. This geometry can generate up to 500 times more current for a given cell area than do flat-plate systems, and high-concentration photovoltaic (HCPV) technology is expected to be economically competitive with flat-plate PV modules.

Five cosponsoring utilities joined the EPRI in this venture: Arizona Public Service, Georgia Power, the Los Angeles Department of Water and Power, Pacific Gas &

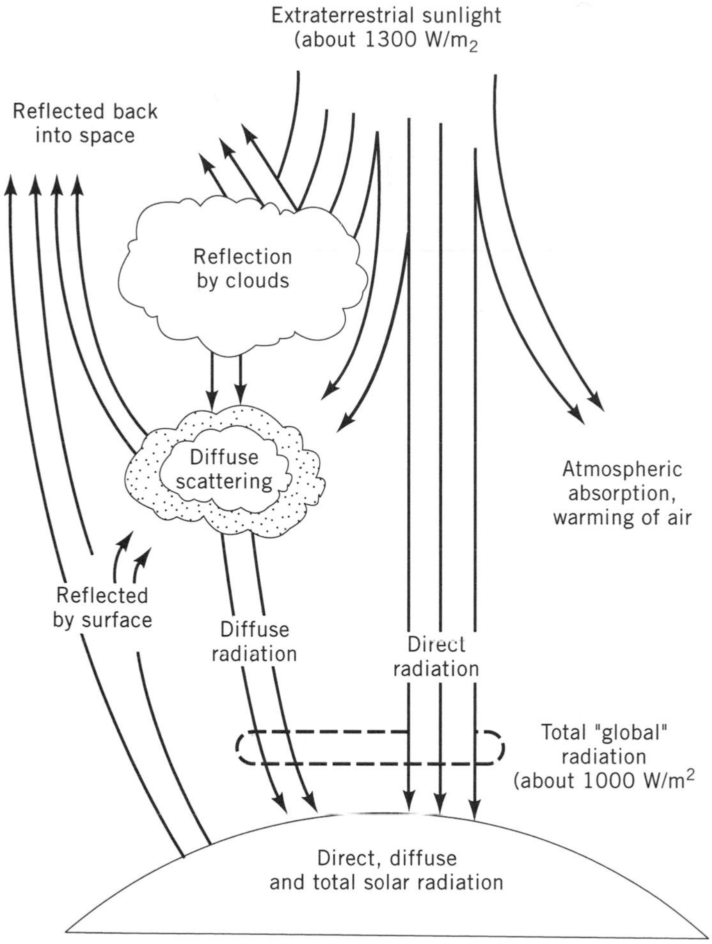

Figure 10. Direct, diffuse, and total solar radiation. Courtesy of NREL.

Electric Company, and Southern California Edison. Other organizations are also active in the concentration approach to PV power.

PHOTOVOLTAICS AND THE ENVIRONMENT

As the environmental costs of conventional sources are factored in, PVs may achieve cost competitiveness with other electric power sources earlier than anticipated. For example, each gigawatt-hour of fossil fuel electric power production puts more than 1000 t of carbon dioxide into the atmosphere. As a result, in 1993 nineteen U.S. states required that environmental issues be considered in selection of power plant fuel. Ten other states were evaluating similar measures.

Depending on market penetration of PVs, the technology can displace the amounts of CO_2 shown in Figure 11. Including energy input, PVs displace about 90% of the CO_2 that would otherwise be produced by today's energy sources.

Environmental Benefits of PVs

The following are environmental benefits of using PVs (10):

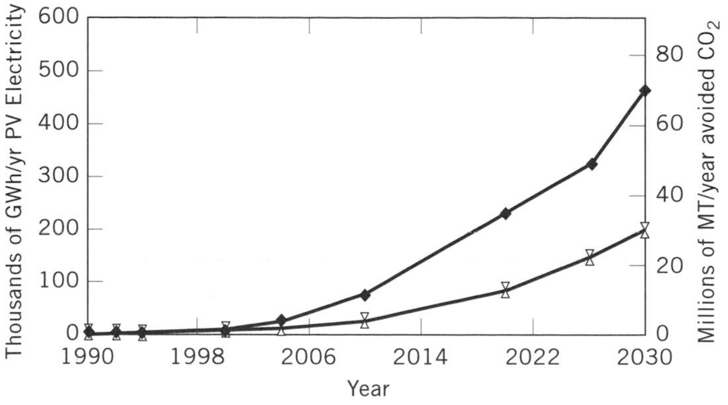

Figure 11. Predicted annual world CO_2 reduction using PV. Courtesy of NREL. A, Accelerated case; B, Base case.

- Environmentally benign
- No fuel costs; thus no risk of future costs hikes
- Low maintenance
- Long life
- Suitability for remote installations
- Systems integrated into buildings or structures have no land costs
- Continuing reductions in cost plus advances in performance
- Modularity permits a wide range of sizes and scaling up as needed
- Short lead times, reducing problems of installing unused capacity
- Solid-state electronics with no moving parts
- National security, industrial growth, and resultant new jobs
- Market competitiveness, expanded economy, and reduced trade deficits.

Potential Environmental Problems with PV Materials

Copper indium diselenide PV production has raised environmental concerns because feedstocks including cadmium and hydrogen selenide are both toxic. However, PV researchers believe that with standard safety precautions the materials need not be a hazard. The Environmental Protection Agency's environmental protection (EP) toxicity test and its even more stringent TCLP test, both of which call for grinding up a CIS module, suspending the particles in various solution and then trying to leach cadmium, selenium, or other toxics from them, have validated this belief (7).

Indium poses not just an environmental hazard but the question of availability of the material for the accelerated production of CIS PV cells. Only about 120 t of this element are mined annually and 50 t of indium is required per gigawatt of PV power, thus limiting annual production to about 2.4 GW. However, since indium is about as abundant as silver and more than 8000 t of silver are mined each year, it is thought that indium production can be increased as needed to match PV needs.

Similarly, a practical solution to the PV manufacturer's problem of disposing of the 225 kg of cadmium waste that might produce in a year would be to recycle the valuable material for use in manufacturing more cells.

AVAILABILITY OF RAW MATERIALS FOR GLOBAL PRODUCTION OF PV CELLS

Serious reservations have also been expressed about the feasibility of producing PV electricity as a global alternative to conventional energy forms. Included among those concerned is the president of a small U.S. PV firm. His thesis assumes a future global population of 10 billion and about 3 kW of electric power needed per inhabitant by the time PV repowering is complete worldwide. For PVs to provide even one-third of global needs would require the installation of about 300,439 km^2 (116,000 square miles) of PV modules using a minimum of 1×10^6 t of PV material. Furthermore, the installed cost for modules must be kept to about 50 cents/W in 40,000 production lines operating 18 hours a day.

Amorphous silicon possesses twin problems of inherent degradation and low conversion efficiency. Copper indium diselenide cells have the disadvantage that there is only about 0.01 Mt of indium available annually. None of the other currently used PV materials are acceptable in the amounts needed for PV globalization because of their toxicity, and the only safe alternative is to investigate and develop a range of new PV materials. Candidate materials include the families of sulfides and phosphides of the abundant metals, comprising at least six semiconductors with bandgaps in the PV range. Zinc diphosphide (ZnP_2) appears to have the greatest potential. These concerns as yet are not shared by many in the industry (12).

This seems to be a minority view, considering the optimistic outlook given in Table 4.

The NREL's Atomic Processing Microscope

A recent landmark invention suggests another possible solution to the PV material supply–demand question. In 1992, NREL scientist Lawrence Kazmerski received a patent for his atomic processing microscope (APM), an innovation making it possible to engineer the fundamental optoelectronic behavior of semiconductor materials. The new instrument was an outgrowth of the scanning tunneling microscope (STM) used in research at the NREL. The STM allowed nanoscale (billionth-of-a-meter) resolution of materials and was invaluable in semiconductor research (13).

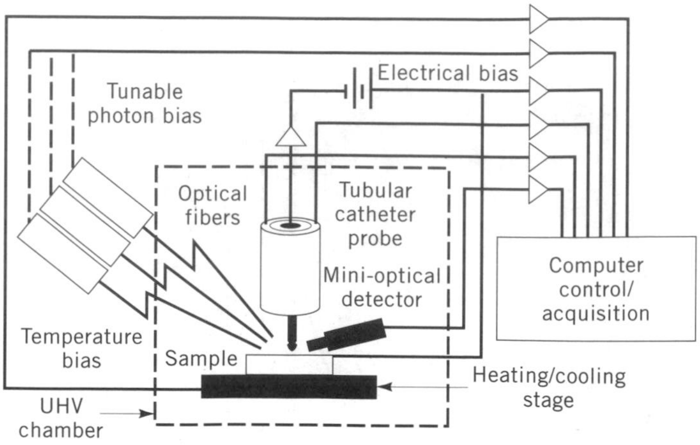

Figure 12. Diagram of atomic processing microscope (APM). Courtesy of NREL.

Kazmerski modified the STM with a mix of electronics, hardware, and computer software to produce the ATM (Fig. 12). This invention received an Industrial Research IR-100 award in 1992.

The APM is unique in that is produces direct images of individual atoms as well as distinguishes between atoms of different types and identifies their bonding characteristics. More importantly, the APM also permits use of single atoms from the surface of a material and their replacement with other atoms with different attributes. This atomic-level legerdemain is accomplished by first cooling the semiconductor material under test to near absolute zero to minimize atomic vibration. The material is then excited with tunable lasers and dc power to supply photons of proper wavelength to gain access to filled or empty electron states. A computer acquires these data, creates a three-dimensional representation of the atoms, assigns colors to different types, and produces an image on a high-resolution printer (13).

Table 4. Availability of Critical Materials

Material	Annual World Production ($\times 10^6$ t)	Amount per GW Peak ($\times 10^6$ t)	Proven Reserves ($\times 10^6$ t)
Indium	120	40	3,600
Tellurium	300	50–150	37,000
Selenium	2,000	60	130,000
Cadmium	20,000	50–150	970,000
Gallium	35	10 (at 1 : 1 Ga : In)	1,000,000

Source: U.S. Bureau of Mines, 1992 Mineral Commodity Summary.

Tailoring the Atomic Structure of PV Cell Materials

Using a computer-controlled probe, a researcher locates the atom of interest, then uses pulsed lasers to move it from its lowest energy state to a particular excited state. Kazmerski has used the APM to remove several selenium atoms from preselected sites to make the test material more n-type and to confirm that selenium vacancies act as donors. He then passivates the region, lending support to the interpretation that oxygen acts as an acceptor.

The APM, equipped with computer programs to rapidly change the characteristics of a candidate material, may move PV technology toward the fabrication, evaluation, and beneficial modification of extremely small devices on the order of a few thousand atoms across and only a few atoms thick. Such devices include PV cells and possibly some with no environmental drawbacks in production or use.

THE FUTURE OF PVs

Global Need for New Electric-generating Capacity

For all its promised potential, it is doubtful that the world can wait for the Solar Power Satellite. By the year 2000, world population is expected to be about 6 billion, a 20% increase in a decade. The DOE's Office of Solar Energy Conversion projects the installation of an additional 1400 MW of PV powerplants by that time, about 900 MW in the United States and 400 MW overseas (3).

Echoing these concerns, the Organization for Economic Cooperation and Development in 1990 estimated that by the year 2030 North America, Western Europe, and the industrialized Pacific will require twice as much electric power as they used in 1985 (Fig. 13). For the developing countries, the factor will be eight times as much. Of the projected 500 GW of new capacity, developing countries will require about 350 GW, China and India using half that amount. Another factor in the power equation is the release of 5 billion tons of CO^2 into the global atmosphere annually for all that time.

Solar Power Satellite

Increasing energy concerns have rekindled interest in a decades-old proposal for solving Earth's energy problems. A tiny military satellite was the first demonstration of PV electricity operating in space. Nikola Tesla experimented with the wireless transmission of electric power more than a century ago but was not successful. In the early 1960s, the U.S. Air force funded development of a microwave-powered helicopter, and in 1965 the Raytheon Company successfully demonstrated the concept (14).

A much different kind of space powerplant was the orbiting Solar Power Satellite (SPS) proposed a decade later in 1968 at an engineering conference in Denver by Peter Glaser, an Arthur D. Little vice-president. Components of this satellite would be placed in low orbit by space shuttles, assembled, and then boosted into an operational 35,000-km synchronous orbit, perhaps using electric power generated by the satellite itself (15).

The SPS was to comprise many square kilometers of PV arrays converting solar radiation to electricity. This output would then be converted to microwave energy and beamed to large rectifying antennae on Earth. Here, the microwave radiation would be converted to alternating current compatible with utility grids. The orbiting PV arrays were expected to function for about 30 years and then would be replaced.

The first International Microwave Power Institute Symposium was held in 1970, attended by representatives of Japan, several European countries, and the Soviet Union. By 1972, the National Science Foundation and the National Aeronautics & Space Administration (NASA) in the United States had prepared a program plan for the SPS, and the first feasibility study was completed for NASA's Lewis Research Center in 1974.

In addition to obvious logistics problems, environmental concerns were raised that the microwave energy flux from the SPS could be harmful to life on Earth. Political complications and the obvious vulnerability of the system to possible military attack were additional problems. (See Microwave Technology.) With these technological problems, plus cost-to-orbit estimates of about $\$10^{12}$ the SPS at the outset seemed a prohibitively expensive solution to the energy problem. But when the 1973 OPEC Oil Embargo hit, more people in high places were willing to listen to Glaser, including NASA, the newly formed Energy Research & Development Agency, and a number of congressmen (16).

Then NASA and the Energy Research and Development Agency (ERDA) became interested enough to begin a 3-year, \$20-million study of the project. By the time the ERDA had become the DOE, a new office was opened to

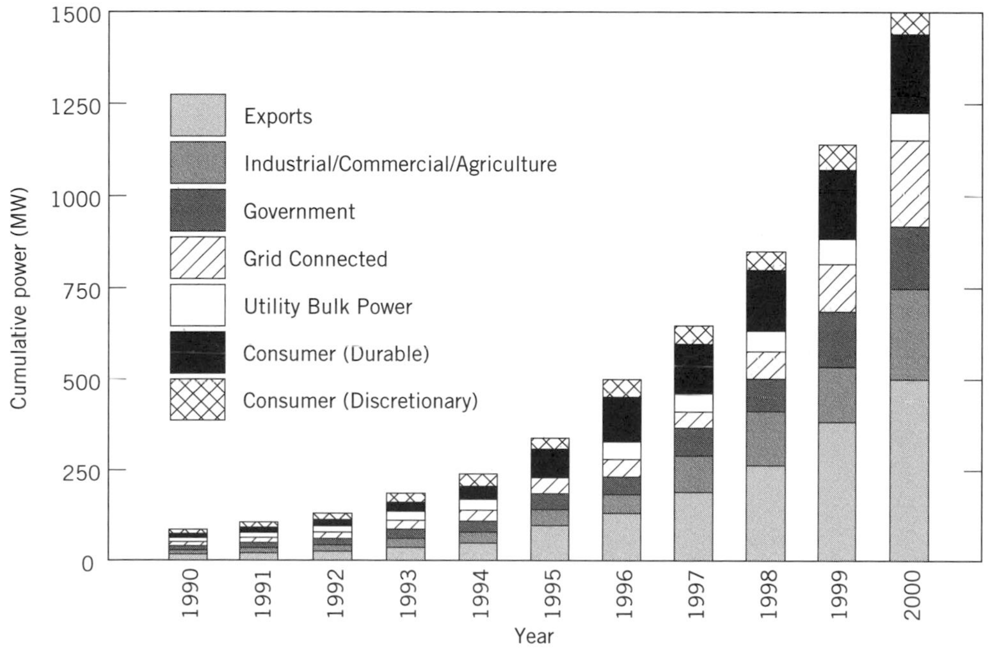

Figure 13. Increased need for PV power in the next century. Courtesy of DOE.

study the SPS in detail. Hundreds of scientists, engineers, and analysts examined the proposal for 3 years and found nearly 50 problems that indicated further research.

No technical problems that could not be handled with enough time and money were found by NASA and the ERDA, but the world was not yet ready for the SPS, especially since cost estimates ran to $100 billion plus over 30 years. The National Research Council (NRC) later put the price tag at $3 trillion over a 50-year period, comparing the SPS with the Manhattan Project, the Panama Canal, and the development of nuclear power.

Nevertheless, in 1977 the DOE began an in-depth study of the engineering, logistic, environmental, and other potential effects of the SPS. By 1980, a combined DOE/NASA report stated that although there were numerous technical problems, none seemed unsolvable. However, it would probably require 30 years to produce the first satellite-rectifying antenna set at a cost of more than $100 billion, with a total cost for the entire SPS array of $781 billion.

Two years later, the American Institute of Aeronautics and Astronautics (AIAA) published a position paper titled "Solar Power Satellites" (17). Even though it conceded the importance of "inexhaustible" energy resources such as the SPS might offer, the position paper concluded with this summary statement:

> In addition to the necessary economic, societal, political, and Earth environmental impact analyses now being conducted under the current DOE program, AIAA recommends that a five-year ground-based technology verification and advancement program be pursued to define costs adequately and to select the preferred system. The cost of such a technology program, including the current DOE effort, is estimated to be of the order of $30 million per year.

In 1981, the National Academy of Science's NRC carried out in-depth studies and hiked the SPS cost estimate to about $3 trillion over a 50-year development period. Furthermore, the NRC recommended that no funds be committed to research for 10 years. With the end of the oil crisis in 1980, the SPS was dropped from the budget. Funding continued for nuclear fusion as a power source, the technology favored by then Energy Secretary James D. Watkins.

France, Japan, West Germany, and the former Soviet Union have proposed similar SPS projects, but only Japan has worked up a detailed program. In the United States the private Space Studies Institute has suggested building the SPS not in space but on the moon, where it would then be more than 20 times as easy to loft into space.

Following the oil crisis in 1973, Japan had implemented its ambitious Sunshine Plan, including the long-range development of a SPS. In 1992, Japanese scientists successfully demonstrated its Microwave Energy Transmission Systems (METS) and a MILAX airplane experiment. In April of 1993, Japan's SPS 2000 study group completed the first phase of the plan and displayed a 1⁄50th-scale model of the SPS. As part of the SPS development, Microwave Country and Microwave Garden projects beamed 10 kW of microwave energy from a tower to study ecological effects and the influence of weather on microwave reflection from the ground. The Microwave Power Transmitting Experiment is planned to beam 100 kW a distance of 12 km to a receiving antenna at the summit of Mt. Fuji.

Japan has announced that in accordance with the New Sunshine Plan a global SPS system is to be in place by the year 2040 and several government agencies and industrial organization are studying the viability of the

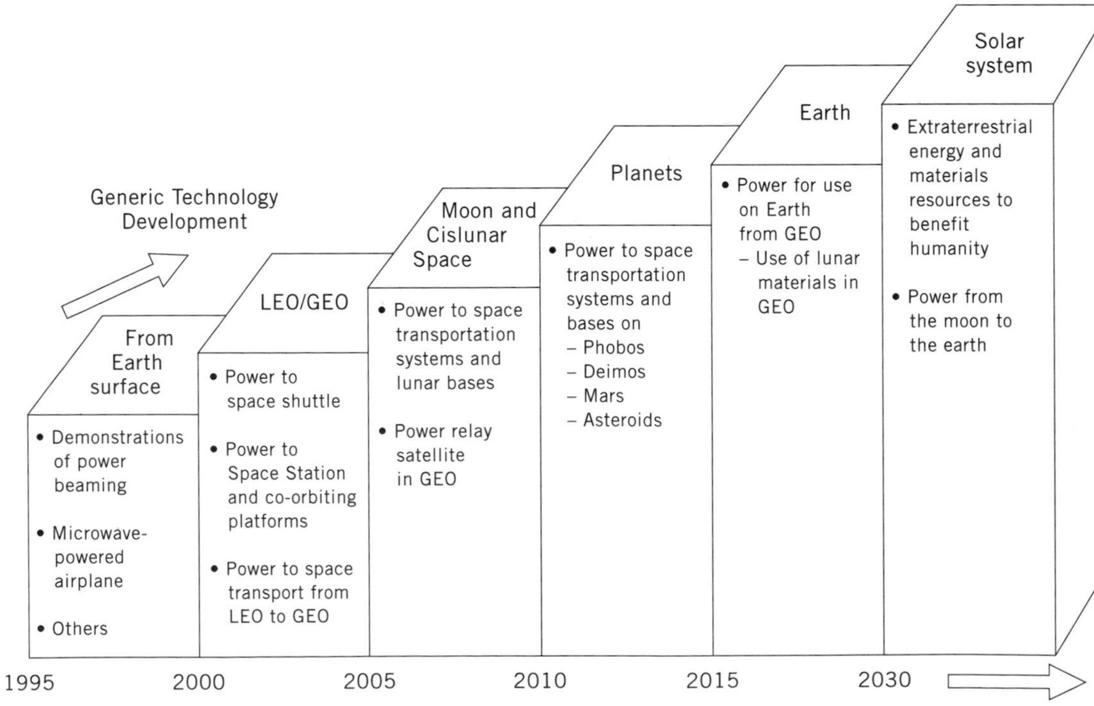

Figure 14. Growth rate for Japan's SPS development. Courtesy of Peter E. Glaser.

project (Fig. 14) (18). Japan presented its new SPS concept at the 44th International Aeronautical Federation Congress of Graz, Austria, in October 1993.

With the developed world expected to require twice as much electric power by the year 2030 and developing nations perhaps eight times as much, the SPS idea has again been brought up in the United states, Peter Glaser suggesting to the NRC that it was time to reevaluate the plan. The idea of multiple orbiting solar satellites had been suggested before, but in light of NASA's Human Exploration Initiative the SPS had taken a new tack: Instead of the original orbiting solar satellite concept, the huge solar collector would be built on a satellite already out there—the moon. Nevertheless, funding has not become available, perhaps because there is not yet a favorable consensus on the SPS in any of its potential forms.

Photovoltaic Hydrogen

In 1990, the DOE's national laboratories estimated that conventional renewable energy strategies could account for only 24 to 34 quads of a total energy use of about 80 quads by the year 2030 (19). Only by being able to store solar energy for later use could we take advantage of intermittent solar energy. Using hydrogen as a storage medium, solar energy could account for another 16–21 quads in that time frame. Following up on this possibility, the EPRI ranked the available storage media for capital costs and hydrogen. Only compressed air is cheaper than hydrogen, and it is not applicable to renewable energy storage.

The Florida Solar Energy Center (FSEC) has been interested in solar hydrogen for a decade, and FSEC Director Dave Block and Kirk Collier, director of FSEC's Photovoltaics and Advanced Technologies Division, have devised a hydrogen production system using parabolic trough concentrators to produce electricity at only 2 V and 200 A, a perfect match for the single-cell electrolyzers. This method emits no carbon dioxide, as does the production of hydrogen from fossil fuels. Hydrogen is then stored in metal hydrides. An analysis of the system will be featured at the 10th World Hydrogen Energy Conference, to be held June 20–24, 1994, in Cocoa Beach, Florida. This work is funded by the NREL, which is also involved in related research on magnesium hydride storage media.

BIBLIOGRAPHY

1. D. Halacy, *The Coming Age of Solar Energy,* Harper & Row, New York, 1963.
2. K. Zweibel, "Thin-Film Photovoltaic Cells," *American Scientist,* (July/Aug. 1993).
3. "Solar 2000: A Collaborative Study," U.S. Department of Energy, Washington, D.C., Feb. 1992.
4. "Photovoltaics Program Plan; 1991–1995" U.S. Department of Energy, Washington, D.C., Oct. 1991.
5. "Photovoltaics for Regulators," National Renewable Energy Laboratory, Apr. 1992.
6. "Photovoltaics: New Opportunities for Utilities," Solar Energy Research Institute, July 1991.
7. H. Kelly, ed., *Renewable Energy,* Island, 1993.
8. *Solar Today* (Sept./Oct. 1993).
9. "Sunlight Brings Electric Power to Rural Brazil," National Renewable Energy Laboratory Press Release, Jan. 26, 1993.
10. T. Surek, "The State of the Art of Thin-Film Photovoltaics," in K. W. Boer, ed., *Advances in Solar Energy,* Vol. 8, Plenum, New York, 1993.
11. T. Moore, "High Hopes for High Power Solar," *EPRI Journal,* (Dec. 1992).
12. T. Mowies, "Photovoltaic Power for the Planet," *Solar Today,* (May/June 1993).
13. L. Kazmerski, National Renewable Energy Laboratory, personal communication.
14. W. C. Brown, *"Experimental Airborne Microwave-supported Platform,"* Technical Report, RADC-TR-65-188, Rome Air Development Center, Rome, N.Y., 1965.
15. P. E. Glaser, "Power from the Sun: Its Future," *Science* **162,** 857–866 (1968).
16. L. Shiner, "300 Billion Watts, 24 Hours A Day, *Air & Space,* (June/July 1990).
17. Technical Committees on Aerospace Power Systems and Space Systems, "Solar Power Satellites," American Institute of Aeronautics & Astronautics, Jan. 1979.
18. M. Nagatomo, *Proceedings of the Twelfth ISAS Space Energy Symposium,* Mar. 10 and 11, 1993, Institute of Space and Astronautical Science (ISAS), Sagamihara, 1993.
19. C. E. Thomas, I. Melody, and J. Ogden. "Solar Hydrogen: A Sustainable Option," *Solar Today,* (Sept./Oct. 1993).

SOLAR COOKING

Dan Halacy
Lakewood, California

About 1% of the 84.32×10^{13} J of energy consumed annually in the United States is used for cooking. This approaches one quad of energy, and from a world perspective the amount must be multiplied many times. The 55 EJ of biomass burned annually accounts for 15% of global energy use, but 38% of that is used in developing countries (1). The sustained growth of biomass does not result in a buildup of carbon dioxide in the atmosphere, because that produced by burning is taken up by new growth. However, the large-scale harvesting of trees is an environmental concern. A continuing and increasing need in the Third World is for firewood or other biomass for cooking, and one solution would seem to be simple, inexpensive solar cookers to replace the biomass resource.

See also Wood for heating and cooking; Energy efficiency; Commercial availability of energy technology.

The sun drying of fruits and vegetables, a very slow form of solar cooking, has long been practiced for preserving foods of various kinds. An area of 1 m^2 normal to the sun receives about 1 kW of heat energy. Because solar cooking is done directly with solar heat, instead of converting thermal energy to electric energy at low efficiency and then back to heat energy; cooking with sunshine was suggested long ago as a practical solution for areas chronically short of firewood or other fuel for cooking.

However, Solar Box Cookers International (SBCI) noted that even though an estimated 25 million people are aware of solar cooking (about 1 in 200), only 220,000–240,000 solar cookers are presently in operation worldwide, including the 10,000 or so in the United States (2). In less-developed countries, the majority of these cookers are box cookers such as those pioneered in Arizona decades ago by Kerr and Cole. A solar box cooker is made with two cardboard boxes, one inside the other with space between them for crumpled newspaper to insulate the inner cooking area. A sheet of glass or plastic over the top of the cooker traps heat inside via greenhouse effect, and an adjustable cardboard reflector covered with aluminum foil adds more solar energy to the cooker. The resulting temperature is sufficient to cook a simple family meal during the day for evening consumption.

THE LONG HISTORY OF SOLAR COOKING

German physicist Tschirnhousen (1651–1708) seemed to have been first to boil water with solar energy concentrated by a large lens but went no farther than that with his solar heating research. The first recorded solar cooking was done in Switzerland in 1767, when Horace de Saussure stumbled onto the technique while measuring the intensity of solar radiation (3).

Saussure was investigating the decrease in air temperature with increased altitude, and his test equipment included an insulated box with several glass covers. This led to his discovery that instead of decreasing, solar radiation was actually more intense at higher altitudes. As a result, he was credited with inventing the hot box, an instrument used by solar researchers for many years. Coincidentally, Saussure noticed that the temperature inside the hot box was sufficient to cook with. Therefore, he built several hot boxes in which he baked various fruits and vegetables; he reported being pleased with the results.

French scientist Ducarla (1738–1816) cooked meat and other foods with a solar cooker; reportedly the food cooked in ca 1 h. In 1834, Sir John Herschel became the first Englishman to cook with the sun. While on a scientific expedition to the Cape of Good Hope, he demonstrated solar cooking.

The first commercially successful solar cooking entrepreneur was the French Augustin Mouchot (1821–1911), who invented and tested many solar devices for cooking. On one occasion, about 1 kg of meat was cooked in 4 h in a "Solar Pot" (see Fig. 1), even though there had been some clouds. He went on to build solar cookers in quantity for Napoleon's armies and received monetary compensation for his efforts. Mouchot's assistant Abel Pifre used solar energy to drive a steam engine and printed the newspaper *Le Journal Solaire*.

While serving in the high court at Bombay, in 1876, William Adams began experimenting with solar cooking. He built an excellent solar concentrating cooker that could boil and bake a variety of foods; he claimed he could cook food for seven soldiers in just 2 h. Adams wrote *Solar Heat*, a book that described the solar cookers.

Charles G. Abbot, who served for many years as secretary of the Smithsonian Institution, experimented with a new kind of solar cooker in 1916. This one used a parabolic trough heat collector located outside of the building. The cooking vessel itself was indoors, allowing more convenient operation. Oil heated by the collector circulated through pipes that went to the kitchen (4). The new solar cooker also tracked the sun with a clockwork mechanism (Fig. 2). Abbot preserved foods, baked bread and meat, and stewed meat and vegetables, often cooking after sunset, using the heat stored in the oil.

MODERN SOLAR COOKERS

India

The world's slow acceptance of solar cookers has not been for lack of need (4). In addition to the quantities of wood burned in cooking, India also burns about 35% of its ani-

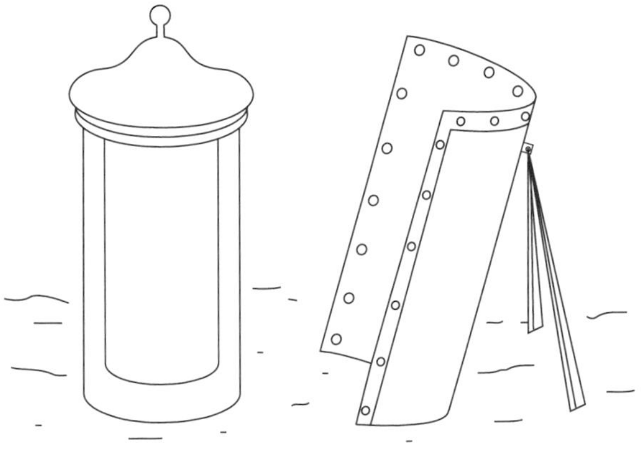

Figure 1. Mouchot's Solar Pot. Courtesy of Richard Pinon.

Figure 2. Charles G. Abbot and his solar cookers. Courtesy of John Yellott.

Figure 3. Japanese solar rice cooker. Courtesy of John Yellott.

mal waste for fuel, impoverishing the soil. The end of World War II saw increased interest in solar cooking in several countries.

India's M. K. Ghosh began working with conventional hot box cookers in 1945 but soon turned to a concentrating type for higher temperatures. By the early 1950s, M. L. Ghai (at India's National Physical Laboratory) had developed and demonstrated parabolic concentrating solar cookers. These high temperature stoves were produced in quantity and sold at a reasonable price by Devidayal Metal Industries (Bombay). They were not favorably accepted, however, and the venture was a commercial failure. Even though the price was about $14 (U.S.), it was apparently more than most potential buyers could afford.

It was many years before serious interest in solar cookers again developed in India. In 1978, the Research Institute in Jodhpur estimated that some Indian villagers were foraging for firewood for up to 10 h daily, several times longer than required only a decade earlier (5). The institute, therefore, warned that the nation's energy crisis was not the increasing price of fossil hydrocarbons but the growing shortage of firewood. As a result, India began promoting solar box cookers with an initial production of 18,000. A year later, an estimated 70,000 had been sold.

In 1990, 15,000 more cookers were reported sold by the Gujarat State Energy Center. But people were not using the cookers because of the limited kinds of cooking that could be done with the solar cookers, the excessively long cooking times that were needed, and the limited time that solar cooking is possible. The limited time that cooking is possible is the most significant problem. Most Third World peoples live in very hot climates and tend to cook meals early in the morning and late in the day, when the sun is not available. Another factor could be cultural; the cooking fire has an almost religious status (6).

There are other reasons why people did not use the cookers. When the Indian Appropriate Technology Development Association reported that the solar box cooker with a single reflector was the most appropriate type for villages, it also recommended that two 100-W light bulbs be included to allow cooking on cloudy days. Concentrating solar cookers did not require supplemental heat but were faulted as being subject to upset by strong winds, requiring frequent adjustments to sun movement, and burning food.

China and Japan

Solar cookers have been much better accepted in China, where three different types have been designed and manufactured. The Academy of Agricultural Research and Planning in Beijing reported that as of 1986 about 100,000 solar cookers had been produced in China and had become the nation's fastest-growing solar technology. China reported a preference for concentrating cookers over the simpler box cookers. In Japan, the Goto Co. manufactured an excellent concentrating solar rice cooker (4).

The United States

Along with more sophisticated solar devices, many cookers and ovens from foreign countries were demonstrated

Figure 4. Commercial solar cooker. Courtesy of *The Pasadena Star News*.

Figure 5. Villager solar oven. Courtesy of Burns-Milwaukee.

at the First World Symposium on Solar Energy, held in Phoenix, Arizona, in 1955 (7). The novelty of cooking with sunshine touched off considerable interest in the United States, and American Maria Telkes and colleagues at New York University (NYU) demonstrated sophisticated concentrating solar ovens that reached temperatures between 150° and 200°C (see Fig. 4).

In 1958, the Curtis-Wright Corp. learned of Telkes's work and hired the NYU solar scientists. The company wanted to manufacture the efficient solar ovens in quantity. Unfortunately, this early solar cooker effort ended in 1960 when a new Curtis-Wright board chairman decided to terminate the project.

From 1950 to 1970, a variety of parabolic reflector solar cookers were produced at the University of Wisconsin by Farrington Daniels and co-workers. These cookers were demonstrated and tested by Native Americans in the U.S. southwest and Coahuila, Mexico. Although the cookers were simple and efficient, they met the same fate as did the similar Ghai reflector cookers in India. The cookers were not used because they were too different from conventional stoves, too hard to learn to use, and too much trouble to keep focused on the sun.

Figure 6. The 1993 U.N. solar box cooker seminar held in Boulder, Colorado. Courtesy of the National Renewable Energy Laboratory.

Colorado solar scientist George Lof adapted the umbrella as a highly portable concentrating solar cooker called the "Umbroiler." This cooker is useful under any weather condition. This device can cook coffee, soup, scrambled eggs, bacon, hotcakes, and even steak (4).

During the 1970s and 1980s, several manufacturers produced and marketed a variety of solar cookers in the United States. These ranged from small cardboard box–aluminum foil reflector cookers to sturdy and efficient concentrating solar ovens that could attain temperatures above 150°C. Sales were modest, however, and the number of commercially available solar cookers has shrunk greatly in recent years.

Commercial solar ovens can reach temperatures of at least 150°C and cook a 5.4-kg turkey in about 3 h on a clear day. The solar oven produced by Burns-Milwaukee is being tested at the Department of Energy's Sandia National Laboratories in Albuquerque, New Mexico, toward further improvements.

AGENCIES INVOLVED IN THE DISTRIBUTION OF SOLAR COOKERS

The Agency for International Development conducted a feasibility study on solar cookers in Haiti, and SBCI has provided many cookers to several less-developed countries, as has Rotary International (5). Canada's Brace Institute provided 20 concentrating cookers of its own design to villagers in Haiti. Unfortunately, 11 of these were soon reported dismantled and abandoned. The remainder were converted to solar water heaters, generally for washing dishes. The Rockefeller Foundation distributed 16 concentrating solar cookers in Mexico but a check after 14 months revealed that only 3 were still operating. Mechanical failure was given as the principal problem.

A Volunteers in Technical Assistance (VITA) member had a similar experience when craftsmen in Oaxaca, Mexico, built a local version of an American concentrating cooker with small glass mirrors glued to the reflector panels. The local glue did not withstand the heat of the sun; when the mirrors fell off, the cookers were discarded and forgotten.

As of 1993, responsibility for fostering widespread use of solar cookers seems to rest largely with SBCI and Solar Box Cookers Northwest in conjunction with the following organizations:

- Academy of Agricultural Research and Planning (Beijing, China).
- Communities in Partnership (British Columbia, Canada).
- Costa Rican Solar Energy Society.
- Energy Development Agency (Gujara State, India).
- Food and Technology branch, the U.S. Department of Agriculture (Washington, D.C.).
- Fraunhofer Institute (Germany).

Freedom from Hunger Foundation.
Free Methodist Church (Bremerton, Washington).
Indo-German Dhauladhar Project (India).
India's Appropriate Technology Development Association.
Indian Institute of Technology (New Delhi).
Medical Exchange International.
Pakistan Council of Appropriate Technology (Islamabad).
Pillsbury's "Plan International."
Rockefeller Foundation.
Rotary International.
SERVE.
SYNOPSIS (France).
Thrasher Research Fund.
United Nations Food and Agriculture Organization (New York).
University of Wisconsin.
VITA.

Tom Burns. Burns, a retired Milwaukee restaurateur, began his solar energy work with hot water heaters and then developed solar cookers at his wife's suggestion. His first cooker was the Sun Oven in 1985, a 9-kg oven that reaches 190°C in about 15 min. Today, it is being used in about 100 countries around the world, thanks to Burns' efforts, Rotary International, and a number of church organizations. There is also the giant Villager model that bakes two dozen loaves of bread at a time and cooks enough beans and rice for a whole village. This large solar oven can also boil hundreds of liters of water for safe drinking and for sterilization purposes. Succeeding where others have failed, Burns-Milwaukee opened an assembly plant near Kingston, Jamaica, in December 1992, with plans to hire 100 Jamaican workers to build ovens for sale in Haiti and other Caribbean countries (8), (see Fig. 5).

Ulrich Oehler. This engineer from Basel, Switzerland, believes "The production of solar cookers began in Khartoum. In Kenya, local craftsmen produce solar cookers for export. Two of our group have produced solar kitchens in Kenya. Some of these are big enough to provide two hot meals a day for 400 students. This year the first solar kitchen has been built in India to feed 100 monks at a Tibetan monastery. And now two kitchens are under construction in two boarding schools in Gujarat. In refugee camps for Afghans in Pakistan, more than 5000 families are using solar cookers."

Didi Contractor. Contractor, a German solar-cooking enthusiast with the Indo-German Dhauladhar Project in India's Himachal Prades State, has helped local people adapt the all-metal cooker designed by the Indian government. Instead of metal, however, these versions use mud and dung or brick, depending on the type of home. The cost is about half that of metal cookers. A total of about 150 were built by 1988; 60% of them were used regularly and another 15%, occasionally.

A FOREST PRODUCTS INDUSTRY'S ATTEMPT TO SOLVE THE PROBLEM

The wood industry in the United States has long been concerned about the world's forests. In 1990, Sy Dimensions Inc. (Healdsburg, California), a large broker of tropical hardwoods, and a group known as the Woodworkers Alliance for Rain Forest Protection commissioned a critical study of Third World Solar cooking (5). The study was to determine the feasibility of Sy Dimensions itself producing and distributing inexpensive solar cookers for use in the Third World. This would aid the people of those countries as well as improve the condition of the forests and consequently the wood industry.

The introduction to the final report (5) stressed the following concerns:

- Approximately 80% of trees felled in developing countries are burned for fuel.
- In Africa and much of Asia, 90% of the cooking fuel is firewood.
- As early as 1979, the United Nations had announced that 33% of the world's population was seriously short of fuel for cooking; by 2003, it will be 50%.
- The U.N. Food and Agricultural Organization reported that for want of sufficient firewood, villagers in many places are obliged to grow what can be eaten raw, thus suffering nutritional consequences.
- There has already been warfare over trees in Natal Province.
- The circle of deforestation around African cities grows so rapidly that residents of Ouagadougou, Burkina Faso, must travel 67.5 km to find firewood.
- In India, 35% of the country's fertilizer is burned for fuel, but attempts at introducing solar cookers continue to fail, despite the fact that the sun delivers the energy equivalent of 250 kg of coal/m^2 of collector per annum.

The report described the impasse succinctly: affordable solar cookers are not acceptable, and acceptable solar cookers are not affordable. Two generalizations about potential markets for commercial solar cookers were made (5):

> *In the higher latitudes*, there are many millions who could afford the relatively sophisticated and expensive solar cookers dictated by the climate. As combustibles become dearer, this market should stimulate lively competition among manufacturers. Some products are already sold in South Asia, the United States, and the European Community. It is likely that in temperate zones most solar cooking technology will be promoted by the private sector, or by subsidized local government programs, not international organizations.
>
> *The tropics* are a different matter. Here are concentrated the millions who suffer most from penury and ignorance. Here as well, simple cookers can work the year around. In a word, there is a high need for low-tech, low-cost solar devices. For humanitarian organizations, the most urgent challenge is in the tropics. In some tropical parts, Jeffersonians (those who favor simple solar technologies) may coax a tolerable cooker from grass roots artisans using locally available mate-

rials, and that may be to the good. In other parts, such as the Sahel, however, there are virtually no materials or disposable income or skills necessary for local manufacture.

In the latter areas, adoption of solar cooking on a significant scale requires the charitable distribution of box cookers mass-produced elsewhere. For private entrepreneurs manufacturing the product, primary customers would be the great governmental and private international welfare organizations—the same people who disdain solar cooking technology and have withheld essential funding for its R&D.

The study also determined whether an American commercial enterprise could contribute significantly to forest protection by promoting solar cooking. The conclusion was no, because of the following reasons (5):

- There is insufficient disposable income for commercial success in the places of most urgent need. Virtually all current manufacture and distribution (of solar cookers) is directed and subsidized by government and NGOs.
- There is insufficient public and institutional awareness of solar-cooking technology's potential to insure its free market success. Indeed, there is significant resistance to it in critical quarters.
- Most potential customers with adequate disposable income do not burn wood or charcoal from endangered forests.
- Private investment in solar cooking promotion will become attractive only when there has been a quantum leap in public awareness of the world's fuel crisis and of solar cooking's potential. Even then, a focus on forest protection will be artificial. This is simply an integral part of many ecological as well as humanitarian advantages to the technology.

THE OUTLOOK FOR SOLAR COOKERS

An international publication is enthusiastic about other uses of solar energy in developing countries, but writes off solar cookers thus (9): "Their high cost, awkward operation, slowness, and inconvenience have prevented this technology from finding a niche. Nevertheless, these devices remain a fixture at research centers and exhibitions, where they regularly amaze visitors. To our knowledge, despite two decades of scattered attempts, there are no examples of the successful introduction of solar cookers."

A report commissioned by the United Nations Solar Energy Group on Environment and Development (1) begins its section on the problems of using biomass for fuel in this way:

> Cooking stoves that burn traditional biofuels are used by half the world's population, yet many are inefficient and hazardous to the health of those who tend them. In recent years, however, a new generation of cook stoves needing less fuel and emitting fewer airborne particulates has emerged. Many of the new designs run on biomass that has been transformed into a liquid, gaseous, or improved solid-fuel form. Alternative cooking systems are compared, and data from cooking trials conducted by the authors in a south Indian village are provided.

Surprisingly, no mention is made of solar cookers, which consume no biomass and emit no particulates, although elsewhere the United Nations continues to stress the need for continued introduction of solar cookers. For example, in May 1993, a U.N. environmental meeting focused on solar cooking as an alternative to conventional fuels and taught several dozen young people, including a number from foreign countries, how to build and use solar box cookers (Fig. 5).

BIBLIOGRAPHY

1. *Renewable Energy: Sources for Fuels and Electricity*, Island Press, 1992.
2. B. Blum, Solar Box Cookers International, personal communication, 1992.
3. R. Pinon, *Passive Solar Energy J.* **2**(2), (1983).
4. D. S. Halacy, Jr., *Fabulous Fireball*, Macmillan, New York, 1957.
5. D. O'Ryan Curtis, *An Appreciation of Solar Cookers*, SY Dimensions, Inc., Healdsburg, Calif., May 1991.
6. J. Zerbe, Forest Products Laboratory, personal communication, 1993.
7. *Proceedings of the World Symposium on Solar Energy*, 1955.
8. S. L. Hickey, *The Rotarian*, July 1993.
9. *Appropriate Technology Source Book*.

SOLAR HEATING

DAN HALACY
Lakewood, Colorado

THE SOLAR RESOURCE

The sun is a fusion reactor at a distance of about 150×10^6 km from the earth (see FUSION ENERGY). Its probable age is 4×10^9 yr and it is expected to radiate energy for another 4×10^9 yr. Solar radiation comprises a spectrum from 10 pm (gamma rays) to 1 km (radio waves). More than 99% of the sun's radiation lies within the optical range of 0.276–4.96 μm. Electromagnetic radiation from the sun amounts to about 3.8×10^{26} W.

Because of the earth's small size and distance from the sun, it receives only a small fraction of total solar radiation; this fraction nevertheless amounts to 1.7×10^{17} W; about half of which is reflected back into space. However, about 13,000 times as much solar energy reaches the earth's surface in a year as the entire world consumes in the form of fossil and nuclear fuels. About 46,400 EJ (44×10^{18} Btu) of solar energy reaches the continental United States annually, compared to the 84 EJ (80×10^{15} Btu) of energy consumed (1 EJ = 0.95×10^{15} Btu or about 1 quad).

Solar energy technology comprises two distinct categories: thermal conversion and photoconversion. Thermal conversion takes place through direct heating, ocean waves and currents, and wind. Photoconversion includes photosynthesis, photochemistry, photoelectrochemistry, photogalvanism, and photovoltaics. Solar radiation is collected and converted by natural collectors such as the atmosphere, the ocean, and plant life, as well as by manmade collectors of many kinds. There are a number of solar technologies (Figs. 1 and 2). (see FUELS FROM BIOMASS).

It has been suggested that by 2050, global carbon dioxide could be reduced to 75% of 1985 levels if renewable

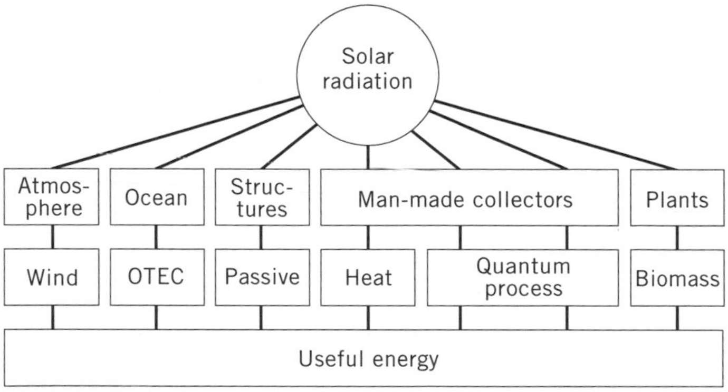

Figure 1. Natural and man-made collectors. OTEC = Ocean Thermal Energy Conversion.

energy and energy efficiency are aggressively adopted. This benefit could be achieved at no extra cost since renewable energy is expected to be cost-competitive with fossil fuels by that time. Furthermore. "In less than 40 minutes, the U.S. receives more energy in the form of sunlight than it does from fossil fuels it burns in a year" (1).

The United States has for many years made use of solar energy in the form of hydroelectric power, presently amounting to about 3 EJ (2.85 × 10^{15} Btu) annually, approximately equal to the present contribution of U.S. nuclear power plants. In addition, about 2 EJ (1.9 × 10^{15} Btu) is derived from the burning of biomass fuels, mostly wood and agricultural wastes. Thus the sun in the aggregate contributes about 6% of total energy used in the United States. The purpose of solar energy development is to appreciably increase this modest use of the vast solar energy resource. The contributions of various energy sources in 1993 are given in Figures 3 and 4.

The surface temperature of the earth is increased by about 250°C over surrounding space by exposure to the sun. Since the sun radiates energy at an effective surface temperature of about 6000 K, terrestrial solar concentrators can capture significant amounts of this thermal energy. A variety of solar thermal collectors provide a temperature range from low-temperature passive residential and commercial building heating through medium temperatures for active water heating, building heating, and industrial process heat, to the much higher temperatures required for power-generating heat engines, and even for operating refractory furnaces and testing materials at temperatures as high as 3,000°C, achieved by concentrating solar radiation by 50,000.

See also AIR CONDITIONING; BUILDING SYSTEMS; THERMOELECTRIC ENERGY CONVERSION SOLAR CELLS ENERGY MANAGEMENT, PRINCIPLES.

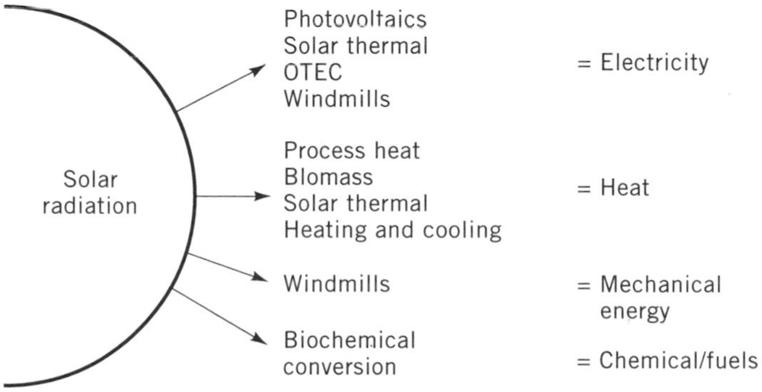

Figure 2. Solar technologies.

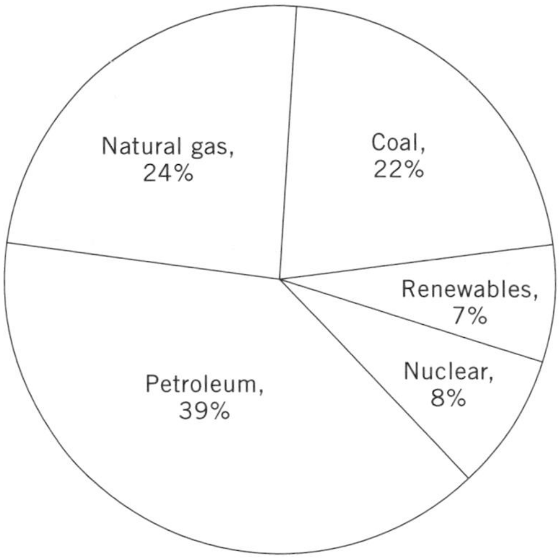

Figure 3. U.S. total energy use, 1990 (84.4 quads). Courtesy of Island Press.

PASSIVE SOLAR TECHNOLOGY

Passive solar technologies include direct and indirect heating as well as cooling.

Heating

As early as the fifth century BC, primitive dwellings in temperate climates were oriented to take advantage of so-

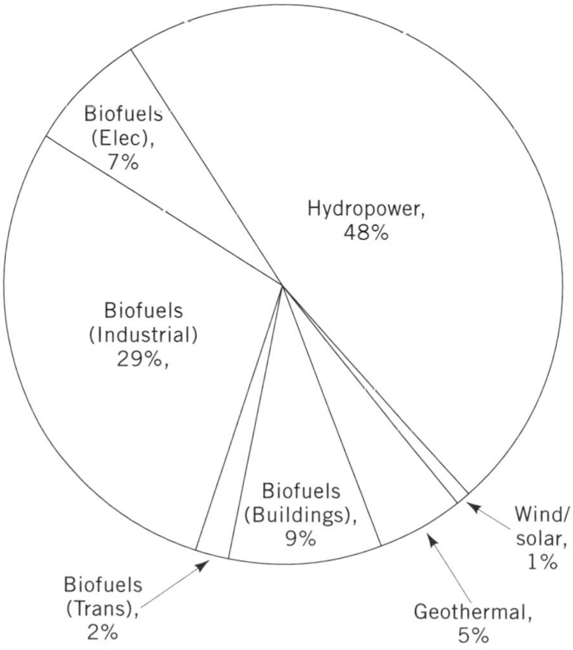

Figure 4. U.S. renewable energy use, 1990 (6.3 quads). Courtesy of Island Press.

lar heat during cold weather. For example, Indians in the American Southwest built passively heated dwellings into southfacing cliffs, which also shaded the structures from summer sun (Fig. 5).

In the past 15 years, about 250,000 passive solar homes and buildings have been constructed or retrofitted in the United States. Passive space heating makes use of solar radiation, with sunlight entering through glazed areas and warming the thermal mass of the building. Pumps or fans are not generally required. Passive technology is used principally in new homes and small commercial buildings. Average passive heating costs range between 4 and 6 cents/kWh, incurred by the extra cost of passive construction (2) (Fig. 6).

In spite of ancient passive heating applications, it is still not widely known that the sun can serve a variety of energy needs. An area of one square meter, normal to the sun, intercepts about 1 kW of solar radiation in clear weather. Thus the southfacing portion of a house roof of 80 m^2 receives more energy than the house consumes for heating, cooling, lights, and appliances. The 90 million or so residences in the United States receive a total of about 6.4×10^6 MW of solar energy annually.

The buildings sector, including homes and commercial buildings, used almost 12 quads of energy from fuels and 7 quads from electricity in 1990 (2). The electrical component required 20 quads of primary fuel, making a total of about 32 quads of the 80 quads or so consumed in the U.S. per year. Energy needs for buildings in the United States are supplied by natural gas (44%), electricity (34%), oil (12%), and other sources including a small amount of renewable solar energy (10%) (2).

Modern society is relearning the ancient passive solar heating techniques and improving them with better design, engineering, and materials. The ancient concept of the building itself as a large-volume solar collector has been revived and is being re-established (see Fig. 6). Proper design combines building orientation with reference to the sun, glazing, shading, insulation, and heat storage. Interest in passive design generally has followed the earlier exploitation of active solar heating.

Even on an overcast day, 50% of solar radiation may reach the earth. Its intensity does not depend on ambient air temperature but on the clarity of the sky and the angle at which the sun's rays are received. A greenhouse or other glazing of 18.6 m^2 facing the sun intercepts up to 422 MJ/d (4×10^5 Btu/d). A well-designed and oriented passively heated building thus receives nearly all the heat it needs from the sun. Surplus heat is stored in floors, walls, rock bins, or water tanks for later use. Incorporation of Trombe walls and/or concrete floors in a passive solar building adds effective heat storage. A Trombe wall is a glazed section over a brick or masonry wall, which supplies the building with heat that has been absorbed on the outside surface of the wall, conducted through the wall and radiated to the room (Fig. 7).

Cooling

Passive cooling techniques include cross ventilation, insulation, shading, evaporation, and the radiation of heat to the clear night sky. Except for natural breezes, however, these are not solar-driven processes. Natural cooling includes that which occurs at night when temperatures are lower, the storing of "coolth" in building mass, and carefully designed shading to minimize solar heat gain in summer.

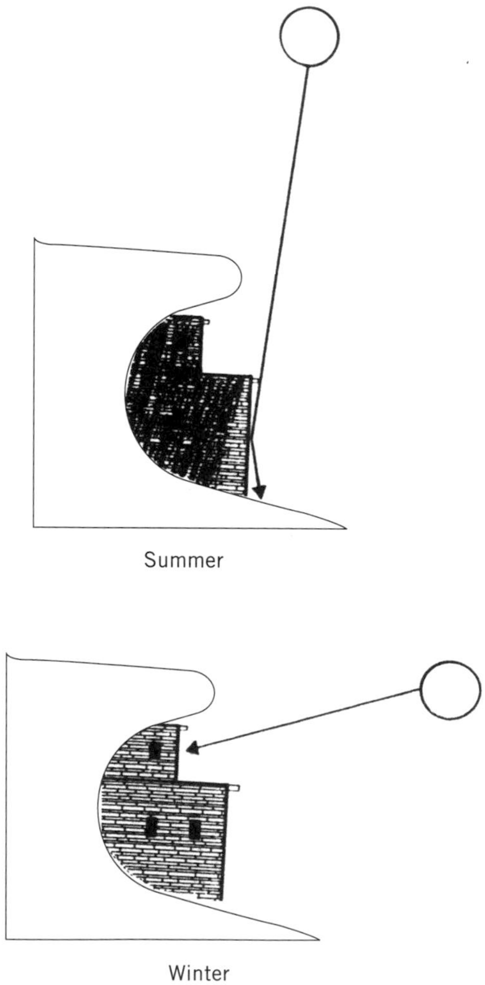

Figure 5. Cliff dwelling passive solar heating.

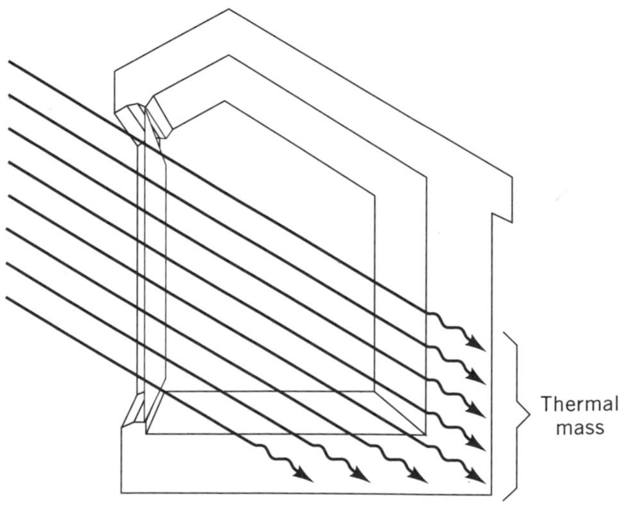

Figure 6. Direct passive heating.

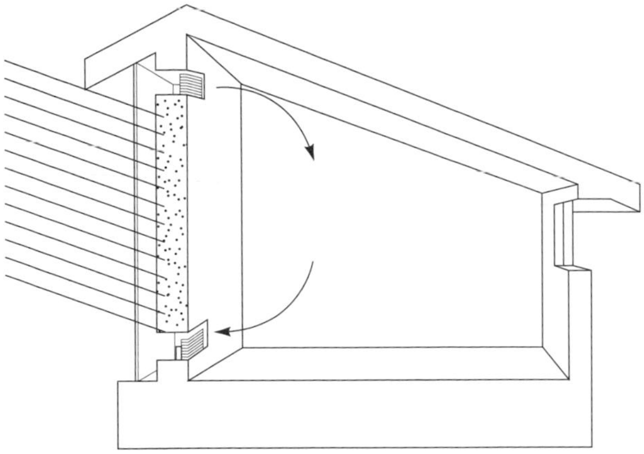

Figure 7. Indirect passive heating system using a Trombe wall. Windows heat the room during daylight hours and the Trombe wall heats it at night.

Daylighting

Daylighting is a passive technique which has become a very effective use of solar energy, not only for homes but schools, stores, hospitals, and commercial buildings as well. In most commercial buildings conventional lighting accounts for 25–30% of total energy cost, and also increases the cost of airconditioning. Daylighting thus reduces lighting and cooling costs as well; and if properly done does not increase the heating load. Well-designed daylighting systems now provide high-quality lighting for 2.5 to 5/kWh; 25–40% less than a decade ago (2).

Passive Solar Homes for National Parks

Early in 1993, The National Renewable Energy Laboratory began working with the National Parks Service in designing passive solar heated and cooled homes to be built in Arizona's Grand Canyon and Yosemite in California (3).

The 1600-square foot homes, which will house Park workers, integrate a range of energy-efficient features including a superinsulated envelope, a Trombe wall, and windows that provide high solar transmission with minimal heat loss through use of special gas fills and coatings. Insulation under the slab minimizes heat loss to the ground and indoor air quality is maintained by continuous forced ventilation. As a result, the Grand Canyon home is estimated to use 90% less energy for space heating and 63% less energy for water heating than conventional housing of the same volume. Electricity use will be minimized by efficient appliances and fluorescent lighting.

The Yosemite home is a passive design expected to cut energy use for summer cooling by 98%. This will be accomplished by prefabricated walls using a foam core, wire framing, and concrete sprayed on both sides to make a durable, well-insulated wall that also meets California's strict seismic standards. Cool night air flowing down from El Portal canyon will cool the building's mass and the house will be closed during daytime, when temperatures reach as high as 43°C. The integrated mechanical system will provide space heating, water heating, and ventilation. Annual energy savings are 73%.

The National Parks solar homes are part of NREL's Exemplary Buildings Program, providing technical assistance to U.S. industry in design evaluation and monitoring of energy-efficient residential and commercial buildings. This is accomplished by conducting industry workshops around the country.

ACTIVE SOLAR TECHNOLOGY

A variety of active solar collectors provide a broad range of applications; including solar water heating, active residential and commercial building heating and cooling.

Water Heating

Solar water heating was common in sunny regions such as California and Florida as early as the 1920s (4). The long popular "Day and Night Water Heater" was so named because its improved design provided solar hot water throughout the 24 hours. Using the principle of thermosiphoning, these flat-plate collectors heated water in elevated storage tanks.

The introduction of cheap natural gas ended this pioneering use of solar water heaters, and then only until a new generation of greatly improved solar water heaters were manufactured in the U.S. during the 1950s and sold in large quantities. Simpler designs were often built by home-owners themselves. The technology is now mature and reliable, heating water to about 60°C with flat-plate, glazed collectors. Most solar water heaters are active types; incorporating pumps but passive, thermosiphoning systems also perform well. Available in a variety of designs, both types comprise combinations of solar collector, storage tank, and a backup of conventional energy of some kind. In cold climates, freeze protection is provided for solar collectors.

About two-thirds of the cost of solar water heaters is in the initial investment. The price of hot water has dropped from 5 to 7 cents/kWh in 1980 to about 3.5 to 5 cents/KWh in 1993 and there are indications that the demand for solar water heaters may increase during the next decade. Earlier interest stemmed in part from utility incentive programs that provided cost reductions through utility/market participation. Manufacturing economies of scale may further reduce solar water heating costs as much as a third by 2000. Solar pool heating also continues to be popular, providing a water temperature of about 27°C. The technology is mature, solar collectors are inexpensive, and pool-heating costs range from 4 to 6/kWh (2).

Space Heating

In active solar heating of a home, a heat-exchange medium is circulated from a solar collector to the space to be heated. The most common solar collectors are adaptations of those long used for domestic water heating. Typically, a shallow rectangular box is insulated on bottom and sides, fitted with a metal absorber plate (having passages for water to flow through) inside the box, and glazed with glass or translucent plastic (Fig. 8). Heat absorbed by the absorber plate is circulated by a heat-exchange medium, eg, water, oil, antifreeze solution, or air. Air systems protect against freezing and no harm results from leaks (except for lower system inefficiency). However, an additional heat exchanger is required if the air collector is to heat water.

The absorber plate of a solar collector is generally painted flat black, or otherwise treated to enhance radiation absorptance. However, a good absorber is generally a good emitter as well. For this reason, selective surfaces are used to obtain high absorption with low emittance. Proportional control varies the speed of heat transfer fluid through a collector, slowing it during periods of weak insolation, whereas a high insolation rate speeds up the pump.

Water systems are more common than air systems. Offering better heat-exchange performance, they do have the disadvantages of potentially damaging fluid leaks and susceptibility to freezing. Because the absorber plate is generally a good radiator as well, water may freeze even at ambient temperatures well above O°C. Protection against freezing includes antifreeze in a separate loop (closed system), a drain-down mechanism, or heating the collector in cold weather either electrically or by pumping in warm water. The antifreeze loop seems to offer the best protection.

Active solar residential heating experienced strong growth during the late 1970s and early 1980s because of a combination of high conventional energy prices and government tax credits for solar energy use. This growth peaked in 1985, by which time about 1 million active solar home-heating systems had been installed in the U.S. Since then, in the absence of most federal and state tax credits, interest in active solar heating has virtually disappeared except for solar water heaters and pool heaters, which continue to experience substantial sales (2).

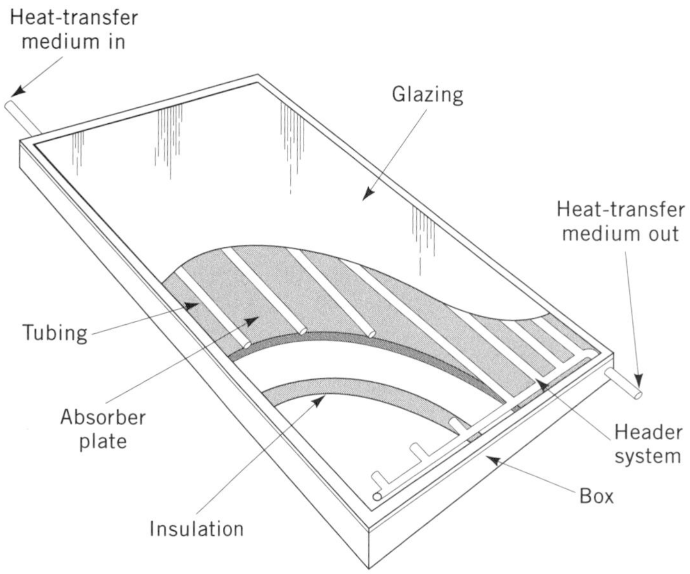

Figure 8. Diagram of a flat-plate solar collector.

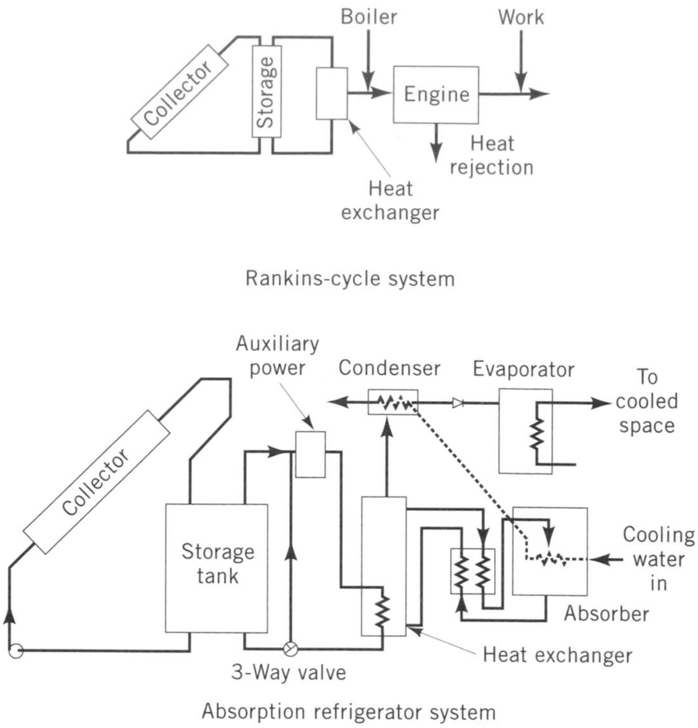

Figure 9. Rankine cycle and absorption solar–refrigeration systems. Courtesy of Marcel Dekker, Inc.

Air-Conditioning

Solar air-conditioning includes solar-powered refrigeration systems of both Rankine-cycle and vapor-compression types, as well as absorption systems (Fig. 9). Other cooling systems include solar-regenerated desiccant cooling. In a system developed by the Institute of Gas Technology, solar heat is substituted for conventional fuel in the desiccant-regeneration cycle.

The Department of Energy operated the National Desiccant Cooling Program in the 1980s using absorption and adsorption technologies, and both open and closed cycles. Open-cycle adsorption systems seemed to offer the best prospects (Fig. 10). In a typical desiccant-cooling cycle, ambient air is adiabatically cooled, dehumidified, cooled both sensibly and evaporatively, and then ducted to the

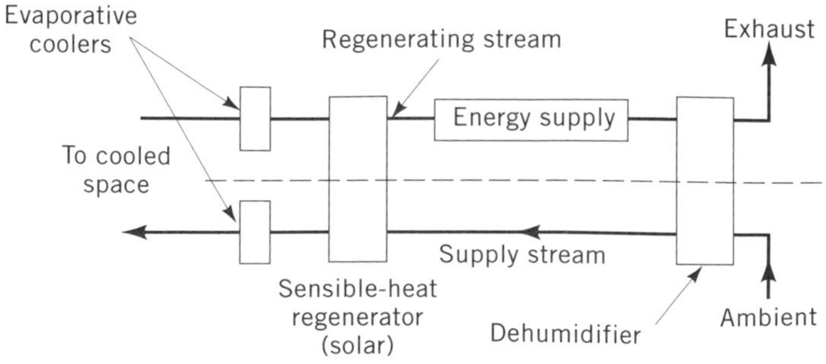

Figure 10. Open-cycle absorption desiccant cooling system. Courtesy of Marcel Dekker, Inc.

living area. In the regenerative stage, air is evaporatively cooled, heated as it cools the supply air stream, heated again by solar collectors, and humidified. Zeolites have been employed in some experimental work, and small solar-refrigeration systems have been built. Simulation and analysis of desiccant cooling systems suggest that solar-regenerated systems can be cost-competitive with conventional vaporcompression or absorption systems, and indicate a 20-yr payback on cost. Desiccant cooling seems best suited for regions with about equal heating and cooling loads and high humidity (see AIR CONDITIONING; REFRIGERATION).

The formation of the Electric Power Research Institute (EPRI) in 1972, an outgrowth of the Edison Electric Institute's Research Committee, was an early factor in utility company interest in developing solar energy applications for heating and cooling. Conservative by nature, the utilities historically have been cautious in developing new and unproved methods of energy production. Solar heating and cooling technologies marked the first utility involvement in using solar energy to supplant conventional energy.

While Detroit Edison, Public Service Electric and Gas of New Jersey, and other utilities had looked into wind energy as an option for producing electric power in the 1930s, and Arizona Public Service participated in the First World Symposium on Solar Energy in Phoenix, Arizona in 1955, most utilities tended to view renewable energy technologies as neither reliable nor economical enough to pursue. Not until the Oil Embargo of 1973 did the utilities begin to look seriously to renewable energy as a possible supplement. Rising electric power rates and concern about pollution of the environment by conventional power generation were other factors that encouraged a reexamination of solar energy in its various forms.

Solar heating and cooling were the applications of choice, because of the broad market and relative simplicity of operation. By 1980, with the utilities now considering solar heating and cooling as demand-side rather than supply-side considerations, EPRI transferred the technologies from its Solar Power Program to a demand-side management program (5).

Hundreds of utilities in all parts of the U.S. participated in thousands of projects involving research and development of solar heating and cooling of homes and buildings. Many lessons were learned in the process. Passive solar heating and cooling were found cost-effective in some locations but operating costs and maintenance of active heating and cooling far exceeded energy savings in many locations.

It is doubtful that the efforts of utilities in DOE's Solar Heating and Cooling program resulted in any great saving of either money or energy. However, many of these programs gave the utilities, the government, and ratepayers as well an increased awareness of the benefits of energy efficiency, conservation measures in all areas of energy use, and an effective understanding of the principles involved in solar heating and cooling toward a cleaner environment.

Although solar water heating and the solar heating and cooling of buildings were the first ventures of most utilities into solar energy use, these technologies are now generally considered a customer information service in the demand-side category. This is particularly true of passive technology, which is considered an energy-conservation measure rather than an electric supply option. Nevertheless, solar heating and cooling technologies are of continuing importance and offer attractive rewards for their wide application.

U.S. DOE PASSIVE SOLAR COMMERCIAL BUILDINGS PROGRAM

This program resulted in the design and construction of 19 commercial and institutional passive solar buildings ranging in size from 70 to 6,000 square meters, in a broad range of building types and climate zones. This effort suggested the potential of the solar heating of buildings.

In each project, the architect–engineer (who often was not familiar with solar heating strategies) was assisted by a team of experts in passive design and construction. The architect–engineer in each case went through one or more redesigns resulting from advice from the experts. Extra design costs incurred were paid by the Department of Energy. In addition to detailed energy calculations usually made with a simulation analysis computer program, two complete cost estimates were made for each base case design. These estimates used two cost databases in common use in the United States.

Following construction, 12 of these buildings were monitored for one year or longer to determine energy use in the four categories of heating, cooling, lighting, and all other uses. Post-occupancy evaluations were made by trained observers on most of the 19 buildings with the following results:

- Actual energy use was 47% less than the base case.
- Total energy costs were reduced by 51%.
- Daylighting was uniform and adequate in the three buildings monitored in detail. Top lighting was bound to be more effective than side lighting.
- Occupants like the buildings and comment favorably on the lighting.

Daylighting was employed in all buildings and was the design driver in most of them.

Following is a comparison of actual construction costs with estimates for the base-case buildings:

- Two buildings cost less than the range of estimates.
- Three of the buildings cost more than the range of estimates
- The cost of the remaining seven buildings fell within the range of estimates

CONCLUSIONS

The results of the DOE program, and many other passive building projects, demonstrates that heating, cooling, and daylighting can be integrated into building design effectively and without conflict. Energy requirements can be reduced in each category, and occupants prefer the solar-heated buildings to their conventional-energy counterparts (6).

SOLAR ENERGY MATERIALS AND EQUIPMENT

Because of the broad range of solar thermal applications and devices, a great variety of materials is used in them. These may be categorized generally as glazing, heat-transfer fluids, and seals and sealants. Criteria for selection include cost, weight, weathering characteristics, toxicity, flammability, outgassing properties, etc.

Glazing. Glazing is the first component of a solar collector to receive incoming solar radiation. Glass remains the material of choice, and plastic substitutes have yet to match its stability, weatherability, and resistance to ultraviolet radiation. Early in its operation, the federal government's Solar Energy Research Institute (now the National Renewable Energy Laboratory) encouraged the manufacture of fusion-drawn glass by Corning Glass Co. The firm produced a highly transmissive, very thin glazing material in an oxidizing atmosphere; containing optically inactive iron (III) and having a transmissivity near the theoretical maximum. The 20-mm thickness also permitted bending for use in linear parabolic mirror applications. The advantages and disadvantages of glass and various plastic glazing materials are given in Table 1.

Reflective Materials. Reflective materials for heliostats or concentrators include silvered glass, polished metals (often anodized or otherwise treated), eg, aluminum such as Alzak or Coilzak, and reflective foils and films. Glass mirrors are subject to breakage, wind loading, scouring by sand and dirt, radiation damage, and degradation by moisture. Metals incur similar damage, and some metallic concentrating collectors used in industry have been covered with plastic reflective films to increase or restore reflectivity. More recently, stretched metallized membranes have also proved efficient and offer lower costs.

Absorber Plates. Absorber plates in solar collectors vary from blackened plastics for low temperature uses, such as pool heating, to metals with selective surface coatings. Nonselective materials may be industrial flat-black or special solar paints with better ratios of absorptance to emittance. Research on selective coatings was pioneered in Israel, where solar water heaters are used extensively. Today, numerous commercial coatings are available (see Table 2).

Absorber plates for low-temperature applications are made from plastics but must resist ultraviolet degradation. Metals are used for higher temperatures and include

Table 1. Advantages and Disadvantages of Glazing Materials

Glazing	Trade Names	Advantage	Disadvantage
Glass		Excellent weathering resistance	Low impact strength High density Poor thermal-stress resistance
Acrylic	Plexiglass Acrylite Flexigard	Good insulator High impact strength Good weathering resistance	Low softening point Distortion when heated Surface abrasion
Fluorinated ethylene–propylene	Teflon	High chemical stability Excellent weathering resistance	High long-wave transmittance
Glass fiber-reinforced polyester	Fiberglass Filon Lascolite	Lightweight High strength High impact resistance High heat resistance Near opacity to long-wave radiation High durability	Poor weathering
Polycarbonate	Lexan Poly-Glaz	Good insulator High initial impact strength of all plastics	Low mar-resistance types do not weather well Surface deterioration by radiation/moisture High cost
Polyester	Mylar	Surface hardness Low cost	High long-wave transmittance UV degradation unless coated
Polyethylene		Light weight Flexible Low cost	Short lifetime High long-wave transmittance Wind and temperature sagging
Poly(vinyl fluoride)	Tedlar	High chemical stability High resistance to abrasion Excellent weathering resistance Low cost	High long-wave transmittance

Table 2. Absorptance and Emittance of Solar Absorber Coatings

Coating	Substrate	Absorptance	Emittance
Black chrome (Cr and CrO)	Nickel-plated metals, copper, steel	0.91–0.96	0.07–0.16
Black copper (CuO and Cu_2O)	Copper, nickel, aluminum	0.81–0.93	0.11–0.17
Black nickel (NiO_2)	Nickel, iron steel	0.89–0.96	0.07–0.17
Iron oxide (Fe_2O_3)	Iron, steel	0.85	0.08
Stainless steel oxide (FeO)	Stainless steel	0.89	0.07
Selective paint	Any	0.90	0.30

aluminum, copper, steel, and stainless steel. Copper and stainless steels have good corrosion resistance, an important characteristic for water-carrying collectors. The absorber is often a composite of aluminum fins bonded to copper-tube water passages. Some collectors are made entirely of aluminum or iron. In such cases, a noncorrosive heat-exchange fluid is used and galvanic action between dissimilar metals must be prevented (see Table 3).

Heat-Transfer Media. Heat-transfer media include liquids and gases; the former are used more commonly. Some solar collectors utilize air; some chlorofluorocarbons, including Freons. The latter may cycle between the liquid and gas phases during operation. Liquid heat-exchange fluids include water, hydrocarbon oils, glycols, and silicones. Choices are made on the basis of cost and physical properties such as heat capacity, specific gravity, thermal conductivity, viscosity, boiling and freezing points, and corrosion resistance (see Table 4).

Water is favored because of its availability, low cost, high specific heat, good thermal conductivity, and low viscosity. Glycols are slightly heavier than water; hydrocarbons and silicones somewhat lighter. Water suffers the disadvantages that it is a liquid over a relatively small temperature range and also is corrosive. However, distilled or deionized water (with appropriate inhibitors) is acceptable in metal passageways. The general pH range is given below.

Metal	pH Range
Aluminum	<5 or >8
Galvanized steel	<7 or >12
Stainless steel	<5 or >12

Copper must be protected from corrosive fluxes and certain chemicals; stainless steels from corrosive fluxes and chloride ions.

Solar Collector Boxes. Collector boxes are made from a variety of materials, including wood, plastic, and metal. However, wood is combustible and also susceptible to rotting, and plastics may not withstand high temperatures. Metals are therefore safer and more durable for solar collectors; aluminum has the added advantage of light weight.

Insulation. Insulation prevents loss of collected heat energy. Operating temperatures and environmental conditions dictate the type and amount of insulation required. Rigid polystyrene foam may be used in a solar swimming-pool heater, which has an upper temperature limit of about 75°C. However, a glazed, medium temperature solar collector can attain very high temperatures and fiberglass or an equivalent insulation is necessary.

Characteristics of various materials are given in Table 5. Because of the outgassing of binder material, some manufacturers use binderless insulation, eg, wool; especially where toxicity cannot be tolerated. Urethane foams are highly toxic at elevated temperatures.

Seals and Sealants. Seals and sealants are used to join materials, accommodate to thermal expansion, prevent leaks, and isolate the collector and system from the environment, as shown in Figure 11. Glazing must be tightly yet flexibly in contact with collector boxes; insulation must adhere to box or absorber; joints must be secure in all weather conditions and against temperature and internal pressure; and pumps must be tightly sealed against leaks. Seal and sealant materials must withstand thermal cycling and undergo minimum outgassing. Elastomers are choice materials because of their rubberlike, elastic qualities. Their temperature limits and compatibility with heat-transfer fluids are given in Tables 5 and 6, respectively.

Table 3. Properties of Heat-Transfer Media

Characteristic	Water	Glycols	Silicones	Hydrocarbons
Heat capacity at 82°C, kJ/(kg·K)[a]	4.18	3.55	1.59	2.00
Specific gravity at 82°C	1.00	1.02	0.93	0.88
Thermal conductivity at 82°C, W/(m·K)	0.64	0.41	0.14	0.12
Viscosity at 82°C, mm^2/s (= cSt)	0.35	1–2	5–10	1–10
Coefficient of thermal expansion, %/°C	0.005	0.017	0.033	0.017–0.044

[a] To convert J to Btu, divide by 1054.

Table 4. Thermal Conductivities and Upper Temperature Limits of Insulation Material

Insulation	Thermal Conductivity, mW/(m·K)	Upper Temperature Limit, °C
Calcium silicate	55	650
Mineral fiber block	36	1040
Mineral fiberboard	55	650
Perlite	69	815
Refractory fiberboard	40	650
Glass fiberboard	36	345
Cellular elastomer	36	105
Polystyrene	28	75
Polyurethane	23	105
Isocyanurate resin	24	120
Phenolic resin	32	135
Urea–formaldehyde resin	35	132

HEAT STORAGE

If solar energy were received continuously and evenly, without differences between winter and summer insolation, the design of solar-energy collectors would be simple. However, insolation varies diurnally as well as seasonally, and it is generally necessary to provide means for storing solar heat for use at night and during cloudy periods. It is seldom cost-effective to design a solar collector for the least amount of energy received. Although such an approach would guarantee some heat even during rare, week-long cloudy periods, capital investment would be prohibitively high. It is thus more economical to employ a conventional energy standby system, with the sun providing some optimum fraction of the needed amount of energy.

Passive solar space heating presents a special example of the need for heat storage. If a sizable greenhouse is attached to a conventional residence, or sufficient south-facing window area is provided, the residence may quickly become very hot by day but lose heat quickly at night. The incorporation of heat-storage materials, eg, masonry floors and wall, rock bins, and water tanks (all commonly called thermal mass) solves the problems of these temperature swings While the sun is shining, the building's thermal mass heats slowly, keeping air temperature moderate. At night, the thermal mass slowly radiates stored heat to keep the house warm (Fig. 12).

Sensible Heat Storage

Heat can be stored in water at about 4 kJ/(kg·K) [1 Btu/(lb–°F)] increase in temperature. Rock also stores heat, although not as much for a given volume. Both these media store sensible heat:

$$Q = MC_pT$$

where Q = heat stored, M = mass of storage material, C_p = specific heat of storage material, and T = temperature change of storage material.

Sensible heat storage in residences is common in Europe, especially in Germany and the United Kingdom. Typical systems use refractory brick for storing heat at temperatures as high as 650°C. Effective and durable insulation is required to keep baseboard units at a safe external temperature of about 60°C. Electric heating rods are also used, and thermostatically controlled fans circulate hot air. Larger, central-heating storage furnaces have refractory-brick or cast-iron cores. Solar energy in residential applications provides only moderate temperatures, and a much larger volume of thermal mass is required.

A heat-storage system called Annual Cycle Energy Storage (ACES) was developed in the late 1970s at Oak Ridge National Laboratories (Fig. 13). 70 m^3 (18,000 gal) of water was stored in the basement of a test house. In

Table 5. Temperature Limits of Elastomers

Elastomer	ASTM Designation	Service Temperature, °C Low	Service Temperature, °C High	Heat Resistance[a]
Acrylate–butadiene	ABR	−30	175	
Butadiene rubber	BR	−75	80	
Styrene–butadiene (buna-S)	SBR	−55	95	F
Isobutene–isoprene (butyl)	IIR	−55	120	G–E
Ethylene–propylene–diene rubber	EPDM	−60	150	G–E
Fluoroelastomer	FKM	−40	230	E
Poly(chloromethyloxirane) (epichlorohydrin polymer)	CO, ECO 494	−40	165	
Chlorosulfonated polyethylene (Hypalon)	CSM 3	−55	135	
Polyisoprene, synthetic	IR 249+	−60	80	
Natural rubber	NR	−60	80	
Polychloroprene	CR	−55	120	G–E
Acrylonitrile–butadiene (nitrile) rubber	NBR	−55	135	G–E
Polyblend (PVC–NBR)	PVC–NBR	−20	120	
Polysulfide (T-13)		−40	110	P—G
Silicone (SI, FSI, PSI 35+, VSI, PVSI)		−95	290	E
Polyurethane (U-138)		−40	120	

[a] P = poor; F = fair; G = good; E = excellent.

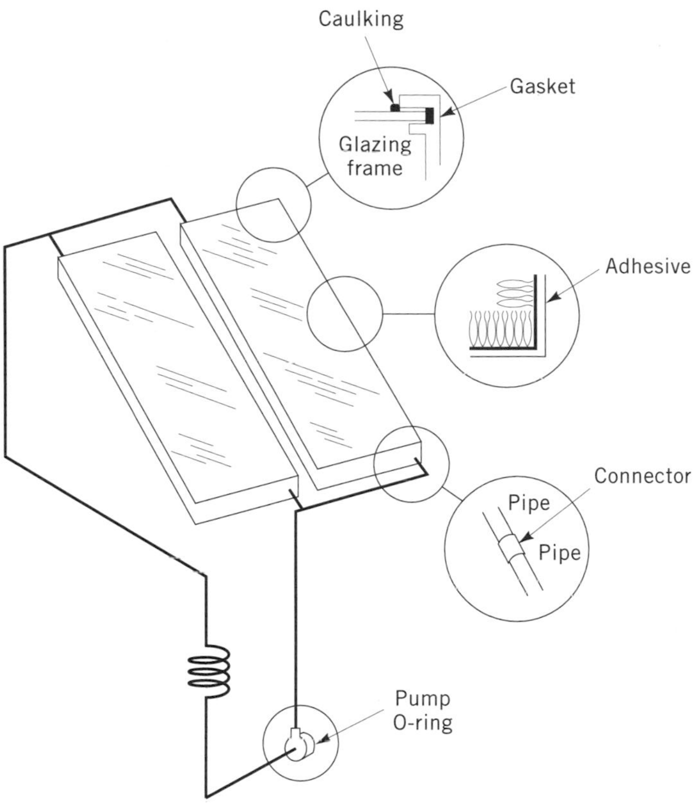

Figure 11. Seals and sealant materials in a solar heating system. Courtesy of *Solar Engineering*.

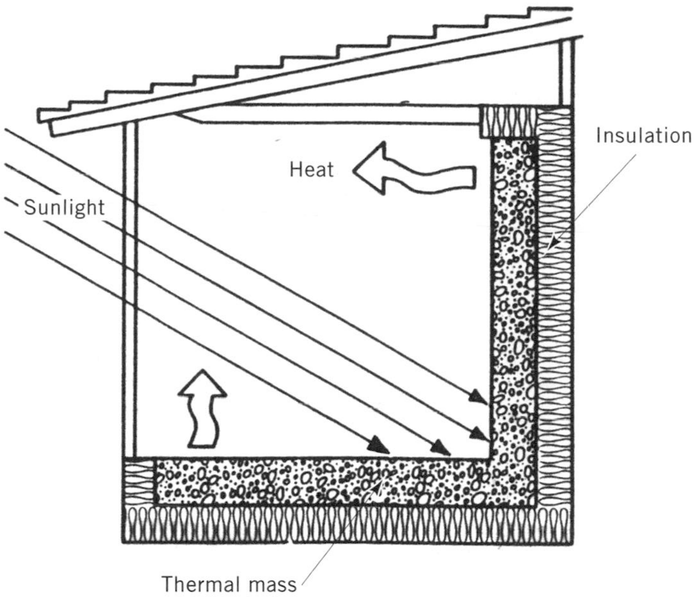

Figure 12. Heat storage in passive solar design.

winter, a heat-pump system heated the house by drawing heat from the stored water. By spring the water had frozen and was then used for cooling the house in summer. Such a system uses both sensible and phase-change heat storage. Analogous large-scale heat-storage systems for communities entail storing heat in aquifers, ponds, earth, and rock. Economies of scale are gained because the ratio of volume to surface area increases with the size of the storage area.

High specific heat is desirable in a storage medium. Furthermore, the thermal mass must provide continuous heat conduction. For example, solid concrete wall provides better thermal storage than a hollow block wall filled with sand. If a block wall is filled with grouting, voids should be prevented. The properties of heat-storage media are given in Table 7.

Latent Heat Storage

Latent heat storage requires a phase change in the storage medium, ruling out materials such as brick, concrete, and rocks. Water is a common example of enhanced heat storage with a phase change. Melting ice absorbs great quantities of heat with no change in temperature. For cooling this property can be put to good use, but the phase-change temperature of water is not suitable for storing heat for later use. A material is needed with a high heat of fusion that changes phase at a higher temperature than water. Desirable characteristics include small change in volume and low vapor pressure. Sodium sulfate decahydrate, commonly called Glaubers salt, was one of the first material used. The properties of typical phase-change material are given in Table 8.

Table 6. Compatibility of Elastomers with Heat-Transfer Fluids[a]

Elastomer	ASTM Designation	Mineral Oils	Aromatic Hydrocarbons		Ethylene Glycol	Silicone Oils
			Oils	Alkylated		
Acrylate–butadiene	ABR	G	G	G	P	G
Butadiene	BR	P	P	P	G–E	F–G
Poly(ethylene-*co*-propylene)	EPM	P	P–F	P–F	E	E
Ethylene–propylene–diene rubber	EPDM	P	P–F	P–F	E	E
Chloroprene	CR	G	P–F	P–F	G–E	G–E
Acrylonitrile–butadiene (nitrile) rubber	NBR	E	F–G	F–G	G–E	G–E
Polysulfide (T-13)		E	E	E	F–E	P–F
Silicone (SI 35+)		P–F	P–F	P–F	F–E	P–F

[a] P = poor; F = fair; G = good; E = excellent.

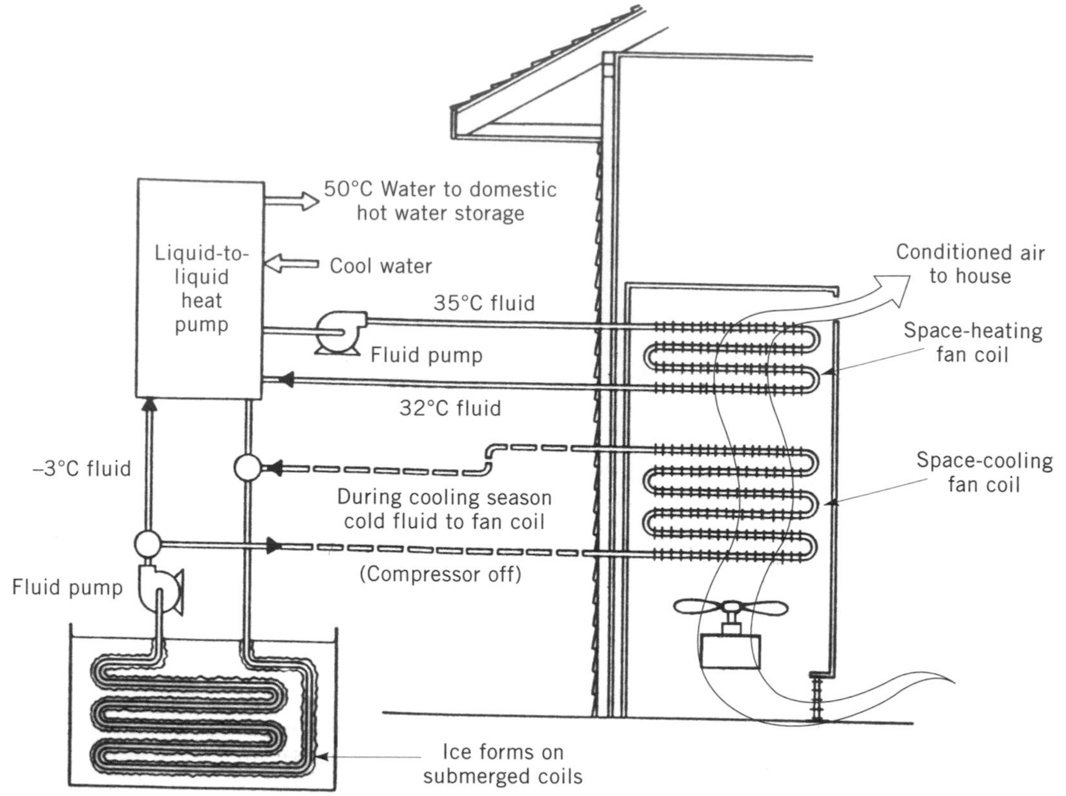

Figure 13. Annual-cycle energy storage system. Courtesy of Oak Ridge National Laboratories.

Storage of one GJ (about 10^6 Btu) at a reasonable space-heating temperature requires about 33 m^3 (1150 ft^3) of rock or 13 m^3 (460 ft^3 or about 3450 gal) water. A phase-changing salt solution, however, can store a GJ (ca 10^6 Btu) in only 2.8 m^3. The problems encountered in phase-change heat storage stem from crystallization and gradual loss of ability to change from liquid to solid after a number of phase-change cycles. Storage in shallow trays, mechanical mixing, rotating drums, or binding agents prevent crystallization or settling out. Compactness is desirable for retrofit heat-storage application. However, care must be taken to prevent undesired interactions with other materials. For example, some PCM materials placed within hollow concrete blocks expand and cause fractures.

Because of the problems presented by phase-changing salts, some researchers have recommended paraffin waxes for solar heating and tetrahydrofuran for air conditioning. However, the Cabot Corporation test house built in the 1950s ago at MIT successfully used a eutectic salt mixture through 3500 cycles without appreciable degradation of performance. Trade journals advertise an energy rod that uses a phase-change compound capable of storing

Table 7. Properties of Various Heat-Storage Media

Storage Medium	Heat Capacity, kJ/(kg·K)	Density, g/cm³	Unit Heat Capacity, J/(cm³·K)	
			No Voids	30% Voids
Water	4.18	1.0	3.6	
Water 30%/glycol 70%	3.34	1.02	3.1	
Scrap iron	0.45	7.82	3.4	2.3
Magnetite	0.67	5.12	3.4	2.3
Scrap aluminum	0.88	2.69	2.3	1.4
Concrete	1.13	2.24	2.3	1.6
Rock (basalt)	0.84	2.88	2.3	1.4
Brick	0.84	2.24	1.8	1.3
Rock salt	0.91	2.17	2.0	1.3
Adobe	0.84	1.92	1.4	
Wood (pine)	2.80	0.49	1.3	
Paraffin oil	2.09	0.78	1.4	

Table 8. Properties of Phase-Change Materials

Material	Mp, °C	Heat of Fusion, kJ/kg[a]	Heat-Storage Density, MJ/m^3[b]	Approx. Cost, $/kg
Sodium sulfate decahydrate	31	251	372	0.02
Sodium thiosulfate pentahydrate	49	209	335	0.16
Barium hydroxide octahydrate	82	265	657	0.66
C_{14}–C_{16} paraffins	2–7	152	119	0.22
Sodium fluoride–magnesium fluoride (1 : 1)	832	625	1370	1.10–0.55
Sodium hydroxide	320	159	284	0.65

[a] To convert kJ/kg to Btu/lb, divide by 2.32.
[b] To convert J to Btu, divide by 1054.

385 MJ/m^3 (10,300 Btu/ft^3) at 27°C, and 485 MJ/m^3 (13,000 Btu/ft^3 at 49°C).

High Temperature Heat Storage

Solar power and industrial process-heat systems call for higher temperature heat storage, ie, >250°C, than residential or commercial applications do. Obviously, more insulation is required to maintain elevated temperatures for long periods. At very high temperatures, for example a turbine driven at 800°C, the containment vessel must be constructed of higher quality material, perhaps stainless steel. The insulating material must not only insulate the heat-transfer medium but also survive the elevated temperatures. High temperature storage media include oil/rock composites and phase-change salts (see Fig. 14).

Reversible Thermochemical Reactions

Heat-storage materials used for thermochemical reactions include hydrated, ammoniated, or methanolated salts, sulfuric acid, hydrogenated metals, and hydrated zeolytes. Functioning as chemical heat pumps, these materials have estimate coefficients of performance about 1.6 in the heating mode, and <1.0 in the cooling mode. Essentially, thermochemical energy is based on the concept: AB = A + B (Fig. 19). It offers high energy density and long-term storage at ambient temperatures. Products and reactants are easily transported, and therefore, endothermic and exothermic reactors can be separated by appreciable distances. However, thermochemical systems are extremely complex compared with sensible and phase-change storage systems.

SOLAR PONDS

Conventional solar collectors are expensive, with typical flat-plate collectors costing $150/$m^2$ (about $14/$ft^2$) and tracking concentrating collectors several times as much. Such high costs for using solar energy make the much simpler solar pond a desirable alternative for some applications.

The phenomenon of nonconvecting or reverse temperature-gradient lakes was discovered at the end of the 19th century at Lake Medve in Transylvania, Romania. Water at the bottom of that lake had a temperature of 65°C, the reverse of the normal situation where warm water rises to the surface and loses its heat quickly. Researchers dis-

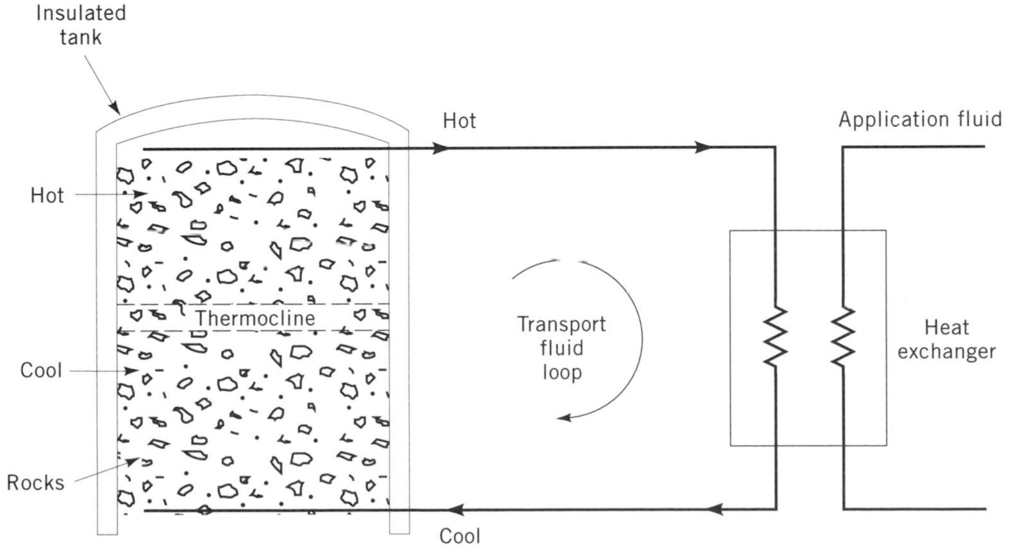

Figure 14. High temperature storage. Courtesy of Marcel Dekker, Inc.

covered stratified, nonmixing water layers of varying density (7). Other natural "reverse-gradient" ponds have since been found in such diversified areas as the state of Washington in the United States, Antarctica, and Venezuela.

Commercial solar ponds were first developed by Israel scientists in the 1950s; stemming from availability of ample solar energy, the Dead Sea as an ideal solar pond, and the new nation's pressing need for alternative energy sources. By 1959, the first man-made, reverse-gradient solar pond was completed in Israel and a specially designed turbine for producing electric power using water at 90°C was installed (8). However, electricity produced by solar ponds could not compete with the very low price of fossil fuels at that time, and the new turbine was later converted for use with fossil fuels.

After 1973 and the growing uncertainty of oil supplies, Israel resumed its work on solar ponds. The new turbines were larger, and in 1979 a 150 kW_e generator began operations at a 7500 m^2 (about 81,000 ft^2) solar pond. This was a test unit; the Israeli program called for a 5 MW turbine/0.25 km^2 pond in 1982, a 20 MW/1 km^2 system in 1983, and a 20 MW/4 km^2 system by 1985. The long-range goal was for 2 GW of power from its solar ponds by the end of the century. The 5 MWe turbine was installed adjacent to a 20-hectare Israeli solar pond in the early 1980s and operated successfully until 1989, when it was shut down for economic reasons.

Solar ponds are constrained by gravity to operate in a horizontal plane, thus suffering appreciable loss of intercepted solar radiation compared with tilted or tracking solar collectors. However, the very low cost of solar pond construction tends to offset this inefficiency. A nonconvecting solar pond typically comprises three layers: a thin, convecting top layer that serves as a buffer for wind; a middle, salt-gradient, non-convecting layer; and a convecting bottom layer that stores most of the heat.

The bottoms of solar ponds are covered with dark, heat-absorbing material. Thin, transparent membranes interposed between layers may be used to prevent mixing. Experimental nonconvecting solar ponds include gelled ponds and viscosity-stabilized ponds; more research is indicated in this area.

The Salton Sea Solar Pond Project

Far exceeding the scope of existing solar pond was the Salton Sea Project proposed in the early 1980s, with a feasibility study beginning in 1981 (Fig. 15). Funded by the Department of Energy, Department of Defense, the State of California, California Edison, and the Israeli Ormat firm that had done most of the work on solar ponds to that time. Proposed to cost only $650,000, the project entailed a Phase 1 feasibility study, a Phase 2 5-megawatt pilot plant, and a Phase 3 600-megawatt electric power plant.

With a surface area of almost 1000 km^2 and a salinity of 4%, the Salton Sea is ideally situated in one of the sunniest regions of the world. Projections including the entire area of the sea suggested a power output of 5 gigawatts. Also considered as sites were San Francisco Bay and Utah's Great Salt Lake; with Salt Lake's potential 22 gigawatts. As yet, however, nothing has resulted from these ambitious plans. Nevertheless, as indicated in Table 9, several relatively small solar ponds have been built in the U.S in recent years.

The First "Progress in Solar Ponds" International Conference was held in Cuernavaca, Mexico from March 29 through April 3, 1987. Seventy attendees participated in the workshop and conference and also visited an Electric Research Institute test solar pond under construction at Zacatapec. The Second International Conference was held in Rome from March 25 to March 30, 1990. Fifty registrants heard a number of presentations, including one from Dr. Harry Tabor, a solar pond pioneer from Israel.

From May 23 to May 27, 1993, the University of Texas at El Paso, hosted the Third International Congress on Progress in Solar Ponds, with delegates representing the United States, China, Switzerland, India, Australia, Israel, Mexico, Italy, and Hawaii. This time, the Bruce Foods solar pond in El Paso was the principal attraction of the conference, and may well have been the reason the meeting was held in Texas (9).

Bruce Foods El Paso Solar Pond

The most successful solar pond in this country as of 1993 was first installed ten years earlier on property of the

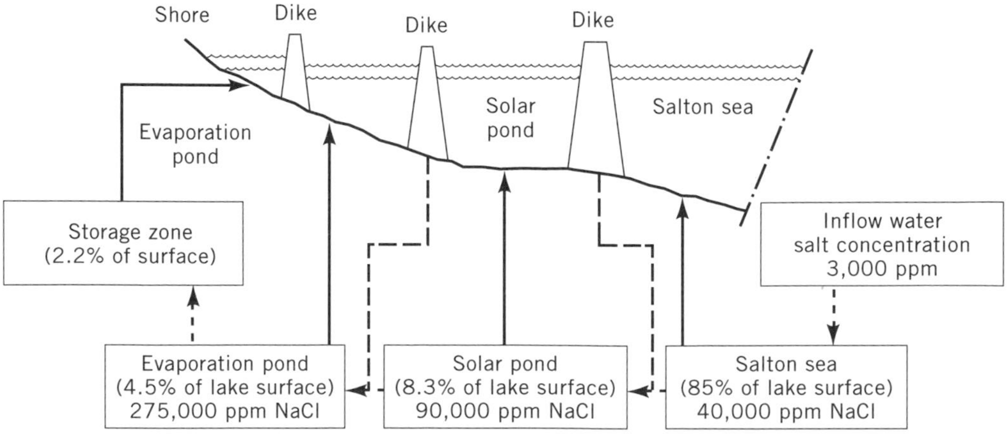

Figure 15. Diagram of Salton Sea Solar Pond.

Table 9. Performance of U.S. Solar Ponds

Application			Pond			Av output, kW (W/m^2)	Energy cost, ¢/(kW · h) ($/$10^6$ Btu)
Process	Location	Latitude	Shape	Dimensions, m	Cost, $/m^2 ($/ft^2)		
Winter space heating	Ohio	40° N	Circular	50 dia	47.50 (4.41)	54 (27.5)	2 (5.86)
Laundry-water preheating	Ohio	40° N	Circular	100 dia	18.33 (1.70)	291 (37)	0.57 (1.67)
Water heating for hotel	Hawaii	20° N	Circular	100 dia	12.00 (1.11)	526 (67)	0.20 (0.59)
Process heat for salt works	Texas	26° N	Rectangular	500 × 2000	5.30 (0.49)	50 (50)	0.12 (0.35)

Ashley Division of Bruce Foods Corporation in El Paso, Texas (10). Initial funding was provided by the U.S. Bureau of Reclamation, the Texas Energy and Natural Resources Advisory Council, and El Paso Electric Company. A 3355 m^2 Hypalon-lined water storage pond that had served as fire protection for a previous tenant was converted into the solar pond by adding an XR-5 liner over the Hypalon layer, with 15 centimeters of sand between liners.

One hundred truckloads of sand obtained from the nuclear-waste storage test facility in Carlsbad, New Mexico was then added to the pond. Adjustable suction and discharge diffusers, an external heat exchanger, and a preheater for feedwater to be used by Bruce Foods were then installed. By late May, 1985, the scanning injection method was used to establish a 100 cm-thick linear gradient zone. The pond began heating on June 12 and process heat was first extracted on July 12 when the pond reached 64°C. However, at 73°C, the gradients began to move apart and the pool rapidly lost heat.

During the winter of 1985–1986, the pond operators' expertise in gradient modification improved. The salinity was increased to 26% (saturated) throughout the pond, and the entire pond was allowed to mix. In May, 1986, a new gradient was established under more controlled conditions and the pond began heating at the rate of 1°C per day and continued at that rate through June, 1986 except for cloudy weather from June 20 to 25. The pond temperature reached 85°C and held at that level. After July 15, heat was delivered to the food-processing plant as preheat for boiler feedwater.

Total energy delivery to September 15 was about 350×10^6 kJ, with the plant receiving 80,000 liters of boiler feedwater. Peak extraction of thermal power for a typical 8-hour working day was on the order of 0.75 MW, with the average being about 0.28 MW/day. A brine flow rate of approximately 1200 liters per minute showed no adverse effect on the pond gradient.

In late July, 1986, a 100 kW organic Rankine-cycle was purchased from Ormat Turbines of Israel, and on September 19, with the solar pond at 85°C, electricity was first generated and grid-synchronized. In its 5 years of operation, this solar pond produced more than 50,000 kWh of electricity. In 1987, a desalination plant was installed for the production of 19,000 liters of drinking water per day from brackish wellwater.

In 1989, the Texas Energy Research in Applications Program established the Texas Solar Pond Consortium, with the University of Texas at El Paso as the lead institution for operational research. Funding from these two agencies has been about $180,000 a year through 1993, with the total amount less than $1 million. In 1992, the El Paso Solar Pond Project successfully generated electric power and also provided the thermal energy to drive the desalting process; the first such demonstration in the world.

West Texas is an ideal site for development of solar ponds, since the area is chronically short of freshwater but has ample supplies of high salinity groundwater. Other proposed Texas applications for solar ponds include oil/brine separation and enhanced oil recovery; wastewater control, mining sodium sulfate, and producing industrial heat for the dying/finishing industry, fish farming, and linking solar pond technology to the production of ethanol to make ethanol cost-competitive.

Shallow Solar Ponds

Shallow, fabricated solar ponds have also been used for industrial process heat, for example those built by Lawrence Livermore Laboratory during the 1980s. Typically 5 × 60 m, and 10 cm deep, the LLL ponds had black, heat-absorbing bottoms, insulated sides, and plastic-glazed tops, as shown in Fig. 16. Water was heated in the ponds during the day and stored at night in underground tanks. Tested in a uranium-ore mining operation at Grants, New Mexico, the ponds were nearly cost- competitive with oil heat.

It has been pointed out that solar ponds are unlikely to generate large amounts of electric power since they are limited because of their significant consumption of water. This is so because water is continuously replenished to compensate for surface evaporation and to recycle salt. However, solar ponds are useful for providing electricity for pumping in desalting applications. This can be useful in remote areas such as coastal deserts or island sites (11).

INDUSTRIAL PROCESS HEAT

In 1990, the U.S. industrial sector used about 21 quads of energy from various fuels, plus an additional 3 quads of

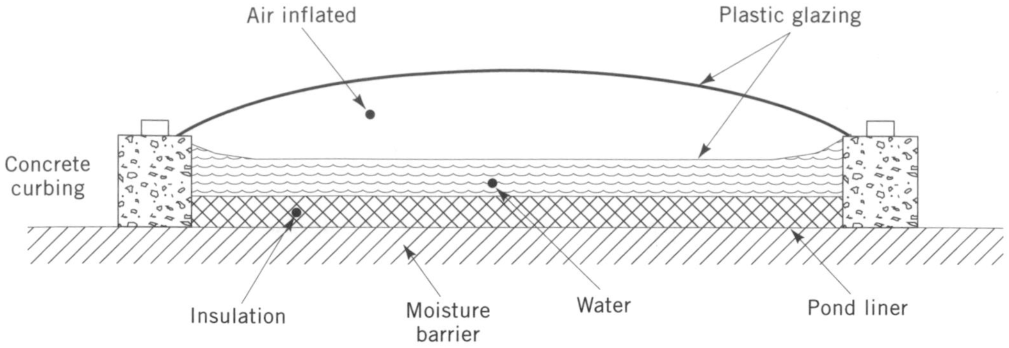

Figure 16. Shallow solar pond. Courtesy of Lawrence Livermore Laboratories.

electricity (2). Since 10 quads of primary energy were consumed in generating the electricity, total energy use by the industrial sector was about 31 quads. This was about the same amount used in 1980, thanks in large part to energy conservation measures put into effect by industry. However, the Energy Information Administration projects an increase of about 20% in industrial energy consumption by the year 2000.

Industrial and agricultural process-heat applications are often amenable to active solar heating. Food dryers and car washes represent the low temperature end of this spectrum, which extends upward to very hot water and even to steam in the food, textile, brewing, and other industries. The best flat-plate collectors produce temperatures up about 100°C, but at considerable loss of efficiency at the high end. Concentrating collectors are therefore preferred in the medium temperature range. Those shown in Figure 17 are capable of producing steam for industrial process heat. Applications include textile processing, irrigation pumps, and domestic and commercial air-conditioning. Approximate percentages for applications and temperatures are given in Tables 10 and 11.

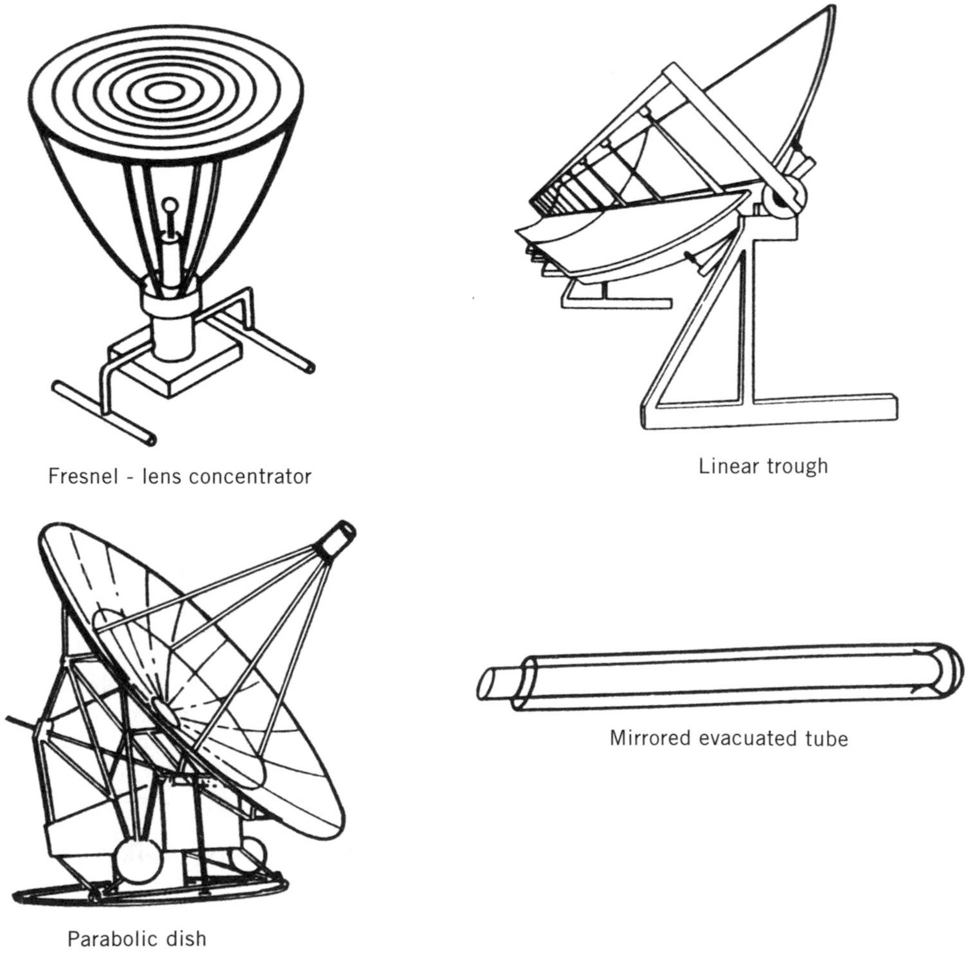

Figure 17. Concentrating heat collectors.

Table 10. Temperature Produced by Various Solar Collectors

Collector Type	Temperature, °C
Solar ponds/flat plate	100
Evacuated glass-tube collectors	50–175
Line-focus concentrators	65–300
Point-focus concentrators	200–800
Central receiver	800–1400

Table 11. Industrial Applications at Various Temperatures

Temperature, °C	Percent of Total
100	10
175	10
175–300	5
300–600	18
600–1100	12
1100	45

Solar collectors shown in Figure 17 are capable of producing steam for industrial process heat. Applications include textile processing, irrigation pumping, and domestic and commercial air-conditioning. Since the manufacturing industry uses about 35% of the nation's energy, it offers a large market for solar-energy systems. Approximate percentages for industrial applications at various temperatures are given in Tables 10 and 11. These are industry-wide averages; requirements within specific industries vary. For example, estimates for the oil-refining industry are: <300°C, 22%; 300 to 600°C, 63%; >600°C, 15%. Some oil companies have investigated the use of solar heat in secondary-recovery operations but found it not cost-effective at prevailing oil or gas fuel prices.

At the other end of the temperature range is the Los Alamos National Laboratory large-area shallow solar-pond collector used in a uranium ore processing demonstration at Grants, New Mexico. This project close to being cost-competitive.

Six industrial process-heat demonstration projects funded by DOE in the 1970s for low-temperature (<38°C) applications included the drying of textiles, lumber, soybeans, and fruit, the curing of concrete block, and can washing (Fig. 18). Both liquid and air were used as working fluids. Installed costs (less prototype design costs) were \$(250–870/m^2 (\$(23–81)/ft^2) of collector area. These high first costs resulted from the fact that each system and application was unique and required special treatment throughout development. A cost of \$(50–100/m^2 (about \$(510)/ft^2) was considered a desirable goal.

As of the early 1980s, several dozen industries in the United States had solar-energy process-heat installations in operation or on the drawing board. Four typical systems, including a brewery, a food-dehydration plant, a natural-gas-processing facility, and a food-processing plant. Operating temperatures ranged up to 300°C. Although these were pilot plants, the economic projections were encouraging. Only one system was ranked as not cost effective; the others estimated payback times ranging from 16 to 21.5 yr with projected fuel cost increases. Typical commercial applications are given in Table 12.

Cost analyses used in the American Solar Energy Society White Paper (2) indicate an optimum size for solar thermal industrial plants between 10 and 100 MW^t. Most

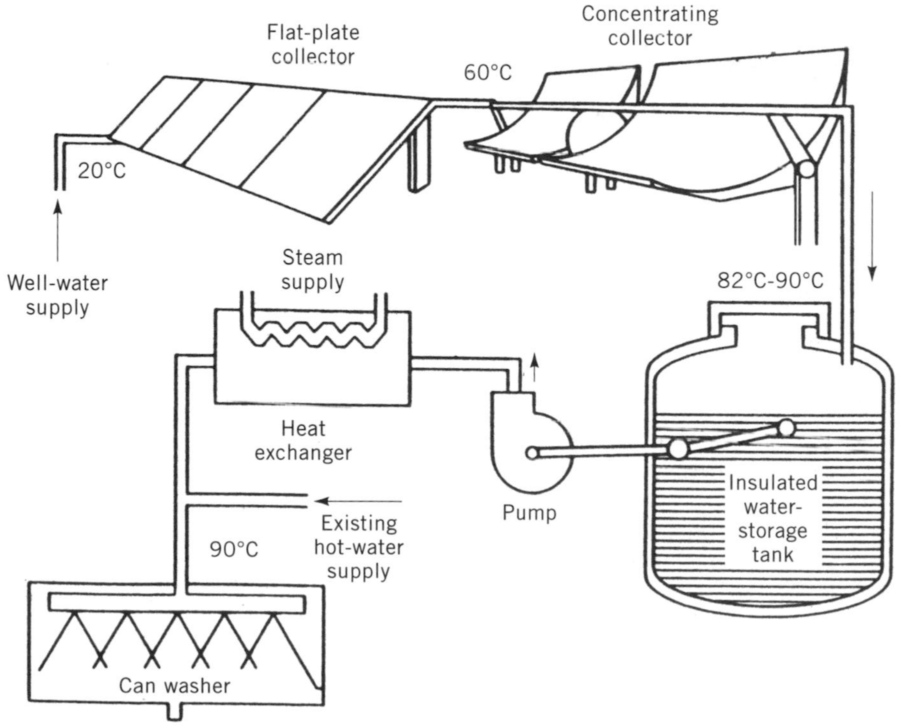

Figure 18. Diagram of solar can-washing system.

Table 12. Early Commercial Applications of Solar Energy

Company	Location	Application	System Type	Energy Saved, GJ/yr[a]
Anheuser-Busch	Jacksonville, Fla.	Pasteurization	Trough	527
Campbell Soup	Sacramento, Calif.	Can washing	Trough	1900
Gold Kist	Decatur, Ala.	Soybean drying	Flat plate	843
J.A. LaCour Kiln Services	Canton, Miss.	Lumber drying	Flat plate	422
Johnson & Johnson	Sherman, Tex.	Bleaching, sterilization	Trough	1580
Lamanuzzi & Pantaleo Foods	Fresno, Calif.	Food drying	Flat plate	1050
Riegel Textile	La France, S.C.	Dyeing	Flat plate	422
Reedy Creek Utility	Lake Buena Vista, Fla.	Air conditioning	Trough	843
York Building Products	Harrisburg, Pa.	Concrete-block curing	Flat plate	422

[a] To convert J to Btu, divide by 1054.

of these systems use simple, one-axis trough concentrators. Two-axis "power towers", or dish arrays are more efficient for higher-temperature applications.

Industrial and agricultural process-heat applications are often suitable for active solar heating. Food dryers and car washes represent the low temperature end of this spectrum, which extends upward to very hot water and other liquids in the food, textile, brewing, and related industries, and even to steam. The best flatplate collectors produce temperatures up to about 100°C, but at considerable loss of efficiency at the high end. Concentrating collectors are therefore preferred for medium and high temperature applications.

Solar thermal process heat demonstrations in a number of varied industries were funded by DOE from the mid-1970s to mid-1980s and generally performed well. However, the loss of "level-playing field" tax credits in the years following, and the problems of effectively integrating solar energy with conventional manufacturing processes caused a return to fossil fuel use in many cases. Land requirements for sufficient insolation, operation, and maintenance costs, and the problems associated with melding the new and unaccustomed solar technologies with long-established fossil fuel heating methods contributed to this shift. However, between 1980 and 1990 the cost per million Btu of industrial solar process heat dropped from \$15–20 to \$5–10. A further reduction to between \$4 and \$8 per million Btu by the year 2000 is expected (2).

There has been a corresponding resurgence of interest in industrial uses of solar heat, resulting in the rehabilitation of many long idle installations. Congress in 1992 allotted \$10 million for solar energy projects in Department of Defense organizations. Such projects include water heating for various uses at March Air Force Base, California, Kirtland Air Force Base, New Mexico, U.S. Army Dugway Proving Grounds, Utah, and construction of new solar plants at Fort Huachuca, Arizona.

A 12,000 ft^2 solar thermal trough space heating system at Luke Air Force Base in Arizona, shut down since 1980, is also being refurbished. A 30,000 ft^2 solar trough hot water system built for a college at Blythesville, Arkansas in 1980 and operated until 1985, is now under consideration for reinstallation at a state prison. The DOE is also considering repowering two 45 MW electric power plants in Albuquerque, New Mexico with solar energy.

In 1991, high winds created operational problems for Industrial Solar Technology's solar trough hot water system at Tehachapi, California. Sandia engineers studied the problem and suggested structural changes in the collector field to prevent further damage. Sandia is also working with Gould Incorporated's Foil Division to upgrade its 60,000 ft^2 solar thermal hot water facility. At the other end of the technology scale, Sandia is testing a commercial concentrating solar oven for campers built by Burns-Milwaukee.

Economic Status and Prospects for Solar Thermal Industrial Heat

Personnel at the National Renewable Energy Laboratory in Golden, Colorado in 1992 (12) projected levelized energy costs (LEC) for industrial process heat equipment used in industrial process heat applications, commercial heating, and commercial cooling. Three families of solar collectors were studied: flat plate concentrators using air or water as the working fluid for temperatures from 40°C to 80°C; parabolic trough technology from 100°C to 350°C, and parabolic dishes to 800°C. Solar furnaces, described later, yield far higher temperatures and innovative approaches to materials processing. Focus of the study was a comparison of annual energy outputs of state-of-the-art parabolic trough systems and enhanced trough systems at three locations:

Location	Phoenix	Bakersfield	Denver
Annual output, GJ	7,720	7,281	6,170
Annual output GJ (enhanced)	12,178	11,557	10,148
Increased output	58%	59%	65%

The LEC study also gave estimates of as much as 5.5 quads of solar energy for the industrial and commercial sectors by the year 2030 if a vigorous R&D program is pursued. This is several orders of magnitude higher than present solar output. A further study showed the effects of increased volume of production of solar collectors, with costs per square meter of collector dropping dramatically from runs of 1000 per year to 100,000 per year (see Table 13).

Table 13. Production Volume Effects on Solar Concentrator Costs

Annual Production Level, m2	Unit Cost $/m2
1,000	188
10,000	99
100,000	70

Solar Heat for Toxic Waste Destruction

U.S. sewers, ponds, and rivers are flooded with some 280 million tons of hazardous industrial wastes every year, more than a ton of it for every American (13). Compounding the problem, most of this is waterborne waste and difficult to get rid of. Nature has long managed to clean up polluted streams with sunlight, and now solar scientists at the the National Renewable Energy Laboratory, Lawrence Livermore Laboratory, and Sandia National Laboratories are adding high solar flux technology to their armament, using a photocatalytic treatment for toxic wastes.

Lawrence Livermore National Laboratory in California was built on the site of a naval aviation base, a training camp, and a maintenance facility. Part of that maintenance involved the cleaning of aircraft engine parts with trichlorethylene (TCE) a highly toxic degreasing agent. In the decades since the war, TCE and other volatile organic compounds have infiltrated local ground water and slowly migrated toward the city of Livermore. In 1987, the contaminated portion of LLNL was designated as another of the many Superfund Sites in the U.S.

Conventional treatment of these sites often simply removed from the water and either dispersed them into the atmosphere or transferred them to carbon filters for disposal in landfills. Solar detoxification, developed by Lawrence Livermore, the National Renewable Energy laboratory in Golden, Colorado, and Sandia National Laboratories in New Mexico, instead breaks down the waterborne contaminants in a one-step process by using concentrated sunlight in combination with a semiconductor photocatalyst such as titanium dioxide.

Contaminated water is carried through glass tubing mounted at the focal point of a large parabolic trough. With energy from the ultraviolet end of the solar spectrum, the photocatalyst produces oxidizing agents in the polluted water and converts contaminants to water, carbon dioxide, and dilute acids that are easily neutralized. This four-month test of solar detoxification en situ rather than the normal process of removing the contaminated water began in 1991. Two strings of concentrating collectors, each 120 feet long, were operated independently and in series during the test. The solar collectors processed about 30 gallons of toxic waste per minute as it flowed through the pipe. Solar radiation reduced the groundwater TCE level from 200 parts per billion to less than 5 parts, the Environmental Protection Agency's standard for safe drinking water.

In a later "one-sun" trial, pumps forced the contaminated water across a flat, catalyst-filled panel and proved that concentrating collectors were not required. While this alternative method was more costly than the concentrating collector approach, researchers anticipate that less-expensive materials may make the one-sun method the standard. Solar detoxification is expected to be a cost effective and fully competitive waste management process by 1995. The process can successfully detoxify 80 toxic chemicals on EPA's list, including dioxins and PCBs.

High-Flux Solar Furnace

In 1991, the National Renewable Energy Laboratory (formerly the Solar Energy Research Institute) offered the use of its new high-flux solar furnace to qualified professionals interested in studying and applying highly concentrated solar energy (14). Using a series of mirrors to concentrate normal sunlight, the long focal length and off-axis design not only provides concentration up to 50,000 suns and temperatures to 3,000°C but greater experimental flexibility and control over the incoming flux as well. The specially designed surfaces of the heliostat and the primary concentrator reflect radiation from the entire solar spectrum of 300 to 2500 nanometers.

In certain applications, surface temperatures exceeding 3000°C are achieved in a fraction of a second. This permits innovative techniques for modifying material surfaces, such as the cladding of expensive materials on inexpensive base material. Metals, ceramics, and composite materials can be treated to yield higher value materials with desirable properties including superconductivity, or greater resistance to corrosion, friction, and oxidation. The liquefaction of coal using concentrated solar flux yields more liquids than do conventional methods. NREL is also experimenting with the manufacture of silicon carbide and other semicrystalline structures, and also seeking a better understanding of the interaction between solar photons and matter. NREL is especially interested in having its high flux furnace used by photochemists to give U.S. industry a competitive edge in high technology markets.

BIBLIOGRAPHY

1. H. Kelly, ed., *Renewable Energy; Sources for Fuel and Electricity,* Island Press, 1993.
2. R. W. Larson, *Economics of Solar Energy Technologies,* American Solar Energy Society, December, 1992.
3. *Lab Talk* **2**(3) (April, 1993).
4. K. Butti and J. Perlin, *A Golden Thread,* Cheshire Books, 1979.
5. K. Boer, ed., *Advances in Solar Energy, 1990,* American Solar Energy Society.
6. J. D. Balcomb, "Integrated Design," *Solar Energy and Buildings Symposium,* Athens, Greece, Dec. 8–10, 1993.
7. *Ann Physics (Leipzig)* **IV**(7), 408 (1902).
8. H. Tabor, *Electron. Power* **10,** 296 (Sept. 1964).
9. *Proceedings of the 3rd International Conference, Progress in Solar Ponds,* University of Texas at El Paso, May 23–27, 1993.
10. A. H. P. Swift and P. Golding, *El Paso Solar Pond Project Fact Book,* University of Texas at El Paso, 1992.

11. S. Folchitto, "A Successful Experience of Establishing and Maintaining Clarity in A Coastal Solar Pond," *Proceedings of the 3rd International Conference Progress in Solar Ponds,* University of Texas at El Paso, May 23–27, 1993.
12. T. Williams and M. J. Hale, "Economic Status and Prospects of Solar Thermal Industrial Heat," *1993 ASME / ASES Joint Solar Energy Conference,* Washington, D.C., April 26–30, 1993.
13. *Lab Talk* **1**(4) (Mar. 1992).
14. *Applications of SERI Science and Technology* (Feb. 1991).

SOLAR THERMAL ELECTRIC

Dan Halacy
Lakewood, Colorado

In the last 15 years, private industry and the federal government's Department of Energy (DOE) have invested more than $2 billion in research and development of solar electric power. The resulting progress to data may be only a beginning, because another 100 GW of new electric power will be needed in the United States alone by the year 2000, with an additional 90 GW during the first decade of the next century. Worldwide, the need will be for another 500 GW by the end of the century. For this reason, the DOE has established the Office of Solar Energy Conversion SOLAR 2000 program, a collaborative effort by the solar industry and government to expedite increased use of solar energy (1).

Fortunately for the United States and the world, a good start toward solar powerplant development has been made. The DOE sees an emerging acceptance by industry, business, and politicians of clean, solar-generated electric power; much of this acceptance is the result of continuing environmental concerns. The resulting political climate has brought about the Clean Air Act Amendments of 1990, plus numerous other environmental legislation measures. The electric power industry itself is the prime mover in the drive toward greater use of clean solar energy (1).

This concern for introducing solar electricity on a broad scale in the United States began more than 15 years ago (2). On May 3, 1978, which had been designated "Sun Day" by President Carter, Public Service Company of New Mexico (PNM) responded appropriately by presenting the solar-minded president with a serious and carefully documented proposal to repower a 50-MW liquid-fuel electric powerplant in New Mexico with solar energy; adding that there were several dozen equally worthwhile utility repowering possibilities elsewhere in the Southwest.

Although the PNM proposal was not funded, many other demonstration solar-thermal-power projects have been built with federal assistance and utility participation; including ownership and operation of a solar power plant or the purchase of electric power from solar plants built by others. There are three major technologies used in the design and construction of solar plants (Figure 1).

By 1980, 54 American utilities had submitted 64 solar-thermal-power-plant proposals, of which the Department of Energy funded 14. All these projects used solar energy to supplant some of the conventional fuels used in existing plants (2). The following are states where power plants were proposed, with number of plants in parentheses: Texas (5); California (3); Arizona (2); New Mexico (2); Nevada (1); and Oklahoma (1).

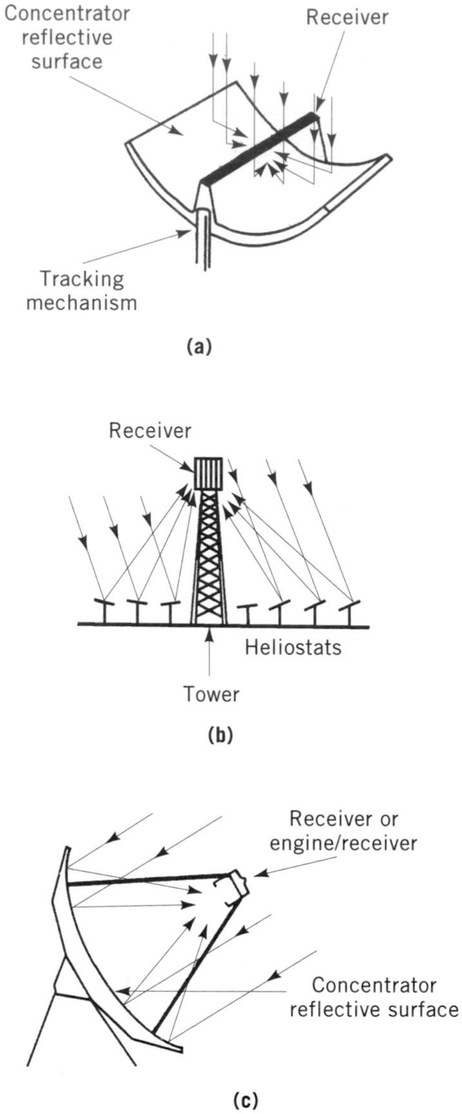

Figure 1. Solar thermal power collectors. (**a**) Parabolic trough; (**b**) central receiver; (**c**) parabolic dish. Courtesy of Island Press.

As Public Service of New Mexico had suggested to President Carter, solar power plants function best in the Southwest, where direct solar radiation is abundant. Average daily solar radiation reaching a 1-m^2 (3.28 ft^2) south-facing heliostat tilted at latitude angle ranges from about 24 MJ in the American Southwest to about 14 MJ in the Northeast. For the same size collector positioned normal to the sun, average daily radiation ranges from about 28 MJ in the Southwest to about 12 MJ in the Northeast. The solar potential is thus enormous, considering the large areas available for solar powerplants: in less than 40 minutes, the United States receives more energy from the sun than it does from the coal, oil, and gas burned in a year. It is estimated that about 20% of the area of Nevada could supply electric power for the entire

Table 1. Average Annual Solar Radiation Available to Three Types of Collector in KWh/m^2

Location	Total, Two-Axis Tracking	Total, Fixed at Latitude	Direct, Two-axis Tracking
Albuquerque, New Mexico	3450	2530	2630
Phoenix, Arizona	3390	2510	2520
Almeria, Spain	3307	2422	2582
Zaragoza, Spain	3293	2437	2552
Denver, Colorado	3100	2280	2340
Sacramento, California	2990	2190	2150
San Diego, California	2720	2110	1860
Honolulu, Hawaii	2580	2000	1610
Madrid, Spain	2549	1782	1887
Austin, Texas	2500	1910	1640
Omaha, Nebraska	2490	1850	1680
Nice, France	2405	1745	1790
Brasilia, Brazil	2397	1877	1649
Miami, Florida	2380	1870	1420
Messina, Italy	2354	1742	1706
Rome, Italy	2288	1677	1664
Athens, Greece	2268	1678	1622
Nashville, Tennessee	2100	1650	1280
Pisa, Italy	2099	1547	1492
Washington, D.C.	2080	1610	1310
Boston, Massachusetts	1920	1470	1170
Manaus, Brazil	1776	1430	1128
Pittsburgh, Pennsylvania	1760	1390	990
Seattle, Washington	1740	1340	1020
Stuttgart, Germany	1729	1276	1167
Zürich, Switzerland	1653	1220	1089
Hamburg, Germany	1497	1083	977

nation. See also THERMOELECTRIC ENERGY CONVERSION; ELECTRIC POWER GENERATION.

Table 1 gives available solar radiation in various locations around the world. Table 2 lists foreign solar power tower plants.

SOLAR THERMAL ELECTRIC POWER

Russian scientists in the 1950s proposed the central-receiver "solar power tower" concept for producing electricity (3). Since that time a number of solar-thermal electric-power facilities have been built, not only in the former Soviet Union but in several other countries as well. In the 1970s the concept was proposed in the United States by Hildebrandt and Hull at the University of Houston. Since 1981, eight such stand-alone solar power plants have been built, including five water/steam plants from 1- to 10-MW capacity in Italy, Japan, Spain, the former Soviet Union, and the United States plus three molten-salt plants in Spain, France, and the United States.

Central receiver solar power plants have the following advantages:

- They collect solar energy optically and transfer it to a single receiver with minimal energy-transport requirements.
- They provide concentration ratios from 300 to 1500, with associated temperatures from 500 to 1500°C, allowing the efficient collection of energy and its conversion to electric power.
- They store solar heat.
- They benefit from economies of scale.

Solar One

More than a decade ago, the Southern California Edison Company helped build, and for four years operated, the largest and most advanced solar-energy power plant to that time. "Solar One" was a 10-MW central-receiver solar thermal powerplant built in the California desert near the city of Barstow. This innovative project was a cooperative venture of the U.S. Department of Energy and a consortium of "The Associates" including Edison, the Los Angeles Department of Water and Power, and the California Energy Commission. Also participating were DOE's Sandia National Laboratories in Albuquerque, New Mexico; McDonnell Douglas Aircraft Corporation; Rocketdyne; Martin Marietta; and Stearns-Roger (4).

Figure 2 is a diagram of Solar One. An array of 1818 mirrored heliostats, each 6.7 m (22 feet) on a side, tracked the sun with computer systems and focused its heat onto the "central receiver" atop a 91.4-m (300-foot) "power tower." Steam at 9,997,750 Pa (1450 psi) and 510°C drove a conventional steam turbine generator to produce a nominal 10 MWe output for the SCE utility power lines. Alternative heat storage was provided by heat-transfer oil, rocks, and sand in a large thermal storage tank for later use at lower temperature and pressure.

Designed primarily to test the feasibility of using solar energy to power an on-line electric utility plant, Solar One produced a considerable amount of power for Southern California Edison during its two years of testing and four years of operation as an electric utility powerplant, and amply demonstrated the feasibility of the power tower concept.

Authorized for construction in 1975, but not completed for almost a decade, Solar One did not begin producing

Table 2. Experimental Foreign Solar Power Tower Plants

Plant	Site	MWe	Receiver Type	Operational	Built By
Eurelios	Sicily	1	Wtr/stm	1981	European Community
SSPS/CRS	Spain	0.5	Sodium	1981	9-Nation Consortium
Sunshine	Japan	1	Wtr/stm	1981	Japan
Solar One	California	10	Wtr/stm	1982	DOE consortium
Themis	France	2.5	Hitec salt	1982	France
CESA	Spain	1	Wtr/stm	1983	Spain
MSEE	New Mexico	0.75	Nitrate salt	1984	U.S.
C3C-5	Crimea	5	Wtr/stm	1985	former Soviet Union

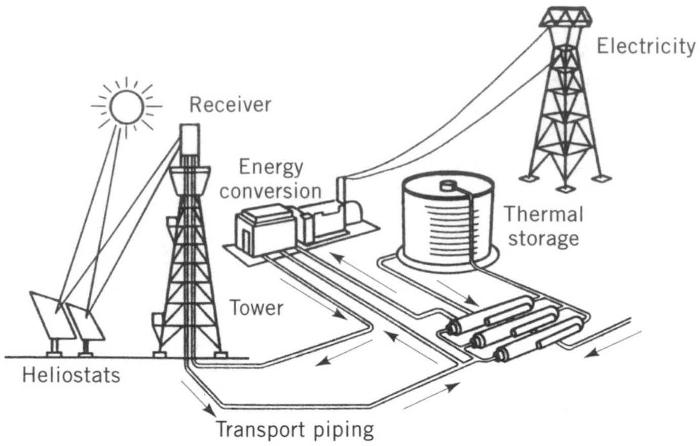

Figure 2. Diagram of Solar One 10-MW electric powerplant. Courtesy of Island Press.

electric power until August 1984. The power plant then operated continuously until 1988, at times reaching an electrical output of 10.8 MWe. Heliostat availability for the four years of operation ranged from 95% to 99%. Plant availability increased from an initial 82% to 96% during the final year of operation. Maximum net daily production varied from 112 MWhe at summer solstice to 48 MWhe at winter solstice.

Original estimates calling for 13% annual efficiency were apparently based on overly optimistic weather data. A revised projection based on 25 years of insolation data for Barstow indicated an annual efficiency of 8.2% with average net electric output about 15,000 MWh per year. Southern California Edison managed the plant and shared its power with the Los Angeles Department of Water and Power. Generator efficiency was about 35% with direct steam, only about 25% with the thermal storage system. Utility effectiveness was of course limited by the fact that Solar One could not collect solar heat at night, and clouds sometimes blocked the sun during the day.

Design and construction costs for Solar One totaled $141.4 million; $14,140 per kW installed and about seven times the installed cost of a conventional 10-MW electric-utility power plant at that time. However, the long-range utility consortium plan included a design for a much larger "Solar 100," whose estimated cost of $4,000 per kW installed was much closer to economic feasibility.

A fair evaluation of Solar One's performance takes into account the fact that it was not a state-of-the-art electric power plant but a very-small-scale pilot plant whose major purpose was to gain information and operating experience applicable to the design of larger, commercial-scale solar power plants. It is a given that future utility use of solar-thermal power plants will be limited to high-insolation areas of the United States. And even in those sites, the central tower concept as executed by Solar One obviously requires further research and development to compete in the utility power market.

Solar Central Receiver Utility Studies Program

In 1985, Southern California Edison, Pacific Gas & Electric, and Arizona Public Service Company cosponsored with the DOE the "Solar Central Receiver Utility Studies." The goal was to select the best solar-thermal electric power-plant design and adapt it for commercialization. In 1989, the Utility Studies group decided on the "Solar One Conversion Project" as the safest approach. A major change would be the use of nitrate salt heat storage rather than steam as in Solar One (3).

The Molten Salt Electric Experiment

Themis, an experimental 2.5-MW central-receiver solar power tower built in France and operated from 1982 to 1986, used molten-salt (Hitec salt) heat storage instead of steam. From May 1984 to June 1985 a Molten Salt Electric Experiment (MSEE) was also conducted by the DOE at its Central Receiver Test Facility (CRTF) at Sandia National Laboratories in Albuquerque, New Mexico. This experiment was designed to evaluate the use of molten salt rather than steam. Molten salt permits considerably higher operating temperatures and increased power output, and MSEE proved that it was also an effective heat-transfer medium; decoupling solar energy from the production of electric power by allowing the solar plant's collection function to follow the availability of solar energy, while power production tracked user demand for electric power.

Solar Two

Pacific Gas & Electric's proposal was for a 30-MW "Carrisa Plains" solar power plant in its own service territory. The design would employ a cavity-type sodium receiver. However, PG&E did not begin the project, nor the much larger 100- to 200-MW plant it later proposed for the Carrisa Plains site. The Arizona Public Service Saguaro Power Solar Plant in Arizona was not repowered with solar electricity either, and Edison withdrew from the remaining proposals because of an excess of generating capacity in its own facilities, lower than anticipated load growth, and reduced conventional fuel prices.

Public Service of New Mexico, which had proposed its modest solar thermal plant to President Carter in vain almost a decade earlier, tried again for government funding of a 25-MW design it believed would cost no more to build than had the 10-MW Solar One Plant at Barstow. Eight other New Mexico utilities were said to be preparing similar proposals, and Texas Utilities teamed with Rockwell International to propose repowering its Permian Basin steam-electric plant near Monahans, Texas, by boosting its output to 115 MW with a solar-thermal power plant (2).

None of these proposals was implemented, and early in 1991 Edison was well along with the design of its "Solar Two," a greatly improved 10-MW follow-on to Solar One incorporating much of that plant's original hardware to reduce construction costs. The U.S. Utility Studies Group was still involved, hoping that the new molten-salt receiver would lead to cost-competitive, 100-MW solar power plants (3).

Solar Two was proposed jointly by a consortium including Southern California Edison, the California Energy Commission, the Electric Power Research Institute, GM Hughes Electronics, Texas Instruments, Idaho Power, Los Angeles Department of Water and Power, Pacific Gas & Electric, PacifiCorp, Sacramento Municipal Utility District, Arizona Public Service Company, the Salt River Project, Sandia National Laboratories, South Coast Air Quality Management District, the Bureau of Reclamation, and the City of Pasadena. The Sacramento Municipal Utility District and the Federal Bureau of Reclamation also expressed interest in building similar new-generation solar power plants.

By using the Solar One site, its 91.4-m (300-foot) tower, heliostats, and as much other original equipment as possible, the consortium estimated that tens of millions in construction costs would be saved. Designed in part as a response to stringent new emissions regulations, the proposed new solar plant could thus be built for a total of only $41 million instead of the $141-million cost of Solar One. Of this, Edison would pay $8 million, other utilities a total of $13.5 million, and DOE $19.5 million. As of late 1992, the Solar Two consortium had secured the $19.5 million for its portion of the project. The Electric Power Research Institute was also a participant, with a $1-million contribution. Efforts were made through Nevada's Senator Bryan to have that state join the consortium, and Sandia was working to include the Indian nations' participation.

Solar 100

In August of 1982, even before completion of Solar One, Southern California Edison, McDonnell Douglas Corporation, and Bechtel Power Corporation had optimistically proposed to DOE a follow-on utility-scale "Solar 100" molten-salt, central-cavity-receiver power plant powered by two large heliostat fields. As its name implied, the new plant was designed to produce 100 MWe (net) with a 60% annual capacity factor. Two large heliostat fields were to be installed sequentially in July 1986 and July 1987 to finance the project more easily (3).

The "Solar Two Participants Agreement," completed in November 1992, also included Pacificorp and Idaho Power. These additions facilitated a cooperative agreement between the Consortium and the DOE toward increased use of solar-thermal electric power and encouraged the building of utility-scale 100-MW and larger solar power plants by the turn of the century. These follow-on plants were expected to produce electricity at a cost between 18 and 36 cents per MJ (5 and 10 cents per kWh). Molten-salt storage raises the capacity factor of solar power towers from about 25% to about 60% annually, and as high as 80% in summer. Southern California Edison believed that the molten-salt Solar Two plant would lead to wide solar–power-plant use (2).

However, the high-temperature salt also creates problems in maintaining receiver piping and equipment at temperatures required for startup. Solar 100 was soon delayed and then deferred indefinitely because of financing, technical risk, and power-marketing considerations, as well as the realization that a larger-scale experiment would be required before construction of a full-scale, molten-salt solar power plant.

By 1985, Edison had decided not to pursue the ambitious Solar 100 venture and instead joined with the DOE, APS, and PG&E in considering eight smaller projects involving the conversion of existing oil or natural-gas electric-utility power plants to solar plants. Instead, work continued on Solar Two, and Bechtel Power Corporation was authorized as constructor and engineer.

Solar One's original computer-driven heliostats and 61-m (200-ft) power tower will be used, but a magnitude 7.5 earthquake in California damaged 4% of the heliostats. These were repaired by riveting, and plans call for the addition of 100 heliostats to the southern portion of the field; parts of these would be obtained from heliostat mirrors used at the now-inactive Carrisa Plains photovoltaic power plant.

Solar Two differs from its forerunner mainly with respect to the heat-storage medium used. Because of stricter emissions regulations promulgated in 1991, the original oil and gravel thermal storage system would be removed and replaced by a state-of-the-art nitrate-salt receiver. This would yield an operating temperature of 565.5°C and allow Solar Two to produce electric power during cloudy periods and at night. Edison expected to complete the engineering, design, procurement, and construction phases of Solar Two by 1995, with demonstration and testing of the new power plant in 1998. The Studies Group believed Solar Two could lead to the building of third-generation 100- to 200-MW solar power plants before the end of the century, producing electric power at levelized energy costs between 18 and 36 cents per MJ (5 and 10 cents per kWh)

These follow-on, large-scale commercial solar power plants will use external cylindrical receiver and generate either 100 or 200 MW of electric power. Annual energy calculations were based on Barstow weather for the year 1977, approximating the region's 25-year average insolation of 9,288 MJ/m^2 (2580 kWh/m^2).

PARABOLIC DISH SYSTEMS

With none of Solar One's follow-on power-tower proposals yet implemented, the second-generation of 1980s DOE-funded solar power plants consisted of small, distributed-receiver types. Prospects for these "dish" systems were then deemed brighter than for power towers because of their high concentration ratios, higher conversion efficiencies, and rapid response times. Dish systems typically have concentration ratios between 600 and 2000, and some achieve temperatures above 1500°C. Modular, they can function singly or in arrays as needed. Each of the four demonstration plants (one a private venture and three of them DOE-funded) used automatic-tracking, parabolic dish collectors (4):

- Georgia Power Company Solar Thermal Total Energy Project at Shenandoah (STEPS)
- LaJet Solar Plant 1 in Warner Springs, California
- Advanco/Vanguard in California
- McDonnell Douglas/United Dish-Stirling-cycle experiment, at Huntington Beach, California.

LaJet Solar Plant 1

This was a privately funded, nominal 5 MWe solar-thermal electric facility in Southern California, using a large collector field of 700 parabolic dish concentrators, each made up of 24 1.5-m stretched-membrane reflective modules. Six hundred of these concentrators generated steam; the others provided steam superheat. A nitrate-salt receiver system was incorporated to provide heat during cloud cover. LaJet sold some electric power to San Diego Gas & Electric beginning in mid-1984, but the plant never reached full-rated operation; this was attributed principally to rapid deterioration of the stretched-membrane heliostats.

STEPS (Solar Total Energy Project, Shenandoah)

This hybrid solar-electric power plant, using 114 large parabolic concentrations plus supplemental gas heating, produced 400 kWe beginning in 1982 and also provided electricity, air-conditioning, and process steam for a textile plant in Shenandoah, Georgia.

The STEPS collectors heated oil to temperatures as high as 360°C. Designed for a thermal-collection efficiency of 62%, the plant produced only about 45% because of serious mismatches in collector output and the textile plant's thermal requirements. As a result, STEPS produced less than 25% of the system's thermal requirements and was decommissioned in 1990.

Vanguard 1

This was a 24-kWe Stirling engine module built by Advanco and sited at Rancho Mirage in California. Vanguard 1 operated for 18 months during 1984 and 1985, and in comparative tests outperformed Solar One with a 2.6-fold higher net conversion efficiency of solar radiation to electric power. Vanguard achieved a record-breaking maximum module net solar-to-electric conversion efficiency of 29.4% and an average daily net efficiency of 22.7%. These results were attributed to the module's high conversion efficiency, low thermal inertia and parasitics, and good part-load performance.

Advantages of Dish Systems

In addition to providing thermal energy for central power generation, dish concentrating systems can also drive Ranking-cycle or Brayton-cycle engines, as well as Stirling-cycle engines. However, tests quickly revealed that organic Rankine-cycle engines were not effective in solar applications. Brayton-cycle engines performed poorly in field tests and were not further developed by DOE. Neither was the sodium heat engine (6).

Analysis at Sandia, based on production of 10,000 dish-Stirling systems a year, projected a levelized energy cost of \$972,000/J (\$0.27/kWh), which is considered competitive for the remote-power applications these systems are designed for. Later testing of three prototype dish-Stirling systems by Southern California Edison indicated a system availability of 50% to 87%, with the majority of power losses caused by circumstances other than hardware failure. Tests were continued through 1993 with the Cummins Power Generation (CPG) dish-Stirling system at CPG's facility in Abilene, Texas, and at the California Polytechnic Institute in Pomona, California. CPG also installed a CPG-460 solar concentrator at Thermacore Industry's test site in Lancaster, Pennsylvania, in August 1992 (6).

In 1993, DOE's Office of Solar Energy Conversion was facilitating the commercialization of 5-kWe and 25-kWe dish-Stirling systems for water pumping and village electrification, with field testing in progress. Nonelectrified rural communities in the developing countries are home to 1.7 billion people, using primary batteries, kerosene for lighting, and gasoline and diesel generators. About 10 million small generators (less than 200 kW) are in use, with another million sold annually. A huge overseas market was foreseen for small-dish electric systems (1).

PARABOLIC TROUGH ELECTRIC POWER SYSTEMS

While electric utilities and the federal government debated between large-scale central-tower solar-thermal power plants and a new generation of sophisticated, high concentration techniques, an ancient solar technology was revived. The technology was the simple, parabolic trough concentrator.

The Shuman/Boys 1912 Solar Thermal Powerplant

In 1912, American engineer Frank Shuman teamed with English physicist C. V. Boys and, with backing from the Eastern Sun Power Company, Ltd., in London, built a sizable solar-steam power plant. Shuman perhaps borrowed the parabolic trough idea from Sweden's John Ericsson, who in 1880 had used such a concentrator to power a hot-air motor of his invention (7).

Shuman's decades-later project was commissioned by a farmer in Meadi, a suburb of Cairo, Egypt, who wanted to use it to pump water from the Nile for irrigating his farm fields. Shuman and Boys built a nominal 100-horsepower steam plant using solar heat collected from an array of seven 31.7-m (104-foot) parabolic trough collectors totaling 3962 m^2 (13,000 square feet) of aperture. Mirrored on their inner surface, these collectors tracked the sun in one plane and reflected its heat onto blackened steam pipes at the focus of the parabola.

On its first test under Egypt's bright sun, the output of the pioneer solar pumping plant ranged between 52.4 and 63 horsepower, and a subsequent cost study predicted a payback for the farmer in only three years. Successful as it was technically, the solar plant was soon shut down by a combination of World War I problems in Egypt and the hostility of Egyptian fellahin, who traditionally irrigated the fields with muscle power. Shuman's pioneering power plant rusted in the desert, a foreshadowing of many solar power experiments to come.

DOE's MISR Program

In the mid-1970s, DOE implemented a development program for parabolic trough concentrators at its Sandia National Laboratories (6). Promising results from this program led to DOE's Modular Industrial Solar Retrofit (MISR) program and, 70 years after Shuman's ill-fated Egyptian solar power plant, DOE sponsored another solar trough-collector irrigation pumping system, this one in the American Desert near Coolidge, Arizona.

The 120-horsepower Coolidge plant delivered about twice the power output of Shuman's pioneering design and used natural gas backup. The demonstration system operated fairly well from October 1979 to September 1981, and dozens of similar trough systems were installed for a variety of industrial uses.

While electric utilities and the federal government debated between large-scale central-tower solar thermal powerplants and a new generation of sophisticated high-concentration techniques, Frank Shuman's ancient solar trough technology was reborn.

The SEGS Power Plants

In 1984, Luz Engineering, an Israeli-based firm, went into the electric-power–generating business near Daggett in the California desert, almost in the shade of Solar One. Its first low profile Solar Energy Generating System (SEGS I), using single-axis trough concentrators, went on line for Southern California Edison in December of that year, producing a modest 14 MWe net. SEGS I produced power for a levelized energy cost of 95.4 cents/MJ (26.5 cents/kWh). The builder was the Kramer Junction Company (3).

SEGS I was followed a year later by an improved SEGS II, a scaled-up, 30-MW module that included a boiler fired by natural gas. Using natural-gas backup, the next five 30-MW plants (SEGS III through SEGS VII) delivered power to SCE for 43 to 61 cents/MJ (12 to 17 cents/kWh), and the SEGS VIII and IX 80-MW plants built in 1989 and 1990 dropped the cost to 36 to 61 cents/MJ (10 to 17 cents kWh). By 1990, 354 MW of electric utility power was being delivered to about 540,000 Edison ratepayers at competitive prices from nine sizable arrays of linear trough-concentrator solar power plants constructed in the California desert.

Federal and state solar energy tax credits were extended only until September of 1990, and Luz was forced to spend an extra $30 million in overtime labor payments trying to complete its SEGS VIII and IX power plants on schedule (see Fig. 3). Compounding this problem, even though both houses of the California Legislature had passed a bill renewing state property-tax exemptions for solar property, the then governor of the state vetoed the bill because he had been erroneously informed it would result in a $60-million deficit in the California budget. The new governor signed an emergency bill after ascertaining that California would instead *receive* about $55 million in extra taxes over the lifetime of the SEGS solar plants. However, the precarious financial situation had by now dried up financial backing for Kramer, and the company filed for bankruptcy.

The nine operating SEGS plants were solar-natural gas hybrid systems built with 1980s power-purchase contracts. However, Luz had tentative contracts for more plants in the United States and Mexico; with Flachglas Solartechnik planning similar projects in Brazil and Morocco. Future parabolic trough development by Luz or a follow-on company will likely be based on the firm's LS-4 collector system. Full-scale LS-4 plants would be built in 200-MWe modules and produce electricity for about 29 to 43 cents/MJ (8 to 12 cents/kWh). Land area of only 2 hm^2/MW of electricity generated is required for SEGS collectors, with the slope not exceeding 1% so that no terracing is needed.

In late 1992, the government of India expressed its interest to the United Nations in a 30-MW Luz trough concentrator solar-electric power plant to be built near Jodhpur (7). The United Nations Development Program in turn requested advice from Sandia National Laboratories in Albuquerque, New Mexico. Sandia officials suggested downsizing the proposed plant to 10 or 15 MW to avoid a large financial loss should the larger plant not prove practical. As of early 1993, the Indian government was considering using the Kramer Junction Company or other U.S. firms toward pursuing the venture.

The demise of Luz Engineering ended the source of commercial-grade cermet selective surfaces for other solar-thermal plant receiver tubes. However, Kramer Junction Company and Sandia worked with Vapor Technology,

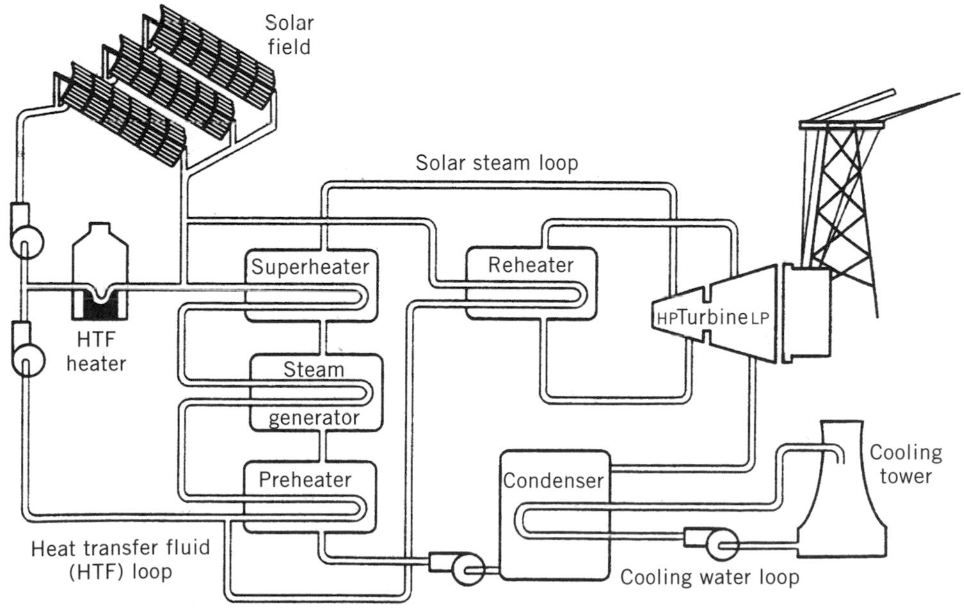

Figure 3. Diagram of LS-4 SEGS Solar Plants VIII and IX. Courtesy of Island Press.

a Boulder, Colorado, firm, to develop a low cost method for application of cermet coating for trough-concentration heat pipes. It is possible that this technique can be applied to the Solar Two receiver tubes as well. The goal is to replicate the optical properties achieved by Luz in Israel, and have results confirmed by optical testing at Sandia and the National Renewable Energy Laboratory. Other SEGS tests have shown that loss of vacuum in receiver tubes causes a 9% loss in heat efficiency, but wind has little effect on collector performance.

Kramer Junction Company also supplied Sandia with seven plant-years of outage records on its trough concentrators. It was found that most such outages should also be applicable to commercial central receivers.

Environmental Effects on Solar Power Plants

Although solar-thermal power plants pose no threat to the environment, the reverse unfortunately is not always true; the environment has occasionally plagued solar-thermal plants. SEGS facilities sustained appreciable damage to their concentrators in an earthquake, requiring Luz to seek government help in replacing them. Assisted by Sandia Laboratories, Industrial Solar Technology (a fabricator of trough concentrators) developed improved replacement mirrors for 16 damaged troughs.

Another act of nature, the volcanic eruption of Mt. Pinatubo in the Philippines caused a significant reduction in solar radiation at the Daggett site, reducing the Kramer Junction Company's revenue by 20%. In the last decade, Pinatubo, and El Chichon in Mexico, reduced solar radiation to the extent that the power output of SEGS plants in California dropped significantly during spring and summer of 1992. Approximately 23.6 million t (26 million tons) of SO_2 were deposited in the stratosphere and converted to H_2SO_4, which mixed with water to produce small aerosols that remain in suspension until removed by raindrops in the troposphere. Researchers at the Atmospheric Sciences Research Center, University of Albany, projected a reduction in insolation between 32°N and 46°N of 3% for summer 1993, 10% to 13% for winter 1993, and 5% to 7% for winter 1994 (8).

Solar 100

The DOE is a member of the U.S. Utilities Studies Group, and in 1993 its Sandia Laboratory issued a contract to Bechtel Power Corporation for an evaluation of the design, costs, and warranty issues related to molten-salt thermal storage and steam-generator designs toward an optimal 100-MWe solar power plant (6).

While the collector field represents the greater part of total plant cost, the receiver is the most difficult engineering challenge. Tubed receivers are prone to constant severe temperature swings because of normal start-up and shut down, as well as clouds obscuring the sun. Solar Two will use nitrate salt as the heat-storage medium and tubed-receivers of Bechtel's design. However, Bechtel has also designed a wire-mesh volumetric receiver for possible future use in later Solar 100 utility plants, and Sandia is testing it.

A volumetric receiver's complexity is evident in the 15 screens of 80-20 nickel chrome resistance wire (41 layers of wire mesh) to make the receiver as "volumetric" as possible; that is, less dense at the opening where the flux is incident, and denser downstream. Bechtel expects thermal efficiency of about 90% at 600°C.

The knit wire mesh for the air receiver was tested in the solar furnace at New Mexico State University for transmissivity and extinction characteristics. Measured extinction coefficients were close to those used in the design of the receiver. Receiver transmissivity may be 1% higher than the design value, affecting performance negatively by 1%. Molten-salt loop testing continues, and the Solar 100 receiver is scheduled to be tested in Almeria, Spain, at the Plataforma Solar as part of the International Energy Agency/Solar Paces Task III. This is the largest solar test facility in Europe.

Sandia is concurrently performing a much simpler analysis for the possible solar repowering of two 45-MW conventional electric power plants near Albuquerque, and at the other end of the power scale is working with the Mexican government on a 30-kWe pilot solar-thermal power plant for Puerto Lobos.

Heliostats

Work at Sandia also includes the development of cheaper heliostats for solar-thermal electric power plants. In 1978, glass-metal cost $1500/m^2. By 1993, costs had dropped to less than $150/m^2, with a goal of $100 to $125/m^2 for a 150-m^2, mass-produced, stretched-membrane unit. Projected levelized energy costs are 43 cents/MJ (12 cents/kWh) for small, solar-only systems and 38 cents/MJ (10.5 cents/kWh) for natural gas/solar hybrids, and only 18 cents/MJ (5 cents/kWh) for 25-kWe dish systems for utility applications. In 1993, Sandia completed the testing and documentation of the prototype 100-m^2, dual-module, stretched-membrane heliostat (4).

Unlike a flat heliostat, a dish collector must accurately track the sun in two axes and is constrained to use a truncated paraboloid reflector. Stretched-membrane reflectors also have inherently short lives and may require frequent replacement at considerable expense and resultant down time. A case in point is the problem experienced with the 1.5-m stretched-polymer reflectors used in LaJet's Solarplant 1. Increasing the size of heliostats also compounds weathering problems.

Three generations of 50-m^2 heliostats have been developed for testing at Sandia, principally by Science Applications International Corporation. Excellent reflective-beam quality was demonstrated in early models at low wind speeds, but gusty winds degraded their performance. Tests at Sandia in 1993 indicated that second-generation technology had corrected this problem; however, silvered polymer reflectors are still easily scratched and have a lifetime estimated at 5 to 10 years. Frequent replacement obviously adds appreciably to solar plant costs. The latest 100^2-m design being tested at Sandia uses twin circular stretched-membrane heliostats attached to a single supporting base.

NEXT-GENERATION SOLAR THERMAL RECEIVERS

Nitrate Salt Direct-Absorption Receivers

The Direct Absorption Receiver (DAR) is a new concept in solar power plant receivers (Fig. 4). In this technology, a

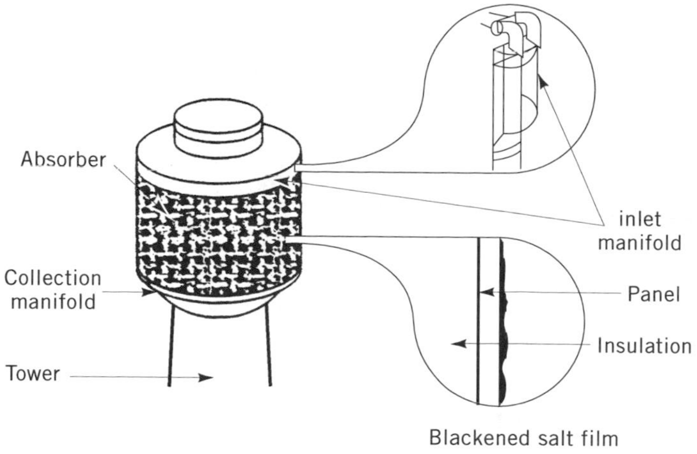

Figure 4. Direct-absorption receiver. Courtesy of Island Press.

film of blackened nitrate salt flows down a near-vertical receiver panel or cylinder (depending on the configuration of the heliostat field) and absorbs concentrated solar radiation. This produces higher temperatures than are available with conventional tubed receivers, making the receiver smaller and less expensive. Studies at the Solar Energy Research Institute (now the National Renewable Energy Laboratory) indicated that a DAR could reduce receiver cost by 50% and also increase the solar power plant's annual output some 16%. However, the technology is not simple, and many improvements have yet to be made.

Air Receivers

Experiments are also being conducted with gaseous transfer media including helium and air. These permit operating temperatures above 1000°C, compatible with higher-efficiency gas turbines or combined Brayton and Rankine-cycle engines. This would permit more efficient production of electricity, and possibly that of synthetic gas and hydrogen as well.

Volumetric Receivers

This type of air receiver uses a porous absorber of metal or ceramic, allowing penetration of sunlight and the gaseous medium into the receiver's interior. Research has been done in the United States and Europe using metal and ceramic absorbers in 200- to 500-kW sizes. However, the technology can operate only at atmospheric pressure unless a transparent window material able to survive a 1000-kW/m^2 solar flux is forthcoming.

Small-Particle Receivers

This concept uses concentrated sunlight focused on an airstream blackened by a suspension of submicron-size particles. Theoretical maximum temperature of the airstream is higher than 2000°C. Such a receiver was built in 1978 and tested at Lawrence Berkeley Laboratory in California. The receiver-outlet air temperature reached about 750°C, considerably lower than predicted.

A windowless "Particle Injection Receiver" was designed and built by Bechtel in 1986. Unlike the LBL small-particle receiver, it operated at atmospheric pressure, using an air curtain rather than a solid window to isolate the darkened air from ambient. Analysis by two-dimensional heat transfer and fluid flow at a 40-MWe power-tower scale indicated a receiver efficiency of 90% and an annual solar-to-electric efficiency of 20% with an overall capacity factor of 26%. Capital costs, largely for heliostats and air-to-air ceramic/metal heat exchangers, were estimated at about $2900 per kilowatt.

Bechtel completed the 2800-hour sample time in its molten-salt corrosion tests in January 1993 and began reviewing reports to make recommendations on steam-generator and thermal-storage systems to be used in Solar 100.

Falling-Sand Receivers

This design concept uses a heat receiver that absorbs solar energy in a curtain of falling sand. Predicted receiver temperature is 1000 to 1200°C, and if attained would drive a gas turbine or combined-cycle power plant directly on solar energy. A number of problems remain, including

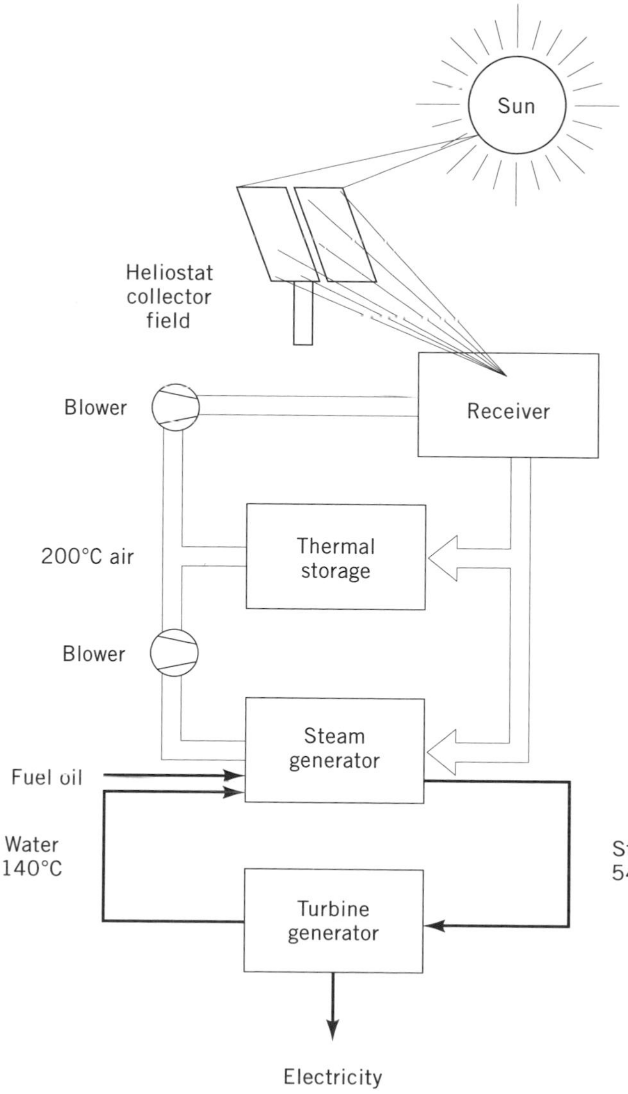

Figure 5. Phoebus air receiver layout. Courtesy of Island Press.

Table 3. Central Receiver Utility Studies Cost Estimates for 100- and 200 MW-Solar Plants[a,b]

Cost Category	Units	200 MW Fifth–Tenth	100 MW First
Capital cost	Million 1987 $	450	295
Annual O&M cost	Million 1987 $	5.6	4.5
Annual net output	GWhe	700	350
Annual capacity factor	Percent	40	40
Levelized energy cost	$/kWhe	0.075	0.10

[a] Courtesy of Island Press.
[b] 1987 costs.

the fusing of sand particles at high temperature and moderate static pressure, as well as receiver durability and efficient solar absorption through repetitive, very-high temperature thermal cycling.

The Phoebus Project

In the early 1990s, Europe's Phoebus Consortium designed a 30-MW hybrid solar/fossil (72% to 28%) air-receiver power plant. This facility was to be built in the desert near Aquaba, Jordan. The north-facing and slightly downward-looking absorber tower would be the focus of a semicircular array of heliostats. Air would be drawn through a metallic wire-mesh receiver at 700°C and flow through a steam generator (see Fig. 5).

Thermal storage would be provided by ceramic bricks. Using weather data from the Solar One experiment at Barstow, Phoebus was calculated to produce about 98,000 MWhe per year. Unfortunately, The Gulf War ended the air-receiver experiment before it could begin. In the interim, a 2.5-MWt air-receiver to be built in 1993 in Almeria, Spain, is expected to serve as proof of concept for the Phoebus receiver design.

The Future of Solar-Thermal Electric Power Plants

Even without the exotic receiver designs just described, cost-effective, utility-scale, solar-thermal power plants seem likely to be built in the near future. The U.S. Utilities Studies Group considers 100- to 200-MW plants with capacity factors from 40 to 60% most economical and estimates that the first central-receiver solar plant in this range will cost from $3000 to $4000 per kWe. The lower estimate assumes a molten-salt receiver and collector system cost of $120 per m^2; the high end allows $175 per m^2 for collectors (3) (see Tables 3 and 4).

The fifth commercial 200-MW central-receiver, with a higher-flux receiver and less-expensive, stretched-membrane heliostats, could be built for an estimated $2225 per kWe and produce power at levelized energy costs between 21.6 and 36 cents/MJ (6 and 10 cents/kWh) by the year 2005. In the following five years, advanced technology receivers mentioned earlier may further reduce costs to less than 18 cents/MJ (5 cents per kWh). It was also noted that central receiver solar power plants require from 2 to 4 hm^2/MW electric power, not considered a limiting factor with low-priced land available. However, cen-

Table 4. Central Receiver Utility Performance Estimates for 100- and 200-MW Solar Plants[a]

Item	Units	200 MW Fifth–Tenth	100 MW First
Land area	km^2	10.0	3.4
Maximum field radius	meters	17,782	1,314
Collector area	m^2	1,818,606	882,690
Number of heliostats		12,235	5,939
Receiver:			
Thermal rating	MWt	936	468
Height	meters	28.4	21.1
Diameter	meters	22.7	19.2
Inlet/outlet temperature	°C	288/566	288/566
Tower height	meters	239	180
Thermal storage capacity	MWht	3,120	1,560
Salt storage tank sizes:			
Hot (H × D)	meters	13.0 × 28.7 (two)	13.0 × 28.7
Cold (H × D)	meters	12.2 × 40.5	12.2 × 28.7
Steam generator			
Rating	MWt	520	260
Outlet temperature	°C	540	540
Outlet pressure	bar	125	125
Turbine gross rating	MWe	220	110

[a] Courtesy of Island Press.

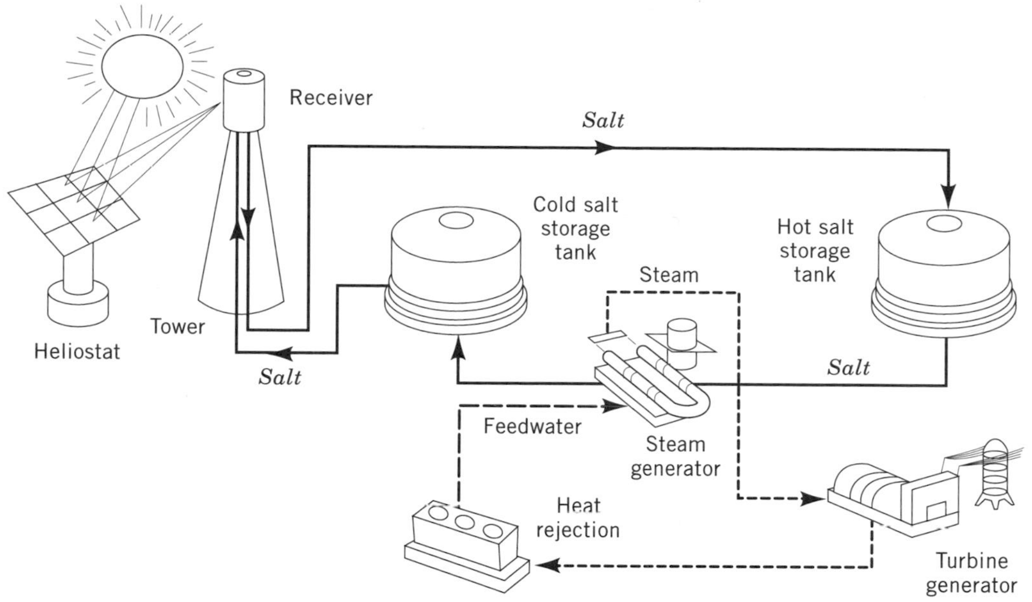

Figure 6. Central-receiver utility studies preferred plant configuration. Courtesy of Island Press.

tral-receiver plant size may be limited to about 1000 MW because at that size the farthest heliostats would be about 2 km (1.24 mi) from the receiver, seriously attenuating the solar energy reaching the receiver (Fig. 6).

Global Prospects for Solar-Electric Power

DOE's Office of Solar Energy Conversion foresees the installation of an additional 930 MW of U.S. solar-thermal systems throughout the world by the year 2000—440 MW in the United States and another 490 MW overseas. This includes construction of one 100-MW solar-thermal electric plant a year beginning in 1998 (1).

Modest penetration of small dish-concentrator solar plants into on-grid and off-grid U.S. markets is also expected in the 1995 to 2000 period. In that same period, an overseas market is foreseen for parabolic-trough, solar–natural-gas hybrid, 80-MW plants and smaller, dish-Stirling engines (see Fig. 7).

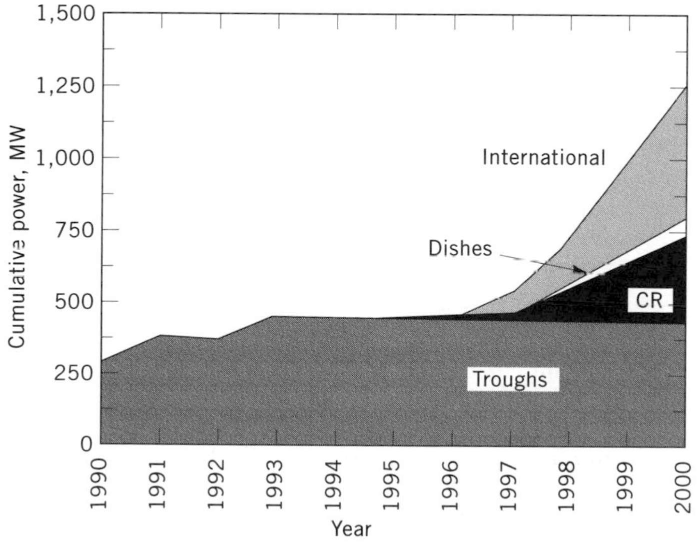

Figure 7. Solar thermal market development projections. Courtesy of DOE Office of Solar Energy Conversion.

OSEC expects the construction of one 100-MW solar-thermal power plant per year following the successful operation of Solar Two, and encourages consideration of 1000 MW "solar parks" of large-scale, solar-thermal central-receiver and parabolic-trough systems situated in sparsely populated regions of western deserts.

OSEC is a partner in the Solar Two Steering Committee, which includes eight utilities, the California Energy Commission, the Electric Power Research Institute, the Department of Energy, and Sandia National Laboratories. It is expected that Solar Two will provide the information base needed for scale up to a 100-MW plant scheduled for completion in 1997. This scale-up is designed to coincide with projections for future capacity needs in the Southwest by the year 2000. Work on Solar Two reduces the technical, construction, and operational risks and address the financial parameters of a utility-scale, 100-MW central receiver.

BIBLIOGRAPHY

1. *Solar 2000. A Collaborative Study*, Office of Solar Energy Technology, United States Department of Energy, Feb. 1992.
2. K. Boer, ed., *Advances in Solar Energy, American Solar Energy Society, 1990.*
3. H. Kelly, ed., *Renewable Energy, Sources for Fuels and Electricity*, Island Press, 1993.
4. *Today's Solar Power Towers*, Sandia National Laboratories, Oct. 1992.
5. *Solar Thermal Power*, Solar Energy Research Institute, Feb. 1987.
6. *The DOE Solar Thermal Electric Program*, Quarterly Progress Report", First Quarter, Fiscal Year 1993. Sandia National

Laboratories, Albuquerque, New Mexico; National Renewable Energy Laboratory, Golden, Colorado.

7. D. S. Halacy, *The Coming Age of Solar Energy*, Harper & Row, New York, 1963.
8. J. J. Michalsky, R. Perez, R. Seals, and P. Ineichen. "Mount Pinatubo and Solar Power Plants," *Solar Today* (July/Aug. 1993).

STEAM TURBINES

J. H. Moore
GE Industrial & Power Systems
Schenectady, New York

STEAM

Modern steam turbines for electrical power generation are the result of nearly a century of engineering development. The first GE production turbine was rated 500 kilowatts (KW) and went into operation in 1901. Just two years later, a unit rated 10 times larger was placed in service at Commonwealth Edison's Fisk St. Station.

Advances in the technology have continued since that time, and today steam turbines are available for both 50 Hz and 60 Hz applications, with ratings from 100 to 1100 megawatts (MW) for fossil-fueled, reheat cycles, and from 600 to over 1300 MW for nuclear applications. This article describes steam turbines for electric power production, utilizing steam from fossil-fuel-fired boilers and nuclear reactors. See also Electric power generation; Energy efficiency—electric utilities.

Ratings and Configurations

Fossil. Steam turbines for fossil-fueled plants accommodate a variety of cycle steam conditions and a large range of ratings. Figure 1 shows the most common cycle steam conditions and the relative effect on heat rate of pressure, temperature and the choice of single or double reheat. Supercritical steam conditions, temperatures greater than 537.8°C (1000°F) and a second stage of reheat are generally associated only with large ratings where they are economically most attractive.

The family of reheat fossil units is illustrated in symbolic fashion in Figure 2 and representative cross sections of two-casing single flow and five-casing six flow steam turbines are shown in Figures 3 and 4. Each turbine type is available for a range of ratings with a selection of steam conditions and exhaust annulus area appropriate to individual technical and economic conditions. The upper limit in rating shown for each configuration is approximate and will depend, in any specific application, on such variables as steam conditions, number of admissions, number and location of extractions, backpressure and whether 3000 or 3600 rpm. The longest last stage bucket offered for 3000 rpm is 1.067 m (42 in.). A 0.016 m (40-in.) titanium bucket is available for 3600 rpm.

At the smallest ratings, a two-casing unit with a single-flow exhaust is available as illustrated in Figure 3. Single-flow designs have the benefit of not requiring an external crossover. Even smaller reheat turbines rated 50 to 75 MW has been built in a single casing more than 30

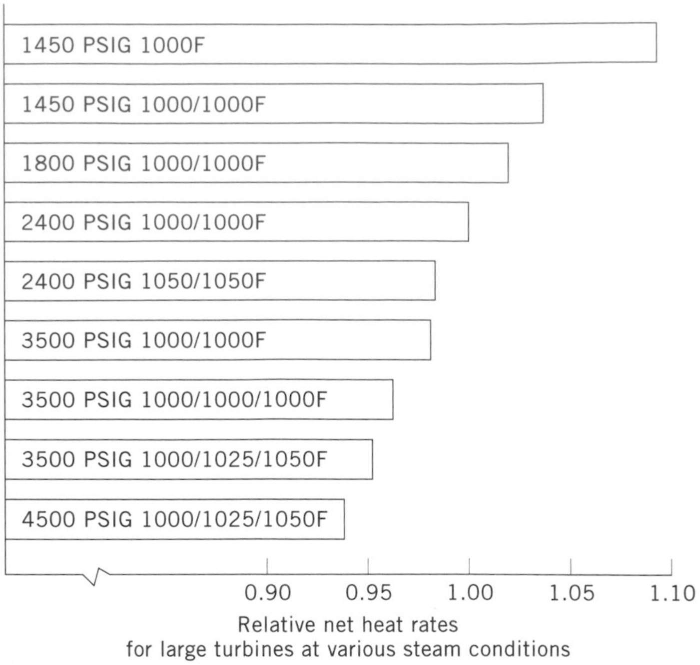

Figure 1. Relative net heat rates for large turbines at various steam conditions.

years ago. However, there is little interest evident today in reheat units smaller than 100 MW. If such a market were to develop, a modern single-casing design could be offered.

The simple two-casing, double-flow arrangement is available with shell-mounted control valves to about 450 MW, and with an off-shell valve chest to about 730 MW, which is the approximate limit for application of the single-span, opposed-flow (HP/IP) section. For the larger ratings in this range, additional annulus area may be appropriate, and a four-flow low-pressure arrangement is required. Larger ratings utilize separate high-pressure and reheat sections, as in Figure 4, with either four- or six-flow low-pressure sections to 1000- or 1100-MW ratings, depending on throttle pressure.

At the larger ratings, highly efficient double-reheat units utilize an opposed-flow, high-pressure first-reheat section with four- or six-flow low-pressure elements. Ratings to 900 MW are available at 3500- and 4500-psig throttle pressure.

Each turbine is designed to meet the individual application requirements as expressed in the final contract heat balances. The designs are tailored to reflect variations in steam extraction for a heater-above-the-reheat-point (HARP), extraction for boiler-feed-pump turbines, and other uses as may be required.

In addition to the designs shown, special designs to meet unusual conditions are available. For example, sites with unusual heat rejection conditions may require a design suitable for exhaust pressure up to 15 inches Hga (50.8 kPa). Specially designed last-stage buckets are available for such applications.

GE steam turbine ratings utilized in these descriptions are based on operation at 11.85 kPa (3.5 in. Hga) exhaust pressure and while heating 3% makeup (MU) flow. Be-

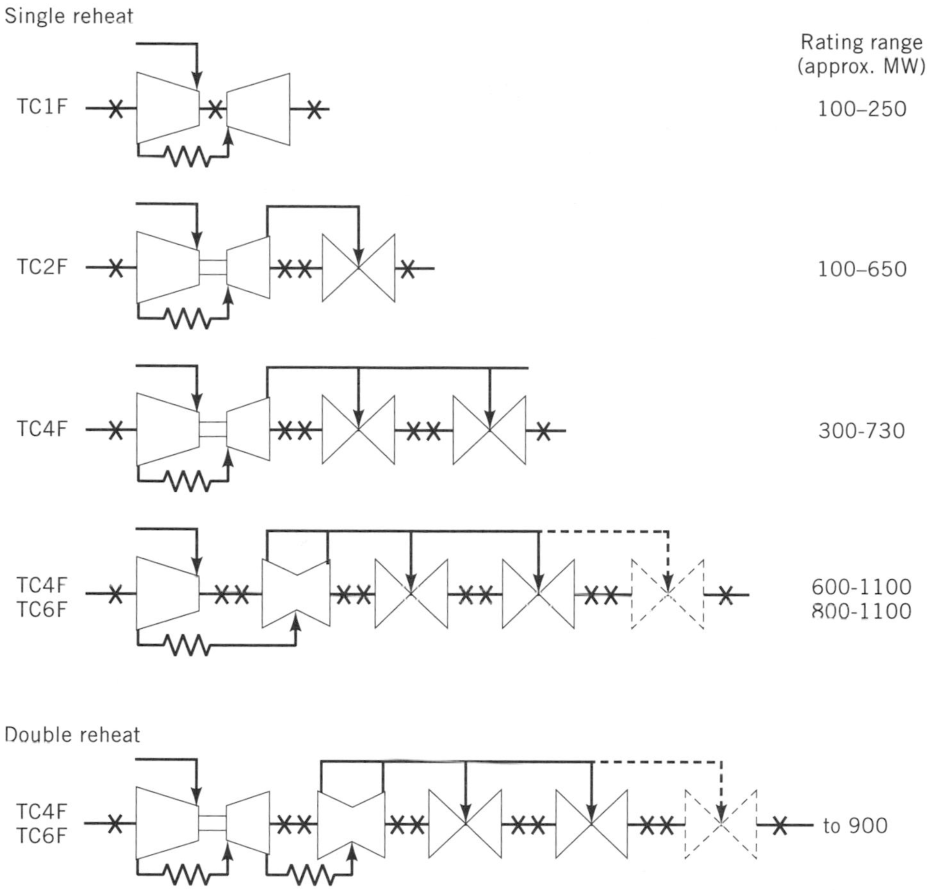

Figure 2. Fossil turbine arrangements.

cause of the effect of variations in flow coefficients from expected values, shop tolerances on drawing areas, etc; the turbine is designed for a flow 5% greater than that required to generate rated output. This flow is called design or valves-wide-open (VWO) flow. To provide plant operating flexibility, the unit is designed for safe operation at VWO flow and throttle pressure 5% above rated pressure. Thus, a 500 MW-rated unit would be guaranteed to produce 500 MW with 11.85 kPa (3.5 in. Hga) exhaust pressure, while heating 3% throttle flow as makeup. When operated at rated throttle pressure with valves wide open, it would be expected to produce about 523 MW (a little less than 5% added output).

It would also be safe to operate at valves wide open with a throttle pressure 5% above rated pressure, where it would be expected to produce about 545 MW. For example, if the exhaust pressure were decreased to 6.77 kPa (2 in. Hga), and there was no makeup, it might produce 550 MW. Thus, units rated in this manner have a maximum expected generation capability about 10% greater

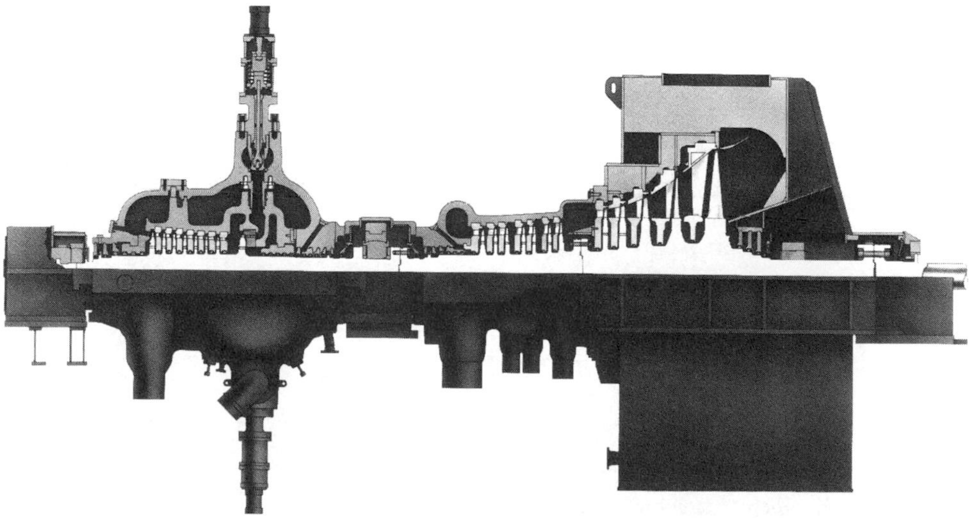

Figure 3. Two-casing, single-flow steam turbine.

Figure 4. Five-casing, six-flow steam turbine.

than the nominal 11.85 kPa (3.5 in. Hga) 3% MU rating.

Nuclear. Although nuclear turbines are available for almost any capacity rating, the licensing requirements for nuclear reactors and the pressures of economy of scale have dictated applications almost exclusively at the larger ratings, utilizing four- and six-flow exhausts. By some projections, however, future reactor designs will more likely be smaller, on the order of 600 MW, and double-flow exhaust turbines with 1.32 m (52-in.) last stage buckets will be suitable. Figure 5 shows the nuclear turbine configurations with moisture separator reheaters (MSRs) between the high pressure and low pressure sections. The cross section typical six-flow design is shown in Figure 6.

As with fossil turbines, each nuclear turbine is designed to meet the individual utility requirements in terms of rating, reactor steam conditions, feedwater temperature and cycle parameters, such as steam reheating, reactor feedpump turbines, and feedwater heater and drain arrangements. Nuclear steam turbines are rated and designed for flow-passing capability in the same manner as fossil turbines. However, they are designed to be suitable for the part-load pressure characteristics of the particular reactor steam supply and are not usually designed for a throttle pressure 5% above rated pressure at valves-wide-open flow.

PRINCIPAL DESIGN FEATURES

Steam turbines across the range of ratings and applications have a number of consistent characteristic features.

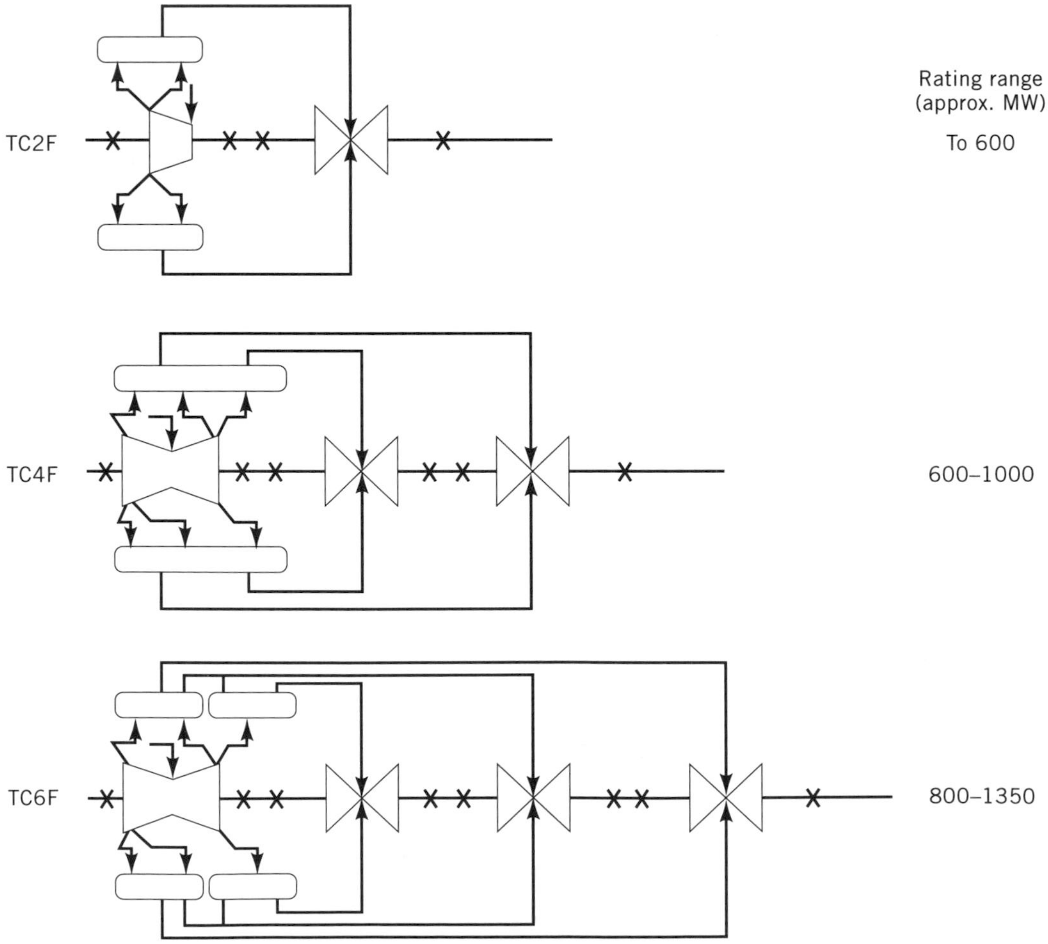

Figure 5. Nuclear turbine arrangements.

Figure 6. Six-flow nuclear steam turbine.

As designs are developed there are almost always conflicting considerations, and major design features result from many years of experience with many units in operation. The reasons a particular feature is adopted over alternative design approaches is not always obvious.

Impulse Staging with Wheel and Diaphragm Construction

The single most important factor relating to design features is the use of impulse stage design, which in turn leads to a construction known as wheel and diaphragm. This design is in contrast to the major alternative technology of reaction stage design with a drum-type rotor and related construction features. GE has developed the impulse design technology after joining forces in 1896 with Charles G. Curtis, who held basic patents.

In a pure impulse stage the entire stage pressure drop is converted into velocity in the fixed nozzles. There is no pressure drop across the moving buckets, which only impose a change in direction of the steam and absorb energy by momentum exchange. In a reaction turbine, some portion of the stage pressure drop, typically 50%, takes place across the moving blades, increasing the velocity of the steam and imparting energy to the blades by reaction, as well as momentum exchange. Peak efficiency is obtained in an impulse stage with more work per stage than in a reaction design (Figure 7), assuming the same diameter. It can be deduced from Figure 7 that a reaction turbine design will require either twice as many stages or 40% greater stage diameters (or some combination thereof) for peak efficiency. GE turbines employ significantly less reaction and have approximately 40% fewer stages in the HP and IP sections than is typical of reaction designs. The contrast is less in the low pressure section where the long bucket length results in a significant increase in velocity of the bucket from the root to the tip. An efficient design requires an increase in the degree of reaction from the root to the tip, and the low pressure stage designs of impulse and reaction turbines tend to be similar.

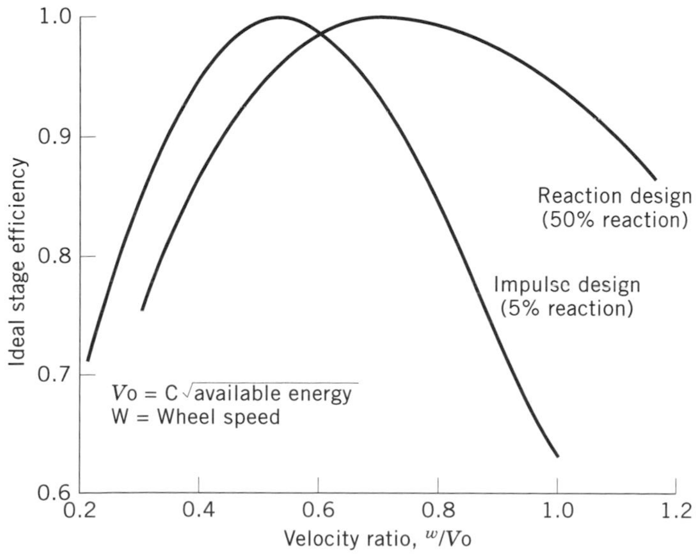

Figure 7. Ideal stage efficiency as a function of velocity ratio for impulse and reaction stage designs.

In the GE stage design, the buckets are mounted on the periphery of wheels and the nozzle partitions are supported in diaphragms, as shown in Figure 8. Because of the relatively large pressure drop that exists across the moving blades in the reaction design, a very high thrust force would exist on the rotor if the blades were mounted on wheels whose faces would also be exposed to the pressure differential. A drum-type rotor is used in reaction-type turbines to avoid excessive thrust.

The significant differences that are associated with these two basic constructions can be separated into those affecting efficiency and those affecting mechanical integrity.

Efficiency. Minimizing stage leakage flows is important to stage efficiency. With less pressure drop across the buckets, the loss due to leakage at the bucket tip is obviously much less for an impulse design than for a reaction design, as shown schematically in Figure 9.

Greater pressure drop exists across the stationary nozzles in an impulse design than in a reaction design; however, the leakage diameter is typically 25% less and, therefore, the cross-sectional area for leakage is correspondingly less. Also, with fewer stages there is sufficient space between wheels to mount spring-backed packings with generous provision for radial movement and a large number of labyrinth packing teeth. In total, the leakage at the shaft packing of an impulse stage is less than that of a reaction stage. The efficiency advantage is even greater than that suggested by the difference in leakage because, the leakage flow in the impulse stage passes through a balance hole in the wheel and does not re-enter the steam path. However, because the construction of a reaction stage precludes the use of balance holes, the leakage flow must re-enter the steam path between the fixed and moving blades causing a disturbance of the main steam flow leading to a significant, additional loss.

In high pressure turbine stages typical of modern designs, tip leakages are two to four times greater and shaft packing flows 1.2 to 2.4 times greater for a reaction design than for an impulse design for turbines of equal rating. The total efficiency loss is even greater due to the re-entry effect of the shaft packing flow inherent with the drum rotor. The effect of leakage losses on stage performance, of course, becomes smaller as the volume flow of the stages increases for both reaction and impulse designs. On a rel-

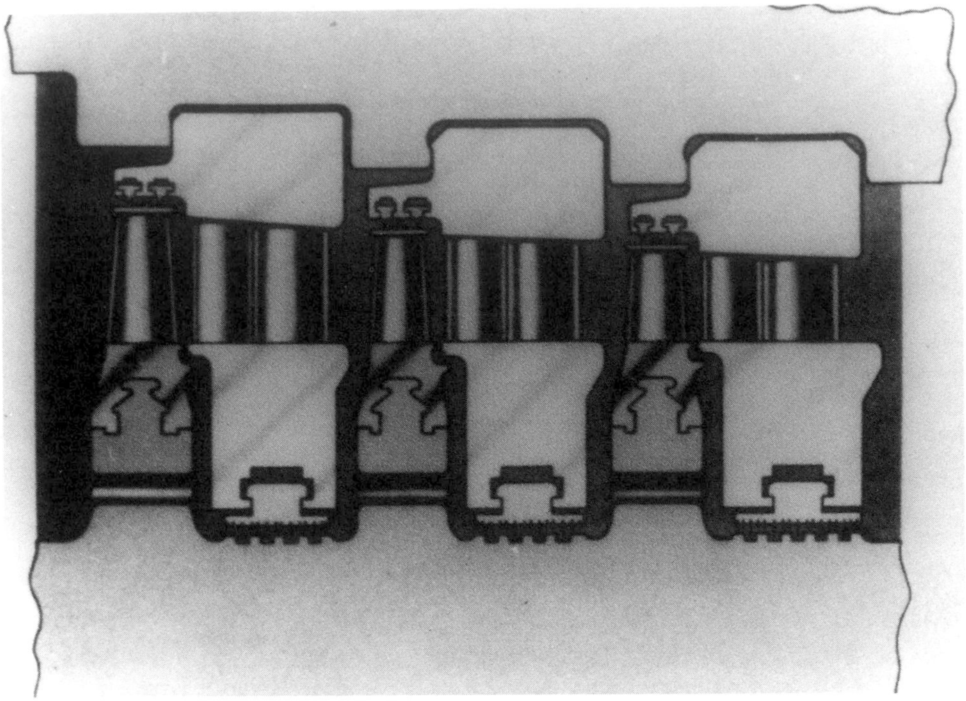

Figure 8. Typical impulse stages, wheel and diaphragm construction.

ative basis, however, the leakage losses on a reaction stage will always be greater than those on an impulse stage designed for comparable application. This situation is also significant from the standpoint of sustained efficiency because the impulse design is less sensitive to the effects of increased packing clearances that might occur in operation.

With more energy per stage, steam velocities in an impulse stage are higher than in a reaction stage. These higher velocities have the potential of resulting in profile losses that could offset the effects of reduced leakage loss if poor nozzle and bucket profiles were used. This loss was a legitimate concern in the early days of steam turbine development with only very simple bucket profiles used. Profile losses, however, are very amenable to reduction with increased sophistication of nozzle and bucket profiles.

With modern computer analysis methods and numerically controlled machine tools, today's turbines have very low stage profile losses as illustrated in Figure 10 in

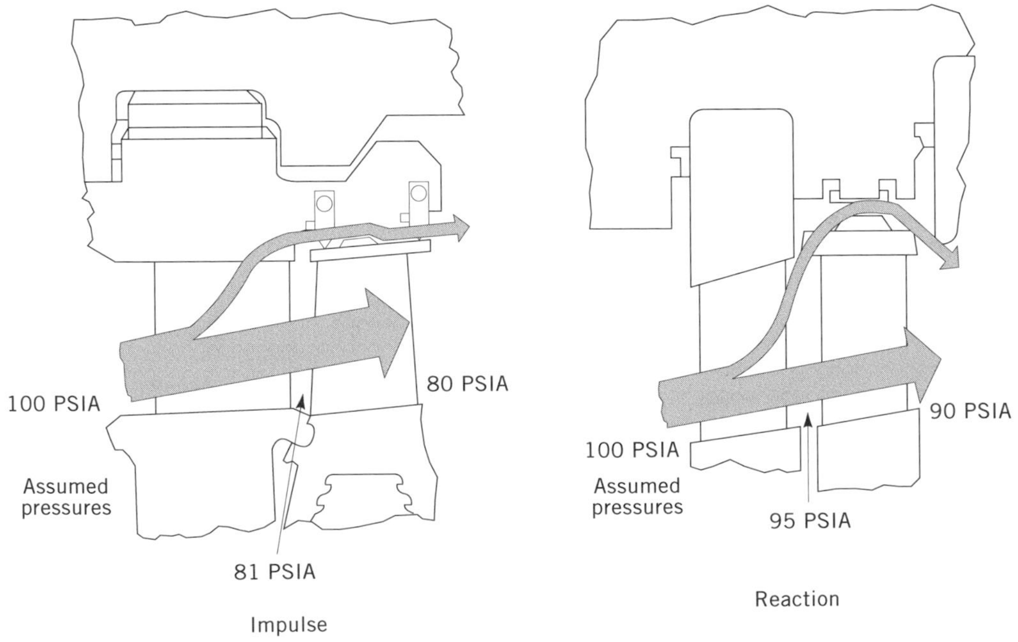

Figure 9. Tip leakage for impulse and reaction stages.

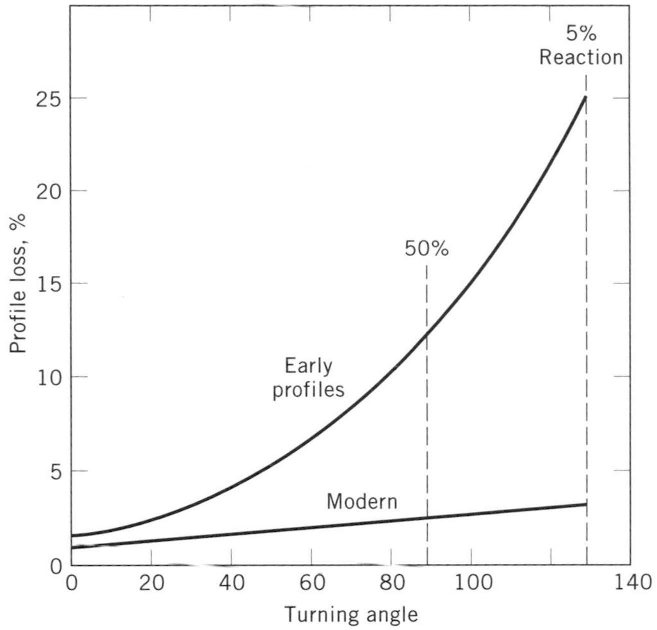

Figure 10. Effect of improved bucket profile on stage profile loss.

which profile loss for early and modern bucket profiles is plotted versus the turning angle of the steam, a parameter inversely related to degree of stage reaction.

With an impulse design, the pressure drop across the diaphragm of the first stage of the reheat and low-pressure sections is high relative to the velocity head of the steam in the inlet pipe, ensuring a uniform flow distribution through the stage. With the lower pressure drop of a reaction stage, poor flow distribution in the first stage of a section can cause performance losses, and complex means such as inlet scrolls are sometimes used to improve flow distribution. Such designs have very little benefit with the impulse design.

Mechanical Design. Impulse stage design with wheel and diaphragm construction lends itself to a rugged, reliable design because the pressure drop occurs across stationary, rather than moving, parts and because the need for fewer stages permits space for sturdy diaphragm design.

Because of the low stage thrust, a balance piston is not required as it is with reaction turbines. Thrust bearings are used with conservative loading without resorting to large sizes.

Thermal stresses in high-temperature rotors limit the rate at which a turbine-generator can change load. These stresses, which are greatest at the rotor surface, depend heavily upon the diameter of the rotor body and the corresponding stress concentration factors. The wheel and diaphragm design results in significantly smaller rotor body diameter and permits ample axial spacing between stages for generous fillet radii at the intersection of the packing diameter and the side of the wheels, resulting in low stress concentration factors at the point of maximum thermal stress. In contrast, the stress concentration factors on drum rotors are relatively high because of the intricate geometry required for blade attachment. The wheel and diaphragm construction, therefore, leads to significantly lower rotor thermal stresses and greater capability for load cycling operation.

Furthermore, the wheel and diaphragm design separates the region of maximum rotor thermal stress from the bucket dovetail region. The dovetail region of the rotor is most likely to be affected by creep resulting from the combination of high temperature, intricate geometry and tensile stress due to the centrifugal load of the buckets. In the drum construction these areas are at the same location, and any creep damage will be additive to low cycle fatigue damage caused by temperature cycling.

One additional advantage of wheel and diaphragm construction arises because high pressure and reheat inner shells with heavy joint flanges tend to undergo distortion due to uneven thermal expansion. Interstage and tip seals are generally supported directly from the inner shells in reaction turbines and distortion of the shell results in movement of the seals, exacerbating the problem of limiting leakage flow. It is GE's practice to support both of these seals from the diaphragm which is unaffected by any distortion of the shell.

Various constructions have been developed with reaction turbines to eliminate or minimize the distorting effect of the joint flange on seal clearances. In one approach, the two halves of the inner shell are held together by a series of rings installed with a shrink fit, creating inward radial forces. This configuration can eliminate the horizontal joint flange, but makes assembly and disassembly difficult. The problem that this design addresses does not exist with GE diaphragm construction.

It is significant to note that although reaction stage design virtually dictates the use of a drum rotor, with an impulse stage design either wheel and diaphragm or drum construction could be used, and that wheel and diaphragm construction is the choice because of its many mechanical and efficiency-related advantages.

Opposed-Flow High Pressure/Intermediate Pressure Design

The single-span, opposed-flow HP/IP design, had been introduced by GE in 1950 in a turbine rated 100 MW. Today there are over 500 turbines with this feature in operation. It is a highly developed design whose maximum rating has been increased over the years to the present limit of approximately 730 MW. This arrangement results in a significantly more compact turbine and station arrangement than that of a unit with the high-pressure and reheat sections in separate bearing spans. There is also one less turbine section to be maintained.

High-pressure steam enters the center of the section and flows in one direction while steam reheated to similar temperature also enters near the center and flows in the opposite direction. This arrangement confines the highest temperature steam to a single central location and results in an even temperature gradient from the center toward the ends, with the coolest steam adjacent to the end packings and bearings.

The opposed-flow design is more compact than a design with separate high-pressure and reheat sections. Tests have shown that this design leads to a lower rate of temperature decay after overnight and weekend shutdowns permitting more rapid restarting.

Although a number of factors affecting performance, including stage packing diameters, shaft-end packing leakages and volume flow effects, are different between the opposed-flow design and one with separate HP and IP sections, the net difference in performance is essentially zero at all ratings.

Reliability statistics on the entire fleet of GE turbines operating in the United States indicate a small but consistent advantage for the opposed-flow design over a design with separate sections at the same rating.

The bearing span for the opposed-flow rotor is greater than the bearing span for either rotor of a unit with separate high-pressure and reheat sections, and the shaft diameter tends to be somewhat larger as a result of designing for similar dynamic characteristics. This design could be a disadvantage at the very largest ratings if the boiler and other plant equipment have a greater capability for rapid starting and loading, and if the unit will cycle frequently. When carefully studied, however, this is seldom found to be the case. In most cases the GE opposed-flow design with wheel and diaphragm construction will have starting and loading capability comparable to a drum-type design with separate high-pressure and reheat sections. Nevertheless, an arrangement with separate HP and IP sections can be provided in the larger ratings when it is believed that the disadvantages are justified by a need for better starting and loading characteristics.

Partial-arc Admission

Except in the very largest ratings, GE turbines are designed for partial-arc admission to improve part-load efficiency. With partial-arc admission, the first-stage nozzles are divided into separate nozzle arcs (typically four for larger units and six for smaller machines), and each arc is independently supplied with steam by its own control valve. For units operating with constant initial pressure, load is reduced by closing these valves in sequence. For smaller units, all four or six valves would close in sequence to give four or six consecutive admissions. For the very largest units in any configuration, three valves would close together and one separately to give two admissions. Intermediate-size units would have two valves closing together with the remaining two closing in sequence to give a three-admission unit.

With a single-admission (or full throttling) machine, load is controlled by throttling on all of the admission valves equally, and all control valves connect into a common chamber ahead of the first-stage nozzles. As load is decreased on the single-admission unit, an increasing amount of throttling takes place in the control valves. In a partial-admission unit on the other hand, less throttling loss occurs at reduced load because the valves are closed sequentially, and only a portion of the steam admitted at any given load undergoes throttling, while the remaining flow passes through fully open valves.

Either free-expanding chests or nozzle boxes are used on multiadmission first stages. Free-expanding chests have removable nozzle plates which facilitate maintenance. The chest, although part of the inner shell, can expand and contract without undue constraints and is well suited to rapid starting and loading.

Nozzle boxes are used on large units. For modern units, these boxes are made from two 180° segments split at the horizontal joint, each containing two separate inlet chambers. A nozzle box for a double-flow first stage consisting of two 180° segments is shown in Figure 11.

It is also possible to change unit output by changing boiler pressure at fixed valve position. Variable-pressure operation is becoming more common, and the question is sometimes raised as to whether the partial-arc admission feature is worthwhile.

If load is reduced by varying pressure with valves wide open, load increase can only be achieved by increasing boiler pressure, which is a relatively slow process. Also, the unit cannot participate in system frequency control. This problem can be solved with a hybrid method of operation in which load is reduced approximately 15% at constant pressure to provide some "throttle reserve" before beginning to reduce pressure. With partial-arc admission it is attractive to fully close one valve and then vary pressure. If a greater capability for rapid load increase is desired, two valves can be closed. In either case partial-arc admission yields a better heat rate than full throttling, even with variable pressure operation.

Most GE steam turbines with partial-arc admission have a feature known as Admission Mode Selection (AMS) which permits all of the valves to be opened and closed together in full-arc admission or sequentially in partial-arc admission. The benefit is that partial-arc admission, with its heat rate benefits, can be used when load is essentially constant, and full-arc admission can be used for starting or making major load changes with a benefit of reduced thermal stress.

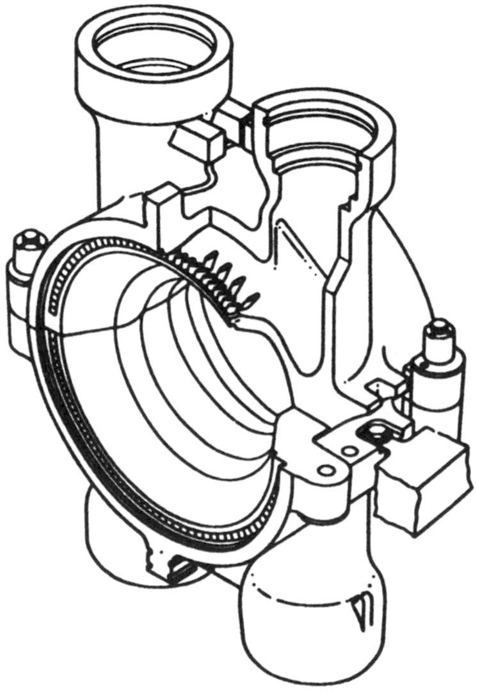

Figure 11. Double flow nozzle box.

Center-Line Support

Turbine components undergo considerable thermal expansion as they undergo changes in temperature. The various stationary components surrounding the rotor in GE turbines are supported at, or very close to, the center line, and are free to expand radially to maintain concentricity. As shown diagrammatically in Figure 12, all diaphragms are positioned by means of radial keys inside inner shells and, in a similar manner, inner shells are positioned inside outer shells, or hoods, by means of radial surfaces at the horizontal joint and at the vertical centerline. Finally, the outer shells are supported by the rotor bearing standards at their true horizontal center lines.

Number of Bearings

Considerable experience exists both for turbine designs employing two bearings per rotor span and with designs that employ fewer bearings. There are advantages and disadvantages to both approaches, but overall the use of two bearings per rotor on large turbines is considered to have sufficient advantage to justify the additional cost and, sometimes, added length. The benefits are less clear, however, on smaller units and current practice is to use three bearings in two-casing machines with single-flow exhausts or small double-flow exhausts.

The use of two bearings per rotor gives the designer flexibility to accurately establish rotor critical speeds by selection of bearing span. It results in shorter bearing span and, therefore, smaller rotor body diameter, which is beneficial to efficiency and starting and loading capability. The shorter, stiffer rotors between bearings and the added damping of the additional bearings reduce susceptibility to rotor instability.

With two bearings per rotor, each rotor can be precision, high-speed balanced on its own journals in the factory. The result is a finely balanced turbine rotor that can be assembled to the other rotors in the field, and in nearly all cases started without additional balancing. If additional balancing is necessary at startup or following a turbine outage, it can easily be accomplished with a minimum of balance shots and downtime because of the relatively small dynamic interaction between adjacent rotors. The imbalance can be located and corrected with a small impact on availability of the unit.

The shared-bearing rotor design results in a rotor system that is more sensitive to imbalance and more difficult to field balance. Turbines with one bearing per span have rotors that are factory balanced using a stub shaft or temporary journal. This procedure creates a difference between the rotor operating conditions during factory balance and the actual conditions when the rotors are fully assembled in the turbine-generator. The result is that ro-

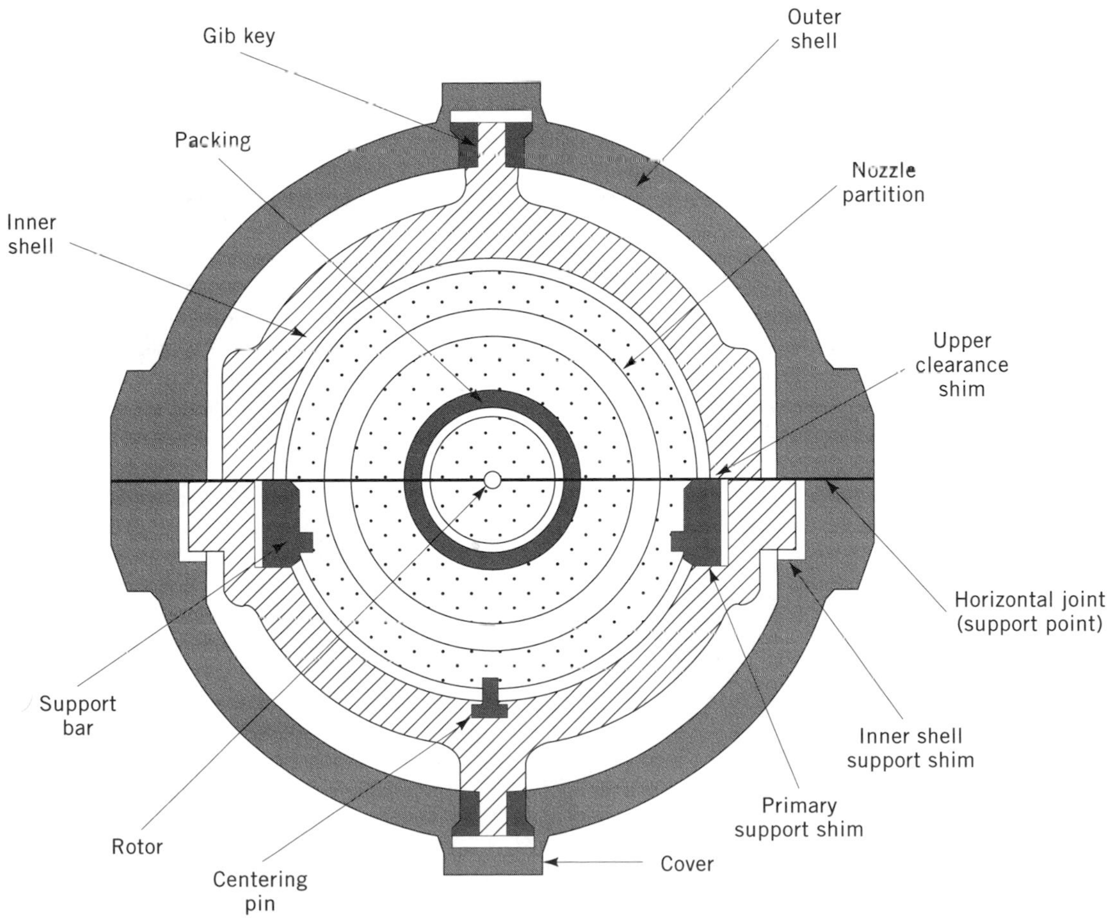

Figure 12. Method of location of stationary components.

tors may require some rebalancing after assembly in the field and, since there is more dynamic interaction between adjacent rotors, balancing is more difficult to accomplish.

General maintenance and bearing inspection are easier with each rotor having two bearings. An auxiliary bearing is not required for support when a coupling is broken and coupling alignment can be more accurately established, further contributing to smooth operating characteristics.

The major benefit of using fewer bearings, other than cost, is that some reduction in overall length of the unit can be achieved. Designing for two bearings per rotor requires some additional length to achieve adequate rotor flexibility between adjacent bearings and thereby, tolerance for misalignment.

Crossover

With fossil turbines, GE uses a single crossover to transport steam from the intermediate-pressure turbine exhaust to the low-pressure turbine inlets. Pressure-balanced expansion joints, as shown in Figure 13, are provided to permit differential thermal expansion between the crossover and the stationary parts it connects without imposing large axial forces due to steam pressure. The stainless steel bellows have high reliability in this application since there is no load imposed in torsion or bending.

Compared to use of a crossaround pipe on each side of the turbine, this design makes for a less-cluttered turbine arrangement with unobstructed access at the floor level. There is also a reliability advantage in that only half as many bellows are required.

In most cases there is also an advantage in reduced bearing span of the IP section and overall length of the machine by making the IP exhaust connection in the upper half and the reheat inlet and extraction connections in the lower half.

Because of the very large volume flow between the high-pressure exhaust and the low-pressure inlets with nuclear turbines, and the presence of moisture separator reheaters between the same points, from four to eight crossaround pipes, are used. The routing is generally three dimensional and, therefore inherently flexible, and the temperatures involved are low, so that expansion bellows are not required.

Solid Particle Erosion Resistance

Carryover of iron oxide particles from boiler superheater and reheater tubes can cause severe erosion to turbine nozzles and buckets. Solid particle erosion (SPE) has a major economic effect in loss of sustained efficiency, and in causing need for longer and more frequent maintenance outages. Extensive efforts to understand the erosive mechanisms in the turbine steam path and develop resistant coatings have led to substantial improvements in the erosion resistance of steam turbines.

Analysis of particle trajectories in steam as a function of density and velocity has led to changes in geometry of nozzle partitions and relative spacing between nozzles and buckets in the first high-pressure and reheat stages that result in dramatic decreases in the rate of erosion.

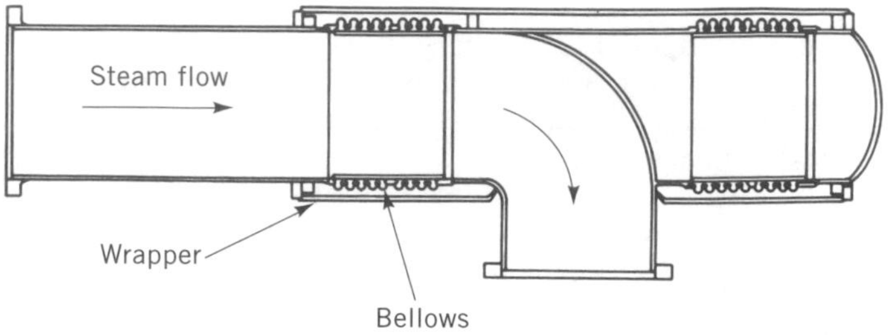

Figure 13. Crossover expansion joint.

These features, along with either plasma spray or diffusion-applied hard coatings in the same regions, have been found so effective that they are provided as standard features on all GE reheat turbines operating with fossil-fuel fired boilers with steam temperature of 537.8°C (1000°F) or greater.

MAINTAINABILITY FEATURES

The extent to which a turbine can be maintained without disassembly and the ease with which it can be disassembled and reassembled for inspection and maintenance, directly impact availability and are important design considerations.

Disassembly of steam turbines is facilitated by a generally uncluttered arrangement, the use of two bearings per span, the use of opposed-flow HP/IP arrangements, to minimize the number of casings, and a minimum number of piping connections to the upper half shell. Any special tools or lifting devices required are provided. Optional features that can be provided include special hydraulically extended coupling bolts and a small jib crane installation for lifting bearing parts without removing the crossover.

Features that reduce the frequency with which major disassembly is required are, if anything, even more important. These include the SPE-resistant features that have been described; provisions for field balancing in all rotors; optional full-flow lube oil filters; positive-pressure variable-clearance packings that provide increased clearance during startup and normal close clearances at load; and access ports for steam path inspection by borescope.

SUPERCARS

AMORY B. LOVINS
Rocky Mountain Institute
Snowmass, Colorado

Since late 1992, a revolution has been rapidly emerging in the conceptual design of cars and other road vehicles (1–4). It artfully combines extremely lightweight and aerodynamically slippery construction with *hybrid-electric drive*. Hybrid drive propels the wheels with motors powered by electricity. However, the electricity is not stored in heavy batteries recharged from the utility grid, but rather is made onboard from fuel, or recovered from braking energy by using the wheel-motors to convert motion back into electricity.

This ultralight–hybrid combination turns out to have a very unusual property. Ultralight construction by itself typically improves fuel efficiency by a factor of about 2.0–2.5. Hybrid-electric drive by itself typically yields only a 1.3–1.5-fold gain. But both together can boost efficiency by about 5–20 times: roughly 5- to 10-fold state-of-the-shelf (using the best commercially available ingredients) and 10- to 20-fold state-of-the-art (still in the laboratory). It is a sort of 2 + 1 = 10 equation. Most remarkably, this dramatic reduction in fuel use, coupled with a reduction in air pollution by one to three orders of magnitude, will apparently involve no trade-offs of other desirable qualities but on the contrary will bring many important new advantages.

Specifically, ultralight family cars with modern hybrid-electric drives could readily achieve combined city/highway fuel efficiency better than 1.6 L of gasoline or equivalent per 100 km (150 mi/U.S. gallon or mpg). This fivefold improvement over typical new U.S. production cars would require only demonstrated and commercially available technologies such as fiberglass-composite construction, high-performance motor systems, conventional buffer batteries, and small gasoline engines. Fuel efficiency of less than 1.0, probably below 0.6, L/100 km (~240–390+ mpg) appears to be achievable with technologies now in the laboratory, such as advanced aerodynamics and materials, monolithic solid-oxide fuel cells, carbon-fiber flywheels, miniature high-speed gas turbines, and small semiadiabatic (insulated) diesel engines. Technical limits may be roughly a factor of two, better still. And far from sacrificing other attributes for efficiency, ultralight hybrids could be more safe, sporty, clean, durable, reliable, quiet, comfortable, and beautiful than existing cars, yet be priced about the same or less.

The key improvements required for at least a fivefold efficiency improvement—chiefly aerodynamic drag and mass somewhat below half those of present U.S. production cars—have been demonstrated, and further reductions of 2–3 times in drag-mass product appear feasible. Net-shape materials, chiefly polymer composites, could do this while cutting production costs through materials savings, hundredfold fewer parts, tenfold less assembly labor and space, and order-of-magnitude lower tooling costs. Epoxy dies, lay-in-the-mold color, and other innovations would permit extremely short product cycles, just-in-time local manufacturing with direct delivery, and onsite maintenance—hence the same retail price even if production costs were considerably higher. This would fundamentally change how cars are made and sold. It could even be the biggest change in industrial structure since the microchip.

Conventional cars, like other technologies, have entered their era of greatest refinement just as they may have become obsolete. Imagine that a seventh of the gross national product (GNP) in, say, the United States were devoted to manufacturing typewriters. The Big Three typewriter manufacturers have gradually moved from manual to electric to typeball models. Now they are making subtle refinements somewhere between a Selectric 16 and a Selectric 17. Their typewriters are excellent and even profitable. People buy over 10 million of them every year. The only problem is that the competition is working on wireless subnotebook computers.

That is where the global auto industry is today—painstakingly refining designs that may soon be swept away, perhaps with terrifying speed, by the integration of very different technologies already in or entering the market, notably in advanced materials, software, electric motors, microelectronics, power electronics, electric storage devices, and computer-aided design (CAD) and manufacturing. This article attempts to sketch the outlines of that potential transformation and of the barriers it faces. Supercars do face serious cultural obstacles in the car industry and institutional barriers in the marketplace. However, their immense societal value merits policy intervention to help speed and smooth this challenging transition, making it less a hardship than a lucrative opportunity. Supercars could also buy time to implement, but cannot replace, fundamental transportation and land use reforms.

Any explanation of how supercars differ from traditional lines of car evolution, how they achieve such exceptional performance, and how they are rapidly emerging from a conceptual into a commercial phase will become out of date in certain details before the ink is dry. Development of this new vehicular design approach has been uniquely rapid: leading designers at many of the world's leading automakers converged on similar logic just in the first half of 1994, and significant new developments are reported almost weekly. However, given the constraints of writing an encyclopedia rather than a newspaper, this article focuses on the basic concepts that will be common to future developments. Later news can be obtained from the Supercar Center (5) and from specialized technical conferences (eg, 6).

See also AUTOMOTIVE ENGINE EFFICIENCY; HYBRID VEHICLES; FUEL CELLS; BATTERIES.

UNITS AND TERMS

- The term *supercar*, promoted by the author since 1992, was changed in November 1994 while this article was being edited. It had already been a term of art for certain kinds of limited-production, high-performance road cars that get 200 mph rather than mpg. It also began to appear in the popular press in early 1994 as a name for the far less efficient (roughly 2.9 L/100 km, or 80 mpg) designs being sought under the U.S. government's 1993 Partnership for a New Generation of Vehicles (originally called the Clean Car Initiative). The resulting confusion made it necessary to switch to a new term, "hypercar," to distinguish advanced ultralight hybrids from these other usages. That is done throughout the rest of this article.
- For simplicity, calendar year is generally used here as a surrogate for automakers' model year.
- Vehicles' fuel efficiency can be expressed either as fuel intensity (L/100 km) or, reciprocally, as fuel economy (mpg). Both kinds of figures are expressed here, unless otherwise stated, in terms of the U.S. Environ-

mental Protection Agency's (EPA's) ratings, weighted 55% for the urban and 45% for the highway driving cycle. To convert from L/100 km to mpg or vice versa, divide by 235.2.

- Both of these federal driving cycles were designed in 1975, are routinely "gamed" by manufacturers in testing, and have come to diverge markedly from typical on-road performance in both efficiency and emissions. By the early 1980s, average well-maintained gasoline-fueled cars' on-the-road fuel economy was typically ~10% lower than rated for city, ~22% for highway, and ~15% for combined driving. By 1990 the combined rating's discrepancy from on-the-road efficiency remained ~15.2% for the car fleet but had widened to ~24.5% for light trucks, with an approximate doubling expected by 2010.
- The EPA gasoline-powered urban fuel economy rating is approximately equivalent to the European urban cycle test (or the Japanese 10-mode test) times 1.12; the EPA highway rating, to the European 90 km/h test (or the Japanese 60 km/h test) times 0.87.
- We adopt here the normal but odd convention of rating fuel efficiency in terms of distance traveled, rather than the product of distance times the passengers or payload carried—like the energy-per-seat-km metric used in analyzing surface mass transit or air travel.
- Also the EPA's convention that a car's "test mass" for purposes of fuel efficiency testing or simulation is 136 kg (300 lb) more than its "curb mass" (vehicle plus required fluids but no passengers or payload).
- All dollars are 1989 U.S.$ (=1.1024 1989 ECU) deflated with the gross domestic product (GDP) implicit price deflator, gallons are U.S. gallons (=3.785 L), and miles are U.S. statute miles (1.609 km). Future costs are levelized or present-valued at a real discount rate of 7% per year.
- Cost of saved energy (CSE)—a Lawrence Berkeley Laboratory method for fair comparisons with the avoided cost of buying the energy being saved—is equal to $Ci/S[1 - (1 + i)^{-n}]$, where C = capital cost, i = annual real interest rate expressed as a decimal (here, 0.07), S = annual fuel savings, and n = lifetime in years. Thus the CSE is capital cost divided by the discounted stream of fuel savings over the car's lifetime. If C includes an appropriate financing charge, the CSE can be compared directly with the levelized price P of delivered motor fuel. Simple payback time in years is then C/SP.

FALLACY OF INCREMENTALISM

Troubled car industries now weaken many national economies, while inefficient light vehicles and their ever-increasing use are major causes of oil dependence, air pollution, noise, climatic threats, and other important social costs. These problems demand transportation and land use innovations, combined with cleaner, more efficient vehicles (7). Yet until about 1993, the conventional wisdom framing the U.S. car efficiency debate was that the doubling of new-car efficiency during 1973–1986 had virtually depleted the "low-hanging fruit"—opportunities for fuel economy consistent with affordability, safety, and performance. Further improvements are widely assumed to be small, incremental, difficult, expensive, and likely to worsen air pollution.

On the contrary, it is turning out that the next doubling of car efficiency will be easier than the first was, because it will come from very different sources: not from incremental refinement of today's cars but from replacing them altogether with a different and functionally superior concept of car design, manufacture, and sales. This concept began to take shape in 1991 (8) and rapidly gained momentum during the following three years (2–4, 6), partly due to the recognition gained through the Nissan Prize awarded at the main European car technology conference, the autumn 1993 ISATA in Aachen, Germany. By mid-1994, the supercar concept had engaged the commercial effort of approximately 20 entities with impressive capabilities, including automakers, large electronics and aerospace companies, and world-class venture capitalists.

This rapidly growing group of supercar entrepreneurs shares a vision of an auto industry transformation that seems technologically plausible and commercially attractive in the 1990s and beyond. It would initially exploit niche markets but has a strong potential to take over virtually the entire industry both for cars and for most other road vehicles, light and heavy. The implications of this transformation are not all welcome, but the issue seems less whether it will happen than who will do it first and best and whether it will be done thoughtfully. To understand these striking and unexpected developments, it is necessary first to explain the historic trends and perceptions from which they emerged.

Historical Trends in Fuel Economy

New U.S.-made cars halved their fuel intensity during 1973–1986, from ~17.8 to a European-like 8.7 L/100 km; ~4% of the savings came from making the cars smaller inside, ~96% from making them lighter and better (9). During 1978–1987 on average, interior volume decreased by <1%, but volume per unit curb weight rose 16% from better packaging, power per unit engine displacement rose 36% through better power plant engineering, and acceleration increased 6% (10). During 1976–1985, weight reduction was the most important (~36%) of the identified causes of improved fuel economy: large amounts of weight turned out to be simply unnecessary, or avoidable by substituting lighter materials.

Unfortunately, not only did weight then increase slightly during 1985–1989, but an important countervailing trend emerged: ~58% of fuel economy gains vanished into more powerful engines to produce ever-faster acceleration (11). Fuel economy is roughly proportional to the square root of acceleration time, both because of increased idling losses with higher displacement engines (10, p. 147) and because of the severe maximum-to-average power mismatch that makes a powerful engine "expensive even when it is not being used" (12). Around

1990, a typical ~1364-kg (~3000-lb) U.S. car had a ~90-kW (~120-hp) engine, sized for ~11-s acceleration from 0 to 97 km/h (0–60 mi/h or mph). That engine was oversized about 6-fold in cruising and 24-fold in city driving, so on average, it would operate at roughly half the efficiency achievable at its optimal point. Such powerful cars have top speeds that average ~206 km/h (~129 mph), twice the maximum U.S. legal limit.

Calculated Scope for Further Improvements

The modest and gradual decoupling of mass from size reached a temporary plateau in the mid-1980s using conventional materials. However, many other kinds of refinements are far from saturated. Further incremental improvements therefore yield a supply curve (Fig. 1) extended 24% from the U.S. Department of Energy's (13) by adding two further measures, idle-off and aggressive transmission management (14). The curve shows cumulative gains in new-car fuel economy, and their empirical marginal costs, from fully deploying a limited list of 17 well-quantified technologies already used in mass-produced platforms, without changing the size, ride, or acceleration of average U.S. 1987 cars. Most of the measures are conventional, such as front-wheel drive, four valves per cylinder, overhead cams, and five-speed overdrive transmissions. Omissions include such simple measures as reducing brake drag to zero or nearly so—commonly done in motorcycles but not yet in cars.

This approach can cut 1987-base fuel intensity in 2000 by ~35%, to 6.99 actual (5.36 rate) L/100 km (14). Such 44-rated-mpg cars would just counterbalance projected U.S. growth in vehicle-km traveled by 2010 (16). Each saved liter would cost on average, $0.14 ($0.53/gal)—less than half today's U.S. gasoline price. At about half that cost, savings ~72% as large are also achievable in U.S. new light trucks (14). Such cost-effectiveness is probably conservative, as illustrated by the *actual* improvements in one subcompact platform (15): the 1992 Honda VX's 56%-improved fuel economy (4.62 L/100 km, 51 mpg) increased its Manufacturer's Suggested Retail Price ("sticker price," normalized for identical cosmetic and safety features) by only ~$650, or $0.18 per saved liter ($0.69/gal)—less than the average-cost supply curve in Figure 1 would predict and far below the marginal cost curve, which is the more appropriate comparison.

A similar but more limited analysis (17), considered authoritative by an official assessment (18), explicitly ignored emerging technologies. These, however, were considered "reasonably certain" over the next 10–15 years, so conservative official findings "should *not* be taken to mean the technological limit of what is possible with the current state of the art" (19): a similar assessment 10–15 years ago would surely have omitted many important

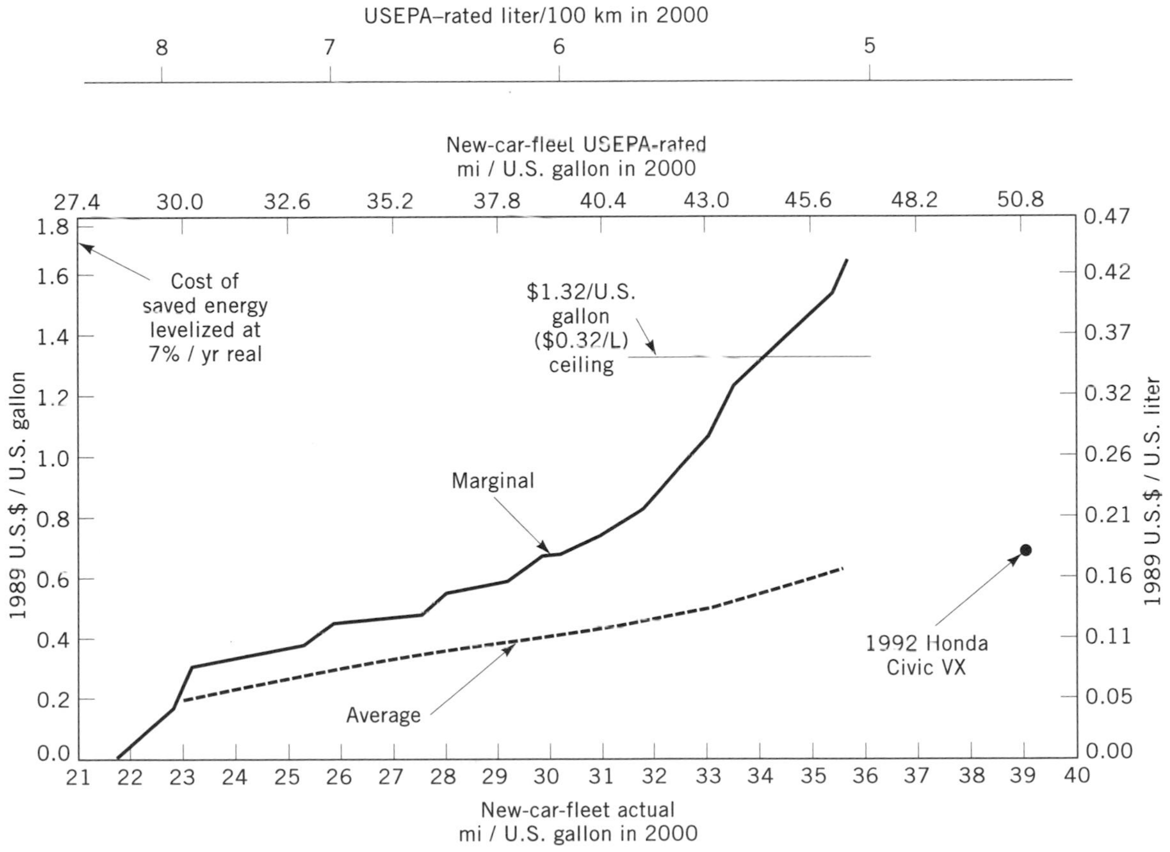

Figure 1. Supply curve for incremental improvements in composite efficiency of average new 1987 U.S. cars as implemented in the year 2000 (14), plus, as an empirical check, the 1992 Honda Civic VX (15).

advances found in today's cars. Confirming how rapidly such estimates are overtaken by events, the 1992 four-passenger Honda VX (15) was 16% more efficient than the U.S. National Research Council's "lower confidence" estimate, published in the same year (19), of what will be technically feasible for a subcompact car *in 2006*.

Curiously, these official assessments barely mentioned certain experimental cars that had already been built and tested. In the mid-1980s, automakers created over a dozen concept cars that combined excellent but fairly conventional components in conventional ways to *demonstrate* doubled or tripled fuel economy (1.7–3.5 L/100 km, 67–138 mpg), often with four- to five-passenger capacity and apparently respectable—in a few cases, superior—safety, emissions, and performance (20). A few of these concept cars had peculiar features, such as sealed windows or rear-facing rear seats, but since these were not shared by others, they were evidently not essential for such good efficiency. At least two versions, from Volvo and Peugeot, would reportedly cost about the same to mass produce as present cars. However, since these concept cars were made in general by European or Japanese firms, they remain little known and are given little official credence in the United States.

The short-term approach universally adopted in U.S. officialdom's pre-1993 car efficiency debate (17–19) is valuable for understanding the potential and limitations of *incremental* improvements to stamped-steel, direct-mechanical-drive, internal combustion, gasoline- or diesel-fueled cars; but it says nothing whatever about what completely *different* designs can do. Attractive though incremental improvements can be, focusing on them diverts attention from a basic challenge to the auto industry: fundamentally redesigning cars and the car business can save much more fuel still, probably cost less, and redefine which firms prosper.

1993 Shift of Attitude

It is no exaggeration to say that designing a car to be three times as efficient as today's U.S. new-car norm of roughly 30 mpg (7.8 L/100 km) is probably *more* difficult than designing it to be 5 or 10 times as efficient. That's because a threefold gain might be achievable by a multitude of very difficult incremental improvements that push traditional methods about to their limit—and therefore require no fundamentally new thinking. Concentration on those limits was the nearly universal attitude of U.S. automakers until autumn 1993 (though not of their suppliers: one survey of 95 such firms, with total revenues of $30 billion a year, found that they strongly favored weight reduction as the key and felt they would "have no problem with a 45 percent improvement" in mandated new-fleet-average fuel economy by 2000 (21)).

The September 29, 1993, agreement between President Clinton and the chief executives of Chrysler, Ford, and General Motors, establishing the Partnership for a New Generation of Vehicles, changed all that. Drawing an explicit analogy to the Apollo program that put people on the moon, it created a joint effort to do something easier and more important: it pledged best efforts, by the Big Three automakers and federal government in cooperation, to produce a tripled-efficiency prototype in 10 years. Setting this goal has already led to two highly desirable results: creating a "leapfrog mentality" that has brought Detroit's most imaginative engineers out of the woodwork, and fostering useful technology transfer from military and National Laboratory practice into the civilian automotive sector.

However, the tripled-efficiency target cuts both ways. Although previously considered far too high for Detroit to imagine, so its adoption represents major progress, it is also just low enough to fall into the trap of continuing incrementalism. Had Vice President Gore been able to get the Big Three to accept a target of improving efficiency by well *over* threefold—a goal his key advisors considered much easier than the automakers did—this would obviously have required a completely different approach, starting with a clean slate. Doing this would have revealed an unexpectedly easy and desirable way to escape from conventional designs' evolutionary trap. That did not occur. Happily, though, some designers both within and outside the formal framework of the partnership have recently recognized this opportunity. It is therefore now plausible that the tripled-efficiency-in-10-years goal will in fact be considerably surpassed, not through any government mandate but through competitive market forces.

Need for a "Leapfrog"

The classical approach to making cars more efficient embodies four key traits:

- thinking about components rather than systems;
- exploring incremental rather than radical changes to those components;
- thinking from the engine toward the wheels, the direction in which the energy flows; and
- supposing that the needed improvements will be mainly in the efficiency of converting fuel into wheelpower via better engines and drivetrains, rather than in improving the basic physics of the platform so it requires less wheelpower.

It is precisely these four traits that have led automaking into its evolutionary trap and blocked major further progress. To understand why, we must consider car physics in its simplest terms—terms so clear and obvious that they are hardly ever stated in the vast literature of car design.

After decades of skillful and devoted effort, automakers have reduced the part of typical U.S. cars' fuel energy that is lost before it gets to the wheels to *only ~80–85%*. This figure is so astonishingly high—perhaps twice what one might expect for an efficient heat engine—largely because of the mismatch of engine size to road load mentioned above. That mismatch, in turn, is due to the habit of building cars out of steel. Steel is heavy; it takes a strong force to accelerate something heavy; therefore accelerating a heavy car sportily enough to satisfy most drivers demands a powerful engine. That engine must be so big—we saw earlier that it uses only about 16% of its power in highway and 4% in city driving—that the underloading roughly halves its average efficiency and increases the difficulty of controlling its emissions. Indeed, matters are even worse, because about 95% of the wheelpower hauls the *car*, not the driver. Thus only about 2% of the input fuel energy ends up providing mobility to the driver.

The most obvious, yet until now most neglected, solution to this conundrum is to work hard on making the *car very light and slippery*, rather than emphasizing tiny incremental improvements to the engine and drivetrain. That is, we must start redesigning the car *from the wheels* (or, more precisely, from the elements that require power to be delivered to the wheels in order to propel the car) rather than from the engine. Because about 5–7 units of fuel are needed to deliver 1 unit of energy to the wheels, *saving 1 unit of energy at the wheels will save (in very round numbers) 5–7 units of fuel energy*. That huge leverage, achieved by avoiding successively compounding losses between fuel and wheels, is a key to supercars' efficiency.

Yet hypercars are more than just an efficient platform. They result from the sophisticated integration of *ultralight construction* with *hybrid-electric drive*. Each of these in turn comprises dozens of modern technologies. Their successful integration requires the very opposite of the traditional design approach. Hypercar designers must think not atomistically, incrementally, and elaborately, but rather holistically, radically, and simply. Successful hypercar design requires whole-system engineering with meticulous attention to detail. It requires designs that are extremely simple (and therefore difficult), obeying Einstein's profound injunction that "everything should be made as simple as possible—but not simpler."

At the root of today's inefficient cars is a deeply rooted error in both design philosophy and choice of materials. As we shall see, a very efficient car cannot be made out of steel, for the same reason that a successful airplane cannot be made of cast iron: it weighs too much. But when we start to design cars less like tanks and more like airplanes, the basic physics of cars magically improves both technical performance and commercial attractiveness. For many car designers, this discovery brings an extraordinary sense of liberation. It is as if a fat pupa, longing for a more graceful form, despairing of shedding weight a milligram at a time, suddenly discovered that it could simply emerge from its chrysalis, spread its luminous new wings, and fly away as a butterfly.

THE ULTRALIGHT STRATEGY

A typical energy budget for a late-1980s U.S. family car (10) shows that after 30% of the input fuel's theoretical potential to do work has been lost because of the irreversibility of combustion, the rest is largely wasted. Of the 100% of combusion heat released in the engine, 47% goes to the cooling system and out the exhaust as waste heat, 17% to friction, 6% to engine auxiliaries and accessories, 13% to torque conversion and transmission, and only 17% to the wheels. What then happens to that 17%?

In urban driving on a level road (22), roughly one-third of wheelpower is used to heat the air that the car pushes aside, one-third to heat the tires and road, and one-third, which previously accelerated the car, to heat the brakes when it stops. Air resistance rises as the square of speed, and the power needed to push through the air rises as the cube of speed, so in highway driving, air drag becomes ~60–70% of total "tractive energy" (wheelpower). Automotive fuel economy, therefore, requires *systematic reductions in all three of these loss terms*. Benefits from improving any one of them are limited, but benefits from improving all together are striking, and they often reinforce each other.

As we shall see, air drag can be reduced by about 2- to 6-fold from today's norms, and road drag by about 3- to 5-fold. Acceleration energy can be reduced roughly 2- to 2.5-fold by mass reduction (the car's mass decreases by 3- to 4-fold, but that reduction is diluted by the unchanged payload), and then at least 70% of the remainder can be recovered by electronic regenerative braking, temporarily stored, and reused. These reductions in wheelpower are then multiplied by the avoided losses in converting input fuel into wheelpower—hence hypercars' astonishing fuel efficiency.

However, the basic parameters of car physics are not equally important. For current U.S. cars, fuel economy is about equally sensitive to reductions in drag and in rolling resistance but is nearly three times as sensitive to reductions in mass (18). Major changes in any of these variables quickly lead to unfamiliar territory (23) where standard coefficients and approximations break down (24): higher fuel economy typically makes aerodynamic drag considerably more important and mass somewhat less, and the speed at which air drag starts to outweigh rolling resistance declines. External factors matter too: drag becomes less important as traffic congestion turns highway driving into stop/go urban-style driving (25). And all the variables interact: mass, for example, becomes less consumption of fuel with low rolling resistance and regenerative braking, while narrower, harder tires (which achieve the former benefit) can help lighter cars to push through puddles without hydroplaning.

In efficient cars, too, previously unimportant terms such as regenerative braking and accessory loads become dominant. Air and road drag become so small that not just one-third but most of urban-driving tractive energy goes to braking and hence becomes available for recovery. Accessory loads (power steering, space conditioning, lights, entertainment systems, etc.) are normally modeled as equivalent to ~9–10% of present average tractive loads, but if unchanged, could use the equivalent of nearly half the wheelpower in a quintupled-efficiency car. Hypercars aim at *total* energy use that approaches the magnitude of accessory loads *alone* in today's cars, so clearly hypercars' accessory loads must be reduced by about an order of magnitude. Fortunately, accessory loads are in fact realistic candidates for such dramatic savings: The engineering concepts and details of what was seen as a minor term have long been appallingly neglected.

Aerodynamic Drag

Aerodynamic drag is proportional to the product of a dimensionless drag coefficient C_D times A (the effective cross-sectional area as seen from the front). Both terms can be markedly improved and already have been in part.

Drag Coefficient. In 1975, when many station wagons had C_D values of 0.6–0.7, a distinguished group of physicists concluded (12) that "about 0.3–0.5 is probably near the minimum for a practical automobile, although even lower values are possible in principle." Now principle has become practice. New U.S. cars' drag coefficient averaged 0.48 in 1979, 0.37 in 1987, and ~0.33 in 1992. Today's sleekest platforms in mass production are ~0.255 (from Opel; among U.S. sedans, 0.29). Volvo's 1993 ECC concept car is 0.23, and the production-engineered 1993 version of GM's electric Impact platform tests at 0.18. Yet 0.18 is also the same value measured by Volkswagen for the 1921 Rumpler Tropfenwagen seven-seater midengine car. Since the mid-1980s, many four-passenger concept cars have achieved <0.2, eg, 0.19 for GM's 1991 Ultralite, 0.186 for Renault's 1987 VESTA II, and 0.137 for Ford's 1985 Probe V. Some LeMans racecars would achieve ~0.1 if not designed to maximize downforce for traction rather than to minimize drag; 1993–1994 unpublished data for the best two- to four-seater road cars are scarcely above this level. (At these low levels, such details as windshield wipers, side mirrors, and door seams become critical.)

The most important step in achieving $C_D < 0.2$ is simply making the bottom of the car as smooth as the top, a task long neglected because nobody could see the bottom. A "needle-nosed" appearance is not required at all, and indeed the car in front can be quite blunt; what matters instead is whether laminar flow adheres to the body or separates from it due to turbulence created by edges and similar discontinuities. Low C_D is easier to obtain in a large than a small car because there is more room to avoid rear-end discontinuities. Contrary to a common belief, the influence of the ground plane ranges from neutral to favorable with good design (26). Even the GM/AeroVironment Sunraycer recumbent-one-seater solar car's 0.125 wind tunnel C_D, considered revolutionary at the time, could have been cut by one-fifth with better wheel-well treatment and other refinements: as noted in a moment, a Swiss team in 1993 actually did cut it by 30%.

For a family car of generally acceptable size and appearance, ultimate practical limits to C_D appear to lie around or somewhat below 0.1, using the following:

- Advanced surfaces with small grooves, dimples, or other textures for passive boundary-layer control, an aircraft and submarine option not previously considered for cars because they were assumed to be made of steel, but easily provided in molded composites or stick-on films.
- Careful design of *internal* air flows for engine cooling and air supply and for passenger ventilation.
- Active (perhaps computer-controlled) suppression of rear-end turbulence.
- Conceivably active control surfaces.
- Cowling the front as well as the rear wheel wells, by, eg, steering via differential speed of the front wheels rather than by conventionally changing their angle. This could be readily accomplished by supercars' special drivemotors but would sacrifice some tire wear and require coordination with active suspension in order to lean properly into turns. It might also require advanced designs to do real-time dynamic simulations incorporating road, load, driver, maneuver, and wind conditions.

As with aircraft, where dozens of small surface refinements can add as much as 180 km/h of air speed, ultralow drag requires unusual care. Joints in the body must be minimized to help laminar flow adhere as far back as possible, a goal facilitated by precise composite body fabrication (which reduces the number and width of seams and the manufacturing variance of geometry, hence of C_D), and perhaps by self-inflating door seals. As mass declines, greater care is also required in designing crosswind response (27), both for stable handling and to avoid the need for parking tiedowns. This art has lately become rather mature, as illustrated by the University of Biel's recumbent-one-seater solar car Spirit of Biel III. In 1993 it achieved a record low C_D of 0.088 (two-thirds body, one-third wheels and canopy), falling to only 0.05 if the airstream is 20° off-axis.

Frontal Area. The other aerodynamic drag factor, frontal area, is also reducible by better packaging and styling. The most important changes come from thin but ultrastrong composite body materials that can make the inside of the cab bigger but the outside smaller. Innovative composite door designs, including versions that rotate and displace laterally, can make doors thinner and simpler, as can the ESORO rotary window design. The same materials' severalfold mass reduction also greatly shrinks the powerplant size needed, again leaving more room inside for people and cargo. More compact powertrain components change not only volume but also form. They permit steeply downsloping (heavily raked) hoods, which in turn permit greater visibility with less glass and hence less mass. Putting ultraminiature headlights (pea-sized metal halides, or fiber-optic ports connected to a central internal light source) inside the bottom edge of the windshield, where they can be cleaned by the same wipers, further lowers front-end height by essentially the height of conventional headlights. It also reduces glare and permits a more crashworthy impact-beam structure at the front corners (and similarly at the rear). Similar, or ultracompact light-emitting diode, options are also available for corner-mounted turn signals and other lights. As these sorts of options lower the front of the car, the hybrid drive's absence of underfloor drivetrain components can also lower the passenger compartment floor while maintaining sufficient ground clearance. All-interior fiber-optic or television rearviews, as in two 1993 Japanese hybrid concept cars (Daihatsu's Dash 21 and Mitsubishi's ESR), can eliminate the frontal area, adjustment and cleaning complexity, and turbulence of side mirrors while providing superior automatic antiglare and contrast adjustments.

The ~2.3-m^2 frontal area typical of new U.S. cars is easily reducible to ~1.9 with no sacrifice of interior roominess; it is, for example, 2.01 in the somewhat boxy 1993 five-passenger Volvo ECC concept car and 1.80 in the 1991 Honda Civic DX. Among four-seater concept cars, GM's 1991 Ultralite achieved 1.71, and Renault's 1987 VESTA II, 1.64. This 1.6–1.7 m^2 range has conventionally been considered a reasonable lower limit, but neither of those cars used a hybrid driveline or exploited most of the additional factors in the previous paragraph. Doing so could probably reduce A to perhaps ~1.5 m^2 while maintaining a spacious interior. After all, two-seaters even with conventional propulsion can yield reasonable comfort with only 1.3 m^2, and their occupants are just as tall as those of four-seaters.

Road Drag

Tire rolling resistance r_0, the dimensionless per-tire ratio of road drag to vertical load, is surprisingly poorly understood and measured, and most tiremakers consider r_0 data confidential. (They tend to be even more reticent about the formal distinction between the low-speed r_0 and high-speed r_1 terms, often lumping them together.) This nonrecoverable loss typically totals ~0.007–0.01 for modern radial tires—half the ~0.02 of 1970-vintage bias-ply tires (20)—and one major automaker reports 0.0062 for good mass-produced tires fully compliant with all U.S. regulations. For a given road surface, tire temperature, pressure, torque, and speed, r_0 depends on energy losses in tire and tread deformation and hence on arcane design details and low-hysteresis materials (28). Considerable progress has been made, though generalizations are difficult because tire and vehicle must be designed together. Improved polymers are an obvious starting point: eg, Venezuelan tests in 1981 suggested a ~15–30% reduction in r_0 and halved tread abrasion from an Austrian liquid-injection-molded aramid/polyurethane tire produced for Soviet offroad vehicles (20). In 1990, Goodyear announced it had cut r_0 in its G-22 4.4-bar concept tires for GM's Impact electric concept car to only 0.0048, rated at 81 km/h (50 mph), with "excellent traction and highway performance." This value has since been further reduced. Comparable developments are underway in Europe and Japan. New concepts also show promise, including variable-camber double-tire and variable-pressure (like Russian truck) configurations and several unusual cross-sectional designs, eg, Scandinavian hollow composite drums with a tread glued onto the outside.

The limits of practical r_0 reduction are now being approached. The Sunraycer's bicycle tire 0.0037 could readily have been cut by one-fifth, but car tires have different requirements: Ride, dry and wet traction, and durability, especially in a light car lightly loaded, cannot be sacrificed in favor of two small r_0. The present car tire art is thus ~0.005–0.006, with advanced concepts approaching ~0.004, a fifth the value of two decades ago. Even more fundamentally, tire rolling resistance is the *product* of r_0 times mass. The lighter the platform and hence the less downforce on each tire, the lower its rolling resistance. Since curb mass, as we shall see, can be reduced by ~3–4 times, and curb mass plus payload therefore by ~2–2.5 times, combining that mass reduction with a reduction of up to half in tire r_0 will reduce the absolute amount of energy lost to road drag by ~3–5 times. In addition, tire designers can usually eliminate the mass and bulk of a spare tire and jack. Self-sealing low-r_0 designs can already cope with ~75% of punctures at the cost of a modest (~8%) increase in r_0V—say, from 0.0038 to ~0.0052.

One might at first suppose that such a light car cannot achieve enough traction for safety, especially on slick surfaces. However, the area of the contact patch will also decrease, yielding higher contact pressure, if the tires are harder and narrower. The combination of this effect with improved tread design, better materials, smart active suspension, and full-time all-wheel drive with superlative antislip traction and antilock braking may well result in a net improvement in the handling elements relevant to safety. At the same time, the harsher ride that could result from harder tires alone can be mitigated by softer suspension without sacrificing precise handling, thanks to the remarkable torsional stiffness of the composite monocoque body described below. Harder tires are not the only option: Ride can be softened, as in GM's Ultralite concept car, with a modified rounded-sidewall profile.

Suspension design is especially challenging in supercars because adequate control, especially in cornering on rough or slippery surfaces, must be maintained while payload varies from roughly 0.1 to nearly 1.0 times curb mass, a far wider range than in heavy platforms. Although a lighter car can grip better in hard cornering—"lopping a thousand pounds off a car chassis does extraordinary things to handling, finesse, acceleration, and economy" (29)—alternative tire profiles may also be desirable. More sophisticated supercar designs are likely to use refined versions of the smart active suspensions now emerging in high-end heavy production cars. These suspensions currently prepare the rear wheels for what the front wheels have just experienced. Supercar versions may also use look-ahead dynamic simulations analogous to those used in high-performance aircraft. Active suspensions using either hydraulic or electrical actuation can probably also recover most of their energy, reducing their net load nearly to zero.

Mass

Mass is doubly important to platform physics, because both rolling resistance and acceleration force are directly proportional to mass. However, mass reductions are even more important than meets the eye, because they snowball or "compound." Saving 1 kg in a conventional steel car is conventionally assumed to save roughly 1.5 kg overall, because less structure and suspension are needed to carry the reduced mass, less driveline and fuel mass to haul it around, etc. However, this "reverse mass compounding" or "mass decompounding" factor appears to be much larger—perhaps around 5—in ultralight cars. The physical explanations for this are complex but include the ability to simplify design by not merely downsizing but by *eliminating* components and systems. For example, a well-designed ultralight car without power steering or power brakes will handle better than a heavy car with these features.

Major mass reductions have already been demonstrated using conventional construction methods and materials. New U.S. production cars' nominal curb mass M averaged 1443 kg in 1990; a light model like a Toyota Tercel LE sedan weighs 946 kg. Yet numerous production and prototype platforms weigh far less: for example (20), VW's five-passenger Auto 2000 (779 kg), Peugeot's five-passenger 205XL (767 kg), Volvo's four-passenger LCP 2000 (707 kg), VW's four-passenger E80 Diesel (699 kg), British Leyland's four-passenger ECV-3 (664 kg), Toyota's five-passenger AXV Diesel (649 kg), Renault's four-passenger VESTA II (475 kg), and Peugeot's four-passenger ECO 2000 (449 kg). This art has been even more highly developed in sports/racing cars: the 1957 two-seater aluminum/steel Lotus Super Seven (later the Caterham 7, still available in kit form) weighs only ~600 kg. Lightweight design has also long been analyzed for family cars. In 1980 the Battelle Memorial Institute in Columbus, Ohio, designed but never built a 545-kg "Pertran" vehicle simulated, with regenerative braking, to achieve composite ratings of 2.76–2.94 L/100 km (80–85 mpg) with a gasoline engine or 2.2–2.4 L/100 km (100–105 mpg) with a diesel engine (20).

All these designs made extensive use of conventional lightweight materials. However, the usual approach—substituting aluminum (plus some magnesium, plastic, and titanium) for steel—reduces weight, increases cost by ~\$2–6/kg saved, offers superplastic molding potential, and often modestly increases fabrication difficulty; in all, it is a useful near-term material option but not an optimal strategic approach. Substituting instead advanced polymer composites and other net-shape materials such as engineering plastics (polyamides, polycarbonates, polyacetals, thermoplastic polyesters, polyphenylene oxides, etc.) can achieve far greater reductions in mass and drag. More importantly, at the same time the substitution of polymers for metal transforms the carmaking process and marketing structure in a way that fundamentally reduces cost—"tunneling through" the cost barrier to conventional mass reduction. Low mass therefore saves not only energy but also production complexity, time, and potentially total vehicle cost, as discussed below.

So far, engineering polymers are used only for specific *components*, where they are rising toward one-fourth of total polymer mass in U.S. cars (30). Polymers of all kinds, mainly ordinary molding compositions but also composites, total ~8% of new cars' mass or ~20–30% of their parts volume, the difference being due to plastics' low density. Although these are significant volumes, most car designers still know steel, think steel, and use comparisons that unduly favor steel. Further metal-vs.-polymer-and-composite substitutions are usually compared with steel only on an incremental, component-specific basis.

Such substitutions often make sense: they save, for example, 60% of mass in leafsprings and 80% in GMC truck driveshafts (31). Much further extensions are technically feasible: the German Aerospace Research Establishment has even made a complete composite powertrain (wrist pins, piston rods, crankshaft, Cardan shaft). Yet component-level comparisons are often piecemeal and unsystematic, and the design often improperly imitative of the original steel part, like the early plastic radio and TV cabinets that were shaped and patterned to look like wood boxes, before designers discovered ergonomic forms. Making the composite part look like the steel part simply does not work: the two materials are so fundamentally different that the composite part must be redesigned from scratch. For example, a composite I-beam with the same dimensions as a steel one may buckle, but a composite box-and-plate beam with the same cross sectional area may be three orders of magnitude stronger (though not stiffer).

Worse, component-level comparisons miss the most important advantage of carrying the substitution *throughout* car design. Most disagreements over the practicality and cost of a massive switch from metal to polymers can be traced back to this difference between a component view and a systems view. This difference is rooted in the cultures of steel and of modern synthetic materials, a gap that is not easily bridged. We shall return to it later.

BEYOND THE IRON AGE: NET-SHAPE MATERIALS

A typical steel part's cost is only ~15% for steel; the other 85% is for shaping and finishing that raw material. Steel is so ubiquitous, and the success of highly evolved steel-car manufacture—one of the most remarkable engineering and managerial achievements in human history—makes its very high design, tooling, fabrication, and finishing costs so familiar that we overlook how they outweigh its cheapness. An electrocoating plant costs a quarter-billion dollars; a classical paint shop a half-billion dollars; and complete tooling for one car model, upwards of one billion (10^9) dollars. Making a steel car requires over a thousand engineers to spend a year designing and a year building a football-field-full of million-dollar steel dies that are used as long as possible (ideally decades), then thrown away.

This inflexible, costly tooling in turn means huge production runs, high risks of stranded investments, and long amoritization times, time to market, and product cycles. These in turn crimp flexibility and innovation. Thus today's most "modern" cars are really the cutting edge of decades-old technology, the highest expression of the Iron Age. Yet new, nonmetallic materials are not just a piece-by-piece substitute for steel, as they have been used so far; they can transform the nature of cars, manufacturing, and marketing. In the process, they also support the ultralight strategy.

GM's Ultralite Example

A striking example of this transformative potential is the Ultralite concept car that ~50 General Motors technologists built in 100 days in 1991 (32–34). It cost ~\$4–6 million, or ~8 hours of GM's 1991–1992 North American losses. This sporty, four-adult, four-airbag car (Fig. 2) achieved $C_D = 0.192$, $A = 1.71\ m^2$, $M = 635$ kg, and highway-speed rolling resistance r_0 ~0.007 at 4.4 bars tire pressure. Its C_DA product (effective drag area) was thus 43%, and its curb mass 44%, of the U.S. norm. These parameters yielded an EPA-rated 3.79 L/100 km (62 mpg), comprising 5.22 L/100 km (45 mpg) city and 2.90 L/km (81 mpg) highway—comparable to the average fuel economy of Volkswagen's relatively conventional IRVW safety test car delivered to the U.S. Department of Transportation 15 years earlier, but this time made in the United States and combining many basic advances in platform physics.

Cruising at 2.35 L/100 km at 81 km/h (100 mpg at 50

Figure 2. General Motors' 1991 Ultralite concept car. (Courtesy of General Motors.)

mph)—enough to go from New York to Los Angeles on 110 L (29 gal) of gasoline—the Ultralite required only 3.2 kW (4.3 hp) of power to the wheels, 71% less than an Audi 100 needs. Efficient, wheels-at-the-corners packaging gave the Ultralite the interior spaciousness of a Chevrolet Corsica, which is twice as heavy and half as efficient, but within the petite outside volume of a Mazda Miata. The Ultralite's 279-cm wheelbase, 66% of its overall length, equaled that of a Lexus LS-400 or nearly of a Buick Park Avenue. The 79-kg rear engine in a removable "pod" was a 1.50-L, 83-kW (111-hp), three-cyclinder-in-line, two-stroke, direct-injection, stratified-charge, all-roller-bearing design driving a slightly modified Saturn four-speed rear-drive transaxle. This nonhybrid, purely mechanical driveline achieved a 218-km/h (135-mph) top speed and accelerated 0–97 km/h (0–60 mph) in 7.8 s. Thus the Ultralite can match the acceleration of a 12-cylinder BMW 750iL while using an engine smaller than a Honda Civic's.

Ultralite's Acceleration/Efficiency Trade-off

For a quick design not yet optimized in either body mass or engine efficiency nor engineered for production, these are impressive achievements. Yet comparisons with Renault's 1987 VESTA II concept car confirms that the Ultralite could be even more efficient with a smaller engine. The VESTA II achieved 64 mpg city, 114 mpg highway, and 86 mpg combined. It has virtually the same C_D as the Ultralite, but 5% smaller A, 160-kg smaller M, slower acceleration, and 36% lower top speed, permitting an engine of only 20 kW (27 hp), *less than one-fourth* the power of the Ultralite's. The VESTA II's reduced engine underloading and slightly better platform physics yielded *two-fifths higher* fuel efficiency.

The sensitivity of fuel economy to acceleration is further illustrated by a calculation (35) that if GM's heavy but sleek and sporty Impact two-seater battery-electric car were powered by a driveline that converted fuel to wheelpower with only 23% efficiency, and if it achieved the same 121-km/h (75-mph) cruising speed with very sluggish acceleration (0–97 km/h in ~32 s rather than the Impact's actual 8 s), then this 682-kg platform with a very small engine (~13 kW, 17 hp) would cruise at <1.04 L/100 km or >225 mpg. Though hardly attractive from a marketing perspective, this is technologically instructive, because with a hybrid-electric drive, such a car could combine powerful acceleration with an engine even smaller than 13 kW, thanks to very powerful electric wheel motors.

A far more marketable actual car, Pininfarina's 1993 two-passenger, thermoplastic-over-spaceframe Ethos II coupe, further confirms that GM's Ultralite has sacrified efficiency for sports car acceleration. The Ethos II weighs more (730 kg), though its identical 0.19 C_D and 10% smaller frontal area (A = 1.53 m^2) yield a 12% lower aerodynamic drag. Its Orbital engine, only half as big and powerful as the Ultralite's, yields an 8% lower top speed and a two-thirds longer acceleration time. Yet by trimming these performance parameters to about the U.S. fleet's average, the Ethos II improves engine optimization enough to yield 2.76 L/100 km at 120 km/h (85 mpg at 74 mph), improving to 2.10 L/100 km at 60 km/h (112 mpg at 37 mph); the 90-km/h cruise rating is 12% more efficient than the Ultralite's. This is not to say that GM's design parameters might not have been a superior marketing decision, but it does illustrate the efficiency/acceleration trade-off.

Can Ultralights be Affordable?

Of the GM Ultralite's more than 100 significant innovations, the most important was its six-piece, 191-kg, ~$12,000-materials-cost monocoque body. Foam was sandwiched between two layers of composite made by applying epoxy resin over carbon-fiber cloth, interwoven with roving (continuous-strand carbon-fiber rope) at stress points. The body was fabricated by Burt Rutan, the maker of the *Voyager* aircraft that his brother then flew round the world on one tank of fuel. Conventional hand-layup methods were used, akin to those common in small-scale, high-end boatbuilding.

The trade and financial press, while admiring the Ultralite's achievements, complained that the miraculously strong black threads of carbon fiber, only 7 μm (0.018 in.) in diameter, cost ~$18–90/kg, two orders of magnitude more than sheet steel at ~$0.55–0.80/kg. But those critics forgot that what matters is not cost per kilogram but cost per *car,* and that there are fundamental differences between steel and composites. Those differences lie at the heart of the economics of producing hypercars:

- Carbon fiber is stiffer and stronger than steel per unit cross section but only one-fourth as dense; it has 2–4 times the stiffness but two-thirds the density of aluminum. Fiber typically occupies ~30–50% of total composite volume in current car practice, ~60–65% in optimal components. Since the rest is a somewhat less dense and considerably less strong epoxy or similar resin, the cross section of a whole composite (not pure carbon fiber) exceeds that of equivalently strong steel, but its mass is roughly two-thirds lower. Thus about one-third to one-half as much mass of carbon is needed as of steel, considering that the fibers can be placed and aligned to match the stress field and interwoven to distribute loads (32), just as a cabinetmaker orients wood grain to stress. The art of composite design is optimal fiber selection and placement to match a controlled load path, so as to capture the material's strengths and compensate for its weaknesses. For example, a body sheet subjected to anisotropic loads can be woven with one axis correspondingly stronger than the other, or with a mixture of materials, to provide just the properties and in the directions required with an elegant frugality of means.

- For many applications, just as serviceable composites can be made from glass fiber as from carbon fiber. Glass is tougher, heavier, more elongating for energy absorption, and ~2–6 times cheaper than 1993 carbon for equivalent strength or stiffness. Specifically, typical 1993 U.S. creel prices for automotive-grade carbon fiber, which is two-fifths costlier than the weaker sporting goods grade, were around \$26/kg. (GM's Pyrograf process should cut this to ~\$4–5/kg but does not yet yield fibers suitable in length for structural applications.) Glass fiber is far cheaper than carbon: \$2/kg for E-glass and ~\$11/kg for S-glass with ~20% higher performance. Simple biaxial cloth in 1993 cost ~\$44/kg for carbon, ~\$31/kg for S-glass, and ~\$8/kg for ordinary E-glass. However, only about half as much carbon as E-glass fiber is needed for equivalent strength and stiffness (but not elongation or toughness) in many applications, so the effective 1993 cost difference was not ~13 times but only ~6 times. (This is because carbon fibers are not only stronger and far stiffer but also smaller, with 2–3 times the surface area of glass, and hence absorb more energy in being separated from the resin matrix.) A composite production car would use a judicious *mixture* of E- and S-glass, carbon, polyaramid, etc., each to do what it does best. Carbon is typically preferred in firewalls, pillars, hardpoints, and similar high-stress structural elements, for lightness in large panels such as roofs and floors, and sometimes for stiffness to protect the passenger compartment (given intrusion protection), while glass is most often used in lower stress areas and for elongation and toughness in front and rear crush zones. Such combinations are used by, for example, U.S. Electricar's Consulier division, which has sold ~100 road-certified monocoque vehicles since 1988. Such sports cars' light weight (the two-seater's basic body weighs only 125 kg out the mold) but normal ~200-hp engine let them outrace any showroom car, even exotics, while achieving 30-mpg average efficiency. However, one Swiss expert, Peter Kägi of ESORO, has cut two-seater body-in-white mass to an impressive 34 kg, or 73% less, using all carbon and polyaramid; without batteries, his whole car weighed only 230 kg.
- Net-shape materials can often be advantageously used in the frame (if any) and other components, not just in the body. Indeed, composite-skin, foam-core materials permit monocoque construction with *no* frame: like an egg, the body *is* the structure. Monocoques' remarkable flexural and torsional stiffness—as solid as a Mercedes at a small fraction of its mass—simplifies many aspects of design and permits softer springing, improving ride and handling.
- Molded composites largely avoid the ~85% of steel parts' total cost that is due to shaping and finishing. Composites emerge from the mold practically ready to use, in complex, sleek, and beautiful shapes unattainable with metal. To be sure, handwork may be required to prepare the materials to be molded and to achieve an exemplary finish (squeegeeing, handling of bleeder cloth and waste resin, etc), but the molded material is already in essentially the shape and finish desired, rather than in a crude form requiring extensive machining, forging, stamping, drilling, etc, plus additional surface finishing operations.
- Composites' molding tolerances are a few tens of micrometers in typical car practice or only a few micrometers in aerospace practice. In contrast, in many car factories one can observe large men with rubber sledges beating steel roof pillars to within 1 or 2 cm of the right position so the roof will fit on. If all parts fit exactly the first time, costly rework is virtually eliminated.
- More importantly, composites' moldability in large, complex units can cut the parts count, in principle, by ~100 times. The body-in-white (basic open-aperture body without trim, chassis, or powertrain) can have not 300–400+ but *only 2–6 parts* that can fit precisely together, slashing the costs of design, tooling, transportation, storage, and inventory.
- Required assembly space and labor drop by ~10 times, largely bypassed by the highly integrated moldings and by simpler design. Total labor probably drops too, but by much less, depending on how much the layup and finish detailing are automated. Although joining is slower than robotized welding, almost none of it is needed because so many parts are integrated together into single moldings (30).
- Components and assemblies and the powertrain elements needed to propel them become much lighter, and hence easier for fewer people to handle with less equipment in a more flexible assembly setup. Volvo concluded in 1985 that its magnesium-intensive LCP 2000 concept car would cost essentially the same as its normal steel cars to mass produce because of its streamlined assembly *alone*, without counting net shape materials and their extra advantages.
- Painting—the costliest, hardest, and most polluting step in carmaking, accounting for nearly 90% of some major automakers' mass of hazardous and toxic releases—can now be eliminated by lay-in-the-mold color that yields a more durable and attractive finish. Painting accounts for about half the finished cost of painted steel car parts and for half the capital cost of assembly shops.
- The coated epoxy dies cost at least 50–60% less per vehicle than tool-steel dies—by some estimates as much as 10-fold less—for three reasons. First, there are far fewer parts. Second, molding each part requires only one die set, not three to seven stepped die sets for successive strikes. It is clearly far cheaper to have one large, complex die for an integrated composite assembly than the hundreds or thousands of dies needed for equivalent separate steel parts. For this

reason, a reaction injection-molded plastic fender may cost the same as a steel fender but have a 70% lower tooling cost (30, 36); substituting polyphenylene oxide for steel in Cadillac parts cut tooling costs by an estimated 74% (31). Third, composite-molding dies are made of a cheap material—typically metal-coated epoxy—that is cheap to fabricate. The epoxy wears out much faster than tool steel but is so cheap to make or recondition that its shorter lifetime hardly matters.

- On the contrary, epoxy dies' shorter life and short development time offer a striking strategic *advantage*. Opportunities for improvement at the time of tooling replacement or refurbishment are far more frequent. New dies can then be made within *days*, not two years, under computer control—even roughed *overnight* directly from the computer design screen by stereolithography, which converts drawings into precise three-dimensional objects. This approach not only eliminates traditional clay modeling and other cumbersome remnants of Victorian manufacturing; it also supports the small design teams, reduced diseconomies of scale, rapid amortization, very short product cycles and time to market, small production runs, high product differentiation, and continuous improvement that market nimbleness demands. This is a revolutionary strategic advantage (37), more important than any direct cost savings: as "a further incentive to adapt the supply of cars rapidly to the evolution of demand, . . . the flexibility introduced in the automobile industry by the technology of synthetic materials has powerful analogies with the flexibility introduced by . . . electronics technology" (30). The resulting flexibility and risk minimization are the strongest motives now driving automakers in the direction of hypercars.

These and other, even more profound, implications for safety, manufacturing, and marketing will be considered below. For the moment, we need only note that the features just listed appear roughly to offset composites' apparent cost disadvantage per kilogram. The materials cost more, but the manufacturing costs less. On balance, production cost *for the complete vehicle* (though not necessary for a given component) should be roughly comparable for composites and may even be lower.

Many examples from other industries confirm that savings on tooling and assembly labor can make seemingly costly molded materials cheaper than steel, not per pound but per finished part or system. For example:

- A 5.9-kg steel tricycle with 126 parts was redesigned to a 1.4-kg, 26-part plastic version at one-fourth the cost (38).
- A mainly brass toilet float/valve assembly was redesigned from 556 to 88 g, 14 parts to 1, and \$3.68 to \$0.58 production cost (38).
- A windshield-wiper arm was redesigned from 49 parts to 1, at lower total cost, even though it was made of ~\$31/kg carbon fiber.
- Chrysler found that composites could cut a steel car's subassembly count by ~75% (saving much assembly labor), plant cost by ~60%, and tooling cost by 50% and its body-in-white parts count by 98–99% (39).

Ultralight composite cars' total production costs remain controversial, but the disagreements appear to come largely from most steel experts' failure to count many of composites' system benefits—and rapidly declining carbon prices. The most pessimistic of the composites experts consulted in our 1993 research estimated mass production cost of ultralight carbon-fiber cars at one or two times that of steel cars today. The most detailed assessments, however, suggested that even carbon-fiber costs could achieve breakeven at fiber costs around the best 1994 creel price of \$11/kg, let alone the astoundingly low \$7/kg quoted in late 1994 (6) for bulk purchases from one major manufacturer's next plant. (This ignores the major price fluctuations that could occur meanwhile if demand temporarily outruns the very limited current supply.) Extensive use of cheaper fibers such as glass, as most designers contemplate, could also offer more robust economic advantages at 1994 prices and at all production volumes.

Most composites-manufacturing experts, including some at major automakers, are convinced that in a "greenfield" (start-from-scratch) comparison, composite cars even at 1993 costs *always* undercut the mass-produced cost of steel cars. Indeed, some practitioners believe that lower tooling cost permits molded 1993 synthetics to beat steel cars' cost at production rates below 30,000 units per year. (For example, Pininfarina has projected a cost below \$20,000 in 1993 dollars for its aluminum/thermoplastic Ethos II at production volume of 10,000 units per year.) Other sources (30) have found the breakeven volume for synthetics has recently doubled, in only a couple of years, to ~50,000–60,000 units per year—a level "slightly more . . . than Jaguar's or Porsche's output." Either of these case-specific levels is ample for exploiting important "boutique" markets—niches large enough to yield attractive production economies and to start moving down the learning curve toward lower costs and larger markets. Add the steep 1994–1995 fall in carbon mixes makes the conclusion even stronger.

Supercars could well launch first in the sporty, high-end luxury-car market. This segment demands exceptional acceleration, handling, freedom from noise and vibration, fit and finish, durability, spaciousness, decoupling of size from mass—exactly the qualities that ultralight composite construction offers—but expects to pay a premium. The relatively high markup of such performance cars also leaves more room for initially higher production costs before manufacturing methods have matured.

Moreover, market segmentation and differentiation, hence more and faster changing models coupled with slower growing aggregate demand, entail "a clear-cut trend towards a decrease in the average production scale. In the long run, this factor could be a powerful force stimulating the diffusion of synthetic materials into the automobile industry" (30). For example, Fiat converted the 12.5-kg hatchback of its Tipo to a molded synthetic material largely because the cheap tooling permitted short production runs and rapid changes (30).

Manufacturing Techniques

The most important, debated, and proprietary area of supercar development is the advanced manufacturing techniques that can bring to composite molding the same degree of simplicity and automation already available from such standard techniques as injection molding for ordinary plastics. Some very interesting and promising techniques are already starting to emerge from the Partnership for a New Generation of Vehicles' unprecedented dialogue between aerospace composites experts, who make high-performance components at roughly a thousand times the cost and one-thousandth the volume needed for automaking, and car experts, who want 100,000 units a year at $9/kg.

In one emerging method, for example, automatically laid-up fibers are wetted with resin as they continuously enter a variable-geometry molding die. As they emerge, intense actinic radiation, such as hard ultraviolet or electron beam, cures the resin about 1000 times faster than heat normally cures thermally activated resins—which is in turn about twice as fast as the heat treating of metal parts. (Actinic-cure resins, now common in dentistry, can be obtained in epoxy, polyester, and other standard formulations; they have the further occupational health and environmental advantage of zero solvents.) Offsetting the "shadow boundary," defined by the sharp edge of the actinic radiation, between the upper and lower sides of the emerging material makes that material hard on one side while it remains soft on the other, allowing it to be bent in three dimensions. Under real-time computer control, such as "electronic sausage machine" could in principle produce large and very complex shapes with high speed, precision, and flexibility. Other methods have also been devised for applying to automaking's complex shapes the established methods that can fabricate simpler composite forms relatively cheaply.

Recycling

Composites are extremely durable—they do not dent, rust, chip, etc—and could last for enough decades to be heirlooms and then, with careful design, be recycled. (Pyrolysis or shredding into short-fiber reinforcement for engineering materials are also possible but do not recover the valuable big-molecule fibers.) Recycling most composites currently requires a brute-force technique such as methanolysis. This appears to be both effective and economically justified by the salvage value of the recovered fibers, which emerge at or very near original strength, and by avoided solid-waste disposal cost. Resin monomers are also recoverable to some degree, though they are much less valuable. However, it would be more elegant to devise resins that unlock with a chemical key at slightly elevated temperature or pressure, simplifying clean fiber recovery. This appears to be feasible, and research on it has recently begun.

The best ways to conserve energy and large molecules in handling large volumes of composites require considerable further research and should not be discouraged by ill-considered regulation. An obvious option is not to sell hypercars but to lease them, so that after perhaps five to seven years they must be returned to the manufacturer. (Such countries as Japan and Sweden already require rapid turnover of their car stocks, and Germany is requiring their return to the automaker for recycling, a huge stimulus to disassembliable design and proper parts labeling.) Otherwise, hypercars' durability will block the introduction of even more advanced models, as durable DC-3s did with later jet aircraft.

Repairs

Composite cars require far fewer repairs than metal cars, because low-speed collisions and minor scrapes do not permanently affect the body; it simply bounces out again. In case of major damage, makers of composite racecars and street-licensed cars (such as Consulier and Horlacher) have long demonstrated that repairs are straightforward, although the required skills are more those of the boatbuilder than of the sheetmetal worker. In essence, the damaged section is defined and cut out, its edges are feathered, and a patch is laid in, resined, cured, and finished. Some issues remain, however, about inspectability for hidden damage and its potential effect on later structural integrity. Composite aircraft wings are routinely repaired to nearly original strength after bird or stone strikes, but such thorough inspection and repair may be difficult for cars. Imbedded conductive or optical fibers may prove useful diagnostic tools, they are already reportedly used in high-performance flight surfaces for military aircraft, some of which are virtually all-composite for radar stealth.

Speed of Transition

As automakers realize the advantages of cars based very largely on composites, a revolution is rapidly emerging in car design, production, and operation that bids to become as profound as the electronics-driven transformation in the 1970s. The challenge to metal might come surprisingly quickly. U.S. passenger cars' bodies are said to have switched from 85% wood-frame in 1920 to over 70% closed steel only 6 years later (40), ensuring the dominance of Henry Ford's assembly line. Mainly in the 1960s, composites rapidly displaced wood and metal in boatbuilding, as they are now doing in aerospace niche markets. Today, the switch to molded synthetic materials could support "a major breakthrough in the technological development of the automobile industry" (30), making it at long last an agile, short-cycle competitor (37).

A perception is growing among the most advanced car designers, too, that when combining ultralight construction with hybrid-electric drive, it makes more sense to start with ultralights and then hybridize them than to start with heavy hybrids and then lighten them, the method of nearly all hybrid car programs until 1993–1994. The reasons for this are both technical and strategic; some will appear further below.

Partly as a result, the previously assumed timescale for a major transition in the auto industry is rapidly shrinking. It has not escaped observers' notice that General Motors created the Ultralite concept car in 100 days; that GM also "productionized" the Impact battery car, starting from a comparable stage of development, in approximately 1 year; and that both these processes involved more the traditional clay-model-and-toolmaking than the virtual-phototyping, CAD-and-stereolithography route. For a firm either big enough to be highly capable or small enough to be equivalently agile, it is not difficult to devise a risk-managed trajectory that combines a number of parallel development paths to reach ambitious objectives quite quickly.

For this reason, because of the readiness of an electronics-conditioned public for all kinds of technological leapfrogs, and because of the competitive imperatives for rapid conversion, the eminent ultralight-vehicle engineer P. MacCready and the equally eminent auto journalist R. Cumberford have predicted that by about 2005, virtually all the cars in American showrooms are likely to have electric propulsion, and most of those will be hybrids. Indeed it seems plausible to suggest that by the mid-1990s there should already be early ultralight–hybrid prototypes (a few experimental ones appeared in 1993–1994); significant early production should start around 1997–1998; and by the end of the 1990s it should be clear that the right parenthesis of the mechanical-drivetrain steel-car era is coming clearly into view.

HYBRID-ELECTRIC DRIVES

Definitions

An ultralight car, even if driven by a conventional driveline (an internal combustion engine mechanically coupled to the wheels), is the first and more important half of the supercar concept. The second half, requiring at least equal technical sophistication, is a carefully integrated hybrid-electric driveline. The term "hybrid" is used here to mean that the wheels are driven mostly or wholly by one or more electric motors but all the electricity is made onboard as needed—or recovered, temporarily stored, and reused to augment the small engine's power for hill climbing and acceleration.

This definition of hybrid-electric ("hybrid" for short) drive is not the official one used by the U.S. Department of Energy (DOE). With emphasis added, the DOE defines hybrids as depending "partially upon *externally* generated electricity for propulsion energy." That approach, typically implemented as a range extender (a battery-electric car with a small engine to recharge the batteries so the car can run longer), is nowadays seldom regarded as very attractive, for the same fundamental reason that limits battery-powered electric cars to niche markets, namely, that batteries inherently contain about 100 times less energy per unit mass than fuel does, so any attempt to get much range from batteries entails a very large and awkwardly compounding battery mass.

The definition used here for hybrids need not exclude all external recharging; one can image hybrids that are able, perhaps on occasion, to draw some recharging power from an external source, but it is not obvious why this is either an important or an attractive option. Recent advances in both electric propulsion and ultralight construction tend to make external recharging at least unnecessary and probably undesirable. (In contrast, as *onboard* photovoltaics become cheap enough, they would probably make an attractive source of nonfueled onboard recharging.) This design approach also finesses difficult technical, economic, and public policy questions about recharging infrastructure and who should pay for it.

Of the many variants on the hybrid driveline concept as defined here, the most common are "series" and "parallel." In a series hybrid, a small onboard engine or other power plant converts fuel into electricity. The electricity drives one or more electric motors to turn the wheels. Electricity generated in excess of instantaneous wheelpower needs, or recovered from electronic braking, is temporarily stored in an onboard battery, ultracapacitor, flywheel, or combination of such storage devices. However, the hybrid's storage is only a buffer, not meant for range, so its energy capacity is an order of magnitude smaller than in a battery-powered car—just a few, say 1–5, kW · h.

In a parallel hybrid, under some conditions, such as running at high speed or up a steep hill, the engine can also provide mechanical power directly to some or all of the wheels. Such designs can, for example, accelerate electrically to cruise speed (taking advantage of electric motors' high torque-to-mass ratio) and then cruise under mechanical engine drive with engine speed varying only ~15% and torque threefold, a small enough performance map to maintain good efficiency. Such parallel designs typically use rear-wheel mechanical drive and front-wheel electric drive, the former with overrunning clutches for reversing. They recover somewhat less braking energy than a series hybrid but can have more graceful failure, since there are two potentially independent power sources. They are also more complex and may weigh and cost somewhat more, depending on design. They can achieve zero tailpipe emissions in the city, where they can run all-electrically, and ultralow emissions under engine or dual power on the highway, so they may qualify for the no-combustion zones being considered by some Western European cities.

The trade-offs between series, parallel, and other hybrid designs—of which there is a nearly infinite variety, many as yet unexplored—are complex. Because of rapid progress in electric propulsion technologies during the 1990s, there is no longer an obvious technical or economic reason to prefer one over the rest. There is also a large spectrum, or rather matrix, of more complex hybrid designs, varying in amount of storage, whether they accept any recharging energy (utility grid or photovoltaic), whether the wheels are ever driven mechanically, and how electric and mechanical torques are combined (to different wheels, in a common planetary gear drive, etc.). Each approach has distinctive features, but none yet seems to offer universally distinctive advantages over the others. Probably a diverse range of designs will be tailored to different market niches, and their comparative advantages will change as one or another technological element advances.

Synergy with Ultralight Construction

The careful combination of ultralight construction with hybrid drive was stated at the beginning of this article to be surprisingly synergistic. There are two main reasons for this:

- Mass savings compound faster with ultralights than with heavy platforms, faster with hybrids than with nonhybrids, and fastest of all with both together.

- Of the ultimate uses for wheelpower—in level urban driving, one-third each to air drag, road drag, and braking—both *irrecoverable* losses (to air and road drag) are reduced by severalfold through ultralight and ultraslippery platform design. But then the only other place the wheelpower can go is into the braking term. That loss is first reduced severalfold by the lower platform mass; then most of the remainder is recovered by the hybrid drive's electronic regenerative braking. Therefore *all three* kinds of wheelpower losses are greatly reduced. Lastly, the wheelpower savings compound by not needing to lose roughly 5–7 units of fuel energy in order to deliver 1 unit to the wheels.

That, in a nutshell, is why well-designed ultralight hybrids can achieve the ~5–20-fold efficiency gains described below.

Unlike today's production cars, hybrids with sufficiently reduced mass and drag will be more efficient and emit less pollution in urban driving than in highway driving. This is mainly because less energy is irrecoverably lost to air drag at low than at high speed. Remember that air drag rises as the cube of speed; road drag, another irrecoverable loss, is also somewhat less at low than at high speed; electric motors have far more torque per unit mass at low speed than do internal combustion engines; and since the brakes are used far more in city driving (the brake lights are on one-third of the time in the EPA urban cycle), there is far more braking energy to recover. Matters are actually far more complex. For example, gradual braking usually recovers less energy than fast braking because there is more time for deceleration to be done partly by irrecoverable air and tire drag. But broadly speaking, ultralight hybrids will perform significantly better in the city than on the highway. Their use could therefore reverse the existing trend of ever-growing divergence between EPA-rated and actual car efficiency and emissions as driving becomes more congested.

The links between hybrid performance and ultralight construction work both ways. Hybrids' design and performance depend critically on mass, drag, and rolling resistance. It is much easier to design a simple, cheap, and extremely clean and efficient hybrid for an ultralight than for a heavy car. Traditional rejection of hybrids because they are allegedly costlier and more complex than nonhybrids—often expressed as "hybrids require two propulsion systems rather than one"—rests on a tacit assumption that the platform is heavy and high-drag. That is neither necessary nor desirable.

Comparison with Battery-Electric Cars

Despite impressive progress since 1990, pure-electric, externally recharged, battery-powered cars work poorly when scaled up to carry four to five passengers rather than one. That is because the battery mass, like any other vehicle mass, compounds: too much energy and power are needed to haul the heavy batteries, requiring heavier batteries to store that extra energy, etc. Battery-electric cars even for two passengers cannot be very big or go very far without excessive battery size or cost. As the Dutch car expert Dr. P.D. van der Koogh notes, battery-electric cars are vehicles that carry mainly butterflies, but not very far and not very fast—otherwise they would need to carry even more batteries. In contrast, electric hybrids scale well to both large and small sizes *of ultralights*. With ultralight construction, the car's size has little to do with its body's mass: going from two to four passengers can add far less than 100 kg, including the extra safety equipment and the incremental mass of the suspension and driveline. With low drag and regenerative braking, too, the energy needed to propel larger vehicles' greater mass is largely recovered, although more powerful and hence typically heavier equipment is needed to accelerate that higher mass.

The two-order-of-magnitude disadvantage of batteries over fuel in specific energy (energy stored per unit mass) cannot in principle, on present knowledge, be overcome by better battery design. Even the most ambitious battery concepts, at least with acceptable life and cost, would have only severalfold higher specific energy than today's lead–acid batteries at ~35 W·h/kg. Fuel is so much lighter, even when coupled with a device to convert it into electricity, that intelligently designed hybrids always beat battery cars on driveline mass, typically by a factor of around 4 for comparable performance. Indeed, for most ultralight-hybrid designs, specific *power* (how rapidly electricity can be added to or recovered from the buffer storage per unit of its mass) is more important than specific *energy*.

Battery cars raise many technical, economic, and policy issues related to their external recharging. The rate at which a gasoline pump supplies energy to the fuel tank of a gasoline-fueled car is equivalent in electric recharging terms to roughly 5–10 MW, which at normal battery voltages would require a cable about as thick as a person's thigh. Some compromise in recharging time thus appears inevitable unless battery banks weighing several hundred kilograms are physically interchanged. (Recharging can be done inductively, but the stray fields may raise biomedical issues.) Providing either battery-changing or battery-charging stations in turn raises significant cost and loadshape issues for electric utilities.

As long as battery cars are few, their loads are of little value or consequence to utilities. However, some utilities have lately been thinking ahead to what would happen if enough battery cars were bought to yield significant reductions in, say, the air pollution in southern California. (That state's Zero Emission Vehicle mandate, discussed below, makes this a far from academic inquiry.) Although circumstances differ widely between utilities and over time, it is a fair generalization that recharging many battery cars at night could strain some utilities' distribution systems (whose investment costs per kilowatt are often at least comparable to generating investments). This problem could be deferred but not wholly avoided by carefully designed timing controls; daytime recharging could incur new generating (and, often, distribution) investments. Arguments in this sphere are complex, varied, and not always internally consistent. For example, one must be alert for simultaneous assertions that battery cars will be recharged only at night (hence will not need new generating capacity) *and* that their limited driving range is not of concern because recharging stations will be conveniently located at workplaces and shopping malls—hardly where one's car is likely to be at 3:00 a.m.

Many electric utilities initially welcomed and promoted battery cars as a new way to sell surplus offpeak (chiefly nuclear) power during the night. Some still do. During 1993–1994, however, more careful analyses both of market competition between battery- and hybrid-electric cars and of potential utility investment burdens have begun to dampen that enthusiasm. Many observers now assert—probably correctly, though only the market will tell—that as hybrid and battery cars compete over the next decade, battery cars are likely to capture niche markets too small to justify investment in public recharging infrastructure. (Hybrids, in contrast, can use the existing fuel distribution infrastructure with little or no modification.) Some utilities have already concluded that battery cars, even if desirable for other reasons, are an unwelcome new load. Bonneville Power Administration has indeed long supported regional research and development (R&D) on hybrid cars as a cost-effective "lost-opportunity resource" (one that is lost if not grasped timely) for avoiding new battery car recharging burdens on the region's power resources.

Onboard Power Plants

Hybrids avoid all these grid-related issues by making their own electricity onboard from fuel as needed. There are five main candidates for performing this conversion:

- A small, low-power, high-efficiency, internal-combustion engine such as an Otto or diesel, preferably optimized for operation at a point or over a small fraction of the traditional engine map. Probably, like aircraft engines, it would be optimized for mass rather than physical size, and hence it would be simpler, lower-stress, cheaper, and more reliable than the ultracompact engines that automakers have developed to squeeze very high shaftpower into tight places.
- An external combustion engine such as a Stirling (or its free-piston Ericsson variant).
- A miniature high-speed gas turbine that, complete with its high-speed generator, could be the size of a champagne bottle (6).
- A thermophotovoltaic converter that optically couples a glowing plate heated by fuel combustion (such as a ceramic gas-fired "waffle" burner) to a nearby photovoltaic array whose bandgap is matched to the peak near-infrared output wavelength, yielding at least 30% conversion efficiency with no moving parts and extremely low emissions.
- Any of the various kinds of fuel cells that convert fuel (typically hydrogen or methane, sometimes more complex hydrocarbons) into electricity with no moving parts and no emissions but water and carbon dioxide.

For present purposes, the choice of power plant is almost immaterial: any of the five candidates, or combined-cycle combinations of them (6), or possibly some other option now emerging, will work, although early versions may use internal combustion engines simply because they are more available and familiar. Most likely these would be advanced stratified-charge engines.

A modern two-stroke Orbital-derivative stratified-charge engine with properly optimized software (the timing is critical) can achieve at least 34% thermal efficiency at 1.2 L (minimum brake-specific fuel consumption ~250 g of low-octane gasoline per mechanical kilowatt-hour) and quite similar values down to ~0.15 L. That is ~5–10% (not percentage *points*) better than many four-valve-per-cylinder four-stroke stratified-charge engines, or ~15% counting indirect effects on vehicle mass and air drag. (The Orbital engine reportedly saves about as much fuel indirectly—from lower A and compounding M savings—as directly.) Even smaller and less sophisticated four-stroke production engines for motorcycles, sized at perhaps an eighth of a liter for a good ultralight hybrid, can produce 80 kW/L at ~10,000 rpm; racing versions, ~192 kW/L at ~20,000 rpm (41). Such small engines might weigh only on the order of 10 kg.

Another attractive alternative, a high-pressure-injection diesel, has achieved 43% efficiency (198 g/kW · h) at 2.46 L (42), a level previously obtained only in large truck engines. But insulated and turbocompounded diesels can do better still, with ~56% in a recent benchscale test of a one-cylinder small diesel. Even higher efficiency may be available with variable-geometry turbocharging, membrane oxygen enrichment, gas bearings, dry ceramic bearings, and other thermal, materials, and tribologic developments. Fuel cells can certainly beat 56%.

Advantages of Hybrid Drive

The simplest (series-hybrid) concept, where the wheels are always driven electrically but the electricity is made onboard as needed by a small powerplant onboard, has four key advantages over direct mechanical drive (23):

- The engine is sized to the *average* load, not the peak load, because a small buffer store, interconnected to the engine-driven generator and the traction motor(s), temporarily stores energy for hill climbing and acceleration. This buffer storage makes the engine ~5–10% the size normal in mechanical-drive heavy cars: typically around 10 kW and possibly less. This is possible because an ultralight platform's "road load" (traction requirement at the wheels) will be only 1–3 kW at cruising speed; accessories will total well under 1 kW under the worst conditions; and "gradeability" (ability to climb hills, typically a 6% grade at cruising speed, fully loaded, against a modest headwind) will add on the order of 8–16 kW. This largest load component can be met partly by the engine and partly by drawing down the precharged buffer store, whose capacity is of course limited, but no hill goes up forever. The engine and storage capacities can be traded off in design, and ultimately a three-axis Global Positioning System-based look-ahead simulator may plan the trip to ensure that the storage is adequately precharged.
- The engine drives a generator, not the wheels, so it runs only at its optimal condition. Just this collapsing of the engine performance map to a point—possible with ultralight but not with heavy hybrids—*doubles* an Otto engine's practical average efficiency and permits simultaneous optimization for emissions. The

scope for such optimization is not yet well known, since little work has been done on point-optimized small engines, but looks large.

- The engine never idles; when not running at its "sweet spot," it turns off. Idle-off in a nonhybrid VW Golf saves 21% of fuel in the European urban test cycle (43). This is broadly consistent with the EPA urban cycle, where the car idles 18% of the time (24). Savings are slightly smaller if, for the driver's peace of mind, idle-off is automatically overridden whenever the turn indicator reports the driver is waiting to turn across traffic.
- Regenerative braking (recovering most of the braking energy into the buffer store) improves conventional platforms' fuel economy by a further one-fourth; in the EPA urban cycle, 23% of the time is spent braking. In principle, with no constraints on safety or driveability, regenerative braking "could recover as much as 70% of (available) kinetic energy in an urban cycle" (44). Wheel-to-wheel ~70% recovery efficiency has in fact been *measured* both in a Western Washington University electric car and in a Brisbane hydraulic-accumulator urban bus (45). (Many projects have achieved values 2–3 times lower because their heavy platforms produced too much power for the battery or inverter to accept, requiring some to be thrown away, or because of suboptimized control algorithms: 70% recovery requires careful engineering.) Recovery around 70% should become the norm with the fine torque/speed control permitted by switched reluctance motors (see below), and recovery approaching 90% should become possible if, as expected, advanced carbon-fiber flywheels enter the market. Regenerative braking can be smooth, be progressive, feel just like mechanical braking to the driver due to software pedal emulation, and be used virtually always, enabling backup friction brake(s) to last as long as the vehicle.

Hybrid Performance

Together, these features permit the fuel-tank/engine/generator to be inherently smaller, lighter, cheaper, and longer lived than the ~300–400 kg of batteries they typically displace in a pure-electric car; those mass savings then compound. This permits far longer range—ultimately across the United States on one tankful—with lower total mass, cost, and refueling inconvenience.

Whatever their design, adding hybrid drive to today's *heavy* cars will typically reduce their fuel intensity by ~10–15% on the highway, ~50% in city driving, and ~35% composite (43,46). Driving a 100-km European test cycle in an experimental VW Golf Diesel hybrid, for example, uses 2.5 L of diesel fuel plus 16.3 kW · h of electricity (47), equivalent in end-use energy terms to ~4.1 L diesel/100 km. Volvo's 1993 ECC parallel-hybrid concept car saves ~38% highway and ~49% city compared with its base model, the automatic transmission Volvo 850, which weighs 134 kg less. (The ECC is rated at 5.2 L/100 km highway, 6.0 city; the 850, at 8.4 and 11.8 L/100 km. The ECC would of course be far more efficient if, despite its greater use of aluminum, it did not weigh 1580 kg, an artifact of a tight project schedule and low budget.) If ultracapacitors were available at reasonable cost, their high specific power could probably boost even heavy hybrids' efficiency, compared with equivalent nonhybrids, by 60% or more (48).

Paper designs with higher drag than the GM Ultralite and 2.5 times as heavy confirm that hybrid drive can cut composite-rated fuel use by up to 60% via regenerative braking plus optimized engine sizing and loading (but not corrected for mass compounding). However, hybridizing lighter, lower drag, lower rolling resistance vehicles will save even more, because a larger share of tractive energy will go into braking, from which it is potentially recoverable. The degree of this gain, which both determines and depends upon driveline mass, may be the biggest single uncertainty in hypercars' ultimate performance limits.

Motor Systems

Many designers add unnecessarily bulky, heavy, costly, inflexible, and hot-running electric drive systems to conventionally heavy platforms. Some designers, notably in the solar car community and in Switzerland, understood early the importance of low mass for electric and hybrid cars, but may still be complicating their task by choosing induction (asynchronous) or direct-current (DC) motors that often require gears.

Electric-hybrid vehicle designers differ on the ideal traction motor system. Impressive progress has enabled both electronically driven induction motors and DC motors to achieve the design goals required for good ultralight hybrids. Yet, especially in the United States, most experts have overlooked the potential further advantages of modern switched reluctance drives, a sophisticatedly simple type developed mainly in England (49,50,51).

These electronically commutated brushless DC machines have a different number of rotor and stator poles, both salient. The rotor is made of laminated iron with no bars, windings, or magnets; it spins around to align itself with the rotating stator field synthesized by power electronics under digital control referred to real-time shaft position sensing. The rotor has low inertia and high strength and runs virtually cold. Fail-safe, soft-start, variable-speed power electronics, driven by sophisticated software and firmware on hybrid power chips, provide optimized stator excitation, with even greater flexibility than a doubly excited DC machine. With possibly only one switch per winding, enhanced torque/ampere, and greatly constrained fault modes, the electronics are cheap, simple, and unusually robust. Noise and torque ripple can be lower than with an induction motor (virtually undetectable torque ripple of 0.05% has been measured at low speeds). Form factor is extremely flexible, and the output shaft can be integrated into the application. Sizes can be milliwatts to megawatts; current designs span five orders of magnitude in speed and eight in torque (49,50).

Recent advances in switched reluctance drives, summarized here for completeness, have reduced decade-old problems to myths (noise, torque ripple, cost, etc) and made switched reluctance drives unusually attractive. A recent review (50) suggests that, for fundamental reasons, properly designed switched reluctance drives can outperform all other types, including electronically commutated permanent-magnet motors, in efficiency, mass, size, versatility, reliability, fault tolerance, *and cost*. Switched reluctance drives' main advantages (49,50,51) include:

- Speed is limited only by rotor bursting strength, with 100,000 rpm readily achieved, far more than gearless cars need but ideal for ungeared gas-turbine generators and free of permanent-magnet generators' high-speed iron loss.
- Starting and low speed torque are uniquely high—typically 4–6+ times higher than for a same-frame induction motor—making the motor typically one to two frame sizes smaller.
- Throughout the very wide speed and torque ranges and in all four quadrants (forward and backward, motoring and generating), the speed/torque curve is entirely under real-time software control; any desired asymmetry can be achieved, including different motoring and braking characteristics.
- Extraordinary overload capacities (often sufficient to absorb all braking energy from an ultralight car) are disproportionately further enhanced by any cooling of the shell, since virtually all heat dissipation is in the stator.
- DC-input-to-shaft efficiency is much higher than for induction or DC systems, eg, 93% for a 2.6-kW drive system built with an amount and quality of copper and iron typical of *standard-* (not premium-) efficiency induction motors, vs. 89–90% for the highly optimized ~1.5–3-kW Sunraycer permanent-magnet drive.
- Exceptional efficiency is also maintained over a far larger operating map.
- Fault-tolerant advantages include "limp-home" if even one pole pair remains energized, whereas induction motors do not run if any pole pair fails. It is partly for this reason that switched reluctance is increasingly used in the most mission-critical aerospace applications and that Ford uses many thousands of units a day for cruise controls.
- The rotor's high strength and small moment permit an angular response of 10^3–10^5 rad/s^2 and control bandwidth of 10^2–10^3 Hz, rivaling the costliest spindle drives or industrial servos.
- Whole-system mass production cost is typically about 15–40% below that of same-torque variable-speed induction systems and usually below that of *constant-*speed induction motors.

These remarkably strong, light servomotors can, but need not, be integrated into each wheel hub, saving net weight. (Suspension engineers prefer to minimize unsprung weight by moving the motors inboard; this requires short axles but improves suspension dynamics and motor protection.) Speed reduction gears could be either omitted or included as dictated by mass, cost, and reliability trade-offs. Depending on failure-mode analysis and the ability of the buffer store to accept high inrush currents, it may be possible to eliminate mechanical brakes except for the emergency handbrake legally required for hill parking. At least in principle, differential wheelspeed, integrated with electronic suspension to lean into turns, may also permit an ultralight car with hard, narrow tires to steer without angling the front wheels, permitting the front wheel wells to be faired against air drag. Switched reluctance drives' only disadvantage is that they are an order of magnitude harder to design than traditional types: Excellent design demands a level of system (especially software) integration and numerical simulation that only a few dozen people have mastered, nearly all of them at a single firm (Switched Reluctance Drives division of Emerson Electric, Leeds, United Kingdom).

Another interesting variant is the recently invented "Electric Wheel" (Town Creek Industries, Leonardtown, MD), which integrates two ring motors, either permanent-magnet or switched reluctance, into a planetary gear system. Independently varying the motors' speeds and directions can smoothly produce an extremely wide range of speed and torque. This approach may save cost and weight compared with conventional motors and gears.

INTEGRATED DESIGN OF ULTRALIGHT HYBRIDS

Redesigning an ultralight-*and*-hybrid car from scratch using aerospace systems concepts could yield an elegantly simply and unusually attractive vehicle (8). A four-passenger family-car version would start with low mass (<600 kg now, <400 kg ultimately) and could achieve high crashworthiness with special materials and design, as discussed below. Like an aircraft, it would be designed for high payload/curb-weight ratio, well above the Peugeot 205XL's 0.56, and would use unusually efficient actuators controlled by fiber optics ("fly-by-light/power-by-wire"). It would combine a drag coefficient of less than 0.2 now and ideally ~0.1 later with smart active suspension and advanced tires. Its hybrid drive would initially use a small internal combustion engine, on the order of 10–15 kW, directly driving a very efficient generator. Buffer storage would be provided initially by just a few kilowatt-hours of improved conventional batteries, such as nickel/metal-hydride or lithium, driving one to four superefficient motors. This design—at least if a series hybrid—eliminates the transmission, driveshaft, universal joints, differential, perhaps axles, and possibly brakes. The entire driveline (propulsion system) would weigh only ~100 kg or less.

Accessory and Auxiliary Loads

Meanwhile, accessory loads would be rigorously reduced, starting with the air conditioner that in a typical U.S. car is now sized to cool an Atlanta house (8). Glazings, which gain ~70% of the unwanted heat, would be very lightweight but spectrally and even angularly selective, perhaps later variable-selectivity. The shell would be light colored (21–24% of new U.S. cars were already white, cutting interior temperatures by ~21°C) or at least low in solar absorptance (which can be done with a wide range of colors) and high in infrared emittance. The roof might be vented like a tropical LandRover's, and any sunroof would be passively gain-controlled with CloudGel, which passively changes from clear to white at a preset temperature. Certain hot interior surfaces, like the deck above the dashboard, could also receive special venting and optical treatments, and would have low heat capacity and perhaps an extended-area texture so as to be less likely to burn the skin at a given temperature. A photovoltaic vent fan and compact firewall and body superinsulation, including radiant barriers, would further cut cooling loads. Air-distributing, minimally insulating ultralight seats, turbulent air movement (especially around the head and neck though not the face), and careful control of the radiant environment would expand the comfort envelope (52). Any remaining cooling would be done by alternative means (52): absorption and/or desiccant devices driven by engine waste heat, or a staged indirect evaporative chiller, or a desiccant/evaporative Pennington cycle, or a very efficient (probably scroll-compressor) heat pump, or an Ericsson heat pump (53), probably with heat-pipe evaporator bypass, economizer, and fuzzy-logic controls. Cooling-system design would separate latent from sensible, average from peak, and shaft from thermal loads in the expectation of vastly reducing shaft loads and hence slicing a substantial layer off the top of the engine map or its hybrid equivalent. Heating would be by passive heat pipe; all ventilation and comfort delivery would be by low-face-velocity coils, high-efficiency variable-speed fans, and low-friction delivery systems (52). This synergistic approach—thermal gain avoidance plus superefficient, largely waste-heat-driven space conditioning—would simultaneously cut mass, drag (via reduced engine compartment bulk), engine loads and sizing, engine performance map range, total cost, emissions, maintenance, chlorinated refrigerants (to zero), and discomfort (8). Astonishingly, the auto industry today uses virtually *none* of these innovations, even though all rest on well-established engineering principles, and their combined application has been shown to reduce buildings' air-conditioning loads by ~90–97% while improving comfort.

Moreover, a single miniature metal halide light source, such as the tripled-efficiency models recently introduced by such firms as Philips, Hella, and General Electric, could provide all exterior and cabin lights via fiber optics and light pipes, excluding heat from the passenger compartment and (with efficient new light-emitting diodes) greatly reducing lighting's energy, maintenance, and mass budgets. Electric loads and mass would be minimized everywhere, from electroluminescent panel lights (used in GM's Ultralite) to speaker supermagnets (used in Ford's Taurus) and from complementary metal–oxide–semiconductor (CMOS) chips to shaft-integrated switched reluctance fans and pumps (which preprogram the device curve onto the motor chips for a perfect match throughout the operating range). Entertainment systems would be as light and power-frugal as the best consumer battery portables. All electrical loads would receive the same power management attention as modern notebook computers. Total accessory loads would drop by half to one order of magnitude, with substantial improvements in comfort and amenity.

All the powertrain friction reductions available, down to the last bearing and advanced lubricant, would be systematically exploited, though scarcely any mechanical powertrain would be left, and power steering and brakes would be as unnecessary as they are in the Ultralite. The pumps and fans in heavy cars' engine auxiliaries would also shrink or disappear. The frustrating ~20–25% irrecoverable loss now typically added to tire rolling resistance to account for losses in wheel bearings, brake drag, etc, thus raising effective total r_0 to nearly 0.008, would be cut by using regenerative electronic braking and lightweight drum brakes, fewer wheel-related bearings and gears, and special, smaller capacity bearings. State-of-the-art all-ceramic bearings have only half the deformation loss of steel bearing elements but retain comparable loading and shock resistance.

Platform Parameters

Such a hypercar's drag-mass product is a rough qualitative indicator, though not a quantitative predictor, of its tractive loads. In an illustrative baseline near-term hypercar, which we might call the "Gaia," this product would fall from the GM Ultralite's 0.19 of the 1990 U.S. new-car mean by a further one-fourth, to ~0.14 of the 1990 norm. Specifically, for model year 1990, average $C_D AM = 0.33 \times 2.3\ \text{m}^2 \times 1443\ \text{kg} = 1095\ \text{m}^2\cdot\text{kg}$. For Gaia, $G_D AM = 0.14 \times 1.9\ \text{m}^2 \times 580\ \text{kg} = 154\ \text{m}^2\cdot\text{kg}$, or sevenfold less.

Gaia could achieve this, for example, with a mass of 580 kg (9% below GM's 1991 Ultralite, or 13% above Renault's 1987 VESTA II), a frontal area of 1.9 m^2 (11% above the Ultralite), and a drag coefficient of 0.14 (2% above Ford's 1985 Probe V). Good $r_0 = 0.007$ tires, halved accessory loads, 90% efficient driveline, and 70% efficient regenerative braking (wheel-to-wheel including storage losses) would also be added. The driveline and accessory assumptions are easily achieved; the 70% regeneration efficiency, though already observed, requires care with configuration, battery choice, other components, and control algorithms, but is consistent with several types of efficient drivesystems now on the market. To power the hybrid drive, we very conservatively assume a 30% efficient gasoline engine. Its minimum brake-specific fuel consumption of 280 g/kW$_{\text{mech}}\cdot$h or 0.46 lb/hp$\cdot$h is typical of today's decent off-the-shelf ~10–20-kW gasoline engines but well below the best. For example, 30% efficiency including powertrain losses (54), or ~33% for just the engine, has been observed with a 70-cm^3, 4-kW 1984 Honda motorscooter engine, so newer and larger designs can do even better.

On these fairly pedestrian assumptions, standard parametric analysis (23) shows a quintupling of fuel economy, to ~146 mpg or ~1.61 L/100 km. Of course, capturing even part of this goal would be richly rewarding; but capturing all of it seems well within reach of technologies already individually proven and only awaiting proper integration. Indeed, Gaia might be considered simply a fairly well optimized GM Ultralite with a quite ordinary gasoline engine hybrid. Simply hybridizing GM's 1991 Ultralite itself, using a 30–35% efficient gasoline engine, would yield efficiencies around 120–190+ mpg, depending on other assumptions. The ESONO H301 confirmed this.

As a sensitivity test, what if we assume completely uncontroversial parameters that are substantially *worse* than the GM Ultralite's but assume a fairly good off-the-shelf gasoline engine to power a hybrid driveline? Suppose $C_D AM = 0.21 \times 1.9\ \text{m}^2 \times 700\ \text{kg} = 279\ \text{m}^2\cdot\text{kg}$: 81% higher than Gaia or 47% higher than the GM Ultralite.

Let's further assume $r_0 = 0.008$ (vs. Ultralite's 0.007), accessory loads 80% those of a normal production car and well above Ultralite's, a more realistic 35% efficient engine, a degraded regenerative efficiency of 60%, and a wholly miserable driveline efficiency of only 75% (or a slightly higher driveline efficiency and equivalently lower engine efficiency). Even these clearly conservative assumptions yield a combined city/highway rating of ~97 mpg or ~2.44 L/100 km—3.2 times as efficient as a typical new car and 21% above the ~80-mpg goal of the Partnership for a New Generation of Vehicles.

As a further illustration of why 80 mpg is hardly an ambitious goal, in April 1994 a student team from Western Washington University in Bellingham, led by the experienced independent designer Prof. Michael Seal, tested its two-seater light hybrid experimental car in Los Angeles traffic more demanding than the EPA urban test cycle. The team's Corvette-sized Viking 21 sports car had $C_DAM = 0.20 \times 1.28\ \text{m}^2 \times 897\ \text{kg} = 230\ \text{m}^2 \cdot \text{kg}$, 21% of the 1990 U.S. norm and only 10% above GM's Ultralite. Viking 21 combined a 1.2-L, ~85-hp motorcycle engine with 8.5 kW · h of NiCd battery buffer storage and a 37-kW electric drive. Its measured fuel efficiency in Los Angeles, including prestored and used electricity expressed as its gasoline equivalent, was *202 mpg* or 1.16L/100 km, ~10 times as efficient as a typical U.S. family car in city driving. (Its highway driving, however, was only on the order of 50–80 mpg; it was not optimized for that mode.) The next version, weighing only 680 kg, will have a better 0.9-L fuel-injected Yamaha engine and improved comfort and safety features, including the ability to pass the 40-mpg barrier crash test. Neither version is meant to be a producible family car, but Viking 21's technical performance certainly commands respectful attention. So does the ~1.6 L/100 km (~147 mpg) combined city/highway efficiency achieved by the ESORO 4-seater H301 parallel-hybrid prototype mentioned below.

Advanced Designs

In practice, our illustrative Gaia model's parameters are substantially *worse* than today's state-of-the-shelf. For example:

- its 580-kg curb mass, though 9% lighter than GM's 1991 Ultralite's, is still well above optimal. For example, two models of ~90% fiberglass ultralight Swiss battery-electric cars with their batteries removed—but still including the extra driveline mass needed to haul their respective 290 and 260 kg of batteries—have curb masses of only 260 kg (for the two-seater Consequento by Horlacher (Möhlin)) and 360 kg (for the two-plus-small-two-seater E301 by ESORO (Oerlikon)). Of the E301's 360 kg, the monocoque body, seats, doors, partial glazing, and surrounding impact beam for crashworthiness total only 100 kg. This rose only to about 120–130 kg for ESORO's November 1994 four-seater H301 prototype, adjusted for its roughly 20–30 kg of excess resin; all the rest of it except the 260 kg of batteries added only another 270 kg. (Note that making the E301 spacious for four adults made it only about 20–30 kg heavier, illustrating the decoupling of size from mass. Both models include electric windows, radio-controlled electric locks, entertainment system, ventilator, and other amenities.) With a lighter-weight driveline and a few kilowatt-hours (~20–40 kg) of NiMH buffer battery, 400 kg would still be a reasonable curb mass assumption *today* for a similarly well engineered 4–5-passenger family car.
- A fairly aggressive $C_DA \sim 0.17\ \text{m}^2$—eg, $C_D = 0.10$ with Ultralite's $A = 1.71\ \text{m}^2$ or $C_D = 0.11$ and (5% below Renault's Vesta II) $A \sim 1.55\ \text{m}^2$ or some equivalent combination—could be combined with excellent 0.005-r_0 tires, a straightforward 70% reduction in today's normal accessory loads, and Gaia's already demonstrated 70% regenerative with 90% driveline efficiencies. With a 50% efficient powerplant—either a rather good diesel (but worse than the 56% already demonstrated) or a fairly mediocre fuel cell—the resulting advanced design, which we might call "Ultima," would average around 376 mpg or 0.63 L/100 km, or over 12 times today's normal new-car efficiency.

Where might the edge of the technological envelope lie? By early in the next decade, some exciting developments now at laboratory bench scale will probably be commercially available. For example:

- Mature versions of the monolithic solid-oxide fuel cells being developed at Idaho National Engineering Laboratory and elsewhere will utimately yield perhaps >1.4 kWe/kg (over twice the power density of an Ultralite engine with generator) and ~2.8 kWe/L (10 kWe from a 15-cm cube). The superinsulated stack would run red hot at ~1000°C, be ~50–60% efficient (more at part load or with a bottoming cycle), and cost perhaps ~$175–260/kWe. It would be self-reforming, could accept a wide range of fuels, and would be reversible (55), eliminating most or all the buffer battery. (Nonreversible proton-exchange-membrane fuel cells, though still relatively heavy, are much closer to marketability, have already been used to run buses, and are projected by some experts, when mass produced, to cost about the same as the driveline components they would displace (16)).
- Miniaturized multifarad ultracapacitors are very costly today but might not remain so. Affordable units could become the peak-power buffer storage devices of choice, as they are very compact, have no moving parts, and can handle extremely high currents. Radical cost reductions, though not yet clearly in view, cannot be ruled out. Even without them, similar buffering tasks can already be assigned to new kinds of high-specific-power batteries (eg, bipolar thin-plate lead-acid) so as to reduce the specific power requirements for the main batteries that store recovered braking energy.
- "Electromechanical batteries" (56)—advanced flywheels with rotors wound from ultrastrong carbon fiber, spinning as fast as 150,000 rpm on almost entirely passive magnetic bearings in a hard vacuum,

with appropriate gimbals and shock mounting—are now the subject of intensive commercial development efforts by at least a dozen organizations. Sophisticated electromagnetic coupling can store energy in and recover it from the flywheel's angular momentum with a round-trip cycle efficiency of ~96–98%. Equally important are flywheels' very high specific power ratings—at least two orders of magnitude above those of ordinary lead-acid batteries—and specific energy ~1–3 times the same batteries'. For example, Oak Ridge National Laboratory in 1985 demonstrated that ordinary IM7 carbon fiber with an ultimate strength of ~4.8 GPa or 0.7×10^6 lb/in.^{2} could store 244 W · h/kg of rotor rim mass before the supporting web failed. Conservatively derating the speed by 15% implies a packaged "can weight" equivalent to ~55–65 W · h/kg; Lawrence Livermore Laboratory, using a lighter containment design, thinks this might approach ~100 W · h/kg with T-1000-class carbon (~6.55 GPa). That is also consistent with the rotorspeed, 49% higher than ORNL's, claimed by SatCon in 1994. (Unlike metal rotors, which can produce shrapnel on failure, composite rotors produce a containable "soft" mixture akin to a hot, swirling maelstrom of sootlike particles and cotton candy.) Specific energy is proportional to fiber strength, which continues to advance: very small quantities of ~10.3-GPa carbon have been made in Japan at ~10^2 times the current IM7 prices, the theoretical limit is severalfold higher, and "buckytubes," a tubular form of buckminsterfullerene, may offer a breakthrough for making even stronger and cheaper carbon fiber. Flywheels using any sort of carbon fiber should have a rundown times of months, be essentially immune to aging and cycle fatigue, and last at least as long as the car. Being made of components similar or identical to those already in mass production for other purposes, they also show promise of reasonable cost. They are also likely to enter mass production as quickly as possible, probably in 1995–1996, for highly competitive uninterruptible power supply and dispersed load management applications; this would rapidly lower their cost for gimballed automotive uses too.

With plausible further progress, an early-next-century hybrid car might, for example, have under its hood:

- a grapefruit-sized fuel cell wrapped in a <40-L superinsulated envelope and user selected to the proper modular size, which could even be temporarily modified for special applications;
- a melon-sized package of power electronics, also modular (plug in an extra "slice" for higher performance, such as towing a holiday trailer lacking its own hybrid drive);
- an orange-sized computer;
- perhaps an optional breadbox-sized space-conditioning package; and
- virtually nothing else.

So why have a hood? Instead, there could be two trunks for extra storage and crush space.

As for what an edge-of-the-envelope "Imagina" car might do, a 0.13-m^2 C_DA (implying passive boundary layer control and other aggressive measures), very advanced $r_0 = 0.005$ tires/bearings, accessory loads one-fourth of today's, 400-kg curb mass, a 56% efficient engine (today's best laboratory diesel, though still worse than many fuel cells), and realistic maximum efficiencies of 96% for the driveline and 88% for regeneration (implying a good flywheel) would together imply an efficiency in excess of 600 mpg (0.38 L/100 km). (By then, of course, fuel-saving returns will have diminished severely, so such innovations would need other motives.) The actual technical limit may well be even higher, because both mass and accessory loads can almost certainly be cut further, and some combined-cycle driveline concepts, such as a miniature gas turbine or high-temperature fuel cell with waste heat recovery, may be capable of efficiencies around 60–70%. However, by the time curb mass fell below, say, 350 kg or so, the payload will be about as heavy as the empty car, and residual accessory loads will probably be the dominant concern.

The nominal parameters just described compare thus with two empirical sets (Table 1).

Flexible Design Options

Infinitely many combinations of parameter values could achieve Gaia- or Ultima-level performance. The values shown in Table 1 are not meant to be optimal, just illustrative. However, some simple sensitivity tests show that the results do not depend on particular assumptions but rather can be achieved with great flexibility.

For example, efficiencies at least as good as Ultima's can be calculated with small adiabatic diesels, already shown to match or exceed the assumed 50% fuel cell efficiency. *Any* 50% efficient power source in the Ultima's low-drag hybrid platform would yield 0.79 L/100 km (298 mpg) even if M were increased to 635 kg, *the same as the Ultralite*, so Ultima's ambitious 400-kg mass target is not essential either (thanks to 70% efficient regenerative braking). That is, a four-passenger hybrid could be ~10 times as efficient as today's production cars if it combined demonstrated power plant and regeneration efficiency, mass, and rolling resistance with excellent driveline efficiency and with aerodynamic drag only one-fifth below the best published empirical levels (VESTA II for A and Probe V for C_D). Such good aerodynamics do not seem necessary either, because modern electrical choices should improve on the assumed driveline and regeneration efficiencies. Conversely, Ultima's 400-kg curb mass (hence 536-kg test mass) would still yield 1.04 L/100 km (225 mpg) even if powered by an ordinary 30% efficient gasoline engine. Thus the efficiency zone claimed for Ultima leaves both technological flexibility and a substantial safety margin, appropriate in view of the many uncertainties. Such performance requires neither the particular mass nor the engine efficiency targets shown, only a suitable combination of these or other parameters. Exact values for a given model of an actual car would of course be optimized for its market segment.

At such low mass, as with 57-kg ultralight aircraft that weigh less than the pilot, payload becomes a large fraction of curb mass. Test mass *including payload* therefore

Table 1. Key Parameters and Composite Fuel Efficiencies of Selected Existing and Hypothetical Four-Passenger Cars[a]

Platform	C_D	A	M	r_0	E_{acc}	C_DAM	Percent Efficiency			Combined City/Highway		
							Engine	Driveline	Regenerative	L/100 km	mpg	Index
Production Cars												
Typical U.S. 1990	0.33	2.3	1443	0.009	1.00	1095	~30	~60	0	8.0	29.4	1.0
Demonstrated Concept Cars												
GM Ultralite 1991	0.192	1.71	635	0.007	?	208	?	?	0	3.79	62	0.5
ESORO H301 1994	0.19–0.22	1.8	680	0.007	?	232–269	?	~90+	?	~1.6[b]	~147[b]	0.2
Hypothetical Ultralight Hybrid Cars Based on Molded Polymer Composites[c]												
"Conservativa," easy	0.21	1.9	700	0.008	0.80	280	35	75	60	2.44	97	0.3
"Gaia," near-term	0.14	1.9	580	0.007	0.50	154	30	90	70	1.61	146	0.2
"Ultima," advanced	0.10	1.7	400	0.006	0.30	68	50	90	70	0.63	376	0.1
"Imagina," ?possible	0.087	1.5	400	0.005	0.25	52	56	96	88	0.38	614	0.05

[a] C_D = drag coefficient, A = frontal area (m^2), M = curb mass (kg), r_0 = tire rolling resistance coefficient including wheel bearing losses, E_{acc} = accessory energy relative to 1990 norm; engine efficiency is fuel to output shaft, driveline efficiency is from that shaft to wheels, regenerative efficiency is from wheels back to wheels (including storage in-out cycle loss and square of driveline efficiency); EPA-rated 55/45 city/highway efficiency estimated by a published approximation (23) rather than by exact simulation; ? = data not publicly available. M for H301 includes 20–30 kg excess body resin and 260 kg (9 kW · h) batteries.

[b] Estimated Swiss combination of ~60% at 10 kW · h/100 km urban battery-electric propulsion (100–120 km range) and ~40% at 2.6 L/100 km (at 90 km/h) internal-combustion propulsion. The C_D range includes both wind-tunnel (0.19) and street (~0.22) values. The 0.125-l, 1-cylinder IC engine (13 kg including clutch) is detuned from 30 to 12 kW; the 55-N · m induction motor (40 kg including controller) produces 40 peak or 21 continuous kW of shaftpower. Top speed is 120 km/h acceleration is ~6 s. Fuel injection will be added in 1995.

becomes critical: a 0.3-L/100-km family car would probably have to leave the family behind. A 300-kg payload reduces the test mass efficiency shown in Table 1 to 2.70 L/100 km (87 mpg) for the Conservativa, 1.82 L/100 km (129 mpg) for Gaia, 0.74 L/100 km (318 mpg) for Ultima, and 0.45 L/100 km (523 mpg) for Imagina. But these are still very respectable levels, and at today's lamentably low load factors (averaging only ~1.2–1.3 adults in central Europe and ~1.1 in North America, not 4+), a supercar fleet's on-the-road fuel economy will typically be better than with a full 300-kg payload.

Fuel Flexibility

A hypercar at roughly an Ultima level of efficiency could comfortably carry a family of four across the lower 48 United States on one small tank of fuel (~30–50 L of gasoline or equivalent). The power plant could burn essentially any desired liquid fuel. But the small mass of fuel required for a long range would also permit the convenient use of gaseous fuels such as compressed natural gas, or other relatively clean and abundant energy gases, without undue tank mass and volume penalty. That is, a small, light, cheap compressed-natural-gas or hydrogen tank would suffice: perhaps even a structurally integrated conformal tank from aerospace practice. (Hollow structural members themselves may make good compressed-gas tanks.) Even relatively costly hydrogen fueling could become attractive, since so little fuel is burned that its cost and that of its storage arrangements become more tractable. (A 10-fold gain in fuel efficiency reduces sensitivity to fuel price by 10-fold.)

If liquid biofuels were used, the small amounts required for such a fleet could be sustainably derived from farm and forestry wastes without requiring special crops or fossil hydrocarbons. The combination of supercars' small fuel requirements, extreme fuel flexibility, "green" market attributes, and insensitivity to fuel price could together produce an unusual "boutique market" in which many small and localized biofuel resources could find sales opportunities unavailable in today's gasoline mass market.

Further, many parts of the country have natural-gas or biogas resources that cannot economically be transported to the pipeline grid. That methane could still be locally compressed and sold to passing cars. This is a novel opportunity to create an oil-free network of literal "gas stations" using a relatively clean and abundant resource that cannot otherwise be sold, that directly displaces oil imported from abroad or from out of state, and that (like local biofuels) captures local jobs and multipliers by keeping more money in the community.

Moreover, onboard energy storage of <5 kW · h (eg, <70 kg of nickel/metal-hydride batteries) would about enable photovoltaics on top of the car to power a typical Southern California commuting cycle just from the energy captured each day from outdoor parking, without ever starting the onboard powerplant, let alone plugging into the grid for recharging. (Weekday driving is about a tenth of daylight hours, the average car is parked ~96% of the time, and monocrystalline silicon cells on a hypercar could collect ~0.8 kW$_e$ in typical daylight.) Even in less favorable climates, a solar boost could improve an ultralight hybrid's fuel economy well beyond the levels shown in Table 1.

Non–Car Vehicles

An early priority should be assessing the transferability of these concepts to vans and light trucks, especially in the United States, whose light trucks are not only a fifth less efficient than cars but are also driven farther for much longer, and so account for upward of half of new light vehicles' projected lifetime fuel consumption. Yet polymer-composite utility vehicles and even buses look very encouraging: experimental models tend to have halved mass, often halved air drag, and doubled fuel economy *before* they are hybridized, implying at least quadrupled efficiency afterward. With such lightweight bodies (a typical composite delivery van weighs less full than a same-sized steel van weighs empty), valuable added height costs very little weight or stability, since it does not materially raise the center of gravity. Van operators can also gain such valuable side benefits as reduced back injuries (since the driver can stand up straight when lifting packages) and handicapped accessibility (required by the Americans with Disabilities Act but not yet provided by any van on the North American market). This suggests important early niche markets in vans and medium vehicles, especially for urban fleets, that could in turn support faster reduction of hypercars' production cost.

Some analogues are also evident for heavy trucks, traditionally inefficient vehicles with efficient and appropriately sized diesel engines. The most fuel-efficient experimental Kenworth or Navistar 36-T (80,000-lb) trucks reportedly test at best values approaching ~25 L/100 km (~9.5 mpg), which for a 909-kg (2000-lb) car would scale to only ~1.2 L/100 km (~200 mpg). A privately designed composite 18-wheel "bullet truck" with $C_D \sim 0.2$ (well above plausible limits) and nearly doubled normal intercity fuel economy is also in model testing (57). Presumably a hybrid version would do even better, especially for hauls with some city or hill driving. The ultralight approach has even been applied to light trains in an Idaho Innovation Center experiment called CyberTran: since about two-thirds of *system* cost is for roadbed and bridges, the light mass permits, among other advantages, a system cost reduction estimated at 5–10 times.

Emissions and Public Policy

Conceptually, hypercars should be from one to three or more orders of magnitude less polluting than today's new cars, because of four effects that are essentially independent and hence multiply:

- The engine is ~10–25 times smaller. This slashes all emissions, although NO_x in particular does not go down linearly with displacement and might even go up in pathological cases if a small engine has poor controls and a very high compression ratio.
- For a given engine, smog-forming emissions per liter of displacement will decline by a substantial factor (Paul MacCready has estimated that it might be on the order of 6 in fairly common circumstances) because of the combined influence of small-map operation, regenerative braking, no idling, and no emission spikes when the car suddenly accelerates (a cause of huge momentary emissions from ordinary cars; for a typical production car, 1 s of wide-open throttle enrichment emits the same CO as ~30 min of part-throttle operation).
- That factor can be substantially increased, probably by at least twofold, by optimizing a series hybrid for single-point operation (which is possible only with ultralights; practical control algorithms in a heavy series hybrid typically require at least a threefold engine range).
- The convenience of inherently clean fuels, such as methane, should be good for at least another twofold improvement, rising toward infinity with solar-produced hydrogen.

Simply multiplying these four terms suggests a total on the order of hundredsfold (much more still with special fuels or with an inherently clean powerplant such as a fuel cell). But even with a conventional gasoline engine, improvement to a mere ~85–100 mpg, around the *low* end of what hypercars can achieve, should yield tailpipe emissions below those of the in-basin power stations that provide a fraction (officially assumed to be about one-third, probably somewhat higher in practice) of the electricity that a battery car in the Los Angeles airshed would need for recharging.

"Virtual" Zero-Emission Vehicles

This logic has interested the powerful California Air Resources Board (CARB) in broadening its definition of so-called Zero-Emission Vehicles (ZEVs) to include any vehicular technology that is at least as clean in the airshed as a battery car *and its associated local powerplants.* Battery cars have no tailpipes, but in reality, as Lee Schipper points out, they are Elsewhere-Emission Vehicles whose pollution is transferred to the (often well-scrubbed) power plant stacks. (Of course, much electricity comes into Southern California from dirty coal plants in the southwestern deserts, but CARB only regulates *California* emissions and hence cannot count what those plants emit.)

CARB launched in 1990, and upheld in May 1994, a controversial set of technology-forcing regulations that require, among other things, 2% of the cars sold by major automakers in California in 1998 to qualify as ZEVs, rising to 10% in 2003. Hitherto ZEVs had been interpreted to mean battery-electric cars exclusively. However, qualifying virtual ZEVs, as now appears likely, would provide an alternative compliance path consistent with hypercars' new technical opportunity to achieve the clean-air goal even more fully. Not only could each hypercar be even cleaner than a battery car, counting both fuel cycles, but being general-purpose rather than limited-range niche vehicles and potentially bigger and cheaper, hypercars should sell better and should relieve automakers' anxieties about whether people will buy, at an acceptable price, the number of battery cars they would otherwise be obliged to sell.

As CARB elaborates the details of virtual ZEVs, it will of course have to compare their full fuel cycle with that of battery cars. This complex exercise will doubtless soon find that the evaporative hydrocarbon emissions from fueling gasoline-powered hypercars has no counterpart in battery cars' fuel cycle and hence could violate the required parity. For this reason, exceptionally, hypercars in the South Coast Air Quality District might have to be fueled with a closed system using compressed methane, hydrogen, or possibly a nonvolatile liquid fuel. (Nonvolatile versions of gasoline that achieve vaporization by other means—preheating the fuel, adding a volatile substance from a separate sealed cartridge, perhaps ultrasonic atomization, etc—may also merit exploration, and not only in these circumstances. It is also conceivable that a technical fix for evaporative emissions, such as sealed cartridge refueling, might be devised when such small volumes of fuel are involved.) But it is much easier to make a compressed-natural-gas fueling infrastructure for cars that refuel very seldom than an electric recharging infrastructure for cars that recharge more or less daily.

This issue may be less important in the 12 northeastern states that in early 1994 petitioned the EPA to adopt California-like standards. But those states would probably achieve far greater reductions in aggregate air pollution from using hypercars than would California. This is because of the "no-third-car" rule: Congress authorized only two kinds of car emission standards, those California sets and the rest, but the distinction is numerical, not descriptive. That is, if California defines a virtual ZEV, it must do so in *numerical terms*—so much of each pollutant per mile—rather than simply as "cleaner than a battery car." But California's electric grid (chiefly renewable, gas, and nuclear) already has far lower air emissions than the power stations prevalent in the northeast. Switching from conventional cars to battery cars in the northeast would therefore tend to release far more powerplant SO_x, NO_x, etc., than in California. But conversely, using hypercars instead of battery cars would avoid bigger power plant emissions in the northeast than in California.

CARB or other authorities may also eventually require or encourage reductions in CO_2 emissions. For battery cars, those obviously depend on who is making the electricity, when, where, and how—a complex calculation. A grid-connected recharger also cannot argue that a particular renewable power source is the marginal resource, since its output could otherwise have been used to displace fossil-fueled generation elsewhere. But if the conversion of fossil carbon into wheelpower is at least as efficient in a hypercar's onboard powerplant as in the utility's central one, corrected for grid and recharging losses, then the hypercar will be less CO_2 intensive, because its driveline will weigh far less than the battery car's; so less energy will be used to haul itself around in order to provide a given amount of mobility. Hypercars should generally be able to meet this test. It also seems plausible that hypercars may be better able than utilities to create niche markets for renewable liquid and gaseous fuels that release no fossil carbon or, in the case of hydrogen, no carbon at all.

CARB's proposed broadening of the ZEV definition to embrace anything cleaner than battery cars would bring full circle the ironic history of CARB's effort to make automakers bring to market what it thought they were capable of. In early 1993, CARB's effort was heavily attacked in major business publications as a quixotic effort to make automakers sell something that was unmakeable, was unprofitable, or people would not want to buy and use. It remains to be seen to what extent this jaundiced view of battery cars proves justified. But there is no doubt that California's ZEV rules launched a global revolution in electric propulsion technology. Nearly all the major progress in this field since 1990 can be directly traced to the market pressures created by CARB's initiative. Unexpectedly, the same body of new technology, when combined with ultralight construction, is now making possible hypercars that can provide all the advantages of electric propulsion without the disadvantages of batteries, and hence can provide the benefits CARB sought for California (clean air, jobs, military conversion, etc) without the concerns about battery cars' range, recharging, cost, and marketability. Just as federal government initiatives earlier created some of the nation's largest industries, such as semiconductors, computers, and aerospace, so CARB's ambitious effort may unintentionally have laid the foundations for an enormous new hypercar industry.

SAFETY, PERFORMANCE, AND AESTHETICS

A common generic objection to fuel-efficient cars is their alleged crash risk. But this confuses fuel economy, mass, size, and design.

Common Misconceptions

Fuel economy and light weight need not compromise safety. There is no correlation, far less a causal relationship, between present cars' crash-test performance and their mass or between their fuel economy and their on-the-road death rate. Although admirably extensive analyses of mass vs. safety have been performed (eg, Ref. 58), the higher crash death rates observed in the *average* of today's light cars (but certainly not in all models) are for a fleet all built with broadly similar methods and materials and hence cannot be used to predict the safety of the completely new kinds of cars proposed here. Historic data before 1985, when the U.S. accident population was only ~15% restrained by belts and airbags (rising to ~50% in 1990 and higher today), are even less meaningful.

The laws of physics, such as momentum transfer, remain valid and require in theory that collisions between two cars identical except in mass will tend to damage the lighter car more. (Practically, this is often incorrect because other, unequal factors such as design dominate. A few years ago, the National Highway Traffic Safety Administration sought to show the danger of light cars via light/heavy crash tests; the light cars reportedly came off better until stronger heavy cars and flimsier light cars were substituted.) But even where other things really are equal, the conclusion is misleading. The idealized theory that heavier means safer leads some to propose that you should always seek to drive a heavier car, thus reducing such collisions' risk to yourself while raising others' risk correspondingly. ("When a crash occurs, other factors being equal [:] The lighter the vehicle, the less risk to other road users. The heavier the vehicle, the less risk to its occupants" (58). The opposite should therefore also be true.)

The right answer is to make *all* cars safe whatever their weight, without putting all the adjustment burden on light cars. The same goes especially for light trucks and sport utility vehicles, which were exempted from U.S. safety requirements, as their crash-test performance clearly shows. Heavier vehicles should be made less "aggressive" (59)—softer, less angular, more absorptive, with bigger ridedown distances—and the road fleet's mass distribution should be further narrowed, eg, by incentives for replacing inefficient with efficient cars. Heavy trucks with slightly relaxed length limits could even be equipped with a highly energy-absorbing structure on the front to help protect any car they might hit; such a structure could easily be an aerodynamic cowling.

It is also correct, as automakers often point out, that better control of destructive driver behavior such as drunkenness is crucial: behavior may be up to a thousandfold more risk determining than the car itself, and only about a twentieth of crashes do not involve driver factors (58). Conversely, light cars may be better able to avoid accidents by stopping sooner and handling more nimbly. Such effects are almost impossible to measure because, for example, larger cars (which tend to be heavier with currently dominant designs) tend to drive more miles, carry more people, and be driven in less urban settings (hence at higher speeds) and in riskier ways, while smaller cars tend to have younger drivers who are more crash prone but survive better (58, pp. 75–76).

Historic data about conventionally designed cars are often used to invite the inference that light cars are unsafe. But those data cannot be used to predict the safety of cars designed very differently. The clearest proof of this is the very large scatter *within* the historic crash-test and death-rate data at any given curb mass. For example, a few years ago Americans could buy a 56-mpg car with a lower death rate than a 23-mpg car; cars with identical efficiencies but over 10-fold-different death rates; and cars at any curb mass that differed in crashworthiness by more than 10-fold. Such comparisons reveal some unusually dangerous cars now on the road at various levels of mass and fuel economy, but they make no case that fuel economy does or must conflict with superior safety. Rather, their high scatter emphasizes the importance of design differences. Indeed, about the only unambiguous correlation that emerges from the historic data is between danger and *acceleration*—a fact unmentioned since Detroit intensified its marketing of muscle cars in the mid-1980s.

Mass and Design

The central conclusion of modern thinking about safety and car mass is that causally *the two are almost completely unrelated* for two reasons:

- Vehicles' design and materials are far more important to their safety than is mere mass.
- The structures and systems needed to protect occupants (and for that matter those outside the vehicle) weigh very little.

Modern designs and materials can do enormously better than Henry Ford had in mind when, in 1926, he said that "a heavy man cannot run as well as a trim man. You do not need weight for strength." In 1991, GM's Ultralite confirmed that mass per unit volume can be cut by more than half below the "steel plateau" level. This decoupling permits fuel-efficient cars to remain ultralight while combining roomy interiors with ample crush length, which appears to improve crash performance somewhat. Yet better materials and design can also substitute for crush length. As a result, fairly well-validated supercomputer simulations show that the Ultralite concept car should be substantially more crashworthy than GM's steel cars weighing more than twice as much. (This is not surprising in view of the more than 10-fold difference in crashworthiness observed between *steel* cars of the *same* weight.)

Strong, Bouncy Materials

Composites and other ultrastrong net shape materials—many stronger than the familiarly durable but lower grade carbon-fiber fishing rods, skis, etc—would dominate the construction of a hypercar. They would bounce without damage in minor fender-bender collisions: Most deformations of composite panels simply pop out again with little or no damage. Under severe loads, composite structures fail very differently than metal, so "totally different design concepts have to be applied," and understanding of failure modes is rapidly evolving but not yet fully mature. However, even under compressive loading, often considered composites' weak point, "composite structural elements . . . show high and in many cases better energy absorption performance than comparable metal structures," nearly twice that of the best aluminum structures and with nearly ideal plastic energy absorption (60). (Compressional strength depends on the specific fibers used: with carbon fibers, which are about equally strong in both modes, it is so great that an unmanned carbon/epoxy minisubmarine broke no fibers at pressures equivalent to 7 km of ocean depth.)

Extensive aerospace experience is available from designing all-composite structures and aircraft (like the Stealth bomber and fighter) to withstand bird and stone strikes, landing stress, etc. This experience confirms such remarkable combinations of strength, durability, and crashworthiness that advanced composite structures are predicted within the next decade to be used "for up to 80 percent of the structural weight of a helicopter" (60). Where interweaving or stacked overlays with tougher fibers such as polyaramid are needed to stabilize brittle-fracturable fibers such as carbon, weight can even go *down,* since, for example, polyethylenes like Dyneema SK60 have specific gravity below unity. Light metals could also be used in hypercars where appropriate, such as in sections of crushable light-metal foam or honeycomb for energy management in a serious crash. These materials, available for two decades (12), have a nearly perfect square-wave response (they squash flat, absorbing enormous energy, before transmitting crash accelerations), making them an ideal substitute for ridedown length.

Special Energy-Absorbing Shapes

Other crash-energy-managing design options include buckling members, down-deflecting heavy driveline components, filament-wound or sheet-and-keel cruciforms (which absorb ~3–4 times the energy per mass of corresponding aluminum structures), and "impact belt" beams. These are foam- or honeycomb-filled, nearly rectangular beams wrapped around the passenger compartments of light cars (500–600 kg curb weight, 2.5–2.8 m overall length). Such a beam, weighing only a few percent as much as the car, can decelerate ~250-kN mean impacts at ~48 mean *g* without intrusion: "Higher impact forces [eg, >0.5 MN] and decelerations can be obtained with small modifications without significant increase of weight" (59). Excellent performance of such very light but effective structures has been proven by actual high-speed head-on and barrier crash tests, chiefly by Horlacher in Switzerland. Some corresponding proprietary U.S. tests have reportedly found that an appropriately designed light composite van could probably be driven away from a 40-mph barrier crash.

Still another example of advanced energy-absorbing structures is the hollow "crush cone," a conical shell developed by Christoff Kindervater at the German Aerospace Research Establishment in Stuttgart. Injection molded of thermoplastic with embedded carbon fibers pointing toward the apex, these 300–400-g cones roughly 15 cm in both length and diameter can be crushed from the apex toward the base only by repeatedly microfragmenting the immensely strong and stiff carbon fibers and by doing the surface work of separating those fibers' very large aggregate surface area from the surrounding matrix. All this takes so much energy that three such cones can smoothly absorb the entire crash energy of a 600-kg car hitting a wall at 40 km/h. In principle, therefore, a few tens of kilograms of such cones could *completely* protect a car from highway-speed crashes at most or all angles.

An ultralight car combining ultrastrong materials, modern airbag and pretensional-belt restraint systems, and crash-energy-managing design can therefore weigh less than half as much as today's platforms (as the Ultralite does) yet be far *safer* than any car now sold. That is why racedrivers are rarely killed nowadays when their lightweight composite cars hit walls at >350 km/h. As tens of millions of Americans saw on their 1992 TV news during the Indy 500 race, the composite car flies to bits, failing at "trigger" sections specifically designed to initiate such breakaway and absorbing extensive crash energy through controlled failure modes, but the "survival capsule" remains intact and the driver generally limps away with perhaps a broken foot. To be sure, ordinary drivers would lack the racedriver's helmet, fitted foam capsule, multipoint restraints, etc, but even in a head-on collision at highway speeds, the energy per unit mass would be only one-fifth the racedrivers', and even if the ordinary drivers were more seriously injured, that would be a great improvement on their fate in today's cars.

Potential Hazards

The main potential safety disadvantages of ultralight hybrids are as follows:

- With their low drag and low or absent engine noise, pedestrians may not hear them coming unless a noisemaker is added that somehow warns without being objectionable.
- Badly designed electrical systems might, after an accident, find dangerous conduction paths via uninsulated carbon fibers.
- Obstacles such as small trees, crash barriers, and lampposts, against which a heavy car can dissipate energy by breaking or deforming them, may instead stop a light car or make it bounce off, increasing deceleration and perhaps bounceback acceleration forces on occupants.

None of these seem either uncorrectable or important.

Potential Advantages

Beyond their general crashworthiness described above, hypercars also offer important and novel safety features:

- The two- or (with series hybrids) four-wheel electric motors, especially if switched reluctance, offer full-time antilock braking and antislip traction, with far greater balance, response speed, sophistication, and effectiveness than today's methods. For example, the algorithms could easily be integrated with a frost sensor, and slippage data from one wheel could instantly be used to generate corrective signals to the other three, consistent with the dynamic simulator's platform stability requirements.
- Hypercars' light weight means potentially faster starts and stops (depending on traction), while their stiff shell means quicker and more precise handling.
- Carbon-fiber designs can be so stiff and bouncy that an ultralight car, if broadsided by a heavy truck, could be designed to go flying, like kicking an empty coffee can. The very unfavorable momentum transfer would go not into mashing the ultralight car but into launching it. Yet occupants restrained by belts, bags, and headrests and protected from intrusion into their protection space might well survive unless accelerated by more than the often survivable ~40–60 g range, in which case they would be dead anyway in any car today, light or heavy, steel or composite.
- In the rare accidents so severe as to crush the composite shell (usually in hammer-and-anvil fashion), the occupants would be far less likely to be injured than by intruding torn metal edges in a steel car; with any potentially intruding carbon-fiber shards overlain by or interwoven into fracture-masking aramid or polyethylene cloth, the crushed composite sections can become relatively innocuous.
- Victims' extrication would be much faster (a crucial element of critical medical care, most victims not dead on the spot can be saved if brought to a hospital within an hour). The doors are likelier to function, and the composite shell can provide easier access. (For example, the Ultralite's thin carbon-fiber door is so strong that it meets the side-impact requirement with no B-pillar, so opening either door gives complete access to the entire passenger compartment.)
- Cutting through the composite body with a rotary wheel is quick and makes no sparks to ignite fuel vapors. Better still, special design arrangements could make the passenger compartment immediately openable without special tools.
- Breakaway energy-absorbing main components would no longer impede access to the passenger compartment.
- The very small fuel volume and the ease of compartmentalizing the fuel (as in some aircraft, using a

spongelike honeycomb or similar structure so that almost no fuel leaks out even in a severe crash) should greatly reduce the risk of postaccident fires. Aircraft-style integral fire suppressants would also become practical.

- Ultralights' decoupling of mass from volume makes it straightforward to maintain a wide track and long wheelbase for rollover resistance, as well as ridedown length.
- Hydroplaning risk should not rise and may fall, because the car weighs less but has narrower, high-pressure tires.
- The small powertrain volume and raked hood are consistent with improved visibility.
- Metal-halide headlamps are exceptionally powerful but optically more controllable, and can cause road markers and certain textiles to fluoresce brightly.
- With careful design, composites', especially foamcore composites', excellent attenuation of noise and vibration could yield an extremely quiet ride—important because road noise is no longer masked by engine noise. Since the drive motors are digitally controlled, electronic antinoise can be cheaply built in. Tires can use noise-canceling tread designs with two out-of-phase halves. There are no body joints to squeak. Wind noise will be virtually absent due to the excellent aerodynamics. The same body foam that provides thermal insulation and structural integrity can also provide spectrally "tuned" noise attenuation, as in some modern aircraft. All these kinds of noise reduction—along with improved thermal comfort—should make driving less fatiguing, potentially boosting driver alertness.
- The whole car is so simple, reliable, corrosion resistant, fault tolerant, and failsafe-designable that dangerous mechanical failures are far less likely.

For all these reasons, the design approach described here could yield substantially improved safety. Hypercars could also offer ample comfort, unprecedented durability and ease of repair, exceptional quietness and freedom from vibration, and beautiful finish and styling while retaining significant stylistic flexibility, impeccable fit and weatherproofness, high performance (light weight means faster acceleration), unmatched reliability, and probably comparable or lower cost.

One caveat is in order, however. Especially in the litigious United States, innovation is deterred by the threat that makers of new and hence initially "unproven" technologies may have to pay or feel compelled to settle damage claims even for accidents in which they are blameless. (In principle, the law does not require best available safety technologies, but anything new may be questioned simply because it cannot yet point to the same broad base of successful experience.) Some experts fear that such potential liability might add exposure up to several thousand dollars per car-year, especially for manufacturers large enough to invite lawsuits but not large enough to defeat them.

Absent tort reform, removing this potential for an important barrier to market entry may require some government indemnity or coinsurance, or possibly preemption of state tort liability, to makers of hypercars meeting a national safety standard, at least until actuarial experience has field-validated their theoretical ability to match or exceed the safety of today's cars. The issue merits careful study. So does the extent to which crashworthiness certification (for which computer simulation can often be substituted) may constitute a barrier to market entry.

Indeed, the whole area of safety regulations that may unnecessarily impede innovation and market entry deserves review. One expert, for example, has remarked that a 1930s requirement still on the books, meant to ensure that runaway cars would drift to a stop, *requires* a certain degree of friction to be built into wheel bearings.

NEW INDUSTRY STRUCTURES, ECONOMICS, AND JOBS

Ultralight hybrids are not just another kind of car. The industrial and market structure they imply is as different as computers are from typewriters, fax machines from telexes, and satellite pagers from the Pony Express. Hypercars imply wrenching changes that may come far more quickly than our ability to manage them. If ignored or treated as a threat rather than seized as an opportunity, these changes are potentially catastrophic for millions of individuals and tens of thousands of companies. The prospect of hypercars can therefore be either devastating or exhilarating. To understand this choice, we must explore how a hypercar industry would differ from today's auto industry in production, sales, and design.

Production

The optimal scale of production may be profoundly different than steel cars' vast tooling and painting investments dictate. It could even be a several-hundred-person franchise operation analogous to a regional soft-drink bottling plant. Today's cars are rapidly built from myriad shipped-in parts on single, extremely costly assembly lines where delay incurs intolerable costs. In contrast, hypercar production's layup/molding/curing operations are slower but fewer than stamping, machining, and welding operations, so total production time per car could be shorter. Moreover, cheap tooling permits hypercar fabrication to run on many parallel lines, reducing holdups, and if done onsite, those lines would occupy most of the space, with little needed for final assembly. Shipments and inventories of parts would be limited because there are relatively few parts, especially for maddeningly complex details like trim and wiring: In-plant skill, not systemwide logistics, would be the key to profitable manufacturing.

The hypercar art will be not in assembly, nor mainly in components, but in design integration. Before this was understood about personal computers, they were misperceived to be "rocket science," a natural monopoly of giant firms like IBM. Then some Texas Instruments engineers set up Compaq, which combined the same Intel chipsets, Seagate drives, and so on, arguably more innovatively. And then firms like Austin and Dell and Gateway 2000 piggybacked on those assemblers' research to substitute their own, bare-bones, low-overhead assembly and marketing operations. As soon as the public learned that mail-order "clone" PCs worked about as well and cost half as much, the original makers' market shares plummeted. Compaq and eventually IBM learned the new rules and now compete head to head with the clonemakers, but the lesson was painful. Now powerful but no-name microcomputers are assembled from the same standardized commodity parts in thousands of basement-scale businesses, an electronic version of piecework home handicrafts, and cutting-edge designs quickly become standard recipes.

The ability to make or buy the basic physical components of hypercars is already potentially widespread: composite parts, sophisticated drivesystems, controllers, and similar elements require great skill to design well but much less skill and little capital to assemble adequately. Specializing their design and greatly expanding their production is a challenging but normal manufacturing task, just like expanding the recently infant microchip, disk-drive, and software industries as demand grew. Hypercars might therefore become surprisingly like PCs if the integration skills also diffused. Hypercars might ultimately become in their turn a virtual mail-order commodity, subject to centralized design testing (like FCC testing for computers' electromagnetic interference) and then made widely to those certified designs. The masters of the essential software and hardware components (the analogues of Microsoft and Intel) might be the big winners, not the final products' assemblers and sellers.

Streamlining Market Structure

Informed by the computer example, rapidly collapsing the levels of the market could be a daring, high-risk, but perhaps high-payoff strategy. Just the way Japanese building agents sell prefabricated-to-order houses today, a hypercar maker could sell cars through CD-ROM demonstrations and test drives provided by a salesperson who visits your home. You choose the options you want; they instantly go by modem to the factory; it makes the car to order; a couple of days later, it is delivered to you. Toyota already does this in Japan; it takes about four days with steel cars and could be far easier with the simpler hypercars.

Once delivered, the hypercar is unlikely to fail (far fewer total parts and moving parts, fewer connections and fasteners, more fault-tolerant electronics and software, cooler running, better preventive diagnostics, etc). But if it does fail, someone from a service company will come fix it; nearly all of its few parts are small, light, readily hand carried, and easily installed. Today, even some photocopiers automatically diagnose themselves and summon technicians by modem; perhaps some handy hypercar owners too, guided by the car's powerful self-diagnostics or remote expert evaluation by modem hookup, might choose to plug in replacement modules automatically express-shipped to their door. Indeed, in the United Kingdom Ford already offers such a 24-hour onsite maintenance service for its cars, whose maintenance is both more difficult and more often needed than it would be for hypercars.

If direct sales, manufacture to order, and onsite maintenance all make sense today for a $1,500 mail-order personal computer, why not for a $15,000 car? Indeed, if Sears could sell a $395 "motor buggy" (then a novel product) by mail order in 1910, why could not a respected automaker do direct mail-order sales of a hypercar in, say, 1998? Such direct marketing and onsite maintenance could transform the economics of hypercars. Today, the suggested retail price of a typical U.S. car, averaging over both economy and performance models, has been marked up by perhaps three-fourths from its bare, no-profit marginal variable cost of production, or by one-half including intended profits, warranty costs, and plant costs. Designs are frozen many years in advance, and the mix of features, colors, etc, is chosen months in advance. Even if demand forecasts turn out wrong, as they often do in today's fickle markets, the undesired products have already been built, so they must be carried and then rebated or discounted until they sell. This adds a severe burden to the already formidable cost of marketing, inventory, selling, and transactions: in the United States, one-fourth more people work for auto dealers than for auto manufacturers.

Conversely, just-in-time assembly-to-order and zero-inventory direct sales would largely squeeze out the causes of the typical ~50% markup. This in turn would presumably enable a hypercar *to sell profitably for the same retail price as now, even if it cost considerably more to produce,* which few composite experts believe it would. That is, streamlining market structure would provide an ample cushion for excess production costs, especially initially when they are higher. This is especially true if introduction starts in high-end segments that normally have higher than average markups.

Market Evolution

If hypercar entrepreneurs were as radical structurally as technologically, they might slash dealer support and eliminate prebuilt inventory, directly costing nearly a million American jobs, many already at risk from the trend toward one-price, no-haggle selling. Their production costs would then be comparable to or lower than for today's platforms and their selling costs would be minimal, so retail car prices well below today's would still support improved gross margins. For those who made the change early, the car business could be enormously profitable, until, as with PCs, competition brought a flood of new market entrants, margins became thin again, and consolidation began. By then, too, some of today's automakers could be out of business—the typewriter and sliderule makers of the 1990s.

Product differentiation and minor support services would become the battleground. Buyers, like buyers of personal computers today, could choose various trade-offs: spend more on composites, less on rust and dents; more on user-friendly diagnostics and plug-in fixits, less on shop repairs; more on safety options, less on hospital bills; more on the car, less on the dealer markup. Even though little of today's markup ends up in the dealer's pocket, reallocating some or all of it offers opportunities for all of a much smaller number of parties to the transaction to be better off.

Jobs and Competitiveness

What about those whom a streamlined hypercar industry may not need? The auto industry and associated businesses employ one in seven American workers (up to nearly two-fifths in some European countries) and represent a tenth of all U.S. consumer spending. They use 40% of the machine tools and 12% of the steel, 20% of the aluminum, zinc, glass, and semiconductors, about 60% of the rubber, carpeting, and malleable iron, and nearly 70% of the lead. David Morris, cofounder of the Institute for Local Self-Reliance, observes: "The production of automobiles is the world's number one industry. The number two industry supplies their fuel. Six of America's ten largest industrial corporations are either oil or auto companies. . . . A recent British estimate concludes that half of the world's earnings may be auto or truck related." Any major structural change in such dominant sectors is bound to be extraordinarily far-reaching.

Hypercars appear to involve up to two orders of magnitude less parts-fabrication work and one order of magnitude less total assembly work than steel cars, though still significant finishwork. Design teams would probably be very small—tens, not thousands—in order to be fast and fully integrated. To be sure, large new industries would spring up, such as composites fabrication to the extent it were not robotized (perhaps using textile technology), advanced motors and electronics, software, etc, but those jobs may go to other people in other places. Petrochemicals and electronics would gain; steel would lose. Computer-aided moldmaking and filament-winding equipment would gain; lathes and milling machines would lose. Diemakers may or may not adapt to computer-aided design and stereolithography. Aerospace firms may or may not adapt to high volumes, low costs, and consumer markets. Startup firms would supplant aging giants. Parts firms and body shops would need new skills but find greatly diminished scope. Auto dealers might become as rare as public stables.

No one has yet begun to assess winners, losers, and relatively graceful transitional patterns for the potentially rapid and traumatic shifts in employment implied by this new type of manufacturing, marketing, and maintenance—a bigger challenge to industrial renewal and retraining policies than any nation has yet faced. But this challenge may be unavoidable, because a country or a company that ducks it may face competitors who feel no such inhibition. It is presumably better to have even a traumatized and diminished auto industry with great new market prospects than to have none. And it is better to have a vibrant, short-cycle, adaptive, keenly innovative, and resilient auto industry than a lumbering, capital-intensive, vulnerable one.

This is also an international issue, a potentially major problem for a country that unwittingly lets others destroy its traditional car industry, and a potentially major opportunity for a country that gets there first. Which countries make, have, and use how many cars could shift rapidly. A very senior Mexican official recently remarked that he wants to make hypercars in Mexico, not only to create a durably competitive car industry, but also to solve domestic air and oil problems. The same thought is already occurring to other developing countries rich in talented low-wage people, increasingly including world-class engineers and software writers, but poor in oil. Those countries include China.

With tooling costs low and design barriers high, the business will flow to integrative talent, not to capital; but with assembly labor reduced, there could be less incentive than now to move assembly offshore. Nonetheless, hypercar jobs, like electronics assembly today, could become a fought-over global commodity much faster than jobs making steel cars, with their huge tooling investments. In car assembly, barriers could become equally low to market entry and to market exit. Automakers wishing to expand into developing countries may find a welcome only for hypercar plants, not for traditional plants. Negative technology transfer is also a risk: if steel cars rapidly became obsolete but their tooling was not scrapped, it could enter a secondary market and be sold cheaply to developing and ex-Socialist countries (as the former East German tooling recently was to the Baltic republics), locking them into economic inferiority and resource waste for even longer.

Capitalization

Then there is the matter of where the car centers of the world (the Detroits, Cowleys, and Wolfsburgs) could get the billions of dollars required to retool to make hypercars, if they were culturally able to do so. Unfortunately, the automakers with the most capital for new ventures, such as Toyota, are the best at steel stamping and hence have the least incentive to change. For other, hungrier firms, however, novel sources of retooling investment might arise. For example, direct project financing by major oil companies could help both sides. After all, the lower 48 United States contain the equivalent of an 8.5-million-barrel-per-day (0.5-TW) oilfield, bigger than the biggest in Saudi Arabia, that is nonpolluting, uninterruptible, and nondepleting. It is the accelerated-scrappage-of-gas-guzzlers oilfield, with substitution by both hypercars and their heavy-vehicle analogues.

Today, oil companies go to the ends of the earth to drill for very costly oil that may not even be there. It would be embarrassing to drill more billion-dollar dry holes while someone *else* found all that cheap "oil" under Detroit. Just as oil majors now hedge upstream/downstream, oil/gas, etc., they could hedge between barrels and "negabarrels": they could project-finance hypercar retooling with upside participation via equity conversion or royalty, so that if the hypercar business proved a great success, they would make less money on oil but more on cars. Preliminary discussions with some cash-rich oil majors have established interest in this investment concept, although it remains to be seen whether Big Oil and Big Cars can put aside their enmity for long enough to collaborate to mutual advantage.

The dynamics of divergent motives and reactions *within* the oil industry, especially between crude-short and crude-long companies, are also complex. Both kinds of firms are becoming intensely interested in hypercars as a nega-OPEC—a potential to save as much oil as OPEC now extracts. Such a frontal assault on the last major nonsubstitutable use of oil (road transport) would presumably more or less permanently crash the oil price. This in turn would have profound and unforeseeable effects on development, geopolitics, and global security (3).

INDUSTRIAL TRANSFORMATION AND CULTURAL CHANGE

Reinventing the automobile is far from the consciousness of most (but not all) of today's automakers. They struggle daily and nightly for next quarter's dividends; the prospect of scrapping their tooling and their mindsets and starting over is another problem they feel they just do not need. The world of hypercars is not only frightening; for many it is so alien as to be hard to conceive at all.

Big automakers start, as Lee Schipper has remarked, with two nearly fatal disadvantages: they are big, and they are automakers. They are dedicated, extraordinarily capable, and often socially aware organizations, but too often their form, style, and speed of learning match the ponderous technologies, vast production runs, and long product cycles inherent in steel cars. Their "productivity improvements have been balanced by a continuous decrease in . . . innovative capability" as ever more highly integrated production processes make innovation more difficult, solidifying a "fluid" industry into a "specific state" (30,61). Automakers have a diemaking and steel-stamping culture, not a composite-molding, electronics, and software culture. They are prisoners of enormous sunk costs which they treat as unamortized assets, substituting accounting for economic principles; hence they go to heroic lengths to adapt four-cylinder engine-block capacity rather than retool. This mindset (61) is a critical obstacle to the transition toward hypercars: new ways cannot diffuse without displacing old ones that resist with distinctive vigor (30).

Many automakers act as if they would rather take writeoffs of their obsolete capabilities later when they do not have a company than now when they do; as if they preferred comfortable obsolescence, even unto bankruptcy, over uncomfortable basic change to ensure long-term profitability. Their strategy appears, at least from the outside, to be to milk old skills and tools for decades, watch costs creep up and market shares down, and postpone any basic innovation until after all concerned have retired, and hope none of their competitors is faster. Yet it only takes one successful hypercar competitor to put them out of business, and they may not even know who it is, for it may be a nonautomaker or even an unheard-of startup company, especially a virtual company.

A different strategy, favored by a growing number of internal policy entrepreneurs and just lately appearing on top management's radar, would enhance automakers' survival prospects. Welcome and capitalize on innovative public-policy instruments that condition the market for hypercars (discussed below); promptly switch to ultralights using net-shape materials in integrated assemblies; refine their design and manufacturing techniques; meanwhile develop sophisticated hybrid drivelines; and then, in one more giant leapfrog, move quickly to electric hybrid drives, managing risk at each step with more conventional fallback positions to cover any temporary technological gaps. If this logic is correct, the first firm that intelligently and aggressively follows it should be able to feel sorry for its former competitors.

Who will that firm be? It might be an automaker or an aerospace company or an electronics giant. It might also be the next Apple or Xerox, a group of smart, hungry, invisible systems engineers in a garage in southern California, eastern Massachusetts, or northern Italy, perhaps even innovators from within the car business but unburdened by its sunk costs and traditional attitudes. The most apt competitors might be smart, agile, unafraid high-technology systems integrators like Hewlett-Packard or Sony, because hypercars are much more a software than a hardware problem: with an order of magnitude more lines of code than today's cars (though much less of it for engine control), a hypercar will be *much more like a computer with wheels than like a car with chips*. Companies that know how to make things like computers, but have nothing to unlearn about cars, may do surprisingly well. And those companies might well be American, because that nation leads in the combination of key technical capabilities needed (systems integration, software, advanced materials, aerospace, and micro- and power electronics) and often in entrepreneurial speed and vision.

The chemical industry may also be a key player. Although the auto industry is woefully undersupplied with people as good at synthetic materials as classical automakers are at steel, some firms are starting to appreciate that molded materials "allow simplification of both cars and productive processes and a more frequent change in the range of models supplied" (30)—factors often even more important than raw cost per part. All the large chemical companies already have "automotive centers" in the Detroit area, and there are analogous European programs. Through this technological fusion, "a new area of research and production, linking the chemical and the automobile industries, is quietly developing," increasing chemical firms' downstream integration while pushing automakers, at least temporarily, toward backwards vertical integration (30).

If the auto industry is to adapt to and grasp the ultralight-hybrid opportunity rather than be run over by others' faster adoption of it, it will need to change its habits:

- It will need to learn in weeks to months, not years to decades. (This speed is what sorts out winning from losing computer companies: Some reports suggest they make 90% of their profit in the first 6 months of a product's lifecycle.) Kelly Johnson's old skunkworks at Lockheed originated certain innovative aircraft from scratch in 4 months. With today's tools, making an innovative car should be faster.
- To achieve this, it will have to keep its work groups lithe, its head counts small, and its bureaucracy suppressed, so as to uplift and liberate its many brilliant individuals.

- It will have to establish a presumption in favor of net-shape and near-net-shape materials in integrated assemblies, defaulting to metal only where temporarily necessary case by case, rather than switching from routine metal to supposedly exotic moldable materials only incrementally as an occasional "frill."
- It will have to redesign components, assemblies, and systems from scratch, using a zero-based mass budget, to exploit the new materials' capabilities: composites are not black steel.
- It will need to determine the best ways to manufacture with net-shape materials, then design cars that best exploit those methods, rather than, as now, designing cars first, like abstract art, and then figuring out the least unsatisfactory way to make them within the constraints of traditional metalforming art.
- It will need to learn that how hard each part is to make and apply is at least as important as how many parts there are.
- It will have to treat temporary uncertainties over the best approaches to recycling composites, field repairs, and certain design and manufacturing techniques as normal problems to be overcome expeditiously, not as reasons to shun net-shape material. The Big Three U.S. automakers are learning only slowly about a possible switch to advanced composites, because they have only a few dozen people exploring it.
- It must involve its workers and suppliers early in thinking through the transition in all its dimensions, from labor flexibility and retraining to occupational health. Waste minimization, recyclability, closed loops, and nontoxic materials will be important when manufacturing with large volumes of composites.
- It will have to pursue ultralight hybrids whether it believes they are the next car or only a niche car. It can assume a small market, easily tested through rental companies, but must stay ready to surge production quickly if the market explodes (as with the Honda CRX, which entered the market almost from bench-scale production). Such flexibility, well clad in transitional risk management, exploits a potential profit opportunity, but unwillingness to try it is a You-Bet-Your-Company decision. In marketing as in invention, chance favors the prepared mind.
- It will have to put more effort into leapfrogging straight to ultralights and then to ultralight hybrids, reaching its objective in only two main retoolings, and less effort into small but very costly marginal refinements in existing platforms with tiny marginal returns. (Of course, each of the two big jumps will include many small improvements, moving from established interim technologies to better ones as they mature, but the nature of the new tooling makes cycle times far shorter than for steel cars.)
- It will need to adopt the X-vehicle philosophy of building and improving rather than studying to death, of going with what we have and then refining it. (Hypercars will need many improvements in manufacturing, repair, recycling, and other techniques, but these are *a need, not a barrier,* they are a normal part of industrial development, not a reason for inaction.) In Paul MacCready's language, it will need not to shoot at the bull's-eye, but to shoot first and then draw the bull's-eye around where the bullet goes.
- It will need a ready-fire-aim culture in which a reasonable number of errors is expected and rewarded and the most successful mistakes are positively celebrated.
- It will have to learn that even if a tactical goal is to improve today's platforms, the strategic goal is to make them obsolete as quickly as possible before competitors do. As *The Wall Street Journal* remarked, surveying the wreckage of the mainframe computer industry (62), "slow reaction stemmed partly from a reluctance to undermine sales of cash-cow large machines. 'You have to face up to the question of destroying your product with new products,' says John Morgridge, chief executive of Cisco, which makes networking hardware. 'If you don't do it, someone else will.'"

Today, most automakers seem far from appreciating the imperative framed by 3M as "We'd rather eat our *own* lunch, thank you." But one way or another, we believe they will learn it—some to their pain, others to their profit. They must choose to be Control Data or Apple, Bull or Dell.

MARKET CONDITIONING AND PUBLIC POLICY

There are compelling public reasons to make cars more efficient, whether incrementally or radically. Cars and light trucks, their efficiency stagnant since 1982, burn ~37% of America's oil, of which ~43% is imported at an annual cost of around $50 billion. The benefits in oil displacement, energy security, international stability, avoided military costs, balance of trade, climatic stabilization, clean air, health and safety, noise, and quality of urban life can hardly be overstated. Just the peacetime readiness cost of U.S. forces earmarked for Persian Gulf intervention plus the direct cost of oil imports is roughly $100 billion per year. In recent years, cars and car parts have accounted for the equivalent of three-fourths of the U.S. trade deficit with Japan.

All told, cars' externalities approach $\$10^{12}$ per year in the United States alone (63,64), many times internal costs (7). But these externalities, perhaps a seventh of the GNP, could be perhaps halved by hypercars, saving several hundred billion dollars a year in pollution, accident, land, noise, vibration, congestion, pavement, military, and climatic-change costs. Indeed, as part of a strategy of industrial regeneration, hypercars could form the centerpiece of a powerful reintegration of the economic, energy, environmental, and military elements of security (65).

Many economists suppose that none of this will happen, because, they assume, efficient cars differ from inefficient cars in only two closely related respects: fuel efficiency and price. But hypercars belie that assumption. They do a complete end run around scholastic debates about how many price elasticities can dance on the head of a pin. People will buy hypercars, *not because they save fuel, but because they are superior cars*—cars alongside which today's most sophisticated steel models might even seem a bit primitive and antiquarian. Hypercars will sell because they work better and redefine the market. Hypercars may well cost less than steel cars too, but it is probably not important whether they do or not, any more than it mattered whether CDs undercut vinyl records or pocket calculators undercut sliderules.

Yet, however ultimately inevitable these competitive factors may make the transition to hypercars, it may if unguided produce two kinds of failures. It may be both unnecessarily disruptive, shattering industrial regions and jobs, and unnecessarily slow and unpredictable in capturing the strategic benefits of saving oil and rejuvenating the economy. Further, many automakers convinced that fuel economy must be antithetical to other marketing factors may resist hypercars for too long and thereby consign themselves and their workers to commercial oblivion. To achieve a relatively smooth transition rapidly and with high confidence may require public-policy interventions in which industrial, oil, security, and environmental imperatives converge: interventions to give automakers strong incentives to pursue the "leap-frog strategy" boldly and to overcome their customers' well-known market barriers to buying fuel-efficient cars.

Weak Price Signals

Thanks to ever-cheaper oil and improving fuel economy, the real 1989-dollar cost of fueling a new American car for 40 km was about \$4 in 1929, \$3 in 1949, \$2 in 1969, and \$1 in 1989 (22). Extrapolation would suggest a cost of zero in 2009, and hypercars will probably make that right within about a nickel, both by using far less gasoline and by crashing the world oil price. In addition, insurance and maintenance costs should be much lower with hypercars than with today's cars, making driving almost a pure capital cost. This greatly reduced variable cost will presumably increase demand for marginal driving.

Both the futures market (which predicts, and can be used to lock in, oil prices) and careful examination of recent technological revolutions in both supply and demand strongly suggest that real oil prices will trend downward for at least the next couple of decades. Although oil prices will doubtless spike occasionally as war or peace breaks out in the Middle East, one cannot count on costly oil to sell fuel-frugal cars. On the contrary, the two are mutually inconsistent: as the 1986 oil price crash proved, efficient cars *prevent* high oil prices.

Are high motor fuel taxes the answer, as most economists prescribe? International comparisons show that motor fuel prices do modestly affect kilometers per year driven but are only weakly related to new-car fuel economy (66). Even in Europe and Japan, with gasoline taxed to ~2–4 times the U.S. prices, new cars are little more efficient than in the United States (66). This is because of dilution by fixed costs, high consumer discount rates (especially if first ownership is customarily short), company car ownership, unusual tax policies, and other distortions that shield drivers from their normal costs (67). Though these factors' relative importance varies by country and over time, collectively they cause a pervasive market failure.

After all, we saw earlier that the efficiency improvement in the 1992 Honda VX, whose fuel economy is markedly higher than the best new-fleet average in Europe, would be like buying gasoline at 69¢/gal, scarcely over half the American gasoline price today, when it is the cheapest in U.S. history, the cheapest in the industrial world, and cheaper than bottled water. Yet most of that car's improvements, and many more it omitted, have not been brought widely to market, because manufacturers have shunned the retooling risk in fear of uncertain market response. OECD on-road fleet-average intensities stayed roughly flat through the 1980s, ranging from nearly 8 L/100 km (Denmark) to nearly 11 L/100 km (Japan), roughly twice the VX's 4.62. Compared with the corresponding previous-year model, the 1991 DX, the VX had 56% higher kilometers per liter, 6% higher interior volume (2.18 m^3), 2% greater length (4.07 m), 10% higher peak torque, 3% lower curb mass (950 kg with airbag, vs. 979 kg with none), and a \$650 higher sticker price. It is a great deal for society but was not widely emulated.

The small fuel price elasticity of new-car efficiency means theoretically that if fuel price were the only instrument available to sway car buyers, extremely high fuel prices would be needed to bring hypercars to market. But hypercars' social value can be signaled and their early production encouraged by other means. For example, the U.S. Corporate Average Fuel Economy (CAFE) standards apparently achieved most or all of the U.S. doubling of new-car fuel economy (68): New cars ended up approximately as efficient as European and Japanese models, perhaps even more so when normalized for size, performance, and accessories. This was despite U.S. gasoline prices so low that fuel is only one-eighth the total cost of driving, so the fuel's price signal is diluted 7:1 by the other costs of owning and running a car.

Standards and Alternatives

CAFE can certainly be improved in many details (18,19). However, performance standards, though a helpful backstop, are not easy to administer, invite gaming, and are technologically static: there is no incentive to do better. A more promising approach would add revenue-neutral "feebates" (a concept developed by R. H. Garwin in the early 1970s, then by both A. B. Lovins and A. H. Rosenfeld, who invented the term a few years ago). Under a "feebate" system, when you buy a new car, you pay a fee or get a rebate; which and how big depends on how fuel-efficient the car is (and perhaps also on how clean and/or safe it is); the fees pay for the rebates. Better still, the rebate for a fuel-efficient new car can be based on its *difference* in fuel economy compared to the old car that is scrapped, thereby getting efficient, clean cars on the road and inefficient, dirty cars off the road faster. (Malfunctioning or ill-maintained "superemitters" are often of 1970s vintage; one-fifth of the U.S. car fleet produces three-fifths of its air pollution.)

Such "accelerated-scrappage feebates" would open large new markets for the auto industry, foster competition, and reward rapid and continuous innovation with market share, potentially without limit. Since the least efficient cars in the United States are now owned more by upper than by lower income people (to whom the more frugal early-1980s cars have now trickled down), scrapping the least efficient cars first would not hurt the poor but on the contrary would reward them by paying them more to scrap and trade up inefficient cars than they would get by trading them in, and would give them economic access to cars that they could then readily afford to fuel.

Many variations on these themes are being considered, including feebates decoupled from separate scrappage rewards, volume normalization to avoid incentives for down-sizing, and rebates paid directly to manufacturers rather than to buyers so as to compound price reductions by reducing markups. Rebates on superefficient cars could be big enough to push the effective retail price below that of a used car, boosting margins, and could even exceed factory prices. The $\sim\$10^{12}$-per-year estimate of U.S. car-related externalities would support marginal feebate slopes on the order of several hundred thousand dollars per gallon per mile (on the order of $\$10^3$ per L/100 km), though much less could prove sufficient. Weak feebates have already been legislated in Maryland (defectively) and the Province of Ontario. A more comprehensive one ("Drive+") was approved by an extraordinary 7:1 margin in the California legislature in 1990, though it was subsequently vetoed; it seems bound to spread to state if not federal agendas. Feebates could command wide consensus and break the political logjam that has long trapped the United States in a sterile debate over higher gasoline taxes vs. stricter CAFE standards, as though those were the only two policy options and small, slow, incremental improvements were the only technical options. Outside North America, governments more used to specific direction of major industries may enjoy even wider policy options.

Negamiles and Negatrips

A successful shift to hypercars, however, will not solve the fundamental problem of too much driving by too many people in too many cars (69) and could worsen by making cars and driving even cheaper and more attractive. If 1-L/100-km, roomy, clean, safe, renewably fueled cars were on the road today, eight million New Yorkers or Los Angelenos or Londoners or one billion hitherto carless Chinese driving them, or just *today's* global car fleet driven ever greater kilometers per year, still would not work; instead of running out of air or oil, we would run out of roads, time, and patience. Avoiding the constraint *du jour* requires far more than extremely fuel-efficient vehicles: they are an essential time-buying step, but no panacea. Rather, hypercars must be *combined* with dramatically decreased automobility (3,4,7), especially because our species is only half as effective at birth control for cars as for people. Global car registrations are growing more than twice as fast as human population, and the 85% of the world's people who own only 24% of its motor vehicles want theirs too.

Sustainable transportation requires designing communities around people, not cars, and rethinking land use so we need not travel so much to get the access we want. This in turn requires an end-use/least-cost access strategy and decision process to foster competition between all modes of access, including those that displace the need for mobility. Such policy innovations can join with hypercars, and their analogues in other modes, to foster global competitiveness and to meet ambitious oil displacement, air, noise, and urban quality, CO_2, equity, and development goals.

This complement to the hypercar revolution requires creative public-policy instruments that introduce market mechanisms to a transportation system still crippled by lopsided subsidies, car dominance, and top-down central planning. In most supposedly market economies, transport planners in good Stalinist fashion decide where the roads, parking, and other car infrastructure will go, socialize its cost through taxes, and generously invite those who prefer alternatives to buy them—and pay for them too. That is hardly a level playing field; it is essentially corporate socialism for cars and free enterprise for most of their competitors.

Happily, the needed innovations are starting to emerge (4). They include congestion pricing of roads and parking, parking feebates, commuting-efficient mortgages, advanced land use planning (70,71), internalization of social costs (63), pay-at-the-pump car insurance (72,73), and making "negamile markets" that maximize competition between all modes of mobility (and ways to get access without mobility, such as telecommuting or being there already, a benign land-use alternative that instead of *solving* the transportation problem simply *avoids* it). How much is it worth paying people to stay off the roads so we need not build and mend them so much? Probably a lot. We should make markets to find out.

Policy innovation will not come a moment too soon (74). In most of the world's cities, cars now drive out street life and the public realm, people know their neighbors mainly through windshields, and social interactions are often reduced, in architect Andres Duany's phrase, to "aggressive competition over squares of asphalt." The privatization of mobility limits opportunity for the one-third of Americans who are too poor, old, young, or infirm to drive, thereby virtually imprisoning them. Unconstrained and fabulously underpriced automobility has indeed created rampant inequity, eroded community, fragmented neighborhoods, devoured land and tranquility, and submerged civilized purpose. As T. S. Eliot warned (7), "A thousand policemen directing the traffic/Cannot tell you why you come or where you go."

None of these challenges and changes will be easy; only easier than not making them. They will take decades, because "the machine that changed the world" (75) has a more formidable momentum than perhaps any other major human achievement. Yet recent industrial history, notably in computing and telecommunications, suggests that the switch to hypercars could be far faster than basic shifts in where people live, work, shop, and recreate. The speed and size of this change could be deeply disruptive and could bring enormous benefits. As with any technological revolution, disruption is inevitable; we can only choose whether to make it hurt or help us. If the technical and market logic outlined here is anywhere near right, we are all about to embark on one of the greatest adventures of our species' industrial history. Ready or not, here it comes.

Acknowledgments

Fuller acknowledgements of many intellectual debts are given in the paper from which this is adapted (1), particularly its coauthors, Dr. John Barnett and L. Hunter Lovins, and our valued collaborators Eric Toler and Ian Jay Czaja. We are grateful for more recent help by Timothy Moore, Michael Brylawski, and David Crammer. The opinions expressed here are personal and should not be attributed to any of our generous teachers, sponsors, or reviewers. The work by Rocky Mountain Institute's Hypercar Center summarized in this article was supported by the Nathan Cummings Foundation, Surdna Foundation, Changing Horizons Charitable Trust, and the Compton, Albert A. List, Max and Anna Levinson, and Florence V. Burden Foundations.

BIBLIOGRAPHY

1. A. B. Lovins, J. W. Barnett, and L. H. Lovins, "Supercars: The Coming Light-Vehicle Revolution," *Procs. 1993 Summer Study of the European Council for an Energy-Efficient Economy* (Rungstedård, Denmark, 1–5 June), Rocky Mountain Institute Publication #T93-10 (5).
2. A. B. Lovins, J. W. Barnett, and L. H. Lovins, "Supercars: The Next Industrial Revolution," RMI Publication #T93-16 (5), Aug. 11 1993.
3. A. B. Lovins, J. W. Barnett, and L. H. Lovins, "Policy Implications of Supercars," RMI Publication #T93-21 (5), Aug. 1993.
4. A. B. and L. H. Lovins, "Reinventing the Wheels," *Atlantic Monthly,* January 1995, RMI Publication #T94-29(S), and supplements in press (9).
5. Publications of The Hypercar Center, Rocky Mountain Institute, 1739 Snowmass Creek Road, Snowmass, Colo. 81654.
6. A. B. Lovins, Chairman, and co-workers, *Proceedings of the Dedicated Conference on Supercars (Advanced Ultralight Hybrids)* 275(1), 75–86. (Aachen, 31 October–2 November 1994), ISATA (International Symposium on Advanced Transportation Applications), Croyden, England.
7. E. W. Johnson, "Taming the Car *and* Its User: Should We Do Both?" *Bull. Amer. Acad. Arts Scis.* **46**(2), 13–29 (1992).
8. A. B. Lovins, "Advanced Light Vehicle Concepts," briefing notes for the Committee on Fuel Economy of Automobiles and Light Trucks, Energy Engineering Board, U.S. National Research Council (Irvine, California, 9 July 1991, updated August 3), RMI Publication #91-20 (5).
9. P. Patterson, "Periodic Transportation Energy Report 1," DOE CE-15, U.S. Department of Energy, Washington, D.C., 202/586-9118, 16 November 1987.
10. M. Ross, "Energy and Transportation in the United States," *Ann. Rev. Energy* **14,** 131–171 (1989).
11. F. W. Westbrook, *Allocation of New Car Fuel Economy Improvements, 1976–1989,* Camden Corp. Report to Oak Ridge National Laboratory, 1991.
12. American Physical Society, *Efficient Use of Energy,* American Institute of Physics Conference Proceedings No. 25, AIP, New York, 1975.
13. C. Difiglio, K. G. Duleep, and D. L. Greene, "Cost Effectiveness of Future Fuel Economy Improvements," *En. J.* **2**(1), 65–83 (1989); see also Energy and Environmental Analysis, *Developments in the Fuel Economy of Light-Duty Highway Vehicles,* August 1988, report to U.S. Congress, Office of Technology Assessment, Washington, D.C.
14. M. Ledbetter and M. Ross, "Supply Curves of Conserved Energy for Automobiles," *Proc. 25th Intersoc. Energy Conversion Engineering Conf.* (Reno, Nevada, 12–17 August 1990), American Institute of Chemical Engineers, New York.
15. J. G. Koomey, D. E. Schechter, and D. Gordon, "Cost-Effectiveness of Fuel Economy Improvements in 1992 Honda Civic Hatchbacks," LBL-32683, Lawrence Berkeley Laboratory, Berkeley, Calif., 1992.
16. H. Kelly and R. H. Williams, "Fuel Cells and the Future of the US Automobile," 7 December 1992, review draft, U.S. Congress, Office of Technology Assessment, Washington, D.C.
17. K. G. Duleep, *Potential to Increase Fuel Economy to 2001,* Energy & Environmental Analysis, Inc., Arlington, Va., 1991.
18. Office of Technology Assessment, *Improving Automobile Fuel Economy: New Standards, New Approaches,* OTA-E-504, U.S. Congress, Washington, D.C., 1991.
19. National Research Council, *Automotive Fuel Economy: How Far Should We Go?* National Academy Press, Washington, D.C., 1992.
20. D. L. Bleviss, *The New Oil Crisis and Fuel Economy Technologies: Preparing the Light Transportation Industry for the 1990s,* Quorum Books, Westport, Conn., 1988.
21. L. Chappell, "Suppliers Say U.S. Car Fleet Can Get 41 mpg in 10 Years," *Automotive News,* 21 August 1991, p. 10.
22. P. B. MacCready, "Further Than You Might Think" and "Electric and Hybrid Vehicles," *Conference on Transportation and Global Climate Changes and Long-Term Options,* Asilomar, California, Aug. 26, 1991.
23. S. M. Rohde and N. A. Schilke, "The Fuel Economy Potential of Heat Engine/Flywheel Hybrid Automobiles," *Supplement to Procs. 1980 Flywheel Technology Symposium* (DOE/ASME/LLL), CONF-801022-Supp (Scottsdale, Ariz.), Oct. 1980, and SAE Special Publication P-91, 1981; see also N. A. Schilke, A. O. DeHart, L. O. Hewko, C. C. Matthews, D. J. Pozniak, and S. M. Rohde, "The Design of an Engine-Flywheel Hybrid Drive System for a Passenger Car," SAE Paper 841306, Society of Automotive Engineers, Warrendale, Pa., 1984.
24. G. Sovran and M. S. Bohn, "Formulae for the Tractive-Energy Requirements of Vehicles Driving the EPA Schedules," SAE Paper 810184, 1981; see also G. Sovran, "Tractive-Energy-Based Formulae for the Impact of Aerodynamics on Fuel Economy over the EPA Driving Cycle," SAE Paper 830304, 1983, and the summary in Ref. 19, App. A.
25. J. D. Maples, "The Light Duty Vehicle MPG Gap: Its Size Today and Potential Impacts in the Future," University of Tennessee Transportation Center, Dec. 1, 1992 draft.
26. P. B. S. Lissaman, "Concepts in the Aerodynamic Drag of Road Vehicles," Lecture 2-1 at pp. 22–23, *GM Sunraycer Case History,* M-101, Society of Automotive Engineers (SAE), Warrendale, Pa., 1988.
27. B. Hibbs, "Sunraycer Aerodynamic Environment," Lecture 2-2, in *GM Sunraycer Case History,* M-101, Society of Automotive Engineers (SAE), Warrendale, Pa., 1988.
28. C. R. Kyle, "The Sunraycer: Wheels, Tires and Brakes," Lecture 2-3, in *GM Sunraycer Case History,* M-101, Society of Automotive Engineers (SAE), Warrendale, Pa., 1988.

29. D. McCosh, "The Promise of Aluminum: The 50-mpg Sedan," *Pop. Sci.,* 76–81 (Feb. 1993).
30. G. Amendola, "The Diffusion of Synthetic Materials in the Automobile Industry: Toward a Major Breakthrough?" *Res. Policy* **19**(6), 485–500 (1990).
31. U.S. Department of Energy, *Lightweight Materials for Transportation: Program Plan,* Office of Transportation Materials, DOE, Washington, D.C., April 1993, §C.7.1, §C.8.2, §C.9.
32. J. Keebler, "GM Builds 100-mpg 'Ultralite' Car," *Automotive News,* 31 December 1991; pp. 1 & 31.
33. D. Sherman, "Using Carbon Fibers to Conserve Hydrocarbons: GM Ultralite," *Motor Trend,* February 1992, pp. 74–76.
34. C. Gromer, "Ultracar," *Pop. Sci.,* April 1992, pp. 33–35 & 125.
35. K. H. Hellman, "Passenger Car Fuel Economy—Beyond 2010," USEPA, 313/668-4246.
36. J. V. Busch, F. R. Field, and J. P. Clarck, "Economics of Automobile Body Panels," MIT Materials Systems Laboratory paper at ATA-CNR Seminar *Innovation in Materials for the Transportation Industry,* Torino, Italy, 4–6 November 1978.
37. J. Romm, "Why America Beat Iraq But Loses to Japan," Rocky Mountain Institute Publication #S91-39 (5), 1991; summarized in "The Gospel According to Sun Tzu," *Forbes,* 9 December 1991, pp. 154–162.
38. R. Seiss, presentation to NRC Irvine hearing (see ref. 8) on behalf of Dow Chemical Co., July 8, 1991.
39. *Automotive News,* "Chrysler Genesis Project Studies Composite Vehicles," 5 May 1986, p. 36.
40. W. S. Abernathy, *The Productivity Dilemma: Roadblock to Innovation in the Automobile Industry,* Johns Hopkins University Press, Baltimore, 1978.
41. S. Yagi, K. Fujiwara, N. Kuroki, and Y. Maeda, "Estimate of Total Engine Loss and Engine Output in Four Stroke S.I. Engines," SAE Paper 910347, 1991.
42. R. Bauder and D. Stock, "The New Audi 5-Cylinder Turbo Diesel Engine: The First Passenger Car Diesel Engine with Second Generation Direct Injection," SAE Paper 900648, 1990.
43. H. Barske, "Rationelle Verwendung von Kraftstoff: Autos mit 3 Liter Benzinverbrauch, eine Utopie?" in *Procs. 6th Ann. Conf., Environment and Renewable Energy II—Objectives 1991 to 2000,* B. Löffler, ed., Basel, 1991.
44. U.S. Department of Energy, "Notes of the Hybrid Planning Workshop" (Dearborn, 22–23 September), DOE, Washington, D.C., 1992.
45. M. K. Vint, "Design and Construction of a Fuel Efficient Braking System," SAE Paper 871233, 1987.
46. J. Delsey, "Environmental Comparison of Electric, Hybrid and Advanced Heat Engine Vehicles," *Proc. Urban Electric Vehicle Conf. (25 May 1992),* OECD/NUTEK/IEA, Stockholm.
47. W. Streicher, "Energy Demand, Emissions and Waste Management of EVs, Hybrids and Small Conventional Cars," *Proc. Urban Electric Vehicle Conf. (25 May 1992),* OECD/NUTEK/IEA, Stockholm.
48. A. F. Burke, "Hybrid/Electric Vehicle Design Options and Evaluations," SAE-920447, 1992.
49. P. J. Lawrenson, "A Brief Status Review of Switched Reluctance Drives," *Eur. Power Electronics and Drives Assn. J.,* September 1992, pp. 71–79.
50. A. B. Lovins and W. Howe, "Switched Reluctance Motor Systems Poised for Rapid Growth," TU-92-4, E SOURCE, Boulder, Colorado, 1992.
51. R. J. Blake and P. J. Lawrenson, "New Applications of Very Large and Small Switched Reluctance Drives," 4.4.1, *Proc. Power Conversion and Intelligent Motion,* Nürnberg, April 1992.
52. D. J. Houghton, R. C. Bishop, A. B. Lovins, B. L. Stickney, J. J. Newcomb, M. Shepard, and B. J. Davids, *The State of the Art: Space Cooling and Air Handling,* E SOURCE (50), Boulder, Colo. 1992.
53. B. Stickney, "Super Efficient Refrigeration Systems," Tech Update TU-92-3, E SOURCE (50), Boulder, Colo. 1992.
54. B. A. P. Remendra, P. B. Hertz, and E. A. Krause, "Comparison, Selection and Modelling of Small Powerplants for the Nexus Vehicle," SAE Paper 880270, 1988.
55. E. Erdle, W. Dönitz, R. Schamm, and A. Koch, "Reversibility and Polarization Behaviour of High Temperature Solid Oxide Electrochemical Cells," *Proc. World Hydrogen Energy Conf. 1990.*
56. W. J. Comfort, S. E. Bumpas, T. A. Edmunds, A. R. Hall, A. D. Lamont, H. K. McCue, and E. Zywicz, "Feasibility Assessment of Electromechanical Batteries for Electric Vehicles," UCRL-ID-109422, May 1992, Lawrence Livermore National Laboratory, Livermore, Calif.
57. L. Weaver, Video and analyses provided 11 May 1992 (14 Home Place, Topsham, Maine 04086).
58. L. Evans, *Traffic Safety and the Driver,* Van Nostrand Reinhold, New York, 1991.
59. R. Käser, "Safety Potential of Urban Electric Vehicles in Collisions," *Proc. Urban Electric Vehicle Conf. (25 May 1992),* OECD/NUTEK/IEA, Stockholm.
60. C. Kindervater, "Composite Structural Crash Resistance," Institute for Structures and Design, Deutsche Forschungsanstalt für Luft- und Raumfahrt (Stuttgart), presented at 3–5 June 1991 DLR Crash User Seminar, Stuttgart.
61. W. S. Abernathy and J. M. Utterback, "Patterns of Industrial Innovation," *Tech. Rev.* **80**(7), 90–98, June–July 1978.
62. G. P. Zachary and S. K. Yoder, "Order from Chaos," *Wall St. J.,* 27 January 1993, p. A1.
63. J. J. MacKenzie, R. C. Dower, and D. D. T. Chen, *The Going Rate: What It Really Costs to Drive,* World Resources Institute, Washington D.C.), 1992.
64. B. Ketcham and C. Komanoff, "Win-Win Transportation, A No-Losers Approach to Financing Transport in New York City and the Region," Transportation Alternatives, New York, 1992.
65. J. Romm and A. B. Lovins, "Fueling a Competitive Economy," *Foreign Affairs* **71**(5), 46–62 (Winter 1992/93); unabridged version. "Making Security Profitable," available from Rocky Mountain Institute (5).
66. L. Schipper, R. Steiner, M. J. Figueroa, and K. Dolan, "Fuel Prices, Automobile Fuel Economy, and Fuel Use for Land Travel: Preliminary Findings from an International Comparison," LBL-32699, Lawrence Berkeley Laboratory, 1992.
67. K. Dolan, B. E. and B. G. Andersson, H. Nishimake, L. Schipper, R. Steiner, and W. Tax, *Fiscal Policies Affecting Automobiles in W. Europe, Japan, and the United States,* LBL draft report originally prepared as an appendix to Ref. 66, 1993.
68. D. Greene, "CAFE or Price?: An Analysis of the Effects of Federal Fuel Economy Regulations and Gasoline Prices on New Car MPG, 1978–89," Oak Ridge National Laboratory/USDOE Office of Policy, Planning and Analysis draft, 10 May 1989.
69. D. Sperling, L. Schipper, and M. DeLuchi, "Is There an Electric Vehicle Future?" *Proc. Urban Electric Vehicle Conf. (27 May 1992),* OECD/NUTEK/IEA, Stockholm.

70. S. Weissman and J. Corbett, *Land Use Strategies for More Livable Places,* Local Government Commission (909 12th St., Suite 201, Sacramento, California 95814), 1992.
71. P. Newman and J. Kenworthy with L. Robinson, *Winning Back the Cities,* Pluto Press Australia Ltd., Leichhardt, NSW, 1992.
72. M. M. El-Gassier, "The Potential Benefits and Workability of Pay-As-You-Drive Automobile Insurance," June 8, 1990 testimony to California Energy Commission (Sacramento), Docket #89-CR-90.
73. A. Tobias, *Auto Insurance Alert! Why the System Stinks, How to Fix It, and What to Do in the Meantime,* Simon & Schuster, New York, 1993.
74. W. Zuckerman, *End of the Road: The World Car Crisis and How We Can Solve It,* Lutterworth Cambridge, Mass., and Chelsea Green, Post Mills, Vt., 1991.
75. J. Womack, *The Machine That Changed the World,* Macmillan/Rawson, New York, 1990, and 1991 paperback edition, HarperPerennial, New York.

SUSTAINABLE RESOURCES, ETHICS

DONALD SCHERER
Bowling Green State University
Bowling Green, Ohio

An article on the ethics of sustainable energy requires elaboration of what sustainability means and clarification of what an ethic is. On these bases an evaluation of energy choices can build.

Heraclitus, the pre-Socratic philosopher, cryptically pronounced that one cannot step into the same river twice. The Heraclitian view has been that, after the step is taken, neither the river nor the one who steps, is ever again the same. Philosophers thereafter have taken his paradoxical pronouncement as a challenge to the concept and criteria of identity or sameness. From Plato's forms to Wittgenstein's aphorism that "same" does not mean the same, philosophers have sought to understand sameness. Heraclitus' remark is pertinent to understanding sustainability. To call an energy *choice* sustainable is to assert that persons like us can make that same energy choice continually and indefinitely. To claim that a *lifestyle* is sustainable is to claim that the impacts of that lifestyle cause no degeneration of the environment incompatible with continuation of that lifestyle. To designate water a sustainable *resource* is to presume its reusability. To affirm that biomass is a renewable and hence sustainable energy *resource* is to assert that, within the limits of regeneration, biomass can continue to be available for human use. In all these formulations sustainability implies at least that what we do will not constrain others like us from carrying on as we have. That human beings ought not act in any such way as to constrain future human beings from acting in that same way is sometimes affirmed as the basic obligation to future generations underlying an ethic of sustainability. Our lives should not diminish theirs. The same choices, the same lifestyle, the same resources ought to be available to them as to us.

See also ENVIRONMENTAL ECONOMICS; LIFE CYCLE ANALYSIS; ENERGY EFFICIENCY; FORESTRY, SUSTAINABLE; RISK ANALYSIS.

Sustainability of Choice

The idea that choices should be sustainable opens this discussion. The most obvious way in which an energy choice might not be sustainable would be if the processes of making energy available consumed materials faster than such materials accumulate. In different parts of the world, for example, people continue to burn wood and other forms of biomass more quickly than the biomass grows, rendering their choice and use of biomass nonsustainable.

Within this mode of thinking, sustainability is conceptually tied to renewability. Through the water cycle, water is transported to higher places so that turbines placed in the mist of falling water can continue to turn, generating electricity. Calling hydroelectric power a sustainable energy choice assumes a minimum rate of flow within a river, the rate sustained in even the driest season. The choice of electric energy generated from falling water will not constrain our successors from making that choice later on, up to the amount generated by that minimum flow.

Scale. An immediate corollary of our example is that even sustainable energy is limited. The sustainability of an energy choice depends on the rate of the renewability of the energy supply. The theoretical limit of recoverable energy from hydroelectric power is set by the operation of the water cycle. Only below that limit is the choice of hydroelectric power sustainable. Similarly, the sustainability of a choice to burn biomass is relative to the size of a population and the efficiency of its energy conversion. Increased use of a forest, resulting from increased human population, increased per person use or decreased efficiency of use, can make a sustainable choice insustainable. Equivalently, decreased use of a forest, resulting from decreased human population, decreased per person use or increased efficiency, can make an insustainable choice sustainable.

Multipurpose Projects. A second implication of our example derives from the multipurpose nature of many human actions, for instance, the construction of large dams. Large dams are typically constructed not only to provide energy but to assure minimal water supply, provide flood protection and create lake recreation. Contention surrounding the issue of whether a choice is sustainability often arises from divergent foci on the purposes the choice serves. Siltation, for instance, may compromise the long-term use of the lake for recreation without compromising minimal water supply. Accordingly, a choice may be literally sustainable and insustainable at once, for different purposes. Whether a present-day choice is sustainable, therefore, often rests on different evaluations about which purposes need to continue to be pursued.

Unintended Consequences. The siltation and salination associated with large dams both illustrate another conceptual complexity of sustainability: the consequences of our choices are often unintended, and some unintended consequences are not benign. Certainly the United States never intended that its construction of dams would increase the salinity of water flowing from the United States into Mexico. Yet by the early 1970s increased salinity threatened Mexican agriculture, whatever the blessings of hydroenergy and an assured minimum water supply. The decision of the United States government to construct water desalinization facilities to clean water entering Mexico reflects the need to evaluate sustainability in terms of actual effects, not merely in terms of the intentions that frame a choice.

A further point emerges from combining considerations of scale, multipurpose projects and unintended consequences. *Thresholds and synergisms,* particularly in light of our *ignorance,* often render the aggregate of our choices insustainable. For example, trace salts, alone and by themselves, seldom affect crop fertility. The fact that the toxicity of a substance is defined as a concentration capable of producing a statistically significant harmful or lethal effect, reflects human awareness that equal increments of an accumulating substance are not equally consequential.

Synergisms are particularly troublesome because of their complexity and therefore because of our more likely ignorance of their workings. It is a standard fact of elementary chemistry that there is no proportion between the effects of relatively benign elements and of the compounds derived from them. When particularly active elements, like chlorine, are components of benign compounds released into an environmental commons, recombination there may produce dangerous compounds. Predicting synergistic effects is extremely difficult, given the lack of laboratory conditions, the novelty resulting from rapidly changing technologies and the variety of harms to which organisms and ecosystems are susceptible. If this complexity were not itself sufficient to engender significant ignorance of synergisms and their ecological effects, public ignorance is regularly increased because the commons contains the emissions and run-off products of many manufacturers, the value of whose manufacturing processes vouchsafes to them, within Anglo–American law, a right to keep secret chemical formulae intrinsic to their manufacture of products. Thus, if synergisms regularly compete with sustainability, the political value of liberty exacerbates public ignorance of their effects and minimizes public prospects for fortifying sustainability.

Negative Fecundity. For a particularly untoward set of consequences Jeremy Bentham (1748–1832) invented the term 'negative fecundity.' An act has negative fecundity when doing the act in order to achieve some effect today increases the difficulty of bringing about the same effect tomorrow (or some later time) through the same action. It is manifest that the increased expense of new generations of pesticides, required to contend effectively against pests selectively evolved for their resistance to older pesticides, has made protection against those pests more difficult. Likewise, the political difficulty of finding adequately safe long-term storage for spent nuclear fuels has increased the difficulty of siting additional fission plants.

From all the above criticisms we may conclude that sustainability is not a simple function of choice. When an individual makes a choice, the sustainability of that choice is a function of the environment in which the choice is made. How often is the choice being made? What other choices, as well as conditions of the physical environment, mitigate that choice? Does making the choice today increase or decrease the costs of repeating it tomorrow? Accordingly, the environment of a sustainable choice includes both the nonhuman environment in which conditions of regeneration occur and the human environment in which further resource-use choices are made.

In summary, then, speaking about sustainable choices is not well advised for, against this understanding, one might urge that a plurality of purposes can make a given choice both sustainable and insustainable, that choices have unintended consequences, and that insustainability, as derived from threshold and synergistic effects, is not directly the product of a given choice.

Sustainability of Lifestyles

It is often thought that the fundamental problem with defining sustainability in terms of choice is that reality is more systemic, less linear, than coheres well with the concept of the consequences of a choice. Defining sustainability in terms of lifestyles, gives sustainability an explicitly systemic character. If the accumulative effects of choices affect sustainability, then the concept of a lifestyle can incorporate the facts that societies are composed of many choice-making individuals, that individuals repeat their choices, and that many effects of choices arise from or are magnified by the organization of the society within which they arise.

A significant advantage of defining sustainability with reference to lifestyle is its focus on social organization. The distancing of residences from workplaces increases energy expended on transportation. A throw-away society creates problems of waste management that impinge its sustainability. The size of residences affects the costs of controlling their temperature. Stereotypes of exercise clothing and dress for work reinforce the inappropriateness of bicycling to work. Ultimately, sustainability, whether of a lifestyle, a resource or a society, is in significant part the sustainability of a pattern of practices. Practices, for example, that reinforce short time lines and impulse buying, thereby reinforce the inefficiency that threatens sustainability. Despite the advantages of holism, defining sustainability as the sustainability of a lifestyle has several disadvantages.

Holism and Choice. Whatever the disadvantages of a definition of sustainability in terms of choice, the existence of options enhances self-realization. Inversely, marketing bundles human choices in packages. On speculation a builder constructs a new home. When a customer buys the home, the customer may feel that the choice made is of the features of the home, not its location. Or the customer may feel that the choice is one of living space, not of insulation and heating system. Yet the builder speculated that the money paid for the package of features enhanced the ability of the new home to be sold. In other words, people's options are typically concrete, encompassing many factors outside the focus of choice. The ability of a product in a market to be sold reflects the holistic appeal of the product. Thus although customer appeal and sustainability both require holistic definitions, nothing guarantees their coherence with one another.

Fascism. The very holism that makes a definition of sustainability in terms of lifestyle attractive carries with it the distinct disadvantage that holism can be antiliberal. If a lifestyle is a style of a society, then maintenance of the lifestyle is compatible with the repulsive treatment of some (dispensable) individuals. Air-pollution-induced emphysema, landfill-leached disease and water-pollution-induced birth defects are all compatible with a society's maintaining a lifestyle. Thus the objection here to defining sustainability in terms of lifestyle is that a lifestyle may produce as its by-product significant hazards for disadvantaged populations. The holism of the definition then encourages a society to look past so basic a right as an individual's right to health, not in the strong sense that others are obliged to provide health care for the needy, but in the very weak sense that others do not have the right to impair or destroy one's health, particularly without one's consent.

Size of Cohort. A lifestyle is not rigidly defined by the number of its participants. Yet the number of participants may be crucial to the sustainability of the lifestyle. The importance of thresholds is that it is often the aggregate concentration of substances that makes them effective. Inversely, some pollutant, benign at a given concentration, may be lethal at another. The greatest mid-range challenge to the sustainability of a western affluent lifestyle will come from the Asian rim, particularly from China, and its attempt to import consumption-driven markets. The adoption of the automobile in Taiwan and Korea and the Chinese (not to mention Eastern European) use of dirty coal flag the problem of cohort-size for sustainability. When 5% of the human population consumes approximately 30% of the human energy supply and emits over 20% of the world's anthropogenic pollutants, the adoption of the lifestyle by increasingly larger cohorts strains thresholds.

Energy Sources. This strain is particularly awkward because of the character of prevalent energy sources. Affluent lifestyles are built largely on nonrenewable energy sources while many marginal lifestyles rely centrally on insustainable use rates of biomass. A nonrenewable energy source is by definition not a sustainable source. To call a lifestyle built on nonrenewable energy a sustainable lifestyle is to postulate that an adequate functional substitute will be found when the nonrenewable resource is exhausted. That is, the lifestyle is sustainable only if the substitute that replaces the nonrenewable energy source continues to allow the energy expenditures the lifestyle requires. To say that calling affluent lifestyles sustainable involves a postulate is a euphemistic way of saying that calling them sustainable involves a perhaps self-indulgent act of faith or, worse, a self-deceptive delusion.

Substitution. Focusing on lifestyle sustainability draws the focus away from the materials and processes that support that lifestyle. These processes may undermine their own continuation, as when rates of use vastly exceed rates of renewability or when pollution emissions destroy resources or undermine health. In such cases lifestyle maintenance is likely to require substitution of materials, the modification of processes and the redistribution of human well-being. A limitation of a definition of sustainability in terms of lifestyle is that it obscures what such adjustments might be and how they take place.

Privileged Persons. Defining sustainability in terms of lifestyle invites the question, "Whose lifestyle?" Advocates will exercise their political ideologies to the extent that a particular lifestyle advantages some race or class or other social grouping. Egalitarian criticisms of traditional treatments of women make points both about women's self-respect and about the costs to human society of the suppression of women. Not only are energies spent reinforcing the suppression but the advantages of the full expression of women's capabilities is lost to the society. Similarly, criticisms arise concerning the developing world. The export of repetitious manufacturing jobs from developed countries, while integrating the workers of developing countries into the world economy, also curtails the roles and opportunities of those workers.

Anthropocentrism. Beyond these issues of intraspecies comparison lies the privileging of the human. Whosever lifestyle the definition addresses is the lifestyle of some cohort of human beings. A definition of sustainability in terms of lifestyle captures some of the holism of human life, but it ignores ecosystemic holism. The sustainability of human energy use, however, rests not simply on lifestyle factors but on renewability rates and the effects of pollutants on nonhuman systems upon which humans ultimately rely. Edward Wilson's hypothesis that the sixth great extinction in the history of life is human-driven casts doubt on the wisdom of defining sustainability in terms of lifestyle.

Sustainability, then, is not well-defined in terms of lifestyle. For such a definition at least controversially, if not illicitly, relies on implicit assumptions about whom the lifestyle privileges. Such a bias makes the use of such a definition of sustainability an often powerful rhetorical ploy. Yet it also undermines its ability to justify the treatment it prescribes for the lives falling outside its privilege. Social and natural feedback systems imply that the long-term consequence of the use of such a definition is both the felt need to suppress the negative feedback the privileging generates and the long-term necessity of adjusting to the rhythms of a more encompassing system than the lifestyle acknowledges.

Sustainability of Resources

When sustainability is understood as the sustainability of neither choices nor lifestyles, discussion may be recast as discussion about the sustainability of a resource. Within the framework a resource is in the broadest sense any material for which a use is established. The material is then a resource for that use. Similarly, a material for which no use is established is at most and at least a potential resource. That is, while the absence of an established use must rather undermine the claim that the material is a resource, the undeniable possibility of devising and establishing a use leaves open the possibility of the material becoming a resource. By the same temporal relatively, much of today's wastes are materials for which uses have ceased being established.

It is typical of human beings that they often take the term "resource" to mean "resource for human beings." Since nothing in the preceding account implies that resources are resources for human beings, the assumption that "resource" is an anthropocentric concept is at once plausible and challengeable. In any event, certain understandings underlie discussion of the sustainability of resources:

Relativity to Use. Sustainability, when glossed as resource sustainability, retains significant relativity. That is, when a material is called a resource, we mean that the material is effective when put to a particular use. The Ollagala Reservoir, beneath much of the Great Plains, is replenished quickly enough to provide drinking water for far more human beings than currently live above it. Yet Great Plains agriculture is arguably unsustainable because of its threatened depletion of the reservoir. Thus, the simple term "resource" can hide from us that a material can be both renewable and nonrenewable, sustainable and unsustainable, as the material is put to different uses.

Even if one confines the view of materials to the energy they can provide, a multiplicity of uses can obscure discussion about sustainability. Energy is used for transportation and for temperature control, for manufacturing and in agriculture. Whether a resource is renewable is a function of the rate of total consumption of the material, not of a particular use to which it is put.

Sustainability of Resources vs Sustainability of Choices. An advantage of defining sustainability as the sustainability of resource uses, rather than the repeatability of choices, arises in the case where a choice may seem sustainable, but the resource may seem unsustainable. Consider fissionable fuels. The choice of fissionable materials as fuels would probably by itself be sustainable for a very long time. While the problem of the continued use of radioactive fuels is in part a practical and political problem of whether safe long-term use can be provided, beyond that problem is the potential problem that fissionable materials potentially refinable into fuels, have other uses, say in armaments. In this way it is possible that a (theoretically sustainable) resource becomes unavailable because a competing and nonsustainable use of the material undermines the market within which the theoretically sustainable resource exists.

Inelasticity of Supply. When it is said that the theoretical limit of hydroelectric power is set by the water cycle, it is often assumed that "the rain falleth where it will," beyond the control of human beings to increase. Likewise, when it is said that the theoretical limit to the use of lumber is the rate at which trees grow, it has similarly been assumed that climatic conditions beyond human control determine tree growth. Contrarily, of course, we have the ability to predict, principally from the age of a tree, its species and the light available to its crown, by how many board feet it will grow in a given time span. Since the mean age of trees in a plot, the species of those trees and the mean light available to each of their crowns are all subject to human calculation and manipulation over time, theoretical limits to the use of lumber have been exceeded in many terrains.

Efficiency of Use. The board feet of timber available from a forest is also relative to the efficiency with which a tree is converted to lumber. The use of computers to calculate the most productive cut of the wood and the development of saw blades that create less saw dust per board foot have both increased production beyond increases in timber supply. The considerable gains in motor efficiency and building insulation that followed the energy crises of the 1970s, notwithstanding, inefficiency of energy use remains a considerable obstacle to sustainability.

Sustainability and Substitutability

Energy from different sources is renewable at different rates. When the rate of consumption vastly exceeds (in orders of magnitude) the rate of renewal, as it does for the common fossil fuels, the energy resource is called nonrenewable. The assumption in this assertion is that efficiencies neither of new technologies nor of social organization will suffice to reduce the resource's rate of consumption to the rate of its renewal.

A significant complexity of the discussion of sustainability has arisen from the reality of substituting one material for another. In the 1840s, for example, whale oil was being used more quickly than whales reproduce. The rate of the resource's use was not sustainable, given the relative constancy of supply. In this circumstance the result was neither the extermination of whales nor the curtailment of oil use. Instead, crude oil was discovered at Titusville, Pa., whaling for whale oil declined, and petroleum use increased. This has been the kind of case in virtue of which economists rest contentions of sustainability not on reuse but on substitution.

Discussion at this point often turns to discounting. Financial or technical resources invested, at a given point in time, in extending the known supply of a form of energy, represent opportunities diverted from other uses of those resources. It is standardly argued that knowing the availability of a form of energy is valuable only because of the time required to devise technologies for accessing that energy and converting it to usable form. In this way, knowing of more than a twenty year supply of petroleum is regularly thought to be worthless. Accordingly, it is inferred that the fact that the known supply of petroleum (at present rates of use) will not last a generation reflects the discounted value of such knowledge, not the absolute shortness of the supply of the energy resource.

The twin arguments of substitutability and discounting are often combined. The argument from discounting is thought to show that supply probably substantially exceeds known supply, and the argument from substitutability is used to conclude that the case in which supply is actually limited to known supply will only occasion the development of new technologies that convert what are now useless materials into tomorrow's resources. In this vein it has even been argued that the present generation's landfills, far from marking the decadence of consumerism, amount instead to the storage of inventoried materials whose recovery will be inexpensive when their future uses become established.

Sustainability and Justice. One of the ethical shortcomings of that argument comes into focus when we shift our attention from cities to rural areas and from countries that use nonrenewable energy resources intensively to countries that rely much more centrally on human labor. When we shift our attention from the center to the peripheries of economic power, we encounter disease, poverty and people who lack opportunities to live in the style either of their parents or of their powerful contemporaries. The likelihood of the siting of hazardous waste in a given area is known to be a function of the race and poverty of its human inhabitants. The effect of a given resource shortage on a particular local human population is a function of the strength of the network of other resources available to that population.

In light of the differential effects of resource shortages on human populations, some rethinking of the concept of sustainability is appropriate. "We should not imperil the availability for future generations of what we have available to us now." When is "now" and who is "us"? Is there a future population that is less deserving of what more affluent societies now have available to them?

Here the ethics of sustainable energy become an ethics of the just use of power. In its starkest form, the danger is that the concept of sustainable energy itself is loaded. The very concept comes under scrutiny from the suspicion that concern about sustainability is no more than a concern of an elite to retain its privileged status.

Justice is an issue of the use of power. Accordingly if sustainability is to escape the label of an ideological concept that imposes the legitimacy of the interests of an established class, then at some level just patterns of resource use or changes in the pattern of resource use must be patterns or changes acceptable to the least powerful. That is, acceptable uses of power among competent adult human beings are largely confined to uses that have the consent of those upon whom power is wielded. Accordingly, there is nothing just about the (forced) imposition of patterns of resource use.

Although this line of argument seems plausible as far as it goes, it leaves significant questions unresolved. Suppose a society establishes a set of conventions or a code of law governing the rightful possession and use of things, in a word, property. Suppose that members of that society are amenable to what they have established, to a set of rules of acquisition and transfer, say. Imagine that those rules include rules governing inheritance. Imagine also that the rules include models of what materials are useful (resources, as we say), models the society outgrows over a period of generations. What facts and principles then ground the justification of the agreement? Who all is bound by this original agreement? Under what conditions, by whom, and to what ends, can it legitimately be modified?

The conservative value of social order is that much human coordination presupposes a large degree, but not a particular definition, of social order. Driving on the right is not better than driving on the left, but having everyone driving on the same side maintains life, health, and property values. Changes in the ongoing definition of an established order always occasions some disruption of coordination mutually beneficial for the parties directly involved. When right turn on red first became the default norm in the United States, pedestrian deaths at intersections rose. At least somewhat contrary to this good, regularly lies the point that certain changes in the ongoing definition of that order would clearly benefit society in general or at least one of the parties directly involved without clear harm, at least in the long run, to anyone. United States pedestrian deaths have since declined and average mileage is increased because time spent idling at intersections is reduced. The point–counterpoint of these assertions defines much of the battle ground for conventional aspects of what justice amounts to.

Justice and the Meaning of Opportunity. The frequent context of these questions has been the still-influential view of John Locke (1632–1702) on property. "In the beginning," Locke said, "all the world was America." By this statement, Locke contrasted England with the new world. In 1661 England had passed the Enclosure Laws that remain to this day the legal foundation of property law in the Anglo–American, liberal tradition. In 1689 when Locke wrote, England was a land parceled into plots to which owners had titles. If one wished to acquire land in England, one's only option was to buy it from its owner. Such was the implication of earlier claims and enacted laws of transfer.

America, as Locke saw it, was a land to which, until recently, no property claims had been made. For in Locke's eye, one can claim previously unworked land as one's own when and only when the value of the land has become, substantially, the product of the labor one has invested in it. Since Locke believed that all American Indians were nomads, he saw them lacking an agriculturalist's investment in, and hence the agriculturalist's claim to, the land. In Locke's eyes, then, America, being inhabited only by nomads (here Locke believed a false premise), was an unclaimed land.

Locke's conclusion was that people come to own previously unowned land by the investment of their labor in the land. Locke reasoned as follows: 90 to 99% of the value of cultivated land has derived from its having been cultivated. Thus, almost nothing of what one claimed after such work would have been available except for one's labor. Indeed inasmuch as the basis of the claim was the creation of value resulting from cultivation, others have been benefited if, because they have also used their labor to create value, they can exchange the value they have created for the value others create and they lack. Therefore, provided that one leaves as much and as good for others (so that they are not deprived of the opportunity to labor and to enjoy the fruits of their labor), the claim of the land as property benefited the claimant (by securing value the claimant has created) without harming anyone else.

This formulation of the view emphasizes Locke's reliance on America as a frontier open for future claimants. The virtual closing of that frontier in the late nineteenth century therefore reopened the question of justice and the meaning of opportunity. The Square Deal, the New Deal, the Fair Deal, and the New Frontier are the prominent slogans of attempts in the United States to address that question.

Of these, Theodore Roosevelt's Square Deal, Truman's Fair Deal, and the Kennedy–Johnson years are central for understanding the meaning 'sustainability' has acquired in the United States. For Roosevelt, the closing of the frontier and the ongoing importance of equal opportunity imply the need for policies that sustain opportunities for future generations. Where Jefferson at the outset of the nineteenth century had inferred from Locke that he ought to buy the Louisiana Territory to secure agricultural opportunities for future generations, Roosevelt inferred at the outset of the twentieth century that efficiency in resource use, through conservation and technological innovation, were required for the United States to succeed in the impending competition among nations. A significant continuing relevance of Roosevelt's view is that the frontiers of energy efficiency and renewable energy remain substantially open.

The relevant plank of Truman's Fair Deal was the GI Bill that created an entitlement to a college education for World War II veterans. The use of this entitlement marked the beginning of mass higher education. Where before the war, the percentage of college educated Americans resembled the percentage of college educated (largely upper class) Britons, enrollments in United Stages colleges after the war began a 500% increase that has made the United States unique in enrolling over half its population in post-secondary education in the second half of the twentieth century. This development created knowledge acquisition as a new frontier, breathing new life into the meaning of equality of opportunity. In the 1980s countries as diverse as Taiwan and Argentina began following the United States in the pursuit of that frontier.

The Kennedy–Johnson years contribute dialectically to the contemporary meaning of sustainability. Outer space has been a new frontier because going to the moon can provide new resources and new platforms on which to develop technology, in a word, new opportunities. The meaning of opportunities, however, has vastly changed. Science education has been seen as vital to prepare people for work on the new frontier so that the frontier will itself become a frontier for the development of further scientific and technical knowledge.

The progressivism of the Roosevelt administration, in other words, has rested on the importance of maintaining equality as an American ideal. New frontiers, whether in education, technology, space, the sea bottom, battery technology or solar energy, are similarly focused on sustaining opportunity by increasing the production of value. Questions of the justice of previous generations' conventions of property are often effectively postponed by the demonstration that their further employment retains promise for the less fortunate.

Sustainability and the Well-Being of Ecosystems. As straightforwardly as these ideas develop Roosevelt's vision, the meaning of 'sustainability' also developed during the Kennedy–Johnson years out of an emerging ecological understanding of both conservation and wilderness preservation. Although late-nineteenth century preservationism had rested on maintaining vestiges of wilderness experience, natural curiosities and the "people's temples" with their remarkable aesthetic–religious appeal, early attempts at applying the preservation ethic in Yellowstone National Park began to convince field biologists that their work required systemic understandings, just as Aldo Leopold reached this conclusion from his early attempts to apply the conservation ethic to forestry and ranching in New Mexico and Arizona. Although Stewart Udall began his tenure as Secretary of the Interior with many older conservationistic and preservationistic attitudes, the Secretary became the champion of more ecological approaches. Since that time "sustainability" has tended to reflect an understanding of the systemic connections that sustain the components of ecosystems we call our resources.

This ecosystemic emphasis, however, does not fit easily with a Lockean conception of property. Most fundamentally, ecology has been called a subversive science because so influential an ecologist as Aldo Leopold seems committed to the view that it makes as much sense to speak of the good of the ecological community inhabiting a mountainside as it does to speak of the good of a human being. An overpopulation of deer, brought about by the human extermination of the deer's predators, harms the mountain, Leopold has urgued, as surely as an injury to one's body harms a human being.

Leopold concluded that we are no more justified in harming a mountainside than Odysseus was in hanging several slave girls on his return from Troy. While neither extant morality nor law gives more status to ecosystems than to slaves, Leopold saw the harm done to mountainsides and slave girls as undermining whatever conventions ignore and justify that harm.

Leopold's observation that extant morality and law did not condemn Odysseus' hanging of his slaves highlights a conflict concerning the property of persons: some hold that value exists precisely because people actually value something. If 1850s Lake Michigan lumbermen saw trees with ten inch trunk diameters as worthless hindrances to the harvesting of the larger-diametered trees Chicago lumberyard dealers were alone willing to buy at that time, then, with no voices raised in dissent, the trees were without value. If 1870s lumber men saw trees of that same size as the largest and most desirable of available trees, then the trees had acquired value. In accord with this mode of thinking, some economists argue that in circumstances of abundance, the destruction of potential resources is justified as often as it is efficient.

The alternative viewpoint is that all organisms (and perhaps ecosystems as well) have goods of their own. A suitable premise here is that organisms are disentropic entities: they absorb energy from their environment; they have capacities, unlike a stone, for maintaining a supply of usable energy, and they have mechanisms within them, unlike an automobile, through which, within limits, they maintain their own functionality. When an entity has both energy-absorbing and self-maintaining capacities it makes sense to speak of the good of the entity. The trees Chicago-based lumbermen saw as worthless in 1850 certainly had a good of their own. The soil that blew off Iowa farms in the 1960s, although not an organism with a good of its own, is a component of a prairie ecosystem that absorbs energy and, in some views, has capacities of self-maintenance. If any entity has a good of its own, then certainly impairing the capacities that underlie that good amount to harming the entity.

Humans continually learn to recognize harms previously invisible to them. The ability to recognize such harms, however, challenges sets of conventions for the acquisition and transfer of property. For "sustainability" within a philosophy oriented to human resources has often meant the sustainability of what the human beings treat as materials to be manipulated by the human beings for their own purposes and therefore independent of any so-called good of non-human entities. In this light, the challenge of Leopold's land ethic, as it is called, is that it is no more right wantonly to ignore the good of an ecosystem than it is to ignore the good of a slave.

The Good of a Person and the Good of an Ecosystem Compared. Maiming, by definition, is injuring in such a way as to cause the loss of some functional capacity. Since a forest, like any ecosystem, has functional capacities, maiming it is a real possibility. A forest has parts that cannot live independently of the forest as a whole anymore than one's leg continues to live after its amputation. The positive good realized in the maintaining of functional capacities implies that something undesirable and, all things being equal, wrong happens when a forest is maimed.

If, however, one is to look for duties of human beings towards ecosystems, those duties must be relative to the entities ecosystems are and thus to what the good of an ecosystem amounts to. The good of an ecosystem is, of course, far different from the good of an enslaved human being. Even if laws made it illegal to teach slaves to read (perhaps on the ground that they would be incapable of learning), human slaves, whose brains evolved as products of cohort-monitoring, are capable of directing their lives autonomously. A slave's degrading treatment by the master may demean the slave. The resultant loss of self-respect may impair the slave's capacities. Those underlying capacities, nevertheless, justify a respect for the slave, laws and cultural mores notwithstanding. Nothing like the centralized self-directing capacities of a human being, however, are present in an ecosystem.

The good of an ecosystem is in its functional capacities. Rather than the thwarting of a will, the thwarting of an ecosystem is the disassembly of its structure, the destruction of its resilience or the degradation of its diversity. It is these function-enabling capacities that justify human respect for them and their well-being. Accordingly, the well-guided tree-hugger hugs the tree not in the misguided conviction that a tree has the capacities and deserves the respect of a "similarly" mistreated human slave. Instead, the clear thinking tree-hugger urges protection of self-sustaining systems.

Both the magnificence of the integrated living system a forest is and the future resource value for human beings of forest components require respect for the integrity of that system. Edward Wilson has recently noted that human beings are dangerously ignorant of how they (frequently) precipitate the collapse of apparently stable but volatile ecosystems. At the same time, not all human use of forests maim them. The regenerative resilience of a forest, whatever its limits, is real. The reality of such resilience implies that the human use of resources is not categorically wrong. One does well to remember that, even in a world where no human beings live, resource relationships are implicit in the concepts of a water cycle or a nitrogen cycle or a food chain. It is not intrinsically wrong for an entity to be a resource. The hope that an encyclopedia of energy is a useful resource is not on its face an immoral hope. The teacher one most greatly respects may be one's mentor and one's most valuable resource. The fact that resources are often treated without respect does not imply that what is one's resource one uses without respect. Respect for an ecosystem is respect for the capacities that give it a good of its own, its structure, its resilience, its diversity.

The ecosystemic concept of sustainability reaches into the discussion of development. Once sustainability is defined ecosystemically, it is clear that many environments are degraded. They are functionally impaired. The question then arises whether the impairment can be corrected. Both the state of technology and the scale of costs immediately enter the discussion. It is, in general, clear that new development conjoined with restoration of the abandoned, previously developed area is more expensive, less feasible and regularly less successful than redevelopment of a previously developed area that leaves undeveloped areas thriving, or, more anthropocentrically, undeveloped. Thus, the sustainability of both ecosystems and of human life regularly favors redevelopment over new.

Sustainability, then, if understood ecosystemically, includes the recognition of goods other than human well-being and the resources that conduce thereto. Once these other goods are recognized, conditions of their stable maintenance exist. The recognition of these goods expands the meaning of 'sustainability' to include the conditions that assure that stable maintenance. Respect for those goods, along perhaps with prudence as well, requires aiming to maintain the conditions that those goods require. Accordingly, human actions thought right or at least permissible, on the grounds that they promote or at least do not harm human well-being, are arguably wrong if and when they contravene broadened, ecosystemic conditions of sustainability.

Ecosystemic Goods as Public Goods. The good of an ecosystem regularly involves public goods for human beings. Clean air and water are the most obvious and universal examples, but there are others. Recently much has been made of the fact that extinction involves loss of genetic information and of the ecosystemic setting in which the emergence of human live was possible. Similarly, the regular flow of energy relative to the rhythms of life promotes the stability and helps to maintain the diversity of complex ecosytems.

A standard social problem about public goods is that individuals will free ride, taking the benefit of the good without contributing to maintaining it, depleting without helping to restore it. Each individual may be tempted to think that an individual defection or contribution is insignificant. Each individual may be concerned that defections by others will leave the contributing individual a hapless sucker. When the good of the individual agent and the good of some larger entity diverge, the agent may act in rationally self-interested fashion. The interest of the larger entity will then move toward managing the options agents encounter.

Managing Common Resources. The management of common resources has sometimes been in government hands (international peace keeping forces, harbors, dams, roadways) and sometimes in the hands of government-regulated monopolies (public utilities, turnpike authorities). This management has grown out of the efficiencies surrounding capital investment for construction and out of the conviction that the community (world, national or local) would benefit, collectively and individually, from a coordinated investment.

What does not follow from these premises is that the resulting management will also be beneficial. To the contrary, a historically prominent result has been that from some egalitarian premise, the further conclusion has been drawn that all individuals are entitled to equal access to the investment and its potential fruits. Accordingly we see user fees in United States National Parks that cover only 7% of the cost of the parks; we see electric utility companies charging customers equally whether the utility's least or most efficient generator produces the customer's energy; we see public tax money subsidizing western cattlemen's access to public land and reservoir water; and we see neither urban companies nor their employees seeking to end rush-hour tie-ups by staggering work days and hours.

Sustainability requires efficiency. The management of common resources is often inefficient. Yet societies often secure individual and collective welfare through institutions that allow nonexclusive access to particular resources. By eliminating the cost of the roadway as part of decisions about travel, the provision of free public roads encourages personal transportation in private vehicles. By leaving unallocated the access to open-water fishing grounds, the provision of public harbors encourages overfishing of targeted areas. Typical features of unsustainably managed resources are that access to them is not borne specially by those who use them and that the benefits of overuse rebound primarily to the user with the costs of overuse distributed primarily to others, especially to future generations, and non-human organisms that suffer largely through habitat degradation, including pollution.

Economists and philosophers have noted, however, that both governmental and private action can effectively structure social environments so that individuals choose to promote, rather than diminish, the public good. The size and community structure of small town neighborhoods has made voluntary curb-side recycling much more successful than was widely predicted as recently as the late 1980s. If energy producers pay full costs on the pollution resulting from the use of some fossil fuel, the producers are induced to avoid the costs through energy production that avoids the pollution. Moreover, such structuring is not aimed merely at preventing losses. While clean air reduces respiratory disease, public goods can be used to generate further public goods. Clean water facilitates an attractive park. Genetic knowledge facilitates health.

The advantage of competition among suppliers is that consumers expand their choice beyond purchasing or not purchasing to a choice of purchases. An upshot of deregulation of the telephone industry, for instance, is that competing companies offer alternative products for alternative prices. Similarly, one might distinguish between the advantages, for the public, of a common provision for universal education and the disadvantages of the noncompetitive expenditure of tax revenue allocated to education.

Public Goods and Governmental Involvement. Public goods create accountability problems because their disrupters are not regularly the victims of the disruption they cause. The run-off from an open strip mine, for example, creates standard upstream/downstream problems. It does not follow, however, that government regulation, permits and fines, are typically required to police disrupters and protect victims. A quarter century of experience has taught the Environmental Protection Agency and environmental cabinet ministries in other countries that heavy regulation is not only prohibitively expensive but also often unnecessary.

The role of government as facilitator of public goods, therefore, takes on different meanings in different cases. (*1*) The years that utilities found capital tied up in unapproved fission plants taught the value of up-front agreements. Foresight about acceptable forms of action avoids both industrial consternation about invested capital and public concern about implementations that require excessive monitoring because of their embodiment of temptations for industry to choose between its own interests and a broader public or environmental interest. (*2*) Sometimes a one-time financial incentive provides the capital base for an ongoing, long-term pay-off. Thus, energy tax credits have been legislated to induce investment in insulation or efficient machinery. (*3*) Sometimes a good name is so significant an asset that support for the public good benefits the benefactor. For such reasons, ARCO, with its technological investments, has found taking a public spirited perspective beneficial. Indeed, certain other companies in the energy industry, lacking similar technologies, have found aping ARCO through token but visible environmental benevolence to be the best strategy available to them with an increasingly environmentally literate public. Clearly, efficiency in social systems argues that people should not fear government involvement per se, especially involvements that surely stabilize the availability and increase of public goods or the decrease or elimination of public bads. Entirely compatible with government involvement for the sake of public goods is the citizen concern of prudence to seek minimal, nonintrusive, one-time or short-term governmental involvements.

In summary, communities and their residents do benefit from the public availability of various goods. Yet common access to those goods often results in inefficient resource use. Therefore, one key to true sustainability lies in alternative management of public goods to avoid the inefficiencies of common access while maintaining the benefits of public availability. While different resources will require different management decisions, the goal of sustainability is to increase the efficiency of public availability. Although a common objection to competitive management is its effects on access for the most disadvantaged, a frequent result of competitive management is increases in efficiency more than offsetting increases in cost. Of course, since this correlation is less rigid than a law of nature, the endeavor to increase efficiency by constraining the terms of public availability must include a caution that constraining the terms of availability without improving efficiency amounts simply to a loss (at least in part) of public goods, typically by poorer individuals.

Proper Resource Pricing. The thesis that a market system produces efficiency involves the assumption that in a market prices reflect costs. Public goods, like clean air and clean water, are threatened and markets become less efficient and less supportive of sustainability when costs are externalized from the market. How, then, are unpriced costs reflected?

The costs of pollution are regularly borne by the polluted. That is, when an Ohio coal-burning electric utility feeds air-conditioning systems by emitting sulfur compounds that contaminate an area two to five hundred miles downwind, it is the lungs of people in New York and Ontario that sustain injury. Similarly, the cost of the emissions from an Alabama plant is a smoggy view of the Shenandoah Valley for summer visitors to the national park.

A "right" to emit pollutants into waterways or the atmosphere would amount to a right to impose costs upon others without their consent. Many of the norms of Anglo–American society, indeed of any human society, function to monitor, constrain and chastise impositions of harm, especially when they are nonconsensual. Without an argument that the harmed eventually benefit from the harm imposed upon them, their innocence would make it extremely difficult to justify any nonconsensual imposition of harm. While the presumption of a frontier society may be that the normal dilution of pollutants will nullify any toxic effect, the present-day intensity and multiplicity or nearby pollutants, especially new, artifactual pollutants, renders that presumption inappropriate.

A small but significant loophole in the preceding argument arises from thresholds. A right to emit pollutants into commons could be a right merely to emit pollutants below the threshold of their harmfulness. If a level of a pollutant can be vouchsafed as harmless to entities whose good would be affected by a greater emission, then the liberty to emit below-threshold amounts of sulfur dioxide and other effluents is a liberty so to act in ways that cause no harm. The Environmental Defense Fund was instrumental in the incorporation of such thinking into the "Emissions Trading" section (Title IV) of the Clean Air Act of 1990.

Justice, Opportunity, and Public Goods. The important relationship between social structure and public goods is typically obscured in frontier situations. Pollution is a central public bad as energy supply is a central public good for an interdependent society. But since pollution is a threshold phenomenon, it is often not critical on a spatial frontier. "Six miles of free-flowing water purifies itself," is an oft-cited rule of thumb. Yet while frontier experiences were forming the national psyche of maximal liberty combined with a minimal conception of harm in late nineteenth century United States, the sophisticated observation of settled life in eastern United States was moving towards different norms.

Ellen Swallow Richards (1842–1911) was at the forefront of much of this movement. Early Massachusetts efforts to establish standards of clean water, early efforts to develop a science of air standards, and early efforts to establish sound practices of public health, glowed from the luster of her pioneering efforts. She matched her keen awareness of physical endangerment of public environmental goods not only with her own tireless and energetic devotion but also with her early mobilization of educated women and of home economists as sources of knowledge and expertise that needed to be structured so as to facilitate public goods. Perhaps through her efforts, more than any one else, there was established a heritage of concern for environmental public goods, a heritage devoted to articulating the elaborate interplay between danger to people, especially less fortunate people, harm to people and harm to non-human life. Great women from Rachel Carson (1907–1964) to Kristin Shrader-Frechette, with her work on siting hazardous wastes, and Theo Colburn, with her work on genetically linked degeneracies arising out of organochlorides in the food chain, have advanced the line Ellen Swallow Richards first developed.

Pollution: Prevention versus Clean-Up. Once pollution is distributed into a fluid medium, isolating and removing pollutants is extremely costly and not very effective. Analyses comparing the costs of preventing the emission of pollutants with the costs of cleaning up after such emissions, regularly conclude the preferability of prevention. The right of the society to avoid, for itself and for its members, the public bad of pollution suggests a rationale for resource pricing that internalizes the costs of appropriate disposal of unwanted by products of production.

For example, an electric utility that has been mandated to pay the cost of disposing of unwanted by-products of its production of electricity. The cost of this disposal now increases the price of the electricity. A monopoly may succeed in passing those costs along to customers, for the monopoly will argue to its regulators that, with the increase of its legitimate costs, it is entitled to an increase in price to sustain its profit margin. If that utility is in a competitive market, however, it must weigh the advantage of alternative means of internalizing its costs. The utility must at least consider the possibilities of cleaning gaseous by-products before emitting them versus compensating victims of long-term emissions.

An enterprising utility will explore further possibilities. Efficiencies will result if cheaper resources can be used to generate electricity. Efficiencies will result if the by-product is eliminated from the generation process. Profits will rise if, instead of the polluting byproduct, a salable product is produced. Imagine, then, a process of converting high sulfur coal in conjunction with such landfilled wastes as automobile tires. Let high temperatures, in the absence of oxygen, reduce complex hydrocarbons in coal to hydrogen and carbon monoxide. Let the by-products of the conversion, rather than any polluting gas, be a solid useful for paving highways and making construction blocks. Imagine that utility companies can burn the produced gases with no need to use post-combustion pollution controls. With an entrepreneurial enterprise like Calderon Energy Company seriously working with utility companies to bring such a process to market, we see fleshed out a promising prospect: even with huge externalized costs in much of the electricity generated from high sulfur coal, a process could be profitable if it reduced what would otherwise become a waste product, to usable forms of energy, created salable byproducts rather than unsalable pollutants and converted otherwise landfilled wastes into components of such salable products. A society that mandated the morally appropriate internalization of the costs of producing electricity from high sulfur coal would thereby speed the technological, and enhance the economic success of such processes.

Even when prices reflect full costs, resources may be improperly priced because different units of the resource are not priced relative to their different costs. For example, the most expensive energy an electric utility generates is that which it generates in middle afternoon on a hot summer weekday. Then, when demand is highest, the company will use its less efficient generators and the average cost per-watt hour will climb. Suppose, however, that pricing does not reflect this cost. For instance, a large user, whose use is increased by air-conditioning demand, may actually pay less if billed at a decreasing block rate. In many other areas flat rates are standard. In any such case, improper pricing is failing to signal to the customer the reasonability of avoiding peak time use. Producers and investors also receive faulty signals from improper price structures.

The most dramatic proof of the efficacy of pricing structures resulted from the "energy crises" of 1973 and 1979. The steeply increased price of energy signaled consumers to improve the efficiency of their energy use. It, "accordingly," signaled producers and investors to develop and market improved efficiency. From insulation to motor efficiency, from temperature controls to construction techniques, long-term efficiencies have resulted.

Sustainability and Reusability

Using a reusable resource is equivalent to making use of it, with the presumption that such use is at least compatible with the reuse of the resource. The presumption in such cases is that using the resource is not equivalent to using it up; that is, transforming it so that it cannot again function as the resource it once was. Water and atmospheric gases are said to be sustainable resources because they are deemed indefinitely reusable.

The value of the reuse of a product, as opposed to recycling of a material, is often noted in this context. The reuse of a cloth shopping bag implies almost no degeneration of the product over time. The recycling of a paper or plastic shopping bag costs tens to hundreds of dollars per ton. Such costs overshadow such cleaning or transportation costs as reuse often involves. Consider water used in cleaning. When water scarcity grows, as in a long-lasting California drought, a processor using water to clean reusable bottles may abandon the practice of disposing of gray water. Rational self-interest demands the reuse of gray water if the scarce supply of water makes cleaning and disinfecting the gray water cheaper than procuring a new supply. Theoretically, the loop of the water, through cleaning bottles, to being cleaned, to cleaning more bottles, is unending.

Similarly, if more complexly, the nitrogen cycle is a cycle of nitrogen from being free in the atmosphere to being fixed in processes of plant growth, perhaps to be tracked into soils, perhaps to be tracked through animals, but eventually to return in free form to the atmosphere, undegraded in its ability to participate in the cycle. Accordingly, one of the worries about green house gases is that oxides of nitrogen will accumulate in the atmosphere. Does such an accumulation interfere with the functioning of the nitrogen cycle? If so, insustainability is threatened in the form of degradation of theoretically reusable resources. Even if not, the concept of global warming rests in part on the assumption that deflecting nitrogen out of the cycle threatens sustainability by violating temperature parameters. Accordingly, we see that the harm in the disruption of reusability may result not only from threatened scarcity of the consumed or diverted resource but also from the transformation of the reusable resource into a pollutant that itself threatens sustainability.

Sustainability and Limits of Reusability Important as resue is to the promotion of sustainability, unending cycles of reuse are of course a theoretical ideal only approachable in practice. The degree to which recycling reduces the average fiber length of recycled paper quantifies the degradation of the paper and thus the limit of its reuse through recycling. In such cases reuse extends the life of a material without indefinitely, much less unendingly, sustaining the material as a resource.

The use of motor oil as a lubricant further illuminates the practical limits of reusability. Just as a scarcity of water may prompt its cleaning and reuse, so the supply of motor oil, in conjunction with the polluting effects of non-reused motor oil, have led to its cleaning and reuse. And, as some nitorgen may be diverted from its cycle into accumulating nitrogen oxides, so some motor oil may be burned or lose viscosity within an engine.

With lubricants, however, a further serious concern is dissipation. Dissipation here is essentially a distributional concept. The distribution of a lubricant throughout a space reduces the amount of lubricant at any one place. The more accessible that space is, the greater the potential for contamination. Both the distribution and the potential contamination increase the cost of recollecting the lubricant for cleaning or for reuse.

Evaluating Whether Resource Loss is Superfluous. These examples of reusability all suggest that actual conditions of use significantly compromise theoretical reusability. The question whether these compromises are significant for the sustainability of resources is often discussed with reference to the example of agricultural practice and soil erosion. Certainly much of Iowa once had more top soil than sustainable agricultural practice requires. The initial abundance of the resource, however, meant that markets did not protect against the early, casual discarding of 'superfluous' materials. Under these conditions loss of some top soil was thought "harmless" on the presumption that the erosive practices would be curtailed and replaced by practices that used top soil up only as fast as new soil accumulates. The advent of low- and no-till plowing has significantly lessened soil erosion since the early 1980s. Familiar, then, is the pattern of resource use by which an abundant and theoretically reusable resource is at first degradingly used and is only later the subject of conservation efforts. Many economists have assumed that, under such conditions, the early unsustainable practice is morally benign.

As degradation of a given resource becomes more threatening to property owners or others who benefit directly from the sustainability of the resource, many have hoped, argued, or trusted that those persons with direct interests will move to ensure reusability. Although the recent history of forestry is often cited as a case in point, real problems remain in replanting some of the steeper harvested slopes or Northwest forest land. Therefore, whether an early "unsustainable" practice is actually morally benign rests upon a falsifiable assumption: that a hypothesized future sustainability will be achievable. Some risk is always involved in such a judgment, a risk likely borne largely by those who are not well off, or freely and easily borne by those who are affluent.

The argument that the loss of Iowa topsoil is superfluous because less topsoil is adequate for agricultural needs also rests on the assumption that the value of Iowa topsoil is confined to its agricultural value for human beings. The "superfluous" loss of materials may be harmful in each of three ways in which this assumption can be challenged. It can be challenged by noting that increased frequency of floods on the Mississippi River between 1950 and 1980 resulted in part from increased siltation of the river itself arising from traditional post-war agriculture. Even when a material is superfluous as a resource, it may be dangerous as a pollutant. The assumption can also be challenged by noting that topsoil may take on nonagricultural value for human beings. Topsoil loss, for example, may curtail shade tree growth. Finally, the assumption can be challenged as anthropocentric. The topsoil is beneficial for the ecosystem it supports. Hundreds of species are benefited by the topsoil and harmed by its loss. A maimed ecosystem is weakened in its diversity and stability.

Resource Reusability and Ecosystem Sustainability. Even though many resources that are theoretically indefinitely reusable are in fact to some extent degraded through use, the question of the sustainbility of those resources remains open. For with respect to many reusable resources, ecosystem sustainability is key to resource sustainability. Slowly, atmospheric pollutants precipitate out, wetlands purify waters, soil accumulates, and trees grow. Therefore, many resources can be refreshed or renewed. Accordingly, the sustainability of many reusable resources rests ultimately on a complex social fact: the relationship between demand, (partially degrading) reuse, and renewal (contravened or facilitated by humans). Lower demand, minimized degradation, and swifter renewal are the friends of sustainability.

The sustainability of a useful material, accordingly, cannot be a function simply of the physical properties of the material and its 'inherent' capacities for reuse or renewal. Contrarily, the concept of sustainability builds on the concept of availability. Rather than a question of theoretical physics, the question of energy sustainability is the question of the human capacity to use some portion of the potential energy in some material. This statement may be either a question of material technology or of social organization. If, for example, food is cooked over a wood stove, the question need not be the technical one of the stove's efficiency in converting the potential energy within the wood into heat. Instead, the question may be to what extent and in what ways the social organization of food preparation detracts from theoretical efficiency.

Distributing Proven Technologies. Social organization can be as important as technology in creating efficiencies. Most obviously, many proven technologies are unavailable to almost all of the approximately 85% of underdeveloped country's populations that are rural. Similarly, efficient technologies are often unavailable to the lower class in advanced societies. Typically, a strong contrast exists between low absolute costs of the technologies and high marginal costs for selected populations. The social value of many of these technologies is two-fold: avoided social costs provide a short-term justifying pay-back for original expenditures, and improved resource efficiency and decreased pollution enhance prospects of sustainability.

Empowering Individuals. The power to reproduce is often the greatest power available to a couple. To the extent that overpopulation is a root cause of poverty and resource degradation, the power to reproduce must be supplemented to avoid poverty and its correlates of disease and violence, and to improve prospects of resource sustainability. Demographic and economic studies concur that literacy and control over some land are changes that by themselves regularly and more effectively than other changes, increase the power and decrease the reproduction of the world's poor. While the trend in rural areas continues toward increased farm size, production per acre is most efficient on small farms. In urban areas worldwide the percentage of illegal aliens undercuts the social practices that would otherwise support individual empowerment.

Reforming Management. The role of the manager is often buffered from economic shocks. If the manager is the decision-maker, the self-interested decision is usually to continue the manager's work. When circumstances change, different management teams, created for different purposes, find their jurisdictions impinging on each other resulting conflicts. Again self-interest may stir parties away from coordination because they insist on their own legitimacy. The social environments in which management will itself be reformed are therefore environments in which managers are not insulated from the effects of their decisions. While older models emphasized the accountability of a common authority (in the business or in the government) to whom conflicting managers would report, newer models emphasize both how stockholders can make managers accountable and how the disintegration of the core corporation is given markets more leverage to make managers accountable.

Coping with Congestion. The plurality of ways in which disruption of the nitrogen cycle might disrupt sustainability occasions a usual schema of environmental degenerations. In addition to resource depletion and pollution accumulation, sustainability can be threatened by congestion, it is often said. The point is that cogestion implies a local exhaustion of resources (road space, living accommodations) or a local accumulation of pollution (climatological inversion, smoke stack emission) that could be remedied by changing the local concentration of resources or pollution. Often the build up of (human) population in a locale and therefore the localization of resource demand and pollution output account for congestion. Congestion, in other words, is a distributional, not a purely aggregative, concept.

This said, one might infer that the category of congestion is theoretically redundant. Still there seem four reasons for retaining the category: first, we must remember that degradation is sometimes a function of the concentration of causes at a site rather than of the absolute amount of the cause. Similarly, the duration of an effect is often a function of the intensity of the cause. Third, "the intensity of the cause" is regularly tied either to the intensity of human resource use or to the sheer prevalence of human beings on the planet. Fourth, remedies for environmental degradation being tied to the above three points, it is salient to focus on actual local conditions of congestion as a means of alleviating degradation and restoring sustainability.

Improved Monitoring. Underlying many of the above problems lie issues of information flow. When information is missing, obscure or difficult and expensive to obtain, persons often proceed in ignorance. Possible efficiencies may go unrealized; realized efficiencies may not be signaled; signaled efficiencies may be obscured by further noise in the social system; market inefficiencies may go undetected; nonmarket coordinations may be obscured: common values may be hidden and common aspirations may be unassumed. What, then, do we know about human beings as monitors?

Every species has evolved in response to vulnerabilities imposed on it and opportunities presented to it by the environment in which its ancestors have lived. The human species is biologically unique because of the ratio of its brain to body size. The human brain is an amazing organ. Requiring an extraordinary amount of continuous energy, the vulnerability the organ brings to the organism argues that only some extraordinary environmental complexity would have stimulated its development. Yet if we begin with the imprecise but consensus view that the hominid line developed on the African savannah, then we have as yet no explanation of why, among the large hunting mammals living on the savannah, brain size increased remarkably only in the hominid line; why when the hominid line proper had a distinctly large brain at its outset at least two million years ago, brains continued to enlarge over the next 1.8 million years.

Accordingly, biologists hold that the complex environment to which hominid's brains have responded is the society of their cohorts. In other words the fundamental problem driving this evolution has been social: how to gain knowledge about and assurances concerning others of one's kind. The reading of facial expressions and vocalizations and the memory of the dispositions of a cohort have been central to the hominid environment for millions of years. Humankind thus evolved in an environment already intricately social.

Human abilities to monitor each other, then, have developed within the traditional social environments of human life. Interpersonal cooperation, nonsymmetrical reliances of individuals on others, possibilities of alliances and possibilities of betrayals have been the stuff of human evolution. The abilities to comprehend gestures and read intentions off their smallest signs, the abilities to assure and to detect phony assurances are the stock in trade of a hominid's mastery of its (social!) environment. Detailed observations of a small number of constant cohorts were the hard work that drove the evolution of the hominid line.

Clearly, however, the social environments of most human beings has altered dramatically in historical times, especially since the industrial revolution. Cohorts of repeated interaction and cohorts of association are much larger than ever in human history. Intricate systems have sprung up for the organization of anonymous, mass society. While in some ways modern life has allowed individuality to flower, in other respects it has required that much human behavior become role-bound and institution-bound so that interactions can proceed easily in the absence of personal histories.

These changes in human society notwithstanding, it is well to focus the environments for which human cognitive and motivational responses are honed. Without affirming the perfection of those responses and without deprecating the capacities of human beings to adapt to social novelty, it is reasonable to believe that maximizing the opportunities of people to respond directly to the voices, faces, gestures, affirmations, assertions and assurances of others improves the quality of the information on the basis of which the humans respond.

Consider the particular success of small communities with curbside recycling. In general transportation expenses make curbside recycling prohibitively expensive in rural areas. Leaving them aside, however, studies find the percentage of participation and the percentage of waste reduction greatest in smaller towns. The apparent explanation of this difference is the nonovert pressure small town people sense to do their part for their town. Part of the confirmation of this explanation is that transients, whom one would expect to identify less with the town, recycle less. Part of the confirmation is that apartment dwellers, whose (non-)recycling is less observable, recycle less. Fluctuating membership of groups and the anonymity of members each detract from community participation.

Similarly, scale is central in delivery systems. Communication theorists have long known the geometrically increasing costs of communication lines in urban areas. Transportation theorists appreciate the complexity and importance of efficiency in delivery routes. These theories have become parts of the basis for movements within large cities toward neighborhood revitalization. The link is that vital neighborhoods contain a multitude of interlinked ties. People, in other words, repeatedly engage a reduced number of cohorts. Given that modern society is defined by a multiplicity of roles, vocational, economic, familial, educational, recreational, religious, etc, a plausible hypothesis is that neighborhoods so organized as to encompass all or most of these roles will so interactively and repeatedly engage persons as to reinforce their concern for each other and for the well-being of their community.

THE GROUNDING OF ETHICS IN NATURAL HISTORY

An ethic of sustainability is an ethic. Humans have not been especially perspicuous in understanding their ethics. Since ethics has something to do with advocacy grounded in the distinction of right from wrong, ethical discussion often focuses on advocacy rather than on understanding the place of ethics in human life. The following is a melded discussion, first of what an ethic is, and then of what norms are appropriate for an ethic of sustainability.

The kind of creature that has an ethic is social and linguistic, for an ethic is a linguistic means of coordination of cohort behavior. A look at the history of the hominid line places what an ethic is and whatever justifiability an ethic may have in focus. The goal of social coordination is to enhance the well-being of individuals without, through coordination, imposing new harms or excessive limits. Eventually norms of social coordination emerge to that end. An ethic is the articulation, justification and criticism of those norms. A slice of history will suggest how society with an ethic emerges.

Four million years ago the climate was getting colder and less rain was falling in the tropical forest, which was consequently shrinking. Some of the primates who were thus losing their habitat for reasons they did not comprehend, our remote ancestors among them, began to move out onto the surrounding savannah. There they encountered many established large mammal species, both herbivorous and carnivorous. Stereoscopic vision they probably had. In the perspective of evolutionary time, they moved quickly, as populations in strange habitats often do, toward a fully upright gait and a loss of brachiation. These animals, already social, also underwent a swift evolution in brain size. Surprisingly, in the line leading to *Homo sapiens,* brain size continued to increase until approximately 100,000 years ago, reaching a point where it requires approximately 20% of an adult's total energy intake and up to 50% of the intake of a human infant.

Evolutionary biologists agree that, however stressful the savannah environment might have been for its mammalian inhabitants, those common stresses cannot explain who one line developed such an extraordinary brain. Instead they find the root of increasing brain size in the extraordinary stress faced by that line arising from the complexity of its intense social living. Exactly how that complexity formed is unknown, but some of its outlines are clear: The lengthening of childhood implied the strengthening of ties among adult support networks. Peer interaction among juveniles promoted the exercising and development of social skills. Squabbles among juveniles provoked partisan adult intervention. The advantages of social harmony promoted adult learning about interventions that avoid not only violence but also hard feelings within support networks. The cessation of estrus, ongoing moderate adult female sexual receptivity, extended relationships between sexual partners, food sharing, and a significant contribution of an adult male's labor to the well-being of his sexual partner(s) and her or their offspring are somehow comingled in the formation of the family as a social unit.

In other words, given that the hominid line lacked the adaptations that secured success for other savannah species, the hominid line relied heavily on strength emerging from numbers and social cohesion. A division of hominid labor, however, unlike a division of bee or ant labor, is not rigidly genetically determined. If instead it is a division to be worked out, then it is a division requiring the likes of trust and assurance. At the same time, the need for trust and assurance amounts to the social possibility of betrayal. For all communities, spread over many generations, circumstances will continually vary between those in which an individual benefits from the existence of mutual aid networks and those in which an individual benefits from striking his (more often than her) own path or ignoring an opportunity for aiding one's cohorts.

The problem then for hominid society becomes centrally social: how to scrutinize the social scene so as to discern the trustworthiness of cohorts. Associated are the individual's problems both of providing the assurance that will make others willing to cooperate and of gaining the freedom to go one's own way or tend one's own concerns without endangering either one's self, one's offspring or the future reliability of one's social network.

In this context the ability to read the motivation of others and the abilities both to show one's genuiness well enough to enlist the aid of others and to conceal one's disposition well enough to avoid becoming other's transparent dupe undergo intense stress. Many coordinations may occur, but breakdowns of coordination will be dangerous and may become violent. It is of course relevant to note how often the advent of modern warfare is interpreted in the light of energy supplies.

Intracohort violence can threaten the security of a social species by reducing numbers, by weakening cohorts to attack from outside the community, and by poisoning the social atmosphere against future cooperation. Intracohort violence, therefore, is dangerous to the survival of a species whose strength emerges from its social ties. At this juncture, the social value of the emotional predisposition toward shame at being implicated in the death of a dear (to one) or valuable (to one) cohort may be that the predisposition may move an animal to shrink in horror from bringing about such a result. Such a predisposition will have the effect of reducing violence and replacing much possible violence with posturing.

The point of speaking of an emotional predisposition is two-fold. First, all social primates have emotional dispositions. Such dispositions are mechanisms much older than language, and consequently, ethics, for regulating social intercourse. Their regulative capacity builds on not only the fundamental social character of those primates' lives but, in a layered fashion, on the fact that behavioral and monitoring problems have become evolutionary problems for earlier generations that had emotional capacities. Thus, emotions respond to the emotions of others and even to the emotions of the self. (One may be frustrated or ashamed that one is angry, for instance.)

Second, a primate emotional disposition is not always overwhelming. Clearly, human behavior does not follow rigidly and inevitably from such dispositions. Circumstances may be so dangerous that expression of the disposition may be pre-empted. Circumstances may be so suspicious that expression of the disposition is checked. In ongoingly oppressive environments the repression of those predispositions may be learned.

In summary, if in social insects social coordination results from a genetically pre-established harmony, in primates the harmony requires significant reading of peer behavior, the stimulus of social emotions and trial-and-error learning about social interaction. Norms of behavior, intact not only with sanctions against violation but also with socially astute tolerances for indiscretions, are already displayed in baboon and chimpanzee societies. Whatever the limitations of such norms, they tend to contribute to social stability and a degree of harmony.

All of the preceding observations and language are among the human tools for the social coordination central to the general animalian capacity for behavior. Animals have evolved from the advantage an organisms may obtain from movement. That is, what is good for an animal regularly requires some behavior by the animal for its realization. The animal does a right, or fitting, or appropriate, thing if the behavior in which it engages is at least behavior that tends to be effective in bringing about or maintaining or restoring some facet of well-being. For social species, a right behavior often does not by itself bring about, maintain, or restore any facet of well-being. If the right behavior effects well-being at all, it does so because it is part of a network of behaviors that effect well-being interdependently.

Whenever language had become a part of human life, all of the preceding was already in the fabric of human society. Actual human practices show us that eventually our social emotions have been supplemented both by articulations of them and by reasonings about the underlying goods and bads of human life and about strategies for promoting the goods and avoiding the bads. Language has become integral to what ethics is. As much as humans may be repulsed by the wrong or tempted to it, they can also articulate what is wrong with it, if it indeed is wrong, and what is attractive about it, if indeed it is attractive. They can reason whether the articulated wrong and the articulated attractiveness are finally incompatible with each other. They can reason about whether their belief that the two are compatible is insight or desire-induced delusion. When they define the bads and who is likely to suffer them, they can devise strategies and new norms for avoiding the bads or for making the suffering fairer. (As such, this article has attempted to define and discuss sustainability as a norm of energy use).

Language is useful directly for improving the social coordination through which a hominid society pursues the good. Language provides its users with means of abstract foci of attention (on the shape or the movement of an object, for instance). Language allows its users to speak of what is not present (the animal that will come to the watering hole) and to use such talk to plan (a sequence of coordinated actions). Language allows its speakers to interpret themselves to others ("What I really want is . . ." "I believe that . . ."), thereby obviating problems of interpretation for trusted or desired allies, while increasing the complexity of problems of interpretation for those who are distrusted and those who are not convinced they are viewed kindly.

Human ethics, then, are not framed simply by emotional dispositions. Part of human intelligence is the ability to observe a recurrence of circumstances and to note that more and less happy effects result from different initiatives. Some behavioral dispositions, human beings may come to believe, are so generally conducive to social facility or harmony that those dispositions are generally or universally to be endorsed, commended, encouraged, and reinforced.

Honesty, courage, and wisdom are often commended as universal human virtues precisely on the argument that the (social) character of human life, regardless of the eccentricities of social organization, is bound to benefit from these dispositions. That is, the veracity of information, resoluteness in the face of danger and discernment of subtle distinctions through which beneficial and enriching pursuits become feasible, is bound to be valuable in human life because of the complex phenomenon that human society is. Because social life is bound to benefit from information-sharing and coordinated planning, for example, honesty and promise-keeping are bound to be excellences; that is, behavioral dispositions such that a society whose members have and act upon, will benefit as a society but whose members lack or do not act upon those dispositions will not.

The Evaluation of Sustainability. The virtue of sustainability is similar; that is, organisms require energy to maintain themselves. Organisms use some of the energy available to them to enhance the assurance of a reliable energy supply. A social animal places its reliance in its creation of a social network. Inconstancy in the supply of energy will mean either that individuals directly suffer a curtailment of energy available to them or that the energy devoted to the social network for assuring energy supplies, is itself curtailed. In the first of these cases, the quality of individual's lives is diminished. In the second, the well-being of a society's future is jeopardized. In this way a social network establishes itself because having the network improves a population's ability to survive (and reproduce). Within primates, such social networks involve patterns of behavior that evolve, are learned by individuals, and are monitored and reinforced by concerned community members. Obviously, then, if an ethic is the articulation, justification, and criticism of a set of social norms, and if the norms themselves are justified by the social well-being they maintain, then sustainability must be a virtue because, regardless of the eccentricities of social organization, the character of human life is bound to benefit if people act not only honestly, courageously and wisely, but also sustainably.

A corollary concerns the place of actions that are not sustainable. If reliance upon those actions for energy supplies, either for individuals or for maintaining the society's assurance structure, is wrong because it is not sustainable, then the more fitting use of non-renewable resources is for boot-strapping. If, having created enhancements of human well-being, the resources needed for the enhancements are no longer needed for its maintenance, then the nonrenewability of the resource is morally unproblematic. To be sure, it is sometimes difficult to foresee whether a present use of a nonrenewable resource will later become superfluous. It is not difficult, however, to observe that many present human uses of resources is the non boot-strapping use of nonrenewables. It is also possible and prudent, when planning the use of a nonrenewable, both to monitor the confinement of its use and to take systematic and diligent steps to outgrow its use.

If some dispositions are universally beneficial, others are less so. In particular, many matters of convention gain the benefit they have, partly or totally, from the establishment of the convention. Property rights in Anglo-American society imply that trespassing is wrong. Property rights in Scandinavian society do not prohibit temporary, nondegrading uses, say of pasture lands. While it is obvious that well-known, widely respected conventions promote the meeting, rather than the frustrating, of expectations, it is not obvious that one of these conventions is better in the sense that a society having it enriches the society or improves its sustainability.

At the same time, it is clear that a convention, established at a given time, is designed to handle a foreseen range of cases. For instance, the so-called taking clause of the Fifth Amendment to the Constitution of the United States prohibits the government's taking of a person's property without compensation. Seizure, eviction, preemptive use of the land for public purposes, all of these were foreseen cases. Case law on the prohibition of coal excavation because of its impact on surface usefulness, on zoning against the downwind effects of smelting and other polluting activities, and on government limitations on the filling of wetlands, have all raised unforeseen issues about when the use of property is rightly restricted to prevent public nuisance or harm versus when an owner has been deprived of economic benefit for the sake of promoting a public good.

Whether, then, a social norm is appropriate, can be evaluated in four distinct ways: to what extent is the norm one of those that reinforces and promotes the dispositions that are valuable in any human society? To what extent does a norm promote the improvement of well-being without, through the use of nonrenewables or the creation of pollutants, substantially endangering well-being elsewhere? To what extent have emerging natural, technical, demographic, or social circumstances skewed the reliance on established norms so that they are not so beneficial or at least neutral as they had been or had been interpreted to have been? And, finally, to what extent is the norm one of those that establishes and facilitates human coordination and, thus, social harmony through means that disrupt goods in either the natural or the social world?

In the light of these criteria, we may evaluate the development of norms for international cooperation regarding environmental concerns. An older practice of norm development for the sake of international cooperation about energy sustainability had been a lowest common denominator standard. From the premise that the international order includes few means of imposing cooperation, it was inferred both that cooperation cannot be constrained and that the only universally achievable standard would be the lowest standard acceptable to any of the nations seeking an agreement.

That practice has been substantially superseded since the early 1980s. The lowest common denominator standard, it has been realized, discouraged nations from seeking mutually beneficial higher standards, misestimated the effect of higher standards on nations obviously falling below those standards, and ignored the often-positive dynamic of higher standards towards the creations of technologies and conventions that facilitate compliance from previously noncomplying countries.

The emerging standard is one of a patch-work of boot-strapping agreements. The dynamics of such agreements not only boast the floor of coordination. Rather, they also encourage new coordinations and additional partnerships. Instead of a norm focused on the avoidance of problems, emerging international norms are focused on the pursuit of goods.

THE ETHICS OF SUSTAINABILITY

Trends in energy conservation, pollution control, population growth, urban planning, and economic development have brought the concept of sustainability to international attention. Yet substantial fuzziness remains in the concept. That fuzziness arises partly from the multiplicity of goods at which actions aim, the multiple uses of resources and the multiple effects of these uses. Different foci then allow the construction of plausible arguments that resources are both sustainable and insustainable and that fuzziness arises significantly from disagreement about what is to be sustained. Set aside the idea that choice is to be sustained and sustainability can either be a matter of lifestyles or of resources. With each of those discussions, one must question how holistically sustainability is to be understood. The key to a non-fascistic but holistic understanding of the sustainability of resources is to conceive resources themselves nonanthropocentrically: all forms of life use resources. From that perspective the functional value intrinsic to an ecosystem provides the only clarity humans presently have about what value sustainability has and why. The most clearly sustainable resources are renewable. The danger that no adequate substitute will be found for some nonrenewable resource grounds the argument that nonsustainable resources, if used, should be used for bootstrapping, not for lifestyle maintenance.

Indeed, the hardest and riskiest way to fulfill the obligation of sustainability to our descendents is to predicate sustainability as the product of an unending substitution of resources. As many substitutions as human history records, the record clearly shows much suffering and nonsustainability for individuals in the midst of substitution. Even the attribution of that suffering to normal market adjustments will not gainsay the moral significance the suffering retains, both because lives are centrally disrupted and because the vulnerability of the disrupted lives underlines the nonconsensual character of their disruption.

The Ethics of Sustainable Energy

The application of these norms to the achievement of energy sustainability, however, requires effort in several areas, for "energy sustainability" continues to have different meanings. It is entirely possible that "energy sustainability" will come to mean the continual availability of adequate energy for human beings of means. If one focuses on the sustainability of human society in the aggregate, then local societies and poorer classes may be dispensible to that sustainability (a matter of fact, true to some degree depending on violence resulting from discontent, etc) and judged and accounted as such (a matter of practice, policy and ethics). It is entirely possible that "energy sustainability" will come to mean the continued availability of adequate energy for human beings if acceptable materials and technologies continue to replace exhausted energy resources. If one focuses on the possible gains of a high-risk energy policy, then a larger group of human beings, perhaps persons of all classes, might suffer substantially from misfortunate calculation. The very suffering that sustainability is primarily directed at averting is made more probable by reliance on scarce, non-renewable energy sources, or on ones the use of which creates significant pollution. It is entirely possible that "energy sustainability" will come to mean making the planet not metaphorically but literally "spaceship earth," a planet composed of almost exclusively, of human beings, materials used as resources for them, and the technologies required for that use. In that case our immense ignorance of ecosystems will assure that before the planet becomes so composed, the human beings will cease to be one of the components.

When Plato advanced the thesis he called the unity of the virtues, the view that each virtue is identical with the several others, he defended it by arguing both that in the ideal case the virtues coincide and that in the total absence of the one, the others are also impossible. Whatever the considerable difficulties of Plato's argumentation, the wisdom of this view is that of cautiously maintaining the coherence that allows several goods to be achieved, or at least approached, simultaneously. Since it is impossible to maximize simultaneously for more than one desideratum, the achievement of several goals requires caution not to make any of them one's sole, or even predominant, guiding light and, indeed, deliberately to pursue each only in ways that facilitate, rather than hinder, the achievement of others.

The outlines of the achievement of energy sustainability are therefore three-fold. First, sufficient poverty exists to drive human population growth, thereby increasing the energy supply required to sustain human life. Honesty requires humans to affirm that the problem is hard enough and that making the problem harder will increase the likelihood that wide-spread social and ecological harm will precede energy sustainability. From this caution must derive the fundamental courage to attack poverty worldwide as a prerequisite of achieving energy sustainability. Otherwise, in wisdom, we discern that the price of sustainability for some will be ongoing social tension, violence, bloodshed, and warfare.

Second, the scale of the operation of maintaining energy supplies for human beings implies a moral demand not only for increasing the efficiency with which the demand is met but, quite significantly, with constraining the pollution of present-day demand meeting. The harms of pollution and their projectable ballooning as developing nations command larger supplies of dirty energy, combine with the unevenness of access to such energy to demand the courage of the immediate pursuit of both the short-term goals of purity and efficiency of using non-renewable energy supplies, and, the longer-term goal of conversion to solar (including wind) and other, hydral forms of renewable energy. Premises about the probable half-century lag time in developing new energy technologies and stabilizing world human population aid discernment of the wisdom of immediate deployment of these strategies. And honesty requires a clear, undiluted statement of the risks, lest people be self-deceptively lulled into ignoring them.

Third, the fuzziness of our understanding of ecosystems and of the myriad species that compose them, argues, in conjunction with the importance of understanding sustainability, for improving human biological knowledge. In our current state of considerable ignorance, the fuzziness of our understanding of ecosystems argues, in light of the value of sustainability, for prudence. Prudence implies relieving the poverty that provides the impetus to human population growth, encouraging healthy, energy-efficient lifestyles chosen informedly in the light of prices that thoroughly reflect costs, and converting energy production to renewable, non-polluting forms. Interestingly, the pursuit of a lifestyle that is not oriented around consumption helps to bring human beings occasions evoking the wisdom involved in discerning and appreciating the good that the integrity of an ecosystem is. Honesty, however, requires the acknowledgment that, like an ecological ethic, a wide-spread ecological consciousness remains no more than what Aldo Leopold called an evolutionary possibility.

BIBLIOGRAPHY

Reading List

Caring for the Earth: A Strategy for Sustainable Living, Earthscan, London, 1991.

A Compendium of Energy Conservation Success Stories, U.S. Department of Energy, Office of Conservation, Washington, D.C. 1988.

Comprehensive National Energy Policy Act of 1992, House of Representatives, One Hundred Second Congress, second session, April 9, 1992, Washington, U.S. Government Printing Office, Supt. of Documents, Congressional Sales Office, 1992.

Controlling Summer Heat Islands: Workshop on Saving Energy and Reducing Atmospheric Pollution by Controlling Summer Heat Islands, Washington, D.C.: U.S. Dept. of Energy, Building Systems Division, Office of Building and Community Systems, 1989.

Energy Efficiency and Renewable Energy Research, Development, and Demonstration, United States Senate, One Hundred First Congress, first session, on S. 488, S. 964 June 15, 1989, Washington, U.S. Government Printing Office, Supt. of Documents, Congressional Sales Office, 1989.

Energy Tax Incentives, United States Senate, One Hundred Second Congress, first session, on S. 26, S. 83, S. 129, S. 141, S. 201, S. 326, S. 466, S. 661, S. 679, S. 731, S. 741, S. 743, S. 992, S. 1157, and S. 1178, June 13 and 14, 1991, Washington, U.S. Government Printing Office, Supt. of Documents, Congressional Sales Office, 1991.

Energy 2000: A Global Strategy for Sustainable Development: A Report for the World Commission on Environment and Development, London, Zed Books, Atlantic Highlands, N.J., 1987.

Global Environment: A National Energy Strategy, House of Representatives, One Hundred First Congress, second session, on H.R. 5521 a bill to establish the global environmental consequences of current trends in atmospheric concentrations of greenhouse gases, September 13, 1990, Washington, U.S. Government Printing Office, Supt. of Documents, Congressional Sales Office, 1991.

Industrial Competitiveness through Energy Efficiency and Waste Minimization, United States Senate, One Hundred Third Congress, first session, April 29, 1993, Washington, U.S. Government Printing Office, Supt. of Documents, Congressional Sales Office, 1993.

Old Growth Forests: Hearing before the Subcommittee on National Parks and Public Lands of the Committee on Interior and Insular Affairs, House of Representatives, One Hundred Second Congress, first session, on H.R. 842, Ancient Forest Protection Act of 1991, H.R. 1590, Ancient Forest Act of 1991, hearing held in Washington, DC, April 25, 1991, Washington, U.S. Government Printing Office, Supt. of Documents, Congressional Sales Office, 1992.

Renewable Energy and Energy Efficiency Technology Competitiveness Act of 1989, U.S. House of Representatives, One Hundred First Congress, first session, May 23, 1989, Washington: U.S. Government Printing Office, Supt. of Documents, Congression Sales Office, 1989.

Renewable Energy Technologies, House of Representatives, One Hundred First Congress, first session, on H.R. 1216, April 26, 1989, Washington, U.S. Government Printing Office, Supt. of Documents, Congressional Sales Office, 1989.

Regulation of Registered Electric Utility Holding Companies, United States Senate, One Hundred Second Congress, second session, on S. 2607, hearing held on May 14, 1992, Washington, U.S. Government Printing Office, Supt. of Documents, Congressional Sales Office 1992.

Report to Congress: Coal Refineries—A Definition and Example Concepts, Washington, D. C., United States Department of Energy, July, 1991, pp. 38–44.

Retiring Old Cars: Programs to Save Gasoline and Reduce Emissions, Washington, D.C., U.S. Government Printing Office, Supt. of Documents, Congressional Sales Office, 1992.

State Initiatives to Promote a Diversified Energy Resource Base, U.S. House of Representatives, One Hundred Second Congress, first session, December 12, 1991, Washingtion, U.S. Government Printing Office, Supt. of Documents, Congressional Sales Office, 1992.

W. Aiken, "Human Rights in an Ecological Era," *Environmental Values,* The White Horse Press, Cambridge, UK, Autumn, 1992.

H. S., Burness, "Risk: Accounting for the Future," *Natural Resources Journal,* (Oct. 1981).

B. R. Barkovich, *Regulatory Interventionism in the Utility Industry: Fairness, Efficiency, and the Pursuit of Energy Conservation,* Quorum Books, New York, 1989.

R., Costanza, ed., *Ecological Economics: The Science and Management of Sustainability,* Columbia University Press, New York, 1991.

H. Daly and J. Cobb, *For Our Common Good,* Beacon Press, Boston, Mass., 1989.

P. L. Fiedler and S. K. Jain, eds., *Conservation Biology: The Theory and Practice of Nature Conservation, Preservation, and Management,* Chapman and Hall, New York, 1992.

W. Fox, *Toward a Transpersonal Ecology: Developing New Foundations for Environmentalism,* Shambhala Publications Inc., Boston, Mass., 1990.

B. Furze, "Ecologically Sustainable Rural Development and the Difficulty of Social Change," *Environmental Values.* The White Horse Press, Cambridge, UK, Summer, 1992.

A. M. Friend, "Economics, Ecology and Sustainable Development: Are They Compatible?" *Environmental Values.* The White Horse Press, Cambridge, UK, Summer, 1992.

J. Garbarino, *Toward a Sustainable Society: An Economic, Social and Environmental Agenda for Our Children's Future,* Noble Press, Chicago, Ill., 1993.

J. P. Gates, *Sustainable or Alternative Agriculture: January 1990–September 1992,* National Agricultural Library, Beltsville, Md. 1993.

D. Hafemeister, H. Kelly, and B. Levi, eds., *Energy Sources: Conservation and Renewables,* American Institute of Physics, New York 1985.

Heraclitus, *Heraclitus: Fragments,* translation with a commentary by T.M. Robinson, Toronto, Ont., and London, University of Toronto Press, 1987.

J. M. Hollander, ed., *The Energy-Environment Connection,* Island Press, Washington, D.C., 1992.

R. B. Howarth, "Intergenerational Justice and the Chain of Obligation," *Environmental Values,* The White Horse Press, Cambridge, Summer, 1992.

R. Howes, and A. Fainberg, eds., *The Energy Sourcebook: A Guide to Technology, Resources, and Policy,* American Institute of Physics, New York, 1991.

F. A. Khavari, *Environomics: the Economics of Environmentally Safe Prosperity,* Praeger, Westport, Conn., 1993.

J. T. MacLean, *Sustainable or Alternative Agriculture, January 1982–December 1989,* National Agricultural Library, Beltsville, Md., 1990.

M. V. Melosi, *Coping with Abundance: Energy and Environment in Industrial America,* Temple University Press, Philadelphia, Pa., 1985.

J. S. Mill "The Stationary State," in *Principles of Political Economy,* 5th ed. Appleton and Company, New York, 1897.

B. G. Norton, "Sustainability, Human Welfare and Ecosystem Health," *Environmental Values,* The White Horse Press, Cambridge, UK, Summer, 1992.

B. G. Norton, *Why Preserve Natural Variety?,* Princeton University Press, Princeton, N.J., 1987.

D. Parfit, "Energy Policy and the Further Future," Center for Philosophy and Public Policy, University of Maryland College, Park, Md., Feb. 23, 1981.

E. Partridge, "On the Rights of Future Generations," in Scherer, D., ed., *Upstream / Downstream Issues in Environmental Ethics,* Temple University Press, Philadelphia, Pa., 1991.

E. Partridge, *Responsibilities to Future Generations,* Prometheus Books, Buffalo, N.Y., 1981, c1980.

D. Pearce, and co-workers, *Blueprint 2: Greening the World Economy,* London, Earthscan Publications in association with the London Environmental Economics Centre, London, 1991.

D. W. Pearce, and J. J. Warford, *World without End: Economics, Environment, and Sustainable Development,* Published for the World Bank [by] Oxford University Press, New York, 1993.

J. Pezzey, "Sustainability: An Interdisciplinary Guide," *Environmental Values,* The White Horse Press, Cambridge, UK, Winter, 1992.

S. L. Pimm, *The Balance of Nature?: Ecological Issues in the Conservation of Species and Communities,* University of Chicago Press, Chicago Ill., 1991.

D. Rothenberg, "Individual or Community? Two Approaches to Ecophilosophy in Practice," *Environmental Values,* The White Horse Press, Cambridge, UK, Summer, 1992.

J. W. Russell, *Economic Disincentives for Energy Conservation,* Ballinger Publishing Co., Cambridge, Mass., 1979.

M. Sagoff, *The Economy of the Earth,* Cambridge University Press, New York, 1988.

M. Sagoff, "Property Rights and Environmental Law," *Philosophy and Public Policy* (Spring 1988).

D. Saxe, "The Fiduciary Duty of Corporate Directors to Protect the Environment for Future Generations," *Environmental Values,* The White Horse Press, Cambridge, UK, Autumn, 1992.

D. Scherer, "A Disentropic Ethic," *The Monist* (Winter, 1988).

D. Scherer, "Towards an Upstream/Downstream Morality for our Upstream/Downstream World," in R. Blatz, ed., *Water Resources, Water Ethics,* University of Idaho Press, 1990.

K. R. Schneider, *Alternative Farming Systems, Economic Aspects: January 1991–January 1993,* National Agricultural Library, Beltsville, Md., 1993.

K. Shrader-Frechette, "Ethical Dilemmas and Radioactive Waste: A Survey of the Issues," *Environmental Ethics* (Winter, 1991).

K. Shrader-Frechette, *Risk Analysis and Scientific Method,* D. Reidel, Boston, Mass., 1985.

K. Shrader-Frechette, *Risk and Rationality: Philosophical Foundations for Populist Reforms,* University of California Press, Berkeley, Calif., 1991.

K. Shrader-Frechette, "Models, Scientific Method and Environmental Ethics," in Scherer, D. ed., *Upstream / Downstream Issues in Environmental Ethics,* Temple University Press, Philadelphia, Pa., 1991.

M. E. Soule, and K. A. Kohm, eds., *Research Priorities for Conservation Biology,* Island Press, Washington, D.C., 1989.

I. F., Spellerberg, *Evaluation and Assessment for Conservation: Ecological Guidelines for Determining Priorities for Nature Conservation,* Chapman & Hall, London, 1992.

C. D. Stone, *Earth and Other Ethics: The Case for Moral Pluralism,* Harper & Row, New York, 1987.

E. O. Wilson, *The Diversity of Life*, Belknap Press of Harvard University Press, Cambridge, Mass., 1992.

L. Wittgenstein, *Philosophical Investigations,* trans. by G.E.M. Anscombe, New York, Macmillan, 1953.

T

TAR SANDS, RECOVERY AND PROCESSING

James Speight
Western Research Institute
Laramie, Wyoming

In addition to conventional petroleum and the heavier crude oils, there remains a class of petroleum that offers some relief to the potential shortfalls in supply: the bitumen found in tar sand deposits (1).

Tar sands, also variously called oil sands or bituminous sand, are a loose-to-consolidated sandstone or a porous carbonate rock, impregnated with bitumen, a heavy asphaltic crude oil with a high viscosity under reservoir conditions.

The hydrocarbons in tar sands are presumed to be the residue of a former oil deposit that is exposed at, or just below, the earth's surface, partly as a result of subsurface migration and partly as a result of erosion of the rocks that once buried the deposit. The volatile substances escaped from the deposit, leaving behind only the heavy hydrocarbon molecules and molecular chains. All major tar sand deposits are in sandstone reservoirs that were deposited by ancient rivers at or near where they drained into the sea.

The bitumen in the tar sand formations is a potentially large supply of energy. However, many of the reserves are exploitable only with some difficulty; optional refinery scenarios will be necessary for conversion of these materials to liquid products because of the substantial differences in character between conventional petroleum and tar sand bitumen (Table 1).

Because of the diversity of available information and the continuing attempts to delineate the various world oil sand deposits, it is virtually impossible to present numbers that accurately reflect the extent of the reserves in terms of the barrel unit. Indeed, investigations into the extent of many of the world deposits are continuing at such a rate that the numbers vary from one year to the next.

The only commercial experience to date—in fact, for almost thirty years (2)—is that of the Alberta, Canada, tar (oil) sand deposits. These deposits (Fig. 1) have played a major role in tar sand science and technology. In the United States, there has been much work carried out on the Utah deposits, but they have not yet been commercially exploited.

See also Energy consumption in the United States; Heavy oil conversion; Petroleum refining; Bitumen.

OCCURRENCE

Tar sand deposits are widely distributed throughout the world (3,4), and the various deposits have been described as belonging to two types: (a) materials that are found in stratigraphic traps; and (b) deposits that are located in structural traps. There are inevitably gradations and combinations of these two types of deposits, and a broad pattern of deposit entrapment is believed to exist. In general terms, the entrapment character of the very large tar sand deposits involves a combination of both the stratigraphic and structural traps.

The largest tar sand deposits are in Alberta and Venezuela. Smaller tar sand deposits are in the United States (mainly in Utah), Peru, Trinidad, Madagascar, the ex–Soviet Union, the Balkan states, and the Philippines (Figure 2). Tar sand deposits in northwestern China (Xinjiang Autonomous Region) are also larger; here the tar oozes to the surface around the town of Karamay.

The potential reserves of hydrocarbon liquids that occur in tar sand deposits have been variously estimated on a world basis as being in excess of three trillion (3×10^{12}) barrels of petroleum equivalent. The reserves in the United States have been estimated to be in excess of fifty million (50×10^{6}) barrels.

That commercialization has taken place in Canada does not mean that commercialization is imminent for other tar sand deposits. There are considerable differences between the Canadian and U.S. deposits (1) that could preclude across-the-board application of the Canadian principles to the U.S. sands.

RECOVERY

The bitumen content of tar sand can range from 0 to 20% w/w but feed-grade material normally runs between 10 and 12% w/w. The balance of the tar sand is composed, on the average, of 5% w/w water and 84% w/w sand plus clay.

The bitumen, which has a gravity on the American Petroleum Institute (API) scale of 8°, is heavier than water and very viscous. Tar sand is a dense, solid material, but it can be readily dug in the summer months; during the winter months, when the temperatures plunge to −45°C, tar sand assumes the consistency of concrete. To maintain an acceptable digging rate in the winter, mining must proceed faster than the rate of frost penetration; if not, supplemental measures such as blasting are required.

Nonmining Methods

In principle, the nonmining recovery of bitumen from tar sand deposits is an enhanced oil recovery technique and requires the injection of a fluid into the formation through an injection well. This leads to the in situ displacement of the bitumen from the reservoir and bitumen production at the surface through an egress (production) well. There are, however, several serious constraints that are particularly important and relate to the bulk properties of the tar sand and the bitumen. In fact, both must be considered in toto in the context of bitumen recovery by nonmining techniques.

For example, the Canadian deposits are largely unconsolidated sands with a porosity ranging up to 45% and have good intrinsic permeability. However, the U.S. deposits in Utah range from predominantly low-porosity,

Table 1. Bitumen and Crude Oil Properties

Property		Bitumen	Conventional
Gravity, °API		8.6	25–37
Distillation	Vol %	°C	
	IBP		
	5	221	
	10	293	
	30	437	
	50	543	
Viscosity, sus. @ 38°C (100°F)		>35,000	<30
sus. @ 99°C (210°F)		513	
Pour point, °F		+50	≤0
Elemental analysis, wt %			
Carbon		83.1	86
Hydrogen		10.6	13.5
Sulfur		4.8	0.1–2.0
Nitrogen		0.4	0.2
Oxygen		1.1	
Hydrocarbon type, wt %			
Asphaltenes		19	≤5
Resins		32	
Oils		49	
Metals, PPM			
Vanadium		250	
Nickel		100	≤100
Iron		75	
Copper		5	
Ash, wt %		0.75	0
Conradson carbon, wt %		13.5	1–2
Net heating value btu/lb		17,500	About 19,500

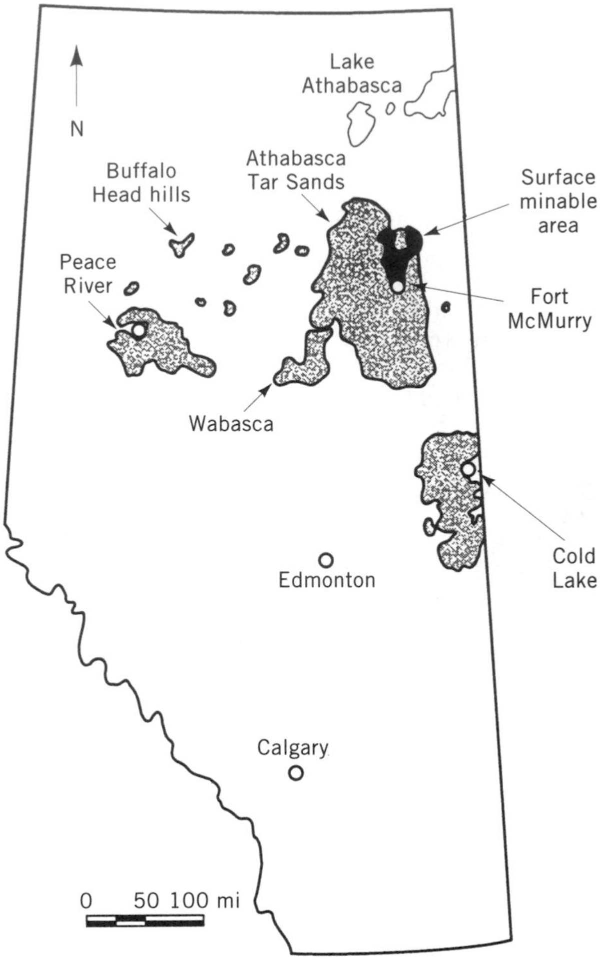

Figure 1. Alberta tar sand deposits.

low-permeability consolidated sand to, in a few instances, unconsolidated sands. In addition, the bitumen properties are not conducive to fluid flow under normal reservoir conditions in either the Canadian or U.S. deposits. Nevertheless, where the general nature of the deposits prohibits the application of a mining technique (as in many of the U.S. deposits), a nonmining technique may be the only feasible bitumen recovery option (5).

The API gravity of tar sand bitumen varies from 5° to ca. 12° depending upon the deposit and viscosities are very high. And, whereas, conventional crude oils may have viscosities of several poise (at 4O°C), the tar sand bitumens have viscosities on the order of 5,000–100,000 Pa.s at formation temperature (ca. 0°C to 10°C), depending on the season. This offers a formidable obstacle to bitumen recovery.

Another general constraint to bitumen recovery by nonmining methods is the relatively low injectivity of tar sand formations. It is usually necessary to inject displacement/recovery fluids at a pressure such that fracturing (parting) is achieved. On one hand, such a technique changes the reservoir profile and introduces a series of channels through which fluids can flow from the injection well to the production well. On the other hand, the technique may be disadvantageous insofar as the fracture occurs along the path of least resistance, giving undesirable (ie, inefficient) flow characteristics within the reservoir between the injection and production wells, that leave a part of the reservoir relatively untouched by the displacement or recovery fluids.

Other variations on this theme include the use of steam and the means of reducing interfacial tension by the use of various solvents. The solvent approach has had some success when applied to bitumen recovery from mined tar sand, but when applied to nonmined material phenomenal losses of solvent and bitumen constitute a major obstacle. This approach should not be rejected out of hand because a novel concept may arise that guarantees minimal (acceptable) losses of bitumen and solvent.

To date, steam methods have been applied almost exclusively in relatively thick reservoirs containing viscous crude oils. In the case of heavy oil fields and tar sand deposits, the cyclic steam injection technique has been employed with some success (1). The technique involves the injection of steam at greater than fracturing pressure 10.3–11.0 MPa for Athabasca sands) followed by a "soak" period, after which production is commenced (6). The technique has also been applied to the California tar sand deposits and in some heavy oil fields north of the Orinoco deposits. The steam-flooding technique has been applied, with some degree of success, to the Utah tar sands and has been proposed for the San Miguel, Texas, tar sands.

Combustion has also been effective for recovery of viscous oils in moderately thick reservoirs where reservoir dip and continuity promote effective gravity drainage, or

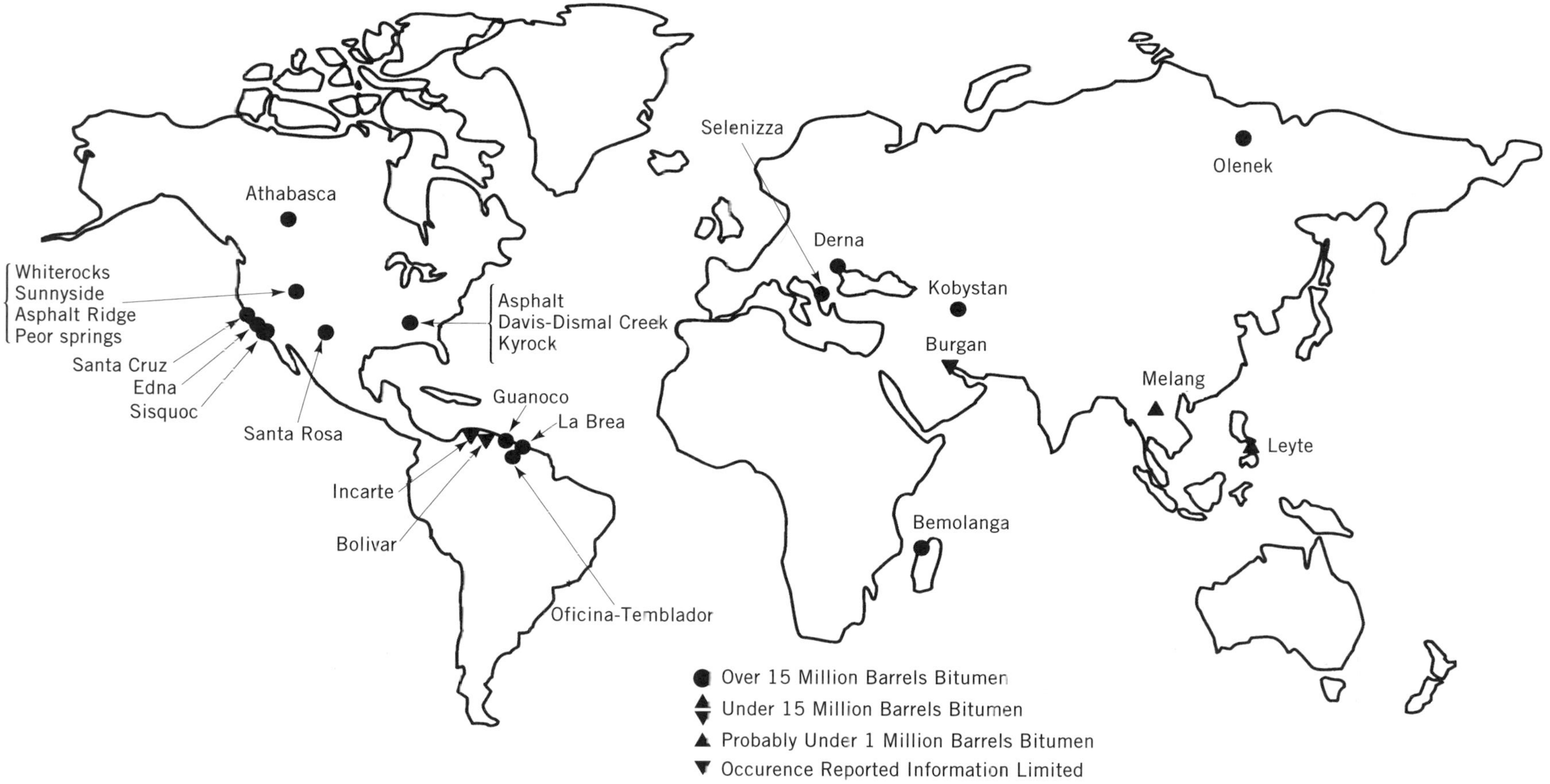
Athabasca
Whiterocks
Sunnyside
Asphalt Ridge
Peor springs
Santa Cruz
Edna
Sisquoc
Santa Rosa
Asphalt
Davis-Dismal Creek
Kyrock
Guanoco
La Brea
Incarte
Bolivar
Oficina-Temblador
Selenizza
Derna
Kobystan
Burgan
Bemolanga
Olenek
Melang
Leyte
Over 15 Million Barrels Bitumen
Under 15 Million Barrels Bitumen
Probably Under 1 Million Barrels Bitumen
Occurence Reported Information Limited

Figure 2. World tar sand deposits.

where several other operational factors permit close well spacing.

Using combustion to stimulate oil production is considered attractive for deep reservoirs and, in contrast to steam injection, usually involves no loss of heat. The duration of the combustion may be short (<30 days) or more prolonged (ca. 90 days) depending on requirements. In addition, backflow of the oil through the hot zone must be prevented or coking will occur.

Both forward and reverse combustion methods have been used with some success when applied to tar sand deposits. The forward combustion method has been used in the Orinoco deposits and in the Kentucky sands. The reverse combustion technique has been applied to the Orinoco deposit (6), Athabasca (7), but the tests were largely unsuccessful owing to the difficulty of controlling the air flow and the risk of spontaneous ignition. However, there has been some success in the application of the reverse combustion technique to the Missouri tar sands, and a modified combustion approach has been applied to the Athabasca deposit. This latter procedure involved a heat-up phase, then a blow-down (production) phase, followed by a displacement phase using a fire-water flood (COFCAW process).

Mining Methods

Oil mining is the term applied to the surface or subsurface excavation of petroleum-bearing sediments for subsequent removal of the oil by washing, flotation, or retorting treatments. Oil mining also includes recovery of oil by drainage from reservoir beds to mine shafts or other openings driven into the oil rock, or by drainage from the reservoir rock into mine openings driven outside the oil sand but connected with it by bore holes or mine wells.

Surface mining consists of strip or open-pit mining. It has been used primarily for the removal of oil shale or bituminous sands lying at or near the surface. Strip mining of shale is practiced in Sweden, Manchuria, and South Africa. Strip mining of bituminous (tar) sand is conducted in Canada.

Oil mining is not new. Mining of petroleum and bitumen occurred in the Sinai peninsula, the Euphrates valley, and in Persia prior to 5000 B.C. In addition, subsurface oil mining was used in the Pechelbronn oil field in Alsace, France, as early as 1735. This early mining involved the sinking of shafts to the reservoir rock, only 30–60 m below the surface, and the excavation of the oil sand in short drifts driven from the shafts. These oil sands were hoisted to the surface and washed with boiling water to release the oil. The drifts were extended as far as natural ventilation permitted. When these limits were reached, the pillars were removed and the openings filled with waste.

This type of mining continued at Pechelbronn until 1866, when it was found that oil could be recovered from deeper, and more prolific, sands by letting it drain in place through mine openings with no removal of sand to the surface for treatment.

The Athabasca Tar Sands deposit in Canada is the site of the only commercial tar sands mining operations. The Suncor operation, located 34 km (20 mi) north of Fort McMurray, Alberta, started production in 1967. The Syncrude Canada project, located 8 km (5 mi) 8 km away, started production in 1978. In both projects, about half of the terrain is covered with muskeg, an organic soil resembling peat moss, which ranges from a few centimeters to (23 ft) in depth. The major part of the overburden, however, consists of Pleistocene glacial drift and Clearwater Formation sands and shales. The total overburden varies from 7 to 40 m (23 to 130 ft) in thickness, and the underlying tar sand strata averages about 45 m (150 ft), although typically 5–10 m (16–33 ft) must be discarded because of a bitumen content below the economic cutoff grade of ca 6% w/w.

Selection of a mining scheme is influenced by the size and life of the ore body, and the geology of the overburden and tar sand strata as dictated by the environments of deposition.

The equipment employed at an oil sands mine is a combination of mining equipment and an on-site transportation system that may (currently) consist of conveyor belts and/or large trucks (Table 2). The mining operation is currently carried out using 8.1×10^6 kgm (8100 t) per hour bucket-wheel excavators and 60 m^3 (80 yd^3) capacity draglines as the primary mining equipment. The mining operation differs in detail depending on the equipment; bucket-wheel excavators "sit" on benches while the draglines "sit" on the surface.

Bucketwheel excavators supplemented by front-end loaders are used to dig overburden and load it through twin chutes onto 15×10^4 kg (150-ton) capacity trucks. Additional equipment is used for maintaining the haul roads and for spreading and compacting the spoiled material. On the other hand, overburden may be stripped with 14 m^3 (18 yd^3) hydraulically operated shovels and a fleet of 15×10^4 kg (150-t) trucks.

Tar sand can be mined with large bucketwheel excavators using units having a 10-m (33-ft) diameter digging wheel on the end of a long boom. Each wheel has a theoretical capacity of 95×10^5 kg/h (9500 t/h) but the average output from digging is about 50×10^5 kg/h (5000 t/h). At the rate of 140×10^6/day (140,000 t/day), tar sand can be transferred from the mine to the plant by a system of 152-cm (60-in.) wide conveyor belts and 183-cm (72-in.) trunk conveyors, operating at 333 m/min (1090 ft/min). Following extraction by a process using hot water, the bitumen is upgraded by coking and hydrogenation to a high-quality synthetic crude.

An even more capital-intensive mining scheme involves large draglines, each equipped with a 71 m^3 (92 yd^3) bucket at the end of a 111-m (364-ft) boom. They can be employed to dig both a portion of the overburden, which is free-cast into the mining pit, and the tar sand, which is piled in windows behind the machine. Bucket-wheel reclaimers, similar to bucket-wheel excavators, load the tar

Table 2. Tar Sand Mining Equipment

Mining	Conveying
1. Draglines	1. Belt conveyors
2. Bucket-wheel excavators	2. 150–299-ton trucks
3. Power shovels	3. Trains
4. Scrapers	4. Scrapers
5. Bulldozer with front-end loaders	5. Hydraulic

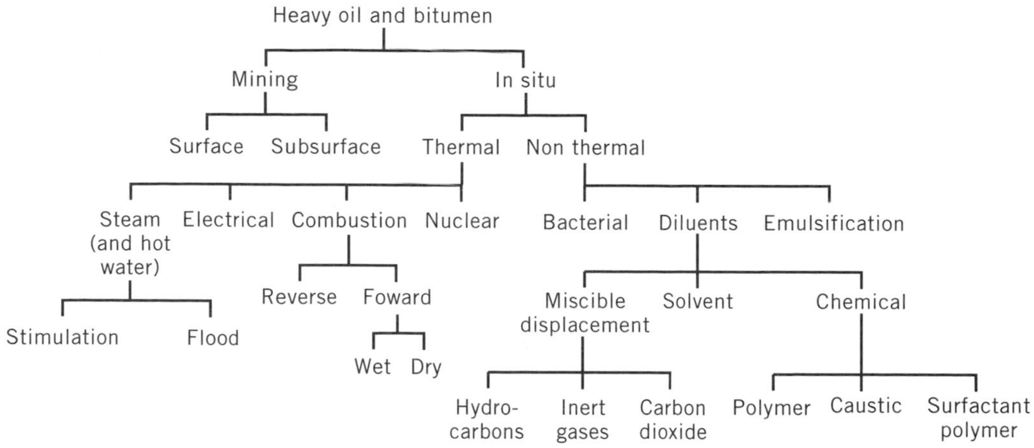

Figure 3. Bitumen recovery processes.

sand from the windrows onto conveyor belts, which transfer it to the plant.

The remoteness of the U.S. tar sands is often cited as a deterrent to development. However, there is more to tar sand development than the accessibility of the location. Topography of the site, overburden-to-ore body ratio, and richness of the ore body are also equally important factors. In the present context of mining the U.S. tar sands, the Utah deposits (Tar Sand Triangle, P.R. Spring, Sunnyside, and Hill Creek deposits) generally have an overburden to net pay-zone ratio above the 0.4–1.0 range with a lean oil content. On the other hand, the Asphalt Ridge deposit is loosely consolidated and could be mined using a ripper/front-end loader (without drilling and blasting) at the near-surface location of the deposit.

One of the major deficiencies in applying mining tech niques to bitumen recovery from tar sand deposits is (next to the immediate capital costs) the associated environmental problems. Moreover, in most of the known deposits, the vast majority of the bitumen lies in formations in which the overburden/pay-zone ratio is too high (6). In addition, another determinant in the economics of a mining operation is the abrasiveness of the sand to the cutting edges of the mining equipment. In-place tar sands are extremely hard and can cause severe "wear and tear" damage to the mining equipment. The northern deposits suffer from temperatures during the winter that range from -10°C to -50°C. This problem may be circumvented to a degree by "loosening" the area to be mined in the autumn season by a series of explosive detonations.

PROCESSING

The Hot-Water Process

The hot-water process is, to date, the only successful commercial process to be applied to bitumen recovery from mined tar sands in North America (1). Many process options have been tested (Figure 3) with varying degrees of success, and one of these options may even supercede the hot water process.

The process (Figure 4) utilizes the linear and the non-

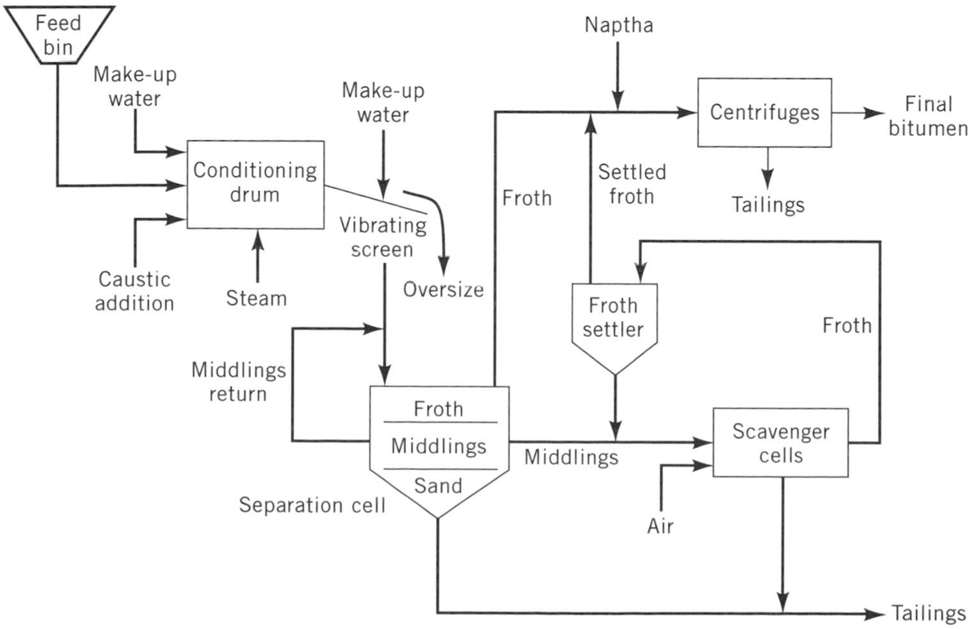

Figure 4. The hot water recovery process.

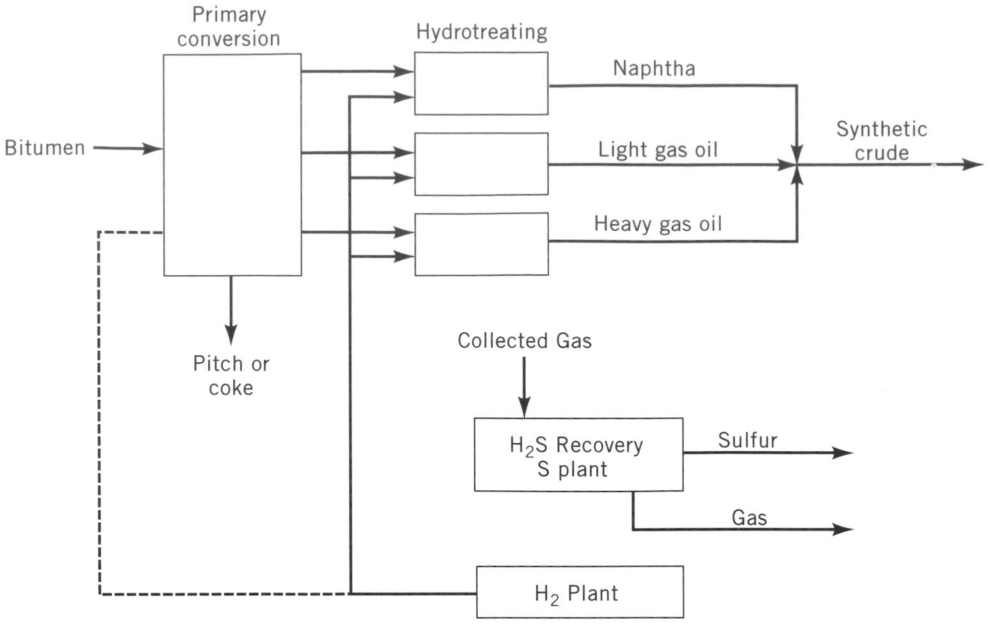

Figure 5. Bitumen conversion.

linear variation of bitumen density and water density, respectively, with temperature so that the bitumen, which is heavier than water at room temperature, becomes lighter than water at 80°C. Surface active materials in the tar sand also contribute to the process (1).

One of the major problems that come from the hot water process is the disposal and control of the tailings. The fact is that each ton of oil sand in place has a volume of ca. 0.45 m^3 (16 ft^3), which will generate ca. 0.6 m^3) (22 ft^3) of tailings, giving a substantial volume gain. If the mine produces ca. 200 × 10^6 kg (200,000 t) of oil sand per day, the volume expansion represents a considerable solids disposal problem.

Environmental regulations in Canada and the United States will not allow the discharge of tailings streams into (9) into the river, (b) onto the surface, or (c) onto any area where contamination of groundwater domains or the river may occur. The tailings stream is essentially high in clays and contains some bitumen; hence the current need for tailings ponds. Removal of the tailings ponds will have to be accommodated at the time of the site abandonment. It is conceivable that the problems may be alleviated somewhat by the development of process options that require considerably less water in the sand/bitumen separation step. Such an option would allow a more gradual removal of the tailings ponds.

The U.S. tar sands have received considerably less attention than the Canadian deposits. Nevertheless, approaches to recover the bitumen from the U.S. tar sands have been made. In the present context an attempt has been made to develop the hot-water process for the Utah sands (8). The process differs significantly from that used for the Canadian sands because the oil-wet Utah sands are different from the water-wet Canadian sands. This necessitates disengagement by hot-water digestion in a high-shear force field under appropriate conditions of pulp density and alkalinity. The dispersed bitumen droplets can also be recovered by aeration and froth flotation (9).

Direct Heating

A process has been developed in Canada that also cracks the bitumen constituents on the sand (10). The processor consists of a large, horizontal, rotating vessel arranged in a series of compartments. The two major compartments are a preheating zone and a reaction zone. Product yields and quality are reported to be high.

Bitumen Conversion

Bitumen conversion, after separation from the associated sand and other materials, is currently carried out by application of coking technologies. The overall conversion of the bitumen is regarded as involving two components: a primary conversion and a secondary conversion (Fig. 5).

In the Suncor operation, bitumen conversion is achieved by use of delayed coking as the primary conversion process, whereas in the Syncrude operation the primary conversion is achieved by application of the fluid coking technology. Bitumen conversion to liquids is on the order of 75%+ with fluid coking giving a generally higher (+1–5%) yield of liquids compared with delayed coking. The remainder appears as coke (ca. 15% w/w) and gases.

Finishing and stabilization (hydrodesulfurization and saturation) of the liquid products is achieved by hydrotreating the liquid streams as two or three separate streams. This is necessary because of the variation in conditions and catalysts necessary for treatment of a naphtha fraction relative to the conditions necessary for treatment of a gas oil (11). It is more efficient to treat the liquid product streams separately and then to blend the finished liquids to a synthetic crude oil. The synthetic crude oil is currently shipped, by pipeline, to a refinery, usually in the Edmonton area.

BIBLIOGRAPHY

1. J. G. Speight, in J. G. Speight, ed., *Fuel Science and Technology Handbook,* Part II, Marcel Dekker Inc., New York, 1990.
2. F. K. Spragins, in G. V. Chilingarian and T. F. Yen, eds, *Bitumens, Asphalts, and Tar Sands,* Elsevier, Amsterdam, 1978, p. 92.
3. P. H. Phizackerley and L. O. Scott, in G. V. Chilingarian and T. F. Yen, *Bitumens, Asphalts, and Tar Sands,* Elsevier, Amsterdam, 1978, p. 57.
4. R. F. Meyer and W. D. Dietzman, In: R. F. Meyer and C. T. Steele, eds., *The Future of Heavy Crude Oil and Tar Sands,* McGraw-Hill, New York, 1981, p. 16.
5. L. C. Marchant and C. A. Koch, *Proceedings, 2nd International Conference on Heavy Crude and Tar Sands,* R. F. Meyer, J. C. Wynn, and J. C. Olson, eds., McGraw-Hill, New York, 1982, p. 1029.
6. J. Burger, In: G. V. Chilingarian and T. F. Yen, eds., *Bitumens, Asphalts, and Tar Sands,* Elsevier, Amsterdam, 1978, p. 191.
7. R. Mungen and J. H. Nicholls, *Proceedings, 9th World Petroleum Congress* **5,** 29 (1975).
8. J. C. Miller and M. Misra, *Fuel Processing Technol.* **6,** 27 (1982).
9. K. E. Hatfield and A. G. Oblad, *Proceedings, 2nd International Conference on Heavy Crude and Tar Sands,* R. F. Meyer, J. C. Wynn, and J. C. Olson, eds., McGraw-Hill, New York, 1982, p. 1175.
10. W. Taciuk, *Energy Proc. Canada* **74(4),** 27 (1981).
11. J. G. Speight, *The Desulfurization of Heavy Oils and Residua,* Marcel Dekker Inc., New York, 1981.

TEMPERATURE MEASUREMENTS

Henry E. Sostmann
Consulting Metrologist
Albuquerque, New Mexico

THE KELVIN THERMODYNAMIC TEMPERATURE SCALE

Temperature is a measure of the hotness of something. For a measure to be rational, there must be agreement upon a scale of numerical values defining hotness and upon devices for realizing and displaying these values. The single temperature scale with an absolute basis in nature is the Kelvin Thermodynamic Temperature Scale (KTTS, or "absolute" scale), which is based on principles that can be deduced from the first and second laws of thermodynamics. [The most commonly used "practical" scale is the International Temperature Scale (ITS), on which temperatures are designated as Celsius degrees (°C).] Since the lower limit of the KTTS is absolute zero and since the scale extends indefinitely upward and is by definition linear, only one nonzero reference point is required to stipulate its slope. The first such reference point was the equilibrium temperature of pure water in its liquid and solid phases at 1 atm pressure, which was assigned the value 0°C, or 273.15 K. In 1954 the present KTTS was established by changing the reference point to more reproducible equilibrium temperature of water in liquid–solid equilibrium under its own vapor pressure, the triple point of water, assigned the value 273.16 K, or 0.01°C. (The unit of temperature on the KTTS is the kelvin, abbreviated K. Since the interval 1 K is identical to the interval 1°C, these symbols may be used interchangeably to indicate an interval but not a temperature. The symbol for a KTTS temperature is T; that for a temperature on any other scale is t).

Most thermometry using the KTTS directly requires a thermodynamic instrument for interpolation. The vapor pressure of an ideal gas is a thermodynamic function, and a common device for realizing the KTTS is the helium gas thermometer. The transfer function of this thermometer may be chosen as the change in pressure with change in temperature at constant volume, or the change in volume with change in temperature at constant pressure. Since rit is easier to measure pressure accurately than volume, constant volume gas thermometry is the usual choice.

A simplified gas thermometer is shown in Figure 1. We will illustrate how it works by using it to show that the temperature of the liquid–solid equilibrium of water (freezing point) under standard pressure is 273.15 K. The figure shows a spherical bulb of constant volume connected by constant-volume tubing to a U-tube manometer. The second connection to the manometer leads to a reservoir of mercury including a means for adjusting the height of mercury in the manometer, in this case a plunger. The bulb and tubing contain an ideal gas. The bulb is first surrounded by an equilibrium mixture of ice and water under a pressure of 1 standard atmosphere, or 101, 325 Pa (C, in Fig. **1a**), and when the gas A is in thermal equilibrium with the water and ice in the bath (at 0°C), the plunger is used to adjust the helium pressure so that both columns of mercury are at the same height, corresponding to the index mark 1.

Next the bulb is surrounded with an equilibrium mixture of liquid and vapor water, ie, the boiling point of water under a pressure of 1 standard atmosphere (D in Fig. 1b). As the gas A is heated to the temperature of the water, its pressure increases, displacing the mercury in both legs of the manometer. The mercury in the left leg of the

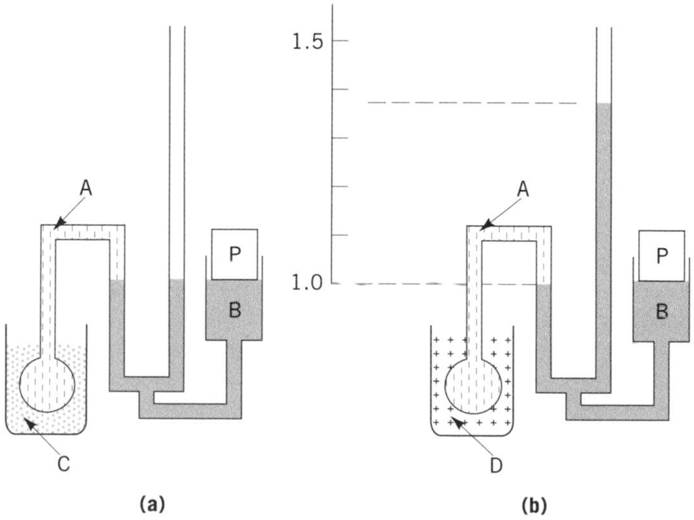

Figure 1. Gas thermometer, A, helium gas; B, mercury; C, water and ice; (0°C); D, Water and steam (100°C); P, plunger for adjusting height of mercury column.

U-tube can be readjusted to the index mark by operating the plunger (restoring the condition of constant volume). However, the mercury in the right leg is now above the index mark. The difference in the heights of the mercury legs, which was zero at 0°C, is the measure of a new pressure which is 1.366099 times the initial or 0°C pressure. From the expression

$$(100°\text{C} - 0°\text{C})/(1.366099 - 1) = 273.15 \quad (1)$$

we can define the interval of the scale as 1/273.15 of the pressure ratio change between 0°C and 100°C. Thus, if the temperature is reduced below 0°C by 273.15 kelvins or Celsius units, an absolute zero is reached (in theory) that is the zero of the KTTS. The zero of the Celsius scale is therefore 173.15 K, and

$$T = t + 273.15 \quad (2)$$

Note in particular that the zero and the interval of the KTTS are defined without reference to properties of any specific substance. (Real measurements with real gas thermometers are much more difficult than the example suggests, and all real gases condense before 0 K is reached.)

See also COMBUSTION MODELING; COMBUSTION SYSTEM MEASUREMENTS; ENERGY MANAGEMENT.

THE FIXED POINTS OF PRACTICAL TEMPERATURE SCALES

Accurate temperature measurements in real-life situations are difficult to make using the KTTS. Most easily used thermometers are not thermodynamic; that is, they do not operate on principles of the first and second laws. Most practicable thermometers depend upon some principle that is a repeatable and single-valued analog of temperature, and they are used as interpolation devices of practical and utilitarian temperature scales which are themselves artifacts. Such principles include the expansion and contraction of liquids and solids, changes in the electrical properties of conductors and semiconductors, and the color and brilliance of light emitted from a very hot source. Any such principles may be used to make a thermometer. Since they are nonthermodynamic, they require construction of a consensus scale to relate the properties of a prescribed interpolation device to the KTTS.

We have described how the KTTS depends upon an absolute zero and one fixed point through which is projected a straight line. Because they are not ideally linear, practicable interpolation thermometers require additional fixed points to describe their individual characteristics. Thus a suitable number of fixed points, temperatures at which pure substances in nature can exist in two or three phase equilibrium, together with specification of an interpolation instrument and appropriate algorithms, define a temperature scale. The temperature values of the fixed points are assigned values based on adjustments of data obtained by thermodynamic measurements such as gas thermometry.

Two-phase equilibria may be solid–liquid, liquid–vapor, or solid–vapor. As is evident from the phase rule of Gibbs, two-phase equilibria are pressure dependent:

$$P + V = C + 2 \quad (3)$$

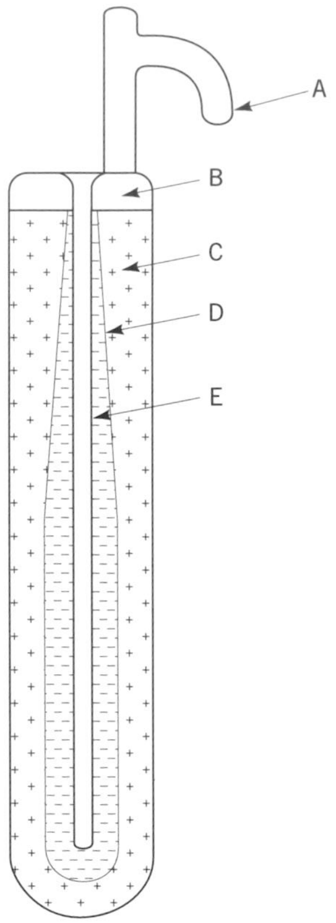

Figure 2. A water triple-point cell. A, handle; B, water vapor; C, liquid vapor; D, solid water (ice mantle): E, thermometer well.

where P is an integer equal to the number of phases present, C is the number of components (for an ideally pure material $C = 1$), and V is an integer giving the number of degrees of freedom. Phase equilibria involving a vapor phase are much more pressure dependent than liquid–solid equilibria. Triple points, equilibria of all three phases, are independent of external pressure.

The triple point of water is the most important of the fixed points, because it is the one point that the KTTS and the ITS have in common, and because it can be realized with great accuracy. A water triple-point cell is shown in Figure 2. The cylindrical envelope is made of borosilicate glass and has a well for a thermometer. The cell is almost completely filled with very pure water of specified isotopic content. The small headspace contains water vapor, whose pressure at the triple-point temperature is about 600 Pa (4.5 torr). The integrity of the cell can be checked by inverting it (carefully). If the cell has retained its integrity, a sharp click will be heard as the water, uncushioned by air, strikes the opposite wall of the cell; also, as the bubble of water vapor is caught in the curved portion of the handle, it is compressed until it is almost invisible. These checks are all that is necessary to confirm the integrity of the cell (although they do not confirm the chemistry of the water). A technical guide and standard for the qualification and use of water triple point cells is found in Ref. 1.

In use, a mantle of ice is frozen onto the outer surface of the thermometer well. A common way to do this is to fill the well with crushed dry ice until the mantle achieves a good thickness. Descriptions of the technique for doing this are given in several publications and in manufacturers' literature. The temperature of the water triple point is 0.01°C, or 273.16 K, by definition. In practice, that temperature can be realized in the cell within ~0.000 15 K of the definition. (In contrast, a bath of ice and water for producing the temperature 0°C is difficult to establish with an accuracy better than 0.002°C.)

Any laboratory requiring precise temperature measurement should be equipped to realize the triple point of water, if no other fixed point. In addition to its status as a primary definition of a temperature, it is a valuable check on the laboratory's thermometers. When thermometers change calibration, they tend to change at all temperatures within their ranges. If the calibration is correct at the water triple point, the probability is very high that it is also correct at any other temperature. The cost of recalibrating one's standard platinum resistance thermometers has become extremely high, and such a check, which may avoid the need for recalibration, may be established for the price of one or two calibrations by the National Institute for Standards and Technology (NIST).

An important class of fixed points is the freezing points of high-purity metals. As solid–liquid equilibria they are pressure dependent, but this dependence is small [for tin, eg, 3.0×10^{-8} K/Pa (0.003 K/atm)] and can be corrected for completely. The metals used in fixed-point cells are better than 99.9999% (6*N*) pure. Figure 3 is one design of such a cell. The metal is contained in a crucible of purified graphite with a graphite cover and sleeve. The crucible is enclosed in an envelope of quartz, the quartz extending into the graphite sleeve to provide a well for the thermometer; thus the graphite assembly is completely enclosed. The cell is charged with pure metal, filled with argon or another inert gas at 101, 325 kPa (1 atm) at the freezing temperature of the metal, and sealed. Thus it is protected against contamination and contains an atmosphere of the correct pressure at the freeze equilibrium temperature. Details of the construction of such cells have been published, and cells are available commercially, as are furnaces specifically designed for melting and freezing the metals. Sealed cells of this sort have shown no measurable change in freezing temperature after 20 years of continuous use. Table 1 shows a list of defining fixed points including the metal freezing points.

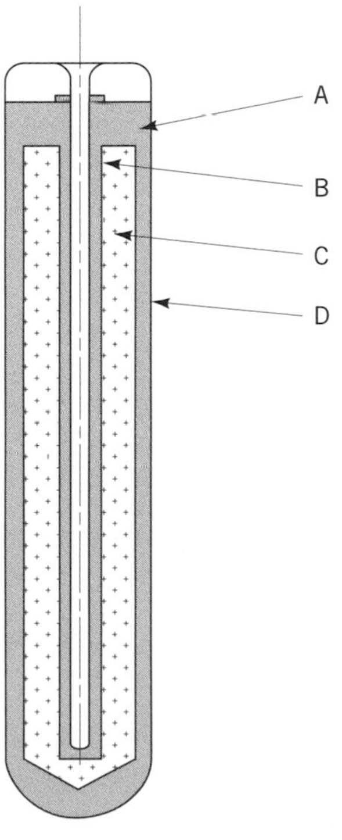

Figure 3. Metal freezing-point cell. A, pure graphite crucible; B, thermometer well; C, highly pure metal; D, quartz envelope.

To perform calibrations using the freezing equilibria of these metals, the cells are heated in furnaces with relatively massive cores to about 10 K above the melting temperature. As the temperature of the metal increases, the onset of melting is indicated when the monitoring thermometer indicates a suddenly constant temperature (the "melt arrest"), as the heat added is used in changing phase rather than changing temperature. When the melt is complete, the temperature increase continues to the furnace setpoint. The furnace control is then reduced to slightly below the melt temperature until the metal is observed to be in freezing equilibrium. The technique for initiating nucleation, managing the supercool, etc., differs for each metal and is described in the literature.

THE ITS AND INTERPOLATION INSTRUMENTS

The ITS is an artifact scale, designed to relate temperature measurements made with practicable instruments as closely as possible to the thermodynamic scale. The scale is established and controlled by the International Committee of Weights and Measures (BIPM) through its Consultative Committee on Thermometry, which was established in 1937. The BIPM itself is established to maintain and implement the Treaty of the Meter, to which most nations of the world subscribe; thus the ITS has not only scientific but legal status in most nations. Within nations, the Temperature Scale is maintained by national standards establishments; eg, in the United States the National Institute for Standards and Technology (NIST); in England the National Physical Laboratory (NPL); and in Germany the Physikalisch-Technische Bundesanstalt (PTB).

Through the years the ITS has been changed a number of times, the better to fit its values to new information regarding the thermodynamic values of the fixed points. In recent history the International Practical Temperature Scale (IPTS) of 1927 has been replaced by the IPTS of 1948, by the IPTS of 1968, (2) and most recently, by the ITS of 1990 (3) effective on January 1, 1990. These changes have included revisions of (a) the range of the scale, (b) the interpolation instruments and their algorithms, and (c) values of the defining fixed points. The changes have been, for most purposes, nontrivial, and it is essential in reading a temperature from the literature

Table 1. Values of Defining Fixed Points of ITS(90) and Some Fixed Points of IPTS(68)[a]

Material	State	T(90) (K)	t(90) (°C)	t(68) (°C)
He	VP	3–5	−270.15 to −268.15	
e-H_2	TP	13.8033	−259.3467	−259.34
e-H_2	VP, CVGT	~17	~−256.15	−256.108
e-H_2	VP, CVGT	~20.3	~−252.85	−252.87
Ne	TP		−248.5939	−248.595*
Ne	BP	54.3584		−246.084
O_2	TP		−218.7916	−218.789
N_2	TP			−210.002*
N_2	BP			−195.802*
O_2	BP	83.8058		−182.962
Ar	TP		−189.3442	
Hg	FP	234.3156		−38.862[b]
Hg	TP	273.16	−38.8344	
H_2O	TP		0.01	0.01
H_2O	BP			100
Ga	MP	302.9146	29.7646	
In	FP	429.7485	156.5985	156.634[b]
Sn	FP	505.078	231.928	231.9681
Zn	FP	692.677	419.527	419.58
Al	FP	933.473	660.323	660.37[b]
Ag	FP	1234.93	961.78	961.93
Au	FP	1337.33	1064.18	1064.43
Cu	FP	1357.77	1084.62	1084.88[b]

[a] Abbreviations: e-H_2, hydrogen with equal distribution of ortho and para states; VP, vapor pressure point; CVGT, constnat volume gas thermometer point; TP, triple point; MP, melting point; FP, freezing point. Note: MP and FP at 101 and 325 Pa ambient pressure.
[b] In the IPTS(68) table designates a secondary reference point.

or in citing a temperature to specify which version of the IPTS or ITS is meant.

The ITS-90 has its lowest point at 0.65 K and extends upward without specified limit. A number of values assigned to fixed points differ from those of the immediately previous scale, IPTS-68. In addition, the standard platinum resistance thermometer (SPRC) is specified as the interpolation standard from 13.8033 K to 961.78°C, and the interpolation standard above 961.78°C is a radiation thermometer based on Planck's radiation law. Between 0.65 and 13.8033 K interpolation of the scale relies upon vapor pressure and constant-volume gas thermometry. The standard thermocouple, which had in previous scales a range between the upper end of the SPRT range and the lower end of the radiation thermometer range, has been deleted.

PLATINUM RESISTANCE THERMOMETER RANGE: −13.8033–1234.93 K

Temperatures on the ITS 90 are expressed in terms of the ratio of the resistance of the SPRT at temperature to the resistance at the triple point of water. The resistance ratio $W(T_{90})$ is calculated as

$$W(T_{90}) = \frac{R(T_{90})}{R(273.16\text{ K})} \tag{4}$$

(Note particularly that in all previous scales the denominator of the ratio W was the resistance at 0°C or 273.15 K.) Restrictions on the SPRT, which are intended to indirectly specify the purity of the platinum from which it is made, are

$$W(302.9146\text{ K}) \geq 1.11807 \tag{5}$$

$$W(234.3156\text{ K}) \leq 0.844235 \tag{6}$$

(302.9146 K is the melting temperature of gallium and 234.3156 K is the triple-point temperature of mercury). An SPRT for use to the silver freezing point has an additional requirement:

$$W(1234.93\text{ K}) \geq 4.2844 \tag{7}$$

where 1234.93 K (961.93°C) is the freezing point of silver. The temperature range over which an SPRT can be used is a function of its design, but no single SPRT can be used over the full range from 13 K to 961°C. As a generality, capsule-type thermometers with a 0°C resistance of about 25.5 Ω may be used from 13 K to about 156°C, 25.5 Ω long-stem thermometers may be used from about 77 K to 660°C, and long-stem thermometers with 0°C resistance of 0.25 or 1 Ω from about 0°C to about 962°C.

The temperature T_{90} is calculated from the resistance–ratio relationship

$$W(T_{90}) - W_r(T_{90}) = \Delta W(T_{90}) \tag{8}$$

where $W(T_{90})$ is the observed value, $W_r(T_{90})$ is a value calculated from a reference function, and $\Delta W(T_{90})$ is the deviation of the observed $W(T_{90})$ of the individual SPRT from the reference function.

There are two reference functions $W_r(T_{90})$, one for the range from 13 K to 0.01°C, another for the range 0.01–962°C. The reference functions represent the calibration of a fictitious thermometer developed from experience in the calibration of many SPRTs over many years. Below 0.01°C, the reference function is

$$\ln[W_r(T_{90})] = A_0 + \sum_{i+1}^{12} A_i \left(\frac{\ln(T_{90}/273.16\ K) + 1.5}{1.5}\right)^i \quad (9)$$

Above 0.01°C the reference function is

$$W_r(T_{90}) = C_0 + \sum_{i=1}^{9} C_i \left(\frac{T_{90}/K - 754.15}{481}\right)^i \quad (10)$$

The ITS-90 provides for 11 SPRT ranges. Three of these are for ranges which extend below −189°C and can be considered to be of interest chiefly to specialists in very low temperatures. The other eight ranges are shown in Table 2. The range should be specified when ordering a thermometer calibration. When the thermometer design permits, a combination of ranges may be specified, e.g., −189 to +420°C. The specification of so many ranges benefits the user of the SPRT over limited ranges, in that (a) accuracy of interpolation may be higher as the range is limited and (b) limited ranges may require fewer fixed-point determinations and consequently substantially lower calibration cost.

Each range has a specific deviation function to be used in Eq. (8). For example, over the range 0–961.78°C, the deviation function is

$$\Delta W(T_{90}) = a[W(T_{90}) - 1] + b[W(T_{90}) - 1]^2 + c[W(T_{90}) - 1]^3 + d[W(T_{90}) - W(933.473\ K]^2 \quad (11)$$

Table 2. Platinum Resistance Thermometer Ranges of ITS(90)

Range	Lower Limit	Fixed Points Required
	Upper Limit of 0.01°C (Eq. 9)	
1	13.8033K	e-H_2 (TP), e-H_2 (VP), e-H_2 (VP), Ne (TP), O_2 (TP), Ar (TP), Hg (TP), H_2O (TP)
2	24.5561	e-H_2 (TP), Ne (TP), O_2 (TP), Ar (TP), Hg (TP), H_2O (TP)
3	54.3584	O_2 (TP), Ar (TP), Hg (TP), H_2O (TP)
4	83.8058	Ar (TP), Hg (TP), H_2O (TP)
	Lower Limit of O°C (Eq. 10)	
5	961.78°C	H_2O (TP), Sn (FP), Zn (FP), Al (FP), Ag (FP)
6	660.323°C	H_2O (TP), Sn (FP), Zn (FP), Al (FP)
7	419.527	H_2O (TP), Sn (FP), Zn (FP)
8	231.928	H_2O (TP), In (FP), Sn (FP)
9	156.5985	H_2O (TP), In (FP)
10	29.7646	H_2O (TP), Ga (MP)
	From −38.8344 to +29.7646°C	
	Reference Function Is Eq. (9) from −38.8344°C to 0.01°C; and Eq. (10) from 0.01°C to +29.7646°C	
11		Hg (TP), H_2O (TP), Ga (MP)

Note: Abbreviations as in Table 1.

where *a, b, c,* and *d* are coefficients derived from calibration of the individual thermometer at specified fixed points–for this range, the water triple point and the freezing points of tin, zinc, aluminum, and silver. Deviation functions for other ranges may be found in reference 3.

THE RADIATION THERMOMETRY RANGE: OVER 1234.93 K

Above the freezing point of silver, T_{90} is defined by the relationship

$$\frac{L\lambda(T_{90})}{L\lambda[T_{90}(X)]} = \frac{\exp\{c_2/\lambda[T_{90}(X)]\} - 1}{\exp[x_2/\lambda(T_{90})] - 1} \quad (12)$$

where $L\ \lambda\ (T_{90})$ and $L\ \lambda\ [T_{90}(X)]$ are the spectral concentrations of the radiation of a blackbody at wavelength (in vacuum) λ at T_{90} and at $T_{90}(X)$, respectively. Here $T_{90}(X)$ refers to the silver freezing point [T_{90}(Ag) = 1234.93 K], the gold freezing point [T_{90}(Au) = 1337.33 K], or the copper freezing point [T_{90}(Cu) = 1357.77 K], and C_2 is the second radiation constant of Planck's radiation formula with the value c_2 = 0.014388 m·K. The usual interpolation instrument is a radiation thermometer, such as an optical pyrometer, which is in itself a thermodynamic device.

RESISTANCE THERMOMETERS

Some of the characteristics of the SPRT, as an interpolation device for the ITS-90, have been discussed above. In order to meet these requirements, the thermometer must be made from almost ideally pure platinum wire mounted in a physical construction which will keep it in a strain-free condition. The conventional resistance 25.5 Ω or some convenient submultiple is historical; over limited ranges it permitted a rough but quick estimate of temperature, since a 1-K change is nearly equivalent to a 0.1-Ω change.

The SPRT is always provided with two current and two potential leads extending from the actual sensing element so that the element can be measured as a four-terminal resistor and the effects of lead resistance eliminated. Capsule thermometers, limited in range from cryogenic temperatures to about 156°C, are usually 6 mm in diameter by less than 6 cm long and are sealed into platinum tubes. They are usually mounted directly in the temperature zone of equipment, where the extension of a long tube from that zone to the external world would represent an unacceptable heat leak. Another common use of capsule thermometers is in calorimeters. The long-stem SPRT with quartz sheaths and an ice point resistance of 25.5 Ω is generally useful from −189 to +660°C. The sheath is commonly 40 cm long and must be immersed at least 15 cm into the medium whose temperature is to be measured. Such a thermometer is shown in Figure 4. High-temperature SPRTs, or HTSPRTs, have been developed specifically to meet the requirements of the ITS-90 and are used from some temperature in the vicinity of 0°C (say −20°C) to an upper limit of the silver freezing point, 962°C. Since platinum has low mechanical strength at these upper temperatures, the winding must be made of large-diameter wire and is of low electrical resistance,

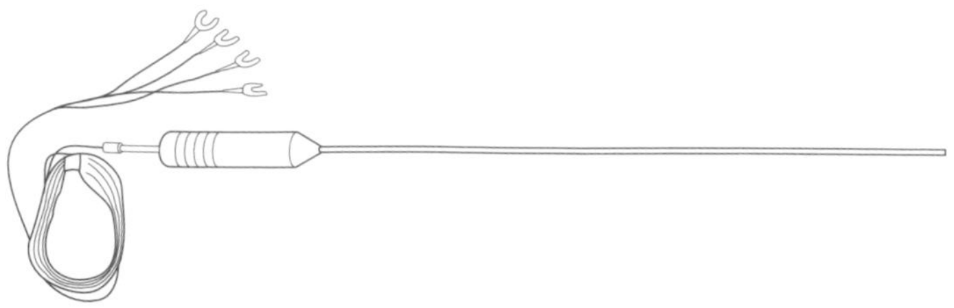

Figure 4. Laboratory standard platinum resistance thermometer (SPRT).

usually 0.25 Ω at 0°C. It is this low resistance and the attendant measurement difficulty which limit the lower end of usefulness.

The SPRTs are devices of superb accuracy and resolution, but they are fragile and can easily be broken. They can also be put out of calibration by strain, induced by even slight mechanical shock or vibration. The major use of SPRTs in science and industry is to maintain the calibrations of working thermometers.

Working-grade metallic resistance thermometers are made of metal wire, usually high-purity platinum, but also (more commonly in past years) copper, nickel, and alloys. The sensitive element is usually protected by a covering of ceramic tubing or ceramic cement, or the individual turns of wire are fixed in place by some means. Glass is occasionally used as a coating, but differences in temperature coefficient of expansion and imperfect insulating properties of glasses at high temperatures limit its use. These forms of protection stabilize the sensing element against shifts due to vibration and shock, while somewhat reducing the temperature coefficient of resistance; consequently they usually do not meet the criteria for interpolation devices of the ITS-90 mentioned in Eqs. (5)–(7).

Working-grade thermometers, conventionally called "industrial" resistance thermometers, are generally smaller than the SPRT element and may be as small as 2.5 mm in diameter and 10 mm in length. They are available in various 0°C resistances, eg, 100, 200, and 500 Ω. They are available as unsheathed elements or in a wide variety of sheaths and enclosures, both standard and custom. They are relatively inexpensive. They are usually made to be interchangeable, without relying on individual calibration, within limits of 0.25 K or closer upon special order. A typical tolerance statement for a precision-class industrial resistance thermometer is

$$\text{Deviation from nominal} \leq (0.1 \pm 0.002t)°\text{C} \qquad (13)$$

In general, manufacturers do not report to the purchaser the calibrations of individual sensors (except upon request) but instead publish tables of resistance versus temperature and tolerance charts for each class. "Deviation" here means departure from a nominal set of values of resistance versus temperature given in a manufacturer's literature.

Industrial resistance thermometers are also the subject of a number of national and international standards, which describe both calibration constants and classes of accuracy and interchangeability. Typical standards are OIML Recommendation of 1985, BS 1904:1984 (England), DIN 43760 (Germany), GOST 6651-84 (USSR), JEMIMA, and SAMA RC4 (1966). With the change from IPTS-68 to ITS-90 and with the political changes in the world (particularly the European efforts at international standardization), the lifetime of any of these standards is probably limited. It is probable that a common European standard will emerge and very quickly be adopted in the rest of the world. It will be adopted verbatim from a standard now in preparation by the International Standards Organization (ISO) to replace ISO publication 751. It will be based on the fixed-point values of the ITS-90 but use the simpler algorithms of the IPTS-48:

$$R(t) = R(0°\text{C})[1 + A_{90}t + B_{90}t^2 + C_{90}t^3(t - 100°\text{C})] \qquad (14)$$

where $C_{90} = 0$ for values of $t \geq 0°\text{C}$.

The following values for the *A*, *B*, and *C* coefficients will be recommended and a table prepared using them:

Coefficient	For ITS-90	Replacing IPTS-68
A_{90}	$3.9083 \times 10^{-3}°\text{C}^{-1}$	$3.90802 \times 10^{-3}°\text{C}^{-1}$
B_{90}	$-5.775 \times 10^{-7}°\text{C}^{-2}$	$-5.802 \times 10^{-7}°\text{C}^{-2}$
C_{90}	$-4.183 \times 10^{-12}°\text{C}^{-4}$	$-4.2375 \times 10^{-12}°\text{C}^{-4}$

These ITS-90 coefficients provide that the α value

$$\alpha = \frac{R(100°\text{C})}{R(0°\text{C})} - 1 \qquad (15)$$

will be $0.00385055°\text{C}^{-1}$, which is very close to the old standard of 0.00385, and are said to permit the calibration of a real thermometer with the same initial sensitivity as the table to within ±0.1°C from −200 to +600°C.

Since the early 1970s, development work has been reported on thick-film, thin-film, and other resistance elements which are made by deposition rather than wound from wire. Such thermometers might have certain advantages of size, response time, cost, etc. To date, they do not have the stability or interchangeability of wrought-wire elements.

Among nonmetallic resistance thermometers, an important class is that of thermistors, or temperature-sensitive semiconducting ceramics. The variety of available sizes, shapes, and performance characteristics is very large. One manufacturer lists in his catalog a choice of characteristics ranging from 100 Ω at 25°C to 1 MΩ at 25°C!

The thermistor material is usually a metal oxide, eg, manganese oxide. Dopants, eg, nickel oxide or copper oxide, may be added to obtain a variety of resistance and slope characteristics. The material is usually sintered into a disk or bead with integral or attached connecting wires. Figure 5 shows a typical series of steps in the production of a disc thermistor.

Bead thermistors are formed by placing two wires, commonly of platinum, in close proximity and parallel to each other and bridging them with a drop of slurry, which is then sintered into a hard bead and encapsulated in protective glass. Such thermistors are quite stable, approaching, over narrow temperature limits, the stability of industrial metallic thermometers. However, the resistance tolerance may vary from unit to unit by as much as ±20%, and matching or interchangeability is usually

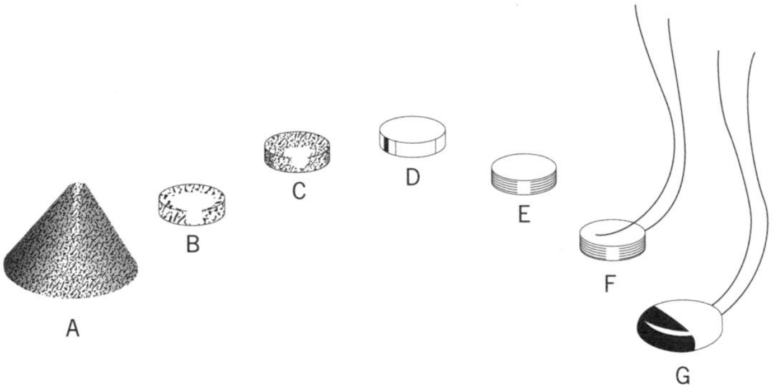

Figure 5. Steps in the manufacture of a disk thermistor. A, ball-milled powder; B, pressed disk; C, sintered disk; D, solvered disk; E, edge-ground disk; F, lead wires attached; G, epoxy-coated.

achieved by selection. Beads can be made quite small, which may allow application in, eg, temperature probes mounted in intravenous needles.

Disc thermistors can be produced to close limits of interchangeability, eg, ±0.1 and ±0.05°C. Disks cannot be made as small as the smallest beads; 2 mm diameter seems an approximate practicable limit. Discs have been historically considered to be less stable than good beads. They are commonly protected with a coating of epoxy resin, which provides less compressive support than the glass coating of bead thermistors. Recent developments have resulted in interchangeable glass-encapsulated disc thermistors which have the stability characteristics of the best beads.

Two characteristics of thermistors are distinctly different from those of metallic resistance thermometers: (1) For all but special types, the temperature–resistance characteristic is negative; that is, resistance decreases with increasing temperature (all metal resistance thermometers exhibit a positive characteristic). (2) The temperature–resistance characteristic is very nonlinear. One typical production thermistor changes resistance 376 Ω/K at 0°C and 13.5 Ω/K at 70°C. Over favorable portions of the nonlinear characteristic, the change in resistance can be more than an order of magnitude larger than that of metallic resistance thermometers. This high dR/dT is a distinct advantage where high sensitivity is required over narrow ranges, as it is in many biological, medical, and environmental situations.

Many attempts have been made to develop an equation based on the semiconducting transfer function of thermistors. These have usually been of the form

$$p^{-1} = f(T)\exp\left(-\frac{\Delta E}{2kT}\right) \tag{16}$$

where E is the energy gap between the valence band energies, T is the kelvin temperature, and k is Boltzmann's constant. None of these efforts has produced a simple law which fits the calibration data. The equation generally used is an empirically derived polynomial (4):

$$T^{-1} = A + B[\ln R(T)] + C[\ln R(T)]^3 \tag{17}$$

known as the Steinhart–Hart equation. Here A, B, and C are derived from calibration at suitable temperatures. It is often useful to have this equation explicit in R, and this may be written as follows:

$$R = e(\exp)\,[(\delta - \tfrac{1}{2}\alpha)^{1/3} - (\delta + \tfrac{1}{2}\alpha)^{1/3})]^{1/2} \tag{18}$$

where $\alpha = (A - 1/T)/C$ and $\delta = (\frac{1}{4}\alpha^2 + \frac{1}{27}\beta^3)^{1/2}$, where $\beta = B/C$ and A, B, and C are the coefficients of Eq. (17).

The nonlinearity of thermistors has been a challenge to a number of designers. With modern circuit components, easily the most flexible way to linearize a curve is with digital active circuitry, breaking the thermistor characteristic at a suitable number of inflection points and assigning an appropriate slope to each line segment between these. Other solutions, not requiring active circuitry but often involving networks of several thermistors combined with several fixed resistors, have also been found. These passive circuits achieve approximate linearity at the cost of sensitivity, the sensitivity being not much higher than that of platinum sensors, which are essentially linear over short temperature intervals.

SEEBECK EFFECT THERMOMETERS (THERMOCOUPLES)

Thermocouples are composed of two dissimilar materials, usually in the form of wires, that accomplish a net conversion of thermal energy into electrical energy with the occurrence of an electrical current. Unlike resistance thermometers, whose response is proportional to temperature, the response of thermocouples is proportional to the temperature difference between two junctions. Figure 6 illustrates such a circuit.

There are three fundamental rules regarding thermocouples: (1) no thermal electromotive force (emf) is produced if heat is applied to a circuit comprising a single homogeneous conductor, (2) the thermal emf of a circuit comprising any number of conductive materials at uniform temperature is zero, and (3) the thermal emf developed by a pair of homogeneous but dissimilar conductors having junctions T_1 and T_3 is the same as the thermal emf developed by an arrangement of the same conductors having junctions at T_1 and T_2 and at T_2 and T_3. This latter principle is illustrated in Figure 7.

It is a common misconception that the generation of the thermal emf is an effect which takes place at the junction between the two wires. Rather it is a bulk effect of the material. Electrons in metals have an electrical charge and a kinetic energy. Along a thermal gradient, charge does not change but kinetic energy does; therefore a transfer of heat must occur. The thermal emf results from a slight rearrangement of electrons in the conductor to respond to thermal transfer. The function of the junction is to assure electrical conductivity between the conductors, and as a consequence, it does not matter how the junction is made (eg, by welding, brazing, clamping, etc.) as long as electrical conductivity is preserved. The assumption of homogeneity of the first and third rules is a condition never actually met in practice and cannot be assumed to be preserved throughout the working life of the thermocouple because of strain, ion migration, etc. For precise work, it is important to verify the performance

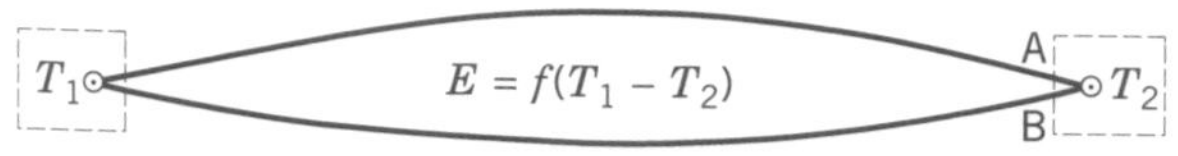

Figure 6. Basic thermocouple circuit. A and B are wires of different materials.

of the thermocouple by calibration at appropriate intervals.

The IPTS-68 stipulated a standard thermocouple whose negative leg was pure platinum and whose positive leg was an alloy of 90% platinum and 10% rhodium as the interpolation standard instrument between the freezing point of antimony (630.74°C) and the freezing point of gold (1064.43°C). (*Note:* These temperatures are given on the IPTS-68; different numbers have been assigned on the ITS-90.) Few sensing elements appear more simple than a thermocouple; few sensors are more subtly and deceptively complex. Errors of the type S Standard thermocouple can be due to accidental alloying of the constituent metals due to metallic vapors in the immediate atmosphere; nonmetallic vapors such as phosphorus, sulfur, arsenic, or easily reduced oxides. The positive element of 90Pt–10Rh can be affected by neutron flux, since rhodium gradually transmuted to palladium. The negative Pt leg forms stable isotopes and is relatively immune to transmutation but may be work hardened by fast neutron bombardment. Calibrations are valid only in the fully annealed state. For these reasons, standard thermocouples increased the uncertainty of realization of the IPTS-68 by a factor of about 10, compared to the SPRT. In the ITS-90, the standard thermocouple has been eliminated by extending the range of the SPRT upward to 962°C and the range of the radiation thermometer downward to this same temperature.

While no longer an interpolation standard of the scale, the thermoelectric principle is one of the most common ways to transduce temperature, although challenged in some disciplines by small industrial platinum resistance thermometers (PRTs) and thermistors. Thermocouple junctions can be made very small, and in almost infinite variety, and for base metal thermocouples the component materials are very cheap. Properties of various types of working thermocouple are shown in Table 3; additional properties are given in reference 5.

The wires of thermocouple pairs must be electrically insulated from each other and protected from the environment. Manufacturers' catalogs indicate the sort of insulation and protection available, including servings of high-temperature ceramic and glass fibers, bare wire in double-bore ceramic beads or tubes, and metal tubing filled with compacted ceramics. Both common and exotic materials are included, and manufacturers' recommendations should be sought. Essentially, the enclosure should provide mechanical protection, but not, through differential expansion, eg, contribute to mechanical strain; provide protection against contamination and not contribute to it via ion migration or evolution of contaminating vapors; and maintain the interconductor insulation. To most potential alterations of state, particularly progressive surface alterations, thick wires may be less susceptible than thin wires.

Another potential source of error, particularly in type K thermocouples between 200 and 500°C, is reversible change of state. Such changes are thought to be the result of lattice ordering–disordering. Reversible changes are particularly subtle, may not be readily discoverable by calibration, and may seriously compromise measurement results. Also, although letter designations by the American Society for Testing and Materials (ASTM) for thermocouple pairs are specific throughout the United States, the nominal calibration designated by the same letters may be different in other countries and in international standards. It is important to determine the specific characteristics of a specific lot of wire, particularly where making replacements.

In industrial applications it is not uncommon that the thermocouple must be coupled to the readout instrument or controller by a long length of wire, perhaps hundreds of feet. It is obvious from the differential nature of the thermocouple that, to avoid unwanted junctions, extension wire be of the same type (ie, for a J thermocouple the extension must be type J). (Where the thermocouple is of a noble or exotic material, the cost of identical lead wire may be prohibitive; manufacturers of extension wire may suggest compromises which are less costly.) Junctions between the thermocouple leads and the extension wire should be made in an isothermal environment. The wire and junctions must have the same electrical integrity as the thermocouple junction. Because the emf is low, enclosure in a shield or grounded conduit should be considered.

Since the thermal emf of the thermocouple is a function of the difference between the hot end and the cold end (usually located at the readout instrument), the measurement can be no more accurate than the isothermality of

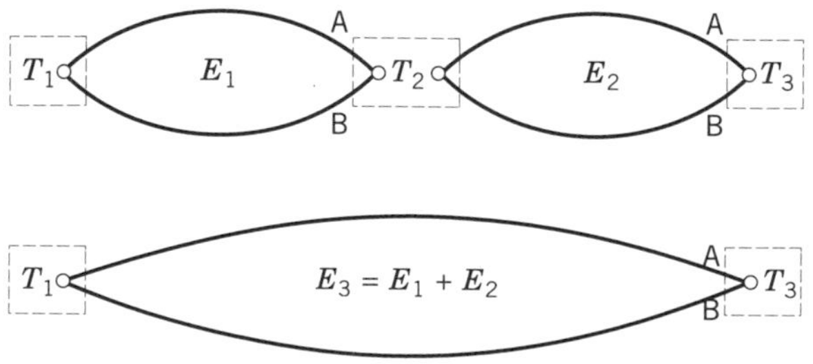

Figure 7. Rule 3 of thermocouples. A and B, the two legs of the thermocouple, are wires of dissimilar materials. Junctions are at temperatures T_1, T_2, and T_3. E_1, E_2, and E_3 are thermal emfs. The thermal emf at T_3 is the same for both circuits.

Table 3. Properties of Various Thermocouple Pairs

ASTM Type	Materials	Range °C	μv/°C[a]
J	Iron-constantan	0–760	64.3
K	Chromel-alumel	0–1260	36.5
R	Pt/87Pt-13Rh[b]	0–1450	13.8
S	Pt/90Pt-10Rh[b]	0–1450	11.8
T	Copper-constantan	−183–375	53.0
E	Chromel-constantan	0–875	78.5
B	70Pt30Rh/94Pt6Rh[b]	870–1700	11.6

[a] Called the Seebeck coefficient. It is given here at the high-temperature end of the range; many thermocouples are nonlinear.

[b] Indicates, eg, one leg of pure platinum, the other of an alloy of 90% platinum and 10% rhodium.

the leads at the cold end and the accuracy with which this temperature is known. Three common methods for addressing these problems are

1. The cold-junction temperature can be fixed by immersing the cold junctions into some known thermal environment: an ice bath or a properly maintained water triple-point cell. A temperature-controlled oven at a temperature above ambient may be used.
2. The cold junction may be attached to or contained within an isothermal mass (eg, a large block of copper, adequately insulated from ambient) whose temperature is measured by some independent means (eg, an embedded thermistor or PRT).
3. Cold-junction compensation can be provided by a network which includes a constant voltage source and a temperature-sensitive bridge to provide an offset voltage which is proportional to the temperature sensitivity of the thermocouple and of opposite sign.

Not all elements of the industrial thermocouple need be wire. For example, if a copper pipe contains a flowing fluid whose temperature is to be measured, a constantan wire attached to the pipe will form a T, or copper–constantan, thermocouple. Such arrangements are difficult to calibrate and require full understanding of the possible inherent problems (eg, Is the copper pipe fully annealed? Homogeneous? Pure, or an alloy? Many ingenious solutions to specific measurement problems are given in reference 6.

MEASUREMENT OF EMITTED RADIATION

Above 962°C, the freezing point of silver, temperatures on the ITS-90 are defined by a thermodynamic function and an interpolation instrument is not specified. The interpolation instrument universally used is an optical pyrometer, manual or automatic, which is itself a thermodynamic device.

The main elements of a disappearing-filament optical pyrometer are shown in Figure 8. There is an optical system for viewing the radiating target, a lamp filament, appropriate filters, an eyepiece, and a calibrated means for varying the lamp current and consequently the brightness of the filament. In the optical path, the filament is visible against the radiating source. The lamp filament current is varied until the image of the glowing filament is of such brilliance that its separate image is indistinguishable from the brilliance of the radiating source. The lamp current is then noted. If a second radiating source also causes the lamp filament to be indistinguishable at that lamp current, the temperature of the second source is said to be the same as that of the first.

Absolute calibrations on the ITS-90 are performed by sighting on a target which is a blackbody at the freezing temperature of silver, gold, or copper. The ideal blackbody is a closed cavity whose walls are at uniform temperature. At every point on the interior wall, radiation is emitted and absorbed, so that the emission and absorption are in equilibrium. If this condition is fulfilled, there is no net loss of energy, and the internal energy is a function of temperature only. For the sake of sighting into such a

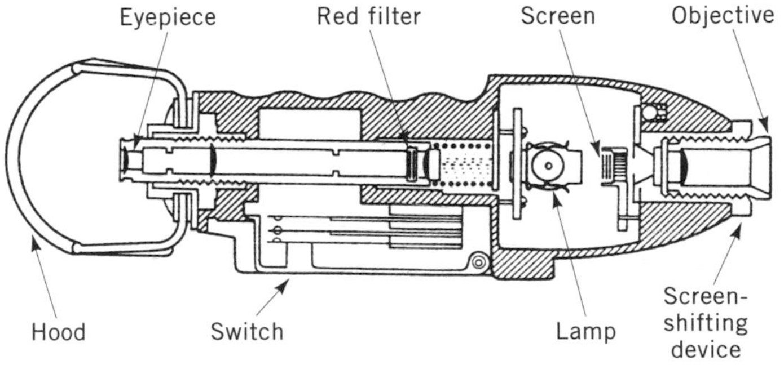

Figure 8. Optical pyrometer.

cavity, a small hole in the wall may be provided which, if it is of proper size and shape, does not disturb the equilibrium condition in a significant way. Under these circumstances the energy emerging from the viewing port has the same energy as the interior of the cavity, which corresponds to the temperature of the blackbody.

A simplified drawing of a blackbody and its furnace is shown in Figure 9. A graphite cavity is surrounded by a graphite crucible, so that the cavity can be surrounded by a shell of silver, gold, or copper. One end of the cavity is open and leads to a conical viewing port. The integral furnace is used to melt and freeze the metal. Such a system, idealized and properly proportioned, may have a temperature reproducibility of 0.02°C and an effective emissivity of 0.99999%; ie, it departs from the ideal by only 10 ppm. The simple blackbody and furnace shown is intended as a calibration device for use by industrial standardizing laboratories.

The total energy radiating from a blackbody of unit area is given by the Stefan-Boltzman law:

$$E = \sigma T^4 \quad (19)$$

where T is the absolute temperature in kelvins and σ is the Stefan–Boltzmann constant, 5.6697×10^{-12} W/cm^{-2} T^{-4}. This expression contains no term for frequency or wavelength, which is distributed over a large spectrum with most of the energy in the infrared. If the spectral radiance N_{bg} is defined as that portion of the blackbody radiance which is at wavelength λ, then

$$N_{b\lambda} = \frac{C_1 \lambda^{-5}}{\exp(C_2/\lambda T) - 1} \quad (20)$$

where C_1 and C_2 are the first and second radiation constants, respectively. The term -1 in the denominator was added by Planck when he observed that Wien's law did not fit experimental data at high temperatures. To explain this stroke of pure intuition, it was necessary for him to invent the quantum theory, for which he received the Nobel Prize in 1918.

This equation permits the evaluation of the energy emitted by a blackbody at a specific wavelength and comparison of it to the energy emitted by a second blackbody at the same wavelength but at a different temperature. Letting T_0 be a reference temperature,

$$N_b(\lambda, T)/N_b(\lambda, T_0) = \exp \frac{C_2}{\lambda T_0} - \frac{1}{\exp(C_2/\lambda T)} - 1 \quad (21)$$

A strip lamp is a convenient means for the calibration of secondary pyrometers (Fig. 10). The notched portion of

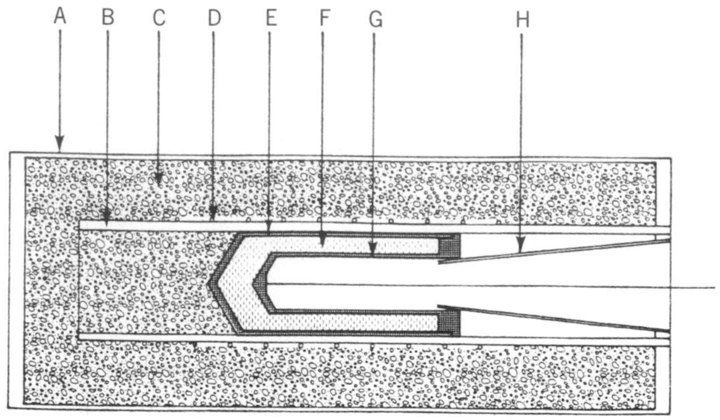

Figure 9. Blackbody and furnace. A, furnace shell; B, furnace tube; C, thermal insulation; D, heater winding; E, outer wall of graphite crucible; F, pure metal (gold or copper); G, inner wall of graphite crucible; H, sighting cone; I, target.

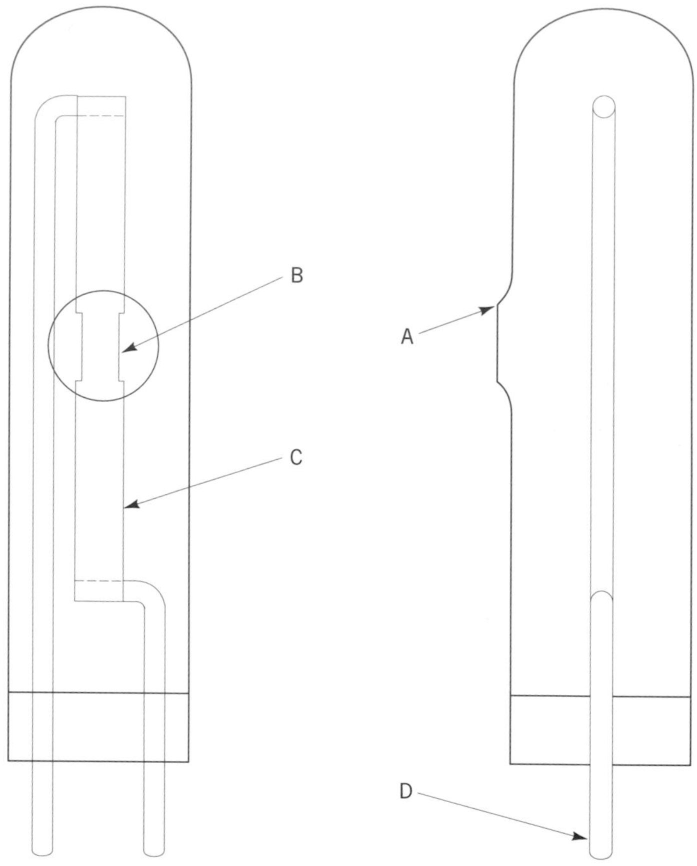

Figure 10. Strip lamp. A, plane window; B, target area of strip; C, strip; D, base pins.

the tungsten strip is the target. A pyrometer which has been calibrated against a radiating blackbody is sighted on the target, and the strip lamp current is adjusted to radiate at the intensity of the blackbody, as transferred by the primary pyrometer. The secondary pyrometer is then substituted for the primary, and the current required to raise the lamp filament to the brilliance of the target of the strip lamp is noted.

MEASUREMENT OF RADIATION FROM REAL OBJECTS

Real objects, when they are in an environment generally hotter or cooler than themselves, radiate, absorb, and reflect energy. The portion radiated is called the emissivity ε. If the portion reflected is r, then

$$\varepsilon + r = 1 \tag{22}$$

No object can radiate more energy than can a blackbody at the same temperature, since a blackbody in equilibrium with a radiation field at temperature T radiates exactly as much energy as it absorbs. Any object exhibiting surface reflection must have emissivity of less than 1. Since pyrometers are usually calibrated with respect to blackbodies, this can cause a serious problem in use. The emissivities of some common materials are listed in Table 4.

The filter and screen of the pyrometer shown in Figure 10 require specific mention. From equation 21 it is evident that the observed radiation must be limited to a narrow bandwidth. Also, peak intensity does not occur at the same wavelength at different temperatures. The pyrometer is fitted with a filter (usually red) having a sharp cutoff, usually at 620 nm. The human eye is insensitive to light of wavelength longer than 720 nm. The effective pyrometer wavelength is $\approx$655 nm.

It is also necessary to reduce the intensity of the radiation admitted into the pyrometer, because pyrometer lamp filaments should not be subjected to temperatures exceeding 1250°C, The reduction is accomplished by a screen or screens; in manually operated secondary pyrometers they are usually neutral-density filters.

Several types of secondary pyrometer are available. In addition to those that measure by varying lamp current, some pyrometers maintain the lamp at constant current but interpose a wedge of graduated neutral density, whose position is a measure of temperature. Also, automatic pyrometers are available in which the eye is replaced by a detector and the measuring element is operated by a servo. In general, the accuracy of the automatic pyrometer is somewhat less than that achieved manually by a skilled operator.

The problem of emissivity from real materials has stimulated the study of pyrometers that measure radiation at two different wavelengths. The principle of the two-color pyrometer is that the energy radiated from a source of one wavelength increases with temperature at a rate different from that radiated at another wavelength. Thus temperature can be deduced from the ratio of the intensities at the two wavelengths, regardless of emissivity. Two-color pyrometers are not widely used.

OTHER ELECTRICAL THERMOMETERS

Many special-purpose electrical thermometers have been developed, either for use in practical temperature measurement, or as research devices for the study of temperature and temperature scales. Among the latter are thermometers which respond to thermal noise (Johnson noise) and thermometers based on the temperature dependence of the speed of sound.

A novel and useful thermometer, based on the change in resonant frequency of a quartz crystal, is produced by

Table 4. Emissivities of Common Materials, = 0.65 μm

Material	Emissivity Solid	Emissivity Liquid
Carbon	0.8–0.9	
Chromium	0.34	0.39
Copper	0.10	0.15
Gold	0.14	0.22
Iron	0.35	0.37
Manganese	0.59	0.59
Molybdenum	0.36	0.37
Platinum	0.30	0.38
Silver	0.07	0.07
Steel	0.35	0.37
Cast iron	0.37	0.40

Hewlett-Packard. The sensor probe contains a precisely cut quartz crystal about 8 mm in diameter. Each quartz sensor has a unique temperature-vs-frequency characteristic, and each sensor is individually calibrated with its operating electronics. Sensitivities of 100 μK are obtained at −50 to 150°C (the effective limits of measurement) and digital readout is provided. The input signal to the conditioner is very low, and the long-term stability is such that frequent adjustment of zero against a reference standard (eg, the melting point of gallium) is required. However, direct readout in units of temperature and great sensitivity make this system a useful and practicable device in many environmental and energy situations, and it is particularly favored by oceanographers.

Yokagawa Electric Works has developed a thermometer based on the nuclear quadrupole resonance of potassium chlorate, usable over the range from −184 to 125°C. This thermometer makes use of the fundamental properties of the absorption frequency of the ^{35}Cl nucleus, and its calibration is itself a constant of nature.

Conventional electrical thermometry is difficult or impossible or involves unacceptable risks in some measurement situations. These include measurements where the process would interfere with the measurement, eg, in microwave and dielectric or inductive heating, high-voltage, high-frequency testing, and studies of the biological effects of electromagnetic radiation; measurements where the presence of an electrical potential is unacceptable, such as where flammable or explosive materials are present; or where the presence of any electrical field would perturb the quantity to be measured. An example is medical patient monitoring during electrosurgery or electrocautery, since the insulated leads of electrical thermometers at these frequencies may become capacitive conductors. A thermometer introduced by Luxtron uses rare-earth phosphors at the end of a thin fiber-optic line and resolves many of these problems. The phosphors are excited by ultraviolet illumination. The fluorescent emissions at different wavelengths have different temperature dependences, but the intensities have a linear dependence upon intensity of excitation. Thus the ratio of emissions is dependent upon temperature but independent of excitation.

NONELECTRICAL THERMOMETERS (LIQUID IN GLASS)

The thermal expansions of liquids are reliable analogues of temperature. Liquids are suitable for use between their freezing and boiling points; mercury and colored alcohols are common materials. The typical liquid-in-glass thermometer includes a thin-walled glass bulb attached to a capillary stem partially filled with a visible liquid and sealed against the environment. The portion of the capillary above the liquid is usually filled with a dry gas under some pressure to avoid column separation. A scale is etched or fused onto the stem. If the range of the thermometer does not include 0°C, a reference mark below the scale at 0°C is usually provided to give a calibration check at the ice point. A contraction chamber may be provided below the main capillary to avoid the need for a long stem before the low end of the calibrated portion of the stem.

There are problems to be considered and avoided when using liquid-in-glass thermometers. One type of these is pressure errors. The change in height of the mercury column is a function of the volume of the bulb compared to the volume of the capillary. An external pressure (positive or negative) which tends to alter the bulb volume causes an error of indication, which may be small for normal barometric pressure variations but large when, for example, using the thermometer in an autoclave or pressure vessel.

Immersion errors are another type. Not only the liquid, but the glass as well, has a thermal expansion coefficient. Three types of thermometer, each having a different immersion requirement, are common. *Complete-immersion* thermometers are calibrated for use with the entire thermometer immersed. *Total-immersion* thermometers are calibrated for use with that portion of the thermometer immersed which contains the liquid column. *Partial-immersion* thermometers are calibrated with the thermometer immersed to a specific index mark on the stem. Most liquid-in-glass thermometers used for accurate measurement are of the total immersion type. However, it may be difficult to read a thermometer whose fluid column is at or below the surface of the medium. In practice, the thermometer is immersed to a depth that allows the meniscus to be seen. The emergent portion of the liquid is then subject to temperature gradients. In this case, and in the case of partial-immersion thermometers, it is necessary to make a correction for the temperature of the emergent stem which contains liquid by measuring the stem temperature with a second thermometer and calculating the correction by means of the equation

$$C_s = KN(t_1 - t_2) \tag{23}$$

where C_s is the correction for stem emergence to be added algebraically to the temperature indicated by the thermometer, K is a coefficient representing the difference in expansion coefficient between the glass and the thermometric liquid (for mercury and usual glass this is 0.00016 for Celsius thermometers), N is the length of liquid column extending above the medium expressed in degrees of the scale, t_1 is the temperature of the thermometer bulb, and t_2 is the average temperature of the liquid column extending above the medium.

Glass-creep errors are also encountered. The liquid-in-glass thermometer should always be used to measure temperatures in ascending order. If the thermometer is stored at room temperature, a temporary ice point depression results, which may be as much as 0.01 K for 10 K of temperature difference, when the thermometer is heated above room temperature. If the thermometer is used to measure a temperature and must then be used to measure a lower temperature, the thermometer should be stored at a still lower temperature for at least 3 days prior to use to assure recovery of bulb dimensions.

Parallax errors are occasioned by failure to read the thermometer from a position exactly normal to its tube axis. Parallax can be reduced by taking the mean of several observations by skilled observers but can be eliminated only by use of a filar telescope mounted so as to move along an axis parallel to the thermometer stem.

Mercury thermometers are subject to separation of the mercury column or to inclusion of bubbles of the fill gas. These may result from shipping and handling and cause a scale offset which can usually be seen upon visual examination and are always recognized by a 0°C verification check. Manufacturers will suggest means by which these temporary defects may be cured.

Frictional errors result from failure of the liquid to perfectly wet the small thermometer bore. They can usually be removed by lightly tapping the thermometer stem, eg, with a wooden pencil, before reading.

Despite these problems, liquid-in-glass thermometers are a (generally) inexpensive means for making temperature measurements of modest accuracy. Many types of liquid-in-glass thermometer are available. Among them are thermometers intended for the performance of specific

Table 5. Temperature-measuring Devices

Device	Manufacturer	Price
Water triple-point cells	Jarrett Instrument Co, Wheaton Md. (USA)	Approx $1000
	Isothermal Technology Ltd Southport, England, and New York	Approx $1,200 with certification
Water triple-point maintenance baths	Isothermal Technology Ltd Southport, England and New York	Approx $4,000 to $15,000
Cells, furnaces, etc, for freezing points of metals	Isothermal Technology Ltd Southport, England and New York	Cells $4000 to $12,000 Furnaces $15,000 to $19,000
Standard platinum resistance thermometers	Isothermal Technology Ltd Southport, England and New York	$2,000 to $5,000 depending upon type and calibration
	YSI Inc. Yellow Springs, Ohio USA	$3,000 to $5,000 depending upon type and calibration
Industrial-grade resistance thermometers	Isothermal Technology Ltd Southport, England and New York	$50 and up, depending upon configuration, etc
	Minco Corporation Minneapolis, Minn. USA	$50 and up, depending upon configuration, etc
	DeGussa AG, Hanau, Germany	$50 up, depending upon configuration
Precision thermistors	YSI Inc Yellow Springs, Ohio USA	Varies with configuration
	Thermometrics, Inc Edison, N.J. USA	Varies with configuration
Standard thermocouples	Isothermal Technology Ltd Southport, England and New York	$500 to $2,000 depending upon configuration and calibration
	Leeds and Northrup Company North Wales, Pa. USA	$500 to $2,000 depending upon configuration and calibration
Industrial thermocouples	Isothermal Technology Ltd Southport, England and New York	Varies with configuration
	Omega Engineering, Stamford, Conn. USA	Varies with configuration
	Yamari Industries Edobori, Osaka, Japan	Varies with configuration
Optical pyrometers	Leeds and Northrup Company North Wales, Pa. USA	≥$5,000
Quartz thermometers	Hewlett-Packard Co Mountain View, Calif. USA	Price on inquiry
NQR thermometers	Yokagawa Electric, Tokyo Japan	Price on inquiry
Fiber-optic thermometers	Luxtron Corp Mountain View, Calif. USA	Price on inquiry
Liquid-in-glass thermometers	Cole-Parmer Co., Chicago, Ill. USA Fischer Scientific Co., Chicago, Ill. USA	Price varies with range, quality, application, certification, etc.

tests, such as ASTM tests; thermometers for very short ranges, eg, the NIST Standard Reference Materials 933 and 934 used in clinical chemistry; thermometers for measuring very small differentials, eg, the Beckman thermometer design; thermometers with fixed or movable electrical contacts for alarm and control functions; and maximum, minimum, or maximum–minimum reading thermometers, which show the extremes of temperatures reached during the period since last reset.

NONELECTRICAL THERMOMETERS (MISCELLANEOUS)

Other nonelectrical thermometers are bimetal, filled-system, and pyrometric cone thermometers. In bimetal thermometers, two strips of metal of differing expansion characteristics are welded together face to face. If one end or such a strip is fixed, the strip bends in response to temperature change as the metal of higher coefficient produces a longitudinal dimensional change. If the strip is formed into a helix, the free end rotates and can actuate a pointer or switch. The filled-system thermometer comprises a rigid bulb, a connecting metal capillary, and a flexible bulb, eg, a metal bellows. The whole system is filled with a liquid or gas of known thermal expansion properties. As temperature change causes the gas to expand, the flexible bulb is displaced and the displacement is sensed as a mechanical motion. Pyrometric cones are geometric shapes, commonly conical, which slump or collapse at some (usually high) temperature. A main use is in estimating peak temperature in kilns.

See Table 5 for a list of temperature-measuring devices.

BIBLIOGRAPHY

1. H. E. Sostmann, *Isotech Journal of Thermometry* **3**(2), 106–113 (1992).
2. H. Preston-Thomas, *Metrologia* **12,** 1 (1976).
3. H. Preston-Thomas, *Metrologia* **27,** 3–10 (1990).
4. J. S. Steinhart and S. R. Hart, *Deep-Sea Research* **15,** 497 (1968).

Reading List

R. P. Benedict, *Fundamentals of Temperature, Pressure and Flow Measurement,* Wiley, New York, 1969.

F. G. Brickwedde (ed.), *Temperature, Its Measurement and Control in Science and Industry,* vol. 3–6, Reinhold, New York, 1962–1992.

H. D. Baker, E. A. Ryder, and N. H. Baker, *Temperature Measurement in Engineering (2 vols.),* Wiley, New York, 1961.

H. B. Sachse, *Semiconductive Temperature Sensors and Their Applications,* Wiley, New York, 1975.

Manual on the Use of Thermocouples, ASTM STP 470, American Society for Testing and Materials, Philadelphia, Pa., 1993.

THERMAL POLLUTION, POWER PLANTS

Charles C. Coutant
Oak Ridge National Laboratory
Oak Ridge, Tennessee

An important by-product of most energy technologies, in the broad sense, is heat. Few energy conversion processes are carried out without heat being rejected at some point in the process stream. Historically, it has been more convenient and less costly to reject such waste heat to the environment than to attempt significant recovery. Although periodic energy scarcities and the high costs have stimulated innovative heat-recovery schemes, the low temperatures of waste heat in relation to process requirements often make reuse impractical and disposal the only attractive alternative.

Heat rejected to the environment by most industries is of little consequence. Cooling flows of air or water are deployed over equipment or through heat exchangers and the relatively small quantities of heat are dissipated to the surrounding air. Small cooling towers, often of the evaporative type, are ubiquitous fixtures of an industrial facility.

Concern over heat rejection arose when the quantities at localized sites rose dramatically as the electric utility industry shifted to water-cooled, thermal-electric generating stations of high unit capacity in the 1950s, particularly in the UK and the United States. Concern was further heightened by a planned shift to nuclear fission reactors as the energy source, which entailed both a further increase in localized generating capacity because of economics of scale and a higher percentage of rejected heat compared to useable electrical energy. In the late 1950s and 1960s, the volumes of cooling water projected to be used by the United States electrical supply system and heated to temperatures 10°C or more above intake temperatures were startling. For example, it was estimated that perhaps 25% of the total available freshwater runoff in the entire nation would be needed for cooling purposes by 1980. The term thermal pollution took on fearsome portents among aquatic scientists, fishery managers, and eventually water-pollution control agencies (1). Directly lethal effects of high temperatures on aquatic life were predicted and, where sublethal temperatures were maintained, effects on reproductive cycles, growth rates, migration patterns, and interspecies competition were hypothesized based on the pervasive nature of temperature as a factor which controls ecological processes (2).

Largely because of these fears and the regulatory attention they attracted, much research and monitoring of thermal effects at power stations was conducted in the decades of the 1970s and 1980s. Both the increased knowledge gained thereby and more stringent regulations led to approaches for using biological requirements of aquatic organisms and local environmental characteristics of the rivers, lakes, and estuaries used for cooling to design nondamaging cooling systems specific for a site. The rate of increase in demand for electricity and new generating stations also diminished, allowing new power plants to be evaluated more thoroughly and located in less susceptible environments. Approaches that evolved by the mid-1980s are still current practice.

See also Electric power generation; Water quality issues; Sustainable resources, ethics.

RISKS

Power-station cooling is fairly straightforward. Generation of electricity by the steam cycle, the most common method regardless of fuel type, ie, coal, oil, gas, nuclear,

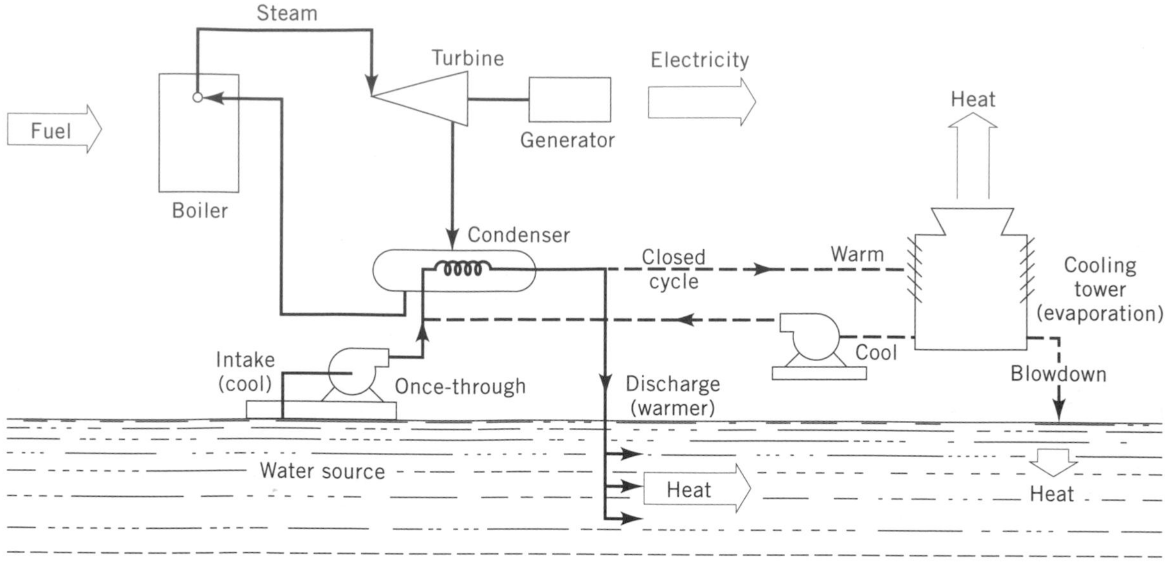

Figure 1. The energy cycle of a thermal electric generating station with two alternative cooling systems: the open-circuit or once-through system (solid lines) and a representative closed-cycle, cooling-tower system (dashed lines). Reproduced by permission from ref. 31.

solid waste, entails production of waste heat (Fig. 1). Although there are some atmospheric losses in the steam cycle, most of the heat is rejected to flowing water through the heat of condensing steam. For a modern 1200 MW (electrical) light-water nuclear power plant, this release is ca 7100 kJ/(kW·h) (1700 kcal/(kW·h)), ie, approximately twice the thermal equivalent of the generated electric energy. The amount of reject heat is proportionately less in fossil-fueled plants, in which up to 40% of the fuel energy can be converted to electricity. The flowing water for the steam condenser was traditionally pumped from a nearby lake or river and returned to it at an elevated temperature in the once-through or open-cycle system; a 1000-MW nuclear power plant with once-through cooling requires ca 49 m^3 of water per second (ca 13,000 gal/s), which is elevated by 10°C. Screens at the water intake prevent entry of objects larger than ca 1 cm. A biocide, usually chlorine, is used to clean heat-exchange surfaces and water conduits.

For purposes of clarifying risks to aquatic life, a clear distinction must be made between heat, which is a quantitative measure of energy that depends upon the mass of an object, in this case the volume of cooling water, and temperature, which is a measure of energy intensity. An amount of heat distributed in two unit volumes of water yields half the elevation of temperature as the same amount of heat distributed in one unit. Although heat is the waste product of electricity generation, temperature is the environmental characteristic to which organisms respond. Thus, the quantity of water used to carry away the load of reject heat is crucial to determining the temperatures created in the environment. A risk is considered here to be the biological or ecological damage that could be done by a certain human alteration of the environment with the likelihood or probability that the specific damaging alteration will occur.

Initially, the source of environmental risk from cooling water was assumed to be the pollutant discharged, ie, heat, in the form of the elevated temperature of the water released from the condensers. Heat is now recognized as only one of several potential risks of power-station cooling (Fig. 2). In this article, the broad view of thermal pollution is taken to encompass each of these ancillary risks.

Generally unheralded until the early 1970s was the physical entrapment and impingement of fish on cooling-water intake screens. Always occurring at chronic low levels at small fossil-fueled plants, the losses of fish and some large invertebrates rose markedly with the startup of new nuclear stations. In particular, plants on the Hudson River began to impinge large numbers of juvenile striped bass, a species of considerable importance locally.

Impingement risks for affected species seem more direct and identifiable than the indirect aspects of temperature change because there are more visible, countable deaths. Even when fish can be removed alive from their attachment to screens, the fatigue and damages suffered generally mean poor chances for survival. The effects of impingement include long periods of futile swimming in screen wells, being held by water currents against the screen mesh (often resulting in suffocation through inability to ventilate the gills), and physical injury from the rotating screen or high pressure water spray used to wash the screen. Beyond risks to individual fish are risks for populations of important species that could be seriously depleted by introducing a new form of chronic mortality into the life cycle. Since impingement acts selectively, affecting some species more than others, ie, generally schooling, pelagic fish, fears of shifts in species composition arose.

Serious analysis of the risks associated with power-plant impingement began when environmental impact statements for nuclear power plants were initiated following the U.S. National Environmental Policy Act of 1969 (NEPA) and the ensuing Calvert Cliffs court decision which assigned responsibility for all cooling-water impacts of nuclear stations to the Atomic Energy Commission (now the Nuclear Regulatory Commission). The analogous problem of impingement of downstream migrating

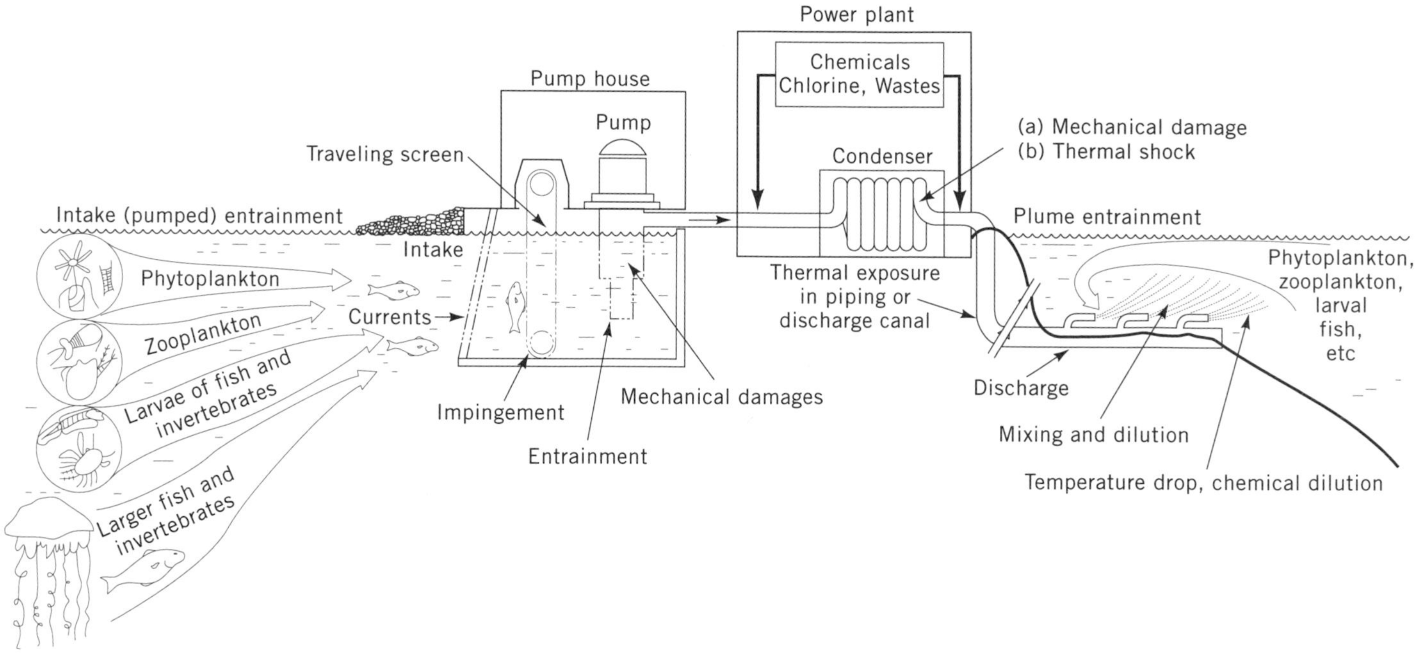

Figure 2. Sources of potential biological damage in the immediate vicinity of a power-station cooling system. Reproduced with permission from Ref. 4.

salmon on screens of irrigation diversion dams had been addressed for many years on the U.S. west coast, but it was not until 1973 that the commonality of both of these problems and of possible engineering solutions was widely recognized (4). Emphasis on solutions was stimulated by the U.S. Federal Water Pollution Control Act Amendment of 1972, which allowed once-through cooling only where intakes minimized losses.

Biocides, principally chlorine, that were used periodically (0.5 hours per day per condenser) for condenser cleaning were identified as toxic risks for organisms in the cooling circuit at the time and for those in the vicinity of the discharge where the biocide dissipates. Concern over residual toxic effects of power-plant chlorination was also largely an outgrowth of NEPA reviews of nuclear stations in the early 1970s, when it coincided with renewed concern over ecological effects of chlorination of treated sewage effluents and evidence of chlorinated organic materials formed in chlorinated water (5,6).

A fourth risk is from combined damages, ie, thermal, physical, and chemical, sustained by small organisms that are pumped through the cooling system, ie, entrainment. These damages were among the earliest to be recognized, but the early emphasis on thermal stresses alone retarded examination of the physical components of the damage, which are now appreciated as the principal risks at many installations (7). During entrainment, any organisms, including phytoplankton, zooplankton, larval fish, invertebrates, and many small fish that cannot swim against the induced current at the cooling-water intake, are drawn into the cooling circuit unless they are large enough to be screened out initially. In the cooling circuit, they receive in rapid sequence (usually <1 min) a series of stresses. These include a pressure drop in front of the pump impeller, risk of physical impact with the impeller or shear stress of a near miss, rapid pressurization downstream of the pump, shear stress as the cooling water is divided among hundreds of condenser tubes ca 2.5 cm in diameter, rapid temperature elevation as heat is transferred to the water through condenser tubes, maintenance of high temperature (usually 8–10°C above ambient) through the discharge system, decreasing pressure in discharge piping (sometimes below atmospheric), followed by turbulent mixing and cooling as the condenser water rejoins the source water body (8). Many entrained organisms do not survive. During periods of biocide treatment to remove heat-transfer-retarding biological slimes from condenser tubes, entrained organisms are also exposed to lethal concentrations of a toxicant, usually chlorine.

Many organisms are exposed to some of the thermal, chemical, and physical stresses of entrainment by being mixed at the discharge with the heated water; this is plume entrainment. The exact number exposed there, which may be many times those pumped through the plant, depends upon the percentage of temperature decline at the discharge that is attributed to turbulent mixing rather than to radiative or evaporative cooling to the atmosphere.

Entrainment was in the forefront of concerns over condenser cooling in the early 1970s as two trends in the planning of power plants developed: gradual decrease in the temperature rise across condensers (ΔT) to reduce thermal effects, which was necessarily accompanied by an increase in volume of water pumped, and increased exploitation by large nuclear stations of estuarine waters, which generally contain far more planktonic eggs and larvae of important fishes and invertebrates than the inland rivers or lakes used previously. Even more than with impingement, the long-term risk with entrainment lies in chronic losses of vulnerable stages to physical stresses which can contribute to instability or decline in populations of valued species. Complex scientific debates over the magnitudes and probabilities of population effects of entrainment mortalities have punctuated power-plant licensing proceedings since 1971, particularly those for the Indian Point Nuclear Power Plant on the Hudson River (9).

Questions also arose over the risks to aquatic life from changes in gas content of the water as a by-product of temperature change; warmer water holds less gas in solution. Dissolved oxygen was hypothesized to decrease below levels necessary for fish by a combination of physical solubility relationships and increased biological demand for oxygen at elevated temperatures. Supersaturation of dissolved gases in water was also viewed, particularly in the northwest United States, as a significant risk arising from heating already saturated river water. Fishery scientists on the Columbia River were especially sensitive to this potential problem, because high gas saturation levels from dam spillways were demonstrably harmful to experimental salmon and trout (10).

Physical changes in habitats near the cooling water intake and discharge structures of power stations were also identified in NEPA analyses as posing some risk or at least potential for change to segments of aquatic communities. Concrete structures, rock jetties, and altered current patterns contribute to habitat modifications that influence suitability of the area for desirable species and can be linked to the overall problem of dissipating rejected heat.

Risks from cooling towers were identified by comparative ecological analyses of cooling-system alternatives and the risks became part of the assessment process. Throughout the 1960s, evaporative cooling towers, whether natural or mechanical draft (see Fig. 1), were seen as the panacea: a closed cooling cycle would replace the transgression of the once-through system on natural waters. Cooling towers operating in the United States and abroad, especially in the UK, seemed to justify their immediate implementation in the United States.

The closed cycle is, however, not really closed. Continuous evaporation would gradually accumulate unwanted salts if there were not some flow through in addition to replacement of cooling water for evaporative losses. This flow through or blowdown carries with it numerous chemicals added to the cooling circuit to prevent corrosion, eg, chromates, zinc, and organophosphorus complexes, to slow mineral scaling of heat-exchange surfaces by, for example, acids or to eliminate biological fouling, eg, by chlorine. Blowdown was traditionally discharged untreated into water bodies that supplied the tower intake, and organisms susceptible to the chemicals were placed at risk. The chemicals also escape the towers in drift in the form of small water droplets and aerosols that are dispersed with air flows to terrestrial surroundings. Cooling-tower sludge, the solid material that accumulates in tower basins, must also be discarded. This sludge includes precipitated chemical elements from the water supply, corrosion inhibitors, and airborne dust, pollen, and tree leaves washed from the tower air flow. At the intake, entrainment and impingement are not eliminated by the addition of cooling towers, although water flows can be reduced to 5–10% of that needed for once-through cooling. The presumed advantage of reduction in water volume for minimizing entrainment risks may be illusory, as organisms entrained into cooling-tower circuits are exposed to lethal conditions, whereas many often survive a rapid once-through passage at well-designed open-cycle plants. Among the first environmental risks recognized from cooling towers were increases in fogging and icing in northern climates.

An alternative closed-cycle system that is used particularly in the Midwest, Southeast, and Texas is the artificial cooling lake. Such lakes exist in a gradient of designs from large multipurpose public reservoirs that provide essentially once-through cooling to privately controlled, diked ponds or canals that rely on evaporative cooling and are often augmented by spray modules. The risks to public resources also vary greatly. Despite the obvious multipurpose benefits of cooling reservoirs where water resources are scarce, these cooling systems were judged in violation of the 1972 Water Pollution Control Act and for several years were not allowed for new power stations. Thus, the number of cooling lakes constructed diminished greatly (11). Although there has been much debate over cooling lakes, numerous water and fish management opportunities from their use has brought them back in favor. Many of the most productive fisheries are fresh water species in lakes also used for power plants.

Changing perceptions of the sources of environmental risk from cooling systems, especially during the period 1968–1973 when thermal concerns were augmented by those of entrainment and impingement, were partly the result of rapid technical improvements in thermal-discharge designs. Spurred by water-temperature standards and limits to sizes of mixing zones imposed by states under the Federal Water Pollution Control Act of 1965, utilities quickly moved away from the traditional discharge schemes. Shoreline canal discharges with plumes of highly elevated temperatures that had been preferred for heat dissipation but which neglected biological requirements gave way to offshore diffusers that exploited turbulent mixing from discharge momentum and buoyancy to quickly diminish temperatures. Such designs had been effective in reducing temperatures rapidly to nondamaging levels at older discharges, such as those in the Columbia River. Risks of thermal damage to aquatic communities were thus diminished, and risks associated with increased cooling-water flow became more apparent.

In the mid-1970s, ecological risks that had commanded the most attention were joined by the risks to human health. The source of risk was the familiar heat additions, but the vector of potential risk, ie, thermophilic microorganisms, was new to most ecological analysts. First to gain attention was the pathogenic amoeba of the genus *Naegleria,* which invades the central nervous system and causes a rapidly fatal meningoencephalitis (12). These protozoans have been isolated from both naturally occurring and artificially heated warm, organic-rich aquatic environments. Evidence is rapidly accumulating that another organism, *Legionella pneumophila,* is also well adapted to warm water and can cause epidemics of pneumonia and general systematic infection, eg, Legionnaires' disease, or a nonpneumonic combination of fever, headache, and myalgia, eg, Pontiac fever (13). Common source outbreaks of Legionellosis have occurred in a pattern that suggests airborne spread of aerosol nuclei generated by cooling towers or evaporative condensers that were documented to contain the organism. Another epidemic occurred among workers cleaning a power-plant condenser. Both amoebae and *Legionella* appear to be basically soil or aquatic-sediment microorganisms for which the ecological niche and dispersal mechanisms to man have been expanded by increasing use of heat-rejection systems, particularly cooling towers. Current practice avoids growths of these organisms by regular cooling tower cleaning and high levels of biocides.

MINIMIZING THE RISKS

Selection of the best cooling system in terms of minimal environmental damages involves matching engineering options to the local aquatic system potentially at risk. General principles of aquatic ecology and of the life histories and environmental requirements of species represented locally can be adapted to local water-resource goals and with detailed understanding of the local aquatic setting to achieve site-specific risk prevention.

Thermal Effects

Temperature is the most all-pervasive environmental factor that influences aquatic organisms. There is always an environmental temperature, whereas other factors may or may not be present. Nearly all aquatic organisms, with the exception of marine mammals, are for all practical purposes thermal conformers or obligate poikilotherms. As such, they are not able to exert significant influence on maintaining a certain body temperature by physiological means. Their body temperatures thus fluctuate in close accord with the temperature of the immediate surrounding water; although especially large, active-moving fish, such as tunas, maintain deep-muscle temperatures slightly higher than the surrounding water. Intimate contact between body fluids and water at the gills and the high specific heat of water assure this near-identity of internal and external temperatures. Behavioral thermoregulation or the control of body temperature by selection of water temperature in natural gradients is, however, a common feature of many fish. Behavioral thermoregulation serves an important ecological role in partitioning aquatic habitats among species, and its understanding is a powerful feature for estimating and ameliorating the impacts of thermal discharges.

Thermal effects on aquatic organisms have been given critical scientific review periodically over the past thirty years or more. In addition, annual reviews of the thermal effects literature have been published since 1968 (14). Water temperature criteria for protection of aquatic life were prepared by the NAS in 1972, and these criteria have formed the basis of the EPA recommendations for establishing water temperature standards for specific water bodies (15,16).

The importance of temperature for organisms lies partly in its fundamental physical–chemical influences. The biochemistry of life is, in general, subject to the basic dictum that a rise of temperature by 10°C results in about a doubling of the rate of chemical reactions. Life also has an upper thermal limit, set partly by the chemical stability of organic molecules, eg, proteins or enzyme systems, and by the balance of food energy inputs and expenditures, that is characteristic of each species and, in some cases, different life stages. Although aquatic organisms must conform to water temperature, many have evolved internal mechanisms other than body-temperature regulation to allow continued functioning as temperature changes occur geographically, seasonally, and daily. Through gradual biochemical adjustments termed acclimation, physiological processes are maintained relatively constant over a prescribed thermal range which is species- or life-stage-specific. Within this range, each species has a zone of optimal physiological performance which, because it tends to coincide with temperatures preferred in gradients, has recently been termed its thermal niche in the environment.

Organisms evolving under annual temperature cycles and in environments with varying temperatures spatially have incorporated thermal cues in reproductive behavior, habitat selection, and certain other features which act at the population level. Thus, the balance of births and mortalities, which determines whether a species survives, is akin to the metabolic balance at the physiological level in being dependent upon the match, within certain limits, to prescribed temperatures at different times of year. At the ecosystem level, relationships among species, eg, predators, competitors, prey animals, and plant foods, are related to environmental temperatures in complex ways. Many of these interactions are poorly understood, and the role of temperature remains speculative in many cases.

Despite the immensity and complexity of known and suspected roles of temperature in aquatic ecosystems, certain thermal criteria have been especially useful in minimizing risks from thermal discharges, at least those short-term ones with which we have the most direct experience. More data have been organized at the physiological level than at higher levels of organization, so that such data have factored prominently in current approaches.

Preventing Mortality

When power stations use small quantities of cooling water, temperatures are raised by large amounts. Organisms entrained in these flows or exposed at the discharge can receive exposures to large temperature rises that, especially in summer, can carry them beyond their upper thermal tolerance. Conversely, organisms that have been nurtured in the warm discharge of a power station in winter can have their cold tolerance exceeded should the power station suddenly shut down. Upper and lower temperature tolerances of aquatic organisms have been well conceptualized over the past 30 yr, and standardized methods are available for determining a species' tolerance ranges under different conditions of thermal history (Fig. 3(a)). On a short time scale, mortality is highly dependent on duration of exposure (Fig. 3(b)), so that brief exposures to potentially lethal temperatures are not actually lethal.

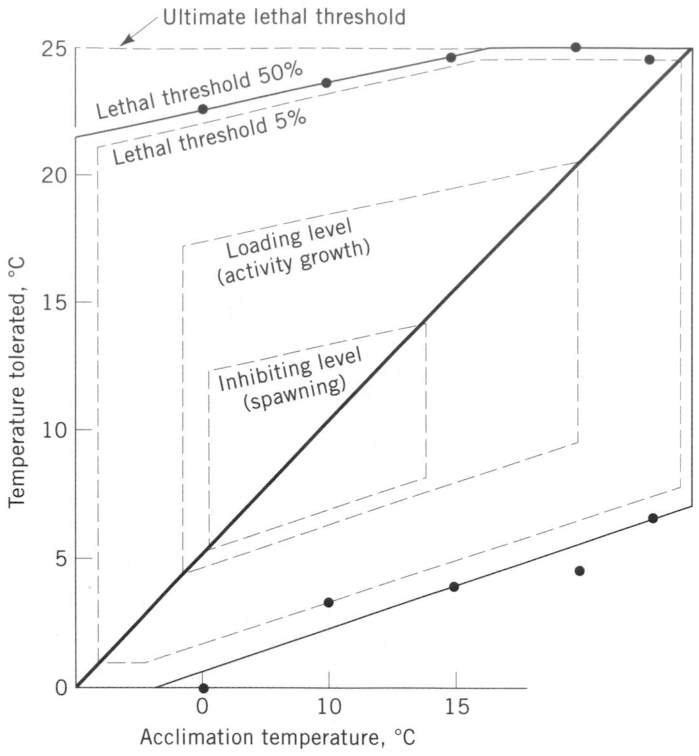

(a)

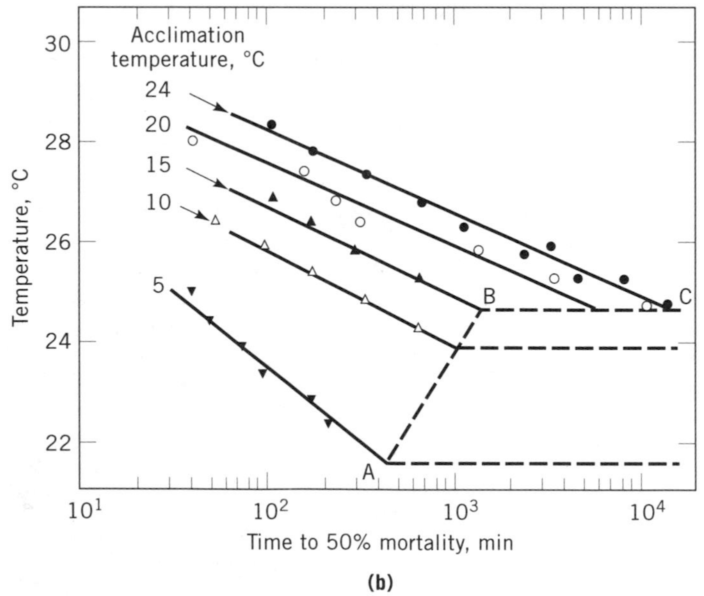

(b)

Figure 3. Lethal temperature thresholds for aquatic species. The patterns are general for all species, but exact temperatures are species-specific. (**a**) Tolerance polygon of upper and lower lethal (50%) temperatures for one-week exposures of an example species (juvenile sockeye salmon) which has been held at the acclimation temperature, with more restrictive thresholds indicated as dashed lines; (**b**) time-dependent mortality (50%) of an example species (juvenile chinook salmon) at temperatures above the one-week lethal threshold after holding at different acclimation temperatures. The dashed line ABC indicates transition to less than 50% mortality at lower temperatures and coincides with the upper lethal threshold of this species' tolerance polygon. Reproduced by permission from ref. 17.

Graphical and mathematical representations of these data for important species at a site can allow a design engineer to tailor temperature elevations and duration of exposure in a plant's piping or in the effluent mixing zone to maximize organism survival (15). This is usually accomplished with detailed mathematical models of cooling-water-effluent dispersion and heat dissipation in the near field where temperatures are highest (18–19). One option used extensively for this purpose in the 1970s was increasing the cooling-water flow and thus decreasing the temperature elevation to below lethal levels. As noted earlier, this tactic had the undesirable result of increasing the numbers of organisms damaged by the physical effects of entrainment and impingement. Thus, there must be a quantitative balance between minimizing direct thermal mortalities and increasing those risks related more to the volume of cooling-water flow.

Preventing Stressful High Temperatures Over Long Periods

For the long term, ie, greater than 1–2 d, simply preventing mortality is insufficient for protecting aquatic species. All of the physiological functions normally performed must be carried out to maintain healthy individuals that are capable of competing in the natural ecosystem. Physiological functions are sufficiently complex and diverse that no single metabolic process has been adequate to define health over a range of temperatures. However, an aggregate measure, growth rate, has proved useful. Growth is an integrator of all physiological functions and some behavioral ones, eg, feeding rate. Growth occurs only if all other metabolic demands are being met and when sufficient food energy is left over for adding biomass. Typically, many physiological functions of well-fed, cold-blooded organisms proceed optimally over a temperature range in which growth rate is maximal (see Fig. 4). Above the temperatures of maximum growth rate, the rate typically declines steeply to a temperature of zero growth, which often occurs 1–2°C below the temperature at which direct mortalities begin. Intuitively, the healthy fish becomes unhealthy as the long-term temperatures it experiences rise from those that yeld maximum growth to that which stops growth. Alternative methods for calculating the upper danger level have been proposed, but each suggests that a long-term decline of growth rate below ca 75% of maximum at high temperatures is unduly risky (15,17).

Standards for upper limits on water temperatures for particular water bodies over periods of about one week or more can be based on species-specific growth rates. With an inventory of important species and life stages in the area during the warm season, the analyst can ascertain the upper temperature limit which does not stress those in the desired aquatic assemblage. Hydrothermal models of heat dissipation in the far field beyond the zone of effluent mixing are important for estimating the zones that may present a long-term risk from elevated temperatures.

Recently, the preferred and avoided temperatures in a gradient have been used as surrogates for optimum growth and upper danger levels for fish. One practical drawback to using growth rate as an index of optimal temperatures is the experimental cost of determining it.

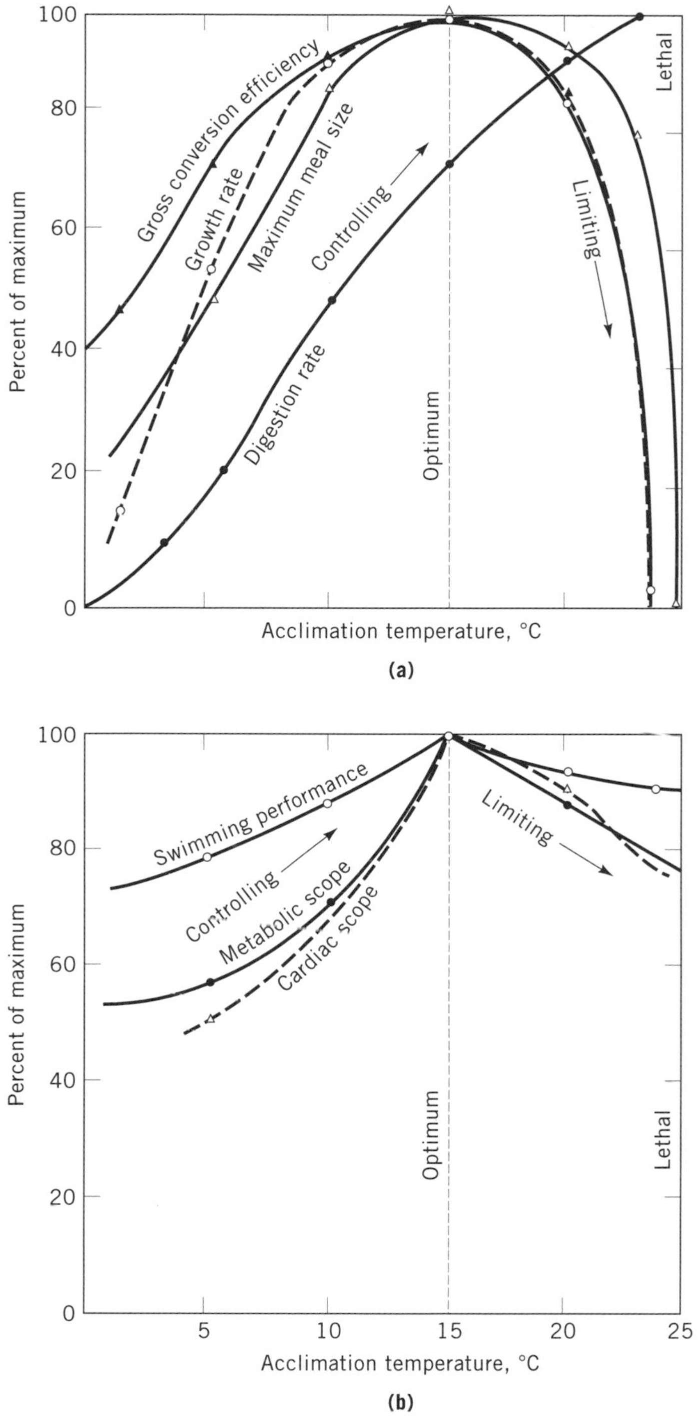

Figure 4. Convergence of numerous physiological functions, including overall growth rate, at an optimum acclimation temperature which is specific for the species, juvenile sockeye salmon. Reproduced by permission from ref. 17.

As more and more data link temperatures of optimal growth rate with preferred temperatures for a species, there is an increasing tendency to conduct only thermal preference tests (20). There is also accumulating evidence that long-term abundance of species at power-plant sites is generally correlated with preferred and avoided temperatures (Table 1).

Preserving Reproduction Cycles

Organisms in the temperate zone have evolved in concert with seasonal temperature variations. Reproductive cycles are generally keyed to the annual cycle, and temperature is often the controlling factor, although changing day length is of equal or greater importance for some species. Reproduction success depends in part on the preservation of an annual temperature pattern, although the precise timing is usually not critical. An analyst can obtain dates and temperatures for spawning of many important aquatic species, and thermal discharges can be designed, usually with the help of far-field mathematical models, to assure the necessary thermal periodicity. Unless the cooling system is a very heavily heat-loaded stream or cooling pond, the thermal output of a power-station complex is rarely sufficient to offset the large natural cooling rates of winter. Thus, annual thermal cycles are generally maintained despite anthropogenic heat rejection.

Maintaining Ecoscystem Structure and Function

Thermal heterogeneity of water bodies is an important structural feature of the environment and plays a large role in determining the composition and functioning of most aquatic systems. Vertical thermal stratification of lakes, reservoirs, and many estuaries in summer (see Fig. 5) segregates an available habitat into discrete zones with differing thermal and water quality characteristics. In the evolution of aquatic species, these discrete zones have been the basis for partitioning the environment, such that different species or life stages within some species occupy discrete portions based, in part, upon temperature selection behavior. This thermal segregation reduces competition for food resources and spawning sites, and often minimizes predation and cannibalism. The disruption of thermal structure by power-station cooling systems can be more damaging to ecosystem composition and function than the direct killing of many individual organisms.

Particularly significant are thermal refuges during periods of extreme high or low temperatures. Aberrant weather or artificial thermal changes can cause large portions of a water body to exceed the physiological limits of some aquatic species for brief periods. The species usually affected are those at the northern or southern limits of their geographic ranges. These species often survive in isolated zones or refuges that retain suitable temperatures, eg, springs where trout can remain cool in the warmest summer months or water bodies where cold-sensitive species, eg, threadfin shad, find relatively warm waters in mid-winter. The demise or survival of such geographically marginal species, which is determined by the availability and sizes of such thermal refuges, can be paramount in establishing the ecological interactions of a water body and ultimately its productivity for sports and commercial fisheries or other uses (22). Power-station discharges can prematurely force desirable cool-water species into refuges in summer and can provide essential warm refuges for desirable warm-water species in winter.

Current thinking is drawing together the previously sketchy and disparate knowledge of thermal heterogeneity of the environment, thermal optima for organism growth rates, and behavioral temperature selection in

Table 1. A Comparison of Preoperational (1968–1971) and Postoperational (1972–1981) Catch Rates of Fish in the Wabash River, Indiana, at the Cayuga Electric Generating Station[a]

Common Name	Species	Preferred Temperature, °C	Avoided Temperature, °C	Percent Change in Catch Rate
Silver redhorse	*Moxostoma anisurum*	24	26	−23%
Shorthead redhorse	*Moxostoma macrolepidotum*	24	26	−17%
Mooneye	*Hiodon tergisus*	24	27	−27%
Golden redhorse	*Moxostoma erythrurum*	24	26	−31%
Sauger	*Stizostedion canadense*	24	28	−22%
Smallmouth bass	*Micropterus dolomieui*	24	27	−9%
Spotted bass	*Micropterus puctulatus*	24	27	−25%
Goldeye	*Hiodon alosoides*	24	27	−48%
Shortnose gar	*Lepisosteus platostomus*		33	−49%
Buffalofish	*Ictiobus* sp.		33	−56%
Blue sucker	*Cycleptus elongatus*			−80%
Skipjack herring	*Alosa chrysochloris*		30	90%
Longnose gar	*Lepisosteus osseus*	30–33	34	155%
Freshwater drum	*Aplodinotus grunniens*	29	31	163%
North River carpsucker	*Carpiodes carpio*		34	178%
Gizzard shad	*Dorosoma cepedianum*		31	217%
White bass	*Morone chrysops*	28–30	31	260%
Carp	*Cyprinus carpio*	29–32	34	498%
Largemouth bass	*Morone salmoides*	27	30	760%
Flathead catfish	*Pyladictus olivaris*		32	1647%

[a] Ref. 21.

ways that are useful for predicting ecosystem effects of thermal alterations (20). Thermal-niche concepts can be used to identify critical habitats for important species in ways that focus the potential influences of power-station discharges (22,23). For example, striped bass (*Morone saxatilis*), an East Coast species introduced on the West Coast and in freshwater reservoirs, partitions a water body in summer along thermal gradients among its age or size classes. Young bass prefer and grow optimally at high temperatures near 24–26°C, subadults prefer ca 22°C, and mature adults select temperatures of 20°C or less. Thermal discharges may benefit growth rates and survival of juveniles, but the adults face a different prospect. Forced to cool water in summer by their genetically based temperature preferences, they may find this habitat severely restricted by thermal additions or compromised by simultaneous depletion of dissolved oxygen as a result, in part, of decomposition of thermally stimulated plankton production. Overcrowding in limited thermal refuges that have sufficient dissolved oxygen has led to starvation, high disease incidence, and abnormally high fishing susceptibility, all of which cause high numbers of deaths.

Altered predator-prey interactions, previously viewed as intractable long-term influences of changing water temperatures, can be estimated at least qualitatively through use of thermal-niche concepts. Adult striped bass that are restricted from access to surface waters by high temperatures seek food among the cool-water species, eg, trout in deep water, rather than among the surface-dwelling shad species which are their normal prey. Trout populations can be decimated, as they were in Lake Mead after

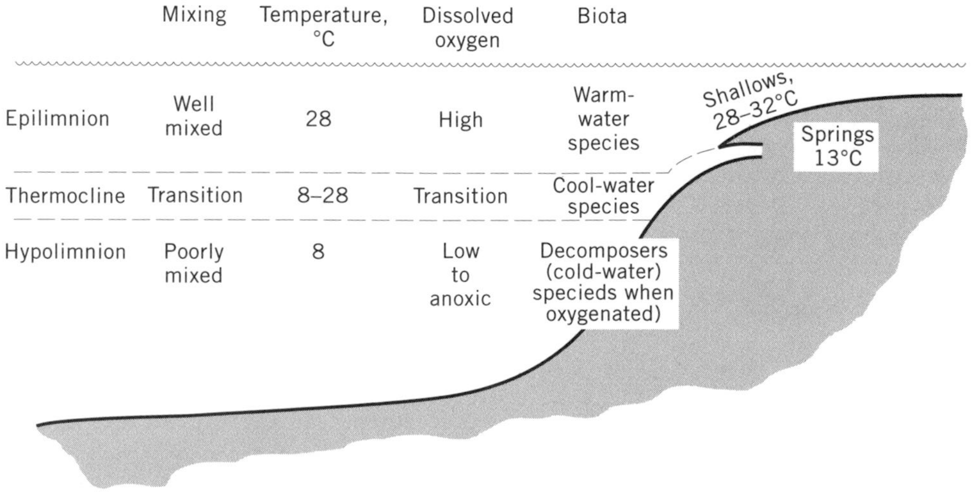

Figure 5. Typical thermal stratification of a lake, reservoir, or poorly mixed estuary in summer which, because of density differences, establishes discrete zones with differing thermal, water quality, and biotic characteristics.

introduced striped bass grew to the size and thermal preference of adults.

Careful planning of thermal additions, including creation of new cooling reservoirs, can yield thermal structures which enhance rather than damage desirable aquatic species. Knowledge of the thermal niches of these species and their potential competitors or predators can permit special provisions for thermal refuges, eg, cool summer zones in a heavily loaded cooling pond. Cooling-water circulation can be designed so that the thermal stratification patterns that are essential for some desired species are maintained. From a different perspective, aquatic species introduced to waters used for power-station cooling can be selected so that their thermal niche matches the thermal structure that the facility creates. Multiple species introductions can be evaluated beforehand to estimate whether available thermal patterns will abnormally aggregate or separate potential predators and prey.

Impingement

An explosion of innovative technology was spawned in the 1970s by the deaths of fish and some invertebrates, eg, blue crabs, on power-station intake screens. Unlike the seemingly intractable complexities of thermal effects in a water body, impingement deaths are clearly related to inlet water velocities, lack of escape routes, and retention on screens either in the water or during the screenwash procedures that were designed for handling nonliving trash. Most of these techniques have been reviewed in a series of conference publications (4,24–25). Repellents, eg, sound, electricity, and light, were used to keep susceptible fish away from intakes with only moderate success. More useful were orientations of the inlet structure such that screens flush with the shoreline allowed lateral escape (Fig. 6(a)). Where this screen orientation was impossible, guidance systems were developed for intake bays to direct fish from the incoming water flows to alternative escape routes (Fig. 6(b)). Guidance devices were also installed at the entrances to unscreened, offshore inlets, where a horizontal velocity cap placed a few feet over a vertical opening proved especially effective in preventing fish entrance (Fig. 6(c)). Small-scale modifications to existing rotating trash screens allow many fish to survive as they are raised from the water and deposited in sluiceways for return to the water. There has been renewed interest in these guidance techniques with expansion of hydroelectric power development.

Experience has also shown that deaths of fish on intake screens, although spectacular, often have minor consequences for local populations of many species (26). Most commonly impinged are open-water, schooling species that have high reproductive potential and a propensity for large natural variations in population numbers. The most common season for impingement mortalities, ie, winter, is a time of natural population reduction because of cold stress for several species. Cold-stressed fish may be impinged in large enough numbers to shut down the facility, but evidence suggests that most are moribund and destined not to survive regardless of cooling-water-intervention.

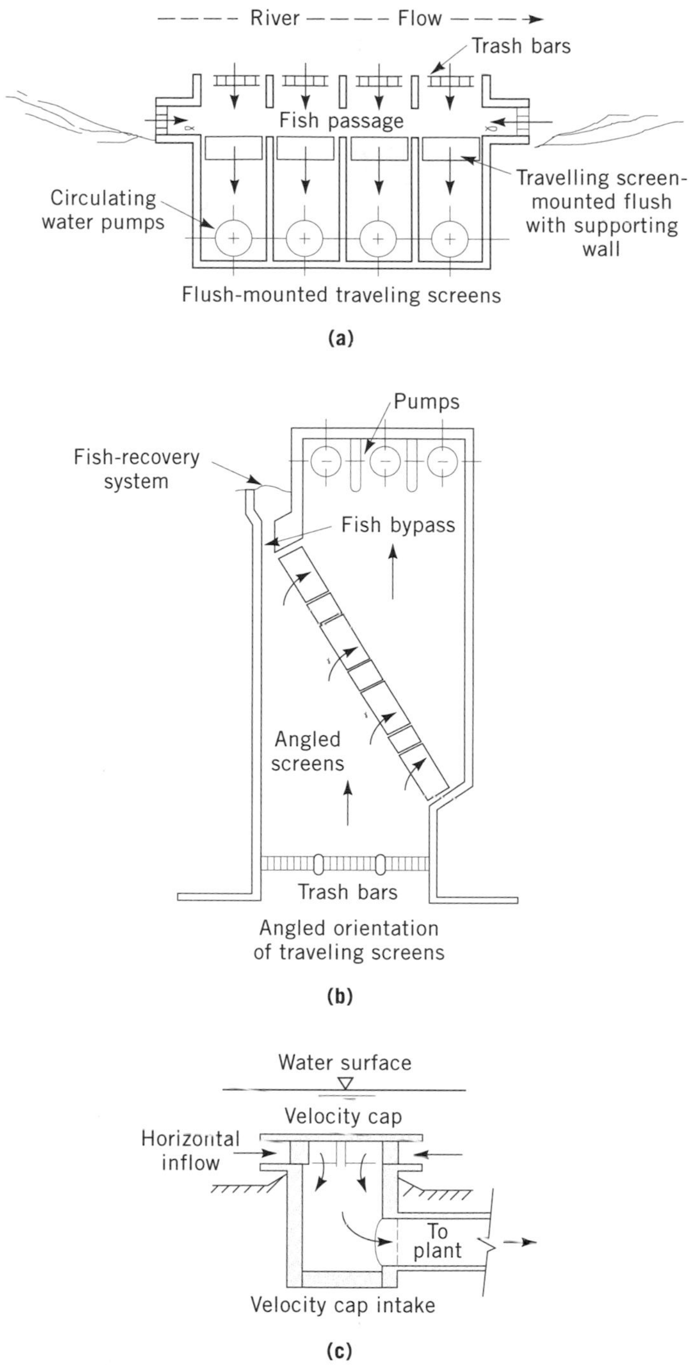

Figure 6. Engineering approaches to minimizing entrapment and impingement of fish and large aquatic invertebrates; eg, blue crabs, on trash screens at intakes. (a) An inlet pump house with vertical traveling screens mounted flush with a river shoreline to minimize obstructions to animal movements parallel flow to direct fish to a recovery chamber that returns to the water body; (c) a velocity cap atop a vertical, offshore inlet; induces a horizontal flow which fish avoid instead of a vertical flow which they do not. Reproduced by permission from ref. 25.

Current perspective on minimizing impingement risks focuses on site-specific analyses of potentially vulnerable species and selection of engineering designs which, within acceptable cost limits, keep impingement deaths few. Successfully accomplishing this objective requires intimate cooperation between biologists and engineers.

Biocides

Chlorine and other biocides are used occasionally in cooling water to kill and dispose of organic growths on heat-exchange surfaces and on piping where water flow could be hampered by such growth. Of necessity, organisms passing through the cooling circuit or residing in the effluent area during periodic chlorine injections experience the potentially lethal exposures. The objective of conscientious plant operators is, however, to maximize the intended kill and minimize extraneous damages, particularly in the receiving water. Methods for accomplishing this goal were developed progressively over the 1970s (27).

Chlorine is a toxicant with a typical dose-response pattern for biota (Fig. 7). There is a time-dependent mortality at high concentrations and a low concentration above which long-term chronic effects are shown. Early methods for using the time-dependent effects of high temperature for predicting safe temperatures and exposure times during cooling-water exposures led to a similar approach for chlorine (3). Because chlorine toxicity data were scant, aggregate graphs for all available species were used to identify safe levels. Different aggregates had to be developed for freshwater and marine assemblages because of the markedly different chlorine chemistry in fresh and salt water. The analytical tools are available for minimizing chlorine damages, although species-specific data are urgently needed.

Entrainment

Stresses to small nonscreenable organisms, eg, fish larvae, during passage through the cooling circuit come from a combination of thermal shock, physical abuses, and periodic injections of biocide. Whereas the thermal and biocide components, ie, the abrupt rise of temperature or chlorine concentration in the condensers followed by a holding period in piping and a decline in the discharge area, are well conceptualized and can be used quantitatively for predicting organism responses, the physical abuses are less well-understood (8). Pressure changes, shear forces, and physical contact with pump impellers or tubing walls are sources of potential damage that vary from system to system. Minor differences in configuration of piping can mean the presence or absence of such devastating features as microcavitation cells. The physical configuration does not lend itself easily to laboratory experimentation, although three such physical simulators have been tried (28–30).

A principal frustration in attempts to minimize entrainment damages has been the contradictory demands of thermal and physical stresses. Thermal stresses can be quantitatively predicted based on dose-response data and minimized by increasing water-flow volumes, which dilute the fairly constant supply of rejected heat. The added volume of cooling water, however, includes proportionately more planktonic organisms, which are subjected to physical stresses. Attempts to minimize thermal damage in new power stations had the demonstrable effect of increasing physical damage. One proposed solution, which is based on the assumption of a high percentage of mortality resulting from physical stresses, is to return to high condenser temperatures at which all of the relatively few organisms entrained would be killed, but those remaining in the waterbody would be assured survival (8).

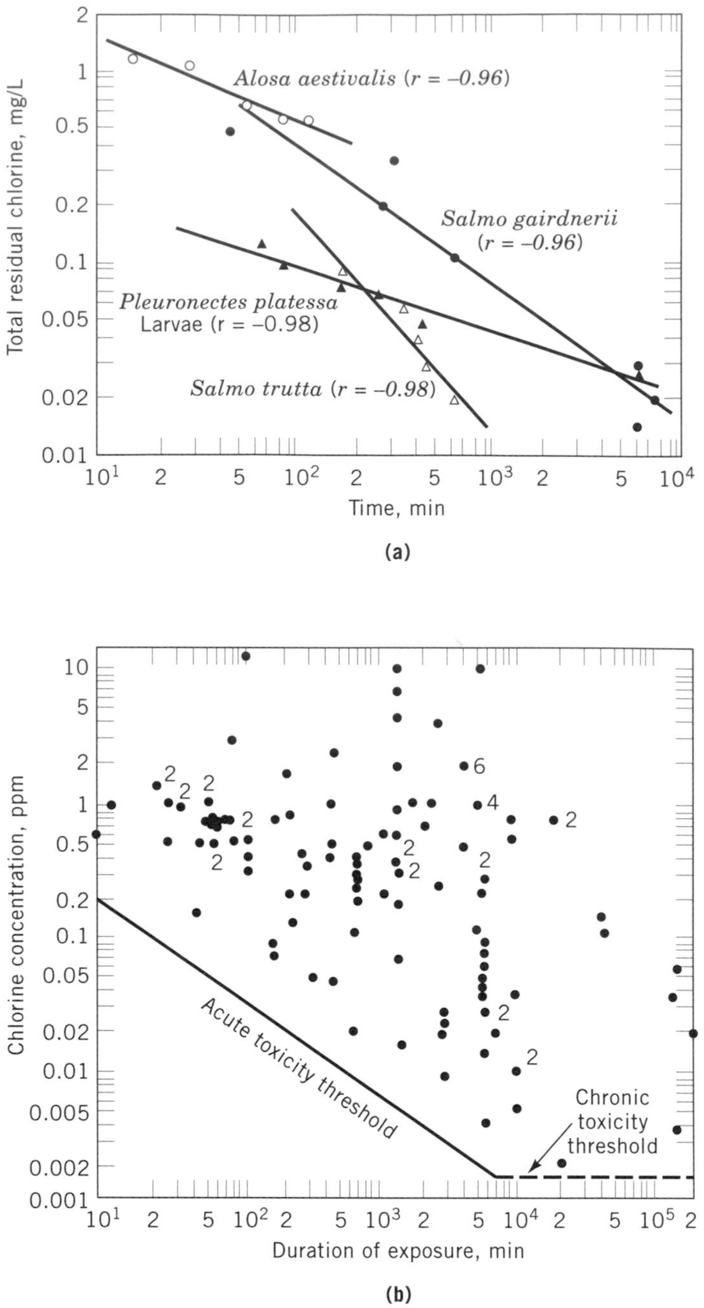

Figure 7. Toxicity of chlorine to aquatic organisms. (a) Time-dependent mortality (50%) of four example species in various levels of total residual chlorine in the laboratory. r = correlation coefficient of the curve; (b) A summary of chlorine toxicity to freshwater species, indicating overall no-effect thresholds for acute and chronic exposures. Numbers indicate where more than one test yielded the same result. A different summary figure applies to marine organisms because of diffierences in the chemistry of chlorine in seawater. Reproduced by permission from ref. 27.

The assumption of high percentage mortality resulting from physical stresses has recently been criticized. New methods of sampling entrained biota, which involve extensive precautions against damaging stresses during the sampling process, indicate a much higher survival of even delicate fish larvae than had been realized with earlier and more crude sampling methods. It appears that ameliorations of thermal stress with flow increases generally are not cancelled by additional physical mortalities except in cases of exceedingly high flows. An optimization procedure, such as that suggested by the Committee on Entrainment, appears fruitful for identifying on a site-specific and seasonally varying basis the most appropriate cooling-water flow regime (8).

There is also a compromise between entrainment and impingement. One of the engineering methods suggested for minimizing entrainment was replacement of 1-cm intake screens with screens of much smaller aperture. Although fewer organists are entrained, most of those saved are, in fact, impinged on the smaller mesh of the screen. The probable new impingement mortality may exceed the risks during entrainment.

Gas Balance

Oxygen in solution does not generally change in a cooling circuit unless there is strong aeration of originally saturated water heated maximally, in which case some dissolved gas is lost. In no case does the loss reduce levels to below that required by aquatic life. When intake water is undersaturated, the agitation of the cooling circuit generally yields an increase. Abundant data support these generalities. Polluted receiving water does, however, have an accelerated microbial oxygen demand because of raised temperature, and this microbial deoxygenation can yield exaggerated dissolved oxygen sag zones downstream. The interaction of thermal effluent and waste stabilization in the river must be considered whenever polluted water is used for cooling, and engineering models for predicting oxygen sag zones are generally capable of incorporating temperature changes.

Damaging supersaturation of dissolved gases has occurred in some cooling-water discharges. Damage has been isolated to circumstances, usually in winter, when temperature elevations above the cool ambient are highest based on the assumption that even a large increase in winter results in temperatures below lethal level or prevalent water quality standards. The practice of winter increases in temperature rise across condensers by cutting back on pumping capacity has either ceased in general practice or the immediate discharge areas have been engineered to prevent long-term residence by susceptible biota. These remedial measures can be completely effective.

Cooling-Tower Chemicals

The new risk posed by changing power-station systems from traditional once-through or open-cycle cooling to cooling towers is from chemicals added to the recirculating water. Blowdown to aquatic systems and drift to the terrestrial landscape carry these chemicals to locations where natural biota can be damaged through direct poisoning or where toxicants can accumulate to potentially detrimental levels by food-chain transfer. Risks from such chemicals can be minimized with fairly straightforward engineering approaches. Airborne drift has been reduced significantly by installation of physical baffles at the air outlets which intercept droplets, coalesce them, and return the water to the cooling-water flow. Blowdown can be treated for removal of chemical constituents, eg, chromates, with the additional benefit of chemical recycling and, thus, cost recovery. Chemical-laden sludges become a more long-term disposal problem; current practices include ponding and landfills. Chemical-recovery processes are also available for sludge treatment. Of significance for all of these processes is the cost which, when added to the initial capital cost of the cooling towers and the operating costs of pumps and fans, can reduce the attractiveness of closed-cycle cooling, especially when less costly mitigative measures exist for the open-cycle system at many locations.

Human Pathogens

Most studies of the risks to humans from pathogens stimulated by environmental conditions in cooling systems, eg, amoebae *Legionella,* have shown that proper designs and good maintenance practices can reduce the potential risks to very low levels. Thermophilic protozoans which thrive in water near human body temperature that is also rich in organic matter find animal tissues similarly favorable, particularly the nasal passages which allow easy penetration to the brain. Field evidence indicates that these organisms are most abundant in stagnant pools and roadside ditches; fatal infections have generally been traced to exposures in such environments. The rapidly flowing water in power-station discharges, although quite warm in summer, generally does not provide either stable substrates or high concentrations of organic material. Additionally, plant security measures generally ensure that bathers do not use the warmest zones. Allowing water contact by the public in discharges, particularly wherever the outlet is a slowly moving canal with abundant vegetation, is risky. The Legionnaires' disease microorganism, identified as a common inhabitant of many cooling-tower systems, especially small ones used for building-temperature control, can be held in check by systematic use of biocides in recirculating cooling water. Contact with the infectious organisms can be minimized if care is taken to isolate inlets for ventilating air from the drift aerosols which are emitted from cooling towers.

MAXIMIZING THE BENEFITS

During occasional episodes of energy scarcity, particularly in the 1970s, attention was given to finding productive uses for power-plant waste heat. The waste of approximately two thirds of the fuel energy during conversion of fossil or nuclear fuels to electricity was viewed as extravagant. Methods were developed for converting the thermal effluents from power stations into useful heat resources. There are potential physical applications of power-plant rejected heat, eg, industrial heating, and biological applications such as fish culture, soil warming, and heating greenhouses and livestock shelters.

The most important factor determining the possible uses of such heat is the temperature of the heat source. There is a threshold between low grade and high grade heat for engineering uses at ca 100°C. Most waste heat from electric power stations is in the form of low grade heat. This is so low in temperature that disposal to the environment has traditionally been considered the only practical alternative. Discharges from once-through cooling systems of thermal power plants have outlet temperatures of 10–40°C, depending upon the season. Circulating water in closed-cycle cooling systems is only slightly warmer (20–50°C). These low temperatures are the result of careful engineering of power stations to extract the maximum amount of electrical energy from the fuel before the rejected heat is dumped to circulating water in steam condensers. Such engineering has produced high efficiencies compared to most other mechanical systems: 33% in nuclear plants to 40% in fossil-fueled plants is the fuel energy converted to electricity.

A more useful, higher grade heat is in a form that has not been degraded to the low temperatures of most power-plant thermal discharges but which remains at 40–200°C following production of initial amounts of electricity. Low pressure steam can be extracted from the turbine system of a power plant before condensation to waste-heat temperature, thus permitting its use for functions requiring higher temperatures. Although this gives better utilization of the energy remaining in the steam, the efficiency of the low pressure turbine for producing electricity is diminished. Condensers can also be operated with smaller cooling-water flows to yield higher discharge temperatures but with some penalty to generation efficiency because of higher turbine back pressures. The overall efficiency of fuel energy use in such multiple-purpose systems can exceed that of electricity production alone, however.

Ideally, a power station could be a potential supplier of electricity, high grade heat, and low grade heat, even though electricity is the main product and most new power stations are optimized accordingly. Depending upon the desired uses at particular sites, future power stations could be designed to alter the ratios of electricity and different grades of heat to produce the most efficient total energy use. Such multiple use was common at early generating stations in the United States and is still common in Europe. Applications in North America are becoming more common again.

Aquaculture

Culture of some aquatic species in essentially unmodified thermal effluents of power stations has been attempted both experimentally and commercially. The principle behind such culture is the temperature-dependence of growth rates discussed earlier. Use of rejected heat to prolong optimal growth temperatures in cool months can significantly increase the sizes attained in a year. Even though power stations rarely provide high enough temperatures in winter for optimal rates, temperatures can usually be maintained at levels that increase growth rates. Maximum use of available temperatures has been made by culturing warm-water shrimp in summer and cool-water trout in winter, each of which can utilize the added temperature imparted by power-station cooling (31). Most thermal aquaculture facilities at power stations have not been economical and have been closed in recent years.

Open-Field Agriculture

Use of warmed water for field crops and orchards has been tested in several studies. Buried pipes can convey thermal-discharge heat to soils where warming aids plant growth and extends the growing season. Soil warming with open irrigation water has also been studied. There are advantages of spraying thermal-effluent water on crops and orchards to prevent freezing of buds and blooms. Whether or not the heat is used, power-plant cooling water can conveniently be combined with irrigation water systems to provide multiple uses of the water. An extensive demonstration project has been established by Electricité de France at their St. Laurent des Eaux power station.

Greenhouse Agriculture

Greenhouse agriculture is well known for its many advantages over open-field agriculture for certain crops. Yields are greater per land area and year-around culture is possible, which allows matching of crop harvest with high demand and price. One drawback is high expense for heating in winter, which can be the most costly part of greenhouse operation. The use of waste heat from steam-electric power plants therefore appears promising as a source of low cost heat for use in greenhouses, especially if the power stations have cooling towers with wintertime operating temperatures of 16°C or higher. The economic analysis is favorable (32).

Experimental greenhouses based on waste heat for temperature control have been operated at several sites and have demonstrated the principle that cooling water can be used to heat a greenhouse in winter. Some have also been based on power-plant effluent for cooling in summer; inexpensive porous packing through which the heated water drips and air circulates is used (Fig. 8). Evaporation provides cooling in summer and sensible heat transfer warms the air in winter. Additional heat and humidity control can be provided with a finned-tube heat exchanger. The French have used pipes and plastic tubes laid in the floor of the greenhouse; warm thermal effluent or water warmed further with heat pumps operating from thermal discharge waters circulates through the pipes and tubes. As with aquaculture systems, greenhouses should affiliate with multiple-unit power stations to ensure continuous warm water supplies.

Animal Shelters

The advantages of temperature control for maximizing weight gain and avoiding animal losses in livestock and poultry are well known. Low grade heat from power station cooling offers the possibility of low cost heating of animal shelters, although few demonstration projects have been developed. Heating animal shelters is a special case of space heating, although capital costs generally must be lower for the system to be economical. The greatest potential for heating animal shelters appears to be in

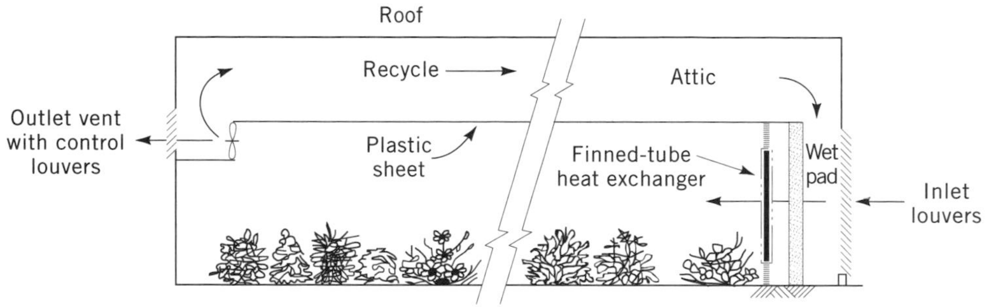

Figure 8. Longitudinal section of an experimental waste-heat greenhouse in which temperature control in all seasons is provided by evaporation and heat transfer as air passes through a fiber pad soaked with power-station cooling water or by heat transfer as air passes through a finned-tube heat exchanger that carries cooling water. A false ceiling provides for recycle of air through the heat-transfer medium. Reproduced by permission from ref. 33.

continental climates where exceedingly cold winters can lead to deaths of livestock.

Space Heating

A large percentage of the energy requirements of most countries in temperate zones is for heating and cooling of living and working spaces and for hot water. The historical use of dual-purpose power plants for electricity generation and central district heating in the United States and their current extensive use in such countries as the Russia, Sweden, and Germany suggests that expansion of this form of waste-heat utilization can contribute significantly to energy conservation and control of concentrated thermal discharges worldwide.

The most economical form of long-distance transport of thermal energy for space heating and cooling by adsorption methods is by heated water rather than steam. Steam had been used from early dual-purpose power plants prior to development of modern water-cooled condensers. It is still used in areas of dense loads, but its range of effective distribution is small because of large pressure drops in distribution systems. Modern experience with dual-purpose power plants (mostly in Europe) and new district heating systems in the United States, eg, in colleges, institutions, and shopping centers, has shown hot water to be superior for dispersed loads, including single-family residences: These dual-purpose stations operate at high thermal efficiency; the Swedish Malmo Plant is researching new technologies for distribution pipes in order to expand economical district heating to dispersed single-family residences.

Economic analyses for the United States indicate that supplying thermal energy to the commercial-residential sector from dual-purpose power stations is more economically competitive in new applications than in cases where existing buildings are to be serviced (34). There is little difference for supplying industrial heating. Such a thermal grid is most competitive with new fossil fuels where there is a high heat-load density and expensive fuel costs.

Industrial Process Heat

Many industries use process heat at 77–110°C. Much of this heat is supplied by combustion of oil and natural gas. Equipment manufacturers are developing industrial heat pumps to capture free industrial-plant waste heat and regenerate it to the desired process-heat temperature, thereby greatly reducing energy costs associated with direct heating. Operations that use heat in the 88–110°C range that can be supplied by new high efficiency heat pumps based on rejected heat sources are washing, blanching, sterilizing, and cleaning operations in food processing; grain drying; metal cleaning and treating processes; distilling operations in the food and petrochemical industries; and industrial space heating. Power-plant rejected heat could be valuable for developing industries that perform these processes.

Cooling Reservoirs

The most extensively developed productive use for power-plant cooling is in multiple-purpose cooling reservoirs. Small impoundments built especially for heat dissipation have been managed for extensive recreational uses as well. Highly productive fisheries for warm-water species, eg, largemouth bass (*Micropterus salmoides*) and channel catfish (*Ictalurus punctatus*), have made cooling reservoirs highly popular. Such reservoirs have been extensively developed in lake-free areas, eg, Texas and Illinois. Broad-scale ecological research on some of these cooling reservoirs has documented the valuable synergistic relationship between cooling-lake fisheries and power-station heat dissipation (35).

Thermal pollution from energy generation is not viewed today as the threat to the environment that it appeared to be in the 1960s and 1970s. This is not because the potential environmental hazards of power station cooling are necessarily any less. The change in perspective has arisen because the hazards have been recognized, the biological and other environmental constraints (and benefits, in some cases) have become understood, and good engineering practice has devised methods to minimize the risks. Where location-specific controversies remain, as they sometimes do, multidisciplinary teams of biologists and engineers can usually design appropriate, site-specific studies to develop the most suitable solutions.

BIBLIOGRAPHY

1. E. W. Moore, *Ind. Eng. Chem.* **50,** 87A (1958).
2. Pennsylvania Department of Health, *Heated Discharges—Their Effects on Streams,* Publication No. 3, Division of Sanitary Engineering, Pennsylvania Department of Health, Harrisburg, Pa., 1962.
3. C. C. Coutant, *Crit. Rev. Environ. Control.* **1,** 341 (1970).
4. L. D. Jensen, ed., *Entrainment and Intake Screening,* Publication No. 74-049-00-5, Electric Power Research Institute, Palo Alto, Calif., 1974.
5. J. A. Zillich, *J. Water Pollut. Control Fed.* **44,** 212 (1972).
6. R. L. Jolley, *J. Water Pollut. Control Fed.* **47,** 601 (1975).

7. S. Markowski, *J. Anim. Ecol.* **28,** 243 (1959).
8. J. R. Schubel and B. C. Marcy, eds., *Power Plant Entrainment—A Biological Assessment,* Academic Press, New York, 1978.
9. S. W. Christensen, W. Van Winkle, L. W. Barnthouse, and D. S. Vaughan, *Environmental Impact Assessment Review* **2,** 63 (1981).
10. W. J. Ebel, *Fishery Bulletin* **68,** 1 (1969).
11. J. Z. Reynolds, *Science* **207,** 367 (1980).
12. S. A. Carter, *Environment* **19,** 16 (1977).
13. D. W. Fraser, D. C. Deubner, D. L. Hill, and D. K. Gilliam, *Science* **205,** 690 (1979).
14. *J. Water Pollut. Control Fed.,* annual review issues.
15. *Water Quality Criteria 1972,* Report No. R-73-033, National Academy of Sciences–National Academy of Engineering, U.S. Environmental Protection Agency, Washington, D.C., 1973, pp. 151–171 and appendix.
16. *Quality Criteria for Water,* U.S. Environmental Protection Agency, Washington, D.C., 1976.
17. C. C. Coutant, *Crit. Rev. Environ. Control* **3,** 1 (1972).
18. D. R. F. Harleman, *Report of a Workshop on the Impact of Thermal Power Cooling Systems on Aquatic Environments,* CONF-750.980, EPRI SR-38, Asildmar, Pacific Grove, Calif., Sept. 28–Oct. 2, 1975, pp. 128–135.
19. H. H. Carter, J. R. Schubel, R. E. Wilson, and P. M. J. Woodhead, *Environ. Manage.* **3,** 353 (1979).
20. J. J. Magnuson, L. B. Crowder, and P. A. Medvick, *Am. Zool.* **19,** 331 (1979).
21. J. R. Gammon, unpublished results, Depauw University, Greencastle, Ind., 1981.
22. C. C. Coutant, *Trans. Am. Fish. Soc.* **114,** 31, 1985.
23. C. C. Coutant, *Striped Bass and the Management of Cooling Lakes, Waste Heat Utilization and Management,* Hemisphere Publishing Corporation, Wash., 1983, pp. 389–396.
24. L. D. Jensen, ed., *Third National Workshop on Entrainment and Impingement,* Ecological Analysts, Inc., Melville, N.Y., 1977.
25. L. D. Jensen, ed., *Fourth National Workshop on Entrainment and Impingement,* Ecological Analysts, Inc., Melville, N.Y., 1978.
26. R. B. McLean, P. T. Singley, J. S. Griffith, and M. V. McGee, *Threadfin Shad Impingement: Effect of Cold Stress,* NUREG/CR-1044, ORNL/NUREG/TM-340, Oak Ridge National Laboratory, Oak Ridge, Tennessee and National Technical Information Service, Springfield, Va., 1980.
27. J. S. Mattice in *Environmental Effects of Cooling Systems,* Technical Report Series 202, International Atomic Energy Agency, Vienna, 1980, pp. 12–26, 148–167.
28. R. J. Kedl and C. C. Coutant in G. W. Esch and R. W. McFarlane, eds., *Thermal Ecology II,* Conf-750425, National Technical Information Service, Springfield, Va., 1976, 394–400.
29. J. M. O'Connor, T. C. Ginn, and G. V. Pase, *The Evaluation and Description of a Power Plant Condenser Tube Simulator,* Report No. 8-0248-956, New York University Medical Center, New York, 1977.
30. J. S. Suffern, *The Physical Effects of Entrainment—Current Research at ORNL,* ORNL/TM-5948, Oak Ridge National Laboratory, Oak Ridge, Tennessee and National Technical Information Service, Springfield, Va., 1977.
31. W. Majewski and D. C. Miller, eds., *Predicting Effects of Power Plant Once-Through Cooling on Aquatic Systems,* Technical Papers in Hydrology 20, United Nations Educational, Scientific and Cultural Organization (Unesco), Paris, 1979.
32. M. Olszewski, S. J. Hillenbrand, and S. A. Reed, *Waste Heat vs. Conventional Systems for Greenhouse Environment Control: An Economic Assessment,* ORNL/TM-5069, Oak Ridge National Laboratory, Oak Ridge, Tennessee and National Technical Information Service, Springfield, Va., 1976.
33. S. E. Beall and G. Samuels, *The Use of Warm Water for Heating and Cooling Plant and Animal Enclosures,* ORNL-TM-3381, Oak Ridge National Laboratory, Oak Ridge, Tennessee and National Technical Information Service, Springfield, Va., 1971.
34. M. Olszewski, *Preliminary Investigations of the Thermal Energy Grid Concept,* ORNL/TM-5786, Oak Ridge National Laboratory, Oak Ridge, Tennessee and National Technical Information Service, Springfield, Va., 1977.
35. W. Larimore and co-workers, *Evaluation of a Cooling Lake Fishery,* EA-1148, 3 Vols., Electric Power Research Institute, Palo Alto, Calif., 1979.

Reading List

Environmental Effects of Cooling Systems, Technical Report Series No. 202, International Atomic Energy Agency, Vienna, 1980.

J. R. Schubel and B. C. Marcy, Jr., eds., *Power Plant Entrainment—A Biological Assessment,* Academic Press, New York, 1978.

G. F. Lee and C. Stratton, *Ind. Water Eng.* **9,** 12–16 (1972).

H. Precht, J. Christophensen, H. Hensel, and W. Larcher, *Temperature and Life,* Springer-Verlag, New York, 1973.

C. C. Coutant, *Crit. Rev. Environ. Control* **3,** 1 (1972).

S. E. Beall, C. C. Coutant, M. Olszewski, and J. S. Suffern, *Ind. Water Eng.* **14,** 8 (1977).

L. B. Goss and L. Scott, *Factors Affecting Power Plant Waste Heat Utilization,* Pergamon Press, New York, 1980.

C. C. Coutant, *Electric Perspectives* **16**(4), 32 (1992).

C. C. Coutant, *Scientific American* **254**(8), 98 (1986).

M. C. Bell, *Fisheries Handbook of Engineering Requirements and Biological Criteria,* U.S. Army Corps of Engineers, North Pacific Division, Portland, Ore. 1991.

L. W. Barnthouse, R. J. Klauda, D. S. Vaughan, and R. L. Kendall, eds. *Science, Law, and Hudson River Power Plants, A Case Study in Environmental Impact Assessment,* American Fisheries Society Monograph 4, Bethesda, Md., 1988.

THERMODYNAMICS

William Kenney
Exxon Research and Engineering
Florham Park, New Jersey

To summarize the critical elements of a science such as thermodynamics, about which many books have been written (see reference list for some of the texts drawn upon for this article), in the space of this article is a formidable challenge. The author has focused on those elements of thermodynamics which are principally involved with energy efficiency and energy conversion. As a result, many of the chemical thermodynamic aspects of the science have been omitted from the article.

See also Energy efficiency; Energy efficiency, calculations; Temperature measurements.

HISTORICAL PERSPECTIVE

Thermodynamics is defined as the science that treats the mechanical action or relations of heat. It was probably first practiced by the cave man when he found that by rubbing two sticks together (the input of work to the process), he could heat them up (the production of heat) to the point where the sticks would ignite kindling wood and start a fire. Of course, what the cave man did by his input of work was to generate enough heat to raise the temperature of his working system to the point where it would initiate a chemical reaction (fire) which liberated the desired heat for cooking or warming his cave. Thus, we have leaped immediately from the simple conversion of work into heat to the more complex process of initiating a chemical reaction to provide a desired commodity. The conversion to heat is one system, and the chemical reaction of the combustion of fuel is another. As we proceed, we shall see that the definition of the boundaries of the system is an important part of the application of thermodynamics to practical problems.

During the course of the next millennia, man struggled to utilize heat and work, but not much progress was made in the scientific understanding of these interrelationships until the middle of the eighteenth century. At that time, the caloric theory of heat was prevalent. Heat was regarded as a fluid that was composed of particles strongly attracted to matter but mutually repulsive of themselves. This fluid was called "caloric." If more calorics were introduced by heating, the repulsive force between the atoms was enhanced and the equilibrium distance between the atoms increased. Thus, matter expanded when its temperature was raised, that is, its density decreased. Latent heat effects, like the melting of ice, were attributed to a chemical reaction between matter and caloric. The temperature would not change until the reaction was over. Careful weighing of objects at different temperatures and of ice before and after melting showed no detectable difference in mass. Therefore, scientists concluded caloric was weightless. Because measurements showed that the same amount of heat was released in freezing the ice as was absorbed in melting it, caloric was a conserved fluid; it could not be created or destroyed. The caloric theory of heat was very plausible at the time because of its ability to provide quantitative descriptions of thermal processes. A number of rather famous physical chemists and mathematicians at the time subscribed to it. These included LaVoisier, Laplace, Priestly, DeLong, Petite, Fourier, and Clapeyron. For all of this, there were many dissidents. For example, Decartes, Boyle, Bacon, Hook, and Newton all believed, but could not prove, that heat was somehow identified with the motions and particles making the matter. This dynamic theory of heat, which followed intuitively from the generation of heat by work through friction, was older than caloric theory but had not been used persuasively to explain thermal phenomena.

One of those to begin the more quantitative definition of the dynamic theory of heat is a man named Benjamin Thomson, who was later to become Count Rumford. Thomson was born in Massachusetts in 1753 but wandered to Europe as a soldier at a relatively young age. He continued his interest in science applied to military problems, and it was in the machining of a canon that the first experiments supporting the dynamic theory of heat were carried out. Rumford put a dull tool in a boring mill and began machining the internal bore of a canon. He succeeded in generating enough frictional heat to boil 12 kg (26 lb) of water into vapor. Only about 255 g (9 oz) of shavings were produces in this process, and Rumford argued that the heat could not have been generated by squeezing or abrading caloric from metal because so little metal was produced. However, devotees of the caloric theory of heat argued that the caloric fluid really weighed very little, and therefore, the experiment proved nothing. In 1799, Humphrey Davie showed that rubbing two pieces of ice together could cause melting. On the basis of careful calorimetry, everyone agreed that the heat content of water was greater than that of ice. However, if caloric was conserved in Davie's experiment, why did the ice melt? The caloric theory offered no satisfactory answer. Even so, the caloric theory continued and was the subject of elegant mathematical formalism evolved by a number of famous scientists at the time.

It was left to an English brewer's son, James Prescott Joule, to demonstrate the exact equivalence of heat and work interactions. This occurred in 1845. Like many scientists of the day, Joule was intrigued by electromagnetic theory and experiments. By using various combinations of electricity and work to make measurements, it was determined that the work equivalent of one British Thermal unit (the amount of heat it takes to raise one pound of water one degree Fahrenheit) was equivalent to 772.5 ft/lb. With the benefit of modern scientific instrumentation, we currently believe that this ratio is 778.28 ft/lb Btu. The precision of Joule's measurements some 50 years after the Rumford experiment managed to put an end to the caloric theory of heat and ultimately led to the first law of thermodynamics. In time, this law, which states that energy is conserved, became so widely recognized that it even influenced thinking about the nuclear processes.

From a practical standpoint, knowing about the equivalence of heat and work can be important. Doing work on a substance will raise its temperature. For example, stirring a pot of liquid will increase its temperature, perhaps to a point where something undesirable happens such as decomposition of the material or accelerated corrosion of the container.

The second law of thermodynamics resulted much less from experimentation and much more from inductive reasoning. Nicholas Leonard Sadi Carnot, a young French military engineer, began to study the maximum work that could be obtained from a given quantity of heat in a heat engine. His paper, entitled "Reflections on the Motive power of Heat and on Machines Needed to Develop that Power," was published in 1824 and illuminated the Carnot principle. He stated that the amount of work which could be obtained from a given amount of heat in a heat engine depended only on the temperatures of the source of heat and the reservoir to which the heat was rejected. His reasoning was based partly on the caloric theory of heat. Lord Kelvin in Scotland and Clausius in Germany independently showed that Carnot's principle was also valid on the basis of the dynamic theory of heat.

At that time, Clausius introduced the function of entropy, which had a far-reaching effect on the quantitative development of the second law. According to Dodge (1), the application of the second law was greatly broadened in subsequent years beyond the field of heat engines and may now be said to underline almost all physical phenomena. J. Willard Gibbs, a mathematical physicist at Yale University, applied second law fundamentals to most of the applications of thermodynamics to chemical systems. He and others developed the concepts of free energy and chemical potential, which are currently used to explain many of the chemical thermodynamic systems we now understand.

The concepts of energy and entropy and the first and second laws of thermodynamics are fundamental to much of modern-day science. Even with our greater understanding of nuclear and stellar phenomena, which resulted in the theory of relativity, the basic assumption that energy is conserved still applies to most industrial processes. The theoretical basis for the neutrino subatomic particle grew from our belief in the conservation of energy.

SOME DEFINITIONS

Thermodynamic principles apply to *systems* which are simply the segment of the real world that is identified for observation and study. Anything outside the system is part of the *surroundings.* Obviously, the two are separated by a *boundary.* A system might be a beaker of water, a collection of pulleys and weights, a heat engine, a gas turbine/waste heat boiler combination, or an entire power plant or oil refinery. Inside the system are working fluids and chemical, mechanical, or thermal *processes,* by which heat, work, and chemical energy are converted to produce the desired results/products.

The quantitative description of a system involves the use of its *properties.* These are measurements of the characteristics of various parts of the system, such as length, mass, volume, temperature, and pressure. The *state* of a system is defined by a particular set of properties. At exactly the same state, the system will have exactly the same property measurements. Thus, 454 g (1 lb) of pure water at atmospheric pressure and 21°C will have exactly the volume (and density) no matter what process was used to bring the water to that state. The water might have been condensed from steam or melted from ice. Clearly, some interaction with the surroundings was required in the process (heat exchange) to establish the water at the desired state.

Some processes are *cyclic,* some are not. In a simple steam/power system, high-pressure water is heated and boiled by heat exchange with the surroundings (a boiler), the resultant steam is expanded through a machine which produces work (which is exported to the surroundings) and then condensed in a heat exchanger (where heat is rejected to the surroundings) and pumped back to high pressure again to repeat the cycle. The initial and final states of the water in this system are the same; therefore, the properties of the water are the same.

Within the steam/power system above are subprocesses which are not cyclic. Take the boiler, for example. High-pressure water is converted to high-pressure steam by the input of heat at a temperature high enough to accomplish the transfer. The water (a working fluid) changes states and properties (density, temperature, physical state) in this process step. The properties of the water do not depend on the path of the process, however. They depend only on the state of the water at any point.

The efficiency of the process (the amount of heat applied) depends on the path of the process and the equipment used. The differences in the properties of the water between the beginning and the end of the process represent the minimum (or *ideal* or *reversible*) amount of heat required for the process step. In real life, more heat is required because inefficiencies are inherent in the process. The second law of thermodynamics addresses some of the reasons for these inefficiencies.

PROPERTIES

In addition to the physical properties of the system and its working parts, there are thermodynamic properties. These permit the quantitative analysis of process steps and systems and have the same characteristics as physical properties; ie, at identical states of a material, the thermodynamic properties are the same.

Temperature is rather a unique property. Intuitively, we think of it as a measure of the "hotness" (or "coldness") of a substance. We know that if we contact a hotter substance with a colder one, "heat " will flow from one to the other until both are equally hot. Measuring (assigning a number to) a temperature is a more complex question.

Fundamental to temperature measurement is a principle that is sometimes called the zero Law of Thermodynamics: Two objects in thermal equilibrium with a third object are in thermal equilibrium with each other. This means that if I put my thermometer in two different pots and get the same reading, then both pots are at this same temperature. One of the problems is that if I use two different types of thermometers, I may get different readings even though both pots are indeed at equal "hotness." This happens because at points apart from the fixed calibration points thermometers using different materials will give different answers. Temperature measurements depend on the properties of the material used to measure them.

Quantitative Definition of Temperature

A quantitative definition of temperature may be based on any property of substances that changes with temperature, such as thermal expansion, electrical resistance, contact electromotive force (emf), or vapor pressure. To establish a numerical scale of temperature, it is first necessary to define the length of a degree. This is commonly done by choosing a fixed and readily reproducible interval of temperature such as that between the freezing point and boiling point of pure water at the standard atmospheric pressure and dividing this fundamental interval into an arbitrary number of degrees (100°C in the case of the celsius scale). The choice of the zero point is also a purely arbitrary matter. Any other temperature is then defined by the following simple equation applying specifically to the Celsius scale:

$$t = \frac{100(G - G_0)}{G_{100} - G_0}$$

where t is any temperature on a Celsius scale, G is any value of the given property at the temperature t, and G_0 and G_{100} are values of the property 0 and 100°C, respectively. This equation merely assumes the simplest possible relationship–a linear one–between the property of a substance and its temperature. There would be no point in (and no basis for) assuming any more complex relationship. The properties most commonly used for the industrial measurement and control of temperatures are as follows:

1. Volume expansion of gases, liquids, and solids at constant pressure
2. Pressure increase of liquids and gases at constant volume
3. Vapor pressure of liquids
4. Electrical resistance of metals (particularly platinum)
5. Electromotive force at the junction of two dissimilar metals

No two scales so defined will agree at any temperatures other than the two fundamental ones of 0 and 100°C. For example, at 50°C, a mercury-filled thermometer will read 50°C, but at 200°C (actual) it will read 202°C. A platinum-rhodium thermocouple will read 50.3°C at 50°C actual, but 222.5°C at 200°C actual. The further one gets from the fundamental calibration interval, the larger the deviations often become.

The potential implications of such deviations are sometimes important. In dealing with heat-sensitive materials a difference of 20 at 200°C can be catastrophic. The properties of many plastics deteriorate very rapidly as temperature rises. A plastic container or pipe might be suitable at 100°C, marginally capable of its task at 150°C, but fail completely at 170°C. Chemical stability can also be critical. Complex molecules often decompose at rapidly accelerating rates as temperature increases. Often there is a threshold temperature below which the heat released by very slow decomposition is dissipiated to the surroundings fast enough to prevent any temperature buildup in the material. Above the threshold temperature heat accumulates, temperatures rise, decomposition rates increase, and the typical exponential cycle begins. In 1985, the east coast from Philadelphia to Bridgeport, Connecticut, was treated to a rotten-egg odor when a chemical plant in New Jersey inadvertantly got a sulfur-containing product too hot by simply stirring a half-full vessel for too long.

To deal with the problems of accurate temperature measurements, a series of international standards of calibration points has evolved and precise measurements are made with special gas thermometers. As described by Fenn (2), gas thermometry is of value because precise measurements of the pressure of a fixed volume of gas can be made. Thus, we base our standard for temperature measurement on the ability to measure two other physical properties: volume and pressure. Gas thermometry is also of value because at very low pressures, readings are independent of which gas is chosen. For example, at essentially zero pressure, the ratio of the pressure in a fixed volume of gas at the boiling point of water to that at the freezing point is exactly 1.36609 ± 0.00004 for all gases. Thus, the material problem is eliminated. It can then be said in general that

$$\frac{t_{\text{steam}}}{t_{\text{ice}}} = \frac{P_{\text{steam}}}{P_{\text{ice}}} = 1.36609$$

in the limiting condition that P_{ice} approaches zero. If one chooses to allow 100 degrees between these two temperatures, as in the Celsius scale, one can calculate that, *on an absolute basis,*

$$t = t_{\text{ice}}\frac{P}{P_{\text{ice}}} = 273.16\frac{P}{P_{\text{ice}}}$$

when the ice point approaches zero. This corresponds to the Kelvin absolute temperature scale used in thermodynamic calculations. Absolute temperatures on the Kelvin scale are normally represented by a capital T and the label K, i.e., 100 K. An absolute temperature scale corresponding to the Fahrenheit scale will be discussed momentarily.

To summarize, to find the temperature of an object with a gas thermometer scale, we find the pressure p exerted by a gas at a given volume when it has been near the object long enough for thermal equilibrium to occur. We also find the pressure p_i exerted by the same amount of gas at the same volume when it is in thermal equilibrium with a mixture of ice and water. To obtain the temperature T of the object, we multiply the ratio of the two pressures, p/p_i, by 273.16. To be precise, we must take the limiting value of the ratio as we decrease the amount of gas in the given volume.

The gas thermometer is not a very convenient device. For everyday use, the Seventh General Congress on Weights and Measures in 1927 adopted a much more convenient reference scale. Revised several times since, it is meant to be as close as possible to the gas thermometer scale and makes use of a number of fixed points that have been carefully measured in laboratories skilled in the art of gas thermometry. The most recent revisions are compared by listing some of the fixed points of each in Table 1 (3,4). The small improvements shown are important to some thermodynamic research and to measurements of properties at very low temperatures such as those involved with superconductors.

In addition to these fixed points, the international scale provides well-defined interpolation procedures for obtaining intermediate temperatures. These procedures make use of platinum resistance thermometers and platinum-rhodium thermocouples. Because it is easy to make

Table 1. Some Fixed Points for International Practical Temperature Scale (1968, 1990)[a]

	At Equilibrium State	
	68 K	90°C
Triple point of equilibrium hydrogen[b]	13.81	13.803
Boiling point of hydrogen at 1 atm	20.28	—
Triple point of neon	24.561	24.556
Boiling point of neon at 1 atm	27.402	—
Triple point of oxygen	54.361	54.358
Boiling point of oxygen at 1 atm	90.188	—
Triple point of mercury	234.308	234.316
Triple point of water	273.16	273.16
Melting point of gallium	302.922	302.915
Boiling point of water at 1 atm	373.15	—
Melting point of indium	429.785	429.749
Freezing point of zinc	692.73	692.677
Freezing point of aluminum	933.61	933.47
Freezing point of silver	1235.08	1234.93
Freezing point of gold	1337.58	1337.33

[a] From Refs. 3, 4.
[b] Triple point is temperature at which solid, liquid, and vapor coexist.

electrical measurements of resistance and potential with greater accuracy, resistance thermometers and thermocouples are widely used as temperature sensors. Very detailed interpretation formulas have been developed to minimize any inaccuracies in interpolation (3,4).

What we have been calling the ice point, T_i, does not appear in today's definition of the gas thermometer scale. In its place is the so-called *triple point* of water, T_{tp} the temperature at which ice, liquid water, and water vapor coexist. The *ice point* proved difficult to reproduce because air has a small solubility in water and any substance dissolved in water will lower its freezing point.

The equivalent absolute Fahrenheit scale is known as the Rankine scale. Temperatures on this scale are identified by the letter R. For example, 491.69 R = 273.16 K. We can define it using the ratio equation and an equation specifying the Fahrenheit degree size:

$$\frac{T_B}{T_i} = \lim_{p \to 0} \frac{P_s}{P_i} = 1.36609$$

and

$$T_B - T_i = 180$$

Here, we find $T_i = 491.68$ and $T_B = 671.68$.

Thus, to convert any Fahrenheit temperature to Rankine, we must add 459.68 (491.68 − 12) because at the ice point temperatures are 0°C and 32°F. To convert any Celsius temperature to Kelvin, we add 273.16. In most practical cases, these are rounded to 460 and 273, but in research at very low temperatures, these and other small inaccuracies can become important.

Thermodynamic Properties

These absolute temperature scales are critical to defining other properties, called thermodynamic properties, of materials which are used in practical calculations.

The first thermodynamic property to be considered is *energy content* (E). Derived from careful experimentation, energy is the basis of the first law of thermodynamics, which states that energy is conserved. In any cyclic process, no matter how complex, a relationship exists between the sum of all work effects and all heat effects.

$$\Sigma W - J \Sigma Q \qquad (1)$$

Where J (for joule) is the constant equivalent of heat and work. Both heat and work are external effects (from/to the surroundings) resulting from changes within a system. They result from changes in energy content within the system and are forms of energy. Energy is something that may either be stored in a system or passed freely from one system to another in the form of heat or work, but it is always *conserved*. In all its transformations, it remains absolutely unchanged in amount. This is the principle of the conservation of energy, which forms the basis of the first law of thermodynamics, and it rests primarily on the experimental facts generalized by Eq. (1).

Energy stored in a system is a definite property of the system and is represented by the symbol E. It is sometimes called "energy content" to distinguish it from energy that may be associated with a system but not a property of it. The energy content E is that due to the configuration of the ultimate particles of the matter composing the system and to their motions. Energy due to mass motion of the system as a whole (dynamic as distinct from static system) or to its position in a gravitational, electric, or magnetic field is not included in the quantity E, as these forms of energy are not properties of the system. A given system under a given set of conditions has a certain energy content; this implies that whenever the system, after a series of changes, returns exactly to its initial condition, the energy content is just the same as it was before. It also implies that under another set of conditions it will have a different energy content. Equation (1) may now be broadened to include these ideas:

$$\Delta E = J \Sigma Q - \Sigma W \qquad (2)$$

where ΔE, the change in energy content of the system, is really defined by this equation. For a cyclic process, $\Delta E = 0$, and the equation reduces to Eq. (1). For a noncyclic process, it simply defines the difference between the heat effects and the work effects as being due to a storage in or depletion of energy from a system, in an amount characteristic of the initial and final states of the system. It should be emphasized that only changes in the amount of energy are being considered here.

For all ordinary phenomena and, more particularly for those involved in engineering work, the first law may be regarded as completely valid within the limits of the most accurate measurements that scientists are able to make. However, it is interesting to note in passing that to explain certain subatomic and also stellar phenomena, it has been necessary to give up the idea of energy conservation and postulate that energy may be created from matter, and vice versa. The relation between them is given by the relativity theory as $E - c^2m$, where c is the velocity of light. Because c^2 is so large, enormous amounts of energy result from the destruction of minute amounts of matter. Thus, 1 g of any substance is equivalent to 9×10^{13} J (25×10^6 kW · hr), the obtaining of which by present methods of power generation would require the burning of about 12,500 t of coal.

Energy is sometimes classified into various forms as expressed by the terms *potential, kinetic, mechanical, chemical, electrical,* and *atomic*. Potential energy is the energy that a system possesses by virtue of its particular configuration. The simplest case would be any object in relation to its position above the earth's surface. Work must be done to elevate any mass above the surface of the earth, and the energy represented by the work done is said to be stored in the system (i.e., the combination of the mass and the earth) as potential energy—"potential" because it is capable of yielding work if the body is allowed to fall to the earth. Like all energy, it has no absolute amount but is purely relative, depending on the choice of a datum plane.

Potential energy of a given system may also be "internal" owing to the particular configuration of the molecules, atoms, electrons, etc., of which it is composed. There are forces of attraction between the molecules of substances; and as the distance between them changes as in the expansion of a gas or the vaporization of a liquid, changes in potential energy occur analogous to that which occurs when a body changes position in the earth's gravitational field.

Kinetic energy is the energy associated with the motion of a body or particle relative to some reference body arbitrarily chosen. It is quantitatively defined for the simple case of a constant velocity by the well-known quotation $K = \frac{1}{2}mu^2$, where m is mass and u is the relative velocity. Kinetic energy may be associated with the motion of large or tangible masses or with the motion of the ultimate particles such as molecules or atoms. All energy that is not kinetic is presumed to be potential.

Energy, whether potential or kinetic, that is associated with relatively large masses (as compared with the ultimate particles of which matter is composed) is generally referred to as mechanical energy. Energy associated with the motion or the relative positions of molecules is frequently called "thermal" or "heat energy." It should be recognized that this is purely a convenient descriptive term referring to any kind of stored energy that can be transferred to or from systems by virtue of temperature differences. It is not to be confused with a heat transfer, Q.

Energy associated with emf's is called *electrical energy* and generally results from movements of the relatively mobile electrons that are present in varying degrees in all substances.

When chemical reactions take place, potential energy is stored or released. The exact form in which this energy is stored is related to the configuration of the atoms and electrons in the molecules. Since it is associated with chemical change, it is conveniently referred to as *chemical energy*.

There is no way of distinguishing the different forms of energy contained within a system. Consequently, all forms of internal energy must be taken together in E. In many practical engineering problems, the chemical and electromagnetic forms of energy play no part. Only the mechanical energy need be considered.

Various forms of energy can be interconverted. Consider a perfectly elastic ball bouncing on a perfectly elastic surface. When the ball is raised above the surface, it has potential energy. When released, this potential energy is converted to kinetic energy by gravity as it accelerates toward the surface. At the moment it strikes the surface, it has no velocity and no elevation, so all of its energy is transferred to elastic strain in the material (like that in a compressed spring). This energy then propels the ball up off the surface and is converted back to kinetic energy, which is converted to potential energy as its elevation increases and velocity decreases due to gravity. The cycle continues forever in the ideal case, but for a finite time in the real world because of friction and the inefficiencies of real energy conversion mechanisms.

The next critical thermodynamic property is *enthalpy* (H) (or heat content). Enthalpy should not be confused with the quantity of heat transferred (Q) in a process step. Enthalpy H is a definite property of a system, whereas Q is not. In most practical processes, enthalpy balances are used to calculate changes in energy and the heat and work interactions with the surroundings.

The second law functions are the *entropy* S and the *work or availability function* A, also called "ideal work" (either maximum or minimum, depending on the process). As stated earlier, other properties/functions which relate primarily to chemical reactions will not be treated here. Entropy was introduced by Clausius in 1851. The function A was originally called "free energy" by Helmholtz.

All the functions mentioned, with the exception of thermodynamic temperature, are extensive properties, i.e., depend on the mass of the system. It will be generally understood that the symbol for each function refers either to a unit mass or to a molal mass. When it is necessary to represent the total value of one of these properties for the system as a whole, the same symbol will be used, but as a capital letter.

Some patience is required in assimilating these definitions and the following section on properties and relationships. Those interested in practical applications of thermodynamics will find them forthcoming, but only when the investment is made in assimilating the theory.

The Energy Content

It was shown earlier that this definition of energy content, which is a definite property of any system, is the essence of the first law of thermodynamics. Equation (2) may now be written in differential form as

$$dE = dQ - dW \tag{3}$$

where dE is a perfect differential since it is the derivative of a point function but dQ and dW merely represent infinitesimal quantities of heat and work and are not derivatives of any functions. This means that if a finite change from state 1 to state 2 is under consideration, we may write

$$\int_1^2 dE = (E_2 - E_1) = \int_1^2 (dQ - dW) \tag{4}$$

In other words, dE can be integrated at once without regard to the nature of the change because E is a property and its value depends only on the initial and final states and in no way on the particular path of the change. The second integration, however, cannot be performed until we know something about the exact way in which the change occurred and so can express dQ and dW in terms of certain properties. In other words, both Q and W depend on the path taken and are not determined by the initial and final states.

We shall deal only with differences in E since nothing is known about the absolute value of this function. For convenience, E may be taken as equal to zero at some arbitrary standard state, but the more usual procedure is to make $H = 0$ at a reference state and determine E from it.

Entropy

The property entropy (s, S) has been the bane of many students of thermodynamics. Various attempts to describe its physical characteristics as was done for energy and enthalpy have been unsatisfactory. Perhaps the best approach is to consider it a property which is useful in analyzing processes and systems thermodynamically and in explaining and applying the second law. The entropy, an exact differential, is defined mathematically by

$$ds = \frac{dq}{T} \tag{5}$$

where dq is a differential quantity of heat and T is the absolute (or thermodynamic) temperature. The definition is true for any system at a constant temperature where changes in state are due to heat interactions alone. Mathematically, it can be shown that for an ideal cyclic process

$$\int_1^2 \frac{dq}{T} = \Delta S = 0 \tag{6}$$

Such processes are called reversible.

A *reversible process* is an idealized process that cannot be realized in practice but represents the best that is theoretically possible. In a reversible process, there is no friction, there are no materials which do not conduct heat, and the forces which drive the process are very small (infinitesimal) in size. As a result, a reversible process proceeds at an infinitesimally slow rate—not very practical or controllable. Its sole purpose is to serve as a standard of comparison against which the inefficiency of practical processes can be measured.

It follows then that the entropy change for practical processes is unequal to zero. In fact, it can be shown that for the whole process system

$$\Delta S > 0 \tag{7}$$

for any practical process.

Availability, Available Energy, Exergy, and Essergy

Materials have different capacities to course change (desired or undesired) depending on their properties/states and compositions. As pointed out by J. Willard Gibbs (5), theoretically, work could be obtained by the diffusion of chemical components from one stream to another across infinitesimal concentration differences as well as from interposing heat or mechanical engines between temperature or force differences.

We can conclude, then, that any material whose temperature, pressure, potential energy, velocity, or composition differs from the general environment has the capacity to cause change if we can devise a process or machine to utilize it. This is the *available energy* of the material. The maximum amount of the desired change will be obtained if a "reversible" process is devised to do the job.

Available energy (denoted by the symbol A) is known by many other names in the literature, some of which have subtly different meanings. Keenan's "availability" and terms involving "work" (available work, ideal work, etc.) originated primarily from mechanical engineering and thermomechanic analysis. These are often perceived as being limited to processes involving mechanical work only, which causes some confusion. However, if one defines the "dead state" (the environment) in terms of composition as well as temperature and pressure, these terms can serve essentially the same purpose as "available energy." The same is true of "exergy," primarily a European term. Evans's (6) "essergy" specifically includes the chemical terms in its definition and, therefore, most precisely represents the full potential of a material not in equilibrium with its environment. The term is not widely used in the literature or in textbooks to date.

Quantitative definitions of A must begin with the definition of "entropy" (B) given above. A reversible and an irreversible process between the same initial and final states for a material will end up with the same values for H, S, A, etc., but the work requirement (or production) will be different for each process path, so the fuel requirements for the process will also differ. Simple evidence of this can be obtained from the steam tables, a tabulation of the properties of water at various temperatures and pressures (7). The property values for 4.1×10^6-Pa (600-psig), 399°C steam are constant whether it is produced in a boiler from fuel or as turbine exhaust from higher pressure steam, but less fuel needs to be burned in the latter case after the fuel attributed to the power generated is subtracted. Therein lies the practical significance of all this theory.

Many investigators have contributed to a definition for available energy which applies to the general case in which work, heat, and mass (composition) come to equilibrium with an environment and cause change in doing so. For a closed system, A_{cs} for a state higher than a "dead state" which is defined by temperature T_0, pressure P_0, and composition μ_{i0} is defined as

$$A_{cs} = (E - E_0) + P_0(V - V_0) - T_0(S - S_0) + \Sigma N_i(\mu_i - \mu_{i0}) \tag{8}$$

where E, V, S, and μ_i represent the internal energy, volume, and entropy of the system and the chemical potential of each component, respectively; E_0. V_0, s_0, and μ_{i0} represent these quantities for the "dead" (or equilibrium) state; and N_i is the number of moles of species i.

In general, the term E should include all elements of the energy system: gravitational (potential), kinetic, magnetic, and electric energy, etc. In industrial processes, these are often neglected and only the thermal-mechanical elements are considered.

The chemical available energy (or diffusion) term is significant only for chemically reactive processes (such as combustion). For other processes, semipermeable membranes would be required to recover work by diffusion. At this writing, such devices do not exist in industry. As a result, the chemical term is often dropped in practical analyses.

For a flow system, the flow work $(P - P_0)V$ must be added to Eq. 8). This results in

$$A = (E - E_0) + (PV - P_0V_0) - T_0(S - S_0) + \Sigma N_i(\mu_i - \mu_{i0}) \quad (9)$$

Noting that $H = E + PV$, Eq. (9) becomes

$$A = (H - H_0) - T_0(S - S_0) + \Sigma N_i(\mu_i - \mu_{i0}) \quad (10)$$

Eliminating chemical terms, where appropriate, leaves

$$A = (H - H_0) - T_0(S - S_0) \quad (11)$$

which is useful in many practical problems.

It should be noted that if only changes in A for two states of the same composition are sought, the dead-state terms H_0 and S_0 drop out, and

$$\Delta A = \Delta H - T_0 \,\Delta S \quad (12)$$

where

$$\Delta H = H_2 - H_1 \quad \text{and} \quad \Delta S = S_2 - S_1$$

This has significance in simplifying many calculations, eg, energy flow in steam systems.

To avoid confusion later, note that A, the available energy of a material, is not necessarily the same as the Gibbs free energy (generally denoted by G) of the material. By definition,

$$\Delta G = \Delta H - \Delta(TS) \quad (13)$$

For an isothermal (constant-temperature) process at the temperature T_0 of the environment, ΔG reduces to (3)

$$\Delta G_{T0} = \Delta H - T_0 \,\Delta S \quad (14)$$

which is identical with ΔA. For the special case in which a process occurs at constant temperature, ΔG is a measure of the work that can be extracted from an isothermal change of state, but ΔA is completely general; it measures the work extractable (or required) from any change of state.

Heat

We stated earlier that the flow of heat into and out of a process (normally designated by Q) is not a property. Heat flows when two (or more) substances at different temperatures are placed in thermal contact. The rate of heat flow is proportional to the temperature difference between the two substances and is important to many thermodynamic calculations. Heat transferred to or from the surroundings is often the variable we manipulate or the product we seek from a practical process.

A key point in understanding heat is to realize that a given amount of heat transferred from substance A to B does not necessarily mean that the temperature changes in A and B will be equal. These temperature changes will depend on the *heat capacities* C of the substances. They will also depend on whether there is a phase change (water to steam or vice versa) in either substance. During a phase change, heat is absorbed or given off with no change in temperature. We call this *latent heat*.

We call heat interactions which do not involve phase changes sensible heat exchange. These are characterized by a temperature change related to the heat capacity of the substance. Heat capacity is defined as the amount of heat needed to change the temperature of a unit mass of a substance by one degree on the temperature scale chosen.

In addition, a measurement of the amount of heat was required. Here again, very specific conditions were needed for a standard because C changes with temperature. Thus, the British thermal unit (Btu) was defined as the amount of heat needed to raise the temperature of one pound of pure water from 59.5 to 60.5°F. The calorie was defined as the amount of heat needed to raise the temperature of one gram of water from 14.5 to 15.5°C. Because the calorie is so small, we often talk in terms of 1000 of these, or kilocalories. (This happens to correspond to the unit of dietary intake.)

The definition of heat capacity is, therefore, given by

$$Q = MC \,\Delta T$$

where M is a unit of mass. Here C has the units of heat per degree per unit mass, typically Btu/lb, °F, or cal/g °C. At the point where the Btu and the calorie were defined, C for water was taken as 1.

Sometimes heat capacities are defined on a molal basis, i.e., the amount of heat needed to raise one mole of a substance by one degree. The unit mass changes so the value of C must be increased by that multiple; i.e., the molal heat capacity of water at 15°C is 18 because the molecular weight of water is 18.

For gases, there are two heat capacities: one at a constant volume, c_v, and one at constant pressure, c_p. The heat capacities at constant pressure (C_p) are easier to measure and generally somewhat higher than those at constant volume. Thus, there are more data available for C_p. For diatomic gases (N_2, O_2, H_2, etc.), C_v at room temperature is essentially the same for each: 4.97 cal/g mol °C. On a weight basis, the C_v for H_2 is, therefore, 2.5 cal/g °C but only 0.155 cal/g °C for oxygen. Tables of constant-pressure heat capacities (C_p) are available on both a weight and molal basis in many references.

Ideal Gas Law

Various researchers observed the behavior of gases and potential laws to describe their observations. In the early 1660s, Robert Boyle proposed Boyle's law, concerning the behavior of a gas at constant temperature as its pressure was changed. Boyle's law says that the volume of the gas will be inversely proportional to the pressure. In the 1787–1802 time frame, Charles and Guy-Lussac asserted that at constant pressure the volume of a gas varied linearly with temperature. Both of these relations lead to the expression

$$PV/T = \text{const} \tag{15}$$

which might be termed the first equation of state.

Shortly after Guy-Lussac, Avogadro proposed that equal volumes of a gas at the same temperature and pressure contained the same number of molecules, regardless of chemical species. Avogadro's law and Avogadro's number have since been verified for all substances at very low pressures. Thus, one gram per mole of any gas at zero Celsius and one atmosphere pressure occupies 22,414 cubic centimeters.

From the above, it follows that the constant in the equation above is the same for all gases. We call it the universal gas constant and label it R. The magnitude of R depends on the units of pressure, temperature, and volume. For example, if pressure is in atmospheres, volume in cubic centimeters, and temperature in kelvin, R = 82.07 cm^3 atm/K. For the units pounds per square foot, cubic feet, and degrees Rankine, R 1545 ft lb force/lb mol R. For gas quantities different than 1 mol, the number of moles present is inserted in the equation so that the ideal gas law becomes

$$PV = nRT \tag{16}$$

This law holds for all gases if the pressure is low enough and is obviously related to the behavior of gases in the gas thermometer. At higher pressures, non-ideal behavior begins to appear and other equations of state are required to describe PVT relationships accurately.

SECOND LAW OF THERMODYNAMICS

The first law says that energy is conserved in one form or another. The second law says that all forms/states of energy are not created equal. Specifically:

1. The quality, or the capacity to cause change, of a material relative to its surroundings (A) varies and is important to consider.
2. This quality will be degraded or destroyed by the steps in potential processes.

The way to measure the inevitable degradation in the capacity to cause change in any process is to calculate the change in the property A across the process. As will be shown later, the first law efficiency of a process is often much higher than the second law efficiency, because the latter measures not only the energy lost in waste streams but the degradation of the quality of the energy that is recovered in useful streams.

The challenge for practical processes as outlined by this author (8) is to use each source of availability (work) appropriately. This means that the capacity to cause change of the source should be matched to the requirements of the task carefully. In many cases high-quality utilities (e.g., electricity or fuel) are used to accomplish thermodynamically menial tasks, such as heating water or domestic living space. The challenge to integrate systems so that the menial tasks are done last with sources of work that have already been used to accomplish more demanding jobs. This is the essence of today's current trend toward "cogeneration."

To understand this concept further, it is useful to consider the hierarchy of energy sources available to practical process users. Electricity is pure work. It can be used for many demanding tasks for which it alone is capable of accomplishing, e.g., driving a motor or an electromechanical process. Next in the hierarchy comes fuel, which is capable of producing heat at very high temperatures (2482°C), but many processes require heat at much lower temperatures (538–1093°C). In addition, the chemical reaction of combustion has large irreversibilities so that about one-third of its capacity to cause change is lost in the burning step.

Next in the hierarchy are various heat/work transfer fluids. Some of these, like high-pressure steam, have the capacity to do work by expansion through an engine as well as provide heat. Others, such as heat transfer oils, mostly move heat at higher temperatures from source to sink. The capacity to do work of these streams varies with their temperature, pressure, and physical state. These in turn are reflected in their enthalpy (H) and available energy (A) contents. For example, steam at 1500 psig and 950°F has an enthalpy content of 3391.8 J/g (1459.2 Btu/lb) and available energy content of 1415.4 J/g (608.9 Btu/lb), while steam at 4.1×10^6 Pa (600 psig) and 393°C has H = 3191.9 J/g (1373.2 Btu/lb) and A = 1194.1 J/g (513.7 Btu/lb). The 4.1×10^6-Pa (600-psig) steam has only 84% the capacity to do work as the 10.3×10^6-Pa (1500-psig) steam (but costs less to produce).

At the lower end of the hierarchy are low-pressure steam and hot water. Atmospheric pressure steam at 100°C has H = 349.6 J/g (150.4 Btu/lb) but A = 474.7 J/g (204.2 Btu/lb) (34% of A for 1500-psig steam). It can still transfer heat but has little capacity to drive machinery.

From these figures it is clear that we should not use electricity or fuel directly to heat our homes or hot water or to provide lower grade services. Not only do these uses cause waste of valuable energy resources, they increase the amount of heat and combustion products rejected to the surroundings, thus exacerbating worries about global warming and heat stress on the environment.

There is a hierarchy of tasks at subambient temperatures as well. Refrigeration requires work to produce. The colder the refrigerant needed, the more work required to produce it. For chilled water at 4°C approximately 71.6 kW of work is needed to produce 1.1×10^9 J (1×10^6 Btu) of refrigeration. At −40°F approximately 171.6 kW is needed, or 2.4 times the work. Thus one should not use cold utilities inappropriately either.

In the real world an entropy change accompanies each of these uses or process steps. The net positive change in entropy for an entire process may include a negative entropy change either in the working fluid or in the surroundings.

The magnitude of the total entropy change is a measure of the inefficiency of the process.

LOOKING AT A PROCESS

We have talked about recovering work or other "capacity to cause change" in a process. Since processes do not always occur spontaneously, available energy must be put in at some point in the process to drive it. From the definition of a reversible process, it should be obvious that ΔA_{rev} measures the minimum input to a process that requires work as well as the maximum output of a process that delivers work.

In real processes, the ΔA required to make the process occur, of course, is larger than ΔA_{min}. Available energy is lost or degraded because of process irreversibilities. Here is where the differences between the first and second laws become apparent; all of the joules (Btu's) entering the process can be accounted for, but their "capacity to cause change" is reduced. Consequently, available energy is consumed in the process. This consumption is what we pay for in fuel costs and what we must strive to minimize. In the literature the consumption of available energy is sometimes termed "lost work." If "work" is interpreted in the general sense discussed earlier, this terminology is appropriate.

Available Energy Balances

Steady-state balances for available energy can be constructed in the same way as the steady-state enthalpy (energy) balance is, with one exception. Whereas the enthalpy is conserved, available energy can be destroyed or degraded, so its balance equation must contain destruction (A_{dest}) and production (A_{prod}) terms. Therefore, per Spiegler (9) and others, the total balance can be written as

$$A_{in} = A_{out} + (A_{prod} - A_{dest}) \tag{17}$$

Since the available energy (A) is a state function, as are enthalpy and entropy, in steady-state cases Eq. (11) can be used to calculate the available energy of any component if H and S are known and an appropriate ambient temperature is established. For example, in a steam turbine the available energy of the inlet steam and the outlet steam can be calculated from the steam tables. At any given turbine efficiency, the work produced by the turbine can be calculated from the Molliere chart, and the available energy destroyed by the inefficiency of the turbine can also be calculated. The sum of these terms must balance. The change in the available energy of the steam in passing through the turbine is then equal to

$$\Delta A = W + A_{dest} \tag{18}$$

where W is the actual work produced by the turbine and A_{dest} the available energy destroyed.

Thermodynamic Cycles

A cyclic process is a series of steps in which the working material starts and ends in the exact same state. For example, in a steam engine water is heated in a boiler where it is vaporized into steam at high pressure. The steam is then expanded to lower pressure through an engine where work is produced. The exhaust steam is condensed in a heat exchanger and is then pumped back to boiler pressure and reintroduced. The water thus undergoes a complete cycle.

Carnot grasped that for the cycle to be complete (i.e., the steam engine to produce work continuously), the working fluid must reject heat to the cold sink as well as absorb heat in the boiler. The production of work required the passage of heat from the high-temperature source to the lower temperature sink. Through intuitive reasoning, Carnot proved that in an ideal (i.e., reversible) cycle, the work produced by the heat engine for a given heat input depended only on the temperature difference between the hot and cold temperatures. This represented the maximum work possible from the cycle, because in any real system irreversibilities would cause further losses due to friction, finite driving forces, etc. Thus, Carnot's principle can be formulated (2) as follows for a reversible process:

$$W = Q\,\frac{T - T_0}{T} \tag{19}$$

where all temperatures are absolute. The work that can be obtained from a heat engine in a cyclic process depends only on the temperatures of the heat source (T) and the heat sink (T_0). In a real process, the actual work obtained will also depend on the path which the process follows, the friction in the engine, the properties of the working fluid, etc., because each of these will introduce irreversibilities.

Considering the Carnot cycle in more detail is instructive. To do so, we need to understand two more terms: adiabatic and isothermal. An *adiabatic* process (e.g., compression) is one that occurs so rapidly or in a system so well insulated from the surroundings that no heat flows across the boundary. An *isothermal* process is one in which heat is exchanged with the surroundings so that the working fluid remains at constant temperature throughout the process step.

The Carnot cycle consists of four steps: (a) constant-temperature expansion, (b) adiabatic expansion, (c) constant-temperature compression, and (d) adiabatic compression. The cycle is pictured schematically on a pressure–volume diagram in Figure 1. It starts at point 1 with the isothermal expansion, and the steps proceed in the order listed. Heat is added or removed in the isothermal steps (a and c), but none is transferred in the adiabatic ones (b and d).

The advantage of showing this cycle on a PV diagram is that the area under the curve for each segment of the cycle represents the amount of work done by or to the working fluid for that segment. Thus, the area between the curve 1, 2, and 3 and the V axis is the work done *by* the working fluid, and that between the curve 3, 4, and 1 is the work done *to* the working fluid. The hatched portion of the curve represents the net work done by the cycle. The ratio of the net work done by the cycle to the heat absorbed in process step a (Q_t) is called the efficiency of the cycle. By experimenting with different temperatures for heat input and rejection, T_1 and T_0 , one can see that

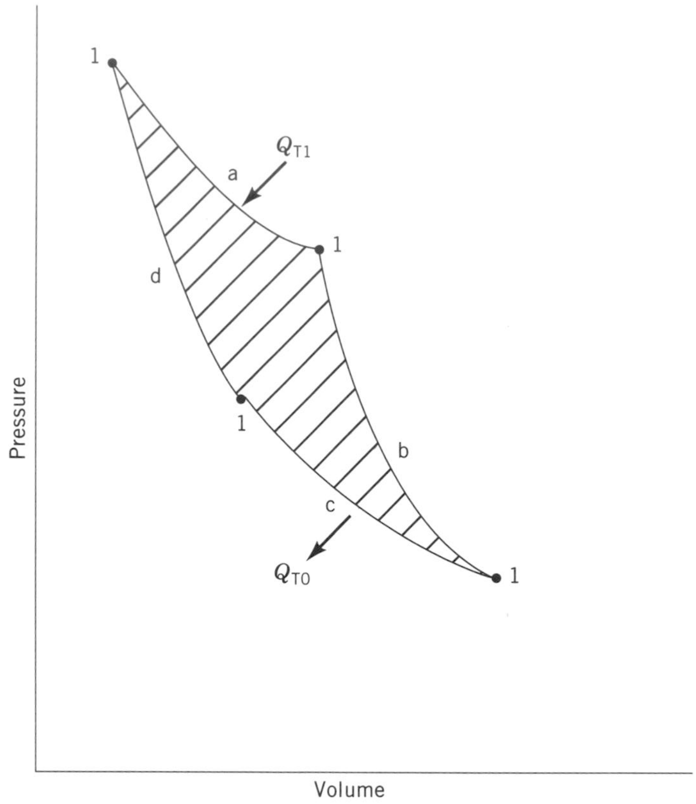

Figure 1. Pressure–volume (*PV*) diagram of the Carnot cycle. Segments *a* and *c* are isothermal; segments *b* and *d* are adiabatic. The hatched area represents the network produced by the working fluid, absorbing heat at T and rejecting it at T_0.

the area enclosed by the cycle path increases with increased difference between T_1 and T_0.

Despite the graphical representation, it was not possible for Carnot to know the paths for the adiabatic steps in the cycle exactly. If he did, it would have been possible to calculate the net work derived from this reversible cycle from the ideal-gas relationships. Even with the handicap, Carnot went on to draw some further conclusions from his observations. He observed that if another engine more efficient than the Carnot engine were possible, it would result in circumstances where heat would have to flow from the lower temperature to the higher. Since this was not possible, he concluded that no engine cycle could have a greater efficiency than the ideal cycle he studied. Thus, the possibility of perpetual-motion machines was addressed and discarded.

If we try to analyze each segment of the Carnot cycle in terms of the ideal-gas relationships, it is possible to prove the Carnot relationship for efficiency. This was presented by Fenn (2). Fundamentally, it is possible to show that the work done and that required in the adiabatic steps (*b* and *d*) cancel. The isothermal work done in step *a* is equal to the heat absorbed at

$$W_a = RT_1 \ln \frac{V_1}{V_2} \tag{20}$$

The work required in step *c* can be calculated similarly but is negative because the system volume decreases. This gives the following relationship,

$$\frac{W_{\text{net}}}{Q_{T1}} = \frac{W_a + W_d}{W_a} \tag{21}$$

By recognizing that T_1 and T_0 are either the initial or final temperature of both adiabatic segments of the cycle, the relationship

$$E \ln \frac{V_3}{V_2} = -C_v \ln \frac{T_0}{T_1} = C_v \ln \frac{T_1}{T_0} \tag{22}$$

can be used to substitute temperatures for the volume terms, leading to the Carnot efficiency equation previously written.

Typically, thermodynamic cycles can also be represented on property diagrams such as the Mollier chart, an enthalpy–entropy (*HS*) diagram for steam. These diagrams can be very intimidating because many other properties—such as temperature, pressure, and percentage of liquid—are plotted as parameters.

Schematically, a simple ideal steam power cycle is shown in Figure 2 on an *HS* diagram. In this process, water is heated, then boiled and then superheated using heat from the surroundings, starting from the lower left of the diagram (point 1 to point 4). The high-pressure superheated steam is then expended through a turbine and work is extracted. This is represented by the enthalpy change from point 4 to point 5 shown in the diagram. The low-pressure steam then enters the condenser where it is condensed to water (5–1). The water is pumped back to high pressure (not shown) and begins the cycle again. Because this is an ideal process, there is no entropy change in the turbine, and the work produced is maximum (W_{ID}).

Though schematic, the diagram is drawn approximately to scale. Note that of the enthalpy put into the process in the boiler, only about one fourth is recovered as work in the turbine. The rest is rejected in the condenser.

The process can be represented on other property diagrams as well, such as pressure–volume (*PV*) and temperature–entropy (*TS*) diagrams, depending on which property changes are most important for the process analyst. Each point in the process shown in Figure 1 has its unique set of properties. While the enthalpy and entropy are shown explicitly on the diagram, the temperature, pressure, and volume at each point around the cycle are also fixed. The work extracted from the steam expansion step is obtained by calculating the enthalpy change (ΔH) for the step (point 4 to point 5). The net heat input to the process from the boiler is shown by ΔH from point 1 to point 4 and the heat rejected in the condenser by ΔH from 5 to 1. Thus, the amount of work produced from a given amount of heat input can be directly calculated from the values in the diagram. Since this is an ideal (reversible) process, the work produced is the difference between the heat input and the heat rejected.

$$W_0 = \Delta H_{1-4} = \Delta H_{5-1}$$

If a temperature–entropy diagram were chosen to represent the process, the heat and work values could not have been calculated directly from the chart, but rather

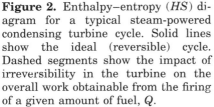

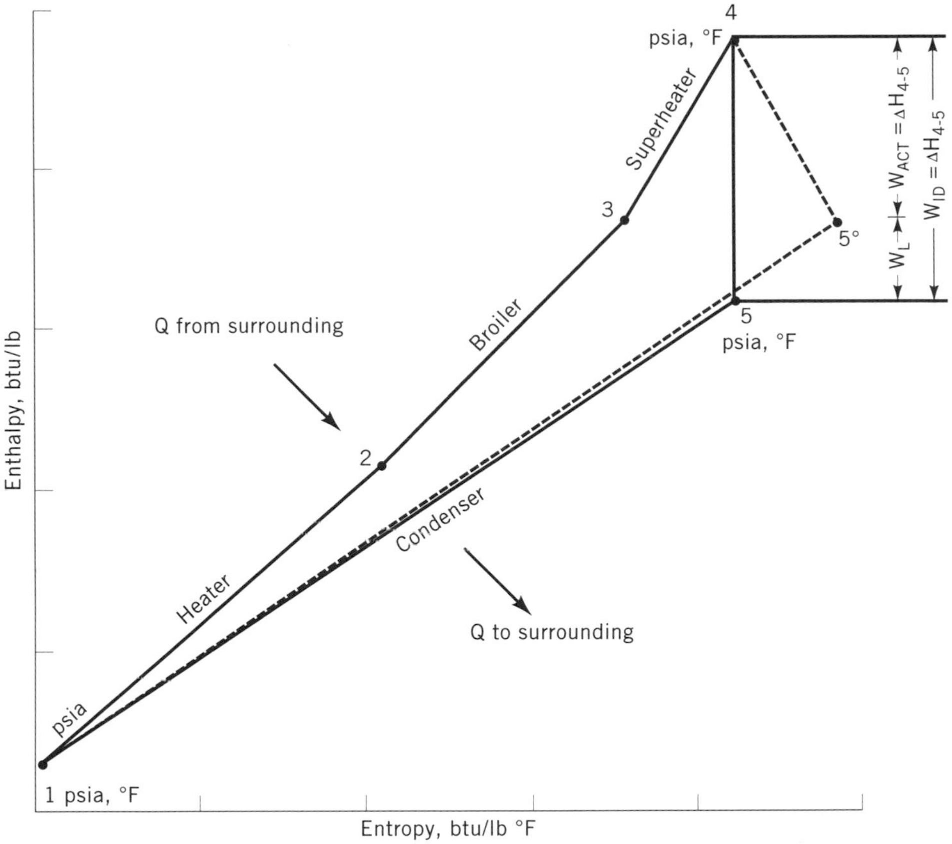

Figure 2. Enthalpy–entropy (*HS*) diagram for a typical steam-powered condensing turbine cycle. Solid lines show the ideal (reversible) cycle. Dashed segments show the impact of irreversibility in the turbine on the overall work obtainable from the firing of a given amount of fuel, Q.

by determining the enthalpy values at each point by calculation from tables of thermodynamic values (i.e., the steam tables).

The impact on the environment can also be seen from these diagrams. Even the ideal process requires an interaction with the environment (heat input and rejection). This heat must be transferred to a cooling medium (usually water or air) which heats up a portion of the environment at large. Heat input is often from a combustion process which consumes fuel and rejects combustion products to the air. Because real processes are less efficient than ideal ones, more heat must be rejected for a given heat input (or work output), and therefore, the lower the efficiency of a process, the greater the environmental impact. Those who would minimize the impact of any process on the environment should focus on reducing the demand for the product (in this case, work) and maximizing the efficiency of the process used to produce the product.

Analysis of Cycles

To analyze process cycles, a systematic, complete-evaluation and an accounting of all the availability which enters and leaves a process are required. Both the first and second laws of thermodynamic are used in this analysis. More is said about the methodology of thermodynamic analysis in the section on thermodynamics in energy systems.

Real processes lose or degrade availability because of the inherent chemical or mechanical irreversibility in the process. This means more work is required in a process which requires it or less is recovered where opportunities for producing work exist in processes. There is also entropy creation due to irreversibility. Irreversibilities occur due to processing requirements for gradients of temperature, pressure, and concentration and for irreversible chemical reactions. These irreversibilities are tolerated because processing tasks have to be accomplished within a finite (usually the smallest) amount of time. All industrial chemical processes are irreversible to a varying degree.

Thermodynamic analysis of process cycles recognizes that *work* (in the most general sense), which is available from either fuel or horsepower, is the most valuable form of energy and that in all real processes the potential to cause change is lost by being degraded to less useful forms of energy, such as low-pressure steam. Thus, many times the terms *work, availability,* and *available energy* are used interchangeably in thermodynamic discussions.

In developing the relationships between availability changes and work, sign conventions need to be addressed. Work done to a process to make it go has a negative sign, the opposite of ΔA. Work done by a process (to the surroundings) is positive.

In a reversible (ideal) process, the total entropy change is zero, so

$$\Delta S \text{ Process} = \Delta S \text{ Surroundings}$$

Neglecting kinetic, potential, and chemical energy changes,

$$\Delta H = \Sigma Q - W_{id} \tag{23}$$

and

$$\Delta S - \Sigma \frac{Q}{T_0} \tag{24}$$

For either the process or the surroundings, the expression for work becomes

$$W_{id} = T_0 \, \Delta S - \Delta H \tag{25}$$

$$W_{id} = - \Delta A_{id} \tag{26}$$

In real processes, work is lost, so

$$W_L = W_{id} - W_{act} \tag{27}$$

For work-producing processes W_{act} is smaller numerically and W_L is a positive number. In processes that require work, both W_{id} and W_{act} are negative, but W_{act} is numerically greater so W_L is positive. Conceptually, the lost work is done to the surroundings and is always positive.

If the total entropy change (process and surroundings) is known, the lost work can be calculated directly:

$$W_L = T_0 \, (\Delta S_{TOT})$$

This corresponds to the term ($A_{produced} - A_{destroyed}$ in the availability balance.

Improvements to a process can result from reductions in either term. Changes in the "ideal work" requirement can only be accomplished by changes in feed or product compositions or conditions. Even small changes in ideal work are worthwhile because they have a geometric impact on "lost work."

Independent of the ideal work, reduction in lost work can be achieved by eliminating sources of irreversibility (e.g., large "delta trees") in the process. In a new plant, these improvements may also reduce net capital investment because smaller boilers, drivers, furnaces, etc., will be required. In general, combustion is a major source of irreversibility because of the chemical changes which take place and because the energy is released at temperatures about 2204°C, but used at relatively low temperatures [eg, 510°C for 10.3×10^6 Pa (1500-psig) steam].

We should note here that the concepts of lost work and lost availability are equivalent in this discussion.

Considering the working fluid cycle in Figure 2, some simple analysis can be instructive. For the ideal process both enthalpy and entropy changes occur in each segment of the process, save one, the turbine. In an ideal turbine, there would be no entropy change; eg, an isentropic expansion would take place.

Starting at point 1, heat comes from the surroundings (the fire box of the boiler) into the working fluid in three stages: sensible heat to the water, latent heat to boil the water, and superheat to the steam. This entire step increases both the enthalpy and entropy of the water at essentially constant pressure. The enthalpy change for each segment represents the net heat input to the process in that segment. These figures are the fundamental heat balance for the process side of the boilers, ie, the first law calculations. The changes in entropy represent the second law considerations.

At the exit of the boiler (point 4), at maximum pressure, temperature, enthalpy, and entropy, the steam enters the turbine where it produces work by expanding to lower pressure and cooling. This work is recovered by mechanical devices which rotate or reciprocate to generate electricity, or do work directly. The amount of work produced is set by the exhaust pressure of the engine and corresponds to the enthalpy change through the machine.

The work recovered in the turbine segment (step 4 to 5) is measured by the enthalpy change in the step. For a given set of steam inlet conditions, this is set by the exhaust pressure of the turbine in the ideal process. In a real process, the efficiency of the engine impacts the exhaust conditions, as shown by the dashed lines in Figure 2. The actual work obtained in exhausting to the same pressure as the ideal process is less. This can only happen if there is an entropy change, as shown in the step 4 to 5′. Thus, the impact of the actual path of the process appears in the diagram.

The efficiency of the actual turbine step is measured by the ratio of enthalpy changes between the actual and ideal processes:

$$\text{EFF} = \frac{\Delta H_{4\text{–}5'}}{\Delta H_{4\text{–}5}}$$

In the analysis of practical processes, the known efficiency of a step is sometimes used to calculate the location of point 5′ on the diagram.

Note that the efficiency of the turbine step affects the duty and efficiency of the condenser as well. Both ΔH and ΔS of the condenser step are larger for the real process than the reversible one. Conceptually, this makes sense because the lost work in the turbine step has to be removed from the process as heat.

So far, we have considered only the thermodynamics of the cyclic working fluid in the steam engine process. The heat-to-work system has other components which are not cyclic. These include the boiler and condenser. The working fluids in these components also undergo thermodynamic changes which interact with the environment.

The condenser is simpler to consider. The cooling fluid is heated and ultimately mixed with the environment directly or indirectly. In Figure 2, the coolant would appear as a line lower than the condenser curve and somewhat parallel to it. This line would be short, however, because each pound of coolant can absorb relatively little enthalpy. Therefore, many pounds of coolant are required. The mixing of this medium with the environment at large (either the cooler ambient air or the cooler source of the cooling water) causes an entropy increase in the receiving body (surroundings). This ultimately may be considered a way to measure environmental impact.

The boiler has a more complex impact on the environment if it is assumed to be a conventional fuel-fired combustion device. The usual device produces a large amount of very hot gas (flue gas) containing CO_2, N_2 and O_2 and very small amounts of CO and oxides of nitrogen. Depending on the composition of the fuel, it may also contain SO_2 and other oxides, elements, or components. The thermal behavior of the flue gas would be a steeply sloped line above the boiler curve. Its behavior is similar to that of the cooling medium but in a reverse direction. Its entropy change is larger and a significant share of its enthalpy is rejected to the air through the flue. The mixing of the hot gas with the cool air causes an entropy change and an environmental impact.

In addition, the combustion process itself is very irreversible. The oxidation reactions of combustion cause a major loss of the availability (10) stored in the fuel because reversible (or nearly reversible) processes to recover the available energy do not exist. For example, the chemical reaction is not carried out in a fuel cell which produces electricity directly. At the stoichiometric mixture of air and fuel, 25–30% of the availability of the fuel is lost for lack of this fuel cell. At this same fuel–air ratio, the temperature of the flue gas is about 2371°C. The highest temperature in the steam cycle is about 538°C so that the available energy captured in the steam is much less than that available from the flue gas even after the chemical losses. Overall, even a modern, efficient boiler captures only 50–60% of the availability of the fuel in the steam.

In addition to the thermal impact, the flue gas also causes chemical change. The CO_2, NO_x, SO_x, etc. must be mixed with the ambient air, which causes an additional entropy change and environmental impact. In these days we know that further chemical and physical effects are also occurring. The NO_x and SO_x react chemically with other atmospheric and biological entities in ways that mostly degrade environmental quality. The CO_2 has mixed effects. Trees and other plants convert CO_2 into plant food and oxygen. The CO_2 also acts as a "greenhouse gas," which reduces the ability of the earth to radiate heat to space (from which earth absorbs radiation) and contributes to fears of global warming.

Thus, we can conclude that for many reasons burning incremental fuel is directionally undesirable. From a strictly thermodynamic point of view, using a resource at 2204°C to heat a working fluid to 538°C is wasteful. The more fuel we burn, the more we use a high-potential resource to do a lower potential job and, therefore, fly in the face of the guidance of the second law. Further, the component interactions of the flue gas with now understood physical and chemical environmental processes are also undesirable.

Thermodynamic efficiency improvement of itself thus serves a number of socially beneficial purposes. Seeking other practical ways to provide the heat needed by society for its various purposes has been sporadic. A generation ago, nuclear power was thought to be the answer. In the author's view, it still has potential (no CO_2, relatively infinite supply, potentially clean) but may have lost any hope for acceptance by society through the industry's lack of responsibility.

Practical applications of "renewable" sources of heat seem to be some time away from economic viability. Perhaps thermodynamics can help to develop directions for these applications.

INTRODUCTION TO EFFICIENCIES

The first law efficiency of a process can be captured by an enthalpy balance: simply calculating the fraction of the input energy which is captured in the working fluid. For the boiler just discussed, the fraction of the combustion heat absorbed in the water equals the first law efficiency of the boiler. For modern boilers, this is typically 85–90% and is controlled by reducing the stack temperature of the exiting flue gas as low as practical (controlled by economic and corrosion). Thus, flue gas at 2204°C has an enthalpy of about 2789.3 J/g (1200 Btu/lb), and when it exits to the atmosphere at 149°C its enthalpy has been reduced to about 232.4 J/g (100 Btu/lb) by exchange with the water and combustion air in the boiler. Obviously, about 90% of the fuel joules (Btu's) has been captured; i.e., the first law efficiency of the boiler is 90%.

The first law efficiency of the turbine segment (for either the reversible or real process) is virtually 100%. The enthalpy in the steam is reduced only by the amount of work extracted (and some very small amount for the friction in the machine). Thus, all of the enthalpy can be accounted for.

Similarly, the efficiency of the condenser is 100% in that all of the heat contained in the low-pressure exhaust steam is captured in the cooling medium. Of course, this represents the heat rejection step so that unless warming the cooling medium is a desired objective, this energy is lost.

If the overall efficiency of the process is considered to be based on the amount of work produced only, the first law efficiency is calculated very simply by dividing the work produced by the heating value of the fuel burned. If the boiler first law efficiency is 90% as discussed, this is calculated as

$$\mathrm{EFF} = 0.9\,\frac{\Delta H_{4\text{-}5}}{\Delta H_{1\text{-}4}}$$

if we ignore the work needed to pump the process water and the cooling medium.

The second law efficiency of the process and its segments is another matter. In a work- (available-energy) consuming process the efficiency is represented by the reversible requirement divided by the actual requirement:

$$\mathrm{EFF} = \frac{W_{\mathrm{id}}}{W_{\mathrm{act}}} = \frac{\Delta A_{\mathrm{id}}}{\Delta A_{\mathrm{act}}}$$

The available-energy input to the process is that contained in the fuel. The product is the work produced in the turbine. While some small differences exist between heating values and availability (energy) values for fuels, the overall second law efficiency of the process is close to the first law efficiency. This occurs because the only useful product considered in both processes is the work produced.

The second law efficiency of the various segments of the process is markedly different from the first law figures. As discussed earlier, the availability recovered in the steam is about 50% of that in the fuel. The work actually produced in the turbine is 75–80% of the ideal. The exhaust steam contains relatively low availability so that the availability rejected to the atmosphere in the condenser is much lower than the enthalpy lost by first law calculation. The second law efficiency of the condenser

Table 2. First and Second Law Efficiency Comparison (%): Simple Steam Cycle

	First Law	Second Law
Overall process	24	24
Furnace and boiler	87	35
Turbine	100	75
Condenser	100	−90

segment is relatively high because temperature-driving forces are low, but not 100%. A conceptual comparison of efficiencies for this simple example is given in Table 2.

A breakdown of the losses in the process from both points of view shows a markedly different picture. This difference in viewpoints provides very different guidance on ways to recover valuable lost energy from the process.

Figure 3 shows the losses calculated by the first and second law for the three major segments of a typical steam/power process cycle. In the first law analysis, some losses are measured in the furnace/boiler segment, but the vast majority occur in the condenser where a large quantity of low-level heat is rejected to the atmosphere. In the second law view, the major losses are in the furnace/boiler segment; some are attributable to the turbine inefficiency and very little availability is lost in the condenser.

An engineer who would improve the process is thus given very different guidance. The second law would say, find ways to reduce fuel fired, insert a topping cycle between the high-temperature flue gas and the steam, increase steam conditions, maximize turbine efficiency, and forget anything around the condenser. The first law would say, find a use for low-level heat and reduce the boiler stack temperature.

From the practical standpoint, finding uses for low-level heat has limited potential. Also, the rejection of low-level heat to the environment has minimum impact if handled properly. To be sure, to the extent that heat sinks are economically viable, they should be used. However, a 5% improvement in turbine efficiency (to 80%) reduces the work lost in the turbine by 7% and reduces the amount of fuel needed by four times the work saved because the overall process efficiency is 24%. Introducing a topping cycle between the flue gas and the steam (mercury boilers were used in the 1930s) would allow more work recovery for the same fuel fired (at increased capital cost).

To summarize, process efficiency is a useful device only when supplemented by further analysis of losses to put possible improvements into perspective. In the separate section on efficiency, more discussion of the differences between first and second law efficiencies and the practical value of such analysis is presented.

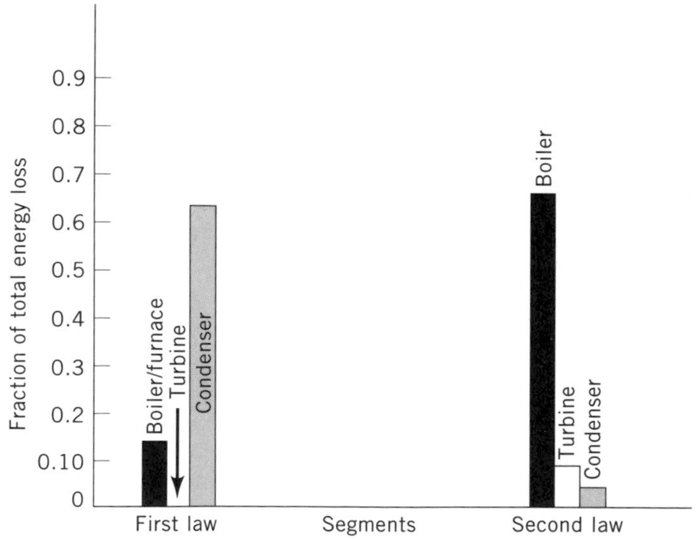

Figure 3. Distribution of losses from steam-powered cycle compared on a first-law and second-law basis. Losses calculated for the different segments of the process vary greatly depending on the viewpoint.

BIBLIOGRAPHY

1. B. F. Dodge, *Chemical Engineering Thermodynamics,* McGraw-Hill, New York, 1944.
2. J. B. Fenn, *Engines, Energy, and Entropy,* W. H. Freeman, New York, 1982.
3. McGlashan M. L., "The International Temperature Scale of 1990 (ITS90), *Journal of Chemical Thermodynamics* **22,** 653–663 (1990).
4. R. L. Rusby, "The Conversion of Thermal Reference Values to the ITS90," *Journal of Chemical Thermodynamics* **23,** 1153–1161 (1991).
5. J. W. Gibbs, *Trans. Conn. Acad. Arts Sci.* **II,** 382 (1973); reedited version of *The Collected Works of J. Willard Gibbs,* 1948.
6. R. B. Evans, "A Proof That Essergy Is the Only Consistent Measure of Potential Work in Chemical Systems," Ph.D. Dissertation, Department of Engineering, Dartmouth College, New Hampshire, 1969.
7. J. H. Keenan and F. G. Keyes, *Thermodynamic Properties of Steam,* Wiley, New York, 1936.
8. W. F. Kenney, *Energy Conservation in the Process Industries,* Academic, Orlando, FL, 1984.
9. K. S. Spiegler, *Principles of Energetics,* Springer-Verlag, Berlin, Germany, 1983.
10. E. P. Gyftopolous and T. F. Widmer, *Potential Fuel Effectiveness in Industry,* Ballanger, Cambridge, MA, 1974.
11. G. N. Lewis and M. Randall, *Thermodynamics* (1923), revised by K. S. Pitzer and E. L. Brewer, 2nd ed., McGraw-Hill, New York, 1961.

Reading List

K. Denbigh, *The Principles of Chemical Equilibriums,* 3rd ed., Cambridge University Press, 1971.

G. N. Hatsopouls and J. H. Keenan, *Principles of General Thermodynamics,* Wiley, New York, 1965. Contains two parts: an introduction for the novice and an advanced approach based on the work of Gibbs.

E. F. Obert, *Concepts of Thermodynamics,* McGraw-Hill, New York, 1960.

O. Redlich, *Thermodynamics,* Elsevier, Amsterdam, 1976. Advanced text with emphasis on unambiguous definitions of fundamentals.

H. T. Odum, "Energy Analysis, Energy Quality and the Environment, in Symposium No. 9, N. W. Gilliland (ed.), *Energy Analysis: A New Public Policy Tool,* American Association of Advanced Science, Washington, DC, 1978.

THERMODYNAMICS, PROCESS ANALYSIS

William Kenney
Exxon Research & Engineering
Florham Park, New Jersey

In most industrial and commercial settings, energy is used in several forms; often, these forms interact. Such interactions take the route of transformations of energy from one form to another, storage and removal from storage, internal recycles, and ultimate rejection to the environment. However, energy is seldom found in the form needed for a particular purpose; it must be transformed (1). Obviously, it is important to get as large a proportion of the original energy as possible into the final desired form, and a large part of an engineer's work is concerned with the factors affecting the efficiency of energy transformations.

The combustion of fuel involves transformations in which stored "chemical energy" is involved. All chemical reactions involve a transfer of energy either from stored chemical energy to the surroundings as heat energy (exothermic reaction) or the reverse (endothermic). In many cases, the transformation is not simply from one form of chemical energy to another, but may also involve mechanical energy and electrical energy as well. Thus, if a chemical reaction takes place in an electrochemical cell, some of the chemical energy may appear as heat emitted to the surroundings, another portion (if there is a volume increase due to the cell reaction) as mechanical work done against the constant pressure of the atmosphere, and the main portion as electrical work.

Energy may pass through several transformations in series before it finally arrives at the desired form. For example, consider the following transformation chain. The chemical energy of a fuel is unlocked by the chemical reaction known as "combustion" and is transformed to heat energy in the products of combustion. This energy is then transferred by conduction, convection, and radiation to energy in steam, which in turn is transformed to mechanical energy (work) in the steam engine or turbine. At this point, there may be only 20% of the original chemical energy in the form of the mechanical energy of large masses in motion. The rest is viewed as representing "losses." Actually, the energy is not lost but merely in another form, and for reasons connected with the second law of thermodynamics may be "unavailable," ie, incapable of being transformed into any useful form.

To continue the chain, the mechanical energy may be transformed to electrical energy by means of a generator, and the electrical energy to chemical energy in a storage battery. Thus, the overall result is a change from chemical energy in one system to chemical energy in another system, and the overall efficiency of the transformation is not over 15%.

See also Energy efficiency; Energy efficiency, calculations; Energy management, principles; Energy management, process.

TYPICAL ENERGY SYSTEM

A typical energy system is shown in Figure 1. Virgin available energy (exergy, availability), ie, fuel and power (work), is introduced at the top of the diagram. Power is

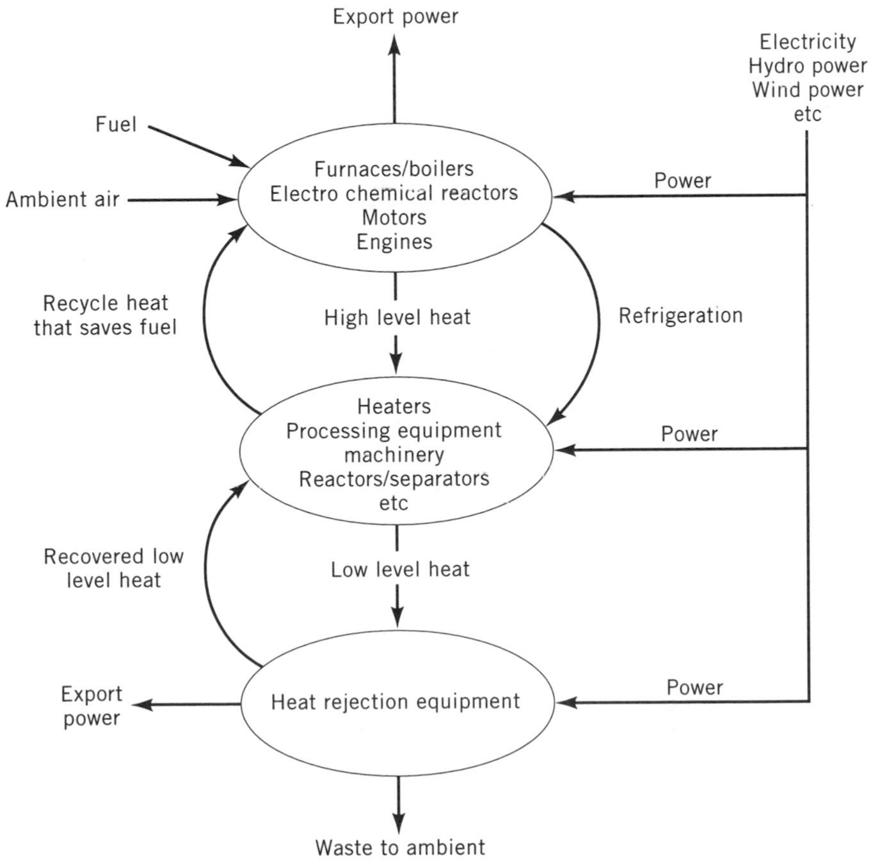

Figure 1. Typical energy system showing hierarchy of uses, virgin availability inputs, and effective recovery paths.

also likely to be needed in other phases of the system, including the heat rejection step at the bottom. In the top segment, "high level" processes take place: combustion of fuel in furnaces, boilers, engines, and in high temperature reactors and processes. Electricity might be used to drive chemical reactions, motors, or electrolytic processes, or to provide high temperature heat as in steel processing. Refrigeration might be produced for use in other parts of the system.

These high level processes often produce high temperature heat for use elsewhere. This might be captured by producing relatively high pressure steam in a waste heat boiler, by exporting the exhaust from an engine (eg, a gas turbine) directly as preheated air to another combustion device, or by making hot flue gases available for some high temperature process use, such as providing heat for a high temperature endothermic reaction (eg, steam cracking (2). Excess power may even be produced and exported outside the site.

The next group of processes typically consume/degrade availability to produce the desired products in the system. Heat and power are degraded/used in various kinds of processing equipment, some of which may recover heat that can be recycled to the high level processes to save fuel or power. For example, it is often possible to recover heat at a moderate level and use it to preheat ambient air to the combustion processes. Power is consumed in this segment to drive pumps, compressors, mechanical devices, material handling equipment, etc.

This group of processes often produces low level heat, which is either rejected to ambient or recovered and recycled. Whether the recovered heat is economic depends on whether the recovery really reduces the input of fuel and power required to produce the desired products. In many cases, the need to cascade any recovered energy back up to the top of the chain is not appreciated. Some mechanism, such as a heat pump or a way to upgrade the value of the low level heat (eg, combustion air preheat), is required to achieve this objective.

In the bottom segment of the diagram, energy that is no longer in a useful form is rejected to the environment. Usually, this is done by exchanging heat with a cooling medium from the environment and/or venting the useless energy to the atmosphere. Energy and investment are required to reject this useless energy to the environment. Pumps and fans for coolants are required as are heat exchangers. Flue gas stacks are needed, as are pressure letdown valves if steam is to be vented. Moreover, once this equipment is installed, it must be operated and maintained.

Within limits, changes in process loads can be accommodated by flexibility within the energy system. For example, if process steam loads are decreased, then a steam turbine driver might be changed for a motor driver, and the firing of the boilers reduced to rebalance the steam system. The question of economics would depend on the amount and price of fuel saved the extra electricity cost incurred.

These interactions lead directly into the consideration of the cogeneration potential at a site. Cogeneration is one way to derive as much useful work as possible from the required available energy inputs in fuel and electricity, before their capacity to cause useful change is exhausted and the depleted energy is rejected to the environment. There are several forms of cogeneration, all of which involve the generation of useful process heat and work in one form or another; they are not limited to the conventional approach where electricity is the work product. In the typical cycle, steam is generated at high pressure and exhausted through a turbine to a pressure where process heat demands exist. The turbine might drive a compressor in the plant or an electric generator. However, internal combustion drivers have an advantage in that less waste heat is produced per horsepower generated than for the exhaust steam from a steam turbine. As a result, a smaller demand for process heat is required to balance the system when internal combustion engines are used.

One approach to characterizing a site's cogeneration potential is to explore the ratio of the amounts of horsepower used at the site to the amount of process heat supplied by steam and the average pressure of steam used for heating (3). (Power that is currently supplied by motors, or can only be supplied by motors for process or mechanical reasons, is excluded from the ratio.) If a site (or a combination of sites) has a ratio of 0.2 or below, it can generally operate completely in a cogeneration mode using steam turbines only. That is, the power needs and the process heat needs can be supplied by running high pressure steam through back pressure turbines to generate the necessary horsepower, with the outlet steam providing the necessary process heat in balance. This critical ratio varies from about 0.1 if the steam is needed at 600 psig, to 0.3 if 15 psig steam is adequate to meet process temperature demands. In this range, overall fuel utilization at the site is maximum and can approach boiler efficiency.

As soon as the critical ratio is exceeded, either the site must use low overall efficiency (25%) condensing turbines to produce the extra power or switch to internal combustion drivers (eg, gas turbines). Using condensing turbines rapidly increases fuel demands at the site. For example, at a ratio of 0.8, 25–30% more fuel is required if condensing turbines are used than if a gas turbine provided the power needed above the critical ratio. At a ratio of 1.4, the extra fuel requirement is about 35%. Thus, understanding and maximizing the site's cogeneration potential significant impacts the fuel demand at the site and the related environmental combustion processes. In some locations, cooperative district heating methods have been developed to take advantage of the potential for grouping process and/or commercial/domestic heat needs. In these methods, excess heat from power generation steps is "exported" to users outside the traditional boundaries of the industrial complex.

In addition to these high level process options, opportunities to recycle heat from otherwise wasted steams exist. As indicated earlier, in most cases, such recycle must displace some virgin availability input to be economic. Sometimes, energy can be "pumped up" in quality via a heat pump to achieve this end. Such heat pumps must be placed appropriately in the system (4). A narrow analysis based on a single unit in the system can produce an expensive error, and unfortunately, few practical applications have been shown to be economic.

The task of the energy systems analyst is to understand the whole system at the site. The goal is to minimize the virgin availability input at the top of the system to minimize costs. Reducing the cost of the necessary rejection of energy at the bottom of the system can often be economic and environmentally beneficial as well. In many cases, seasonal and diurnal cycles must be considered in any optimization scheme; no longer is it competitive to consider only an average steady-state energy balance in analyzing an energy system. Also, below-capacity operations and irregular alignments may provide opportunities to reduce the required availability input, at least part of the time. This is the challenge of the energy systems analyst.

Energy Conversion

A number of energy conversions are indicated in Figure 1, including combustion, transfer of high temperature heat from flue gas to process or steam, the generation of power from fuel or by the expansion of steam from high pressure to low, and the use of power to move fluids, form metal, raise weights, and produce chemical change. These conversions individually and collectively have efficiencies and impacts on the overall system efficiency. The objective of the energy system engineers is to define the optimum combination of availability inputs and outputs to minimize the real consumption (and cost) of the various forms of energy required to make the desired products from each process in the system. In general, this will correspond to the lowest entropy change in the system. For new facilities, this may also correspond to lower capital costs than seen in nonoptimum systems as well.

Thermodynamic Analysis

To analyze energy systems, a systematic approach involving both first and second law considerations is needed. This approach involves a number of steps, which are briefly described below (2):

1. *Obtain a Consistent Heat and Material Balance for the Process.* Care should be taken in using a consistent set of numbers, because errors in heat and material balances are magnified in availability calculations.
2. *Determine the Temperature of the Surroundings.* The chosen temperature is generally that at which heat is rejected to the atmosphere; the temperature can either be the air temperature, the cooling water temperature, or an arbitrary temperature at which heat is judged to have no value. In general, a small difference in the specific temperature chosen will not have a significant effect on the results of the analysis.
3. *Determine the Overall Boundary for the Process System.* This involves defining the process limits, the feeds, the products, and the utility usage; the portions of the process included or excluded from the boundary may have a large effect on the analysis.
4. *Calculate the Required Thermodynamic Properties.* The availability function for each stream that crosses a boundary is required; entropy and enthalpy data are required for the calculation of availability function. The availability of shaft work or electricity that crosses a boundary is equal to the energy crossing the boundary and must be considered as a stream in the analysis.
5. *Determine the Reference State.* The reference state represents the temperature and pressure that are used for the calculation of thermodynamic properties. Thermodynamic properties from multiple sources should be corrected for any differences in reference state. The most common reference temperature is 298.180°K; two useful data sources are available (5,6).
6. *Compute Minimum Availability Required for Process.* The minimum availability required by the process is defined as availability of product streams minus the availability of raw material streams. If the process produces steam, electricity, or work, the availability of these products is included in the minimum availability. Availability gain by cooling water is excluded unless the purpose of the process is to heat water or air.
7. *Compute Actual Availability Consumed.* Availability is supplied to a process in terms of fuel, steam, electricity, work, or heat. Before proceeding with the analysis, sign conventions must be established; if heat transferred into a process is considered positive, then work transferred out of a process must also be positive. In effect, the process converts a certain share of the input heat to output work and thus, the following sign conventions are usually used:
   1. Q_{in} and W_{out} are positive.
   2. Q_{out} and W_{in} are positive.
   3. W_{lost} is positive as it is done on the surroundings.
8. *Calculate Lost Availability.* This quantity is simply the difference between the actual availability consumed and the minimum required by the process.
9. *Determine the Boundaries for the Smaller Parts of the Process.* This involves defining boundaries of the smaller processing sections to locate the main availability losing sections of the process; the analysis is usually repeated on consecutively smaller sections of the process until the segments or equipment contributing most to the loss of availability are identified.

The energy systems engineer is still faced with using the data generated by the above to achieve the objective of reducing the input of availability (and, hence, the energy cost) of the system. This effort focuses on the implementation of two strategies:

1. Reducing the ideal work (availability change) for each desired change.
2. Reducing lost work (availability), ie, the entropy change (ΔS) for each step of each process included in the overall system.

The approaches have different characteristics and limitations.

Changing the ideal work of a process can involve only a limited number of variables. Theoretically, the ideal work is independent of the process chosen and depends only on the initial and final states of the raw materials and products. Thus, the variables to be manipulated are confined to the initial and final conditions, and the initial and final compositions. In some cases, changes in the mass flow rates are more related to ideal work changes than they are to lost work changes, and might be considered to belong in the ideal work category.

By their nature, reductions in ideal work requirements are generally strategic in nature, result from an overall view of the process, and require less data than studying reductions in lost work. However, they are also more difficult to find and generally require more radical departures from existing technology, resulting in a difficult "sell" to company management. These difficulties should not inhibit the search for improvements. Relatively little work is required to carry out the analysis, and even small changes generally produce significant results. Many commercial processes have overall thermodynamic efficiencies in the range of 10 to 20%; as a result, small improvements in ideal work requirements have large multipliers back through the energy system.

Reducing the lost work (availability) of a process provides a broader field for creative thinking because many more variables can be changed to achieve results. For example, one can improve the efficiency of a piece of equipment, match a utility better to a service, reduce internal recycle, and so forth. These improvements all involve reducing the net entropy change ΔS of the total system by improving steps in the process path; however, these improvements all also require more detailed analysis to define the appropriate steps.

For some applications, less complicated forms of analysis may be useful. To optimize heat exchanger systems, the heat exchanger network technology might be used. This is often termed "pinch" technology, which attempts to eliminate the need to calculate entropy changes but still to achieve benefits in more efficient use of heat exchanger trains (7). This is done by analyzing the temperature—enthalpy (TQ) curves for both the heating and cooling streams, and matching sources and sinks in a network of heat exchangers that maximizes the amount of heat recovered economically. By minimizing the number of exchangers through more sophisticated network analysis, economics can be enhanced in new facilities. The thermodynamic principles of closely matching the quality of the energy used to that required for a given process effect produces the minimum entropy change for the step (8). Researchers are working to expand use of the technology approach to heat pumps and other process equipment (9).

For example, Linnhoff has established a cooperative research institute at the University of Manchester in the United Kingdom; this work has produced software to assist engineers in optimizing heat exchanger networks. Other industrial organizations and technical software companies have also developed software to assist in the synthesis of optimum configurations. Although these tools are useful in applicable situations, they lack the broad applicability of the fundamental thermodynamic analysis approach.

Returning to the concept of reducing lost work, if the lost work of a refrigerated separation process is reduced as to reduce refrigeration requirements, in effect, the ideal work of the refrigeration section has been reduced. This is true because the only product of the refrigeration section is to supply the available energy needed to cool the process in question. The ideal work of the refrigerant compressor is reduced and with it the lost work encountered in that step. Working backward through the power supply train, the actual work for the compressor becomes the ideal work for the turbine that drives the compressor. Similarly, the actual work for the turbine is the ideal work for the boiler and steam system. All along the chain, any change in ideal work results in a change in lost work in the preceding segment of the energy supply chain. The magnitude of each lost work change will depend on the efficiency of the particular step. Compressors and drivers have high thermodynamic efficiencies, but boilers do not. Thus, a small change in refrigeration load can be magnified substantially at the boilers. Note that the real source of the improvement was a reduction in the work lost in a process segment, which is where most improvements will be found.

Thermodynamic analysis is a systematic way of characterizing what happens to the available energy of the materials in a process. It is based on the usual heat and material balance for the process but extends it to encompass entropy changes and the attendant losses of available energy. Even a simplified analysis can assist in identifying fundamental process improvements.

Once sufficient breakdown of the losses in each process step have been developed, the location of significant losses will be clear (10,11). The challenge is to identify ways that will allow improvement of the efficiency of the overall system by reducing the losses in a given process step. The most simple improvement is to eliminate the process step entirely, which in some cases is possible. For example, consider the case where a hot intermediate product is cooled, sent to storage, and then returned to the next step in the process through a heater. Even if this step is accomplished through a hot product—cool feed heat exchanger, some losses occur due to the temperature difference in the exchanger. Why not bypass the tank and go directly to the next process step, using the storage tank only when process imbalances or interceptions occur?

A second approach is to realign the available energy sources to better match the needs of consumers and maximize the conservation of availability. For example, the intermediate pressure levels in a multi-level steam system might be adjusted to maximize power recovery between levels (12). The system shown in Figure 2 is a relatively simple steam system, generating power and supplying heat to two process users, identified as reboilers for simplicity. The steam system is not in balance with 4,536 kg/h 10,000 lbs/h) of 190 psia steam being vented. This might occur only during part of the year, eg, in summer when steam demand is low but power load remains constant or even grows due to air-conditioning or heavier refrigeration loads. A qualitative inspection of the system shows a number of availability losses. The most obvious is the vent. However, the exhaust pressure at the turbine is apparently set by heat exchanger design of reboiler A with a 14°C temperature driving force (ΔT). This may or may not have been optimum at the time of the design, but

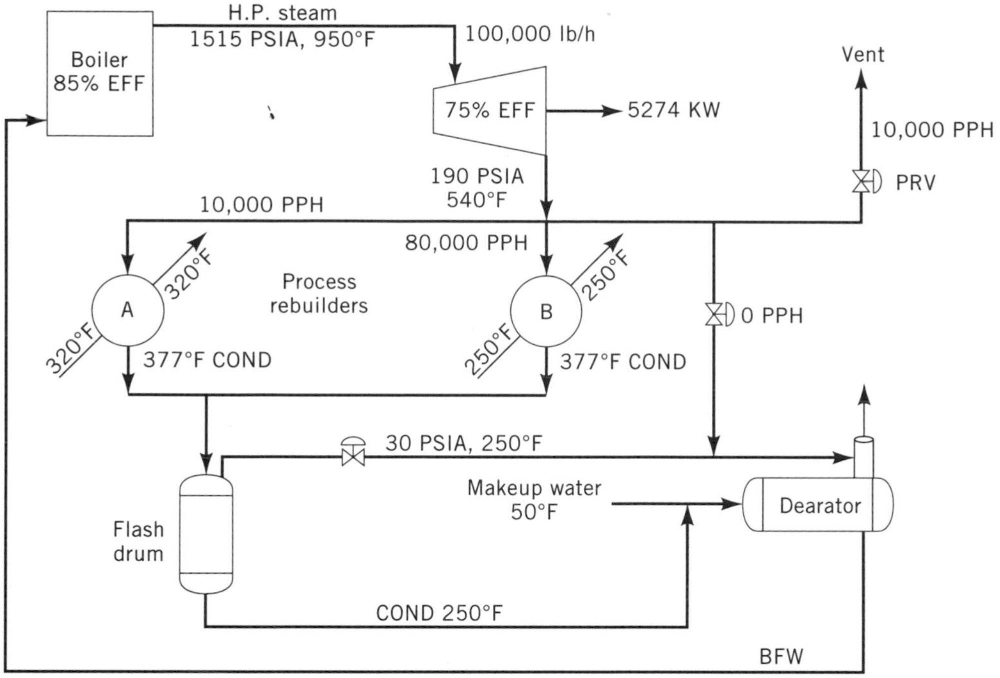

Figure 2. Simple steam system, showing power generated, process heat demands, and overall steam balance.

using 190 psia steam (192°C saturation temperature) to drive the larger reboiler B (53°C ΔT) is wasteful. Any letdown across the valve pictured would also be wasteful.

One approach to reduce lost availability at the cost of increased heat exchanger surface would be to reduce the exhaust pressure of the turbine. This would reduce the required steam flow for the necessary power production, thus reducing the venting but decreasing the temperature driving force at both reboilers. If, for example, the exhaust pressure were reduced to 125 psia, only 7.48 kg (16.49 lbs) of steam would be required to produce a kilowatt of power vs the 8.60 kg (18.96 lbs) required at 190 psia exhaust pressure. Thus, the total steam flow needed is reduced from 45,360 kg/hr (100,000 lbs/hr) by the ratio 7.48/8.60, which results in a steam flow of 39,463 kg/hr (87,000 lbs/hr). This would seem to be a too large reduction because 40,824 kg/hr (90,000 lbs/hr) are apparently required for the process heat demand. However, the latent heat of vaporization increases with lower pressure. In this case, the valve increases from 847 BTU/lb at 190 psia, to 875 BTU/lb at 125 psia, which makes the balance close.

The saturation temperature for 125 psia steam is 173°C vs the 192°C for 190 psia. This decreases the driving force on reboiler A from 14°C to -4°C, clearly requiring more surface. In reboiler B, the ΔT is reduced from 53°C to 34°C, also a significant change. Before rushing out to buy more heat exchangers, however, it would be wise to reassess the operating pressures of the equipment being reboiled by A and B. If these could be lowered, then the required process temperatures would also be reduced, perhaps restoring some of the lost capacity in B. Further, if these reboilers are driving fractionation equipment, lower pressure usually improves the relative volatility of the separation, making less reflux (and, therefore, less heat) necessary. This example proves the point about the interactions depicted in Figure 1.

The economics of this change are fuel savings of 13%. To produce 45,360 kg/hr (100,000 lbs/hr) in an 85% efficient boiler requires firing of 1.54×10^{11} j/hr (1.46×10^{8} Btu/hr), plus the requirements of the boiler itself. At 39,463 kg/hr (87,000 lbs/hr), this is reduced by 2.85×10^{10} J/hr (27 BTU/hr). At $2/MBTU fuel price, this amounts to $38/hr or about $300,000/yr for a full-year operation, a savings usually worthy of further study, especially as emissions from the stack are also reduced.

If this system were revamped in an existing plant, there would be no investment savings to offset the need for heat exchanger additions, unless a boiler expansion were being contemplated for other reasons (3,13). In a new plant design, however, one could balance boiler savings vs exchanger cost and likely find that a lower investment option was possible because boilers generally cost more per unit capacity than heat exchangers. As indicated above, one could go further into optimizing the process system by coordinating the process (energy) requirements with the utility system for an overall optimum in both energy efficiency and investment. This is rarely done because of the complexity encountered. A number of researchers were working on global thermoeconomic optimizer systems in the early 1980s; so far, few practical applications of these efforts have been reported (14–18).

The next step in the analysis of an energy system improvement is to perform a more detailed analysis on the entire system that would be affected by the change. This would include not only the steam system but also those segments of the process system that interact with it (eg, the equipment driven by reboilers A and B and related systems). Lost work (availability) analysis might be a useful tool, as the system gets more complex, to quantify the relative magnitudes of the losses in the process and to verify the qualitative observations already made. The more detailed analysis should only be applied where needed to provide additional insight, not indiscriminately.

Another form of lost work (availability) that often merits investigation is pressure drop. Here, designers often balance the size of pipe ducts and other conduits against the investment and operating costs of the pumps and compressors needed to transport the material through them. These trade-offs are well-known, but often the efficiency of the pump or compressor and driver is not evaluated in as much detail as might be beneficial. Again, higher efficiency means lower power requirements. Also, the choice of driver and its impact on the overall energy system should be taken into account when pertinent.

In some cases, pressure drop can be decreased while obtaining improved process performance. The use of new packings in fractionation services as opposed to the tried-and-true sieve trays is one example. There is evidence that packings are practical in many applications traditionally served by trays. The resulting tower investments are lower as well as the pressure drops through the system in grass-roots situations. As indicated earlier, these benefits can also reduce both the quantity and the temperature level of the energy source needed to drive the process step.

SHORTCUTS/CHECKLISTS

As should be clear from these examples, detailed thermodynamic analysis is not always required to increase efficiencies, but a thorough understanding of the principles of the second law always is. In addition, a systems viewpoint is necessary. It is possible to capture the essentials of second law thinking in checklists of varying detail to serve as a trigger for the imagination of those interested in improving energy efficiency. Some items on a checklist can be drawn from the foregoing discussion:

- Minimize the temperature driving forces in heat exchange and optimize these forces as a function of temperature levels.
- Minimize combustion, electricity, and other virgin availability inputs, and maximize appropriate energy reuse.
- Maximize the efficiency of engines, pumps, compressors, and other machinery (often by using the maximum number of stages).
- Use expanders to generate power from pressure letdowns instead of throttle valves.
- Select drivers and conditions for process heat demand on the basis of integrated system analysis, and maximize cogeneration potential.
- Ensure that any process step is necessary before trying to improve it.
- Minimize the downgrading and loss of availability.
- Minimize pressure drops.
- Carry out separations at conditions that maximize separation potential and avoid mixing anything that later will have to be separated.

The logic of this list leads to other suggestions that are related to second law principles:

- Carry out exothermic reactions at maximum temperature to allow heat recovery at maximum availability levels.
- Conversely, carry out endothermic reactions at the lowest possible temperature to minimize the availability level of the energy source.
- Minimize the use of diluents, solvents, etc, and maximize reaction yields.
- Maximize the number of stages in separations and provide distributed energy input/rejection; both steps minimize driving forces.

The final caveat is that greater integration in any system often decreases its flexibility. At some point, the basic operability and safety of the system may be compromised. Clearly, no efficiency improvement will be economic if it creates a significant risk of an incident, whether it be an explosion or a release to the environment.

ENERGY STORAGE

Earlier, one approach to providing energy storage was discussed: the generation of power by a conventional combustion process and its storage in batteries for later use. The net efficiency of the many transformations in this chain was quite low, yet the idea of energy storage can be both economically viable and environmentally conservative.

Cyclical loads provide most of the opportunities, with the economics usually driven by the investment needed to handle peak demands vs that required for more consistent loads. In hydropower systems, "pumped storage" is sometimes used to provide a better match of the relatively continuous power generation from the water flow and the typical cycle of residential/commercial power demand. During the valley in the consumer load cycle, excess power is used to pump water to an elevated storage place. Later, when demand peaks, it is run downhill through a turbine to generate peak electricity. Some environmentalists object to the construction of these facilities, but positive trade-offs are achieved by not having to build other peak power generators (usually relatively inefficient gas turbines), by not restricting the flow of the primary source of water flow, and by not causing possible user inefficiencies (or substitutes) related to power restrictions.

Commodity storage can also level power demand and production investment in cyclical load situations. The most obvious are the large gas holders on utility gas systems and the elevated head tanks on municipal water systems. High pressure reservoirs in compressed air systems also fit this description. The storage of liquid oxygen and nitrogen for later use as a gas is also a form of energy storage. These elements are separated from air by the application of energy to liquify them in a fractionation tower. Conserving this energy by storing the liquified product makes economic and environmental sense.

Storing electricity in batteries is no always wasteful. Storage batteries involve direct-current electricity, whereas, most generation currently is alternating-current. Batteries thus represent just another transformation of the energy as well as a storage mechanism. In some cases, it is the transformation that drives the application.

The storage of both heat and mechanical energy has also been practiced in other forms. Flywheels are a traditional way of storing and extracting mechanical energy by accelerating and decelerating a rotating mass. This can be a relatively efficient process as the changes in speed can be slow, implying small entropy change. Such systems are often very massive, requiring large investment. In high temperature processes, hot off-gases are sometimes used to heat refractory masses which, in turn, are switched to discharge this heat to cool feed streams. This approach is often forced on the energy conserver when direct feed to effluent heat exchange is precluded by fouling or materials considerations. The approach is relatively inefficient because two temperature-driving forces are required: one to heat the refractory, and one to discharge the heat to the cooler stream.

Attempts to capture solar heat are inherently cyclical. The direct generation of electricity must end when the sun disappears. However, applications such as capturing the sun's heat in domestic hot water, in a massive stone wall, by melting a solid salt, etc, for later use in space heating at night for hot water supplies has proven economic. Recovering such heat via absorption refrigeration to air-conditioning spaces has also been used and has the merit of direct coupling in time to the load. All of these suffer from the "double ΔT" inefficiency and significant investment requirements.

Changing the physical or chemical form of an energy source to make it more transportable or more suitable for a specific application has been done. It is often successful when the energy source is in a remote location and would otherwise be wasted, eg, flared or vented to the atmosphere. Natural gas associated with remote oil fields is often a candidate for such processes. Sometimes, such gas is merely condensed and transported to market. Obviously, a significant energy input is required to achieve the condensing. Recent studies are considering changing the chemical form of remote natural gas to convert it to heavier liquid molecules. Not only is the new material more transportable, it is also suitable to a wider range of end uses than simple combustion. It might serve as feedstock to a variety of petroleum or chemical processes as well as vehicle fuel. Again, the efficiency of current processes is relatively low, about 50%. Fundamental analysis of proposed processes at an early stage might identify areas of research that would reduce enthropy generation and reduce waste in new processes (19). Again, there is some potential for using thermodynamic analysis as a means to prevent pollution.

The storage of agricultural products and fibrous garbage also present opportunities. Corn has long been stored as feed for animals, a form of energy storage, and can also be converted to ethanol and used as a motor fuel. Of course, vast areas of corn fields and huge silos would be required to contribute significantly to the vehicle fuel pool. Also, the net efficiency of converting the solar energy to the motor fuel form via the agricultural chain is currently low.

Anerobic fermentation of garbage and domestic sewage is a source of energy that is relatively untapped in the industrialized world (although practiced in various forms in less advanced societies). Yet, garbage is a serious social concern. Large amounts of garbage are stored, and fermentation goes on to some extent unbidden. Refuse to energy plants are built over the protests of environmentalists and truly concerned neighbors, although these plants are generally well-designed and operated. Whether the reaction to the logistics, collection, distribution, and in-home use problems of fermentation gas would be any different is problematical. Still, both approaches have the advantage of starting with a currently wasted source and reducing some currently perceived problems. At the moment, the efficiency of both approaches is likely to be low, but as more knowledge about biological processes, is gained, the future outlook may be more attractive.

In summary, cyclical energy demands offer some opportunities for energy storage and reuse. The economics of any particular application are driven either by the capital cost of providing fixed facilities equal to peak demands, or the potential for using otherwise wasted sources. In the longer run, finding inexpensive ways of using solar energy, already accumulated masses of wastes, or of conversion of future wastes into useful energy forms is a part of the challenge.

BIBLIOGRAPHY

1. B. F. Dodge, *Chemical Engineering Thermodynamics,* McGraw-Hill, New York.
2. W. F. Kenney, "Combustion Air Preheat Saves Energy in Olefins Production," *Oil and Gas Journal,* 130–132 (Oct. 17, 1983).
3. W. F. Kenney, *Energy Conservation in the Process Industries,* Academic Press, Orlando, Fla., 1984.
4. B. Lennhoff and B. W. Townsend, "Designing Total Energy Systems," *CEP* 72–80 (July 1982).
5. J. H. Keenan, F. G. Keyes, G. Hill, and J. G. Moore, *Steam Tables,* John Wiley & Sons, Inc., New York, 1969.
6. D. R. Stull, E. F. Westrum, and G. C. Sinke, *The Chemical Thermodynamics of Organic Compounds,* John Wiley & Sons, Inc., New York, 1969.
7. B. Lennhoff and D. R. Vredeveld, "Pinch Technology Has Come of Age," *CEP,* 33–40 (July 1984).
8. T. Umeda, T. Harada, and K. Shiroko, *Comp. Chem Eng.* **3,** 273–282 (1979).
9. B. Lennhoff and W. D. Witherell, "Pinch Technology Guides Retrofit," *Oil and Gas Journal,* 54–65 (April 7, 1986).
10. W. R. Ovllette and A. Bejan, *Energy* **5,** 587–596 (1982).
11. D. Poulikakos and A. Bejan, *Trans. ASME* **104,** 616–623 (1982).

12. W. Lockett and J. J. Guidl, "Site Energy Surveys," *Proceedings of the Industrial Energy Conservation Technology Conference,* Houston, Tex., 1981.
13. E. Gyftopoulas and M. Benedict, "Air Separation Process," *ACS Symp. Ser.* **122,** Washington, D.C., 1980.
14. M. Tribus and Y. M. El Sayed, *Thermoeconomic Analysis of an Industrial Process,* MIT Center for Advanced Engineering Study, Massachusetts Institute of Technology, Cambridge, Mass., Dec. 1980.
15. Y. El Sayed, "The Exergy Concept Beyond the Academic Environment," *Proceedings of the Industrial Energy Technology Conference,* Houston, Tex., 1988.
16. R. B. Evans, P. V. Kadaba, and W. A. Hendrix, *ACS Symp. Ser.* **235,** 239–261 (1983).
17. B. B. Moore and W. J. Wepfer, *ACS Symp. Ser.* **235,** 289–306 (1983).
18. G. Wall, *Energy (Oxford)* **11,** 957–967 (1986).
19. T. C. Vogler and W. Weissman, *Chem. Eng. Prog.* **84,** 35–42 (1988).

THERMOELECTRIC ENERGY CONVERSION

E. A. Skrabek
Fairchild Space & Defense Co.
Germantown, Maryland

Thermoelectric energy conversion is the science of the interchange of thermal and electrical energy in simple solid state devices, generally, heavily doped semiconductors. Figure 1 shows schematically two thermocouples connected in series. If thermal energy is applied at the top of the device from some heat source, and removed at the bottom by a heat sink, an electrical potential will appear across the device as indicated by the polarity signs. The size of the ΔT and the basic material properties will determine the voltage across the couple (usually of the order of tenths of a volt per couple). Practical working voltages are obtained by connecting a large number of these thermocouples in series. The amount of current is dependent upon the cross-sectional area and length of the legs.

See also Commercial availability of energy technologies.

If, on the other hand, an electric current is passed through the device from an outside source, heat will be absorbed at one junction and released at the other, causing the device to operate as a solid state heat pump. Thus, thermoelectric devices can be used to generate electrical power, refrigerate or transfer large amounts of heat without the use of freons or compressors or any rotating machinery. However, the overall efficiency of these devices is somewhat lower than other systems and the initial cost is often higher. The economics of mass production and newer design and manufacturing techniques will undoubtedly decrease this difference in the future; in addition, the fact that the operating life of many of these solid state devices is extremely long (decades or more) should allow the recovery of the costs.

Segmenting of legs (Fig. 2) to take advantage of the higher performance of certain materials in a given temperature range has been used to increase the conversion efficiency to some extent. However, this technique is limited since all of the thermal and electrical currents must flow through each segment. Every material has its own optimum I/q ratio, so in a given segmented couple it is normally impossible to have each segment operating at its own optimum efficiency unless the couple is used at precisely the design hot and cold junction temperatures.

In thermoelectric cooling applications, extensive use has been made of cascaded systems to attain very low temperatures, but since the final stage is so small relative to the others, the thermal flux is limited. This is illustrated in Figure 3. The relative sizes of the stages are adjusted to obtain the maximum ΔT. Then the way to get higher cooling capacity is to increase the size of each stage while maintaining these area ratios.

Since Joffe's pioneering treatise in 1957 (1), a number of good books have been published on this topic. For a really thorough survey of the field and recent information on materials properties, the CRC Handbook is recommended (2). Varying approaches to the analysis, design,

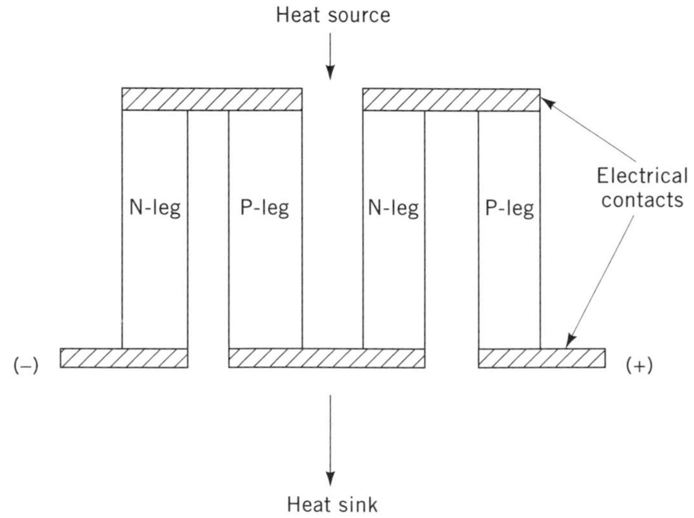

Figure 1. Simple thermoelectric device with two thermocouples in series.

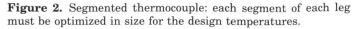

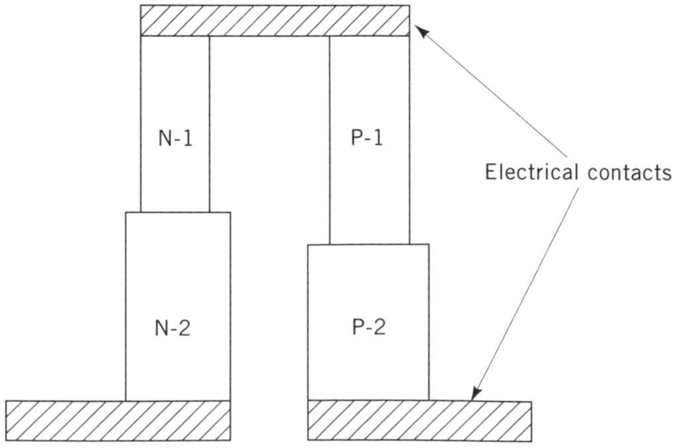

Figure 2. Segmented thermocouple: each segment of each leg must be optimized in size for the design temperatures.

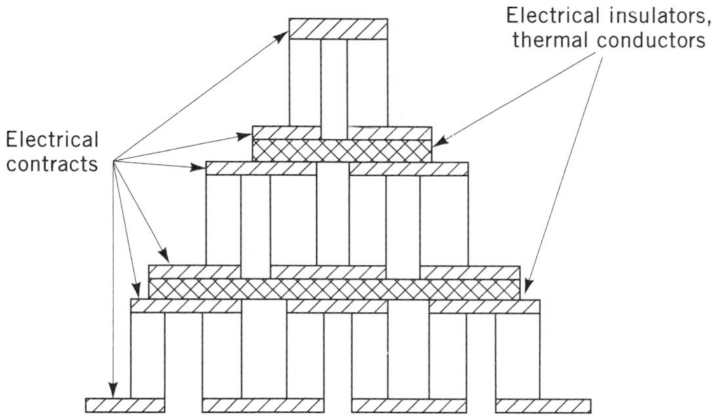

Figure 3. Three stage Cascaded Thermopile. The relative sizes of the stages must be adjusted to obtain the maximum delta T.

and construction of thermoelectric devices can be found in other works (3–6). Most of the current materials research and device design work is presented annually in two conferences:

1. The International Conference on Thermoelectrics, sponsored by the International Thermoelectrics Society and one or more organizations in the current year host country. (Contact: Professor K. R. Rao, The University of Texas at Arlington, Box 19016, Arlington, Tex., 76019, U.S.).
2. The Intersociety Energy Conversion Engineering Conference, hosted on a rotating basis by seven cooperating technical societies: ACS, AIAA, AIChE, ANS, ASME, IEEE, and SAE. The work on thermoelectrics is only a few sessions in this large conference, but it is generally on the latest operating systems.

HISTORICAL BACKGROUND

The basic ideas of thermoelectricity have been known for nearly two centuries, but until well after the Second World War the primary use was for temperature measurement with metallic wires. Then with the improvements in semiconductor technology, electrical power generation and refrigeration came under serious consideration.

In 1822 Seebeck began a study of the "magnetic polarization of metals and ores produced by a temperature difference." His goal was to prove that the earth's magnetic field arose from temperature differences between the equator and the poles. The extensive data that he collected laid the foundation for later developments even though he steadfastly refused to believe that the magnetic fields he observed in his experiments were due to the electric current flowing as a result of the electrical potentials generated by the temperature differences in his circuits of dissimilar materials. It is tempting to speculate on the course of events had he built a power converter with the best materials he had studied. It would have had a conversion efficiency of about 3%, which was the best that a steam engine could do in those days.

The second major thermoelectric effect was discovered by Peltier in 1834. He reported on the production or absorption of heat at the interface of two dissimilar conductors when an electric current was passed through them. Interestingly, Peltier too misinterpreted the significance of his discovery. He thought he had shown that Joule heating (I^2R) only occurred with strong electrical currents. However, in 1838 Lentz gave the correct interpretation of the Peltier effect. He provided a phenomenological demonstration by placing a drop of water at the junction of bismuth and antimony rods and caused it to freeze or melt depending upon which direction an electrical current was made to flow through the rods.

Finally in 1857 W. Thomson (Lord Kelvin) placed the whole field on firmer footing by using the newly developing field of thermodynamics to clarify the relationship between the Seebeck and Peltier effects. He also discovered the Thomson effect, a much weaker thermoelectric phenomenon which causes the generation or absorption of heat (other than Joule heat) along a current carrying conductor in a temperature gradient.

In 1919 Altenkirch and others attempted to build electrical power generators using these effects. However, they used metallic components causing the efficiencies to fall well below 1%, and they were not economically justifiable.

With the rapid growth of semiconductor theory around the middle of this century, interest in these devices revived and practical power generators with 6 to 11% efficiency were developed. These generators were especially useful for remote terrestrial and space applications. Indeed radioisotope thermoelectric generators proved to be an enabling technology for the deep space exploration programs to the outer planets and beyond.

THERMOELECTRIC EFFECTS

The primary thermoelectric phenomena considered in practical devices are the reversible Seebeck, Peltier, and, to a lesser extent, Thomson effects, and the irreversible Fourier conduction and Joule heating. The Seebeck effect causes a voltage to appear between the ends of a conductor in a temperature gradient. The Seebeck coefficient is given by:

$$S = \frac{\delta V}{\delta T}$$

For metals, S generally varies between about 0 and 40 μV/K, while semimetals and heavily doped semiconductors usually have values up to several hundred μV/K, and low conductivity semiconductors and insulators have values of S up to 1000 μV/K and higher.

Many articles refer to the Seebeck voltage being generated only at the junction of dissimilar conductors in an electrical circuit, but consideration of the Thomson effect leads to the conclusion that the Seebeck voltage is generated along the entire thermoelectric element in a temperature gradient.

The Thomson effect is the generation or absorption of heat (other than I^2R heat) in a current-carrying conductor subjected to a thermal gradient.

$$q_\tau = \tau I$$

The Thomson coefficient τ is given by $\tau = -T\dfrac{\delta S}{\delta T}$

and, since the third law of thermodynamics requires $S = 0$ at absolute zero

$$S = -\int_0^T \frac{\tau}{T}\,dT$$

This gives us a way to determine the absolute value of the Seebeck coefficient for a material without the added complication of a second conductor.

The Thomson heat is generally in an order of magnitude less than the Peltier heat and is often neglected in device design calculations.

The other major thermoelectric phenomenon is the Peltier effect, which is the generation or absorption of heat at the junction of two different conductors when a current flows in the circuit. The amount of heat involved is determined by the magnitude of the current and the Peltier coefficients, π, of the materials.

$$q_p = \pi_{A,B} I$$

Whether the heat is evolved or absorbed is determined by the direction of the current flow.

Kelvin showed the interdependence of these phenomena by thermodynamic analysis by assuming that the irreversible processes were independent of the reversible ones. This approach was later proved theoretically sound by the use of Onsager's concepts of irreversible thermodynamics (7).

$$\pi_{A,B} = S_{A,B} T$$

$$\tau_A - \tau_B = S_{A,B} - \frac{2\pi_{AB}}{2T}$$

The irreversible phenomena represent entropy gain through irrecoverable heat losses.

$$q = I^2R \quad \text{electrical (Joule heating)}$$

$$q = \lambda \frac{A}{l}\Delta T \quad \text{heat flow (Fourier conduction).}$$

Taking all of the above into account, one can show that the efficiency η of a thermoelectric power generator is given by

$$\eta = \frac{P_{out}}{Q_{in}} = \frac{I^2R_0}{K\Delta T + ST_H I - \frac{1}{2}I^2R} \quad \text{(neglecting Thomson heat).}$$

Optimizing for both leg geometry and load resistance and using average values for S, ρ, and λ

$$\eta = \frac{\Delta T}{T_H}\left(\frac{\sqrt{1+ZT_{av}} - 1}{\sqrt{1+ZT_{av}} + T_c/T_H}\right)$$

where $$Z = \frac{(|S_A| + |S_B|)^2}{(\sqrt{\lambda_A\rho_A} + \sqrt{\lambda_B\rho_B})^2}\ \text{deg}^{-1}$$

The first term in the efficiency equation is simply the Carnot efficiency for a reversible heat engine and the second term is the reduction in efficiency due to the irreversible effects. The important factors can easily be seen to be Z (the figure of merit) and ΔT. Increasing Z or T_H or decreasing T_c all increase the efficiency in general. However, since S, ρ, and λ are all temperature dependent, changing T will also change the value of these properties making the prediction of the overall effect less easy to estimate without detailed calculations using actual material properties.

For insulators, Z is very small due to ρ being very high (little electrical conduction); for metals, Z is very small due to S being very low. Z peaks for semi-conductors at $\sim 10^{19}\ \text{cm}^{-3}$ charge carrier concentration (about 3 orders of magnitude less than for free electrons in metals). Thus for electrical power production or heat pump operation the optimum materials are heavily doped semiconductors.

Operating in the opposite mode, ie, causing an electric current to flow through the device from an outside source rather than impressing a thermal gradient on it, will result in cooling or heating systems. In thermoelectric cooling applications, the effectiveness of a device is measured in terms of the coefficient of performance, β, rather than efficiency. This is the ratio of the heat removed from the cold reservoir to the electrical power input.

$$\beta = \frac{q_c}{P_e} = \frac{ST_cI - \frac{1}{2}I^2R - \lambda\Delta T}{SI\Delta T + I^2R}$$

When optimized for geometry and current this becomes

$$\beta = \frac{Tc}{\Delta T} = \left(\frac{\sqrt{1+ZT_{av}} - T_H/T_c}{\sqrt{1+ZT_{av}} + 1}\right)$$

Under these conditions the maximum heat pumping rate, q_{cm}, the maximum temperature difference across the module, ΔT_m, and the minimum temperature attainable on the cold side, $T_{c\min}$, are given by the following relations:

$$q_{cm} = \frac{S^2T_c^2}{2R} - \lambda\Delta T$$

$$\Delta T_m = \frac{S^2T^2}{(\sqrt{\rho_n\lambda_n} + \sqrt{\rho_p\lambda_p})^2} = \tfrac{1}{2}T_c^2 Z_m$$

$$T_{c\min} = \frac{\sqrt{1+2ZT_H}}{Z}$$

Currently available devices are generally capable of cooling rates of 2–3 W/cm² with electrical current densities of 1 amp/mm^2. Typical temperature drops are about 30°C (86°F) across a single stage. But much larger ΔTs can be obtained by cascading.

CURRENT TECHNOLOGY

Materials

Since the Seebeck coefficient, the electrical resistivity and the thermal conductivity of thermoelectric materials all vary with temperature, then no single material can function well over the entire temperature range in which thermoelectric devices are used. Table 1 indicates the generally accepted operating range categories and the primary thermoelectric materials used in each range. Many other materials have been or are currently being developed to either obtain higher conversion efficiencies or less expensive manufacturing (2).

The physical form of the thermocouples varies significantly depending upon the applications. Most spacecraft power supplies utilize separate thermocouples which can be checked for performance at successive stages of manufacturing and be replaced if necessary. This approach fits in very well with the extremely high reliability requirements imposed on such systems. In terrestrial systems where such individualized attention is not economically feasible, modular assemblies are generally used which can contain tens to hundreds of couples in a single unit.

Since most thermoelectric materials are very strong in compression but rather weak in tension and shear, most low and moderate temperature devices are either spring loaded or are made integral with the heat exchangers to avoid stress problems and to assure good thermal contact for conductive coupling with the heat source and sink. The high operating temperature of silicon germanium devices and the physical ruggedness of these materials have allowed cantilever mounting of the thermocouples from the cold surface of the generator housing and the use of radiant heating to supply heat to the hot shoes.

Thermoelectric Power Generation

Thermoelectric devices still represent niche markets, but as economic and environmental conditions continue to change, they appear poised to advance into more common use. TE power generators are in use in many areas, including satellites, deep space probes, remote area weather stations, undersea navigational devices, military and remote area communications, and cathodic protection.

Table 1. Accepted Operating Range Categories and the Primary Thermoelectric Materials Used in Each Range

Temperature °C			
Hot	Cold	Predominant Materials	Operating Mode
200	−130	Bismuth telluride; bismuth antimony telluride	Cooling or power
600	100	Lead telluride; silver antimony germanium telluride	Power
1000	300	Silicon germanium	Power

Undoubtedly, the most exciting application of thermoelectric power has been in the radioisotope thermoelectric generators (RTG) of the U.S. space programs. These began very modestly with two SNAP-3B three watt demonstration units flown as proof of principle tests on the mainly solar powered Transit 4A and 4B navigational satellites in 1961. These were followed by the 27 watt SNAP-9A design on the Transit 5BN-1 and 5BN-2 satellites in 1963. These were the first spacecraft to be fully powered by radioisotope power sources. These plus all of the other U.S. RTG powered missions are summarized in Table 2. The power levels have steadily increased over the years up to the 850 watts scheduled to fly on the Cassini mission which will send a probe into the atmosphere of Saturn's moon Titan and then spend four years orbiting Saturn and making multiple encounters with its other moons. The launch for this spacecraft is scheduled for October 1997.

These thermoelectric generators were recognized as an enabling technology making possible the extended missions in which spacecraft have explored the giant outer planets and even the regions beyond the far reaches of the Solar system in the case of the two Pioneer and two Voyager missions. The Pioneer spacecraft have been operating over twenty-two years in the frigid reaches of outer space and have survived extremely high radiation fields on very close flybys of Jupiter and are still sending back data from over six billion miles away from the sun. None of the missions have encountered any failures in their thermoelectric power systems, although several have experienced problems in their on board electronics systems and radio systems (8).

More mundane, but in the long run, much more important economically and environmentally are the terrestrial systems for power generation and cooling. A number of these systems are briefly described below. The primary focus of the power generating devices is to recapture some of the waste heat now being dumped by commercial processes and vehicle engines, or to make use of the "free" heat from geothermal sources. In cooling devices the emphasis is on eliminating the use of refrigerant gases which could be environmentally harmful, and in providing very small systems for specialized uses.

The California Energy Commission, Energy Technology Advancement Program and the U.S. Department of Energy have sponsored a project whose purpose is to reduce diesel engine fuel use and NO_x padiculates (9). The current engine driven electrical generator uses up 3 to 5 HP from the engine drive shaft, while the TE generator uses only exhaust heat and coolant from the radiator with very little degradation of the engine power output. The unit is about 1 meter long and about 25 cm in diameter. It uses Bismuth Telluride Thermoelectric modules. Initially more expensive than the engine driver alternator, it could recover the cost differential in 2 years in U.S. or 6–8 months in Europe at current fuel costs. It is believed that on-board electrical usage in trucks for powering computers NO_x reduction systems, particulate trapping, etc, will

Table 2. Summary of Radioisotope Thermoelectric Generators Successfully Launched by the United States (1961–1990)

Power Source	Initial Average Power, W	Spacecraft	Mission Type	Launch Date	Orbital Lifetime, Years
SNAP-3B7	2.7	Transit 4A	Navigational	6/29/61	500
SNAP-3B8	2.7	Transit 4B	Navigational	11/15/61	1200
SNAP-9A	>25.2	Transit 5BN-1	Navigational	9/28/63	1900
SNAP-9A	26.8	Transit 5BN-2	Navigational	12/5/63	1800
SNAP-19B3	56.4	Nimbus III	Meteorological	4/14/69	3600
SNAP-27	73.6	Apollo 12	Lunar	11/14/69	On lunar surface
SNAP-27	72.5	Apollo 14	Lunar	1/31/71	On lunar surface
SNAP-27	74.7	Apollo 15	Lunar	7/26/71	On lunar surface
SNAP-19	162.8	Pioneer 10	Planetary	3/2/72	Beyond solar system
SNAP-27	70.9	Apollo 16	Lunar	4/16/72	On lunar surface
Transit-RTG	35.6	Triad	Navigational	9/2/72	137
SNAP-27	75.4	Apollo 17	Lunar	12/7/72	On lunar surface
SNAP-19	159.6	Pioneer 11	Planetary	4/5/73	Out of solar system
SNAP-19	84.6	Viking 1	Mars Lander	8/20/75	On Martian surface
SNAP-19	84.2	Viking 2	Mars Lander	9/9/75	On Martian surface
MHW-RTG	307.4	Les-8	Communications	3/14/76	1,000,000
MHW-RTG	308.4	Les-9	Communications	3/14/76	1,000,000
MHW-RTG	477.6	Voyager 2	Planetary	8/20/77	Out of solar system
MHW-RTG	470.1	Voyager 1	Planetary	9/5/77	Out of solar system
GPHS-RTG	576.8	Galileo	Planetary	10/18/89	Jupiter encounter
GPHS-RTG	~288.	Ulysses	Planetary	10/6/90	solar-polar orbit

only increase in the future making the gains from the TE system even better.

The U.S. Coast Guard has been experiencing very high fuel and maintenance costs on their Diesel motor-generator powered Major Aids to navigation's systems, and are therefore sponsoring work on such systems as photovoltaics, wind, or wave power generators to float charge large battery packs. Since each of these new power supplies is subject to significant down times due to adverse weather conditions, a very reliable low maintenance back-up power supply is required. For this a 1.5 W TE generator burning diesel fuel is being investigated. It uses a segmented lead telluride/bismuth telluride module.

As an offshoot of the Saudi–German HYSOLAR program that uses photovoltaics to produce hydrogen by electrolysis of water, giving in effect "storable solar energy," the hydrogen can later be burned at 200–700°C using catalytic combustion (10). This procedure will provide the heat to a modified Teledyne Energy Systems TE converter that was originally designed for natural gas, propane or butane fuels. The H_2 is both storable and transportable and the combustion product is water.

Osaka University and the University of Wales are collaborating on a research project aimed at the recovery of electrical power from commercial processes waste heat (11). This research is a low temperature (80°C, 176°F) low ΔT (~60°C, ~140°F) process which makes use of very thin thermoelements and very large area heat exchange devices. The initial costs could be quite high, but the low temperature operation should result in extremely long life times to depreciate the costs effectively. Since this system is using only waste heat from another process, electrical power is gained with no additional pollution of any sort. The efficiency of these systems is rather low compared to other devices operating over wider temperature ranges, but since it is piggybacking on other processes that take care of the economic investment, it is a promising approach to alleviating the energy crunch in the future.

Cold climate electrical heaters for trucks in far northern areas of Canada, Northern Europe, and Russia are very important. They are used when the trucks are idle and provide heat to the engine coolant and the cab without idling the engine for extended periods (12). This device both minimizes engine wear and reduces pollution as well as provides significant economic savings. Many jurisdictions now prohibit extended idling of engines. In extremely cold environments, engines can rather quickly become difficult to very nearly impossible to start. If ordinary gasoline or diesel oil fired heaters are used, the coolant circulation pump, air fan, etc, must be powered from the vehicle's batteries, thus curtailing the time the system can be used, especially at very low temperatures when it is needed the most. By adding PbTe thermoelectrics to such heater systems, about 2% of their thermal output can be turned into electricity to run the heater's electronics, fuel pump, combustion fan and coolant circulation pump, with sufficient power left over to keep the vehicle's battery fully charged. The market for such units is in the hundreds of thousands if manufacturing costs can be reduced.

Thermoelectric Cooling

In the 1960s there were a number of attempts to make thermoelectric cooling a major industry. Although all of the systems worked very well, they could not compete economically with the low energy costs then in effect, but the groundwork was laid and in today's situation, these areas are being revisited.

A corporate headquarters building in Wisconsin was fitted with a thermoelectric air conditioning system by Carrier Corporation. It consisted of 30 decentralized units and was still operating well over 10 years later (13). However, Carrier withdrew from the business.

Small refrigerators were developed by several companies and some were even installed in hotel rooms in Chicago. Borg-Warner and other companies produced many compact systems for laboratory uses (14). Air-Industry in France built an air conditioning system for a passenger railway coach that was still in daily use after 10 years of operations without a single thermoelectric failure (15).

With the recently concluded international agreements to ban chlorofluorocarbons by the year 2000, the efficiency of refrigeration systems using substitute working fluids has moved much closer to that of the thermoelectric cooling systems. Considering the economics of scale, it now seems possible to market thermoelectric refrigerators at only 10 to 20% above the cost of similar sized compressor units. The quieter operation and much longer life of thermoelectric units could make them competitive in the near future (16).

The U.S. Army is sponsoring work on a modular device to be used in protecting personnel in severe conditions. This is a thermoelectric cooling device designed for mounting in a 2 man army vehicle and to provide a controlled working atmosphere for two personnel wearing Nuclear, Biological, and Chemical (NBC) protective clothing (17). It consists of a thermoelectric cooling unit, a multiple filter pack, and a blower assembly. The unit can supply 24 CFM of cooled, filtered air with 250 watts of cooling to each man. The unit operates off of the vehicle's 28 volt electrical system and can easily maintain an effective working climate inside the suits over an external ambient range from 35°C (95°F) 85% RH to 49°C (120°F) 3% RH.

BIBLIOGRAPHY

1. A. F. Ioffe, *Semiconductor Thermoelements and Thermoelectric Cooling,* Infosearch Limited, London, UK, 1957.
2. D. M. Rowe, ed., *CRC Handbook of Thermoelectrics,* CRC Press, Inc., London, UK, 1993.
3. R. R. Heikes and R. W. Ure, Jr., *Thermoelectricity: Science and Engineering,* Interscience Publishers, N.Y., 1961.
4. S. W. Angrist, *Direct Energy Conversion,* Allyn and Bacon, Inc., Boston, Mass., 1965.
5. F. J. Blatt, P. A. Schroeder, C. L. Foiles, and D. Greig, *Thermoelectric Power of Metals and Alloys,* Plenum Press, N.Y., 1976.
6. T. C. Harmon and J. M. Honig, *Thermoelectric and Thermomagnetic Effects and Applications,* McGraw-Hill Book Co., Inc., N.Y., 1967.
7. C. A. Domenicalli, *Rev. Mod. Phys.* **26,** 237 (1954).
8. E. A. Skrabek, "Performance of Radioisotope Thermoelectric Generators in Space," *Proceedings of the 7th Symposium on Space Nuclear Power Systems,* Albuquerque, N. Mex., Jan. 1990.
9. J. C. Bass and N. B. Elsner, "Current Thermoelectric Programs at Hi-Z Technology, Ins.," *Proceedings of Eleventh International Conference on Thermoelectrics,* Arlington, Tex., Oct. 7–9, 1992, pp. 1–3.
10. M. Al-Garni and M. Bajunaaid, "Design and Test of a Thermoelectric Power Generator Fueled by Hydrogen," *Proceedings of Eleventh International Conference on Thermoelectrics,* Arlington, Tex., Oct. 7–9, 1992, pp. 4–9.
11. K. Mastsuura, D. M. Rowe, K. Koumoto, G. Min, and A. Tsuyoshi, "Design Optimization for a Large Scale, Low Temperature Thermoelectric Generator," *Proceedings of Eleventh International Conference on thermoelectrics,* Arlington, Tex., Oct. 7–9, 1992, pp. 10–16.
12. A. G. McNaughton and G. J. McBride, "Electrically Independent Heaters Using Thermoelectric," *Proceedings of Eleventh International Conference on Thermoelectrics,* Arlington, Tex., Oct. 7–9, 1992, pp. 17–20.
13. J. G. Stockholm, "Modern Thermoelectric Cooling Technology," *Proceedings of IX International Conference on Thermoelectrics,* Arlington, Tex., Mar. 1990, pp. 90–108.
14. A. B. Newton, "Designing Thermoelectric Air Conditioning Systems for Specific Performance," ASHRAE Transactions, July, 1965, pp. 133–147.
15. L. Stockholm, Pujod–Soulet, and P. Sternat, "Prototype Thermoelectric Air Conditioning of a Passenger Railway Coach." *Proceedings of IV International Conference on Thermoelectrics,* Arlington, Tex., Mar., 1982.
16. P. M. Schliklin, "Thermoelectricity, A Possible Substitute for the CFCs," *Proceedings of IX International Conference on Thermoelectricity,* Pasadena, Calif., Aug., 1990, pp. 381–395.
17. P. Hanan and B. Mathiprakasam, "Development of Two-Man Thermoelectric Micro-climate Conditioner for use in Army Ground Vehicles," *Proceedings of Eleventh International Conference on Thermoelectrics,* Arlington, Tex., October 7–9, 1992, pp. 181–184.

TIDAL POWER

Robert H. Clark
Consulting Engineer
Sidney, British Columbia
Canada

Tidal power is often thought of as a relatively new source of energy, yet it has been available as long as the oceans have existed. Tides originate with the gravitational forces of the Moon and Sun and offer a wide range of advantages over other forms of energy suited to conversion into electricity. They are a freely available source that is of indefinite duration, nonpolluting, and accurately predictable. It has been roughly estimated that the total annual tidal energy theoretically available could be in the range of 2,000 to 3,000 TWh. For more than a millennium, a minute portion of this natural energy has been harnessed and put to good use. Several patents have been filed to protect devices and concepts for extracting this energy (1). However, there is no longer any technological barrier to the commercial conversion of tidal energy into electrical energy. There are three commercial tidal–electric plants in successful operation, one of which has been operating since the late 1960s.

See also Renewable resources; Oceanography; Hydropower.

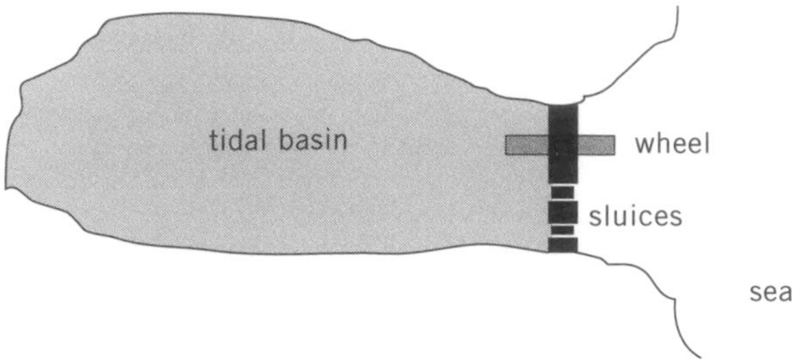

Figure 1. A tidal mill layout.

HISTORICAL PERSPECTIVE

Inhabitants of the coastlines, particularly those bordering the North Atlantic Ocean, installed simple tidal mills on the shores of Gaul, Andalusia, and England, at least as early as the Middle Ages, to carry out some of the more tedious tasks of society at the time, such as the grinding of corn.

The tidal mill was a version of the conventional waterwheel mill but required extra skill and care to operate. The plan of a typical tidal mill layout is shown in Figure 1.

The dam, closing off a small bay or mill pond from the main estuary or the sea, was equipped with sluice or flap gates. These mills used the ebb tide for the generating cycle. The gates were designed to open automatically with the tidal current during flood tide, as illustrated in Figure 2**a**, and to close with the commencement of the ebb tide to trap the water in the mill pond, as in Figure 2**b**; the water was then allowed to escape from the pond back to the sea by driving a wooden paddle wheel (Fig. 2**c**). A sluice gate was provided to control the amount of water fed back to the sea through the waterwheel when the tide was ebbing. An ungated relief weir or spillway in the dam ensured that any flood waters flowing into the tidal basin or mill pond from upstream could be safely discharged from the pond under high tide conditions to reduce flooding possibilities and disruption of milling operations.

Many mills were automated to the extent that they could be operated by one person having a series of ropes and pulleys by which any part of the mill's machinery could be set in motion. At locations with semidiurnal tides, operation was possible twice daily for periods of about five hours each, and since the tidal cycle and solar cycle are out of phase in most locations around the world, the miller had to mill when the tidal conditions were right.

Tidal mills were being operated in the eleventh century, and many were constructed during the ensuing centuries along the Atlantic Coast of Europe, mostly in Great Britain, France, and Spain. One such installation, in the Deben Estuary of England, was mentioned as early as 1170 and although the present building dates from the eighteenth century, the mill was in operation until 1957. A tide mill at Eling near Southampton, U.K., was restored in 1980 and is the only known tidal mill in western Europe that regularly produces wholemeal flour. Figure 3 provides a cut-away view of the mill to illustrate the transition from hydraulic energy to mechanical energy which is then transferred through gears, levers, and shafts to the milling operation. The site is thought to have been in more or less continuous use for well over 900 years, although the fabric of the mill has been renewed many times (2). Fourteen tidal mills are reported to have been in operation in the bays or coves of La Rance estuary, and

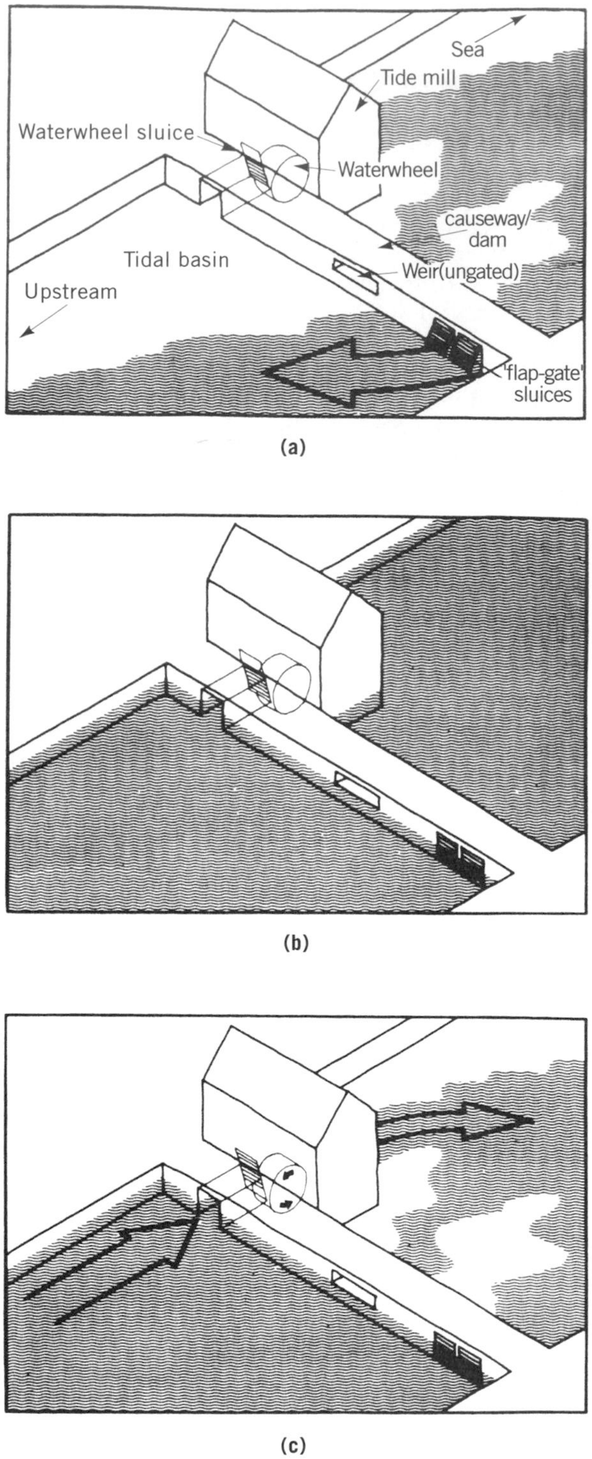

Figure 2. Operation of a tidal mill development. **(a)** Tide rising; **(b)** high tide; **(c)** tide falling.

Figure 3. Eling tide mill near Southampton. Courtesy of the Tide Mill trust.

one or two were still functioning in the 1950s during the investigations leading up to the construction of France's 240 MW tidal electric generating station in the Rance estuary.

Tidal energy was exploited as early as 1607 in North America when a grist mill was constructed at Port Royal, Nova Scotia. A tidal mill, which developed about 35 kW, was built in 1734 at Chelsea, Massachusetts, to grind spices and at least two tidal grist mills were in operation in Passamaquoddy Bay prior to 1800. Tidal power installations were also developed for pumping; huge wooden, tide-generated waterwheels, installed under London Bridge in 1580 as part of the London water supply system, were still in operation in 1824 and an installation for pumping sewage was still in use in Hamburg in 1880. A concise review of the early history of the exploitation of tidal energy is available (3).

Early tidal mills could not be constructed to extract more than a minute portion of the energy potential of a site, perhaps the equivalent of 20–50 kW in mechanical energy, for use locally. The Industrial Revolution (late eighteenth century) increased large-scale power and energy demands and the technology of river hydro and steam-electric plants responded. Today the cost of energy is only about 5%, in terms of labor hours, what it was at the beginning of the eighteenth century. Although tides as a source of energy lost appeal and tidal power technology was neglected in the nineteenth century, the latter part of the twentieth century has seen a renewed interest in exploiting the energy of tides. Three modern, commercial, tidal electric plants of 3, 20, and 240 MW capacity were in operation by 1984 and feeding into electrical utility systems. In-depth investigations have been carried out at a number of potential sites around the world with the conclusion that development of several of the sites are economically viable. Tidal energy is renewable, pollution-free, relatively inflation-free, and competely predictable and its exploitation would reduce environmental problems associated with fossil fuels conversion.

TIDES AND ENERGY GENERATION

Tidal Characteristics

Ocean tides are the result of gravitational forces and centrifugal action caused by the Earth's rotation, and by changes in the relative positions of the Earth, Moon, and Sun (4). Despite its comparatively small size, the Moon exerts a much greater influence on the tides than does the Sun because it is so much nearer Earth. Theoretically, the solar force is about 46% of the lunar. The average period of the lunar tide is about 12 hours and 25 minutes, ie, the time between successive high waters, and there are two lunar tides in each lunar day of 24 hours and 50 minutes. High water, however, does not directly follow the variation in the tide-producing differential force but lags behind it due to the viscosity of the water, the friction of the boundaries of the water mass, and the characteristics of the coastline. The excess of 50 minutes over the solar day results in maximum water level occurring at progressively different times on successive days; thus tidal power cannot be used regularly to meet peak load demand which occurs at about the same period every day. In light of this characteristic, tidal power development is generally assessed for its energy output only.

Since the orbit of the Moon is an ellipse, its distance from Earth varies during the course of a lunar month, and its effect on the ocean tides is maximum when it is closest to Earth. Similarly, due to the elliptical orbit of Earth around the Sun, the tide-producing force of the Sun varies. Maximum tidal ranges, ie, spring tides (the higher tides) are produced every two weeks when the Sun and Moon are in alignment (Fig. 4). The highest spring tides occur during the equinoxes in March and September

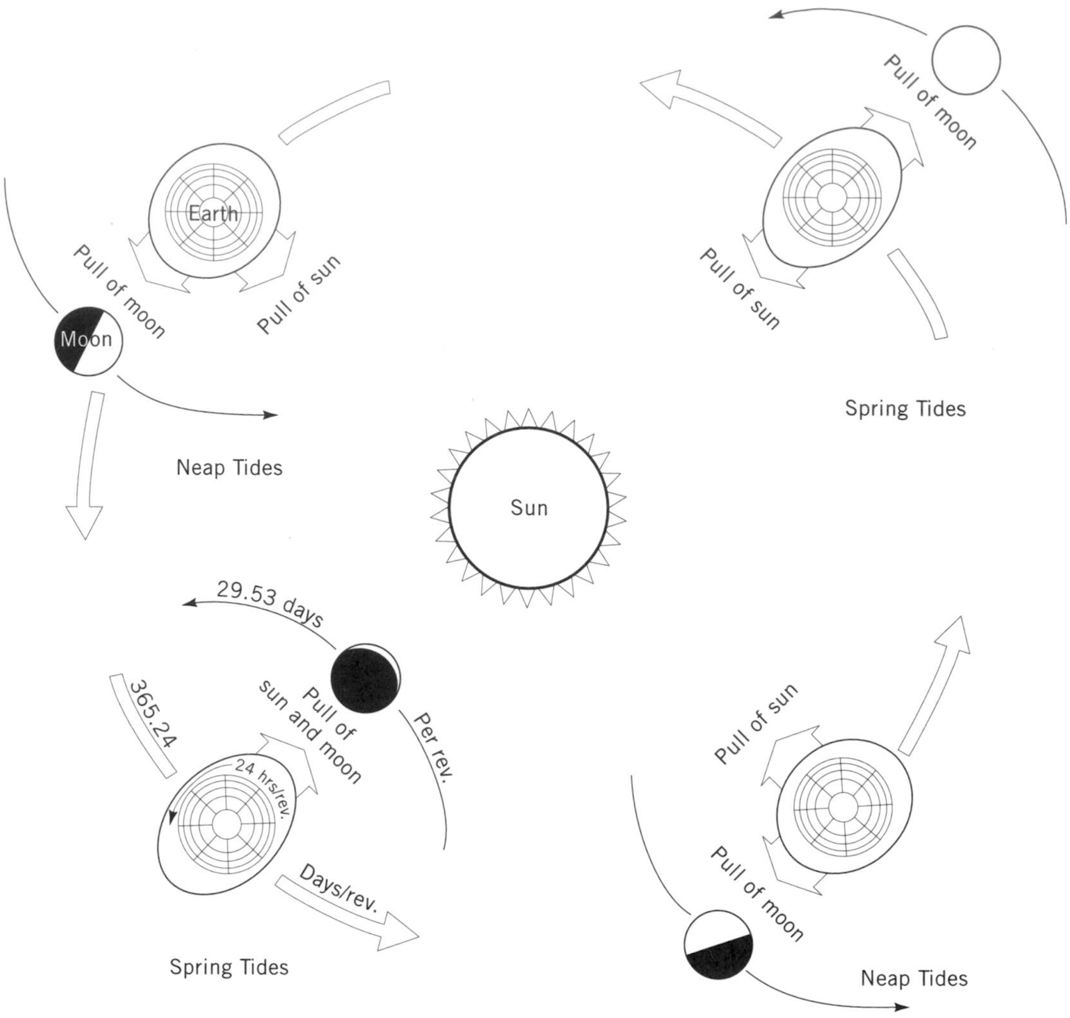

Figure 4. Origin of the tides.

when the Moon, in alignment with Earth and the Sun, is also nearest Earth. Minimum tidal ranges, ie, neap tides (the lower tides), occur when the Moon is in quadrature with the Sun so that they partially neutralize each other's gravitational influence. The lowest neap tides occur during the solstices in June and December when the Moon, in quadrature with the Sun, is farthest from Earth. In some locations, the spring tidal range may be as much as 12 times the neap range; such differences have an important bearing on the design of a tidal power plant. On the other hand, there are many locations where the range of the tides is significantly lower, only a few centimeters at Nantucket Island, for example. The island of Tahiti is in a unique location. It seems to be at the node of a lunar-influenced oscillating basin in the Pacific Ocean, so that the Sun has full control of the small tides, producing high tides regularly at noon and midnight and low tides at 0600 and 1800 hours.

The varying gravity and centrifugal forces react on one another to produce three principal types of ocean tides: (*1*) the semidiurnal type which is represented by two high and two low waters each day with the morning and afternoon tides being very nearly equal in magnitude; (*2*) the diurnal type which has only one high and one low water during a day; and (*3*) the mixed type which results from a combination of the semidiurnal and diurnal type and contains two high and two low waters each day that differ in form and height.

Amplification of Tidal Ranges. Throughout the ocean, the tide is almost always a symmetrical undulation having the same form as the long swell of the ocean the summit causing high water and the trough low water. The ranges of these ocean tides are generally less than 1 m. In the Atlantic Ocean, for example, the oceanic tide runs up from between South America and Africa to its north end between Canada and Europe. When the ocean tidal undulation enters straits, inlets, bays, and estuaries, it is subject more to the laws of hydraulics and wave motion than to its astronomical influences and undergoes the same kind of modifications that affect waves in shallowing water near the shore. Tidal effects may be amplified appreciably. The observed sea level is raised and depending on the configuration of these inlets, the amplitude of the wave or undulation may be increased many times. Such a wave tends to move along as estuary with a speed given by $\sqrt{gd}$, where d = mean depth. If the mean depth is 30 m, then the wave length L = wave period × speed = approx. 767 km. The period of a semidiurnal ocean tide is long enough to make the wave length much longer than the physical length of an estuary. However, an estuary with a wave length that is about one-quarter or one-half the wave length of the tidal undulation imparts a resonant response so that the amplitude of the tidal undulation could increase along the estuary by factors of four or more. The Bay of Fundy provides a good example of this effect. The narrowing and shelving of the Bay brings the whole system from the Continental Shelf through the Gulf of Maine to the upper reaches of the Bay, to near resonance with the semidiurnal tides. Thus, along the edge of the Continental Shelf, the oceanic tide has a mean range of about 1 m. At the mouth of the Bay of Fundy, the range has increased to about 3.5 m; in Cobequid Bay, at the top end of Fundy, the range is about 13 m (Fig. 5).

Effects of Barrage on Tidal Regime. A barrier or barrage across any part of an estuary affects the resonant response of that estuary. However, it must be emphasized that a tidal barrage is not the impermeable barrier of a river dam; it must be capable of not only storing water but also discharging the tidal flow, as appropriate, under the influence of the natural tidal currents. The tidal range at the barrage could be increased because the barrage would bring the foreshortened estuary closer to resonance or the range could be reduced materially if the natural resonant conditions were upset. The extraction of energy also affects the tidal range since the time phase of flows past the barrage site are changed. Since the energy output of a tidal power station is very sensitive to the available tidal range at the barrage site, predictions of the effect on this range caused by the construction and operation of a tidal electric plant are essential for purposes of design, optimization, and economic analysis. Moreover, both the amplitude and currents in the vicinity of the site affect the construction techniques to be adopted, as well as the design of civil works and thus the cost of the project.

Any modification of the tidal range and water movement pattern in the estuary also affects other important socio-economic and environmental factors, such as navigation, land drainage, sediment patterns and movements, as well as biological aspects. Also, any divergence from the natural state always raises questions about damage to coastal installations or impairment of their usefulness.

Developments in hydrodynamic numerical modeling along with advances in computer science have provided the means to simulate the physical system of an estuary. As a result, enormous progress has been made in the understanding of problems caused by the existence of a barrage and the calculation of its effect on the tide itself, which is a fundamental problem in tidal power investigations. For example, it has now been confirmed (5) that the Bay of Fundy–Gulf of Maine region (Canada–United States) out to the Continental Shelf represents a single resonant system with a period close to that of the 12 hour and 25 minute period of the semidiurnal tide. This, of course, is the reason for the very high tides of the order of 13 m at the head of the Bay of Fundy. The operation of a tidal electric plant in the upper reaches of Fundy could, by decreasing the effective length of the system, alter the tidal ranges at the barrage and affect the regime of the Bay and Gulf. Numerical model results developed for Fundy investigations have shown only minor reductions in tidal ranges for plants operated at the economically feasible sites, although there would be slight increases in tidal ranges in the Gulf of Maine. The Severn Estuary

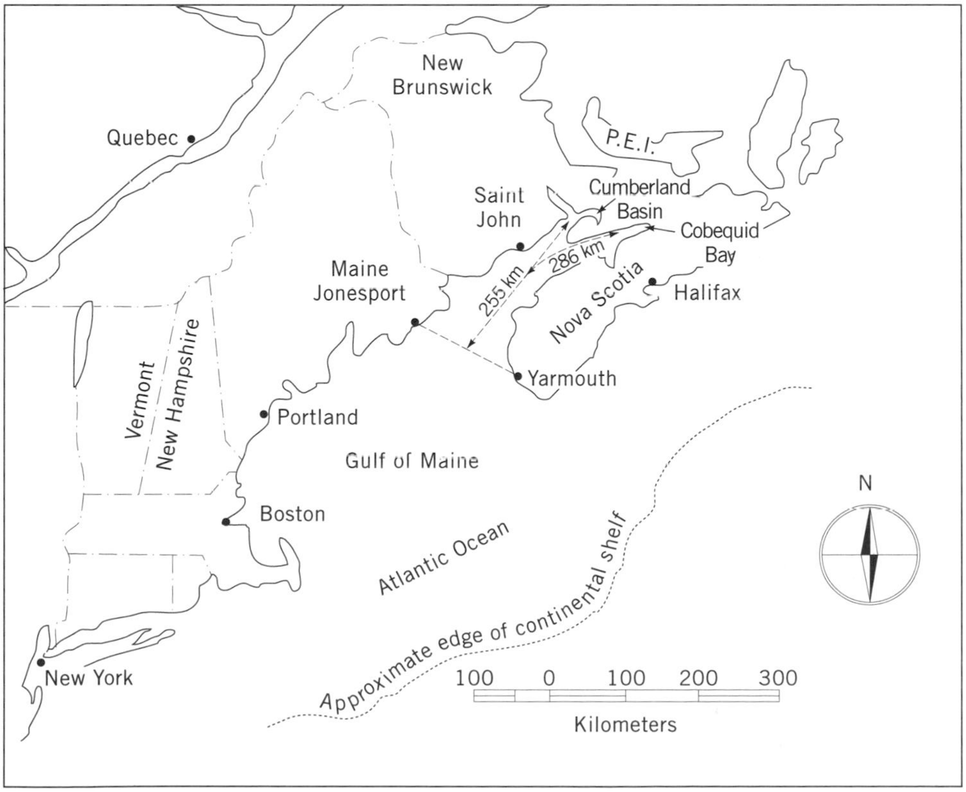

Figure 5. Tidal resonance of the Bay of Fundy.

(U.K.) studies also concluded that the proposed Severn barrage would decrease the tidal range slightly but the effect would fall off rapidly with distance seaward (6). Obviously, the calibration of such models requires tidal data not only in the estuary but seaward, out to the Continental Shelf.

Over such large areas, two-dimensional, finite-difference models provide sufficient accuracy. A three-dimensional hydrodynamic model is necessary to describe the tidal and current patterns in the estuary for both the natural regime and the regime modified by the installation of a tidal power station. In addition, it is likely that hydraulic models would be employed in pre-investment investigations of a potential development because they are better tools to resolve problems associated with situations involving considerable stratification, complicated boundaries and flow patterns, significant bottom effects, scouring and sedimentation problems, and random or breaking waves. It is also possible to simulate the Coriolis force (the effect of Earth's rotation on ocean levels) on physical models by means of a large diameter, slowly rotating table. Such models were constructed for investigations of the various Chausey projects in Brittany, France, in the early 1950s before significant advances in mathematical modeling and computers. A relatively new technique, hybrid modeling, links the advantages of both hydraulic and mathematical models. Hybrid modeling is particularly attractive when studying the effects of construction and plant operation in large estuaries where the open boundary is far from the site so that it is impractical to include it in the physical model (7). Thus, the hydraulic model could be constructed at the appropriate scale factor to minimize distortions and interfaced dynamically with a numerical model which could be extended as far afield as necessary to ensure effective dissipation of reflections at its outer boundary. Information would be exchanged at the interface so that the two models would run as one model of the whole area, with the hydraulic model handling all the nonlinear problems for which it was designed, and the numerical model handling the dissipation of reflections.

Predictability of Tides. Regardless of local variations, tides occur in an orderly fashion and are predictable. Although there may be other elements than the tide-producing forces, such as friction, barometric pressure and wind, the shapes of the continents, and depths of the oceans and water passages, it is possible to predict accurately the basic astronomical tides. In locations where tidal ranges are suitably large to justify power development, climatic influences may have some significant effects. Both the periodicity and predictability of the tides are characteristics with a significant bearing on the utilization of this phenomenon as an electrical energy source. Tides at any location tend to repeat themselves, almost identically, about every 19 years. The variations from year to year, however, are relatively small so that the available energy is practically the same year after year which is in contrast to river hydroelectric energy with its attendant, unpredictable wet and dry cycles. Moreover, a tidal power plant does not have to cope with wet or dry years nor with large seasonal variations in flows. Thus, except for the minor influences that meterological disturbances exert upon the water surfaces, with tidal energy it is possible to know how much energy will be available in the next week, the next month, or the next year.

Factors Influencing Approach to Development

Although both low-head river hydroelectric plants and tidal electric plants utilize the same turbine–generator machinery to convert potential energy into electrical energy, the unique aspects of the tides require a significant difference in approach to the development and utilization of the latter power resource. Water utilized by a tidal power plant travels in and out of a basin twice a day, at least for the common semidiurnal tides. The major elements of a tidal power development then include a barrage, a tidal basin, and a transmission link to the electrical utility system grid. The barrage is formed by the powerhouse, the sluiceway section, and dyke sections (inactive portions of the barrage) to the shore to complete the cut-off from the main estuary or ocean and provide a controlled tidal basin. Water is captured in the controlled basin on the flood tide and held there until a sufficient head is developed between the basin and the ebbing tide to turn the turbines and generate electricity.

A tidal electric plant is essentially a low-head development with the head varying continuously with the tidal fluctuations and the fluctuation in basin levels resulting from the operation of the plant. The output depends not only on this varying head but also on the area of the tidal basin, the capacity of the sluiceways for filling and emptying the basin, the capacity of the generating units, as well as on the mode of plant operation. As a result, the installed capacity of a tidal plant is determined on the basis of economic considerations rather than on the available head and flow regime, the latter factors being the primary considerations for determining the capacity of a hydroelectric plant. Similarly, the sluiceway is designed to allow for timely filling and emptying of the basin to enable optimization of energy production, rather than to pass a hypothetical "probable maximum flood" which is the sluiceway capacity criterion for major hydroelectric developments. Moreover, since the energy available in a tidal cycle varies from zero to a maximum, the latter available for a short period only, the selection of the generating unit size requires a rather more complex economic analysis than for a hydroelectric plant.

The energy produced by a tidal power plant is proportional to the volume of the controlled basin and the range of the tide or, alternatively, to the area of the basin and the square of the tidal range. A favorable site, therefore, should have a basin as large as possible and a high tidal range, taking into account the following considerations:

- the number and size of turbogenerators that can be installed in a given length of barrage is limited so that the entrance to the basin should be wide enough to accommodate the appropriate number of generators to optimize the development
- the sluice gates connecting the sea and basin must be capable of passing large flows under small differences in level and, therefore, have large dimensions

the cost of closure dykes increases rapidly in proportion to the depth of soil to rock or suitable foundation material on which they must be constructed.

A number of sites worldwide can meet the foregoing technical considerations, but at this point only a few could provide competitive power and energy.

Development Concepts

The simplest operating cycle for a tidal power plant is that which generates electricity on the ebb tide, referred to as ebb-generation. In the first phase of this cycle, the turbines are stopped and the sluice gates are opened to allow the water to flow into and fill the basin as the tide rises. When the basin is full, the sluice gates are closed, initiating the second or waiting phase which lasts until the sea level ebbs to below that in the basin to provide sufficient head to activate the turbines. This initiates the third or production phase which terminates when the head becomes too low for generation. The sluice gates are then opened for the first phase of a new cycle. The result of ebb or single-effect generation is two slugs of energy per day, of about five hours duration each. Energy can also be generated on the flood tide, known as flood generation, but the output is not as great as ebb generation since the latter mode has the advantage of being able to discharge greater volumes in the upper levels of the basin under higher heads.

The civil works for a single-basin scheme are the least complex and therefore are likely to be the least costly, but because this mode of operation produces energy only, it cannot produce continuous power and so cannot be considered a dependable peak power source. Periods of generation are dictated by the times of the tides (which vary from day to day); thus there would be many occurrences when energy was produced at times of low demand and only periodically at times of maximum demand. This moving cyclical output can be managed to some extent by double-effect generation, i.e., generating on both the ebb and flood cycles. More complex and expensive generating equipment, in addition to more costly civil works would, however, be required. Fortunately, with the sophisticated turbogenerator sets now available it is possible to generate on both the flood and ebb cycles. Sluicing capacity is enhanced due to the generator's capability to pump in both directions and act as orifices in the two directions. However, the ability to generate in both directions, ie, double-effect generation, would effect an increase in total output of only about 5% because some of the ebb-generation period would have to be curtailed to permit a more effective flood generation.

If the turbines were also designed to operate in a pumping mode when driven by the generator operating as a motor powered from the electrical system, the pumping operation would go into effect at or near high tide while the basin was still being filled through the sluices and for a short time after they had been closed. Relatively little energy is needed to lift the water through the small difference in level between the sea and the basin near high tide. However, during the subsequent generating cycle, ie, ebb tide, this additional water would fall through a greater head and produce more energy than was consumed in pumping, yielding a pumped storage efficiency greater than 100%. Similarly, on emptying the basin, the level could be lowered further by pumping from basin to sea, at or near ebb tide, creating additional filling capacity for a double-effect scheme. Such pumping would usually be carried out with the relatively low-cost energy from the power system with which the tidal plant is integrated.

However, the double-effect capability combined with the ability to pump in either direction enhances the operating flexibility of the plant and enables its output to be adjusted to some extent to meet load demands of the system with which the plant is connected. A schematic arrangement of a single-basin, double-effect plant and its possible outputs to provide energy or a peaking capability is provided in Figure 6. Although such operating flexibility adds a degree of complexity to the plant, recent conclusions support the economic justification for pumping, at least during the slack water period at the end of the flood cycle (6), since the net energy output may be increased in the order of 10%.

A double-effect, single-basin development is capable of many modes of operation between that of energy production and dependable peak production. In any power system the value of energy varies so that a single-basin, double-effect tidal plant can be operated to maximize the revenue from the production of its energy throughout the solar day. However, such flexibility can only be achieved with the sacrifice of the site potential and at an increased capital cost.

The problem of continuity of output has been one of the main disadvantages of tidal energy. One means of providing continuous output is by hydraulically linking two tidal basins as shown by Figure 7. This type of scheme was proposed for exploiting the tidal energy of Passamaquoddy and Cobscook Bays as an international project (refer Fig. 16). The water level in one basin is maintained high and that in the other, low, with generation always in the one direction. The high basin is filled during the flood tide through a set of sluices and the low basin emptied during ebb tide through its sluiceways. Although such a development would provide continuous output, the energy production would be lower than that from a single-basin scheme using either of the basins. Moreover, such a linked-basin scheme would require an interconnecting waterway in which to locate the power plant, as well as dykes and sluiceways to provide for the control of each basin. The results of the in-depth studies by the Bay of Fundy Tidal Power Review Board (5) and by the Severn Barrage Committee (8) have shown that the single-basin scheme with either single- or double-effect turbogenerators offers the lowest unit cost of energy and can extract closest to the maximum developable potential of a site.

Assessing Energy Ouput

The operation of a tidal electric project is significantly more complex than the operation of a hydroelectric project. Figure 6 illustrates the continuously changing water levels acting on the turbines as a result of the natural

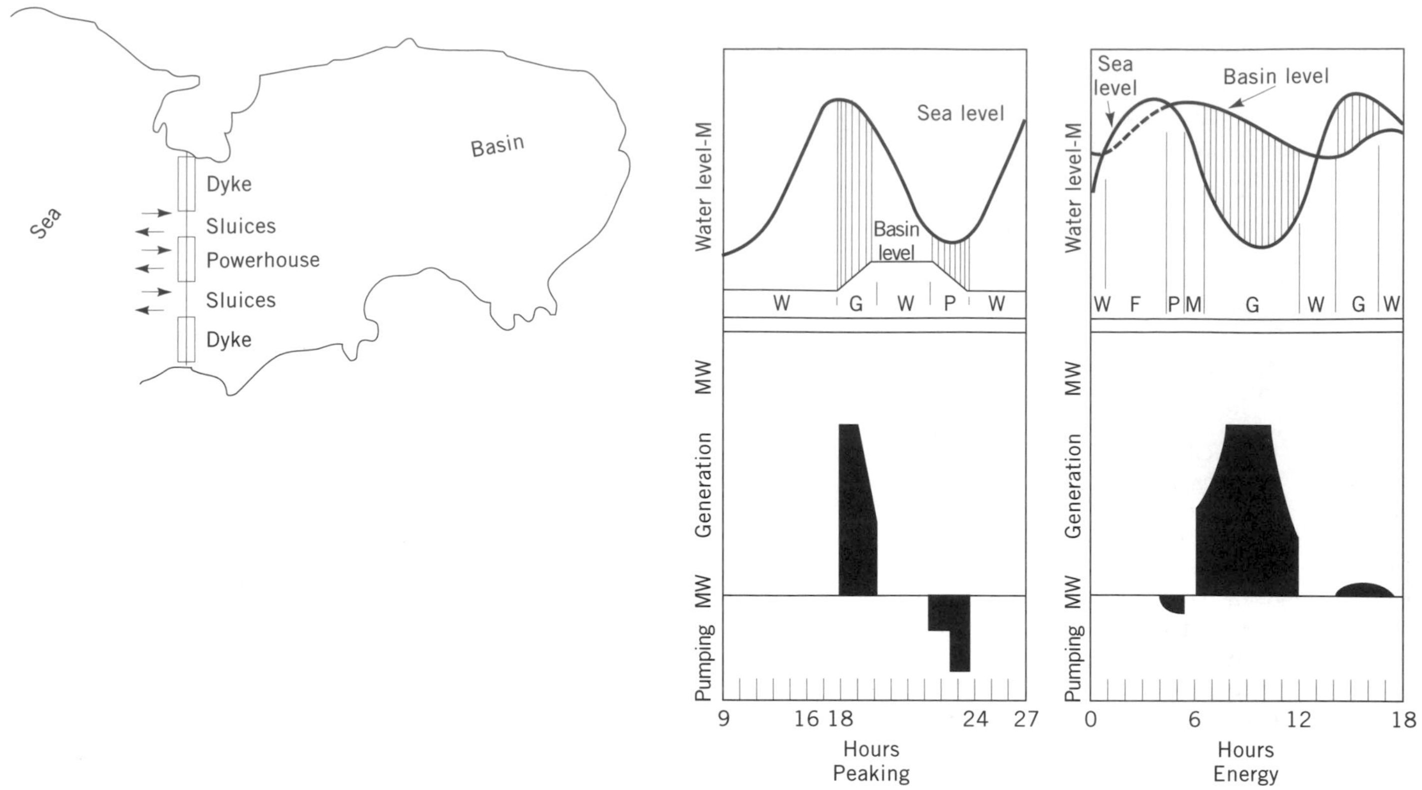

Figure 6. Form of energy output with double-effect generation. G, generating; W, waiting; P, pumping; F, filling.

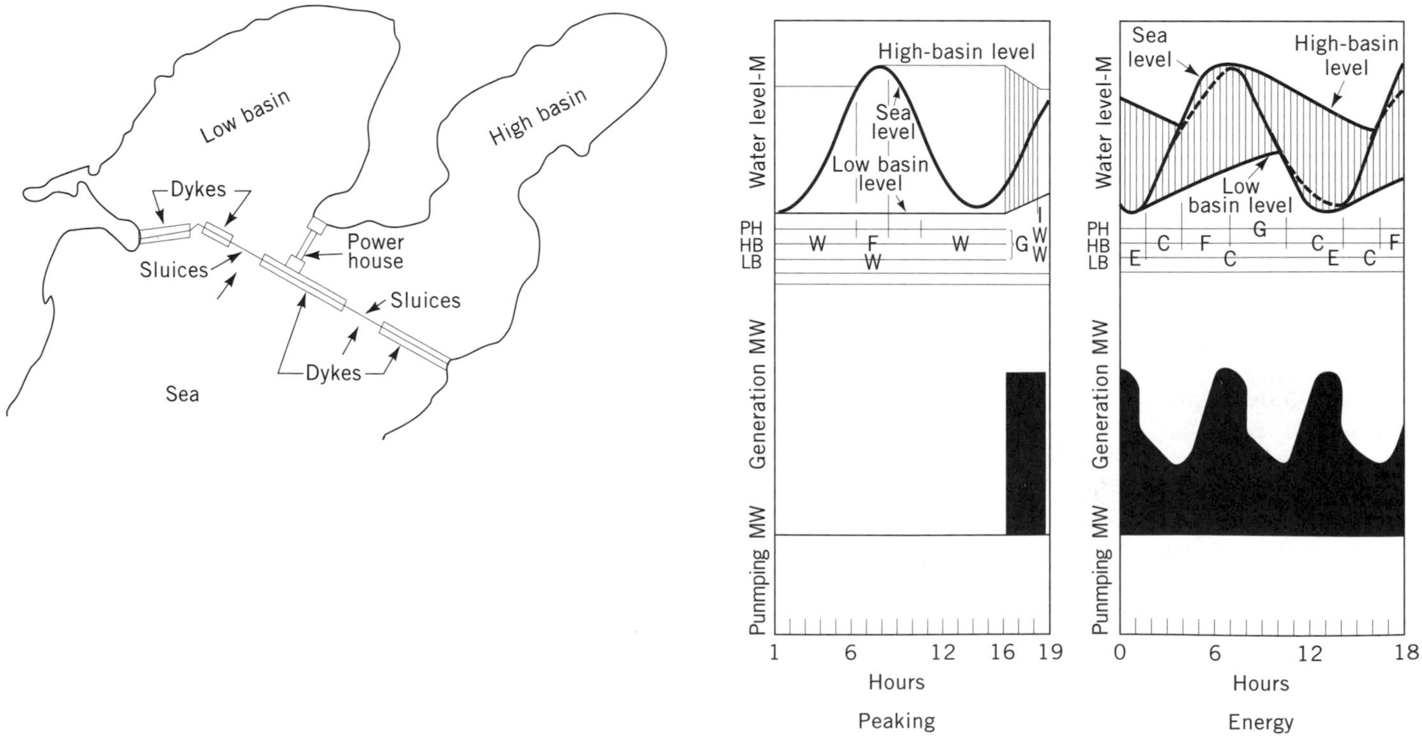

Figure 7. Form of output from a double-basin scheme. G, generating; F, filling; I, idle (no generation); W, waiting, E, emptying (closed basin).

cycle of the tide seaward of the barrage, and the operational decisions of opening sluices and running turbines either as generators or pumps. Analyzing the effect which the many interacting variables will have on the output is complex and requires computer-based mathematical modeling (9,10).

Retiming Tidal Output

The phenomenon of the successive daily advance in time of the high and low tidal cycles by approximately 50 minutes poses difficulties in meeting the peak energy demands of an electrical utility system. Moreover, it must be borne in mind that the amount of power that can be produced by any tidal power plant, regardless of its mode of operation, is directly related to the daily tidal range. In some locations, the range of the spring tidal cycle may be twice that of the neap tidal cycle.

The ability of an electrical power system to absorb raw tidal energy depends on several factors, such as the capacity of the tidal plant in relation to the total system capacity and the relative amount of energy and capacity from the various types of generation sources making up the system supply. Since a single-basin tidal electrical development cannot provide dependable capacity, because it is not able to operate during certain periods of the day, supplemental generation may be installed to firm up the power so as to provide a more valuable form of energy. During the day, demands for power peak during certain hours, eg, from 1600 to 1900 hours, and demand is relatively low during the night and on weekends. To meet the short period load demands, gas turbines are generally used to supply extra power during peaking hours since they can be started and fully loaded in a very short time and can be shut down quickly when demand decreases.

Large-scale energy storage facilities are similar to hydroelectric facilities adapted to short-term cycling. Energy can be stored by pumping water to an elevated storage reservoir, using low-value, off-peak energy, and using that water during peak load demand periods for generation under normal hydroelectric operations (11). About 20–25% of the energy thus delivered to a storage site is lost in pumping and generating, yielding a net energy storage efficiency of 75–80%, less any transmission losses to obtain the total system energy loss.

Tidal energy may also be stored in underground caverns in the form of compressed air. The first such commercial project, the 260-MW Huntorf compressed air energy-storage plant (CAES), built in northwest Germany in the 1970s, has proved a useful adjunct to the electrical power system (12). The plant stores off-peak energy by compressing air at high pressure (70 atmospheres) into caverns leached in salt domes. At time of peak demand, the air is released to charge an oil- or gas-fired expansion turbine driving an electrical generator.

Compressed-air energy storage or conventional pumped-storage facilities may be a valuable adjunct to the exploitation of tidal energy, but this can only be determined by economic assessment in relation to particular system conditions. A study of the economic viability of a CAES facility to retime the ouput from a proposed Bay of Fundy development at Site A8 (refer Fig. 15) provided a prefeasibility level of assessment of this retiming procedure. The result showed that at this time and with the current mix of generating facilities in the Fundy region, such a facility would not be economically viable (13).

With today's very large power systems, continuity of supply can be assured by the diversity of power sources with which they integrate. Such systems have the capability for absorbing energy from intermittent generating sources, particularly when the output is entirely predictable in magnitude and time. Each system must be assessed separately. All plants must assume the most cost-effective position on the system load curve and this approach should be maintained throughout the lifetime of each contributing station. Tidal energy is sufficiently flexible to be exploited in different ways.

CIVIL WORKS AND CONSTRUCTION

Tidal power developments are characterized by ocean and coastal sites, usually requiring lengthy barrages to close off an estuary or inlet. More often than not, foundation conditions involve deep overburden and silt layers and sites often have deposits that are moved by current and tidal effects. These will be affected by changed conditions during construction and during operation of the tidal scheme.

As noted previously, the main components of a development are the powerhouses, the sluiceways, and the dykes or inactive sections of the barrage isolating the estuary or basin from the sea. Many construction considerations and techniques which apply to conventional hydroelectric power installations also apply to tidal electric projects. Nevertheless, there are some aspects of tidal barrages that are unique and require special construction techniques. For example, the techniques must take into account the twice-daily fluctuations in tidal flow ranging from a maximum into the basin to a maximum in the opposite direction. Moreover, there is also a problem common to many of the large potential sites of the relatively great water depths at such sites and, in many cases, the deep bed of sediment in the bottom of the estuary.

Hydroelectric dams are normally built within cofferdams to provide a dry foundation. This was the technique followed in 1961 in constructing the Rance tidal power barrage. In this case, the cofferdams accounted for about 30% of the civil works costs of the project. The construction time with this method is longer which substantially increases the cost of the final project, and closing an estuary completely could have devastating effects on the ecosystems of the estuary. Therefore, this technique is not suitable for the construction of tidal electric developments in large estuaries, such as the Severn and the upper embayments of the Bay of Fundy.

The alternative approach, currently applied more frequently for various projects, is construction in the wet. The objective is to design the barrage so that it can be constructed in modular units which can be cast as caissons at an off-site dry dock and subsequently towed to the site and sunk onto a prepared foundation. This technique was first used in large scale in the 1940s for the Mulberry Harbor and since then has found application in the construction of underwater tunnels in the United States, Canada, Germany, and Italy; the closure of the estuaries in the Netherlands' Delta Works Plan; the siting of oil production platforms in the North Sea; and the construction during the 1980's of the Storm Surge Barrier to protect St. Petersburg, Russia (14).

An incentive for using this technique is that it permits the construction of the major civil works to be carried out under protected in-shore conditions. The equipment may be placed and fitted at the shore facility, or the turbine-generator assemblies can be barged to the caisson which has already been positioned and fixed in its place on the barrage line on prepared foundations. The large floating lifting derricks now available are capable of transferring such equipment to the caisson and setting it up in position.

In 1968, one of the main objectives for siting the small, experimental tidal electric plant at the entrance to Kislaya Guba, an estuary off the Barents Sea was to test the application of this float-in method of construction under the rigorous conditons of the northern coastline of Russia. Russian engineers were searching for reductions in construction time and costs to open the way for exploitation of the tidal potential on the northern coasts. The powerhouse module was placed within a few millimeters of the correct vertical position and only a few centimeters from the proposed horizontal position (15).

In the Bay of Fundy studies, the Management Committee assessed the comparative economics of the alternative construction concepts for the single-effect schemes at the three major sites under consideration. It was concluded that construction behind cofferdams would cost from 30 to 100% more, depending on the site. However, there are locations, particularly for small estuaries, where the site configurations favor construction in the dry, such as in the case of the 20-MW Annapolis tidal electric plant However, for larger estuaries, the state-of-the-art in marine construction clearly favors the float-in technique which provides the following advantages in constructing a barrage: (*1*) it minimizes environmental impact no only with respect to sediment movement during construction but also to the myriad of ecosystems depending on the twice-daily water exchange in an estuary; (*2*) it permits the powerhouse and sluiceway-unit caissons to be manufactured in several locations remote from the barrage thus decentralizing the resources required for construction; and (*3*) it significantly accelerates the installation program yielding significant economies as a result of the reduction in the construction schedule.

This float-in technique has also already been used in the construction of low-head hydroelectric plants. In one case, a French company prefabricated a steel, three-unit turbine-generator caisson in its St. Nazaire shipyard which was transported across the Atlantic ocean and floated up the Mississippi and Ohio rivers to its destination at the Greenup Lock and Dam. The technique has been suggested for installing powerhouses at navigation dams, currently without powerhouses, in the United States (16). In another recent application of this technique, a prefabricated steel caisson containing eight 24-MW generators driven by geared-type bulb turbines, was constructed in New Orleans, Louisiana, and floated to its final destination near Vidalia, into a newly excavated channel between the Mississippi River and an existing outflow channel into the Red/Achafalaya river basins. The eight-unit structure was floated and set between concrete buttress dam sections and on foundations prepared in the dry. The power station barrage was then secured to the adjacent concrete buttress dam sections, filled with concrete, and grouted to the concrete base (17).

The advantages of the float-in technique for tidal energy exploitation become evident when it is realized that the barrages of the proposed large-scale tidal power developments, subjected to intensive study since the early 1980s, could have lengths of up to 16 km with perhaps 200 turbogenerators and 150 sluiceways. Some of the potential developments in northern Russia would be significantly larger. A powerhouse caisson may contain up to four turbine-generators and the sluiceway caissons, up to three sluices, depending on the local topographical and tidal conditions. The access and closure dykes, or nonactive embankments of the barrage, are, for the studies carried out to date, assumed to be constructed using the conventional procedure with bottom-dump barges, a walking platform surmounted by a large crane, and end dumping. The use of blind caissons for closure could prove more economical and also yield a quicker closure sequence, but the extension of the floated-in technique to the closure dyke or other inactive sections of the barrage depends entirely on the site conditions and configurations.

MECHANCIAL AND ELECTRICAL EQUIPMENT

Tidal power development requires large volumes of seawater to pass through turbines under relatively low pressure head. The highest heads, the difference of water levels upstream and downstream of the development, at accessible sites around the world may be of an average of 12 meters or so. The form and layout of the required hydraulic turbine equipment has a strong influence on the overall design of the tidal power plant. The electromechanical equipment for a tidal electric plant is essentially the same as that used in low-head, run-of-river hydroelectric projects. Because of the significantly lower operating heads available at tidal sites, the turbines must be capable of discharging large volumes of water in order to develop the available power potential. These physical factors, along with economic considerations, prescribe the largest diameter turbines that can be supplied and installed.

Any major tidal electric development includes an unusually large number of turbogenerating units; for example, 128 units are proposed for the Minas Basin site in the Bay of Fundy, and 216 will be required for the Severn Barrage project. It will be necessary to make provision to operate them in blocks of six to ten units in order to simplify the automatic control procedures involved in starting and stopping of units, as well as providing for joint load sharing. Since the machines will be operating in a salt water environment, all submerged hydraulic equipment must be protected against corrosion. The cathodic protection system has been shown to perform extremely well (18). Algae growth may also be a potential problem so antifouling paints may need to be applied to some submerged metal surfaces.

Turbine-Generators

Early studies of tidal power developments, such as the International Passamaquoddy Tidal Power Project, were based on vertical axis Kaplan turbines which were frequently used in hydroelectric developments under modest heads (Fig. 8). These conventional, vertical shaft units have variable pitch runner blades and operate with a distributor having adjustable wicket gates or guide vanes, but require a deep elbow draft tube and a wide intake section both of which contribute to higher powerhouse costs. Moreover, losses are incurred in deflecting the water through the right-angled, draft-tube bend.

The bulb turbine, which is a horizontal shaft, axial-flow turbine, eliminates the extra excavation required for the elbow draft tube as well as the hydraulic losses in the bend. The generator is direct driven and housed in an underwater bulb (Fig. 9). This turbine type was developed for application in the Rance tidal electric plant and is now used extensively in modern, large, low-head hydroelectric schemes. By allowing flow through the barrage without change of direction; the powerhouse structure is more adaptable to the modular caisson construction. As a result, the dimensions of these structures are determined by stability criteria rather than by water passage hydraulics, as is the case for the vertical axis turbine.

One of the main disadvantages of the bulb turbogenerator is its low rotational inertia value because the diameter of the generator is kept as small as possible to minimize the size of the bulb housing and the hydraulic losses it causes. The generator diameter increases as the design rotational speed is reduced and the speed must be kept low to minimize cavitation damage. Because the inertia of the generator is important in maintaining stable operation and since the generator defines the diameter of the bulb in the upstream water passage, an increase in the generator size requires an increase in the length of the convergent section of the waterway if higher energy

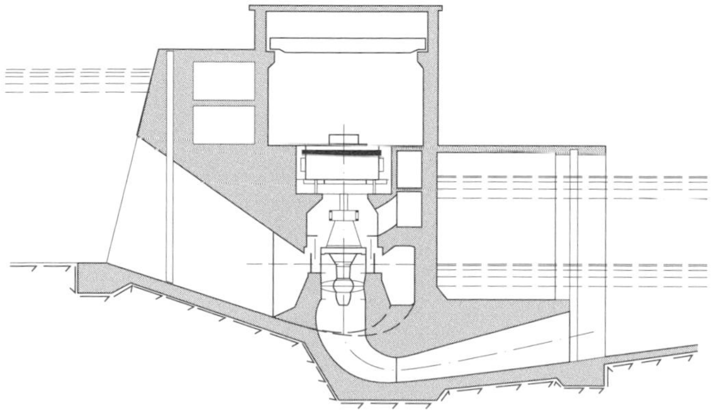

Figure 8. Sectional elevation of a powerhouse housing a vertical-shaft Kaplan turbine.

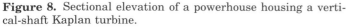

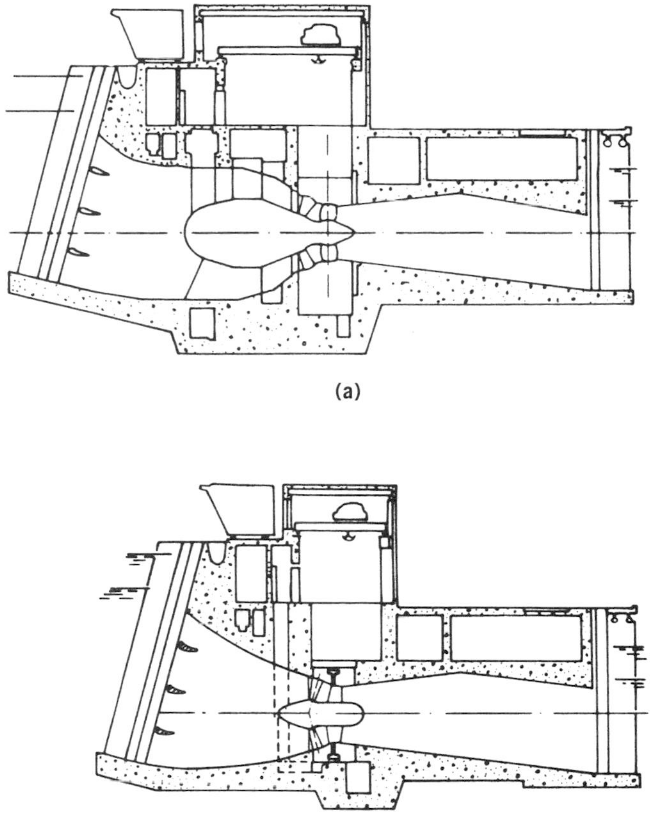

Figure 9. Comparison of **(a)** bulb and **(b)** Stratflo turbine powerhouses for identical output under same head.

losses, due to the sharper change in direction of flow past the bulb, are to be minimized.

This problem can be overcome by the straight-flow turbine in which the propeller blades are used as a spider to support the generator rotor, doing away with the generator bulb and straightening the water passages (19). The concept of attaching the generator rotor to the outer ends of the turbine blades was first applied in small units of Escher Wyss manufacture, installed in low-head, run-of-river plants on the Lech and Iller rivers in Bavaria around the 1940s. The units in these plants are about 1.5–2 m in diameter and develop up to about 2 MW under heads of about 7.5 m. With the upsurge in interest in tidal energy exploitation in the 1960s, Escher Wyss resumed development work to apply the concept to larger machines and thus evolved the Straflo turbine (Fig. 9). The key to the successful design was the development of sealing devices which could operate on relatively large diameters and at high surface velocities. Ten units for hydroelectric plants in Europe were in service in the early 1980s with runner diameters from 3 to 7 m and capacities ranging from 1.5 to 8 MW. The Annapolis tidal electric station comprising one, 20-MW Straflo turbine with a diameter of 7.4 m was commissioned in 1984 and relatively few problems have been encountered, none of which has appreciably affected the operation of the plant (20).

The main characteristic of the Straflo is that the generator forms a rim attached to the periphery of the turbine runner. To prevent water getting into the generator, seals are placed between the flanks of this rim and the sides of the water passages upstream and downstream from it. With low heads, low peripheral velocities, and only small axial variations, flexible lip seals can be used. With high peripheral velocities and higher pressures, as in big turbines, and in watercourses having high concentrations of abrasive solids, continuous self-cleaning, fluid-film hydrodynamic seals are used with no direct contact. This enables metal-to-metal surfaces to be used on rotational-to-fixed seals. In low-head units, the turbine-generator assembly is light and the generator can be supported, like the other moving parts of the unit, by the shaft bearings, via the runner blades. With higher heads, ie, in the range where vertical Kaplan turbines are now used the power developed requires much heavier generators that cannot be borne by the relatively flexible turbine blades. Therefore, high-capacity Straflo turbines are combined with generators whose rotors are supported and centered by hydrostatic bearings.

The concept of the rim-type generator is attractive because it operates in the dry outside of the turbine water passages and does not have the electrical problems associated with the long-core design required in bulb turbine-generators. The rim generator operating characteristics are more closely comparable to the conventional, vertical-shaft generator and its inertia is several times greater than that of the direct-driven, bulb turbine. Thus, the rotational stability is much improved and with the removal of the bulb the water passage can be made shorter and, therefore, less expensive and more efficient. Fig. 9 illustrates the comparative dimensions of powerhouses for bulb and Straflo turbine installations for the same power output under an identical head. One inherent handicap of the Straflo turbine is that adjustable blades are not yet feasible with the large diameter models.

As already noted, a tidal electric plant can be designed to generate on the ebb tide only, ie, single effect, or it may be designed to generate on both the ebb and flood, ie, double effect. Economics of the electrical system influence the choice of operation mode. Since relatively wide ranges of net operating head and flow must be accommodated over a tidal cycle, turbine runners with variable pitch blades provide good efficiency throughout the operating zone. The bulb turbine with its variable pitch runner blades is generally employed under such conditions because the blades are regulated to give good efficiency under the continuously varying head. When the wicket gates are also adjustable, double regulation is possible and an even higher efficiency can be maintained. Such turbines can also pump in two directions so that a greater range between high and low water in the basin can be achieved (Fig. 6.). However, when pumping would be desirable under certain situations, single regulation, ie, with adjustable runner blades only, provides almost as much energy as double regulation, ie, with adjustable runner blades and adjustable wicket gates or guide vanes, and eliminates the operating mechanism for guide vane adjustment which results in a more compact machine, more readily adaptable to caisson installation (6).

Although each different turbine type and configuration offers certain advantages, the choice of turbine type for a tidal electric development depends on site characteristics, optimum plant operating strategy, as well as competitive prices.

Sluice Gates

Tidal power plants operate on the continuously varying difference in level between the water in the basin and that in the sea. The basin must, therefore, be filled from the sea or emptied to the sea as required by the operating regime of the power plant in order to coordinate the output with the load curve of the system with which it is connected. Such operation requires sluiceways equipped with gates that can be opened and closed frequently, quickly, and reliably, and be as free as feasible from maintenance and operating problems. The gates must withstand the impact of storm-engendered waves and, in cold regions, the masses of ice carried by the flow into and out of the basin, freezing of the operating mechanisms, as well as damage due to the corrosive nature of the marine environment. In northern climates, submerged intake structures in hydroelectric power plants are relatively free from ice problems. In sharp contrast to the operation of a hydroelectric development where the sluice gates may be operated only several times annually, the sluice gates of a tidal electric plant are required to open and close each tidal cycle, resulting in 705 complete operations annually for single-effect generation or more than 1400 complete operations for double-effect generation.

For single-effect ebb generation the total sluiceway capacity, combined with the flow through the generating units (acting as orifices), must be sufficient to fill the basin during the flood tide so that the water level in the basin stands at as high a level as possible for the next generating cycle. Thus an optimum relationship between the cost of providing sluiceway capacity and the energy output is formed. For double-effect operation, the total capacity of the sluiceway is even more critical if full advantage is to be taken of this mode of operation. In this case, the optimum relationship between sluiceway capacity and power output is somewhat more complex. The various comprehensive investigations of tidal sites completed to date have selected either vertical lift gates with hydraulic hoists or a radial-type gate incorporating buoyancy chambers.

EXISTING TIDAL ELECTRIC PLANTS

Since the 1950s, there have been a number of advances in technology, notably in hydraulic turbine-generator units, marine construction, and mathematical understanding of tidal cycle variations. The technological problems inhibiting large-scale tidal power development have now been resolved.

La Rance

The initial design in 1951 for this development was based on classical vertical shaft generators using Kaplan turbines (variable pitch blades and variable guide vanes) which had been highly successful in exploiting low-head hydro sites. However, it became apparent that a completely new hydraulic turbine and generator set had to be developed that would be better suited to the special operating conditions of a tidal power plant, ie, turbine and pumping operations in both directions of flow as well as being able to assist sluicing operations. Moreover, because of the large number of turbine-generators required to take advantage of the potential of La Rance estuary, it was necessary to install them much closer together than would be the case in a conventional hydroelectric station. The horizontal shaft bulb unit was the outcome of the effort to solve the problem. After prototype turbine-generator sets had been installed and operated at several hydroelectric plants in the 1950s, as well as in an abandoned harborlock at Saint Malo subject to tidal fluctuations, a final design was adopted and construction authorized in 1961. The general layout of the development is shown in Figure 10. First energy was delivered in 1966 and the installation completed in 1967 with 24 10-MW turbine-generator units, comprising double-regulated reversible bulb turbines capable of turbining and pumping in both directions. These were the largest diameter bulb units (5.35 m) designed up to that time and since they proved so successful, substantially larger units have been built and operated successfully.

The construction of the locks, powerhouse, and sluiceway sections were carried out in the classical manner of hydroelectric plant construction, ie, in the dry, behind cofferdams. However, using this method, the estuary was cut off from the sea for three years during the powerhouse construction period. As a consequence, salinity fluctuations, heavy sedimentation, and accumulation of organic matter in the basin led to the almost total disappearance of marine flora and fauna. When the powerplant started operation and tidal flows resumed, a new biological equilibrium was reached within a decade, although not the same equilibrium that had prevailed before the project, and aquatic life flourished (21).

After two decades of exposure to marine attack, the reinforced concrete structure was considered perfect. The impressed current cathodic protection system has proven effective in preventing corrosion of the stainless steel runner blades and removable draft tube adjacent to the runner, as well as normal steel used for the bulb, the draft tube liner, and other submerged metal parts. A sophisticated paint system had also been used on the normal steel components, but this additional protection is no longer considered necessary. La Rance proved to be a source of valuable lessons as well as an economically attractive power station and in 1976, the standard cost of tidal energy generated was equal to the average cost of nuclear and thermal energy in the Electricité de France system (18). Advances in construction techniques since the 1960s have resulted in significant cost reductions, and the successful operation of even larger bulb turbines has provided further reductions in cost per MW of output. An

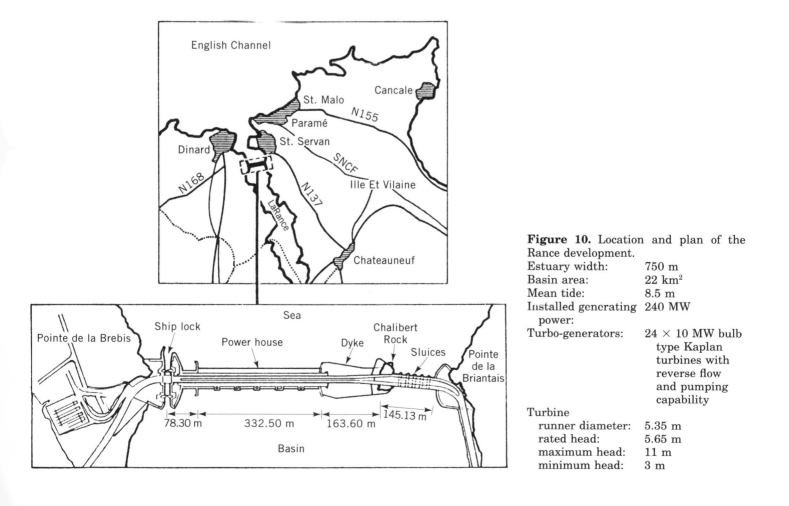

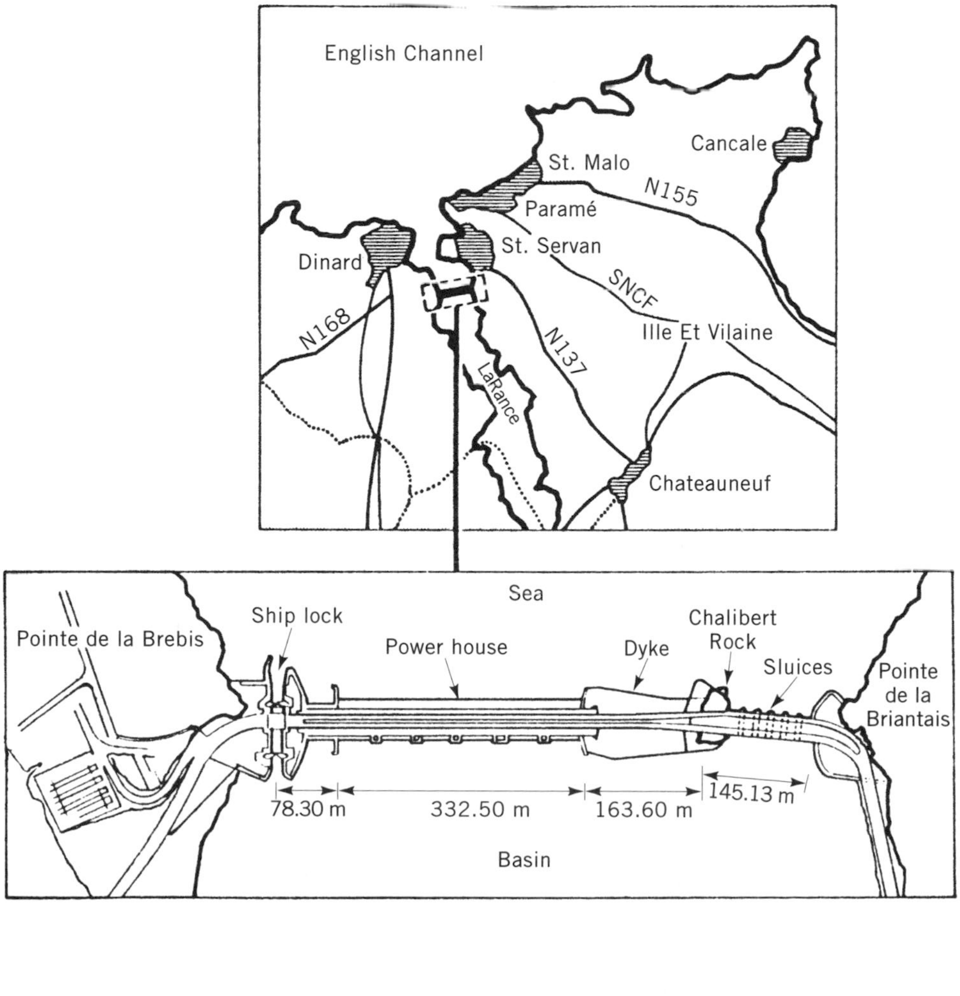

Figure 10. Location and plan of the Rance development.

Estuary width:	750 m
Basin area:	22 km^2
Mean tide:	8.5 m
Installed generating power:	240 MW
Turbo-generators:	24 × 10 MW bulb type Kaplan turbines with reverse flow and pumping capability
Turbine	
runner diameter:	5.35 m
rated head:	5.65 m
maximum head:	11 m
minimum head:	3 m

overview of the project and its operation over a 20-year period is provided (22).

Annapolis

In the late 1970s, although there were many straight-flow turbines in operation, mainly on European rivers, all were of relatively small diameter, on the order of one to two meters. The Straflo turbine design appeared to have cost advantages for low-head hydro applications and so could improve the economics of harnessing undeveloped, low-head hydro resources in Canada as well as the prospects of major tidal energy sites in the Bay of Fundy. However, there was some concern that unforeseen defects or design handicaps might emerge in a scale-up from existing machines; it was decided to demonstrate the commercial operation of a large Straflo turbine. The site (Fig. 15) for the installation and testing of the prototype was a causeway that had been constructed across the mouth of the Annapolis River in 1963 to provide better drainage and protection for the dyked lands surrounding Annapolis Basin and to carry a roadway to replace an old bridge. The site chosen required minimum capital outlay for civil works but provided conditions appropriate for installation and testing operations. The 7.6-m diameter runner provides a true prototype for future large-scale applications. This is the size, for example, that has been proposed for a large-scale tidal plant at Site A8 in Cumberland Basin (Fig. 15).

Taking into account the nature of the site, it was not necessary to consider caisson construction. A pit was dug in a strip of land separating the Annapolis River estuary and the sea, and the powerhouse constructed in the pit. After backfilling, the intake and discharge canals were excavated to connect the powerhouse to the river estuary, which served as the tidal basin, and the sea. The 20-MW plant was commissioned in 1984 (23). Its single Straflo turbine was set as shallow as possible to expose any tendency of the runner to cavitate. No such damage has occurred. Active cathodic protection from corrosion was provided for the wetted surfaces of turbine and draft tube and has proved adequate.

The machine operates unattended at most times, controlled by optimization programs in the powerhouse computer. A system control and data acquisition terminal transmits status information to the Western Zone control center (~100-km) of Nova Scotia Power Incorporated and permits the Zone operators to control the machine at will. The powerhouse computer also receives information on river flow, basin level, and sea level and controls the sluice gates and the turbine in a manner that ensures no environmental damage can occur due to excessive water elevations. The annual output of the plant has been less than the 50-GWh design estimate due to several factors: the basin level has been kept lower than design level; the intake and discharge canals were not excavated to design dimensions resulting in higher hydraulic losses; and tidal ranges have been less than expected. Since the mean tidal range at Annapolis is only about 7 m, even small head losses have an appreciable effect on output. It has been concluded that the Straflo turbine is eligible for consideration in large tidal power developments even though it cannot be constructed with adjustable blades (20). The choice of turbine type for a tidal electric development depends on site characteristics, optimum plant operating strategy, as well as competitive prices.

Jiangxia

China began the exploitation of tidal electric power in the mid-1950s and a 40-kW, single-effect tidal electric plant was completed in 1959 in Zhejiang Province; the plant was subsequently enlarged to 200-kW capacity. Another single-effect plant, with a capacity of 155 kW, was built on Jingang Creek in Shandong Province in 1970. It has three, 55-kW turbine-generators and supplies electricity to nine villages. In the 1970s, there was a total of six tidal plants with 19 turbines aggregating more than 1900 kW of installed capacity in Zheiang, Jiangsu, and Shandong Provinces. The maximum ranges at these sites vary from 3.5 to 7.8 m and the installed unit capacities range between 75 and 160 kW.

In April 1980, the Jiangxia tidal electric power plant in Wenling County was commissioned. The complete installation is to include five bulb units with a total capacity of 3200 kW and one Straflo unit of 700 kW capacity, for a plant capacity of 3.9 MW (24). The bulb units operate in the double-effect mode and are expected to deliver 10,700 GWh annually to a small grid supplying villages about 20-km distance through a 35-kVA transmission system.

Kislaya Guba

Russia and Siberia have considerable tidal energy potential along their northern shores amounting to an estimated 10% of the developable global tidal energy resources. These developable sites are located on the coasts of the Barents Sea, the White Sea, and the Sea of Okhotsk. In 1962, a decision was made to construct the Kislaya Guba station as an experimental plant to verify the application, under the particular physical and climatic conditions of the region, of the float in place method of construction, ie, the use of caissons. The application of this method to the rocky coastlines having difficult land approaches and a bleak climate between 65 and 70°N latitude seemed to offer considerable advantages, particularly since the station or floating components could be built in the favorable conditions of a seacoast industrial center, towed in a completed state (with equipment already assembled) to the site, and sunk onto prepared foundations. Because the solutions to problems that would arise during construction of this experimental plant could be of value for hydraulic construction in general, and under the rigors of the far north in particular, a broad program of comprehensive investigations related to civil works was an integral part of the project, such as the testing of frost-resistant reinforced concrete containing special additives to provide strength and water tightness (15).

The location of this scheme was selected so minimum work to close off the bay, which would become the tidal basin for the station, would be required. The neck connecting the Kislaya Gulf to the Barents Sea is 50 m wide. Although the tides there only have a range between 1.3

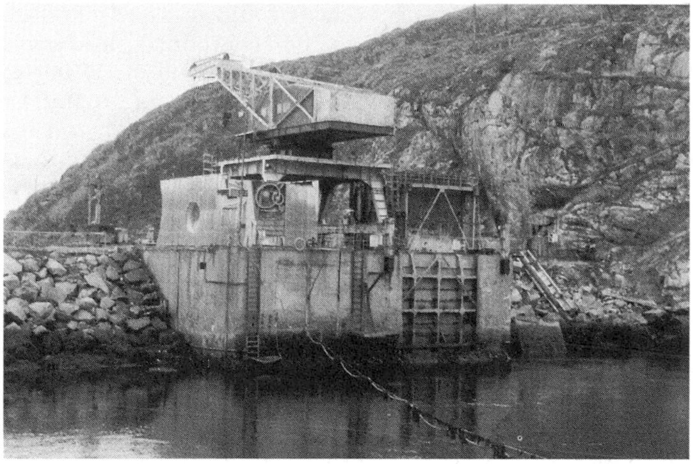

Figure 11. Kislaya Guba experimental tidal power station (view from the sea side).

and 3.9 m, the site is about 100 km from the cities of Murmansk and Kola, the latter being where the caisson was fabricated and equipped. However, because of the relatively low range, it was possible to carry out investigations of the power characteristics of the installation under fairly complex conditions. The experimental station (Fig. 11) commissioned in 1968, has yielded much valuable information regarding construction and placement techniques, including foundation preparation and concrete durability under harsh climatic conditions (25).

DEVELOPMENT POSSIBILITIES

Since the early part of the twentieth century, serious and sometimes quite elaborate schemes have been advanced for the exploitation of the tidal energy at potential sites, particularly in an attempt to overcome the discontinuity of output. The overall level of interest has fluctuated with time as a result of national and international economic and political fortunes. Currently, the interest in renewable energy resources is high.

In addition to a sufficiently high tidal range, a tidal power site should include a natural bay with an adequate area and volume, and be so situated that the operation of a plant will not significantly reduce the tidal range. There are few such locations worldwide to justify consideration of overall tidal power development. Figure 12 identifies major sites where tidal ranges have been considered suf-

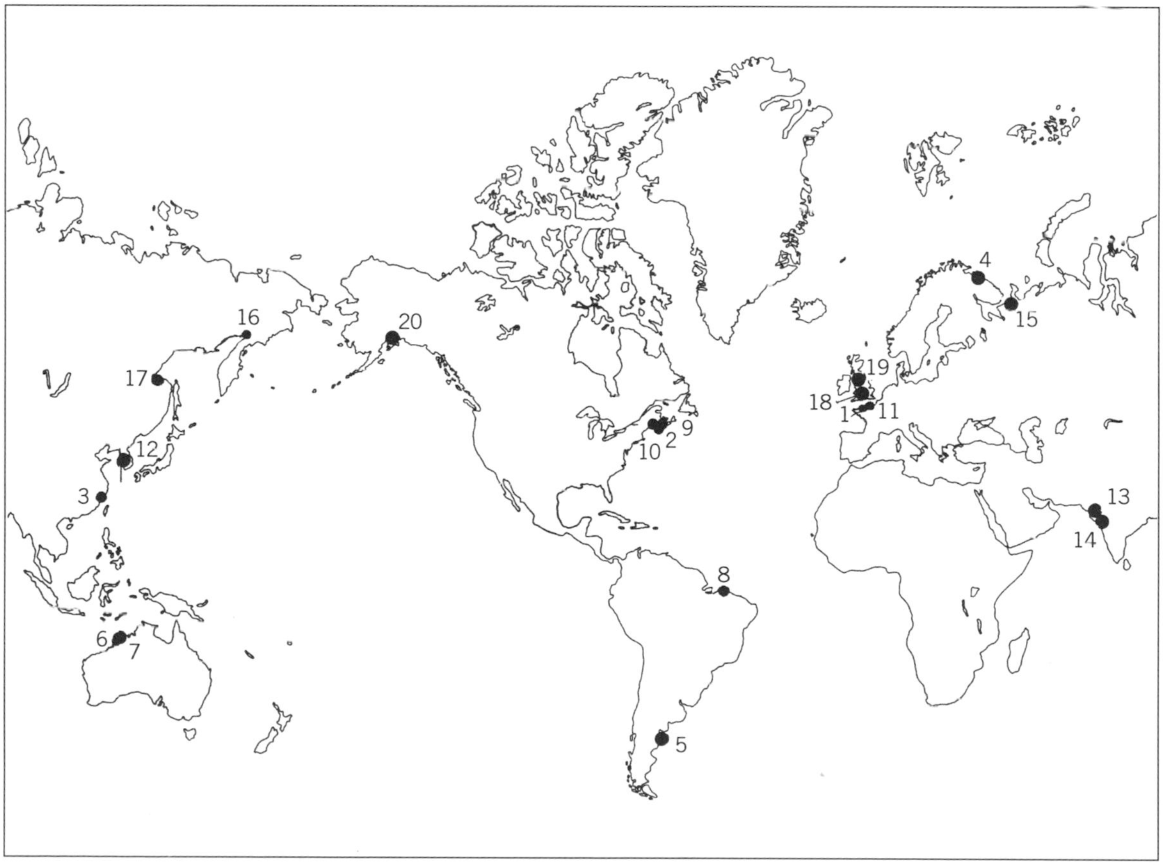

Figure 12. Location of operating tidal power plants (locations 1–4) and principal potential sites. 1. La Rance, 2. Annapolis, 3. Jiangxia, 4. Kislaya Guba, 5. Gulf of San José, 6. Secure Bay, 7. Walcot Inlet, 8. Northeast Coast Brazil, 10. Passamaquoddy/Cobscook Bays, 11. Chausey Island, 12. Garolim Bay, 13. Gulf of Kachchh, 14. Gulf of Cambay, 15. Gulf of Mezen, 16. Gulf of Penzhinsk, 17. Gulf of Tugursk, 18. Severn Estuary, 19. Mersey Estuary, and 20. Cook Inlet.

ficiently large and where the physiography appears to posses promising physical characteristics to support potential tidal power projects. Some sites tend to recieve concentrated attention for a short while and then disappear from the literature. Some sites which have undergone fairly intensive scrutiny or study since the 1970s, such as on the west coast of South Korea and on the Kimberley coast of Australia, hardly warranted mention in a previous text (26) which devoted considerable time to sites in northern Germany and the Netherlands which are now seldom, if ever, mentioned. The Severn, Fundy, and Passamaquoddy projects have been a perennial focus in this field of renewable energy resources. Results of investigations of some of the major locations with tidal power potential provide an insight to the possibilities for further exploitation of this renewable resource. Obviously, not all of the sites have been investigated to the same level of detail and proposals for some sites are based on very meager tidal data and other physical information.

Argentina

There are very high (~13.3 m) tides near the Magellan Strait which decrease progressively northward in the direction of the consumption centers. The coastline configuration creating the Valdez Peninsula (Fig. 13) and the phase difference in the tides, close to a one-half wave period between the two sides of the 6–8-km wide isthmus, has periodically drawn interest in development of tidal electric potential available. The Valdez Peninsula is an extensive projection into the Atlantic Ocean that causes tide delays of a half period between the northern coast of the isthmus, ie, the San José Gulf, and the southern coast or Nuevo Gulf. A level difference of almost 16 m results. It is this drop that could be used to produce electricity by dredging a canal through the isthmus and installing a double-effect tidal electric plant with a possible installation of between 600 and 1200 MW at the canal's end in the San José Gulf. As yet, the tidal phenomenon exhibited here has not been fully investigated to determine if such massive use might cause it to be greatly modified. Substantial tide and current data will have to be collected in order to understand the mechanics of the phenomenon.

However, assuming a single-basin development only based on the San José Gulf, it has been roughly estimated (27) that a near optimal installation would require about 180 generating units for a total installed capacity of 5040 MW with a mean annual output of 9400 GWh. The total cost of such a single-basin development, based on caisson construction, would make the cost of tidal energy higher than the long-term marginal values of Argentina's electric system.

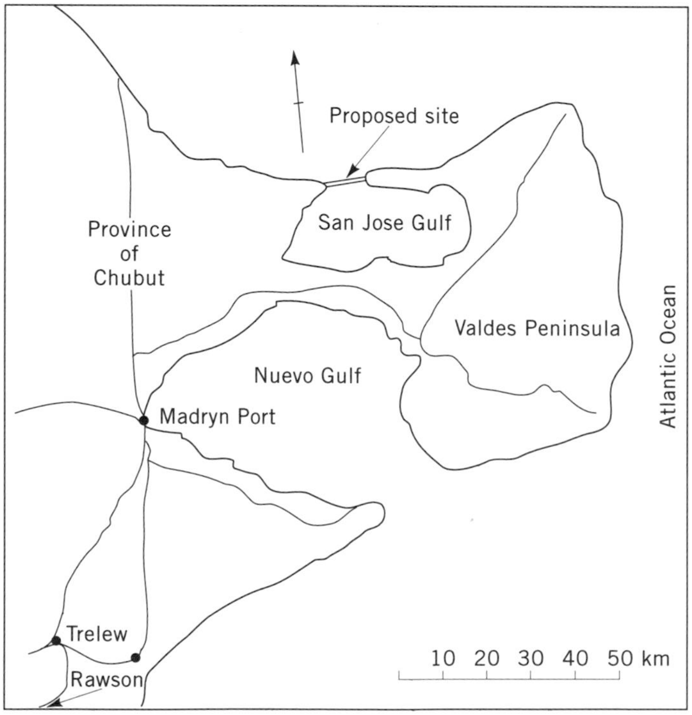

Figure 13. Location of proposed tidal power project for Gulf of San José.

Australia

There are a number of bays and estuaries on the rugged northern coastlines of western Australia where the mean spring tide ranges from about 5 to 11 m. Two adjacent estuaries, Walcott Inlet and Secure Bay off Collier Bay, offer favorable configurations with mean spring tidal ranges of about 11 m (Fig. 14). Prefeasibility investigations carried out in 1976 for the State Energy Commission of Western Australia were based on the use of caissons and bulb turbines and indicated the possibility for an installed capacity of 570 MW at Secure Bay with 30 turbines providing an annual energy output of 1650 GWh (28). The surface area of Secure Bay at high tide is about 180 km^2. At Walcott Inlet, which has a considerably larger surface area of about 415 km^2, the installed capacity could be 1259 MW yielding an annual energy output of 3940 GWh. It was concluded at that time that tidal power could not compete, by a considerable margin, with fossil-fuel generation in the Australian power market. A pumped-storage element could be incorporated with a Secure Bay development by taking advantage of a high level basin adjacent to the Bay so as to provide a firm power output. There is also a possibility that these two inlets could be developed as a double-basin scheme, using Walcott Inlet as the high basin, in order to provide firm power. The area is virtually uninhabited so there is no established local energy market. Further prefeasibility investigations would be required to justify a costly, definitive feasibility study of any of these possibilities.

In its final report to the Legislative Assembly of Western Australia in 1991, the Select Committee on Energy and the Processing of Resources made several recommendations regarding the tidal resources in the Kimberley region. These included the need to prepare long-term plans to harness Kimberley tidal power as a renewable energy resource that will reduce greenhouse gas emissions, and ensure that any future natural gas pipelines from the northwest of Australia to markets in the south or east be designed to allow the later transmission of hydrogen produced by electrolysis from tidal power in the Kimberley (29).

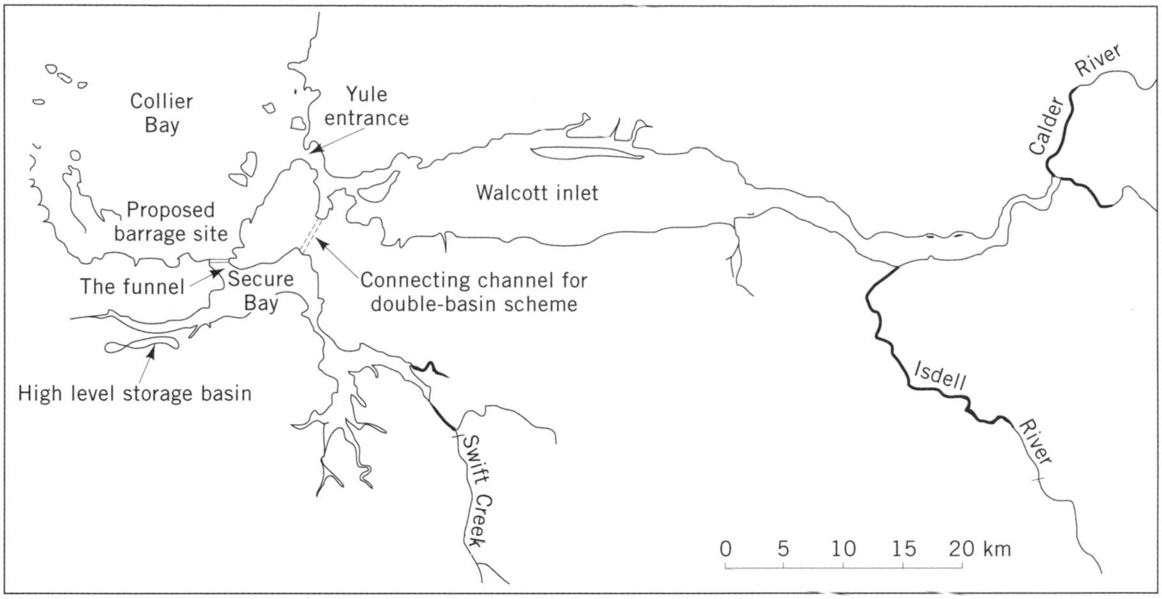

Figure 14. Proposed tidal power sites for Secure Bay and Walcott Inlet.

Brazil

Along the northeastern coast of Brazil, mean tidal ranges vary from about 5 m near the estuary of the Parnahyba River to about 8 m near the border with French Guiana. A maximum range of 13.5 m has been recorded in Maraca Harbor, near Montenegro, Amapa Province. In 1979–1980, Eletrobras, Brazil's Central Electricity Authority, commissioned a survey of this tidal power potential (30). A large number of sites were identified and, based on field reconnaissance, 12 were selected for preliminary study. A design was also prepared for a pilot plant which would make use of an existing dam spillway in the Bacanga estuary near Sao Luis. The capacity of the proposed pilot installation had not been selected but could be fixed at either 5, 7, or 9 MW provided by two bulb turbines. There has been little further activity in tidal energy exploitation in this region.

Canada

Bay of Fundy. Although a power project involving the Petitcodiac and Memramcook estuaries in the upper reaches of the Bay of Fundy was outlined in 1919, it was not until 1944 that the first major study of tidal possibilities was made (31). The Governments of Canada and New Brunswick authorized the investigation of a double-basin scheme linking these two estuaries. That study (32) concluded that such a scheme could produce about 1.3 TWh annually but could not be economically justified. Subsequently, several investigations of limited scope were undertaken at other possible sites during the next two decades but none was sufficiently definitive to establish the merits of development.

With the construction of La Rance almost completed, the Governments of Canada, New Brunswick, and Nova Scotia, in August 1966, jointly initiated a comprehensive investigation of the feasibility of large-scale tidal power developments in the Bay of Fundy. The details of the investigation and conclusions are set forth in the October 1969 reports of the Atlantic Tidal Power Programming Board (ATPPB) and of its Engineering and Management Committee (33) which was set up to carry out the technical and economic studies and surveys. It was found that, although technically feasible, it was not possible to exploit tidal energy at a cost competitive with energy from alternative sources. However, in concluding its studies, the Programming Board recommended that further studies should be authorized when (*1*) the interest rate on money drops sufficiently to suggest the possibility of an economic tidal power development; (*2*) a major breakthrough in construction costs or in the cost of generating equipment suggests the possibility of designing economic tidal power development; (*3*) pollution abatement requirements substantially magnify the cost of using alternative sources of power; or (*4*) the alternative sources of more economic power supply become exhausted.

Subsequent to the submission of that report, not only did significant cost increases occur in thermal energy sources but a greater emphasis was placed on environmental considerations, tending to further increase the costs of all types of generating plants. Accordingly, the Governments of Canada and of the provinces of New Brunswick and Nova Scotia established the Bay of Fundy Tidal Power Review Board in February 1972 to undertake a critical review of the findings of the Programming Board in the light of the prevailing and projected conditions and to recommend procedures for a reassessment of feasibility, as appropriate. The Review Board's findings indicated that the economic position of tidal power had improved significantly since 1969, taking into account the significant increase in the international price of fuel oil and the current trends in system power development. In addition, the Review Board pointed out additional merits encouraging the development of tidal power: (*1*) corresponding reductions in fossil-fired energy production would result in reduced atmospheric and water pollution; and (*2*) reduction in demand on increasing oil and gas reserves which could be diverted to more critical uses. Although not specifically assessed in economic terms, the Board also recognized that the transmission facilities which would be necessary with a tidal power project would undoubtedly offer additional opportunities for interchange of power within and between power systems of Eastern Canada. Acting on the recommendations of the Review Board, the three governments agreed in December 1975 that further investigation was desirable and in the public interest in view of the distinct possibility that tidal energy be shown as an economic resource in the Atlantic region, and authorized the reassessment of Fundy tidal power.

The agreement established a Management Committee to carry out the study program under the general direction of the Review Board. The Committee contracted the work with a number of Canadian consulting firms. The results of the investigation are set out in the reports of the Bay of Fundy Tidal Power Review Board and Management Committee (5). Thirty sites (Fig. 15), including the 23 sites examined in the previous study, were evaluated in terms of approximate development costs and outputs. Successive screenings yielded three sites, confirming the final site selection by the Programming Board, for further, more detailed studies. The results of these studies, which incorporated the use of floated-in powerhouse and sluiceway caissons, showed conclusively the fundamental economic feasibility of tidal power and the technical and economic feasibility of its integration into the existing generation supply systems. The benefit–cost ratios for developments at sites A8 and B9, which the Programming Board had found to be less than unity, were now assessed at 1.2 with a break-even period of 30 to 35 years. The Review Board also concluded that site A8, the Cumberland Basin site, would be the preferred candidate for development and recommended that funding be provided to complete detailed investigations and definitive designs for the development. In a subsequent updating study (34), the benefit–cost ratios had nearly doubled.

In 1985, a further study was financed by the Governments of Canada and of the province of Nova Scotia and the results of that study confirmed that site A8 was favored as the target site for development. The mode of construction would be heavily dependent on placement of float-in caissons, not only accommodating the turbogenerators and sluices but also the access and closure structures. The barrage length of 2560 m would accommodate 42 turbines with an installed capacity of 1428 MW at a rated head of 9.1 m.

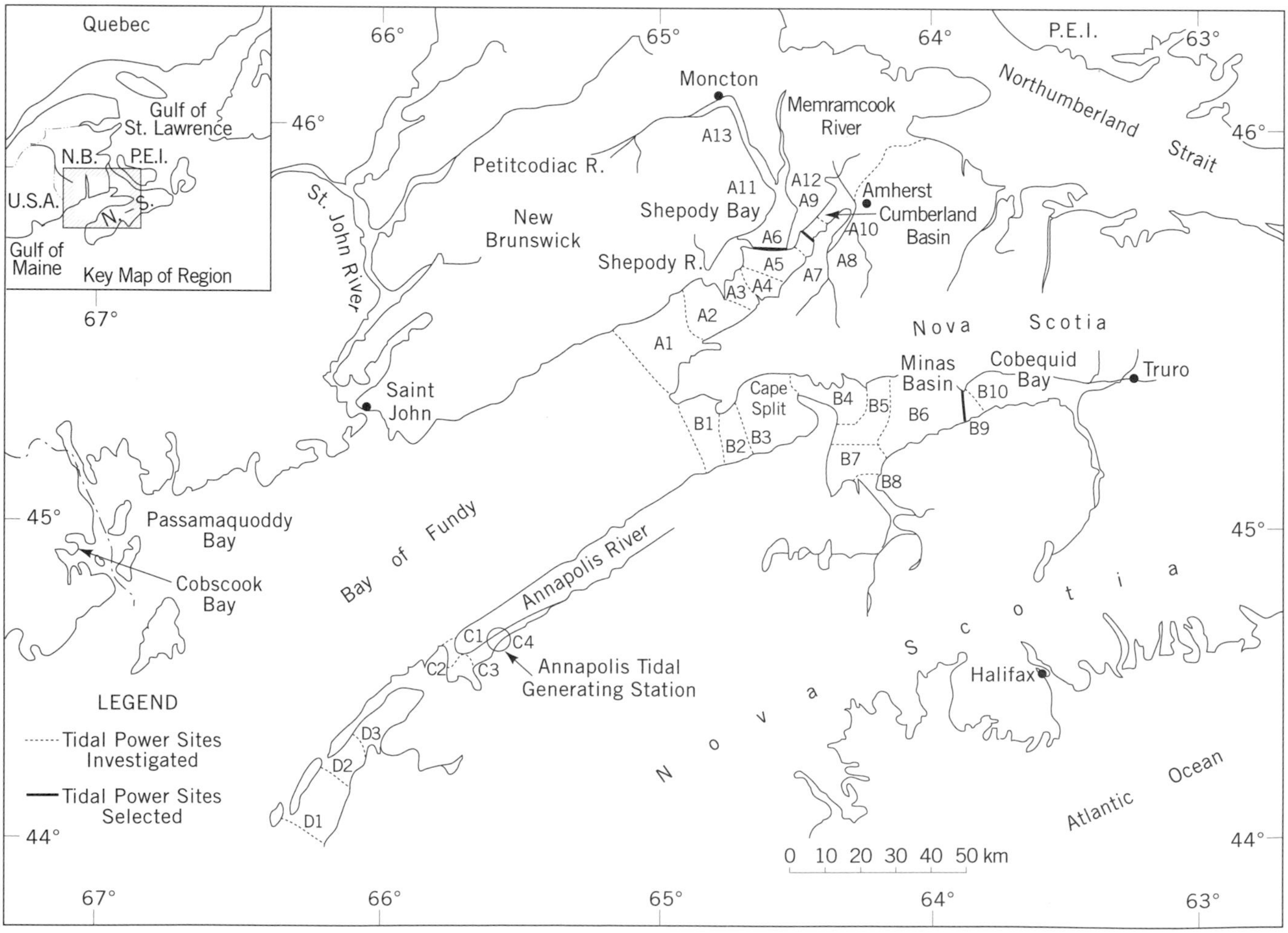

Figure 15. Bay of Fundy and possible sites for tidal energy development.

Other Possible Canadian Sites.

In addition to the Bay of Fundy, there are several potential tidal power sites on the western and northern coasts of Canada with possibilities for development. The British Columbia Power Authority made a cursory examination (35) of 13 sites on that province's coast where the mean tidal range varies from 3.5 to over 5 m. Their potential annual energy outputs would be in the range of 100 to 2200 GWh. However, the unit cost of energy was estimated to be about an order of magnitude greater than other generation options available to the Authority.

The geographical configuration of Ungava Bay along the Hudson Straits, with its mean tidal ranges of 7 to 8.5 m, offers sites with a total potential installation of about 5500 MW (36). The main impediment, even before contemplating the potential developments themselves, is the distance exceeding 1500 km to major load centers and the lack of communication and development in this hinterland of Canada.

Canada–United States

Plans were promoted in the early 1940s by United States interests for a two-basin scheme involving Passamaquoddy Bay and Cobscook Bay near the entrance to the Bay of Fundy. The major portion of Passamaquoddy Bay lies in Canadian territory while Cobscook Bay is entirely within the United States (Fig. 16). Eventually, the possibilities were investigated, at the request of Canada and

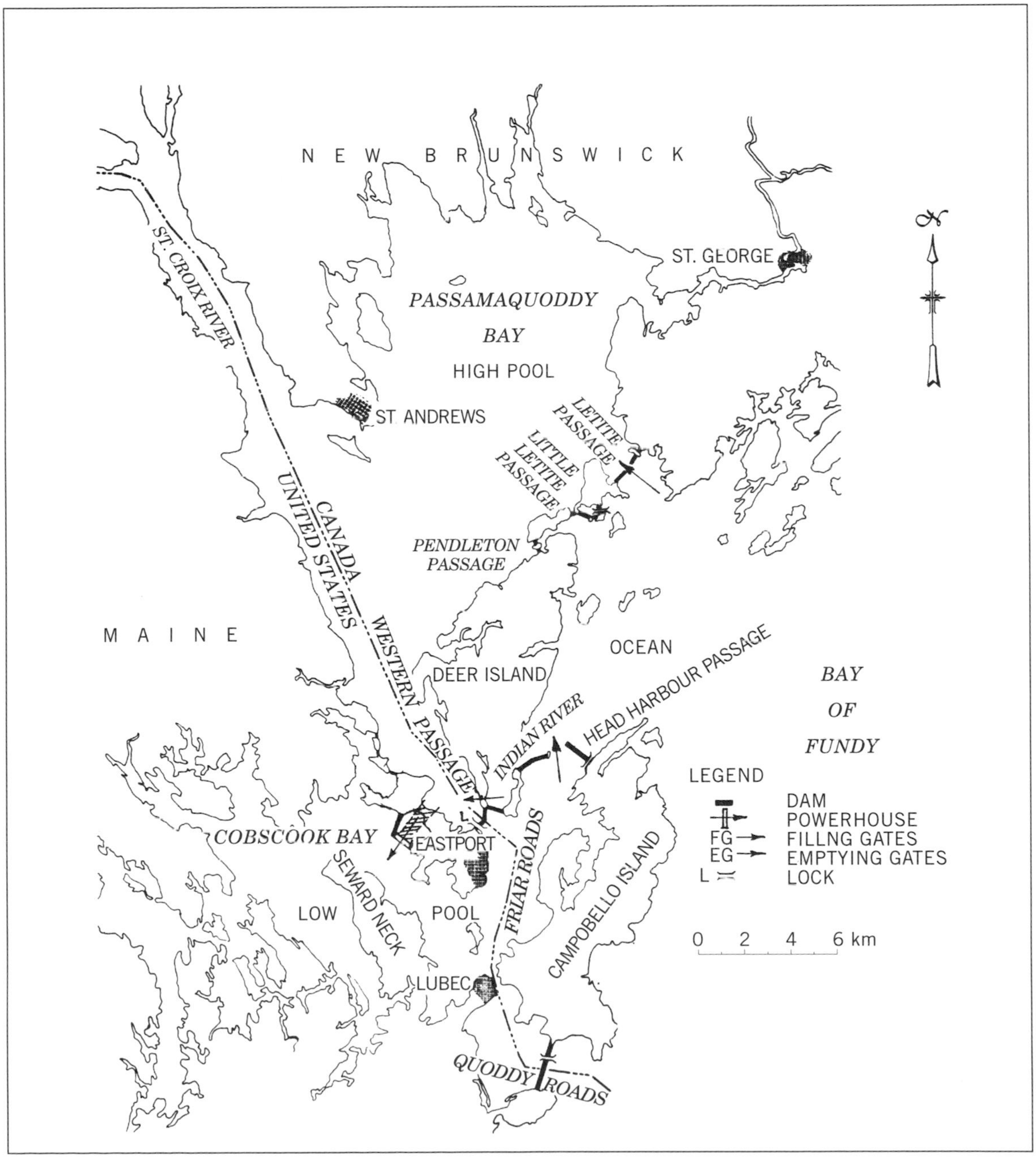

Figure 16. Layout of International Passamaquoddy tidal power scheme.

the United States, by the International Joint Commission established by the Boundary Waters Treaty of 1909 between the two countries. The Commission's report of 1961 (37) concluded that hydraulically linked, two-basin scheme with an installed capacity of 300 MW and a dependable capacity of 95 MW was not economical; in fact, the benefit–cost ratio was substantially below unity. Although the studies have been updated periodically by the United States Corps of Engineers, the conclusion remains (38).

France

Following the commissioning of La Rance project in November 1966, interest turned again to the major project embracing Chausey Island and Mont St. Michel Bay which had been under study for many years and for which the La Rance tidal project was considered, at the time, as a "scale-model."

Electricité de France has studied several development arrangements, one of which would include a power station of 300 units of 40 MW each. Such a project, which would generate about 25,000 GWh annually, would effect the tides and currents in the English Channel, and perhaps beyond. Another project, proposed in 1971, would provide for a power plant with 200 30-MW units with two basins, each 1100 km^2 in area, 35 km of dykes, and 50 km of sluice gates. Such a plant could generate about 34 TWh per year with an assured output of 3900 MW (39). In addition to altering the established ecosystems in the Bay of St. Michel and adjacent coastal waters and changing the landscape, it would also affect a large population making a living from the sea. Studies to assess environmental impacts and effects on regional economic development have been outlined.

Korea

The indigenous energy resources of Korea are quite limited and the major source of energy is oil which must be imported. Exploitation of the potential energy offered by the many bays and inlets along the west coast of Korea, notably Incheon, Asan, Garolim, and Cheonsu, where mean tidal ranges are about 5.7, 6.1, 4.8, and 4.5 m, respectively, has been of continuing interest for many years. In 1977, a Phase 1 study was commissioned to evaluate the tidal power potential on the west coast of Korea to determine the priority of sites and to recommend a site for detailed feasibility study for Phase 2 of the tidal power program undertaken by the Government of Korea. A tidal power station was included in the Korean five-year plan that commenced in 1982. The report of the Phase 1 study, which concentrated on tidal power aspects only, ignoring multipurpose considerations, recommended Asan Bay in which two sites were proposed for further detailed study. The smaller of the projects in this Bay was shown to be marginally economic at the time with an installation of about 450 MW and an annual energy output of 1345 GWh. Subsequently, it was decided that the Phase 2 technical and economic feasibility studies would be carried out for a tidal electric plant at the mouth of Garolim Bay. The study was completed in 1982, and a scheme was recommended for a tidal electric plant equipped with 24 turbogenerators of 20 MW each in a barrage which would control a basin of about 100 km^2. The barrage would have a total length of about 2.1 km with the powerhouse and sluiceway sections being constructed of precast, floated-in caissons.

The tides in Garolim Bay are semidiurnal having a large diurnal inequality and spring and mean tide ranges of 7.9 and 4.8 m, respectively. A reassessment study (40) carried out in 1986 undertook further oceanographic surveys and essentially confirmed the alignment, layout, etc., developed for the 1982 report with several refinements such as a re-optimized installation of 20 turbine-generators at 20 MW each yielding about 836 GWh annually. However, the project still could not be justified economically because of the high discount rate. It has been noted that the project could support recreational facilities, marine aquaculture, a small harbor, and dock marina, in the basin whose benefits could bring the tidal project to economic feasibility. Such a multipurpose development is expected to be studied in the near future (41).

India

There are relatively large tidal power potentials available in the gulfs of Kachchh and Cambay in Gujarat Province and possibilities of small-scale developments in the Sunderbans area of West Bengal. These locations are in regions of India where energy dependence rests on coal imported from East Bengal. In 1975, a preliminary assessment of the technical and economic feasibility of developing tidal power projects at suitable locations was sponsored by the United Nations. The results of that assessment suggested two single-basin schemes to develop the energy potential of these two estuaries. The size of an installation in the Gulf of Cambay, where the mean tidal range is about 7 m, could range from 5000 to 7000 MW, but the intermittent output from such a large, single-basin scheme could pose a significant problem for integration into India's Western Grid. A feasibility study for the Gulf of Kachchh, with a mean tidal range of 5.45 m was launched in 1982 and completed in 1987. The current proposal envisages an installed capacity of about 900 MW (42).

A power barrage would be constructed at the narrows, east of Navlahki Port, with two small subsidiary structures to prevent impounded waters in the basin from flowing through Kandla creek. About 40 km of low embankment dykes would be required to provide a basic area of about 260 km^2. The barrage would be formed by powerhouse and sluiceway caissons and a rock-fill section. Locks would also be required to ensure the continuation of navigation to Navlahki Port. Changes in patterns of sedimentation in this estuary caused by the barrage could have an impact on the project. Interest in development of the tidal potential at Navlakhi in the Gulf of Kachchh is maintained by the Tidal Power Cell of India's Central Electricity Authority. The integration of the output of this tidal power project into the highly, thermal-intensive Western Grid system is being given careful consideration.

Russia

Coastal configurations with fairly high tidal ranges along the northern coasts of Russia present opportunities for very large-scale tidal power developments. Assessment of this potential has resulted in the selection of three favorable sites: Mezensk on the White Sea coast and Penzhinsk and Turgursk on the coast of the Sea of Okhotsk (43).

At the Mezensk site, where the mean tidal range is 5.66 m, a powerhouse with 800 19-MW bulb units of 10 m diameter would form about 22% of the 85.6 km-long barrage. The remainder of the barrage would be constructed of rock and earth fill with a sand core. The annual output of this 15,200 MW plant could average about 45.4 TWh.

The Penzhinsk site is a very remote possibility, not only because of its potential magnitude of 21.4 GW of installation for an average annual output of 71.4 TWh, but also because of its remoteness, ie, about 2000 km from other large generating sources with which power exchanges would be required in order to utilize the tidal output. Also, a principal problem with the construction of such a plant would be the adverse climate where the annual average temperature is -6.5°C and the minimum is −50°C.

The effectiveness of a Tugursk tidal electric plant presupposes the development of those power systems where the whole of the peak load and some fraction of the base load will be supplied with the aid of the existing hydroelectric stations as well as those under construction and design. The total capacity of these stations is estimated (43) to be 2000 MW by the turn of the century and about three times that capacity by the year 2010. As currently contemplated, the Tugursk tidal electric plant would consist of 420 bulb units, each of 16.2-MW capacity, for a total installation of 6804 MW. The annual average output of such a plant is estimated at 16,200 GWh. The powerhouse would be formed by 105 floated-in caisson units each containing four turbine-generators, for a total length of 10.5 km. The mean tidal range at the site is 5.63 m and the area of the tidal basin is 1120 km^2. The benefit-to-cost ratio would be less than unity (ca 1994). The report on a prefeasibility study of the Tugursk potential development is under preparation.

It has been recognized that before embarking on the construction of 6000+ GW installations, long-term, detailed investigation, research, and perhaps the construction and testing of the operation of a prototype development with 10-m diameter bulb turbines, would be required. Such a prototype development is contemplated by the proposal for construction of the Kolskaya pilot-commercial plant on the Barents Sea coast near Murmansk. The plant would contain four bulb units with runner diameters of 10 m. The project would be operated in conjunction with a river hydroelectric plant which has adequate storage for the cyclical output of the tidal electric plant.

United Kingdom

Severn Barrage. Original interest in the possibility of developing tidal power from the Severn Estuary began early in the twentieth century. The tides in the estuary have a mean range of 8.5 m. In 1978, renewed national interest resulted in the establishment, by the Secretary of State for Energy, of the Severn Barrage Committee to assess the feasibility of development. The work of the Committee concentrated attention on an Inner Barrage scheme with an installed capacity of 7200 MW and an annual average output of 12,900 GWh (6). This energy output equals that from consumption of about five million tons of coal equivalent per year and, from the 1981 perspective, economics of development were to a considerable extent dependent on future balance of nuclear and coal produced electricity.

Following publication of the Committee's report, a consortium of industrial interests known as the Severn Tidal Power Group (STPG) published its report in 1986 (44) which endorsed the barrage alignment from Cardiff to Weston. Although the scheme was considered insufficiently attractive for private investment funding, the benefits of diversification, regional, and nonenergy benefits produced a recommendation from the STPG to undertake further work. The STPG, the former Department of Energy, and the former Central Electricity Generating Board decided that the Severn Barrage Development project should be initiated with equal funding from each party. This report, published in 1989, concluded that the development would consist of 216 turbine-generators, 9-m diameter bulb turbines with a rating of 40 MW each, to provide an average annual output of 17,000 GWh equivalent to 7% per of the electrical consumption in England and Wales (6). The barrage would have a length of 15.9 km and would be constructed from a series of prefabricated concrete, floated-in caissons, which would house either turbine-generators or the 116 required sluices, and form blank sections of the barrage. A shiplock would also be included to allow access to upstream ports. It was estimated that the barrage would take nine years to build but would be able to generate electricity after seven years.

Environmental concerns have been and continue to be of deep concern to the proponents and government jurisdictions involved around the Severn Estuary. A further, more detailed investigation is underway to evaluate the impact of construction and long-term operation of a Severn Barrage on the Severnside region, regional infrastructure, and local ports.

Mersey. A tidal power barrage in the Mersey Estuary was proposed in the early 1980s and a private sector group formed the Mersey Barrage Company to carry out the necessary feasibility studies and promote the project. As presently conceived (45), the project would consist of 28 geared and pit-mounted Kaplan units, 8 m in diameter. The turbine-generators will be housed in concrete caissons with four units per caisson. Although the turbines and sluices will be floated to the site in caissons, the required ship locks will be constructed *in situ* within cofferdams. The annual energy output should be about 1400 GWh. If the CO_2 reduction initiatives and the value of the barrage as an additional Mersey road crossing were factored into the economic analysis, the net present value of the project would be positive.

Small-Scale Sites. There are several estuaries around the United Kingdom that possess the main requirements for a small-scale tidal scheme and several of these have been subjected to prefeasibility studies or are under investigation. The results of the investigations of the Conwy (46) and Wyre (47) estuaries indicate that the annual energy output from a development at the former site would be about 60 GWh and from the latter, about 130 GWh. The mean spring tidal ranges at these locations are 6.5

and 8.6 m, respectively. At the 1994 pool price for electricity in the United Kingdom, these schemes would not be economical.

United States

Cobscook Bay. The United States government has explored the feasibility of exploiting the energy of the tidal potential of Cobscook Bay only. This Bay (see Fig. 15) has an average tidal range of about 5.4 m and an area of about 98 km^2, and could support an installed capacity of about 250 MW. Because of the many islands and their locations within the Bay, double-basin schemes have also been considered in the exploitation studies (38).

Cook Inlet. Cook Inlet, Alaska, also offers tidal energy exploitation possibilities. A preliminary assessment of the tidal power potential and characteristics of Cook Inlet was carried out under the authority of the Governor of the State of Alaska and completed in 1981 (48). Sixteen sites within the Inlet were examined, and on the basis of the assessment the two most attractive sites were across either Knik Arm or Turnagain Arm (Fig. 17). It was also concluded that studies should proceed for the site above Eagle Bay on Knik Arm. A tidal plant at this location could have an estimated capacity of 1440 MW from 60 bulb turbogenerator sets and 36 sluices for an annual energy output of about 4037 GWh. The barrage would be formed by prefabricated powerhouse and sluiceway caissons. The Eagle Bay site could be developed in a manner which would be consistent with the most likely energy demand growth in the region, provided the energy could be retimed economically. A major field investigation program will be necessary to remove or reduce uncertainties arising from the limited current data base and environmental concerns.

ENVIRONMENTAL PERSPECTIVE

Ecosystems unaffected by humans, do not remain static but evolve as their environments gradually change over time. Constructing a barrage composed of sluices and powerhouses across a tidal estuary will produce environmental changes much more rapidly. Ecosystems adjust to changes and continue to function in a modified form, but consequences are far-reaching and difficult to predict. Some environmental changes may affect natural resources over a region larger than the resulting tidal basin while others may affect the operation of the tidal power station itself. On environmental grounds, tidal power offers a very significant advantage over conventional, fossil-, and nuclear-fueled electricity generating systems; the exploitation of tidal energy does not produce harmful by-products which can cause emission or storage problems. Environmental alterations expected to occur as a result of the construction and operation of a tidal power station clearly must be considered along with the economic and engineering factors in the decision-making process. An environmental assessment of a tidal electric development is a complex and difficult task because it must deal with a dynamic estuarine environment on a site-specific basis.

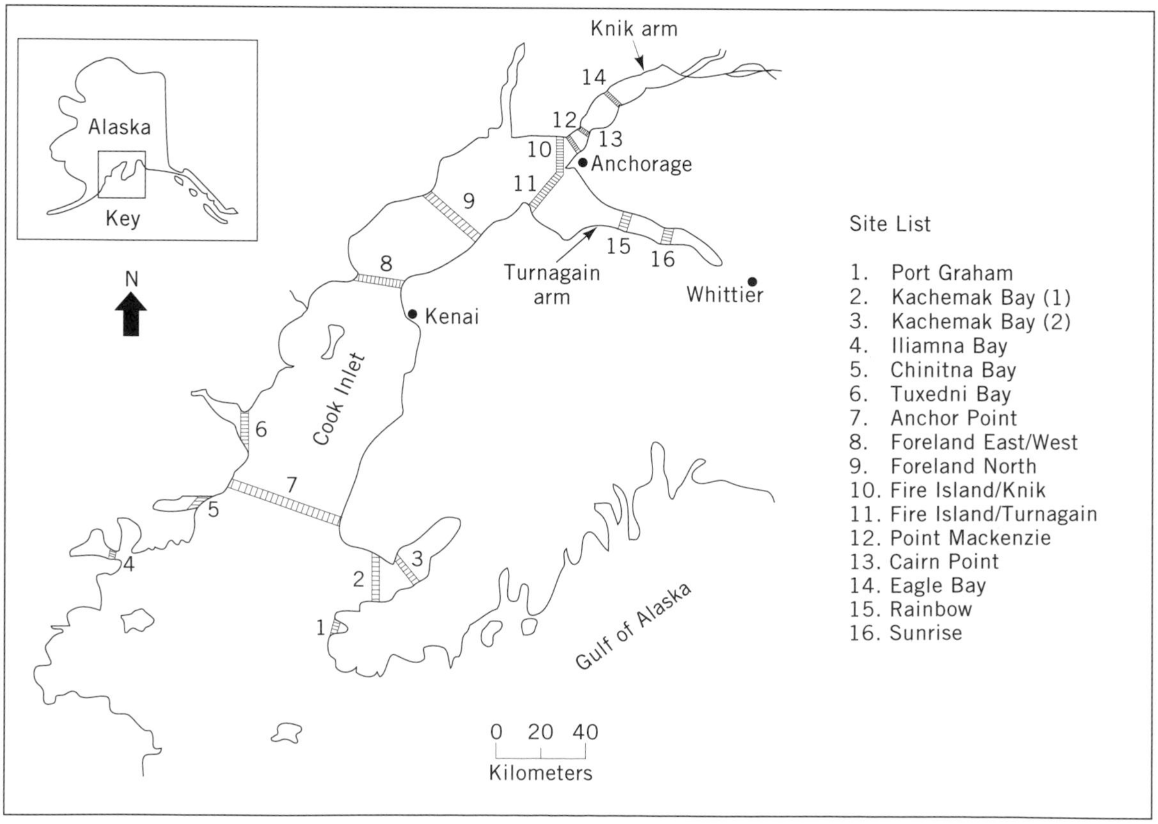

Figure 17. Location of potential power sites in Cook Inlet.

Physical Effects on Natural Resources

Although many environmental consequences will be common to all proposed tidal power developments, the relative importance of these and other changes is site specific. An account of the probable consequences of such developments can only be drawn up with reference to a specific location, and would not likely be transferable from site to site without substantial modification. Many effects of a tidal electric development would be essentially the same as those experienced with the construction of a major marine project, such as dredging and filling operations, site access, and traffic disruptions during construction. Additionally, however, a tidal electric plant will produce unique and important changes stemming from its location and the nature of its operation, ie, from its physical presence.

Physical Changes. As already described, the suitability of sites for tidal energy exploitation is limited to those estuaries with a large tidal range and favorable topographical characteristics. Large tidal ranges are accompanied by strong tidal streams and combined with wave action result in significant sediment movement. A direct effect of constructing a tidal electric station would be to alter the range and pattern of sea level fluctuations and the properties of tidal currents, ie, the tidal regime, not only within the controlled or tidal basin but also seaward of the development.

One principal effect of a project will be to alter the water level in the tidal basin. In a state of nature, lands bordering the estuary are alternatively flooded and exposed with each tidal cycle. There would also be land bordering the basin which, under extreme low tide, would only occasionally be exposed and drained and, on the other hand, there would be land only rarely submerged during periods of high tide. The operation of a tidal electric station under the ebb-generation mode, for example, would permanently alter this natural regime. Since the level of a tidal basin would normally not fall as low nor probably rise as high, the average basin level would be maintained above the natural mean tide level as illustrated by Figure 18. The result would be a decrease in the area of the intertidal zone which would change the habitat for wild fowl, shore birds, and miscellaneous species.

Such physical changes would indirectly affect the wave climate, temperature and salinity characteristics, stratification, density currents, sediment deposition and ice formation, and movement. These effects would be felt within the basin as well as seaward of the development. The ground and surface freshwater processes in the controlled estuary or tidal basin could also be directly affected as a result of flooding, draining of marsh lands, and salt-water intrusion, for example.

Climate. The climate regime in the vicinity of a tidal power station could be affected by such changes in the physical oceanographic regime as water temperature and the extent and duration of an ice cover. Such changes could, in turn, influence local air temperature, cloud cover, and the amount and distribution of precipitation which may impact farm and forest land.

Water Quality. Changes in the physical regime would affect important chemical processes such as the cycling and availability of inorganic nutrients essential for the growth of aquatic organisms.

Geological Characteristics. A tidal electric station would alter the current velocities and wave action with consequential effects on the pattern of shoreline erosion and sediment regime. Such changes not only have implications for the biological processes within their influence, but sediment transportation and deposition patterns have an important bearing on the design and operational aspects of a tidal electric generating facility. The combined action of tidal currents and waves is largely responsible for determining sediment features. River flows into a tidal basin will also be important where they are of sufficient magnitude to influence the regime resulting from the tides and waves. In summary, the operation of a tidal electric development will produce the following changes in the physical characteristics of an estuary with their consequential impacts on society and the environment:

Causes—changes in water levels and water flow, velocity and sedimentation patterns

Impact on society—agriculture, land drainage and flooding, water quality, opportunity for estuary crossing, employment, industry, electrical energy with no atmospheric pollution, amenity, recreational opportunities, ports and navigation, and shore protection

Impact on the environment—migratory birds and pelagic fish, aquatic mammals, marginal wetlands, and the estuarine ecosystem balance.

The foregoing potential impacts are not listed in order of importance since they would have different values and influences in different estuaries. Most of the potential impacts arise from the altered tidal regime, ie, the range and pattern of water level fluctuations and properties of tidal currents, and would, therefore, be felt in relatively close proximity to the barrage. However, depending on the scale of development and its position relative to the resonant characteristics of the estuary, the tidal regime on coast lines relatively remote from the development could be affected. For example, although the operation of a tidal plant at the mouth of Cumberland Basin (see Fig. 15) would have no measurable effect on the tide levels at Boston, Massachusetts, about 650 km south along the Atlantic coast, the larger development in Minas Basin (Site B9) could increase the average tidal range there by about 13 cm (49). This would not change the mean tide level. The global-mean sea level is expected to increase by about 18 cm during the period 1985–2030 under the "business-as-usual" scenario (50).

Ecological Effects

The physical phenomena of tidal regime, water circulation, velocities, turbidity levels, and mud flats will be dominant in an estuary with potential for energy exploitation, and quantification of these processes is necessary to understand the biological processes since the two are closely linked within an estuary. The distribution, abundance, and productivity of living organisms will be affected by the foregoing physical changes. Marine ecosys-

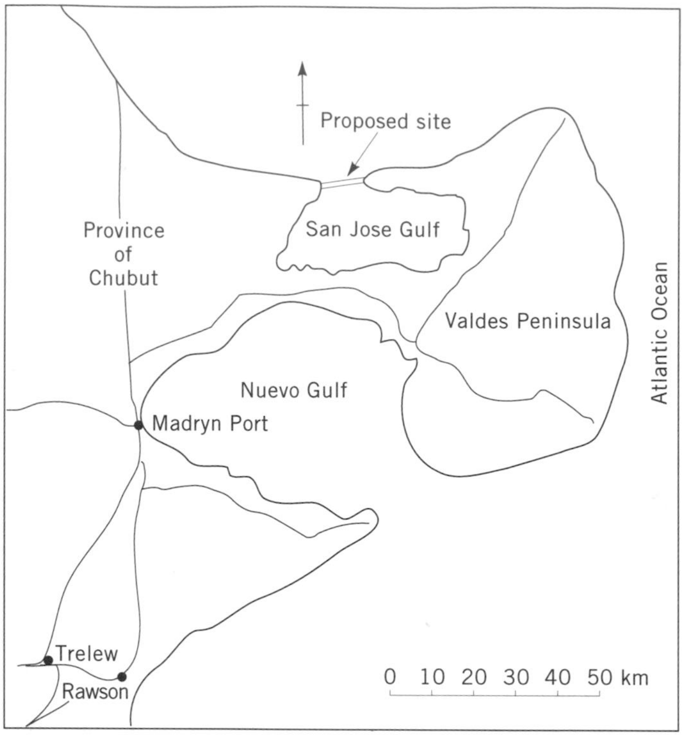

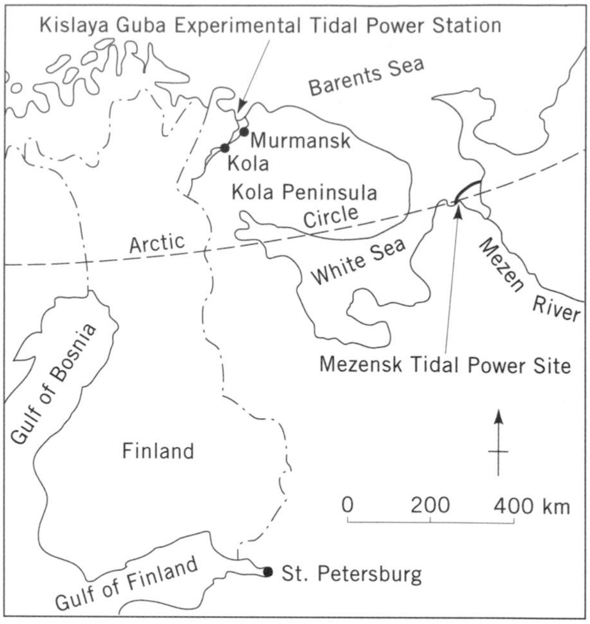

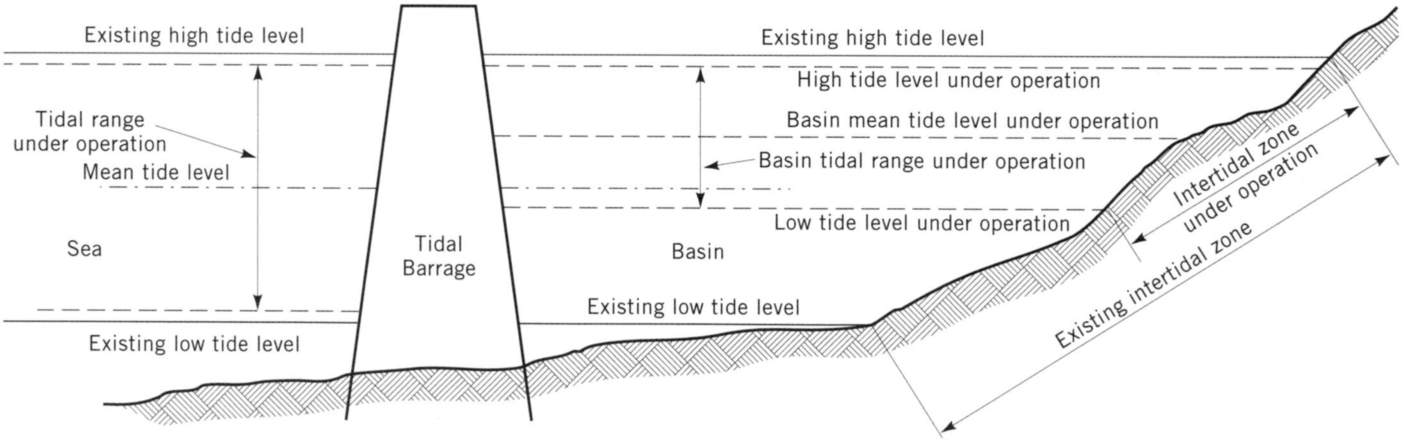

Figure 18. Relative position of tidal range in the basin for ebb-generation mode.

tems in estuaries with high tidal ranges suitable for energy exploitation will be highly stressed, but, having adjusted to the physical regime, will be productive. A high-stress environment is one in which there is significant variation in the parameters, such as salinities and temperatures, essential to the support of life forms living in the water. For example, in an estuary that receives large seasonal freshwater flows, the salt water in the estuary will be forced seaward during those seasons. High suspended sediment concentrations not only lowers the quality of water for growth of zooplankton but also severely limits the penetration of light necessary for primary production of phytoplankton, the basis of the estuarine food chain. The result would be that planktonic and benthic environments would be characterized by low biological productivity.

The intertidal regions are additionally stressed by large tidal ranges, waves, and alternating currents which continuously inundate and drain the land at the water's edge and scour, erode, and deposit the transitional sediments. Mud flats, visible at low tide but submerged at high tide, can be very productive. Reducing these areas in an estuary, which would result with the operation of a tidal station at its mouth, is likely to reduce the overall productivity of the intertidal zone. The dominant species of organisms might also be changed as a result of silting and changes in turbidity. Reduction of the productivity and nature of the fauna of intertidal flats could have a substantial influence on bird migrations. For example, the upper reaches of the Bay of Fundy are important fueling areas for the southerly migration of the birds passing along the east coast of North America.

On the other hand, there may not be a reduction in productivity. For example, a study carried out on the Severn estuary (51) suggested that there would be an increase in productivity of invertebrates so that, despite a reduction in the intertidal area that would result from the proposed Severn tidal power barrage, the population of most species would be maintained. Under current conditions, production of this life in the highly turbid waters is almost certainly light limited, but the greater water clarity expected after barrage construction, because of reduced tidal resuspension, could increase planktonic production and offset the reduction anticipated by the smaller intertidal area. New dominant organisms would likely develop under the changed regimes. Experience has shown that shore birds are seldom food limited and will crowd together on mud flats in response to a loss of habitat and show considerable flexibility in habitat selection.

Fishery concerns may be expressed in three areas: the passages of fish from sea to basin to sea; fish production in the tidal basin; and effects on fisheries seaward of the development (52). A tidal electric plant would create a barrier to the normal passage of fish in either direction past the plant location. Those fish species frequenting the vicinity of a tidal plant as part of international stocks which frequent the vicinity of a plant, would be likely to make multiple passes through the barrage. For example, shad, while feeding in the upper estuaries of the Bay of Fundy, are observed to drift in and out with the tides. Therefore any given fish would probably pass one of the proposed barrage sites about 15 times during its seasonal visit (53). With regard to the inward migration of spawning fish, potential problems would arise as a result of possible delays, disorientation, and failure to reach spawning grounds.

Expected impacts on fish have, so far, been based on experience with conventional river hydro plants for which published mortality rates vary from 0 to 60%. For example, the environmental assessment for the Annapolis project estimated negligible mortality for fish of 0.5 m length (shad), rising to a 10% average mortality and a 25% maximum mortality for fish of 1 m length. However, results on fish mortality at the Annapolis Tidal Generating Station has indicated modality rates somewhat higher (54). This result is no doubt due to the fact that all conclusions about turbine fish mortality have been based on research involving salmonids, whereas the exploitation of tidal energy involves the exposure of a wide range of anadromous and catadromous species to the hazards of turbine passage. Pressure changes and shear during such passage may be a significant source of modality in tidal plants. Such plants, of necessity, operate at low heads and low efficiencies at the beginning and end of each operating cycle; shear potential increases as turbine efficiency decreases. The use of asynchronous machinery, as compared to the synchronous turbines used at the Annapolis and La Rance stations, could reduce mortality since higher efficiencies and smoother flow conditions would be maintainable at the lower heads.

In view of the multiple-passage problem, it will likely be necessary to divert fish from turbine passages to other routes past the barrage by means of either sluices or fishways. This may be best accomplished by behavioral devices, such as light, sound, etc., combined with attraction flows which have been successful at many river hydro plants. Further research and experiment will be necessary to develop an effective solution.

Summary. Table 1 (48) presents a matrix of the elements of the environment that may be affected by the construction of a tidal power station and its operation. Obviously, the various impacts and their causes are, to a large extent, interrelated. A major environmental field data collection and analysis program is necessary in order to define environmental impacts with some precision, although in this field there will always remain a number of uncertainties. There is, as yet, no simple mechanism for taking environmental factors into account when considering the real cost of electricity. The conclusions reached on environmental impact will almost inevitably be less certain than those reached on technical and economic feasibility. It may not be possible, nor indeed desirable, to quantify some impacts in economic terms.

With regard to environmental assessment, a tidal electric generating facility should be considered as only one prospective component of the future generation mix of the regional utility system. Thus, the assessment of environmental implications of a tidal project should be placed in perspective through comparison of system impacts, with and without a tidal electric generating facility. In so doing, judgments of nonquantifiable aspects, such as the reduction of acid-forming gases and other pollutants, may be approached on a comparative basis and lead to a better appreciation of the relative importance of potential impacts, both beneficial and adverse.

FUTURE PROSPECTS

The economics of exploiting tidal energy depend primarily on site-specific factors, such as tidal range, basin area, and the required barrage length and height. The civil works component of a tidal electric power project amounts to about 55–60% of the capital cost of a project, according to the results of the detailed studies for the Severn Estuary (6) and Bay of Fundy (5) projects. The electromechanical component, including the transmission facilities to shore, approximates 35% of the capital cost with the remainder allocated to feasibility and model studies, engineering and management, and miscellaneous costs. The

Table 1. Potential Interaction of a Tidal Power Development and Elements of the Environment

Construction Activities and Plant Operation	Topography/Bathymetry	Mineral Resources	Soil	Drainage and Surface Runoff	Gound Water	Tidal Regime	Fresh/Salt Water Interface	Physical Oceanography	Sedimentation/Erosion	Water Chemistry	Aquatic Ecosystems	Migrating Aquatic Species	Benthic Ecosystems	Intertidal Zone Productivity	Wetland Vegetation	Wetland Wildlife Habitat	Upland Vegetation	Upland Wildlife Habitat	Climate	Air Quality	Noise and Vibration	Ice Formation	Surface Water Hydrology
Construction activities																							
Site Development, Land-Based																							
Clearing, grading, surface excavation, building structures, material storage	•		•	•	•				•	•	•				•	•	•	•		•	•		•
Road rail spur construction	•		•	•	•				•	•							•	•		•	•		
Excavation for abuttments	•		•	•	•				•	•				•	•	•	•	•		•	•		
Material placement	•		•	•	•				•	•				•	•	•	•	•		•	•		
Operate land-based marine equipment	•		•	•	•			•			•		•	•	•	•	•	•			•	•	
Worker facilities and use	•		•	•	•					•						•	•	•					
Site Development, Marine																							
Pile driving	•		•					•	•		•	•	•	•	•	•		•			•	•	
Intertidal construction zone	•		•	•				•	•	•	•	•	•	•	•	•	•	•		•	•	•	
Dredging	•		•					•	•	•	•	•	•	•	•	•					•	•	
Mattress/dike placement	•		•					•	•	•	•	•	•	•	•	•					•		
Tug and barge operation	•									•	•	•								•	•	•	
Caisson storage and transport	•							•	•		•	•	•	•							•	•	
Caisson installation	•							•	•		•	•	•								•	•	
Stationary marine equipment	•							•	•	•	•	•	•	•							•	•	
Mechanical/electrical equipment installation																				•	•		
Site Accessibility																							
Road, rail transport of personnel materials or equipment			•														•	•		•	•		
Marine transport of personnel, materials or equipment								•		•	•			•	•	•					•	•	
Remote Construction Facilities																							
Construction material source areas	•	•	•	•	•												•	•		•	•		
Dredge disposal sites, upland	•		•	•	•												•	•		•	•		
Dredge disposal sites, marine	•							•	•	•	•	•	•	•									
Operation of Permanent Facilities																							
Access and closure dike (presence)	•			•	•	•	•	•	•	•	•	•	•	•	•	•	•	•				•	•
Physical estuary barrier	•				•	•	•	•	•	•	•	•	•	•	•	•			•			•	•
Powerhouse and sluiceway (presence)						•	•	•				•	•									•	
Turbine operation						•	•	•	•	•		•									•	•	•
Sluiceway operation	•					•	•	•	•	•		•									•	•	•
Power facilities (presence)	•			•													•	•				•	
Switchyard operation																					•		
Drydock and dock facilities (presence)	•			•					•	•	•		•	•	•	•						•	
Long-term operation										•	•			•		•			•		•	•	
Impoundment (presence)	•			•	•	•	•	•	•	•	•	•	•	•	•	•			•			•	•
Water level fluctuation	•		•	•	•	•	•	•	•	•			•	•	•	•			•			•	•
Locks (presence)													•									•	
Operation								•	•			•	•								•	•	
Site access (presence)	•			•										•	•	•	•	•					•
Road, rail spur use																	•	•		•	•		
Marine use								•		•	•			•	•	•				•	•	•	
Worker facilities (presence)	•			•								•					•	•					
Use of workers					•							•									•		

Column heading: Elements of the environment[a]

[a] • indicates potential for interaction betweeen environmental elements.

advances in marine construction techniques and in turbogenerator design over the past several decades have been instrumental in substantially reducing project costs and in improving energy output. Since the technology is capital intensive, the cost of electrical output will be very sensitive to the discount rate for the capital invested. However, real, nonenergy benefits to the region and the environment will arise from the construction of a barrage.

The technology for exploiting the energy of the tides is now relatively mature. However, there undoubtedly will be further improvements in both overall design and individual components although the gains in cost-effectiveness will likely be relatively small. As a capital-intensive development, one potential area for cost reductions could be in shortening the construction time which currently is in the order of 10 years or more, depending on the magnitude of the development. This could involve, for example, refinements of caisson design to provide for the installation of the overall prime-mover units, prior to the towing of the caisson to the site. The larger proposed projects would likely require the manufacturing capabilities of a number of countries to minimize construction time.

The determination of the economic feasibility of tidal power, ie, its competitiveness over the long term with conventional sources of energy used by electrical utility systems, involves the interaction of many factors under future conditions and trends which cannot be predicted with certainty. The primary objective, then, should be to identify the generation program providing the required service to the system at the least cost, ie, the objective of minimum cost with maximum efficiency and minimum wastage. Such an approach entails comparing the cost of the electric utility operation with and without a tidal power development and requires the optimization of utility generation expansion plans for the two cases and simulation of hour-by-hour operation. It also requires assumptions regarding future load growth and the shape of future load curves, construction costs, fuel costs, and inflation and interest rates.

Real interest rates are undoubtedly one of the two crucial parameters of competitiveness. For long lead-time, capital intensive energy projects, the risks of adverse changes in future inflation, interest rates, or fuel costs is a serious deterrent to embarking on the construction of a tidal electric station. A real interest rate of 5%, a level now being achieved, would likely make economically viable several of the projects that have been investigated in some detail and a few would undoubtedly become competitive with fossil-fired alternatives. The second crucial parameter relates to the extent to which social costs of power generation from hydrocarbon fuels are internalized. Social and environmental costs are significant and must be fully exposed in any choice between alternatives. This means that the full social and environmental costs of conventional alternatives must be internalized, a principle inherent in the resolutions of the 1992 Earth Summit.

The magnitude of some potential schemes and their high capital costs and long lead times present a related problem and will undoubtedly require the active support and intervention of governments to proceed. However, once completed, a tidal power project represents an insurance against the inflation of fuel prices. In this regard, the useful life of the project will be about three times longer than fossil-fueled generating stations. This aspect is undervalued by present methods of analysis, in part, because of unduly high interest rates.

In summary, the tidal energy source is completely predictable, inflation proof, and has no atmospheric pollution. With the return to real interest rates in the 4–5% range, the near future will likely see the installation of several, commercial tidal electric plants. In fact, the World Energy Conference (55) estimates that the total annual energy output from tidal electric plants by the year 2020 will be 81 TWh, provided all costs are fully internalized, or about half that amount under a scenario based on present economic conditions and treatment of social costs.

BIBLIOGRAPHY

1. R. Gibrat, *L'Energie des Marées,* Presses Universitaires de France, Paris, 1966.
2. M. Southgate, *The Old Tide Mill at Eling,* Eling Tide Mill Trust Limited.
3. R. H. Charlier, *Tidal Energy,* Van Nostrand Reinhold, New York, 1982.
4. R. G. Dean, "Tide and Harmonic Analysis," in A. T. Ippen, ed., *Estuary and Coastline Hydrodynamics,* Engineering Societies' Monographs, McGraw-Hill, New York, 1966, Chapt. 4.
5. *Reassessment of Fundy Tidal Power,* reports of Bay of Fundy Tidal Power Review Board and Management Committee, Ottawa, Canada, Nov. 1977.
6. *The Severn Barrage Project,* general report by the Severn Tidal Power Group, Energy Paper No. 57, Department of Energy, London, 1989.
7. V. Argintaru and D. H. Willis, "Simulation of the Tidal Propagation in the Bay of Fundy Using a Hybrid Model, Xth Congress," *International Association for Mathematics and Computer Simulation,* Montreal, Canada, 1982.
8. *Tidal Power from the Severn Estuary,* report by the Severn Barrage Committee, Energy Paper No. 46, Department of Energy HMSO, 1981.
9. N. Birkett, B. M. Count, and N. K. Nichols, "Optimal Control Problems in Tidal Power," *Water Power and Dam Construction,* 37–42 (Jan. 1984).
10. S. T. Y. Lee and C. Deschamps, "Mathematical Model for Economic Evaluation of Tidal Power in the Bay of Fundy," *IEEE,* Summer Power Meeting, Mexico City, 1977.
11. *Proceedings of the International Conference on Pumped Storage Development and Its Environmental Effects, University of Wisconsin, Sept. 19–24, 1971,* American Water Resources Association, Urbana, Ill., 1971.
12. J. L. Haydock and J. G. Warnock, "Compressed Air Energy Storage and Tidal Power," *Proceedings of the Conference on New Approaches to Tidal Power,* Bedford Institute of Oceanography, Halifax, Nova Scotia, June 1982.
13. *Assessment of Retiming Tidal Power from Bay of Fundy Using Compressed Air Energy Storage,* SSC File No. 48SZ23283-7-613, Energy Mines and Resources of Canada, Ottawa, May 1988.
14. Y. Sevenard, "Leningrad Flood Protection," paper presented to the *Conference of the International Association for Bridge and Structural Engineering,* Nyborg, Denmark, 1991.

15. L. B. Bernshtein, "Kislaya Guba Experimental Tidal Power Plant and Problem of the Use of Tidal Energy," in T. J. Gray and O. K. Gashus, eds., *Tidal Power,* Plenum Press, New York, 1972.
16. G. A. Makela, "Float-in Powerhouses," *J. Energy Eng.* ***109***(2), 43–59 (June 1983).
17. B. Hadley and L. E. Lindestrom, *The Sidney A. Murray Jr. Hydro Power Project,* Institute of Mechanical Engineering seminar, Apr. 1990.
18. J. Cotillon, "La Rance Tidal Power Station, Review and Comments," *Proceedings of the Colston Symposium 30, Tidal Power and Estuary Management, Apr. 1978, University of Bristol,* Scientechnica, 1979, pp. 49–66.
19. M. Braikevitch, "Straightflow Turbine," in T. J. Gray and O. K. Gashus, eds., *Tidal Power,* Plenum Press, New York, 1972.
20. R. G. Rice and G. C. Baker, "Annapolis: The Straflo Turbine and Other Operating Experiences," *Proceedings of the Conferrence on Tidal Power,* London, Mar. 19–20, 1992, Institute of Civil Engineers, London, 1992.
21. J-P. Frau, *Usine Marémotrice de la Rance: Quel impacts sur son environnement après 23 ans de functionnement,* Congrès de L'Union des Oceanographes de France, Paris Nov. 20–23, 1989.
22. P. Hillairet, "Vingt ans après: La Rance, une expériencee marémotrice," *La Houille Blanche,* No. 8, 1984.
23. R. P. DeLory, "The Annapolis Tidal Generating Station," *Water for Energy,* 3rd International Symposium on Wave, Tidal, OTEC and Small Scale Hydro Energy, Brighton, England, May 14–16, 1986.
24. "Jiangxia Power Station," *SHP News,* newsletter of the Asia–Pacific Regional Network for Small Hydro Power, No. 2, 1985; issued by the Hangzhou Centre for Small Hydro Power.
25. J. N. Usachov, "Studies at Kislogubskaya TPP," *Proceedings of the International Symposium on Tidal Power Stations, Murmansk, Sept. 18–23, 1991,* Hydroproject Institute, Ministry of Fuel and Energy, Russian Federation, Moscow, 1992, pp. 233–236.
26. L. B. Bernshtein, *Tidal Energy for Electric Power Plants,* translated from 1961 Russian text by I. Program for Scientific Translation, Jerusalem, 1965.
27. G. Aisiks, "Tidal Power in Argentina and Its Potential Development," in Ref. 25, pp. 73–86.
28. P. Ackers, "Tidal Power Projects in Australia," in Ref. 18, pp. 108–114.
29. E. T. Haws, N. Reilly, and P. Wood, "Prospective Tidal Power Projects in the Kimberley Region of Western Australia," in Ref. 20.
30. M. L. de Carvalho, *Tidal Power Generation Potential of the Maranhao-Para-Amapa Coast,* internal report of Centrais Electricas Brazileiras S.A.—Eletrobras, Rio de Janeiro, June 1982.
31. R. H. Clark, *Tidal Power and Canada—A Review,* report to Department of Energy, Mines and Resources, Ottawa, Canada, 1987.
32. *Report on Tidal Power, Petitcodiac and Memramcook Estuaries,* Department of Mines and Resources, Ottawa, Canada, 1945.
33. *Feasibility of Tidal Power Development in the Bay of Fundy,* reports of the Atlantic Tidal Power Programming Board and the Engineering and Management Committee, Ottawa, Canada, Oct. 1969, 6 vols.
34. *Fundy Tidal Power, Update '82,* Tidal Power Corporation, Halifax, Nova Scotia, 1982.
35. *Tidal Power in British Columbia,* British Columbia Hydro and Power Authority, Vancouver, June 1979.
36. G. Godin, "The Power Potential of Ungava Bay and its Hinterland," *Water Power,* **26,** 167–171 (1974).
37. International Joint Commission, *Report on Investigations of the International Passamaquoddy Tidal Power Project,* Docket 72, Washington, D.C., and Ottawa, Canada, Apr. 1961.
38. W. W. Wayne, *Final Report on Tidal Power Study for the United States Energy Research and Development Administration,* Contract No. E(49-18)-2293 with Stone and Webster Engineering Corp., Boston, 1977.
39. R. Bonnefille, "Historical Sketch of the French Chausey Islands Tidal Power Project Plant," *Proceedings of the International Symposium on Korean Tidal Power,* Seoul, Korea, Nov. 14–15, 1978, pp. 27–41.
40. *Korea Tidal Power Study—1986,* report to Korea Electric Power Corp., Sept. 1986.
41. W-O. Song, *A Review of Korean Tidal Power Studies,* in Ref. 25.
42. H. R. Sharma, "Kachchh Tidal Power Project," *Proceedings of the Symposium on Water for Energy,* Brighton, England, May 14–16, 1986.
43. L. B. Bernshtein, "Soviet Tidal Power Projects (Mezen, Tugar, Kolskaya) and Their Realization," in Ref. 25.
44. *Tidal Power from the Severn,* report of the Severn Tidal Power Group, 1986.
45. B. I. Jones, C. D. I. Morgan, D. Phillips and M. W. Pinkey, "The Mersey Barrage—Civil Engineering Aspects," in Ref. 20.
46. M. E. Matthews and R. M. Young, "Feasibility of a Conwy Barrage," in Ref. 20.
47. M. E. Matthews and R. M. Young, "Tidal Energy from the Wyre," in Ref. 20.
48. *Preliminary Assessment of Cook Inlet Tidal Power, Phase 1 Report,* prepared for the Office of the Governor, State of Alaska, by Acres American Inc., Columbia, Md., Sept. 1981.
49. G. C. Baker, "Current Status of Tidal Power in the Bay of Fundy," in Ref. 20.
50. R. A. Warrick and J. Oerlemans, "Sea Level Rise," in *Scientific Assessment of Climatic Change,* report of the Intergovernmental Panel on Climate Change to World Meteorological Organization and United Nations Environmental Programme, Geneva, June 1990.
51. S. J. Muirhead, "The Environmental Effects of Tidal Energy," in Ref. 20.
52. M. J. Dadswell, R. A. Rulifson, and G. R. Daborn, "Potential Impact of Large-Scale Tidal Power Developments in the Upper Bay of Fundy on Fisheries Resources of the Northwest Atlantic," *Fisheries,* **11**(8), 26–35 (1986).
53. M. J. Dadswell, G. D. Melvin, P. J. Williams, and G. S. Brown, "Possible Impact of Large-Scale Tidal Power Developments in the Upper Bay of Fundy on Certain Migratory Fish Stocks of the Northwest Atlantic," in D. C. Gordon and M. J. Dadswell, eds., *Update on the Marine Environmental Consequences of Tidal Power Development in the Upper Reaches of the Bay of Fundy,* Canadian Technical Report of Fisheries and Aquatic Sciences, No. 1256, Bedford Institute of Oceanography, Dartmouth, Nova Scotia, 1984, pp. 577–600.
54. G. C. Baker, *Environmental Impacts of Large-Scale Tidal Power Development in the Bay of Fundy,* Tidal Power Corp., Halifax, Nova Scotia, May 1987, 29 pp.
55. *Renewable Energy Resources, Opportunities and Constraints, 1990–2020,* World Energy Conference, London, U.K., 1992.

TRANSPORTATION FUELS—AUTOMOTIVE GASOLINE

Lewis M. Gibbs
Chevron Research and Technology Co.
Richmond, California

Gasoline is defined as "a volatile, highly flammable, colorless liquid mixture of hydrocarbons produced by the fractional distillation of petroleum and used chiefly as a fuel in internal combustion engines" (1). An expanded definition, based on state petroleum inspection laws in the United States (U.S.), is "a volatile mixture of flammable, liquid hydrocarbons derived from petroleum, natural gas, shale oil, or coal, boiling within the range 25 to 225°C (77 to 437°F) containing small amounts of additives, which is suitable for use as a fuel in spark-ignition, internal combustion engines" (2).

Composition

As indicated above, gasoline is not a single chemical compound like water. It is a mixture of compounds called hydrocarbons, which are composed of the elements carbon and hydrogen. As many as 3000 individual hydrocarbons could be in gasoline, but only 200 to 300 hydrocarbons can be readily identified through analysis of any blend of gasoline. The hydrocarbons found in gasoline belong to one of four principal groups: paraffins, olefins, naphthenes, or aromatics. The differences between the groups lie in the ratio of hydrogen to carbon atoms and in how the atoms are arranged. Gasoline contains hydrocarbons ranging from butane (a paraffin with only four carbon atoms in its structure) to methyl naphthalene (an aromatic containing eleven carbon atoms). Oxygenated gasolines contain alcohols and/or ethers in addition to hydrocarbons. The characteristics of a particular fuel are strongly influenced by the types and relative amounts of hydrocarbons and oxygenates it contains. Refiners select the mix of components obtained from catalytic crackers, catalytic reformers, alkylation units, ether plants, and other processing units, to blend products with the required properties.

TYPES

There are several types of gasoline wherein the type descriptor is based on the end use of the fuel. The most widely used type of gasoline is automotive or motor gasoline. It is marketed for use in vehicles powered by spark-ignition internal combustion engines (automobiles, trucks, buses, motorcycles, and motor scooters). It is also used in motorboats (both four-stroke cycle inboards and two-stroke cycle outboards) and small nonautomotive power equipment (lawn mowers, garden tools, chain saws, and snowmobiles). Another type is aviation gasoline, which is marketed for use in light aircraft. Aviation gasoline is also used in some commercial aircraft powered by spark-ignition engines. Racing gasoline is a type of gasoline that is used in high speed race cars, off-road vehicles, and boats.

Leaded and Unleaded Gasoline. From 1926 until recently, nearly all automotive gasoline in the U.S. has contained a lead (Pb) antiknock additive (3). Until 1970, the principal exception was a premium unleaded gasoline sold by American Oil Company (now Amoco Oil Company) on the East Coast and in the Southeast of the U.S. since 1915. Beginning in 1970, unleaded gasoline was introduced throughout the U.S. and by 1981 the sales of unleaded gasoline surpassed those of leaded gasoline. As of 1994, leaded gasoline is still available in the U.S., but its sales are dropping off rapidly because most cars built since 1975 and all light trucks built since 1979 must use unleaded gasoline. The phasedown of lead in leaded gasoline began in 1979. The average quarterly lead level as required by U.S. Environmental Protection Agency (EPA) regulations reached 0.1 g Pb/gal (0.03 g Pb/L) maximum in 1988 (4,5). The Clean Air Act Amendments of 1990 require that lead antiknock compounds be eliminated from reformulated gasoline in 1995 and from all U.S. highway gasoline by 1996 (6).

Oxygenated Gasoline. Beginning in 1978, a new class of automotive fuel was widely marketed in the U.S. The fuel was a blend of 10 vol % ethanol (ethyl alcohol) in gasoline (7). This ethanol blend was popularly called "gasohol," and also was known as gasoline–ethanol blend. Ethanol is classified as an oxygenate. Oxygenates are defined as oxygen-containing, ashless compounds, such as alcohols and ethers. As will be discussed later, gasoline–oxygenate blends, or oxygenated gasoline, were required beginning in 1992 in 39 specified areas of the U.S. during the wintertime. Further, in 1995 oxygenates will be required in a modified fuel called "reformulated gasoline."

Although the American Society for Testing and Materials (ASTM) uses the term "automotive spark-ignition engine fuel" to describe both gasoline and gasoline–oxygenate blend, the term "gasoline" will be used throughout this article. When the performance is different for oxygenated gasoline and hydrocarbon gasoline, the specific oxygenated gasoline composition (gasoline–alcohol blend, gasoline–ethanol blend, gasoline–ether blend, etc) will be indicated.

History

Gasoline, which was considered a worthless by-product of petroleum, entered commercial usage no later than 1863. It is difficult to establish who discovered it, but Joshua Merrill may have isolated gasoline in Boston as a result of his efforts to further refine kerosene, the primary petroleum product at that time (8). Gasoline was first used in air–gas machines to produce fuel that could be piped and burned in gaslights to illuminate mills and factories. Gasoline was the fuel used in 1876 in the first four-stroke cycle engine built by Nicolaus Otto in Germany. As the demand increased for gasoline during the years 1900 to 1920, gasoline ceased to be a by-product and the lower boiling (more volatile) portion of the kerosene fraction was diverted to gasoline in order to meet the requirements. As demand increased, a process called thermal cracking was developed in 1913 to increase the amount of gasoline. To increase octane quality, thermal reforming was introduced in the 1930s. Then in 1936, catalytic cracking was developed, followed by catalytic reforming in 1940. Hydrocracking (1959) is similar to catalytic cracking, but it takes place in a hydrogen atmosphere. Isomerization (1943) is a process which upgrades octane quality by forming isomers from normal (straight-chain) paraffins. Alkylation (late 1938) and polymerization combine small, gaseous hydrocarbons to form larger, high octane, liquid hydrocarbons. Coking (1930s) can produce gasoline by pyrolysis of heavy, low value petroleum fractions. As a result of these processing developments, gasoline today is the principal, high quality product of U.S. crude oil refining (9–11) (see REFINING for more processing details).

Properties

Gasoline is blended to provide good performance in motor vehicles under a variety of ambient operating conditions. Good fuel performance means that the vehicle starts easily, warms up rapidly, runs smoothly, does not stall under hot or cold conditions, accelerates without surge or hesitation, achieves good fuel economy, does not knock, and forms minimal engine deposits. In addition to performance factors, the impact of fuel properties on exhaust and evaporative emissions has become an important consideration. The following sections discuss the significance of the important physical and chemical properties of gasoline.

Antiknock Quality. Combustion in a spark-ignition internal combustion engine under normal conditions is initiated by a spark. The flame front fans out from the spark plug and travels across the combustion chamber rapidly and smoothly until the fuel–air mixture is consumed and the piston is driven down the cylinder by expanding gases. If the antiknock rating is insufficient, the last portion of the unburned mixture ahead of the flame front will ignite spontaneously (autoignition) and then burn very rapidly. This abnormal combustion causes the pressure in the cylinder to rise rapidly, resulting in a characteristic knocking or pinging sound.

Octane Scale. Antiknock, or octane rating, is a measurement of a gasoline's resistance to knock (autoignite) and is reported in terms of octane numbers. The octane number scale is an arbitrary one which was established in 1929. On this scale, two pure paraffin hydrocarbons were selected as primary references. One is isooctane (2,2,4-trimethylpentane), which has a high resistance to knock. It is assigned the value of 100. The other one, with a low resistance to knock, is normal heptane. It is assigned a value of zero. The octane number of the reference fuel is defined as the volume percentage of isooctane in the blend of isooctane and normal heptane. For example, a blend of 90 vol % isooctane and 10 vol % normal heptane by definition has an octane number of 90. To produce reference fuels with octane numbers above 100, the lead antiknock compound, tetraethyllead, is added to pure isooctane according to a specified formula (12).

Octane Measurement. Two standard laboratory engine test methods are used to determine octane ratings of gasoline. They are ASTM D 2699, Test Method for Knock Characteristics of Motor Fuels by the Research Method, and ASTM D 2700, Test Method for Knock Characteristics of Motor and Aviation Fuels by the Motor Method (12). Both test methods employ a standard single-cylinder variable compression ratio test engine. The Motor method engine operates at a higher speed and higher fuel-air mixture temperature than does the Research method. To determine an octane number of an unknown gasoline sample, the compression ratio and fuel–air ratio are varied to produce a standard knock intensity. A primary reference fuel blend (mixture of isooctane and normal heptane) is then found which matches the knock intensity of the sample under the same compression ratio conditions. The octane number of the reference fuel blend is the octane number of the sample. The higher the octane number, the greater is the fuel's resistance to knock. If the test is conducted in the Research method engine, it is reported as the Research octane number. Since the Motor method conditions are more severe, the Motor octane number of commercial gasoline is lower than the Research octane number. The arithmetic difference between Research and Motor octane numbers is called "sensitivity."

Antiknock Index. Both Research and Motor octane number are commonly reported to define the antiknock performance of a gasoline. Antiknock Index is defined as the average of the Research and Motor octane ratings of a fuel, or (Research + Motor)/2 which is commonly referred to as (R + M)/2. In the U.S., federal regulations require the posting of the Antiknock Index on the fuel dispensing pump. This index is the currently accepted method of relating Research and Motor octane numbers to actual antiknock performance in vehicles. However, the exact relationship depends on vehicle design, transmission type, and operating conditions.

Road Octane Number. The octane numbers determined in single-cylinder laboratory engines cannot completely predict the performance of a fuel in multicylinder engine vehicle service. Road octane number is determined in a fleet of 10 to 15 vehicles using the Coordinating Research Council (CRC) Modified Uniontown Procedure F-28 (13). The Road octane number of a fuel is defined as the octane number of a primary reference fuel blend that produces the same knock intensity as the test fuel while operating under the same spark timing and acceleration conditions. The Road octane number of a commercial gasoline lies between the Research and Motor octane numbers of that fuel. The difference between Research octane number and Road octane number is called "depreciation." The Road octane number of a fuel will vary depending on the group of vehicles in which it is measured.

Octane Number Requirement. The octane number requirement (ONR) of an engine is defined as the octane number of a reference fuel that will produce trace knock under the most severe speed and load conditions on a level road or chassis dynamometer. Trace knock is the knock intensity that is just audible to the trained technician. The most severe conditions may be under full-throttle or part-throttle operation. The reference fuels used to determine ONR may be either the primary reference fuel blends used in the Research and Motor methods, or a full-boiling series developed from commercial gasoline components. ONR is determined using the CRC E–15, Technique for Determination of Octane Number Requirement of Light Duty Vehicles (14).

The ONR of an engine changes with its operating conditions (15). It increases with decreasing absolute humidity and with increasing temperature of the ambient air and/or engine coolant. ONR decreases as altitude increases (barometric pressure decreases). However, the effects of altitude are much smaller in late model cars equipped with barometric compensation for spark timing and fuel-air ratio. Advancing either the basic or automatic spark advance system increases ONR. Improperly functioning ignition or emission control systems can cause either an increase or decrease in the ONR.

Octane Requirement Increase. The build-up of carbonaceous deposits in the engine's combustion chamber can also increase the ONR. This increase is called octane requirement increase (ORI). The build-up of combustion chamber deposits in a new engine rapidly increases ONR over the first few thousand miles of service and then approaches an equilibrium level after 10,000 to 15,000 miles (16,000 to 24,000 km). ORI can range from 1 to 13 numbers with the average effect being around 5 numbers. The equilibrium ONR is affected by engine design, mileage accumulation conditions, fuel composition, and engine lubricant composition and consumption. Although these deposits raise compression ratio, the more important effect is that they insulate the combustion gases from the engine cooling system, trap heat, and transfer it to the incoming fuel–air mixture. This increases combustion chamber temperatures and promotes knock (10,15).

Vehicle Performance. If the antiknock rating of the fuel is below the antiknock requirement of the engine, knock occurs under these operating conditions. Knock that is just audible may irritate the driver, but rarely causes any engine damage. Loud knocking over extended periods of time may cause a loss of power, overheating of the engine, and damage to pistons or other engine parts.

Using fuels with antiknock ratings higher than that required to prevent knock does not improve vehicle performance. However, in vehicles equipped with knock limiters that retard the spark timing when knock is detected, a fuel with a higher antiknock rating may improve performance. This is especially true in turbocharged vehicles (16).

Although Antiknock Index is widely used to correlate with vehicle performance, the performance of some individual vehicles is best predicted by Research octane number, whereas the performance of others may best be predicted by Motor octane number. The importance of Research and Motor octane number differs, from model to model and from year to year and reflects changes in engine and transmission design. In general, Research octane number correlates best with low speed, mild knocking conditions; Motor octane number correlates best with high speed and high temperature road conditions and with part-throttle operation. Research octane number also correlates best with a condition called "run-on," "after-running,'" or "dieseling."

Run-on, after-running, or dieseling is a phenomenon during which the engine continues to run after the ignition switch is turned off. This condition usually can be corrected by using a fuel with a higher Research octane number or by removing combustion chamber deposits. Another important factor that affects run-on is high idle speed. This is why many carbureted engines are equipped with an idle stop solenoid.

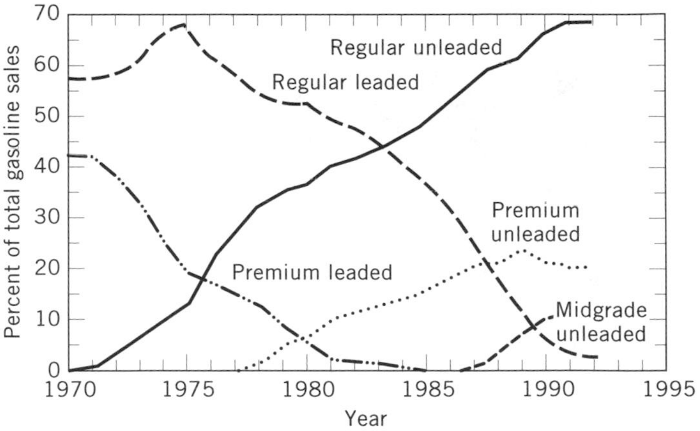

Figure 1. U.S. market share of gasoline grades: national average.

Grades of Gasoline. Gasolines are sold by octane grades in the U.S. Most areas offer three grades of unleaded gasoline. They are regular, midgrade, and premium. At lower altitudes, the Antiknock Indexes of the three grades are nominally 87, 89, and 92, or 93 (depending on the area). At high altitude [large areas above 762 meters (2,500 feet)], the Antiknock Index is nominally 86, 87.5, and 91 for the same three grades. There still are a few areas in the U.S. where a regular leaded gasoline (nominally 88.5 Antiknock Index) is marketed instead of the midgrade unleaded gasoline. In 1992 California banned the sale of leaded gasoline. Federal regulations will ban the sale of leaded gasoline in 1995 for reformulated gasoline areas and for all areas of the U.S. in 1996 (see Clean Air Act Section).

Figure 1 shows the changes in the sales of the various grades of gasoline with time. Figure 1 indicates when the three unleaded grades were introduced in the U.S. and the phaseout of the leaded grades.

Antiknock Index Trends. Figure 2 shows the history of average U.S. gasoline Antiknock Index for the various leaded and unleaded grades (17–20). Antiknock Index increased rapidly as lead usage increased concurrently until World War II when a brief downward dip in both occurred. After the war, the so-called "octane race" occurred whereby increases in Antiknock Index and increases in engine compression ratio leapfrogged each other. By 1965, the Antiknock Indexes of both leaded grades had stabilized. The Antiknock Index actually dropped during the 1973 oil embargo and the 1979 energy crisis.

Leaded premium gasoline essentially disappeared in 1981 and the unleaded premium that replaced it has shown an 1.2 number increase in octane level from 1981 to 1994. The Antiknock Index of unleaded regular gasoline has decreased about 0.1 number from 1981 to 1994, as shown in Figure 2. Since midgrade unleaded gasoline was widely introduced in mid-1988, its Antiknock Index has remained essentially constant.

Volatility. Gasoline must vaporize in order to burn. Therefore volatility, or the tendency to vaporize, is one of the more important physical properties. Gasoline must vaporize more easily in cold weather so that a vehicle will

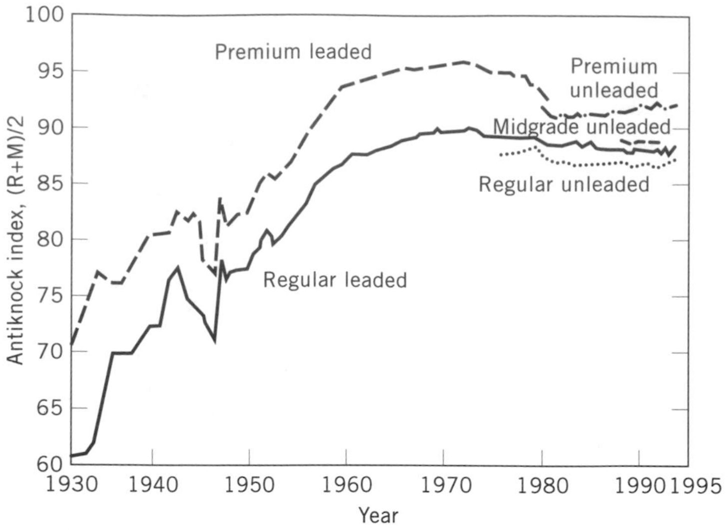

Figure 2. U.S. gasoline national average antiknock trends (17,20). Courtesy of the Society of Automotive Engineers.

start easily and run smoothly until it is warmed up and not cause dilution of the engine oil by unburned fuel. In warm weather, gasoline must be less volatile to prevent the formation of too much vapor in pumps, fuel lines, carburetors, or fuel injectors, which can result in loss of power or stalling due to vapor lock. Under certain atmospheric conditions, fuels that vaporize too easily can cause icing in the carburetor which can result in rough idling and stalling. This problem has mostly been eliminated with the introduction of port fuel injection systems and intake air heating for emission control. Gasoline volatility is changed seasonally to provide good cold-start and warmup driveability and warmed up and hot-start driveability at the anticipated minimum and maximum ambient temperatures and altitude for each locality. Driveability is an assessment of a vehicle's actual performance relative to its anticipated response to the driver's movement of the accelerator pedal. Terms used to describe driveability malfunctions are stall at idle or while driving, backfiring, rough idling, stumble, surge, hesitation, and starting time. Volatile gasolines also emit more evaporative, refueling, and running loss emissions than less volatile fuels.

Gasoline volatility is primarily defined by boiling range (distillation curve), vapor pressure, and vapor-liquid ratio. Some fuel producers use indexes, such as Driveability Index (DI) and Front End Volatility Index (FEVI) to control volatility.

Distillation. Since gasoline is a blend of a variety of hydrocarbons, each having a different boiling point, it has a boiling range rather than the single boiling point that a pure compound would have. This temperature range is measured by ASTM D 86, Test Method for Distillation of Petroleum Products (21), which determines the temperatures at which given percentages of the gasoline sample are evaporated under the conditions specified by the test. These temperatures have been correlated with engine performance and are one of the bases for controlling the volatility characteristics of gasolines during blending.

Figure 3 is a stylized ASTM D 86 distillation curve which shows the specific vehicle performance factors affected. A gasoline distillation curve is divided into three parts: front-end volatility, mid-range volatility, and tail-end volatility. Each of these divisions of the distillation curve affects certain performance characteristics (10,15,22).

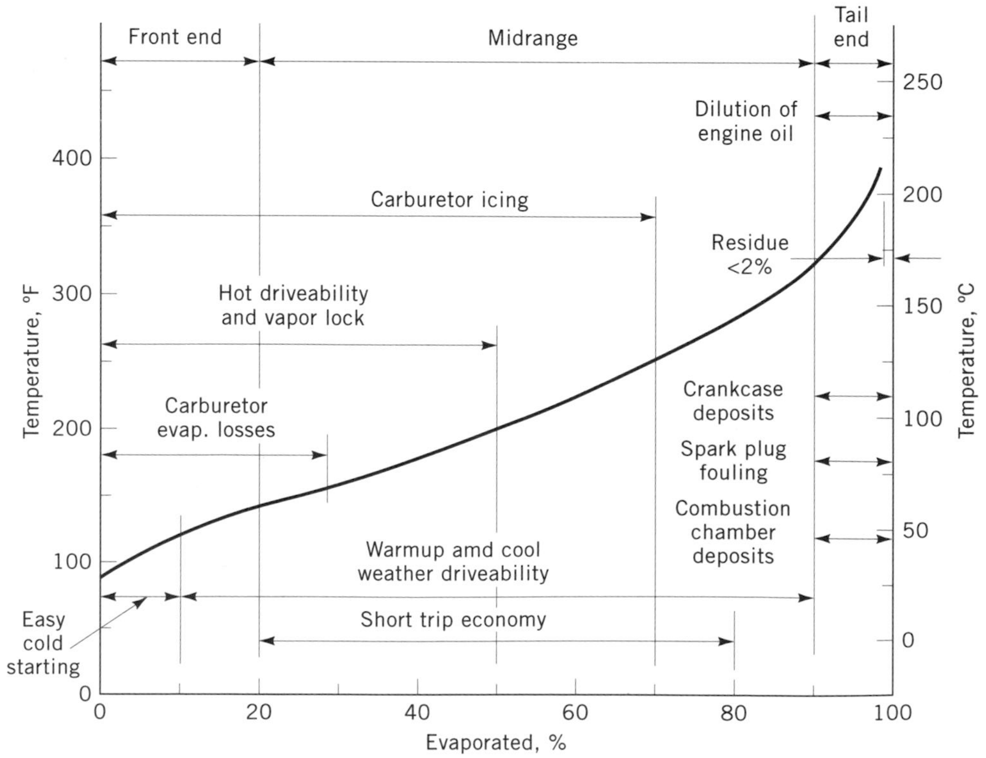

Figure 3. Significance of ASTM D 86 distillation curve. Courtesy of Chevron Research and Technology Company.

Front-end volatility must be balanced to provide (1) easy cold starting, (2) easy hot starting, (3) freedom from vapor lock, and (4) low evaporative and running loss emissions.

Mid-range volatility must be adjusted to give (1) rapid warmup and smooth running, (2) good short-trip fuel economy, (3) good power and acceleration, and (4) protection against carburetor icing and hot stalling.

Tail-end volatility must be balanced to provide (1) good fuel economy after engine warmup, (2) negligible engine deposits, (3) minimal dilution of the oil by unburned fuel, and (4) minimal hydrocarbon exhaust emissions.

Driveability Index. Although each area of the distillation curve is important, the entire curve is what the engine has to distribute, vaporize, and burn. To predict cold-start and warmup driveability, a Driveability Index (DI) (23) has been developed using common distillation points as follows:

$$\text{DI} = 1.5 \times (10\% \text{ evaporated point}) + 3.0 \times (50\% \text{ evaporated point}) + 1.0 \times (90\% \text{ evaporated point})$$

Lower values of DI equate to better cold-start and warmup driveability. The 50% evaporated point has the greatest effect as indicated by its having the largest coefficient in the DI equation. Within the limits on each of the percent evaporated points provided in ASTM D 4814, Specification for Automotive Spark-Ignition Engine Fuel (24), a fuel blender can trade an increase in one of the percent evaporated values by reducing one or both of the other values according to the DI equation. The equation was developed using data obtained on hydrocarbon gasolines and has not been proven to be applicable to oxygenated gasoline. Work continues to improve the DI equation and to make it universally applicable.

Reducing DI has been shown to reduce exhaust hydrocarbon emissions (25). However, the weighting factors for 10%, 50%, and 90% evaporated points of the DI equation probably are not optimized for exhaust emissions predictions. Results from the Auto/Oil Air Quality Improvement Research Program (AQIRP) have indicated that the 90% evaporated point is most important (26–28).

Distillation Trends. Figure 4 presents the trends in distillation properties (10%, 50%, and 90% evaporated) for regular leaded gasoline from 1906 to 1981, and for unleaded regular gasoline from 1981 to 1994 (18–20,29). From the advent of the automobile as a principal consumer of gasoline in 1908, the 50% and 90% evaporated points initially increased rapidly as the more volatile portion of kerosene was added to the gasoline pool to meet the increasing demand. The importance of volatility and its effects on driveability was first acknowledged in 1929 with the general introduction of seasonal gasoline (29). As shown in Figure 4, there has mostly been a steady decrease in all three distillation parameters since that time. However, from 1981 to 1994, the 90% point of unleaded gasoline has shown a slight increase. There also has been a summertime rise in 10% and 50% evaporated points which corresponds to the implementation in 1989 of the EPA Phase I vapor pressure regulations which lowered vapor pressure for much of the nation. This regulation was implemented to reduce refueling, evaporative, and running losses of hydocarbons (see discussion in Clean Air Act Section). The effect on both 10% and 50% points of the EPA vapor pressure regulations is primarily caused by the removal of butane.

A review of premium unleaded gasoline data shows that the 10% evaporated point is at the same level and has a similar trend as those shown for unleaded regular in Figure 4. The 50% point is about 8°C (15°F) higher for

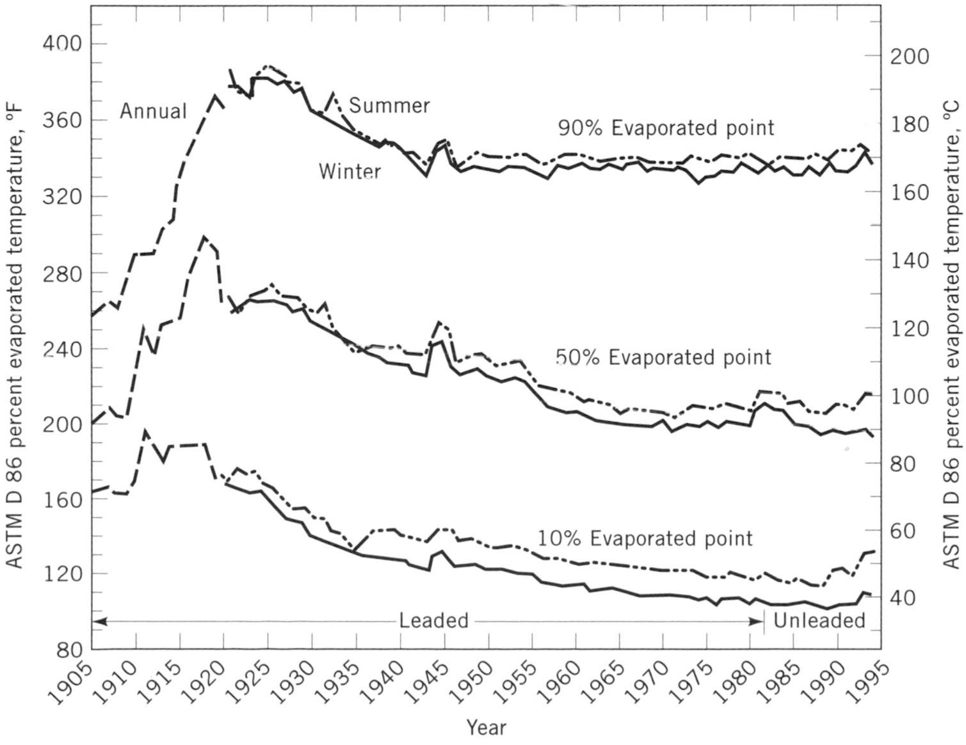

Figure 4. U.S. regular grades nation average distillation trends (19,20,29). Courtesy of the Society of Automotive Engineers, Inc.

premium and there is no downward trend. The 90% point of premium unleaded averages about 6°C (10°F) lower, but shows the same slightly upward trend as unleaded regular gasoline.

Since about 1978, oxygenates have been blended into gasoline. The National Institute for Petroleum and Energy Research (NIPER) gasoline survey reports, which were used to develop Figure 4, have included ether-containing fuels in the average, but have reported the results for alcohol-containing fuels separately. Including the gasoline–alcohol blend data in the data base to develop the average values would lower the 50% evaporated point less than 2°C (4°F) and have a negligible effect on the 10% and 90% points.

Vapor Pressure. One gasoline front-end volatility parameter is vapor pressure, which is the force per unit area exerted on the walls of a closed container by the vaporized liquid at a specified temperature. Historically, vapor pressure has been measured at 37.8°C (100°F) by ASTM D 323, Test Method for Vapor Pressure of Petroleum Products (Reid Method) (21). The results from this test are reported as Reid Vapor Pressure. Because trace amounts of water are present in the test method, it can only be used for gasoline and gasoline–ether blends. The result obtained for a gasoline–alcohol blend will be incorrectly low. To avoid this problem, a similar method using the same apparatus and a procedure modified to keep the fuel and apparatus dry was developed as ASTM D 4953, Test Method for Vapor Pressure of Gasoline and Gasoline-Oxygenate Blends (Dry Method) (24). Advances in instrumentation have resulted in two additional all-inclusive test methods; ASTM D 5190, Test Method for Vapor Pressure of Petroleum Products (Automatic Method) and ASTM D 5191, Test Method for Vapor Pressure of Petroleum Products (Mini Method) (24). The results are reported as vapor pressure without reference to temperature.

Driveability. The vapor pressure of a fuel must be sufficiently high to provide easy cold starting. However, if the vapor pressure is too high, vapor locking can occur. Vehicles equipped with carburetors and throttle body-fuel injection have shown poorer cold start driveability as vapor pressure was reduced (30). However, lowering the 50% evaporated point helped reduce the negative effects of lowering vapor pressure. This loss in driveability was not found in vehicles equipped with port fuel injected engines.

Evaporative Emissions. It has long been recognized that evaporative emissions increase with an increase in vapor pressure (31,32). Fuel tank evaporative emissions (diurnal or daily losses) correlate with vapor pressure while carburetor evaporative emissions correlate with the percent evaporated at 71°C (160°F) (32). For modern cars with evaporative emission control systems, vapor pressure is the fuel volatility parameter that correlates best with both diurnal and hot soak evaporative losses (33). In 1988, a comprehensive test program to investigate the effects of vapor pressure at varying ambient conditions confirmed that lowering vapor pressure reduces evaporative diurnal and hot soak emissions, and showed that the greatest effects occurred at the highest ambient test temperature of 27°C (80°F) (34). Reducing vapor pressure not only lowers vehicle evaporative emissions, but also lowers losses from fuel storage tanks and vehicle refueling. Vapor pressure controls are generally in effect during the summer months when ozone concentrations are the highest. In the cooler months, vehicle cold start and warmup driveability may be adversely affected if the vapor pressure is too low.

Exhaust Emissions. Lowering vapor pressure can affect exhaust emissions as well as evaporative emissions. At the 27° C (80°F) test condition, exhaust hydrocarbon emissions decrease with reductions in vapor pressure (34). However, at 12.8°C, (55°F) reducing vapor pressure increased exhaust hydrocarbon emissions, whereas at 1.7°C (35°F), changes in vapor pressure had no significant effect on exhaust hydrocarbon emissions. Changes in vapor pressure had no significant effect on carbon monoxide emissions at 1.7°C (35°F) and 12.8°C (55°F). However, lower vapor pressure did cause a reduction in carbon monoxide exhaust emissions at 26.7°C (80°F). Vapor pressure changes have shown no effect on oxides of nitrogen exhaust emissions (34). In another study, the Auto/Oil AQIRP showed that lowering vapor pressure reduced exhaust hydrocarbon and carbon monoxide emissions and had no effect on oxides of nitrogen emissions (26,35).

Safety. There is a safety concern associated with the delivery of low vapor pressure fuels by tank truck and storage of the fuel in vehicle fuel tanks at low ambient temperatures. Normally, gasoline vapor in the vapor space of fuel tanks is too rich to burn under typical ambient conditions. As vapor pressure is reduced, less vapor is formed, and the vapor space can fall within the flammability limits at ambient temperatures. This potential safety problem becomes most acute during the spring phase-in of low vapor pressure summertime gasolines because the ambient temperatures may still be low and the vapor pressure may be especially low in order to blend down existing high vapor pressure inventories. A further concern for fuel tanks is that as the vehicle is driven, fuel weathers (loses the more volatile components) to an even lower vapor pressure, thus aggravating the situation (36).

Vapor Pressure Trends. The average vapor pressures for regular grade gasolines in the U.S. (leaded 1942–1981, unleaded 1981–1994) are shown in Figure 5 (18–20). For both summer and winter gasolines there was a steady upward trend until 1989 when there was a sudden drop due to the EPA Phase I vapor pressure regulations. The slight reduction in summertime vapor pressure in 1971 might be a result of California regulations which limited vapor pressure to 62 kPa (9.0 psi) maximum. Beginning in 1992, California further limits vapor pressure to 53.8 kPa (7.8 psi) maximum. California accounts for about 11% of the U.S. gasoline market.

For 1990 and later surveys, including the data for gasoline–alcohol blends would increase the average vapor pressure by less than 0.7 kPa (0.1 psi).

Vapor–Liquid Ratio. Another parameter for expressing front-end volatility is vapor-to-liquid ratio (V/L) at a particular temperature. ASTM D 2533, Test Method for Vapor–Liquid Ratio of Spark-Ignition Engine Fuels (21), is a method for determining V/L across a range of temperatures. In this test method, either glycerin or mercury can

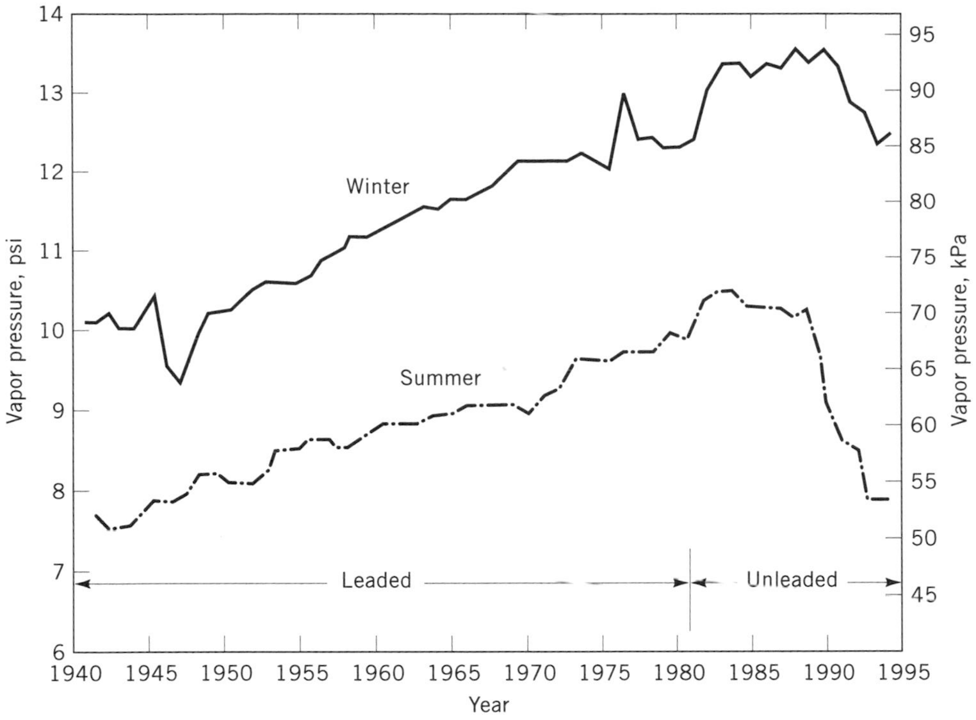

Figure 5. U.S. regular grades national average vapor pressure (18–20).

be used to confine the vapors. Glycerin can only be used for testing hydrocarbon gasoline because alcohols and ethers are soluble in glycerin. Mercury can be used for any fuel, but mercury requires special handling because of its toxicity. In 1992, a new V/L test procedure, ASTM D 5188, Test Method for Vapor–Liquid Ratio Temperature Determination of Fuels (Evacuated Chamber Method) (24), became available. This method does not use a confining fluid and thus is applicable to all fuels. ASTM D 4814, Specification for Automotive Spark-Ignition Engine Fuel (24), contains three methods for calculating the temperature for a V/L of 20. Although these calculation methods are stated as not applicable to gasoline–oxygenate blends, recent studies suggest that the calculation methods are acceptable for gasoline-ether blends.

Temperature for a Vapor–Liquid Ratio of 20. The temperature for a V/L of 20 is the best single property for predicting a gasoline's tendency to vapor lock (38). Vapor lock is the condition where the fuel supply is reduced or interrupted by the formation of excessive vapor in the fuel system (fuel pump, fuel line, carburetor, or fuel injector). Vapor locking causes a lean fuel–air ratio which may result in a loss of power, surging, backfiring, and engine stoppage with difficulty in restarting. A rule of thumb is that the hottest and thus most critical portion of the fuel supply system operates at about 17°C (30°F) higher than ambient temperature. Therefore, the minimum temperature for a V/L of 20 for a given marketing area is set at 17°C (30°F) above the highest anticipated 90th percentile maximum daily temperature. This correlation was developed for vehicles equipped with suction-type fuel pumps and carburetors. This control rationale may not be applicable to late-model pressurized fuel injector systems which are less prone to vapor lock (37,38).

Front-End Volatility Index. An alternative to using V/L is to use a parameter which is a combination of vapor pressure and some distillation characteristics. The Front-End Volatility Index (FEVI) is such a parameter. It is calculated from the following equation where vapor pressure is expressed in psi (37):

$$FEVI = \text{vapor pressure} + 0.13 \times (\text{percent evaporated at } 158°\text{F})$$

In Europe, a similar parameter, Vapor Lock Index (VLI), is used. It is calculated from the following equation where the vapor pressure is expressed in bar:

$$VLI = 14.5 \times (\text{vapor pressure}) + 0.16 \times (\text{percent evaporated at } 70°\text{C})$$

Seasonal Volatility Control. As previously discussed, gasoline volatility must be changed with the seasons (ambient temperatures) to provide good cold-start and warmup driveability and to prevent hot fuel handing problems under warmed up conditions. Further, volatility has to be controlled to prevent excessive evaporative and running loss emissions. To accomplish these goals in the U.S., ASTM D 4814, Specification for Automotive Spark-Ignition Engine Fuel (24), contains two tables. One table specifies vapor pressure and percent evaporated limits for six volatility classes (AA through E), as shown in Table 1. The second table (Table 2) specifies V/L requirements for five volatility classes (1 through 5). To specify a volatility requirement, a combination of a letter from Table 1 and a number from Table 2 is used (for example, C-3). An alphanumeric designation is specified for each month and for each state or portion of state based on anticipated

Table 1. ASTM D 4814 Vapor Pressure and Distillation Class Requirements[a]

Vapor Pressure/ Distillation Class	Vapor Pressure, max, psi (kPa)	Distillation Temperatures, °C(°F), at % Evaporated					Distillation Residue, vol %, max
		10 vol %, max	50 vol % min	50 vol % max	90 vol %, max	End Point, max	
AA	54 (7.8)	70 (158)	77 (170)	121 (250)	190 (374)	225 (437)	2
A	61 (9.0)	70 (158)	77 (170)	121 (250)	190 (374)	225 (437)	2
B	69 (10.0)	65 (149)	77 (170)	118 (245)	190 (374)	225 (437)	2
C	79 (11.5)	60 (140)	77 (170)	116 (240)	185 (365)	225 (437)	2
D	93 (13.5)	55 (131)	77 (170)	113 (235)	185 (365)	225 (437)	2
E	103 (15.0)	50 (122)	77 (170)	110 (230)	185 (365)	225 (437)	2

[a] Ref. 24. Courtesy of ASTM.

temperatures and the requirements of EPA Phase II volatility regulations.

Sulfur Content. Sulfur is a naturally occurring constituent of crude oil. During the refining of crude oil to produce gasoline, the level of sulfur is greatly reduced. One refining process, sweetening, converts mercaptans into disulfides to improve the stability of gasoline, to reduce its corrosive effects on some metallic fuel system components, and to improve its odor. Other sulfur compounds that can be present in gasoline are thiophenes, benzothio-Phenes, polysulfides, sulfides, thiophenols, and even elemental sulfur.

Total sulfur content can be determined using ASTM D 1266, Test Method for Sulfur in Petroleum Products (Lamp Method) (21), ASTM D 2622, Test Method for Sulfur in Petroleum Products by X-Ray Spectrometry (39), or ASTM D 3120, Test Method for Trace Quantities of Sulfur in Light Liquid Hydrocarbons by Oxidative Microcoulometry (39). Mercaptan sulfur level can be determined by ASTM D 3227, Test Method for Mercaptan Sulfur in Gasoline, Kerosene, Aviation Turbine, and Distillate Fuels (Potentiometric Method) (39). Mercaptans can be determined qualitatively by the Doctor Test, ASTM D 4952, Test Method for Qualitative Analysis for Active Sulfur Species in Fuels and Solvents (Doctor Test) (24). The presence of elemental sulfur or reactive sulfur compounds can be detected by ASTM D 130, Test Method for Detection of Copper Corrosion from Petroleum Products by the Copper Strip Tarnish Test (21).

The average sulfur content of U.S. gasolines in 1994 is 0.034 wt % for regular unleaded and 0.014 wt % for premium unleaded. However, there still are a few fuels being marketed with sulfur levels in excess of 0.1 wt % (20). From the 1940s to mid-1950s, sulfur content averaged about 0.09 wt % for regular leaded gasoline and about 0.08 wt % for premium leaded (18). By the 1970s, average sulfur level had been reduced to about 0.04 wt % for regular leaded and to about 0.025 wt % for premium leaded. Doctor "Sweet" gasolines contain less than 10 to 15 parts per million by weight (ppm) mercaptans.

Table 2. ASTM D 4814 Vapor Lock Protection Class Requirements[a]

Vapor Lock Protection Class	Vapor/Liquid Ratio (V/L) Test Temperature, °C(°F)	V/L, max
1	60 (140)	20
2	56 (133)	20
3	51 (124)	20
4	47 (116)	20
5	41 (105)	20

[a] Ref. 24. Courtesy of ASTM.

In the U.S. there are no federal limits on maximum sulfur content. ASTM D 4814 sets a maximum sulfur limit of 0.10 wt % for unleaded gasoline and 0.15 wt % for leaded gasoline. California since 1975 has limited the maximum sulfur level in unleaded gasoline to 0.030 wt % (300 ppm).

Sulfur oxides formed during combustion may be converted into acids that promote corrosion of engine parts and exhaust systems. However, lowering sulfur levels from the present averages will provide only marginal improvement in engine and exhaust system life. Sulfur dioxide and sulfur trioxide (sulfate) emissions are undesirable atmospheric pollutants. However, the contribution of today's low-sulfur fuels to sulfate emissions is very small. A recent concern is the deactivation by sulfur of three-way catalysts used in exhaust emission control systems. An AQIRP study indicated changing fuel sulfur level has an immediate and reversible effect on exhaust emissions. Testing showed that hydrocarbon, carbon monoxide, and oxides of nitrogen tailpipe emissions from recent technology vehicles were reduced when sulfur content was lowered (40).

Existent Gum. During storage, gasoline may be slowly oxidized by air to form undesirable products such as peroxides and/or gum. These oxidation products are usually soluble in the gasoline, but the gum may be deposited as a sticky residue when the gasoline evaporates. ASTM D 381, Test Method for Existent Gum in Fuels by Jet Evaporation (21), is used to determine gum content. The test method uses a heptane wash to remove high boiling, oil soluble materials. When the wash step is omitted, the residue is reported as unwashed gum. The unwashed gum determination can be used to detect polymeric additives and carrier oils present in the gasoline.

The existent gum or washed gum content of U.S. gasolines averages about 2 mg/100 mL (20). The ASTM D 4814 limit is 5 mg/100 ML maximum.

The insoluble oxidation products can clog fuel filters. The soluble products can contribute to deposits on the surfaces of carburetors, fuel injectors, intake manifolds, ports, valves, and valve guides, and combustion chambers.

Stability. The oxidation stability of gasoline is an indication of its resistance to gum formation during extended storage. ASTM D 525, Test Method for Oxidation Stability of Gasoline (Induction Period Method) (21), is used to indicate the long-term oxidation stability of gasoline. The results are expressed as a time delay (induction period) before rapid oxidation begins. Gasolines readily exceed the 4 hours minimum induction period specified in ASTM D 4814 and most exceed 24 hours. Gasolines stored for more than one year are considered suspect and should be inspected for Existent Gum content and particulate contamination before using.

The correlation between the ASTM D 525 induction period and the formation of gum during long-term storage will vary depending on the storage conditions, fuel composition, and additives used (15,22). A combination of four-hours oxidation in the ASTM D 525 apparatus, filtration of the insoluble gum, and determination of the gum after induction by ASTM D 381 is believed to be a better method for predicting long-term storage. This procedure is used to measure a parameter called potential gum. ASTM D 873, Test Method for Oxidation Stability of Aviation Fuels (Potential Residue Method) (21), is a similar procedure except induction takes place for 5 or 16 hours. Oxidation inhibitors (antioxidants) are frequently evaluated by using a 12 weeks at 43°C (110°F) storage stability test. Aging for one week by this procedure is generally accepted to being equivalent to approximately one month of storage at ambient conditions.

Hydrocarbon Composition. The four major types of hydrocarbons in gasoline are aromatics, olefins, naphthenes, and paraffins. Naphthenes and paraffins are grouped together as saturates. ASTM D 1319, Test Method for Hydrocarbon Types in Liquid Petroleum Products by Fluorescent Indicator Adsorption (21), determines the amounts of aromatics, olefins, and saturates present in a gasoline. This method does not detect oxygenates and in fact provides the same result after the addition of an oxygenate as before. Therefore, the results must be corrected downward for the amount of oxygenates present. Many laboratories use nonstandardized gas chromatographic test methods to determine the individual hydrocarbon species present. Olefins content can be determined from an analysis by ASTM D 1159, Test Method for Bromine Number of Petroleum Distillates and Commercial Aliphatic Olefins by Electrometric Titration (21). This test method does not fully respond to the dilution effect of adding oxygenates, and thus Bromine Number is not reduced as much as expected. The amount of benzene present can be determined using ASTM D 4053, Test Method for Benzene in Motor and Aviation Gasoline by Infrared Spectroscopy (39), and by ASTM D 3606, Test Method for Benzene and Toluene in Finished Motor and Aviation Gasoline by Gas Chromatography (39). ASTM D 3606 suffers from interferences when methanol and ethanol are present.

The concentrations of hydrocarbons in commercial U.S. gasolines as determined by ASTM D 1319 fall into the following ranges: aromatics from 12 to 54 vol %, olefins from 0 to 37 vol %, and saturates from 35 to 85 vol %.

Olefins are generally the most reactive type of hydrocarbons in the formation of ozone. They have been controlled in Los Angeles County since 1959 by setting a maximum Bromine Number limit of 30. In 1971, the control on Bromine Number was extended to the South Coast Air Basin. Many U.S. gasoline manufacturers limit the benzene content to less than 5 wt % because federal toxic substances regulations require special labeling if that level is exceeded. The Clean Air Act Amendments of 1990 will limit benzene content and the toxics reduction will control aromatics content beginning in 1995, and California regulations will limit olefins, benzene, and aromatics contents beginning in 1996 (see Clean Air Act Section). The Auto/Oil AQIRP testing showed that lowering olefins content increases hydrocarbon and lowers oxides of nitrogen exhaust emission in both current and older model vehicles. The study also showed that lowering aromatics content reduces hydrocarbon and carbon monoxide emissions from current model vehicles. Lowering aromatics increases hydrocarbon and lowers oxides of nitrogen exhaust emissions from older model vehicles (26,27,41). Lowering aromatics content only marginally increased oxides of nitrogen exhaust emissions in current model vehicles (27).

Density. The mass per unit volume or density of gasoline is not a property that is controlled by specification. However, it is a widely used and reported property. The density will depend on the chemical composition of the fuel. At a given carbon number (number of carbon atoms in the molecule), aromatics are most dense, then olefins, and then paraffins with the lowest density of the hydrocarbon types. Density is reported at a reference temperature, such as 60°F in the U.S. or 15°C in Europe. Relative density is the ratio of the mass per unit volume of the fuel at a given temperature to the mass of an equal volume of water at the same temperature. The U.S. oil industry generally uses the term "API Gravity at 60°F," which is an arbitrary hydrometer scale related to relative density or specific gravity as follows:

$$\textit{API Gravity}^\circ = (141.5/\text{specific gravity } 60/60^\circ\text{F}) - 131.5$$

API Gravity is an inverse scale with higher numbers representing less dense petroleum products. AFI Gravity of gasoline ranges from about 50 to 70° API (relative density 0.78 to 0.70). Density, relative density, and API Gravity can be determined using ASTM D 1298, Test Method for Density, Relative Density (Specific Gravity) or API Gravity of Crude Petroleum and Liquid Petroleum Products by Hydrometer Method (21), or by ASTM D 4052, Test Method for Density and Relative Density of Liquids by Digital Density Meter (39).

Volume Correction. Gasoline is frequently wholesaled on the basis of volume corrected to 60°F (15.6°C). To determine the volume at 60°F (15.6°C), the API Gravity, density, or relative density is determined at the bulk product temperature. The volume at 60°F (15.6°C) is calculated from the volume and density measured at the bulk temperature and the Volume Correction Factor from the appropriate tables in ASTM D 1250, Petroleum Measurement Tables–Volume Correction Factors (42). The coefficients of thermal expansion for oxygenates are greater at the same density than those of hydrocarbons. Therefore, ASTM D 1250 overcorrects the volume for pure oxygenates when the bulk temperature is above 60°F (15.6°), and undercorrects for bulk temperatures below 60°F (15.6°) up to 0.5% within normal storage temperature ranges. There will be less than a 0.1% error for gasoline–oxygenate blends.

Energy Content. There is a gross empirical correlation between volumetric heating value of gasoline and its API Gravity or relative density. A fuel with a greater density has a higher heat content.

Heating Value. Heating value can be determined directly by ASTM D 240, Test Method for Heat of Combustion of Liquid Hydrocarbon Fuels by Bomb Calorimeter (21), or by ASTM D 2382, Test Method for Heat of Combustion of Hydrocarbon Fuels by Bomb Calorimeter (High Precision Method) (21). Fuel economy is directly related to the heating value of the fuel (43). The volumetric lower or net (liquid fuel–water vapor product) heating values for U.S. gasolines range from 109,000 to 119,000 Btu/gal (30 400 to 33 200 kJ/L). Because winter gasolines contain more volatile and less dense hydrocarbons, their heating values on average are 1.5% lower than summer gasolines. In general, premium unleaded gasolines contain more aromatics, which are the most dense hydrocarbons, and on average have a heating value 0.7% higher than regular unleaded gasolines. Heating value is a property which is not usually limited by specification.

Rust and Corrosion. Copper corrosion is determined by ASTM D 130 as discussed in the Sulfur Section. Ferrous corrosion (rusting) is determined for gasoline using a modification (NACE TM-01-72) of ASTM D 665, Test Method for Rust-Preventing Characteristics of Inhibited Mineral Oil in the Presence of Water (21). While this test method is accepted for predicting pipeline corrosion, no correlation has been developed for predicting vehicle component corrosion. Automobile manufacturers generally use full-scale fuel pumping rigs to evaluate fuel system durability.

Workmanship. At the point of marketing, gasoline is expected to be visually free of undissolved water, sediment, and suspended matter. It should be clear and bright at the ambient temperature or 21°C, whichever is higher. Solid and liquid contamination can lead to restriction of fuel metering orifices, corrosion, fuel line freezing, gel formation, and filter clogging. Refiners, distributors, and marketers must maintain constant vigilance to control contamination (15,22).

Additional Properties. There are a number of additional properties of gasoline which are not controlled by specification and are rarely measured. However, these properties are of interest to some users, equipment designers, and researchers. These properties are summarized for a typical gasoline in Table 3 (15,44).

Table 3. Additional Properties of Gasoline[a]

Property	Value
Autoignition temperature, °C (°F)	260 (500)
Coefficient of thermal expansion at 15.6°C, per °C (per °F)	0.00105–0.00133 (0.00058–0.00074)
Composition, wt %	
Carbon	85–88
Hydrogen	12–15
Electrical conductivity, mhos/cm	1×10^{-14}
Flammability limits, vol % vapor in air	1.4 to 7.6
Flash point, closed cup, °C (° F)	−43 (−45)
Freezing point, °C (° F)	<−40 (<−40)
Latent heat of vaporization, Btu/gal (kJ/L)	900 (251)
Molecular weight, average	100–105
Refractive index, n_D at 20°C (68° F)	1.4–1.5
Specific heat, Btu/lb-°F (kJ/kg-°K)	0.48 (2.01)
Surface tension, N/m	20×10^{-3}
Viscosity, mm/s	
20°C (68° F)	0.5–0.6
−20°C (−4° F)	0.8–1.0
Water solubility, ppm	100–200

[a] Refs. 15, 44.

ADDITIVES

An additive, by definition, is "a substance added to another in small quantities to produce a desired effect . . . as an antiknock added to gasoline, etc" (1). Additives are added to gasoline to provide improved performance or to correct deficiencies. Additives frequently can provide a benefit at lower cost than through refinery processing, or they can provide one that is not obtainable by processing. Additives have been used commercially in gasoline in the U.S. since 1923. As time has passed, new additives have appeared and old additives have disappeared.

The principal types of gasoline additives being used today are listed in Table 4 (17). Also shown for each additive is the general chemical type along with the primary function of each. The concentrations vary greatly among and within types, as shown in Figure 6 (17). The concentrations are shown both in ppm and the usual refinery units of pounds per 1000 barrels (lb/1000 bbl). The amounts and types of additive used in a given gasoline depend on the properties of the base gasoline and on the performance desired by the marketer. With the exception of lead and manganese antiknocks and some antioxidants, standardized test methods are not available to determine the type and concentration of specific additives.

Additives initially were put into gasoline at the refinery and many still are. In the refinery, antioxidants and metal deactivators are injected at processing units. Dyes and antiknocks are added at in-line gasoline blenders or into finished gasoline tanks. Demulsifiers may be added at the in-line blenders or to refinery or terminal tankage. Pipeline companies inject corrosion inhibitor additives. Detergents were initially injected at the refinery, but now both detergents and the broader-performing deposit control additives are usually injected at pipeline or marketing terminals.

Table 4. Current Typical Gasoline Additives[a]

Additive	Type	Function
Oxidation inhibitors (antioxidants)	Aromatic amines and hindered phenols	Inhibit oxidation and gum formation
Corrosion inhibitors	Carboxylic acids, amides, and amine salts	Inhibit corrosion of iron
Metal deactivators	Chelating agent	Inhibit oxidation and gum formation catalyzed by certain metals, particularly copper
Carburetor/injector Detergents	Amines and amine Carboxylates	Prevent and remove deposits in carburetor throttle bodies and port fuel injectors
Deposit control	Polybutene amines, polyether amines	Prevent and remove deposits throughout carburetor, fuel injectors, intake manifold, and intake ports and valves
Demulsifiers	Polyglycol derivatives	Improve water separation
Antiknock compounds	Lead alkyl and organo-manganese compounds	Increase octane number
Antiicing	Surfactants, alcohols, and glycols	Prevent icing in carburetor and fuel system
Dyes	Azo and other-oil soluble compounds	Identification

[a] Ref. 17. Courtesy of the Society of Automotive Engineers, Inc.

Effects on Emissions

Some additives have direct effects on emissions and emission control systems as follows:

Lead Antiknocks. On December 9, 1921, Thomas Midgley, Jr. and Thomas A. Boyd of General Motors Research Corporation discovered that tetraethyllead (TEL) was a very effective antiknock additive (17). Gasoline containing TEL was first marketed on February 1, 1923. TEL was the only lead alkyl antiknock that was commercially available until 1960 when Standard Oil Company of California (now Chevron Corporation) began to use tetramethyllead (TML). Physical and chemically reacted mixtures of TEL and TML were also offered. Laboratory and Road octane responses to lead antiknock additives depend on which lead additive is used and on the composition of the gasoline. Refiners attempt to use the most cost-effective lead antiknock additive in their gasoline compositions. Lead has always been a very cost-effective antiknock, which is the reason why it was so successful. In the U.S. its usage peaked in the late 1960s. Regular leaded gasoline contained an average of about 2.3 grams of lead per U.S. gallon (0.6 gram per liter). Premium

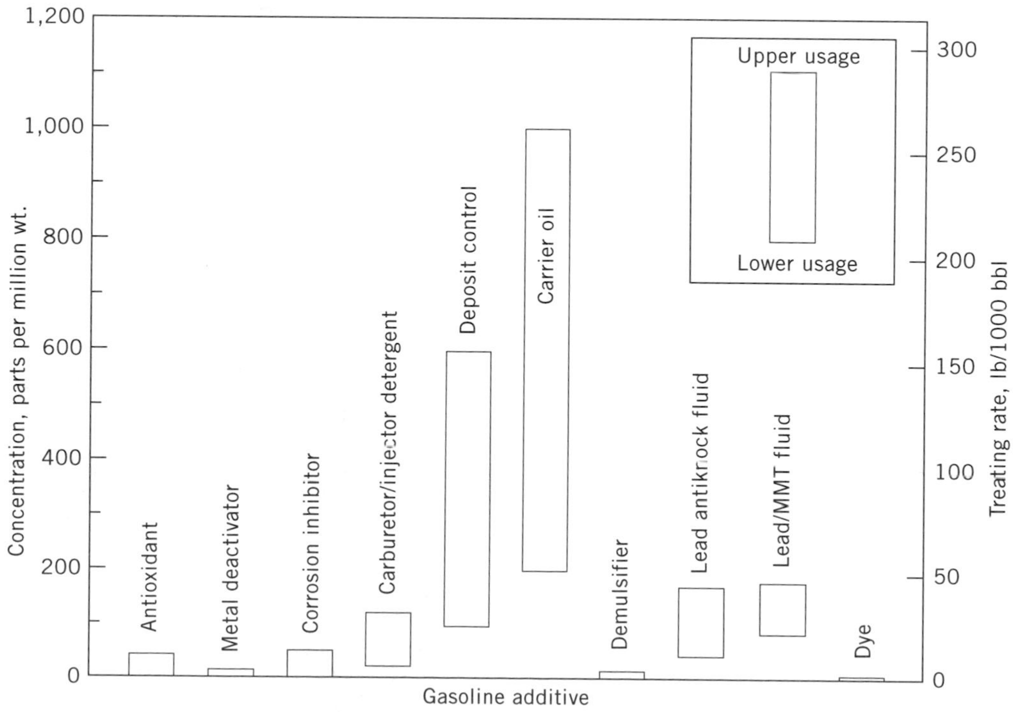

Figure 6. Current gasoline additives' dosage ranges (17). Carrier oil used with some deposit control additives. Courtesy of the Society of Automotive Engineers, Inc.

leaded gasoline contained an average of about 2.8 grams of lead per U.S. gallon (0.74 gram per liter).

The lead content in leaded gasoline was regulated to protect public health because lead is a known toxic material and one that is especially damaging to children. The federal maximum average limit of 0.10 gram per U.S. gallon (0.025 gram per liter) became effective in 1986. In 1995 and 1996 lead will be eliminated from highway gasoline under federal regulations (see Clean Air Act Section). California eliminated leaded gasoline in 1992.

The maximum lead content for unleaded gasoline is 0.05 gram per U.S. gallon (0.013 gram per litre). This federal regulation became effective in July 1974 and was adopted to provide adequate protection for catalytic emission control devices. Concurrently, phosphorus was limited to 0.005 gram per U.S. gallon (0.0013 gram per liter). Deactivation of catalysts by lead or phosphorus results in substantial increases in tailpipe emissions of hydrocarbons, carbon monoxide, and oxides of nitrogen (45).

Manganese Antiknocks. Besides lead, manganese is an effective antiknock element. In 1959, Ethyl Corporation commercialized a manganese antiknock compound, methylcyclopenta dienyl manganese tricarbonyl (MMT). MMT is also marketed in an antiknock additive package with TEL. As lead began to be phased down in the late 1970s as a result of federal regulations, the use of MMT in combination with lead increased.

With the introduction in 1974 of the 1975-model year exhaust catalyst-equipped cars, which required unleaded gasoline, MMT began to be used alone as an antiknock. Additive concentrations up to 0.125 gram manganese per U.S. gallon (0.033 gram per liter) were recommended. After the passage of the Clean Air Act Amendments of 1977, MMT could not be used in unleaded gasoline unless a waiver was obtained from the EPA. Four waiver requests have been submitted, but none have been granted because of concern about increased emissions of exhaust hydrocarbons and the effect of the additive on future low emission vehicles. Manganese is still being used in Canada in unleaded gasoline at concentrations up to 0.063 gram manganese per U.S. gallon (0.017 gram per liter).

Carburetor Detergents. The introduction in 1954 of a carburetor detergent additive by the Standard Oil Company of California was the first use of surface-active materials to alleviate problems caused by deposits in engine induction systems (17). This first additive was used to solve a carburetor gumming problem that was prevalent at that time, particularly in urban fleets. The use of carburetor detergents had a resurgence in 1985 after problems with port fuel injector deposits surfaced and automobile manufacturers requested that additives be used to keep injectors clean. However, to obtain the necessary cleanliness performance, nearly double the carburetor detergent treatment rate was required. In many cases, the higher treatment rates caused increases in intake valve deposits (46–48).

Deposit Control Additives. The carburetor detergents were effective in preventing and in many cases removing deposits from carburetor throttle bodies. However, they were not effective in handling deposits in other parts of the carburetor, such as the air bleeds, or in the rest of the engine intake system (10). The introduction of emission control systems in the 1960s and 1970s, such as positive crankcase ventilation (PCV), exhaust gas recirculation (EGR), and intake air heaters, increased deposit levels in the whole intake system. As a result, carburetor detergents were not as effective as they were in the simpler 1950s vehicles.

The first deposit control additive, which cleaned up and kept clean all areas of the carburetor (and later fuel injection systems), intake manifold hot spot and runners, intake ports, and the difficult-to-control intake valves, was commercialized in 1970 by Standard Oil Company of California. This additive was a polybutene amine and was used in conjunction with a carrier oil.

The use of oil as a gasoline additive was not new. Oils were added to provide lubrication for top piston rings and intake valve guides. At 1000 ppm to 3000 ppm levels, they showed a modest benefit in helping to prevent intake valve deposits. However, their use in conjunction with a polymeric surface-active additive provided both cleanup and keep-clean performance. Vehicle tests showed that a combination of deposits in the carburetor and/or PCV system produced enrichment of the fuel-air ratio supplied to the engine. This usually causes an increase in exhaust hydrocarbons and carbon monoxide. Because the additive can remove these deposits, hydrocarbon and carbon monoxide exhaust emissions are reduced and fuel economy improves (49,50).

The original purpose of using deposit control additives was to keep induction systems of carbureted engines clean so that exhaust emissions stayed well below the level of emissions generated with fuels containing no additive. When the port fuel injector deposit problem appeared in 1985, deposit control additives as well as carburetor/injector detergents were shown to effectively clean up and keep clean port fuel injectors (47,48).

It has long been recognized that heavy deposits on intake valves can reduce maximum engine power by throttling the intake charge (51). Experiments have shown a link between intake valve deposit weight and oxides of nitrogen emissions. The results suggest that the increase can be caused by the deposits affecting the combustion rate or the level of residual gases in the combustion chamber (51). Tests have shown that the use of deposit control additives can prevent increases in oxides of nitrogen exhaust emissions (52).

Fleet tests in which intake valve deposits were removed manually showed that deposits increased carbon monoxide and oxides of nitrogen tailpipe emissions. An increase in exhaust hydrocarbons was observed, but it was not statistically significant. Also, it was observed that as intake valve deposit level increases, tailpipe emissions increase proportionally (53).

At a 1989 CRC workshop, a number of automobile manufacturers reported cold-start driveability problems with low levels of intake valve deposits. Problems appeared to be aggravated by low-volatility fuel (high 50% evaporated point). It was also shown that the use of effective deposit control additives could prevent poor cold start driveability caused by intake valve deposits (54). A CRC study showed that the extent that intake valve deposits affected cold-start driveability depended on engine design and emission control calibration (55).

There are no current federal requirements for the use of deposit control additives, but such requirements become effective in 1995 (see Clean Air Act Section). Beginning in 1992, California required the use and certification of deposit control additives.

OXYGENATES

As discussed earlier, the Nebraska Gasohol Committee in 1978 began a program of adding 10 vol % ethanol to gasoline and marketing the fuel as gasohol as a means to increase fuel availability during the oil embargo. Although low levels of alcohols, such as isopropyl alcohol, had been used since 1950 as an antiicing additive and Atlantic-Richfield Company (ARCO) had used gasoline-grade *tert*-butyl alcohol (GTBA) since 1969, the ethanol program was the start of the widespread use of oxygenates.

Oxygenate Regulations

In the U.S., the Clean Air Amendments of 1977 control the use of oxygenates in unleaded gasoline. The regulations developed to implement this law prohibited the introduction or increase in concentration of a fuel or fuel additive which is not substantially similar to any fuel or fuel additive utilized in the certification of any 1975 or subsequent model year vehicle. A waiver to the regulation could be obtained on proving that the candidate fuel or fuel additive will not cause or contribute to a failure of any emission control device or system. If the EPA Administrator does not act to grant or deny an application within 180 days, a waiver is treated as granted. The regulations do not apply to leaded gasoline.

Waivers. In 1978 the EPA, on failure to take action within 180 days, granted a waiver to Gas Plus, Inc. for ethanol at 10 vol % by operation of the statute. This waiver is popularly called the "gasohol waiver." In 1982 EPA issued an interpretive ruling that the waiver also applied to concentrations of ethanol less than 10 vol %. ARCO obtained an EPA waiver in 1979 to use GTBA at up to 7.0 vol % in unleaded gasoline. Also in 1979, Sun Oil Company received an EPA waiver that allowed the use of 2.75 vol % methanol along with 2.75 vol % GTBA in a blend with unleaded gasoline. ARCO, in 1981, received a waiver for an oxygenate blend that contained equal amounts of methanol and GTBA. This mixture could be used at concentrations that resulted in up to 3.5 wt % oxygen content (about 9.5 vol % total alcohol). Later in 1985 and 1988, waivers were also issued to Du Pont and Texas Methanol (OCTAMIX) for mixtures of methanol and higher alcohols (cosolvents) with a 3.7 wt % maximum oxygen content limit along with a maximum methanol content of 5.0 vol % and a minimum cosolvent content of 2.5 vol %. Depending on the cosolvent used, these two waivers allowed up to 10 vol % of the alcohol mixture and required the addition of specific corrosion inhibitor additive packages. ARCO obtained a waiver in 1979 to add up to 7.0 vol % methyl *tert*-butyl ether (MTBE) to gasoline. In 1988 Sun Oil obtained a waiver that increased the maximum MTBE limit to 15 vol % (56). All of the waivers, except the one for ethanol, require that the finished fuel meet ASTM volatility requirements. For the latest status on waivers, the EPA should be contacted.

Substantially Similar. The EPA issued the "substantially similar" ruling in 1981. This rule allowed fuels containing aliphatic alcohols (excluding methanol) and/or ethers to be blended into gasoline up to a concentration that would result in 2.0 wt % oxygen in the blends. It also allowed the use of 2.75 vol % methanol along with 2.75 vol % GTBA. In 1991 EPA revised the "substantially similar" rule by increasing the maximum oxygen content of gasoline-oxygenate blends (excluding methanol) to 2.7 wt %. The "substantially similar" rule requires the finished fuel to possess all the physical and chemical characteristics of an unleaded gasoline specified by ASTM D 4814-88 for at least one of the seasonal and geographic volatility classes specified in the standard.

In order to assure an adequate supply of base gasoline for ethanol blending, EPA has ruled that gasolines containing up to 2.0 vol % MTBE subsequently blended with 10 vol % ethanol would not violate the ethanol waiver.

None of the above uses of oxygenates should be considered as additive usage, which is the addition of small quantities of materials. At the levels used, alcohols and ethers are more properly defined as blending components. ASTM D 4815, Test Method for Determination of MTBE, ETBE, TAME, DIPE, Tertiary-Amyl Alcohol, and C_1 to C_4 Alcohols in Gasoline by Gas Chromatography (24), is used to determine the type and quantity of oxygenates present in gasoline–oxygenate blends.

Chemical and Physical Properties.

Chemical and physical properties of some commonly used oxygenates and oxygenates that are under consideration are shown in Table 5 for alcohols and Table 6 for ethers (44,56). Figure 7 graphically shows for an average gasoline the nominal maximum volumetric amounts of various oxygenates and combination of oxygenates permitted under the "substantially similar" rule and EPA waivers. When the limit is specified as a maximum oxygen content, the actual volume permitted will depend on the density of the base gasoline. The amount permitted under "substantially similar" also is indicated in Figure 7 for waivered oxygenates. For ethanol, propanols, and butanols the waiver limit is higher than that permitted under "substantially similar." The controlling maximum limits for propanols and butanols are actually from waivers of mixtures containing methanol, but since there is no minimum methanol requirement, the total amount allowed can be all cosolvent alcohol at concentrations that provide up to 3.7 wt % oxygen.

Effects on Physical Properties

Adding oxygenates to hydrocarbon gasoline changes the properties of the fuel. The effect depends on the type and amount of oxygenate. As a class, ethers have less of an effect than do alcohols (44).

Table 5. Properties of Alcohols[a]

Property	Methanol	Ethanol	Isopropyl Alcohol	*n*-Butanol	Gasoline-Grade *t*-Butanol
Autoignition temperature, °C (° F)	464 (867)	423 (793)	399 (750)		478 (892)
Blending (R + M)/2[b]	119	115	106	86	97
Blending vapor pressure, kPa (psi)[c]	276 (40)	124 (18)	97 (14)	62 (9)	62 (9)
Boiling point, °C (°F)	65 (149)	78 (173)	82 (180)	118 (244)	80–83 (176–181)
Chemical formula	CH_3OH	C_2H_5OH	$(CH_3)_2CHOH$	C_4H_9OH	$(CH_3)_3COH$
Coefficient of thermal expansion at 15.6°C (60°F), per °C (°F)	0.0012 (0.00067)	0.0011 (0.00062)			
Composition, wt %					
Carbon	37.49	52.14	59.96	64.82	65.0
Hydrogen	12.58	13.13	13.42	13.60	13.7
Oxygen	49.93	34.73	26.62	21.58	21.3
Electrical conductivity, mhos/cm	4×10^{-7}	1.35×10^{-9}			
Flammability limits, vol % vapor in air	7.3 to 36.0	4.3 to 19.0	2.0 to 12.0	1.4 to 11.2	2.4 to 8.0
Flash point, °C (°F)	11 (52)	13 (55)	12 (53)	29 (84)	11 (52)
Heating value, lower, kJ/L (Btu/gal)	15,800 (56,800)	21,200 (76,000)	24,400 (87,400)	27,000 (96,800)	26,300 (94,100)
Latent heat of vaporization, kJ/L (Btu/gal)	931 (3340)	663 (2378)	585 (2100)	475 (1700)	474 (1700)
Molecular weight	32.04	46.07	60.09	74.12	73.5
Refractive index, n_D at 20°C (68°F)	1.3286	1.3614	1.3772	1.3993	1.3838
Relative density 15.6/15.6°C (60/60°F)	0.7963	0.7939	0.7899	0.8137	0.7810
Specific heat, kJ/kg-°K (Btu/lb-°F)	2.51 (0.60)	2.39 (0.57)	2.55 (0.61)	2.35 (0.56)	3.82 (0.72)
Stoichiometric air-fuel ratio, wt	6.45	9.00	10.3	11.1	11.1
Viscosity, mm/s					
20°C (68°F)	0.74	1.50	3.01	3.54	7.4
−20°C (−4°F)	1.44	3.58	7.43		Solid
Water solubility, 21°C (70°F)					
Fuel in water, vol %	100	100	100	100	100
Water in fuel, vol %	100	100	100	100	100

[a] Refs. 44, 56.
[b] In 87 (R + M)/2 typical composition unleaded gasoline.
[c] At nominally 10 vol %.

Volatility. Adding low molecular weight alcohols to hydrocarbon gasoline depresses the boiling temperature of individual hydrocarbons. The effect is greater for aliphatic hydrocarbons than for aromatic hydrocarbons (44). The addition of ethers produces the same effect as adding a hydrocarbon of equal volatility (57). The effect on the shape of the distillation curve of adding equal volumes of ethanol, methyl *tert*-butyl ether (MTBE), and ethyl *tert*-butyl ether (ETBE) to hydrocarbon gasoline is shown in Figure 8 (57). The addition of alcohols highly distorts the shape of the distillation curve. Studies have shown that as the molecular weight of the alcohol increases, the distortion of the curve becomes less and moves up the curve (44,58).

The effect on vapor pressure of adding various oxygenates to hydrocarbon gasoline is illustrated in Figure 9 (57,58). Adding small amounts of low molecular weight alcohols produces large increases in the vapor pressures of the mixtures. Then as more alcohol is added, the vapor pressure levels out in a nonideal (nonlinear) blending relationship (44). Blending ethers is like blending hydrocarbons with similar volatility properties (57). Figure 9 shows that for a 62 kPa (9 psi) base fuel, the addition of 10 vol % ethanol increases the vapor pressure about (6.9 kPa (1.0 psi) while the addition of a MTBE has little effect on vapor pressure. Adding ETBE lowers the vapor pressure. The increase in vapor pressure observed with the addition of ethanol depends on the vapor pressure of the base gasoline. As shown in Figure 10, the increase is less as the vapor pressure of the base fuel increases.

Tables 5 and 6 show the blending vapor pressure of numerous oxygenates used at 5 to 20 vol % in a 48.3 -103 kPa (7.0 to 15.0 psi) base gasoline.

To protect against vapor lock and to provide good cold and hot start driveability, the same volatility limits (vapor pressure, distillation properties, and temperature for a vapor–liquid ratio of 20) apply to gasoline and gasoline–oxygenate blends. For all oxygenate waivers, except the one for ethanol, the EPA requires the finished blend to meet ASTM volatility requirements. For Phase II volatility rules, the EPA permits blends containing between 9 and 10 volume % ethanol to have 6.9 kPa (1.0 psi) higher

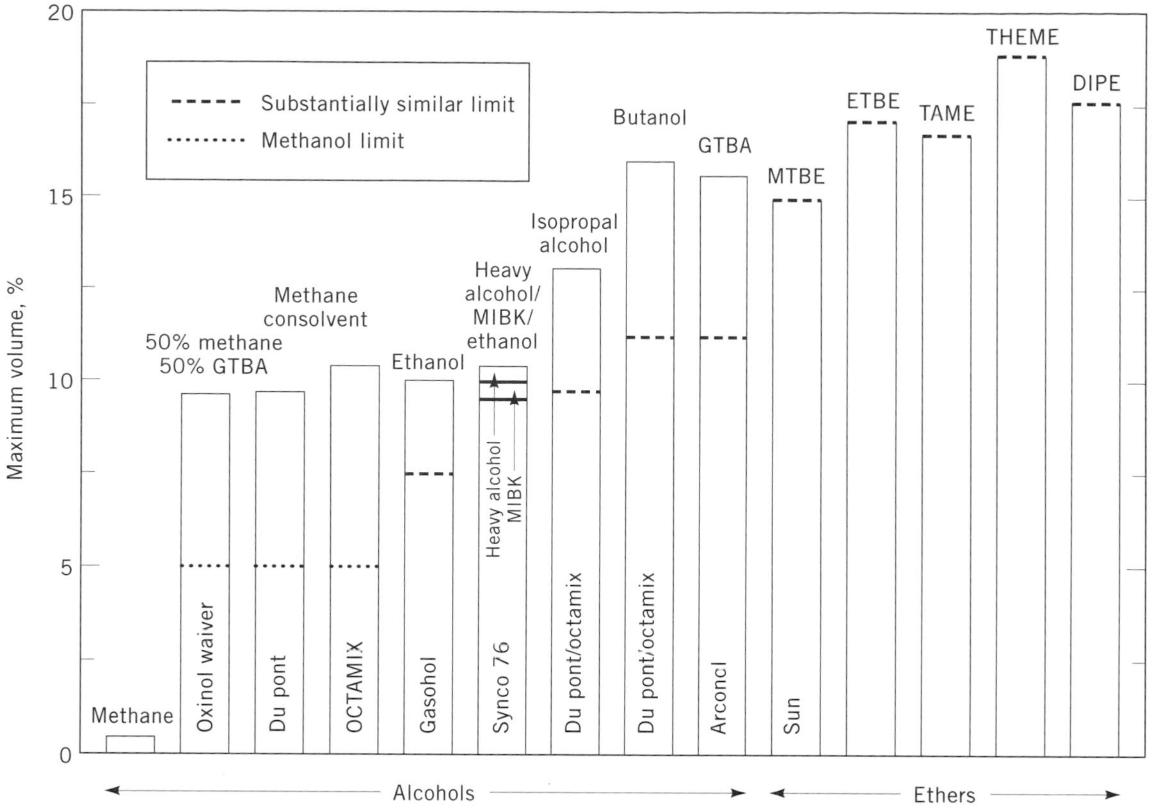

Figure 7. Maximum volumes of U.S. EPA approved oxygenates for unleaded gasoline. Average relative density (0.74) gasoline. GTBA = gasoline grade *t*-butanol; MTBE = methyl *t*-butyl ether; TAME, *t*-amyl methyl ether; ETBE = ethyl *t*-butyl ether; THEME, *t*-hexyl methyl ether; DIPE, diisopropyl ether. .– –., substantially similar limit; . . . . , methanol limit.

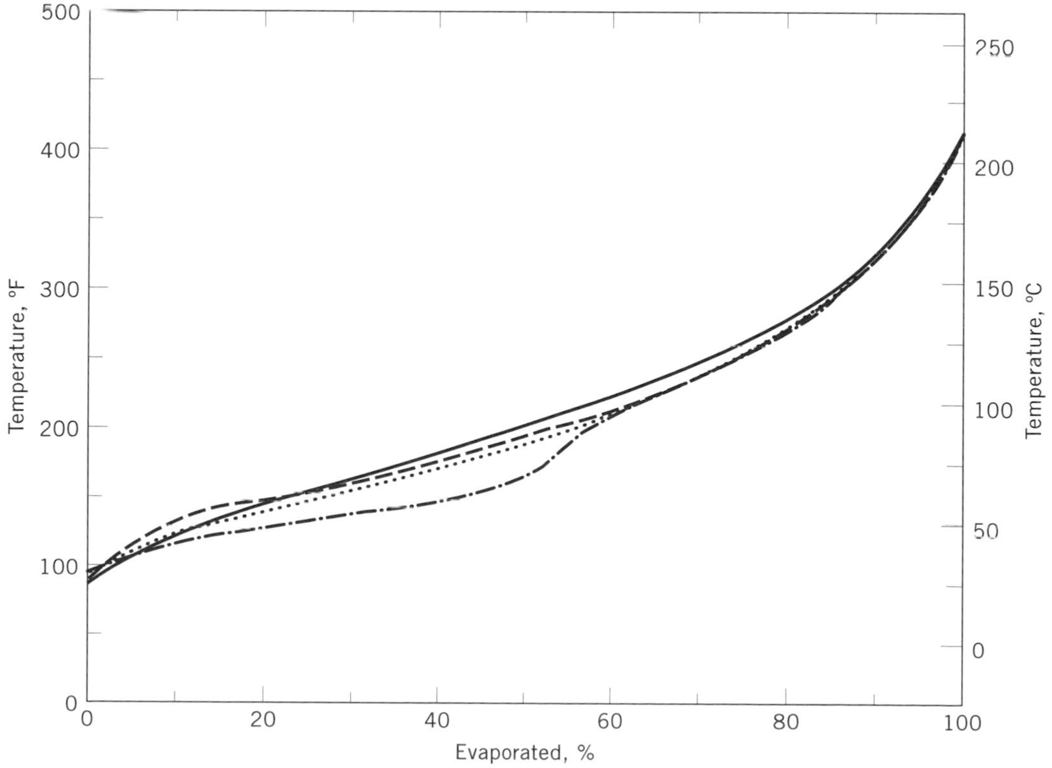

Figure 8. Effect of equal volumes of oxygenates on the distillation curve. ___, gasoline; – – – –, ethyl *t*-butyl ether, 10 vol %; . . . . methyl *t*-butyl ether, 10 vol %; ·—·—· ethanol, 10 vol %. Data courtesy of General Motors Research (57). Courtesy of Society of Automotive Engineers.

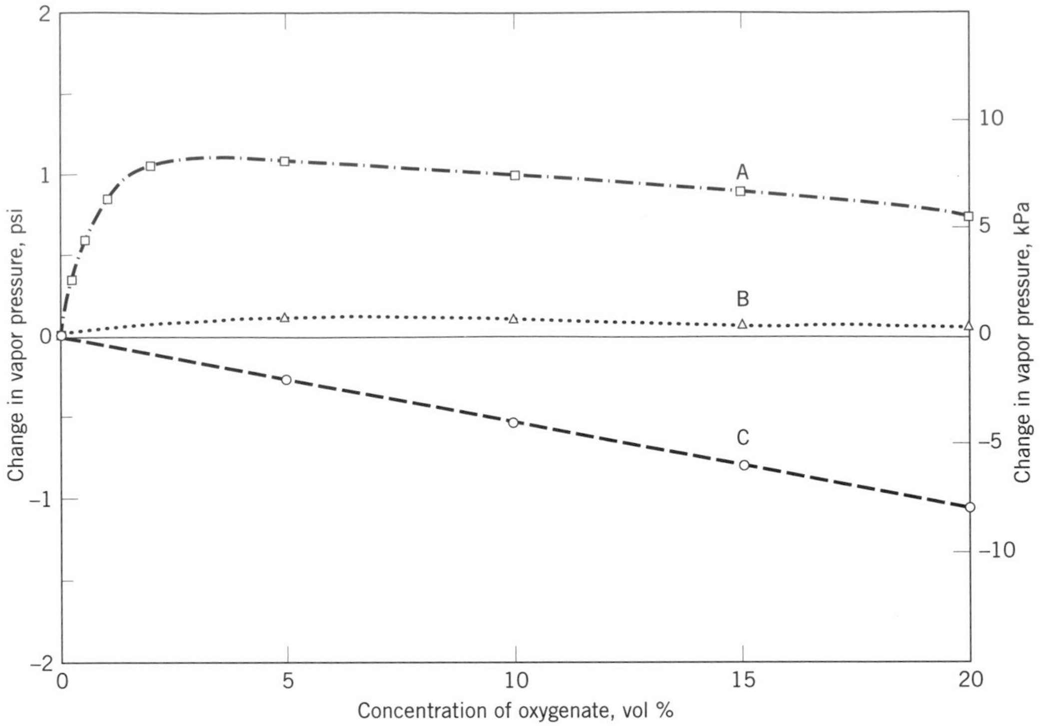

Figure 9. Effects of equal volumes of oxygenates of vapor pressure of 62 kPa (9 psi) gasoline. Data courtesy of General Motors Research (57). A, ethanol; B, methyl *t*-butyl ether; C, ethyl*t*-butyl ether. Courtesy of the Society of Automotive Engineers.

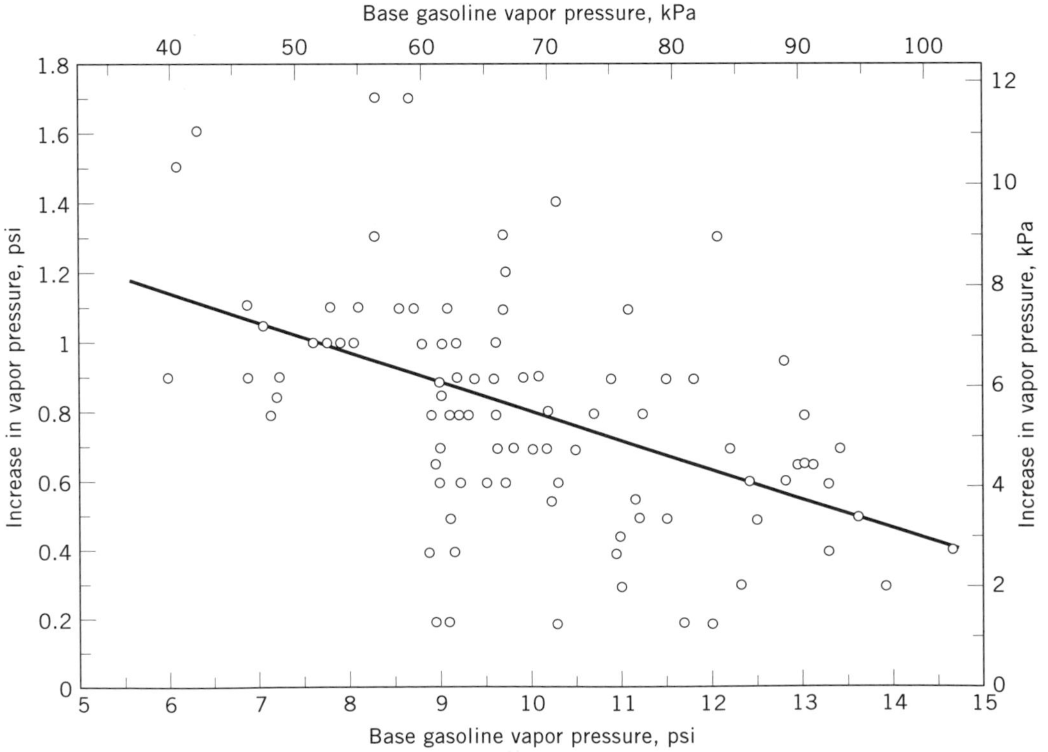

Figure 10. Increase in vapor pressure from addition of 10 vol % ethanol. Analysis of 101 data points from multiple sources.

Table 6. Properties of Ethers[a]

Property	Methyl Tert-Butyl Ether	Tert-Amyl Methyl Ether	Tert-Hexyl Methyl Ether	Ethyl Tert-Butyl Ether	Tert-Amyl Ethyl Ether	Diisopropyl Ether
Autoignition temperature, °C (°F)	435 (815)					443 (830)
Blending (R + M)/2[b]	110	105	95	111	100	107
Blending vapor pressure, kPa (psi)	55 (8)	14 (2)	7 (1)	28 (4)	14 (2)	5 (0.7)
Boiling point, °C (°F)	55 (131)	86 (187)	~110 (~230)	73 (163)	101 (214)	68 (155)
Chemical formula	$(CH_3)_3COCH_3$	$(CH_3)_2(C_2H_5)COCH_3$	$(CH_3)_2(C_3H_7)COCH_3$[c]	$(CH_3)_3COC_2H_5$	$(CH_3)_2(C_2H_5)COC_2H_5$	$(CH_3)_2CHOCH(CH_3)_2$
Coefficient of thermal expansion at 15.6°C, per °C (°F)	0.00143 (0.00078)			0.00140 (0.00077)		0.00144 (0.00080)
Composition, wt %						
Carbon	68.13	70.53	72.35	70.53	72.35	70.53
Hydrogen	13.72	13.81	13.88	13.81	13.88	13.81
Oxygen	18.15	15.66	13.77	15.66	13.77	15.66
Electrical conductivity, mhos/cm						
Flammability limits, vol % vapor in air	1.6 to 8.4	1.0 to 7.1		1.0 to 6.8		1.4 to 7.9
Flash point, °C (°F)	−26 (−14)	−11 (11)		−19 (−3)		−12 (9)
Heating value, lower, kJ/L (Btu/gal)	26,000 (93,500)	28,000 (100,600)		27,000 (97,000)		27,900 (100,000)
Latent heat of vaporization, Btu/gal (kJ/L)	241 (863)	243 (870)		232 (830)	228 (816)	251 (900)
Molecular weight	88.15	102.18	116.20	102.18	116.20	102.18
Refractive index, n_D at 20°C (68°F)	1.3689	1.3760		1.3885	1.3912	1.3681
Relative density, 15.6/15.6°C (60/60°F)	0.7460	0.7758		0.7452	0.7705	0.7289
Specific heat, kJ/kg-°K (Btu/lb-°F)	2.09 (0.50)	2.18 (0.52)		2.13 (0.51)	2.09 (0.50)	
Stoichiometric Air-Fuel Ratio, wt	11.7	12.1	12.5	12.1	12.5	12.1
Viscosity, mm/s						
20°C (68°F)	0.47					
−20°C (−4°F)	1.44					
Water solubility, 21°C (70°C)						
Fuel in water, vol %	4.8	1.15		1.2	0.4	
Water in fuel, vol %	1.5	0.6		0.5	0.2	

[a] Ref. 44, 56.
[b] In 87 (R + M)/2 typical composition unleaded gasoline.
[c] Most prevalent of four isomers.

vapor pressure than the specified limit for other summertime fuels. If the vapor pressure is not adjusted or the distillation curve modified, the addition of ethanol will lower the temperature for a vapor–liquid ratio of 20 (44,57) and increase the chances of encountering vapor lock. Cold-start and warmup driveability testing in older vehicles has shown poorer driveability performance for gasoline-alcohol blends. Testing in more recent vehicles showed that Driveability Index-type equations underestimate the demerits for gasoline–alcohol blends. In 1991 the CRC conducted a large customer/rater driveability program at intermediate temperatures. The study compared the performance of hydrocarbon and oxygenated gasolines. The results showed that consumers were less satisfied with ethanol blends than with hydrocarbon gasoline and MTBE blends. This dissatisfaction was greatest in carbureted vehicles. Hesitation was associated with the ethanol blend dissatisfaction, while surge and stumble were the major complaints for hydrocarbon gasoline (59). The poorer driveability performance is generally attributed to the leaning effect resulting from the addition of oxygen (44,60).

When ethanol or MTBE is added to gasoline to meet mandated oxygenated gasoline requirements, the 50% evaporated point in some cases is lowered below the minimum specified by ASTM D 4814. At the request of ASTM, the CRC initiated a study in 1992 to determine if the minimum 50% point limit could be reduced without encountering hot-start driveability problems. The test results showed that lowering the 50% point increased driveability problems for carbureted and fuel injected vehicles when the vapor pressure was low [51.7 kPa (7.5 psi)], but not when it was high [82.7 kPa (12.0 psi)] (61). Based on these findings, ASTM is considering reducing the 50% point only for volatility Classes D and E which have high vapor pressure limits.

Water Tolerance. Water tolerance is the term used to indicate the ability of a gasoline-oxygenate blend to dissolve water without phase separation. Water is only slightly soluble in ethers (Table 6) and is even less soluble in hydrocarbon gasoline (Table 3). When a gasoline–ether blend contacts more water than it can dissolve, two distinct layers will be present–a gasoline–ether layer that contains traces of dissolved water and a lower layer that is essentially water. This is not substantially different from when hydrocarbon gasoline contacts water. Usually, this is not a problem.

Gasoline–alcohol blends behave differently. Water and low molecular weight alcohols are completely miscible (Table 5). When a gasoline–alcohol blend contacts water, it will dissolve a significant amount of water. However, if there is too much water for the gasoline–alcohol blend to dissolve, it will separate into two layers–a gasoline-rich upper layer and an alcohol-rich aqueous layer on the bottom. The aqueous layer can be corrosive to metals and the engine cannot operate on it (22,44,56).

As the molecular weight of alcohols increases, so does their water tolerance. Increasing the concentration of the alcohol and/or the aromatics content of the blend improves water tolerance. However, the most important factor affecting water tolerance, aside from the quantity of water contacted, is the fuel temperature. The solubility of water increases as the temperature increases. At room temperature a 10 vol % ethanol blend will separate into two phases after contacting only about 0.7 wt % water even in a highly aromatic fuel (44).

Because it is difficult and expensive to keep pipelines, tankers, barges, and refinery and terminal storage tanks free of water, alcohols are usually blended with gasoline at terminals into tank trucks ("splash blending"). Water tolerance problems can usually be avoided by using dry fuels and taking care to prevent contact with water during distribution and storage. Sources of water include condensation out of humid air and contaminants in the fuel distribution system.

ASTM D 4814 provides limits and a test method for water tolerance where the fuel is cooled under specified conditions to the temperature at which it is expected to be used.

Deposits. The addition of alcohol may reduce the effectiveness of some carburetor detergents and corrosion inhibitors. As a result, higher concentrations of these additives may be needed to provide performance equivalent to hydrocarbon gasoline with normal additive use levels (17). Part of the poorer carburetor detergent performance observed may be because the addition of some alcohols increases the amount of deposits formed in carburetor throttle bodies. Ethers have not shown a similar effect Studies of port fuel injector deposits did not show any impact on deposit formation with the addition of alcohol. It has been shown that alcohols can contribute to intake valve deposits, which cause cold-start driveability problems (17).

Antiknock Index. The addition of oxygenates increases the Antiknock Index of the base gasoline. The blending Antiknock Index values for various oxygenates when added to a base 87 (R+M)/2 gasoline are shown in Tables 5 and 6. The blending values decrease as the Antiknock Index of the base gasoline increases and as the base gasoline becomes more aromatic.

Heat Content. As shown in Tables 5 and 6, the heating values of oxygenates are less than that of hydrocarbon gasoline (see Heating Value section). This is because the molecule contains oxygen. The heating value of the blend is reduced in proportion to the amount of oxygenate present. For example, adding 10 vol % ethanol reduces the heating value about 3.3%. The reduction in heating value will increase fuel consumption. The loss in fuel economy will depend on the design of the emission control system and fuel metering calibration. For ethanol, a study covering many model years of automobiles showed an overall loss of about 1.7% (44).

Fuel System Materials Compatibility. Fuel system metals, plastics, and elastomers must be compatible with the fuels that will be contacting them.

Unless phase separation occurs, metal corrosion is generally no worse for gasoline–oxygenate blends than that encountered with gasoline, especially if corrosion inhibitors are used. When gasoline–alcohol blends contact water, the aqueous phase which separates out can be particularly aggressive toward metals (44). Gasoline–methanol blends are less tolerant of water and are most corrosive.

Since the 1980s, most fuel system elastomers and plastics have been designed to be compatible with the commonly used concentrations and types of oxygenates. Older vehicles and equipment designed for hydrocarbon gasoline have potential incompatibility problems with oxygenates. Methanol blends are of greatest concern to equipment manufacturers and some do not accept them and most limit their use. In addition to the type and concentration of oxygenate, the aromatics content of the fuel has a significant effect on the performance of the elastomer or plastic.

Volume Growth. A small but measurable growth in volume occurs when alcohols, particularly methanol and ethanol, are added to hydrocarbons. The addition of 10 vol % ethanol can cause an increase in volume ranging from 0.2% for a low relative density gasoline (0.70) to 0.14% for a high relative density gasoline (0.78) (44).

Cooling Effect. Another phenomenon occurs when alcohols, particularly low molecular weight alcohols, are added to gasoline. The temperature of the mixture is decreased a few degrees due to the negative heat of solution.

Effects on Emissions

One air quality related reason for adding oxygenates to a fuel is to increase the oxygen content of the fuel. This reduces carbon monoxide exhaust emissions (62). The amount of the carbon monoxide reduction is a function of the emission control technology used on the vehicles. The greatest reduction is obtained in carbureted vehicles equipped with oxidation-catalysts. The closed loop vehicles without adaptive learning show a lesser response to oxygenate addition and the adaptive learning vehicles exhibit a small reduction in carbon monoxide emissions (62).

Adding oxygenates also has effects on other regulated emissions. Typically oxygenates also provide a modest reduction in exhaust hydrocarbons and a small increase in oxides of nitrogen (26,35,62). The magnitude of these effects appears to be dependent on the quantity of oxygenate in the fuel and on the technology used to control exhaust emissions. The reduction in exhaust hydrocarbons may be a result of the reduced 50% evaporated point rather than an oxygen effect (25).

When ethanol, methanol, or methanol with cosolvent alcohols are added to gasoline to provide the oxygen, there can be a significant increase in vapor pressure (57,58). Also, when an alcohol-containing fuel is mixed with gasoline of the same vapor pressure, the resulting mixture will have a higher vapor pressure than either of the two components (63). The amount of the increase will depend on the ratio of the two components with the maximum increase occurring when the ratio is 80 vol % gasoline to 20 vol % alcohol blend containing about 10 vol % alcohol. When a 10 vol % ethanol blend is mixed with gasoline and both have a 62 kPa (9 psi) vapor pressure, the maximum increase will be about 5.5 kPa (0.8 psi). If the ethanol blend is mixed with an MTBE blend rather than hydrocarbon gasoline, the increase may be as much as 40% less.

As discussed earlier, increasing vapor pressure results in increased evaporative emissions. This is also true when the vapor pressure increase is a result of alcohol addition or mixing alcohol-containing fuels with hydrocarbon gasoline (34,60,63). This increase in evaporative emissions is generally not considered a serious problem in the wintertime low ozone season. However, increased vapor pressure in the summertime can contribute to increased hydrocarbon emissions and more ozone production.

CLEAN AIR ACT AMENDMENTS OF 1990

In the U.S., federal regulations to control gasoline properties did not exist until 1974 when a 0.05 gram per U.S. gallon (0.013 gram per liter) maximum lead limit became effective for unleaded gasoline (45). Previously, gasoline properties were controlled by individual state regulations. Except in California, the primary purpose was to protect consumers from fraud or misrepresentation. California has controlled gasoline properties since 1959 in an effort to improve air quality. Beginning in January 1988, Colorado became the first state to mandate the use of oxygenated gasoline in certain areas around Denver. Later oxygenated gasoline programs were implemented in areas of Arizona (Maricopa County, 1989; Pima County, 1990), Nevada (Clark County and Washoe County, 1989), New Mexico (Albuquerque/Bernalillo County, 1989), and Texas (El Paso, 1991). Starting in the spring of 1989, the EPA Phase I summertime volatility regulations limited specified areas to maximum vapor pressures of 62 kPa (9.0 psi) 65.5 kPa (9.5 psi), and 72.4 kPa (10.5 psi).

National Gasoline Requirements

The portions of the Clean Air Act Amendments of 1990 that were related to fuels provided laws which regulated vapor pressure in 1992 and will have significant impacts on fuel properties in 1995 (6).

Vapor Pressure. The first requirement of the law, which became effective in 1992 for the 48 contiguous states and the District of Columbia, set maximum limits on vapor pressure. Alaska and Hawaii were exempted from the vapor pressure regulations. The basic limit is that all gasolines at refineries and terminals (May 1 through September 15) and at service stations and fleet dispensers (June 1 through September 15) must not exceed 62 kPa (9.0 psi) vapor pressure. In addition, from June 1 through September 15, there is a 53.8 kPa (7.8 psi) maximum limit for those EPA Class B areas not in ozone attainment. Nonattainment areas are those areas that do not meet statutory primary ambient air quality standards (in this case the ozone standard). Class B areas correspond generally to the southern states. California has its own 53.8 kPa (7.8 psi) maximum limit which applies to the whole state and over a longer summertime period than the federal regulations.

Under federal regulations, fuels containing 9 to 10 vol % ethanol are allowed to have a vapor pressure 6.9 kPa (1.0 psi) higher than the applicable limit during the summertime control periods. Gasolines blended with other oxygenates must meet the required limits.

Deposit Control Additives. To prevent the accumulation of intake system deposits the Clean Air Act Amendments of 1990 require that all gasolines must contain a deposit control additive beginning in 1995. As of early 1994, the regulations have not been finalized, but engine test procedures are to be used to certify additives at a specified level of performance.

Lead Antiknocks. Beginning on January 1, 1996, gasoline for highway use cannot contain any lead or lead antiknocks. Thus, only limited quantities of leaded gasoline blended for off-highway use, such as aviation and farming, will be permitted.

Carbon Monoxide Nonattainment Areas

The law requires that states with carbon monoxide nonattainment areas must revise their state implementation plan to require the use of oxygenated gasoline. To assist the states in the development of their plans, the EPA developed guidelines for 39 carbon monoxide nonattainment areas. The program began on November 1, 1992. Some states allow averaging of the oxygen content across all fuels marketed in a given nonattainment area. The average oxygen level required is 2.7 wt % with a minimum oxygen requirement for each gallon of 2.0 wt %. By allowing averaging, the EPA is encouraging the trading of oxygen credits. If averaging is not used, then the minimum oxygen content is 2.7 wt % for all gasoline.

The oxygenates used to meet the requirements must either fall under the "substantially similar" rule or have been issued an EPA waiver. Since the minimum oxygen requirement for oxygenated gasoline is the same as the maximum allowed under "substantially similar," EPA modified the maximum oxygen content to 2.9 wt % for "substantially similar" oxygenates used in the 39 areas. This allows for some margin of error in blending oxygenates and in measuring their oxygen content. When EPA waivers are used for oxygenates, the maximum limits specified in the waivers apply.

The law also permits the waiving, in whole or in part, of the oxygenated gasoline requirements if a state can demonstrate that the use of oxygenated gasoline would prevent or interfere with the attainment of air quality standards for any air pollutant other than carbon monoxide. California set its own oxygen limit at 1.8 wt % to 2.2 wt % because of concern that the higher levels of oxygen required by the federal law will increase the oxides of nitrogen emissions. Such an increase could adversely impact the attainment of ozone, nitrogen dioxide, and particulate air quality standards. California has petitioned EPA to obtain approval for its lower oxygen level.

Ozone Nonattainment Areas

The Clean Air Act Amendments of 1990 also address ozone nonattainment areas. The law applies to the nine worst nonattainment areas and is effective January 1, 1995. It also allows other less severe ozone nonattainment areas the option of choosing (opt in) to have the requirements apply to their areas. Permission is granted on petition by the governor of their state. As of early 1994, several areas in the U.S. northeast and Dallas/Fort Worth, Texas have opted in.

For the period from 1995 through 1999, the law requires during the high ozone season that the volatile organic compounds (VOC) emissions be reduced the greater of at least 15% or of that provided by a fuel formula specified by Congress. The formula is 1.0 vol % maximum benzene content, 25.0 vol % maximum aromatics, 2.0 wt % minimum oxygen, lead free, and the required use of a deposit control additive. Subsequent assessments of the formula fuel indicate that it does not provide the required 15% reduction.

The law also requires during the entire year that toxics emissions [benzene, 1,3-butadiene, acetaldehyde, formaldehyde, and polycyclic organic matter (POM)] be reduced the greater of at least 15% or of that provided by the formula fuel during the period 1995 to 1999. Assessment of the formula fuel showed it would provide more than a 15% reduction in toxics and thus this new level became the requirement. As in the VOC season, specific fuel requirements are a maximum benzene content of 1.0 vol %, minimum oxygen content of 2.0 wt %, prohibition of heavy metals, no increase in oxides of nitrogen emissions, and no dumping into areas not in the program.

The heavy metals ban specifically listed lead and manganese, but did permit metals other than lead to be used if the metal was shown not to increase toxic (mass or cancer-risk basis) vehicle emissions. The antidumping provision was designed to prevent blending of the high volatility components, benzene, aromatics, etc, into fuels for use outside the nonattainment areas if these components were rejected from the refinery pool for use in ozone nonattainment areas.

A fuel that complies with the requirements described above is defined in the law as a "reformulated gasoline."

Reg Neg. The Clean Air Act Amendments of 1990 are the law. However, they do not provide the detailed regulations required to implement and enforce the provisions of the law. The issuing of the regulations is the responsibility of the EPA. Usually, the EPA develops the regulation and then publishes it as a Notice of Proposed Rulemaking in the *Federal Register*. A public hearing on the regulation is held and written comments are accepted. A revised regulation may then be published as a Supplemental Notice of Proposed Rulemaking in the *Federal Register*. Another public hearing may be held and written comments are accepted. The final rule is then published in the *Federal Register*.

Because of the diversity of the interested parties, (EPA, Department of Energy, automobile industry, petroleum industry, fuel marketers, environmentalists, state and local air pollution regulators, driving public, and oxygenate producers), the EPA used a regulatory negotiation (Reg Neg) process to develop reformulated gasoline regulations. They believed that it would be more expeditious and more effective in preventing future litigation than the process of notice and comment rulemaking described above. The process is to convene a committee that represents a balance of interested parties to negotiate in good faith a consensus rule within a fixed period of time. After a consensus agreement is reached, EPA publishes in the *Federal Register* a proposed rule based on the agreement. After publication, participants and the general public will be provided time for comments. However, the participants are bound not to challenge the basic rule.

The Agreement in Principle was signed by the Reg Neg participants on August 16, 1991. Table 7 summarizes the Reg Neg agreement for reformulated gasoline which developed the "simple model" for predicting vehicle VOC and toxics reductions relative to 1990 baseline gasoline in the 1990 vehicle fleet. The "simple model" contains only two variables, vapor pressure and oxygen content, to reduce VOC. Two options are provided for benzene control: (1) a maximum of 1.0 vol % benzene for each gallon, or (2) an average of 0.95 vol % benzene and maximum 1.3 vol % benzene for any given gallon. At the 1.0 vol % benzene level, the vehicle toxics emission reductions shown in Table 7 will be achieved by blending reformulated gasoline containing less than about 27 vol % aromatics. Oxygen content may be controlled for each gallon or on an average basis as shown in Table 7. Vapor pressure is controlled by region. Region 1 (formerly EPA Class B) areas generally are southern states while Region 2 (formerly EPA Class C) areas are northern states. Again in this case either an averaging or an each gallon basis is provided as shown in Table 7. The averaging periods are summarized in Table 7.

Table 7. Reg Neg Agreement 1995–1997 Reformulated Gasoline Regulations

Property	Flat Limit Per Gallon	Average[a]	Cap for Average
Reid vapor pressure, kPa max., June 1–September 15 at retail[b]			
Region 1	49.6 (7.2)	49 (7.1)	51.0 (7.4)
Region 2	55.8 (8.1)	55.2 (8.0)	57.2 (8.3)
Benzene, vol % max.[c,d]	1.0	0.95	1.3
Oxygen content, wt %, min.[d]	2.0	2.1	1.5
Toxics, % Reduction Min.[c]			
Summer	~21	At 1.0 vol % benzene, limits aromatics to ~27 vol % maximum	
Winter	~13.5		
Annual	~18.5		
Sulfur and olefins contents, and 90% evaporation point	≤1990 average levels		

[a] More stringent requirement for averaging if any supplied area is out of compliance with each gallon limits.
[b] Averaging period for vapor pressure and VOC: June 1 to September 15.
[c] Averaging period for benzene, toxics, and NOx: calendar year.
[d] Trading allowed for oxygen and benzene.

The average limits become more restrictive if any supplied area is out of compliance. The vapor pressure limits also apply to ethanol-containing blends and the previously discussed 6.9 kPa (1.0 psi) allowance does not apply; otherwise the VOC requirement would not be met. Refiners and marketers are allowed to trade oxygen and benzene credits. A further requirement for reformulated gasoline is that the sulfur and olefins contents and the 90% evaporated point must be equal to or less than the 1990 levels for the supplied areas.

Complex Model. There also is a "complex model" for predicting VOC, oxides of nitrogen, and toxics reductions. The final model is a result of considerable interfacing of the oil and automobile industries and the EPA. In addition to the vapor pressure, oxygen content, and benzene content terms of the "simple model," the "complex model" contains sulfur, olefins, and aromatics contents, oxygenate type, and distillation parameter terms. The distillation terms are the percents evaporated at 200°F (93°C) and 300°F (149°C) which are generally referred to as E200 and E300, respectively. The percent evaporated at given temperatures terms are more linear functions than the more commonly used temperature for a percent evaporated which for 90% evaporated is generally referred to as T_{90}.

Antidumping. The Reg Neg agreement also contains a section on antidumping regulations. The purpose of this regulation is to prevent blending in other areas with high volatility components, benzene, aromatics, etc, which were rejected from the refinery pool as unsuitable for blending reformulated gasoline. The antidumping baseline is based on 1990 gasoline properties. To meet the "no increase in benzene requirement," either the "simple model" or the "complex model" may be used. If the "simple model" is used, the additional limitation that sulfur content, 90% evaporated point, and olefins content may not exceed 125% of the 1990 averages is required.

Presidential Intervention. In October 1992, then the President, George Bush announced that the EPA was to modify the Reg Neg agreement and establish new volatility requirements to encourage the use of ethanol. The maximum vapor pressure for all reformulated gasoline in Region 2 (northern states), except for ethanol blends, is to be reduced from 55.8 kPa (8.1 psi) as shown in Table 7 to 53.8 kPa (7.8 psi). Then, gasoline–ethanol blends, up to 30% market share of reformulated gasoline, could be marketed with a 6.9 kPa (1.0 psi) higher vapor pressure in Region 2. States could choose to permit a larger market share for ethanol blends, but they would have to reduce the vapor pressure level of reformulated gasoline more to compensate for the increased emissions of the incremental ethanol blends. In Region 1 (southern states), the states may choose to participate in a reformulated gasoline with ethanol program that reduces the vapor pressure from 49.6 kPa (7.2 psi) to 48.2 kPa (7.0 psi) maximum and allows 20% market share for gasoline-ethanol blends which will have a 6.9 kPa (1.0 psi) higher maximum limit.

It is the former President's position that the lowering of the maximum vapor pressure of reformulated gasoline without ethanol and of the base stock used for ethanol blending will offset the increased hydrocarbon emissions caused by the vapor pressure increase that results from ethanol addition. Thus, the program, according to President Bush, will be neutral with respect to ozone formation.

Final Reformulated Gasoline Rule. The EPA issued a proposed rule to implement the Bush plan, but it received many negative comments. A big complaint was that the proposed rule invalidated the Reg Neg agreement and would increase evaporative emissions. A further important point was that the proposed rule was so complex that it was unworkable. When the final rule was issued on December 15, 1993, the special treatment for ethanol

Table 8. California Phase 2 Gasoline Regulations

Property	Flat Limit	Average	Cap for Average or Alternative Formulation
Reid vapor pressure, kPa (psi) max.[a]	48.3 (7.00) Summertime		48.3 (7.00) Summertime
Sulfur, wt % max.	0.0040	0.0030	0.0080
Benzene, vol % max.	1.0	0.8	1.2
Aromatics, vol % max.	25	22	30
Olefins, vol % max.	6.0	4.0	10.0
90% Evaporated point, °C (°F) max.	149 (300)	143 (290)	166 (330)
50%, Evaporated point, °C (°F) max.	99 (210)	93 (200)	104 (220)
Oxygen content, wt %	1.8–2.2		1.8–2.7 Wintertime

blends was eliminated and the Reg Neg agreement reinstated. However, the EPA announced that new rules would be proposed to require that 30% of the oxygen in reformulated gasoline must come from renewable oxygenates. Further, only renewable ethers (ETBE and TAEE) would be permitted in the summertime to prevent increased evaporative emissions. In the wintertime, renewable alcohols or renewable ethers would be permitted.

The final rule involves Phases I and II. Phase I is implemented effective January 1, 1995, and will reduce both ozone forming VOC and air toxics emissions from vehicles 15 to 17%. Oxides of nitrogen emissions must not increase. The Reg Neg "simple model" will be used through 1997. The "complex model" must be used after January 1, 1998, but can be used earlier at the option of the producer. Phase II becomes effective on January 1, 2000, and will reduce ozone forming VOC emissions from vehicles 25 to 29%. Air toxics emissions will be reduced 20 to 22%. The EPA is also requiring that oxides of nitrogen vehicle emissions be reduced 5 to 7%. The "complex model" will be required for Phase II. However, the weighting for exhaust emissions becomes greater relative to evaporative emissions in 2000.

CALIFORNIA PHASES 1 AND 2 REGULATIONS

Phase 1

As discussed previously, California has led the nation in implementing regulations to help reduce air pollution. The California Phase 1 gasoline regulations became effective in January 1992 (64). The vapor pressure limits are the same as those for federal Class B ozone nonattainment areas, except that they apply to all areas of the state and for a longer time period (as long as from April 1 through October 31, depending on the area). California requires that all gasolines must contain a deposit control additive and that the fuel effectively prevents port-fuel injector and intake valve deposits as determined by defined vehicle test procedures. The California procedures are based on the CRC 15 minute run/45 minute hot-soak portfuel injector test procedure and the Southwest Research Institute/BMW 16,000 km (10,000 mile) intake valve keep-clean test procedure (64–66). California also requires port fuel injector deposit cleanup performance.

Leaded gasoline can no longer be marketed in California, but fuel that contains manganese antiknock additive or does not meet the other requirements of unleaded gasoline can be dispensed through wide nozzles into vehicles designed to use leaded gasoline. Effective on January 1, 1994, only unleaded gasolines can be marketed.

Phase 2

California has a Phase 2 gasoline regulation program which becomes effective on March 1, 1996. Table 8 shows the limits for vapor pressure; oxygen, sulfur, benzene, aromatics, and olefins contents; and 90% and 50% evaporated points. The limits for benzene and oxygen are about the same as those of federal reformulated gasoline. The California aromatics limit is a specific value while the federal limit is dictated by the 15% toxics reduction requirement. The California limit for vapor pressure is more restrictive than federal limits. The 50% evaporated point is not controlled by federal regulations. The federal controls on sulfur content, olefins content, and 90% point are not fixed values for all gasolines in that they are based on refinery or company-wide averages. The California sulfur and olefins limits are substantially below almost all current refinery or company averages. The California flat limit applies to each gallon of reformulated gasoline when averaging is not used. Table 8 also shows the average limit and the corresponding maximum value for any given gallon of gasoline. California has developed a Predictive Model, which is like the federal "Complex Model," but does not use the same terms or coefficients.

BIBLIOGRAPHY

1. *Webster's New World Dictionary*, Third College Edition, Simon and Schuster Inc., New York, 1988.
2. *Digest of State Inspection Laws—Petroleum Products*, API Publication 926, American Petroleum Institute, Washington, D.C., 1992 Edition.
3. C. Roberts, *Ethyl: A History of the Corporation and the People Who Made It*, University Press of Virginia, Charlottesville, Va., 1983.
4. *Code of Federal Regulations*, Title 40, Part 80, Section 20 (40 CFR 80.20), July 1, 1979 through July 1, 1989.
5. Environmental Protection Agency, *Federal Register*, **44**(178), 53144 (Sept. 12, 1979).

6. Title I—Provisions for Attainment and Maintenance of National Ambient Air Quality Standards, *Public Law 101-549*, U.S. Government Printing Office, Washington, D.C., Nov. 15, 1990.
7. J. K. Paul, *Ethyl Alcohol Production and Use as a Motor Fuel*, Noyes Data Corporation, Park Ridge, N.J., 1979.
8. H. F. Williamson and A. R. Daum, *The American Petroleum Industry—The Age of Illumination 1859–1899*, Northwestern University Press, Evanston, Ill., 1959, pp. 234–238.
9. L. Raymond, "Today's Fuels and Lubricants—and How They Got That Way," *Automotive Engineering*, **88**(10), 27 (Oct. 1980).
10. "Motor Gasolines," *Technical Publication*, Chevron Research and Technology Company, Richmond, Calif., July 1990.
11. W. F. Bland and R. L. Davidson, *Petroleum Processing Handbook*, McGraw-Hill, New York, 1967.
12. *1994 Annual Book of ASTM Standards*, Section 5, 05.04, ASTM, Philadelphia, Pa., 1994.
13. *CRC F-28-75 Modified Uniontown Technique*, Coordinating Research Council, Atlanta, Ga., 1975.
14. *CRC E-15-93 Technique For Determination of Octane Number Requirement of Light Duty Vehicles*, Coordinating Research Council, Atlanta, Ga., 1993.
15. "SAE J312, Automotive Gasolines Recommended Practice," *SAE Handbook*, Vol. 3, Society of Automotive Engineers, Warrendale, Pa, 1994.
16. M. J. McNally and co-workers, "The Effects of Gasoline Octane Quality on Vehicle Performance—A CRC Study," *SAE Paper 912394*, SAE, Warrendale, Pa., Oct. 1991.
17. L. M. Gibbs, "Gasoline Additives—When and Why," *SAE Paper 902104*, (SAE Transactions **99**), Oct. 1990.
18. E. M. Shelton, M. L. Whisman, and P. W. Woodward, "Trends in Motor Gasolines: 1942–1981," *DOE / BETC / RI-82 / 4*, Bartlesville Energy Technology Center, Bartlesville, Okla., June 1982.
19. C. L. Dickson, P. W. Woodward, and P. L. Bjugstad, "Trends of Petroleum Fuels, 1987," *NIPER-309*, National Institute for Petroleum and Energy Research, Bartlesville, Okla., Dec. 1987.
20. C. L. Dickson and P. W. Woodward, *Motor Gasolines, Summer 1993, NIPER-183 94 / 1*, Bartlesville, Olka., July 1994.
21. *1994 Annual Book of ASTM Standards*, Section 5, 05.01, American Society for Testing and Materials, Philadelphia, Pa., 1994.
22. "Automotive Gasolines," G. V. Dyroff, ed., MNLI, American Society for Testing and Materials, Philadelphia Pa., 1993.
23. D. A. Barker, L. M. Gibbs, and E. D. Steinke, "The Development and Proposed Implementation of the ASTM Driveability Index for Motor Gasoline," *SAE Paper 881668*, (*SAE Transactions* **97**) Oct. 1988.
24. *1994 Annual Book of ASTM Standards*, Section 5, 05.03, American Society for Testing and Materials, Philadelphia, Pa., 1994.
25. J. A. Gething, "Distillation Adjustment: An Innovative Step to Gasoline Reformulation," *SAE Paper 910382*, SAE, Warrendale, Pa., Feb. 1991.
26. Auto/Oil Air Quality Improvement Research Program, SAE SP-920, SAE, Warrendale, Pa., Feb. 1992.
27. "Initial Mass Exhaust Emissions Results From Reformulated Gasoline," *Technical Bulletin No. 1, Auto / Oil Air Quality Improvement Research Program*, Coordinating Research Council, Atlanta, Ga., Dec. 1990.
28. "Preliminary Results—Phase II Heavy Hydrocarbon (T90) Study, Matrix 1, Auto/Oil Air Quality Improvement Research Program," Coordinating Research Council, Atlanta, Ga., Dec. 7, 1992.
29. J. O. Esinger and D. P. Barnard, "A Forgotten Property of Gasoline," *SAE Paper 350106, SAE Journal*, **37**(2), 293, (Aug. 1935).
30. J. P. Graham, B. Evans, R. M. Reuter, and J. H. Steury, "Effect of Volatility on Intermediate-Temperature Driveability with Hydrocarbon-Only and Oxygenated Gasolines," *SAE Paper 912432*, SAE, Warrendale, Pa., Oct. 1991.
31. D. T. Wade, "Factors Influencing Vehicle Evaporative Emissions," *SAE Paper 670126,* SAE, Warendale, Pa., Jan. 1967.
32. M. W. Jackson and R. L. Everett, "Effect of Fuel Composition on Amount and Reactivity of Evaporative Emissions," *SAE Paper 690088*, SAE, Warrendale, Pa., Jan. 1969.
33. T. H. DeFries, S. Kishan, and R. F. Klausmeier, "Relative Importance of 22 parameters to Evaporative Emissions—A Sensitivity Analysis of EVAP 2.0," *SAE Paper 881594*, Warrendale, Pa., Oct. 1988.
34. *Effect of Fuel RVP and Fuel Blends on Emissions at Non-FTP Temperatures.* Vol. 1: Interim Summary Report, American Petroleum Institute, Washington, D.C., July 1991.
35. Emission Results of Oxygenated Gasoline and Changes in RVP," *Technical Bulletin No. 6, Auto / Oil Air Quality Improvement Research Program*, Coordinating Research Council, Atlanta, Ga., Sept. 1991.
36. W. F. Marshall and G. A. Schoonveld, "Vapor Space Flammability of Automobile Tanks Containing Low RVP Gasolines," *SAE Paper 902096*, SAE, Warrendale, Pa., Oct. 1990.
37. "Evaluation of Expressions for Fuel Volatility," *CRC Report No. 403,* Coordinating Research Council, Atlanta, Ga., 1967.
38. "1983 CRC Two-Temperature Vapor Lock Program Using Gasoline-Alcohol Blends, *CRC Report No. 550*, Coordinating Research Council, Atlanta, Ga., October 1986.
39. *1994 Annual Book of ASTM Standards*, Section **5**, **05.02**, American Society for Testing and Materials, Philadelphia, Pa., 1994.
40. "Effects of Fuel Sulfur on Mass Exhaust Emissions, Air Toxics, and Reactivity," *Technical Bulletin No. 8, Auto / Oil Air Quality Improvement Research Program,* Coordinating Research Council, Atlanta, Ga., February 1992.
41. "Mass Exhaust Emissions Results From Reformulated Gasolines in Older Vehicles," *Technical Bulletin No. 4, Auto / Oil Air Quality Improvement Research Program*, Coordinating Research Council, Atlanta, Ga., May 1991.
42. *ASTM D 1250 Petroleum Measurement Tables—Volume Correction Factors*, American Society for Testing and Materials, Philadelphia, Pa., 1980.
43. "SAE J1498, Heating Value of Fuels Information Report," *SAE Handbook*, Vol. 3, Society of Automotive Engineers, Warrendale, Pa., 1993.
44. *Alcohols and Ethers—A Technical Assessment of Their Application as Fuels and Fuel Components,* API Publication 4264, 2nd ed., American Petroleum Institute, Washington, D.C., July 1988.
45. Environmental Protection Agency, *Federal Register,* **38**(6) 1254, (Jan. 10, 1973).
46. R. C. Tupa and C. J. Dorer, "Gasoline and Diesel Fuel Additives for Performance/Distribution/Quality—II," *SAE Paper 861179*, (*SAE Transactions*, **95**) Sept. 1986.
47. B. Y. Taniguchi, R. J. Peyla, G. M. Parsons, S. K. Hoekman,

and D. A. Voss, "Injector Deposits—The Tip of Intake System Deposit Problems," *SAE Paper 861534*, Oct. 1986.
48. G. T. Kalghatgi, "Deposits in Gasoline Engines—A Literature Review," *SAE Paper 902105*, (*SAE Transactions* **99**), Oct. 1990.
49. D. W. Houser, R. F. Irwin, L. J. Painter, and G. H. Amberg, "Field Tests Show Gasoline Deposit Control Additives Effective in Emission Reduction," *APCA Paper 71-83*, June–July 1971.
50. D. W. Hall and L. M. Gibbs, "Carburetor Deposits—Are Clean Throttle Bodies Enough?" *SAE Paper 760752,* SAE, Warrendale, Pa., Oct. 1976.
51. J. A. Gething, "Performance-Robbing Aspects of Intake Valve and Port Deposits," *SAE Paper 872116*, SAE, Warrendale, Pa., Nov. 1987.
52. R. A. Lewis and co-workers, "A New Concept in Engine Deposit Control Additives for Unleaded Gasoline," *SAE of Japan Paper 830938*, Nov. 1983.
53. K. R. Houser and T. A. Crosby, "The Impact of Intake Valve Deposits on Exhaust Emissions," *SAE Paper 922259,* SAE, Warrendale, Pa., Oct. 1992.
54. *Proceedings of the CRC Workshop on Intake Valve Deposits*, Coordinating Research Council, Atlanta, Ga., Aug. 22–24, 1989.
55. J. P. Graham and B. Evans, "Effects of Intake Valve Deposits on Driveability," *SAE Paper 922220*, SAE, Warrendale, Pa., Oct. 1992.
56. "SAE J 1297, Alternative Automotive Fuels," *SAE Handbook,* Vol. 3, Society of Automotive Engineers, Warrendale, Pa., 1994.
57. R. L. Furey and K. L. Perry, "Volatility Characteristics of Blends of Gasoline with Ethyl Teritary-Butyl Ether (ETBE)," *SAE Paper 901114,* SAE, Warrendale, Pa., May 1990.
58. R. L. Furey, "Volatility Characteristics of Gasoline-Alcohol and Gasoline Ether Fuel Blends," *SAE Paper 852116*, SAE, Warrendale, Pa., Oct. 1985.
59. "Assessment and Correlation of Customer and Rater Response to Cold-Start and Warmup Driveability," *CRC Report No. 585,* Coordinating Research Council, Atlanta, Ga., Oct. 1993.
60. L. M. Gibbs, "The Impact of State Air Quality and Product Regulations on Current and Future Fuel Properties," in K. H. Strauss and W. G. Dukek, eds., *The Impact of U.S. Environmental Regulations on Fuel Quality,* ASTM STP 1160, American Society for Testing and Materials, Philadelphia, Pa., Jan. 1993.
61. "Effects of RVP, T_{50}, and Oxygenates on Hot-Start and Driveability Performance at High and Low Altitude," *CRC Report No. 584*, Coordinating Research Council, Atlanta, Ga., June 1993.
62. "Wintertime Exhaust Emissions at High Altitude and Sea Level Using Gasoline/Oxygenate Blends," *CRC Report No. 572*, Coordinating Research Council, Atlanta, Ga., Oct. 1992.
63. R. L. Furey and K. L. Perry, "Vapor Pressure of Mixtures of Gasolines and Gasoline-Alcohol Blends," *SAE Paper 861557,* SAE, Warrendale, Pa., Oct. 1986.
64. *California Code of Regulations, Title 13, Chapter 5, Article 1, Sections 2250–2257*, Barclays Law Publishers, South San Francisco, Calif., 1992.
65. R. C. Tupa, B. Y. Taniguchi, and J. D. Benson, "A Vehicle Test Technique for Studying Port Fuel Injector Desposits—A Coordinating Research Council Program," *SAE Paper 890213*, SAE, Warrendale, Pa., Feb. 1989.
66. W. H. Bitting and co-workers, "Intake Valve Deposits Fuel Detergency Requirements Revisited," *SAE Paper 872117,* SAE, Warrendale, Pa., Nov. 1987.

TRANSPORTATION FUELS—ELECTRICITY

Mark Mills
Mills McCarthy & Assoc.
Chevy Chase, Maryland

As the twentieth century comes to a close, few citizens would deny the profound personal and economic benefits of the transportation revolution. Never before in history have so many had so much mobility and personal freedom. At the same time, rising affluence and steadily declining real costs of transportation have made flying and driving so popular that in many places the severe smog and congestion have become almost as much of a defining experience as a first solo drive in a car or flight on an airplane. In addition, the transportation sector's appetite for fuel has raised concerns about the nation's dependence on oil imports. Two-thirds of all oil is used for transportation. Energy consumption for transportation also fuels the debate over global warming. Transportation carbon dioxide emissions have grown three times as much as emissions from all other sources over the past two decades.

There are some striking parallels between the state of transportation in America today and the state of affairs at the end of the nineteenth century. The burgeoning industrial revolution brought increasing population and transportation demands in the late 1800s. The horse-drawn cart, which for centuries served people well, was the primary form of transportation. It was abundantly clear that the option was no longer up to the task, nor was the horse a clean motive source at the levels of demand experienced in the late 1800s. By the time the 1890s came to a close, the problems sounded, at least in their broad impact, remarkably like those of today.

The United States of 1880 had over 100,000 horses and mules pulling 18,000 cars and depositing in excess of one million pounds of manure each day–which attracted flies until it dried up and was blown around by the wind (1). By the dawn of the twentieth century, there were over three million horses in U.S. cities. Traffic jams were commonplace, and were so bad that police often had to break them up using billy clubs. There was more to the transportation problem than emissions from the horse power in those days. Typically, 15,000 horse carcasses had to be removed (not always expeditiously) from the streets of New York City each year (2). The pollution problems of transportation forced planners and policy makers to recognize a connection between waste in the cities and health, and in fact spawned the era of sanitation laws.

The solution to the nineteenth century transportation congestion and pollution problem was dramatized at the 1884 Chicago World's Fair. The Fair featured a variety of electric people movers; an electrified moving sidewalk, battery driven launches on the lagoons, electric elevators, and an electric trolley (1). The electric trolley took U.S. cities by storm and came to dominate urban transportation in the early part of the twentieth century, literally transforming the landscape. The electric trolley was not, however, a new idea at the time. There had been some four decades of experimentation with the technology before it was adopted (3).

The transportation revolution which began with the electric trolley has typically been viewed in technical terms. But as David Nye, American history professor at the Copenhagen University, in his insightful chronicle of the dawn of the electrical revolution noted, its impact was far broader:

> Just as the electric light for too long has been understood merely in terms of its practicality, so too electric traction has been studied primarily as a form of transportation, without recognizing that it also became a vehicle of political ideologies, or seeing that it altered the city's image (1).

Planners and prognosticators are once again reviewing options and preparing plans to alleviate transportation problems. The landscape of the clean transportation debate strongly resembles that of a century ago. The electric trolley truly revolutionized the city. But it had built-in limits and was displaced almost entirely within 50 years by the automobile. The cycle continues. Professor Nye's observation about planners' prescience, or lack thereof, may be a useful caution for today's planners:

> Planners were also wrong about the automobile. Just as few imagined the demise of the streetcars, at first few thought the automobile would serve as mass transportation.

Today's debate over future clean and convenient transportation options is focused on the congestion and emissions caused by both the personal automobile and heavily traveled intercity air traffic routes. And while the debate has largely been dominated among the oil, automotive, and aviation industries, new players are emerging: the electricity and natural gas industries, and their suppliers and allies. At the moment many favor natural gas as the fuel of choice to alleviate at least the emissions part of the transportation problem. But the time frames and scope of the problem require a broader look at where the nation could be 20 years from now.

While many fuels could replace gasoline, natural gas and electricity are the clear leaders. There is no evidence that biomass-based ethanol, methanol, coal, or hydrogen are remotely competitive using foreseeable technologies or resources. In each of these cases, the net energy consequences on a life cycle basis are worse, and the net emissions impacts are frequently worse. Both options use domestic resources, although significant reliance on natural gas vehicles (NGVs) would likely result in substantial increases in natural gas imports.

NGVs have a clear advantage now; they are already practical. Electric vehicles (EVs) for road use, on the other hand, have a way to go. EVs are currently practical for a variety of nonroad applications ranging from forklifts and airport tugs, to golf carts and various types of materials handling vehicles as well as a wide range of nonroad mobile sources. Nonroad EVs can offer significant potential for near-term emissions reductions of smog precursors and CO_2. But, when EVs get there, how much better than NGVs will they be? In addition, EVs can replace more than gasoline ground transportation. For example, magnetic levitation trains (the only means for effectively electrifying airplanes) could be practical replacements for high density domestic intercity travel corridors.

NGVs already offer an environmentally superior alternative to gasoline vehicles. Therefore, the issue for policy makers and parties interested in committing public and private resources to develop practical electric vehicles is: are EVs enough better than NGVs to justify substantial investment in their development?

BACKGROUND

The driving forces in the search for cleaner and more convenient alternatives to the current oil-dominated transportation system are primarily environmental and economic. Market barriers to new transportation technologies are frequently cited; in particular, existing vehicle infrastructure for sales, fueling, and maintenance. Far from being barriers these characteristics of the existing transportation infrastructure have set the standards against which any alternative will be held. New transportation options have little chance of significant penetration if they do not replicate or generally improve on existing economic, convenience, and performance characteristics of gasoline vehicles.

Indeed, it is the relative convenience and low cost of today's automobile and air travel that fuel the rapid increase of travel. Most projections do not envision a diminution in the growth in travel over the next two decades. Figure 1 shows that passenger travel has grown about 60% over the past two decades, and will grow yet another 50% over the next two. Similarly, as Figure 2 illustrates, domestic air travel is rising rapidly and will have grown about 8-fold between 1970 and 2010.

As the rest of the marketplace has switched away from the direct use of combustible fuels, the transportation sector has come to account for an ever larger share of total U.S. oil consumption (6). As Figure 3 shows transportation now consumes two-thirds of the nation's oil, up from 50% two decades ago. Two decades from now, unless there is a change transportation will consume about 70% of all oil use.

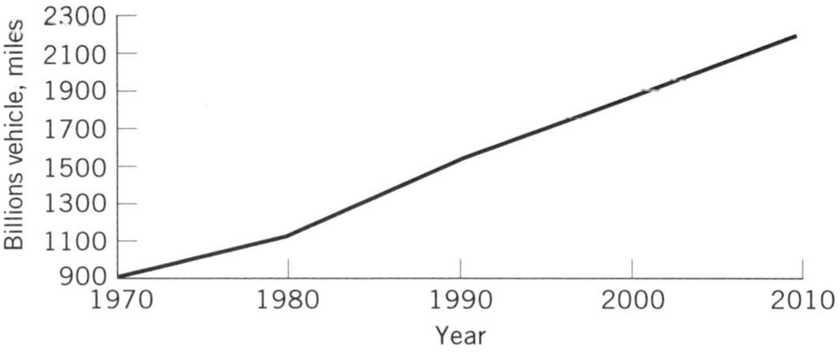

Figure 1. Total U.S. passenger car travel, historic and projected trends (4,5).

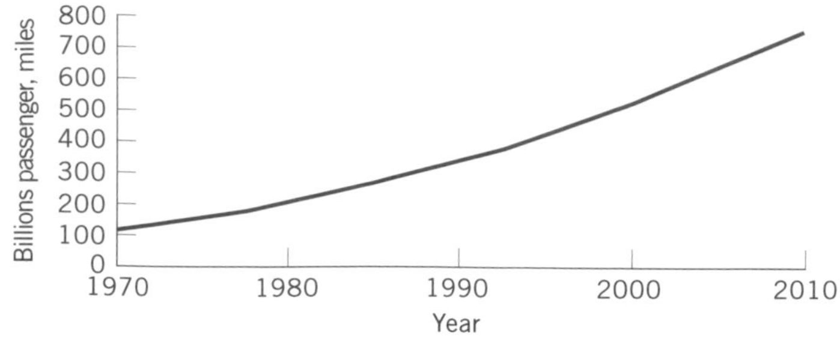

Figure 2. Total domestic air travel, historic and projected trends (4,5).

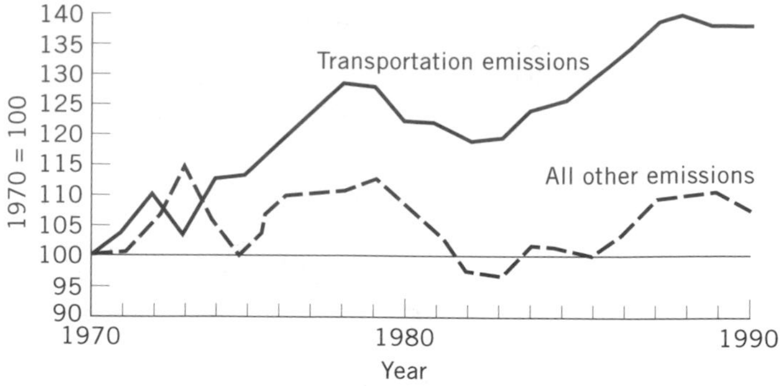

Figure 4. Historic carbon dioxide emissions: transportation vs. all other.

Numerous environmental issues cannot be addressed without considering transportation's unique characteristics. At least 50% of the smog problem in U.S. cities arises from emissions by cars, trucks, buses, and aircraft. With respect to the continuing debate over global warming, plans to minimize future carbon dioxide (CO_2) emissions must contend with the fact that transportation CO_2 emissions have grown at a pace some three times greater than all other sources of emissions, as shown in Figure 4. Projections (see Fig. 5) show that the historic trend is expected to continue; transportation sector CO_2 emissions are expected to grow far more rapidly than all other sources combined.

The transportation sector is made up of passenger as well as freight travel. In addition, transportation occurs by car, truck, bus, rail, and water. The focus of this analysis is on passenger transportation on the ground and by air. Ground transportation can be converted from gasoline to natural gas or electricity, and to a significant extent air transportation could be displaced by high speed electric maglev trains. Air and ground passenger transportation account for two-thirds of all transportation energy use and CO_2 emissions: ground passenger transportation is 50% of the total transportation sector, and passenger air travel 16%. All other forms of transportation, both for people and freight, marine, rail, truck, bus, mass transit, total the balance of 34% (4).

METHODOLOGY

The methodology utilizes a full fuel-cycle evaluation, and a technology-balanced criteria. Fuel consumption and emissions with all aspects of the fuel cycles are included: refining and/or conversion, transportation/distribution, and end-use conversion.

The analysis is based on a comparison of similar 25 mpg and 40 mpg vehicles. The technology of a 25 mpg vehicle is taken as typical for 1995 and the 40 mpg vehicle for 2010 respectively. The reduced weight, rolling and wind resistance (and similar mechanical gains) associated with 40 mpg cars are assumed to be symmetrically available to both NGVs and EVs.

The analysis compares the emissions from NGVs and gasoline vehicles and EVs that employ commercially viable propulsion systems. Of course, EVs are not today commercially available. While on-road prototypes, such as the Chrysler TEVan, are important developmental vehicles, they do not represent the type of commercially viable product that would be in dealers' showrooms post-1995 and ca. 2010. There is no relevance to comparing the environmental impact of today's EV prototype with a commercial gasoline or NGV. Only commercially viable vehicles will ever compete (3).

The principal energy/emissions difference between prototype and practical EVs is the amount of electricity required per mile of travel. Current prototypes use at least twice as much electricity as will future commercially available vehicles. Analyses based on current prototypes will grossly understate the environmental benefits of EVs. For example, the NESCAUM analysis of electric vehicles used the current 0.5 kwh/mile of travel as a starting baseline, with 0.25 and 0.1 identified as the most efficient projected for 2015 (3). EVs that are less electricity-intensive

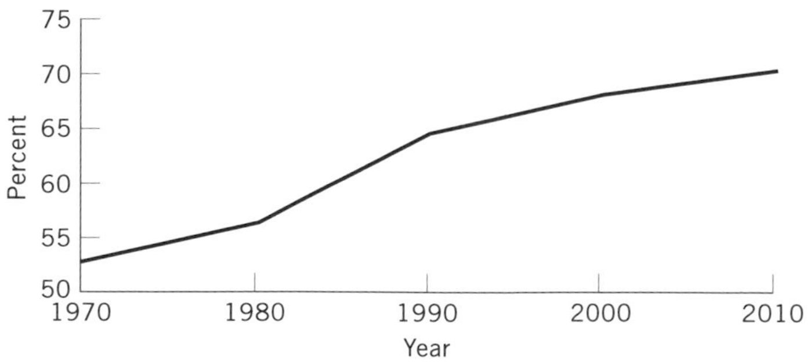

Figure 3. Share of total U.S. oil consumption used for transportation (4,6).

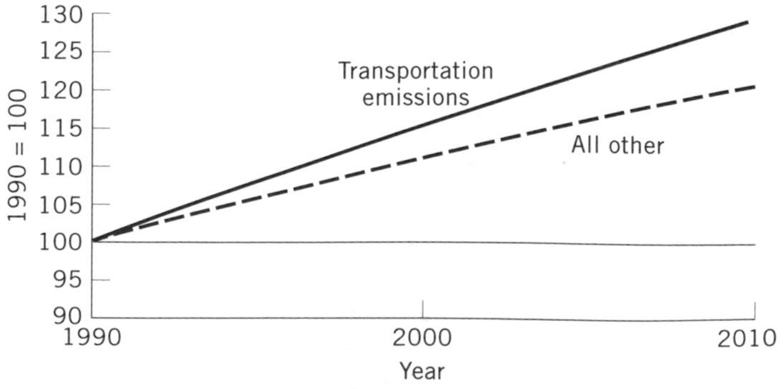

Figure 5. Projected carbon dioxide emissions: transportation vs. all other.

than today's prototypes are a prerequisite for achieving practical range.

The financial impediments to EVs are principally the absence of sufficient demand to mass produce practical vehicles and batteries; neither will be achieved until the obvious technology or technologies of choice are finally identified and proven. Without economic viability no alternative will penetrate. Therefore, this analysis is an environmental assessment based on achievable technological targets.

The greatest uncertainties in this type of analysis fall into four categories:

Future Price of Natural Gas. Comparisons of operating costs are frequently based on the assumption that users will pay the current industrial price for natural gas (7). The current industrial price is about one-half the current residential price for natural gas. In addition, the price of natural gas in the year 2010 (the relevant time for price comparisons), is projected to be substantially higher than today, possibly 1.5 times higher (8,9). Doubling the price of natural gas could almost wipe out the operating cost advantage of NGVs over gasoline vehicles.[9]

Future Price of Batteries Improved battery performance is prerequisite for practical EVs. There is no experience with probable future battery costs since little is known about the likely technology-of-choice and costs of volume production. Nonetheless, it is safe to say that they will be significantly cheaper than today.

Methane Leakage. Methane is one of the principal gases implicated in the global warming theory. Methane is the principal component of natural gas. Burning methane results in about one-half as much CO_2 per BTU as does burning coal. However, unburned methane, ie, leaking natural gas, is at least 20 times more powerful in global warming terms than is CO_2. In recent years, significant research has been undertaken to document methane leaks from the existing natural gas network (10). There appears to be precious little information available regarding the potential for methane leaking from a widespread commercial system that involves compressing natural gas, filling the high pressure tanks in NGVs, and operating and maintaining high-pressure fuel systems on NGVs. There is a similar paucity of data on potential leaks from the highly pressurized fuel system or fugitive leaks from the engine during vehicle operation. Because of the powerful effect of methane, extremely small leak rates can substantially erode, and could entirely erase the greenhouse gas advantage of NGVs over gasoline vehicles.

Advancing Gasoline Vehicle Technology. Proponents of NGVs and EVs frequently make the implicit assumption that technology is fixed regarding the future performance of gasoline-powered vehicles. The magnitude of technological inertia associated with the internal combustion engine suggests that there are still technology surprises in store. Some potential improvements in internal combustion engine technology have the potential to significantly improve the overall energy efficency, and therefore the emissions, associated with the old-fashioned car engine.

Diamond-coated engine components almost eliminating friction (an emerging technology with rapidly declining costs), ceramic engines, and improved plastics are only a few examples of the rapidly evolving area of materials science for automotive engines. Improvements in vehicle weight, rolling resistance, and wind resistance benefits NGVs and EVs as much as gasoline vehicles. However, research in ceramic turbocharged engines suggests that doubling the thermodynamic efficiency of an internal combustion engine may be possible; new computer-controlled camshafts may boost fuel efficiency as much as 20% (11).

There is a probability that gasoline vehicles will be substantially more efficient and cleaner in the future. This presents a moving target for competing fuels, and indicates that a competitive technology will need to be dramatically better than gasoline vehicles.

OIL: THE BASELINE

There are a myriad incremental engineering finesses which can reduce emissions from gasoline engines, eg, catalytic converters, multiple valves, and computer-controlled fuel-injection. Acoustical mufflers, for example, offer a very intriguing electrotechnology already available (from Noise Cancellation Technologies) to reduce fuel consumption. The principal: detect noise at the engine and in the exhaust pipe, generate an equal and opposite noise via speakers placed as a collar around the exhaust pipe, thereby cancelling the exhaust noise. Removing the baffles in a muffler is now possible and results in reduced backpressure, thereby increasing fuel efficiency by several percent. Environmental impacts go down when the amount of energy needed to carry a person one mile goes down; ie, the most energy-efficient transportation mode is virtually always the one with the smallest environmental impact.

There are, of course, many criteria determining how energy-efficient a transportation mode can be. Similarly, there are many criteria determining which mode is chosen. Moving very large quantities of freight across the ocean is most economical and practical on a ship. This also happens to be rather energy-inefficient. The most energy-efficient mode of transportation ever invented is the bicycle. It is even 30% more energy-efficient than walking. (Although here the fuel energy is derived from food, of course, but in terms of the physics, it is still energy.) But, both walking and bicycling, while frequently enjoyable, have obvious limitations for many transportation applications.

Table 1 below summarizes the range of energy efficiencies associated with the principal means of moving people (12). The old-fashioned bus is a surprise performer for many people. Close on its heels in energy efficiency is the twenty-first century alternative to flying between cities in the United States, the 300 mph magnetically-levitated train, or maglev. A New York to Washington D.C., or Los Angeles to Sacramento maglev option, comparably priced to flying, would in all likelihood be heavily used, and would provide quicker door-to-door service than flying, while also saving energy.

Transportation modes are chosen for reasons largely

Table 1. Energy Efficiency of Transportation

Mode of Transport	No. of Passengers/Vehicle	Passenger mpg
Bicycle	1	650 (equivalent)
Bus (intercity)	45	225
Maglev train	140	200
Subway	1000	150
Auto (high-occupancy)	4	120
A300 aircraft	267	80
Auto (typical)	1	30
Executive jet	8	16
Snowmobile	1	12
Ocean liner	2000	10

unrelated to energy efficiency. The transportation modes of choice are, in effect, revealed in the shares of total transportation energy used by each technology (13) (see below).

Mode of Transport	Share of Total Transport Energy, %
Auto	48
Truck	37
Air (commercial)	8
Pipelines	3
Rail (freight)	2
Bus	1
Air (other)	1
Water	1
Rail (passenger)	<0.1
Mass transit	<0.1

The above data suggest that reducing the environmental emissions associated with cars would provide the single largest benefit. As a practical matter the next largest opportunity is associated with commercial air travel via electric airplanes, ie, maglev trains. There is no realistic comparable NGV opportunity in regard to air travel. Extracting hydrogen from natural gas for fueling conventional aircraft is theoretically feasible, but would result in significantly greater expense and overall energy consumption.

The emissions baseline that NGVs and EVs must compete against are shown in Table 2. Emissions directly from the vehicle arise from the engine's combustion process. Fuel cycle emissions arise from processing and distributing the gasoline. Total emissions are shown for both current 25 mpg cars and future (large) 40 mpg cars.

NATURAL GAS VEHICLES

There are in excess of 30,000 NGVs in operation, principally by large truck fleet operators such as Brooklyn Union Gas (15). This is a sufficient number to establish commercial viability. Compressed natural gas filling stations are still rare, however. Amoco, perhaps the biggest booster of NGVs, still has only 12 compressed natural gas filling stations anywhere in the country, compared to nearly 10,000 gasoline filling stations (16).

In order to properly evaluate the net environmental impact of switching from gasoline to NGVs, there are four important factors (in addition to basic emissions data).

Energy Efficiency. Retrofitting a gasoline vehicle to burn compressed natural gas can reduce the overall energy efficiency of the engine by as much as 10%. On the other hand, building a vehicle specifically to take advantage of the properties of natural gas would permit an improvement in efficiency of up to 10% (17). In this analysis, it is assumed that purpose-built NGVs are used, thus gaining the full 10% energy benefit.

Emissions. Previous analyses summarized in Table 3 have included in the emissions tally the various parts of the fuel cycle to provide the natural gas to the vehicles. Emissions from the fuel cycle arise during the extraction, processing, transportation, and distribution of natural gas.

Compression Emissions. In order to hold a sufficient quantity of natural gas in a vehicle, the gas must be compressed and stored at very high pressures. Compression is accomplished using electrically driven pumps. While the cost of purchasing electricity is counted as an operating cost for NGVs, the emissions associated with the electricity used in compression is often overlooked. AGA cites a \$0.67/mile cost to compress the natural gas, and notes that this is a slow-fill-mode requiring less compression energy, a mode that fleet operators may tolerate. However, the 5-minute quick-fill will be, according to AGA, more costly, ie, require more electricity.

Table 2. Typical Gasoline Vehicle Emissions[a] (grams/mile)

Emission Type[b]	From Vehicle at 25 mpg	From Fuel Cycle at 25 mpg	Total at 25 mpg	Total at 40 mpg
CO_2	345	55	400	250
NO_x	0.40	0.24	0.64	0.4
CO	3.4	0.17	3.57	2.2
VOC	0.40	0.15	0.55	0.3

[a] Ref. 7,14.
[b] CO = carbon dioxide
NO_x = nitrogen oxides
CO = carbon monoxide
VOC = volatile organic compounds

Table 3. Unadjusted NGV Emissions[a,b] (grams/mile)

Emission Type	From Vehicle at 25 mpg	From Fuel Cycle at 25 mpg	Total at 25 mpg	Total at 40 mpg
CO_2	235	11	246	154
NO_x	0.40	0.15	0.55	0.34
CO	0.31	0.02	0.33	0.21
VOC	0.06	0.01	0.07	0.04

[a] Ref. 7.
[b] Emissions from the American Gas Association (AGA) data have been reduced by 10% since the AGA data does not assume a purpose-built vehicle with a 10% efficiency gain that is possible; emissions are linearly pro-rated down for the improved efficiency associated with a 40-mph vehicle.

On the order of 2 kwh of electricity is required to compress the natural gas equivalent of a gallon of gasoline. AGA (17) allows, for example, for a \$0.67/mile for compression cost which translates, at \$0.074/kwh, into about 2 kwhr/gallon equivalent. The amount of NO_x, CO, and VOC emissions associated with 2 kwhrs is negligible, and does not significantly change the total fuel cycle impact for those emissions from an NGV. However, post-1995 U.S. average CO_2 emissions levels for the national electric generating system are projected to be 640 grams/kwhr. An NGV with a vehicle efficiency of 40 mpg, would require 2 kwhrs, or 1,280 grams of CO_2 for 40 miles of driving or 32 grams of CO_2/mile. At a 25 mpg efficiency, the CO_2 emissions rate from compression would be 51 grams/mile.

Methane Emissions. Natural gas is comprised of methane, a more powerful greenhouse gas than carbon dioxide, as is now widely acknowledged. Although the warming impact of methane in the atmosphere declines much more rapidly than does the essentially permanent effect of CO_2, it is nonetheless significant for those concerned with global warming. CO_2 does not change over time in terms of its atmospheric warming potential. Methane, on the other hand, starts out for the first two decades after release with 63 times greater warming potential (pound for pound) than does CO_2. This greater impact declines to 20 times more than CO_2 over a 100-year period, and about 9 times over a 500-year period. Most analysts used the 20-fold additional warming impact as the average.

There is inevitably some leakage of natural gas from the entire system of distribution and consumption starting at the well and ending with the customer. There has been considerable uncertainty and some debate over the magnitude of total system methane leakage associated with using natural gas, and whether or not the impact is significant.

Gas industry studies claim a total system leakage of about 0.35% (18). Separately and more recently, the U.S. Department of Energy (DOE) and the Environmental Protection Agency (EPA) undertook studies identifying total natural gas system leakage ranging from 1 to 1.5% (19). For example, a Pipeline System Inc. 1989 study found withdrawal and field separation leakage of 0.34%, processing and transmission of 0.48%, and distribution at 0.13%. While it is possible to minimize leakage from high pressure pipelines, clearly the challenge is greater in the more diverse, lower pressure local distribution pipelines (20).

Methane is also associated with virtually all fuel cycles: it is not unique to natural gas. The mining of coal, extraction, and combustion of oil, and use of biomass, all lead to methane emissions. On a global basis, current natural gas operations and coal operations emit comparable amounts of methane, while biomass burning emits about 30% more and oil combustion about two-thirds less. Globally the emissions are about 35 m t/yr from natural gas storage, transmission, and distribution, 38 m t/yr from coal mining, 55 m t/yr from biomass burning, and 10 m t/yr from oil combustion (20).

The identified amount of methane leaking from underground coal mining translates into the energy equivalent of about a 1% methane leak rate (20). The coal-seam leakage issue is clearly relevant since about 55% of the nation's electricity is coal-fired. Coal-seam methane leakage suggests that a methane leakage penalty for natural gas is comparable to coal-seam methane emissions. It is arguably easier to capture and use methane from coal mines than to minimize methane leaks from a labyrinthine pipeline system (21).

This analysis assumes that the impact of methane leakage up-to-the-customer, is a wash for the coal-dominated electricity system and the natural gas distribution system. This is probably a generous assumption. AGA's identification of a 0.35% total system leakage is lower than the DOE and EPA studies findings of 1% to 1.5%. And while a 1% equivalent leak rate applies to underground coal mining, coal accounts for 55% of electricity generation, and not all U.S. coal comes from underground mines. Altogether, the methane equivalent leak rate associated with the electric sector may be more like 0.5%, and the natural gas to consumer system may well have a leak rate two to three times greater.

There is more to the NGV fuel cycle than delivering the gas to the filling station. There are potential leaks associated with compressing natural gas to high pressures needed for storage on board a vehicle. There is also leakage from the activity of purging and plugging in a high pressure line to fuel a vehicle, as well as fugitive leakage from the vehicle's engine and fuel system during operation. Such sources of leakage are unique to the NGV and are not to be found in the evaluations made thus far regarding NGV global warning impacts. A similar potential for small leakage at the point-of-use with appliances in homes should also be counted. The level of leakage that constitutes a significant increase in global warming impacts is substantially lower than the level of leakage that constitutes an explosive hazard in a home. Indeed, there appears to be a paucity of data on this issue. The potential for methane leakage has prompted some to raise cautionary notes regarding the global warming benefits of NGVs.

> Methane is a potent greenhouse gas, and no one knows how much might leak into the atmosphere. There may be anywhere from a 25% decrease in greenhouse gas emissions to an 11% increase. In short, there is no guarantee that switching to compressed natural gas vehicles would reduce global warming gases (20).
>
> However, this favorable [global warming] comparison for compressed natural gas vehicles assumes that there are no unburned methane emissions or leakage during vehicle operation. The widespread use of compressed natural gas vehicles for alleviating the greenhouse problem would require the development of several technologies. The most important of these is the ability to fuel and operate the vehicle without leakage of methane to the atmosphere. The leakage of only 1.4 to 1.8 grams of methane per kilometer would negate the carbon dioxide advantage of compressed natural gas as compared to gasoline (21).

A simple calculation reveals that the global warming advantage of NGVs over gasoline vehicles would be entirely eliminated at a 1.5% total methane leak rate from the compressor through to and including the filling opera-

Table 4. Total NGV Emissions (CO_2 and CO_2-Equivalent), grams/mile[a]

Emission Type	From Vehicle at 25 mpg	From Fuel Cycle to Station	Electricity for Compress. 25/40 mpg	Methane Leakage 25/40 mpg	Total at 25 mpg	Total at 40 mpg
CO_2	235	11	50/31	22/14	318	199
NO_x	0.40	0.15	n	n	0.55	0.34
CO	0.31	0.02	n	n	0.33	0.21
VOC	0.06	0.01	n	n	0.07	0.04

[a] n = negligible.

tion itself and the vehicle storage system and engine. A 25 mpg car emits about 400 grams/mile (gpm) of CO_2, compared to 236 gpm CO_2 for the NGV, an advantage of about 150 gpm. At 1.5% leakage during compression/filling/operation, the 7.5 gpm of methane leaking is equal (in warming terms) to 150 gpm of CO_2. Assuming a 1% total leak rate for compression, filling and operations. There is a CO_2 equivalent penalty of 22 gpm for the 25 mpg vehicle and 14 gpm for the 40 mpg vehicle.

Table 4 shows the emissions for NGVs counting both emissions associated with electricity for compression, and CO_2-equivalent leakage of methane.

ELECTRIC VEHICLES

The body of research on electric vehicles is daunting. However, the presence of research, instead of direct experience with practical vehicles is, in effect, an indictment when it comes to marketplace. The issue is not whether there exists a practical EV today (there doesn't) but if one were developed and used, would there be substantial environmental benefits, taking into account the entire fuel cycle? This issue has relevance both for smog precursors to help cities clean the air, and for the global climate change debate.

The introduction of the GM Impact spawned a flurry of popular media coverage for EVs, and the reappearance of EV development programs (22,23). The reality of the types of vehicles that are feasible with current battery and propulsion technology has, on the other hand, kept auto makers from even making vague predictions about possible introduction dates for electric cars. The automotive press is generally scornful of the state of EVs (24). The principal problem is widely recognized and has been called the great battery barrier (25). To overcome this barrier, industry and government have formed the U.S. Advanced Battery Consortium. The goal of the consortium is to develop a low cost battery that has almost three times the energy storage ability (along with faster recharging) when compared to today's best. Ultimately, it may require the type of serendipitous break-through that is frequently characteristic of science-based engineering solutions; there is already encouraging evidence that the Consortium may reach its goal.

The basic issue, however, is whether or not the environmental benefits of a hypothetical, practical EV are substantially better than either gasoline vehicles, or NGVs. If the benefits are substantial, one can argue that the pursuit of a practical vehicle is worthwhile.

In Table 5 are summarized the emissions from the entire fuel cycle of an electric vehicle. Once again, we consider both 25 mpg and 40 mpg equivalent vehicles. The data in Table 5 do not incorporate improved batteries; such efficiencies will be essential for practical EVs, and they will result in even lower emissions because fewer total kwhrs will be required to make up for reduced battery losses.

In various previous comparisons of EVs and NGVs, the existing prototype EVs are compared to road-ready NGVs. A more valid comparison is the future, road-ready EV and future purpose-built NGV. (The purpose-built NGV, with its energy efficiency gain over a retrofit NGV, was assumed in the previous section.) The prototype Chrysler TEVan consumes 0.5 kwhr/mile of travel. Practical EVs will need to be in the 0.25 to 0.15 kw·h/mile range shown to be possible in engineering feasibility studies. The GM Impact consumes about 0. 1 kwhr/mile (3). In Table 5, the weight reduction, rolling, and wind resistance features of the 25 and 40 mpg vehicles are assumed to be in place for the EV (as they were for the gasoline and NGV). In addition, the data below incorporates a 0.3 kw·h/mile fuel consumption for post-1995 vehicles and 0.2 kw·h/mile for 2010 vehicles.

Post-1995 power plant emissions are assumed in calculating Table 5 since any comparison of vehicles prior to that period is meaningless (14). Electric Power Research Institute data for power plant emissions from a 16 mpg-equivalent EV are used and extrapolated to 25 and 40 mpg-equivalent EVs (14).

Improved Power Plant Efficiency. The marginal supply of electricity over the next 20 years will be met by new or repowered old power plants that will be significantly more efficient than today's. Table 5 is based on today's average of a 35% power plant efficiency. This understates the probable net emissions benefits of EVs as new power plants ranging from 40 to 48% efficiency are now practi-

Table 5. EV Emissions, grams/mile

Emission	Total for 25 mpg Equiv. 0.3 kW·h/mi	Total for 40 mpg Equiv. Vehicle 0.2 kW·h/mi
CO_2	192	120
NO_x	0.21	0.12
CO	0.3	0.02
VOC	<0.01	<0.01

Table 6. Comparative Year 2010 Emissions for Typical Vehicle, grams/mile

Emission Type	From Gasoline	From NGV	From EV	NGV Emissions as % Gasoline	EV Emissions as % Gasoline
CO_2	250	199	120	80	48
NO_x	0.4	0.34	0.12	85	30
CO	2.2	0.21	0.02	10	1
VOC	0.3	0.04	<0.01	13	<1

cal. The net effect of using such power plants to fuel EVs will be to reduce emissions by at least another 30% over those used here.

SO_x and Solid Waste

Some fuel cycle analyses include the emissions of SO_x and solid waste from coal-fired power plants since coal provides 55% of the nation's electric supply. Since there are essentially none of either for NGVs, this is one area in which NGVs have a clear win. However, the solid waste issue, while a legitimate concern, is not relevant insofar as urban smog and global warming are concerned.

With respect to SO_x: under the Clean Air Act Amendments, utilities will not be permitted to increase SO_x emissions. Thus any emissions associated with EVs will, *a priori*, need to be eliminated. While the net effect may place a greater challenge on utility SO_x reduction programs, (i.e., more scrubbers, coal cleaning, purchase of clean coal, or emissions credits) there will be no net societal increase in SO_x. For this reason, it makes no sense to counting SO_x since there will, under the law, be no increase in SO_x for any reason, regardless of the use of EVs or any other electric technology.

Table 6 compares the results contained in Tables 4 and 5 for NGVs and EVs respectively. And, with regard to just greenhouse gas emissions, the EV fuel cycle offers substantially larger reductions, as shown below

Fuel	Percent Change in Greenhouse Gas Emissions Compared to Oil
EVs at current power mix	−43
NGV	−16
Methanol from natural gas	8
Gasoline from oil	0

Air Travel: Maglev

The 300 mph magnetically levitated train is, at least technically, a viable alternative to air travel for many city-to-city domestic routes (26). Door-to-door transit time will be quicker via maglev than via airplane for cities closer than 600 miles. Some projections put future maglev passenger fares below that of air travel, suggesting that once built, maglev would be heavily used. The primary benefits of a maglev system would appear to be greater convenience for travelers and reduced air travel congestion in the busiest corridors.

Because maglev (and steel-wheeled high speed rail) are so much more energy efficient than either aircraft or automobiles, there will be a substantial net reduction in energy, smog emissions, and CO_2 associated with passengers switching to maglev. On average, there is a four-fold reduction in primary energy needed for a passenger-mile on a maglev compared to an aircraft (again, this takes into account fuel use at electric power plants). This translates into a reduction of about 170 gram CO_2 per passenger-mile of travel. Assuming 3,000 BTU/passenger-mile for aircraft and 750 for maglev. Savings of 2,250 BTU/passenger-mile result in average emissions of 174 gams CO_2. The calculation assumes CO_2 emissions/BTU of oil for the aircraft and CO_2 emissions from a blended national average fuel mix of fuels for electric generation which is comparable to that of oil combustion alone. The potential CO_2 benefits of maglev can be quickly assessed against the fact that the 10 busiest intercity corridors where the distances are less than 600 miles account for about 20 billion passenger-miles of air traffic.

Maglev would also significantly reduce emissions of NO_x in dense urban areas. NO_x emissions from Los Angeles International airport alone are equivalent to more than one million passenger cars (27). There is also evidence to suggest that there is a greater smog impact from the NO_x emissions injected into the atmosphere by aircraft engines during take-off.

Other Transportation Electrification Opportunities

There are a multitude of other opportunities for electricity to replace gasoline in internal combustion engines. These include subways, light rail, off-road vehicles (ranging from golf carts to airport tugs), and a group of engines classified as utility engines by EPA under the mobile sources of emissions category. These utility engines include all manner of lawn and garden equipment, of which lawn mowers account for the largest component. It appears that in all of these cases, the net energy and net emissions benefits from using electric drive instead of burning gasoline are at least as great or greater than using electric on-road vehicles. The net emissions reductions will typically be greater for off-road sources since that class of internal combustion engines are not currently regulated and thus emit much more per horsepower-hour of operation than do on-road engines.

In fact, electrifying just one class of utility engines, lawn mowers, with battery-powered cordless technology would confer remarkable energy savings, CO_2 emissions reductions, and air quality benefits. By the mowing season of 1994, there were at least five manufucturers of cordless electric mowers; Black & Decker, with a product entering its third mowing season, Ryobi, MTD, and at least two European manufucturer looking for U.S. market entry. Clearly, only a cordless electric mower is a practical replacement for a gasoline mower. The Black & Decker mower will mow a 1/4 acre lawn, which Black & Decker notes accounts for about one-half of all U.S. residental lawns, on a single charge. Preliminary research shows that if one-half of the population that uses gasoline mowers converted to battery-powered electric mowers, there would be a net emissions improvement equal to taking nearly three million cars off the road, and about one million tons of CO_2 would be eliminated per year (net of power plants.)

In addition, the practical prospects for, and emissions benefits of electrifying many of the vehicles at airports is being actively explored. Overall, there has been remarkably little study of the potential emissions benefits via electrification in the entire range of nonroad engines, a surprising circumstance considering that current battery technology makes many of these applications practical now.

COMPARATIVE IMPACTS: NGV AND EV

For a number of analysts, the conclusions regarding NGVs versus EVs are clear. In testimony before the U.S Congress, the Environmental Defense Fund concluded (28)

> Of the fuels we examined, electricity is the most promising for achieving significant greenhouse gas reductions. The use of an average electricity fuel mix would result in a 27% reduction in greenhouse gas emissions (GHG) versus gasoline. . . . The use of compressed natural gas in vehicles results in only modest GHG benefits at best (0 to −15%).

In an important study of EVs, the Northeast States For Coordinated Air Use Management (NESCAUM) concluded (3):

> The results of this analysis demonstrate that battery-powered electric vehicles represent the cleanest motor vehicle technology likely to be available to the Northeast within the next six to eight years. . . . Replacing internal combustion engines with battery-powered electric vehicles will also reduce the amount of CO_2, greenhouse gas, emitted by the motor vehicle fleet.

In yet another analysis (29), where methane-equivalent leakage from the NGV fuel cycle was incorporated, the conclusions shown above were reached for total global warming impacts.

While there are variations in the assumptions that different analysts use to reach conclusions on the overall greenhouse benefits of electric transportation, their conclusions are all remarkably similar. Correct accounting reveals that EVs offer substantially more reductions in emissions than do the other alternatives.

Table 6 summarizes the data developed in this analysis, for vehicles likely to be available in 2010:

- 40 mpg gasoline car
- 40 mpg-equivalent NGV purpose-built with a 10% efficiency advantage over the gasoline vehicle and a realistic compression/filling/operating system in the vehicle (with total leakage of 1%);

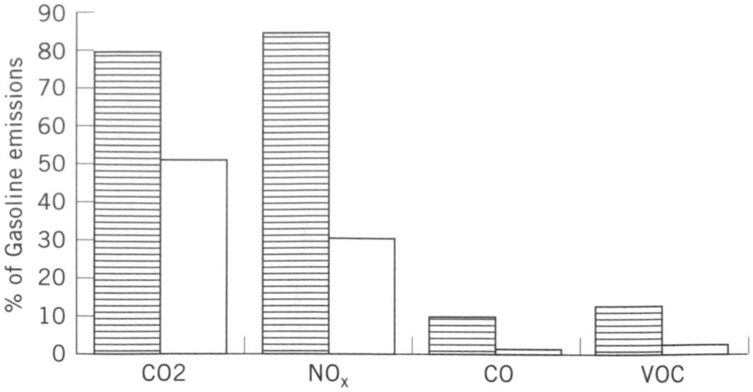

Figure 6. Year 2010 comparative total fuel-cycle emissions; NGV and EV as % of gasoline. ▤, NGV; □, EV.

- state-of-the-art EV consuming 0.2 kw · h/mile (coal 55% of total electric supply)

Table 6 summarizes the results of the analysis for a typical year 2010 high efficiency vehicle. In all three important categories of emissions for cities, the EV will lead to 10 to 35% of the emission level of an NGV. And for the greenhouse gas, CO_2 total emissions are about 65% of an NGV. Clearly, switching to EVs will achieve greater emissions reductions than switching to NGVs. As a practical matter, the switch to NGVs can begin immediately since there are already many practical applications today. The magnitude of the additional emissions reductions expected to arise from EVs suggests that substantial effort is worth devoting to the development of a practical EV.

BIBLIOGRAPHY

1. D. Nye, *Electrifying America: Social Meanings of a New Technology,* MIT Press, Cambridge, Mass., 1990, p. 86.
2. M. Melosi, *Garbage in the Cities,* Texas A&M University Press, College Station, 1981.
3. *Impact of Battery-Power Electric Vehicles on Air Quality in the Northeast States,* Northeast States for Coordinated Air Use Management (NESCAUM), Feb. 24, 1992.
4. *Annual Energy Outlook 1993 With Projections to 2010,* DOE/EIA-0383(93), U.S. Energy Information Administration, Washington, D.C., Jan. 1993.
5. *Statistical Abstract of the United States 1992,* U.S. Dept. of Commerce, Washington, D.C.
6. U.S. Energy Information Administration, *Annual Energy Review,* Washington, D.C., 1992.
7. *Natural Gas and Electric Vehicles: An Economic and Environmental Comparison With Gasoline Vehicles,* American Gas Association, Policy & Analysis Issues, Ap. 16, 1991.
8. Ref. 4, p. 91.
9. *1992 Edition of the GRI Baseline Projection of U.S. Energy Supply and Demand to 2010,* Ap. 1992, p. 140.
10. *California Energy Markets,* 3 (Mar. 12, 1993).
11. S. Nadis and J. Mackenzie, *Car Trouble,* World Resources Institute, Beacon Press, 1993, p. 46.
12. Howes and Fainberg, *The Energy Sourcebook: A Guide to Technology, Resources and Policy,* American Institute of Physics, 1991, p. 360.
13. *National Transportation Statistics Annual Report.*

14. *Electric Van and Gasoline Van Emissions: A Comparison,* technical brief, Electric Power Research Institute, 1989.
15. *Fortune,* 21 (Mar. 22, 1993).
16. *Wall Street Journal,* B1 (Jan. 5, 1993).
17. D. Sperling, *New Transportation Fuels: A Strategic Approach to Technological Change,* University of California Press, 1988.
18. *Natural Gas Transmission and Distribution Methane Emissions,* engineering technial note, American Gas Association, Nov. 1989.
19. *Limiting Net Greenhouse Gas Emissions in the U.S.,* Vol. 2, DOE/EIA, Washington, D.C., Sept. 1991.
20. M. Wallis, *New Scientist,* 62 (Nov. 10, 1990).
21. G. MacDonald, "The Near and Far Term Technologies, Uses and Future of Natural Gas," *IAE / OECD Expert Seminar on Energy Technologies for Reducing Emissions of Greenhouse Gases.* Ap. 12–14, 1989.
22. M. Fischetti, *Smithsonian,* (Ap. 1992).
23. "Electric Vehicles: Special Report," *IEEE Spectrum* (Nov. 1992).
24. P. Bedard, *Car and Driver,* (May 1992).
25. The Great Battery Barrier," *IEEE Spectrum,* (Nov. 1992).
26. *In Pursuit of Speed: New Options for Inter-city Passenger Transport,* Transportation Research Board, National Research Council, Nov. 1991.
27. R. Price, *Strategic Planning for Energy and the Environment,* 37 (Spring 1992).
28. D. Fisher, "Comments on the Greenhouse Gas Impacts of Alternative Transportation Fuels," Environmental Defense Fund, before the Subcommittee on Energy and Power, Committee on Energy and Commerce, U.S. House of representatives, Washington, D.C., June 19, 1991.
29. D. Sperling, *New Transportation Fuels: A Strategic Approach to Technological Change,* University of California Press, 1988, p. 326.

TRANSPORTATION FUELS—ETHANOL FUELS IN BRAZIL

SERGIO C. TRINDADE
SE²T International, Ltd.
Scarsdale, New York

WALTER VERGARA
The World Bank
Washington, D.C.

The largest alternative transportation fuels program in the world today is Brazil's Proalcool Program About 7.6 million metric tons of oil equivalent (MTOE) of ethanol, derived mainly from sugar cane, have been consumed as transportation fuels in 1992 (equivalent to 151,000 barrels of crude oil per day). The total primary energy supplied to the Brazilian economy in 1992 has been 140.8 million MTOE, and approximately 4.3 million vehicles–about one third of the total vehicle fleet or about 40% of the total car population–run on hydrous or "neat" ethanol at the azeotropic composition (96% ethanol, 4% water, by volume). Additional transportation fuels available in the country are diesel and gasoline, the latter of which is defined by three grade: Gasoline A (regular, leaded gas) has virtually been replaced by gasoline C, a blend of gasoline and up to 22% anhydrous ethanol by volume, and gasoline B (premium gasoline) has been discontinued as a result of neat ethanol market penetration.

THE PROALCOOL PROGRAM

There have historically been a number of stakeholders in the Brazilian transportation/fuel industry: the government; ethanol and automobile manufacturers and distributors; the petroleum/natural gas and sugar cane industries; and Brazilian and international consumers. The economic interests of these stakeholders are often intertwined both in the national and international marketplaces.

For example, ethanol has been used as an automotive fuel in Brazil since the 1920s. But, up until 1975, when Proalcool was launched, ethanol was only used occasionally in blends of varying composition with gasoline to protect the important sugar industry from the vagaries of the international market. The production of ethanol absorbed part of the sugar production that was not economically feasible for export, and the government thereby guaranteed a market for ethanol.

By 1957, transnational corporations had begun local car production, and Brazil's automobile industry had since grown, contributing in the range of 8–12% of the industrial GNP and taking its place as one of the 12 largest auto industries in the world.

Stakeholders and Government Policy

The introduction of the Proalcool Program in 1975 coincided with a convergence of the interests of the key stakeholders in Brazil's transport fuel economy. The program was originally conceived to implement the new policy of promoting the market penetration of ethanol as a transportation fuel via gasoline C, thus displacing gasoline A. The rationale behind the policy was the concern over prices and availability of liquid fuels for transportation in connection with the first oil crisis of 1973.

These concerns were exacerbated in 1979 with the onset of the Iranian oil crisis, and the Iran-Iraq conflict, which led the Brazilian government to implement policies supporting a more radical approach to promote ethanol, in the form of neat ethanol, as a transportation fuel, in addition to gasoline C. Ethanol moved from its earlier role of gasoline extender to being a fuel in its own right.

At the root of the Proalcool initiative were government concerns with the security of supply of transportation fuels refined in Brazil from predominantly imported crude oil. Furthermore, the oil price hikes of 1973 and 1980 put considerable pressure on the country's balance of payments. Oil imports' share of total imports jumped from 9.8% in 1973 to 43.3% in 1985. (Tables 1 and 2 provide a more comprehensive picture of the impact of oil imports on Brazil's international trade and finances.) Ethanol has been produced from domestic resources paid for in local currency. Its use therefore has saved precious foreign exchange.

In founding Proalcool, the government was convinced that the expansion of the ethanol industry could also help Brazil's economic growth. It would provide the automobile industry with a more reliable source of domestic fuel

Table 1. Brazilian Petroleum Imports, 1973–1992

Year	Avg. Oil Price U.S. #/Barrel	Volume 1000 bbl/d	Value million, $	Share of Total Imports, %
1973	2.54	652.8	605.2	9.8
1975	10.53	703.5	2704.1	22.2
1979	16.83	1019.7	6263.5	34.6
1980	29.98	886.0	9372.4	40.8
1985	27.24	545.1	5418.0	43.3
1989	15.70	592.0	3390.0	20.6
1990	20.89	571.0	4354.0	27.9
1991	18.26	507.0	3370.0	19.3
1992 est.	16.30	532.2	3166.0	21.7

based on the well-established sugar cane agriculture. Direct and indirect job creation connected with ethanol production would offer growth opportunities for the rural areas and allow the reduction of economic imbalances between the diverse regions of such a vast country as Brazil.

Background of the Key Stakeholders

At its peak in the mid 1980s, Brazil had some 600 distilleries capable of producing about 16 million m^3 (4.2 billion gallons) of ethanol. On an energy content basis, ethanol's share of liquid fuels used for ground transportation (which employs both Diesel and Otto engines) increased from 0.5% in 1973 to a peak of 22% in 1988 and 1989. The pace of market penetration of ethanol fuels can be better illustrated however, by the displacement of gasoline from the Brazilian domestic market, as both ethanol and gasoline have been used primarily in Otto engines. On an energy basis, the share of ethanol in this market increased from 0.9% to a peak of 52% during the period from 1975 to 1988 while the market for diesel fuel practically doubled in size from 8.1 to 15.9 million MTOE as the ratio of diesel fuel to Otto fuels moved from 73% to 141%. Meanwhile, Otto fuels (ethanol and gasoline) grew only 2.0% to 11.3 million MTOE. Since 1989, Diesel fuel has continued to grow, gasoline regained market lost to ethanol, and ethanol stabilized (see Table 3).

The large market expansion of ethanol as a transportation fuel stimulated the development of technologies aimed at improving sugar cane yields, more efficient conversion from sugar cane juice and molasses into ethanol, and higher engine efficiency. After all, the ethanol production technology available at the inception of Proalcool was geared to the low volume, high value-added, and traditional potable ethanol market. Historically, Otto engine technology, although originally developed with alcohols (ethanol, methanol) in mind, was later optimized for hydrocarbon fuels.

To expedite the launching of Proalcool, the government resorted to subsidies for the development of sugar cane agribusiness and ethanol production. These subsidies have practically been removed. Meanwhile, Brazil has been undergoing an acute inflation for over a decade. While inflation has not entirely impeded economic growth, it has severely disorganized both public and private finances and distorted relative prices in the economy.

As a result, Petrobras–the main ethanol distributor to all marketers and a company with a former longstanding tradition of profitability in its 40 years of existence–has suffered substantial losses in the distribution of ethanol.

Under inflationary conditions and with prices for ethanol ex-distillery and ethanol CIF-retailers set by different government agencies at different times, the distributor often ends up buying ethanol from producers at a higher price than it can sell to retailers. The slow adjustment of oil product prices under the inflationary regime of the past decade has also compounded Petrobras's losses.

On the other hand, the sugar and ethanol industry has grown extraordinarily with Proalcool and has a large stake in the program's further growth. Yet, the industry has suffered from the price-setting and commercial mech-

Table 2. Trade Balance and Foreign Debt, 1973–1992, Billion U.S. $

Year	Exports	Imports	Trade Balance	Net Foreign Debt
1973	6.2	6.2	0	6.2
1975	8.7	12.2	− 3.5	17.1
1979	15.2	18.0	− 2.7	40.2
1980	20.1	23.0	− 2.8	46.9
1985	25.6	12.2	+13.4	84.2
1989	34.4	18.3	+16.1	89.6
1990	31.4	20.7	+10.7	87.5
1991	31.6	21.0	+10.6	82.7
1992 est.	34.9	20.0	+14.9	N.A.

Table 3. Transportation Fuels Consumption in Brazil, 1973–1991

Year	Diesel Fuel[a]	Gasoline[a]	Ethanol[a]	Total Otto Fuels[a]	Ratio Diesel/Otto, %
1973	6.4	10.4	0.2	10.6	60.4
1975	8.1	11.0	0.1	11.1	73.0
1979	11.9	10.0	1.2	11.2	106.3
1980	12.4	8.7	1.4	10.1	122.8
1985	13.5	5.9	4.1	10.0	135.0
1989	16.4	6.4	6.3	12.7	129.1
1990	16.5	7.3	5.7	13.0	126.9
1991	17.2	7.9	6.9	13.8	124.6

[a] Million MTOE

anisms of ethanol sales, and often finds itself at odds with Petrobras, the main ethanol buyer. And on recent occasions of higher international sugar prices, the industry has switched to a considerable extent, to more sugar and less ethanol output, as in the years before Proalcool. During such occasions, some ethanol producers who do not make sugar, shut down production for lack of economic incentive.

Currently, the automobile industry is relatively indifferent to the type of fuel available–neat ethanol or gasoline–as its large gasoline engine and vehicle export program make it possible to supply vehicles for both fuels to the domestic market. Vehicle owners, however, are very sensitive to retail fuel price differences between ethanol and gasoline, and consider total vehicle operation costs. In addition, consumer perception of fuel availability is crucial in determining type of vehicle purchase. In this latter context the opinion of taxi drivers all over the country is crucial.

The above issues have led to a situation that challenges the continuation of Proalcool. The reintroduction of gasoline to the market has alleviated the excess of refinery output relative to domestic demand. Occasions of short domestic supply of ethanol have led Petrobras to import methanol and ethanol to fill the gap. A new transportation fuel–ethanol/methanol/gasoline blend (60, 33, and 7% by volume, respectively)–has appeared on the Brazilian market in early 1990.

As of September 1992, retail prices for automobile fuels in Brazil at the official commercial exchange rate were U.S. $0.35 per liter (U.S. $1.32 per gallon) for gasoline C, U.S. $0.26 per liter (U.S. $0.98 per gallon) for neat ethanol and U.S. $0.26 per liter (U.S. $0.98 per gallon) for Diesel fuel. International comparisons with Brazilian domestic prices expressed in U.S. $ can be tricky, given, inter alia, that inflation and currency devaluation vary at different rates.

GOVERNMENT MACROECONOMIC OBJECTIVES

Since 1975, Proalcool has generally met government objectives, although not necessarily by its own devices. Concerns over the security of supply and high prices of fuels have abated due to the relaxation of international oil markets and to the increase in domestic crude production, which nearly quadrupled from 1975 to 1987, reaching a plateau thereafter, and has resulted in over 50% oil self-sufficiency.

The ensuing decrease in imported oil price and the overall export drive of Brazil has considerably improved the trade balance, but unfortunately, the parallel increase in foreign debt and debt service has offset the trade surpluses. The growth in oil prices since 1973 is one of the causes of the outstanding Brazilian debt.

The rate of economic growth has fluctuated since 1975. Nevertheless, most analyses indicate that agricultural employment and industrial activities connected with ethanol fuel production, distribution, and utilization benefited substantially from Proalcool.

Economics as if the Environment Mattered

Despite significant technological developments in ethanol production, by the end of 1985 the best ethanol producer in Brazil was barely cost-competitive with gasoline at a crude price of U.S. $28 per barrel. On the other hand, straight economic comparisons do not take into consideration: (1) the value of ethanol in reducing oil imports in a country as highly indebted as Brazil; (2) the higher value of ethanol as an octane booster; and (3) ethanol superior environmental characteristics. As the long-term marginal cost of oil is likely to increase, which will be translated into market economic terms, ethanol can become a rational economic option for some countries.

The environment has not been an explicit concern in the formulation of Proalcool, but as the program evolved, the costs and benefits to the environment have become evident. The distillation of ethanol produced from the fermentation of sugar cane juice and/or molasses, yields a large volume of liquid effluent (stillage) with a high biological oxygen demand/chemical oxygen demand (BOD/COD) content. Each volume of ethanol yields 13 to 15 volumes of stillage. At the outset of Proalcool, it has been traditional to dispose of raw stillage in bodies of water. This process has become untenable as ethanol output expanded. The dominant approach today is to return stillage to the sugar cane fields for fertilization and irrigation purposes. In this application, stillage is valued for its potassium content, which is usually sufficient to pay back the required investment within a commercially acceptable time.

Ethanol burning in Otto engines has a generally beneficial effect on emissions. The actual effect depends on a variety of factors including electronic injection, compression ratio, tuning, ethanol content in the blend, catalytic processing of tail pipe off-gases, and driving cycle. The major negative effect is the increase in aldehyde emissions (particularly acetaldehyde) in relation to gasoline.

As of November 1991, lead compounds were completely phased out of Brazilian gasolines. No addition of aromatics was necessary to compensate for the loss of octane, as the ethanol content of up to 22% by volume, in gasoline, brings in the required octane rate to a minimum of 80 Motor Octane Number (MON).

The addition of ethanol to gasoline cuts down the blend content of sulfur, aromatics and olefins, thereby reducing emissions of sulfur and toxic and photochemically reactive substances into the atmosphere. The leaning effect of ethanol in the gasoline blend cuts down the emission of carbon monoxide and gum deposits in the engines. In the U.S., ethanol and other oxygenates are being added to gasoline blends for the latter purpose, in conformity with the 1990 amendments of the Clean Air Act.

Aldehyde emissions of neat ethanol-fueled vehicles are typically three times larger than gasoline-fueled vehicles. Ethanol related aldehyde emissions are dominated by acetaldehyde (85%), whereas gasoline related aldehyde emissions are principally composed of formaldehyde, a much more toxic substance. Formaldehyde is a known carcinogen and has a photochemical reactivity of 70% higher than acetaldehyde, a feature directly related to the build-up of photochemical smog. It is interesting to note that the aldehyde emissions of diesel-fueled vehicles is on the same order of magnitude of neat ethanol-fueled vehicles. Diesel related aldehyde emissions are more toxic than ethanol related emissions, and contribute to the unpleasant aroma of diesel exhaust gases.

Aldehyde emissions of neat ethanol vehicles have been gradually decreasing in recent years. Typically for 1992 models, aldehyde emissions are 0.03 g/km for neat ethanol vehicles and 0.01 g/km for gasoline blends containing 22% ethanol by volume. Improved neat ethanol vehicles with catalytic convertors and/or electronic injection have aldehyde emissions of the same level as conventional gasoline vehicles (with carburetor, no catalytic convertor), and lower than diesel fueled vehicles of similar power.

A 1990 study of the Sao Paulo State environmental agency (CETESB) has confirmed that acetaldehyde makes about 70% of the aldehydes measured in the atmosphere of the City of Sao Paulo – a sprawling megalopolis with over 10 million inhabitants. Acetaldehyde concentrations have varied in the range of 4–47 ppb (parts per billion) in open air. The maximum observed concentration was measured inside a major tunnel, at 132 ppb. As a reference, the World Health Organization acetaldehyde limit of occupational tolerance has been 1000 ppb.

Furthermore, emissions of unburned fuel of neat ethanol vehicles (basically ethanol) is much less toxic and less photochemically reactive than unburned hydrocarbons. Neat ethanol combustion yields much less soot than hydrocarbon burning, which causes less respiratory troubles in the population and provides for better atmospheric visibility.

Regarding nitrogen oxide emissions, the advantages of ethanol's large latent heat of vaporization and higher thermodynamic efficiency are offset by the higher compression ratio of neat ethanol vehicles, thus resulting in NO_x emissions similar to gasoline vehicles, but much lower than diesel vehicles.

Studies with guinea pigs and rats conducted by the University of Sao Paulo have demonstrated that ethanol vapors and tailpipe gases are less toxic than those of gasoline, in terms of acute toxicity. Although there are presently no data on chronic toxicity of ethanol vapors and combustion gases, over a decade of extensive utilization of ethanol fuels has not revealed evidence of negative health effects, even with gas station workers, who are more intensively exposed. Studies by CETESB showed that the maximum concentrations of ethanol in the city of Sao Paulo reach no higher than 2.3 ppm, which compares well with the WHO limit on occupational tolerance at 1000 ppm.

The recent concerns over the net contribution of fossil fuels to the level of carbon dioxide in the atmosphere and the ensuing greenhouse effect on climate change, have given sugar cane ethanol an unexpected environmental value. Today sugar cane ethanol is the only commercial transportation fuel that can claim a zero, or almost zero, net contribution of carbon dioxide to the atmosphere.

CONCLUSIONS

Government policymakers, in Brazil and elsewhere, must arbitrate inherent conflicts between the main stakeholders in any alternative transportation fuels program. In the Brazilian case the relevant production stakeholders and their positions are: Petrobras, which produces domestic oil, refines the totality of the crude, transports and distributes ethanol and petroleum products; ethanol producers, which take most of the value added generated by the Proalcool program, but are exposed to sudden bankruptcy as a result of uneven pricing policies or severe inflation; and automakers, who depend upon the availability of various fuels.

The reintroduction of gasoline into the domestic market beginning in 1989, after years of gasoline market erosion, reflected an attempt to avoid expensive investment in refining to match market demand for crude slate under a policy of self-sufficiency in refining. The automobile industry output of gasoline-fueled vehicles has been growing fast since 1989 in response to this change in policy. Ethanol capacity is likely to remain at the current level in the foreseeable future, although actual ethanol output will depend on the reaction of producers to the pricing policies of the government and the practices of Petrobras in commercializing ethanol. Ethanol and methanol imports, by Petrobras, are already filling the gap left by domestic ethanol producers who are unwilling to sell at prices they do not consider remunerative-particularly those who have the option of switching to other products, such as sugar.

A new consensus among key stakeholders has emerged to replace the understanding achieved during 1975–1980. There is a new transportation fuels market emerging in Brazil, and the gasoline C share is fast increasing. How-

Table 4. Brazilian Vehicle Sales in the Domestic Market, 1973–1992, % by Fuel Type

	Cars			Light Commercial			Heavy Commercial		
Year	Gasoline	Ethanol	Diesel	Gasoline	Ethanol	Diesel	Gasoline	Ethanol	Diesel
1973									
1979	100.0			99.5		0.5	36.4		63.6
1981	99.7	0.3		82.6	0.9	16.5	1.9	0.0	98.1
1985	71.3	28.7		37.5	11.1	51.4	0.1	1.6	98.3
1989	4.0	96.0		4.8	68.5	26.7	0.1	3.0	96.9
1990	39.0	61.0		29.0	39.3	31.8	0.0		100.0
1991	86.8	13.2	0.	62.5	9.1	28.4	0.2	0.0	98.8
1992	77.8	22.1	0.1	56.2	16.9	26.9	0.2	0.0	99.9
est.	71.2	28.1	0.1	52.0	24.6	23.4	0.2	0.0	99.8

ever, this new consensus, as its predecessor, is subject to dynamic change, and it remains to be seen how much value can be attributed to the carbon dioxide recycling feature of sugar cane ethanol—or biomass-derived ethanol in general, as well as the net positive environmental qualities of ethanol fuels.

Other countries can learn from the Brazilian Proalcool experience that a delicate balance between key stakeholders is germane to the successful implementation of alternative transportation fuels programs. Other lessons include:

- Decisions on alternative transportation fuels must be seen in the long-term perspective.
- Domestic production of alternative fuels cannot be entirely dependent on uncertain and unpredictable oil prices. (It may be better and safer to spend two dollars in local currency at home than one dollar on imports, given heavy indebtedness and soft currency.)
- Consumers respond most favorably to: economic incentives which mitigate cost of motor vehicle ownership and operation; consistent and farsighted policies; and the initial reputation of vehicles powered by alternative fuels. The chaotic transient of the initial introduction of neat ethanol engines into the market, their dominance of the market throughout the 1980s and their later retreat in the late 1980s, shown in Table 4, illustrate the point.
- Stakeholders' consensus is critical for the stable implementation of an alternative fuels program. However, consensus achieved at program launch must be checked and renegotiated periodically throughout the life of the program.
- Air quality can be improved and a total, or almost total, recycle of carbon dioxide emissions can be achieved.
- Nevertheless, under current and foreseeable economic supply conditions, very few countries in the world should actually embark on an ethanol fuels program. The optimal candidates are countries with biomass surpluses and energy deficits that are landlocked and without their own oil resources. Examples are: Paraguay and Northern Argentina in South America; Kenya, Uganda, Zimbabwe, Malawi, Zambia in Africa. This scenario offers a more viable economic prospect for the penetration of ethanol in the transportation fuel market. In the U.S. and Europe, the emerging market for oxygenated gasolines, built on the demand for a better environment and for octane boosters, provides an opportunity for ethanol to compete with other oxygenates, such as methanol and methyl ter-butyl ether—MTBE. One possibility for the penetration of ethanol in transportation fuels markets, is in the form of ETBE, ethyl ter-butyl ether. The total, or quasi-total, carbon dioxide recycle feature of biomass-derived ethanol, adds a new dimension to the future prospect of this non-fossil transportation fuel.

Acknowledgements

This article draws heavily from Reference 7. The authors wish to thank Arnaldo Vieira de Carvalho Jr. of Promon, Rio de Janeiro; Alberto Mortara of Sao Paulo; Julio Borges of Copersucar, Sao Paulo; Luiz Celso de Castro Leal of Dyna, Rio de Janeiro, for providing most of the data contained in the tables and to Ake Brandberg, of Eco-Traffic, Stockholm, for his comments on an earlier draft. The detailed information on the air quality impact of ethanol fuels is due to Alfred Szwarc of CETESB, Sao Paulo.

Disclaimer

The findings, interpretations and conclusions expressed in this article are entirely those of the authors and should not be attributed to the world bank or to members of its board of executive directors.

BIBLIOGRAPHY

1. *Politica de Combustiveis Liquidos Automotivos* (*Automotive Liquid Fuels Policy*), CNE–Brazilian National Energy Council, Brasilia, Brazil, 1988.
2. D. Sperling. "Brazil, Ethanol and the Process of System Change," *Energy* **12**(1) 11–23 (1987).
3. S. C. Trindade, and A. Vieira de Carvalho, Jr., "Transportation Fuels Policy Issues and Options: The Case of Ethanol Fuels in Brazil," in D. Sperling, ed., *Alternative Transportation Fuels–An Environmental and Energy Solution,* Quorum, New York, 1989, pp. 163–186.
4. S. C. Trindade, *Oxygenated Transport Liquid Fuels: The Total System,* World Energy Council Monograph, London, UK, 1989.
5. S. C. Trindade, *Brazilian Alcohol Fuels: An International Multisponsored Program,* Rio de Janeiro, Brazil, 1984.

6. *Brazilian Automotive Industry Statistical Yearbook, 1957–1991.* ANFAVEA, Sao Paulo, Brazil, 1992.
7. S. C. Trindade, "Nonfossil Transportation Fuels: The Brazilian Sugar Cane Ethanol Experience," in J. Tester, and coworkers, eds., *Energy and the Environment in the 21st Century,* MIT Press, Cambridge, Mass., 1991, pp. 277–284.
8. V. Yang, S. C. Trindade and J. R. Castello Branco, "How Brazil Grows Motor Fuel," *CHEMTECH* **11**(3), 168–172 (1981).

TRANSPORTATION FUEL—HYDROGEN

T. NEJAT VEZIROGLU AND FRANO BARBIR
Clean Energy Research Institute
University of Miami,
Coral Gables, Florida

Hydrogen energy is not a new concept. Actually, the basic technologies for hydrogen production, storage, transportation, and utilization are already technically feasible. Over the last two decades, there have been increasing research efforts to investigate the various aspects of the hydrogen energy system and technologies. Most of the results have been presented in the proceedings of the Hydrogen Economy Miami Energy (THEME) Conference (1) and in those of the 10 World Hydrogen Energy Conferences (2–11) held to date. Books and reports by Bockris (12), Veziroglu (13), Ohta (14), Williams (15), Skelton (16), and Winter and Nitsch (17) cover the hydrogen energy system, hydrogen production methods, storage, transportation, and utilization in some depth. Figure 1 shows the hydrogen energy system and the technologies involved (18). The following sections will emphasize some recent developments in hydrogen energy technologies with respect to early implementation and commercialization.

Hydrogen Production

Hydrogen is currently being produced mainly by steam reforming of natural gas, and also by partial oxidation of heavy oil and by coal gasification. Different methods of hydrogen production based on renewable energy sources have been or are being developed. They are direct thermal decomposition or thermolysis, thermochemical processes, electrolysis and photolysis. It appears that so far water electrolysis is the only method developed that can be used for large-scale hydrogen production. Today, available advanced electrolyzers are more than 90% efficient, and research efforts are aimed toward new advanced concepts in electrode design and material improvement in order to reduce the costs, increase reliability, and extend the lifetime. Electrolysis can effectively be used in combination with photovoltaic (PV) cells. The results of an experimental 10-kW PV–electrolysis plant in Germany have indicated a low specific energy consumption (3.84 kWh/Nm^3), good gas purities, fast dynamic response, and a wide range of operation (19). These results, and the results of other testing facilities, indicate that the PV–electrolysis systems have the potential to become available for large-scale hydrogen production as well as for individual stand-alone applications. Electrolysis can also be used with hy-

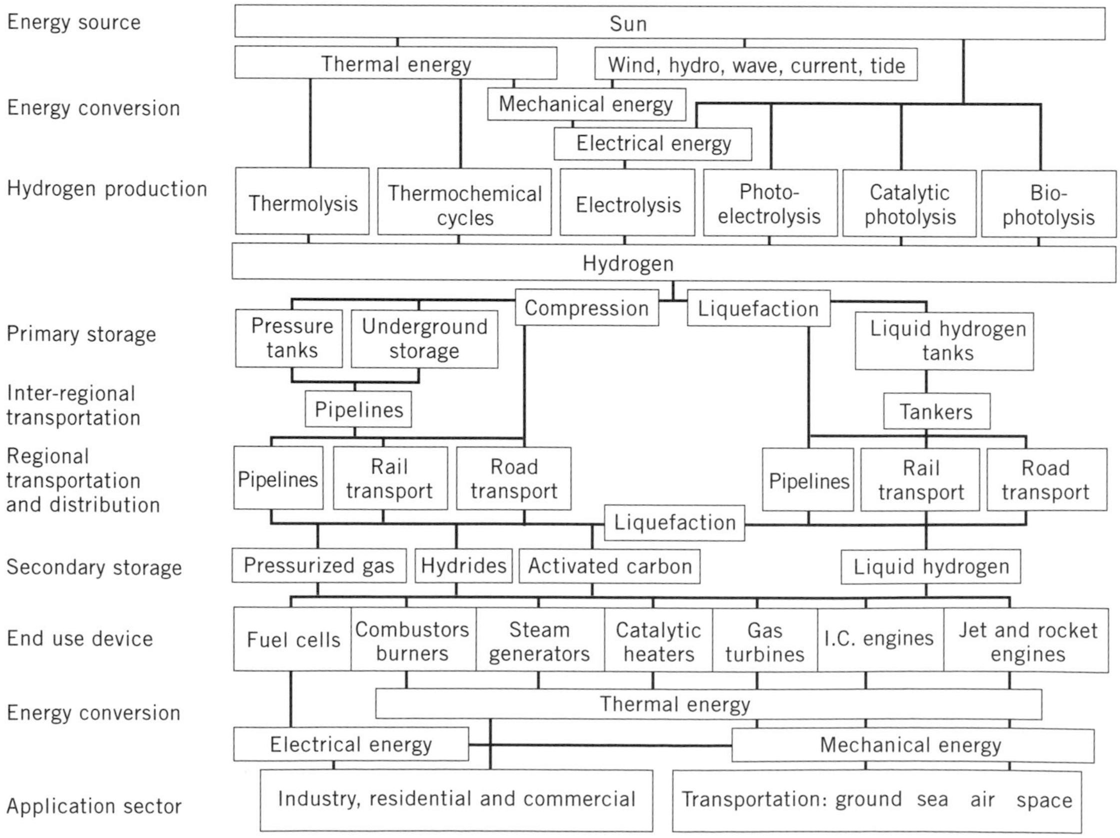

Figure 1. Solar hydrogen energy system block diagram.

dro, wind, wave, current, tide, and ocean-thermal produced electricity. The concept of a hydro-hydrogen clean energy system is under investigation in a 100-MW_e international project (20). The goal of this investigation is to prove the feasibility of conversion of the Canadian hydropower into hydrogen, the maritime transport of liquid hydrogen or methylcyclohexane to Europe, and the storage, distribution, and end use there.

Photolysis, or direct extraction of hydrogen from water using only sunlight as an energy source, can be accomplished by photobiological systems, photochemical assemblies, or photoelectrochemical cells (21). Intensive research activities open new perspectives for photoconversion, where new redox catalysts, colloidal semiconductors, immobilized enzymes, and selected microorganisms could provide the means of large-scale solar energy harvesting and conversion into hydrogen.

Hydrogen Delivery

Hydrogen must be stored to overcome daily and seasonal discrepancies between energy source availability and demand. For large-scale storage, the recommended method is the underground storage of hydrogen in aquifers, depleted petroleum and natural gas fields, and man-made caverns resulting from mining and other activities. The latter method is already successfully being utilized in some countries.

In the hydrogen energy system it is envisaged that from the production plants and/or storage, hydrogen will be transported to consumers by means of underground pipelines and/or supertankers (22). To match consumption demand, hydrogen can be regionally transported and distributed, both as a gas and as a liquid by pipelines or in insulated containers by road and rail transportation. Depending on the end use, hydrogen has to be stored in stationary or mobile storage systems at the consumer end as a pressurized gas, as a liquid, or using some of its unique physicochemical properties, in metal hydrides or in activated carbon. Because of hydrogen's low density, this secondary storage could be a problem, especially where space is limited (eg, in automobiles). For some applications (air and aerospace transportation) hydrogen must be used in liquid form (or even in a form of slush hydrogen, which is a mixture of liquid and solid hydrogen). Since liquefaction of hydrogen occurs at 23 K, large amounts of energy (about 35–40% of its energy content) are required to produce liquid hydrogen and to keep it liquid. Therefore, novel liquefaction methods are being investigated in order to reduce the energy requirements and the cost. It has recently been announced that the engineers and scientists at the Astronautics Technology Center in Madison, Wisconsin, were designing and building a revolutionary magnetic liquefier that is expected to offer higher efficiency than conventional equipment as well as reduced size and cost (23).

Hydrogen Utilization

Hydrogen as an energy carrier has many possible applications. It can be used instead of fossil fuels for virtually all purposes: as a fuel for surface and air transportation, as a fuel for heat production, and even as a fuel for production of electricity directly (in fuel cells) or indirectly (through gas- and steam-turbine-driven generators).

Large amount of research work has been done on the use of hydrogen as a fuel for cars, trucks, and buses. They have been mainly aimed toward conversion of the existing internal combustion engines to run on hydrogen (and solving problems related to hydrogen combustion characteristics) and studying the problem of hydrogen storage in vehicles. Two concepts based on liquid hydrogen storage and hydride storage have been developed, tested, and successfully demonstrated (24). Problems of on-board hydrogen storage can be avoided if engine efficiency is improved. The hydrogen/air fuel cells in conjunction with an electric motor are about two times as efficient as internal combustion engines, and do not generate any emissions (except water). They are considered to be ideal for the new generation, zero emission vehicles. Energy Partners, of West Palm Beach, Florida, has already developed and demonstrated a fuel cell powered prototype passenger vehicle (25).

Liquid hydrogen has numerous advantages as a fuel for commercial subsonic and especially for supersonic aircraft. In April 1988, the flight of a commercial airliner (*Tupolev 155*) fueled with liquid hydrogen was demonstrated in the former USSR. The Germans are carrying out preliminary work on a project for a hydrogen-fueled commercial airliner (*Airbus*) (26). The former Soviet Union and Germany signed an agreement to jointly develop hydrogen-powered propulsion technology for civilian aircraft (27). All the aerospace programs, including the U.S. National Aero-Space Program, are based on liquid or slush hydrogen as fuel.

For domestic applications, the Fraunhofer Institute for Solar Energy Systems, in Germany, has developed, designed, and tested appliances based on the principle of catalytic combustion of hydrogen, which are now deliverable to customers (28). These appliances are a catalytic hydrogen stove and an absorption refrigerator with catalylic H_2 eliminator systems. Catalytic combustion of hydrogen is advantageous with respect to efficiency and emissions.

Other possible hydrogen applications in homes and commercial buildings are heat (space heating and air conditioning, water heating) and electricity generation. Hydrogen fuel cells are very efficient devices (up to 80% combined electric and thermal efficiencies), which can produce both electricity and heat. The Toshiba Corp. of Japan, jointly with the International Fuel Cells Corp. (IFC) of the United States, is now planning to sell on-site 200-kW fuel cell power plants (29). The 4.5-MW fuel cell plant, designed by United Technologies Corp. for Tokio Electric Power, has operated flawlessly and to date has logged hundreds of hours at peak power. Another, an 11-MW plant, began operating in December 1990 for the same utility company (30).

Efficient hydrogen use in electricity generation will become very important for solar power plants, where hydrogen serves as an energy storage medium. During periods when solar energy availability is higher than energy demand, surplus energy can be used in electrolyzers to produce hydrogen, and during the periods when demand is higher than availability or the sun is not available, electricity can be produced from hydrogen via fuel cells.

The German Aerospace Establishment (DLR) has developed a hydrogen/oxygen steam generator which is extremely efficient (almost 100%), compact, and relatively inexpensive (31). This steam generator has been developed for spinning reserve in power plants, but it can also be used for peak-load electricity generation, in industrial steam supply networks, and as a micro–steam generator in medical technology and biotechnology (24).

Standards for Hydrogen Energy Technology

International standards are needed to facilitate the economic and safe production, storage, transport, and utilization of hydrogen as an environmentally compatible energy carrier and feedstock and to enable international development and exchange of hydrogen energy technology. The International Standards Organization (ISO) has established a technical committee (TC-197) which will work on international standards for hydrogen technology (32). This is a very important activity and will lay the foundations of the hydrogen energy system.

CRITERIA FOR FUEL CHOICE

As we approach the end of the fossil fuel era, it has become clear that we shall have to manufacture our future fuel, since there are no natural energy sources or fuels having the convenience of utilization of the fossil fuels, especially that of the fluid fossil fuels. This in fact provides humankind with a great opportunity to decide on the best possible fuel to manufacture.

The following is a list of criteria that must be considered in selecting the future fuel:

- Transportability
- Versatility
- Utilization efficiency
- Environmental compatibility
- Safety
- Economics (effective cost)
- It must be a convenient fuel for transportation.
- It must convert with ease to other forms of energy at the user end.
- It must have high utilization efficiency.
- It must be compatible with the environment.
- It must be safe to use.
- It must be inexpensive.

Table 1. Energy Content (HHV) of Fuels

Fuel	Chemical Formula	Energy per Unit Mass (MJ/kg)	Energy per Unit Volume (GJ/m^3)	Motivity Factor Φ_M
		Liquid Fuels		
Fuel oil	$C_{20\leq}H_{42\leq}$	45.5	38.65	0.78
Gasoline	$C_{5-10}H_{12-22}$	47.4	34.85	0.76
Jet fuel	$C_{10-15}H_{22-32}$	46.5	35.30	0.75
LPG	$C_{3-4}H_{8-10}$	48.8	24.40	0.62
LNG	$\sim CH_4$	~50.0	~23.00	0.61
Methanol	CH_3OH	22.3	18.10	0.23
Ethanol	C_2H_5OH	29.9	23.60	0.37
LH_2	H_2	141.9	10.10	1.00
		Gaseous Fuels		
Natural gas	$\sim CH_4$	~50.0	0.040	0.75
GH_2	H_2	141.9	0.013	1.00

Transportation Fuel

Surface vehicles and airplanes must carry their fuel for a certain distance before replenishing their fuel supply. In the case of space transportation, the space vehicles must carry their fuel, as well as the oxidant, necessary for their

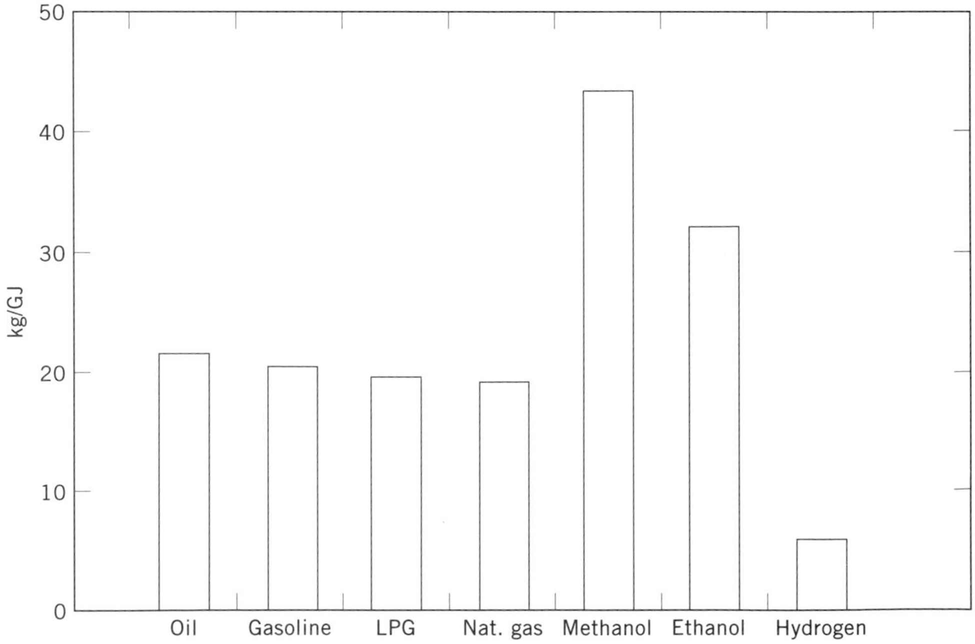

Figure 2. Mass per unit energy (HHV) generated for various fuels.

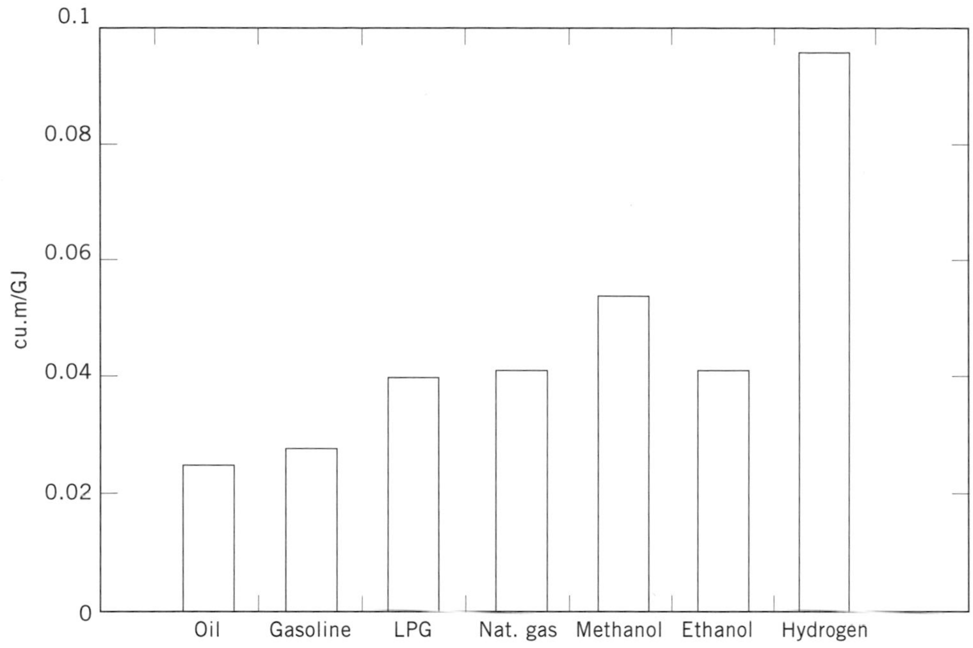

Figure 3. Volume per unit energy (HHV) generated for liquid fuels.

scheduled range. Therefore, it is important that the transportation fuel be as light as possible and also take as little space as possible. Table 1 presents the pertinent information, ie, energy per unit mass and energy per unit volume, for various fuels that are presently used and/or under consideration for future use. Figures 2–4 present the same information in bar chart form.

From the tables and figures, it can be seen that hydrogen is the lightest fuel for a given amount of energy, methanol the heaviest, and the others in between. When liquid fuels are compared, on an energy per unit volume basis, oil and gasoline are on the lower side and hydrogen occupies more space than any other fuel for a given amount of energy. If the utilization efficiency advantage of hydrogen is taken into account, there is some 36% reduction in the volume of hydrogen. When the gaseous fuels are compared on an energy per unit volume basis, natural gas occupies about 32% of the volume needed by hydrogen for the same amount of energy. If the utilization efficiency of hydrogen is taken into consideration, then the volume required by hydrogen becomes about double that of natural gas.

In order to combine the mass and volume characteristics of fuels, we could carry out the following simple analysis: Mass per unit energy for a given fuel must be as small as possible, since the acceleration force is proportional to mass. This can be expressed as

$$F_a \propto \frac{M}{E} \tag{1}$$

where F_a is the acceleration force needed, M is the mass of the fuel, and E is the energy generated by the same fuel. The volume of the fuel affects the drag force in proportion to the surface area of the volume it occupies. This could be expressed as

$$F_d \propto \left(\frac{V}{E}\right)^{2/3} \tag{2}$$

where F_d is the drag force and V is the volume of the fuel. For the best possible transportation fuel, both F_a and F_d must be as small as possible. Consequently, their sum should be as low as possible, and also their product would be expected to be low. In order to avoid the necessity of having dimensional compatibility between Eq. (1) and (2), the latter criteria will be used, ie,

$$F_a F_d \propto \frac{M}{E}\left(\frac{V}{E}\right)^{2/3} \tag{3}$$

should be as low as possible. Conversely, for a good transportation fuel, the inverse of the product $F_a F_d$ must be as high as possible. We shall call a normalized form of this inverse product (normalized with respect to the properties

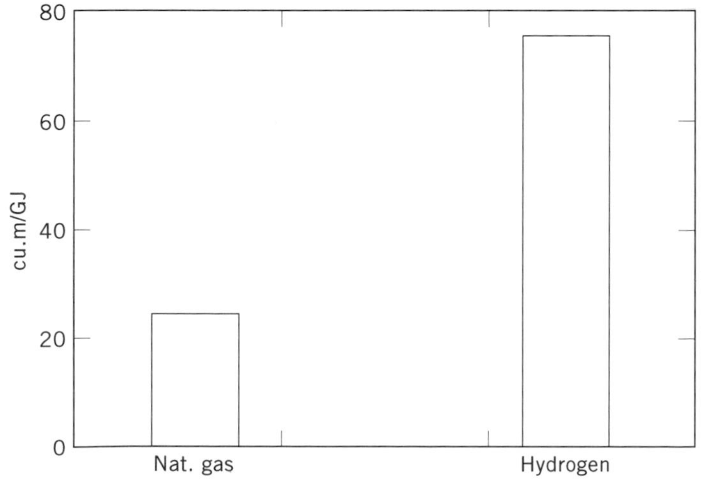

Figure 4. Volume per unit energy (HHV) generated for gaseous fuels.

of hydrogen) the *motivity factor,* ie,

$$\Phi_M = \frac{(E/M)(E/V)^{2/3}}{(E_h/M_h)(E_h/V_h)^{2/3}} \quad (4)$$

where the subscript h refers to hydrogen. Table 1 also lists the motivity factor for various fuels. Figure 5 presents the same information in bar chart form. It can be seen that among the liquid fuels, LH_2 has the best motivity factor, while methanol has the lowest motivity factor. Among the gaseous fuels, GH_2 has the best motivity factor. Consideration of the utilization efficiency advantage of hydrogen further improves hydrogen's standing as the best transportation fuel. Of course, this is one of the reasons why hydrogen is the fuel of choice for the space programs around the world, even though presently it is more expensive than fossil fuels.

Versatility

At the user end, all the fuels must be converted through a process (such as combustion) to other forms of energy, eg, thermal energy, mechanical energy, and electrical energy. If a fuel can be converted directly to various forms of energy at the user end, it becomes more versatile and more convenient to utilize. Table 2 lists various fuels and processes by which they can be converted to other forms of energy at the user end. It can be seen that all the fuels, except hydrogen, can be converted through only one process, that of combustion. Hydrogen, however, can be converted to other forms of energy in five different ways; ie, in addition to flame combustion, it can be converted directly to steam, converted to heat through catalytic combustion, act as a heat source and/or heat sink through chemical reactions, and converted directly to electricity through electrochemical processes (33).

Table 2. Versatility (Convertibility) of Fuels

Conversion Process	Hydrogen	Fossil Fuels
Flame combustion	Yes	Yes
Direct steam production	Yes	No
Catalytic combustion	Yes	No
Chemical conversion (hydriding)	Yes	No
Electro-chemical conversion (fuel cells)	Yes	No

The applications of hydrogen flame combustion ($H_2 + \frac{1}{2}O_2 \rightarrow H_2O$ + heat are as follows:

- IC engines
- Diesel engines
- Gas turbines
- Rocket engines
- Heating
- Cooking

It covers all the applications of fossil fuel flame combustion; ie, it can be used as fuel in all the engines as well as for heating and cooking.

Combustion of hydrogen with pure oxygen produces pure steam at high temperatures (about 300°C) By adding water to this high-temperature steam, temperature is reduced to levels permissible for containment in metallic structures, and the amount of steam is increased. In the so-called Aphodid steam generators, efficiency is high (up to 99%) and the size of the generator is very much smaller than the conventional boilers. If hydrogen is produced by electrolysis, then the oxygen needed would be readily available. Applications of hydrogen-produced steam [$H_2 + \frac{1}{2}O_2 + n\text{-}H_2O \rightarrow (n+1)\text{-}H_2O$ "steam"], which ranges

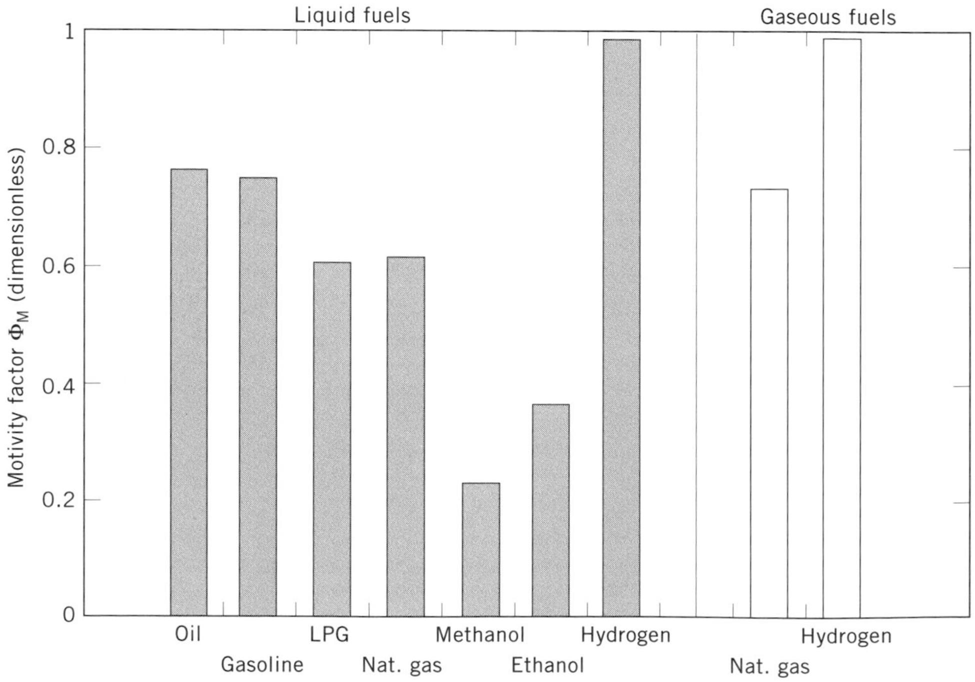

Figure 5. Motivity factor for various fuels.

from steam for power plants to steam for heating, are as follows:

- Steam turbines
- Spinning reserve
- Industrial steam
- Heating steam

In the presence of some catalysts, such as platinum, rhodium, and palladium, hydrogen will combine with oxygen (pure oxygen or oxygen from the air) at lower temperatures than those of flames without a flame, producing water vapor and heat. Temperatures can be controlled by adjusting the hydrogen flow, and the efficiencies are high. There is little or no NO_x production. Catalytic combustion of hydrogen is especially suitable for cooking, space heating, and appliances requiring heat. The possible applications of the catalytic combustion of hydrogen ($H_2 + \frac{1}{2}O_2 +$ catalyst $\rightarrow H_2O +$ heat) are as follows:

- Cooking
- Water heating
- Space heating
- Heat for absorption cooling

When a hydride is formed by the chemical reaction of hydrogen with an element, a metal, or an alloy, heat is generated. In other words, the charging or absorption process is exothermic. Conversely, the discharging or desorption process is endothermic; ie, heat must be supplied to a hydride in order to liberate hydrogen from it. The reactions given below sum up the reactions described above:

$$m\text{-M} + nH_2 \rightarrow M_mH_{2n} + \text{heat}$$

$$M_mH_{2n} + \text{heat} \rightarrow m\text{-M} + n\text{-}H_2$$

and the applications of hydrogen chemical conversion (hydriding) are as follows:

- H_2 storage
- H_2 purification
- D_2 separation
- Compressing
- Pumping
- Heat pumping
- Refrigeration
- Air conditioning
- Electricity generation

The heat generated during the charging process and the heat needed for discharging are functions of the hydriding substance, the hydrogen pressure, and the temperature at which the heat is extracted or supplied. Using different metals and by forming different alloys, different hydriding characteristics can be obtained. In other words, it is possible to make or to find hydriding substances which are more suitable for a given application. There is a wide range of potential applications, from hydrogen storage to deuterium separation and from pumping to electricity generation (see the preceding list).

Hydrogen is the most suitable fuel for electricity generation. It can directly be converted to direct current (DC) electricity in fuel cells, which are not subject to Carnot cycle limitation, with efficiencies higher than those for any other fuel. Fuel cells are similar to electrolyzers but operate in reverse, ie, generate DC electricity using hydrogen and oxygen, instead of producing hydrogen and oxygen using DC electricity, as is the case with electrolyzers.

Fuel cells consist of two electrodes with an electrolyte solution between them. The electrolyte can be a basic solution of sodium or potassium hydroxide, an acidic solution (eg, phosphoric acid), or one of the various ceramic or polymeric solids that carry electric current in the form of hydroxyl ions or hydrogen ions. At the anode, hydrogen gas reacts to produce hydrogen ions (H^+) and electrons. The electrons are driven through the external circuit, producing DC electricity, where they can do work, and finally to the cathode, where they react with oxygen (from air or from an oxygen supply) and the electrolyte to produce hydroxyl ions (OH^-). In the electrolyte, the hydroxyl ions and the hydrogen ions react to produce pure water. The water is then removed from the cell. In addition to the electricity, the water produced can be used, eg, as potable water. In spacecraft, fuel cells are used to produce electricity, and the water produced is used by the astronauts as drinking water. Table 3 presents information on the different types of fuel cells, their operating temperatures, efficiencies, and developers and/or manufacturers.

It is clear from the above discussion that at the user end the fossil fuels can be converted to other forms of energy by flame combustion alone, while hydrogen can be converted by five different processes including flame combustion. In order to quantize versatility, hydrogen will be assumed to have the maximum *versatility factor* of unity, ie,

$$\Phi_{Vh} = 1 \qquad (5)$$

while the fossil fuels will have a versatility factor equivalent to the fraction of the energy utilization field which should normally be satisfied by the flame combustion of hydrogen. In the hydrogen energy system, it is expected that because of the utilization efficiency advantage all of the electric power (30% of the field) will be generated by fuel cells; two-thirds of the transportation sector (30% of

Table 3. Direct DC Electricity Generation through Electrochemical Conversion ($H_2 + \frac{1}{2}O_2$ + fuel cell $\rightarrow$ DC electricity + H_2O + heat)

Type	Operating Temperature (°C)	Efficiency	Developer and Manufacturer
Alkaline	50–90	50–70	UTC, Elenco Siemens
Phosphoric acid	180–210	40–55	UTC, Toshiba, IFC Fuji
Solid polymer	50–125	60–75	Ballard, IFC, EP
Molten carbonate	600–800	55–75	UTC, ERC Hitachi, Toshiba
Solid oxide	900–1200	60–80	Westinghouse

the field) will be energized by flame combustion; and one-half of the heat generation field (40% of the field) will be energized by flame combustion. Consequently, the flame adds up to 0.30(2/3) + 0.40($\frac{1}{2}$) = 0.4 of the total energy utilization field, and the fossil fuel versatility factor becomes

$$\Phi_{Vf} = 0.4 \tag{6}$$

From the foregoing, it is clear that hydrogen is the most versatile fuel with many energy conversion possibilities for a given application.

Utilization Efficiency

In comparing the fuels, it is important to take into account the utilization efficiencies at the user end. For utilization by the user, fuels are converted to various energy forms, such as mechanical, electrical, and thermal. Studies show that in almost every instance of utilization, hydrogen can be converted to the desired energy form more efficiently than other fuels (34).

Thermal Energy Generation. In many industrial, commercial, and residential applications, fuels are converted to thermal energy (eg, for space heating, water heating, cooking, steam generation, and direct heat in industrial processes). Hydrogen is burned in large amounts in the chemical industry to produce heat, especially when it is not economical to purify it. However, some design changes in the existing burners are required to match hydrogen-burning properties, which are somewhat different than those of methane and other gaseous fuels, but the efficiency is about the same as for fossil fuel combustion.

Catalytic combustion of hydrogen is more efficient than flame combustion. In some applications, eg, for space heating, catalytic combustion can be up to 99% efficient, since all the heat of the catalytic reaction remains inside the heated space; there are no exhaust gases. Catalytic burners can also be used in kitchen ranges.

In the Aphodid steam generators hydrogen burns with oxygen, producing pure steam at quite high temperatures. Then an appropriate amount of water is added in order to bring the temperature down. Such a device, named the H_2/O_2 steam generator, has been developed and tested at the DLR. This steam generator is very simple, compact, and extremely efficient (up to 99%). Conventional steam generators are usually about 75–80% efficient.

Electrical Power Generation. Hydrogen can be converted to electricity in fuel cells with much greater efficiencies than those possible in thermal power plants using fossil fuels. While conversion efficiencies for the latter are in the range of 35–38%, practical efficiencies in hydrogen fuel cells are in the range of 50–70%. In the advanced hydrogen fuel cells now being developed, even higher efficiencies (up to 85%) can be expected. This is another important, unique property of hydrogen, which can also increase the conversion efficiencies in transportation vehicles, and consequently reduce transportation fuel consumption. Hydrogen fuel cell/electric motor combinations would yield more than two times greater conversion efficiencies than an internal combustion engine running on a gasoline or diesel engine.

Transportation. Hydrogen-powered vehicles with internal combustion engines have been proven to be more efficient than gasoline vehicles. Hydrogen can be considered more thermally efficient than gasoline, primarily because it burns better in excess air and permits the use of a higher compression ratio. Data from engine tests indicate that hydrogen combustion is 15–50% more thermally efficient. The overall fuel efficiency of a hydrogen vehicle, which takes into account the thermal efficiency as well as the weight of the vehicle, is also better when compared with a gasoline vehicle. On average, hydrogen vehicles are 22% more efficient. As hydrogen can burn in lean fuel–air mixtures as well as in rich mixtures, it can cause large improvements in fuel use efficiencies in stop–start city driving.

Investigations show that for a given number of passengers and a given payload, a subsonic jet passenger airplane would use 19% less energy if it were to use liquid hydrogen instead of fossil-based jet fuel. In the case of a supersonic jet plane, the efficiency advantage of hydrogen is even greater; it is 38% better than jet fuel.

The above-discussed hydrogen utilization efficiencies are summarized in Table 4 as the *utilization efficiency factor,* defined as the ratio of the fuel utilization efficiency to the hydrogen utilization efficiency for a given application. Figure 6 presents the same information in bar chart form. It can be seen that hydrogen is the most efficient fuel. This results in conservation of resources in addition to conserving energy.

ENVIRONMENTAL COMPATIBILITY

Since the utilization of fuels can adversely affect the environment, it is important to examine the environmental compatibility of fuels. Throughout the process of fossil fuel consumption (extraction, transportation, processing, and particularly their end use, combustion), there are harmful impacts on the environment, which cause direct and indirect negative effects on the economy. Excavation

Table 4. Utilization Efficiency Advantage of Hydrogen

Application	Utilization Efficiency Factor, $\Phi_U = \eta_F/\eta_H$
Thermal energy	
Flame combustion	1.00
Catalytic combustion	0.80
Steam generation	0.80
Electric power, fuel cells	0.54
Surface transportation	
IC engines	0.82
Fuel cells/EM	0.40
Subsonic jet transportation	0.84
Supersonic jet transportation	0.72
Weighted average	0.72
Hydrogen utilization efficiency factor	1.00
Fossil fuel utilization efficiency factor	0.72

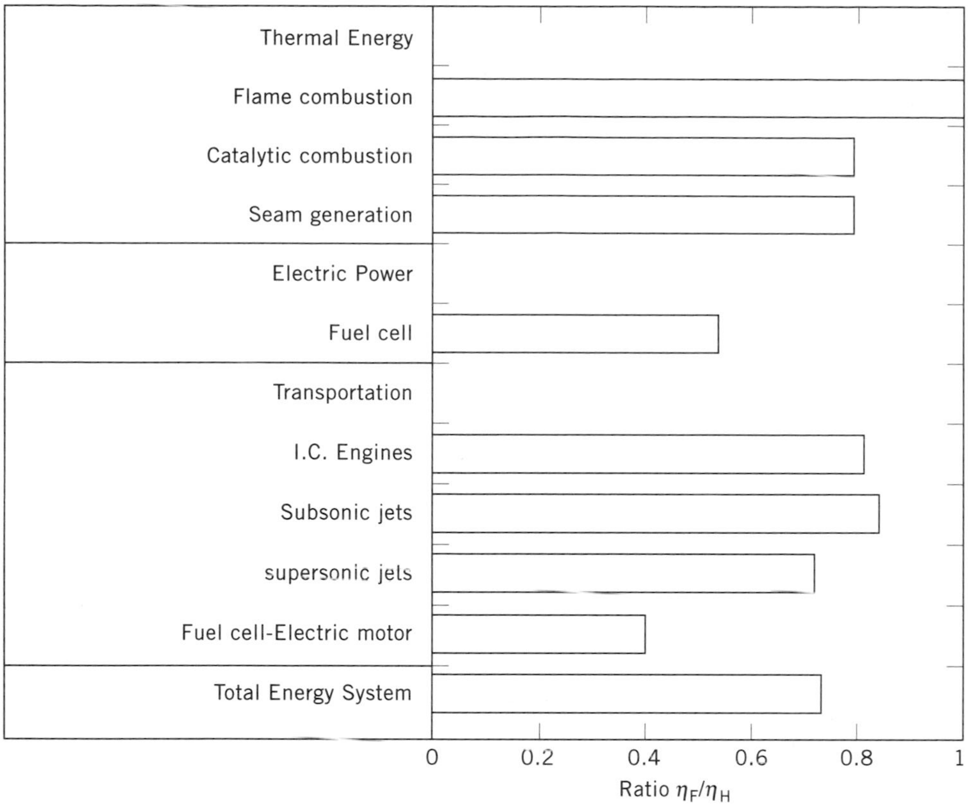

Figure 6. Ratios of fossil fuel utilization efficiency (η_F) to that of hydrogen (η_H) for various applications.

of coal devastates the land, which has to be reclaimed, and is out of use for several years. During the extraction, transportation, and storage of oil and gas, spills and leakages occur, which cause water and air pollution. Refining processes also have negative environmental impact.

Most of the fossil fuel environmental impact occurs during the end use. Their end use involves combustion, irrespective of the final purpose (ie, heating, electricity production, or motive power for transportation). The main constituents of fossil fuels are carbon and hydrogen, but also some other ingredients, which are originally in the fuel (eg, sulfur) or are added during refining (eg, lead, alcohols). Combustion of the fossil fuels produces various gases (CO_x, SO_x, NO_x, CH), soot and ash, droplets of tar, and other organic compounds, which are all released into the atmosphere and produce the air pollution. Air pollution causes damage to human health, animals, crops, and structures and reduces visibility.

Once in the atmosphere, triggered by sunlight or by mixing with water vapor and other atmospheric compounds, the primary pollutants may undergo chemical reactions, change their form, and become secondary pollutants, like ozone, aerosols, peroxyacyl nitrates, various acids, etc. Precipitation of sulfur and nitrogen oxides, which have dissolved in clouds and in rain droplets to form sulfuric and nitric acids, is called acid rain. Acid deposition causes soil and water acidification, resulting in damage to the aquatic and terrestrial ecosystems, affecting humans, animals, vegetation, and structures.

The remaining products of combustion in the atmosphere, mainly carbon dioxide, together with other so-called greenhouse gases (methane, nitrogen oxides, and chlorofluorcarbons) result in thermal changes by absorbing the infrared energy Earth radiates back into the atmosphere and reradiating it back to Earth partially, causing global temperature increases. The effects of the temperature increases are melting of the icecaps, sea level rise, and climate changes, which include heat waves, droughts, floods, stronger storms, more and bigger wildfires, etc.

Taking into account various reported damages, Table 5 has been prepared (34,35). It summarizes the damages attributed to the fossil fuels. Environmental damage due to coal utilization is the highest, \$9.82/GJ; then comes petroleum, with \$8.47/GJ; and natural gas is the least harmful among fossil fuels, with the environmental damage estimated at \$5.60/GJ.

The cost of the above-described negative effects are not included in the market price of fossil fuels and can be considered external costs. These costs are paid by the society and/or eventually will be paid by the society, since in the long term any disturbed ecosystem will affect the human society, its environment, and its economy.

Environmental damage is not the only external cost of fossil fuels. In a complete economic study, one must take into account other external costs, such as the military costs for protecting oil and gas supplies. It is estimated that the U.S. military expenditures for maintaining bases to secure oil supplies from fragile regions such as Middle East are about \$60 billion a year, or \$1.7/GJ of petroleum consumed in the United States.

Flame combustion of hydrogen will produce some NO_x, about the same as the flame combustion of fossil fuels. However, it will not produce other pollutants, acid rain

Table 5. Environmental Damage Due to Fossil Fuel Consumption (1990 $/GJ)

Type of damage	Coal (E_{coal})	Petroleum (E_{pet})	Natural Gas (E_{gs})
Effect on humans	3.48	2.83	2.09
Effect on animals	0.51	0.42	0.30
Effect on plants and forests	1.35	1.09	0.81
Effect on aquatic ecosystems	0.18	1.05	0.11
Effect on man-made structures	1.12	0.90	0.67
Other air pollution costs	0.98	0.79	0.59
Effect of strip mining	0.49	—	—
Effect of climatic changes	1.39	1.13	0.84
Effect of sea level rise	0.32	0.26	0.19
Total environmental damage	9.82	8.47	5.60

ingredients, or greenhouse gases. In the hydrogen energy system, energy conversion will not be achieved by flame combustion alone, but there will be employed conversion processes such as catalytic combustion, direct steam generation, and chemical and electrochemical processes, which do not produce NO_x. Hence, NO_x production in the hydrogen energy system will be relatively small.

In order to compare the overall environmental effects of various fuels, it will be better to compare the energy systems based on these fuels. We shall therefore consider the following three possible scenarios: (1) the fossil fuel energy system as it is today; (2) the coal and coal-based synthetic fuels system; and (3) the hydrogen system based on renewable energy sources (36).

Fossil Fuel System

In this scenario, it is assumed that the present fossil fuel system will be continued as it is today. Therefore, it is assumed that 40% of primary energy (in fossil fuel equivalent units) will be used for thermal energy generation, 30% for electric power generation, and 30% for transportation (two-thirds for surface transportation and two-thirds for air transportation). Energy supplied by hydropower and nuclear power plants (mostly in the form of electric power) and by other non–fossil fuel sources has not been taken into account, since it is assumed that it will be the same for all three scenarios.

In the fossil fuel system, it is assumed that one-half of the thermal energy will be supplied by natural gas and the rest by petroleum fuels (fuel oil and residual oil) and coal. Coal is assumed to be the main energy source for electricity generation, gasoline for surface transportation, and jet fuel for air transportation. This is of course a simplified version of the fossil fuel energy system, but it is close enough to the present patterns.

Coal/Synthetic Fossil Fuel System

In this case, it is assumed that the present fossil fuel system will be continued by the substitution with synthetic fuels derived from coal wherever convenient and/or necessary. Patterns of energy consumption are also assumed to be unchanged. Coal will be used extensively for thermal power generation and for electric power generation, because it is much cheaper than synthetic fuels. However, some end uses require fluid fuels. Therefore it has been assumed that synthetic natural gas (SNG) will be used for thermal energy generation (primarily in the residential sector) and also as a fuel for surface transportation, where it will share the market with synthetic gasoline. Synthetic jet fuel will be used in air transportation.

Solar–Hydrogen Energy System

In this case, it is assumed that conversion to hydrogen energy will take place, and one-third of the hydrogen needed will be produced from hydropower and two-thirds by direct and indirect (other than hydropower) solar energy forms. The same percentages of energy demand for various sectors as the above scenarios will be assumed. It will further be assumed that one-half of the thermal energy will be achieved by flame combustion, one-quarter by steam generation with hydrogen/oxygen steam generators, and the last quarter by catalytic combustion; electric power will be generated by fuel cells; one-half of the surface transportation will use gaseous hydrogen burning internal combustion engines and the other half will use fuel cells. In air transportation (both subsonic and supersonic), liquid hydrogen will be used.

Table 6 lists the pollutants for the three energy systems described above. Figure 7 presents the same information in bar chart format. It can be seen that the coal/synthetic fossil system is the worst from the environmental point of view, while the solar–hydrogen energy system is the best. The solar–hydrogen system will not produce any CO_2, CO, SO_x, hydrocarbons, or particulates, except some NO_x. However, the solar–hydrogen-produced NO_x is much less than that produced by the other energy systems for the reasons given earlier.

Table 7 presents the environmental damage per gigajoule of the energy consumed for each of the three energy systems considered and also for their fuel components in 1990 U.S. dollars as well as *environmental compatibility factors,* defined as the ratio of the environmental damage due to the hydrogen energy system to that due to a given energy system. Figure 8 shows similar information for the three energy systems in bar chart format. The environmental damage for the solar–hydrogen energy system is due to the NO_x produced. It can be seen that the solar–hydrogen energy system is environmentally the most compatible system.

It should be mentioned that hydrogen also has the answer to the depletion of the ozone layer, mainly caused by chlorofluorocarbons. Refrigeration and air-conditioning systems based on the hydriding property of hydrogen do

Table 6. Pollutants Produced by Three Energy Systems (kg/GJ)

Pollutant	Fossil Fuel System	Coal/Synthetic Fossil System	Solar–Hydrogen System
CO_2	72.40	100.00	0
CO	0.80	0.65	0
SO_x	0.38	0.50	0
NO_x	0.34	0.32	0.10
HC	0.20	0.12	0
PM	0.09	0.14	0

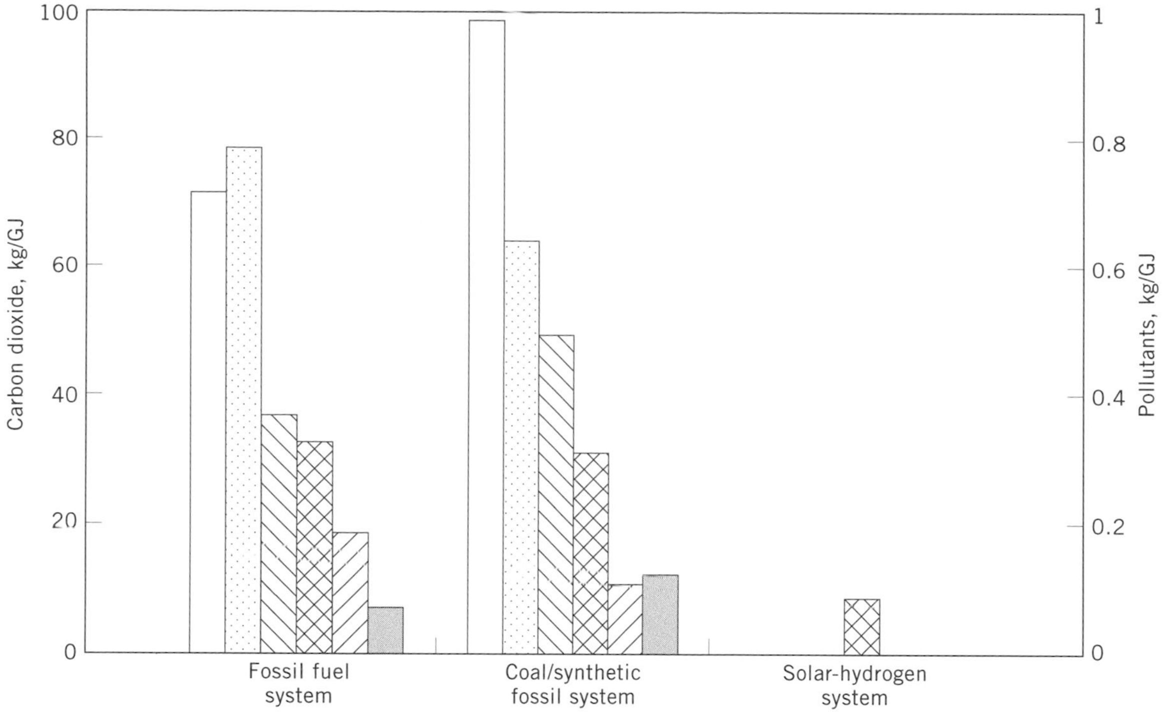

Figure 7. Pollutants produced by three energy systems. □, CO_2; □, CO; ▧, SO_x; ▩, NO_x; ▨, CH_x; ■, PM.

not need chlorofluorocarbons but need hydrogen, and any hydrogen leak would not cause ozone layer depletion. Such refrigeration systems are also very quiet, since they do not have any moving machinery.

Vapor Generation

When environmental compatibility of fuels are examined, the amount of water vapor generated must also be considered. When combusted by flame combustion, all the fuels produce water vapor via the oxidation of hydrogen. In the case of the hydrogen energy system, obviously there will be more water vapor generation as compared to the fossil fuels. Global warming, which is caused by the utilization of the fossil fuels, also causes an increase in water vapor generation. Assuming that Earth's mean temperature has increased by 0.5°C since the beginning of the Industrial Revolution (37), this additional water vapor generation and that produced by the combustion of fuels have been calculated. The results are presented in Table 8 and Figure 9. It can be seen that (1) the two fossil fuel systems generate much more additional (above natural) water vapor than the solar–hydrogen energy system, (2) the additional water vapor generated by global warming is much greater than that produced by the combustion of fuels, (3) the amount of water vapor generated by fuels is minimal compared to that generated naturally, and (4) the solar–hydrogen system causes the smallest increase in vapor generation. Again, when the additional vapor generation is considered, the solar–hydrogen system becomes environmentally the most compatible system.

Table 7. Environmental Damage and Environmental Compatibility Factors

Energy System and Fuel	Environmental Damage (1990 \$/GJ)	Environmental Compatibility Factor, Φ_E
Fossil Fuel	8.44	
Coal	9.82	0.047
Oil	8.47	0.054
Natural Gas	5.60	0.082
Coal/Synthetic Fossil	10.47	
Syn-Gas	13.77	0.033
SNG	9.13	0.050
Solar–hydrogen	0.46	
Hydrogen	0.46	1.000

SAFETY

The safety aspects of fuels involve their toxicity on one hand and the fire hazard properties on the other. In addition to the toxicity of their combustion products, the fuels themselves can be toxic. The toxicity increases as the carbon-to-hydrogen ratio increases. Hydrogen and its main combustion product, water or water vapor, are not toxic. However, NO_x, which can be produced through the flame combustion of hydrogen (as well as through the combustion of fossil fuels) displays toxic effects. When the amounts of toxic pollutants produced per unit of energy consumed are considered, it is clear that hydrogen is the safest fuel, followed by methane and gasoline in that order (see Tables 5 and 6).

Table 9 lists the characteristics of fuels related to fire hazards. Lower density makes a fuel safer, since it increases the buoyancy force for speedy dispersal of the fuel in case of a leak. For the same reason, higher diffusion

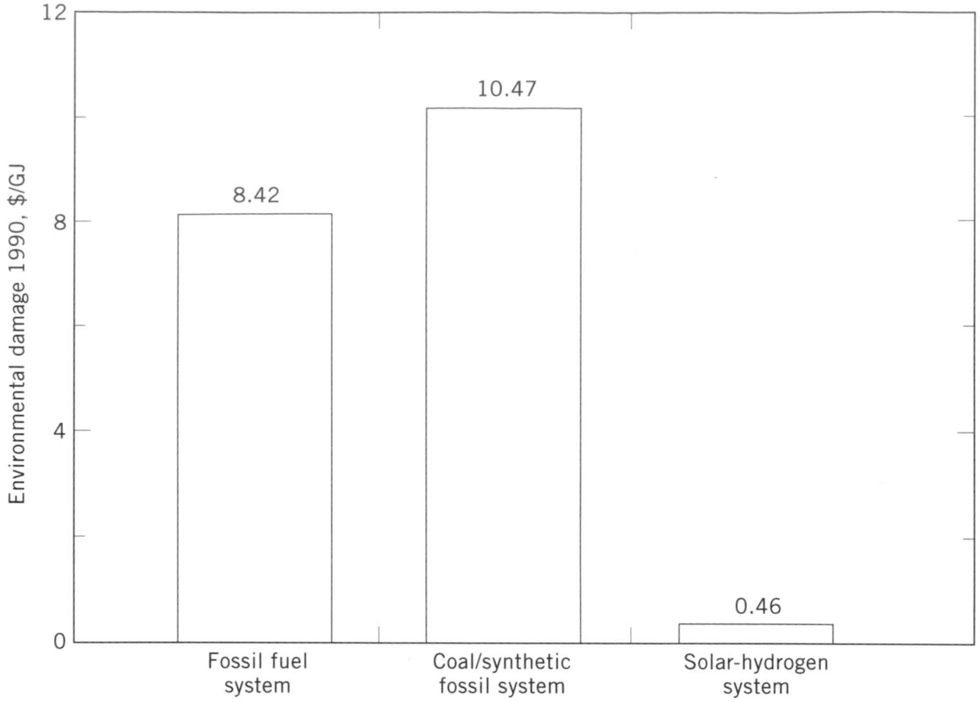

Figure 8. Comparison of environmental damage by three energy systems.

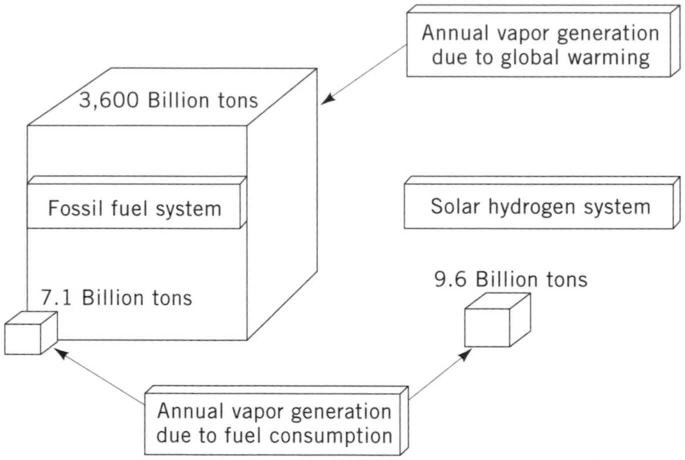

Figure 9. Vapor generation by energy systems for 1990 energy consumption (vapor generated due to global warming has been calculated for 0.5°C global temperature increase).

coefficients are helpful. Higher specific heat causes a fuel to be safer, since it slows down the temperature increases for a given heat input. Wider ignition limits, lower ignition energies γ, and lower ignition temperatures make the fuels less safe, as they increase the limits in which a fire could commence. Higher flame temperature, higher explosion energy, and higher flame emissivity make a fuel less safe as well, since its fire would be more damaging.

Table 10 compares the safety of fuels. For each of the toxic element and fire hazard characteristics, it ranks the fuels from 1 to 3, 1 being the safest and 3 the least safe. These rankings have been summed up for each fuel in order to arrive at an overall ranking. The total rankings have been prorated to obtain the *safety factors,* defined as the ratio of the total ranking for hydrogen to that of a given fuel. It can be seen that hydrogen becomes the safest fuel, while gasoline is the least safe, methane being in between the two.

Table 8. Comparison of Vapor Generation by Three Energy Systems for 1990 Energy Consumption

Item	Unit	Fossil Fuel System	Coal/Synthetic Fossil System	Solar–Hydrogen System
Annual vapor generation by energy system	10^{12} kg	7.1	7.4	9.6
Annual vapor generation due to global warming	10^{12} kg	3600	3600	0
Total vapor generation due to energy system and global warming	10^{12} kg	3607	3607	9.6
Total vapor generation as fraction of that produced naturally	%	0.721	0.721	0.002

Note: Annual vapor generation due to solar heating: 5×10^{17} kg.

Table 9. Characteristics Related to Fire Hazard of Fuels

Property	Gasoline	Methane	Hydrogen
Density[a] (kg/m^3)	4.40	0.65	0.084
Diffusion coefficient in air[a] (cm^2/sec)	0.05	0.16	0.61
Specific heat at constant[a] pressure (J/g K)	1.20	2.22	14.89
Ignition limits in air (vol %)	1.0–7.6	5.3–15.0	4.0–75.0
Ignition energy in iar (mJ)	0.24	0.29	0.02
Ignition temperature (°C)	228–471	540	585
Flame temperature in air (°C)	2197	1875	2045
Explosion energy[b] (g TNT/kJ)	0.25	0.19	0.17
Flame emissivity (%)	34–43	25–33	17–25

[a] At normal temperature and pressure.
[b] Theoretical maximum; actual 10% of theoretical.

ECONOMICS

The economical comparison between competing fuels and/or energy systems should be based on the effective costs of the services these fuels provide. The effective costs include the utilization efficiency, the cost of the fuel, and the costs associated with fuel consumption but which are not included in its price (so-called external costs). External costs include the costs of the physical damage done to humans, fauna, flora, and the environment duc to harmful emissions, oil spills and leaks, and coal strip mining as well as governmental expenditures for pollution abatement and expenditures for military protection of oil supplies.

In economic considerations, it is also important to compare the future costs of hydrogen (which will be considerably lower than they are today because of assumed market and technology development) with the future costs, both internal and external, of fossil fuels (which will unavoidably be higher than today's prices due to depletion, international conflicts, and environmental impact).

The effective cost of a fuel can be calculated using the following relationship:

$$C_r = (C_i + C_e)\frac{\eta_{fk}}{\eta_{sk}} \quad (7)$$

where C_i is the internal cost or the conventional cost of the fuel, C_e the external cost including the environmental damage caused by the fuel, η_{fk} the fossil fuel utilization efficiency for application k, and η_{sk} the synthetic fuel (including hydrogen) utilization efficiency for the same application or the end use.

In order to evaluate the overall cost (C_0) to the society, the three scenarios considered earlier will be used. This cost can be calculated from the relationship

$$C_0 = \sum_{n=1}^{n} \alpha_n C_{rn} \quad (8)$$

where α_n is the fraction of energy used by the energy sector n, such as electricity generating, heat producing, surface transportation, subsonic air transportation, and supersonic air transportation. Since α_n is a fraction, their sum is

$$\sum_{n=1}^{n} \alpha_n = 1 \quad (9)$$

Substituting Eq. (7) into Eq. (8), one obtains

$$\sum_{n=1}^{n} \alpha \left[(C_i + C_e)\frac{\eta_{fk}}{\eta_{sk}} \right]_n \quad (10)$$

Using Eqs. (7)–(10), Tables 11–13 have been prepared for the three energy scenarios, ie, the fossil fuel system, the coal/synthetic fossil fuel system, and the solar–hydrogen energy system, for the year 2000 in 1990 U.S. dollars. Table 14 presents the summaries of the effective costs for the three energy systems, their fuel components, and the *economy factors* (Φ_c), defined as the ratio of the overall effective cost for hydrogen to that of a given fuel. Figure 10 shows the overall results in bar chart format. Comparing the results, it becomes clear that hydrogen is the most cost-effective fuel and results in the lowest overall effective cost to society.

Table 10. Safety Ranking of Fuels

	Fuel Ranking[a]		
Characteristic	Gasoline	Methane	Hydrogen
Toxicity of fuel	3	2	1
Toxicity of combustion products (CO, SO_x, NO_x, HC, PM)	3	2	1
Density	3	2	1
Diffusion coefficient	3	2	1
Specific heat	3	2	1
Ignition limit	1	2	3
Ignition energy	2	1	3
Ignition temperature	3	2	1
Flame temperature	3	1	2
Explosion energy	3	2	1
Flame emissivity	3	2	1
Totals	30	20	16
Safety factor Φ_s	0.53	0.80	1.00

[a] 1, safest; 2, less safe; 3, least safe.

Table 11. Effective Cost of Fossil Fuel System (1990 U.S.$/GJ)

Application	Energy Consumption Fraction	Fuel	Effective Cost	Fraction × Cost
Thermal energy	0.20	Natural gas	11.82[a]	2.36
	0.10	Petroleum fuels	18.66[a]	1.87
	0.10	Coal	12.02	1.20
Electric power	0.30	Coal	11.68	3.50
Surface transportation	0.20	Gasoline	21.40	4.28
Air transportation	0.10	Jet Fuel	17.59	1.76
Total	1.00			
Overall effective cost			14.97	

[a] Average for residential and industrial sector.

Table 12. Effective Cost of Coal/Synthetic Fossil System (1990 U.S.$/GJ)

Application	Energy Consumption Fraction	Fuel	Effective Cost	Fraction × Cost
Thermal energy	0.30	Coal	12.02	3.61
	0.10	SNG	24.81	2.48
Electric power	0.30	Coal	11.68	3.50
Surface transportation	0.10	SNG	24.81	2.48
	0.10	Syn-gasoline	34.97	3.50
Air transportation	0.10	Syn-Jet	30.77	3.08
Total	1.00			
Overall effective cost			18.65	

Table 13. Effective Cost of Solar–Hydrogen Energy System (1990 U.S.$/GJ)

Application	Energy Consumption Fraction	Fuel[a]	Effective Cost	Fraction × Cost
Thermal energy				
Flame combustion	0.20	GH_2	17.63	3.53
Steam generation	0.10	GH_2	14.10	1.41
Catalytic combustion	0.10	GH_2	14.10	1.41
Electric power				
Fuel cells	0.30	GH_2	9.52	2.86
Surface transportation				
IC engines	0.10	GH_2	14.46	1.45
Fuel cells	0.10	GH_2	7.05	0.71
Air transportation				
Subsonic	0.05	LH_2	17.78	0.89
Supersonic	0.05	LH_2	15.24	0.76
Total	1.00			
Overall effective cost			13.02	

[a] It has been assumed that $\frac{1}{3}$ of hydrogen will be produced from hydropower and $\frac{2}{3}$ from solar.

Table 14. Overall Effective Costs and Economy Factors

Energy System and Fuel	Effective Cost (1990 U.S.$/GJ)	Economy Factor, Φ_c
Fossil Fuel	14.97	
Gasoline	21.40	0.61
Natural gas	11.82	1.10
Coal/synthetic fossil	18.65	
Syn-Gas	34.97	0.37
SNG	24.81	0.52
Solar–hydrogen	13.02	
Hydrogen	13.02	1.00

OVERALL COMPARISON OF FUELS

In order to compare the merits of fuels, six properties have been identified and each has been quantized by a dimensionless factor: motivity factor φ_M, versatility factor φ_v, utilization efficiency factor φ_U, environmental compatibility factor φ_E, safety factor φ_s, and economy factor φ_c. Factors have been normalized with respect to hydrogen properties so as to make them vary between zero and unity, unity being the most desirable factor. We could obtain an overall merit factor (Φ_0) for each of the fuels by

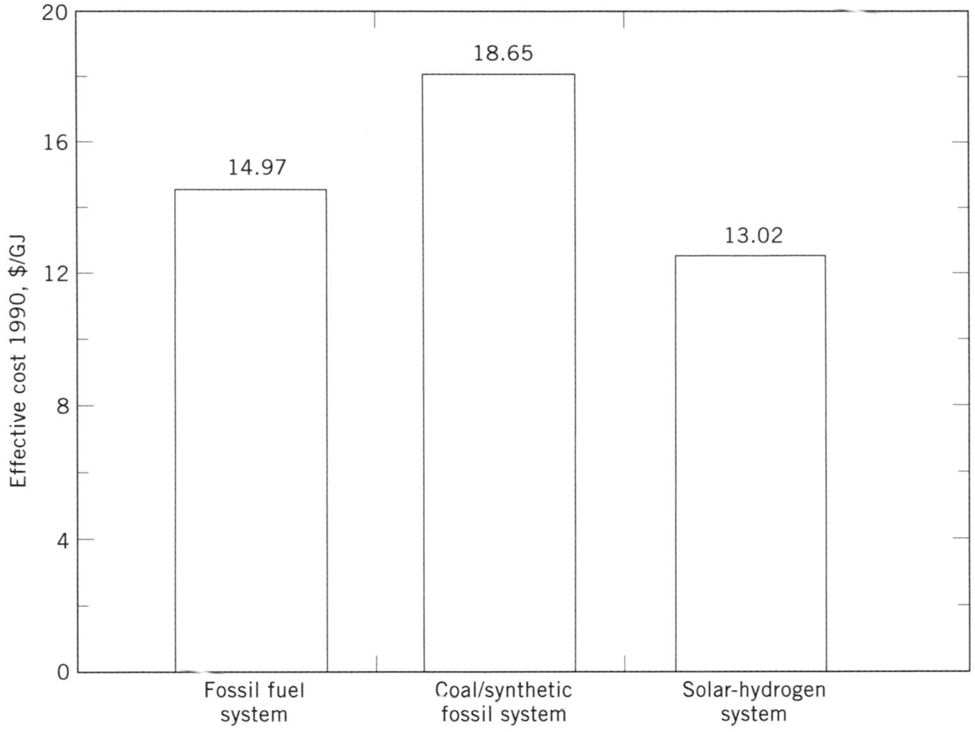

Figure 10. Overall effective costs of three energy systems (year 2000).

defining a weighted average as follows:

$$\Phi_0 = \frac{\sum_{j=1}^{6} W_j \Phi_j}{\sum_{j=1}^{6} W_j} \qquad (11)$$

where W is the weighting coefficient for a given property and j is the property. Assuming each property has the same weighting coefficient, Eq. (11) becomes

$$\Phi_0 = \frac{1}{6} \sum_{j=1}^{6} \Phi_j \qquad (12)$$

Using Eq. (12), Table 15 has been prepared. The overall merit factors are presented in Figure 11 in bar chart format. It can be seen that hydrogen has the best rating for each criteria as well as the best overall rating. Natural

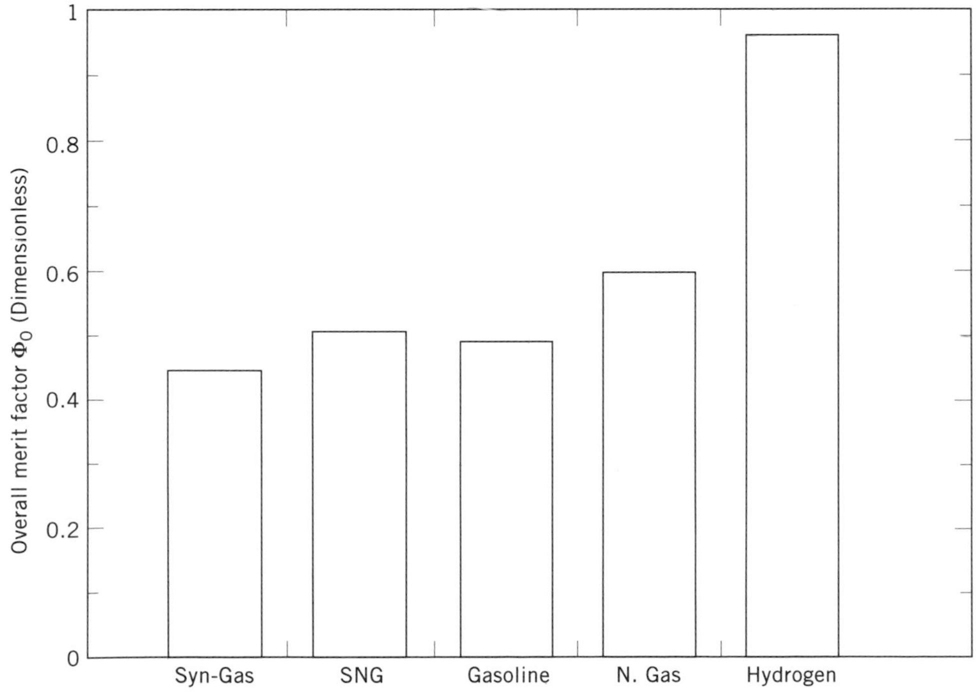

Figure 11. Overall merit factors (Φ_0) of fuels.

gas comes the second best, followed by synthetic natural gas, gasoline, and synthetic gasoline.

INITIATION OF HYDROGEN ENERGY SYSTEM

Although a hydrogen energy system based on renewable energy sources has many advantages when compared with the existing energy system mainly based on utilization of fossil fuels, it will be very difficult to establish it without major governmental and international actions. There are many obstacles for establishing such an energy system, such as economics, lack of infrastructure, and lack of public acceptance.

Energy policies in developed as well as in developing countries should be so planned as to dictate the best possible ways to a sustainable energy system. In order to make the market work, external costs have to be internalized, because only in that way can the clean technologies compete. One way of achieving this would be the introduction of the so-called carbon tax or environmental surcharge on the polluting fuels.

When hydrogen's higher utilization efficiency and its minimal environmental costs are taken into account, hydrogen will become a cost competitive fuel, probably as soon as at the turn of the century.

Hydrogen and electricity are convenient for both centralized and decentralized production, which makes the future energy system more flexible. The most probable and least painful path toward such an energy system would be the gradual implementation of hydrogen production, storage, distribution, and utilization facilities until the hydrogen energy system is established. This process would take 50–60 years to complete.

Natural gas could be the link between today's and the future energy systems. At present, natural gas is the dominant source for hydrogen production. It is also an established energy source and currency with established infrastructure and with basically no public acceptance problem. In the interim, a further market penetration of natural gas prepares infrastructure and end-use technologies for an eventual transition to hydrogen. Most importantly, natural gas buys time for the sustainable energy system to develop and to mature and therefore represents the best energy source and currency for the transition period. Also, as the natural gas supplies decrease, hydrogen could be mixed and transported with the natural gas to make up for the gaseous fuel shortage, since such mixtures can be used in most of the natural gas utilization systems with small adjustments. Therefore, the lack of infrastructure can be relatively easily overcome by gradual implementation and by relying on natural gas in the transition period.

Finally, any new energy system will be publicly accepted as it proves to be better than the old one. For this reason, demonstration or pilot projects can be extremely useful. Not only will they prove hydrogen energy feasibility, but also they will enable evaluations and improvements to be carried out as well as provide training for operating personnel.

POSSIBLE PILOT PROJECTS

There is no reason to delay the actions for transition toward the hydrogen energy system. Since the technology for utilization of renewable energy sources in general is still not competitive, it is recommended to start with smaller projects, which should be subsidized by governmental and international organizations and perhaps by energy companies (including petroleum, natural gas, and coal companies) planning to be involved in the future energy system based on renewable energy sources. The initiation actions should be directed toward establishing pilot and/or demonstration projects for hydrogen production, storage, and utilization. These projects should be designed primarily to satisfy the need for energy (in the form of fuel and electricity) at selected locations and in that way to demonstrate feasibility.

Technologies Involved

Technologies involved in the pilot projects (and for that matter in the hydrogen energy system in general) can be grouped according to their functions: primary energy conversion system, hydrogen production systems, electricity production systems, hydrogen delivery systems, hydrogen utilization systems, and hydrogen-utilizing industries. Table 16 presents the three systems for primary energy conversion: solar radiation, wind energy, and hydropower.

Water electrolysis is the only commercially developed technology today suitable for hydrogen production from renewable energy sources. However, electricity for water electrolysis can be produced employing different technologies and different energy sources, eg, hydropower, wind energy, or solar energy. Hydropower is already a mature technology for electricity production and the main characteristic is relatively constant availability. Hydrogen can also be produced using off-peak electricity at very competitive costs and used as a fuel in homes, transportation, and industry.

Table 15. Overall Ranking of Fuels

Merit Factor	Syn-Gas	SNG	Gasoline	Natural Gas	Hydrogen
Motivity, Φ_M	0.76	0.64	0.76	0.64	1.00
Versatility, Φ_V	0.40	0.40	0.40	0.40	1.00
Utilization efficiency, Φ_U	0.72	0.72	0.72	0.72	1.00
Environmental compatibility, Φ_E	0.033	0.050	0.054	0.082	1.00
Safety, Φ_S	0.53	0.80	0.53	0.80	1.00
Economy, Φ_C	0.37	0.52	0.61	1.10	1.00
Overall merit, Φ_0	0.47	0.52	0.51	0.62	1.00

Table 16. Primary Energy Conversion Systems for Pilot Projects

System	Subsystem and/or Component
1. Solar	Photovoltaic array
	Power conditioning subsystem
	Control subsystem
2. Wind	Wind turbine–generator sets (AC and/or DC)
	Power conditioning subsystem
	Control subsystem
3. Hydropower	Water turbine–generator set(s) (AC and/or DC)
	Power conditioning subsystem
	Water intake and discharge subsystem
	Control subsystem

Solar energy is characterized by diurnal and seasonal variations and without energy storage is incapable of meeting the demand. Hydrogen can be produced during the day, stored, and used for electricity generation (when needed) and/or as a fuel for other applications. Solar thermal energy conversion can be cost effective at certain locations, but PV solar energy conversion might be more appropriate because of its simplicity (no moving parts, easier maintenance), although it is presently more expensive. However, it is expected that the costs of PV cells and arrays will decline continuously from today's \$4–\$5/W_p to \$0.2–\$0.4/W_p over a period of 10–15 years. Photovoltaics are already cost effective for small power generation in remote locations, but at \$0.2/W_p they will become competitive with any electricity generation technology, Consequently, hydrogen produced will also become cost competitive. There are already several PV–electrolysis pilot and demonstration plants (eg, SWB Bavaria, Germany, and Hysolar, Germany–Saudi Arabia joint project), which will provide useful information for further development of this technology (38).

Technology for wind energy conversion is already developed and commercially feasible at some locations. There are already about 2000 MW of installed wind turbines around the world (most of them in California), and many countries (such as India and China) are developing wind energy programs. Wind energy is also characterized by diurnal and seasonal variations (however, these variations are less predictable than those of the solar energy). Similarly, hydrogen could be produced during windy periods, stored, and used for electricity generation and/or as a fuel.

Table 17. Hydrogen Production System for Pilot Projects

Subsystem	Type
Feedwater supply and treatment	Ion exchange, reverse osmosis, electrodialysis, distillation
Power conditioning	Rectifier
Electrolyzer	Alkaline: monopolar, bipolar, bipolar high pressure; membrane (SPE); high-temperature steam
Cooling system	Open cycle, closed cycle
Gas separation and purification	Gas separator, scrubber, washer, cooler, catalyzer
Inert gas supply	Pumps/compressors
Control system	Mechanical/electrical

Table 18. Electricity Supply Systems for Pilot Projects

System	Type
Primary energy conversion system	Photovoltaics, wind energy converters, hydropower converters
Fuel cells	Alkaline, phosphoric acid (PAFC), molten carbonate, solid oxide (SOFC), solid polymer electrolyte
Power distribution	AC, DC

Table 17 presents the subsystems needed in an electrolysis plant for the production of hydrogen and oxygen.

With the exception of hydropower, the primary energy source (solar and wind) would be available intermittently and with variable intensity. Hence, in these cases there must be an additional system to produce electricity whenever needed using the stored energy, ie, hydrogen. Table 18 shows the continuous electricity production systems for all the three primary energy sources.

Table 19 presents the hydrogen and oxygen storage and delivery systems and options. It should be noted that although the oxygen produced could be released into the atmosphere, it may be more convenient and more economical to store the oxygen and utilize it in fuel cells and/or industry.

Table 20 presents the various hydrogen-utilizing appliances and transportation vehicles and the options available, and Table 21 presents some industries where hydrogen and/or oxygen are used as energy carriers or chemical feedstock.

Project Types and Locations

It would be very useful to be able to use all or most of the options available as to the primary energy sources and as

Table 19. Hydrogen/Oxygen Delivery Systems for Pilot Projects

System	Type
Gaseous hydrogen storage	High-pressure vessel(s), high-pressure compressor, pressure gauges, valves, pressure regulators, safety valves
Hydride storage	Hydride vessel, compressor, heat exchangers, temperature gauges, pressure gauges, valves, pressure regulators, safety valves
Hydrogen distribution	Pipelines, valves, flowmeters, safety valves, pressure gauges, pressure regulators, car refueling station
Gaseous oxygen storage	Pressure vessel, oxygen compressor, oxygen distribution system
Safety and fire fighting	Hydrogen detectors, water sprinklers

Table 20. Hydrogen/Electricity Utilization Systems (Residential, Commercial, Transportation) for Pilot Projects

Sector	Device	Type
Residential and commercial	Cooking range/oven	Flame, catalytic, electric, hybrid
	Refrigerator	Absorption with flame burner, absorption with catalytic burner, electric, hydride
	Water heater	Solar, flame, catalytic
	Air conditioning (heating/cooling)	Electric, catalytic heater, heat pump, absorption heat pump, evaporative cooler, hydride
Transportation	Cars, buses, trucks, tractors, fork-lifts	IC engine, fuel cell/electric

Table 21. Hydrogen-utilizing Industries for Pilot Projects

Industry	Description
Any industry which needs process steam (food, chemical, textile, etc.)	Hydrogen–oxygen steam generator
Fertilizer	Ammonia synthesis
Chemical	Hydrogenation reactions for production of organic chemicals and products, alcohol production, hydrogenation of aromates, synthesis of amines and hydrogen peroxide
Food	Hydrogenation of natural fats and oils
Metallurgy	Direct reduction of iron ore, reduction gas for production of nonferrous metals (tungsten, nickel, molybdenum), reduction component in powder sintering processes, reduction gas in heat treatment of steels
Electronics	Reduction gas in silicon processing
Glass	Reduction gas in float glass production

to the hydrogen energy technologies in the pilot projects, so as to be able to evaluate them and decide on the best possible options for future applications. Therefore, three types of pilot projects are proposed, each based on different primary source: solar, wind, and hydro.

Figure 12 presents the block diagram of the pilot project based on solar energy. In this case the PV-produced electricity is partially used for hydrogen (and oxygen) production and partially to meet the electricity needs during the sun's availability. Hydrogen and oxygen produced electrolytically are stored and then used for electricity production when the sun is not available. Hydrogen is also used as fuel and as a chemical feedstock if necessary. The remaining oxygen can be used in industry or elsewhere (eg, in hospitals) or could be released into the atmosphere.

Figure 13 presents the block diagram of the pilot project based on wind energy. Since the wind is also intermittently available, the block diagram is similar to the one based on direct solar energy.

Figure 14 presents the block diagram of the pilot project based on hydropower. Since hydropower would be

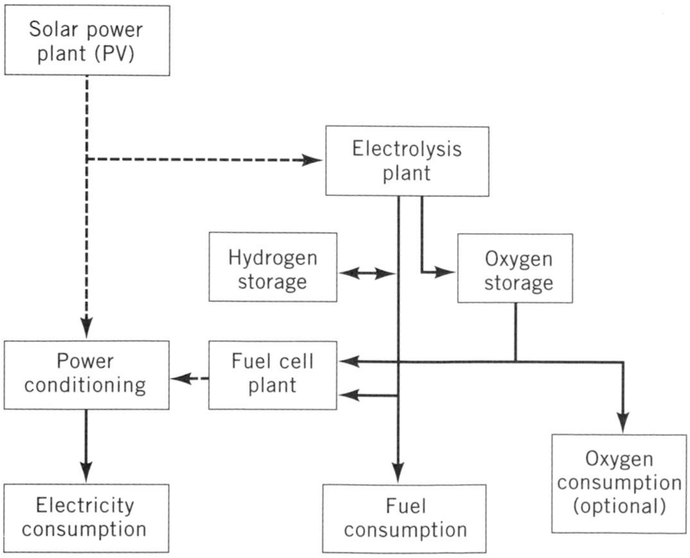

Figure 12. Block diagram of pilot project based on solar energy.

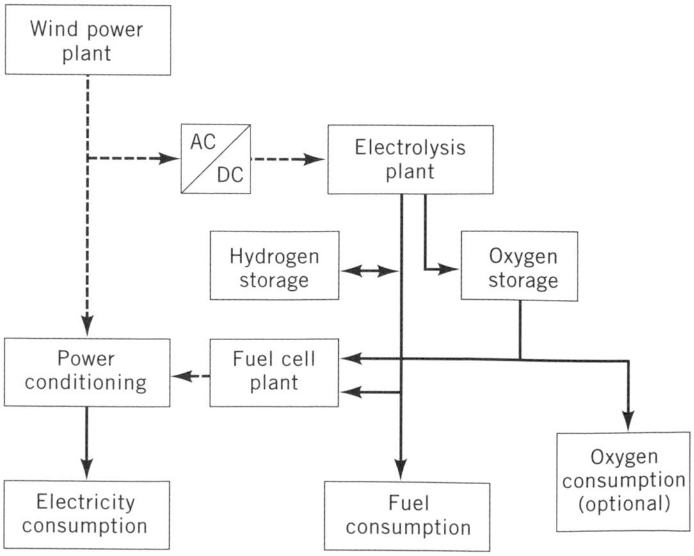

Figure 13. Block diagram of pilot project based on wind energy.

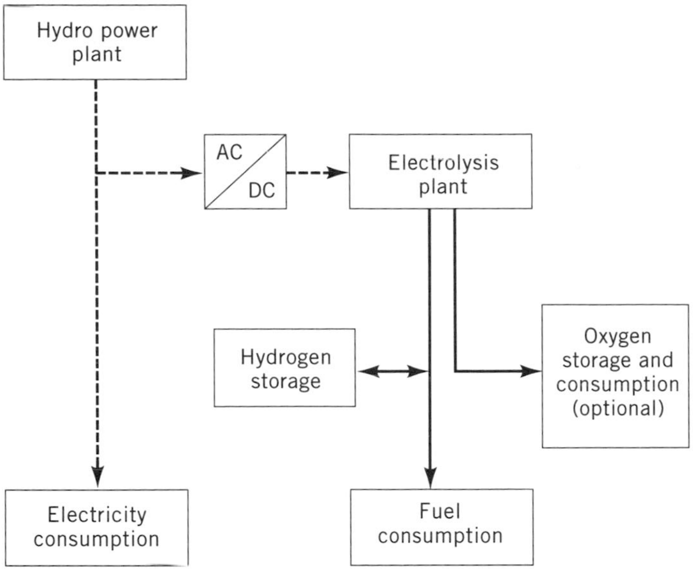

Figure 14. Block diagram of pilot based on hydropower.

available at all times, there is no need for a fuel cell to produce electricity. With this exception, the block diagram is similar to that for wind energy.

It is recommended to establish the first three pilot projects at locations where it will be relatively easy to demonstrate feasibility. The main criteria for the location selection should be:

- Resource availability (water, solar energy, wind, hydro potential)
- Need for energy (both in the form of electricity and fuel)
- Relative vicinity of some major academic institution
- Low environmental impact
- Investment subsidy availability

Project Costs and Comparisons

The project cost mainly depends on the technologies involved and the project size. A pilot project size is limited by at least two factors: (i) a too big project will be prohibitively expensive and (ii) a too small project would not prove feasibility. Therefore, it appears that a midsize project, which could provide electricity and hydrogen for a community of up to 1000 inhabitants, would be optimal for the given function. Precise capacity for each of the facilities will be determined with regard to the quantity and patterns of energy need which should be satisfied. It would therefore very much depend on the individual location.

As the annual energy consumption of a community of up to 1000 people, we have assumed a hydrogen consumption of 2500 GJ/yr and electricity consumption of 1,000,000 kW·h/yr. About half of these would be used by the residents and the other half by the local industry.

The cost estimates for the pilot projects based on solar energy, wind energy, and hydropower are shown in Table 22. It should be emphasized that the above-stated costs are the costs for the pilot projects. Once large-scale conversion to the hydrogen energy system starts and the market grows, there would be large reductions in the estimated costs.

Table 22. Preliminary Cost Estimate of Pilot Projects

	Cost (million U.S.$)		
Item	Solar	Wind	Hydro
Electricity generation	5.5	1.5	0.5
Hydrogen production	2.0	2.0	1.5
Hydrogen storage	1.5	1.5	1.0
Hydrogen utilization	4.0	4.0	3.0
Subtotal (equipment)	13.0	9.0	6.0
Construction and installation	1.5	1.5	2.5
Infrastructure	3.0	2.5	2.0
Design and engineering	2.5	2.0	1.5
Total	20.0	15.0	12.0

Note: Total electricity generation capacity: 1 MW (for solar and wind) and 0.25 MW (for hydro); hydrogen to consumer: 2500 GJ/yr; electricity to consumer: 1 GW·h/yr.

CONCLUSIONS

Hydrogen is by far the best fuel with many unique and unmatched qualities as it is the *best transportation fuel,* the *most versatile fuel,* the *most efficient fuel,* the *environmentally most compatible fuel,* the *safest fuel,* and the *most cost-effective fuel to society.* It is obvious that hydrogen should be the fuel of choice for the post–fossil fuel era, when we shall have the option to manufacture the fuel of our choice.

BIBLIOGRAPHY

1. T. N. Veziroglu, ed., *Hydrogen Energy,* Parts A and B, Proc. Hydrogen Economy Miami Energy Conf. (THEME), Plenum Press, New York, 1975.
2. T. N. Veziroglu, ed., *Proc. 1st WHEC* (3 vols.), Clean Energy Research Inst., University of Miami, Coral Gables, FL, 1976.
3. T. N. Veziroglu and W. Seifritz, eds., *Hydrogen Energy System* (t vols.), Proc. 2nd WHEC, Pergamon Press, Oxford, 1979.
4. T. N. Veziroglu, K. Fueki, and T. Ohta, eds., *Hydrogen Energy Progress* (4 vols.), Proc. 3rd WHEC, Pergamon Press, Oxford, 1981.
5. T. N. Veziroglu, W. D. Van Vorst, and J. H. Kelley, eds., *Hydrogen Energy Progress IV* (4 vols.), Proc. 4th WHEC, Pergamon Press, Oxford, 1982.
6. T. N. Veziroglu, and J. B. Taylor, eds., *Hydrogen Energy Progress V* (4 vols.), Proc. 5th WHEC, Pergamon Press, Oxford, 1984.
7. T. N. Veziroglu, N. Getoff, and P. Weinzierl, eds., *Hydrogen Energy Progress VI* (3 vols.), Proc. 6th WHEC, Pergamon Press, Oxford, 1986.
8. T. N. Veziroglu and A. N. Protsenko, eds., *Hydrogen Energy Progress VII* (3 vols.), Proc. 7th WHEC, Pergamon Press, Oxford, 1988.
9. T. N. Veziroglu and P. K. Takahashi, eds., *Hydrogen Energy Progress VIII* (3 vols.), Proc. 8th WHEC, Pergamon Press, Oxford, 1990.
10. T. N. Veziroglu, C. Derive, and J. Pottier, *Hydrogen Energy Progress IX,* (3 vols.), Proc. 9th WHEC, International Association for Hydrogen Energy, Coral Gables, FL, 1992.

11. D. L. Block and T. N. Veziroglu, *Hydrogen Energy Progress X,* (3 vols.), Proc. 10th WHEC, International Association for Hydrogen Energy, Coral Gables, FL, 1994.
12. J. O'M. Bockris, *Energy: The Solar-Hydrogen Alternative,* Australia and New Zealand Book Co., Sydney, 1975.
13. T. N. Veziroglu, "Solar Hydrogen Energy System and Hydrogen Production," in *Heliotechnique and Development,* Development Analysis Associates, Cambridge, MA, 1975.
14. T. Ohta (ed.), *Solar-Hydrogen Energy System,* Pergamon Press, Oxford, 1978.
15. L. O. Williams, *Hydrogen Power: An Introduction to Hydrogen Energy and Its Applications,* Pergamon Press, Oxford, 1980.
16. L. W. Skelton, *The Solar-Hydrogen Energy Economy: Beyond the Age of Fire,* Van Nostrand Reinhold, New York, 1984.
17. C. M. Winter and J. Nitsch, *Hydrogen as an Energy Carrier,* Springer-Verlag, Berlin, 1988.
18. H. J. Plass, Jr., H. Miller, and F. Barbir, "Hydrogen Systems Application Analysis, Task A," in *Solar Hydrogen Energy System,* T. N. Veziroglu (ed.), Report to DOE/SERI #XL-9-19168-1, Clean Energy Research Institute, Coral Gables, FL, 1989.
19. W. Hug, et al., "High Efficient Advanced Alkaline Electrolyzer for Solar Operation," in *Hydrogen Energy Progress VIII,* Vol. 2, T. N. Veziroglu and P. K. Takahashi (eds.), Pergamon Press, Oxford, 1990, pp. 681–690.
20. R. Wurster and A. Malo, "The Euro-Quebec Hydro-Hydrogen Pilot Project," in *Hydrogen Energy Progress VIII,* Vol. 2, T. Veziroglu and P. K. Takahashi (eds.), Pergamon Press, Oxford, 1990, pp. 59–70.
21. I. Willner and B. Steinberger-Willner, "Solar Hydrogen Production through Photobiological, Photochemical and Photoelectrochemical Assemblies," *Int. J. Hydrogen Energy* **13,** 593–604 (1988).
22. T. N. Veziroglu, "Hydrogen Technology for Energy Needs of Human Settlements," *Int. J. Hydrogen Energy* **12,** 99–129 (1987).
23. *Hydrogen Lett.* **V**(12), (1990).
24. U. Sprengel and W. Hoyer (eds.), *Solar Hydrogen—Energy Carrier for the Future,* DLR, ZSW, and Ministry of Economic Affairs and Technology for the State of Baden-Wurttemberg, Stuttgart, Germany, 1990.
25. M. Nadal and F. Barbir, Development of a Hybrid Fuel Cell/Battery Powered Electric Vehicle, in D. L. Block and T. N. Veziroglu, eds., *Hydrogen Energy Progress X,* Vol. 3, Int. Assoc. Hydrogen Energy, Coral Gables, Fla., 1994, pp. 1427–1440.
26. *Hydrogen Lett.* **VI**(9), (1991).
27. *Hydrogen Lett.* **V**(6), (1990).
28. K. Ledjeff, "New Hydrogen Appliances," in *Hydrogen Energy Progress VIII,* Vol. 3, T. Veziroglu and P. K. Takahashi (eds.), Pergamon Press, Oxford, 1990, pp. 1429–1444.
29. *H_2 Digest* **1**(10), (March 1990).
30. J. H. Hirschenhofer, "International Developments in Fuel Cells," *Mech. Eng.,* 78–83, (August 1989).
31. H. J. Sternfeld and P. Heinrich, "A Demonstration Plant for the Hydrogen/Oxygen Spinning Reserve," *Int. J. Hydrogen Energy* **14,** 703–716 (1989).
32. G. R. Grob, "Implementation of a Standardized World Hydrogen System, in *Hydrogen Energy Progress VIII,* Vol. 1, T. N. Veziroglu and P. K. Takahashi (eds.), Pergamon Press, Oxford, 1990, pp. 195–200.
33. T. N. Veziroglu, "Hydrogen Technology for Energy Needs of Human Settlements," *Int. J. Hydrogen Energy* **12**(2), (1987).
34. F. Barbir, T. N. Veziroglu, and H. J. Plass, Jr., "Environmental Damage Due to Fossil Fuel Use," *Int. J. Hydrogen Energy* **15**(10), (February 1990).
35. A. H. Awad and T. N. Veziroglu, "Hydrogen versus Synthetic Fossil Fuels," *Int. J. Hydrogen Energy* **9**(5), (1984).
36. F. Barbir and T. N. Veziroglu, "Effective Costs of the Future Energy Systems," *Int. J. Hydrogen Energy* (1992).
37. T. N. Veziroglu, I. Gurkan, and M. M. Padki, "Remediation of Greenhouse Problem Through Replacement of Fossil Fuels by Hydrogen," *Int. J. Hydrogen Energy* **14**(4) (April 1989).
38. C. J. Winter and M. Fuchs, "HYSOLAR and SOLAR-WASSERSTOFF-BAYERN: Research, Development and Demonstration of Two Solar Hydrogen Energy Projects," *Int. J. Hydrogen Energy* **16**(11), 723–734 (1991).

TRANSPORTATION FUELS—NATURAL GAS

Frank Stodolsky
Kenneth Wilund
Argonne National Laboratory
Washington, D.C.

The primary impetus for the use of nonpetroleum transportation fuels in the United States, which includes natural gas and propane, is the desire to reduce environmental damage (including mitigation of global warming) and the desire to reduce U.S. dependence on imported oil for national security reasons. The use of natural gas and propane in light- and heavy-duty vehicles is discussed, compressed natural gas (CNG), liquefied natural gas (LNG), and propane (or liquefied petroleum gas, LPG) are examined. Natural gas can be also used to make methanol, additives for gasoline such as the oxygenate methyl tertiary butyl ether (MTBE), or alkylates (which are octane enhancers).

HISTORICAL PERSPECTIVE

1900–1990

LPG and CNG have been used as engine fuels since the early 1900s. LPG has been used in vehicles since 1913, and natural gas vehicles were being tested in Europe during World War I and in the United States by the mid-1930s (1). Low pressure gas vehicles developed in the UK during World War I had fabric bags attached to the roof of the vehicle for storing the gas. These fabric containers ranged from 30 to 220 m^3 in volume, about as large as the vehicle to which it was attached. Enough gas was stored in this arrangement to provide a range of 24–40.2 km between refueling (2). It was not until World War II that the widespread use of LPG and CNG for automobiles began in Europe as a means of alleviating fuel shortages. In 1940, there were 80 gas filling stations in London alone. Germany had gone the furthest in the development of gas vehicles during this period. In 1939, Germany had six types of bottled gas on the market, including CNG, LPG, butane, city gas, coke oven gas, and *Ruhrgasol.* They had more than 50 filling stations and about 75,000 vehicles using compressed gases of various types, with CNG and city gas more widely available than LPG or butane. Natural gas fields, coal hydrogenation, coal carbonization, and oil refineries provided most of the methane available at this time. Several European countries were also using methane derived from the fermentation of city sewage. The sewage gas, which was typically 64% methane and 35% carbon dioxide, was normally purified before being used. During World War II, the Italian government prohibited the use of natural gas for any purpose other than motor fuel. By 1947 there were about 25 natural gas filling stations in operation in various places in northern Italy, including 5 in Milan alone. Most cars were equipped with three cylindrical steel bottles. Common bottles weighed 41 kg each and had a capacity of 0.43 m^3. The bottles were normally filled to a pressure of 20.3 MPa so that each bottle held 108 m^3 of natural gas. This would allow motorists to travel from 240 to 257 km. Italy was also using methane gas, compressed to 30.4 MPa to run a railway locomotive between Rovigo and Venice (3).

Table 1 shows the number of vehicles in various European countries that had been converted to compressed gases in 1941. Besides the countries listed in the table, the Soviet Union, France, and Denmark had also developed compressed gaseous fuels at that time (2). In the United States, experimental cars and trucks utilizing LPG were being tested in the 1930s and 1940s. However, the first widespread use of LPG use in motor vehicles was in tractors. In 1941, Minneapolis-Moline introduced the first LPG tractor as a regular factory option. Beginning in the early 1940s until the mid-1960s, LPG gained in popularity to the point where it became a significant part of tractor production and all large tractor manufacturers provided LPG fuel systems as an option. By 1953, 135,000 tractors had been converted to LPG. However, the use of LPG tractors declined in the late 1960s as turbocharged diesel engine tractors began to take over (4).

Standard model forklift trucks powered by LPG entered the U.S. market in the early 1950s (5). By 1957, LPG industrial lift trucks were the fastest-growing portion of the LPG market (6). While most uses of LPG in internal combustion engines declined in the 1970s, the use of LPG for forklifts increased slightly (7). In the late 1960s and early 1970s the use of LPG in internal combustion engines was at its peak. In 1969 alone, there was a 20% growth in the production of engines capable of burning natural, liquefied, or condensed petroleum gas. In 1972, internal combustion usage of LPG reached 5.6×10^6 m^3 about 7% of total LPG sales. But by 1978, the market declined to 3.8×10^6 m^3, which represented only 4.5% of the total LPG sold (7). Forklifts were the only vehicles continuing to use LPG to a significant degree in the late 1970s and it was not until the 1980s that the conversion of heavy-duty engines to LPG began to occur.

LPG is the predominant alternative transportation fuel. Although its share of the market has been slowly declining. Up to the late 1980s, CNG was used extensively as a transportation fuel only in Italy (8). CNG is used extensively only in Italy (8). Table 2 shows the number of gaseous fueled vehicles in selected countries in 1987 (9).

Table 1. Gaseous-fueled Vehicles in Europe, 1941[a]

Country	Number of Vehicles
Germany	75,000
UK	10,000
France	500
Sweden	1,367
Denmark	358
Italy	20,000
Total	*107,225*

[a] Ref. 2.

1990–Present

In the early 1990s, interest in "clean fuels" to mitigate environmental damage and to diversify energy sources has spurred interest in natural gas and LPG as transportation fuels. In 1990, the 101st Congress enacted the Clean Air Act Amendments (CAAA), which provides new incentives for the production of natural gas-fueled vehicles. Under the CAAA, the sale of "clean fuel vehicles" is required. California has implemented its own standards, which are more stringent than those under the CAAA. The Energy Policy Act of 1992 sets the stage for alternative fuel use for transportation by providing various incentives. In addition, an executive order signed in 1993 calls for federal government purchases of up to 15,000 alternative fueled vehicles in 1995.

Largely in response to the government initiatives for alternative fueled vehicles, approximately 7,000 natural gas vehicle conversion systems were installed in 1992, and between 18,000 and 23,000 units were expected to be installed in 1993 (10). There were more than 700 operational natural gas vehicle refueling stations operating in mid-1993. In 1994, more than 40,000 natural gas conversions are predicted. LPG highway vehicles number more than 330,000 (11). In 1987, the U.S. LPG fleet used 1.7×10^6 m^3 of LPG.

FUEL PROPERTIES AND STORAGE CHARACTERISTICS

Properties and storage characteristics of natural gas and LPG are shown in Table 3. CNG is normally stored at 20.7 MPa at room temperature as a gas, whereas LNG is stored cryogenically as a liquid at 138 kPa and −150°C. LPG is stored at 1.4 MPa at room temperature, at which it is a liquid. CNG, LPG, and LNG all have lower volumetric energy densities than gasoline and diesel fuel. Therefore, the range of a natural gas or LPG vehicle be-

Table 2. Gaseous-Fueled Vehicle Population, 1987[a]

	CNG	LPG
Japan		1,400,000–1,700,000
France	2,000	43,000–53,000
Italy	300,000	700,000–850,000
FRG		12,000–15,000
Canada	15,000	130,000
United States	30,000	300,000–370,000

[a] Ref. 9.

Table 3. Natural Gas and LPG Fuel Properties and Fuel Storage Characteristics Compared with Gasoline[a]

Parameter	Fuel			
	Gasoline	LPG	CNG[b]	LNG[c]
Density, g/cc	0.74	0.50	0.17	0.41
Energy content[d]				
MJ/kg[e]	44.0	46.3		50
compressed			50	
adsorbent			50	
MJ/L[f]	32.7	23.1		20.3
compressed			8.4	
adsorbent			4.3	
Relative fuel volume, approximate[g]	1.0	1.4		1.5[h]
compressed			3.9	
adsorbent			7.6	
Relative fuel tank external volume, approximate[i]	1.0	1.4		1.8 [h]
compressed			4.3 [j]	
adsorbent			8.4	
Relative full fuel tank weight, approximate[i]	1.0	1.2 [k]		1.4[h,k]
compressed			3.7	
adsorbent			7.4	
Vehicle range, same tank volume, km[l]	483	338		298
compressed			129	
adsorbent			72	

[a] Ref. 18.
[b] For steel cylinders at 20.7 MPa.
[c] Saturated liquid at 138 kPa and −114°C.
[d] Lower heating value; heat of reaction at constant pressure or at a constant volume at standard temperature, excluding the heat of condensation of water vapor. Lower heating value is used for heat engine fuels because heat engines do not normally extract energy from water vapor in exhaust gases.
[e] To convert MJ/kg to Btu/lb, multiply by 430.
[f] To convert MJ/L to Btu/gal, multiply by 3589.
[g] For equivalent range; all engines assumed to have equal thermal efficiency.
[h] Need for vapor space in "full" cylinder is neglected. Vapor space is always required to allow for vaporization and liquid expansion that occurs as the temperature and pressure increase because of warming of the cylinder over time. The vapor space could add 5–10% to the cylinder volume.
[i] For typical 1136-L medium-duty truck tank. Relative volumes expected to be similar for passenger car size tanks (typically 57-L).
[j] For single cylinder made of steel; volume for multiple cylinders would be approximately 40% higher than the value indicated here.
[k] Preliminary estimates.
[l] Assumes a 483-km range between refueling for a gasoline vehicle, all fuel tanks are the same size, and thermal efficiency is the same for each fuel.

tween refueling stops is much lower than when operating on conventional fuels. To achieve a range comparable to conventional fuels, natural gas and LPG vehicles must have substantially larger and heavier tanks, which could limit passenger or cargo space and reduce acceleration. Limited range between refueling is especially acute for CNG vehicles. Limited range is a principal stumbling block to consumer acceptance of natural gas and LPG vehicles. There are no feasible technical solutions to this tank size–range problem.

Gas composition has a significant effect on vehicle performance and emissions. Typical natural gas and LPG compositions are shown in Table 4. Surveys of natural gas supplies have shown a wide variation in composition. Nonmethane constituents can vary between 3% (mol) and 18% (mol), according to one U.S. survey (12). Fuel composition affects the lean-flammability limit of the fuel–air mixture. A change in the lean-flammability limit can result in misfire and engine damage. Fuel composition also affects knock resistance. Knock, the name for the sound transmitted through the engine structure, occurs when a portion of the end gas (the fuel, air, residual gas mixture) spontaneously ignites ahead of the propagating flame in a spark-ignition engine. Resistance to knock is desired to

Table 4. Typical Composition of Natural Gas and LPG Fuel (vol %)[a]

Component	CNG	LNG	LPG
Methane	92.29	94.00 (min)	
Ethane	3.60	2.75 (max)	
Propane	0.80	0.90 (max)	92.5 (min)
Butanes	0.29	nil	2.5 (max)
Propylene			5.0 (max)
Pentanes	0.13	nil	
Hexanes	0.08	nil	
CO_2	1.00	nil	
Nitrogen	1.80	0.50	
Water	0.01	nil	

[a] Refs. 12, 18, and 19.

avoid piston ring damage, cylinder head gasket failure, and erosion of the piston and cylinder head surfaces. Resistance to knock is important in natural gas engines whose compression ratios have been increased to take full advantage of the gas. Heavier hydrocarbons such as ethane and propane lower knock resistance, while inerts such as nitrogen and carbon dioxide improve knock resistance. LNG contains almost no water and carbon dioxide, because these constituents must be removed to accommodate liquefaction. Because the noncombustible constituents have been removed, LNG will typically have a greater heating value (on a mass basis) than CNG, but a lower knock resistance.

The knock resistance of natural gas is measured differently from that for gasoline. Knock resistance of a spark-ignition engine fuel is commonly measured by the octane scale. Two methods are used in the United States to evaluate resistance to knock. They are the research octane number (RON) method (13) and the motor octane number (MON) method (14). The octane scale has not been adapted for use in rating natural gas fuels, because the upper limit of this scale is at the point where most natural gas octane numbers would be expected to fall. The methane number (MN) scale has been developed to overcome this limitation. Correlations between octane number and methane number have been developed. A MON of 100 is equivalent to a MN of about 40; a MON of 130 is equivalent to a MN of about 85 (15).

LPG characteristics are more consistent because gas composition standards have been developed for vehicular use (16). Saturated compounds such as propane and butane decrease the tendency to knock, whereas high levels of unsaturated compounds such as propylene and *n*-butenes will increase the tendency to knock. ASTM has a MON standard for measuring knock resistance for LPG, which permits direct comparison with gasoline (17). The MON for LPG is typically 90–94. A RON standard has not yet been developed for LPG. Propane, with a boiling point of 6.7°C gives LPG its excellent cold-weather performance. LPG with high levels of butane perform more poorly in cold weather because butane has a lower boiling point (−10°C) than propane.

Although natural gas composition varies widely, there is a high degree of consistency of the Wobbe number and heating value among natural gas supplies in the United States, which is desirable from an engine perspective. (The Wobbe number is a parameter that determines the chemical energy that will flow past an orifice with a given pressure drop.) The variation in Wobbe number for natural gas in the United States will effect, at most, a ±3% change in the equivalence ratio (defined as the ratio between the actual fuel : air ratio and the stoichiometric fuel : air ratio) of an uncontrolled CNG engine (12). Heating value (the amount of energy contained in a stoichiometric mixture of natural gas and air) was observed to vary only ±1%. Composition of LNG can change during the course of vehicle operation ("enrichment"), which could affect performance and emissions. *Enrichment* (or weathering) and *fractionation* are terms used to describe the boiling of the lighter methane fraction from the heavier ethane, propane, and butane fractions during short trips. When the engine is first started, methane vapor fuels the engine to maintain the cylinder pressure within limits. The remaining liquid becomes enriched with the heavier hydrocarbons over time and the engine could experience an abrupt change in fuel composition when the liquid–vapor control system switches to liquid during operation. LNG requires the use of either nearly pure methane or additional engine control systems to reduce the risk of engine detonation due to enrichment caused by short-trip driving patterns (18).

VEHICLE SYSTEMS

Light-duty Engines

Spark ignition engines can be modified to run on CNG, LNG, and LPG (CNG and LNG enter the combustion chamber as a gas, and therefore, combustion is essentially the same). A conventional spark ignition gasoline-engine vehicle can be converted to dual-fuel natural gas–gasoline by modifying the fuel delivery system and advancing the ignition timing to allow for the lower flame speed of natural gas. This type of engine can operate on only one fuel at a time. It is an established technology; in 1992, more than 7000 vehicles were converted to dual fuel (10). The compression ratio cannot be raised in the dual-fuel spark ignition engine to improve thermal efficiency, because gasoline cannot tolerate the high compression ratios that natural gas can tolerate. Power output is usually reduced between 10 and 20%, because of the lower mixture density and reduced inlet air flow (20). Fuel economy of a natural gas–gasoline dual-fuel vehicle is about 4% lower than a gasoline vehicle having the same overall weight and acceleration characteristics (21). Given drawbacks in power output and thermal efficiency, converted spark ignition gasoline engines are seen as an interim step toward utilization of natural gas as a vehicular fuel. For optimum performance and efficiency, increased compression ratios and improved mixing must be designed into the engine. In one test, an increase in compression ratio from 9 : 1 to about 12.8 : 1 with advanced timing allowed the natural gas engine to match power and efficiency of its gasoline counterpart (22).

Both mechanical and electronic fuel delivery systems have been developed for natural gas engines. The fuel system for LPG is similar to a natural gas fuel system. Mechanical systems are common; electronic systems employing fuel injection are not yet fully commercial. Mechanical systems employ a pressure reducer and either a venturi or a variable restriction carburetor. In the venturi carburetor, the fuel is supplied in the throat section. An adjustment screw is located betwene the reducer and the venturi, which allows the air : fuel ratio to be set. The variable restriction carburetor employs a membrane, which is activated by the pressure drop due to the air flow. The membrane is connected to a gas metering valve, which controls the rate of natural gas flow. The mechanical systems are limited in thier ability to maintain a given air : fuel ratio. Changes in air : fuel ratio occur due to variation in natural gas composition and temperature. Some mechanical systems use electronic trimming of the air : fuel mixers. The signal is provided by an exhaust oxygen sensor and computer control system. An example of this kind of system is shown schematically in Figure 1. Electronic fuel injection systems for natural gas engines have been developed over the past decade and are almost commercial. The natural gas injector is a valve with electromagnetic actuation like gasoline injectors. However, there are problems that have not been completely solved in these systems. The low gas density requires larger injectors than those used by gasoline engines. Therefore, the larger size and mass reduces maximum operating frequency, which limits top-end engine speed below that of a similar gasoline engine. Problems with sealing, wear, and air–fuel mixture homogeneity have also been observed (20).

Both stoichiometric and lean-burn combustion techniques are being developed for low emission natural gas engines. Stoichiometric combustion, whereby the air : fuel ratio is at or near the theoretical value for combustion, is typically used on gasoline engines converted to dual-fuel or dedicated natural gas. This approach provides the maximum power output for a given engine displacement. Stoichiometric engines use a three-way catalytic converter to control nitrogen oxides (No_x), carbon monoxide (CO) and hydrocarbons (HC). For natural gas engines, the efficiency of conversion is strongly dependent on equivalence ratio (actual fuel : air ratio to stoichiometric fuel : air ratio). Therefore, the engine must operate in a narrow window of equivalence ratio, requiring closed-loop electronic control systems. One drawback of natural gas engines with three-way catalytic converters is the inability to sufficiently oxidize methane especially at the lower exhaust temperatures characteristic to natural gas combustion (12). Methane is a greenhouse gas and may contribute to global warming (see below).

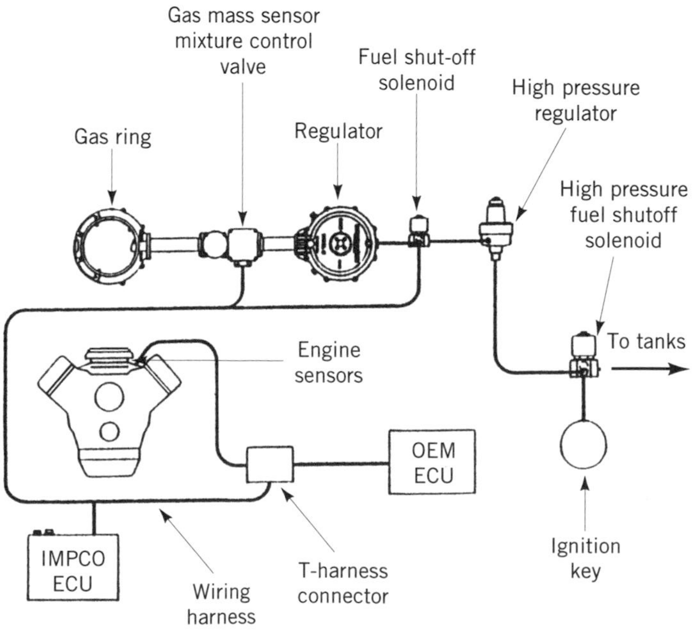

Figure 1. General Motors natural gas fuel metering system. Courtesy of Natural Gas Vehicle Coalition, Arlington, Virginia.

Lean-burn combustion techniques for gasoline engines are being applied to natural gas engines. Natural gas is inherently attractive for lean-burn technology because of its tolerance to diluted combustion as well as the high homogeneity of fuel–air mixture that can be achieved. An optimized lean-burn natural gas engine would employ a high compression ratio (ranging from 11 : 1 to 15 : 1), a high energy ignition system, and a high turbulence combustion chamber. A turbocharger is required to maintain a satisfactory power output because the energy content of the fuel–air mixture is substantially reduced using lean-burn technology. Emissions from lean-burn engines would be controlled differently from those for stoichiometric engines since three-way catalysts are not efficient in a lean (oxidizing) environment. Lean-burn engines would require a suitable catalyst to simultaneously oxidize unburned methane and reduce NO_x emissions.

Heavy-duty Engines

Heavy-duty natural gas engines are typically derived from diesel engines. Typically, heavy-duty diesel engines start in the 150 kW range for highway trucks and buses and go up to more than 2980 kW for locomotives. In diesel engines, natural gas can either be fumigated into the combustion chamber and compressed as a premixed charge, or injected directly with high pressure injectors. The premixed charge method is proven technology, while direct injection of natural gas is still experimental. In the premixed charge method, an external ignition source such as diesel fuel or spark is needed to combust the natural gas because of its high autoignition temperature. When designed for spark ignition operation, spark plugs and a high energy ignition system are used in place of diesel fuel injectors. Because fuel is present in the charge, compression ratio must be lowered from about 19 : 1 to between 10 : 1 and 15 : 1 to prevent knock. Diesel engines converted to run on natural gas can operate at higher compression ratios than natural gas spark ignition engines because they are designed to tolerate higher pressures.

The term *dual-fuel engine* is used when diesel fuel is used with natural gas. Dual-fuel natural gas engines can run on pure diesel fuel or varying amounts of natural gas replacing diesel fuel, up to about 90% at high loads (23). Conventional fuel injector design makes it impossible to reduce pilot fuel further. However, "micro" pilot systems have been tested that limit the pilot diesel fuel to less than 1% of full-load energy consumption (24).

In the direct injection system, natural gas is not premixed with air but injected at a high pressure near top dead center. The absence of premixing prevents knock, but makes control of NO_x more difficult. Efficiency comparable to diesel-fueled engines can be achieved using direct injection of natural gas.

As in lean-burn light-duty engines, suitable catalysts must be developed to reduce emissions of unburned methane and NO_x under lean conditions.

Fuel Storage and Refueling

The weight and bulk of fuel cylinders is a concern both from the standpoint of overall vehicle weight (and hence vehicle acceleration and fuel economy) and space available for cargo and passengers. Unlike gasoline tanks, cylinders do not conform to existing vehicle packaging. All vehicle cylinders must comply with American Society of Mechanical Engineers (ASME) or U.S. Department of Transportation (DOT) specifications for pressurized cylinders.

Several combinations of materials are typically used for natural gas cylinders. Steel, hoop-wrapped steel, hoop-wrapped aluminum, full-wrapped aluminum, and all-composite cylinders have been developed. A comparison of fuel tank weights are shown in Figure 2. All-composite cylinders afford a significant weight reduction, but even the best designs still have a weight disadvantage compared to a gasoline or diesel fuel tank. Measured on a basis of weight per liter of diesel equivalent, an all-composite CNG cylinder would weigh about three times more than a tank of diesel fuel. Recently, cylinders containing a molecular sieve or activated carbon adsorbent have been developed for low pressure storage of natural gas 2.1–6.2 MPa. Low pressure allows the use of lower cost cylinder materials and single-stage compressors for refueling. However, these systems cannot store as much natural gas as compressed gas cylinders. Therefore, this technology is considered an interim solution for natural gas vehicles. Adsorbent storage technology is based on the principle that the amount of natural gas stored in a pressurized cylinder is greatly enhanced by placing an activated carbon substrate inside the cylinder. This principle is valid up to about 12.4 MPa, beyond which the carbon becomes a greater impediment to storage capacity than is capable through adsorption. The performance of the adsorbent material is a function of its methane adsorption capacity, packing density, and thermal conductivity.

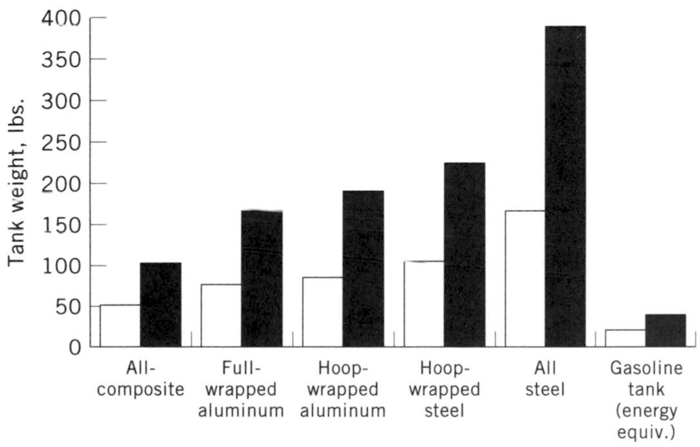

Figure 2. Estimated weights of different CNG cylinders compared with gasoline tank with same volume excluding fuel weight. Adapted from Refs. 25 and 26. ■ 12.7 m^3 tank, yields a range of approximately 177 km; □ a 31.2 m^3 tank yields a range of approximately 402 km.

For refueling, natural gas at line pressure (typically 34.4 kPa) must be pressurized to the cylinder pressure (20.7–24.8 MPa) using natural gas compressors. CNG vehicles can be filled using either fast-fill or slow-fill methods using these compressors. With the slow-fill method, there is a direct connection between the vehicle and the natural gas compressor. The volume delivery rate is limited to the discharge rate of the compressor itself. A typical slow-fill system requires filling overnight. In the fast-fill method, the compressor is not directly linked to the vehicle fuel tank being filled. Instead, the fast-fill method uses a bank of high pressure cylinders that have been filled previously by the compressor. The bank of cylinders are controlled so that the lowest pressure cylinder is used first and an automatic or manual sequential valve system switches to the higher pressure banks as the pressure between the cascade and the vehicle fuel tank is equalized. Fleet operations would probably use slow-fill, and commercial fueling stations would be exclusively fast-fill. Currently, all public natural gas refueling stations are the fast-fill type.

LNG tanks for vehicles are not yet commercial. Much of the technology is based on larger LNG storage applications. LNG must be kept at cryogenic temperatures. Therefore, double tank construction with insulation between the inner and outer shell is required. LNG operates at lower pressures, requiring less demanding materials for the cylinders themselves. However, insulation and the tank-in-a-tank design add to the tank volume, thereby reducing the benefit of the higher volumetric energy density of LNG compared with CNG. Insulation used in LNG tanks are typically multiple layers of reflectors (such as aluminum foil) and reflector separators (such as glass fiber felt) contained in an evacuated ($<1.3 \times 10^{-2}$ Pa) annular space. Even with high efficiency insulation, heat influx occurs that evaporates the liquid natural gas. Because expansion occurs on evaporation, a pressure-relief system is needed. Storage times before overpressurization due to heat influx in excess of 10 days appears feasible. While typical practice requires that an overpressurized system be vented, approaches have been explored to limit or manage these evaporative emissions. Concepts include fueling the vehicle directly from the tank gas, use of inline adsorbents (similar to gasoline charcoal canisters), higher operating tank pressures, and catalytic incineration of vent gas streams. Smaller LNG tanks inherently transfer more heat per unit volume than larger tanks given the same insulation capacity. Therefore, one large tank is preferable to two smaller tanks. Smaller tanks as required in passenger cars would be more susceptible to heat influx than larger tanks, which would be used on heavy-duty vehicles and freight trains.

LNG dispensing technology for vehicles is less developed than CNG dispensing technology. LNG must be actively pumped from storage tanks maintained at approximately −162°C during vehicle refueling. The fuel storage tank must be jacketed containing a vacuum and insulation. The LNG pump must be submerged in the LNG and operated at the same temperature as the tank to avoid cavitation. Because LNG vaporizes during refueling and vapor is expelled from the vehicle tank during refueling, some sort of natural gas vapor recovery will probably be required. Some argue that the LNG entering the tank will collapse the vapor, therefore eliminating the need for vapor recovery. However, refueling nozzle designs have emerged that employ simultaneous liquid inflow and vapor recovery. Like CNG systems, LNG delivery nozzles would incorporate a quick disconnect, which would also seal the delivery system from leaks. Vapor collected by the system must be refrigerated to convert it back into a liquid.

LPG cylinders are common and require no new technology for use in vehicles. LPG cylinder diameters range from 30.5 to 61 cm, with lengths ranging from 76 to 165 cm (27). They are designed to withstand pressures of 1.7 to 2.1 MPa. LPG cylinder size is limited to 757 L (water capacity, WC) for passenger vehicles and 1135.5 L WC for larger highway vehicles, according to DOT specifications. These limitations are designed to minimize the fire hazard in case of an accident. Fuel tanks for passenger cars are typically 76–151 L. For pickup trucks, cylinder size ranges from 95 to 303 L. To allow for expansion, LPG cylinders are normally filled to about 80% of their water capacity. When a cylinder is mounted inside a vehicle, or in the trunk, a vapor barrier is required to prevent propane vapors from seeping into the passenger compartment in the event of a leak. A sealed trunk or the sealing of fittings with a gas-tight enclosure can serve as a vapor barrier. A relief valve is required on the cylinder in the event of overpressurization.

LPG refueling does not require high pressure compressors as does natural gas. Small LPG fuel facilities use tank pressure to deliver fuel to the vehicle. However, in large facilities, a constant displacement pump is used for metering and to provide a constant flow rate. No vapor recovery system is needed because the storage and dispensing system is sealed, but delivery lines must be equipped with excess flow check valves that shut flow off when it exceeds design values. Delivery lines must also contain check valves that shut off flow in case of insufficient backpressure, or emergency shut off valves.

EMISSIONS

From an environmental perspective, tailpipe (exhaust) and evaporative emissions, ozone formation potential, and global warming effects are of interest. The primary environmental benefit expected from the use of nonpetroleum fuels is reduced ozone levels in urban areas. NO_x and HC emissions are critical to ozone formation. Hence, much of the emphasis is on reducing emissions of these gases. Evaporative HC emissions from LPG and CNG are limited to refueling losses, which are expected to be minor because of the sealed fueling system. LNG evaporative HC emissions are a function of the boil-off rate, tank level, and length of time between vehicle startups. LNG refueling emissions were discussed above. Below emissions and standards for natural gas and LPG relative to gasoline are discussed, including tailpipe emissions and emission standards, ozone forming potential, and emissions of gases suspected to contribute to global warming.

Light-duty Vehicle Tailpipe Emissions and Standards

In general, emissions of CO, nonmethane hydrocarbons (NMHC), and air toxics (defined in the CAAA as 1,3 butadiene, polycyclic organic matter, benzene, formaldehyde, and acetaldehyde) are lowered using natural gas in light-duty vehicles compared with gasoline, but NO_x emissions could be higher or about the same as gasoline. Natural gas vehicles emit higher levels of unburned methane than gasoline vehicles. Data on LPG emissions are scarce. LPG-fueled vehicles seem to produce levels of emissions less than gasoline but greater than methanol (28). While LPG and natural gas can generate significant levels of aldehydes, levels are below that of gasoline.

Table 5 shows selected 1994 model year and later U.S. emission standards (28,29). For natural gas and LPG vehicles, it was EPA's intent to apply emission standards that are numerically equivalent to those applicable to other fuels, where appropriate. For LPG, EPA directly applied total hydrocarbon (THC) and NMHC emission

Table 5. Light-duty Vehicle Emission Standards (g/km)[a,b]

	Model Year				
	1994			1996 (Phase I)	2001 (Phase II)
				Clean Fuel Vehicle Programs	
Pollutant[c]	Gasoline	Natural Gas[d]	LPG[d]		
THC	0.26		0.26		
NMHC	0.156	0.156	0.156		
CO	2.1	2.1	2.1	2.1	2.1
NO_x	0.25	0.25	0.25	0.25	0.25
PM		0.05	0.038		
NMOG				0.018	0.047
HCHO				0.009	0.009
Evap. HC (quantity/test)[d]		1.25	1.25		

[a] Refs. 28 and 29.
[b] Applicable for the first 80,450 km or 5 yr (whichever occurs first); other standards (not shown) apply to 160,900 km or 10 yr.
[c] *PM,* particulate matter; *NMOG,* nonmethane organic gas; *HCHO,* formaldehyde; *Evap. HC,* evaporative hydrocarbons.
[d] Applies to 40% of manufacturers 1994 model year vehicles, 80% of 1995 vehicles, and 100% of 1996 and later vehicles.

Table 6. 1994 California Light-duty Vehicle Standards (g/km)[a,b]

Pollutant	1994 Vehicles[c]	TLEV[d]	LEV[d]	ULEV[d]	ZEV[d]
NMHC	0.156				
NMOG		0.078	0.047	0.025	0.00
CO	2.1	2.1	2.1	1.1	0.00
NO_x	0.25	0.25	0.13	0.13	0.00
HCHO		0.009	0.009	0.005	0.00

[a] Ref. 28.
[b] *TLEV,* transitional low emitting vehicle; *LEV,* low emitting vehicle; *ULEV,* ultra low emitting vehicle; *ZEV,* zero emitting vehicle.
[c] Vehicles that are not part of the low emission vehicle phase-in schedule.
[d] Phased in between 1994 and 2003.

standards for petroleum-fueled vehicles, based on the similarity of their emission characteristics to those of petroleum-fueled vehicles. High methane emissions observed from natural gas vehicles would make it unfeasible for current technology natural gas vehicles to comply with the THC standard for petroleum fuels. Therefore, THC standards were deemed not feasible until adequate catalyst technology is developed.

By 1998, centrally fueled fleets in several classes of ozone nonattainment areas (outside of California) must begin to purchase clean fuel vehicles that meet the gasoline vehicle emission standards for model year 2001 (Phase II standards). In addition, the CAAA contains a "California Pilot Program" where fleets and private vehicles must meet the Phase I standards starting in 1996. California has implemented its own standards, which are more stringent than those under the CAAA. California emission standards are shown in Table 6. The concept of "equivalent stringency" has been developed for other size classes of vehicles. Generally, the emission standards for heavier light-duty trucks (>1701 kg) and medium-duty trucks are derived from light-duty vehicle emission standards, corrected for vehicle weight and frontal area. To reduce the complexity of the discussion, these standards are not included here.

Generally, natural gas vehicles appear to meet the NMOG and CO standards down to the ULEV level, but cannot yet meet the 0.13 g/km NO_x standard at the LEV and ULEV level (28). Some prototype natural gas vehicles have met the California ULEV standards at low mileage; emissions durability over longer periods of operation (160,900 km) is still unproven. It is believed that LPG vehicles could meet all the California standards through the LEV level. Air toxics are of less concern for natural gas and propane than for gasoline (30).

Heavy-duty Engine Tailpipe Emissions and Standards

Compared with diesel vehicles, the natural gas fueled heavy-duty vehicle offers lower smoke, particulates, and NO_x emissions. NO_x emissions can be reduced in dual-fuel engines by limiting the amount of diesel pilot fuel to about 1% of full-load fuel consumption (24). HC emissions from dual-fuel engines tend to be higher than those of comparable spark ignition lean-burn engines. Experimental direct-injected natural gas diesels have shown higher NO_x levels than in premixed charge operation but much lower NO_x levels than those of an uncontrolled diesel-fueled engine. Some direct-injected diesels have shown high HC emissions due to leakage of natural gas at the injector tip.

Heavy-duty vehicle emission standards are based on a per-unit power output (g/kWh) basis and not a per-kilometer basis (g/km), because heavy-duty vehicles are normally used for hauling freight or performing lifting or other nonpassenger transportation functions. The heavy-duty emission standards are shown in Table 7. Their stringency cannot be compared with the light-duty vehicle standards, as the heavy-duty engine test cycle is different than the light-duty vehicle test cycle.

Ozone Formation

Reactivity of a gas is a measure of how much ozone will be formed. The "Carter reactivities" of some of the exhaust

Table 7. Emission Standards for Heavy-duty Vehicles (g/kWh)[a]

Pollutant	1994 Standards		California Light-heavy			Federal Clean Fuel Vehicle
	Federal	California	1995	LEV	ULEV	
THC	1.7					
NMHC		1.5				
NO_x	6.7	6.7				
NMHC + NO_x			5.2	4.7	3.3	4.2
CO	20.8	20.8	19.3	19.3	9.7	20.8
PM	0.13[b]	0.13[b]	0.13	0.13	0.07	0.13

[a] Ref. 31.
[b] 0.07 g/kWh for urban bus engines.

components of gasoline, natural gas, and LPG are shown in Table 8. The term NMOG includes all NMHCs as well as ethane and aldehydes. NMOG is usually used when discussing ozone-forming potential. Methane has a low reactivity, therefore, it is excluded from the definition of NMOG. For emissions that evaporate from the fuel system (evaporative emissions), the individual emission components must be multiplied by the relative reactivities shown in Table 8. Because fuel systems for natural gas and LPG are sealed, most nontailpipe emissions are a result of leakage during refueling and not from the engine bay or tank itself.

California has proposed a set of reactivity adjustment factors (RAFs) for vehicles running on alternative fuels. RAF values represent the relative amount of ozone formed by a given mass of NMOG emissions compared to gasoline. NMOG emissions from natural gas and LPG combustion are less reactive than NMOG emissions from gasoline (Table 9).

On a reactivity-equivalent basis, EPA projects that CNG dual-fuel vehicles will emit 36–47% less VOC-equivalent emissions than typical future gasoline vehicles, while optimized dedicated CNG vehicles are projected to emit 80–93% less than typical future gasoline vehicles (22). For heavy-duty vehicles, EPA projects that NMHC emissions from heavy-duty CNG engines will be between 67 and 80% lower than heavy-duty diesel NMHC emissions (23).

Fuel Cycle Energy Conversion Efficiency and Global Warming

As with conventional fuels, the production and combustion of natural gas and LPG generate carbon dioxide (CO_2) and other "greenhouse" gases that could contribute to global warming. Global warming potential of gases is still under active discussion and research; while the understanding of global warming has advanced significantly in the past decade, there remains significant uncertainty as to the relative impact of anthropogenic and natural sources of greenhouse gases. CO_2 emissions are a function of the vehicle energy consumption rate, carbon content of the fuel, and the fate of the carbon in the fuel (eg, complete oxidation or partial oxidation to CO or unburned fuel). Unburned methane, other organic compounds, CO, NO_x, and nitrous oxide (N_2O) are also greenhouse gases and are more stable than CO_2. Therefore, a unit mass of these compounds could contribute more to warming than a unit mass of CO_2. The Intergovernmental Panel on Climate Change (IPCC) publishes indices to estimate the direct effects of greenhouse gases on warming potential over different time horizons. The global warming potential (GWP) consists of direct and indirect radiative forcing that could contribute to global warming. Direct forcing occurs when the gas itself is a greenhouse gas; indirect forcing occurs when the original gas produces a gas or gases that themselves are greenhouse gases. Typically, emissions are expressed in CO_2-equivalents (ie, the GWP for CO_2 is set at 1) and only include the direct GWP. The warming potential of the gases decrease over time; therefore, values are usually expressed in terms of 20-yr, 100-yr, and 500-yr time horizons. The IPCC estimates of direct GWP for 20- and 100-yr time horizons are, respectively, methane 35 and 11 and N_2O 260 and 270. Indirect GWP of methane and N_2O is uncertain; as an approximation, the indirect GWP for methane is positive and could be comparable in magnitude to the direct value (32). Indirect GWP of N_2O is considered much smaller and is expected to have a negligible effect on total greenhouse gas emissions.

Table 8. Reactivities of Organic Emissions From Gasoline, Natural Gas, and LPG[a]

Component	Reactivity, g ozone/g NMOG	
	MIR[b]	MOR[c]
Gasoline NMOG[d]	3.44	1.30
Methane	0.01	0.01
Propane	0.48	0.31
Butane	1.02	0.66
Acetaldehyde	5.51	2.17

[a] Ref. 30.
[b] Maximum incremental reactivity (MIR) is used when the reaction is limited by organic content. CARB recommends the use of MIR values in California.
[c] Maximum ozone reactivity (MOR) is used when the reaction is limited by NO_x concentration.
[d] Gasoline contains components with MIRs ranging from less than 1 to about 10, and MORs up to about 3. Numbers are for exhaust components, not from evaporative emissions.

Table 9. Exhaust Emission Reactivity Adjustment Factors for CNG and LPG Compared with Gasoline[a]

Method	Reactivity Adjustment Factor		
	CNG	LPG	Gasoline
MIR	0.18	0.50	1.00
MOR	0.24	0.59	1.00

[a] California Air Resources Board in Ref. 30.

For most fuels, CO_2 comprises the bulk of greenhouse gas emissions, and most of it is generated when the fuel is consumed to power the vehicle (33). When looking at the merits of alternative fuels from a global context, it is customary to estimate the energy conversion efficiency over the complete fuel cycle. The fuel cycle includes feedstock recovery, feedstock transport, fuel production, fuel distribution, compression or liquefaction, and end use in the vehicle. Estimates of the energy efficiency (J out/J in) for each step in the fuel cycle for CNG, LPG, and gasoline for passenger cars are shown in Table 10. CNG, LPG, and gasoline have similar fuel cycle energy efficiencies, even though the efficiency in the individual process steps are different.

Natural gas and LPG have lower carbon : hydrogen ratios than gasoline, methanol, and ethanol (Fig. 3). Therefore, these fuels have an inherent advantage over gasoline when used in engines. When CO_2 emissions from other parts of the fuel life cycle and unburned methane emissions are included, the advantage of natural gas and LPG is not as clear.

Because methane has a GWP per unit mass at least an order of magnitude greater than CO_2, estimates are

Table 10. Fuel Cycle Energy Efficiency Estimate for CNG, LPG, and Gasoline Use in Passenger Cars

Conversion Step	CNG	LPG	Gasoline
Fuel[a]			
Feedstock production	0.96	0.96	0.97
Feedstock transport	1.00	1.00	0.98
Preparation	0.97	0.89	1.00
Conversion and refining	1.00	1.00	0.91
Fuel transport	0.95	0.97	0.98
Distribution	0.89	0.99	0.99
Vehicle[b]	0.32[c]	0.32[c]	0.30
Total Fuel Cycle Energy Efficiency	*0.25*	*0.26*	*0.25*

[a] Estimates in Ref. 34.
[b] Estimates in Ref. 33.
[c] Assumes engine is optimized to run on natural gas or propane and filled fuel tank weighs approximately the same as the gasoline vehicle to yield similar acceleration performance, at the expense of range between refueling.

extremely sensitive to assumptions about methane emissions. Assumptions about the GWP value, tailpipe methane emissions rate, and fuel source (oil or natural gas wells) greatly affect the results. For example using methane that would otherwise be vented or flared at oil wells would result in a large net methane reduction and would far outweigh any benefits of substituting one fuel for another (35). However, if the natural gas emitted during extraction of oil is not used (36), then the methane emissions attributable to gasoline production would be high.

Greenhouse gas emissions from LPG depend on whether the LPG comes from petroleum refining or natural gas product. Tailpipe methane emissions are not a concern for LPG as it is for natural gas. Overall, LPG is generally recognized as having a smaller direct GWP than gasoline or natural gas.

COSTS

The conversion of gasoline-fueled vehicles to dual-fuel (gasoline–natural gas) is an established technology, while fully optimized dedicated natural gas vehicles are just now becoming commercial. Conversions of existing gasoline vehicles involves "underhood" modifications such as installation of a carburetor designed for use with natural gas, pressure-reducing valves, shut-off valves, gauges,

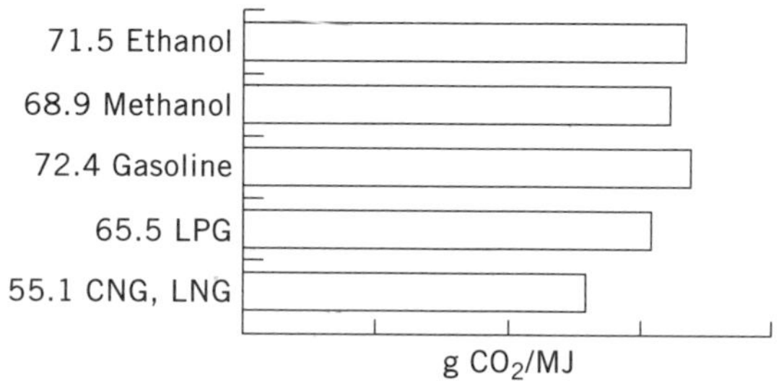

Figure 3. CO_2 emissions from complete combustion of selected fuels.

and fuel-selector switches. Conversions also involve the replacement of the gasoline tank with gas cylinders and supporting structures. Cost estimates for conversions dependent on quantity of vehicles, and size, pressure, and material of construction of the gas cylinders. In 1989 dollars, conversion of a passenger car to dual-fuel use is estimated to cost between $2550 and $3250 using 20.7 MPa steel cylinders, and between $1650 and $2250 using 16.5 MPa steel cylinders (22). For comparison, LPG retrofit costs between $700 (1989 dollars) (37) and $1000 per vehicle (11). The lower cost of LPG conversion is mainly due to the fact that the cylinders are smaller and carry the fuel at a lower pressure. For high production rates of CNG vehicles, conversions have been estimated at $1600 (23).

Cost estimates for dedicated CNG vehicles are less certain and depend on the degree of engine optimization, cylinder materials, and production volume. Estimates of the incremental costs for a dedicated, optimized CNG passenger vehicle produced by a manufacturer in large volumes range between $500 and $1100.

Costs to convert an existing diesel engine to dual-fuel methane-diesel fueled operation range between $3100 and $5600 (1989 dollars). This estimate does not include costs of cylinders and fuel-handling components. For an urban transit bus tank designed for 224 m^3 net increase in fuel cylinder costs are estimated to range between $7400 and $9900 (1989 dollars) when fiber-wrapped steel cylinders are used.

The cost to retrofit a gasoline service station to fast-fill CNG refueling has been estimated to range between $225,000 (990 m^3 storage capacity) and $400,000 (2830 m^3 storage capacity) (1989 dollars) (22). About half of the cost is for an electric compressor motor having an output of 5.7 to 8.5 m^3/min. at 24.8 MPa. About 25% of the total cost is for the gas storage cylinders that are arranged in cascades. Each cylinder is typically 7 m long and 69 cm in diameter and stacked. Costs do not include CNG piping to the service station.

Most studies on LNG for transportation have focused on heavy-duty applications and fleet automobile applications that could be refueled at facilities owned by the fleet operator. The incremental cost for large-scale bus conversions has been estimated to be $40,000 per bus (1991 dollars), or about 20% more than the price of a new bus (18,23). Cost estimates of LNG, LPG, and diesel fuel technology for transit fleet operations are shown in Figure 4.

HEALTH AND SAFETY

Although the primary interest of natural gas and LPG fuels revolves around their potential to reduce vehicle emissions and ozone-forming compounds, consideration must be given to health and safety. Exposure to transportation fuels could be through inhalation, skin contact, and ingestion. Natural gas and LPG appear to present a low health risk from direct exposure in normal dilution concentrations in air. Refueling of CNG could pose risk of physical injury from cryogenic burns caused by gas cooled by rapid expansion during refueling. Safety concerns as-

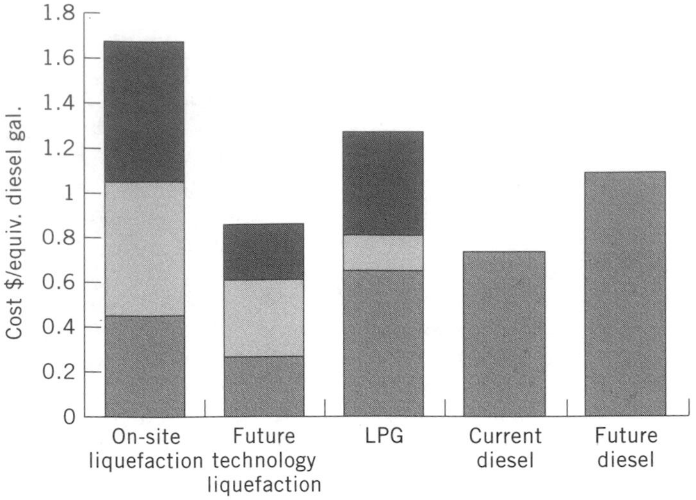

Figure 4. Fuel-specific cost estimates for bus transit fleet operations (18,29). Natural gas prices assumed to be \$2.50/MCF and \$1.50/MCF for future technology. LPG price assumed to be \$0.42/gal. ■ fuel cost; □ liquefier, fueling facility costs; ■ incremental vehicle maintenance costs.

sociated with alternative fuels involve production, distribution, and refueling activities and vehicle operation. Chemical and physical properties that affect the safety of alternative fuels are (*1*) potential damage from detonation, (*2*) potential for combustion of vapors in the fuel tank, (*3*) potential for combustion of vapors in open spaces, (*4*) potential for combustion in restricted spaces, and (*5*) hazard if a fire occurs. LPG is more likely to detonate from a spill than either CNG or LNG. LNG is more likely to detonate in a spill than CNG, because a spill of LNG is more likely to generate concentrations above the lower detonation limit than a CNG leak. Spills of CNG, LNG, and LPG would not result in contamination of soil or groundwater like gasoline or diesel fuel. Combustion of vapors in the tank are of no concern for the fuels. Combustion in open spaces is a concern for LPG because of its low flammability limits and moderate ignition temperature. LPG is heavier than air; therefore, it does not disperse as quickly as methane, which is lighter than air. LNG is only a concern during large spills when the gas cannot disperse quickly because the spilled fuel cools surrounding air, which results in a dense mixture; otherwise, LNG is similar to CNG where only a slight concern exists for combustion in open spaces. For all the fuels, there are significant concerns about combustion in restricted spaces. Methane from a leaking CNG cylinder can build up in closed spaces like a parking garage and can form combustible mixtures. Pressure venting of LNG as a result of warming of the cylinder can also result in combustible mixtures. LPG does not disperse quickly so there is a high concern for the potential for gas buildup in closed spaces from leaking tanks. In the event of a fire, methane burns with a relatively cool flame compared with gasoline, and a CNG cylinder will not explode in the presence of a fire. LNG tanks could pressure vent when heat from a fire contacts the tank. LPG fires are more difficult to put out than CNG and LNG fires because of extreme heat, and heat can cause pressure venting.

Safety risks as a result of vehicular accidents involve fire, physiological damage from fire, explosion, and overall risk. CNG and LPG vehicles in use up to now show good safety records. As of 1991, CNG vehicles have shown a lower injury and death rate than gasoline vehicles (39). No deaths have been attributed to fuel system failures. There are no safety data available for LNG vehicles, because they are more experimental. Overall, safety of CNG and LPG vehicles in collision scenarios appear to be between that of gasoline vehicles (less safe) and diesel vehicles (safer) (23). Currently, various organizations are developing safety standards. CNG, LNG, and LPG tanks must be designed according to ASME pressure vessel codes, a variety of Compressed Gas Association codes, Canadian Gas Association codes, U.S. Code of Federal Regulations, and American Petroleum Institute (API) standards. The U.S. Code of Federal Regulations describe DOT standards, which describe cylinder performance requirements, testing requirements, and allowable contaminants in the gas to prevent corrosion. Standards for compressed gas cylinders used in vehicles are now being developed.

FUTURE VEHICLE DEVELOPMENTS

The use of natural gas and LPG as a transportation fuel could lower NO_x and VOC emissions, thereby reducing ozone formation. Use of these fuels could lower the amount of oil imported. However, full benefits from the use of natural gas as a transportation fuel will not be realized until vehicles are designed specifically for natural gas and a distribution and refueling infrastructure is established. The engine, fuel cylinder, and fuel delivery system should be optimized to take advantage of the low tendency of natural gas to knock. Optimized engines would include the use of lean-burn technology, optimized compression ratios, improved valve materials, quick-response electronic fuel injection systems, closed-loop feedback control systems for air : fuel ratio and spark-timing control, and improved catalysts to reduce methane and other HCs in the exhaust. It is generally recognized that catalytic reduction of NO_x in lean-burn natural gas engines needs a breakthrough.

Thermal efficiency of natural gas engines could be about 6% higher than a similar gasoline engine assuming that engine components are optimized; efficiency on LPG would be slightly less. However, a CNG vehicle is less efficient overall because natural gas cylinders are heavier than a filled gasoline tank, assuming equal energy content. The development of lighter, stronger fuel cylinders would greatly improve the range of natural gas vehicles. However, even with the most advanced cylinder construction (all composite), weight of CNG storage is about three times the weight of an equivalent tank of gasoline, on an energy basis.

LNG tanks require improvements to maintain cryogenic temperatures and to prevent overpressurization especially during periods of nonuse.

Diesels operating on natural gas would require improved lubricants to reduce particulate emissions and improved valve seals to reduce oil consumption (20). Also, high pressure direct natural gas injection needs to be refined for diesel applications. Given the inherent range limitation, natural gas fuels require the necessary infracture to ensure that refueling outlets match range limitations of the vehicle. In general, early targets for developing an alternative fuel infrastructure are centrally fueled fleet operations for which vehicles return to a common point at the end of the day. Finally, better understanding of emissions of natural gas and LPG fuels, including understanding of global warming concept and atmospheric chemistry would provide a more sound scientific basis for introducing alternative fuel vehicles.

BIBLIOGRAPHY

1. "Automobile Operated with Natural Gas Demonstrated in Minneapolis as Feature of Mid-West Convention," *Gas Age*, 488 (April 25, 1936).
2. G. Egloff and P. M. Van Arsdell, *Chem. Eng. News*, 649–658 (May 25, 1942).
3. R. A. Ransom, *Gas Age*, 28–34 (Nov. 27, 1947).
4. R. N. Coleman and K. W. Burnham, in *An Historical Perspective of Farm Machinery,* Society of Automotive Engineers, Inc., Warrendale, Pa., 1980.
5. *Mill and Factory,* Conover-Most Corp., New York, Aug. 1953, pp. 246–248.
6. G. R. Benz, W. F. DeVoe, and A. F. Dyer, *Am. Gas Assoc. Monthly*, 42–48 (Jan. 1963).
7. J. Nichols, *Butane Propane News*, 26–27 (May 1980).
8. U.S. Department of Energy, *Assessment of Costs and Benefits of Flexible and Alternative Fuel Use in the U.S. Transportation Sector, Progress Report One: Context and Analytical Framework,* DOE report DOC/PE-0080, Washington, D.C., Jan., 1988.
9. U.S. Department of Energy, *Assessment of Costs and Benefits of Flexible and Alternative Fuel Use in the U.S. Transportation Sector, Progress Report Two: The International Experience,* DOE report DOC/PE-0085, Washington, D.C., Aug., 1988.
10. *Status of the Natural Gas Vehicle Conversion and Refueling Infrastructure,* AGA EA 1993-2, American Gas Association, Arlington, Va., Aug. 13, 1993.
11. *Using LP-Gas as an Alternative Fuel,* National Propane Gas Association, Lisle, Ill., undated.
12. W. E. Liss and W. H. Thrasher, *Natural Gas as a Stationary Engine and Vehicular Fuel, SAE paper no. 91234,* Society of Automotive Engineers, Inc., Warrendale, Pa., 1991.
13. *ASTM Standard D2699–92, Test for Knock Characteristics of Motor Fuels by the Research Method,* American Society for Testing and Materials, Philadelphia, Pa, 1993.
14. *ASTM Standard D2700–92, Test for Knock Characteristics of Motor Fuels by the Motor Method,* American Society for Testing and Materials, Philadelphia, Pa, 1993.
15. J. Kubesch, *Effect of Gas Composition on Octane Number of Natural Gas Fuels, Gas Research Institute Topical Report GETA 92-01,* Chicago and Southern California Gas Co., Los Angeles, May 1992.
16. *ASTM Standard D1835–91, Specification for Liquefied Petroleum Gases,* American Society for Testing and Materials, Philadelphia, Pa, 1993.
17. *ASTM Standard D2623–86, Test for Knock Characteristics of Liquified Petroleum (LP) Gas by the Motor (LP) Method,* American Society for Testing and Materials, Philadelphia, Pa., 1993.
18. C. Powars, D. Luscher, C. Moyer, and L. Browning, *A White Paper: Preliminary Assessment of LNG Vehicle Technology, Economics, and Safety Issues* (Revision 1), Gas Research Report no. GRI 91/0347.1, Gas Research Institute, Chicago, Jan. 10, 1992.
19. K. Owen and T. Coley, *Automotive Fuels Handbook,* Society of Automotive Engineers, Inc., Warrendale, Pa., 1990.
20. A. Unrich, R. M. Bata, and D. W. Lyons, *Natural Gas: A Promising Fuel for I. C. Engines,* SAE paper no. 930929, Society of Automotive Engineers, Inc., Warrendale, Pa., 1993.
21. R. I. Bruetsch, *Emissions, Fuel Economy, and Performance of Light-Duty CNG and Dual-Fuel Vehicles,* Environmental Protection Agency technical report EPA/AA/CTAB-88-05, Environmental Protection Agency, Ann Arbor, Mich., June 1988.
22. U.S. Environmental Protection Agency, *Analysis of the Economic and Environmental Effects of Compressed Natural Gas as a Vehicle Fuel,* Vol. I, *Light-Duty Vehicles,* Office of Mobile Sources Special Report, Ann Arbor, Mich., Apr. 1990.
23. U.S. Environmental Protection Agency, *Analysis of the Economic and Environmental Effects of Compressed Natural Gas as a Vehicle Fuel,* Vol. II, *Heavy-Duty Vehicles,* Office of Mobile Sources Special Report, Ann Arbor, Mich., Apr. 1990.
24. C. S. Weaver and S. H. Turner, *Dual-fuel Natural Gas/Diesel Engines: Technology, Performance, and Emissions,* SAE paper no. 940548, Society of Automotive Engineers, Inc., Warrendale, Pa., 1994.
25. Brunswick Composites, *Brunswick's All-Composite NGV Fuel Containers,* Brunswick Corp., Lincoln, Nebr., 1993.
26. M. A. DeLuchi, *Emissions of Greenhouse Gases from the Use of Transportation Fuels and Electricity,* Argonne National Laboratory report ANL/ESD/TM-22, Vol. 1, Argonne, Ill., Nov., 1991.
27. *Installation and Service Guidebook: LP-gas Tanks, Cylinders, and Equipment,* National Propane Gas Association, Lisle, Ill., 1989.
28. D. E. Gushee, *Alternative Fuels for Automobiles: Are They Cleaner Than Gasoline?* Congressional Research Service report 92-235 S, Library of Congress, Washington, D.C., Feb. 27, 1992.
29. U.S. Environmental Protection Agency, *Federal Register,* **57**(215), 52912–52928 (Nov. 5, 1992).
30. M. E. Pitstick, *Emissions from Ethanol and LPG Fueled Vehicles,* Transportation Research Board paper no. 930527, Transportation Research Board, Washington D.C., 1993.
31. G. F. Baker, J. A. Draves, and R. F. Klausmeier, *Assessment of Environmental, Health, and Safety Issues Related to the Use of Alternative Transportation Fuels,* Gas Research Institute report GRI-91/0291, Coerr Environmental Consulting, Chapel Hill, N.C., Feb. 10, 1992.
32. J. T. Houghton and co-workers, *Climate Change 1992,* Cambridge University Press, Cambridge, UK, 1992.
33. F. Stodolsky and D. J. Santini, *Chemtech,* **23**(5), 54–59 (Oct. 1993).
34. S. P. Ho and T. A. Renner, *Global Warming Impact of Gasoline vs. Alternative Transportation Fuels,* SAE paper no. 901489, Society of Automotive Engineers, Inc., Warrendale, Pa., 1990.
35. D. J. Santini, M. A. DeLuchi, A. Vyas, and M. Walsh, *Greenhouse Gas Emissions from Selected Alternative Transporta-*

tion Fuels Market Niches, paper presented at the American Institute of Chemical Engineers Summer National Meeting, Aug. 20–23, 1989.

36. S. Unnasch, B. Moyer, D. D. Lowell, and M. D. Jackson, *Comparing the Impact of Different Transportation Fuels on the Greenhouse Effect,* California Energy Commission Consultant Report, Sacramento, Calif., Apr. 1989.

37. R. Moreno Jr and D. G. Fallen Bailey, *Alternative Transport Fuels from Natural Gas,* World Bank technical paper no. 98, The World Bank, Washington, D.C., 1989.

38. Acurex Environmental Corp., *GRI Workshop on LNG Vehicle Technology, Economics, and Safety Issues: Focus Group Recommendations Summary,* Gas Research Institute report GRI 92-0330, Gas Research Institute, Chicago, July 7, 1992.

U

UNDERGROUND GASIFICATION

James Speight
Western Research Institute
Laramie, Wyoming

Underground gasification of coal (also called in situ gasification of coal) is the gasification of coal in place (ie, without mining) by interaction of the coal with air to yield a low-heat content (low Btu gas), or with a mixture of steam and oxygen to produce an intermediate heat content (intermediate Btu) gas (1).

See also Coal gasification; Fuels, synthetic—gaseous fuels; Manufactured gas; Fischer-Tropsch process and products.

The concept of the underground gasification of coal dates to Britain (1868) and Russia (1888) with extensive development occurring in Russia from 1933 onward. The United States and Canada have been interested in coal gasification since the 1950s, although development of the technology did not formally begin until the early 1970s.

Underground coal gasification technology has been tested in the United States, but in terms of commercial development the former Soviet Union has been the most active country (1).

The fundamental principles of underground gasification are similar to those of above-ground gasification with the exception that the reactor is the coal seam.

The gasification of coal (be it above-ground or underground) is essentially the conversion of coal (by any one of a variety of processes) to produce combustible gases which may be of low, medium, or high Btu content depending upon the defined use. High-Btu gas consists predominantly of methane with a heating value of approximately 37.3×10^3 kJ/m^3 (1000 Btu/ft^3) and is compatible with natural gas insofar as it may be mixed with, or substituted for, natural gas.

On the other hand, medium-Btu gas consists of a mixture of methane, carbon monoxide, hydrogen, and various other gases. The heating value of medium-Btu gas usually falls in the range of $11–26 \times 10^3$ kJ/m^3 (300–700 Btu/ft^3) and is suitable as a fuel for industrial consumers. Finally, low-Btu gas (the gas usually produced by underground gasification of coal) consists of a mixture of carbon monoxide and hydrogen (H_2), and has a heating value of less than 300 Btu/ft^3 (11 kJ/m^3). This gas is of interest to industry as a fuel gas or even, on occasion, as a raw material from which ammonia, methanol, and other compounds may be synthesized.

The importance of coal gasification as a means of producing fuel gases for industrial use cannot be underplayed. But all coal gasification systems have environmental issues that must be addressed; a range of undesirable products are also produced which must be removed before the desirable ones are used to provide fuel and/or to generate electric power.

Coal gasification involves the thermal decomposition of coal, the reaction of the carbon in the coal, and other pyrolysis products with oxygen, water, and hydrogen to produce fuel gases, such as methane, either directly

$$[C]_{coal} + [H]_{coal} \rightarrow CH_4$$

or through the agency of hydrogen (which may be generated during the reaction)

$$[C]_{coal} + 2\,H_2 \rightarrow CH_4$$

although the reactions are more numerous and more complex.

The presence of oxygen, hydrogen, water vapor, carbon oxides, and other compounds in the reaction atmosphere during pyrolysis may either support or inhibit numerous reactions with coal and with the products evolved. The distribution of weight and chemical composition of the products are also influenced by the prevailing conditions (ie, temperature, heating rate, pressure, residence time, etc) and, last but not least, the coal feedstock.

If air (as is usually the case underground) is used as the combustant, the product gas will have a heating value of $5.6–11.2 \times 10^3$ kJ/m^3 (150–300 Btu/ft^3) and will contain undesirable constituents, such as carbon dioxide, hydrogen sulfide, and nitrogen. The use of pure oxygen, although expensive, can result in a product gas having a heating value of 300–400 Btu/ft^3 (11.2–14.9 MJ/m^3) with carbon dioxide and hydrogen sulfide as by-products.

If a high-Btu gas (33.6–37.3 kJ/m^3; 900–1000 Btu/ft^3) is required from the products of an underground reactor, efforts must be made to increase the methane content of the gas by further conversion above ground:

$$CO + H_2O \rightarrow CO_2 + H_2$$

$$CO + 3\,H_2 \rightarrow CH_4 + H_2O$$

$$2\,CO + 2\,H_2 \rightarrow CH_4 + CO_2$$

$$CO + 4\,H_2 \rightarrow CH_4 + 2\,H_2O$$

The reaction rates for these reactions are relatively slow with negative heats of formation and catalysts may be necessary for complete reaction.

PRIMARY AND SECONDARY GASIFICATION

In general terms, the gasification of coal (be it above ground or underground) involves both primary and secondary gasification.

Primary gasification involves thermal decomposition of the raw coal and many schemes to produce mixtures containing various proportions of carbon monoxide, carbon dioxide, hydrogen, water, methane, hydrogen sulfide, nitrogen, and typical products of thermal decomposition, such as tar, oils, and phenols. A solid char product may also be produced, and often represents the bulk of the weight of the original coal. This type of coal being gasified determines (to a large extent) the amount of char produced and the analysis of the gas product.

Secondary gasification usually involves gasification of the char from the primary gasification stage. This is usually done by reaction of the hot char with water vapor (generally produced from the coal or, on occasion, injected into the underground reactor) to produce carbon monoxide and hydrogen:

$$[C]_{char} + H_2O \rightarrow CO + H_2$$

The gaseous product from the underground gasifier generally contains large amounts of carbon monoxide and hydrogen, plus lesser amounts of other gases. Carbon monoxide and hydrogen (if they are present in the mole ratio of 1:3) can be reacted in the presence of a catalyst to produce methane:

$$CO + 3\,H_2 \rightarrow CH_4 + H_2O$$

For the initiation of the process and to maintain the reaction in the seam, air (or oxygen) and steam are injected into one of two wells drilled into a coal seam. After ignition, the coal is partially combusted to a synthesis gas mixture plus tar, which is recovered through the second well.

Low heat content gas (1.9–10.4 MJ/m^3; 50–280 Btu/ft^3) is generated, the range of heat content varying with the oxidants used. Air, for instance, yields the lowest Btu product (1.9–5.2 MJ/m^3; 50–140 Btu/ft^3), whereas an air–steam cycling sequence produces gas having a much higher heat content (9.3–10.4 MJ/m^3; 250–280 Btu/ft^3) (2).

The success of underground coal gasification depends in large part on effective contact between the coal and the injected gases. Thus, a primary limitation to efficient underground gasification is the low permeability of coal seams. In some instances large shafts are drilled into the seam, and after some mining, panels of coal are sectioned off by brickwork. This is called the chamber or warehouse method (Fig. 1). It is costly in terms of the labor required to construct the chambers and to break into or bore into the panels to create passageways for the oxidants.

Borehole Gasification

In a borehole gasification scheme, a set of parallel boreholes (galleries) are drilled into the seam and are connected by a series of horizontal channels. Air is forced into the galleries and remote electric ignition initiates combustion (Fig. 2). The borehole method is most appropriate for level seams. For seams with a large dip, inclined galleries, called streams, are sunk into the seam and are connected at the bottom by a horizontal "fire drift" channel. The gasification is initiated in the fire drift channel and the vertical openings carry inlet and product gases. This method uses the process called reverse combustion, ie, the combustion front moves toward the source of injected oxidant (Fig. 3). In the chamber and borehole methods, combustion and inlet gas move in the same direction (forward combustion). Shaftless methods for *in situ* gasification include the percolation method. After an array of paired boreholes is drilled, the combustion gases are forced into one hole of the pair and product gases are released through another, depending on the permeability of the coal. The combustion zone can be forward or reverse and the process is repeated sequentially throughout the array (Fig. 4).

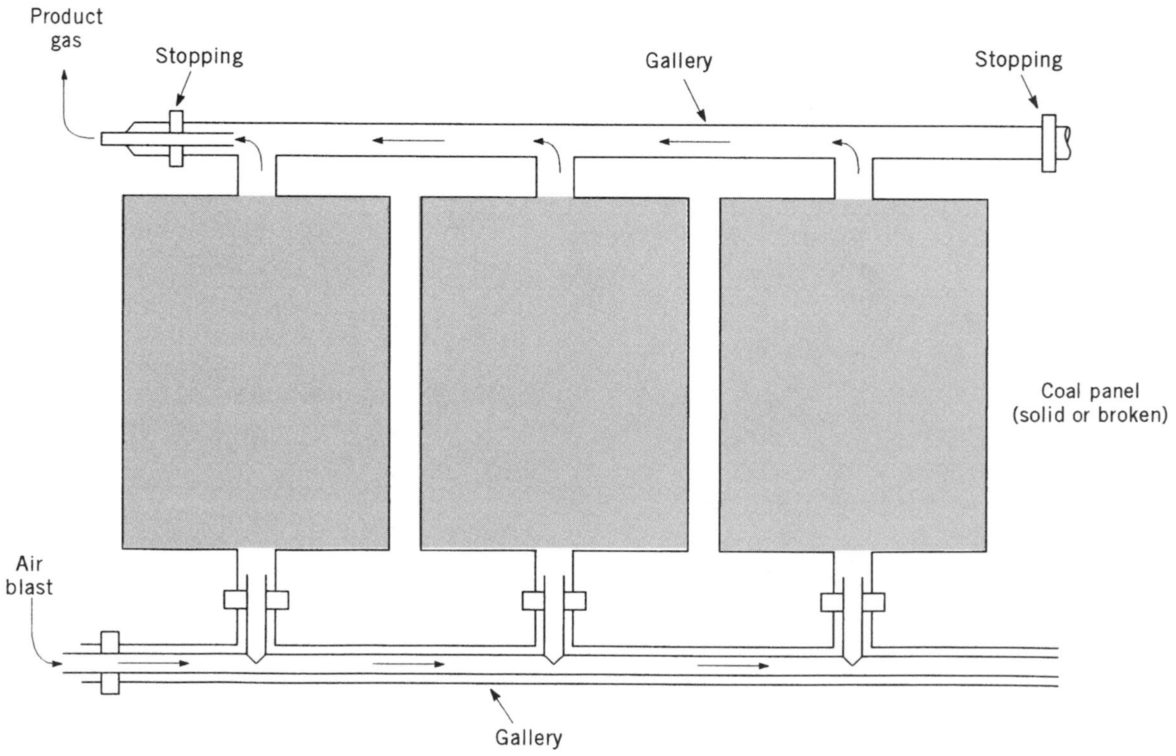

Figure 1. The chamber method of underground coal gasification.

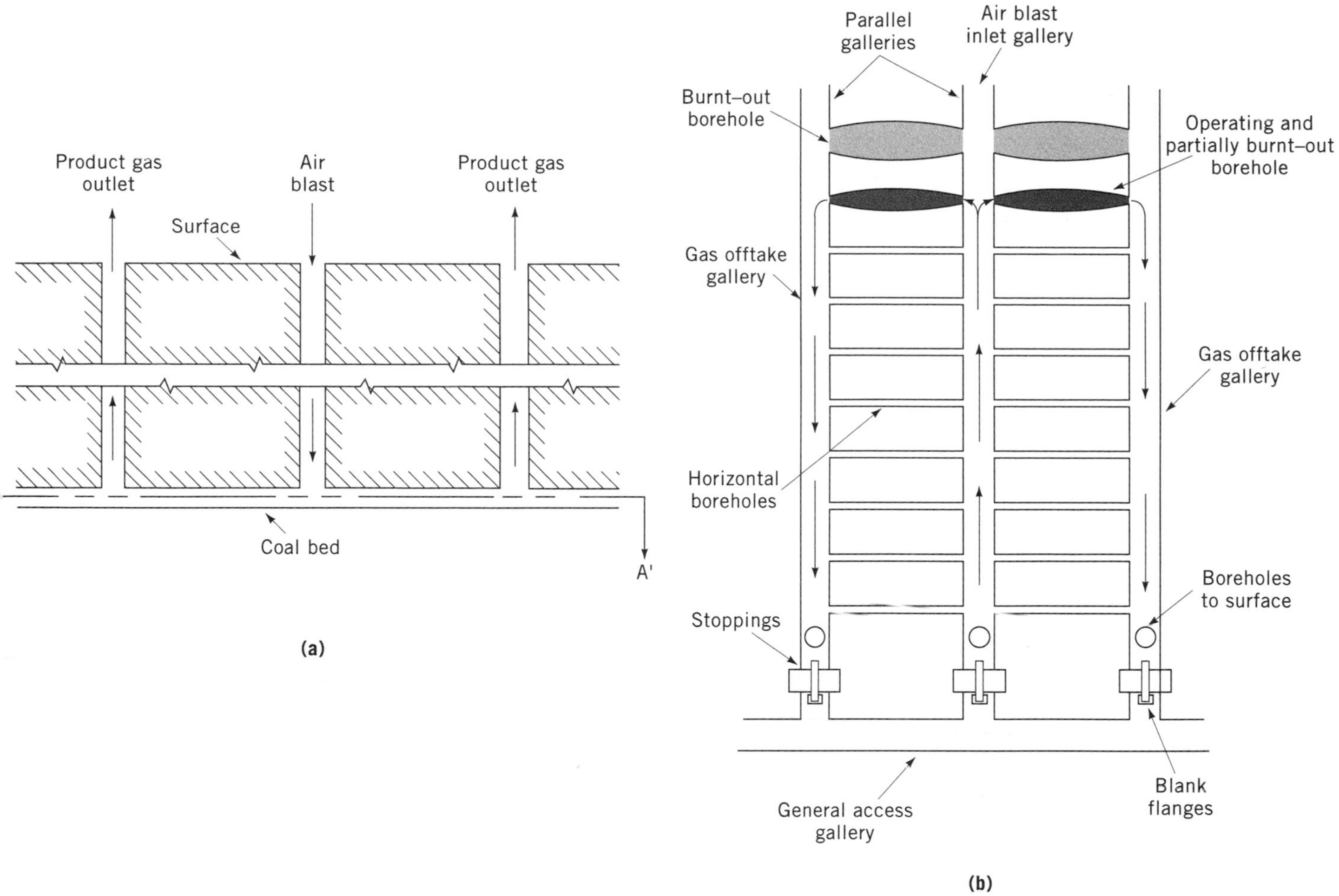

Figure 2. The borehole method of underground coal gasification.

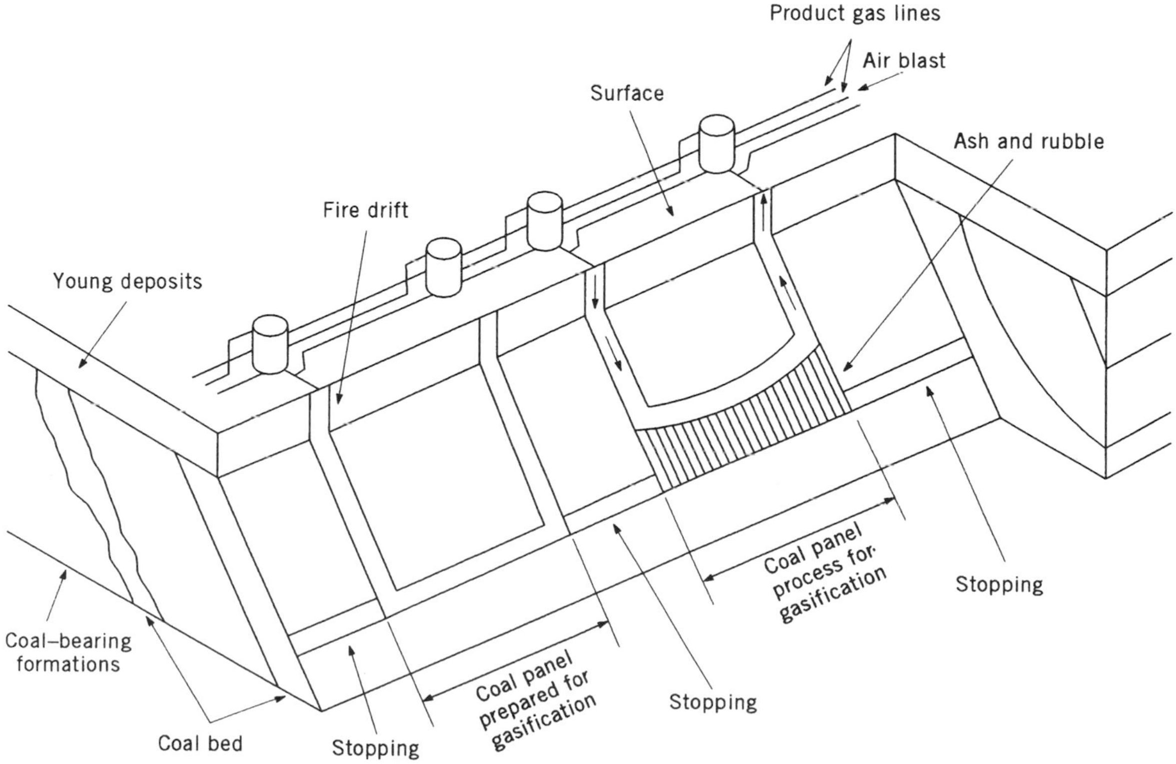

Figure 3. The stream method of underground coal gasification.

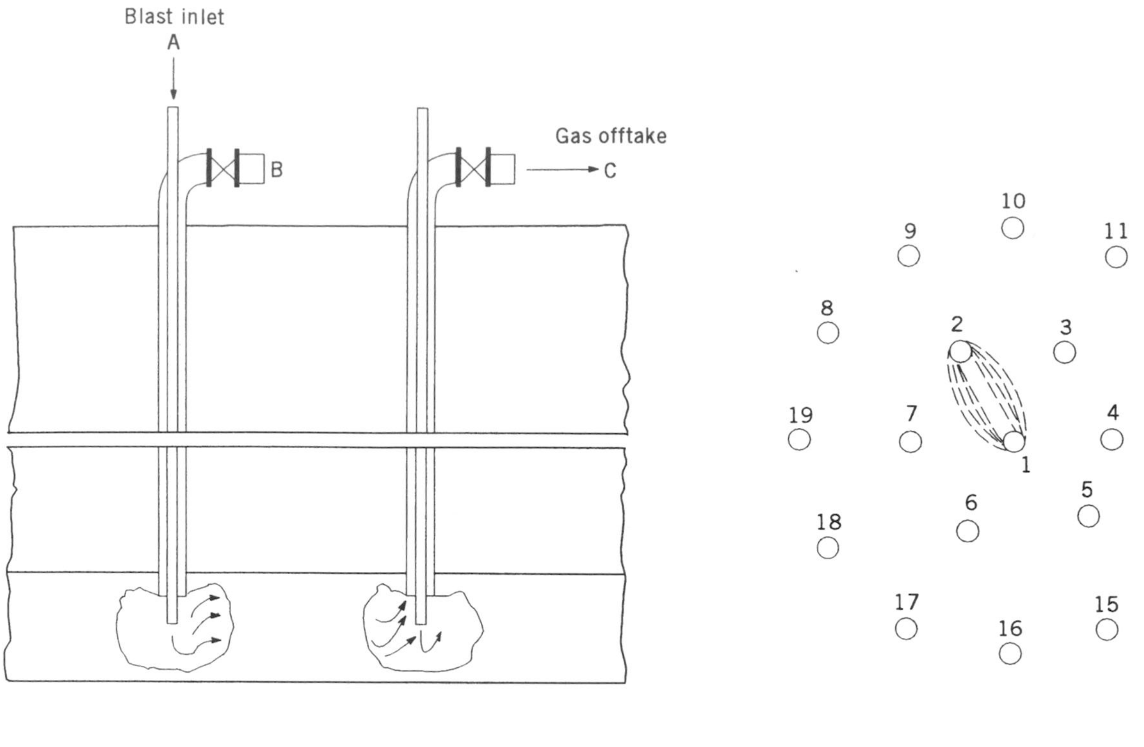

Figure 4. The percolation method of underground coal gasification.

In a design intended to produce high heat content (high Btu/ft^3, high MJ/m^3)-Btu gas from underground seams up to 1000 m below the surface, a scheme was devised that first fractures the coal using explosives after which boreholes can be drilled and cased. Combustion with oxygen and steam is proposed in order to produce a gas rich in methane (Fig. 5). Cleanup and additional methanation could be carried out at the surface.

Currently, there is no commercial underground coal gasification facility operational in the United States. However, a test at Rawlins (Wyoming), sponsored by the U.S. Department of Energy, the Gas Research Institute, Amoco, the Electric Power Research Institute, and Union Pacific Resources, was completed in the later 1980s with environmental monitoring continuing into the early 1990s.

In this test, two commercial-sized underground gasification modules of different configurations were simultaneously operated to produce medium heating value gas. The extended linked well (ELW) module consists of a horizontal borehole connected to vertical injection wells. The product gases were produced out of the horizontal well.

Controlled Retracting Injection Point (CRIP)

The other module configuration tested was the (*CRIP*) (controlled retracting injection point) system (Fig. 6). In UCG operations, gasification efficiency generally declines with time due to increased heat losses to the overburden. More overburden is exposed as the UCG module matures. The CRIP system permits the movement

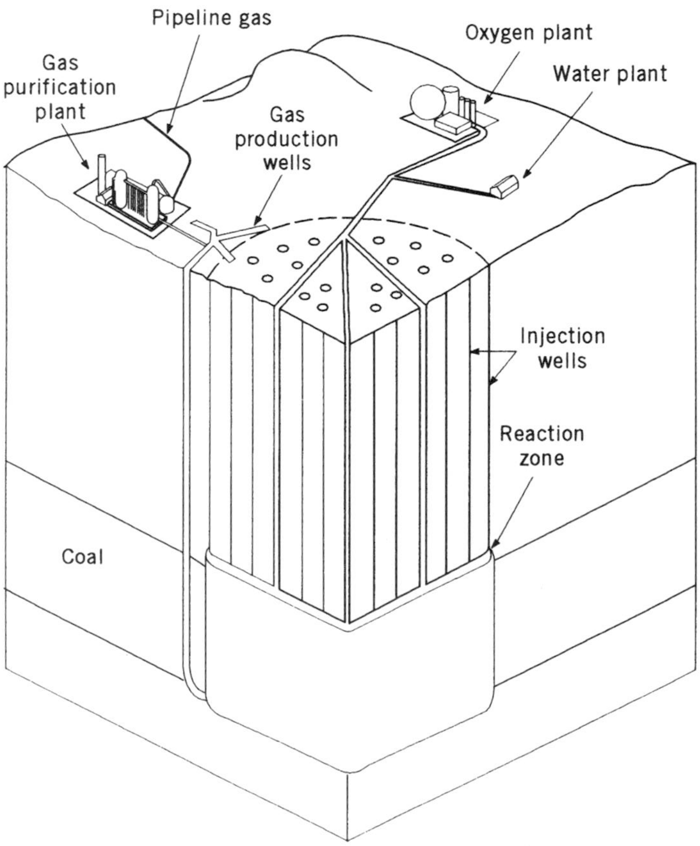

Figure 5. A method for the generation of high heat content gas.

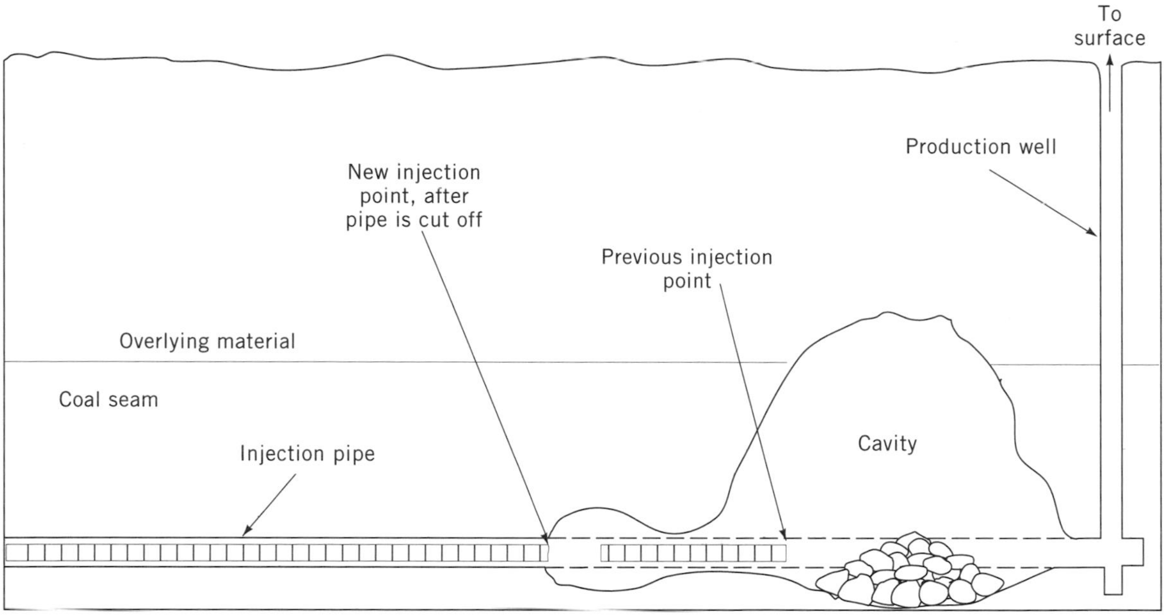

Figure 6. The CRIP method for underground coal gasification.

of the injection point away from the existing cavities to expose fresh coal to the processs, and reduces heat losses to exposed overburden. In the CRIP module, a liner is inserted in a horizontal borehole in the coal seam. Steam and oxygen are injected through the liner in order to react with the coal and its products, fueling the gasification reactions. When the gasification efficiency declines to some economically determined value, "CRIP maneuver" is performed. A burner in the liner is retracted to a new location where the liner is burned off. The coal is ignited at this point, initiating the development of a new gasification cavity.

Operations of both UGG modules were very successful; however, the CRIP module performed better. The CRIP module was operated for approximately 100 days, consumed 9000 of coal, and produced an average gas heating value of 295 Btu/ft^3 (11.0 MJ/m^3). Three CRIP maneuvers were accomplished which in each case improved the heating value of the product gas. The ELW module was operated for approximately 60 days, consumed 4000 t (4×10^6 kgm) of coal and produced an average gas heating value of 275 Btu/ft^3 (10.2 MJ/m^3).

One of the most important aspects of the RM1 UGG test is that it appears to have been conducted with minimal environmental impact. Previous UGG tests have raised concerns about the environmental acceptability of the UGG process, particularly with respect to the groundwater.

Based on evaluation of past UGG tests and laboratory research, the Western Research Institute, with the support of the Department of Energy and the Gas Research Institute, developed a set of environmental control methods to minimize or eliminate groundwater contamination associated with UCG. These methods were demonstrated at the RM1 test site and preliminary results have shown that the methods were successful in reducing the generation, deposition, and transport of contaminants.

The potential for any product is determined by economics. UCG must be shown to be profitable before any commercial development will be attempted. With the current domestic energy oversupply situation, the risks of an unproved technology in UCG may inhibit commercialization. However, a return to higher natural gas and oil prices will make UCG more attractive.

Many potential uses for UCG product gases are possible. There is very little difference between product gas composition from surface gasifiers and that from UCG reactors. Therefore, in many applications, gas produced by UCG can be substituted for gas produced by surface gasifiers. These uses include methanol and synthetic gasoline production, feed for combined cycle power plants, and ammonia production.

BIBLIOGRAPHY

1. J. G. Speight, J. G. Speight, ed., in *Fuel Science and Technology Handbook,* Marcel Dekker Inc., New York, 1990, p. 726.
2. R. M. Nadkarni, C. Bliss, and W. I. Watson, *Chem. Tech.* 230 (1974).

V

VAPOR LOCK

Ramon Espino
Exxon Research and Engineering
Annandale, New Jersey

One of the most common hot weather driveability problems in gasoline engines used to be vapor lock. In modern engines and with high quality gasoline, the possibility of experiencing vapor lock problems is more remote. Vapor lock occurs when the gasoline vaporizes prematurely in the feed line and prior to being delivered to the carburetor or the fuel injectors. In a mild case of vapor lock, the engine experiences hesitation during acceleration but in extreme cases the engine can stall.

From the point of view of automobile design, the most critical factor is the location of the fuel pump(s). If at least one of the pumps is submerged in the fuel tank, the gasoline will be delivered to the carburetor or injectors under pressure and therefore it will be less likely to vaporize. It is also important to design the engine for minimum heating of the fuel lines. However, some of the new automobile designs have very low air-drag and consequently very little cooling air flows by the engine compartment. This increases the temperature under the hood including that in the fuel lines. Turbocharged engines also tend to increase the temperature under the hood and the tendency of the fuel to vaporize in the lines. Even the fuel in the tank can get hot, particularly if the exhaust system is positioned directly under the tank.

Obviously, the ambient temperature is very important in causing hot weather driveability problems. High altitude driving can increase the likelihood of vapor lock too. The gasoline being under less pressure has a greater tendency to evaporate, but even more important is the fact that at high altitude the engine compartment tends to get very hot since the engine is frequently operating under a high-load, low-speed driving mode.

The properties of the fuel are also critical in preventing vapor lock problems. The "front-end" volatility of the gasoline is the most important property impacting on hot weather driveability problems, including vapor lock. The temperature at which the gasoline yields a ratio of vapor to liquid volume (V/L) of 20 is widely used to measure front-end volatility. Clearly, the higher the temperature at which this volume ratio is reached, the less the front-end volatility of gasoline. An alternative measure to the V/L ratio is to determine the Reid Vapor Pressure (RVP) of the gasoline and the volume of gasoline that distills at 70°C. This technique is widely used in the United States to measure the hot weather driveability performance of gasolines. The volume of gasoline that distills at 70°C is measured using ASTM D86 as the distillation method. The Reid Vapor Pressure and the volume distilled at 70°C (E70°C) is combined in an expression RVP + n E70°C. This is called the Vapor Lock Index (VLI) or Front-End Volatility Index (FEVI). The constant n is 0.13 if the RVP is measured in psi. The term E70°C has units of % volume. The term n has a value of 9 if the RVP is measured in milibars. For European and Japanese cars, the value of n that correlates best with vehicle performance lies between 5 and 7 (RVP measured in milibars).

The driving cycle, the vehicle itself as well as the gasoline quality, determine the hot weather performance of a vehicle. The driver is also an important factor too. It has been observed that professional testers of driveability problems are more critical than the consumer public in detecting problems. In the United States, minimum quality standards have been established for different geographic regions and depending on climatic conditions. The process to set these specifications is based on measuring the level of satisfaction of drivers with a given gasoline. A representative population of vehicles and drivers is used for predetermined geographic regions and climate conditions. Gasoline marketers should not exceed the front-end volatility specification for that region. The specification can either be the temperature for a V/L ratio of 20 or Vapor Lock Index (VLI).

The introduction of oxygenated fuels in gasoline has brought about renewed concerns about front end volatility and potential vapor lock. Oxygenates are mandated to be used in gasoline in certain regions of the country where the amount of carbon monoxide in the atmosphere exceeds local or federally mandated standards. Fortunately, carbon monoxide limits are exceeded mainly in the winter months and thus the greater tendency for alcohols to evaporate does not cause hot weather driveability problems since gasolines are only required to contain oxygenates in the winter. However, oxygenates are also being considered as gasoline components year around to reduce tail pipe emissions of hydrocarbons and the corresponding high ozone levels that is experienced in the Los Angeles area and other large metropolitan areas of the country. Gasolines with high levels of oxygenates may bring an increased number of vapor lock problems in the summer months. Vehicle design changes and stringent front-end volatility limits are expected to keep this problem to a minimum.

BIBLIOGRAPHY

K. Owen and T. Coley, *Automotive Fuels Handbook,* Society of Automotive Engineers, Inc.

E. F. Obert, *Internal Combustion Engines and Air Pollution,* Harper and Row, Publishers, New York.

VISBREAKING

James Speight
Western Research Institute
Laramie, Wyoming

Visbreaking (viscosity reduction, viscosity breaking) is a relatively mild thermal cracking operation used to reduce the viscosity of residua.

Residua are sometimes blended with lighter heating

oils to produce fuel oils of acceptable viscosity. By reducing the viscosity of the nonvolatile fraction, visbreaking reduces the amount of the more valuable light heating oil that is required for blending to meet the fuel oil specifications. The process is also used to reduce the pour point of a waxy residue.

Although the viscosity of residua drops dramatically with temperature, their viscosity is very high at ambient temperatures. This very mild thermal cracking process is aimed at reducing the viscosity of residua or heavy and extra heavy crude oils so that they meet specifications desirable for fuel oil applications and for pipeline transportation.

The visbreaking process uses the approach of mild thermal cracking as a relatively low-cost and low severity approach to improving the viscosity characteristics of the residue without attempting significant conversion to distillates. Low residence times are required to avoid polymerization and coking reactions, although additives can help to suppress coke deposits on the tubes of the furnace.

Visbreaking conditions range from 455–510°C at a short residence time and from 344×10^3–344–2068 kPa (50–300 psi) at the heating coil outlet. It is the short residence time which brings to visbreaking the concept of being a mild thermal reaction. This is in contrast to, for example, the delayed coking process where residence times are much longer and the thermal reactions are allowed to proceed to completion. The visbreaking process uses a quench operation to terminate the thermal reactions. Liquid-phase cracking takes place under these low severity conditions to produce some naphtha, as well as material in the kerosene and gas oil boiling range. The gas oil may be used as additional feed for catalytic cracking units, or as heating oil.

In the process, a crude oil residuum is passed through a furnace where it is heated to a temperature of 480°C with a residence time of 1–3 minutes and under an outlet pressure of about 690 kPa (100 psi) (see Fig. 3 in HEAVY OIL CONVERSION). The heating coils in the furnace are arranged to provide a soaking section of low heat density, where the charge remains until the visbreaking reactions are completed. The cracked products are then passed into a flash-distillation chamber. The overhead material from this chamber is then fractionated to produce a low quality gasoline as an overhead product and light gas oil as bottoms. The liquid products from the flash chamber are cooled with a gas oil flux and then sent to a vacuum fractionator. This yields a heavy gas oil distillate and a residual tar of reduced viscosity. A quench oil may also be used to terminate the reactions (1,2).

A 5–10% conversion of atmospheric residua to naphtha is usually sufficient to afford at least a ca 5-fold reduction in viscosity. Reduction in viscosity is also accompanied by a reduction in the pour point.

An alternative process design uses lower furnace temperatures and longer times, achieved by installing a soaking drum between the furnace and the fractionator. The disadvantage of this approach is the need to decoke the soaking drum.

The main limitation of visbreaking, and for that matter all thermal processes is that the products can be unstable. Thermal cracking at low pressure gives olefins, particularly in the naphtha fraction. These olefins give a very unstable product, which tends to undergo polymerization reactions to form tars and gums.

The reduction in viscosity of distillation residua tends to reach a limiting value with conversion, although the total product viscosity can continue to decrease. The minimum viscosity of the unconverted residue can lie outside the range of allowable conversion if sediment begins to form. When pipelining of the visbreaker product is the process objective, addition of a diluent such as gas condensate can be used to achieve a further reduction in viscosity.

BIBLIOGRAPHY

1. J. G. Gary and G. E. Handwerk, *Petroleum Refining: Technology and Economics,* Marcel Dekker Inc., New York, 1984.
2. J. G. Speight, *The Chemistry and Technology of Petroleum,* 2nd ed., Marcel Dekker Inc., New York, 1991.

W

WASTE MANAGEMENT PLANNING

LOCH MCCABE
Resource Recycling Systems, Inc.
Ann Arbor, Michigan

The purpose of this chapter is to provide the reader with a conceptual understanding of a proven methodology for integrating waste management practices in a manner that minimizes costs to the community. The planning methodologies suggested here combine basic planning theory with standard economic modeling practices. The use of economic modeling is critical because the comparison of solid waste management strategies is essentially the evaluation of costs and benefits over time given a variety of political, regulatory, social, and fiscal constraints.

Community planners and decision-makers are faced increasingly with the challenging task of developing an integrated solid-waste management strategy that best meets the community's short- and long-term waste recovery and disposal needs while minimizing costs. This task requires a balancing of different waste management approaches, including recycling, composting, landfilling, and incineration. Achieving this balance in a technically and economically optimal manner requires the planner to undertake four steps; ie, determine alternative waste management options within each strategy, maximize the cost-effectiveness of each option, maximize the cost-effectiveness of each strategy, and choose the most cost-effective strategy.

Upon completing this decision-making process, planners should have the information they need to build the type, size, and number of waste recovery/disposal collection and processing systems that are likely to be the most cost-effective in the short and long run for that community.

The planning experiences of a hypothetical county called Green County will be used to facilitate understanding of the these concepts and how they may be applied. Green County is determining whether it should: develop an Integrated Waste Management Strategy that emphasizes source reduction/reuse, recycling, and composting, or continue with the status quo strategy of landfilling nearly all the county's waste.

Given the county's fiscal constraints, the Integrated Waste Management Strategy would need to be at least 10% more cost-effective over a 10-year period to be adopted. See also FUELS FROM WASTE; INCINERATION; RECYCLING; COGENERATION.

DETERMINING ALTERNATIVE WASTE MANAGEMENT OPTIONS

Given a certain volume and composition of solid waste generated in the community, the planner may select alternative waste management options that meet all or part of the community's waste management needs. Some of the most common options, and their components, are listed in Table 1.

The strategies selected will generally be chosen based on a variety of regulatory, technical, political, and financial criteria as required by the community. Source reduction/reuse options tend to be favored in communities that have high waste disposal costs and have already achieved high waste recovery through recycling and composting. Recycling and composting tend to be preferred by communities that have high waste disposal costs or there is significant community resistance to building additional waste disposal capacity nearby. Incineration- and landfilling-based strategies are typically adopted by communities with relatively low waste disposal costs and little internal or external pressure for additional waste recovery through recycling and composting.

Green County, which is experiencing both rising waste disposal costs and proenvironmental community pressure, would thus choose to emphasize the source reduction/reuse, recycling, and composting components of its Integrated Waste Management Strategy. This strategy also provides for the landfilling of all remaining refuse.

MAXIMIZING THE COST-EFFECTIVENESS OF EACH OPTION

After selecting a waste management option to explore, it is necessary to determine how it might be implemented most cost-effectively. That is, to identify the optimal operating level and scale for each option that minimizes the net costs of that option independently of other options.

If the planner is developing a recycling option, the entire waste stream (W) may divided into recyclables (q) and non recyclables (*1*). First, the planner would determine the portion of the waste stream to be recycled (q^*). This quantity is typically based on regulatory requirements such as recycling "goals." Then evaluate the cost-effectiveness of different scales (sizes) of recycling collection and processing systems to handle the desired volume of recyclables (q^*). The planner might avoid choosing a scale that is too small ("undersizing"), or a scale of production that is too large ("White Elephant" syndrome). It is most desirable to choose an "optimal" scale, one that will produce the appropriate quantity of recycling at the lowest price.

For example, Figure 1 indicates that the medium-scale recycling effort will be the most cost-effective scale to achieve the recycling of q^* tons. The small-scale recycling effort could also recover q^* tons, but at a higher average cost.

The actual costs for recycling will depend on several variables including the specific number and types of materials collected, the level of source separation or presorting done by generators (residents and businesses), the quantity of material handled (ie, the scale), and the efficiency and effectiveness of the program designers and operators. These variables notwithstanding, typical costs to

Table 1. Typical Waste Management Options and Option Components

Waste Management Options	Option Components
Source reduction/reuse	The collection and disposal/recovery of many packaging materials such as boxes, wrapping, pallets, and other dunnage may be avoided through source reduction and reuse activities.
Recycling	Some paper, glass, metal, plastic, and other materials (including textiles, batteries, motor oil, appliances, and tires) may be collected and recovered at appropriate residential and commercial recycling facilities.
Composting	Leaves, grass, brush, wood waste, and even food waste may be collected and composted at appropriate facilities. Many of these materials may be home composted as well.
Incineration	Disposing of solid waste through an appropriate incineration facility with possible energy recovery. Ash residue would be disposed of at an appropriate landfill within or outside of the community.
Landfilling	Disposing of solid waste at an appropriate landfill within or outside of the community.

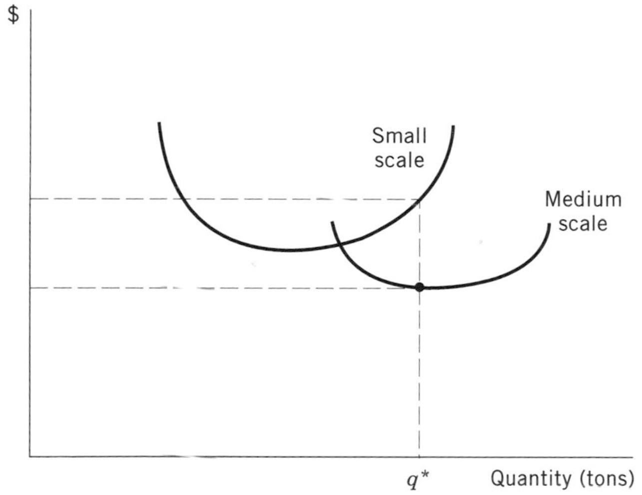

Figure 1. Cost effectiveness of two scales of recycling q^* tons of material in Community A.

both collect and process the standard set of residential and commercial recyclables are shown.

	Collection Costs $/t	Processing Costs $/t
Residential Recyclables	90–140	30–50
Commercial Recyclables	20–70	15–30

These cost estimates may differ and do not include material revenues, which vary widely depending on the particular materials marketed and their quality.

Taking the next step, similar analysis would be used to determine the most cost-effectiveness scale for composting other materials or landfilling residual waste. For Green County, the cost-effectiveness of composting will depend largely on the types of materials handled, the level of site preparation and maintenance required, the level of regulation the compost site must follow, and the markets for the resulting compost. Landfilling costs will be highly affected by the distance to the landfill, the landfill's regulatory requirements, its remaining available capacity, and competition from other landfills.

MAXIMIZING THE COST-EFFECTIVENESS OF EACH STRATEGY

Once individual options within strategies have been optimized, it is necessary to link those options together to evaluate the cost-effectiveness of the entire strategy as a single unit. If the strategy being evaluated include only recycling and landfilling and the most cost-effective scales of recycling and landfilling the residual (1^*) have been identified, these options would be combined together to determine an overall cost for the strategy (Fig. 2).

There is, of course, an interrelationship between the quantity and costs of recycling and the quantity and costs of landfilling. Thus it is often desirable for the planner to conduct several iterations of the analysis to determine the *mix* of options to ensure that the overall waste stream is managed in the most cost-effective manner given the community's political, social, and financial goals and constraints.

Attaining the most cost-effective strategy will often require the planner to evaluate different scales for each option within the strategy as well as evaluating different existing and potential financing and regulatory mechanisms to implement and fund each option.

In Green County, it has been determined that the Integrated Waste Management Strategy could be implemented most cost-effectively residential and business sec-

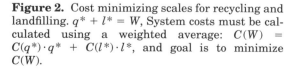

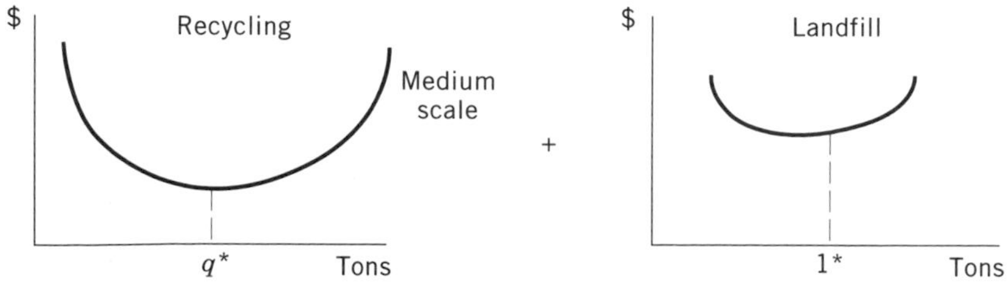

Figure 2. Cost minimizing scales for recycling and landfilling. $q^* + l^* = W$, System costs must be calculated using a weighted average: $C(W) = C(q^*)\cdot q^* + C(l^*)\cdot l^*$, and goal is to minimize $C(W)$.

tor waste is reduced by 3% through education and similar waste reduction efforts, recycling collection and processing systems recover 29% of the current residential and business sector waste, composting collection and processing systems recover 13% of the current residential and business sector waste, and a transfer station is built to haul the remaining waste to an adjacent county for disposal in that county's landfill.

Through this combination of waste recovery systems, it was estimated that the Integrated Waste Management Strategy would effectively divert 45% of the waste that would otherwise go to the landfill, as shown in Figure 3.

Additional analysis reveals that the Integrated Waste Management Strategy is highly sensitive to the quantity of waste material diverted from landfill. The cost-effectiveness of the Integrated Waste Management Strategy improves significantly as the quantity of waste recovered through source reduction, reuse, recycling, and composting increase.

The county also has determined that its Status Quo Strategy could be implemented most cost-effectively if the county used the current county landfill until it closed the next year; built a transfer facility to transfer waste to a landfill in an adjacent county until new in-county landfill was built; and built and operated a new in-county landfill within seven years.

The projected total nominal system costs of this combination are shown in Figure 4. These per-ton costs include the costs to collect, transfer, and dispose of waste from Clark County.

Maximizing the cost-effectiveness of each option must be evaluated in one time period, but must be evaluated over time. Projecting costs will depend on a range of variables including: changing waste generation rates and composition characteristics, changing waste recovery goals, improvements in technology, the inflation of operating costs, and the anticipated impacts of changing regulations. Such cost projections should also include costs for landfill closure, equipment salvaging, and other end-of-project costs that will be incurred. Typically, cost projections are made for 5, 10, and sometimes 20 years.

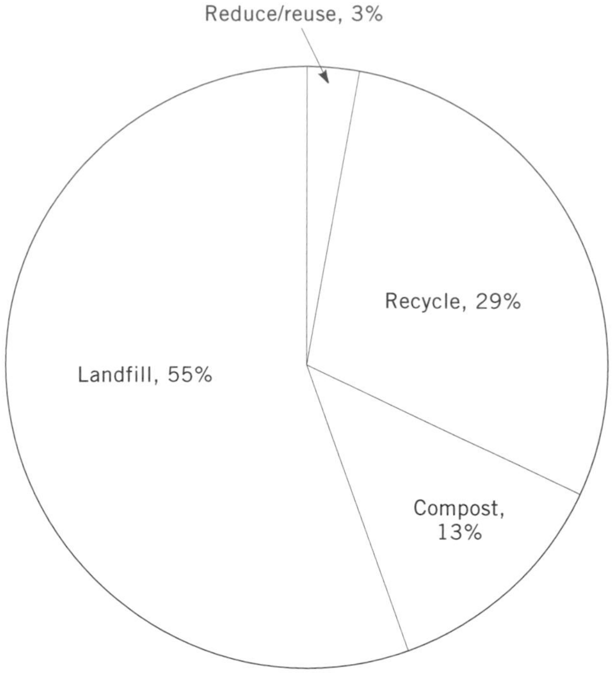

Figure 3. Mix of Waste management options for Integrated Strategy.

Green County planners projected nominal per-ton system costs (collection + processing/transferring + marketing/disposal) for each strategy. They determined that the Integrated Waste Management Strategy would incur higher waste recovery than landfill costs per ton in the short run, and relatively lower waste recovery costs

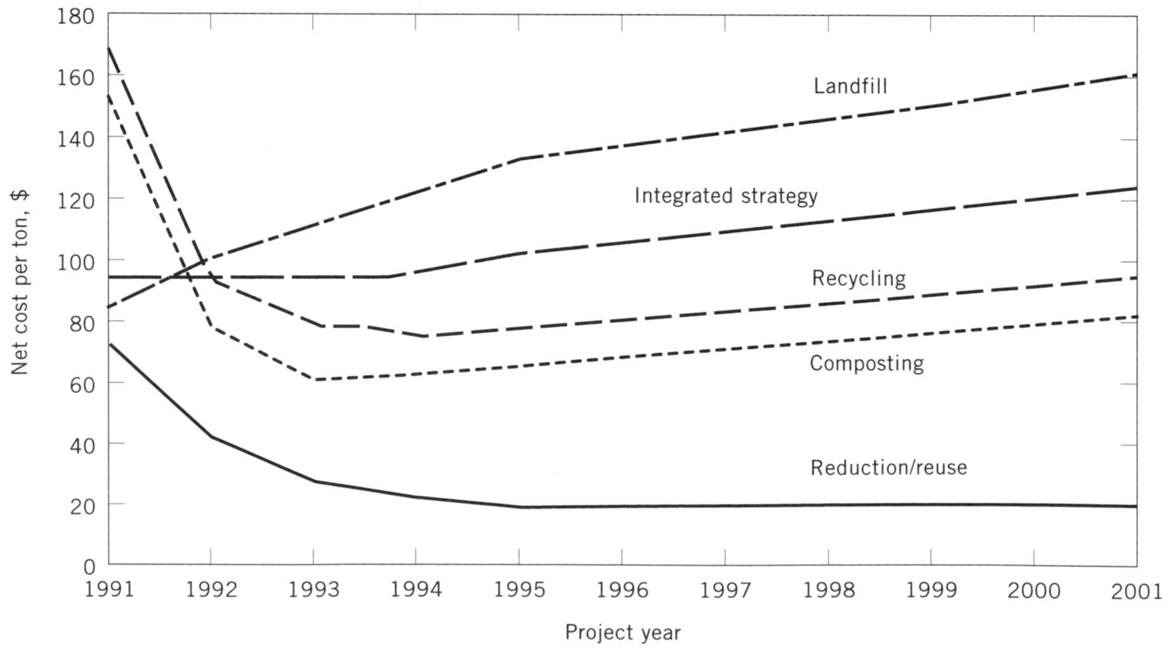

Figure 4. Nominal net costs for components and whole of Integrated Strategy.

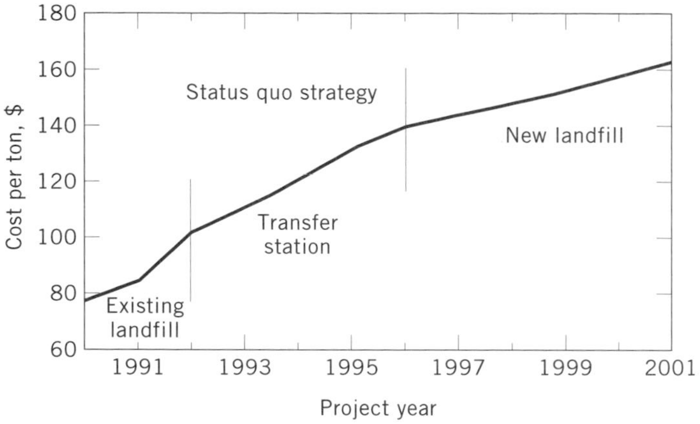

Figure 5. Projected nominal costs per ton for refuse collection and landfill disposal under Status Quo strategy.

per ton in the longer term. Landfill costs would continue to rise during the 10-year period considered.

With this information, the weighted average of options per-ton costs of the Integrated Strategy were projected over the planning period. These trends are reasonably typical, particularly because they reflect the cost-effectiveness gained as recycling, composting, and waste reduction/reuse quantities increase.

Green County planners also estimated costs per ton to implement the Status Quo Strategy over the same period, as shown in Figure 5.

The projected rapid rise in waste disposal cost because of projections could decrease to some extent if landfill operators lower tipping fees because of (a) unforeseen competition from other waste disposal sites, or (b) significantly lower quantities of refuse disposed because of recycling and composting. Waste disposal costs could increase if the new landfill is not built, or if additional regulatory requirements add to landfill costs.

CHOOSING THE MOST COST EFFECTIVE STRATEGY

Finally, the projected costs of the competing strategies can be compared over the same time period (Fig. 6).

It is likely that the relative costs of the strategies will shift over time. If this is true (as it is in Figure 6), the planner should determine the net present value (NPV) of each strategy over the time period. Net present value analysis allows the planner to directly compare net costs in Year 1 with net costs in later years through the equalization of the value of money over time. Once this is done, the planner may then choose the strategy with the greatest net present value.

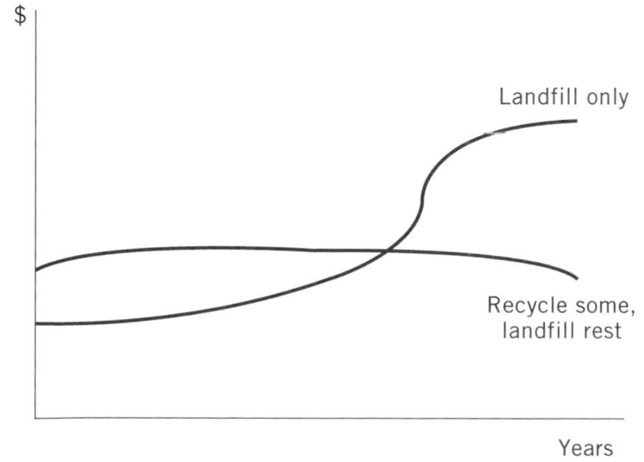

Figure 6. Changing nominal costs for two strategies.

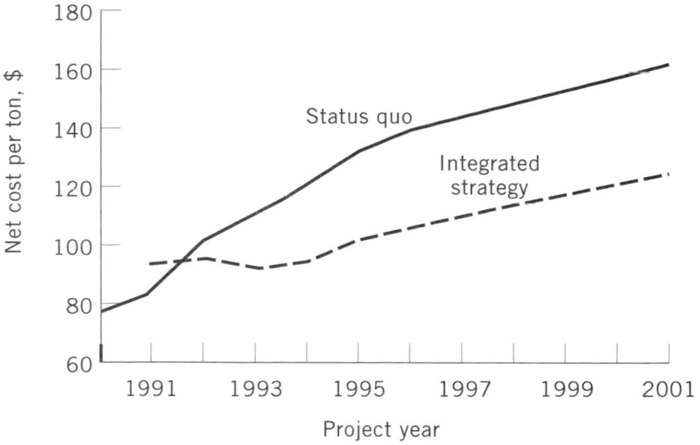

Figure 7. Nominal net costs for the status quo and Integrated Strategies.

When planners in Green County compared the Integrated Waste Management Strategy with the Status Quo Strategy, it was not absolutely clear which strategy should be chosen (Fig. 7).

It is hypothesized that the Integrated Strategy would be at least 10% less costly in constant dollar terms than the Status Quo Strategy, in part because of the higher initial costs of the Integrated Strategy. To confirm this hypothesis, the planners conducted a net present value analysis ($r = 10\%$) for both strategies over the project period (1991–2001). The NPV results of the analysis were Status Quo Strategy, $91,394,730; Integrated Strategy, $75,383,058. Thus the hypothesis was confirmed. The Integrated Waste Management Strategy was projected to eventually save Green County $16 million dollars, or 17% in present-value dollars over the project period.

CONCLUSION

Using basic planning theory and standard economic models, the planner can conclusively lend critical assistance to the solid waste management decision-making process. With such models, community planners, managers, and politicians may effectively consider a range of interdependent waste management alternatives and then focus on the optimal combination of waste management programs that best meet the community's needs.

BIBLIOGRAPHY

Reading List

Biocycle Journal of Composting and Recycling, JG Press Inc. Published monthly by JG Press Inc. (215-967-4135).

Recycling Today; The Business Magazines for Recycling Professionals, GIE Inc. Published monthly by GIE Inc. (1-800-456-0707).

Resource Recycling Magazine; North America's Recycling Journal. Published monthly by Resource Recycling, Inc. (503-227-1319).

Solid Waste and Power; The Magazine of Waste Management Solutions. Published bimonthly by HCI Publications, Inc. (816-931-1311).

Waste Age, published monthly by the National Solid Waste Management Association.

U.S. Congress, Office of Technology Assessment. *Facing America's Trash: What's Next for Municipal Solid Waste.* OTA-O-424 U.S. Government Printing Office, Washington, D.C., Oct. 1989.

G. Tchobanoglous et al, *Integrating Solid Waste Management Engineering Principals and Management Issues,* McGraw-Hill, New York, 1993.

WASTE-TO-ENERGY TECHNOLOGIES

MARY K. WEES
HDR Engineering, Inc.
Omaha, Nebraska

Waste-to-energy, a general term for the combustion of municipal solid waste (MSW) with energy recovery, is also referred to as municipal waste combustion (MWC) or resource recovery. MSW includes garbage, trash, refuse, and yard debris that has been discarded as being spent or worthless and is destined for disposal from households, commercial establishments, and institutions. Waste-to-energy is one of several MSW management and disposal alternatives available (1). The benefit of waste-to-energy is that it reduces the volume of MSW requiring land disposal by approximately 90%. Waste-to-energy is generally more capital intensive than other solid waste management alternatives, but the value of the recovered energy is used to offset a portion of the expenses. Waste-to-energy technologies must provide for both environmentally sound MSW disposal and efficient energy production.

When considering the technical aspects of recovering energy from MSW, it is essential to recognize the differences between MSW as a fuel source and other typically used fossil fuels. MSW is a heterogeneous mixture of materials. The primary MSW components are paper, plastics, textiles, wood, rubber, glass, ceramics, ferrous metals, aluminum and other nonferrous metals, various chemicals and solvents, and organic materials in the form of food wastes and yard debris. Generally, the combustible portion makes up 70–80% of the waste stream (2). The composition of MSW changes constantly and is affected by general economic conditions, consumer habits, seasonal variations, and the recovery of selected constituents through recycling endeavors.

Recovery of energy from MSW was developed in the 1930s and 1940s in Europe: Today, waste-to-energy is used in Europe, the Far East, and the United States for MSW disposal. The varying nature of MSW has led to a wide diversity in the techniques for energy recovery. Many techniques are adaptations from combustion of the more conventional fossil fuels. Because of the unique characteristics of MSW, an exact analysis of the combustion process is virtually impossible. Consequently, there are few standards in waste-to-energy, and waste-to-energy is often referred to as an "art."

This article focuses on the most commonly used waste-to-energy technologies and design features. Technologies that hold promise for future development are also identified. The two generic waste-to-energy technologies are mass burn and processed waste combustion. These technologies are discussed below, together with the by-products of waste-to-energy: the energy recovered, air emissions, and remaining ash and residue.

See also AIR POLLUTION; FUELS FROM WASTE; INCINERATION; FUELS, SYNTHETIC; COGENERATION.

MASS-BURN TECHNOLOGY

Mass burn refers to the combustion of MSW as received, with little or no processing or removal of materials. Combustion occurs in furnaces that have been specifically designed to accommodate the heterogeneous characteristics of the waste stream. Mass burn is the waste-to-energy technology most widely used worldwide. The principal components of a mass-burn plant include the waste storage and feeding systems, combustion grates, and combustion units (furnace and boiler).

Waste Storage and Feeding Systems

Mass-burn systems typically have a large pit, or bunker, to store the MSW that has been collected from waste generators and delivered in a variety of collection vehicles. An alternative floor storage system, used primarily by processed waste combustion systems and some smaller mass-burn systems, is discussed below under Processed Waste Technology. The capacity of the storage pit is primarily a function of the disposal capacity (generally referred to as the throughput capacity) of the waste-to-energy plant. Disposal capacity is expressed in tons per day (tpd). The storage pit capacity is determined in view of the delivery schedule of collection vehicles (normally 5–6 days/week and 8–10 h/day, with no deliveries on holidays); the continuous nature of the combustion process (24 h/day, 365 days/year); seasonal variations in the generation of MSW; and the scheduled maintenance and forced outages of the combustion units. Typically, the storage pit capacity is designed to accommodate 3–5 days of MSW storage at the rated plant disposal capacity. The storage pit design takes into consideration a variety of factors, including the type of collection vehicles that will use the plant, the number of spaces allocated for vehicles to simultaneously discharge their waste loads (which influences traffic congestion at the plant), the MSW density, and protection against groundwater infiltration into the storage pit. The storage pit area, and often the adjacent vehicle maneuvering area, is fully enclosed, and a slight negative pressure is maintained to minimize the release of odors. Combustion air is drawn from the storage pit area to produce this negative-pressure condition. An overhead traveling bridge crane system spans the entire length of the pit. The crane system is used to mix the waste in the pit to provide more uniformity and manage storage space by stacking waste against the pit walls. It also removes large, bulky, or noncombustible materials such as furniture or appliances.

Most mass-burn plants use a gravity feed chute to introduce MSW into the combustion unit. The crane system removes MSW from the storage pit and deposits the MSW into the gravity feed chute. The feed system must be designed to prevent burnback from the furnace and provide an adequate air seal for the furnace. Typically, the chute is water cooled and the chute and hopper design configuration prevents the MSW from bridging or jamming. The bottom of the feed chute often has a cycling ram to control the waste feed rate onto the combustion grates. Automation of the ram feed system helps maintain efficient combustion.

Combustion Grates

Because MSW is heterogeneous, its combustion results in an uneven heat release. This affects the efficiency of the combustion system, which in turn impacts air emissions and energy recovery. To minimize the impacts of a heterogeneous fuel, various waste-to-energy combustion grate systems have been specifically designed to agitate the MSW during the combustion process and thereby provide a more uniform and thorough burnout. Combustion grates are usually inclined and are designed to tumble, turn, and move the waste through the combustion chamber of the furnace. Many mass-burn system designs incorporate various combustion zones in the furnace for drying, volatization, ignition, and burnout. Varying the grate speed and combustion air distribution within these zones enhances combustion control.

Combustion grate designs commonly used in mass-burn systems include reciprocating grates, rocking grates, step grates, cascading grates, reverse reciprocating grates, drum or roller grates, and rotary kilns (3). In addition to agitating the waste, the combustion grate should be designed to provide a relatively uniform depth of bed and combustion surface area. A hydraulic system is typically used to provide grate motion. Combustion air is introduced through the grates as underfire air in a manner that evenly distributes combustion air through the bed and cools the grates. Grates may be made from special alloys with heat-resistant properties. Potential problems in the design of combustion grates include blockage of the air openings by small ash particles or melted metals, abrasion of moving grate parts, and wear of hydraulic mechanisms.

Combustion Units

Waste-to-energy combustion units generally fall into two categories: field-erected and modular systems. Field-erected systems are used primarily for large-scale facilities [226.8 t/day (250 tpd) and larger], whereas modular systems are used primarily for smaller scale facilities and are normally available in module sizes of 22.3–113.4 t/day (25–125 tpd). As the names imply, field-erected systems are built on site, but modular systems are shop assembled and shipped to the construction site as modules for installation.

Field-Erected Combustion Units. There are two basic designs for field-erected combustion units: refractory systems, which have a refractory furnace chamber and a separate waterwall boiler, and waterwall systems, which have an integrated refractory-lined furnace and waterwall boiler. The primary difference between the two is the boiler location. Refractory systems have a boiler located downstream of the combustion chamber; waterwall units have a furnace–boiler combination constructed with water tube membrane walls and boiler radiant sections that are an integral part of the combustion chamber walls. Waterwall systems typically have a slightly higher thermal efficiency than refractory systems.

Refractory is provided in areas where the furnace is subject to flame impingement. In refractory systems, the refractory insulates the furnace and limits heat loss. In waterwall systems, the refractory controls corrosion and erosion of the waterwall tubes in the furnace section. Refractory is sensitive to abrasion and chemical attack. Slagging can be controlled through proper air flow within the furnace.

Combustion takes place under stoichiometric conditions in the furnace; typically, 80–100% excess air is used. Key considerations in the design of combustion units include the fuel composition and combustion volume required, the interaction of the flame and hot gases within the combustion chamber, the physical configuration of the combustion chamber, and the quantity and distribution of combustion air. The design must provide for the maintenance of adequate combustion temperatures and mixing of gases. Failure to maintain combustion temperatures of at least 760°C throughout the furnace may result in cold spots within the furnace and incomplete combustion, leading to the formation of carbon monoxide and unburned hydrocarbons. Auxiliary fuel is used for startup as well as for maintaining control of furnace temperature. Combustion air is also introduced above the fuel bed as overfire air and is often preheated. The quantity and distribution of the overfire air assist in turbulent mixing of the combustion gases while also promoting complete combustion and even temperature distribution. Failure to maintain proper combustion conditions such as even distribution of fuel or air results in oxygen-deficient burning and a reducing atmosphere. This reducing atmosphere in combination with the presence of chlorine and other chemicals, particularly metal salts, results in highly corrosive conditions. Because of the corrosive nature of the combustion gases, refractory coatings and special alloy materials are used for the water tubes (3).

Hot flue gases from the furnace flow through the boiler section of the combustion unit. Heat is absorbed by the waterwalls and boiler tube bundles in the convective passes of the boiler. One or more empty passes facilitate heat transfer and reduce the flue gas temperature. Depending on the boiler design, an air-preheating or economizer section may be located near the boiler exit and a superheater section may be located near the boiler entrance. Screen tubes may also be used to protect the superheater tubes from corrosion and erosion caused by high flue gas temperatures and velocities. Because of the heavy ash loading in the flue gases, tube spacing and gas velocities must be carefully analyzed to minimize ash fouling and erosion of tube surfaces. Sootblowing or rapping is commonly done to clean the tube surfaces and prevent bridging.

Modular Combustion Units. Modular combustion units are basically refractory-lined furnaces with two combustion chambers: primary and secondary. Some modular system designs operate in excess-air conditions in the primary chamber; others use starved-air (less than stoichiometric) conditions in the primary combustion chamber. The secondary chamber completes the combustion of the gases, normally with the assistance of auxiliary fuel. Heat recovery occurs after the primary and secondary chambers in a waste heat recovery boiler. Energy transfer is by convection only, without the benefit of radiant heat recovery in the furnace. Consequently, the thermal efficiency of modular systems is typically lower than that of field-erected systems. Because modular systems are often batch fed rather than the continuous-feed operation of the field-erected systems, they require different waste feed systems. Also, frequent startup and shutdown can result in rapid deterioration of refractory, primarily from temperature variations and reducing atmospheres.

PROCESSED WASTE TECHNOLOGY

The basic goal of the processed waste technology is to reduce the size of the MSW and remove noncombustible materials in order to provide a more homogeneous, easier-to-handle solid waste fuel. The main product of processed waste systems is a prepared fuel commonly referred to as refuse-derived fuel (RDF). The more uniform nature of the RDF leads to better control of heat release during combustion and greater thermal efficiency than achieved by mass-burn systems. This provides the potential to use existing conventional fuel combustion systems and thereby achieve cost savings. RDF can be used either as a supplemental fuel in existing industrial and utility combustion systems that have been modified or as the primary fuel in a dedicated combustion unit. The requirements of the selected combustion system will determine the physical size and characteristics of the RDF.

Processed waste technologies involve mechanical, and sometimes manual, separation and sizing of MSW. This provides an opportunity to recover valuable or reusable materials such as aluminum, ferrous metals, and glass from the MSW during processing. Processed waste technologies have also been advocated as a means of removing waste constituents that can appear in air emissions or in the ash and residue. For example, battery removal can reduce the amount of mercury or cadmium in the waste stream.

Processed waste systems serve two separate functions: waste processing and RDF combustion. These activities may be located together at one site or waste processing may occur at a site separate from the RDF combustion system. The following describes various processing and RDF combustion systems.

Processing Systems

The basic processing techniques include some type of size reduction or shredding of the MSW to produce a more homogeneous, relatively uniform, and small particle size. Screening or sorting techniques are used to separate combustible and noncombustible materials, resulting in a concentration of the combustible fraction. MSW processing systems can vary in complexity and in the degree of processing performed. The level of processing is a function of the RDF combustion system requirements. To provide an industrywide standard for the classification of various types of RDF, the American Society for Testing and Materials has developed RDF classifications (4). The degree of processing is measured in part by the percentage of RDF passing through various sizes of mesh screens. The two principal classifications are a coarse RDF (designated as RDF-2) and a fine RDF (designated as RDF-3) with most of the metals, glass, and other inorganics removed. In terms of weight, 95% of the RDF-2 passes through a 38.7-cm^2 (6-$in.^2$) mesh screen, and 95% of the RDF-3 passes through a 12.9 cm^2 (2-$in.^2$) mesh screen.

Various arrangements for MSW storage and retrieval have been developed for processed waste systems. The most common is a large, enclosed tipping floor rather than a storage pit and crane system as used in mass-burn systems. Collection vehicles discharge waste loads onto the tipping floor. This allows visual inspection of the MSW and removal of bulky or oversized materials that may not be processible in the system or could be a potential hazard during waste processing, as in the case of explosive materials, for instance. The tipping floor capacity is based on some of the same considerations as apply to storage pit sizing with respect to the frequency of waste deliveries, design capacity of the processing system, and vehicle unloading requirements. Front-end loaders transfer the material from the tipping floor to the processing system equipment. They also move and stack MSW for storage management during peak delivery periods.

There are three general categories of waste processing systems: dry processing, wet processing, and biodegradation. These categories are discussed below.

Dry Processing. Dry processing usually involves various stages of waste sizing and sorting, with different sequences of mechanical equipment in a processing line arrangement. Multiple processing lines provide redundancy in the event of mechanical failure and facilitate equipment maintenance. Dry processing systems are typically sized based on an operating schedule of less than 24 h/day and less than 7 days/week. Conveyors transfer the MSW from the tipping floor and between various pieces of processing equipment. Considerations in conveyor design include maintaining a level waste depth, minimizing transfer points to avoid excessive spillage, and providing for heavy-duty operation. Conveyors also allow for visual inspection of the MSW.

Generally, the first activity in the processing line is to open any bags or bundles. Bag breaking can be done by a number of techniques, including manual slitters positioned along the infeed conveyors, spikes in a trommel, rotating knives, or flail shredders. The next step is generally a size separation process designed to separate the small noncombustible content (grit, sand, glass, and metal containers) from the remaining waste stream. Early removal of small noncombustible material reduces abrasion and wear on subsequent processing equipment. The size separation equipment may also occur at additional places in the processing line after size reduction, depending on the RDF requirements. Size separation equipment may consist of trommels, disc screens, vibrating screens, air density separation, or a combination of these approaches.

Oversized materials are generally conveyed to a shredder for size reduction. Because of the potential for explosive materials such as pressurized canisters or gas cans to be present in the waste stream, the shredding operation generally includes explosion suppression and containment systems. At various points along the process line, magnetic separation is generally incorporated to remove ferrous metals from the RDF fraction. Throughout the processing system, there are dust collection devices to control airborne particulates. The undersized fraction from the sizing processes may be subjected to various material recovery processes. These may include electrostatic separation, eddy current separation, vibrating tables, optical sorting, froth flotation, and heavy media separation (5).

The sized and separated RDF fraction requires storage prior to burning. This storage may take the form of bins, bunkers, or large storage buildings using pneumatic systems, conveyors, or cranes to remove the material. The RDF storage and retrieval system must take into consideration the low bulk density of the RDF and its tendency to compress, making it difficult to remove.

RDF may also be densified using various types of pelletizing, cubing, extrusion, or briqueting machines. The resulting fuel is commonly referred to as d-RDF. In the densification process, the RDF may be subjected to high pressure and sometimes high temperature with or without a binding medium. The benefits of densification include easier storage and handling. In addition, the higher bulk density and the shape of the densified fuel pellets allow the d-RDF to be used as a supplemental fuel in conjunction with certain existing coal handling equipment.

Wet Processing. Wet processing systems are an adaptation of the hydrapulping techniques used in the pulp and paper industry. Wet processing involves several preliminary dry processing activities to screen the MSW initially. The screened fraction, containing the concentration of organics, is introduced together with water into the hydrapulper, which functions like a blender. When the material is pulverized, the resulting slurry passes through perforated bottom extraction plates and centrifuges to separate the organic and nonorganic fractions. Wet processing reduces the potential for explosions during shredding, and the fuel is cleaner and more homogeneous than RDF produced from dry processing systems. However, the moisture inherent in a wet processing system must be removed before the fuel is burned, and the heat required to dry the fuel lowers the net energy recovery efficiency. This processed waste system is no longer used in the United States.

Biodegradation. Biodegradation is a biochemical process that converts the organic fraction of MSW into methane and other biogases. A dry separation process is used to produce a finely shredded, concentrated fraction of biodegradable materials. This fraction is then mixed with nutrients, recycled wastewater from the process, and makeup water as necessary to achieve a specific proportion of solids to liquids. Sewage sludge is often used in the process. The slurry is fed to an anaerobic digester, where bacteria break down and decompose the organic matter, producing a biogas. The biogas is saturated with moisture and contains both methane and carbon dioxide, thus requiring conventional gas scrubbing techniques prior to use as a fuel gas. The bacteria used in the process are anaerobic and require air-tight reaction vessels. Factors influencing the system include proper nutrients and pH, absence of toxins, proper temperature control, and adequate residence time for the bacteria to function. Although several anaerobic digestion technologies have been tested on a demonstration level, biodegradation processes using MSW are not in commercial use at this time (6).

RDF Combustion Systems

RDF combustion systems basically fall into the following categories: spreader-stoker firing, suspension firing, fluidized-bed incineration, and pyrolytic conversion.

Spreader-Stoker Firing. Though typically used for the dedicated combustion of coarse RDF, spreader-stoker firing is also applicable when a coarse RDF is cofired with coal or other fuels. In spreader-stoker firing, RDF is fed into the combustion unit through air-swept spouts or pneumatic distributors. Some of the RDF is burned in suspension, while the heavier particles fall onto the grate where combustion is completed. Uniform distribution of the RDF into the combustion unit and onto the grate is required for efficient combustion. Standard traveling, reciprocating, or vibrating grates may be used. Most spreader-stoker combustion units have a waterwall design, with refractory installed in the area above the grates to protect the tubes from corrosion and erosion. Conventional boiler design is appropriate, with combustion air distributed both as overfire air above the fuel bed on the grate and as underfire air through the grate. The overfire air provides the required turbulence and mixing of combustion gases to ensure burnout of suspended particles and volatile matter. Approximately 40–50% excess air is required for combustion. As with mass-burn combustion systems, special alloys are used in the water tubes along the grates to resist the corrosive effect of the combustion gases, and the boiler section must have adequate provisions for ash cleaning to reduce plugging and fouling of the tubes. Older, existing fossil fuel combustion units can be modified to accept RDF as a fuel. The use of existing combustion units may provide an opportunity for cost savings.

Suspension Firing. Suspension firing involves the cofiring of RDF in suspension with pulverized coal. RDF normally supplies 10–20% of the total heat input. Suspension firing requires a more highly refined and processed RDF than spreader-stoker firing. Although there are a variety of fuel feed techniques, the RDF is typically introduced into the combustion unit by a pneumatic system with a surge bin. Either heated combustion air or ambient air is used to blow the RDF into the boiler at various levels around the pulverized coal burners. A dump grate is required in the bottom of the boiler to allow any RDF not completely combusted in suspension to burn out before being discharged. Combustion air may be added above and below the grates to ensure complete combustion of the heavier particles. Suspension-fired systems have more efficient air distribution than spreader-stoker systems, and excess air requirements are in the range of 10–20%. Because of the high gas velocities in the combustion unit, suspension-fired systems have a higher percentage of ash entrained in the flue gas. Similar to the case of spreader-stoker firing, the boiler design must take into consideration the corrosion, erosion, and ash fouling and plugging. Suspension firing is most applicable when a market exists for a supplemental fuel to be burned in an existing boiler designed for suspension firing of coal. Typically, utility plants are such a market.

Fluidized-Bed Combustion. Fluidized-bed combustors have a bed of thermally inert material, such as sand and/or limestone, which is kept suspended in the combustion unit through the action of fluidizing air distributed below the bed. RDF can be introduced into or on top of the bed, mixed by the fluidizing air and bed material, and combusted in the turbulent bed of inert material. The turbulence of the fluidized bed allows the combustion to take place at a lower temperature than in conventional combustion systems without a significant effect on the combustion efficiency. This is beneficial in reducing certain air emissions, particularly sulfur dioxide and nitrogen oxides. RDF is typically not fired alone in fluidized-bed combustion systems; coal or wood are cofired with the RDF to help in maintaining stable firing conditions. Although fluidized-bed combustion has been used for a variety of fuels, experience with RDF is limited. The level of processing required may vary depending on the design characteristics of the specific fluidized-bed technology used. Problems may occur if the RDF contains excessive amounts of glass, which can cause agglomeration of bed particles. Accumulation of noncombustible objects may lead to defluidization of the bed. Ash and residue from the RDF may also change the physical and chemical composition of the bed, thus requiring continuous bed replacement (7).

Pyrolysis. Pyrolysis systems convert organic materials in an oxygen-deficient environment into a combustible gaseous or liquid product and a solid, carbon-rich residue. Although unprocessed MSW may be subjected to pyrolytic conversion, a more homogeneous feedstock is necessary to improve operation efficiency and product quality. MSW must generally be finely shredded and screened to produce a high-quality RDF material that has the majority of glass, sand, grit, and metals removed. The composition and yield of the pyrolysis products can be varied by controlling operating parameters such as feedstock composition, pressure, temperature, time, and feedstock particle sizing and by using catalysts or cofired auxiliary fuels. The variable nature of RDF composition complicates control of the specific chemical reactions that take place. Because of the problems associated with controlling the chemical reactions and maintaining product quality, the marketing of gaseous and solid pyrolysis products has not been successful, and pyrolysis is not in commercial operation in the United States (8).

ENERGY RECOVERY

Energy from mass-burn systems or RDF combustion in conventional boilers is generally recovered in the form of hot water, steam, or electricity. Hot water is used primarily in district heating applications; steam may be used in various industrial processes (including heating and cooling) or passed through a conventional turbine-generator to produce electricity. There is limited experience with the energy recovered from biological and chemical processes such as anaerobic digestion or pyrolysis. It is generally anticipated, however, that the nature of the RDF fuel stock would require some type of fuel cleaning or scrubbing to provide commercial-grade products. The following discussion focuses on the energy recovered by mass-burn and RDF combustion systems.

The amount of energy recovered from MSW is a function of the energy content of the incoming waste and the recovery efficiency of the combustion technology. The energy content of MSW will vary based on the individual waste components and their moisture content. As a rule of thumb, the higher heating value of unprocessed MSW will vary from 3800 to 5500 Btu/lb (8833–12,784 J/g) of MSW, with an average higher heating value of 4500 Btu/lb (10,460 J/g). MSW processing will result in removal of a portion of the waste materials (ranging from 20–40%, depending on the level of processing) and subsequent increase in the higher heating value of the remaining RDF. The higher heating value of RDF can vary from 5200 to 7500 Btu/lb (6718–9681 J/g) of RDF.

MSW combustion technologies have varying levels of energy recovery efficiency. Field-erected mass-burn units have an expected thermal efficiency of 65–74% for waterwall systems and a slightly lower range of 60–70% for refractory systems. Modular mass-burn units vary widely, with energy recovery efficiencies ranging from 40–60%. Dedicated RDF combustion units can be expected to have an energy recovery efficiency of 70–75%. Typically, 0.9–1.4 kg (2–3 lb) of steam per pound of unprocessed MSW can be generated in mass-burn and processed waste systems. Steam conditions are limited by flue gas temperatures and corrosion concerns. Steam conditions can range from 315°C, 42 kg/cm^2 (600 psig), for modular mass-burn systems to 454°C, 60 kg/cm^2 (850 psig) for field-erected mass-burn systems.

The net electrical output is affected by in-plant power consumption. Mass-burn systems have steam or electrical demands to run motors, fans, and other plant equipment. Process waste systems use various degrees of electricity, depending on the specific equipment and the degree of processing. Plant-specific factors include the use of various air pollution control technologies and the use of air-cooled versus water-cooled condensers. Given the fact that energy recovery will vary widely, the net electrical production for various technologies ranges from 300 to 400 kW·h/ton of incoming MSW for modular mass-burn systems, from 400 to 600 kW·h/ton for field-erected mass-burn systems, and from 420 to 525 kW·h/ton for processed waste systems (5,9).

AIR EMISSIONS

Combustion of MSW results in the release of products of combustion, some of which are considered air pollutants. If uncontrolled, these pollutants can pose a potential risk to human health (10). The primary air pollutants of concern for waste-to-energy technologies are particulate matter, acid gases, oxides of nitrogen (collectively referred to as NO_x), carbon monoxide (CO), volatile organic compounds (VOCs), polychlorinated dibenzo-*p*-dioxins (PCDDs) and polychlorinated dibenzofurans (PCDFs) (collectively referred to herein as dioxins and furans), and toxic metals. Waste-to-energy air emissions are controlled by using well-designed and efficient combustion equipment, installing air pollution control equipment, and employing good combustion practices during operation. In the United States, waste-to-energy air emission concentrations are governed by the Clean Air Act of 1970, as amended, and by regulations promulgated by the U.S. Environmental Protection Agency (EPA) and state environmental agencies (11). Environmental regulations over the past 20 years have led to increasingly stringent emission limits for various pollutants.

Air Pollution Control Equipment

Equipment for controlling most air emissions is located downstream from the boiler in the combustion units. State-of-the art equipment arrangements include spray dryer absorbers and dry scrubbers in combination with baghouses, which are also referred to as fabric filters, or electrostatic precipitators (ESPs); ESPs in combination with wet scrubbers; in-duct sorbent injection in combination with a baghouse; and furnace injection in combination with a baghouse or ESP. The most commonly used air pollution equipment in new waste-to-energy plants in the United States is a spray dryer scrubber and baghouse arrangement. Gases leaving the air pollution control equipment are dispersed through a tall stack. The stack height is determined primarily by building configuration (height and width) to keep the emissions plume from becoming entrained within air flow over the building. Stack height may be optimized by mathematical modeling of dispersion that takes into consideration prevailing wind, surrounding topography, ambient air conditions, and the amount of mixing that will occur between the stack exhaust and the surrounding atmosphere.

Particulate Matter

Uncontrolled particulate matter is made up of completely and incompletely burned materials suspended as fly ash and condensible particles formed from the volatilization and recondensation of certain chemical constituents in the MSW. In waste-to-energy combustion systems, the use of underfire and overfire air to promote complete combustion results in the entrainment of fly ash. RDF combustion that occurs in suspension tends to entrain more particulate matter in the flue gas than does a mass-burn system. Proper design of the furnace and grate system is important in reducing fly ash entrainment; however, it is difficult to maintain exact control over all areas of the furnace and grate system at all times, and particulate generation is unavoidable.

Particulates can be controlled through ESPs, baghouses, and wet scrubber systems. ESPs are generally reliable pieces of equipment and can accommodate flue gas temperatures in the general range of 121–454°C. ESP removal efficiency is a function of the collection plate area, gas velocity through the collectors, number of fields, particle size distribution and properties, and in-field resident time. Baghouses are not as sensitive to small variations in gas volume, particle size, and composition as are ESPs and can accept surges in gas flow and particulates with no significant increase in particulate emissions. Baghouses also have a greater control efficiency for submicrometer particle sizes, which assists in the control of trace metals and complex organics. In baghouses, problems can occur with metal corrosion and sparks that cause fires. These problems can be mitigated by employing a dry scrubber upstream of the baghouse to effectively control acid gases and reduce flue gas temperatures. Wet venturi scrubbers can also collect particulates but have generally not been applied to waste-to-energy in the United States because of their high energy consumption, large quantities of wastewater, and potential problems with acid corrosion. Particulate control exceeding 99% removal efficiency has been demonstrated, and particulate emissions below the federal regulatory limit of 0.015 grains per dry standard cubic foot for new waste-to-energy facilities are readily achievable.

Acid Gases

Acid gases resulting from the combustion of MSW include sulfur dioxide (SO_2), hydrogen chloride (HCl), and hydrogen fluoride (HF). Emission rates are a function of the concentrations of sulfur, chlorine, and fluorine in the MSW and of the combustion system design. Air pollution control equipment in the form of wet scrubbers or spray dryers is the conventional means of reducing acid gas emissions from waste-to-energy plants. A spray dryer is generally located immediately after the boiler exit. Spray dryers in waste-to-energy applications generally use lime as the reagent in either a dry power or a wet slurry. Removal efficiencies exceeding 80% for SO_2 and 95% HCl for new waste-to-energy plants are readily achievable. In addition, acid gas emissions can be lowered by removing MSW components with high sulfur, chlorine, and fluorine contents prior to combustion. This can be done by recycling prior to MSW collection and by various waste processing activities.

Nitrogen Oxides

NO_x emissions from waste-to-energy are the result of oxidation of nitrogen compounds that are in the combustion air at high temperatures (sometimes referred to as thermal NO_x) and in the fuel at lower temperatures (fuel

NO_x). Formation of NO_x is highly dependent on temperature and residence time in the combustion unit. Lower, uniform combustion temperatures and uniform mixing of air and fuel generally eliminate high oxygen concentration gradients and sharp temperature gradients that are conducive to NO_x formation. However, most waste-to-energy systems are designed for complete combustion of organic gases using high temperatures and longer residence times. Combustion control techniques for limiting NO_x formation include control of the grate speed, waste feed, and underfire and overfire air supply. Flue gas recirculation designs generally suppress thermal NO_x formation by reducing the temperature in the flame zone. In addition, NO_x emissions can be reduced by removing certain nitrogen-rich components of the MSW, particularly the organic fraction from yard wastes.

There are several air pollution systems that are effective in reducing NO_x emissions: selective catalytic reduction, selective noncatalytic reduction, electron beam irradiation, and copper oxide processes. Good combustion control and selective noncatalytic reduction, which involves injecting ammonia or urea into the furnace, are the methods commonly used to control NO_x in waste-to-energy plants in the United States. Operating data on the effectiveness of NO_x controls indicates reductions on the order of 40% can be maintained.

Carbon Monoxide and Volatile Organic Compounds

Carbon monoxide is the result of incomplete combustion of carbon in the fuel. Similarly, VOCs are emitted as a result of incomplete combustion of various forms of organic materials. Some VOCs are known carcinogens. In waste-to-energy systems, incomplete combustion may be caused by introduction of an overly wet or large load of waste in the furnace, inadequate air supply or improper air distribution, low combustion temperatures, or inadequate residence time. Control of CO and VOCs is a function of good furnace design, good combustion practices, and proper monitoring of combustion. Design features that mitigate the formation of CO and VOCs include optimal placement of overfire air jets, use of high-pressure overfire air to promote turbulence and complete mixing of gases, and control of the combustion air to maintain a minimum furnace temperature of approximately 982°C. Continuous monitoring of CO, carbon dioxide, oxygen, and NO_x is an important factor in providing proper combustion controls. Air pollution control equipment specifically designed for the control of CO and VOCs has not been incorporated into waste-to-energy plant design.

Dioxins and Furans

Dioxins and furans are generally formed during the combustion process but may also be present in certain constituents of MSW. Dioxins and furans can form due to incomplete combustion. Thus, they may occur in a waste-to-energy system where furnace designs allow combustion gases to follow cool pathways and avoid the turbulent combustion zones or during transient upset conditions caused by fluctuations in the MSW composition or loading. Dioxins and furans are effectively destroyed at high temperatures, generally above 815–982°C. The common practice for controlling dioxins and furans in waste-to-energy systems is to maintain a combustion temperature of 982°C for a residence time of 1–2 s. In addition, providing adequate combustion air for the control of CO serves to control dioxin and furan formation. Because dioxins and furans may condense onto fine particulate matter at lower temperatures, dry scrubbers with a baghouse also tend to enhance removal.

Toxic Metals

A portion of the lead, mercury, beryllium, cadmium, and various other metals present in trace quantities in MSW will be volatized and exist in the flue gas. Volatile trace metals in the flue gas can be expected to condense on smaller particles in the exhaust stream and be collected by the air pollution devices used to control particulates. The exception is mercury, which tends to remain in the vapor phase at flue gas temperatures above 200°C. Various techniques are being tested for their effectiveness in controlling mercury emissions from waste-to-energy systems. These techniques include wet scrubbing and injection of activated carbon-based adsorbents in the flue gas. In addition, removal of MSW components with high mercury contents, such as fluorescent lamps and mercury–zinc batteries, will reduce mercury emissions (12).

ASH AND RESIDUE

As a result of MSW combustion, two residuals are present; fly ash and bottom ash. Ash quantities and characteristics are a function of the MSW composition, waste-to-energy technology, operating conditions, and type of air pollution control equipment. In field-erected mass-burn plants, nearly all of the material received is processed through the combustion system. Combustion results in a 90% reduction by volume. The total ash product ranges from 15–25% of the incoming waste on a dry weight basis, or 20–40% on a wet weight basis. Bottom ash accounts for approximately 70–85% by weight of the total ash product, and fly ash quantities range from 15–30% by weight of the total ash product. Because modular mass-burn technologies typically experience higher levels of unburned material in the ash, the total ash product can be expected to range from 15–35% by dry weight. Processed waste systems will remove many of the noncombustible materials that would ultimately become bottom ash in a mass-burn plant. The total ash from an RDF system can be expected to range from 8–20% by dry weight per ton of RDF, depending on the level of processing performed. Spent reagent from dry scrubbers may increase the amount of ash by an additional 1–3% on a dry weight basis.

The chemical composition of ash and residue consists primarily of silicon, aluminum, iron and calcium oxides, sulfate, and chloride ions. Minor constituents include organic and inorganic compounds. The actual ash and residue composition can vary considerably, particularly for mass-burn systems. Heavy metals in the ash and residue include lead found in batteries, solder in cans, paints, ink pigments, and metal scrap; cadmium from coating and plating of metal equipment parts, electronics, and rechargeable batteries; and mercury from disposable batteries, calculators, thermometers, and cameras. Processed waste systems and certain recycling efforts can reduce the concentration of metals in the ash and residue. The fly ash component usually contains much higher concentrations of metals and dioxins and furans than the bottom ash. Generally, dioxins and furans and semivolatile compounds are relatively immobile in the environment, whereas lead and cadmium have the highest potential for leaching of any of the metal constituents (13).

Waste-to-energy ash- and residue-handling systems generally use water for cooling the ash and residue. Some system designs totally immerse the ash and residue in an enclosed water-filled quench tank. Inclined drag chain conveyors or ram ejectors continuously remove the ash and residue from the quench tank. Dry systems use a water spray to cool the ash and residue and reduce dust. Ash and residue may be processed to recovery materials or aggregate appropriate for road-building materials. Processing techniques can include screening, trommeling, crushing, magnetic separation, and heavy media separation to recover various materials. Processing reduces the amount of material requiring land disposal. Additional ash and residue treatment has the goal of immobilizing metals contained in the ash for environmentally safe disposal. Such treatments include encapsulation, cementation, vitrification, and fixation to solidify the ash and residue. Stabilization techniques include metal extraction and sorbent addition.

BIBLIOGRAPHY

1. U.S. Environmental Protection Agency (USEPA), *Decision-Makers Guide to Solid Waste Management,* USEPA, Office of Solid Waste and Emergency Response, EPA/530-SW-89-072, Washington, D.C., November 1989.
2. U.S. Environmental Protection Agency (USEPA), *Characterization of Municipal Solid Waste in the United States, 1992 Update,* USEPA, Office of Solid Waste and Emergency Response, Washington, D.C., July 1992.
3. W. D. Robinson (ed.), *The Solid Waste Handbook, A Practical Guide,* Wiley, New York, 1986.
4. American Society for Testing and Materials (ASTM), "Standard Definitions of Terms and Abbreviations Relating to Physical and Chemical Characteristics of Refuse Derived Fuel," in *1993 Annual Book of ASTM Standards,* Vol. 11.04, ASTM, Philadelphia, PA, 1993.
5. HDR Engineering, *Solid Waste Technology Assessment for City Public Service of San Antonio, Texas,* HDR Engineering, Omaha, Nebr., June 1990.
6. SRI International, *Data Summary of Municipal Solid Waste Management Alternatives,* Vol. X: *Appendix H–Anaerobic Digestion of MSW,* National Renewable Energy Laboratory, NREL/TP-431-4988J, Golden, Colo., October 1992.
7. SRI International, *Data Summary of Municipal Solid Waste Management Alternatives,* Vol. V: *Appendix C–Fluidized-Bed Combustion,* National Renewable Energy Laboratory, NREL/TP-431-4988E, Golden, Colo., October 1992.
8. SRI International, *Data Summary of Municipal Solid Waste Management Alternatives,* Vol. VI: *Appendix D–Pyrolysis and Gasification of MSW,* National Renewable Energy Laboratory, NREL/TP-431-4988F, Golden, Colo., October 1992.
9. SRI International, *Data Summary of Municipal Solid Waste Management Alternatives,* Vol. III: *Appendix A–Mass Burn Technologies,* National Renewable Energy Laboratory, NREL/TP-431-4988C, Golden, Colo., October 1992.
10. "Incineration 2000 Phase II Report: An Analysis of the Health Risks Associated with Emissions from Solid Waste, Sewage Sludge, Hospital Infectious Waste and Hazardous Waste Incinerators in the New York/New Jersey Metropolitan Region," New York State Department of Environmental Conservation, Division of Air Resources, Albany, N.Y., January 1993.
11. Clean Air Act, enacted by Public Law 90-148, as amended, 42 U.S.C. 7401.
12. Solid Waste Association of North America, *Mercury Emissions from Municipal Solid Waste Combustors: An Assessment of the Current Situation in the United States and Forecast of Future Emissions,* National Renewable Energy Laboratory, NREL/TP-430-5399, Golden, Colo., May 1993.
13. U.S. Environmental Protection Agency (USEPA), *Methodology for Assessing Environmental Releases of and Exposure to Municipal Solid Waste Combustor Residuals,* USEPA, Office of Research and Development, EPA/600/8-91/031, Washington, D.C., April 1991.

WASTE-TO-ENERGY ECONOMICS

E. Larry Beaumont
Beacon Tech Net, LLC
Jeffrey F. Clunie
R. W. Beck and Associates
Denver, Colorado

Waste-to energy (WTE) economics are complex, and the list of variables is long: availability of waste supply, heating value of the waste fuel, size of facility, annual tonnage throughput, type of technology, thermal efficiency of the technology, capital cost, operating and maintenance (O&M) cost, type of energy produced, market price of energy, type of material recovery from plant waste streams, type of financing, credit worthiness of the participants, and so on. Because of this complexity, there are few economic rules that can be universally applied to the wide range of WTE facilities. Each facility must be considered and evaluated on its own particular merits. Understanding the interdependence of the variables involved is the key to judging the economic merits of any proposed project.

WTE facilities are characterized by relatively high capital and O&M costs when compared to other solid-waste processing technologies, such as material recovery (recycling) and composting facilities, but they provide the greatest benefit if measured in terms of preserving landfill capacity. That is, for every ton of waste processed, WTE facilities have smaller volumes of residue after processing that require landfilling, with the reduction in volume typically approaching 90%. The ability to finance WTE facilities is generally enhanced when they have executed contracts for (1) long-term energy sales to a creditworthy customer(s), because marketing of other recovered products from the wastestream is more volatile; and (2) long-term "put or pay" waste disposal agreements with public and private entities responsible for waste disposal.

WTE facilities can be difficult to implement on a standalone, nonrecourse basis (a project financing basis) because of the potential for complexity of contractual arrangements and perceived financial risk on the part of the potential bond purchaser. However, they have served well as flagships of overall integrated solid-waste management systems (financed on the basis of a system financing) dedicated to landfill capacity preservation at the least total system cost (1). The basic difference between a project financing and a system financing is that in project financing, only the revenues of the WTE facility are pledged to repay bondholders. Thus, if the WTE facility fails to operate properly, the bondholders have no other source of payment. In a system financing, the revenues of the entire system, including WTE, recycling, composting, and landfill, are pledged to repay the bonds.

The following discussion addresses WTE economics on a project financing basis, but the effects of the economic variables on integrated systems, which feature recycling, composting, WTE, and landfill as complementary tools, should also be considered. See also Air pollution; Fuels from waste; Incineration; Fuels, synthetic; Cogeneration.

WTE ECONOMICS OVERVIEW

Given the complexity of issues affecting WTE economics and for ease of explanation, this discussion has been organized around the following two equations:

$$NDC = (OM + DS - PR)/TP$$

where NDC is net disposal cost (\$/ton); OM is operating and maintenance cost (\$/year); DS is debt service (gross) (\$/year); PR is product revenues, eg, energy, recovered materials, interest earnings on reserve funds, etc (\$/year); and TP is annual throughput (tons/year). The second equation:

$$RTF = NDC - OR$$

where RTF is required tip fee (\$/ton) and OR is other revenues, eg, subsidies, utility bills, tax payments, etc, to support the tip fee (\$/ton).

THROUGHPUT (TP)

The annual solid-waste throughput (TP) of a WTE facility, expressed in tons per year (tpy), is a function of the size of the facility, the annual capacity factor (ACF) (sometimes incorrectly referred to as "availability" if the WTE is operated at less than its maximum continuous rating or MCR), and the amount of waste actually delivered to the facility.

The size of a facility is usually expressed in terms of tons per day (tpd) of throughput capacity. Because the primary function of a WTE facility is to process and reduce the volume of solid waste, with energy generation as a valuable by-product, most WTE facilities are intended to operate 24 hours per day, 7 days per week; therefore, daily throughput capacity is usually expressed in terms of tons per 24-hour operating day. The capital cost of a WTE facility is often expressed as the cost per daily throughput capacity, or \$/tpd. Therefore, the estimated capital cost of a WTE facility must be greater if the proposed plant size is larger than needed to process the quantity of currently available waste (to provide for future growth, for example); or is lesser if the facility is smaller than the current size of the wastestream (to minimize capital cost and/or to provide incentive for increased recycling).

The ACF is expressed as a percent of annual throughput capacity (ATC), which is daily TP at MCR times 365 days per year. The difference between ATC and ACF is that ACF accounts for the amount of downtime required during a year for scheduled and unscheduled outages. For example, if a WTE facility size is 1,000 tpd, then

$$ATC = 1{,}000 \times 365 = 365{,}000 \text{ tpy}$$

This assumes that the WTE runs at 100% of its MCR for every hour of the year, without any downtime for scheduled maintenance or forced outages.

However, if the assumed ACF is 85% (after taking into account the time necessary for scheduled and unscheduled downtime), then annual throughput actually is

$$TP = ATC \times ACF = 365{,}000 \times 0.85 = 310{,}250 \text{ tpy}$$

In this example, as long as more than 310,250 tpy of waste are delivered to the facility, the TP is calculated as shown above. However, if total waste available is less than 310,250 tpy, then the annual TP will also be less than 310,250 tpy. The annual TP will then be equal to that amount of waste actually delivered to the WTE facility, as the facility is capable of processing more than is available.

Confusion often surrounds the term "availability" when discussing the TP capacity of a WTE facility; the concept of ACF should not be confused with the electric utility term "availability." The availability of an electric utility generating unit is defined as the hours per year available for operation, even if it is not actually operating. Therefore, an 85% availability in the electric utility industry does not necessarily mean that a facility can actually produce 85% of annual production capacity. For WTE facilities, ACF is the better measure of actual facility processing capability.

OPERATING AND MAINTENANCE (O&M) COSTS

Operating and maintenance (O&M) costs for a WTE facility include salary costs and benefits; outside contract maintenance services; office furnishings and supplies; spare parts; consumables; residue, bypass, and ash disposal; testing and lab fees; insurance; site lease payments; utilities; host community fees; permit renewals; administrative expenses; and taxes. Many, but not all, elements of O&M cost should be considered as having both fixed and variable components, such that annual O&M costs are the sum of fixed costs plus variable costs. For example, salary and benefits can be calculated as a fixed cost in dollars per year, based on the guaranteed annual TP of the facility (usually referred to as the guaranteed annual tonnage, or GAT). However, there may be a variable component of salary cost if the facility receives and processes more than its GAT, eg, employees working overtime in order to process additional quantities of waste. Variable costs are a function of varying throughput, so total salary cost might be expressed as

$$SC = FC + (VCR)(ITP)$$

where SC is salary cost (\$/year); FC is fixed salary cost (\$/year); VCR is variable salary cost rate above GAT (\$/ton); and ITP is incremental throughput above GAT (tons/year).

Fixed and variable O&M cost analysis is critical to understanding WTE economics because total annual O&M cost is a significant portion of net disposal cost (NDC). As a rough estimate, a $1 million per year O&M cost component is equivalent to a $10 million capital cost in terms of the annual cost of debt service over a 20-year amortization period at recent interest rates. For example, if the price per ton of lime in a typical air pollution control device should increase by $30 per ton, due to a change in market conditions, and the scrubber consumes 2,637 tons per year of lime in a 1,000 tpd facility (based on 17 lb of lime per ton of solid waste TP and 85% ACF), the increase in NDC is

$$\frac{(\$30/\text{ton})\,(2{,}637\ \text{tons lime/year})}{(1{,}000\ \text{tons MSW/day})\,(365\ \text{days/yr})\,(0.85\ \text{ACF})} = \$0.26/\text{ton}$$

If the base net disposal cost was $60 per ton, this one change would have caused a 0.4% increase in NDC.

The same potential impact on NDC is true for other elements of O&M cost. An unexpected increase in quantities of residue, ash, or bypass waste, or a significant change in disposal costs of these materials, has the potential to cause a significant change in NDC. The economics of WTE depend on carefully assessing the potential change in all variable expenses and their effects on NDC. Even when a contract includes a guaranteed O&M price, there are often "passthrough" costs, above the guaranteed amount, which can dramatically affect NDC as described above. Typical passthrough costs may include the cost of utilities, insurance, and residue disposal.

In addition, operating costs may also be impacted by future changes in law, such as a new requirement for additional pollution control equipment or changes in the disposal of residue ash. Such increases are typically passed along contractually to the municipal users of the facility.

PRODUCT REVENUES (PR)

WTE facilities usually produce energy in the form of electricity and/or steam; hot water is also generated in some countries for district heating. The resulting unit measures of energy production are kilowatt hours (kWh) of electricity generated per ton of waste processed, and thousand pounds (Mlb) of steam generated per ton of waste processed. Applying those units of measure to annual TP yields projected annual energy generation. The quantity of energy generation is in turn applied to the selling price of energy (cents per kWh of electricity or dollars per Mlb of steam) to determine annual energy revenues.

Most energy sales contracts are complex and are based on a wide variety of factors, including avoided production costs, long-range avoided demand and energy costs, on-peak and off-peak energy production, and least-cost alternative means of generation. For example, electric utilities will often differentiate the price they pay for energy between on-peak hours (often 8:00 a.m.–8:00 p.m. on weekdays) and off-peak hours, or during on-peak and off-peak months of the year. The electric rates may also be based on a fixed price, or they may fluctuate with the electric utility's avoided energy costs as established by the state energy regulatory commission. Variations in the price of electricity can significantly impact energy revenues and, therefore, on NDC. Regardless of the complexity of the energy sales contract, many agreements can be reduced to capacity and energy components for the purpose of computing energy revenues. Capacity generally refers to the capital cost a utility incurs to construct generating capability. Energy, on the other hand, refers to the operating expenses incurred in producing electricity from that installed capacity.

For example, if a utility has adequate or excess generating capacity for the foreseeable future, it may pay only for the energy component (kWh). Conversely, if generating capacity will be needed during the near future, the utility may also pay for the avoided capacity (expressed in kW) which does not need to be built because of the existence of the WTE facility. For electric generation, capacity refers to the size of the generating unit in kW, and energy refers to the total energy production in kWh. So the total annual energy revenue in dollars per year is

$$(\text{Capacity Payment} \times \text{Size}) + (\text{Energy Payment} \times \text{Production})$$

These factors can then be combined into a total unit energy revenue, expressed in dollars per kilowatt hour.

Depending on the terms and conditions of the power sales agreement, the magnitude of these payments may change for each year of the contract. Thus, an attractive net energy revenue formula in the early years of the project may be frontloaded with higher payments in the earlier years and decreasing revenue payments during the latter years of the project, at the time when O&M expenses may be increasing due to wear and tear. Similar concepts apply for steam sales. For contracts which are frontloaded, the energy customer will usually develop a "bank account" whereby the additional energy payments made by the energy customer in the early years of the project are "loaned" to the project but must be "repaid" in the latter years.

An important aspect of energy sales contracts, which may also impact O&M cost, is that whereas WTE facilities produce energy for sale when in operation, they also purchase power from local utilities for periods when they are not running. The net energy revenue, therefore, must take into account the capacity and energy charges which the WTE facility pays for purchasing power, which in some cases, may be quite high. If a facility does not achieve its guaranteed ACF, there can be a substantial penalty for increased purchased power caused by frequent WTE facility outages.

Energy revenues are also a function of the thermal efficiency of the WTE facility. For each ton of waste processed, there is an optimum level of energy production (usually expressed in kWh/ton of waste processed). Generally, larger WTE facilities are able to operate at higher steam temperatures and pressures, and therefore will often generate more electricity per ton of waste processed. If combustion operating parameters such as excess air and furnace exit gas temperature are not controlled properly, or if proper maintenance to keep heat transfer surfaces relatively clean is not performed, energy production can drop significantly.

Many WTE contracts involving operation of the facility include provisions for sharing of energy revenues between the owner of the facility and the operator of the facility, or between a governmental entity and the operator. If a governmental entity owns the facility but contracts out the operation, the contractor is often given a share of the revenues as an incentive to operate the facility as efficiently as possible. Electricity revenues are often shared on the basis of 10–20% to the operator, with the balance to the owner. If a private company owns and operates the facility, the host government may either be paid a host community fee based on a contracted-for dollars per ton of waste received at the facility, or the host community may be given a share of the electricity revenues.

Revenues from the recovery of other products will depend on the type of technology employed at the facility. Mass-burn facilities may realize revenues from the recovery of ferrous metal from the residue ash. Refuse-derived fuel facilities offer the potential for increased revenues from the recovery and sale of ferrous metal, nonferrous metal (primarily aluminum), and corrugated cardboard. If a WTE facility is to be linked with a material recovery facility, other materials such as glass, plastic, and newspaper may be recovered for sale. WTE facilities also have the technical capability to sell residue ash for use in the manufacture of asphalt or cement, or for use as a roadfill material, although the actual implementation of such programs has not yet been significantly realized. These opportunities for revenue enhancement depend on obtaining approval for use by state regulatory agencies.

In each case, the incremental capital and O&M costs that would be incurred to recover such additional materials must be analyzed in comparison to the potential revenues to be received from the sale of recovered materials with variable market value. (In some cases, the avoided disposal costs of the components as part of the residue stream may justify their recovery.) Some facilities have developed markets for enhancing recovered product streams to improve marketability; others have invested in ash treatment systems which produce aggregate or building blocks, but approval for commercial use of these products is still pending. Such approval by regulating authorities would be a major boost to WTE economics. Finally, in certain instances, decisions have been made to recover additional materials in order to meet a state recycling goal, in spite of the fact that it may not have been economical to do so.

FINANCING AND DEBT SERVICE (DS)

Annual debt service (DS) is a function of the amount borrowed, interest rate, term of loan or bonds, construction schedule, and contingency funding. Various financing mechanisms are available for the financing of WTE facilities and their impact on DS (2). However, such mechanisms will work and be available only if a logical basic underlying structure of the transaction exists and if the various potential areas of risk associated with the WTE facility have been properly addressed. Examples of potential areas of risk in the financing of a WTE facility and how they might be addressed are summarized in Table 1. Attempts are often made to address these areas of risk by utilizing some type of credit enhancement. However, WTE facilities in the greatest need of credit enhancement, because of some fatal flaw in the financing structure of the project, will unlikely be able to issue revenue bonds or to secure bond insurance or a letter-of-credit. On the other hand, WTE facilities that have been developed where creditworthy entities have assumed the risk associated with their areas of responsibility, have been able to avail themselves of the various financing mechanisms described above.

Method of Ownership

One of the initial financing decisions involves the method of ownership. Will the owner of the WTE facility be the public entity responsible for solid-waste management in an area, such as a municipality, county, or regional authority (public ownership); or will the WTE facility be owned by a for-profit private company that intends to operate the WTE facility (private ownership)? The method of ownership determines the types of financing mechanisms available.

Public Ownership. The types of financing available if the WTE facility is publicly owned include those discussed below and those listed in Table 2.

1. *Payment from the General Funds of the Public Sponsor.* It is possible, although unlikely, that the sponsor of the WTE facility may decide to pay for the capital cost of the WTE facility. This is unusual for two reasons: (1) the capital cost of a WTE facility usually exceeds the unencumbered balance then currently available in most general funds; and (2) moneys in general funds are being used to pay for other projects such as schools, public buildings, and roads; projects that do not lend themselves to the creation of a separate enterprise fund.

2. *General Obligation Bonds.* Communities may issue general obligation notes or bonds to pay for the capital cost of their WTE facilities. This financing mechanism generally represents the lowest level of financial risk to an investor and, therefore, should enjoy the lowest interest rate. It is also usually the easiest type of bond to issue in terms of issuance costs and disclosure requirements.

Table 1. Components of a WTE Facility Financing That Must Be Addressed

Item of Risk	How to Address
Waste or recovered material supply	Under long-term contract on a put-or-pay basis with a creditworthy entity
Competitive cost with alternative means of same service	Flow control; subsidy
How revenues are collected	Tipping fees; utility bill; property tax bill
Construction cost overrun	Guaranteed capital cost; parental guarantee; performance bond
Operating and maintenance cost overrun	Guaranteed operating expense; passthroughs; parental guarantee
Useful life of the WTE facility	Utility grade equipment; operate at manufacturers' recommended level; preventive maintenance; scheduled renewals and replacements
Loss of market for recovered materials	Formula for payment based on meeting revenue requirement
Loss of environmental operating permit	Contractual requirement with private owner/operator
Future change in law	Usually paid by the party receiving the service; formula for payment based on meeting revenue requirement
Change in composition of waste stream	Flexibility in design of the WTE facility and formula for payment based on meeting revenue requirement
Performance guarantees: Throughput Recovery rates Utility consumption Residue Quality	Contractual requirement with private owner/operator

The reasons why general obligation bonds are not used more often for the financing of WTE facilities include: (1) most communities have limitations on the amount of general obligation bonds that they are allowed to issue and often use their allotment for other capital projects, which do not have the potential to be supported by an enterprise fund; (2) the cost of the solid-waste management service is supported by the entire taxpayer base on the basis of a tax assessment, rather than by the actual users of the solid-waste management system on the basis of the amount of solid waste requiring disposal; and (3) some public entities responsible for solid-waste management, such as regional solid-waste authorities, are not authorized to issue general obligation bonds.

3. *Revenue Bonds.* The issuance of tax-exempt revenue bonds is one of the more traditional means of financing publicly owned solid-waste management facilities, particularly WTE facilities, although revenue bonds will typically result in a higher interest rate on the bonds than if general obligation bonds had been issued by the same public entity. The advantages of revenue bonds, however, include the following: (1) the use of revenue bonds for solid-waste management projects will not decrease a public entity's general obligation bonding capacity and will allow general obligation bonds to continue to be available for schools, roads, etc; (2) depending on the method of revenues collection, it allows the users of the WTE facility to be charged on the basis of the amount of

Table 2. Financing Mechanisms for Publicly Owned WTE Facilities

Mechanism	Advantages	Disadvantages
Payments from general funds	Lowest financing cost; easiest to implement	Funds needed for other purposes that cannot be financed with revenue bonds
General obligation bonds	Lower interest rates	Bonding capacity required for other purposes
Revenue bonds: Project revenue bonds System revenue bonds	Users of the system pay for the service provided; preserves general obligation bonding capacity	Higher interest rates
Municipal bond insurance	Higher credit rating and lower interest rates	Bond insurance premiums increase cost; difficult to obtain if WTE facility not structured properly
Performance bond	Assures completion of construction through mechanical completion	Does not guarantee the performance level of the WTE facility
Subordination of operating fee	Improves debt service coverage	May provide incentive for operator to forego preventive maintenance and making of renewals and replacements
Certificates of participation	Avoids the use of general obligation bonding capacity	Requires annual appropriation

material that requires processing and/or disposal; and (3) it allows for the development of an enterprise fund established for solid-waste management services rather than having such costs paid from the general fund of a public entity. Revenue bonds can further be categorized as follows:

a. *Project Revenue Bonds.* Project revenue bonds for a WTE facility would be secured only by the revenues which the WTE facility is able to realize, such as disposal fees, revenues from the sale of recovered materials, and interest income on the reserve funds. There is generally no pledge of other sources of revenues from other parts and components of the WTE system, such as revenues from tipping fees at a transfer station or a landfill. Similarly, there is typically no pledge of the general tax revenues of the public entity to repay the revenue bonds issued for the WTE facility. The commitment on the part of the public entity to pay whatever is required to meet the revenue requirement of the WTE facility and to keep the bondholder whole must be relatively unconditional. Thus, a project revenue bond takes on certain characteristics of a general obligation bond, because a prospective investor is looking past the WTE facility for repayment of the project revenue bonds and, instead, is looking at the creditworthiness of the public entity that has pledged to pay whatever is required.

b. *System Revenue Bonds.* A more common financing mechanism for a WTE facility is the issuance of system revenue bonds, where the revenues of the entire solid-waste management system are pledged to repay the bonds. Obviously, this particular mechanism will work only when other components of a solid-waste management system exist, and such components are able to realize revenues. Examples of components of a solid-waste management system that can collect revenues include a transfer station, a WTE facility, and a landfill.

Because system revenue bonds can call upon a broader base of revenues for repayment of the bonds, they have become an increasingly popular means of financing WTE facilities. The prospective bond investor no longer has to be concerned with the technical aspects of the WTE facility in order to assess whether the bonds will be repaid. Rather, the investor will view the creditworthiness of the entire solid-waste management system. Typical issues analyzed in connection with the financing will include solid-waste flow control, tipping fees charged at competing facilities, changes in environmental regulations, and the method of billing and collecting for solid-waste management services.

4. *Municipal Bond Insurance.* The financing mechanism of municipal bond insurance assures the bondholders that the bonds will be repaid regardless of the performance and operating results of either the WTE facility or the solid-waste management system. Because of this reduced risk to the investor, the rate that will be paid on the bonds will be reduced because investors are then looking through the WTE facility or the system to the creditworthiness of the bond insurance company. This financing mechanism can increase the rating of a bond from A or less to AAA. However, caution must be exercised when viewing this financing mechanism.

First, not all WTE facilities or projects will be able to obtain municipal bond insurance; only 3 out of 10 proposed bond issues are actually approved for municipal bond insurance. Therefore, this particular financing mechanism will be available only for those WTE facility financings that have properly addressed the potential technical and economic risk; attempts to shift risk from a public entity to a bond insurance company will meet with failure. Second, the benefit of a lower interest rate associated with having municipal bond insurance is offset by the insurance premium. An economic evaluation will have to be performed for each WTE facility financing, which will calculate the potential interest savings during the term of the bonds and will compare them with the insurance premiums that will be paid upfront at the time of financing. This will typically involve a life-cycle cost analysis that will take into account the current value of future savings. The insurance premium will vary, depending on the perceived risk associated with the WTE facility and on the competitive forces in the bond insurance marketplace.

Several years ago, when interest rates were significantly higher than their current levels, the spread, or difference between interest rates, between insured and uninsured bonds was much greater. With the recent decrease in interest rates, that spread has narrowed, impacting the benefits associated with bond insurance. Only a detailed analysis will determine whether this financing mechanism will provide economic benefits.

5. *Performance Bond.* Whereas not a financing mechanism in and of itself, many investors prefer to see evidence of a performance bond as part of the overall financing package for a WTE facility. The term performance bond is somewhat misleading, because the bond will not guarantee that the WTE facility will actually perform as proposed. Rather, a performance bond assures that the WTE facility will be constructed, the walls of the building will stand, the equipment will be in place, and the WTE facility will be capable of achieving mechanical completion. The ability of the WTE facility to process material will not be addressed by performance bond. Investors prefer to see a performance bond for a WTE facility because it indicates that a bonding company has reviewed the proposed construction of the WTE facility and has determined that the WTE facility is capable of being constructed; this, therefore, reduces some portion of the construction risk.

6. *Subordination of Operating Fee.* The operating expense that a private operator will charge the public owner of a WTE facility typically includes an operating fee. One potential method to improve debt service coverage on bonds is to make the payment of the operating fee to a private operator, which is intended to cover the operator's administrative costs and provide some profit, subordinate to the payment of debt service on the bonds issued by the public entity. The subordination of this payment provides additional incentive for the private operator to manage the WTE facility as efficiently as possible. The public owner should ensure, however, that in establishing such

Table 3. Financing Mechanisms for Privately Owned WTE Facilities

Mechanism	Advantages	Disadvantages
100% equity investment	Lowest financing cost; easiest to implement	No opportunity to leverage the investment
Tax-exempt revenue bonds	Lower interest rates	Limitation on application of bond proceeds; requires a bond allocation from the state
Taxable revenue bonds	Can be used in conjunction with other financing mechanisms	Higher interest rates due to taxable interest
Letter-of-credit	Lower interest rates	Bank imposes requirements
Efficacy insurance	Guarantees technical performance; pays debt service and operating expenses	Not widely available

an arrangement, the private operator does not have the opportunity or financial incentive to forego undertaking routine maintenance of the WTE facility or to avoid making required renewals and replacements to the WTE facility. The short-term benefit of subordinating the operating fee could be lost in the long-term cost of improper maintenance.

Private Ownership

WTE facilities that are privately owned will have a somewhat different menu of financing mechanisms available. Like the public ownership options, however, the ability to obtain financing for most privately owned WTE facilities will heavily depend on the strong support of the public sector. This is particularly true for those WTE facilities that will receive material collected by, or on behalf of, a public sector entity. A discussion of the financing mechanisms available for privately owned WTE facilities is presented below and listed in Table 3.

1. *One Hundred Percent Equity Investment.* Most privately owned WTE facilities will include some equity participation by the owner. The amount of equity could theoretically vary from 100 to 0%. An equity investment in the range of 20–30% is not unusual for the financing of many WTE facilities. Generally, the greater the amount of equity invested in a WTE facility, the higher the overall cost of capital, due to an equity investor's demand for a higher return on what is perceived to be a riskier investment. Also, a private company must pay federal and state income taxes on equity earnings, further increasing the cost of owning the WTE facility.

A privately owned WTE facility financed with 100% equity is somewhat similar to its publicly owned counterpart financed from the general funds of the public sponsor, in that both are being paid for with available cash instead of utilizing other financing mechanisms. Given the alternative financing mechanisms that are usually available, it is extremely unusual for a WTE facility to be financed with a 100% equity investment because it offers no opportunity to leverage the investment.

2. *Tax-Exempt Revenue Bonds.* Privately owned WTE facilities that process solid waste are eligible for financing a major portion of the capital cost of the project as long as they are receiving solid waste and meet the Internal Revenue Service's (IRS's) definition of being a solid-waste disposal facility. The portion of the WTE facility that is eligible for tax-exempt financing ends at the point in the process where the solid waste has been converted to a useable product. This is usually considered to be at the point where steam has been generated and is available at a piping header terminal point. The balance of the cost of the WTE facility, usually the turbine generator and auxiliaries, would be financed by equity provided by the owner and/or the issuance of taxable revenue bonds. The availability of tax-exempt revenue bonds for the financing of privately owned WTE facilities depends on receiving an allocation from the state's private activity bond volume cap.

The prospective investors in a privately owned WTE facility, who would be purchasing tax-exempt revenue bonds, or taxable revenue bonds for that matter, will have the same concerns regarding risk as was previously discussed regarding publicly owned WTE facilities financed with revenue bonds. An investor will be looking for some guarantee of payment of the bonds, from either a creditworthy private company or from a public entity, as part of a service agreement with the private owner of the WTE facility. Usually, prospective revenue bond purchasers, who are generally being asked to make a long-term investment of 15–20 years or more, will not be willing to accept the market risk associated with short-term contracts for the delivery of materials or the sale of energy. Any type of revenue bond financing of a privately owned WTE facility will have to be supported by a contract(s) with creditworthy entities that will agree to pay (most likely on a put-or-pay basis during the term of the bonds) for the solid-waste management services being provided by the WTE facility. The lack of those strong underlying contracts would probably preclude the use of any type of revenue bond for the financing of a privately owned WTE facility.

3. *Taxable Revenue Bonds.* Taxable revenue bonds can be used either in conjunction with tax-exempt revenue bonds to help pay for the cost of those portions of the privately owned WTE facility that are not eligible to be financed with tax-exempt revenue bonds, or in conjunction with some form of equity investment to pay for a privately owned WTE facility that cannot be financed with tax-exempt revenue bonds. Because the interest on the bonds is taxable to the bondholder, this type of financing mechanism is expected to bear a higher interest rate than the tax-exempt revenue bonds. However, there are fewer restrictions on the application of the bond proceeds, and they are generally easier and less expensive to issue in terms of legal expenses.

4. *Letter-of-Credit.* The letter-of-credit is often used and is provided by a bank, usually rated as an A credit or better. The letter-of-credit may guarantee the repayment of the bonds in the event of a future problem, or a letter-of-credit (or line-of-credit) may pay for some future, unforeseen expense, such as increased operating expenses or the need to modify a major item of equipment. The letter-of-credit is used to strengthen the overall credit of the WTE facility financing and, if obtained, should allow the owner of the WTE facility to realize a lower interest rate on the bonds issued to pay for the cost of construction. If the bonds are supported by a letter-of-credit, then prospective bond purchasers will often look past the WTE facility and to the creditworthiness of the bank in evaluating whether or not to invest in the WTE facility. In that sense, the letter-of-credit is somewhat similar to the municipal bond insurance for revenue bonds described previously.

A letter-of-credit is also similar to municipal bond insurance in that a letter-of-credit is neither inexpensive nor easily obtained. The WTE facility must appear to be a sound financial investment before the bank will agree to provide the letter-of-credit. The bank will normally undertake an in-depth review of the WTE facility, including those risk items identified in Table 1. Despite the experiences of some savings and loan institutions during the 1980s, it is incorrect to think that a letter-of-credit bank is willing to accept items of risk that a bondholder will not find acceptable.

If the WTE facility is able to qualify for a letter-of-credit, the owner of the WTE facility needs to determine if the savings in interest payments associated with the lower interest rates are great enough to pay the letter-of-credit fees, which are normally expressed as a percentage of the principal amount of the bonds being issued. Because the letter-of-credit bank is, in effect, becoming the guarantor of the bonds, the letter-of-credit bank may request changes in the underlying credit structure of the transaction, including the creation of various reserve funds, changes in contract provisions, and meeting certain debt service coverage requirements before there is any payment to an equity investor. If the letter-of-credit bank is to participate, it will do whatever it feels is appropriate to assure that its investment will be repaid.

5. *Efficacy Insurance.* Various forms of efficacy insurance have been available in the solid-waste management area during the last 10–15 years. In the early 1980s, it was possible to obtain a single efficacy insurance policy that would pay out for three types of coverage: (1) the capital cost to undertake any repairs or modifications to a facility that were required to allow that facility to be operated at its guaranteed operating capacity; (2) the operating and maintenance expenses of the facility during the period that the repairs and modifications were being undertaken; and (3) the debt service on the bonds during the period that the repairs and modifications were being undertaken. This particular type of insurance policy, in effect, shifted the technical risk of a solid-waste management facility from the bondholder to the insurance company, and was viewed as the financing mechanism that would allow a number of solid-waste management projects to be financed.

Unfortunately, several solid-waste management projects that obtained this type of insurance faced major technical problems, and the insurance companies were called on to make significant payments. As a result, this type of efficacy insurance basically disappeared by the mid-1980s. During the past several years, variations of efficacy insurance have re-emerged and are again being considered as a possible financing mechanism for the development of WTE facilities and other solid-waste management projects.

6. *Leverage Lease.* Leverage lease financing is a mechanism that is theoretically available but, in reality, has seldom been used since the Tax Reform Act of 1986. This type of mechanism usually involves the leasing of equipment, rather than land or a building. Its lack of use in the financing of WTE facilities is due to two reasons. First, the Tax Reform Act eliminated investment tax credit. Second, the IRS has determined that it will not allow "double dipping" of tax benefits and does not permit charging depreciation expense and having tax-exempt bonds. The implementation of the alternative minimum tax (AMT) requirements necessitates that the issuance of private activity bonds for solid-waste projects is motivated by considerations other than reducing tipping fees by providing tax incentives.

Summary

As indicated above, there are a number of financing mechanisms available to finance a WTE facility. However, most of them represent a variation on the theme that the WTE facility must be based on a firm foundation, where items of risk have been allocated to the appropriate parties who are in the position both to be responsible for managing that item of risk and to have the financial capability to pay the resulting cost if the item of risk should actually occur. It is unrealistic to expect that the utilization of some type of financing mechanism will somehow allow the risk associated with the WTE facility to be shifted to an investor, because investors will not accept it.

Like any situation involving a wide array of options, there is no simple answer; one particular financing will not be appropriate for all cases. Situations vary from state to state, jurisdiction to jurisdiction, system to system, and project to project. All solid-waste management officials should review the circumstances surrounding their particular WTE facilities with their bond counsel, financial advisor, and investment banker, to review options and to determine what is most appropriate for their particular situation. The time to begin considering various financing mechanisms for the financing of a WTE facility is after: completion of a needs-assessment for the type of services the WTE facility will provide; policy decisions have been made regarding the method of ownership and operation; a technically sound and proven technology has been selected; permits have been applied for; and contracts with creditworthy entities have been developed to guarantee revenue payments adequate to pay operating expenses and service financing charges.

NET DISPOSAL COST (NDC)

Net disposal cost (NDC), expressed in dollars per ton, can be calculated for each year of the facility's operation once each of the following input variables has been computed: construction period, term of bonds or loan, interest rate, principal amount, construction drawdown of schedule, interest rate earned on funds, reserve fund amounts, and equity. To understand the economics of WTE completely, numerous sensitivity cases should be run to identify the possible range of NDC for variations in each input variable. Whereas the "base case" typically has contractually guaranteed quantities, variations above and below these levels should be studied, in order for the municipal entity, which is guaranteeing to pay the NDC, to understand fully the potential range of tipping fees it may have to pay due to decreases in energy revenues, increases in operating expenses, decreases in waste quantities, and so forth.

If a WTE facility is developed on a project finance basis, where a guaranteed supply of waste can be provided and where tip fees at the gate are calculated in accordance with a formula (a formula tipping fee) that is intended to cover NDC, then NDC will equal the facility tip fee. However, WTE facilities developed in this manner rarely exist in a market vacuum, and a guaranteed supply of waste cannot be assumed, as there is usually considerable competition for solid waste. For these facilities, NDC may be higher than competing tip fees at other waste management facilities in the area, particularly landfills, especially if competing tip fees at area landfills are being artificially depressed in order to draw waste away from the WTE facility. Under these conditions, it may be necessary to obtain other sources of revenue (OR), from either subsidy payments from the general tax coffers for publicly owned facilities, or from additional equity payments from the owner for privately owned facilities.

In order to obtain long-term financing, in the absence of guaranteed waste supply at a tip fee which equals NDC, OR must be pledged either by a governmental entity or a creditworthy private company (depending on the method of ownership), such that the resulting tip fee at the WTE facility is competitive with surrounding facilities. This is generally referred to as "economic flow control" and it differs from the concept of "legal flow control," where ordinances are passed requiring the waste to be delivered to the WTE facility regardless of cost. The resulting WTE facility tip fee is then

$$TF = NDC - OR$$

WTE SENSITIVITIES

Variations in capital cost, O&M cost, and energy price all impact NDC; most variations in WTE project economics can be reduced to one of these variables. All costs and revenues in this discussion are expressed in 1994 U.S. dollars. Considering the range of capital cost from \$40,000 to \$80,000 per tpd solid-waste TP in the 100–500 tpd-size range, and \$70,000–110,000 per tpd TP in the larger size ranges; O&M cost range of \$30–45 per ton solid-waste TP; and energy price range of \$0.02–0.08 per kWh, the estimated NDC of a potential WTE project can vary by 100% or more, depending on the assumptions used for economic analysis or on the quality of proposals received (3).

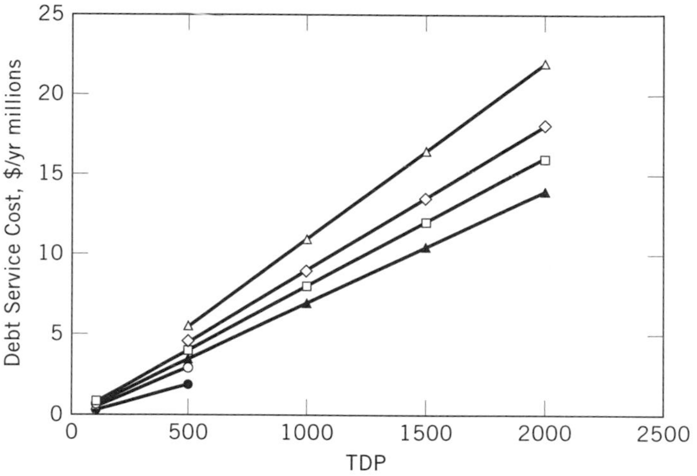

Figure 1. Debt service cost versus tons per day (tpd), based on 10% capital recovery factor. (**a**) 40,000 \$/tpd; (**b**) 60,000 \$/tpd; (**c**) 70,000 \$/tpd; (**d**) 80,000 \$/tpd; (**e**) 90,000 \$/tpd; (**f**) 110,000 \$/tpd.

Figures 1–3 show examples of translating WTE facility size into annual cost or revenue components of debt service (as a function of capital cost), O&M cost, and energy revenues, which can then be inserted into the basic formula for NDC previously presented. Figure 1 shows that for a 1,000-tpd WTE facility, debt service can range from \$7–11 million per year, depending on the quality of the technology chosen. For this example, debt service is calculated based on a capital recovery factor of 0.10, which is equal to a 20-year revenue bond issue at about 8% interest. In calculating the capital cost of a project and thus the total debt service, the total capital cost includes not only the cost of design, construction, and startup, but also financing, legal, and development fees, that is, the total amount borrowed.

Figure 2 presents annual O&M costs as a function of throughput, for a typical range of unit O&M costs in dollars per ton. Typically, WTE technologies that are more

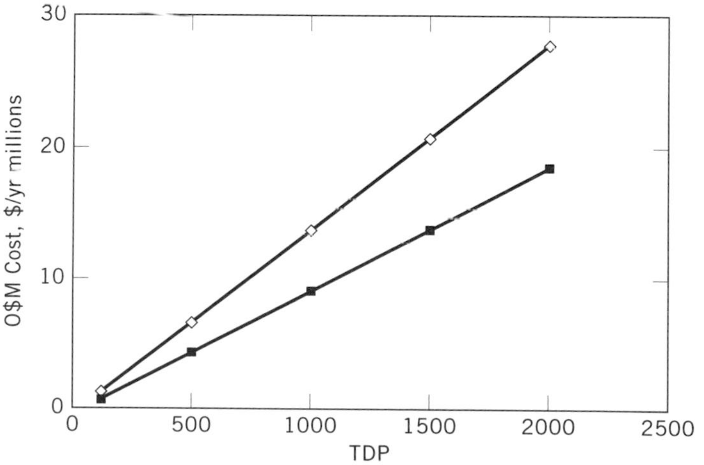

Figure 2. Operating and maintenance (O&M) costs versus tons per day (tpd), based on 85% annual capacity factor. (**a**) 30 \$/ton; (**b**) 45 \$/ton.

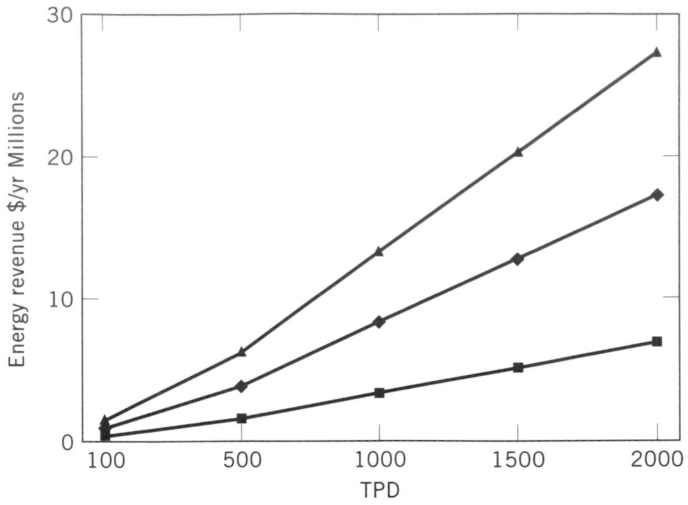

Figure 3. Energy revenues versus tons per day (tpd), based on 85% annual capacity factor. (**a**) 0.02 $/kWh; (**b**) 0.04 $/kWh; (**c**) 0.08 $/kWh.

expensive in terms of capital cost can be expected to yield O&M costs that are relatively lower than low capital-cost technologies. For a typical 1,000-tpd facility with the characteristics identified on Figure 2, O&M costs would range from $9.3 to $14 million per year. Similarly, energy price sensitivities for typical WTE facilities are shown on Figure 3. Assuming net energy production of 550 kWh per ton TP for sale, energy revenues would range from $3.4 to $13.4 million per year for a 1,000-tpd facility.

Using the lowest value of each of these variables listed above, NDC would be calculated as follows for a 1000-tpd facility:

$$NDC = (OM + DS - PR)/TP$$
$$= (\$9{,}307{,}500 + \$7{,}000{,}000 - \$3{,}412{,}750)/310{,}250$$
$$\text{tpy} = \$41.56 \text{ per ton}$$

Using the highest value of each variable,

$$NDC = (\$13{,}961{,}250 + \$11{,}000{,}000 - \$13{,}651{,}000)/310{,}250 \text{ tpy} = \$36.46 \text{ per ton}$$

As a further sensitivity, if O&M cost could be reduced from $45 per ton to $35 per ton because high quality equipment has been selected, then

$$NDC = (\$10{,}858{,}750 + \$11{,}000{,}000 - \$13{,}651{,}000)/310{,}250 \text{ tpy} = \$26.46 \text{ per ton}$$

This example demonstrates the importance of considering WTE economics on a case-by-case basis. There are no set equations that can accurately predict NDC without considerable analysis of the wide array of variables affecting final disposal cost.

Other examples of sensitivity cases that impact NDC and DS coverage and are thus often considered by prospective bond purchasers include the following:

1. Increase in operating expenses
2. Decrease in waste throughput for technical reasons
3. Decrease in waste availability due to recycling
4. Decrease in the price of electricity
5. Change in the assumed rate of inflation
6. Future changes in environmental laws requiring increased capital and operating costs
7. Decreases in prices paid for recycled materials

BIBLIOGRAPHY

1. E. L. Beaumont, "How Can Communities Judge the True Cost of Solid Waste Management?" *Solid Waste & Power* **3**(5) (Oct. 1989).
2. J. Clunie, "Mechanisms for the Financing of Material Recovery Facilities," *Proceedings from SWANA 31st Annual International Solid Waste Exposition,* San Jose, Calif., Aug. 2–5, 1993.
3. E. Berenyi and R. Gould, *1991 Resource Recovery Yearbook,* Governmental Advisory Associates, Inc., New York, 1991.

WATER CONDITIONING

JAMES ROBINSON
Betz Laboratories
Trevose, Pennsylvania

Abundant supplies of fresh water are essential to the development of industry. Enormous quantities are required for the cooling of products and equipment, for process needs, for boiler feed, and for sanitary and potable water supply. See also WATER QUALITY ISSUES; WATER QUALITY MANAGEMENT; PETROLEUM REFINING—EMISSIONS AND WASTES.

WATER IMPURITIES

Water impurities include dissolved and suspended solids. Calcium bicarbonate is a soluble salt. A solution of calcium bicarbonate is clear, because the calcium and bicarbonate are present as atomic-size ions that are not large enough to reflect light. Suspended solids are substances that are not completely soluble in water and are present as particles. These particles usually impart a visible turbidity to the water.

CLARIFICATION

Suspended matter in raw water supplies is removed by various methods to provide a water suitable for domestic purposes and most industrial requirements. The suspended matter can consist of large solids, settleable by gravity alone without any external aids, and nonsettleable material, often colloidal in nature. Removal is generally accomplished by coagulation, flocculation, and sedimentation. The combination of these three processes is referred to as conventional clarification.

Coagulation involves neutralizing charged particles to destabilize suspended solids. In most clarification processes, a flocculation step then follows. Flocculation starts when neutralized or entrapped particles begin to collide and fuse to form larger particles. This process can occur naturally or can be enhanced by the addition of polymeric flocculant aids.

Table 1 lists a number of common inorganic coagulants. Typical iron and aluminum coagulants are acid salts that lower the pH of the treated water by hydrolysis. Depending on initial raw water alkalinity and pH, an alkali such as lime or caustic must be added to counteract the pH depression of the primary coagulant. Iron and aluminum hydrolysis products play a significant role in the coagulation process, especially in cases in which low turbidity influent waters benefit from the presence of additional collision surface areas.

With aluminum sulfate, optimum coagulation efficiency and minimum floc solubility normally occur at pH 6.0–7.0. Iron coagulants can be used successfully over the much broader pH range of 5.0–11.0. If ferrous compounds are used, oxidation to ferric iron is needed for complete precipitation. This may require either chlorine addition or pH adjustment.

Polyelectrolytes refers to all water-soluble organic polymers used for clarification, whether they function as coagulants or flocculants. Water-soluble polymers may be classified as follows:

Anionic. Ionize in water solution to form negatively charged sites along the polymer chain.

Cationic. Ionize in water solution to form positively charged sites along the polymer chain.

Nonionic. Ionize in water solution to form slight negatively charged sites along the polymer chain.

Polymeric primary coagulants are cationic materials with relatively low molecular weights (under 500,000). The cationic charge density (available positively charged sites) is high. Polymeric flocculants or coagulant aids may be anionic, cationic, or nonionic. Their molecular weights may be as high as 50,000,000. Table 2 describes some typical organic polyelectrolytes.

The use of organic polymers offers several advantages over the use of inorganic coagulants:

- The amount of sludge produced during clarification can be reduced by 50–90%; the approximate dry weights of solids removed per pound of dry alum and ferric sulfate are approximately 0.11 and 0.23 kg, respectively.
- The resulting sludge contains less chemically bound water and can be more easily dewatered.
- Polymeric coagulants do not affect pH; therefore, the need for supplemental alkalinity, such as lime, caustic, or soda ash, is reduced or eliminated.
- Polymeric coagulants do not add to the total dissolved solids concentration, eg, 1 ppm of alum adds 0.45 ppm of sulfate ion (expressed as $CaCO_3$); the reduction in sulfate can significantly extend the capacity of anion exchange systems.
- Soluble iron or aluminum carryover in the clarifier effluent may result from inorganic coagulant use; therefore, elimination of the inorganic coagulant can minimize the deposition of these metals in filters, ion exchange units, and cooling systems.

In certain instances, an excess of primary coagulant (whether inorganic, polymeric, or a combination of both) may be fed to promote large floc size and to increase settling rate. However, in some waters, even high doses of primary coagulant will not produce the desired effluent clarity. A polymeric coagulant aid added after the primary coagulant may, by developing a larger floc at low treatment levels, reduce the amount of primary coagulant required.

Generally, very high molecular weight anionic polyacrylamides are the most effective coagulant aids. Nonionic or cationic types have proven successful in some clarifier systems. Essentially, the polymer bridges the small floc particles and causes them to agglomerate rapidly into larger, more cohesive flocs that settle quickly. The higher molecular weight polymers bridge suspended solids most effectively. Coagulant aids have proven quite successful in precipitation softening and clarification to

Table 1. Common Inorganic Coagulants

Name	Typical Formula	Typical Strength	Typical Forms Used in Water Treatment	Density	Typical Uses
Aluminum sulfate	$Al_2(SO_4)_3 \cdot 14$ to $18H_2O$	17% Al_2O_3	lump, granular, or powder	60–70 lb/ft^3	primary coagulant
Alum		8.25% Al_2O_3	liquid	11.1 lb/gal	
Aluminum chloride	$AlCl_3 \cdot 6H_2O$	35% $AlCl_3$	liquid	12.5 lb/gal	primary coagulant
Ferric sulfate	$Fe_2(SO_4)_3 \cdot 9H_2O$	68% $Fe_2(SO_4)_3$	granular	70–72 lb/ft^3	primary coagulant
Ferric-floc	$Fe_2(SO_4)_3 \cdot 5H_2O$	41% $Fe_2(SO_4)_3$	solution	12.3 lb/gal	primary coagulant
Ferric chloride	$FeCl_3$	60% $FeCl_3$, 35–45% $FeCl_3$	crystal, solution	60–64 lb/ft^3 11.2–12.4 lb/gal	primary coagulant
Sodium aluminate	$Na_2Al_2O_4$	38–46% $Na_2Al_2O_4$	liquid	12.3–12.9 lb/gal	primary coagulant; cold–hot precipitation softening

Table 2. Common Organic Polyelectrolytes

Polymer Type	Typical Formula	Typical Molecular Weight	Available Forms	Typical Uses
Nonionic	Polyacrylamide $[-CH_2-CH(C=O)(NH_2)-]_n$	1×10^6 to 2×10^7	powder, emulsion, solution	flocculant in clarification with inorganic or organic coagulants
Anionic	Hydrolyzed polyacrylamide $[-CH_2-CH(C=O)(NH_2)-]_n[-CH_2-CH(C=O)(ONa)-]_y$	1×10^6 to 2×10^7	powder, emulsion, solution	flocculant in clarification with inorganic or organic coagulants
Cationic	Poly(DADMAC) or poly(DMDAAC) polymers $[-CH_2-CH-CH-CH_2-$ with CH_2 and CH_2 joining to $N^+(CH_3)(CH_3)]_n + Cl^-$	250 to 500×10^3	solution	primary coagulant alone or in combination with inorganics in clarification
Cationic	Quarternized polyamines $[-CH_2-CH(OH)-CH_2-N(CH_3)(CH_3)-]_n$	10 to 500×10^4	solution	primary coagulant alone or in combination with inorganics in clarification
Cationic	Polyamines $[-CH_2-CH_2-NH_2-]_n$	10^4 to 10^6	solution	primary coagulant alone or in combination with inorganics in clarification

achieve improved settling rates of precipitates and finished water clarity.

PRECIPITATION SOFTENING

Precipitation softening processes are used to reduce raw water hardness, alkalinity, silica, and other constituents. This helps prepare water for direct use as cooling tower makeup or as a first-stage treatment followed by ion exchange for boiler makeup or process use. The water is treated with lime or a combination of lime and soda ash (carbonate ion). These chemicals react with the hardness and natural alkalinity in the water to form insoluble compounds. The compounds precipitate and are removed from the water by sedimentation and, usually, filtration. Waters with moderate to high hardness and alkalinity concentrations (150–500 ppm as $CaCO_3$) are often treated in this fashion.

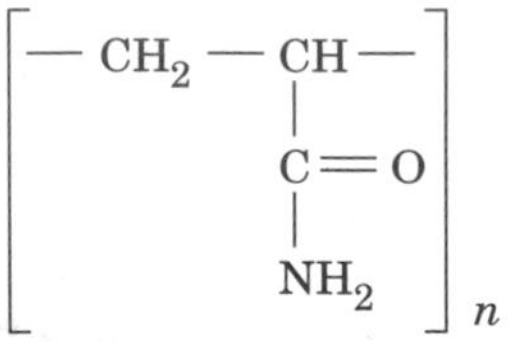

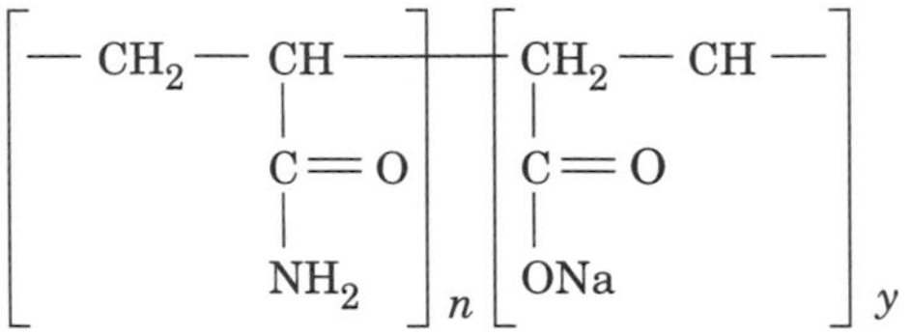

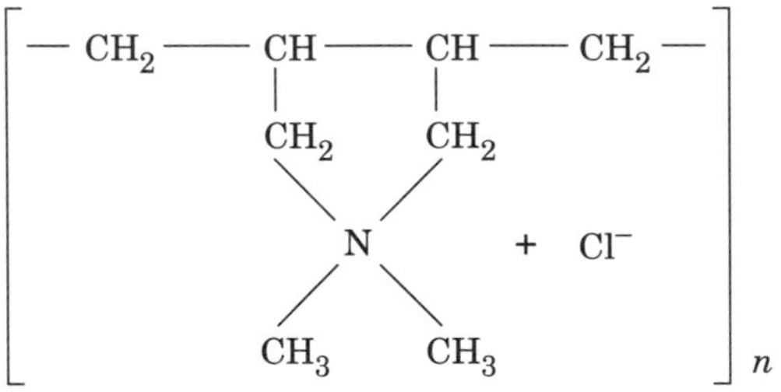

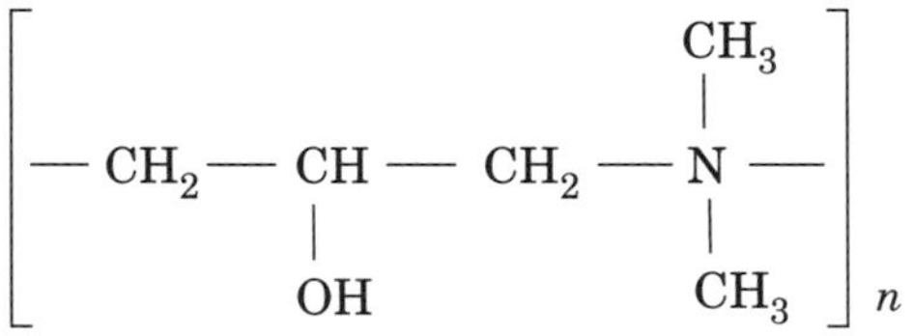

Cold Lime Softening

Precipitation softening accomplished at ambient temperatures is referred to as cold lime softening. When hydrated lime, $Ca(OH)_2$, is added to the water being treated, the following reactions occur:

$$CO_2 + Ca(OH)_2 = CaCO_3 + H_2O$$

$$Ca(HCO_3)_2 + Ca(OH)_2 = 2CaCO_3 + 2H_2O$$

$$Mg(HCO_3)_2 + 2Ca(OH)_2 = Mg(OH)_2 + 2CaCO_3 + 2H_2O$$

Noncarbonate or permanent calcium hardness, if present, is not affected by treatment with lime alone. If noncarbonate magnesium hardness is present in an amount greater than 70 ppm and an excess hydroxyl alkalinity of about 5 ppm is maintained, the magnesium will be reduced to about 70 ppm, but the calcium will increase in proportion to the magnesium reduction.

To improve magnesium reduction, which also improves silica reduction in cold process softening, sodium aluminate may be used. The sodium aluminate provides hydroxyl ion (OH^-) needed for improved magnesium reduction, without increasing calcium hardness in the treated water. In addition, the hydrolysis of sodium aluminate results in the formation of aluminum hydroxide, which aids in floc formation, sludge blanket conditioning, and silica reduction.

Warm Lime Softening

The warm lime softening process operates in the temperature range of 49°–60°C. The solubilities of calcium, magnesium, and silica are reduced by increased temperature. Therefore, they are more effectively removed by warm lime softening than by cold lime softening.

Hot Process Softening

Hot process softening is usually carried out under pressure at temperatures of 108°–116°C. At the operating temperature, hot process softening reactions go essentially to completion. This treatment method involves the same reactions described above, except that raw water CO_2 is vented and does not participate in the lime reaction. The use of lime and soda ash permits hardness reduction down to 0.5 g/gal, or about 8 ppm, as calcium carbonate. Magnesium is reduced to 2–5 ppm, because of the lower solubility of magnesium hydroxide at the elevated temperatures.

Silica Reduction

Hot process softening can also provide good silica reduction. The silica reduction is accomplished through adsorption of the silica on the magnesium hydroxide precipitate. If there is insufficient magnesium present in the raw water to reduce silica to the desired level, magnesium compounds (such as magnesium oxide, magnesium sulfate, magnesium carbonate, and dolomitic lime) may be used.

Alkalinity Reduction

Treatment by lime precipitation reduces alkalinity. However, if the raw water alkalinity exceeds the total hardness, sodium bicarbonate alkalinity is present. In such cases, it is usually necessary to reduce treated water alkalinity in order to reduce condensate system corrosion or permit increased cycles of concentration.

Treated Water Quality

Predicted analyses of a typical raw water treated by various lime and lime–soda softening processes are presented in Table 3. Treatment by lime converts the sodium bicarbonate in the raw water to sodium carbonate as follows:

$$2NaHCO_3 + Ca(OH)_2 = CaCO_3 + Na_2CO_3 + 2H_2O$$

Calcium sulfate (gypsum) may be added to reduce the carbonate to required levels. The reaction is as follows:

$$Na_2CO_3 + CaSO_4 = CaCO_3 + Na_2SO_4$$

FILTRATION

Filtration is used in addition to regular coagulation and sedimentation or precipitation softening for removal of solids from surface water or waste water. This prepares the water for use as potable, boiler, or cooling makeup. Waste water filtration helps users meet more stringent effluent discharge permit requirements.

Table 3. Typical Softener Effluent Analyses

Factor	Raw Water	Removal of Calcium Alkalinity Cold-Lime	Lime–Soda Softening (Cold)	Lime–Soda Softening (Hot)[a]	Lime Softening (Hot)[a]
Total hardness (as $CaCO_3$), ppm	250	145	81	20	120
Calcium hardness (as $CaCO_3$), ppm	150	85	35	15	115
Magnesium hardness (as $CaCO_3$), ppm	100	60	46	5	5
"P" alkalinity (as $CaCO_3$), ppm	0	27	37	23	18
"M" alkalinity (as $CaCO_3$), ppm	150	44	55	40	28
Silica (as SiO_2), ppm	20	19	18	1–2	1–2
pH	7.5	10.3	10.6	10.5	10.4

[a] Removal of SiO_2 by the hot process, to the levels shown, may require the feed of supplemental magnesium oxide. Sludge recirculation is necessary. All raw water constituents will be diluted by the steam used for heating by approximately 15% if the process is hot.

Filtration does not remove dissolved solids, but may be used together with a softening process, which does reduce the concentration of dissolved solids. For example, anthracite filtration is used to remove residual precipitated hardness salts remaining after precipitation softening.

In most water clarification or softening processes in which coagulation and precipitation occur, at least a portion of the clarified water is filtered. Clarifier effluents of 2–10 NTU may be improved to 0.1–1.0 NTU by conventional sand filtration. Filtration ensures acceptable suspended solids concentrations in the finished water even when upsets occur in the clarification processes.

ION EXCHANGE

Ion exchangers exchange one ion for another, hold it temporarily, and then release it to a regenerant solution. In an ion exchange system, undesirable ions in the water supply are replaced with more acceptable ions.

Ionizable groups attached to the resin bead determine the functional capability of the resin. Industrial water treatment resins are classified into four basic categories:

Strong Acid Cation (SAC)
Weak Acid Cation (WAC)
Strong Base Anion (SBA)
Weak Base Anion (WBA)

SAC resins can neutralize strong bases and convert neutral salts into their corresponding acids. SBA resins can neutralize strong acids and convert neutral salts into their corresponding bases. These resins are used in most softening and full demineralization applications. WAC and WBA resins are able to neutralize strong bases and acids, respectively. These resins are used for dealkalization, partial demineralization, or (in combination with strong resins) full demineralization.

Sodium Zeolite Softening

Sodium zeolite softening is the most widely applied use of ion exchange. In zeolite softening, water containing scale-forming ions, such as calcium and magnesium, passes through a resin bed containing SAC resin in the sodium form. In the resin, the hardness ions are exchanged with the sodium, and the sodium diffuses into the bulk water solution. The hardness-free water, termed soft water, can then be used for low to medium pressure boiler feedwater, reverse osmosis system makeup, some chemical processes, and commercial applications, such as laundries.

Demineralization

Softening alone is insufficient for most high-pressure boiler feed waters and for many process streams, especially those used in the manufacture of electronics equipment. In addition to the removal of hardness, these processes require removal of all dissolved solids, such as sodium, silica, alkalinity, and the mineral anions (Cl^-, SO_4^{2-}, and NO^{3-}).

Demineralization of water is the removal of essentially all inorganic salts by ion exchange. In this process, strong acid cation resin in the hydrogen form converts dissolved salts into their corresponding acids, and strong base anion resin in the hydroxide form removes these acids. Demineralization produces water similar in quality to distillation at a lower cost for most fresh waters.

The standard cation–anion process has been modified in many systems to reduce the use of costly regenerants and the production of waste. Modifications include the use of decarbonators, weak acid and weak base resins. Several different approaches to demineralization using these processes are shown in Figure 1.

Condensate Polishing

Ion exchange can be used to purify or polish returned condensate, removing corrosion products that could cause harmful deposits in boilers. Typically, the contaminants in the condensate system are particulate iron and copper. Low levels of other contaminants may enter the system through condenser and pump seal leaks or carryover of boiler water into the steam. Condensate polishers filter out the particulates and remove soluble contaminants by ion exchange.

Most paper mill condensate polishers operate at temperatures approaching 93°C, precluding the use of anion resin. Cation resin, which is stable up to temperatures of over 132°C, is used for deep bed condensate polishing in these applications. The resin is regenerated with sodium chloride brine, as in a zeolite softener. In situations in which sodium leakage from the polisher adversely affects the boiler water internal chemical program or steam attemperating water purity, the resin can be regenerated with an ionized amine solution to prevent these problems.

MEMBRANE PROCESSES

In recent years, membrane processes have been used increasingly for the production of "pure" waters from fresh water and seawater. Membrane processes are also being applied in process and wastewater systems. Although often thought to be expensive and relatively experimental, membrane technology is advancing quickly, becoming less expensive, improving performance, and extending life expectancy.

Common membrane processes include ultrafiltration (UF), reverse osmosis (RO), electrodialysis (ED), and electrodialysis reversal (EDR). These processes (with the exception of UF) remove most ions; RO and UF systems also provide efficient removal of nonionized organics and particulates. Because UF membrane porosity is too large for ion rejection, the UF process is used to remove contaminants, such as oil and grease, and suspended solids.

Reverse Osmosis

Osmosis is the flow of solvent through a semipermeable membrane, from a dilute solution to a concentrated solution. This flow results from the driving force created by the difference in pressure between the two solutions. Osmotic pressure is the pressure that must be added to the concentrated solution side to stop the solvent flow through the membrane. Reverse osmosis is the process of reversing the flow, forcing water through a membrane

System	Application	Typical Effluent	Advantages & Limitations
SA → WB	Silica and CO_2 are not objectionable	Conductance: 10–40 intro silica unchanged	Low equipment costs low regenerant costs
SA → SB	Lower alkalinity, raw water, silica, and CO_2 removal required	Conductance: <15 intro silica: 0.02–0.10 ppm	Low equipment costs Medium regenerant costs
SA → D → SB	High alkalinity, raw water, silica, and CO_2 removal required	Conductance: <15 intro silica: 0.02–0.10 ppm	Low regenerant costs Repumping required
SA → D → WB → SB	High alkalinity, chloride and sulfate, raw water, silica, and CO_2 removal required	Conductance: <15 intro silica: 0.02–0.10 ppm	Higher equipment cost lowest regenerant cost repumping required
WA → SA → D → WB → SB	High hardness, alkalinity, chloride, and sulfate, raw water, silica, and CO_2 removal required	Conductance: <10 intro silica: 0.02–0.06 ppm	Higher equipment cost lowest regenerant cost repumping required
CF → D → SB	High sodium, raw water, low leakage required	Conductance: <10 intro silica: 0.02–0.06 ppm	Medium equipment cost lower acid cost for leakage obtained
SA → SB → SA	High sodium, raw water, existing 2-bed system, low leakage required	Conductance: <5 intro silica: 0.02–0.06 ppm	Easy to retrofit system danger of acidic water on anion breakthrough
MB	Low solids, raw water, high purity required	Conductance: <1 intro silica: 0.01–0.05 ppm	Low equipment cost high chemical cost, higher attention required
SA → SB → MB	High solids, water, high purity required	Conductance: <1 intro silica: 0.01–0.05 ppm	Medium equipment cost high chemical cost, higher attention required

SA Strong acid cation exchanger; SB Strong base anion exchanger; D Degasifier; MB Mixed bed

WA Weak acid cation exchanger; WB Weak base anion exchanger; CF Counterflow cation

Figure 1. Demineralizer systems consist of various unit processes arranged to meet the system needs.

from a concentrated solution to a dilute solution to produce pure water. Figure 2 illustrates the processes of osmosis and reverse osmosis.

Reverse osmosis is created when sufficient pressure is applied to the concentrated solution to overcome the osmotic pressure. This pressure is provided by feed-water pumps. Concentrated contaminants (brine) are removed from the high pressure side of the RO membrane, and pure water (permeate) is removed from the low pressure side. Membrane modules may be staged in various design configurations, producing the highest quality permeate with the least amount of waste.

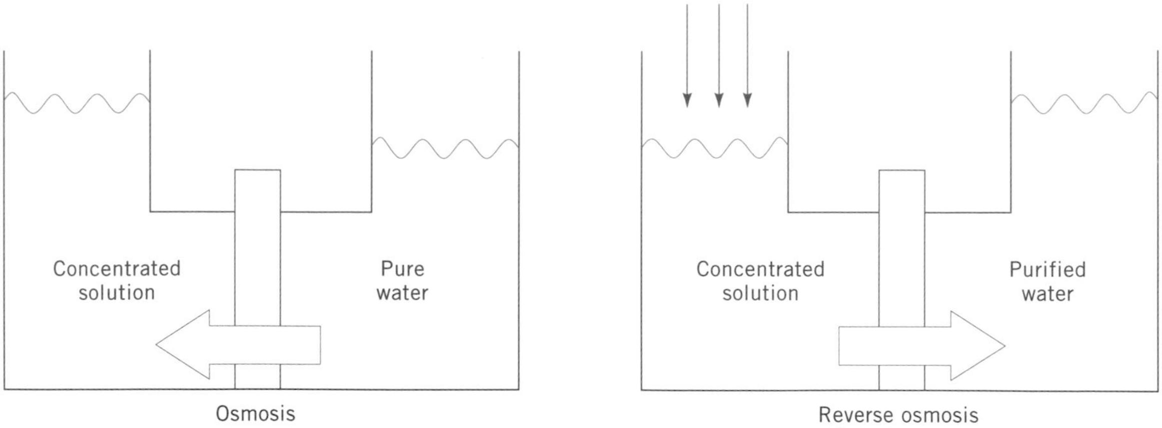

Figure 2. In the osmosis process, water flows through a membrane from a dilute solution to a more concentrated solution. In reverse osmosis, applied pressure causes water to flow in the opposite direction.

Typically, 95% of dissolved salts are removed from the brine. All particulates are removed. However, due to their molecular porosity, RO membranes do not remove dissolved gases, such as Cl_2, CO_2, and O_2.

Electrodialysis

Electrodialysis processes transfer ions of dissolved salts across membranes, leaving purified water behind. Ion movement is induced by direct current electrical fields. A negative electrode (cathode) attracts cations, and a positive electrode (anode) attracts anions. Systems are compartmentalized in stacks by alternating cation and anion transfer membranes. Alternating compartments carry concentrated brine and purified permeate. Typically, 40–60% of dissolved ions are removed or rejected. Further improvement in water quality is obtained by staging (operation of stacks in series). ED processes do not remove particulate contaminants or weakly ionized contaminants, such as silica.

Electrodialysis Reversal

Electrodialysis reversal processes operate on the same principles as ED; however, EDR operation reverses system polarity (typically three to four times per hour). This reversal stops the buildup of concentrated solutions on the membrane and thereby reduces the accumulation of inorganic and organic deposition on the membrane surface. EDR systems are similar to ED systems, designed with adequate chamber area to collect both product water and brine. EDR produces water of the same purity as ED.

BOILERS

Boiler System Corrosion

The dissolved gases normally present in water cause many corrosion problems. For instance, oxygen in water produces pitting that is particularly severe because of its localized nature (Fig. 3). Carbon dioxide corrosion is frequently encountered in condensate systems and less commonly in water distribution systems. Water containing ammonia, particularly in the presence of oxygen, readily attacks copper and copper-bearing alloys. The resulting corrosion leads to deposits on boiler heat transfer surfaces and reduces efficiency and reliability. In addition to the gases, corrosion can also be caused by the concentration of caustic or acidic species in the feed water.

Caustic Corrosion. Concentration of caustic (NaOH) can occur either as a result of steam blanketing (which allows salts to concentrate on boiler metal surfaces) or by localized boiling beneath porous deposits on tube surfaces.

Caustic corrosion (gouging) occurs when caustic is concentrated and dissolves the protective magnetite (Fe_3O_4) layer. Iron, in contact with the boiler water, forms magnetite and the protective layer is continuously restored. However, as long as a high caustic concentration exists, the magnetite is constantly dissolved, causing a loss of base metal and eventual failure (Fig. 4).

Steam blanketing is a condition that occurs when a steam layer forms between the boiler water and the tube wall. Under this condition, insufficient water reaches the tube surface for efficient heat transfer. The water that does reach the overheated boiler wall is rapidly vaporized,

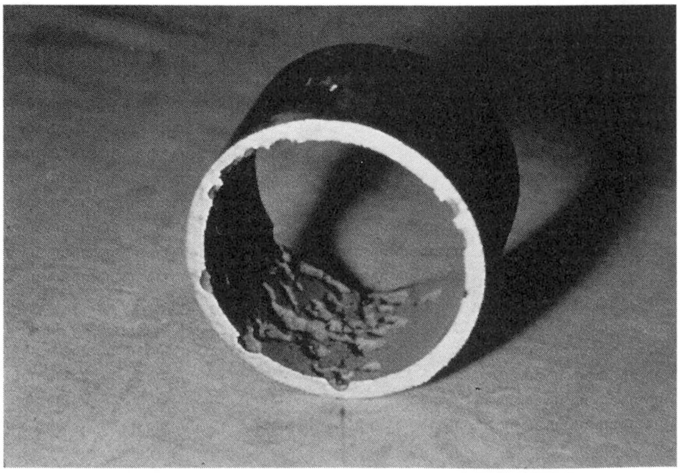

Figure 3. Tube severely damaged by oxygen.

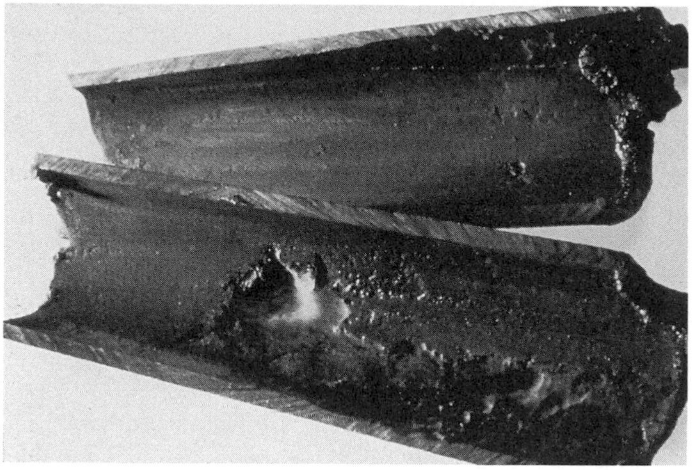

Figure 4. Caustic gouging caused failure of this boiler tube.

leaving behind a concentrated caustic solution, which is corrosive.

Porous metal oxide deposits also permit the development of high boiler water concentrations. Water flows into the deposit and heat applied to the tube causes the water to evaporate, leaving a concentrated solution. Again, corrosion may occur. Caustic attack creates irregular patterns, often referred to as gouges. Deposition may or may not be found in the affected area.

Acidic Corrosion. Low makeup or feed-water pH can cause serious acid attack on metal surfaces in the preboiler and boiler system. Even if the original makeup or feed-water pH is not low, feed water can become acidic from contamination of the system. Common causes include improper operation or control of demineralizer cation units, process contamination of condensate (eg, sugar contamination in food processing plants), and cooling water contamination from condensers. Acid corrosion can also be caused by chemical cleaning operations. Overheating of the cleaning solution can cause breakdown of the inhibitor used, excessive exposure of metal to cleaning agent, and high cleaning agent concentration. Failure to neutralize acid solvents completely before startup has also caused problems. In a boiler and feed-water system, acidic attack can take the form of general thinning, or it can be localized at areas of high stress such as drum baffles, U bolts, acorn nuts, and tube ends.

Hydrogen Embrittlement. Hydrogen embrittlement is rarely encountered in industrial plants. The problem usually occurs only in boilers operating at or above 689.5×10^3 Pa. Hydrogen embrittlement of mild steel boiler tubing occurs in high pressure boilers when atomic hydrogen forms at the boiler tube surface as a result of corrosion. Hydrogen permeates the tube metal, where it can react with iron carbides to form methane gas, or with other hydrogen atoms to form hydrogen gas. These gases evolve predominantly along grain boundaries of the metal. The resulting increase in pressure leads to metal failure.

The initial surface corrosion that produces hydrogen usually occurs beneath a hard, dense scale. Acidic contamination or localized low pH excursions are normally required to generate atomic hydrogen. In high purity systems, raw water in-leakage (eg, condenser leakage) lowers boiler water pH when magnesium hydroxide precipitates, resulting in corrosion, formation of atomic hydrogen, and initiation of hydrogen attack.

Stress Corrosion Cracking. Stress corrosion cracking occurs from the combined action of corrosion and stress. The corrosion may be initiated by improper chemical cleaning, high dissolved oxygen levels, pH excursions in the boiler water, the presence of free hydroxide, and high levels of chlorides. Stresses are either residual in the metal or caused by thermal excursions. Rapid startup or shutdown can cause or further aggravate stresses. Tube failures occur near stressed areas such as welds, supports, or cold worked areas.

Boiler Deposits

Deposition is a principal problem in the operation of steam generating equipment. The accumulation of material on boiler surfaces can cause overheating and/or corrosion. Both of these conditions frequently result in unscheduled downtime. Common feed-water contaminants that can form boiler deposits include calcium, magnesium, iron, copper, aluminum, silica, and (to a lesser extent) silt and oil. Most deposits can be classified as one of two types: scale that crystallized directly onto tube surfaces or sludge deposits that precipitated elsewhere and were transported to the metal surface by the flowing water.

Boiler feed water pretreatment systems have advanced to such an extent that it is now possible to provide boilers with ultrapure water. However, this degree of purification requires the use of elaborate pretreatment systems. The capital expenditures for such pretreatment equipment trains can be considerable and are often not justified when balanced against the capability of internal treatment.

The quality of feed water required depends on boiler operating pressure, design, heat transfer rates, and steam use. Most boiler systems have sodium zeolite softened or demineralized makeup water. Feed-water hardness usually ranges from 0.01 to 2.0 ppm, but even water of this purity does not provide deposit-free operation. Therefore, good internal boiler water treatment programs are necessary.

Boiler Water Treatment

Oxygen Control. To meet industrial standards for both oxygen content and the allowable metal oxide levels in feed water, nearly complete oxygen removal is required. This can be accomplished only by efficient mechanical deaeration supplemented by an effective and properly controlled chemical oxygen scavenger.

To deaerate the boiler feedwater, water is sprayed into a steam atmosphere. This heats the water to within a few degrees of the temperature of the saturated steam. Because the solubility of oxygen in water is low under these conditions, 97–98% of the oxygen in the incoming water is released to the steam and is purged from the system by venting. Although the remaining oxygen is not soluble under equilibrium conditions, it is not readily released to the steam. Therefore, water leaving the heating section of the deaerator must be scrubbed vigorously with steam to maximize removal.

In addition to mechanical deaeration, chemical oxygen scavengers are used to remove any remaining oxygen. The oxygen scavengers most commonly used in boiler systems are sodium sulfite, sodium bisulfite, hydrazine, catalyzed versions of the sulfites and hydrazine, and organic oxygen scavengers, such as hydroquinone and ascorbate.

It is of critical importance to select and properly use the best chemical oxygen scavenger for a given system. Principal factors that determine the best oxygen scavenger for a particular application include reaction speed, residence time in the system, operating temperature and pressure, and feed-water pH. Interferences with the scavenger–oxygen reaction, decomposition products, and reactions with metals in the system are also important factors. Other contributing factors include the use of feed water for attemperation, the presence of economizers in the system, and the end use of the steam. Chemical oxygen scavengers should be fed to allow ample time for the scavenger–oxygen reaction to occur. The deaerator storage system and the feed-water storage tank are commonly used feed points.

Scale and deposits are controlled through the use of phosphates, chelants, and polymers. Phosphates are precipitating treatments, and chelants are solubilizing treatments. Polymers are most widely used to disperse particulates but they are also used to solubilize contaminants under certain conditions.

Phosphate Treatment. Calcium phosphate is virtually insoluble in boiler water. Even small levels of phosphate can be maintained to ensure the precipitation of calcium phosphate in the bulk boiler water, away from heating surfaces. Therefore, the introduction of phosphate treatment eliminates the formation of calcium carbonate scale on tube surfaces. When calcium phosphate is formed in boiler water of sufficient alkalinity, a particle with a relatively nonadherent surface charge is produced. This does not prevent the development of deposit accumulations over time, but the deposits can be controlled reasonably well by blowdown.

In a phosphate precipitation treatment program, the magnesium portion of the hardness contamination is precipitated preferentially as magnesium silicate. If silica is not present, the magnesium will precipitate as magnesium hydroxide. If insufficient boiler water hydroxide is being maintained, magnesium can combine with phosphate. Magnesium phosphate has a surface charge that can cause it to adhere to tube surfaces and then collect other solids. For this reason, alkalinity is an important part of a phosphate precipitation program.

Phosphate–Polymer Control. Phosphate treatment results are improved by organic supplements. Naturally occurring organics such as lignins, tannins, and starches were the first supplements used. The organics were added to promote the formation of a fluid sludge that would settle in the mud drum. Bottom blowdown from the mud drum removed the sludge.

There have been many advances in organic treatments. Synthetic polymers are now used widely, and the emphasis is on dispersion of particles rather than fluid sludge formation. Although this mechanism is quite complex, polymers alter the surface area and the surface charge to mass ratio of the boiler solids. Many synthetic polymers are used in phosphate precipitation programs. Most are effective in dispersing magnesium silicate and magnesium hydroxide as well as calcium phosphate. The polymers are usually low in molecular weight and have numerous active sites. Some polymers are used specifically for hardness salts or for iron; some are effective for a broad spectrum of ions.

Chelant Control. Chelants are the prime additives in a solubilizing boiler water treatment program. Chelants have the ability to complex many cations (hardness and heavy metals under boiler water conditions). They accomplish this by locking metals into a soluble organic ring structure. The chelated cations do not deposit in the boiler. When applied with a dispersant, chelants produce clean waterside surfaces.

Chelant–Polymer Control. Iron oxide is of particular concern in today's boiler water treatment programs. Deposition from low (less than 1.0 ppm) hardness boiler feed water is eliminated with chelant programs and can be reduced by up to 95% by a good polymer–phosphate treatment program. Iron oxide is an increasingly significant contributor to boiler deposits, because of the virtual elimination of hardness deposits in many systems and because the high heat transfer rates of many boilers encourage iron deposition.

A chelant–polymer combination is an effective approach to controlling iron oxide. Adequate chelant is fed to complex hardness and soluble iron, with a slight excess to solubilize iron contamination. Polymers are then added to condition and disperse any remaining iron oxide contamination.

A chelant–polymer program can produce clean waterside surfaces, contributing to much more reliable boiler operation. Out-of-service boiler cleaning schedules can be extended and, in some cases, eliminated. This depends on operational control and feed-water quality. Chelants with high complexing stabilities are "forgiving" treatments; they can remove deposits that form when feed-water quality or treatment control periodically deviates from standard.

Phosphate–Chelant–Polymer Combinations. Combinations of polymer, phosphate, and chelant are commonly used to produce results comparable to chelant–polymer treatment in boilers operating at 4137×10^3 Pa or less. Boiler cleanliness is improved over phosphate treatment, and the presence of phosphate provides an easy means of testing to confirm the presence of treatment in the boiler water.

Polymer-only Treatment. Polymer-only treatment programs are also used with a degree of success. In this treatment, the polymer is usually used as a weak chelant to complex the feed-water hardness. These treatments are most successful when feed-water hardness is consistently low.

High Pressure Boiler Water Treatment. High pressure boilers usually have feed water composed of demineralized makeup water and a high percentage of condensate returns. Because of these conditions, high pressure boilers are prone to caustic attack. Low pressure boilers that use demineralized water and condensate as feed water are also susceptible to caustic attack.

There are several means by which boiler water can become highly concentrated. One of the most common is iron oxide deposition on radiant wall tubes. Iron oxide deposits are often quite porous and act as miniature boilers. Water is drawn into the iron oxide deposit. Heat applied to the deposit from the tube wall generates steam, which passes out through the deposit. More water enters the deposit, taking the place of the steam. This cycle is repeated and the water beneath the deposit is concentrated to extremely high levels. It is possible to have 100,000 ppm of caustic beneath the deposit while the bulk water contains only about 5–10 ppm of caustic.

Boiler feed-water systems that use demineralized or evaporated makeup or pure condensate may be protected from caustic attack through coordinated phosphate and pH control. Phosphate buffers the boiler water, reducing the chance of large pH changes due to the development of high caustic concentrations. Excess caustic combines with disodium phosphate and forms trisodium phosphate. Sufficient disodium phosphate must be available to combine with all of the free caustic in order to form trisodium phosphate.

Disodium phosphate neutralizes caustic by the following reaction:

$$Na_2HPO_4 + NaOH = Na_3PO_4 + H_2O$$

This results in the prevention of caustic buildup beneath deposits or within a crevice where leakage is occurring. Caustic corrosion (and caustic embrittlement, discussed later) does not occur, because high caustic concentrations do not develop.

Figure 5 shows the phosphate–pH relationship recommended to control boiler corrosion. Different forms of phosphate consume or add caustic as the phosphate shifts to the proper form. For example, addition of monosodium phosphate consumes caustic as it reacts with caustic to form disodium phosphate in the boiler water according to the following reaction:

$$NaH_2PO_4 + NaOH = Na_2HPO_4 + H_2O$$

Conversely, addition of trisodium phosphate adds caustic, increasing boiler water pH:

$$Na_3PO_4 + H_2O = Na_2HPO_4 + NaOH$$

Control is achieved through feed of the proper type of phosphate either to raise or to lower the pH while maintaining the proper phosphate level. Increasing blowdown

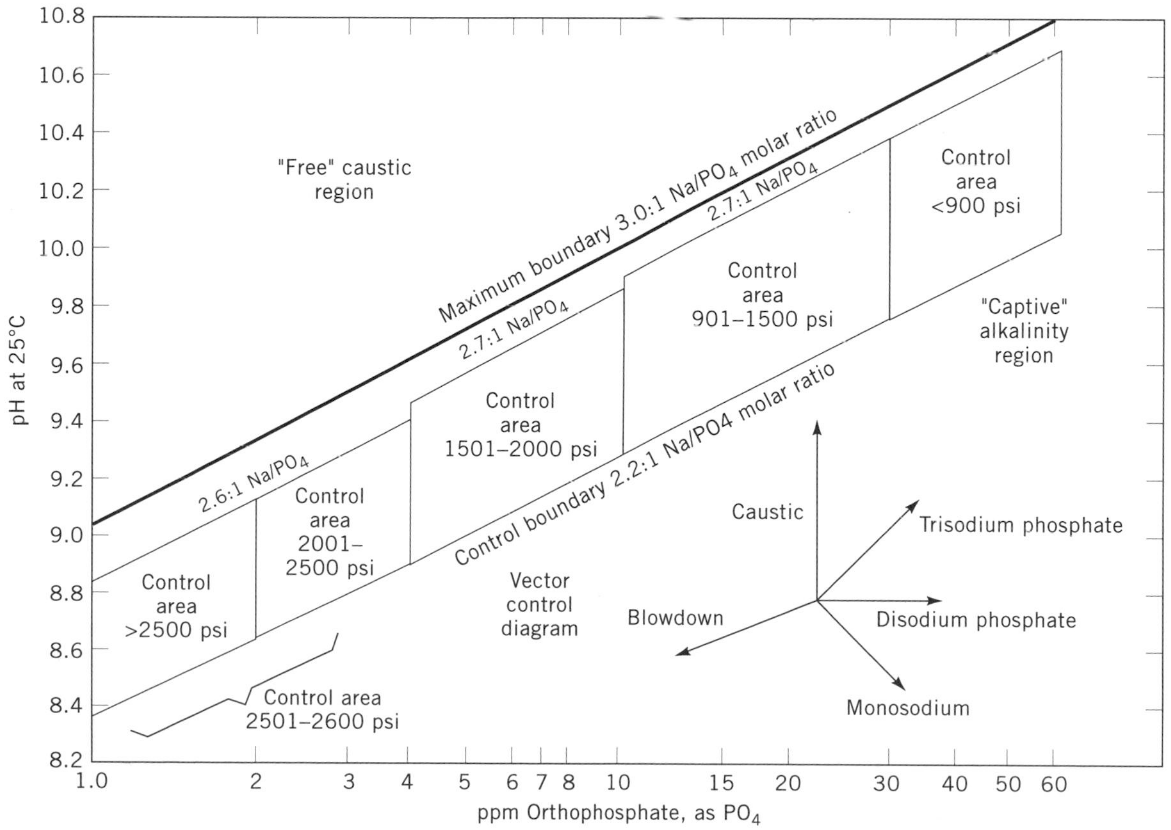

Figure 5. Coordinated phosphate–pH control avoids both acid and caustic corrosion.

lowers both phosphate and pH. Therefore, various combinations and feedrates of phosphate, blowdown adjustment, and caustic addition are used to maintain proper phosphate–pH levels.

Elevated temperatures at the boiler tube wall or deposits can result in some precipitation of phosphate. This effect, termed *phosphate hideout,* usually occurs when loads increase. When the load is reduced, phosphate reappears.

Clean boiler water surfaces reduce potential concentration sites for caustic. Deposit control treatment programs, such as those based on chelants and synthetic polymers, can help provide clean surfaces.

When steam blanketing is occurring, corrosion can take place even without the presence of caustic, due to the steam–magnetite reaction and the dissolution of magnetite. In such cases, operational changes or design modifications may be necessary to eliminate the cause of the problem.

If deposits are minimized, the areas where caustic can be concentrated is reduced. To minimize the iron deposition in 6.895–12.07 $\times$ 10^6 Pa boilers, specific polymers have been designed to disperse the iron and keep it in the bulk water. As with phosphate precipitation and chelant control programs, the use of these polymers with coordinated phosphate–pH treatment improves deposit control.

Supercritical boilers use all-volatile treatments, generally consisting of ammonia and hydrazine. Because of the extreme potential for deposit formation and steam contamination, no solids can be tolerated in supercritical once-through boiler water, including treatment solids.

Steam Purity

Boiler water solids carried over with steam form deposits in nonreturn valves, superheaters, and turbine stop and control valves. Carryover can contaminate process streams and affect product quality. Deposition in superheaters can lead to failure due to overheating and corrosion, as shown in Figure 6.

Superheated steam turbines are particularly prone to damage by carryover. Sticking of governor and stop valves due to deposits can cause turbine overspeed and catastrophic damage. Solid particles in steam can erode turbine parts, while deposition on turbine blades can reduce efficiency and capacity. Losses of 5% in turbine efficiency and 20% in turbine capacity have occurred due to deposition. When large slugs of boiler water carry over with steam, the resulting thermal and mechanical shock can cause severe damage.

Steam can be contaminated with solids even when carryover is not occurring. Contaminated spray attemperating water, used to control superheated steam temperature at the turbine inlet, can introduce solids into steam. A heat exchanger coil may be placed in the boiler mud drum to provide attemperation of the superheated steam. Because the mud drum is at a higher pressure than superheated steam, contamination will occur if leaks develop in the coil.

CHEMICAL TREATMENT OF CONDENSATE SYSTEMS

Condensate systems can be chemically treated to reduce metal corrosion. Treatment chemicals include neutralizing amines, filming amines, and oxygen scavenger-metal passivators.

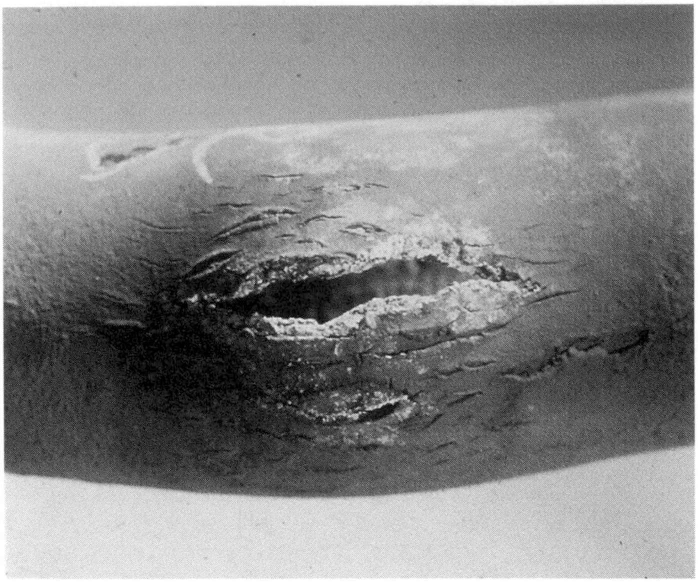

Figure 6. Boiler water contamination of the steam caused superheater deposits, which led to tube metal overheating and failure.

Neutralizing Amines

Neutralizing amines are used to neutralize the acid (H+) generated by the dissolution of carbon dioxide or other acidic process contaminants in the condensate. These amines hydrolyze when added to water and generate the hydroxide ions required for neutralization:

$$R—NH_2 + H_2O \quad R—NH_{3+} + OH^-$$

The overall neutralization reaction can be written as shown:

$$R—NH_{3+} + OH + H_2CO_3 = R–NH_{3+} + HCO^{3-} + H_2O$$

By regulating the neutralizing amine feed rate, the condensate pH can be elevated to within a desired range (eg, 8.8–9.2 for a mixed copper-iron condensate system).

Filming Amines

Another approach to controlling condensate system corrosion is the use of chemicals that form a protective film on metal surfaces (Fig. 7). This approach has come into widespread use with the development of suitable products containing long-chain nitrogenous materials. Filming amines protect against oxygen and carbon dioxide corrosion by replacing the loose oxide scale on metal surfaces with a thin amine film barrier.

Advances have been made in formulating filming amine treatments. Straight filming amines containing one ingredient, such as octadecylamine, are effective but often fail to cover the entire system and can produce fouling. Emulsifiers and, in some cases, small amounts of

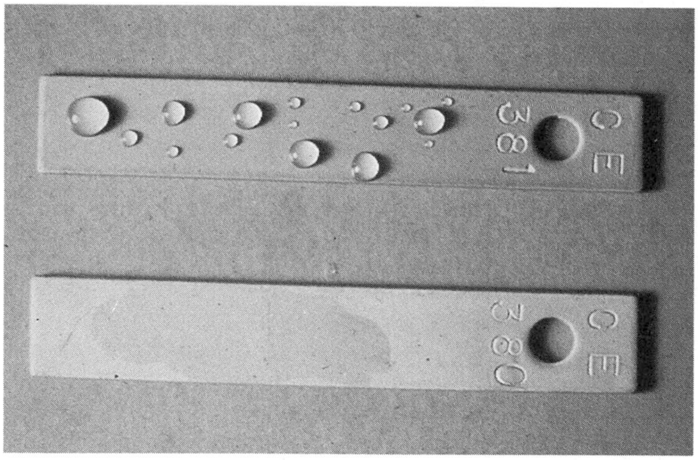

Figure 7. Test specimens illustrate the nonwettable surface produced by a filming amine (381) compared with an untreated surface (380).

neutralizing amines can be added to improve film distribution by providing more uniform coverage. This increases system protection and reduces the fouling potential. Application experience has shown that combination amines (filming and neutralizing amines with dispersant aids) provide a superior film bond, reduce deposit problems, and provide better system coverage and thus provide more complete and economical corrosion protection.

Oxygen Scavenging and Metal Passivation

Where oxygen invades the condensate system, corrosion of iron and copper-bearing components can be overcome through proper pH control and the injection of an oxygen scavenger. One important factor to consider in choosing an oxygen scavenger for condensate treatment is its reactivity with oxygen at the temperature and pH of the system. A scavenger that removes oxygen rapidly provides the best protection for the condensate metallurgy. Hydroquinone has been shown to be particularly effective for most systems.

The use of neutralizing amines in conjunction with an oxygen scavenger–metal passivator improves corrosion control in two ways. First, because any acidic species present is neutralized and pH is increased, the condensate becomes less corrosive. Second, most oxygen scavenger–passivators react more rapidly at the mildly alkaline conditions maintained by the amine than at lower pH levels. For these reasons, this combination treatment is gaining wide acceptance, particularly for the treatment of condensate systems that are contaminated by oxygen.

COOLING SYSTEMS

Cooling System Corrosion

Corrosion can be defined as the destruction of a metal by chemical or electrochemical reaction with its environment. In cooling systems, corrosion causes two basic problems. The first and most obvious is the failure of equipment with the resultant cost of replacement and plant downtime. The second is decreased plant efficiency due to loss of heat transfer, the result of heat exchanger fouling caused by the accumulation of corrosion products.

Corrosion occurs at the anode, where metal dissolves. Often, this is separated by a physical distance from the cathode, where a reduction reaction takes place. An electrical potential difference exists between these sites, and current flows through the solution from the anode to the cathode. This is accompanied by the flow of electrons from the anode to the cathode through the metal (Fig. 8).

For steel, the typical anodic oxidation reaction is

$$Fe = Fe^{2+} + 2e^-$$

This reaction is accompanied by the following:

$$Fe^{2+} + 2OH^- = Fe(OH)_2$$

The ferrous hydroxide then combines with oxygen and water to produce ferric hydroxide, $Fe(OH)_3$, which becomes common iron rust when dehydrated to Fe_2O_3.

The primary cathodic reaction in cooling systems is

$$O_2 + H_2O + 2e^- = 2OH^-$$

The production of hydroxide ions creates a localized high pH at the cathode, approximately 1–2 pH units above bulk water pH. Dissolved oxygen reaches the surface by diffusion, as indicated by the wavy lines in Figure 8. The oxygen reduction reaction controls the rate of corrosion in cooling systems; the rate of oxygen diffusion is usually the limiting factor.

Another important cathodic reaction is

$$2H^+ + 2e^- = H_2$$

At neutral or higher pH, the concentration of H^+ ions is too low for this reaction to contribute significantly to the overall corrosion rate. However, as pH decreases, this reaction becomes more important until, at a pH of about 4, it becomes the predominant cathodic reaction.

TYPES OF CORROSION

The formation of anodic and cathodic sites, necessary to produce corrosion, can occur for any of a number of rea-

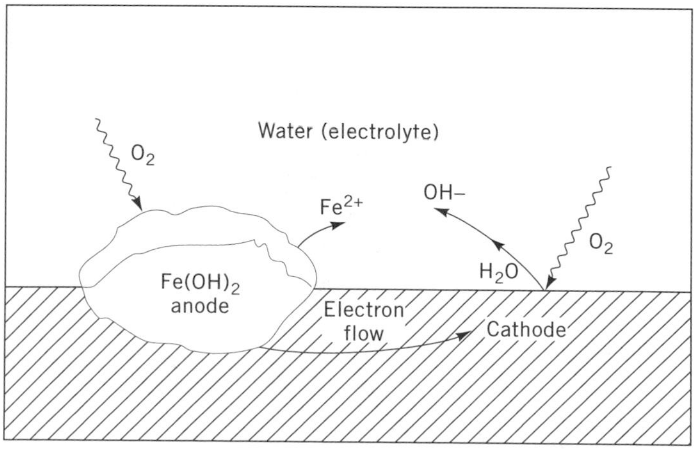

Figure 8. A corrosion cell.

sons: impurities in the metal, localized stresses, metal grain size or composition differences, discontinuities on the surface, and differences in the local environment (eg, temperature, oxygen, or salt concentration). When these local differences are not large and the anodic and cathodic sites can shift from place to place on the metal surface, corrosion is uniform. With uniform corrosion, fouling is usually a more serious problem than equipment failure.

Localized corrosion, which occurs when the anodic sites remain stationary, is a more serious industrial problem. Forms of localized corrosion include pitting, selective leaching (eg, dezincification), galvanic corrosion, crevice or underdeposit corrosion, intergranular corrosion, stress corrosion cracking, and microbiologically influenced corrosion. Another form of corrosion, which cannot be accurately categorized as either uniform or localized, is erosion corrosion.

Pitting

Pitting is one of the most destructive forms of corrosion and also one of the most difficult to predict in laboratory tests (Fig. 9). Pitting occurs when anodic and cathodic sites become stationary due to large differences in surface conditions. It is generally promoted by low velocity or stagnant conditions (eg, shellside cooling) and by the presence of chloride ions. Once a pit is formed, the solution inside it is isolated from the bulk environment and becomes increasingly corrosive. The high corrosion rate in the pit produces an excess of positively charged metal cations, which attract chloride anions. In addition, hydrolysis produces H^+ ions. The increase in acidity and concentration within the pit promotes even higher corrosion rates, and the process becomes self-sustaining. Inhibitors can be used to control pitting, but they must be applied correctly.

Selective Leaching

Selective leaching is the corrosion of one element of an alloy. The most common example in cooling systems is dezincification, which is the selective removal of zinc from copper–zinc alloys. The conditions that promote the pitting of steel also promote the pitting of brass, which in cooling systems usually occurs by dezincification. Low pH conditions (<6.0) and high free chlorine residuals (>1.0 ppm) are particularly aggressive in producing dezincification. The dezincification resistance varies with the alloy. For example, 70–30 brass is less resistant than admiralty brass (70–30 brass plus 1% tin), which is less resistant than inhibited Admiralty brass (Admiralty brass plus a small amount of arsenic, antimony, or phosphorus).

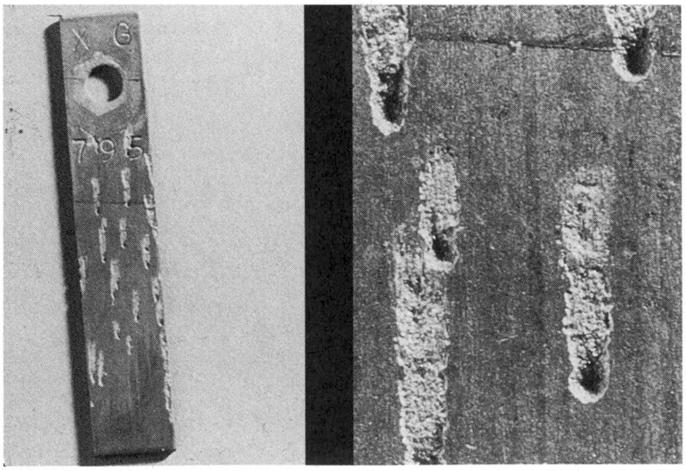

Figure 9. Pitting corrosion is damaging because it can lead rapidly to equipment failure.

Galvanic Corrosion

Galvanic corrosion occurs when two dissimilar metals are in contact in a solution. The contact must be good enough to conduct electricity, and both metals must be exposed to the solution. The driving force for galvanic corrosion is the electric potential difference that develops between two metals. This difference increases as the distance between the metals in the galvanic series increases.

Table 4 shows a galvanic series for some commercial metals and alloys. When two metals from the series are in contact in solution, the corrosion rate of the more active (anodic) metal increases and the corrosion rate of the more noble (cathodic) metal decreases.

Galvanic corrosion can be controlled by the use of sacrificial anodes. This is a common method of controlling corrosion in heat exchangers with Admiralty tube bundles and carbon steel tube sheets and channel heads. The anodes are bolted directly to the steel and protect a limited area around the anode. Proper placement of sacrificial anodes is a precise science.

The most serious form of galvanic corrosion occurs in cooling systems that contain both copper and steel alloys. It results when dissolved copper plates onto a steel sur-

Table 4. Galvanic Series of Metals and Alloys[a]

Corroded End
(anodic, or least noble)

Magnesium	Inconel (active)
Magnesium alloys	Hastelloy A
Zinc	Hastelloy B
Aluminum 2S	Brasses
Cadmium	Copper
Aluminum 17ST	Bronzes
Steel or iron	Copper-nickel alloys
Cast iron	Titanium
Chromium-iron (active)	Monel
Ni-Resist	Silver solder
18-8-Cr-Ni-Fe (active)	Nickel (passive)
15-8-3′-Cr-Ni-Mo-Fe (active)	Inconel (passive)
Hastelloy C	Chromium-iron (passive)
Lead–tin solders	18-8 Cr-Ni-Fe (passive)
Lead	18-8-3 Cr-Ni-Mo-Fe (passive)
Tin	Silver
Nickel (active)	Graphite

Protected End
(cathodic, or most noble)

[a] Courtesy of International Nickel Company, Inc.

face and induces rapid galvanic attack of the steel. The amount of dissolved copper required to produce this effect is small and the increased corrosion is difficult to inhibit once it occurs. A copper corrosion inhibitor is needed to prevent copper dissolution.

Crevice Corrosion

Crevice corrosion is intense localized corrosion that occurs within a crevice or any area that is shielded from the bulk environment. Solutions within a crevice are similar to solutions within a pit in that they are highly concentrated and acidic. Because the mechanisms of corrosion in the two processes are virtually identical, conditions that promote pitting also promote crevice corrosion. Alloys that depend on oxide films for protection (eg, stainless steel and aluminum) are highly susceptible to crevice attack because the films are destroyed by high chloride ion concentrations and low pH. This is also true of protective films induced by anodic inhibitors.

The best way to prevent crevice corrosion is to prevent crevices. From a cooling water standpoint, this requires the prevention of deposits on the metal surface. Deposits may be formed by suspended solids (eg, silt, silica) or by precipitating species, such as calcium salts.

Intergranular Corrosion

Intergranular corrosion is localized attack that occurs at metal grain boundaries. It is most prevalent in stainless steels that have been improperly heat treated. In these metals, the grain boundary area is depleted in chromium and, therefore, is less resistant to corrosion. Intergranular corrosion also occurs in certain high strength aluminum alloys. In general, it is not of significance in cooling systems.

Stress Corrosion Cracking

Stress corrosion cracking (SCC) is the brittle failure of a metal by cracking under tensile stress in a corrosive environment. Failures tend to be transgranular, although intergranular failures have been noted. Commonly used cooling system alloys that may crack due to stress include austenitic stainless steels (300 series) and brasses. The susceptibility of stainless steels to SCC increases significantly as the temperature is increased. Most laboratory stainless steel SCC testing is done at about 150°C, because it is difficult to promote cracking at temperatures below 93°C. For this reason, SCC of stainless steels has not been widely observed in cooling systems.

Chloride is the main contributor to SCC of stainless steels. High chloride concentrations, resulting from high chloride levels in the makeup water and/or high cycles of concentration, will increase susceptibility. Although low water temperatures generally preclude cracking, SCC of stainless steels can occur in cooling systems.

For brasses, the ammonium ion is the principal cause of SCC. Few service failures have been reported when ammonia is not present.

The most likely places for SCC to be initiated are crevices or areas where the flow of water is restricted. This is due to the buildup of corrodent concentrations in these areas. For example, chloride can concentrate from 100 ppm in the bulk water to as high as 10,000 ppm (1%) in a crevice. Deposits are initiating sites because of crevices formed beneath them. The low water velocities in shell-side cooling are also detrimental.

The most effective way to prevent SCC in both stainless steel and brass systems is to keep the system clean and free of deposits. An effective deposit control treatment is imperative. A good corrosion inhibitor is also beneficial. Chromate and phosphate have each been used successfully to prevent the SCC of stainless steel in chloride solutions.

Microbiologically Influenced Corrosion (MIC)

Microorganisms in cooling water form "biofilms" on cooling system surfaces. Biofilms consist of populations of sessile organisms and their hydrated polymeric secretions. Numerous types of organisms may exist in any particular biofilm, ranging from strictly aerobic bacteria at the water interface to anaerobic bacteria such as sulfate-reducing bacteria (SRB) at the oxygen-depleted metal surface. The presence of a biofilm can contribute to corrosion in three ways: physical deposition, production of corrosive by-products, and depolarization of the corrosion cell caused by chemical reactions.

As discussed above, deposits can cause accelerated localized corrosion by creating differential aeration cells. This same phenomenon occurs with a biofilm. The nonuniform nature of biofilm formation creates an inherent differential, which is enhanced by the oxygen consumption of organisms in the biofilm.

Many of the by-products of microbial metabolism, including organic acids and hydrogen sulfide, are corrosive. These materials can concentrate in the biofilm, causing accelerated metal attack. Corrosion tends to be self-limiting due to the buildup of corrosion reaction products. However, microbes can absorb some of these materials in their metabolism, thereby removing them from the anodic or cathodic site. The removal of reaction products, termed *depolarization,* stimulates further corrosion. Figure 10 shows a typical result of microbial corrosion. The surface exhibits scattered areas of localized corrosion, unrelated to flow pattern. The corrosion appears to spread in a somewhat circular pattern from the site of initial colonization.

Erosion Corrosion

Erosion corrosion is the increase in the rate of metal deterioration from abrasive effects. It can be identified by grooves and rounded holes, which usually are smooth and have a directional pattern. Erosion corrosion is increased by high water velocities and suspended solids. It is often localized at areas where water changes direction. Cavitation (damage due to the formation and collapse of bubbles in high velocity turbines, propellers, etc) is a form of erosion corrosion. Its appearance is similar to closely spaced pits, although the surface is usually rough.

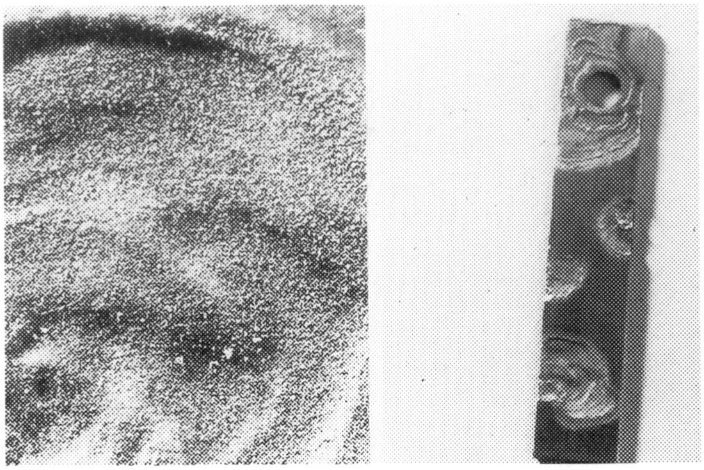

Figure 10. Microbiologically influenced corrosion (MIC).

CONTROL OF COOLING SYSTEM CORROSION

Corrosion control requires a change in either the metal or the environment. The first approach, changing the metal, is expensive. Also, highly alloyed materials, which are resistant to general corrosion, are more prone to failure by localized corrosion mechanisms such as stress corrosion cracking.

The second approach, changing the environment, is a widely used, practical method of preventing corrosion. In aqueous systems, there are three ways to effect a change in environment to inhibit corrosion: (*1*) form a protective film of calcium carbonate on the metal surface using the natural calcium and alkalinity in the water, (*2*) remove the corrosive oxygen from the water, either by mechanical or chemical deaeration, and (*3*) add corrosion inhibitors.

Calcium Carbonate Protective Scale

The Langelier Saturation Index (LSI) is a useful tool for predicting the tendency of a water to deposit or dissolve calcium carbonate. Work published in 1936 deals with the conditions at which a water is in equilibrium with calcium carbonate. An equation developed by Langelier makes it possible to predict the tendency of calcium carbonate either to precipitate or to dissolve under varying conditions. The equation expresses the relationship of pH, calcium, total alkalinity, dissolved solids, and temperature as they relate to the solubility of calcium carbonate in waters with a pH of 6.5–9.5:

$$pH_s = (pK_2 - pK_s) + pCa^{2+} + pAlk$$

where pH_s is the pH at which water with a given calcium content and alkalinity is in equilibrium with calcium carbonate, K_2 is the second dissociation constant for carbonic acid, and K_s is the solubility product constant for calcium carbonate. These terms are functions of temperature and total mineral content. Their values for any given condition can be computed from known thermodynamic constants. Both the calcium ion and the alkalinity terms are the negative logarithms of their respective concentrations. The calcium content is molar, while the alkalinity is an equivalent concentration (ie, the titratable equivalent of base per liter). The calculation of the pH_s has been simplified by the preparation of various nomographs. A typical one is shown in Figure 11.

The difference between the actual pH (pH_a) of a sample of water and the pH_s ($pH_a - pH_s$) is called the Langelier Saturation Index. This index is a qualitative indication of the tendency of calcium carbonate to deposit or dissolve. If the LSI is positive, calcium carbonate tends to deposit. If it is negative, calcium carbonate tends to dissolve. If it is zero, the water is at equilibrium.

The LSI measures only the directional tendency or driving force for calcium carbonate to precipitate or dissolve. It cannot be used as a quantitative measure. Two different waters, one of low hardness (corrosive) and the other of high hardness (scale-forming), can have the same saturation index.

The Stability Index developed by Ryzner makes it possible to distinguish between two such waters. This index is based on a study of actual operating results with waters having various saturation indexes.

$$\text{Stability Index} = 2(pH_s) - pH_a$$

Where waters have a Stability Index of 6.0 or less, scaling increases and the tendency to corrode decreases. Where the Stability Index exceeds 7.0, scaling may not occur at all. As the Stability Index rises above 7.5 or 8.0, the probability of corrosion increases. Use of the LSI together with the Stability Index contributes to more accurate prediction of the scaling or corrosive tendencies of a water.

A uniform coating of calcium carbonate deposited on the metal surfaces physically segregates the metal from the corrosive environment. To develop the positive LSI required to deposit calcium carbonate, it is usually necessary to adjust the pH or calcium content of the water. Soda ash, caustic soda, or lime (calcium hydroxide) may be used for this adjustment. Lime is usually the most economical alkali because it raises the calcium content as well as the alkalinity.

Theoretically, controlled deposition of calcium carbonate scale can provide a film thick enough to protect, yet thin enough to allow adequate heat transfer. However, low temperature areas do not permit the development of sufficient scale for corrosion protection, and excessive scale forms in high temperature areas and interferes with heat transfer. Therefore, this approach is not used for industrial cooling systems. Controlled calcium carbonate deposition has been used successfully in some waterworks distribution systems where substantial temperature increases are not encountered.

Mechanical and Chemical Deaeration

The corrosive qualities of water can be reduced by deaeration. Vacuum deaeration has been used successfully in once-through cooling systems. When all oxygen is not removed, catalyzed sodium sulfite can be used to remove the remaining oxygen. The sulfite reaction with dissolved oxygen is

$$Na_2SO_3 + O_2 = Na_2SO_4$$

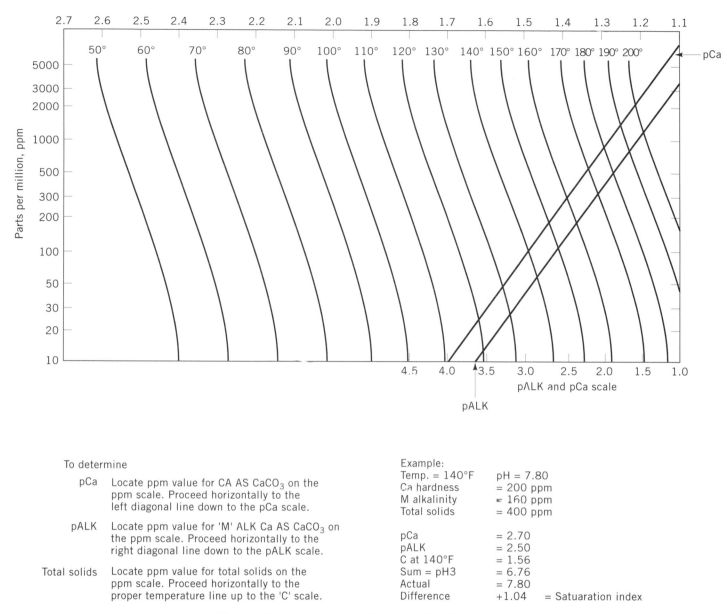

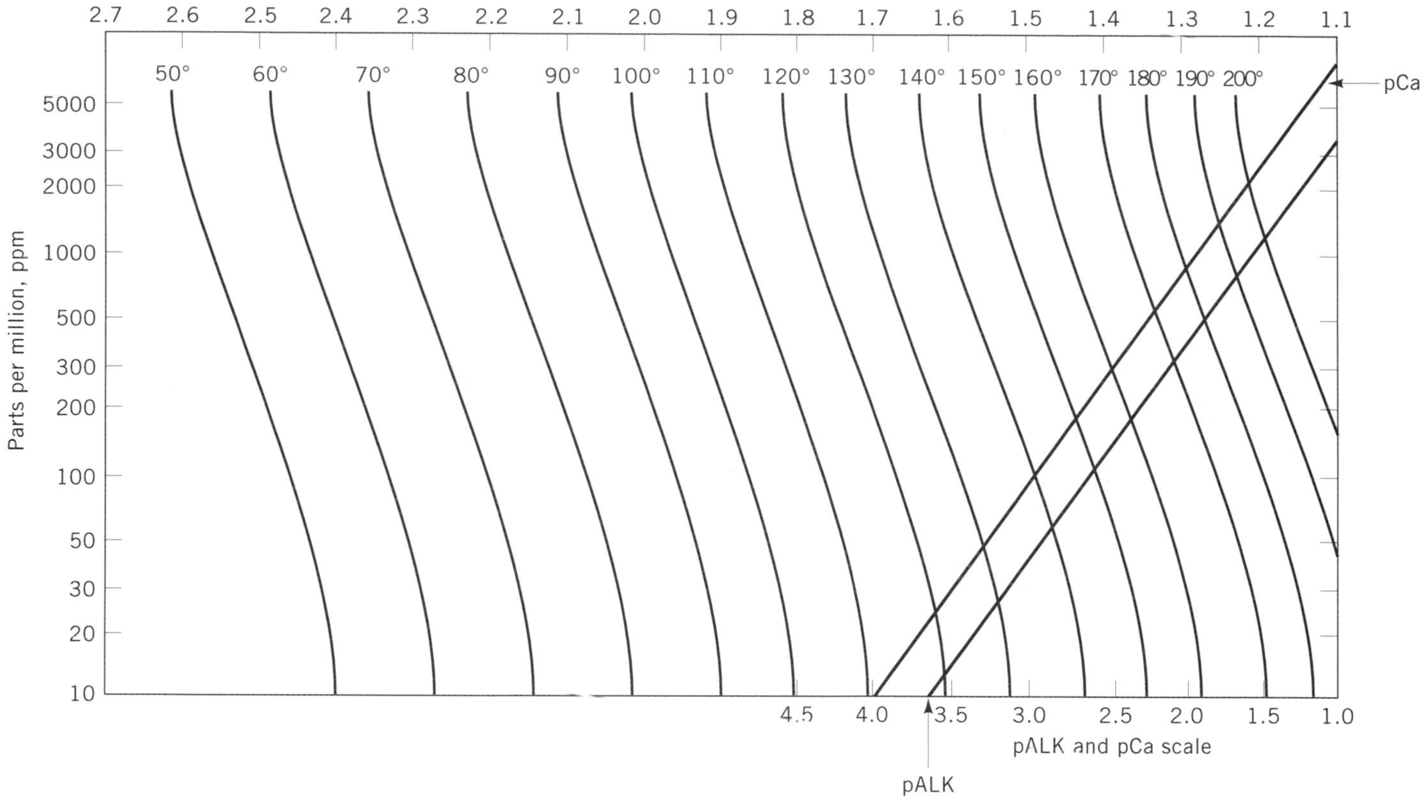

To determine

pCa	Locate ppm value for CA AS $CaCO_3$ on the ppm scale. Proceed horizontally to the left diagonal line down to the pCa scale.
pALK	Locate ppm value for 'M' ALK Ca AS $CaCO_3$ on the ppm scale. Proceed horizontally to the right diagonal line down to the pALK scale.
Total solids	Locate ppm value for total solids on the ppm scale. Proceed horizontally to the proper temperature line up to the 'C' scale.

Example:

Temp. = 140°F	pH = 7.80	
Ca hardness	= 200 ppm	
M alkalinity	= 160 ppm	
Total solids	= 400 ppm	
pCa	= 2.70	
pALK	= 2.50	
C at 140°F	= 1.56	
Sum = pH3	= 6.76	
Actual	= 7.80	
Difference	+1.04	= Satuaration index

Figure 11. A typical Langelier Saturation Index chart.

The use of catalyzed sodium sulfite for chemical deaeration requires eight parts of catalyzed sodium sulfite for each part of dissolved oxygen. In certain systems where vacuum deaeration is already used, the application of catalyzed sodium sulfite may be economically justified for removal of the remaining oxygen. The use of sodium sulfite may also be applicable to some closed loop cooling systems. In open recirculating cooling systems, continual replenishment of oxygen as the water passes over the cooling tower makes deaeration impractical.

Corrosion Inhibitors

A corrosion inhibitor is any substance that effectively decreases the corrosion rate when added to an environment. An inhibitor can be identified most accurately in relation to its function: removal of the corrosive substance, passivation, precipitation, or adsorption.

Deaeration (mechanical or chemical) removes the corrosive substance–oxygen.

Passivating (anodic) inhibitors form a protective oxide film on the metal surface; they are the best inhibitors because they can be used in economical concentrations and their protective films are tenacious and tend to be rapidly repaired if damaged.

Precipitating (cathodic) inhibitors are simply chemicals that form insoluble precipitates that can coat and protect the surface; precipitated films are not as tenacious as passive films and take longer to repair after a system upset.

Adsorption inhibitors have polar properties that cause them to be adsorbed on the surface of the metal; they are usually organic materials.

Passivation Inhibitors. Examples of passivators (anodic inhibitors) include chromate, nitrite, molybdate, and orthophosphate. All are oxidizers and promote passivation by increasing the electrical potential of the iron. Chromate and nitrite do not require oxygen and thus can be the most effective. Chromate is an excellent aqueous corrosion inhibitor, particularly from a cost perspective. However, due to health and environmental concerns, use of chromate has decreased significantly and will probably be outlawed soon. Nitrite is also an effective inhibitor, but in open systems it tends to be oxidized to nitrate.

Both molybdate and orthophosphate are excellent passivators in the presence of oxygen. Molybdate can be an effective inhibitor, especially when combined with other chemicals. Orthophosphate is not really an oxidizer per se, but becomes one in the presence of oxygen. If iron is put into a phosphate solution without oxygen present, the corrosion potential remains active and the corrosion rate is not reduced. However, if oxygen is present, the corrosion potential increases in the noble direction and the corrosion rate decreases significantly.

A negative attribute of orthophosphate is its tendency to precipitate with calcium hardness found in natural waters. In recent years, deposit control agents that prevent this deposition have been developed. Due to its relatively low cost, orthophosphate is widely used as an industrial corrosion inhibitor.

Precipitating Inhibitors. As discussed earlier, the localized pH at the cathode of the corrosion cell is elevated due to the generation of hydroxide ions. Precipitating inhibitors form complexes that are insoluble at this high pH (1–2 pH units above bulk water), but whose deposition can be controlled at the bulk water pH (typically 7–9 pH). A good example is zinc, which can precipitate as hydroxide, carbonate, or phosphate. Calcium carbonate and calcium orthophosphate are also precipitating inhibitors. Orthophosphate thus exhibits a dual mechanism, acting as both an anodic passivator and a cathodic precipitator.

Copper Corrosion Inhibitors. The most effective corrosion inhibitors for copper and its alloys are the aromatic triazoles, such as benzotriazole (BZT) and tolyltriazole (TTA). These compounds bond directly with cuprous oxide (Cu_2O) at the metal surface, forming a “chemisorbed” film. The plane of the triazole lies parallel to the metal surface, thus each molecule covers a relatively large surface area. The exact mechanism of inhibition is unknown. Various studies indicate anodic inhibition, cathodic inhibition, or a combination of the two. Other studies indicate the formation of an insulating layer between the water surface and the metal surface. A recent study supports the idea of an electronic stabilization mechanism. The protective cuprous oxide layer is prevented from oxidizing to the nonprotective cupric oxide. This is an anodic mechanism. However, the triazole film exhibits some cathodic properties as well.

In addition to bonding with the metal surface, triazoles bond with copper ions in solution. Thus dissolved copper represents a “demand” for triazole, which must be satisfied before surface filming can occur. Although the surface demand for triazole filming is generally negligible, copper corrosion products can consume a considerable amount of treatment chemical. Excessive chlorination will deactivate the triazoles and significantly increase copper corrosion rates. Due to all of these factors, treatment with triazoles is a complex process.

Adsorption Inhibitors. Adsorption inhibitors must have polar properties to be adsorbed and block the surface against further adsorption. Typically, they are organic compounds containing nitrogen groups, such as amines, and organic compounds containing sulfur or hydroxyl groups. The size, orientation, shape, and electrical charge distribution of the molecules are all important factors. Often, these molecules are surfactants and have dual functionality. They contain a hydrophilic group, which adsorbs onto the metal surface, and an opposing hydrophobic group, which prevents further wetting of the metal.

Glycine derivatives and aliphatic sulfonates are examples of compounds that can function in this way. The use of these inhibitors in cooling systems is usually limited by their biodegradability and their toxicity toward fish. In addition, they can form thick, oily surface films, that may severely retard heat transfer.

Silicates. For many years, silicates have been used to inhibit aqueous corrosion, particularly in potable water systems. Probably due to the complexity of silicate chemistry, their mechanism of inhibition has not yet been firmly established. They are nonoxidizing and require oxygen to inhibit corrosion, so they are not passivators in the classical sense. Yet they do not form visible precipitates on the metal surface. They appear to inhibit by an adsorption mechanism. It is thought that silica and iron corrosion products interact. However, recent work indicates that this interaction may not be necessary. Silicates are slow-acting inhibitors; in some cases, 2 or 3 weeks may be required to establish protection fully. It is believed that the polysilicate ions or colloidal silica are the active species and these are formed slowly from monosilicic acid, which is the predominant species in water at the pH levels maintained in cooling systems.

COOLING SYSTEM DEPOSITS

Deposit accumulations in cooling water systems reduce the efficiency of heat transfer and the carrying capacity of the water distribution system. In addition, the deposits cause oxygen differential cells to form. These cells accelerate corrosion and lead to process equipment failure. Deposits range from thin, tightly adherent films to thick, gelatinous masses, depending on the depositing species and the mechanism responsible for deposition.

Deposit formation is influenced strongly by system parameters, such as water and skin temperatures, water velocity, residence time, and system metallurgy. The most severe deposition is encountered in process equipment operating with high surface temperatures and/or low water velocities. With the introduction of high efficiency film fill, deposit accumulation in the cooling tower packing has become an area of concern. Deposits are broadly categorized as scale or foulants.

Scale

Scale deposits are formed by precipitation and crystal growth at a surface in contact with water. Precipitation occurs when solubilities are exceeded either in the bulk water or at the surface. The most common scale-forming salts that deposit on heat transfer surfaces are those that exhibit retrograde solubility with temperature.

Although they may be completely soluble in the lower temperature bulk water, these compounds (eg, calcium carbonate, calcium phosphate, and magnesium silicate) supersaturate in the higher temperature water adjacent to the heat-transfer surface and precipitate on the surface.

Scaling is not always related to temperature. Calcium carbonate and calcium sulfate scaling occur on unheated surfaces when their solubilities are exceeded in the bulk water. Metallic surfaces are ideal sites for crystal nucleation because of their rough surfaces and the low velocities adjacent to the surface. Corrosion cells on the metal surface produce areas of high pH, which promote the precipitation of many cooling water salts. Once formed, scale deposits initiate additional nucleation, and crystal growth proceeds at an accelerated rate.

Scale control can be achieved through operation of the cooling system at subsaturated conditions or through the use of chemical additives. The most direct method of inhibiting formation of scale deposits is operation at subsaturation conditions, where scale-forming salts are soluble. For some salts, it is sufficient to operate at low cycles of concentration and/or control pH. However, in most cases, high blowdown rates and low pH are required so that solubilities are not exceeded at the heat transfer surface. In addition, it is necessary to maintain precise control of pH and concentration cycles. Minor variations in water chemistry or heat load can result in scaling (Fig. 12).

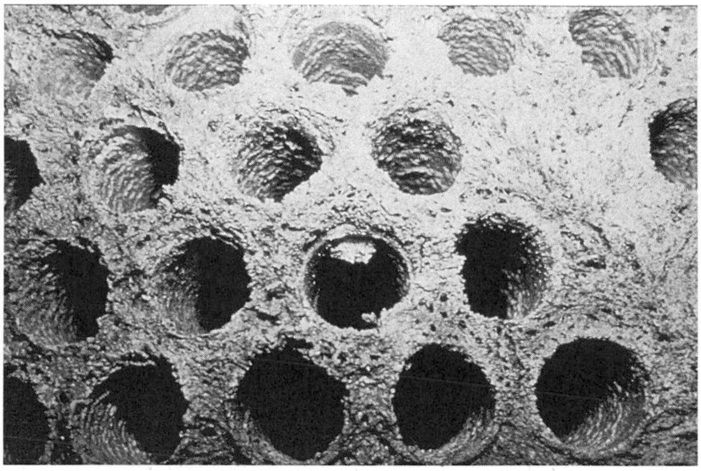

Figure 12. Calcium carbonate scaling of a surface condenser due to poor pH control.

Threshold Inhibitors. Deposit control agents that inhibit precipitation at dosages far below the stoichiometric level required for sequestration or chelation are called *threshold inhibitors.* These materials affect the kinetics of the nucleation and crystal growth of scale-forming salts and permit supersaturation without scale formation. Threshold inhibitors function by an adsorption mechanism. As ion clusters in solution become oriented, metastable microcrystallites (highly oriented ion clusters) are formed. At the initial stage of precipitation, the microcrystallite can either continue to grow (forming a larger crystal with a well defined lattice) or dissolve. Threshold inhibitors prevent precipitation by adsorbing on the newly emerging crystal, blocking active growth sites. This inhibits further growth and favors the dissolution reaction. The precipitate dissolves and releases the inhibitor, which is then free to repeat the process.

Threshold inhibitors delay or retard the rate of precipitation. Crystals eventually form, depending on the degree of supersaturation and system retention time. After stable crystals appear, their continued growth is retarded by adsorption of inhibitor. The inhibitor blocks much of the crystal surface, causing distortions in the crystal lattice as growth continues. The distortions (defects in the crystal lattice) create internal stresses, making the crystal fragile. Tightly adherent scale deposits do not form, because crystals that form on surfaces in contact with flowing water cannot withstand the mechanical force exerted by the water. The adsorbed inhibitor also disperses particles, by virtue of its electrostatic charge, and prevents the formation of strongly bound agglomerates.

The most commonly used scale inhibitors are low molecular weight acrylate polymers and organophosphorus compounds (phosphonates). Both classes of materials function as threshold inhibitors; however, the polymeric materials are more effective dispersants. Selection of a scale control agent depends on the precipitating species and its degree of supersaturation. The most effective scale control programs use both a precipitation inhibitor and a dispersant. In some cases this can be achieved with a single component (eg, polymers used to inhibit calcium phosphate at near neutral pH).

Fouling

Fouling occurs when insoluble particulates suspended in recirculating water form deposits on a surface. Fouling mechanisms are dominated by particle–particle interactions that lead to the formation of agglomerates. At low water velocities, particle settling occurs under the influence of gravity. Parameters that affect the rate of settling are particle size, relative liquid and particle densities, and liquid viscosity. The relationships of these variables are expressed by Stokes' Law. The most important factor affecting the settling rate is the size of the particle. Because of this, the control of fouling by preventing agglomeration is one of the most fundamental aspects of deposition control.

Foulants enter a cooling system with makeup water, airborne contamination, process leaks, and corrosion. Most potential foulants enter with makeup water as particulate matter, such as clay, silt, and iron oxides. Insoluble aluminum and iron hydroxides enter a system from makeup water pretreatment operations. Some well waters contain high levels of soluble ferrous iron that is later oxidized to ferric iron by dissolved oxygen in the recirculating cooling water. Because it is insoluble, the ferric iron precipitates. The steel corrosion process is also a source of ferrous iron and, consequently, contributes to fouling.

Both iron and aluminum are particularly troublesome because of their ability to act as coagulants. Also, their soluble and insoluble hydroxide forms can each cause precipitation of some water treatment chemicals, such as orthophosphate. Airborne contaminants usually consist of clay and dirt particles but can include gases such as hydrogen sulfide, which forms insoluble precipitates with many metal ions. Process leaks introduce a variety of contaminants that accelerate deposition and corrosion.

Foulants, such as river water silt, enter the system as finely dispersed particles, which can be as small as 1–100 nm. The particles carry an electrostatic charge, which causes similarly charged particles to repel each other, favoring their dispersion. The net charge a particle carries depends on the composition of the water. Cycling of cooling water increases the concentration of counter-charged ions capable of being electrostatically attracted to and adsorbed onto a charged particle. As counterions adsorb, the net charge of the particle decreases. Particles begin to agglomerate and grow in size as their repulsive forces are diminished.

Settling occurs when the energy imparted by fluid velocity can no longer suspend the particle, due to agglomeration and growth. After particles have settled, the nature of the deposit depends on the strength of the attractive forces between the particles themselves (agglomerate strength) and between the particles and the surface they contact. If attractive forces between particles are strong and the particles are not highly hydrated, deposits are dense and well structured; if the forces are weak, the deposits are soft and pliable. Deposition continues as long as the shear strength of the deposit exceeds the shear stress of the flowing water.

Removal of Particulate Matter

The amount of particulate entering a cooling system with the makeup water can be reduced by filtration and/or sedimentation processes. Particulate removal can also be accomplished by filtration of recirculating cooling water. These methods do not remove all of the suspended matter from the cooling water. The level of fouling experienced is influenced by the effectiveness of the particular removal scheme employed, the water velocities in the process equipment, and the cycles of concentration maintained in the cooling tower.

High Water Velocities

The ability of high water velocities to minimize fouling depends on the nature of the foulant. Clay and silt deposits are more effectively removed by high water velocities than aluminum and iron deposits, which are more tacky and form interlocking networks with other precipitates. Operation at high water velocities is not always a viable solution to clay and silt deposition because of design limitations, economic considerations, and the potential for erosion corrosion.

Dispersants

Dispersants are materials that suspend particulate matter by adsorbing onto the surface of particles and imparting a high charge. Electrostatic repulsion between like-charged particles prevents agglomeration, which reduces particle growth. The presence of a dispersant at the surface of a particle also inhibits the bridging of particles by precipitates that form in the bulk water. The adsorption of the dispersant makes particles more hydrophilic and less likely to adhere to surfaces. Thus dispersants affect both particle-to-particle and particle-to-surface interactions.

The most effective and widely used dispersants are low molecular weight anionic polymers. Dispersion technology has advanced to the point at which polymers are designed for specific classes of foulants or for a broad spectrum of materials. Acrylate-based polymers are widely used as dispersants. They have advanced from simple homopolymers of acrylic acid to more advanced copolymers and terpolymers. The performance characteristics of the acrylate polymers are a function of their molecular weight and structure, along with the types of monomeric units incorporated into the polymer backbone.

Surfactants

Surface-active or wetting agents are used to prevent fouling by insoluble hydrocarbons. They function by emulsifying the hydrocarbon through the formation of microdroplets containing the surfactant. The hydrophobic (water hating) portion of the surfactant is dissolved within the oil drop, while the hydrophilic (water loving) portion is at the surface of the droplet. The electrostatic charge imparted by hydrophilic groups causes the droplets to repel each other, preventing coalescence. Through a similar process, surfactants also assist in the removal of hydrocarbon-containing deposits.

Cooling water systems, particularly open recirculating systems, provide a favorable environment for the growth of microorganisms. Microbial growth on wetted surfaces leads to the formation of biofilms. If uncontrolled, such films cause fouling, which can adversely affect equipment performance, promote metal corrosion, and accelerate wood deterioration. These problems can be controlled through proper biomonitoring and application of appropriate cooling water antimicrobials.

BIOFOULING

Microbiological fouling in cooling systems is the result of abundant growth of algae, fungi, and bacteria on surfaces. Once-through and open or closed recirculating water systems may support microbial growth, but fouling problems usually develop more quickly and are more extensive in open recirculating systems. Once-through cooling water streams generally contain relatively low levels of the nutrients essential for microbial growth, so growth is relatively slow. Open recirculating systems scrub microbes from the air and, through evaporation, concentrate nutrients present in makeup water. As a result, microbe growth is more rapid. Process leaks may contribute further to the nutrient load of the cooling water. Reuse of waste water for cooling adds nutrients and also contributes large amounts of microbes to the cooling system.

In addition to the availability of organic and inorganic nutrients, factors such as temperature, normal pH control range, and continuous aeration of the cooling water contribute to an environment that is ideal for microbial growth. Sunlight necessary for growth of algae may also be present. As a result, large, varied microbial populations may develop.

The outcome of uncontrolled microbial growth on surfaces is "slime" formation. Slimes typically are aggregates of biological and nonbiological materials. The biological component, known as the biofilm, consists of microbial cells and their by-products. The predominant by-product is extracellular polymeric substance (EPS), a mixture of hydrated polymers. These polymers form a gel-like network around the cells and appear to aid attachment to surfaces. The nonbiological components can be organic or inorganic debris from many sources that have become adsorbed to or embedded in the biofilm polymer.

Slimes can form throughout once-through and recirculating systems and may be seen or felt where accessible. In nonexposed areas, slimes can be manifested by decreased heat transfer efficiency or reduced water flow. Wood-destroying organisms may penetrate the timbers of the cooling tower, digesting the wood and causing collapse of the structure. Microbial activity under deposits or within slimes can accelerate corrosion rates and even perforate heat exchanger surfaces.

Microorganisms

The microorganisms that form slime deposits in cooling water systems are common soil, aquatic, and airborne microbes. These microbes may enter the system with makeup water, either in low numbers from fresh water sources or in high numbers when the makeup is waste water. Significant amounts may also be scrubbed from the air as it is drawn through the cooling tower. Process leaks may contribute microorganisms as well.

Bacteria. A wide variety of bacteria can colonize cooling systems. Spherical, rod-shaped, spiral, and filamentous forms are common. Some produce spores to survive adverse environmental conditions such as dry periods or high temperatures. Both aerobic bacteria (which thrive in oxygenated waters) and anaerobic bacteria (which are inhibited or killed by oxygen) can be found in cooling systems.

Fungi. Two forms of fungi commonly encountered are molds (filamentous forms) and yeasts (unicellular forms). Molds can be quite troublesome, causing white rot or brown rot of the cooling tower wood, depending on whether they are cellulolytic (attack cellulose) or lignin degrading. Yeasts are also cellulolytic. They can produce slime in abundant amounts and preferentially colonize wood surfaces.

Algae. Algae are photosynthetic organisms. Green and blue-green algae are common in cooling systems (blue-green algae are now classified with the bacteria and are called cyanobacteria). Various types of algae can be responsible for green growths that block screens and distribution decks. Severe algae fouling can ultimately lead to unbalanced water flow and reduced cooling tower efficiency. Diatoms (algae enclosed by a silicaceous cell wall) may also be present but generally do not play a significant role in cooling system problems.

Biofilms

Microbiologists recognize two different populations of microorganisms. Free-floating (planktonic) populations are found in the bulk water. Attached (sessile) populations colonize surfaces. The same kinds of microorganisms can be found in either population, but the sessile population is responsible for biofouling.

When fouling occurs, even mechanical cleaning does not remove all traces of the biofilm. Previously fouled and cleaned surfaces are more rapidly colonized than new surfaces. Residual biofilm materials promote colonization and reduce the lag time before significant fouling reappears.

Biofilms on heat exchange surfaces act as insulating barriers. Heat exchanger performance begins to deteriorate as soon as biofilm thickness exceeds that of the laminar flow region. Microbes and hydrated biopolymers contain large amounts of water, and biofilms can be more than 90% water by weight. As a result, biofilms have thermal conductivities close to that of water, and in terms of heat transfer efficiency, a biofilm is the equivalent of a layer of stagnant water along the heat exchange surface.

Biofilms can promote corrosion of fouled metal surfaces in a variety of ways. This is referred to as microbially influenced corrosion. Microbes act as biological catalysts promoting conventional corrosion mechanisms: the simple, passive presence of the biological deposit prevents corrosion inhibitors from reaching and passivating the fouled surface; microbial reactions can accelerate ongoing corrosion reactions; and microbial by-products can be directly aggressive to the metal.

Oxidizing Antimicrobials

The oxidizing antimicrobials commonly used in industrial cooling systems are the halogens, chlorine and bromine, in liquid and gaseous form; organic halogen donors; chlorine dioxide; and to a limited extent, ozone. Oxidizing antimicrobials oxidize or accept electrons from other chemical compounds. Their mode of antimicrobial activity can be direct chemical degradation of cellular material or deactivation of critical enzyme systems within the bacterial cell. An important aspect of antimicrobial efficiency is the ability of the oxidizing agent to penetrate the cell wall and disrupt metabolic pathways.

The relative microbiological control ability of typical halogens is as follows:

$$HOCl \geq HOBr \geq NH_xBr_y >> OCl^- > OBr^- >>> NH_xCl_y$$

Cooling water pH affects oxidizing antimicrobial efficacy. The pH determines the relative proportions of hypochlorous acid and hypochlorite ion or, in systems treated with bromine donors, hypobromous acid and hypobromite ion. The acid forms of the halogens are usually more effective antimicrobials than the dissociated forms. Under some conditions, hypochlorous acid is 80 times more effective in controlling bacteria than the hypochlorite ion. Hypochlorous acid predominates below a pH of 7.6. Hypobromous acid predominates below pH 8.7, making bromine donors more effective than chlorine donors in alkaline cooling waters, especially where contact time is limited.

Antimicrobial efficacy is also affected by demand in the cooling water system, specifically demand exerted by ammonia. Chlorine reacts with ammonia to form chloramines, which are not as efficacious as hypochlorous acid or the hypochlorite ion in microbiological control. Bromine reacts with ammonia to form bromamines. Unlike chloramines, bromamines are unstable and reform hypobromous acid.

Most microbes in cooling systems can be controlled by chlorine or bromine treatment if exposed to a sufficient residual for a long enough time. A free chlorine residual of 0.1–0.5 ppm is adequate to control bulk water organisms if the residual can be maintained for a sufficient period of time.

Continuous chlorination of a cooling water system often seems most prudent for microbial slime control. However, it is economically difficult to maintain a continuous free residual in some systems, especially those with process leaks. In some high demand systems it is often impossible to achieve a free residual, and a combined residual must be accepted. In addition, high chlorine feed rates, with or without high residuals, can increase system metal corrosion and tower wood decay. Supplementing with nonoxidizing antimicrobials is preferable to high chlorination rates.

Sodium hypochlorite and calcium hypochlorite are chlorine derivatives formed by the reaction of chlorine with hydroxides. The application of hypochlorite to water systems produces the hypochlorite ion and hypochlorous acid, just as the application of chlorine gas does.

Halogen donors are chemicals that release active chlorine or bromine when dissolved in water. After release, the halogen reaction is similar to that of chlorine or bromine from other sources. Solid halogen donors commonly used in cooling water systems include 1-bromo-3-chloro-5,5-dimethylhydantoin, 1,3-dichloro-5,5-dimethylhydantoin, and sodium dichloroisocyanurate.

Chlorine dioxide, ClO_2, is another chlorine derivative. This unstable, potentially explosive gas must be generated at the point of application. The most common method of generating ClO_2 is through the reaction of chlorine gas with a solution of sodium chlorite.

Ozone

Ozone is an allotropic form of oxygen, O_3. Because it is an unstable gas, it must be generated at the point of use. Ozone is an effective, clean oxidizing agent possessing powerful antibacterial and antiviral properties.

Nonoxidizing Antimicrobials

Nonoxidizing antimicrobials usually control growths by one of two mechanisms. In one, microbes are inhibited or killed as a result of damage to the cell membrane. In the other, microbial death results from damage to the biochemical machinery involved in energy production or energy utilization.

Quaternary ammonium compounds (quats) are cationic surface-active molecules. They damage the cell membranes of bacteria, fungi, and algae. As a result, compounds that are normally prevented from entering the cell are able to penetrate this permeability barrier. Conversely, nutrients and essential intracellular components concentrated within the cell leak out. Growth is hindered, and the cell dies. At low concentrations, quats are biostatic because many organisms can survive in a damaged state for some time. However, at medium to high concentrations, quats can control the organisms.

Many antimicrobials interfere with energy metabolism. Because all microbial activity ultimately depends on the orderly transfer of energy, it can be anticipated that interference with the many energy-yielding or energy-trapping reactions will have serious consequences for the cell. Antimicrobials known to inhibit energy metabolism include organotins, bis(trichloromethyl) sulfone, methylenebis-(thiocyanate) (MBT), β-bromo-β-nitrostyrene (BNS), dodecylguanidine salts, and bromonitropropanediol (BNPD). All of these compounds are effective when applied in sufficient concentrations. Dodecylguanidine salts also have surfactant properties, which probably contribute to their effectiveness.

MACROFOULING ORGANISMS

Fouling caused by large organisms, such as oysters, mussels, clams, and barnacles, is referred to as macrofouling. Typically, organisms are a problem only in large once-through cooling systems or low cycle cooling systems that draw cooling water directly from natural water sources. Water that has been processed by an influent clarification and disinfection system is usually free of the larvae of macrofouling organisms.

Macrofouling has always been a concern in certain regions of the United States, especially in coastal areas. However, in the last 10 yr, the incidence of problems in the United States caused by macrofouling has increased dramatically. This is due primarily to the "invasion" of two organisms that were accidentally introduced to this country: the Asiatic clam and the zebra mussel. Both organisms have flourished and represent a significant threat to system reliability. Adding to the problem is the decreased use of chlorine and heavy metal antimicrobials, which permits the infiltration and growth of macrofouling organisms in plant water systems.

Asiatic Clams

Asiatic clams are freshwater mollusks. They probably originated in China or eastern Asia and were introduced into North America and Europe in the past century. They were originally found in warm water but their territory now extends to Minnesota. They have not yet been seen in Canadian rivers or lakes.

Asiatic clams do not attach to surfaces but burrow into sediments in their natural environment. Larvae and juvenile clams easily pass through intake screens (Fig. 13) and settle in low flow areas. Within 6 months to 1 yr, the clams grow to 1.5–2.54 cm in size. When a clam dies, the shell gapes open. Shells of living or dead organisms are carried by water flow and can wedge in condenser or heat exchanger tubes. Once a shell is wedged in a tube, other shells and debris collect and plug the tube further (Fig. 14). The Asiatic clam reaches adulthood in about 1 yr and reproduces in warm months, releasing thousands of larvae into the system.

Thermal Treatment (Thermal Backwash)

The organisms that cause macrofouling can be killed by heated water. Some systems are designed to allow the heated water from the outlet of the condenser to be recirculated back to the intake. As the water recirculates, it is heated and improves macrofouling control. A 15- to 60-min exposure to water at 40°C or higher has effectively controlled zebra mussels. Thermal treatment is not used extensively because most systems are not designed to recirculate water. Also, when heated water is recirculated the system cooling capacity is greatly diminished.

Oxidizing Antimicrobials

The application of an oxidizing antimicrobial such as chlorine for the control of undesirable organisms is a well-known and long-practiced procedure. Chlorine is toxic to all living organisms from bacteria to humans. However,

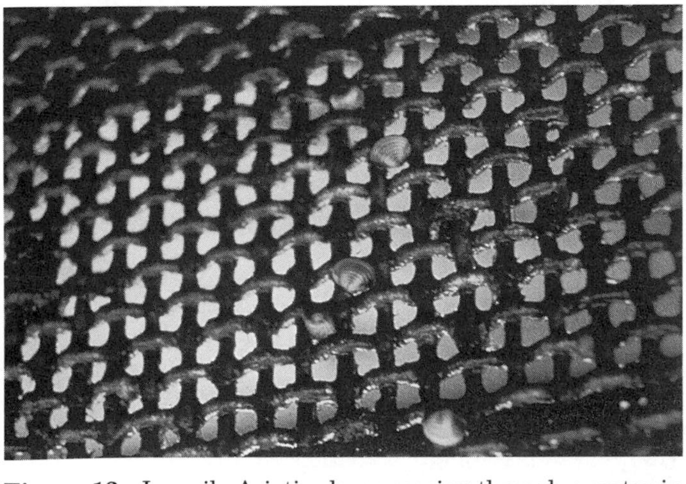

Figure 13. Juvenile Asiatic clams passing through a water intake screen.

Figure 14. Surface condenser infested with adult Asiatic clams.

in the case of hard-shelled creatures, including some mollusks and crustaceans, exposure is not easily accomplished. Some mollusks (eg, oysters, blue mussels, Asiatic clams, and zebra mussels) and crustaceans (eg, barnacles) have sensitive chemoreceptors that detect the presence of oxidizing chemicals such as chlorine (hypochlorite), bromine (hypobromite), ozone, and hydrogen peroxide. When oxidizers are detected at life-threatening levels, the animal withdraws into its shell and closes up tightly to exclude the hostile environment. Animals like oysters and mussels can remain closed for days to weeks if necessary. There is evidence to suggest that, during extended periods of continuous chlorination, the creatures may eventually die from asphyxiation rather than chlorine toxicity.

Even when their shells are closed, the animals continue to sense their environment, and as soon as the oxidant level decreases, they reopen and resume siphoning. Continuous chlorination often fails to eradicate these macrofouling creatures because of interruptions in the feed, which can occur for various reasons, such as chlorine tank changeover or plugging of feedlines. If the interruption lasts long enough (1 h or possibly less), the animals have time to reoxygenate their tissues between the extended periods of chlorination. Any oxidant, such as chlorine, bromine, or ozone, elicits the same response from these creatures. Therefore, only continuous, uninterrupted applications are successful.

Nonoxidizing Antimicrobials

There are several categories of nonoxidizing antimicrobials that have proven to be effective in controlling macrofouling organisms. Quaternary amine compounds and certain surfactants have been applied to infested systems for relatively short intervals (6–48 h). These compounds do not trigger the chemoreceptors of the mollusks. The mollusks continue to filter feed and ingest a lethal dosage of the antimicrobial throughout the exposure period. These compounds produce a latent mortality effect; the mollusks may not die until several hours after the antimicrobial application. Cold water temperatures may extend this latent mortality effect and may also require slightly higher feed rates and longer feed durations due to slower organism metabolism. The advantages of a nonoxidizing antimicrobial program include ease of handling, short application time, and relatively low toxicity to other aquatic organisms. In addition, some of these compounds can be readily detoxified.

WATER EFFICIENCY

Scott Chaplin
Jim Dyer
Andrew Jones
Richard Pinkham
Rocky Mountain Institute–Water Program,
Snowmass, Colorado

Water efficiency is just now receiving the attention that energy efficiency has enjoyed for the last 15 years. While the similarities between the two resources are significant, the connections are perhaps even more profound. Saving energy by using water more wisely has been one of the primary reasons for the development of water-efficient technologies and management schemes. This is most apparent in the development of technologies to save hot water and water that is pumped for irrigation. Water utilities are also beginning to view water efficiency as a way to save energy used in the transport and treatment of water. This entry provides several examples of how energy utilities are saving energy by promoting water efficiency, often by forming partnerships with water utilities. See also Energy efficiency; Energy efficiency, calculations; Energy efficiency, electric utilities.

DEFINITIONS AND TERMINOLOGY

The term *water efficiency,* as it is now used by water utilities, consultants, and other professionals in the water supply field, means providing the same or better water services using fewer resources. This is sometimes referred to in both the water and energy fields as "demand-side management" (see below). Water services include taking showers, flushing toilets, cleaning cars, irrigating crops, etc. Water efficiency connotes not only saving water but also the energy, chemicals, and other inputs used to heat, treat, and transport water both before and after use. This term differs somewhat from the more common term *water conservation,* which also includes curtailment and restrictions and perhaps a loss in the quality of water services provided.

Demand-side management, a term used commonly in the energy efficiency field to refer to adjustments in the amount or timing (peak versus nonpeak periods) of energy required for satisfaction of end uses, is also used in the water efficiency field. In both cases, savings in demand, or customers' consumption, can produce "new" supplies of the resource that are as useful as, and comparable to, water or energy provided by new infrastructure projects.

With both energy and water, it is often less expensive to save than it is to secure new supplies. Energy utilities often refer to the *cost of saved energy,* which is the cost of an efficiency program to a utility in a given year divided by the annual energy savings of efficiency measures installed in that year, levelized at a given interest rate over the lifetime of the measures. It provides a dollar figure for economic comparison of efficiency programs with the levelized cost of traditional supply projects such as power plants. Water utilities are just now beginning to use the same strategy to calculate the *cost of saved water.*

STATE-OF-THE-ART WATER-EFFICIENT TECHNOLOGIES

In recent years, manufacturers have made tremendous improvements in the water efficiency of a wide variety of water-consuming technologies. Toilets made before 1980, for example, often consumed 18.9–26.5 L (5–7 gal) per flush. In the late seventies and early eighties, 13.2 L (3.5 gal) per flush toilets became the norm. Today, most toilets made in the United States are designed to use less than 6.1 L (1.6 gal) per flush. In other countries, toilets using less than 3.8 L (1 gal) per flush are common. For some situations, improved models of the waterless, composting toilet, first introduced in the 1960s, may be suitable.

To a large extent these improvements have been spurred by water supply and treatment capacity shortages. Some of these improvements are now incorporated into building and manufacturing standards, including the Energy Policy Act passed by the U.S. Congress in late 1992. Figure 1 compares the efficiency of several pre-1980 water technologies available in the United States with those that are available today. (This figure includes technologies identified in ref. 1. The national plumbing standards are included in ref. 2. The figure also includes the use of waterless, or composting, toilets.) In addition, the 1992 U.S. standards are shown. As the technologies continue to improve, it is likely that these standards will become more stringent.

Perhaps the two best examples of technologies that can increase both water and energy efficiency are high-efficiency showerheads and faucets. Over 40 different models are on the U.S. market, all of which use less than 9.5 L/min (2.5 gal/min). Surveys of users show consumer acceptance of the technologies to be high–spray styles range from misty to tingly to blasting and designs run from mundane to hi-tech to elegant. Replacement of the average showerhead (which flows at a rate of 15.1–22.7 L/min (4–6 gal/min) in the United States), with a high-efficiency model would save significant quantities of water and energy. Replacement of faucets would have a similar, although not as large, effect. In addition to reducing water and energy use, such replacements would also reduce pollution and save consumers money. For example, retrofitting one U.S. household's showerheads and faucets (assuming an electric water heater) would result in the annual savings shown in Table 1.

Manufacturers have made similar improvements in other types of water-consuming technologies. Commercial laundry systems are now available that reuse rinse water, reducing water consumption by up to 75%. Recirculation cooling systems for commercial and industrial applica-

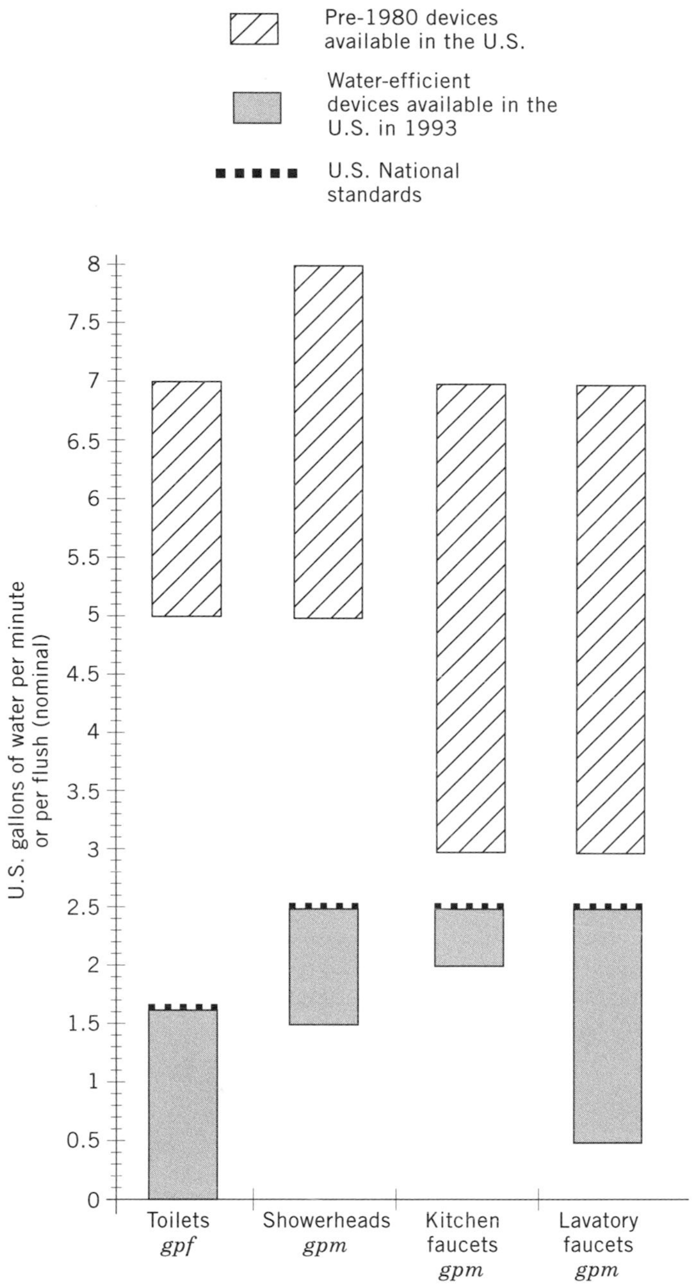

Figure 1. Water used by U.S. faucets, showerheads, and toilets. A comparison of the approximate range of water used by pre-1980 devices, and the water-efficient technologies available in 1993. (**a**) Pre-1980 devices available in the U.S. (**b**) U.S. National Standards, water-efficient devices available in the U.S. in 1993.

Table 1. Annual Savings from Replacing Showerheads and Faucets in an Average U.S. Home

Water	5,600–15,000 gallons
Electricity	590–1,700 kilowatt hours
CO2 emissions	700–1,800 pounds
NOx emissions	1.2–3.0 pounds
SOx emissions	2.5–6.4 pounds
Utility bills	\$25–\$170 savings

Source: Reference 3.

tions are replacing once-through systems. In landscaping, the use of low-water-consuming native plants and high-efficiency irrigation systems are becoming increasingly popular. As with high-efficiency showerheads, each technological improvement that saves water also saves energy. In all cases energy savings are available directly from reduced needs for pumping, treating, and/or heating. In some cases energy is saved indirectly, from reduced heating loads by decreasing the amount of cold water brought into a heated building.

While there is a high potential for water savings in municipal uses of water, the greatest opportunities are in agriculture. Farmers can increase irrigation water efficiency in a variety of ways. Careful monitoring of soil moisture and use of localized data on crop evapotranspiration (plant water demand and evaporation from soil) permit better scheduling of irrigation, avoiding wasteful application of water. Field modifications such as land leveling using laser-guided earth-moving equipment and soil management techniques like furrow diking and conservation tillage improve water distribution, infiltration, and retention. Changing irrigation equipment can also provide significant water savings while maintaining or increasing yields. Surge valves for furrow irrigation systems, low pressure sprinklers, drip and trickle systems, and subsurface irrigation are a few of the technologies available to irrigators. Decreasing the amount of water needed will lead to significant energy savings primarily from reduced pumping needs. Energy savings of 10–25% are easily achievable, and much greater savings can sometimes be obtained. For many farmers, the energy savings are far more significant financially than the water savings.

ENERGY SAVINGS FROM INCREASED MUNICIPAL WATER EFFICIENCY

Financing water and energy efficiency has often been difficult for both municipal water and energy utilities. Most investments made by utilities are large long-term investments in facilities that they will own and operate. The potential for increased efficiency, however, lies primarily in changing technologies that customers own and operate as well as the management practices of customers. Increasingly, many water and energy utilities are considering savings obtainable at the customer level from increased efficiency and shifting of loads to nonpeak periods, often called "demand-side management," to be a "new" supply that is often less costly than traditional supplies. Because of this, innovative water and energy utilities are now pooling their finances and management skills to promote municipal water efficiency. These win–win partnerships save money for the utilities by avoiding expensive supply-side options and cutting program implementation costs. Water utilities save water normally wasted in showers and toilets. Energy utilities save part of the electricity and gas used to heat, treat, and pump water. By forming partnerships to combine water and energy audits and retrofit programs, utilities can reduce distribution and installation costs and make significant advances in saving money, saving resources, and preserving the environment (1).

Most partnerships are primarily financial, as utilities split the cost of a water efficiency program. One utility may provide the products, the other, installation. Even better, utility planners are pooling their skill in creating more effective demand-side management programs. The energy utilities have more experience in marketing efficiency, while the water utilities have greater understanding of water systems. Working out the partnership contract between the various entities has been the greatest challenge, as different regulations and procedures govern energy and water utilities. For example, the Metropolitan Water District of Southern California (MWD), as a water wholesaler, cannot deal directly with the people who use its water, so MWD's involvement in efficiency programs must be channeled through a water retailer. Utilities also are accountable to different regulatory groups such as public utility commissions or state departments of water resources, groups that may be relatively unfamiliar with such innovative programs.

Another challenge for the utilities has been budget constraints. While water efficiency programs are modular and can be implemented in pieces, they still require an outlay of capital that may be scarce. Conversely, this very challenge has been a motivation for the partnerships. When the capital investment is split, each utility gets all its savings for one-third to one-half of the cost. Furthermore, the consumer benefits as better funded projects allow for more efficiency measures to be included in the retrofit package. Acting separately, the utilities might only be able to undertake a few inexpensive measures, "skimming the cream" off the top of the range of available savings and thereby perhaps making future additional measures less cost-effective.

Interested utilities have little trouble accepting the concept that both water and energy savings will result from water efficiency. The precedent for energy savings from reduced water heating loads is strong in many utilities across the country. Savings of pumping energy are well demonstrated in California, with its many aqueducts, but few electric utilities across the country have focused on capturing the savings in that sector. Even more tenuous, and yet untapped, will be the energy savings in wastewater and drinking-water treatment plants. Although savings have not yet been well documented, treatment plants should use less energy as a result of reduced flows. For that reason, it will be increasingly important to involve the local wastewater treatment districts in partnerships.

There are two keys to successful partnerships. One is an open, cooperative dialogue between the parties. Contracts have been most readily signed between groups who have come to the table with a willingness to listen that is as great as their desire to be involved. A second key to success is a long-term commitment to water efficiency. With many efficiency improvements such as repair of leaks, regular attention is required to ensure a steady supply of new water and energy savings. The following cases illustrate some of the more innovative municipal programs capturing the rewards of such cooperation and long-term planning.

Northeast Utilities

Connecticut Light and Power Co. (CL&P) has teamed up with three water utilities, Connecticut-American Water Company, Connecticut Water Company, and Metropolitan District Commission (MDC), to promote residential water and electric efficiency. Contractors visited 1100 homes to install resource-efficient hardware, including water heater and pipe wraps, high-efficiency showerheads, faucet aerators, compact fluorescent lamps, and other devices. Beyond the hardware distribution, the maintenance component of the retrofit included cleaning refrigerator coils. The electric utility paid for the energy-efficient devices and the water utility paid for the water-efficient devices, but they shared the installation cost in an innovative way: if the home had an electric water heater, CL&P paid for the labor, and if the home had a gas water heater, the two shared the labor costs.

The utilities shared the costs of installing water efficiency retrofit kits for elderly and handicapped people who requested assistance with installation. Over 30,000 households have accepted kits, which include high-efficiency showerheads, faucet aerators, toilet tank dams, and toilet leak detection tablets. Overall investment in 1991 was approximately $50,000 by the energy utility and $25,000 by the water utilities.

Southern California Water and Energy Conservation Partnership

Southern California Gas (SoCal Gas), Southern California Edison, Metropolitan Water District (MWD), Los Angeles Department of Water & Power, and Pasadena Water & Power Department have joined forces to develop long-term water and energy conservation programs inspired by the realization that the uses of water, gas, and electricity are interrelated. The first program was a pilot project to install 16,000 water-efficient toilet flappers in low-income homes. By the end of 1991, 10,000 flappers had been installed. The program was funded by dividing the cost of the flappers and the installation between Edison and MWD. Following the first several months of installation, a survey was taken to determine overall satisfaction with the flappers. Results showed that 71% of the flapper recipients were satisfied or very satisfied. The only source of dissatisfaction resulted from some owners of 13.2-L (3.5-gal) toilets who discovered that the flapper did not work well in that size tank. The program was revised to include installation only in 18.9-L (5-gal) tanks. The utilities are now implementing a toilet replacement program, which includes funding from an electric utility that saves energy by not pumping as much water. The partnership is considering a program to encourage the use of horizontal-axis washing machines that are both energy efficient and water efficient.

San Diego Area Utilities

San Diego Gas and Electric and San Diego County Water Authority have teamed up to promote residential water and energy efficiency. The utilities have organized a door-to-door installation of high-efficiency showerheads–13,800 in 1990 and 70,800 in 1991–1992. The water utility funded the technology kits while the energy utility paid for installation. As the program matured, the utilities expanded the retrofits to include other energy-saving devices, including 9500 compact fluorescent lamps.

Seattle/King County Area

In 1992, The Seattle Water Department and its 27 wholesale purveyors, Puget Sound Power and Light, and Seattle City Lights ran the largest collaborative residential water and energy saving program in the United States, with sponsorship also from Washington Natural Gas and Metro (the regional sewer utility). They went door to door, providing conservation kits to over 300,000 residences in Seattle and the nearby suburban areas. For electric hot water customers, Puget Power and Seattle City Light paid for the products, distribution, and marketing. For customers with nonelectric hot water, the Seattle Water Department (with support from the gas and sewer utilities) paid for all toilet retrofits, products, distribution, and marketing. Bonneville Power Administration reimbursed 75% of City Lights' costs and supported the program with assistance in purchasing, evaluation, and marketing. The program budget for all partners was approximately $3.2 million.

ENERGY SAVINGS FROM INCREASED WATER EFFICIENCY IN AGRICULTURE

Wherever irrigation water is pumped, irrigating efficiently–applying as little water as needed to obtain desired crop yields–saves water *and* saves energy by reducing pumping. The potential to save energy is becoming a powerful incentive for energy utilities and farmers, as well as irrigation water providers, to collaborate on programs to increase the efficiency of water use in irrigation. Farmers benefit from lower energy bills, and the energy savings can be part of an energy utility's demand-side management strategy. Increasing irrigation water and energy efficiency can benefit irrigation districts that face large energy bills for delivering water to farmers. Efficiency programs have wider benefits as well. Better water management may improve a farmer's ability to adapt to drought and contributes to improved management of soil nutrients and applied fertilizers. The water savings can also reduce water diversions from rivers, lakes, and wetlands, a development increasingly welcomed by society (5).

Improved irrigation practices and equipment can save large amounts of water, and correspondingly large amounts of pumping energy, but these changes require an investment of a farmer's time and money. Programs that educate farmers, assist them with technical matters, and support them financially can spark action. Energy utilities are an increasingly important source of such programs. They gain low-cost energy supplies by assisting farmers with water management. Many energy utilities now help farmers test, maintain, and improve the energy efficiency of irrigation pumps and shift irrigation energy loads from peak-use periods, such as hot afternoons, to evenings and other off-peak times. But these programs typically do not reduce water use very much. Savvy energy utilities are now promoting programs that will save additional energy by saving water. Irrigation water management programs at some such utilities have demonstrated easily obtainable savings in both water and energy of 10–25% per irrigation system. Savings up to 40% are possible in some cases. They have shown that the cost of saved energy achieved by encouraging better irrigation water management is often well below 1 cent/kW·h, *much* lower than the cost of energy obtained by building new plants and usually significantly lower than the generating costs at existing plants. Some utilities benefit additionally if water saved in irrigation is left instream, where it can contribute to hydroelectric power production. Significant savings in natural gas, an important energy source in some irrigated areas, are also achievable.

Energy utilities are helping farmers save water in several ways: (1) assisting farmers in the purchase of water-saving equipment by providing rebates and low-interest loans; (2) providing technical assistance through trained staff or consultants to help farmers with irrigation scheduling, equipment evaluations, system analyses, and recommendations for system modifications or conversions; (3) educating farmers through workshops and on-farm training as well as technical publications; and, (4) aiding farmers indirectly by providing financial support for programs and services–such as weather stations, irrigation scheduling services, and water management seminars–run by government agencies, Cooperative Extension offices, or private firms. Because of the low cost of these programs relative to the savings, it is likely that energy utilities will become increasingly active in promoting irrigation water efficiency in coming years.

Existing programs provide four lessons for success. First, utilities must utilize experienced field personnel who are familiar with local crops, irrigation systems, and the needs of farmers. Second, approaching farmers respectfully enhances success. Most farmers understand quite clearly the impact of energy costs on their profitability. From credible field personnel, they welcome assistance with improving their equipment and practices. Third, success requires a long-term effort as well as program modification as experience reveals the most successful and cost-effective ways to reach farmers. Fourth, cooperation between wholesale energy providers, retail energy utilities, government agencies such as the Soil Conservation Service and Cooperative Extension, and other organizations increases the chances of success. (For additional information on successful irrigation efficiency implementation techniques, see ref. 6.)

As energy utilities become more involved in irrigation water management, some of the creative approaches to financing, marketing, and implementing programs developed for residential, commercial, and industrial demand-side management will no doubt be applied to serving the needs of irrigators. Collaboration between energy utilities and water providers, increasingly common for municipal water and energy efficiency programs (see previous section of this entry), will also develop. These organizations can explore the potential for mutually beneficial sharing of the management and costs of irrigation water efficiency programs. Approaches that might not be as attractive for either water or energy savings alone may indeed be very cost-effective when both are considered.

To begin taking advantage of linked water and energy savings in irrigation, potential partners might consider expanding an existing energy program to include water-saving measures that save energy as well. Energy utilities can look for water efficiency programs offered by irrigation water providers or other agricultural organizations that could benefit from partnering with organizations interested in energy savings. Farmer involvement in program development is another way to get started. Ultimately they are the ones whose decisions will determine a program's success. Finally, the ideas and experiences of managers of existing programs are an excellent resource for program design. Clearly, the link between energy and water savings in irrigation provides opportunities for collaborative efforts to achieve greater efficiency in the use of both important resources. The following case studies illustrate some successful efforts by energy utilities to link water and energy efficiency in agriculture.

Western Area Power Administration: Eastern Colorado

Western Area Power Administration (WAPA) wholesales power, largely from federal hydroelectric projects, to municipal utilities, rural electric cooperatives, irrigation districts, and investor-owned utilities who in turn provide retail electric service to customers in 15 western and midwestern states. Federal law requires utilities purchasing power from WAPA to prepare integrated resource plans for their energy supplies. WAPA aids their efforts with a number of energy efficiency programs. In cooperation with the Colorado State Soil Conservation Board, the Colorado Office of Energy Conservation, and the U.S. Soil Conservation Service, WAPA's Loveland, Colorado, office supports the *Pump Testing and Irrigation Efficiency Program,* carried out by three field teams in the Ogallala Aquifer region of eastern Colorado. Pump efficiency tests help the teams get a "foot in the door" with farmers. In addition to the pump tests, the teams check the water application uniformity of sprinkler systems. They then suggest equipment improvements and work with interested farmers on conversions to medium-pressure and low-pressure systems, including LEPA (Low-Energy Precision Application) systems. The teams also teach farmers to monitor soil moisture, and in some cases they install gypsum block moisture sensors. They help farmers adopt conservation tillage techniques for retaining soil moisture and occasionally assist in the installation of surge valves for some surface systems. Working together, the teams and farmers customize systems for the particular requirements of agricultural production in eastern Colorado.

Since beginning their efforts in 1986, the teams have assisted nearly 2000 farms, leading to installation of over 300 LEPA systems and many other medium- to low-pressure systems. Energy savings for improvements made from 1986 to 1992 now total over 33,000,000 kW·h per year. Water savings for all measures implemented over the program's history now exceed 55,000 acre-ft per year. The program's operating costs have typically been about $120,000 per year. The cost of saved energy to the program funders ranged from 0.37 to 1.49 cents/kW·h over the 4 years from 1988 to 1991 (assuming a 5% discount rate and a 4-year measure life) (7). WAPA has historically contributed over 60% of the funding. The program is now moving toward greater self-sufficiency by charging $100 per pump test.

Some of the local utilities in eastern Colorado provide support for the WAPA program. The K.C. Electric Association, a rural electric cooperative, provides the team based in Burlington, Colorado, with $350 per year for the purchase of gypsum blocks (devices used to monitor soil moisture) that are installed in fields of customer farmers. Better water management may initially reduce K.C.'s revenues but will help individual farmers stay in business and maintain the groundwater resource of the region, thereby assuring the long-term viability of the cooperative. The Burlington, Colorado, office of People's Natural Gas, a multistate, privately owned utility, pays the $100 pump test fee for customers and supplies one set of gypsum blocks per center pivot system. In 1992, People's spent roughly $3500 on these measures, an investment in good customer relations, promotion of natural gas, and the long-term viability of their customer base and local economy. The Morgan County Rural Electric Association provides farmers participating in the pump testing and irrigation efficiency program a $50 rebate. This cooperative supports the program in order to reduce demand charges farmers and the utility must pay to acquire summertime power and to maintain load by keeping marginal wells in operation.

Besides saving energy and water, the WAPA program has created three full-time and several part-time field staff jobs and has generated new business for local agricultural equipment suppliers. Irrigation water efficiency also contributes to management of groundwater quality problems in eastern Colorado by reducing leaching of nitrates. Keys to success have been the technical skills of the field teams, their extensive one-on-one work with farmers, and the ability of the teams to serve as liaisons between farmers and the Soil Conservation Service, other government agencies, utilities, Colorado State University, farm equipment dealers, and manufacturers.

Pacific Gas & Electric: Central and Northern California

Due to provisions of California's energy utility regulations, this investor-owned utility can profit from investing in demand-side management. It plans to meet future growth in demand with renewable energy sources and cost-effective savings from existing customers. Pacific Gas & Electric's (PG&E's) agricultural demand-side management efforts include pump tests, energy audits, irrigation system analyses, and two popular rebate programs.

The rebate programs focus on energy savings, but some of the supported measures also save water. Farmers can receive rebates through the *Retrofit Express* program for water-saving equipment, such as time clocks with battery or spring-wound back-up ($50 per time clock), surge valves ($450 per valve), and low-pressure sprinkler nozzles (50 cents per nozzle). Besides these standardized rebates, which a farmer can obtain by submitting a simple form and receipts, the *Retrofit Customized* program provides rebates for farmers with special savings opportunities, such as changing from flood to more efficient drip or sprinkler irrigation. Regulatory scrutiny requires that the rebates produce clear and cost-effective energy savings, so PG&E carefully tracks expenses and savings. In the 1992 *Express* program alone, PG&E spent $119,000 on rebates for the three water-saving measures, resulting in annual energy savings of over 3,900,000 kW·h and demand savings of 789 kW. PG&E does not track the water savings from these measures. However, the program has been very popular due to California's recent drought and reductions in water deliveries from state and federal water projects. Farmers clearly recognize the potential water savings from these measures. One key to the success of this program has been the enthusiasm and skill of the PG&E agricultural representatives, who have established long-standing relationships with area farmers.

Pacificorp: Northern California

Pacificorp is an investor-owned electric utility with operations in seven northwestern states. In 1992, it began a pilot irrigation efficiency program, *Irrigation FinAnswer,* in northern California. Irrigation system analyses are provided free of charge to interested farmers. Consultants use computer software to model a farmer's irrigation system equipment and develop recommendations for optimizing its performance. They also make recommendations on irrigation scheduling. As part of this program, Pacificorp provides publicity and marketing for irrigation water management seminars put on by county extension agents. Pacificorp also reimburses farmers the $50 fee for the two-day course.

To encourage farmers to replace energy-inefficient equipment, the *Irrigation FinAnswer* program offers loans, with a novel twist. Pacificorp finances, at the prime interest rate, energy-saving measures that are cost-effective for the company and funds at prime plus 3% measures that are cost-effective business investments for the farmers but provide only marginal demand-side resource returns for the utility. In this way, farmers can put together a package of improvements that together will save energy and meet other goals, such as managing water, increasing yields, reducing labor, and otherwise improving profitability. Farmers pay the loan back via an energy service charge on their electric bills, in months that they specify according to their expected cash flow. Supported measures include changing sprinkler systems from high-pressure "big guns" to much lower pressure pivots or side-roll wheel lines, replacement of worn sprinkler heads and nozzles, resizing of distribution pipe, and changing or repairing pumps to match system flow and pressure requirements. Pacificorp believes the program produces water as well as energy savings.

Bonneville Power Administration: Pacific Northwest

A power wholesaler for federal hydroelectric projects in the Pacific Northwest, Bonneville Power Administration (BPA) funds a variety of irrigation efficiency programs. It supports operation and maintenance of the *AgriMet* network of automated weather stations operated in the region by the U.S. Bureau of Reclamation. These stations feed data to a Bureau of Reclamation computer that provides farmers and irrigation scheduling services with information on crop water demands. In its *WaterWise* program, BPA currently pays several dozen local utilities for the costs of pump tests and system analyses for interested farmers. BPA then channels rebates, through the local utilities, to farmers who install recommended, cost-effective measures. Rebate amounts are based on the particular hardware or the size of the farm and the potential savings. BPA also reimburses local utilities for some rebate-related administrative costs. Most of the supported measures impact energy use only, but some reduce water use as well.

Beginning with the 1993 irrigation season, BPA expanded *WaterWise* to support local utility involvement in irrigation water management. Participating utilities receive reimbursements from BPA for providing farmers one or more of the following services: (1) information on weather and crop water demands, disseminated through local media; (2) on-site, "hands-on" training in water management techniques; (3) computer-based irrigation scheduling, accessed through a farmer's phone and modem (see the case study below for Umatilla Electric Cooperative Association); and (4) local utility-designed programs.

The water management training feature of *WaterWise* grew out of a BPA-funded pilot program operated in 1991 and 1992 by one of BPA's utility customers, the Central Electric Cooperative in Redmond, Oregon, with the help of Northwest Irrigation Utilities, a regional utility consortium. The aim of the *Irrigation Management Pilot* program was to reach small- and medium-sized irrigators in the area, to test the voluntary adoption of conservation measures after an educational visit, and to quantify the potential energy and water savings of a simple, "low-tech," site visit approach. In 1992, an agricultural consultant visited 73 irrigators in the Central Electric Cooperative's service area, spending approximately 3 hours at each farm. The consultant surveyed irrigation equipment and practices and suggested simple changes in practices and equipment (such as renozzling sprinklers) to save water and energy. He gave away soil probes, taught basic soil moisture monitoring techniques, and showed farmers how to use the *AgriMet* weather station data. The consultant found that most of the irrigators were overwatering, and he suggested appropriate reductions in set times. He also referred some farmers to the *WaterWise* program for audits and possible retrofits.

Results from the pilot project indicated easily achievable savings for both energy and water of 10–20% and up to 40% in some instances. A follow-up survey showed that most of the suggested efficiency improvements were voluntarily implemented by farmers after the initial visit. Based on the consultant's on-farm analyses and recommendations and the follow-up survey of implementation rates, this $19,850 program in 1992 produced energy savings of 406,000 kW·h/year from reduced set times and installation of new sprinkler nozzles and water savings of 1818 acre-ft/year from the reduced set times alone. The cost to the utilities of the saved energy was 0.71 cents/

kW·h, assuming a 3.5% discount rate and an eight-year measure lifetime (7). (Costs do not include some of the utilities' overhead, which was not quantified. The savings would remain attractive even if overhead were included.) Using the same assumptions and allocating all program costs to the water savings, the cost of saved water amounted to $1.58 per acre-ft, indicating that simple changes can produce large water savings at low cost. The "low-tech" approach of this pilot project succeeded because of its simplicity and the skill and local reputation of the consultant, a retired county extension agent. Local farmers trusted his evaluations and suggestions. The *WaterWise* program now offers funding to any BPA customer utility that offers similar services to its farm customers.

Umatilla Electric Cooperative Association: Oregon

Faced with rapidly increasing wholesale power costs in the early 1980s, local energy utilities in the Pacific Northwest scrambled to find new ways of servicing the needs of their customers. Umatilla Electric, one of the region's largest irrigation power providers, put together an "Agribusiness Task Force" of farmers, food processors, extension personnel, and others. The task force recommended that the utility hire an irrigation specialist to develop irrigation energy demand-side management programs.

Besides implementing irrigation system audits and other programs directly focused on energy savings, the specialist convinced the utility that water efficiency was also important. He developed a computer and phone network to provide weather forecasts, crop evapotranspiration data, and irrigation recommendations, based on data from Bureau of Reclamation *AgriMet* weather stations. The computer service is menu driven; farmers can choose information for their local area and for irrigation systems of different types and water delivery rates. There is no charge to use the system, but farmers must have their own computer and modem. Known as the *Northwest Irrigation Network,* the system is now used to schedule irrigation on well over half of the 200,000 irrigated acres Umatilla Electric serves. Farmers using the scheduling information have achieved 20–25% water and energy savings, on average.

Attention to the needs of local irrigators has led to the program's success. As a rural electric cooperative, Umatilla Electric's customers are also its owners. The cooperative's manager believes that demand-side management efforts are often the utility's best energy buy, and at the same time these efforts strengthen its customers' operations. Concern for the impact of rising electric rates on customer farmers, who support much of the local economy, drove the formation of the Agribusiness Task Force. From the very beginning, the cooperative's board of directors provided sustained financial support for the computer network.

Beginning in 1990, Bonneville Power Administration (BPA) began a three-year pilot program to offer the service to selected farmers in other local utility territories in the Pacific Northwest. The pilot indicated that computer-based irrigation scheduling could be successfully implemented in other parts of the region. BPA's retail utility customers with *WaterWise* contracts can now obtain funding to offer similar services (see the BPA case study above).

FUTURE TRENDS

Improvements in plumbing technologies and water management techniques will continue to increase the potential to save water in the municipal sector. Some of these improvements will focus on the technologies themselves, while others will reduce the need for water indirectly. Increasing building insulation, for example, will indirectly lead to water and energy savings in areas where evaporative cooling systems are used. The use of graywater and rainwater for nonpotable needs will also indirectly reduce the need for energy used to treat water.

In irrigated agriculture, future savings in energy through increased water savings will be substantial. As the case studies above show, much can be done already, at relatively low cost, with existing management techniques and currently available irrigation equipment. The most critical frontier is in developing new financing mechanisms and implementation techniques. Innovative partnerships between progressive farmers, energy utilities, and irrigation water providers will go a long way. One area ripe for partnerships is the piping of irrigation water delivery systems, which can save water by eliminating ditch seepage and save energy by producing gravity pressure sufficient to run sprinklers and other water application equipment. New technologies are also arising. Computerized on-farm irrigation systems with automated soil moisture sensors are one area of significant research at present. Advances in drip, trickle, and subsurface irrigation systems are also being continually made.

SOURCES FOR ADDITIONAL INFORMATION

The following are publications and organizations that may be of interest to those seeking more information about water efficiency:

Publications

Proceedings of Conserv 93: The New Water Agenda, December 1993, American Water Works Association, 6666 W. Quincy Ave., Denver CO 80235. Proceedings of one of the largest conferences on water efficiency.

F. van der Leeden et al., *The Water Encyclopedia,* 1990, Lewis Publishers, 121 S. Main St., Chelsea, MI 48118. Provides useful facts and figures on water supply, use, and quality, primarily for the United States.

A. Vickers, *Handbook of Water Conservation,* 1994, Lewis Publishers, 121 S. Main St., Chelsea, MI 48118. The most comprehensive document on water efficiency available. Includes both municipal and agricultural water uses.

Water: Conservation and Reclamation, 1991, The Global Cities Project, 2962 Filmore St., San Francisco, CA

94123. A useful reference for local governments. Contains a wide variety of municipal water efficiency legislation.

Organizations

American Water Works Association (AWWA), 6666 West Quincy Avenue, Denver, CO 80235. Phone: (800)559-9855.

Energy Efficiency and Renewable Energy Clearinghouse (EREC) P.O. Box 3048, Marrifield, Va. 22116 (800)523-2929.

National Small Flows Clearinghouse, 258 Stewart Street, Morgantown, WV 26505. Phone: (800)624-8301.

Rocky Mountain Institute–Water Program, 1739 Snowmass Creek Road, Snowmass, CO 81654-9199. Phone: (303)927-3851.

BIBLIOGRAPHY

1. C. Laird, *Water-Efficient Technologies: A Catalog for the Residential / Light Commercial Sector,* Rocky Mountain Institute, Snowmass, Colo., 1991.
2. "Energy Policy Act of 1992," *Congressional Record,* vol. 138, no. 142, part V, U.S. Government Printing Office, Washington, DC, October 1992.
3. A. P. Jones, *High-Efficiency Showerheads and Faucets,* Rocky Mountain Institute, Snowmass, Colo., 1993.
4. A. P. Jones and J. Dyer, *Water and Energy Utility Partnerships,* Rocky Mountain Institute, Snowmass, Colo., 1992.
5. R. Pinkham and J. Dyer, *Linking Water and Energy Savings in Irrigation,* Rocky Mountain Institute, Snowmass, Colo., 1993.
6. C. Laird and J. Dyer, *"Feedback and Irrigation Efficiency,"* Rocky Mountain Institute, Snowmass, Colo., 1992.
7. *Western Area Power Administration, Pump Testing and Irrigation Efficiency,* profile no. 40, The Results Center, Aspen, Colo., 1992, pp. 14–15.
8. *Final Report: 1992 Waterwise Irrigation Management Pilot Program,* report by W. Trimmer and Northwest Irrigation Utilities to Bonneville Power Administration, November 1992, p. 6.

WATER QUALITY ISSUES

JOHN D. HEM
U.S. Geological Survey
Water Resources Division
Menlo Park, California

The physical effectiveness of water moving through the hydrologic cycle, as a means of modification of surface topography, is impressively demonstrated in such geographic erosional features as the Grand Canyon of the Colorado River. Besides the physical movement of erodible material, however, the chemical attack of circulating water on rock minerals over time brings into solution a wide variety of elements that are transported to the oceans by rivers.

In the early years of development of the science of geochemistry there was a strong interest by investigators Goldschmidt and Vernadski in Europe and Clarke in the United States in the significance of chemical composition of river water and how the observed composiiton was reached through water-rock chemical reaction. There also were direct practical motivations of studies of river water composition. For example, during the nineteenth century the settlement of the interior of the United States was continuous. The growing major cities commonly were located on navigable streams, or at the shoreline of other surface water bodies, that could be used both as sources of water supplies and for disposal of wastes as well as for transportation. That these uses were commonly not compatible became clear as populations of cities increased.

Evaluation of water for its fitness to be used as a public supply, the procedures generally termed "sanitary analysis," involve such determinations as dissolved oxgyen, biochemical oxygen demand, testing for presence of various forms of bacteria, and related evaluations, are described in "Standard Methods" (1) and are discussed only briefly in this article. The evaluation of river water for possible industrial uses is generally concerned with determination of major dissolved constituents in filtered sample aliquots. Samples must be taken at intervals frequent enough to show the degree of variability of these concentrations over a range of flow rates.

This article considers three general aspects of surface water chemistry: (1) a brief overview of natural geochemical control factors, (2) evaluation of circulation rates of essential plant nutrient elements coupled to the hydrologic cycle, and (3) evaluation of historical trends in stream water quality in four large U.S. river basins. The manner in which these trends are related to human activities is then discussed. In addition, literature references that review ongoing studies in the field of specific industrial and sanitary sewage pollution problems are given and various research studies relating to roles of sediment and stream biota to river water quality are cited and briefly described.

Ideally the historical record of stream water quality would extend back to a time when human activities in the drainage basin had no significant effects. This "pristine" condition had probably already passed in most U.S. rivers before any organized water quality studies were made, as concern about apparent stream pollution was commonly a motivating factor in starting such studies.

See also OIL SPILLS; WATER QUALITY MANAGEMENT.

FACTORS AFFECTING STREAM WATER QUALITY

General Principles of Stream Water Geochemistry

The composition of pristine stream water can range from nearly that of rainwater, a very dilute solution containing up to a few tens of milligrams per liter of dissolved material, to concentrations of as much as several thousand milligrams per liter, which mainly are found in semiarid regions where saline springs contribute chemical constituents derived from readily soluble rock strata. In general, however, the dissolved-solids concentrations of principal streams in humid and subhumid regions of the conterminous United States have a narrower range. Analyses showing maximum and minimum concentrations for a recent 3-year period for 29 stream-sampling sites in the United States recently were published (2). Seven of these sites had maximum total dissolved-solids concentrations greater than 1000 mg/L; these high concentrations are attributable to natural saline inflows, although human activities probably intensified their effects.

Chemical analyses of stream water that have been published since the early years of this century generally include determinations for four positively charged ions (cations)—calcium (Ca^{2+}), magnesium (Mg^{2+}), sodium (Na^+), and potassium (K^+)—and five negatively charged ions (anions)—bicarbonate (HCO_3^-), sulfate (SO_4^{2-}), chloride (Cl^-), fluoride (F^-), and nitrate (NO_3^-)—and uncharged dissolved silicic acid (generally reported in terms of silica, SiO_2). These are the major constituents in most natural stream water and are those given principal attention in this article. Minor constituents, present at concentrations substantially less than 1 mg/L, include a wide variety of inorganic and organic constituents, and although these constituents commonly are the result of human activities, they are not considered in detail in this article because few reliable measurements were made until the 1970s. Short-term trends for these constituents have been evaluated using statistical techniques (3).

Chemical Reactions that Govern Stream Composition. Weathering is a general term for mechanical and chemical alteration of rock minerals that are exposed to the atmosphere and circulating water at and near the land surface. Chemical reactions that occur during weathering produce both water-soluble and non-water-soluble products; those that are water soluble are transported from the reaction site in surface runoff or in moving groundwater. To a certain extent, at least, the concentrations of dissolved elements would be expected to reflect the relative abundance of the elements in the rocks exposed at the reaction site. Such a broad generalization has some validity for silicon (Si) and the four elements that form the major cations of most natural stream water. These five elements are among the eight most abundant elements, with oxygen being the most abundant, in igneous and sedimentary rocks of the Earth's outer crust. The other two of the most abundant elements—aluminum (Al) and iron (Fe)—form oxides or hydroxides of very low solubility during normal rock weathering and, therefore, generally are not present in large amounts in stream water. On the other hand, major anions in stream water display a more complex relation to rock composition. In the average stream water sample, the five most abundant anions represent the nonmetallic elements carbon, sulfur, chlorine, fluorine, and nitrogen. Also, oxygen is included in three of the anions of these elements.

Oxygen is by far the most abundant element in crustal rocks, composing 46.6% of the lithosphere (4). In rock mineral structures, the predominant anion is O^{2-}, and water (H_2O) itself is almost 90% oxygen by weight. The nonmetallic elements fluorine, sulfur, carbon, nitrogen, chlorine, and phosphorus are present in lesser amounts in the lithosphere. These elements all play essential roles in life processes of plants and animals, and except for phosphorus and fluorine, they commonly occur in earth surface environments in gaseous form or as dissolved anions.

In a very broad general sense, then, the major cationic constituents of stream water tend to reflect the composition of associated rocks and the relative resistance of the rock minerals to weathering. The anions, which must be present in these water solutions in electrochemical balance with the cations, tend to reflect the influence of various chemical and biochemical processes that have broken down the rock minerals as well as the chemical, biochemical, and physical processes taking place in the aquatic and surrounding environments. The predominance of bicarbonate anions in most stream water is related to cycling of carbon dioxide (CO_2) from air and to biological processes in soil. Dissolution of carbon dioxide in water produces carbonic acid (H_2CO_3) that attacks rock minerals. Bicarbonate anions are formed in solutions participating in such reactions in amounts equivalent to the amount of cations that are released.

Sulfur and nitrogen participate in biologically mediated oxidation reactions producing hydrogen ions (H^+) that become available for weathering of rock minerals. For example, pyrite (FeS_2) can be converted to dissolved ferrous iron and sulfate as a result of oxidation of sulfur by dissolved oxygen, and the hydrogen ion is a major byproduct. Carbonate or sulfate in sedimentary rocks (eg, limestone and gypsum) can be taken directly into solution and can add substantially to the bicarbonate and sulfate contents in water that is in contact with such rocks.

Chlorine plays a less significant role in chemical weathering processes than do sulfur and carbon. Most geochemists believe that much, or most, of the chloride in stream water in coastal areas is derived from sea salt that is carried landward or deposited by rainfall. Farther inland, however, a major part of the chloride loads in streams is the result of human activities.

The final composition of stream water is the product of the weathering reactions and related processes outlined above. However, the chemical processes are influenced and controlled by an intricate combination of environmental factors that are characteristic for each drainage system. Therefore, the composition of the bedrock in an area and the residual material left at the surface as soil and subsoil exert a strong influence on the chemical composition of runoff from the area. The reactions of water with this material are the ultimate geological control and are the source of soluble weathering products.

Most igneous and metamorphic rocks are composed predominantly of aluminosilicate minerals, including feldspar such as albite ($NaAlSi_3O_8$) or anorthite ($CaAl_2Si_2O_8$) and crystalline forms of silica such as quartz (SiO_2). Various mixed metal-plus-silicon oxides such as olivine [$(Mg,Fe)_2(SiO_4)$] and pyroxene [$Mg_2(SiO_3)_2$] can be major constituents in darker colored igneous rocks that are relatively low in total silicon.

Rocks that were deposited as sediment can consist of unaltered fragments of a precursor rock body and are represented by sandstone and conglomerate. The finer grained sedimentary rocks, such as shale or siltstone, also can contain some unaltered particles but also usually have high proportions of slightly soluble alteration products, such as clay minerals, formed during weathering of resistant silicate minerals. Such rocks are classified as hydrolyzates. Another class of sedimentary rock of major importance is the precipitates, such as limestone and dolomite, which are predominantly composed of calcium carbonate and calcium plus magnesium carbonate, respectively. Evaporites are sedimentary rocks produced by extensive evaporation of water from weathering solutions. Common examples are gypsum and anhydrite, primarily composed of calcium sulfate, or halite (rock salt), primarily composed of sodium chloride. Obviously the more readily soluble minerals of evaporite or precipitate rocks can dissolve rapidly when exposed to circulating water. Carbonates also can act as cementing material between the mineral grains of resistate and hydrolyzate rocks.

The extent to which minerals are attacked and dissolved from igneous and metamorphic terranes depends in large part on the availability of reaction sites on solid surfaces and the length of time the solution–solid contact is maintained. The effects of weathering are controlled by kinetic factors, such as the rates of the chemical reactions and the general rates of water and sediment movement, and biological factors that include the effects of biotic growth in the weathering zone. Also, the hydrologic properties of the drainage system (precipitation, evaporation, runoff, slope of the area), the relative permeability of rocks and soils, and the degree to which the surface drainage system is coupled to the groundwater reservoirs are important modulating forces. In rocks that contain more soluble minerals such as calcite, the degree to which solids are dissolved and carried off in the runoff is more likely to be governed by chemical thermodynamic factors, and in carbonate systems a state of chemical equilibrium could be closely approached.

Mineral dissolution reactions of importance generally require a continuous supply of hydrogen ions in the incoming solution. To some degree, reacting hydrogen ions are supplied from the water itself, which always includes, to some extent, water molecules that have broken apart (dissociated) into hydrogen and hydroxide ions. Under standard conditions (25°C and 1 atm pressure), the effective concentration (activity) of hydrogen ions in pure water is $10^{-7.00}$ mol/L. (A mole of a chemical element is a quantity in grams numerically equal to the atomic weight. For hydrogen, $10^{-7.00}$ moles per liter is equivalent to 0.1 microgram per liter.)

The pH scale commonly used to express acidity is defined as the negative base-10 logarithm of the activity of hydrogen ions in a solution. At neutrality under the conditions defined in the preceding paragraph, the pH therefore will be 7.00. A change of 1 pH unit represents a 10-fold change in hydrogen ion activity. The activity of a dissolved ion is exactly equal to its concentration only in very dilute solutions. This topic has been discussed and explained more extensively elsewhere (5).

In natural systems the most effective sources of hydrogen ions generally are chemical reactions involving dissolved constituents. An important source is carbon dioxide gas, which is present in weathering solutions as a result of contact with air; it is produced in larger quantities by plant root respiration and decay of soil organic matter and by the metabolic processes of various organisms in water and sediment. Equation (1) shows that some of the carbon dioxide that dissolves forms carbonic acid (note that in the following equations the arrows indicate the direction in which the reaction normally proceeds; reactants are on the left side and products on the right; double arrows indicate that reactions can proceed in either direction):

$$\underset{\text{carbon dioxide}}{CO_2} + \underset{\text{water}}{H_2O} \rightleftarrows \underset{\text{carbonic acid}}{H_2CO_3} \tag{1}$$

Equation (2) shows that the acid dissociates to form bicarbonate ions and hydrogen ions:

$$\underset{\text{carbonic acid}}{H_2CO_3} \rightleftarrows \underset{\text{bicarbonate}}{HCO_3^-} + \underset{\text{hydrogen}}{H^+} \tag{2}$$

and carbonate anions can be formed in a second dissociation step:

$$\underset{\text{bicarbonate}}{HCO_3^-} \rightleftarrows \underset{\text{carbonate}}{CO_3^{2-}} + \underset{\text{hydrogen}}{H^+} \tag{3}$$

The H^+ that is supplied by reactions in Eqs. (2) and (3) can react with silicate minerals such as the sodium-bearing form of feldspar:

$$\underset{\text{albite}}{2NaAlSi_3O_8} + \underset{\text{hydrogen}}{2H^+} + \underset{\text{water}}{9H_2O} \rightarrow \underset{\text{kaolinite}}{Al_2Si_2O_5(OH)_4} + \underset{\text{silicic acid}}{4H_4SiO_4(aq)} + \underset{\text{sodium}}{2Na^+} \tag{4}$$

to produce the clay mineral kaolinite, undissociated silicic acid (H_4SiO_4 which also can be written as SiO_2), and sodium ions. This is essentially an irreversible process (note the single arrow) in that albite is not readily synthesized from the reaction products under ordinary natural weathering conditions, and the reaction will continue to proceed to the right as long as reactants are available.

The hydrogen ion flux that is provided by carbonic acid dissociation also can attack calcite ($CaCO_3$):

$$\underset{\text{calcite}}{CaCO_3} + \underset{\text{hydrogen}}{H^+} \rightleftarrows \underset{\text{bicarbonate}}{HCO_3^-} + \underset{\text{calcium}}{Ca^{2+}} \tag{5}$$

This reaction is relatively fast and readily reversible so that in drainage basins in carbonate-dominated terranes the stream water commonly will have near-equilibrium concentrations of hydrogen, bicarbonate, and calcium ions. At equilibrium, the rates of forward and reverse processes represented in Eq. (5) are equal.

In effect, the forward progress of reactions such as is shown in Eqs. (4) and (5) will be controlled by the availability of hydrogen ions, but the final result, as indicated by the chemical composition of stream water from any given drainage basin, will be influenced by a complicated set of interrelated physical factors that influence the volume and rate of water movement, ecologic and climatic factors that control the type and density of plant and bacterial growth and soil development, and the human development of water and land resources. Thus, the geochemistry of stream water in any given drainage basin is unique to that basin. The historical record of water chemistry in a specific basin cannot be interpreted without giving proper attention to the way various hydrologic and other environmental factors in the basin have influenced the chemical composition of the water. An extensive body of research on the topic of global stream water geochemistry is summarized in Ref. 6, who agree with the need expressed in this paragraph for a broad consideration of cause and effect when evaluating stream water chemistry.

Effects of Human Activities. A considerable part of the currently existing motivation for organized long-term water quality studies has been public concern that human activities in many drainage basins have induced destructive changes in stream water quality. From examples cited in this article, such effects can indeed be documented. Also, in some basins water quality management has succeeded in correcting some of the human-caused deterioration and is substantially restoring the quality of the water. The concept of sustainability is relevant, and the development goal for drainage systems is to maintain suitable water quality while permitting levels of water use that will sustain the basin's existing and reasonable future economic development.

Human activities that alter stream flow characteristics and thus cause water quality changes include the building of structures that impound or regulate rates of stream flow, diversion of water from one drainage basin to another, irrigation of land adjacent to streams, and lowering of tributary groundwater tables by pumping from wells. Waste disposal, directly or indirectly, into streams also influences water quality by adding chemicals and suspended matter. Disposal of untreated organic waste into streams was common in urban and rural areas of the nation until the early twentieth century. Besides pathogenic bacteria in the waste, large amounts of organic chemicals and suspended material depleted the dissolved oxygen of receiving waters and killed much of the aquatic biota in some streams. Thus, the concentration of dissolved oxygen in stream water also is considered a contamination index. Normally, the dissolved-oxygen content is near the saturation level that can be calculated for water that is in contact with air at ambient tempertaure, but oxidizable material in solution, especially organic waste, can substantially deplete the dissolved-oxygen content. Additionally, phosphate (PO_4^{-3}) concentrations are indicative of contamination from waste sources. Phosphate is a constituent of domestic and industrial waste, in part, because of the widespread use of phosphate compounds as detergent additives.

Land use changes and related developments also can affect stream water quality. Examples include urbanization, clearing of forests, various agricultural practices, such as use of fertilizers and pesticides and return flow of drainage water from irrigated fields, and industrialization. Urbanization and industrialization lead to various side effects. Mining for coal and metals generally contaminates water during and after the mining activity. The smelting of ores to recover metals and the burning of coal to generate power release pollutants into the air, and eventually some of these pollutants find their way into water supplies. Metal ores, coal, and other organic fuels commonly contain, or are associated with, reduced sulfur. Oxidation of the sulfur by burning the fuel and smelting the ores and oxidation in the mines or in waste dumps when sulfides are exposed to air constitute major sources of sulfur in stream water. High metal concentrations commonly are found in streams draining metal-mining areas, and sodium chloride and calcium chloride are dispersed widely by salt and sand mixtures used to melt ice from highways.

Circulation Rates of Elements

The concept of cycling of individual elements, in part coupled to the hydrologic cycle, has been developed and quantified over the past half century. Besides the total quantities of the elements present in various reservoirs—bedrock (the lithosphere), soils, all forms of living matter (the biosphere), the oceans, the atmosphere, and fresh water—the rates of exchange and mechanisms of movement from one reservoir to another are considered in the cycle. This concept is highly relevant in developing a frame of reference for evaluating possible environmental effects of energy technology.

A simplified diagram representing the various reservoirs and transport mechanisms and pathways involved in the cycles of nutrient elements at and above the surface of the Earth is given in Figure 1. The processes are those considered to be the most important in the context of this article, but others of lesser significance can be postulated. For some of the elements, notably carbon, sulfur, chlorine, and nitrogen, considerable research has been done to evaluate (quantitatively) the amount of the various elements in the reservoirs and the rates of transfer. Each of these elements is expanded on in the following discussions.

Carbon. Most of the Earth's supply of carbon is stored in carbonate rocks in the lithosphere. Normally the circulation rate for lithospheric carbon is slow compared with that of carbon between the atmosphere and biosphere. The carbon cycle has received much attention in recent years as a result of research into the possible relation between increased atmospheric carbon dioxide concentration, most of which is produced by combustion of fossil fuel, and the "greenhouse effect," or global warming. Extensive research has been done on the rate at which carbon dioxide might be converted to cellulose and other photosynthetically produced organic compounds by various forms of natural and cultivated plants. Estimates also have been made of the rate at which carbon dioxide is released to soil under optimum conditions by various kinds of plant cover, such as temperature-zone deciduous forests, cultivated farm crops, prairie grassland, and desert vegetation.

The efficiency of the weathering of rocks in using carbonic acid produced in the carbon cycle is affected by various hydrologic, environmental, and cultural controls. The fact that the principal anion in fresh surface water worldwide almost always is bicarbonate attests to the overrid-

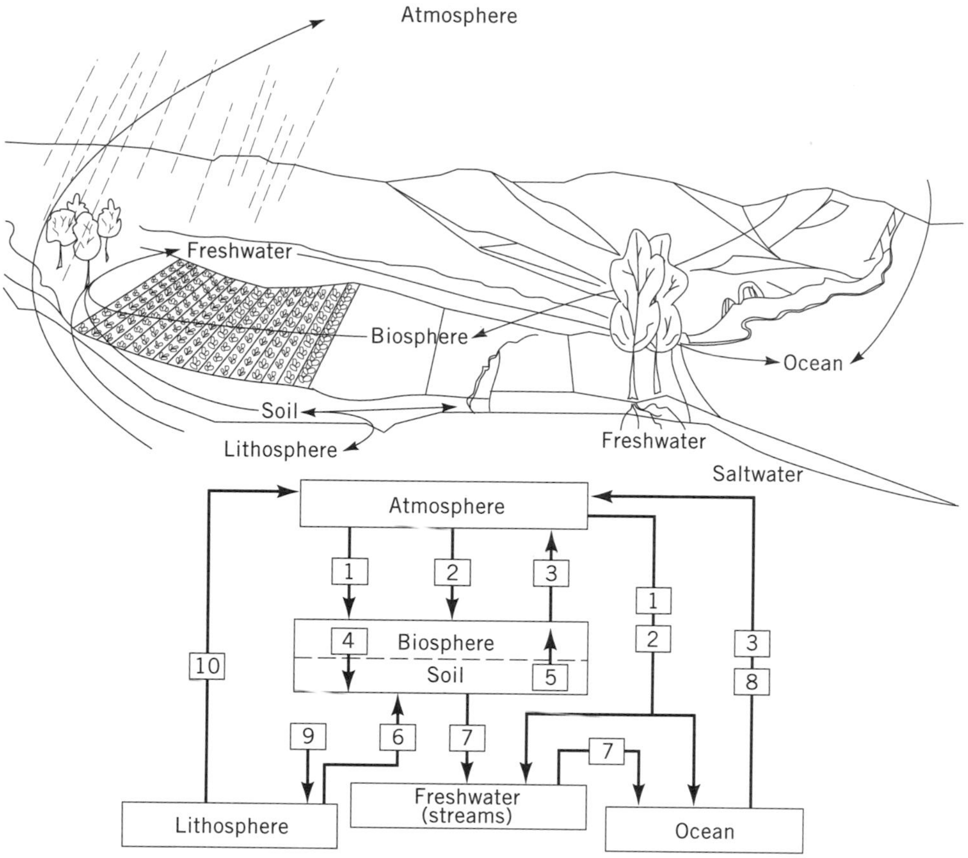

Figure 1. Generalized cycle of the various reservoirs and transport mechanisms and pathways involved in the circulation of nutrient elements. The numbered arrows represent processes by which elements transfer among the reservoirs. Processes shown are those considered to have the most important influence on stream water quality. (Modified from ref. 7.)

ing importance of this process. Exceptions are systems in which evaporite minerals are available for dissolution by groundwater or where human activities are major sources of sulfate or chloride inflow.

Quantitative estimates of the magnitude of the carbon cycle, based mainly on a compilation in Ref. 8, suggest that the total amount of carbon dioxide converted to organic matter during a year might be as great as 7–9 tons/acre in midlatitude deciduous forests and about a tenth that amount in native grassland. Cultivated farm crops presumably convert somewhat more carbon dioxide per unit area than might be expected in native grassland. Respiration of carbon dioxide by plant roots is at a rate equal to about 25% of the total carbon dioxide used by the plants, and the potential supply of carbon dioxide from vegetation for weathering rock is about 1.65 tons/acre/yr on the average, over the whole land area of the Earth (8). This is equivalent to about 1060 tons/mi^2 yr of carbon dioxide. If all this carbon dioxide were dissolved in water and the resulting carbonic acid reacted with silicate minerals to maintain near-neutral pH, the annual bicarbonate yield could be as great as 1500 tons/mi^2. A drainage basin in a limestone terrane theoretically could produce bicarbonate at an even higher rate, as the reaction between dissolved carbon dioxide and solid carbonate minerals produces two bicarbonate ions per reacting dissolved carbon dioxide molecule.

As might be expected, the average rates of bicarbonate yield in drainage basins of the world are far below these numbers, as only a relatively small proportion of the carbon dioxide released in soil is likely to participate in rock weathering. A general indication of the role of dissolved carbon dioxide in weathering processes in streams can be obtained by calculating the average annual bicarbonate yield from average concentration of dissolved bicarbonate, average water discharge, and drainage basin area. Examples of this type of calculation can be obtained from published literature. Ref. 9 studied 56 drainage basins in the conterminous United States and Hawaii to determine the relative importance of stream temperature, human population density, atmospheric precipitation, and predominant type of bedrock (limestone, sandstone, or crystalline) as controls of constituent yield. Of the 19 limestone basins, the Kalamazoo River at Saugatuck, Michigan, had the maximum annual bicarbonate yield, which was about 324 tons/mi^2. However, in the other basins, annual bicarbonate yields were substantially lower—15 had less than 225 tons/mi^2, and the average yield for the 19 limestone basins was 114 tons/mi^2.

In general, bicarbonate yields reported in Ref. 9 were much smaller for basins in crystalline (noncarbonate) rock terranes than for those in limestone or sandstone terranes. Other investigators have reported relatively large bicarbonate yields for certain igneous terranes. For example, Ref. 10 reports large bicarbonate yields from igneous drainage basins in the Pacific Northwest. Among studies referred to is a study for a basin having an active glacier in the northern Cascade Mountains of Washington (11). In this basin, the annual bicarbonate yield was near or slightly greater than 100 tons/mi^2.

The high bicarbonate yield in some of the Pacific Northwest streams could be the result of glacial action and mechanical erosion of rock material in basins having steep slopes that produce large amounts of relatively freshly fractured and fine-grained sediment that would participate readily in dissolution and alteration reactions. The rate at which such reactions would proceed is a function of net surface area of sediment available per unit volume of runoff and the length of time such material was exposed to water in the runoff process. Therefore, the efficiency of solvent erosion under conditions in the Pacific Northwest region could be considerably increased by such factors.

The chemical analyses tabulated in this article identify "alkalinity" as a property of the water rather than a simple constituent. Alkalinity has been more broadly defined as "capacity for acid neutralization" (12,13). Common practice in water analysis is to report alkalinity in terms of bicarbonate and carbonate concentrations, although other ionic species also may contribute by reacting with the titrating acid.

Sulfur. The cycle of sulfur in weathering environments is affected by a more diverse set of reactions than for carbon. As is the case with carbon, most of the Earth's supply of sulfur is stored in the lithosphere (5). Although some sulfur actually is taken up by vegetation, the growth and decay of plants does not tie up large proportions of the total sulfur supply. The element's geochemical behavior in the environment is summarized by Nriagu (14). In igneous and metamorphic rocks, sulfur generally is present in the chemically reduced sulfide form (S^{2-}) and commonly is associated with metals. Polysulfide minerals in which the nominal valence of sulfur is between -2 and 0 also are common. An example is pyrite (FeS_2), which commonly occurs in association with coal seams and other sediments laid down under conditions where free oxygen was not available.

In weathering environments where oxygen is continuously available, negatively charged reduced forms of sulfur are converted to positively charged oxidized forms such as sulfate (SO_4^{2-}) in which the sulfur is in the fully oxidized (S^{6+}) state. Oxidation and reduction reactions of sulfur commonly are bacterially mediated. The calcium sulfate minerals gypsum ($CaSO_4 \cdot 2H_2O$) and anhydrite ($CaSO_4$) are common constituents of evaporite rocks, and in semiarid regions where such rocks are near the land surface, stream water can contain substantial concentrations of sulfate. However, streams in humid regions generally carry relatively low concentrations of sulfate unless human activities have intervened. Weathering reactions that involve oxygen are important in the development of soils, especially in humid and subhumid climates. Commonly, oxygen is in relatively short supply at shallow depths in the soil zone, having been depleted by oxidation of organic matter. In wetland soil and submerged sediments, oxygen can be in very short supply and sulfur is in reduced form. However, as noted in Refs. 15 and 16, the sulfur in aerated soil is principally a component of organic compounds. In any event, any reduced species of sulfur may be converted to sulfate by oxidation when wetlands are drained or soils are converted from their natural state to agricultural cropland. All these processes can increase the availability of sulfate for transport by streams and cause increased sulfate loads in areas affected by development.

Geochemical studies of sulfate in streams have been approached in various ways. For example, Ref. 17 is a review of literature on sources of sulfate in the dissolved loads of streams, especially in areas where development effects are likely to be strong. It was concluded that for North America as much as 60% of the average yield of sulfate carried to the ocean is related to human activities. This study assigned a higher proportion of the total yield of sulfate in European streams to human sources, but for other continents it was thought the effects of human activities to be relatively minor.

Some of the compounds produced in the sulfur cycle are gases. For example, combustion of fossil fuels, especially coal, produces sulfur dioxide gas (SO_2), which is further oxidized in the atmosphere to sulfur trioxide (SO_3) that combines with water to form sulfuric acid (H_2SO_4). Reduction of sulfate in anaerobic soils and sediment produces hydrogen sulfide (H_2S) gas that also is reoxidized to sulfate in the presence of air. As a result, precipitation from the atmosphere is a major source of sulfate in streams in parts of North America and Europe. The estimate for human sources includes both sulfur dioxide and hydrogen sulfide reaction pathways (17). A compilation by the U.S. Environmental Protection Agency (18), summarizing research on acidic deposition, quoted estimates and measurements of "excess sulfate" yields in precipitation and in lakes and streams in the northeastern United States and southeastern Canada. Among measured data cited was a wet-plus-dry annual sulfate deposition rate for a site in the Adirondacks that ranged from about 8.2 to 16 tons/mi^2. Although a wide range of values was observed, these rates probably can be considered reasonable for much of the eastern half of the United States. The term "excess sulfate" is defined as the amount "over and above that supplied by sea salt cycling." Ref. 9 estimates that, on average, 60% of the sulfate yield observed for the stream basins he studied was assignable to atmospheric sources.

Another source of sulfur in the global hydrologic cycle that is not related to human activities is gaseous emission of hydrogen sulfide and sulfur dioxide from volcanoes and other geothermal sources. Although effects of these emissions can be locally intense, they generally are thought to be much less significant on a global scale than the human sources.

Sulfate concentration in streams and changes over time are discussed later in this article as one of the principal indices of human influences on stream water composition. Also, it will be shown that differences and similarites in sulfate yields help in attaining a reasonable perspective on the importance of various hydrologic and geochemical characteristics of individual drainage systems.

Chlorine. Nearly all chlorine compounds are readily soluble in water. As a result, the major reservoir for this element in Figure 1 is the ocean (5). Chloride, as noted earlier, is naturally present at low levels in rain and snow, especially over and near the oceans. Widespread increases in chloride concentration in runoff in much of the United States can be attributed to the extensive use of sodium chloride and calcium chloride for deicing of streets and highways. Ref. 19 points out the importance of the increased use of deicing salt as a cause of increased chloride concentrations in streams of the northeastern United States and the role of this factor in the chloride trends in Lake Ontario. Increases in chloride concentration also can occur as a result of disposal of sewage, oil field brines, and various kinds of industrial waste. Thus, chloride concentration trends also can be considered as an index of the alteration of streamwater chemistry by human development in the industrialized sections of the world. Although chlorine is an essential element for animal nutrition, it is of less importance for other life forms.

Nitrogen. About three-fourths of the Earth's nitrogen is present in the atmosphere as nitrogen gas. Because of its importance as an essential element in plant and animal nutrition, nitrogen, in its various oxidation states and its yield and concentration, is of considerable interest in studies of human influences on stream water composition. Certain small- and medium-sized streams in the intensively developed agricultural areas of the United States have been strongly affected by nitrogen-bearing runoff from fertilized soil. However, because of its use by aquatic vegetation, the amount and form of nitrogen tend to be seasonally variable, and especially for larger streams, it is more difficult to use than sulfate or chloride as an index of human effects on water composition. Although nitrate concentrations are reported in some of the streams evaluated in this article, no attempt at interpretation of the nitrogen chemistry is made here; however, Ref. 20, using mainly U.S. Geological Survey (USGS) data, evaluates nitrogen trends since 1905 in the Mississippi River at St. Francisville, Louisiana.

STREAM WATER QUALITY TRENDS IN FOUR DRAINAGE BASINS

Data Available

The first comprehensive nationwide study of the chemical composition of surface water in the United States was begun by the USGS in 1905. This program received cooperative support from state agencies in Illinois, Minnesota, Kansas, California, Oregon, and Washington and entailed collection of daily samples of water at or near gaging station sites on principal streams for a period of about 1 year. The daily samples were combined into 10-day composites, which were filtered to remove suspended material and analyzed for principal dissolved constituents using standard "wet-chemical" procedures. As a result of decreased funding and related factors, this work was substantially curtailed after 1907; however, the studies did produce compilations of analyses for about 100 stream-sampling sites east of the 100th Meridian (21) and for about another 55 sites in the western part of the country (22). Somewhat more detailed reports for each of the six cooperating states also were issued in the USGS Water-Supply Paper series, and Clarke (23) used many of the analyses from the program in his summary of the composition of river and lake waters of the United States.

From about 1914 to 1940 studies of river water quality were carried on at only a few sites in the United States. A revival of interest began in the 1940s, however, and by the 1960s, a much larger number of sampling sites were in operation than had existed at the height of the 1905–1907 program. The available historical data base for most U.S. sampling stations has no water quality measurements for the 1910–1940 period.

Stream water quality commonly varies greatly in response to water discharge; thus, a single year of record is not adequate for reliable extrapolation, and in any exacting comparison of historical data, this factor needs to be taken into account. From the beginning, it has been a general policy in the USGS surface water quality program to locate sampling sites at or near gaging stations where records of stream flow are obtained. Until about 1970, many of the USGS water quality records were based on daily sampling, generally with determinations of specific electrical conductance (an indicator of total cation and anion concentration) on each sample, but with extensive analyses performed only on composited daily samples. Composites usually contained 10 daily samples. However, where stream discharge and other factors caused substantial day-to-day changes in specific conductance, the composite period was shortened to prevent mixing of chemically dissimilar samples and to give a clearer indication of the stream chemistry variability. Annual averages of these analyses, weighted by time or discharge, were used to summarize the records. After 1970, complete analyses were done on single samples collected at various time intervals ranging from semimonthly to quarterly, and analyses of composite samples were no longer made. Analytical procedures changed from time to time as improved instrumentation and techniques became available.

In order to identify and consider the importance of the various sources and controlling factors that operate in specific drainage basins in the United States to produce water having hydrochemical properties that are characteristic of each basin, four drainage basins having long-term hydrologic and water chemistry records were chosen for study in this article—the Great Lakes—Upper St. Lawrence River at and near Ogdensburg, New York, and Cornwall, Ontario, Canada; the Columbia River at and upstream from the Dalles, Oregon; the Allegheny River upstream from Pittsburgh, Pennsylvania; and the lower Mississippi River at and upstream from New Orleans, Louisiana.

The historical records mentioned above are compared with more recent data for each of the drainage basins in order to detect and explain major differences between water composition observed early in the twentieth century and that observed more recently. Sulfate, as mentioned earlier, is one of the principal indicators of the effects of human activities on stream water composition, and it is the principal indicator used in the following discussions. The records selected for study display general trends in concentrations and yields that appear to be well-enough defined to outweigh the influences of different sampling frequencies and changing compositing practices. Trends that can be detected in the data collected in the past 20 years generally are too subtle to be evaluated closely by the methods used in this article and are more appropriately studied by the more sophisticated procedures which are described in Refs. 3, 24, and 25.

Selected analytical and related data for the drainage basins are given in Tables 1–6. The location of these basins is shown in Figure 2. Apparent trends with time are evaluated by various means described in the text. The annual minimum and average constituent concentrations and annual yields of sulfate in tons per unit area of drainage basin are the focus of the discussion.

Descriptions of sampling and analytical procedures for the early USGS data suggest that, in general, the results for major constituents should be comparable in precision and accuracy with those given in later records. However, in some respects the procedures used in the early work may have introduced inaccuracy and bias. The daily samples were collected in 120-mL capacity soft-glass bottles, which limited the volume of water for analysis. The rather low sensitivity of some of the procedures available at that time probably affected the analytical accuracy for constituents present at small concentrations. Analyses reported (22) for stations in the West give only bicarbonate, chloride, and total dissolved-solids concentrations for the individual samples and weekly composites. Methods used to prepare composite samples for more complete analysis for these stations were inconsistent and introduced substantial bias; hence cationic compositions given for these waters probably are not geochemically useful.

Analyses (21) represent composites of equal volumes of daily samples. Generally there were three composites for each month. The complete analyses for these stations are much more reliable than those given in Ref. 22 as indicators of the actual water composition, but some bias is introduced by compositing equal volumes for each day when stream discharge was variable.

Other Constituents of Stream Water

The records reported in Refs. 21 and 22 were obtained for the primary purpose of evaluating the suitability of surface water resources of the United States for utilization by industry and for irrigation of agricultural lands in the western part of the country. These stream waters also provide public water supplies for many municipalities. Evaluations of water quality for the latter purpose emphasize constituents that were not given detailed consideration in Refs. 21 and 22 summaries, although there are references in Ref. 21 to work done in various state health laboratories and municipal treatment plants.

Suspended-Material Separation. Water samples collected for the studies (21,22) were filtered before the analyses were begun. Procedures at that time generally used shredded and washed asbestos mats in Gooch crucibles.

Difficulties were reported in removing suspended sedi-

Figure 2. Drainage basins and related data of the four water quality stations from which data were obtained for determining long-term water quality trends. Drainage area and discharge from refs. 26–30.

ment from the composite samples before the analyses of dissolved concentrations were started in both Refs. 21 and 22. It is evident from the reported concentrations that the various filtration and pretreatment methods used sometimes left particulate material in the samples and caused improbably high determined concentrations for iron, silica, and possibly some other solutes. Since about 1960 plastic membrane filters having pore diameters of 0.45 μm have been specified for removal of suspended material from water samples.

Suspended Particulate Concentrations. Tables of analyses include suspended-sediment concentrations based on dried and weighed residues left on the filters (21,22). The sample collection procedures, however, were not directed toward obtaining sediment concentrations representative of those in the stream cross section, and the resulting sediment data are of limited significance. For some stations, there are reported values for "turbidity," a measure of the blocking of transmitted light by the water sample compared with standards containing various concentrations of suspended Fuller's earth.

In subsequent USGS stream quality studies made up through the 1940s the emphasis continued to be on the major dissolved constituents. However, improvements in sampling techniques and analysis were made in work in the Colorado River basin in the 1920s (31) and subsequently through interagency studies in the 1940s to 1960s sampling equipment and techniques were developed so that loads of sediment moved by streams could be accurately determined (32–34).

The mineralogy, surface chemistry, and other properties of sediment particulates tends to be a principal factor in controlling concentrations of dissolved trace metals and organic compounds in associated river water, and with improved analytical instrumentation during the 1960–1980 period a need for more detailed study of suspended material in stream water had arisen. This brought about extensive research that is still ongoing. Some estimates of elemental fluxes associated with stream sediments made by (35) indicate that of 48 elements investigated only 7 were transported in solution in river water at rates greater than they were transported in sediment associated with the stream flow, and for most other elements the sediment-associated fluxes were more than 10 times the dissolved fluxes. (Totals of elements in sediment were determined by procedures that included all solid-phase forms.)

Organic Particulates. Much of the particulate material contributed by human activities is organic. It may include various forms of microorganisms and nutrients for aquatic vegetation that bring about conditions of eutrophication—an increasing seasonal growth of aquatic biota that uses the nutrient supply and then dies back. The decay of the dead biota can lower the dissolved oxygen in the water to levels that may interfere with propagation and growth of larger aquatic life forms such as game fish. Such conditions have commonly occurred downstream from sewage treatment plants and various types of industrial outfalls. Equipment has been recently developed and refined for collecting integrated representative samples of suspended and dissolved materials in cross sections of large streams. In an ongoing study described in Ref. 36 equipment is carried in a research vessel that has been used in a series of cruises to sample the Mississippi River at a series of cross sections from the upper river to its mouth below New Orleans. Some results of the work related to occurrence of agrochemicals in the river were given (32), and concentrations of an extensive suite of trace metals were reported (38). Some general aspects of the pollution problems resulting from industrial and sanitary wastes at specific locations on eastern U.S. rivers are reviewed in the proceedings volume for the 13th Mid-Atlantic Industrial Waste Conference (39).

Surface Water Ecology

Ecologic studies of contaminated and uncontaminated streams have been the subject of extensive research. Those streams that have been subjected to contaminant additions through disposal of industrial and other forms of waste generally show substantial decreases in the diversity and abundance of instream biota. These effects are often first observed in decreases of native game fish such as brook or rainbow trout, but more extensive research has shown many more subtle changes. A compilation (40) includes summaries of ecologic studies in 12 river basins in North America.

The drainage basin of the Clark Fork, an eastern tributary of the Columbia River, has been affected for long distances by the copper mining and ore treatment processes that were conducted at Butte and Anaconda, Montana, for about 125 years, ending in 1982. Conditions in this stream have been described (41). A more recent paper (42) pointed out that many streams in mineralized areas had relatively high metal concentration prior to the beginning of mining activity.

Means for Summarizing Dissolved-Material Contents

Water analyses traditionally have included a summarizing value termed "total dissolved solids." A determined value specified as "residue on evaporation" (ROE) is obtained by evaporating a measured volume of the water sample to dryness in a dish whose weight is known exactly, drying the residue at a specified temperature (in USGS practice at 180°C for 1 h), and weighing the dish and its contents after cooling in a moisture-free atmosphere. The more modern analyses also include electrical conductance, which is related to the total concentration of dissolved cations and anions, and electronic determination of pH, which is a measure of the acidity or hydrogen ion (H^+) activity. Although the ROE procedure appears to be simple and direct, it has complicating factors. The bicarbonate in solution in the original sample is converted to carbonate by drying, with loss of an equivalent amount of carbon dioxide gas. However, some types of water, such as those having high concentrations of dissolved calcium and sulfate, can deposit residues that contain water incorporated in crystalline minerals such as gypsum ($CaSO_4 \cdot 2H_2O$). To avoid this problem, a calculated total dissolved solids (SUM) commonly is reported for water containing more than about 1000 mg/L total constituent concentration. The SUM calculation is made by adding reported concentrations for all the major constituents, with a correction factor converting dissolved bicarbonate to an equivalent carbonate content of the dried residue.

Great Lakes–Upper St. Lawrence River

The St. Lawrence River, at the outlet of Lake Ontario, represents the drainage from the Great Lakes basin and has a relatively constant chemical composition, owing to the large storage capacity of the lakes and their efficiency in mixing of inflow. The water surface of the Great Lakes makes up a substantial fraction of the basin area. The basin also has a relatively constant water discharge. These characteristics make it a good site for study of water quality trends. Data from two water-quality-sampling and stream-flow-gaging stations—Cornwall, Ontario, Canada, and Ogdensburg, New York (40 miles upstream from Cornwall)—are used to show constituent change over time. Although the two stations are 40 miles apart and there is a small amount of tributary inflow between them, the chemical composition of the water at the sites does not differ significantly and thus can be used for comparison of data over time. At Cornwall, the drainage area of the St. Lawrence River is 298,800 mi^2 and the long-term average discharge is 245,000 ft^3/s, as reported by the U.S. Geological Survey (26).

Data from these stations are summarized briefly in Table 1, where each of the columns represents data for a particular year progressing from 1906–1907 to 1990. The data show that principal dissolved-ion concentrations in the river were nearly constant during the period 1969–1980 and that most of the change occurred between 1907 and 1956. Data for some major ions for samples from Lake Ontario that were presented (46) indicate that relatively continuous increases occurred until the 1960s. An illustration in Ref. 2 based on a graph from Ref. 47 suggests that sulfate concentration in Lake Ontario steadily increased from 12–14 mg/L in 1906 to 28–30 mg/L in the 1970s. However, during the 1980s, the sulfate concentration in the lake as represented by the St. Lawrence River at Cornwall did not show any further upward trend, and it seems likely that the sulfate concentration had stabilized. Samples taken at 1- or 2-month intervals during water years 1981–1985 had sulfate concentrations of 28 mg/L or more on 11 occasions. During water years 1986–1990 no sulfate concentrations greater than 27 mg/L were observed at Cornwall.

Behavior of chloride concentration during the period of record was somewhat similar to that of sulfate—about a threefold increase compared to the doubling of sulfate. Data in Table 1 indicate that a long-term decrease in bicarbonate of 10–20 mg/L is a reasonable estimate.

Table 1. Chemical Composition of Water and Related Data for St. Lawrence River, Representing Outflow from Lake Ontario, Selected Years (1906–1990)

	Ogdensburg, N.Y.,	Cornwall, Ontario, Canada		Ogdensburg, N.Y.		Cornwall, Ontario, Canada		
Constituent Property, and Related Data	1906–1907[a] (1)	1935[b] (2)	1940[b] (3)	1956[c] (4)	1969[d] (5)	1977[b] (6)	1980[b] (7)	1990[e] (8)
Silica (SiO_2)	6.6	6.9	4.1	3.5	0.6	0.2	0.5	—
Calcium (Ca)	31	33	36	36	39	38	36	—
Magnesium (Mg)	7.2	7.9	8.5	8.2	7.3	8.0	7.8	—
Sodium (Na)	{ 6.3[f]	7.3[f]	7.5[f]	9.7	12	13	13	—
Potassium (K)				1.5	1.3	1.6	1.4	—
Alkalinity as bicarbonate(HCO_3) (property)	122	110	97	112	110	110	99	—
Sulfate (SO_4)	12	19	21	25	28	27	26	27
Chloride (Cl)	7.7	15	16	21	26	27	26	—
Fluoride (F)				0.1	0.1	0.1	0.2	—
Nitrate (NO_3)	0.3	0.6	1.0	1.0	0.3	0.53	0.49	—
Dissolved solids	134	156	158	179	169	169[g]	160[h]	—
Specific conductance, μS at 25°C (property)				299	314	370	270	300
pH units (property)		8.0	8.0	6.8–7.9	7.3–8.0	7.8	7.1	—
Drainage area, mil^2	295,200	298,800	298,800	295,200	295,200	298,800	298,800	298,800
Average discharge, ft^3/s	242,300	192,000	226,000	257,000	270,000	250,000	296,000	264,800

Note: Concentration values are in milligrams per liter unless otherwise noted. Data from the Ogdensburg, N.Y., and Cornwall, Ontario, Canada, stations are considered equivalent. *Sources*: column 1, ref. 21; columns 2, 3, ref. 43; column 4, ref. 44; column 5, ref. 45; columns 6, 7, ref. 2; column 8: U.S. Geological Survey, New York District office written communication, 1991.

[a] Average of 11 monthly samples, Sept. 1906–Jan. 1907, March–Aug. 1907.

[b] Single sample collected on Aug. 30, 1935 (col. 2), Aug. 22, 1940 (col. 3), June 27, 1977 (col. 6), and Sept. 29, 1980 (col. 7).

[c] Average of 10-day composite samples collected during water year 1969.

[d] Average of 10 individual monthly samples collected during water year 1969.

[e] Average of 5 bimonthly samples collected between October 1989 and July 1990.

[f] Sodium and potassium not determined separately; value reported is total Na + K expressed as an equivalent amount of sodium.

[g] Maximum total dissolved solids for period 1977–1980.

[h] Minimum total dissolved solids for period 1977–1980.

Refs. 47 and 48 attribute the historical changes in water composition of the Great Lakes to human activities, which have increased as the population of the drainage basins of the lower four lakes has increased since the beginning of the twentieth century. Industrial waste and incompletely treated sewage are thought to be major sources of sulfate and chloride. Other nonpoint sources are agricultural fertilizers and fossil fuel combustion and urban runoff that add sulfate and sodium and calcium chlorides used for deicing highways. The decrease in alkalinity of the water leaving Lake Ontario can possibly be explained as a direct consequence of the low pH of rain that falls on all the lakes. However, because the loss of bicarbonate alkalinity is only about two-thirds as great as the gain in sulfate, the observed changes cannot be simply assigned to sulfuric acid in this rainfall. It seems more likely that some of the acidity in the rain that falls over the tributary drainage is neutralized by reaction with sedimentary rock minerals, as calcium and sodium show increases.

During the 1960s there was much concern about the increasing rate of eutrophication of Lake Erie. The limiting nutrient for aqueous microbiota in the Great Lakes was perceived to be phosphorus (48), and household synthetic detergents containing sodium phosphate were thought to be an important source. Consequently, a concerted effort was made to decrease the sale of detergents containing phosphate and to improve sewage treatment processes. By the 1980s, some declines in phosphate concentration had occurred in Lake Erie and Lake Ontario (48).

An accompanying effect of eutrophication that is more readily observable in Table 1 is a decrease in silica concentration in Lake Ontario. Some decline in dissolved silica apparently has occurred in all of the lakes except Lake Superior. This decline is brought about by the growth of diatoms, a species of aquatic microorganisms in the upper layers of lake water that is widespread in all types of water impoundments where the water is clear and exposed to the sun. The silica is used by these microorganisms to form their skeletons and is later precipitated and becomes part of the bed sediment.

Silica determinations on the 1906–1907 St. Lawrence samples (Table 1, col. 1) might have been affected by inadequate filtration, and possibly some silica was dissolved from the glass sample bottles during storage before analysis. Thus, the average silica concentration of 6.6 mg/L might be too high. However, a rather well-defined downward trend is observable in more recent silica concentration records for the St. Lawrence at Ogdensburg and Cornwall that has greater experimental certainty. Determinations of dissolved silica on individual samples collected during water year 1956 ranged from 1.1 to 6.8 mg/L (44). A record of once-monthly samples from water year 1969 (45) shows silica concentration that ranged from a low of 0.0 (below the minimum reporting limit) to a maximum of 1.9 mg/L. Samples taken during water years 1977–1980 show concentrations of 0.2 and 0.5 mg/L (Table 1, cols. 6, 7).

The Lake Superior drainage basin is underlain predominantly by metamorphic and igneous rocks, whereas most of the Great Lakes drainage area eastward to the outlet of Lake Ontario is underlain by carbonate-rich sediment. It would be expected, therefore, that the water of Lake Superior would have different chemical characteristics than those of the other Great Lakes because it has a different set of geochemical controls. Also, the major ion composition of water in Lake Superior has not changed significantly during the period of record.

Additional insight into changes in dissolved material flowing from the lakes over time can be obtained by calculating average annual yields of major constituents observed in the Great Lakes–St. Lawrence basin (Table 2). Ref. 9 contains calculations of the annual yields of principal constituents for 56 drainage basins and evaluated the degree of correlation with four environmental factors—bedrock type, annual precipitation, population density, and average stream temperature. The study shows that in basins dominated by limestone bedrock, the average annual yield for calcium was 36.0 tons/mi^2 and for bicarbonate was 114.2 tons/mi^2. Calcium yields listed in Table 2 for the 1956–1980 period are close to the averages reported by Peters and are significantly higher than yields for 1906–1907. However, the bicarbonate yields indicated in Table 2 are lower than the average reported by Peters (9) and show no well-defined trend. Eight of the 56

Table 2. Estimated Annual Yield of Constituents (t/mi^2), St. Lawrence River Outflow from Lake Ontario, Selected Years (1906–1990)

Constituent or Property	Ogdensburg, N.Y., 1906–1907 (1)	Cornwall, Ontario, Canada, 1937 (3)	Ogdensburg N.Y. 1956 (5)	Ogdensburg N.Y. 1969 (6)	Cornwall, Ontario, Canada 1977 (7)	Cornwall, Ontario, Canada 1980 (8)	1990 (9)
Calcium (Ca)	25.0	24.2	30.9	35.1	31.3	35.1	NA
Sodium (Na)	5.1[a]	6.7[a]	8.3	10.8	10.7	12.7	NA
Bicarbonate (HCO_3)	98.6	75.9	96.0	99.0	90.6	96.5	NA
Sulfate (SO_4)	9.7	14.8	21.4	25.2	22.2	25.4	23.6
Chloride (Cl)	6.2	10.8	18.0	23.4	22.2	25.4	NA

Note: Calculated for data in Table 1 using the equation tons per square miles = annual average constituent concentration in milligrams per liter × annual average stream discharge in cubic feet per second × conversion factor 0.9844 divided by drainage area in square miles. Column numbers are those used in Table 1. NA, data not available for calculation.

[a] Equivalent amount of sodium from combined sodium and potassium.

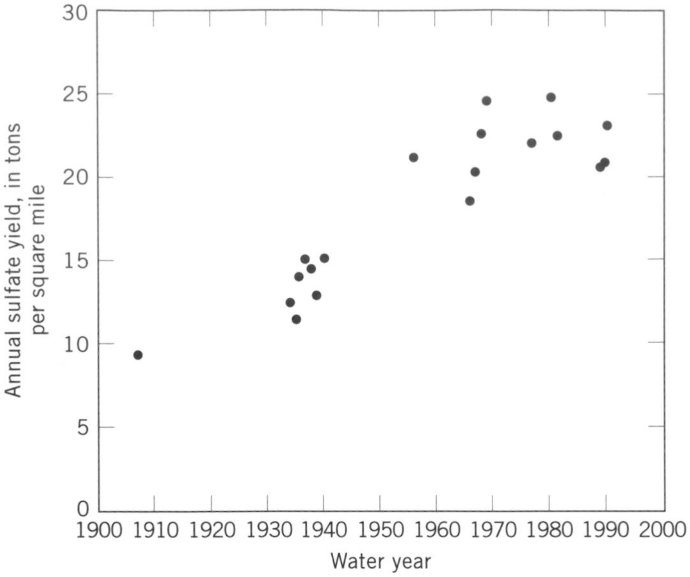

Figure 3. Annual sulfate yields of the St. Lawrence River at Cornwall, Ontario, Canada, 1906–1990. (Calculated from data in U.S. Geological Survey water-resources records and ref. 43.)

streams studied by Peters are tributaries of the Great Lakes–St. Lawrence system, and the average of the annual sulfate yield reported for these eight streams was 28.5 tons/mi^2. This finding agrees approximately with the sulfate yield of 25.4 tons/mi^2 given in Table 2 (col. 8) for Cornwall during water year 1980, considering the rather wide range of annual sulfate yields in the eight basins that were selected.

Although the substantial effects of human activities on the composition of the Great Lakes water is well documented by the data cited here, the system as a whole in recent years has shown a substantial ability to maintain a relatively high degree of chemical stability. The effect of acidic precipitation, for example, has not significantly depleted the capacity of the system to neutralize acid as has occurred in smaller lakes and streams in noncarbonate terranes in other parts of the northeastern United States. Although further increases in sulfate concentration in the lower Great Lakes have been predicted by various investigators (48), the concentrations of sulfate and chloride in the St. Lawrence River just downstream from the outlet of Lake Ontario appear to have remained the same or perhaps declined slightly since about 1980 (25).

The annual yields of sulfate during 1906–1990 for the St. Lawrence–Great Lakes basin (Fig. 3) are influenced considerably by differences in water discharge from year to year. The maximum values of about 25 tons/mi^2 in 1969 and 1980 and 23 tons/mi^2 in 1990 represent discharges substantially above the long-term average. The minimum yield for 1960–1970, 18.8 tons/mi^2, occurred in 1966, a year that had below-normal discharge. However, the yield more than doubles between 1906 and 1956 and has an upward trend through about 1970.

Columbia River Upstream from The Dalles, Oregon

Another major river drainage system shared by the United States and Canada is that of the Columbia River. The total drainage area of the Columbia at its mouth is 258,200 mi^2, and its long-term average discharge is 281,200 ft^3/s (49). Six stations—four on the Columbia and two on the Snake River, a major tributary to the Columbia—are used in this discussion to show constituent changes over time. On the Columbia River, data from stations at Northport, Washington, and Cascade Locks, The Dalles, and Rufus, Oregon, are used; the data from the three Oregon stations are considered equivalent. On the Snake River, the data from Clarkston and Ice Harbor in Washington are considered equivalent. The major downstream station used in this discussion is The Dalles, which is about 80 miles upstream from the city of Portland and has a continuous water quality record from 1950 to 1982. However, in some years only specific-conductance data were obtained. Additionally, about 40 miles downstream from The Dalles about 2 years of intensive sampling was done at Cascade Locks, near and upstream from the present site of Bonneville Dam, during 1910–1912 (50). The long-term average discharge at The Dalles is 192,000 ft^3/s (27), and the drainage area upstream from the sampling station is 237,000 mi^2 (Fig. 2).

Although the drainage basin of the Columbia is very different in many respects from that of the St. Lawrence above Cornwall, the basins are similar in size, discharge, and mean annual runoff. The area drained by the St. Lawrence above Cornwall is about 1.26 times greater than that of the Columbia above The Dalles, and the average discharge of the St. Lawrence River at Cornwall is about 1.27 times greater than that of the Columbia at The Dalles (Fig. 2). Hence, the runoff per square mile in the two systems is nearly the same.

Within the United States, the station farthest upstream on the Columbia is at Northport, Washington, about 8 miles from the United States–Canada International Boundary. Water quality records for this station cover most of the period 1952–1990, although for some years the information obtained was minimal. In addition, Van Winkle (50) includes complete analyses for major components in 10-day composites of daily samples collected from February 1, 1910, to January 31, 1911. The drainage area of the Columbia at Northport is 59,700 mi^2. However, not all of this area is in Canada, as the Pend Oreille and Kootenai rivers that drain more than 25,000 mi^2 of western Montana and northern Idaho join the Columbia a few miles north of the International Boundary. The average discharge of the Columbia at Northport for 54 years of record ending in 1991 is 99,870 ft^3/s (51).

The water contributed by the Snake River, the principal tributary in the reach between Northport and The Dalles, is about 54% of the gain in discharge of the Columbia in that reach. The remaining inflow is contributed by smaller tributaries such as the Okanogan, Spokane, Yakima, Umatilla, John Day, and Deschutes Rivers. Irrigation is extensive in most of the tributary basins, especially along the Snake River. Much of the area of the Columbia drainage basin south of the International Boundary, however, contributes very little direct surface runoff. In Idaho, for example, large areas are underlain by basalt, and although this terrane transmits groundwater readily, many of the stream drainage patterns are poorly developed. Also, precipitation at lower elevations

Table 3. Chemical Composition of Water and Constituent Annual Yields and Related Data for Columbia River at and near Northport, Washington, and The Dalles, Oregon, and the Snake River Near its Mouth, Selected Years (1910–1968)

Constituent, Property, and Related Data	Columbia River						Snake River, Clarkston, Wash., 1954 (4)		Columbia River				Snake River, Ice Harbor, Wash., 1968 (7)		Columbia River, The Dalles, Oreg., 1968 (8)	
	Northport, Wash., 1910–1911 (1)		Cascade Locks, Ore., 1910 (2)		Northport, Wash., 1954(3)				Rufus, Oreg., 1954 (5)		Northport, Wash., 1968 (6)					
	Concentration	Yield	Concentration	Yield	Concentration	Yield	Concentration	Yield	Concentration	Yield	Concentration	Yield	Concentration	Yield	Concentration	Yield
Silica (SiO_2)	8.7	15.9	13	11.5	6.3	13.6	20	9.4	11	9.6	4.8	8.1	16	6.0	11	7.6
Calcium (Ca)	18	32.9	16	14.2	21	45.1	22	10.3	19	16.5	22	37.0	21	7.9	20	13.8
Magnesium (Mg)	4.7	8.6	4.2	3.7	5.0	10.8	7.9	3.7	4.9	4.3	4.4	7.4	7.6	2.8	5.1	3.5
Sodium (Na)	{ 4.7[a]	8.6[a]	7.1[a]	6.3[a]	1.6	3.4	18	8.4	6.3	5.5	1.7	2.9 }	19[a]	7.1[a]	6.3[a]	4.3[a]
Potassium (K)					.8	1.7	2.3	1.1	2.1	1.8	.7	1.2 }				
Alkalinity, as bicarbonate (HCO_3) (property)	73	134	67	59.3	75	161	103	48.2	78	67.7	74	124	2.5	.94	1.6	1.1
Sulfate (SO_4)	12	22.0	13	11.5	12	25.8	27	12.6	14	12.2	14	23.5	106	39.6	78	53.8
Chloride (Cl)	0.6	1.1	2.0	1.8	0.9	1.9	9.5	4.5	3.2	2.8	0.4	0.67	27	10.1	19	13.1
Fluoride (F)	NA	NA	NA	NA	NA	NA	NA	NA	0.2	NA	0.2	NA	8.8	3.3	3.9	2.7
Nitrate (NO_3)	0.23	NA	0.43	NA	0.7	NA	1.3	NA	1.0	NA	0.5	NA	0.4	NA	NA	NA
Dissolved solids	84	154	89	78.7	85	183	159	74.4	103	89.5	88	148	1.1	NA	NA	NA
Specific conductance, μS at 25°C (property)	NA	NA	NA	NA	147	NA	247	NA	167	NA	149	NA	160	59.8	102	70.3
Drainage area, mi^2	59,700		237,000		59,700		103,200		237,000		59,700 252	 NA	167	NA		
Average discharge, ft^3/s	111,000[b]		213,000		130,400		49,060		209.100		102,000 108,500		237,000 41,210		166,600	
Dissolved solids load, million tons per year	9.2		18.7		10.9		7.69		21.2		8.84		6.50		16.6	

Note: Concentration values are in milligrams per liter unless otherwise noted; annual yields are in tons per square mile. Data from the Cascade Locks, Rufus, and The Dalles, Oreg., stations on the Columbia River are considered equivalent, as are data from Ice Harbor and Clarkston, Wash., stations on the Snake River. Yields are calculated from concentration values and discharge data. NA, data not available or not completed. *Sources:* Concentration values in columns 1, 2, ref. 50, columns 3–5, ref. 53; columns 6–8, ref. 54.

[a] Sodium and potassium not determined separately; value reported is total Na + K expressed as an equivalent amount of sodium.

[b] Avearage discharge was estimated by assuming that the ratio of the 1910 value to the long-term average discharge was the same for the river at Northport as it was at The Dalles.

in most of eastern Washington and Oregon is too low to contribute much runoff because moisture in air masses from the west is removed by their passage over the mountains in western Washington and Oregon (52).

A general indication of the chemical characteristics of the stream water in the Columbia River system is shown in Table 3. Data for Northport (Table 3, cols. 1, 3, 6) for several different years show that all constituents with the exception of silica are at similar concentrations, although the discharge in 1968, the year of lowest discharge, was about 78% of that in 1954 (53,54). Records for this station show that although the annual maximum discharge during the 1950s and 1960s commonly was as much as 10 times as great as the annual minimum discharge, the maximum annual sulfate concentrations were only about twice as great as the annual minimum sulfate concentrations. These relations suggest that strong homogenizing forces are at work. Three such forces are (1) the mixing effect of a large volume of naturally stored surface water, such as in the Great Lakes, can dampen out fluctuations in concentration related to dilution by runoff or snowmelt; (2) regulation of flow by storage reservoirs impounded by dams can contribute to decreases in the range between yearly maximum and minimum concentrations, as demonstrated by records for the Colorado River below Hoover Dam (2, p. 227); and (3) the chemical interaction between runoff and surficial material, especially where there is a large volume of stream flow of which a considerable part is contributed by groundwater, can prevent extensive chemical variations. The third factor probably is important in the area of the Columbia basin upstream from Northport.

Changes in concentration between Northport and The Dalles (as determined at Rufus, Oregon, about 20 miles upstream from The Dalles) that occurred in water year 1954 are shown in columns 3 and 5 of Table 3. The greatest increases in concentration are in silica, sodium, and chloride. The increased concentrations observed at The Dalles (Table 3, col. 8) possibly are the result of the relatively large silica and sodium concentrations in the Snake River as determined at Clarkston, Washington (Table 3, col. 4), which is the farthest downstream station on the Snake River. Annual yields of the major constituents also are given in Table 3 for each of the sampling stations; however, the yields at Northport in 1910–1911 (Table 3, col. 1) are based on estimated discharge inasmuch as no measured discharge data are available for that year. The yields for the Cascade Locks station (Table 3, col. 2) are much lower than those for the Northport station, which reflects the fact that much of the tributary drainage area between the stations is effectively noncontributing. Water quality data for the Columbia at The Dalles after the early 1970s are sparse for most years.

Maximum and minimum sulfate concentrations observed at Northport from 1910 to 1985 are plotted in Figure 4. It is apparent that these sulfate values do not have the well-defined increasing trend with time shown for the St. Lawrence in Figure 3. Least-square fitted lines shown in Figure 4 were drawn using data from 1910 and 1952–1970. The data collected after 1970 are less complete owing to decreased sampling frequency. The significance of the upward trend was tested by using only the 1952–1970 data and extrapolating the resulting straight line to 1910. The line for the maximum sulfate values intercepted the 1910 coordinate at 16 mg/L compared with the observed value of 15 mg/L. However, a similar calculation for the minimum sulfate data gave a poorer agreement. The existence of an upward trend in the sulfate concentrations at Northport is obviously questionable.

The historical increases in sulfate concentration shown in Figure 4 are much smaller in an absolute as well as in a relative sense than the increases observed in the St. Lawrence from 1906 to the 1970s. The average sulfate concentration in the St. Lawrence increased from 12 to 28 mg/L between 1906 and 1969 (Table 1), an increase of more than 130%. For the Columbia at Northport, the increase in average sulfate concentration was from 12 mg/L in 1910 to 14 mg/L in 1968 (Table 3), an increase of about 17%.

It is evident that the annual sulfate yields (Fig. 5) for the drainage basin of the Columbia River at Northport do not show any clearly defined trend. However, the sulfate yields have fluctuated between about 18 and about 26 tons/mi^2 for the years of record up to 1981, and this fluctuation could be interpreted as representing a rather poorly defined decreasing trend from 1960 to the 1980s.

Allegheny River Near Pittsburgh, Pennsylvania

The Allegheny River drains an area of about 11,500 mi^2 in southwestern New York and western Pennsylvania, and the long-term average discharge is 19,680 ft^3/s (Fig. 2). At Pittsburgh, the Allegheny joins the Monongahela River to form the Ohio River. Four stations are used in this discussion to show constituent changes over time. In downstream order these are Kittanning, which is on the

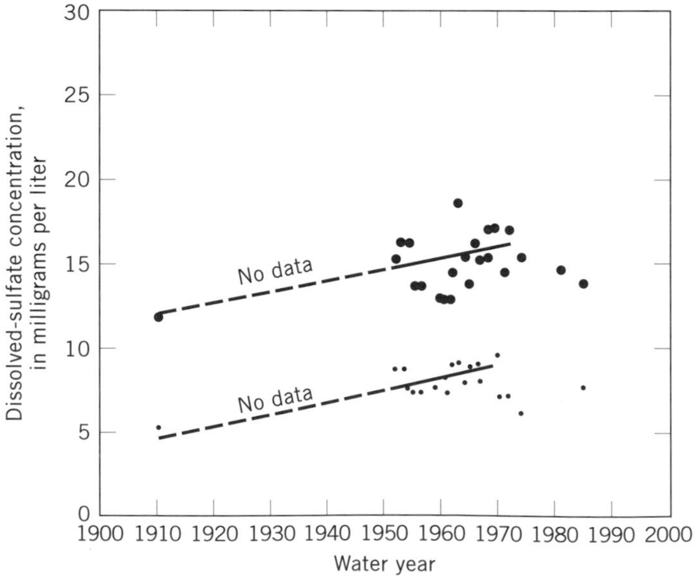

Figure 4. Annual maximum and minimum dissolved sulfate concentrations of the Columbia River at Northport, Wash., 1910–1985. (From U.S. Geological Survey records of water-quality data.)

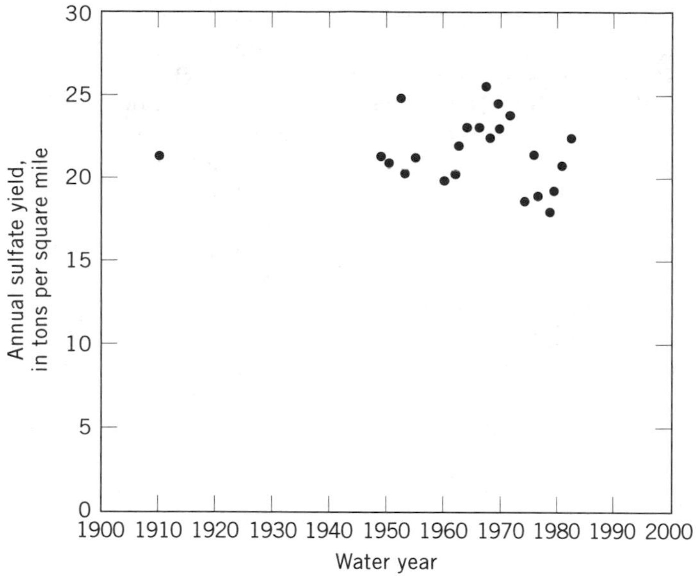

Figure 5. Annual sulfate yield of the Columbia River at Northport, Wash., 1910–1981. (Calculated from U.S. Geological Survey water-resources records.)

Allegheny about 40 miles upstream from Pittsburgh and about 12 miles upstream from the confluence of the Allegheny and the Kiskiminetas River; Vandergrift on the Kiskiminetas about 12 miles upstream from its confluence with the Allegheny, which is about 28 miles above the confluence of the Allegheny and the Monongahela; and New Kensington and Sharpsburg, data from which are equivalent, downstream from the confluence of the Kiskiminetas with the Allegheny.

In the Allegheny drainage basin, the bedrock is sandstone, shale, and limestone and contains extensive deposits of bituminous coal (55). Exploitation of petroleum and natural gas has produced some stream contamination through release of saline groundwater associated with the petroleum. For about 70 miles upstream from Pittsburgh, the Allegheny receives drainage from many active and abandoned coal mines that has been affected by oxidation products of pyrite and related sulfide minerals that accompany the coal. This drainage is a solution of sulfuric acid and commonly carries high concentrations of iron and other cations. The Allegheny provides a part of the water supply for the city of Pittsburgh, and the deterioration of water quality in the river has been a matter of considerable concern for many years.

The station at Kittanning, Pennsylvania, has the only analytical data for the Allegheny River recorded during the early 1900s (21). As shown in Table 4 (col. 1), the time-weighted average sulfate concentration for 1906–1907 was 17 mg/L. In contrast, in a study of water quality records of the Allegheny and its tributaries up to about 1965 reported that the average sulfate concentration for water year 1962 was 48 mg/L (55). Although this is a substantial increase, even larger sulfate and hydrogen ion contributions to the river are received farther downstream. Among other aspects of the water quality problems in the lower part of the Allegheny basin, McCarren (55) noted the large inflow of acid mine drainage carried

Table 4. Chemical Composition of Water and Related Data for Allegheny River at Downstream from Kittaning, Pennsylvania Selected Years (1906–1989)

	Allegheny River			
			At/Near New Kensington	
Constituent, Property, and Related Data	At/Near Kittanning, 1906–1907[a] (1)	At/Near Sharpsburg, 1947[b] (2)	1975[c] (3)	1989[c] (4)
Silica (SiO_2)	7.9	NA	4.1	3.5
Calcium (Ca)	14	NA	33	38
Magnesium (Mg)	3.0	NA	7.3	10
Sodium (Na)	{11[d]	NA	14	23
Potassium (K)		NA	2.6	2.3
Alkalinity, as bicarbonate (HCO_3) (property)	38	0	24	40
Sulfate (SO_4)	17	193–200	110	110
Chloride (Cl)	14	46–49	14	24
Fluoride (F)	NA	NA	0.4	0.2
Nitrate (NO_3)	0.7	4.1–5.4	2.2	2.2
Dissolved solids	87	NA	224	232
Specific conductance, μS at 25°C (property)	NA	609–625	350	NA
pH units (property)	NA	4.0–4.2	6.7	7.5
Drainage area, mi^2	8,973	11,410	11,500	11,500
Average discharge, ft^3/s	NA	3,700	3,740	3,580

Note: Concentration values are in milligrams per liter unless otherwise noted. Data from the New Kensington and Sharpsburg stations are considered equivalent. NA, not available. *Sources*: column 1, ref. 21; column 2, ref. 56, column 3, ref. 57; column 4, ref. 20.

[a] Time-weighted average.

[b] Range of values for four cross-sectional samples on Nov. 10, 1947.

[c] Single sample collected on Aug. 25, 1975 (col. 3) and Sept. 5, 1989 (col. 4).

[d] Sodium and potassium not determined separately; value reported is total Na + K expressed as an equivalent amount of sodium.

by the Kiskiminetas River. He mentioned that a plan for improvement of the water quality in the lower Allegheny River was being implemented. In general, the flood control and multipurpose reservoirs that had been constructed, or were planned for construction in the Allegheny basin, were to be operated in such a way as to increase water discharge rates in the lower Allegheny at times of natural low flow. By providing dilution water for such times, the effect of acid mine drainage from streams such as the Kiskiminetas River would be eased. At the same time, efforts to clean up abandoned mines and waste disposal sites and decreasing contamination from other sources would continue.

Some effects of flow management and other contamination abatement measures on water composition in the Kiskiminetas River are indicated by data in Table 5. The composite sample for November 11–20, 1947 (Table 5, col. 1), represents an average discharge about three-fifths as great as that for the 1976 sample (Table 5, col. 2). The smaller sulfate concentration in the 1976 sample was mainly the result of dilution, as the daily sulfate loads were about the same. However, the pH values show a hundredfold decrease in acidity in the 1976 sample. This demonstrates the fact that the diluting water contained bicarbonate alkalinity that reacted to neutralize the acid, and the effect is not simply a matter of dilution.

In addition to the 1906–1907 data mentioned above, Table 4 gives analyses for three samples from the Allegheny below the Kiskiminetas—one from Sharpsburg (November 10, 1947, col. 2) and two from New Kensington (August 25, 1975, col. 3; September 5, 1989, col. 4). Water discharge rates were about the same for all three samples. However, the proportion of flow coming from the Kiskiminetas probably was greater at the time of the 1947 sampling than it was for the later sampling dates because flow control became more extensive after 1950 (55, p. 66). The water quality records for New Kensington indicate, in general, an upward trend in alkalinity.

Figure 6 shows the dissolved-sulfate concentration and pH of samples collected from the Allegheny River below the Kiskiminetas at low-flow stages during the period of record for water years 1948–1989. The values represent single samples that were collected about once a month, and the sample used for each year was the one having the lowest reported stream discharge. No special effort seems to have been made to collect samples at minimum-flow stages. However, for nearly all the years of record, a reasonably representative sample was obtained for low-flow conditions. In the 1948–1952 period, the stream dis-

Table 5. Chemical Composition of Water and Related Data for Kiskiminetas River at Vandergrift, Pennsylvania, at Low Flow (1947, 1976)

	Kiskiminetas River at Vandergrift, Pa.	
Constituent, Property, and Related Data	1947[a] (1)	1976[b] (2)
Silica (SiO_2)	NA	NA
Calcium (Ca)	NA	43
Magnesium (Mg)	NA	16
Sodium (Na)	NA	NA
Potassium (K)	NA	NA
Alkalinity, as bicarbonate (HCO_3) (property)	0	0
Sulfate (SO_4)	360(1027)[c]	225(1086)[c]
Chloride (Cl)	12	21
Fluoride (F)	NA	0.2
Nitrate (NO_3)	NA	NA
Dissolved solids	NA	325
Specific conductance, μS at 25°C (property)	981	700
pH units (property)	3.2	5.2
Drainage area, mi^2	1825	1825
Average discharge, ft^3/s	1058	1790

Note: Concentration values are in milligrams per liter unless otherwise noted. NA, not available. *Sources*: Column 1, ref. 56; column 2, ref. 58.
[a] Composite sample, Nov. 11–20, 1947.
[b] Single sample, Jan. 22, 1976.
[c] Sulfate load, in tons per day.

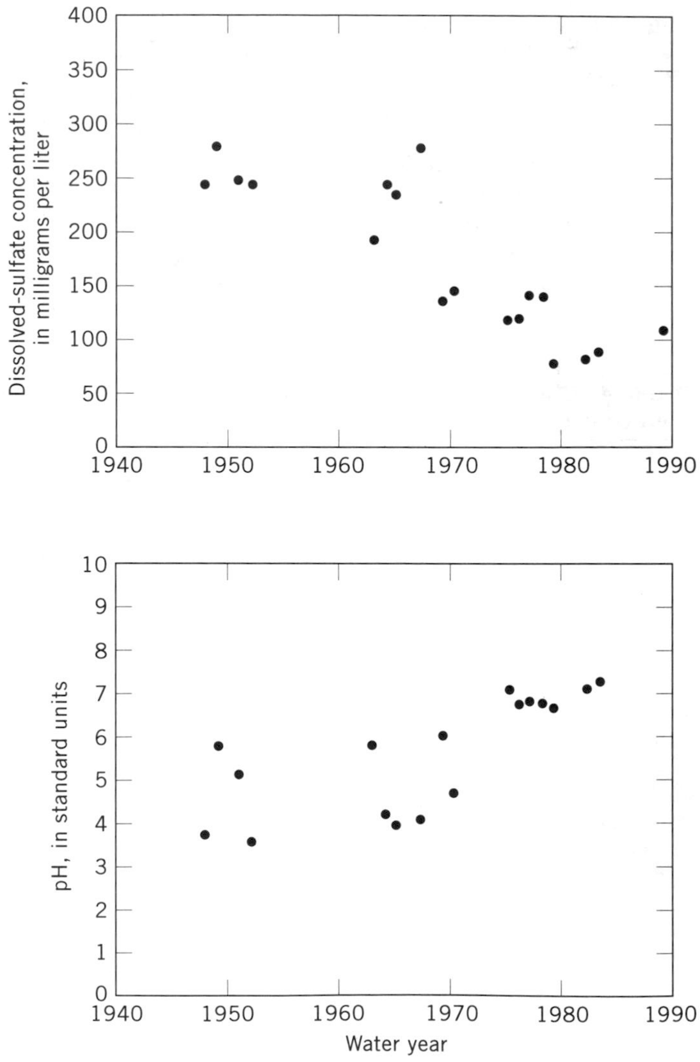

Figure 6. Dissolved sulfate concentration and pH of the Allegheny River at New Kensington, Pa., 1948–1989. Annual values shown were measured in samples collected at the lowest discharge rate for each water year. (From U.S. Geological Survey records of water-quality data.)

charges for samples represented in Figure 6 ranged from 1580 ft^3/s in 1948 to 2500 ft^3/s in 1949. After 1965, the discharges for most of the plotted data values were considerably higher, although in 1967 the flow was 2960 ft^3/s and in 1982 was 2920 ft^3/s. The pH values measured in monthly samples from New Kensington before 1970 were as high as 6.2 but only when stream discharges exceeded 5000 ft^3/s. In the 1980s, the pH of the water was above 6.5 at all sampling times, although discharges reported at those times were as low as 2920 ft^3/s.

The improvement in water quality in the Allegheny River at low flow during the time period covered by Figure 6 is rather dramatic. Data in Table 4 indicate a nearly 50% decrease in sulfate concentration at low flow in the 1975 and 1989 data compared to the 1947 data. The fact that discharge rates were about the same for all three sampling dates means that the actual sulfate loads at low flow in 1975 and 1989 were only about half that observed in 1947.

Lower Mississippi River, Louisiana

The Mississippi River is the largest North American river in terms of both drainage area and discharge, and the possible existence of long-term historical trends in the chemical composition of the river is of great interest. The water composition of the lower Mississippi near its mouth reflects the effects of human activities in the entire Mississippi drainage basin, and data collected since 1954 at the St. Francisville, Louisiana, water quality sampling station (Fig. 2), along with data collected in 1905–1906 near New Orleans, Louisiana, are used in this discussion. The contributing drainage area of the river near St. Francisville is about 1,125,300 mi^2, and the average discharge, which was determined from the nearest upstream gaging stations (Tarbert Landing, Miss., and Red River Landing, La.), is 514,000 ft^3/s for the years 1973–1982 (29).

The earliest water quality records for the Mississippi near its mouth (21) are for the New Orleans public water utility intake for the period April 29, 1905, to April 28, 1906; a shortcoming of the New Orleans record is a lack of river discharge data (Table 6). Detailed chemical quality sampling on the lower Mississippi was not undertaken by the USGS until 1954. The sampling station is located at a highway ferry crossing 2 miles southwest of St. Francisville, about 35 miles upstream from Baton Rouge and more than 160 miles upstream from the New Orleans sampling station. Beginning in 1957, these records were

Table 6. Chemical Composition of Water and Related Data for Mississippi River at New Orleans, Louisiana (1905–1906), and near St. Francisville, Louisiana (water years 1966, 1989)

	New Orleans Water Works Intake 1905–1906[a] (1)			Mississippi River near St. Francisville					
				Water Year 1966[d] (12)			Water Year 1989[g] (3)		
Constituent, Property, and Related Data	July 30–Aug. 5, 1905[b]	Feb. 4–10, 1906[c]	Annual Average	Dec. 1–10, 1965[e]	Feb. 12–20, 1966[f]	Annual Average	Nov. 14, 1988[h]	Mar. 14, 1989[i]	Annual Average
Silica (SiO_2)	13	7.8	11	7.8	8.4	7.6	4.1	5.4	5.0
Calcium (Ca)	33	28	32	44	36	43	40	24	33
Magnesium (Mg)	8.5	6.8	8.4	22	9.3	13	19	6.6	11
Sodium (Na)	21[j] (Na + K)	7.8[j] (Na + K)	13[j] (Na + K)	28	17	21	34	NA	22
Potassium (K)				3.4	2.0	3.1	4.1	NA	3.0
Alkalinity, as bicarbonate (HCO_3) (property)	116	91	111	168	116	145	162	77	115
Sulfate (SO_4)	36[k]	17[l]	24	87[k]	38[l]	59	76[k]	29[l]	44
Chloride (Cl)	10	6.3	9.7	28	21	22	27	9.2	19
Fluoride (F)	NA	NA	NA	0.4	0.2	0.3	NA	NA	0.2
Nitrate (NO_3)	2.7	2.5	2.5	1.8	3.6	1.7	2.6	4.2	3.9
Dissolved solids	195	130	166	320	201	250	NA	NA	200
Specific conductance, μS at 25°C (property)	NA	NA	NA	511	337	413	507	228	349
pH units (property)	NA	NA	NA	7.6	7.6	7.4	7.7	7.5	7.6
Drainage area, mi^2	—	1,243,700	—	—	1,125,300	—	—	1,125,300	—
Average discharge, ft^3/s	NA	NA	NA	224,000[m]	601,000[m]	337,000[m]	145,000[m]	1,120,000[m]	569,000[m]

Note: Concentration values are in milligrams per liter unless otherwise noted. NA, data not available. *Sources*: Column 1, ref. 21, column 2, ref. 59, column 3, ref. 60.

[a] Analyses of weekly composites of daily samples, April 29, 1905–April 28, 1906.
[b] Composite sample having the maximum dissolved sulfate concentration during the 1905–1906 period.
[c] Composite sample having the minimum dissolved sulfate concentration during the 1905–1906 period.
[d] Analyses of 10-day composites of daily samples, water year 1966.
[e] Composite sample having the maximum dissolved sulfate concentration during water year 1966.
[f] Composite sample having the minimum dissolved sulfate concentration during water year 1966.
[g] Analyses of single monthly samples, water year 1989.
[h] Monthly sample having the maximum dissolved sulfate concentration observed during water year 1989.
[i] Monthly sample having the minimum dissolved sulfate concentration observed during the water year 1989.
[j] Total sodium plus potassium as equivalent amount of sodium.
[k] Maximum dissolved sulfate for the year.
[l] Minimum dissolved sulfate for the year.
[m] Discharge at Tarbert Landing, Miss., for sampling dates.

supplemented by a daily sampling program at the Luling-to-Destrehan ferry crossing, 17 miles west of and upstream from New Orleans. However, the difference in drainage agrea between the St. Francisville station and the Luling-to-Destrehan ferry crossing is considered to be insignificant, and therefore, except for possible effects of industrial activity and waste disposal at and near Baton Rouge, the records for New Orleans and the Luling-to-Destrehan ferry crossing should be essentially the same as those for St. Francisville. Because the data from the St. Francisville station are more complete, they are used here as the basis for investigating historical changes in water composition.

As with all delta formation, the discharge channels of the Mississippi River delta continually migrate. As sediment builds up in one channel, it eventually forces the river to abandon that route and find another. An example of the process can be seen in the development of the Atchafalaya River. This stream probably was first formed by ancient floods near the confluence of the Red and Mississippi Rivers, and by the nineteenth century discharge from the Mississippi River through the Old River into the Atchafalaya River, later aided by navigation channel-clearing improvements and flood control activities, had become significant (Shlemon, 61). To prevent the abandonment of the main Mississippi channel through New Orleans, which had been predicted as a possibility by 1975 (Wells, 62, p. 6), the U.S. Army Corps of Engineers designed and, by 1963, had installed the Old River Control Structure. Since July 1963, the Old River Control Structure has been operated to divert variable amounts of flow from the Mississippi into the Atchafalaya. The amount of this diversion is required to be at least 30% of the total flow of the Atchafalaya, which also carries the entire discharge of the Red River from a drainage area of 87,570 mi^2 (63). As a result of the diversion by the control structure, the average flow of the Mississippi at Tarbert Landing is decreased by about 20% compared to the discharge upstream from the diversion point (62). This would not be expected to influence significantly the dissolved-sulfate or chloride concentrations in the river near St. Francisville, but it does influence calculation of yields of dissolved constituents because the effective drainage area upstream from Tarbert Landing cannot be exactly determined due to the variable amounts of flow routed down the Atchafalaya River.

Many other factors complicate the interpretation of changes that have taken place in water composition above New Orleans since 1906. In addition to many cultural changes, agricultural and industrial development changes, dams and reservoirs, channel controls, the effects of variable rainfall and runoff in the basin upstream, and the possible effects of the Atchafalaya diversion, there have been various changes in sampling protocol over the years of record. A once-daily sampling schedule prevailed in New Orleans in 1905–1906, but each composite covered only 1 week. In the years from 1955 to 1968, daily sampling continued but generally with a standard 10-day equal-volume composite period. Some composites, however, contained a month of daily samples if day-to-day variation was small. Beginning in 1969, complete analyses were made on single larger samples collected during flow measurements in a cross section according to USGS procedures to ensure reliable representations of conditions. The sampling frequency generally was about once a month, but in some years it was two or three times a month, and in a few years it was less than once a month.

For most years, daily average values of specific conductance and temperature were taken from on-site recorded data, and for some years other properties or constituents, including sulfate and chloride, were measured daily. The number of determinations of sulfate available per year has ranged from 365 to as few as 6.

Because of these factors, a simplified method of approximating sulfate trends in the lower Mississippi that was used earlier (2) was adapted for preparation of Figure 7. From the data for each year of record the analyses showing the maximum and minimum dissolved-solids concentrations were selected, and these were used to obtain a maximum and minimum sulfate concentration for that year. In addition an average sulfate concentration weighted by discharge was calculated for each year.

A linear least-squares regression line was fitted to each of the three sets of data points, and by using different time periods, it was found that the lines with the largest correlation coefficients showed an increasing trend between about 1906 and 1970, with no well-defined upward or downward trend during the subsequent 20 years. (Fig. 7). The 1906 data point was used as the point of origin for each line.

Some evidence of the composition of the Mississippi River water at New Orleans during the long gap between 1906 and 1955 can be obtained from reports (64,65) on quality of public water supplies in the United States. The data given by Lohr and Love (65) are for the chemical

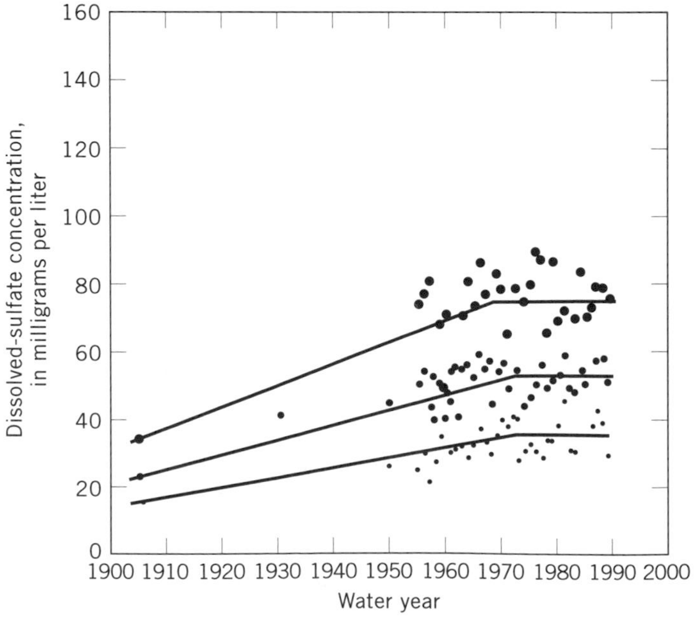

Figure 7. Annual maximum, average, and minimum dissolved sulfate concentration of the Mississippi River at New Orleans, La. (1905–1906), and near St. Francisville, La. (1955–1989). (From U.S. Geological Survey data compilations and refs. 2 and 21.)

quality of the Mississippi River at the New Orleans water supply intake and include maximum, minimum, and average total hardness values for the year 1950. A rather well-defined linear relation between sulfate concentrations and hardness can be developed from the analytical data for St. Francisville in the 1950s over the range in total hardness from 80 to 125 mg/L. If applied to the minimum and average hardness values reported by Lohr and Love (65) for the untreated water at New Orleans during 1950, calculated sulfate concentrations for 1950 are 30 mg/L for the minimum hardness and 44 mg/L for the average hardness. These values are very close to the trend lines in Figure 7. The sulfate concentration given in Ref. 64 for the 1931 sample furnished by the New Orleans Water Utility is 42 mg/L. This probably is not an average value for the entire year, and it lies somewhat above the average line in Figure 7.

The most likely cause of the upward sulfate concentration trend from 1906 to 1967 in the lower Mississippi River would seem to be the contamination of the river and its tributaries by human activities, as it is similar to the upward trend in sulfate in the St. Lawrence River and the lower Great Lakes during the same period. However, factors other than simple waste disposal practices might be responsible. For example, the records for the Allegheny River above Pittsburgh demonstrate effects of human activities that modified flow patterns by releasing, during periods of low flow, water stored in reservoirs during high-flow stages in order to dilute what would normally have been a low flow having a high sulfate concentration. Some of the sulfate behavior in the Mississippi could possibly be ascribed to the holding back of dilute flood water in storage facilities upstream and releasing them at later times. The general effect of storage on water quality at downstream points is to decrease the variability of constituent concentrations.

Table 6 gives an indication of the overall change in composition of the water of the lower Mississippi during the 60 years following the 1905–1906 record (21). As noted in the discussion of Figure 7, the slopes of the fitted straight-line trends appear to change at or shortly after 1967.

In general, the historical data for the Mississippi near St. Francisville show that minimum values for sulfate and chloride concentrations occur during high-flow periods and maximum values occur at low flow, but the relation of discharge and concentration is poorly defined. In summary, a very large amount of data that suggests water quality changes have occurred in the lower Mississippi is available. However, interpretation of these data in order to closely define the causes of the trends is still rather uncertain.

Sensitivity of River Systems to Basin Development and Related Waste Disposal Practices

The sensitivity of a drainage system to human-induced water quality degradation is a function of numerous physical properties of the system relating to climate, bedrock geology, and rate of water circulation or residence time. The interaction of these factors from the time water reaches the land surface as precipitation until it flows as runoff into a drainage channel leaves a characteristic imprint on the amount and chemistry of the stream water.

The data presented for the four drainage basins demonstrate how the impacts of various human activities have changed sulfate ion concentrations and yields, and it is instructive to review the similarities and differences between water quality records that were observed and the hydrologic and other factors that can be postulated as sensitivity controls.

St. Lawrence River Basin. The St. Lawrence River at the outlet of the Great Lakes system shows a well-defined upward trend in sulfate concentrations between 1906–1907 and 1980 (Fig. 3 and Table 1). The trend is less masked by seasonal or short-term variability than for most other streams in the conterminous United States because of the large volume of water stored in the lakes and the effectiveness of seasonal mixing that occurs in all of them. Also, the sedimentary rocks and the glacial drift derived from them that underlie the four lowermost lake basins provide carbonate to neutralize acid precipitation, and ample precipitation sustains a large outflow.

The hypothesis that the doubling of the sulfate concentration in the St. Lawrence River near the outlet of Lake Ontario since 1906–1907 was caused primarily by increased atmospheric sulfur contributions to the Great Lakes drainage basin appears to be reasonable. The increase in annual sulfate yield between 1906 and the average value for the 1969–1980 period is about 14.3 ± 1.5 tons/mi^2 (Table 2). Although some of the sulfate yield in 1906 could have been caused by atmospheric sulfur, the concentrations of sulfate in Lake Ontario that were determined at that time (12 mg/L) were thought (48) to represent the natural condition extending back into the nineteenth century.

The compilation of data by the EPA (18) on quantities of sulfate brought to the land surface in rain and snow and dry fallout shows a rather wide range from year to year and from one measuring site to another. However, within the St. Lawrence–Great Lakes drainage system and much of the northeastern United States, the sulfate fallout and rainout rate that is attributable to human activities probably is in the range of 8.2 to about 16 tons/mi^2. Studies in geographically adjacent areas (66,67) show sulfate deposition rates that are within this range. Both studies observed some downward trends in sulfate deposition with time during the period from about 1970 through the 1980s. The pH of rainfall, however, did not generally increase during this period.

The study (66) was conducted in the Hubbard Brook Experimental Forest in New Hampshire beginning in 1963. This area is underlain by metamorphic rocks with glacial till derived from these rocks overlying them. This material reacts very slowly with acid deposited by precipitation and streams and lakes have water with pH values less than 5.0. The study (67) covers the period 1965–1979 and has data for streams draining areas where the bedrock is of sedimentary origin. The decreased sulfate loading can probably be explained as a result of regionally extensive efforts to decrease sulfur species emissions from power plants and other human-related sources. Some increase in fallout of oxidized nitrogen species has been observed in the areas studied, which also can produce acid precipitation.

From the time the sulfate fallout rate stabilized and began to decline to the time sulfate yields of the St. Lawrence at Cornwall approximately stabilized would appear from plotted points in Fig. 3 to have been about 10 years. This may not be an unreasonable lag time, as the total quantity of water storage in the Great Lakes is very large compared to the annual water quantities discharged by the St. Lawrence River.

Columbia River Basin. The average annual discharge for the Columbia River at The Dalles, Oregon (Fig. 2), is about the same as that of the St. Lawrence River at Cornwall, Ontario, Canada, but many of the other physical, hydrologic, and geochemical characteristics of the two systems are very different. A large part of the Columbia drainage in eastern Washington and Oregon and in adjacent areas of Idaho has a semiarid climate and contributes little direct runoff to the Columbia. Although the Columbia and its tributaries have some large natural lakes and many reservoirs created by dams, the area of water surface in the Columbia River basin is not nearly as large as the area of the Great Lakes.

Sulfate contributions from the atmosphere over the Columbia drainage basin are relatively minor compared to those in the Great Lakes–St. Lawrence and are more likely to be intermittent and of natural origin, as for example the 1980 eruption of Mount Saint Helens. Ref. 68 notes some relatively small sulfate contributions of human origin in runoff from the North Cascade Range, which is drained by streams flowing toward Puget Sound. It is rather surprising to note, however, that the sulfate yield for the Columbia River at Northport (Table 3, Fig. 5) is in the range of about 18 to about 27 tons/mi^2 for each year of record, nearly the same as the yields for the Great Lakes–St. Lawrence system during the 1970–1990 period. Possible human-related and natural sources for the high sulfate yield can be discerned.

A substantial part of the discharge of the Columbia at Northport is contributed by the two main tributaries that drain areas of western Montana and northern Idaho—the Clark Fork–Pend Oreille and the Kootenai. The Kootenai River also drains a substantial area in southeastern British Columbia. A considerable area of land is irrigated by diversion of water from the two tributaries. In addition, the Clark Fork has been affected by copper mining and ore treatment processes in the Butte-Anaconda area (41). The Clark Fork discharges into Lake Pond Oreille in northern Idaho and the lake discharges through the Pond Orielle River into the Columbia just north of the International Boundary. It seems likely that there are significant sulfate contributions from human sources from the two tributaries.

A daily sampling record for the Kootenai River at Porthill, Idaho, for water year 1950 (69) gave a calculated annual sulfate yield of 17.7 tons/mi^2. A calculation for the Pend Oreille at Metaline Falls, Washington, based on daily sampling during the same water year (69) gave a sulfate yield of 14.2 tons/mi^2. The average water discharge for these two tributaries in 1950 was 60,900 ft^3/s, which is more than one-half of the 1950 average discharge of the Columbia at Northport, Washington (111,000 ft^3/s). Water quality records were not obtained at that station in 1950. However, the sulfate yield in water year 1952 was 21.8 tons/mi^2 (70) and, as shown in Figure 5, generally has ranged from about 22 to about 25 tons/mi^2. In order for annual sulfate yields to be this high at Northport, the annual sulfate yield for the Columbia basin upstream from the Kootenai and Pend Oreille sampling stations must be substantially greater than 30 tons/mi^2. Using the 1950 and 1952 data quoted above, the calculated annual sulfate yield for the 20,800 mi^2 of the Columbia basin that is upstream from the two tributaries is 33.8 tons/mi^2. Records for 1980–1989 are incomplete, but generally they support this estimate.

Mineral and thermal springs are among possible natural sources of sulfate in the Canadian part of the Columbia drainage basin. As noted by Van Everdingen (71), some of the springs in this region yield water having a large enough sulfate concentration to approach saturation with respect to gypsum. Extensive exposures of sedimentary rock in the southern Rocky Mountains of Canada (72), could also yield sulfate to weathering solutions.

Measurements of sulfate concentrations at four sampling stations on the Columbia River (73) from 1985 to 1991 provide some additional insights into sulfate sources. The data for a station that is upstream from natural lakes and artificial storage reservoirs show that annual minimum sulfate concentrations near 10 mg/L are reached in most years during June or July, when flow is augmented by melting snow. During the fall and winter seasons, sulfate concentrations commonly are in the range of 25–40 mg/L, and a maximum concentration of 53.5 mg/L was reported on February 11, 1986.

These sulfate concentrations are substantially more variable and reach higher values than any at downstream stations on the Columbia. The highest sulfate concentration observed at The Dalles in the 1951–1974 period was 34 mg/L, which occurred in 1972. The higher maximums observed at the Canadian site suggest that the high sulfate yield could come from groundwater inflows and dissolution of sedimentary rock minerals in the upper part of the Columbia drainage basin.

Flow control structures for aid in navigation and generation of electric power are numerous on the Columbia main stem. These have had some notable effects on the ecology of the river, especially on fish populations. These aspects of human-caused changes are discussed in Refs. 40 and 74. Ref. 40 contains descriptive articles on stream ecology in 12 North American river basins.

Allegheny River Basin. The average runoff of the Allegheny River basin above New Kensington is 1.72 ft^3/s mi^2, which is nearly the same as the water yield of the Columbia basin above Northport. However, the average discharge of the Allegheny is only about 19% of that of the Columbia at Northport. Hence, one could predict the higher sensitivity to point sources of contamination in the Allegheny that is demonstrated by the data for various sampling stations in the Allegheny basin. The average of the data published (21) for the Allegheny shows a relatively pristine condition in 1906–1907 at the Kittanning, Pennsylvania, sampling station (Table 4). Minimum and maximum sulfate concentrations reported by Dole for the year were 10 and 35 mg/L, respectively. McCarren's (55) analytical data for that station for 1961–1962 show a range of sulfate concentrations from a minimum of 32 mg/L to a maximum of 67 mg/L. These data are substantial evidence of sulfate contamination at Kittanning, and probably much of the contamination can be attributed to atmospheric fallout. Ref. 55 indicates that most of the coal-mining activity was downstream from the Kittanning station.

Major sources of mine drainage and industrial waste contamination of the Allegheny River immediately upstream from Pittsburgh are located on the Kiskiminetas River. As shown in Table 5, during the period November 11–20, 1947, the sulfate load in the Kiskiminetas was 1027 tons/day at an average discharge of 1058 ft^3/s. On January 22, 1976, at an average discharge of 1790 ft^3/s, the sulfate load was 1086 tons/day. The similarity of sulfate loads for these two sampling periods contrasts with the decrease in sulfate concentration from 360 mg/L in 1947 to 225 mg/L in 1976. This decrease in concentration can be attributed to dilution brought about by release of water from upstream storage reservoirs. The hundredfold decrease in hydrogen ion activity indicated by the concurrent increase in pH in the 1976 sample indicates the diluting water had a substantial acid neutralization capacity.

An indication of the relative contamination sensitivity of the Allegheny compared to the Columbia can be gained by considering daily sulfate loads. The Columbia at Northport in 1968 (Table 3, col. 6) had an average sulfate concentration of 14 mg/L and an average discharge of 102,000 ft^3/s, which represents an average daily sulfate load of 3850 tons. If the 1027-tons/day sulfate load equal to that of the Kiskimetas was added to the Northport sulfate load at a discharge of 1058 ft^3/s, the average sulfate concentration at Northport would increase to 18 mg/L. The complete analytical data for 1968 for this station (54) show a minimum average discharge of 40,500 ft^3/s for the composite sample representing January 1–20 and a sulfate concentration of 19 mg/L. Here, the addition of 1058 ft^3/s of discharge carrying 1027 tons/day of sulfate would have given a final sulfate concentration of 28 mg/L. Although the effect on sulfate concentration at low flow would be significant, the effect at the annual average discharge rate is only a little greater than the analytical uncertainty in the sulfate determinations.

A much more obvious effect of discharge on sulfate concentration can be seen for the Allegheny River. From Tables 4 and 5 it can be calculated that the sulfate concentration in the Allegheny upstream from the Kiskiminetas at low flow in November 1947 would have been about 136 mg/L. As the data in column 2 of Table 4 show, the sulfate concentration was about 200 mg/L after receiving discharge from the Kiskiminetas. At a near-average discharge of 19,640 ft^3/s and a typical sulfate concentration of 40 mg/L in the Allegheny upstream from the Kiskiminetas, the effect of the inflow at the rate of 1058 ft^3/s and a sulfate concentration of 360 mg/L would be to increase the sulfate concentration of the Allegheny below the river confluence to 57 mg/L.

Additional data in Table 4 (cols. 2–4) show that for similar flow stages in the Allegheny below the Kiskiminetas, there was indeed a substantial improvement in water quality between November 1947 and September 1989. The sulfate concentration decreased by about one-half and pH increased from near 4.0 to 7.5. Data in Figure 6 indicate that most of the decrease in sulfate and increase in pH at low flow had taken place by 1975. Flow augmentation by timed reservoir releases and related remedial measures have had an obvious beneficial effect on the quality of the Allegheny at New Kensington.

Lower Mississippi River Basin. In the lower Mississippi River, data on concentrations of sulfate and changes between 1905–1906 and 1989 (Fig. 7) show an approximate doubling of annual minimum, maximum, and average concentrations. Most of the change occurred before 1980. The calculation of sulfate yield for the whole Mississippi drainage basin is fraught with many uncertainties, owing to the complex nature of flow control and water diversion efforts and incomplete discharge and water quality data. However, a rough approximation of sulfate yield can be made using discharge data from the Tarbert Landing station and the rather arbitrary contributing area assigned in USGS water data compilations of 1,125,300 mi^2 for the Tarbert Landing station. The results obtained using the annual average sulfate concentration for each year indicate that for most of the years from 1964 to 1989 the annual yield was between 20 and 25 tons/mi^2; the 5 years in which the yield was below 20 tons/mi^2 were years of low mean discharge.

These estimated annual sulfate yields are similar in magnitude to those calculated for that general time period for the Great Lakes–St. Lawrence basin (Table 2). However, if it could be estimated, the gain in yield between 1905–1906 and the 1964–1989 period is a more meaningful figure. The discharge of the river at the New Orleans sampling site during the 1905–1906 period was not measured. A further uncertainty relates to the amount of flow that might have left the main river at the site of the present Old River Control Structure. If it is assumed this was a minor effect in 1905–1906, the long-term average discharge (about 570,000 ft^3/s) given by Wells (62) for the measuring point at Vicksburg, Mississippi, about 140 river miles upstream from the Old River Control Structure, could perhaps be assumed to approximate the 1905–1906 average for the Mississippi at New Orleans. The average sulfate concentration of 24 mg/L given for that period leads to a calculated annual yield of 12 tons/mi^2 of sulfate for the drainage basin.

Although this estimate has a substantial uncertainty, the indicated increase in annual yield of sulfate since 1906 is about 9–14 tons/mi^2. This increase is equivalent to the increase calculated for the Great Lakes–St. Lawrence system from 1906 to the 1970s and 1980s (Fig. 3) and could be considered as possibly explainable by an increase in atmospheric contributions of sulfate. It also appears from the relatively minor changes that have occurred during the 1970s and 1980s that the river has reached an approximate steady state with respect to present-day sulfate contributions from the drainage basin and effective transport of sulfate out of the basin to the ocean. Continuing efforts to control sulfur releases to the atmosphere and hydrosphere in the drainage basin could eventually decrease the sulfate concentrations and yields observed at St. Francisville to an extent that will be more obvious.

None of the four major river systems that are described in this article display a high degree of sensitivity to point sources of contamination. Also, the systems that receive significant contributions from "acid rain" have soils and bedrock that neutralize the acid effectively. This is in contrast to the strong effects observed by researchers at the Hubbard Brook site in New Hampshire (66) where streams and lakes generally have pH values of 5.0 or less. The soils and bedrock at the Hubbard Brook site are dominated by silicate minerals that react very slowly with acid. This is an important factor in producing the high sensitivity of surface water systems to human-related pollution.

SUMMARY

Under pristine conditions, that is, in the absence of human civilization and development, the chemical composition of inland stream and lake waters is, ideally, controlled by the alteration of rock minerals through chemical weathering processes, which liberate soluble products. These processes in turn are controlled or influenced by climatic factors such as rainfall, air temperature, and evaporation and by associated biological or biochemical processes, such as photosynthesis and transpiration by plants, decay of vegetative debris, and the effects of aquatic life processes. Circulation of essential nutrient elements, including carbon, sulfur, chlorine, and nitrogen, generally is bound to elemental oxygen from the atmosphere and provides most anionic species occurring in natural water, such as bicarbonate, sulfate, chloride, and nitrate. Other constituents of natural surface waters, including calcium, magnesium, sodium, and potassium, can be correlated in general with the chemical composition of rocks and soils in a given drainage basin and are found as principal cationic species and are in electrochemical balance with anions in these waters.

The influence of human activities in a stream drainage basin can be relatively simple and direct, as in the disposal of soluble organic and inorganic waste, or more subtle and complex, as in the conversion of prairie or forest land to agricultural use. Such effects can be expected to increase as population density and agricultural, industrial, and mining activities increase. For the detailed comparison of stream water chemistry in the early years of the twentieth century with more modern conditions, four drainage basins having diverse hydrologic characteristics were selected for study. They were the Great Lakes–Upper Saint Lawrence River basin (near the outlet of Lake Ontario), the Columbia River basin (upstream from The Dalles, Oregon), The Allegheny River basin (upstream from Pittsburgh, Pa.), and the Mississippi River basin (upstream from New Orleans, La.). Principal attention was given to sulfate concentrations and annual yields in tons per square mile, the reason being that major quantities of sulfate are released to the atmosphere by fossil fuel combustion and various industrial activities, and the sulfate is brought to the land surface in rain and dry fallout.

In the Great Lakes–Upper St. Lawrence River basin, for example, the yield of sulfate in tons per square mile per year in the St. Lawrence River nearly doubled between 1905 and 1956 and continued to increase, but at a lesser rate, until about 1970, when the yield leveled off or perhaps even declined slightly. The continuing yield of between about 19 and 25 tons/mi^2 indicates that the basin may have reached a steady state between the natural and human-induced loading of sulfur to the basin and its removal by the St. Lawrence River. In contrast, sulfate concentrations in the upper Columbia River basin at Northport, Washington, show less clearly defined trends in sulfate concentrations and yields during the century. These data indicate that human-induced effects are largely masked by the large amount of runoff available in the Columbia River basin and by the effects of storage and mixing in lakes and reservoirs.

Coal mining was extensive in the Allegheny River drainage basin in Pennsylvania in this century, and sulfate concentrations in the river near Pittsburgh increased substantially between the early 1900s and 1962 as a result of drainage from many active and abandoned coal mines. The operation of flood control and multipurpose reservoirs in the basin has caused a nearly 50% decline in sulfate concentration in the Allegheny River near Pittsburgh at low flow, as shown in analysis of samples collected in 1947 and compared to samples collected in 1975 and 1989. This flow augmentation by timed reservoir releases has had a beneficial effect on the quality of the Allegheny River.

The Mississippi River drains more than 1,125,000 mi^2 of the conterminous United States and integrates the effect on stream water quality of a large range of human activities across a large continental area. The calculation of sulfate yield in the basin has many uncertainties because water quantity and water quality data are incomplete, and the effects of flow control and water diversions are difficult to measure. However, from the available data it can be estimated that sulfate concentrations in the lower Mississippi River at and upstream from New Orleans, Louisiana, probably have doubled between 1905–1906 and 1989; most of this increase seems to have occurred before 1980. Estimates of the increase in annual sulfate yield due to human activities since 1905 ranging from about 9 to about 14 tons/mi^2 are consistent with the increase in sulfate yields in the Great Lakes–St. Lawrence River system. In both instances, yields seem to have leveled off around 1970 or 1980 and to have remained fairly stable since. Possibly, both drainage systems have reached a steady state, and the natural and human-induced sulfate loading to the basins, much of it from atmospheric deposition, is now stable. Data for total SO_2 emissions for each state from 1965 to 1980 (75) show that in most states adjacent to the Great Lakes the emission rates decreased significantly after 1970.

BIBLIOGRAPHY

1. American Water Works Association, American Public Health Association, and Water Environment Foundation, *Standard Methods for the Examination of Water and Wastewater*, 18th ed., American Public Health Assc., Washington D.C., 1992.
2. J. D. Hem, A. Demayo, and R. A. Smith, "Hydrogeochemistry of Rivers and Lakes," in *Surface Water Hydrology*, V O-1, *The Geology of North America*. The Geological Society of America, Boulder, Colo., 1990, pp. 189–231.
3. R. A. Smith, R. B. Alexander, and M. G. Wolman, *Analysis and Interpretation of Water-Quality Trends in Major U.S. Rivers*, 1974–81, U.S. Geological Survey Water-Supply Paper 2307, U.S. Geological Survey, Reston Va., 1987.
4. V. M. Goldschmidt, *Geochemistry*, Clarendon, Oxford, England, 1954.
5. J. D. Hem, *Study and Interpretation of the Chemical Characteristics of Natural Water*, 3rd ed., U.S. Geological Survey Water-Supply Paper 2254, U.S. Geological Survey, Reston, Va., 1985.
6. M. Meybeck and R. Helmer, "The Quality of Rivers—From Pristine Stage to Global Pollution," *Palaeogeogr. Palaeoclimatol. Palaeoecol.* (*Global Planet. Change Sec.*) **75,** 283–309 (1989).
7. M. Meybeck, "Atmospheric Inputs and River Transport of Dissolved Substances," in Symposium on Dissolved Loads of Rivers and Surface Water Quantity/Quality Relationships, August 1983, Hamburg, Germany, *Int. Assoc. Hydrol. Sci. Bull.* **141,** 173–192 (1983).
8. H. Lieth, "The Role of Vegetation in the Carbon Dioxide Content of the Atmosphere," *J. Geophys. Res.* **68**(13), 3887–3898 (1963).
9. N. E. Peters, *Evaluation of Environmental Factors Affecting Yields of Major Dissolved Ions of Streams in the United States*, U.S. Geological Survey Water-Supply Paper 2228, U.S. Geological Survey, Reston Va., 1984.
10. D. P. Dethier, "Weathering Rates and the Chemical Flux from Catchments in the Pacific Northwest, U.S.A.," in S. M. Colman and D. P. Dethier (eds.), *Rates of Chemical Weathering of Rocks and Minerals*, Academic, Orlando, Fla., 1986, pp. 503–530.
11. R. C. Reynolds, Jr. and N. M. Johnson, "Chemical Weathering in the Temperate Glacial Environment of the Northern Cascades Mountains," *Geochim. Cosmochim. Acta* **36,** 537–554 (1972).
12. J. R. Kramer, "Alkalinity and Acidity," in R. A. Minear and L. H. Keith (eds.), *Water Analysis*, vol. 1: *Inorganic Species*, part 1. Academic, New York, 1982, pp. 85–135.
13. W. Stumm and J. J. Morgan, *Aquatic Chemistry*, 2nd ed., Wiley-Interscience, New York, 1981.
14. J. O. Nriagu (ed.), *Sulfur in the Environment, Part II: Ecological Impacts*, Wiley-Interscience, New York, 1978.
15. M. Nyborg, "Sulfur Pollution in Soils," in J. O. Nriagu, ed., *Sulfur in the Environment*, Part II: *Ecological Impacts*, Wiley-Interscience, New York, 1978, pp. 359–390.
16. J. W. Fitzgerald, "Naturally Occurring Organosulfur Compounds in Soil," in J. O. Nriagu, ed., *Sulfur in the Environment, Part II: Ecological Impacts*, Wiley-Interscience, New York, 1978, pp. 391–443.
17. R. A. Berner, "Worldwide Sulfur Pollution of Rivers," *J. Geophys. Res.* **76,** 6597–6600 (1971).
18. U. S. Environmental Protection Agency (EPA), *The Acidic Deposition Phenomenon and Its Effect*, vol. 2: *Effects Sciences*, Critical Assessment Review Papers, Report EPA 600/9-83-016AF, EPA, Washington, D.C., 1984, pp. 4–11.
19. R. C. Bubeck, W. H. Diment, B. L. Deck, A. L. Baldwin, and S. D. Lipton, "Runoff of Deicing Salt—Effect on Irondequoit Bay, Rochester, N.Y.," *Science* **172,** 1128–1132 (1971).
20. R. E. Turner and N. N. Rabalais, "Changes in Mississippi River Quality in This Century—Implications for Coastal Food Webs," *Bio-Science* **41**(3), 140–147 (1991).
21. R. B. Dole, *The Quality of Surface Waters of the United States—Part 1, Analyses of Waters East of the One-Hundredth Meridian*, U.S. Geological Survey Water-Supply Paper 236, U.S. Geological Survey, Washington D.C., 1909.
22. H. Stabler, *Some Stream Waters of the Western United States, with Chapters on Sediment Carried by the Rio Grande and the Industrial Application of Water Analyses*, U.S. Geological Survey Water-Supply Paper 274, U.S. Geological Survey, Washington D.C., 1911.
23. F. W. Clarke, *The Composition of the River and Lake Waters of the United States*, U.S. Geological Survey Professional Paper 135, U.S. Geological Survey, Washington, D.C., 1924.
24. D. R. Helsel, "Statistical Analyses of Water Quality Data," in R. W. Paulson, E. B. Chase, J. S. Williams, and D. W. Moody, eds., *Compilers, National Water Summary 1990–91*, U.S. Geological Survey Water-Supply Paper 2400, U.S. Geological Survey, Reston Va., 1993, pp. 93–100.
25. R. A. Smith, R. B. Alexander, and K. J. Lanfear, "Stream Water in the Conterminous United States—Status and Trends of Selected Indicators During the 1980's," in R. W. Paulson, E. B. Chase, J. S. Williams, and D. W. Moody, eds., *Compilers, National Water Summary 1990–91*, U.S. Geological Survey Water-Supply Paper 2400, U.S. Geological Survey, Reston Va., 1993, pp. 111–140.
26. U.S. Geological Survey, *Water Resources Data for New York*, vol. 1: *Eastern New York Excluding Long Island—Water Year 1988*, U.S. Geological Survey Water Data Report NY 88-1, Springfield, Va., National Technical Information Service, 1989.
27. U.S. Geological Survey, *Water Resources Data for Oregon*, vol. 1: *Eastern Oregon, Water Year 1990*, U.S. Geological Survey Water Data Report OR-90-1, Springfield, Va., National Technical Information Service, 1991.
28. U.S. Geological Survey, *Water Resources Data for Pennsylvania*, Vol. 3: *Ohio and St. Lawrence River Basins, Water Year 1990*, U.S. Geological Survey Water Data Report PA-90-3, Springfield, Va., National Technical Information Service, 1991.
29. U.S. Geological Survey, *Water Resources Data for Louisiana, Water Year 1983*, U.S. Geological Survey Water Data Report LA-83-2, Springfield, Va., National Technical Information Service, 1984.
30. U.S. Geological Survey, *Water Resources Data for Louisiana Water Year 1990*, U.S. Geological Survey Water Data Report LA-90-1, Springfield, Va., National Technical Information Service, 1991.
31. C. S. Howard, *Suspended Matter in the Colorado River in 1925–28*, U.S. Geological Survey Water-Supply Paper 636B, U.S. Geological Survey, Washington D.C., 1929, pp. 15–44.
32. Interagency Committee on Water Resources, Federal Interagency Sedimentation Project, "Laboratory Investigation of Suspended Sediment Samplers," Interagency Report 5, Iowa University Hydraulics Lab., Iowa City, Iowa, 1941.
33. B. R. Colby, "Fluvial Sediments—a Summary of Source, Transportation, Deposition and Measurement of Sediment Discharge," *U.S. Geol. Surv. Bull.* **1181A** US Geological Survey, Washington D.C., (1963).

34. Interagency Committee, "A Summary of the Work of the Federal Interagency Sedimentation Project," Interagency Report S, St. Anthony Falls Hydraulics Lab., Minneapolis, Minn., 1963.
35. J-M. Martin and M. Meybeck, "Elemental Mass-Balance of Material Carried by World Major Rivers," *Marine Chem.* **7,** 173–206 (1979).
36. R. H. Meade and H. H. Stevens, Jr., "Strategies and Equipment for Sampling Suspended Sediment and Associated Toxic Chemicals in Large Rivers—with Special Emphasis on the Mississippi River," *Sci. Total Environ.* **97/98,** 125–135 (1990).
37. W. E. Periera, C. E. Rostad, and T. J. Leiker, "Synthetic Organic Agrochemicals in the Lower Mississippi River and Its Major Tributaries—Distribution, Transport, and Fate," *J. Contam. Hydrol.* **9,** 175–188 (1992).
38. H. E. Taylor, J. R. Gabarino, and T. I. Brinton, "Occurrence and Distribution of Trace Metals in the Lower Mississippi River and Its Tributaries," *Sci. Total Environ.* **97/98,** 369–384 (1990).
39. C-P. Huang (ed.), *Industrial Waste—Proceedings of the 13th Mid-Atlantic Conference, June 29–30, 1981*, Ann Arbor Scientific, Ann Arbor, Mich., 1981.
40. C. D. Becker and D. A. Neitzel (eds.), *Water Quality in North American River Systems*, Battelle, Columbus, Ohio, 1992.
41. J. N. Moore and S. N. Luoma, "Hazardous Wastes from Large-Scale Metal Extraction," *Environ. Sci. Technol.* **24,** 1278–1285 (1990).
42. D. O. Runnells, T. A. Shepherd, and E. E. Angino, "Metals in Water—Determining Natural Background Concentrations in Mineralized Areas," *Environ. Sci. Technol.* **26,** 2316–2323 (1992).
43. H. A. Leverin, *Industrial Waters of Canada—Report of Investigations 1934 to 1940*, Bureau of Mines Report 807, Canada Department of Mines and Resources, Ottawa, Ont., 1942.
44. U.S. Geological Survey (USGS), *Quality of Surface Waters of the United States 1956*, Parts 1–4, U.S. Geological Survey Water-Supply Paper 1450, USGS, Washington D.C. 1960.
45. U.S. Geological Survey (USGS), *Quality of Surface Waters of the United States 1969*, Parts 4–5, U.S. Geological Survey Water-Supply Paper 2144, USGS, Reston, Va., 1974.
46. R. R. Weiler and V. K. Chawla, "Dissolved Mineral Quality of Great Lakes Waters," in *Proceedings of the 12th Conference on Great Lakes Research, Ann Arbor, Mich.*, International Association for Great Lakes Research, 1969, pp. 801–818.
47. H. F. H. Dobson, "Principal Ions and Dissolved Oxygen in Lake Ontario," in *Proceedings of the 10th Conference on Great Lakes Research, Toronto, Canada, 1967, Ann Arbor, Mich.*, International Association for Great Lakes Research, 1967, pp. 337–356.
48. R. R. Weiler, "Chemistry of the North American Great Lakes," *Verhandlungen der Internationalen Vereinigung fur Theoretische und Angewandte Limnologie*, **21,** 1681–1694 (1981).
49. D. K. Leifeste, *Dissolved Solids Discharge to the Ocean from the Conterminous United States*, U.S. Geological Survey Circular 685, U.S. Geological Survey, Reston, Va., 1974.
50. W. Van Winkle, *Quality of the Surface Waters of Washington*, U.S. Geological Survey Water-Supply Paper 339, U.S. Geological Survey, Washington D.C., 1914.
51. U.S. Geological Survey, *Water Resources Data for Washington, Water Year 1991*, U.S. Geological Survey Water Data Report WA-91-1, Springfield, Va., National Technical Information Service, 1992.
52. U.S. Geological Survey (USGS), *National Water Summary 1988–89 Hydrologic Events and Floods and Droughts*, U.S. Geological Survey Water-Supply Paper 2375, USGS, Reston, Va., 1991.
53. U.S. Geological Survey (USGS), *Quality of Surface Waters of the United States 1954*, Parts 9–14, U.S. Geological Survey Water-Supply Paper 1353, USGS, Washington D.C., 1959.
54. U.S. Geological Survey (USGS), *Quality of Surface Waters of the United States 1968*, Parts 12–16, U.S. Geological Survey Water-Supply Paper 2100, USGS, Reston, Va., 1973.
55. E. F. McCarren, *Chemical Quality of Surface Water in the Allegheny River Basin, Pennsylvania and New York*, U.S. Geological Survey Water-Supply Paper 1835, U.S. Geological Survey, Washington D.C., 1967.
56. U.S. Geological Survey (USGS), *Quality of Surface Waters of the United States 1948*, Parts 1–6, U.S. Geological Survey Water-Supply Paper 1132, USGS, Washington D.C., 1953.
57. U.S. Geological Survey, *Water Resources Data for Pennsylvania*, vol. 3: *Ohio River and St. Lawrence River Basins, Water Year 1975*, U.S. Geological Survey Water Resources Data Report PA-75-3, Springfield, Va., National Technical Information Service, 1976.
58. U.S. Geological Survey, *Water Resources Data for Pennsylvania*, vol. 3: *Ohio River and St. Lawrence River Basins, Water Year 1976*, U.S. Geological Survey Water Data Report PA-76-3, Springfield, Va., National Technical Information Service, 1977.
59. U.S. Geological Survey (USGS), *Quality of Surface Waters of the United States 1966*, Parts 7–8, U.S. Geological Survey Water-Supply Paper 1994, USGS, Reston, Va., 1971.
60. U.S. Geological Survey, *Water Resources Data for Louisiana, Water Year 1989*, U.S. Geological Survey Water Data Report LA-89-1, Springfield, Va., National Technical Information Service, 1990.
61. R. J. Shlemon, *Development of the Atchafalaya Delta-Hydrologic and Geologic Studies of Coastal Louisiana*, Louisiana State University, Coastal Research Unit, Baton Rouge, La., 1972.
62. F. C. Wells, "Hydrology and Water Quality of the Lower Mississippi River," Louisiana Department of Transportation and Development, Office of Public Works, Baton Rouge, La., Water Resources technical report no. 21, 1980 (in cooperation with the U.S. Geological Survey).
63. U.S. Geological Survey, *Water Resources Data for Louisiana*, vol. 2: *Southern Louisiana, Water Year 1980*, U.S. Geological Survey Water Data Report LA-80-2, Springfield, Va., National Technical Information Service, 1981.
64. W. D. Collins, W. L. Lamar, and E. W. Lohr, *Industrial Utility of Public Water Supplies in the United States, 1932*, U.S. Geological Survey Water-Supply Paper 658, U.S. Geological Survey, 1934.
65. E. W. Lohr and S. K. Love, *Industrial Utility of Public Water Supplies in the United States 1952*, Part 2: *States West of the Mississippi River*, U.S. Geological Survey Water-Supply Paper 1300, U.S. Geological Survey, Washington D.C., 1954.
66. C. T. Driscoll, G. E. Likens, L. O. Hedin, J. S. Eaton, and E. H. Bormann, "Changes in the Chemistry of Surface Waters—25 Year Results at the Hubbard Brook Experimental Forest, *N.H." Environ. Sci. Technol.* **23,** 137–143 (1989).
67. N. M. Peters, R. A. Schroeder, and D. E. Troutman, *Temporal Trends in the Acidity of Precipitation and Surface Waters of New York*, U.S. Geological Survey Water-Supply Paper 2188, U.S. Geological Survey, Reston Va., 1982.
68. D. P. Dethier, "Atmospheric Contributions of Stream Water Chemistry in the North Cascade Range, Washington," *Water Resourc. Res.* **15,** 787–794 (1979).
69. U.S. Geological Survey (USGS), *Quality of Surface Waters of*

the United States, 1950, Parts 9–14, U.S. Geological Survey Water-Supply Paper 1189, USGS, Washington, D.C., 1954.
70. U.S. Geological Survey (USGS), *Quality of Surface Waters of the United States 1952*, Parts 9–14, U.S. Geological Survey Water-Supply Paper 1253, USGS, Washington D.C., 1957.
71. R. O. Van Everdingen, *Thermal and Mineral Springs in the Southern Rocky Mountains of Canada*, Environment Canada, Water Management Service, Ottawa, Canada, 1972.
72. B. S. Norford, *Ordovician and Silurian Stratigraphy of the Southern Rocky Mountains*, Bulletin 176, Geological Survey of Canada, Ottawa, Ont., 1969.
73. P. H. Whitfield, Environment Canada, Aug. 1991, personal communication.
74. Q. J. Stover and R. E. Nakatani, "Water Quality and Biota of the Columbia River System," in C. D. Becker and D. A. Neitzel, eds., *Water Quality in North American River Systems*, Columbus, Ohio, Battelle, 1992, pp. 53–83.
75. R. A. Smith and R. B. Alexander, *Evidence for Acid-Precipitation Induced Trends in Stream Chemistry at Hydrologic Bench-Mark Stations*, U.S. Geological Survey Circular 910 U.S. Geological Survey, Reston, Va., 1983.

WATER QUALITY MANAGEMENT

W. W. Eckenfelder, Jr.
Eckenfelder, Inc
Nashville, Tennessee

Over the past decade, water pollution control has progressed from an art to a science. Increased emphasis has been placed on the removal of secondary pollutants, such as nutrients and refractory organics, and on water reuse for industrial and agricultural purposes. This in turn has generated both fundamental and applied research, which has improved both the design and operation of wastewater treatment facilities.

Solving water pollution problems today involves a multidisciplinary approach in which the required water quality is related to agricultural, municipal, recreational, and industrial requirements. In many cases, a cost–benefit ratio must be established between the benefit derived from a specified water quality and the cost of achieving that quality.

Wastewaters emanate from four primary sources:

1. municipal sewage,
2. industrial wastewaters,
3. agricultural runoff, and
4. stormwater and urban runoff.

Estimating municipal wastewater flows and loadings can be done in one of several ways, based on knowledge of past and future growth plans for the community, sociological patterns, and land use planning. Two possible ways are as follows:

1. *Population Prediction Techniques.* Several mathematical techniques are available for estimating population growth. Caution should be employed in the use of these procedures, particularly in areas subject to rapid industrial expansion, rapid suburban development, and changing land use patterns.
2. *Saturation Population from Zoning Practice.* Percentages of a saturation population can be estimated for fully developed areas based on zoning restrictions (single-dwelling residential, multiple-dwelling residential, commercial, etc.).

Provisions should be included for infiltration in the case of separate sewers as well as storm flows in the case of combined sewers.

As municipal and industrial wastewaters receive treatment, increasing emphasis is being placed on the pollutional effects of urban and agricultural runoff. The range of concentration of pertinent characteristics in these wastewaters is given in Table 1. Present research on stormwater treatment considers large holding basins in which the stormwaters are treated in the municipal facility after the storm (an in situ treatment by screening, sedimentation, chlorination, etc.). In the future, water quality management in highly urbanized areas will have to consider stormwater as a major pollutant.

Agricultural runoff is a major contributor to etrophication in lakes and other natural bodies of water. Effective control measures have yet to be developed for this problem. Runoff of pesticides is also receiving increasing attention.

See also Air pollution; Air pollution control methods; Exhaust control, industrial.

WATER QUALITY STANDARDS

Water quality standards are usually based on one of two primary criteria, stream standards or effluent standards. Stream standards are based on dilution requirements for the receiving water quality based on a threshold value of specific pollutants or a beneficial use of the water. Effluent standards are based on the concentration of pollut-

Table 1. Pollution from Urban and Agricultural Runoff

Constituent	Urban Runoff[a] (Stormwater)	Agricultural Runoff[b]
Suspended solids, mg/L	5–1200	—
Chemical oxygen demand (COD), mg/L	20–610	—
Biological oxygen demand (BOD), mg/L	1–173	—
Total phosphorus, mg/L	0.02–7.3	0.10–0.65
Nitrate nitrogen, mg/L	—	0.03–5.00
Total nitrogen, mg/L	0.3–7.5	0.50–6.50
Chlorides, mg/L	3–35	—

[a] From Ref. 1.
[b] From Ref. 2.

ants that can be discharged or on the degree of treatment required.

Stream standards are usually based on a system of classifying the water quality based on the intended use of the water.

Although stream standards are the most realistic in light of the use of the assimilative capacity of the receiving water, they are difficult to administer and control in an expanding industrial and urban area. The equitable allocation of pollutional loads for many industrial and municipal complexes also poses political and economic difficulties. A stream standard based on minimum dissolved oxygen at low stream flow intuitively implies a minimum degree of treatment. One variation of stream standards is the specification of a maximum concentration of a pollutant (ie, the BOD) in the stream after mixing at a specified low-flow condition.

Note that the maintenance of water quality and hence stream standards are not static, but subject to change with the municipal and industrial environment. For example, as the carbonaceous organic load is removed by treatment, the detrimental effect of nitrification in the receiving water increases. Eutrophication may also become a serious problem in some cases. These considerations require an upgrading of the required degree of treatment.

Effluent standards are based on the maximum concentration of a pollutant (mg/L) or the maximum load (lb/day) discharged to a receiving water. These standards can be related to a stream classification.

In 1972 the U.S. Legislature passed Public Law 92-500, which requires certain levels of treatment for industrial wastewater discharges. Effluent guideline criteria (expressed as kilograms pollutant per unit of production) have been developed for each industrial category to be met by specified time periods.

The BPT is defined as the level of treatment that has been proven to be successful for a specific industrial category and that is currently in full-scale operation. Sufficient data exist for this level of treatment so that it can be designed and operated to achieve a level of treatment consistently and with reliability. For example, in the pulp and paper industry, BPT has been defined as biological treatment using the aerated lagoon or the activated sludge process with appropriate pretreatment.

The BAT is defined as the level of treatment beyond BPCTCA that has been proven feasible in laboratory and pilot studies and that is, in some cases, in full-scale operation. BAT in the pulp and paper industry may include such processes as filtration, coagulation for color removal, and improved in-plant control to reduce the wasteload constituents.

In general, effluent guidelines are developed by considering an exemplary plant in a specific industrial category and multiplying the wastewater flow per unit production by the effluent quality attainable from the specified BPT process to obtain the effluent limitation in pounds or kilograms per unit of production. The effluent limitations consider both a maximum 30-day average and a 1-day maximum level. In general, the daily maximum is two to three times the 30-day average. For example, the average wastewater flow from an exemplary plant is 30,000 gal/ton of production and the average effluent BOD is 30 mg/L.

The effluent limitation can then be computed:

$$(30{,}000 \text{ gal/ton}) \times (8.34 \times 10^{-6}) \times (30 \text{ mg/L}) = 7.5 \text{ lb/ton}$$

It is recognized that the wastewater volume and characteristics from a specific industrial category will depend on such factors as plant age, size, raw materials used and in-plant processing sequences.

The U.S. Environmental Protection Agency (EPA) has also developed pretreatment guidelines for those industrial plants which discharge into municipal sewer systems. In general, compatible pollutants such as BOD, suspended solids, and coliform organisms can be discharged providing the municipal plant has the capability of treating these wastewaters to a satisfactory level. Noncompatible pollutants, such as grease and oil, heavy metals, etc., must be pretreated to specified levels. Rigid limitations have been developed for the discharge of toxic substances to the nation's waterways.

In several cases, such as shellfish areas and aquatic reserves, the usual water quality parameters do not apply because they are nonspecific as to detrimental effects on aquatic life. For example, COD is an overall measure of organic content, but it does not differentiate between toxic and nontoxic organics. In these cases, a species diversity index has been employed as related to either free-floating or benthic organisms. The index indicates the overall condition to the aquatic environment. It is related to the number of species in the sample. The higher the species diversity index, the more productive the aquatic system. The species diversity index K_D is computed by the equation $K_D = (S - 1)/\log_{10} l$, where S is the number of species and l the total number of individual organisms counted.

Regulations establishing effluent limitations guidelines, pretreatment standards and new source performance standards for the organic chemicals, plastics, and synthetic fibers (OCPSF) were promulgated in 1987. In these regulations, specific organic chemicals are defined by the EPA as priority pollutants:

1. Acenaphthene
2. Acrolein
3. Acrylonitrile
4. Benzene
5. Benzidine
6. Carbon tetrachloride (tetrachloromethane)

Chlorinated benzenes (other than dichlorobenzenes)

7. Chlorobenzene
8. 1,2,4-Trichlorobenzene
9. Hexachlorobenzene

Chlorinated ethanes (including 1,2-dichloroethane, 1,1,1-trichloroethane, and hexachloroethane)

10. 1,2-Dichloroethane
11. 1,1,1-Trichloroethane
12. Hexachloroethane
13. 1,1-Dichloroethane

14. 1,1,2-Trichloroethane
15. 1,1,2,2-Tetrachloroethane
16. Chloroethane (ethyl chloride)

Chloroalkyl ethers (chloromethyl, chloroethyl, and mixed ethers)
17. Bis(chloromethyl)ether
18. Bis(2-chloroethyl)ether
19. 2-Chloroethyl vinyl ether (mixed)

Chlorinated napthalene
20. 2-Chloronapthalene

Chlorinated phenols (other than those listed elsewhere; includes trichlorophenols and chlorinated cresols)
21. 2,4,6-Trichlorophenol
22. *para*-Chloro-*meta*-cresol
23. Chloroform (trichloromethane)
24. 2-Chlorophenol

Dichlorobenzenes
25. 1,2-Dichlorobenzene
26. 1,3-Dichlorobenzene
27. 1,4-Dichlorobenzene

Dichlorobenzidine[a]
28. 3,3′-Dichlorobenzidine

Dichloroethylenes (1,1-dichloroethylene and 1,2-dichloroethylene)
29. 1,1-Dichloroethylene
30. 1,2-*trans*-Dichloroethylene
31. 2,4-Dichlorophenol

Dichloropropane and dichloropropene
32. 1,2-Dichloropropane
33. 1,2-Dichloropropylene (1,2-dichloropropene
34. 2,4-Dimethylphenol

Dinitrotoluene
35. 2,4-Dinitrotoluene
36. 2,6-Dinitrotoluene
37. 1,2-Diphenylhydrazine
38. Ethylbenzene
39. Fluoranthene

Haloethers[a] (other than those listed elsewhere)
40. 4-Chlorophenyl phenyl ether
41. 4-Bromophenyl phenyl ether
42. Bis(2-chloroisopropyl) ether
43. Bis(2-chloroethoxy) methane

Halomethanes[a] (other than those listed elsewhere)
44. Methylene chloride (dichloromethane)
45. Methyl chloride (chloromethane)
46. Methyl bromide (bromomethane)
47. Bromoform (tribromomethane)
48. Dichlorobromomethane
49. Trichlorofluoromethane
50. Dichlorodifluoromethane
51. Chlorodibromomethane
52. Hexachlorobutadiene
53. Hexachlorocyclopentadiene
54. Isophorone
55. Naphthalene
56. Nitrobenzene

Nitrophenols[a] (including 2,4-dinitrophenol and dinitrocresol)
57. 2-Nitrophenol
58. 4-Nitrophenol
59. 2,4-Dinitrophenol[a]
60. 4,6-Dinitro-*o*-cresol

Nitrosamines[a]
61. *N*-Nitrosodimethylamine
62. *N*-Nitrosodiphenylamine
63. *N*-Nitrosodi-*n*-propylamine
64. Pentachlorophenol[a]
65. Phenol[a]

Phthalate esters[a]
66. Bis(*e*-ethylhexyl) phthalate
67. Butyl benzyl phthalate
68. Di-*n*-butyl phthalate
69. Di-*n*-octyl phthalate
70. Diethyl phthalate
71. Dimethyl phthalate

Polynuclear aromatic hydrocarbons (PAHs)[a]
72. Benzo(*a*)anthracene (1,2-benzanthracene)
73. Benzo(*a*)pyrene (3,4-benzopyrene)
74. 3,4-Benzofluoranthene
75. Benzo(*k*)fluoranthene (11,12-benzofluoranthene
76. Chrysene
77. Acenaphthylene
78. Anthracene
79. Benzo(*ghi*)perylene (1,12-benzoperylene)
80. Fluorene
81. Phenanthrene
82. Dibenzo(*a*,*h*)anthracene (1,2,5,6-dibenzanthracene)
83. Indeno(1,2,3-cd)pyrene(2,3-*o*-phenylenepyrene)
84. Pyrene
85. Tetrachloroethylene[a]
86. Toluene[a]
87. Trichloroethylene[a]
88. Vinyl chloride[a] (chloroethylene)

Pesticides and metabolites
89. Aldrin[a]
90. Dieldrin[a]
91. Chlordane[a] (technical mixture and metabolites)

DDT and metabolites[a]
92. 4-4′-DDT
93. 4,4′-DDE (*p*,*p*′-DDX)
94. 4,4′-DDD (*p*,*p*′-TDE)

Endosulfan and metabolites[a]

95. α-Endosulfan-alpha

96. β-Endosulfan-beta

97. Endosulfan sulfate

Endrin and metabolites[a]

98. Endrin

99. Endrin aldehyde

Heptachlor and metabolites[a]

100. Heptachlor

101. Heptachlor epoxide

Hexachlorocyclohexane (all isomers)[a]

102. α-BHC-alpha

103. β-BHC-beta

104. γ-BHC (lindane)-gamma

105. δ-BHC-delta

Polychlorinated biphenyls (PCBs)[a]

106. PCB-1242 (Arochlor 1242)

107. PCB-1254 (Arochlor 1254)

108. PCB-1221 (Arochlor 1221)

109. PCB-1232 (Arochlor 1232)

110. PCB-1248 (Arochlor 1248)

111. PCB-1260 (Arochlor 1260)

112. PCB-1016 (Arochlor 1016)

113. Toxaphene[a]

114. 2,3,7,8-Tetrachlorodibenzo-*p*-dioxin (TCDD)[a]

These chemicals are regulated as a concentration level in the effluent. In most cases, these levels are in the microgram-per-liter range.

Recent air pollution regulations limit the amount of volatile organic carbon (VOC) that can be discharged from wastewater treatment plants. Benzene is a particular case in which air emission controls are required if the concentration of benzene in the influent wastewater exceeds 10 mg/L.

The water quality criteria for various industrial uses are summarized in Table 2. The surface water quality criteria for public water supplies have been summarized in reference 4. Color should not exceed 75 units and odors should be virtually absent. Ammonia nitrogen, nitrite nitrogen, and nitrate nitrogen should not exceed 0.5, 1, and 10 mg/L, respectively. Chloride should not exceed 250 mg/L, and pH value should be between 5.0 and 9.0. Sulfate should not exceed 250 mg/L, and the geometric means of fecal coliform and total coliform densities should not exceed 2000/100 mL and 20,000/100 mL, respectively.

For freshwater aquatic life, the pH should be between 6 and 9 and the alkalinity should not be decreased more than 25% below the natural level. For most fish life, the dissolved oxygen should be in excess of 5.0 mg/L. For wildlife, the pH should be between 7.0 and 9.2, and the alkalinity should be between 30 and 130 mg/L.

For irrigation, the pH should be between 4.5 and 9.0, and the sodium adsorption ratio should be within the tolerance limits determined by the U.S. Soil Salinity Laboratory Staff. For continuous use on all soils, the metal contents for aluminum, cadmium, chromium, cobalt, copper, iron, lead, and zinc should be no more than 5.0, 0.01, 0.1, 0.05, 0.2, 5.0, and 2.0 mg/L, respectively.

Federal water quality-based criteria (WQBC) are ambient concentrations of a chemical which, if not exceeded, will protect the designated uses of a water body. These differ from technology-based criteria, which are based on the achievable concentrations of a chemical when treated by the best available treatment technology for a given industrial category. Neither type of criteria are enforceable limits and are intended as guidance values to assist regulations in protecting water resources. However, WQBC are often used by states to establish enforceable standards when setting effluent discharge limits.

WQBC are designed to be protective of a water body's designated uses such as potable water supply and propa-

Table 2. Industrial Water Quality Limits[a]

Industry	Turbidity (units)	Color (units)	Hardness	Temperature (°C)	pH	TDS	SS	S_iO_2	Fe	M_aCl	Cl	SO_4	Alkalinity
Textiles (SIC 22)	0.3–5	0–5	0–50	—	—	100–200	0–5	25	0–0.3	0.01–0.05	100	100	50–200
Pulp and paper (SIC 26)													
Fine paper	10[b]	5	100	—	—	200	—	20	0.1	0.03	—	—	75
Kraft paper													
Bleached	40[b]	25	100	—	—	300	—	50	0.2	0.10	200	—	75
Unbleached	100[b]	100	200	—	—	500	—	100	1.0	0.50	200	—	150
Groundwood papers	50[b]	30	200	—	—	500	—	50	0.3	0.10	75	—	150
Soda and Sulfite paper	25[b]	5	100	—	—	250	—	20	0.1	0.05	75	—	75
Chemicals (SIC 28)	—	500	1,000	[c]	5.5–9.0	2500	10,000	[c]	10	2.00	500	850	500
Petroleum (SIC 2911)	—	25	900	—	6.0–9.0	3500	5,000	85	15	—	1,600	900	500
Iron and steel (SIC 33)	—	—	—	38	5.0–9.0	—	100[d]	—	—	—	—	—	—
Food canning (SIC 2032, 2033)	—	5	250	—	6.5–8.5	500	10	50	0.2	0.20	250	250	250
Tanning (SIC 3111)													
Tanning processes	[e]	5	150	—	6.0–8.0	—	—	—	50	—	250	250	[c]
Finishing processes	[e]	5	[f]	—	6.0–8.0	—	—	—	0.3	0.20	250	250	[c]
Coloring	[e]	5	[e]	—	6.0–8.0	—	—	—	0.1	0.001	—	—	[c]
Soft drinks (SIC 2086)	—	5	[g]	—	[g]	[g]	—	—	0.3	0.05	500	500	85

[a] From ref. 3; units are mg/L unless otherwise specified. Abbreviations: TDS, total dissolved solids; SS, suspended solids.
[b] Units in mg/L as SiO_2.
[c] Not considered a problem of concentrations encountered.
[d] Settleable solids.
[e] Not detectable by test.
[f] Lime softened.
[g] Controlled by treatment for other constituents.

gation of fish and wildlife. As such, WQBC must protect both "human health" and "aquatic life" uses. For example, the freshwater aquatic life chronic criterion for cyanide is 5.2 μg/l where the criterion for the protection of human health is 700 μg/l. Depending upon the designated uses of the water body, the lowest criterion can be considered by the regulatory agency in establishing discharge limits for cyanide.

POLLUTIONAL EFFECTS IN NATURAL WATERS

The movement and reactions of waste materials through streams, lakes, and estuaries is a resultant of hydrodynamic transport and biological and chemical reactions by the biota, suspended materials, plant growths, and bottom sediments. These relationships can be expressed by a mathematical model that reflects the various inputs and outputs in the aquatic system. Considering the oxygen balance, the general relationships for the oxygen-sag curve are given as

$$\frac{\partial C}{\partial t} = \varepsilon \frac{\partial^2 C}{\partial X^2} - U\frac{\partial C}{\partial X} \pm \sum S \qquad (1)$$

where C = concentration of dissolved oxygen
t = time at a stationary point
U = velocity of flow in X direction
ε = turbulent diffusion coefficient
ΣS = sources and sinks of oxygen
X = distance downstream

Equation (1) assumes that the concentration of any characteristic is uniform over the stream cross section and that the area is uniform with distance. If this is not the case, Equation (1) must be suitably modified.

The sources of oxygen are

1. incoming or tributary flow,
2. photosynthesis, and
3. reaeration.

The sinks of oxygen are

1. biological oxidation of carbonaceous organic matter,
2. biological oxidation of nitrogenous organic matter,
3. benthal decomposition of bottom deposits,
4. respiration of aquatic plants, and
5. immediate COD.

Sources of Oxygen

The quantity of oxygen in the incoming or tributary flow is considered as an initial condition in Equation (1). The dissolved oxygen present in waste discharges should also be considered if the waste flow is large relative to the stream flow.

Photosynthesis is the production of oxygen from the growing of green plant life in water courses from sunlight, carbon dioxide, and other stream nutrients. The degree of photosynthesis depends upon sunlight, temperature, mass of algae and rooted plants, and available nutrients. It will exhibit a diurnal variation.

Reaeration. By the process of natural reaeration, oxygen is added to the water body. Reaeration is primarily related to the degree of turbulence and natural mixing in the water body (high in sections of rapids, low in impounded areas).

Sinks of Oxygen

In the biological oxidation of carbonaceous organic matter, the rate of removal is related to the amount of unstabilized organics present.

When unoxidized nitrogen is present in the wastewater, nitrification will result with time of passage or distance downstream. The rate of nitrification is less when untreated wastes are discharged to the stream and the concentration of nitrifying organisms is low, and the rate increases in the presence of well-oxidized effluents with a high seed of nitrifying organisms.

Nitrifying organisms are sensitive to pH and function best over a pH range of 7.5–8.0. The rate of nitrification decreases rapidly at dissolved oxygen levels below 2.0–2.5 mg/L, so that at low oxygen levels in the water body little or no nitrification will occur. Denitrification has been observed to occur in stretches of zero or near-zero oxygen concentration.

Benthal Decomposition of Bottom Deposits. In polluted streams, the river bottom may be covered by active biological materials, such as sludges and slimes. The growth and accumulation of these materials results from deposition of suspended organics and/or the transfer of soluble organics to the flowing slimes. The BOD is metabolized to new cell growth and aerobically or anaerobically decomposed. Deposition occurs at low velocities. At high velocities (above 0.3 to 0.45 m/s) deposited materials may be resuspended and cause a secondary increase in BOD. In areas where bottom deposits occur, oxygen will be used by diffusion into the upper layers of the deposit (the rate will increase in the presence of worms, which increases the porosity of the deposit) and from the diffusion of organic products of anaerobic degradation into the flowing stream water, which will increase the soluble organic oxygen demand of the water, as shown in Figure 1.

Respiration of Aquatic Plants. Oxygen will be removed from the water body by the respiration of aquatic plants. (Oxygen may also be contributed by photosynthesis.) It has been shown that photosynthesis can provide as much as two-thirds of the dissolved oxygen present in the water during the day, while at night respiration can deplete the oxygen resources.

Many industrial wastes contain chemicals that will exert an immediate oxygen demand, such as sulfites. This will usually be immediately apparent in the oxygen sag curve and the BOD curve in the stream.

Salinity in a river or estuary causes a depression in the oxygen saturation level. As an example, at 25°C during summer critical conditions, if the salt concentration is increased to 10,000 mg/L, the oxygen saturation value is reduced to 7.56 mg/L from 8.18 mg/L, which is about 8%. This phenomenon frequently occurs in tidal streams and waters that receive industrial wastewater such as oil field

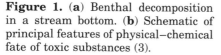

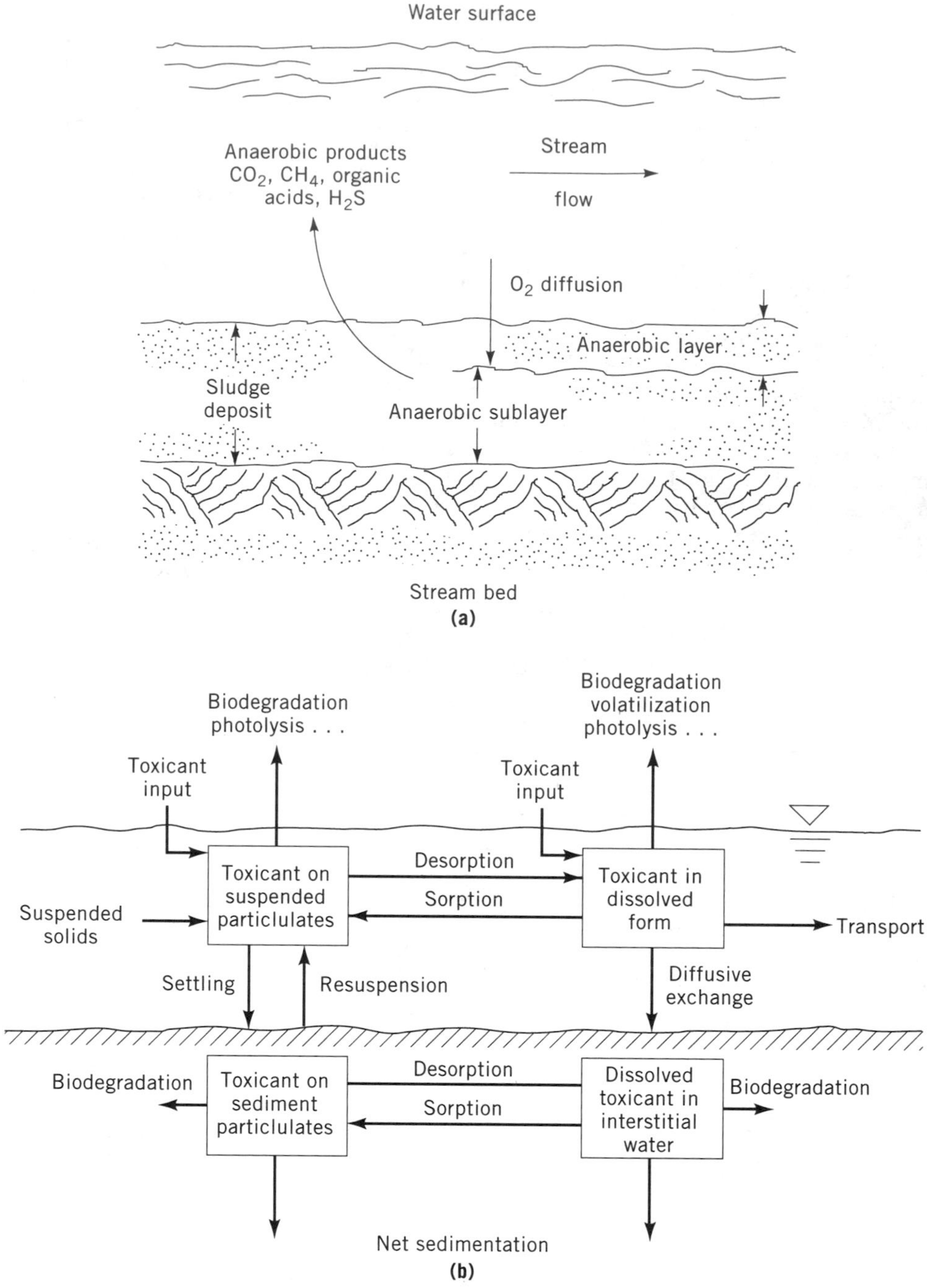

Figure 1. (**a**) Benthal decomposition in a stream bottom. (**b**) Schematic of principal features of physical–chemical fate of toxic substances (3).

brines or cucumber brine from a pickling process. The salinity problem is also observed in many southwestern rivers and streams through irrigation practices in the United States.

In the case of a stream or river, Equation (1) can be expressed as the simplified Streeter–Phelps model (Fig. 2):

$$\frac{D}{dt} = K_1L - K_2D \tag{2}$$

where D = oxygen deficit (saturation minus dissolved oxygen)
L = organic concentration
K_1 = organic degradation rate coefficient
K_2 = oxygen reaeration coefficient

The effects of photosynthesis, benthal demand nitrification, and immediate oxygen demand must be appropriately introduced into the equation. The mathematical manipulation of this and similar relationships are available in references 3 and 5–7.

Considerable attention is now being placed on the concentration of toxic chemicals in the sediment. Many chemical constituents tend to attach or sorb to the solids. The implication for wastewater discharges is that a substantial fraction of some toxic chemicals is associated with the suspended solids in the effluent. In accordance with adsorption equilibrium, these toxics will dissolve in the water with time and could cause serious water pollution problems. The fate of toxic materials relative to sediments is shown in Figure 1**a**.

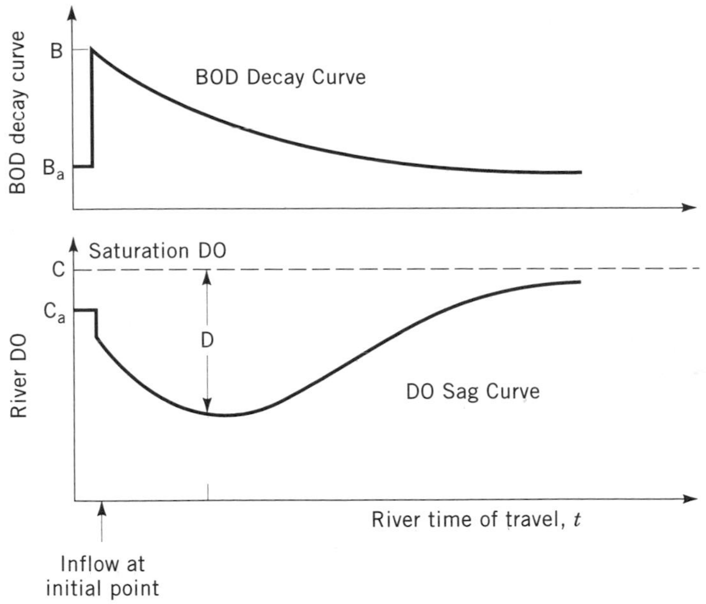

Figure 2. Typical Streeter–Phelps model.

WASTE MINIMIZATION

Before end-of-pipe wastewater treatment or modifications to existing wastewater treatment facilities to meet new effluent criteria, a program of waste minimization should be initiated.

Reduction and recycling of waste are inevitably site and plant specific, but a number of generic approaches and techniques have been used successfully across the country to reduce many kinds of industrial wastes.

Generally, waste minimization techniques can be grouped into four major categories: inventory management and improved operations, modification of equipment, production process changes, and recycling and reuse. Such techniques can have applications across a range of industries and manufacturing processes and can apply to hazardous as well as nonhazardous waste.

Many of these techniques involve source reduction—the preferred option on the EPA's hierarchy of waste management (8). Others deal with on- and off-site recycling. The best way to determine how these general approaches can fit a particular company's needs is to conduct a waste minimization assessment, as discussed above. In practice, waste minimization opportunities are limited only by the ingenuity of the generator. In the end, a company looking carefully at bottom-line returns may conclude that the most feasible strategy would be a combination of source reduction and recycling projects.

Waste minimization approaches as developed by the U.S. EPA are as follows:

Inventory Management and Improved Operations

- Inventory and trace all raw materials.
- Purchase fewer toxic and more nontoxic production materials.
- Implement employee training and management feedback.
- Improve material receiving, storage, and handling practices.

Modification of Equipment

- Install equipment that produces minimal or no waste.
- Modify equipment to enhance recovery or recycling options.
- Redesign equipment or production lines to produce less waste.
- Improve operating efficiency of equipment.
- Maintain strict preventive maintenance program.

Production Process Changes

- Substitute nonhazardous for hazardous raw materials.
- Segregate wastes by type for recovery.
- Eliminate sources of leaks and spills.
- Separate hazardous from nonhazardous wastes.
- Redesign or reformulate end products to be less hazardous.
- Optimize reactions and raw material use.

Recycling and Reuse

- Install closed-loop systems.
- Recycle on site for reuse.
- Recycle off site for reuse.
- Exchange wastes.

In order to implement the program, an audit needs to be made, as described in the following:

Phase I: Preassessment

- Audit focus and preparation.
- Identify unit operations and processes.
- Prepare process flow diagrams.

Phase II: Mass Balance

- Determine raw material inputs.
- Record water usage.
- Assess present practice and procedures.
- Quantify process outputs.
- Account for emissions: to atmosphere, to wastewater, and to off-site disposal.
- Assemble input and output information.
- Derive a preliminary mass balance.
- Evaluate and refine the mass balance.

Phase III: Synthesis

- Identify options: identify opportunities, target problem areas, and confirm options.
- Evaluate options: technical, environmental, and economic.
- Prepare action plan: waste reduction plan, production efficiency plan, and training.

Pollution reduction can be directly achieved in several ways. The five major ways are as follows (9):

Table 3. General Analysis of In-Plant Modifications

Type of Modifications	Description	Percent of Total RWL Reduction
Equipment revision and additions	Self-explanatory	25
Unit shutdowns	Shutdowns due to the age of the unit or the product. These shutdowns are not a direct result of pollution considerations, but they are somewhat hastened by these considerations.	10
Scrubber replacement	Replacement of scrubbers associated with amine production by burning of the off vapors.	3
Segregation, collection, and incineration	Of specific concentrated wastewater streams	35
Raw material substitutions	Self-explanatory	3
Reprocessing	Collecting tail streams from specific processes, then putting these streams through an additional processing unit to recover more product and concentrate the final wastestream	3
Miscellaneous small projects	A variety of modifications which individually do not represent a large reduction in RWL	21

1. *Recirculation.* In the paper board industry, white water from a paper machine can be put through a saveall to remove the pulp and fiber and recycled to various points in the paper-making process.
2. *Segregation.* Clean streams can be separated for direct discharge. Concentrated or toxic streams may be separated for separate treatment.
3. *Disposal.* In many cases, concentrated wastes can be removed in a semidry state. In the production of ketchup, the kettle bottoms after cooking and preparation of the product are usually flushed to the sewer. The total discharge BOD and SS can be markedly reduced by removal of this residue in a semidry state for disposal. In breweries, the secondary storage units have a sludge in the bottom of the vats which contains both BOD and SS. Removal of this as a sludge rather than flushing to the sewer will reduce the organic and solids load to treatment.
4. *Reduction.* It is common practice in many industries, such as breweries and dairies, to have hoses continuously running for clean-up purposes. The use of automatic cutoffs can substantially reduce the wastewater volume.

   The use of drip pans to catch products, in such cases as in a dairy or ice cream manufacturing plant, instead of flushing this material to the sewer considerably reduces the organic load. A similar case exists in the plating industry, where a drip pan placed between the plating bath and the rinse tanks will reduce the metal dragout.
5. *Substitution.* The substitution of chemical additives of a lower pollutional effect in processing operations, eg, substitution of surfactants for soaps in the textile industry.

To date, the discussion has considered in-plant measures to reduce wastewater volume and strength that do not involve major capital expenditures or process modifications. It is possible in many cases to markedly reduce the raw waste load by in-plant process changes. In general, this involves an economic trade-off between in-plant changes and end-of-pipe treatment. Some examples of this follow.

In the organic chemicals industry, an example might be drawn from the Union Carbide plant in South Charleston, West Virginia (10). This plant is reasonably representative of what might be expected from a major multiproduct chemicals complex. A detailed study of waste load reduction by in-plant changes showed that the present plant flow of 11.1×10^6 gal/day and 55,700 lb BOD/day could be reduced to a flow of 8.3×10^6 gal/day and a BOD to 37,100 lb/day. The ways of achieving these reductions are shown in Table 3. Equipment revision and additions and unit shutdowns refers to those units that would be replaced. Incineration involves the option of taking the more concentrated wastewater streams and rather than discharging to the main sewer, segregating them with incineration. Reprocessing refers to taking the tank bottoms that still have product present and reprocessing them for further product recovery. The results of a waste minimization program in three other industries are shown in Table 4.

CHARACTERIZATION OF WASTEWATERS

A comprehensive analytical program for characterizing wastewaters should be based on relevancy to unit treatment process operations, the pollutant or pollutants to be removed in each, and effluent quality constraints. The

Table 4. Source Management and Control

Case Studies	Before	After
1. Chemical industry		
a. Volume (m^3/d)	5000	2700
b. COD (t/d)	21	13
2. Hide and skin industry		1800
a. Volume (m^3/d)	2600	2 ×
b. BOD (t/d)	3 × 6	6
c. TDS (t/d)	20	10
d. SS (t/d)	4 ×	3 ×
	83	7
3. Metal preparation and finishing		
a. Volume (m^3/d)	450	270
b. Chromium (kg/d)	50	5
c. TTM (kg/d)	180	85

qualitative and quantitative characteristics of waste streams to be treated not only serve as a basis for sizing system processes within the facility, but also indicate streams having refractory constituents, potential toxicants, or biostats. Such streams are not amenable to effective biological treatment, as indicated by the characterization results, and require treatment using alternative processes.

It should be recognized that the total volume of wastewater as well as the chemical analyses indicating the organic and inorganic components are required with statistical validity before conceptualizing the overall treatment plant design. The basis parameters in wastewater characterization are summarized as follows (11):

1. Source information for the individual points of origin: waste constituents (specific compounds or general composition), discharge rate (average and peak), batch discharges, and frequency of emergency discharges or spills.
2. Chemical composition

   Organic and inorganic constituents

   Gross organics: COD, total organic carbon (TOC), BOD, and extractables

   Toxics, hazardous compounds, priority pollutants

   Gross inorganics: TDS

   Specific inorganic ions; As, Ba, Cd, CN, Hg, Pb, Se, Ni, Sg, nitrates

   pH, acidity, alkalinity

   Nitrogen and phosphorus

   Oil and grease

   Oxidizing reducing agents (eg, sulfides)

   Surfactants

   Chlorine demand
3. Physical properties: temperature range and distribution, particulates (colloidal, settleable, and flotable solids), color, odor, foamability, corrosiveness, and radioactivity.
4. Biological factors: BOD, toxicity (aquatic life, bacteria, animals, plants), and pathogenic bacteria
5. Flow characteristics: average daily flow rate, duration and magnitude of peak flow rate, maximum rate of change of flow rate, and stormwater flow rate (average and peak).

The causes of variability in waste characterization are as follows:

changes in production rate,
variations in plant product mix,
batch operations,
variations in efficiencies of production units,
changes in raw materials,
upsets in production processes,
maintenance (equipment shutdown and cleanout),
miscellaneous leaks and spills, and
contaminated drainage and runoff from rainstorms.

Industrial Wastewater Flow

The design flows for industrial complexes generally consist of the following:

1. base process flows resulting from normal production operations,
2. sanitary sewage,
3. contaminated storm runoff, and
4. other sources (extraordinary dumps, tank draining, ballast discharge, etc.)

The base flow and sanitary contribution can be measured in open channels or closed conduits using a variety of methods such as automatic metering devices, weirs, or less sophisticated devices. Care should be taken to ensure flows are measured during workday and weekend operations, during different work shifts, and over a sufficiently long period of time to reflect statistical reliability.

Within the last decade, contaminated storm runoff has become an area of increasing concern within industrial complexes. Storm flow is intermittent and unpredictable in nature, and little data have been collected to typify its characteristics. The level of flow and degree of contamination varies within an installation as it has its own geometric characteristics which influence patterns of surface runoff.

DEFINITION OF WASTEWATER CONSTITUENTS

Parameters used to characterize wastewaters can be categorized into organic, inorganic, and biological constituents. The important general contaminants of concern are identified in Table 5. The organic content of wastewater is estimated in terms of oxygen demand using BOD, COD, or total oxygen demand (TOD). Additionally, the organic fraction can be expressed in terms of carbon using TOC. It should be recognized that these parameters do not necessarily measure the same constituents. Specifically, they reflect the following:

1. BOD: biodegradable organics in terms of oxygen demand.
2. COD: organics amenable to chemical oxidation as well as certain inorganics, such as sulfides, sulfites, ferrous iron, chlorides, and nitrites.
3. TOD: all organics and some inorganics in terms of oxygen demand.
4. TOC: all organic carbon expressed as carbon.

The oxygen demand and organic carbon of selected industrial wastewaters are shown in Table 6. In addition to the total organic content, it is important to identify VOC and the presence of specific priority pollutants, which were identified earlier.

The inorganic characterization schedule for wastewaters to be treated using biological systems should include those tests which provide information concerning the following:

1. potential toxicity, such as heavy metal, ammonia, etc.;

Table 5. Important Contaminants of Concern in Wastewater Treatment[a]

Contaminants	Reason for Importance
Suspended solids	Suspended solids can lead to the development of sludge deposits and anaerobic conditions when untreated wastewater is discharged in the aquatic environment.
Biodegradable organics	Composed principally of proteins, carbohydrates, and fats, biodegradable organics are measured most commonly in terms of BOD and COD. If discharged untreated to the environment, their biological stabilization can lead to the depletion of natural oxygen resources and to the development of septic conditions.
Pathogens	Communicable diseases can be transmitted by the pathogenic organisms in wastewater.
Nutrients	Both nitrogen and phosphorus, along with carbon, are essential nutrients for growth. When discharged to the aquatic environment, these nutrients can lead to the growth of undesirable aquatic life. When discharged in excessive amounts of land, they can also lead to the pollution of groundwater.
Priority pollutants	Organic and inorganic compounds selected on the basis of their known or suspected carcinogenicity, mutagenicity, teratogenicity, or high acute toxicity. Many of these compounds are found in wastewater.
Refractory organics	These organics tend to resist conventional methods of wastewater treatment. Typical examples include surfactants, phenols, and agricultural pesticides.
Heavy metals	Heavy metals are usually added to wastewater from commercial and industrial activities and may have to be removed if the wastewater is to be reused.
Dissolved inorganics	Inorganic constituents such as calcium, sodium, and sulfate are added to the original domestic water supply as a result of water use and may have to be removed if the wastewater is to be reused.

[a] From ref. 12.

2. potential inhibitors, such as TDS and chlorides;
3. contaminants requiring specific pretreatment, such as pH, alkalinity, acidity, suspended solids, etc.; and
4. nutrient availability.

Aquatic toxicity is becoming a permit requirement on all discharges (14). Aquatic toxicity is generally reported as an LC_{50}, the percentage of wastewater which causes the death of 50% of the test organisms in a specified period, ie, 48 or 96 h or as a no observed effect level (NOEL) in which the NOEL is the highest effluent concentration at which no unacceptable effect will occur, even at continuous exposure.

Toxicity is also frequently expressed as toxicity units (TU), which is 100 divided by the toxicity measured:

$$TU = \frac{100}{LC_{50} \text{ or NOEL}}$$

Table 6. Oxygen Demand and Organic Carbon of Selected Industrial Wastewater[a]

Waste	BOD_5 (mg/L)	COD (mg/L)	TOC (mg/L)	BOD/TOC	COD/TOC
Chemical[b]	—	4,260	640	—	6.65
Chemical[b]	—	2,410	370	—	6.60
Chemical[b]	—	2,690	420	—	6.40
Chemical	—	576	122	—	4.72
Chemical	24,000	41,300	9,500	2.53	4.35
Chemical refinery	—	580	160	—	3.62
Petrochemical	—	3,340	900	—	3.32
Chemical	850	1,900	580	1.47	3.28
Chemical	700	1,400	450	1.55	3.12
Chemical	8,000	17,500	5,800	1.38	3.02
Chemical	60,700	78,000	26,000	2.34	3.00
Chemical	62,000	143,000	48,140	1.28	2.96
Chemical	—	165,000	58,000	—	2.84
Chemical	9,700	15,000	5,500	1.76	2.72
Nylon polymer	—	23,400	8,800	—	2.70
Petrochemical	—	—	—	—	2.70
Nylon polymer	—	112,600	44,000	—	2.50
Olefin processing	—	321	133	—	2.40
Butadiene processing	—	359	156	—	2.30
Chemical	—	350,000	160,000	—	2.19
Synthetic rubber	—	192	110	—	1.75

[a] High concentration of sulfides and thiosulfates. From ref. 13.

Table 7. Chemical Waste Treatment

Treatment Method	Type of Waste	Mode of Operation	Degree of Treatment	Remarks
Ion exchange	Plating, nuclear	Continuous filtration with resin regeneration	Demineralized water recovery, product recovery	May require neutralization and solids removal from spent regenerant
Reduction and precipitation	Plating, heavy metals	Batch or continuous treatment	Complete removal of chromium and heavy metals	One day's capacity for batch treatment, 3-h retention for continuous treatment, sludge disposal or dewatering required
Coagulation	Paperboard, refinery, rubber, paint, textile	Batch or continuous treatment	Complete removal of suspended and colloidal matter	Flocculation and settling tank or sludge blanket unit, pH control required
Adsorption	Toxic or organics, refractory	Granular columns of powdered carbon	Complete removal of most organics	Powdered activated carbon (PAC) used with activated sludge process
Chemical oxidation	Toxic and refractory organics	Batch or continuous ozone or catalyzed hydrogen peroxide, advanced oxidation processes	Partial or complete oxidation	Partial oxidation to render organics more biodegradable

In which the LC_{50} or the NOEL is expressed as the percentage of effluent in the receiving water. Therefore, an effluent having an LC_{50} or 10% contains 10 TU.

Effluent toxicity can also be defined as a chronic toxicity in which the growth or reproduction rate of the species is affected.

WASTEWATER TREATMENT

In order to meet present requirements, existing plants need to be retrofitted and new plants incorporate advanced wastewater treatment technology. A substitution flow sheet showing available technology is shown in Figure 3. Physical-chemical treatment alternatives are shown in Table 7. Biological treatment alternatives are shown in Table 8. Options for meeting restrictive requirements are tertiary treatment following biological treatment, source treatment of toxic or refractory wastewaters, or modifications of the existing biological technology incorporating PAC. The applicable technologies to meet present effluent guidelines and standards for various industrial categories are shown in Table 9. These technologies are discussed in detail in the text.

Table 8. Biological Waste Treatment

Treatment Method	Mode of Operation	Degree of Treatment	Land Requirements	Equipment	Remarks
Lagoons	Intermittent or continuous discharge, facultative or anaerobic	Intermediate	Earth dug; 10–60 days retention		Odor control frequently required
Aerated lagoons	Completely mixed or facultative continuous basins	High in summer, less in winter	Earth basin, 2,44–4.88 m (8–16 ft) deep; 8.55–17.1 $m^2/(m^3/d)$ [8–16 acres/(mil/gal/d)]	Diffused or mechanical aerators, clarifier for sludge separation and recycle	Excess sludge dewatered and disposed of
Activated sludge	Completely mixed or plug flow, sludge recycle	>90% removal of organics	Earth or concrete basin; 3.66–6.10 m (12–20 ft) deep; 0.561–2.262 $m^3/(m^3/d)$ [75,000–350,000 ft^3/(mil gal/d)]	Diffused or mechanical aerators, clarifier for sludge separation and recycle	Excess sludge dewatered and disposed of
Trickling filter	Continuous application, may employ effluent recycle	Intermediate or high, depending on loading	5.52–34.4 $m^2/10^3$ [225–1400 ft^2/(mil gal/d)]	Plastic packing 6.10–12.19 m (20–40 ft) deep	Pretreatment before POTW or activated sludge plant
Rotating biological contactor (RBC)	Multistage continuous	Intermediate or high		Plastic disks	Solids separation required
Anaerobic	Complete mix recycle; upflow or downflow filter, fluidized bed; upflow sludge blanket	Intermediate		Gas collection required, pretreatment before POTW or activated sludge plant	
Spray irrigation	Intermittent application of waste	Complete, water percolation into groundwater and runoff to stream	6.24×10^{-7}–4.68×10^{-6} $m^3/(s \cdot m^2)$ [40–300 gal/(min·acre)]	Aluminum irrigation pipe and spray nozzles, movable for relocation	Solids separation required, salt content in waste limited

Notes:

$$\text{ft} = 0.305\ \text{m}$$
$$\text{acre/(mil gal}\cdot\text{d)} = 1.07\ \text{m}^2/(\text{m}^3\cdot\text{d})$$
$$\text{ft}^3/(\text{mil gal}\cdot\text{d}) = 7.48 \times 10^{-3}\ \text{m}^3/(\text{thousand m}^3\cdot\text{d})$$
$$\text{ft}^2/(\text{mil gal}\cdot\text{d}) = 2.45 \times 10^{-2}/(\text{thousand m}^3\cdot\text{d})$$
$$\text{gal/(min}\cdot\text{acre)} = 1.56 \times 10^{-8}\ \text{m}^3/(\text{s}\cdot\text{m}^2)$$

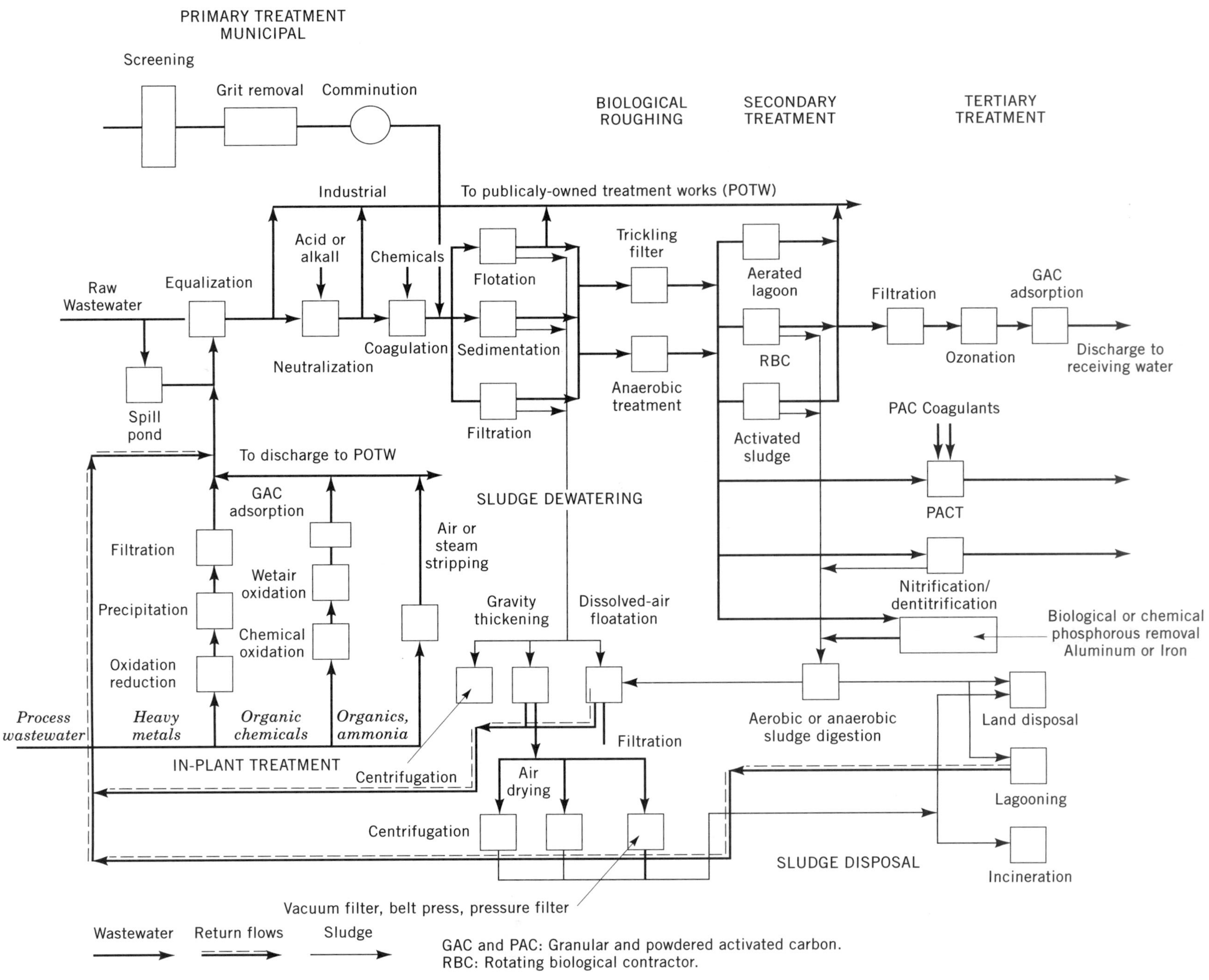

Figure 3. Alternative wastewater treatment technologies.

Table 9. Applicable Technologies to Meet Present Guidelines and Standards

Process Industry	Effluent Guidelines and Standards[a]: Reference[a]	Effluent Guidelines and Standards[a]: Parameters Regulated	Applicable Treatment Technologies: Neutralization	Sedimentation and/or Filtration	Biological Treatment	Chemical Precipitation	Chemical Oxidation/ Reduction	Air and/or Steam Stripping	Activated Carbon Adsorption	Resin Adsorption	Ion Exchange	Ultrafiltration and/or Reverse Osmosis	Flotation/ Phase Separation
Dairy products	405	pH, BOD, TSS		X	X								
Grain mills	406	pH, BOD, TSS		X	X								
Canned/preserved fruits and vegetables processing	407	pH, BOD, TSS		X	X								
Canned/preserved seafood processing	408	pH, BOD, TSS, O&G		X	X								X
Sugar processing	409	pH, BOD, TSS, O&G, temperature, fecal coliform	X	X	X		X						
Textile mills	410	pH, BOD, TSS, O&G, COD, sulfide, phenol, chronium	X	X		X							X
Cement manufacturing	411	pH, TSS, temperature	X	X									
Feedlots	412	BOD, fecal coliform			X		X						
Electroplating	413	pH, TSS, metals, cyanide, total toxic organics (TTO)	X	X		X	X				X	X	
Organic chemicals, plastics, and synthetic fibers	414	pH, BOD, TSS, metals, cyanide, volatiles, semivolatiles	X	X	X	X	X	X	X	X	X	X	
Inorganic chemicals manufacturing	415	pH, TSS, COD, O&G, metals, chlorine, TOC, ammonia, cyanide	X	X		X	X	X			X	X	X
Soaps and detergents	417	pH, BOD, TSS, O&G, COD, surfactants (MBAS)	X	X	X		X					X	X
Fertilizer manufacturing	418	pH, BOD, TSS, phosphorus, ammonia, organic nitrogen, nitrate, fluoride	X	X		X	X	X			X		
Petroleum and petroleum refining	419	pH, BOD, TSS, O&G, COD, sulfide, chromium, hexavalent chromium, phenolic compounds (4AAP), ammonia	X	X	X	X	X	X	X	X	X		X
Iron/steel manufacturing	420	pH, TSS, O&G, metals, cyanide, phenols (4AAP), ammonia, naphthalene, benzene, tetrachloroethylene, benzo(*a*)pyrene, chlorine	X	X		X	X	X			X	X	X

Table 9. *(continued)*

	Effluent Guidelines and Standards[a]		Applicable Treatment Technologies										
Process Industry	Reference[a]	Parameters Regulated	Neutralization	Sedimentation and/or Filtration	Biological Treatment	Chemical Precipitation	Chemical Oxidation/ Reduction	Air and/or Steam Stripping	Activated Carbon Adsorption	Resin Adsorption	Ion Exchange	Ultrafiltration and/or Reverse Osmosis	Flotation/ Phase Separation
Nonferrous metals	421	pH, TSS, O&G, benzo-(*a*)pyrene, metals, cyanide, ammonia	X	X		X	X	X			X	X	X
Phosphate manufacturing	422	pH, TSS, phosphorus, fluoride	X	X		X					X		X
Steam electric power generating	423	TSS, O&G, metals, chlorine, priority pollutants	X	X		X	X		X		X	X	X
Ferroalloy manufacturing	424	pH, TSS, metals, cyanide, phenols, ammonia	X	X		X	X	X			X	X	
Leather tanning and finishing	425	pH, BOD, TSS, O&G, chromium, sulfide	X	X	X	X	X						X
Glass manufacturing	426	pH, BOD, TSS, Oil, COD, lead, fluoride, phenol, phosphorus, ammonia	X	X		X	X	X			X	X	X
Asbestos manufacturing	427	pH, TSS, COD	X	X									
Rubber processing	428	pH, BOD, TSS, O&G, COD, metals	X	X	X	X							X
Timber products	429	pH, BOD, TSS, O&G, phenols, chromium, arsenic, copper	X	X	X	X	X						
Pulp, paper, and paper board mills	430	pH, TSS, BOD, zinc, pentachlorophenol	X	X	X	X			X				
Builders paper and board mills	431	pH, BOD, TSS, settleable solids, pentachlorophenol	X	X	X				X				
Meat products	432	pH, BOD, TSS, O&G, fecal coliform, ammonia	X	X	X								X
Metal finishing	433	pH, TSS, TTO, metals, cyanide	X	X		X	X				X		
Coal mining	434	pH, TSS, iron, manganese	X	X		X					X	X	
Offshore oil and gas extraction	435	O&G											
Mineral mining and processing	436	pH, TSS, flouride	X	X		X					X	X	X
Pharmaceutical manufacturing	439	pH, BOD, TSS, COD, cyanide, TSS, settleable solids	X	X	X		X		X	X			

Ore mining and dressing	440	pH, TSS, settleable solids, metals, ammonia	X	X		X					X		
Paving and roofing materials	443	pH, BOD, TSS, O&G	X	X									X
Paint formulating	446	No process water discharge										X	
Ink formulating	447	No process water discharge										X	
Gum and wood chemicals manufacturing	454	pH, BOD, TSS	X	X	X								
Pesticide chemicals manufacturing	455, pending	pH, BOD, TSS, COD, organic pesticide chemicals	X	X					X	X	X	X	
Explosives manufacturing	457	pH, BOD, TSS, COD, O&G	X	X					X	X		X	X
Carbon black manufacturing	458	O&G		X									X
Photographic processing	459	pH, cyanide, silver	X	X			X				X	X	
Hospitals	460	pH, BOD, TSS	X	X	X								
Battery manufacturing	461	pH, TSS, COD, O&G, metals	X	X		X					X	X	X
Plastics molding and forming	463	pH, BOD, TSS, O&G	X	X	X	X						X	X
Metal molding and casting	464	pH, TSS, O&G, TTO, phenols, copper, zinc, lead	X	X		X	X				X	X	X
Coil coating	465	pH, TSS, O&G, metals, cyanide, TTO	X	X		X	X				X	X	X
Porcelain enameling	466	pH, TSS, O&G, metals	X	X		X					X	X	X
Aluminum forming	467	pH, TSS, O&G, metals, cyanide, TTO	X	X		X	X				X	X	X
Copper forming	468	pH, TSS, O&G, metals, TTO	X	X		X	X				X	X	X
Electrical and electronic components	469	pH, TSS, metals, TTO	X	X		X	X				X	X	X
Nonferrous metals forming and metal powders	471	pH, TSS, O&G, metals, cyanide	X	X		X	X				X	X	X
Waste treatment	437, pending	Proposal pending	X	X	X	X	X	X	X	X	X	X	X
Metal products and machinery	438, pending	Proposal pending	X	X		X	X				X	X	X
Industrial laundries	441, pending	Proposal pending	X	X		X					X	X	X
Transportation equipment cleaning	442, pending	Proposal pending	X	X									X

[a] Code of Federal Regulations, Title 40, Parts 405–471.

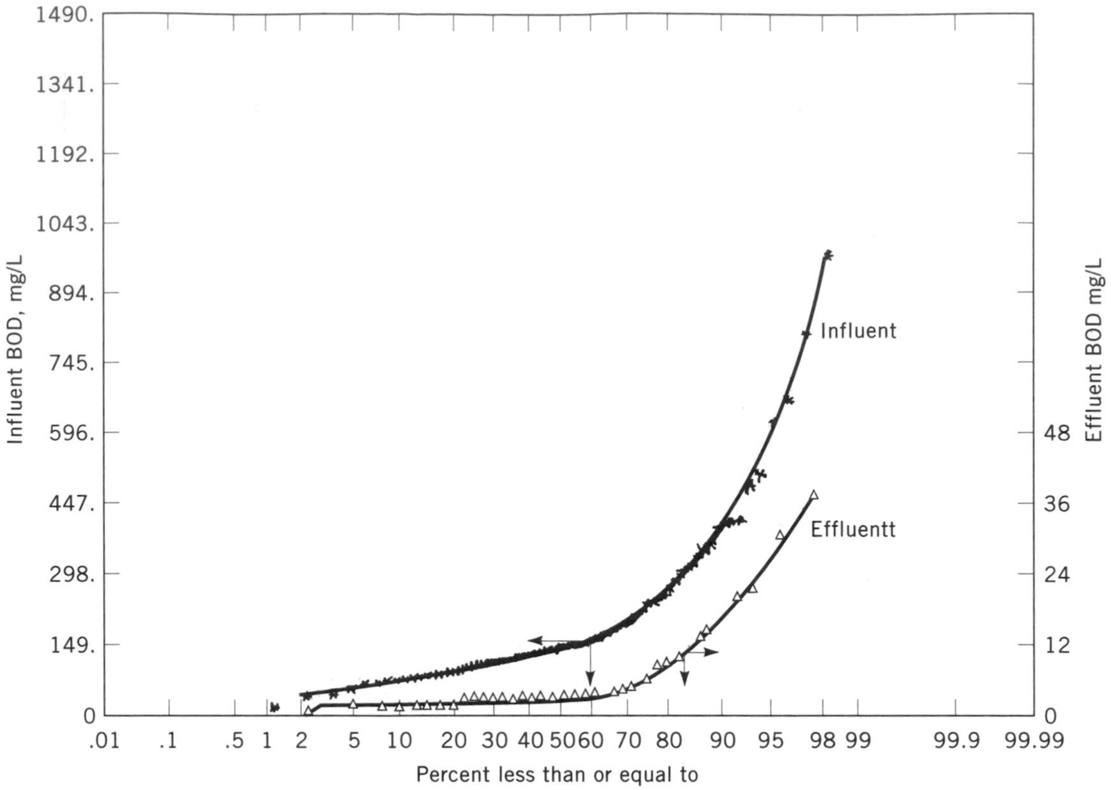

Figure 4. Influent and effluent variability for a petroleum refinery.

Equalization

Equalization is employed to reduce variability in flow and/or strength of industrial wastewaters so they can be successfully treated in either biological or physical-chemical treatment processes. In most cases, the effluent concentration from a wastewater treatment process will be proportional to the influent concentration. This is shown in Figure 4 for a petroleum refining activated sludge effluent for 24-h daily composite values. It is therefore necessary to control the influent concentration in order to achieve a predictable effluent concentration.

If the wastewater flow is fairly constant (as in a pulp and paper mill), a constant-volume basin can be employed in which wastewater strength is equalized.

If the wastewater and strength are highly variable (as in a batch process chemical plant), a variable volume basin with a variable inflow and a constant outflow is employed to equalize both flow and strength.

If the wastewater is readily degradable (as in a brewery), aeration is provided in the equalization basin to avoid septicity and the generation of odors.

Equalization basin types are shown in Figure 5.

In order to handle accidental spills or overflows, a spill basin may be provided in which flow is diverted when the concentration of a particular constituent exceeds a predetermined value. When the spill is contained, the wastewater flow is diverted back to the equalization basin. The contents of the spill basin are then pumped at a constant controlled rate to the equalization basin, as shown in Figure 6.

Neutralization

Wastewater discharge usually requires a pH between 6 and 9. Exceptions are a biological process in which microbial respiration degrades acidity (acetic acid is oxidized to CO_2 and H_2O) or the CO_2 generated by microbial respiration neutralizes caustic alkalinity (OH^-) to bicarbonate HCO_3.

Neutralization usually follows equalization so that acidic and alkaline streams can be partially neutralized in the equalization basin. If the wastewater is always

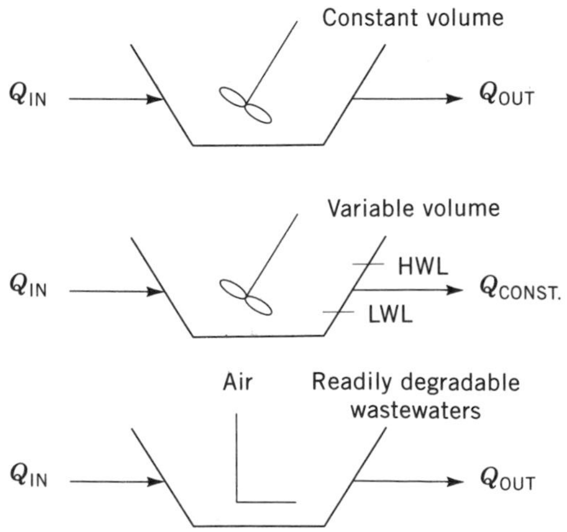

Figure 5. Equalization alternatives.

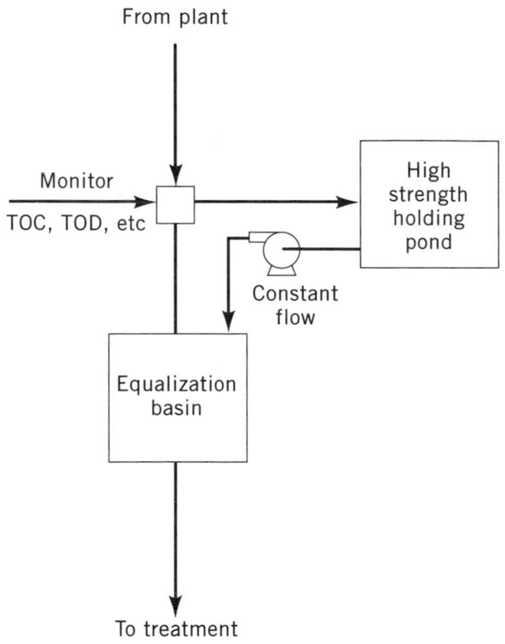

Figure 6. Control of spills and shock loads.

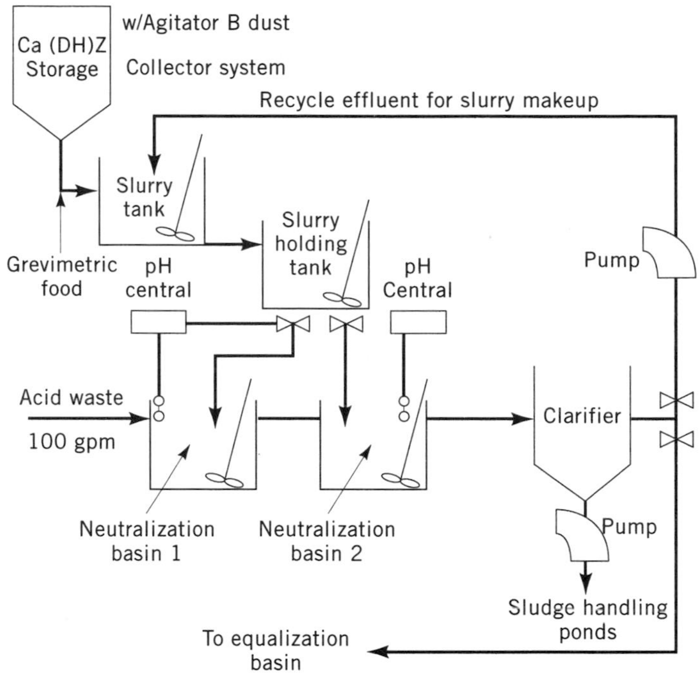

Figure 7. Schematic diagram of lime neutralization system.

acidic, neutralization may precede the equalization basin to minimize corrosion in the equalization basin.

Acidic wastewaters can be neutralized with lime, magnesium hydroxide, caustic, or limestone.

Lime is preferred over caustic because it is cheaper in cost and usually produces a more dewaterable sludge. Magnesium hydroxide will eliminate the precipitation of $CaSO_4$.

A limestone bed is simple to operate and is applicable to moderately acidic wastewaters.

Highly acidic wastewaters require a two-stage process because of the logarithmic nature of pH, as shown in Figure 7. The first stage will adjust the pH to 3–3.5 and the second stage trim the pH to 6.5–7.5.

Alkaline wastewaters can be neutralized with H_2SO_4 or HCl or using flue gas (CO_2).

Removal of Oil and Grease

High concentrations of oil and grease can be removed in a gravity separator where the lighter oils and greases float to the surface where they are skimmed off.

The API gravity separator removes oil globules 0.015 cm or greater and can achieve an effluent oil content of less than 50–100 mg/L.

The corrugated plate separator with a narrow separation space can remove oil globules 0.01 cm or greater. As

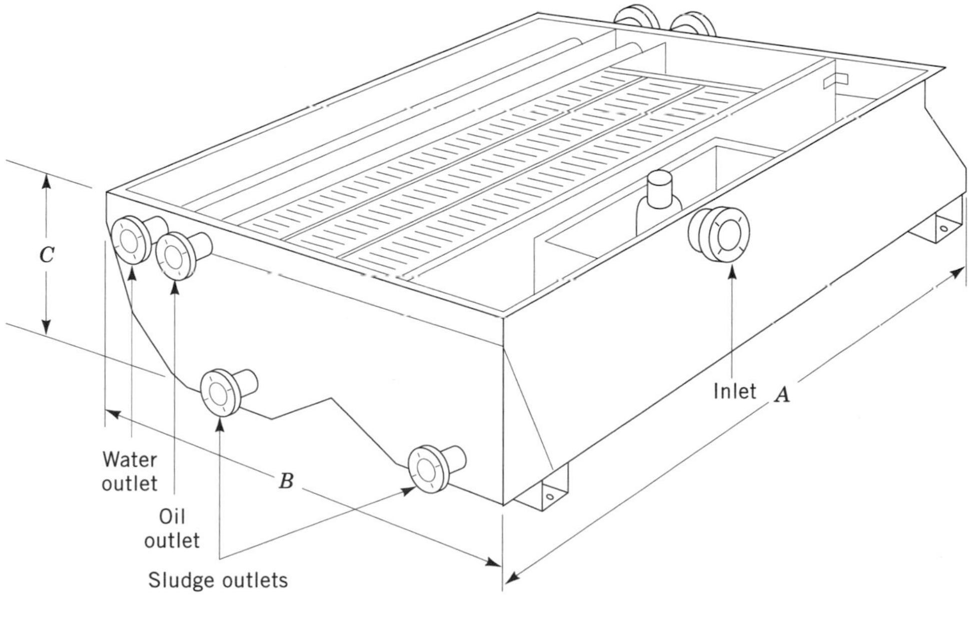

Figure 8. Corrugated plate separator.

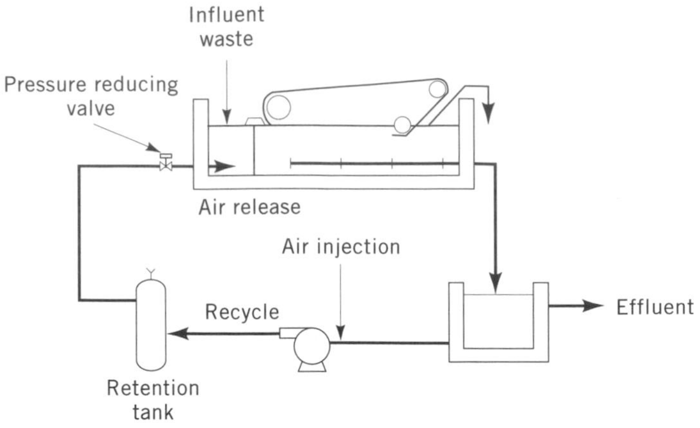

Figure 9. Pressurized dissolved air flotation system.

a result, effluent oil concentrations as low as 10 mg/L are achievable (Fig. 8).

Low concentrations of oil can be removed by dissolved air flotation (DAF). In this process, an effluent recycle is pressurized in the presence of excess air, causing additional air to go into solution (in accordance with Henry's law). When this water is discharged to the inlet chamber of the flotation unit (at close to atmospheric pressure), the dissolved air comes out of solution in the form of tiny air bubbles which attach themselves to and become enmeshed in suspended solids and oil globules, as shown in Figure 9. The performance of DAF for the treatment of several wastewaters is shown in Table 10. In cases where the oil globules are of a very small size, a coagulant, usually alum, and a polymer are added to flocculate the particles, thereby enhancing bubble attachment and flotation.

Alternatively, induced air flotation (IAF) can be employed in which air bubbles are generated through an inductor, as shown in Figure 10. The removal mechanism is the same as the DAF.

Emulsified oil contains a liquid film so that it will not separate by gravity without first breaking the emulsion. This is achieved by adding surfactants, emulsion-breaking polymers, or coagulants. After the emulsion is broken, the conventional technologies described above are applicable.

SUSPENDED SOLIDS REMOVAL

Depending on the concentration and characteristics of the suspended solids, they can be removed by filtration, flotation, or sedimentation.

Coarse solids are removed by screening.

Settleable suspended solids are removed in a clarifier which may be circular or rectangular. The efficiency of solids removal is a function of the overflow rate (m^3/m^2·day or gal/ft^2·day). Typical clarifiers are shown in Figure 11. The relationship between overflow rate and suspended solids removal for a pulp and paper mill wastewater and municipal sewage is shown in Figure 12.

Removal of dispersed solids requires the addition of a coagulant (alum, iron, or a polymer) to flocculate the solids so they will settle in the clarifier. A sludge blanket unit with sludge recirculation will frequently result in lower coagulant requirements and enhanced clarification, as shown in Figure 13.

The application of coagulation for removal of dispersed and colloidal solids of several industrial wastewaters is shown in Table 11.

Filtration is employed when the suspended solids concentration is less than 100 mg/L and high effluent clarity is required. Finely dispersed suspended solids will require the addition of a coagulant prior to filtration.

Filters most commonly used in wastewater treatment are a dual media (anthrafilt and sand) or a moving bed or continuous backwash sand filter, as shown in Figures 14 and 15. Performance data for the tertiary filtration of municipal and industrial wastewater are shown in Table 12.

HEAVY METALS REMOVAL

Heavy metals should be removed prior to biological treatment or other technologies which generate sludges to avoid comingling metal sludges with other nonhazardous sludges.

Technologies Available

- *Precipitation of Metals as Hydroxide.* In this case, lime or caustic is added usually to the pH of minimum solubility. This process has limitations in cases where multiple metals are present with a pH range of minimum solubility, as shown in Figure 16.

Table 10. Air Flotation Treatment of Oily Wastewaters

Wastewater	Coagulant (mg/L)	Oil Concentration (mg/L)		
		Influent	Effluent	Percent Removal
Refinery	0	125	35	72
	100 alum	100	10	90
	130 alum	580	68	88
	0	170	52	70
Oil tanker ballast water	100 alum + 1 mg/L polymer	133	15	89
Paint manufacture	150 alum + 1 mg/L polymer	1900	0	100
Aircraft maintenance	30 alum + 10 mg/L activated silica	250–700	20–50	90+
Meat packing		3830	270	93
		4360	170	96

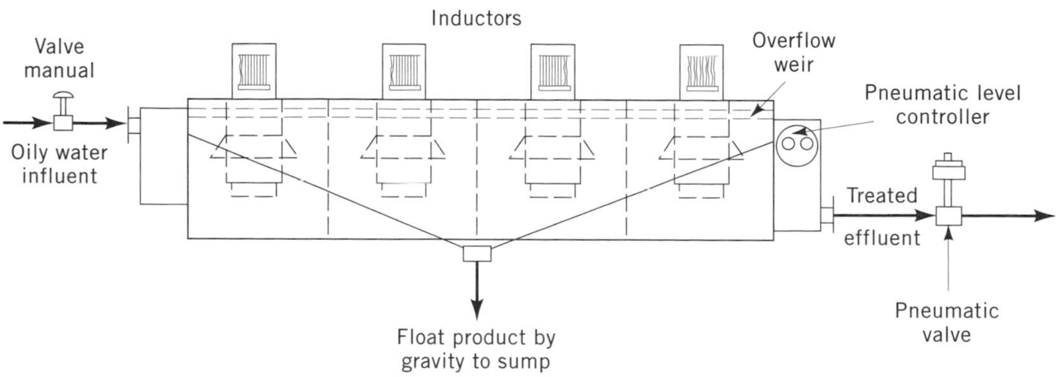

Figure 10. Induced air flotation unit.

- *Precipitation as Carbonate.* This follows the same relations as the hydroxide.
- *Precipitation as Sulfide.* This has the advantage of having a wide range of minimum solubility. The disadvantage is that a poorer thickening sludge is generated and sulfide presents a potential health hazard.

As a result, sulfide precipitation is frequently used as a polishing step following hydroxide precipitation:

- Coprecipitation: Many metals such as arsenic will adsorb on alum or iron flocs and be effectively removed over a near-neutral pH range. The disadvantage of coprecipitation is the generation of large quantities of sludge.
- New chelating ion exchange resins are able to selectively remove many heavy metals. This process is suitable for end-of-pipe polishing and for metal concentration and recovery. Ion exchange is also employed for metal recovery and water reuse.
- Many heavy metals are removed on activated carbon. A primary mechanism is sulfide precipitation on the carbon.
- Reverse osmosis (RO) can be employed to remove and recover heavy metals, particularly nickel.

Achievable effluent concentrations of heavy metals is summarized in Table 13 (15).

Table 11. Coagulation of Industrial Wastewaters

Wastewater	Parameter	Influent (mg/L)	Effluent (mg/L)
Board mill	BOD	127	68
	TSS	593	44
Tissue mill	BOD	140	36
	TSS	720	15
Ball bearing	TSS	544	40
	O/G	302	28
Laundry	ABS	63	0.1
	BOD	243	90
Latex	COD	4,340	178
	BOD	1,070	90
	Total Solids	2,550	446

REMOVAL OF VOLATILE ORGANICS

Volatile organics (benzene, toluene, etc.) should usually be removed prior to biological treatment.

- *Air Stripping.* Removal of volatile organics by air stripping is accomplished in packed or tray towers in which air is introduced to the bottom of the tower counterflow to the liquid passing down the tower. Removal of volatiles is a function of Henry's constant, the air–liquid ratio, and the transfer efficiency of packing. High concentrations of volatiles require treatment of the off gas through vapor phase carbon or combustion (16).

Volatile removal as a function of the air–liquid ratio is shown in Figure 17. Typical stripping towers are shown in Figure 18.

- *Steam Stripping.* Steam introduced to a packed tower will cause volatiles to be removed in the vapor phase. An azeotropic mixture is formed resulting in a separation of the volatiles from the water. An effluent recycle is usually employed to reduce volatiles in the liquid effluent. A steam stripping tower is shown in Figure 19.
- *Carbon Adsorption.* Most volatile organics will adsorb on activated carbon in the liquid state.
- *Biologically Active Carbon.* Low concentrations of biodegradable volatiles will be removed by adsorption and biodegradation on activated carbon.
- *Chemical Oxidation.* Many volatiles will be chemically oxidized using conventional or advanced chemical oxidants.

NUTRIENT REMOVAL

In many locations, nitrogen and phosphorus must be removed in order to meet effluent limitations.

Nitrogen (17)

- Nitrogen is most commonly removed through the process of biological nitrification and denitrification, as

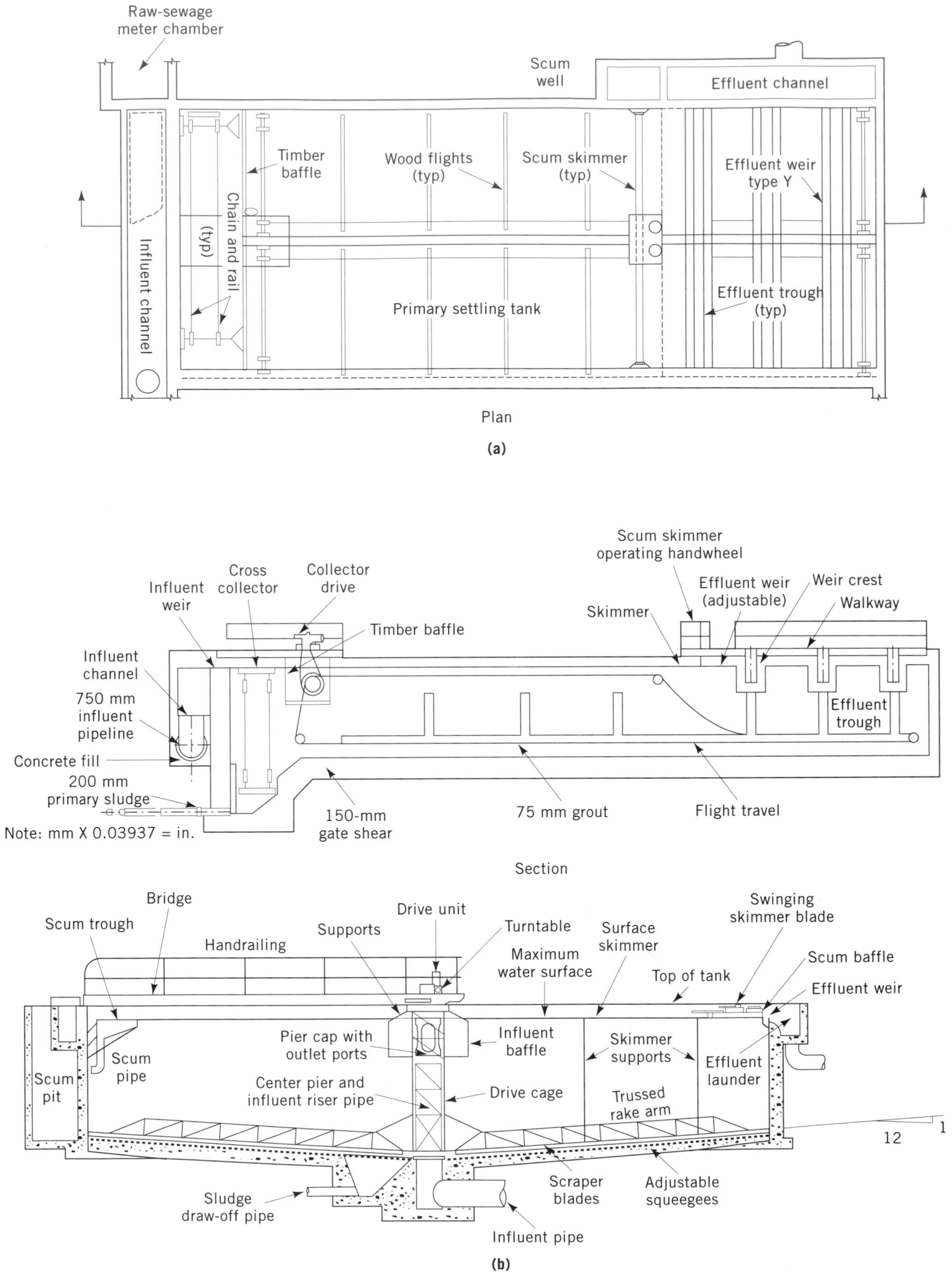

Figure 11. (**a**) Typical rectangular sedimentation tank. (**b**) Typical circular sedimentation tank.

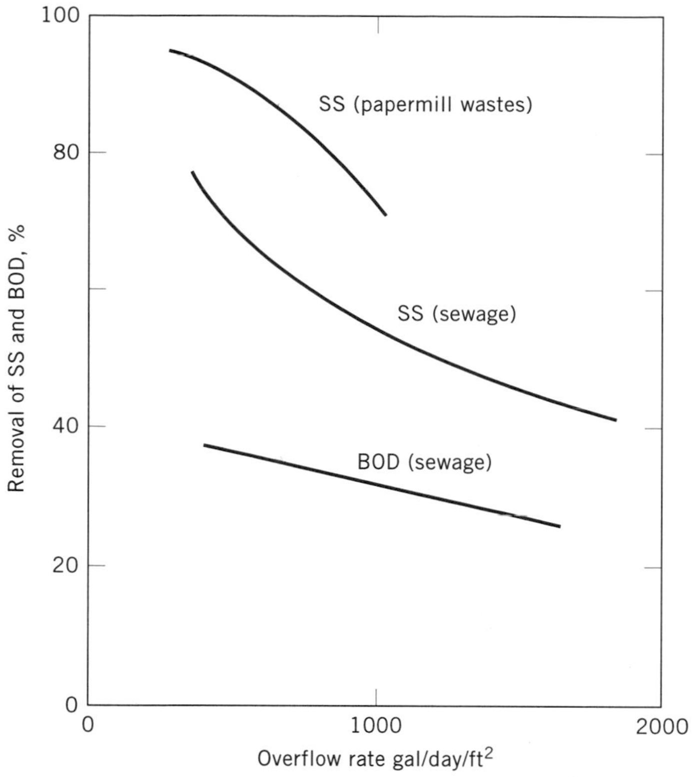

Figure 12. Suspended solids and BOD removal from domestic sewage and paper mill wastes by sedimentation.

shown in Figure 20. In this process, organic nitrogen is hydrolyzed to ammonia. Ammonia, in turn, is oxidized to nitrite through the action of a specific organism, *Nitrosomonas*. This reaction also generates two hydrogen ions for each nitrogen oxidized so that alkalinity must be available to neutralize the acidity.

The nitrite, in turn, is oxidized to nitrate through the action of a specific organism, *Nitrobacter*. Since *Nitrobacter* has a much higher growth rate than *Nitrosomonas*, the critical design parameter is that the solids retention time (SRT) or sludge age in the process must exceed the growth rate of *Nitrosomonas* or they will be washed out of the system.

The nitrifying organisms are relatively sensitive to many toxic organics so that the treatment of industrial wastewaters requires special attention to the presence of toxics. Denitrification is a process in which facultative organisms will reduce nitrate to nitrogen gas in the absence of molecular oxygen. This consequently results in the removal of BOD_5. The denitrification process also generates one hydroxyl ion so that alkalinity requirements are reduced to half when both nitrification and denitrification are practiced.

Phosphorus Removal (18)

Phosphorus can be removed from wastewater either chemically or biologically:

Chemical Removal

Phosphorus can be precipitated with lime to form $Ca_3(PO_4)_2$. The actual composition of the precipitate is a complex compound called apitate. Achieving minimum phosphorus concentrations requires a pH in excess of 10.5.

Alum or iron will precipitate phosphorus as $AlPO_4$ or $FePO_4$. This procedure is generally employed in conjunction with the activated sludge process in which the coagulant is added at the end of the aeration

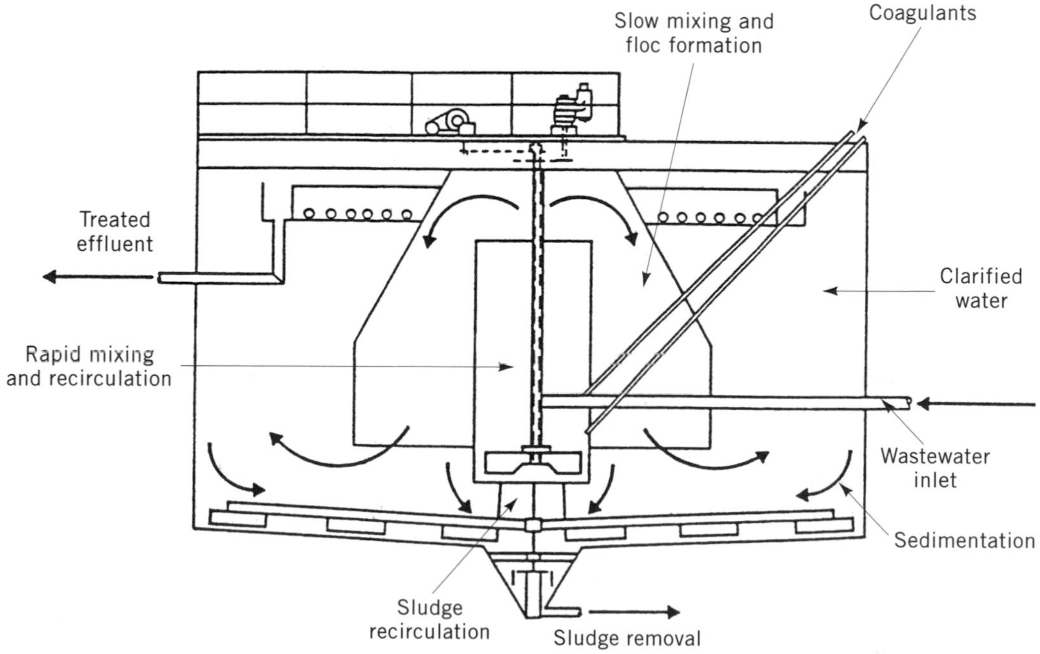

Figure 13. A reactor clarifier designed for both coagulation and settling.

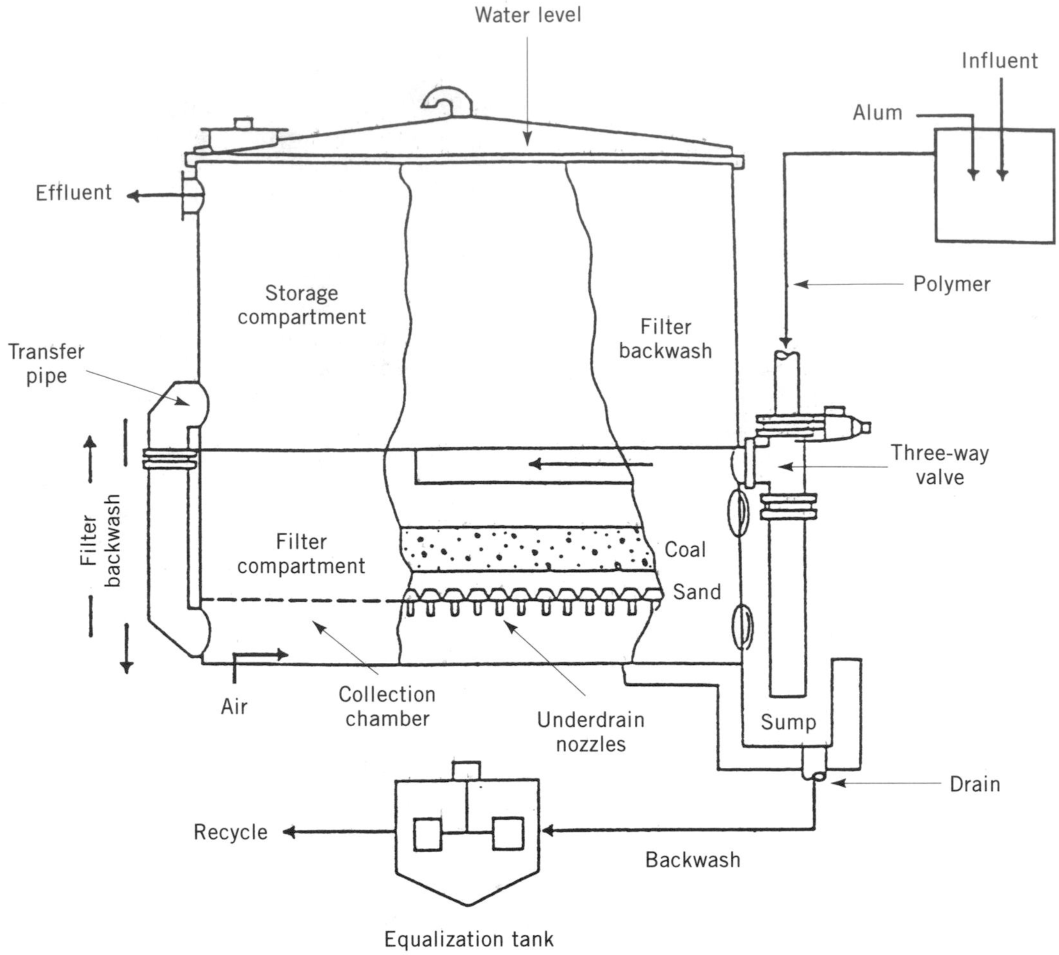

Figure 14. Typical automatic dual-media filter.

basin or between the aeration basin and the final clarifier.

Biological Removal

Certain organisms normally present in activated sludge have the ability to store phosphorus. The process configuration for bio-P removal involves an anaerobic step in the presence of volatile fatty acids principally acetic in which phosphorus is released and acetate taken up by the bio-P organisms. Under proper operating conditions, soluble effluent phosphorus levels of 0.1 mg/L are achievable from mu-

Table 12. Filtration Performance

Filter Type	Wastewater	Filter Depth (ft)	Hydraulic Loading [gal/(min·ft²)]	Percent Removal		Effluent (mg/L)	
				SS	BOD	SS	BOD
Gravity downflow	TF effluent	2–3	3	67	58	—	2.5
Pressure upflow	AS effluent	5	2.2	50	62	7.0	6.4
Dual media	AS effluent	2.5	5.0	74	88	4.6	2.5
Gravity downflow	AS effluent	1.0	5.3	62	78	5	4
Dynasand	Metal finishing	3.3	4–6	90	—	2–5	—
	AS effluent	3.3	3–10	75–90	—	5–10	—
	Oily wastewater	3.3	2–6	80–90[a]	—	5–10[a]	—
Hydroclear	Poultry	1	2–5	88	—	19	—
	Oil refinery	1	2–5	68	—	11	—
	Unbleached kraft	1	2–5	74	—	17	—

Note:
ft = 0.305 m
gal/(min·ft^2) = 4.07 × 10^{-2} m^3/(min-m^2)
[a] Free oil.

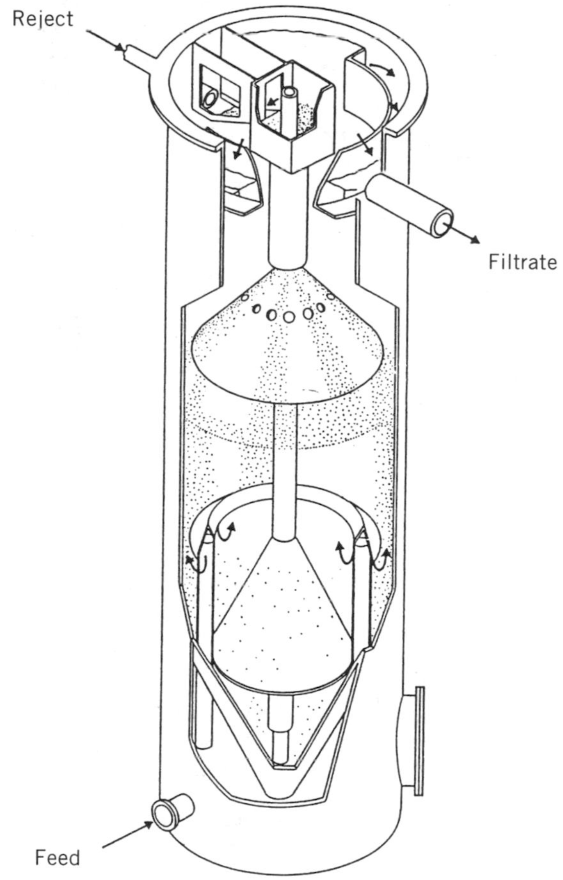

Figure 15. Dynasand filter (DSF). (Courtesy of Parkson Corporation.)

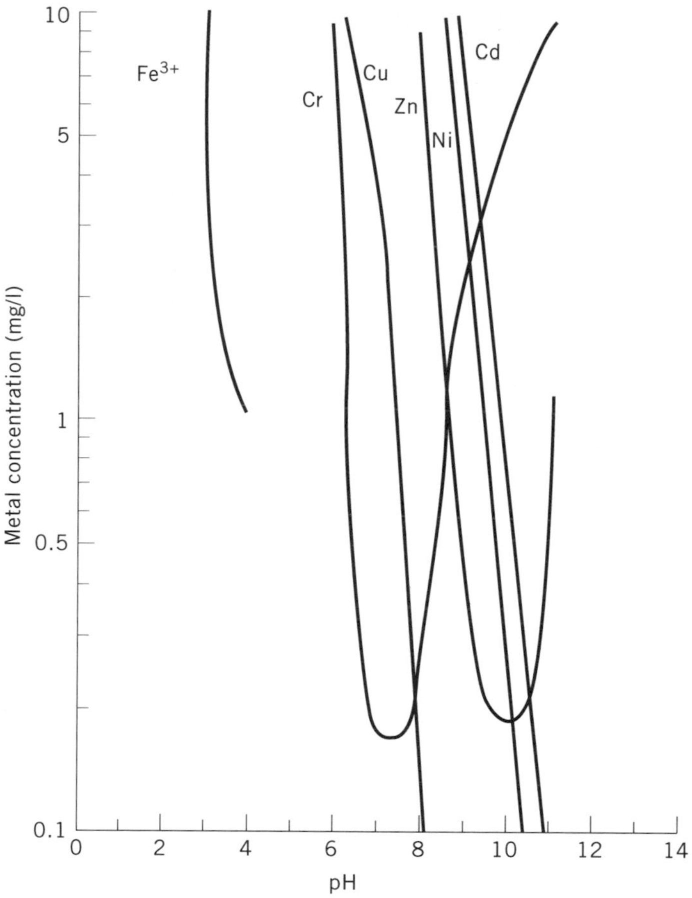

Figure 16. Solubility of metals versus pH.

Table 13. Effluent Levels Achicvable in Heavy Metal Removals

Metal	Achievable Effluent Concentraton (mg/L)	Technology
Arsenic	0.05	Sulfide precipitation with filtration
	0.06	Carbon adsorption
	0.005	Ferric hydroxide coprecipitation
Barium	0.5	Sulfate precipitation
Cadmium	0.05	Hydroxide precipitation at pH 10–11
	0.05	Coprecipitation with ferric hydroxide
	0.008	Sulfide precipitation
Copper	0.02–0.07	Hydroxide precipitation
	0.01–0.02	Sulfide precipitation
Mercury	0.01–0.02	Sulfide precipitation
	0.001–0.01	Alum coprecipitation
	0.0005–0.005	Ferric hydroxide coprecipitation
	0.001–0.005	Ion exchange
Nickel	0.12	Hydroxide precipitation at pH 10
Selenium	0.05	Sulfide precipitation
Zinc	0.1	Hydroxide precipitation at pH 11

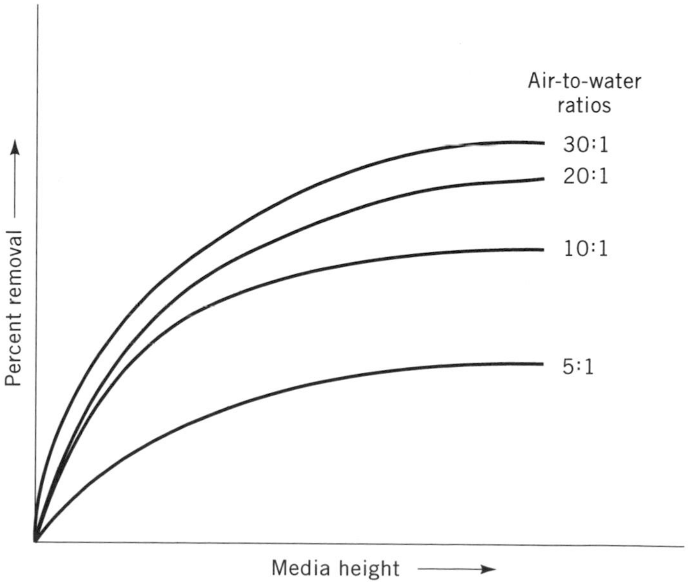

Figure 17. Illustration relationships of air stripping removal efficiencies to media height and air-to-water ratios.

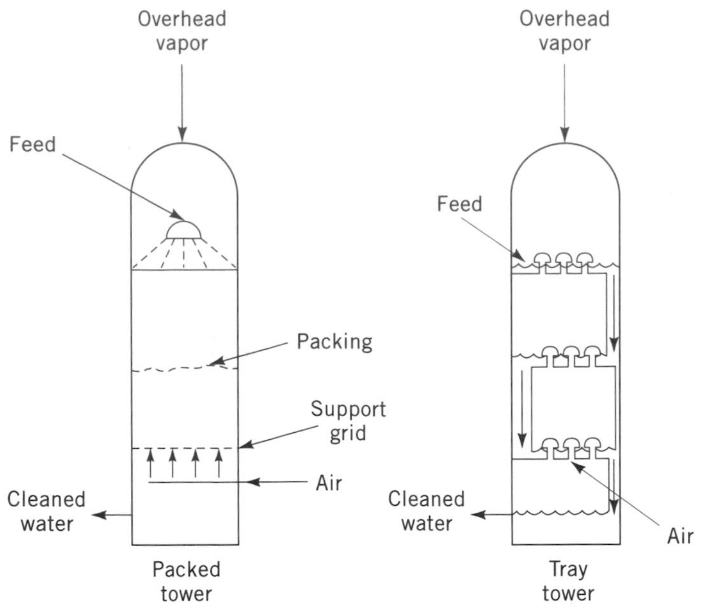

Figure 18. Air stripping towers.

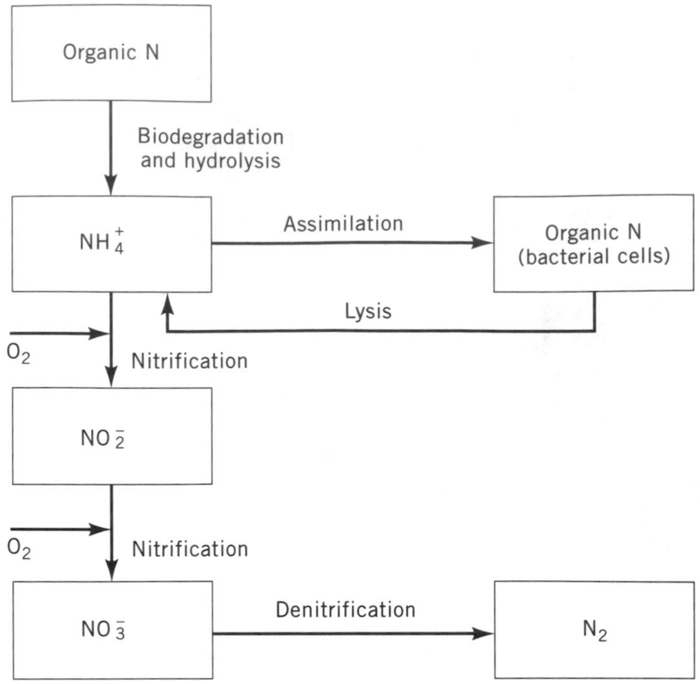

Figure 20. Nitrogen transformation.

nicipal wastewater. This process is schematically shown in Figure 21.

BIOLOGICAL TREATMENT

Aerobic treatment is generally applied to lower strength wastewaters while anaerobic treatment is employed as a pretreatment for high-strength wastewaters. The objective of biological treatment is to remove biodegradable organics.

Biological treatment can be aerobic or anaerobic. In an aerobic biological-treatment process the mechanisms of removal are stripping, sorption, and biodegradation.

Stripping (14)

- Degradable VOC (ie, benzene) will both biodegrade and strip from the solution.
- The percentage stripped will depend on the power level in the aeration basin and/or the type of aeration equipment (ie, enhanced stripping with surface aerators).
- The biodegradation rate will decrease with increased halogenation and hence the percentage stripped will increase, as shown in Figure 22.

Sorption

- Most organics sorb to a very small degree on the biofloc, ie, <2%.
- An exception is the nondegradable pesticide Lindane, other pesticides, and PCBs.
- Heavy metals will complex with the cell wall and precipitate within the floc. Metal accumulation will increase with increasing sludge age.

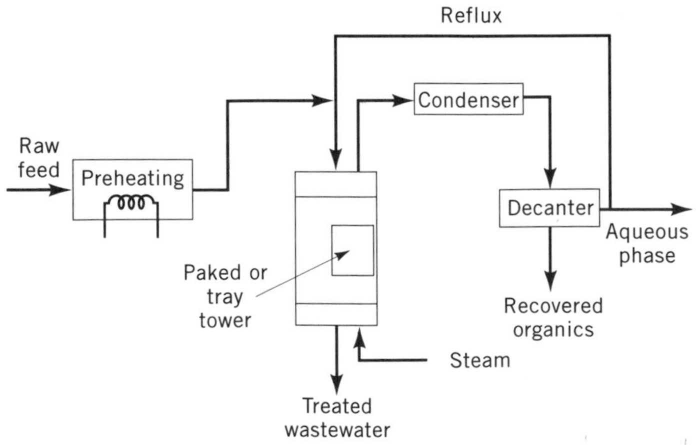

Figure 19. Typical steam stripping process.

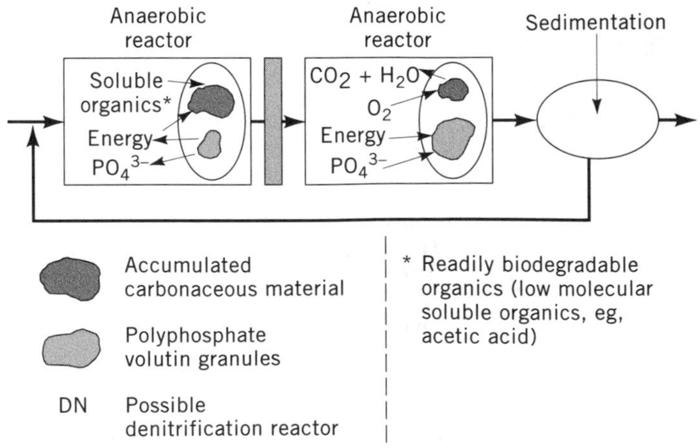

Figure 21. Mechanics of biological phosphorous removal.

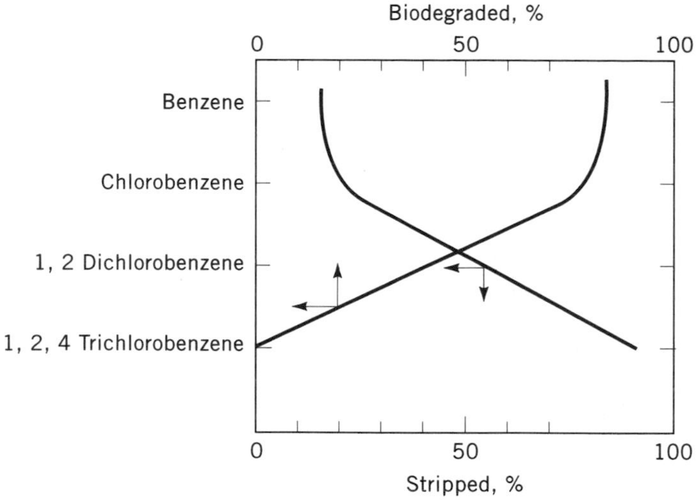

Figure 22. Effect of chlorination on biodegradation and stripping.

Biodegradation (9,12,19)

- Approximately one-half of the organics removed are oxidized to CO_2 and H_2O and one-half synthesized to biomass.
- Three to 10% of the organics removed result in SMPs (soluble microbial products). The SMP is significant because it causes aquatic toxicity.
- Nitrogen and phosphorus are required in the reaction at an approximate ratio of BOD–N–P of 100:5:1. Nitrogen and phosphorus are amply available in municipal wastewaters but frequently deficient in industrial wastewaters.

The reactions in an aerobic biological process are

$$\text{Organics} + O_2 + N + P \xrightarrow[K]{\text{cells}} \text{new cells} + CO_2 + H_2O + \text{SMP}$$

where K is a reaction rate coefficient which is a function of the degradability of the wastewater and SMP is the nondegradable soluble microbial products and

$$\text{Cells} + O_2 \rightarrow CO_2 + H_2O + N + P + \text{nondegradable cellular residue} + \text{SMP}$$

In the activated sludge process, performance is related to F/M (food to microorganism ratio) and refers to the pounds of BOD applied per day per pound of VSS, where VSS is equated to the biomass concentration or to the sludge age, which is the average length of time the organisms are in the process:

$$\text{Sludge age} = \frac{\text{mass of organisms under aeration}}{\text{mass wasted/day}}$$

The performance, therefore, is related to F/M and the degradability (K), as shown in Figure 23. The biodegradation characteristics of various organics is shown in Table 14.

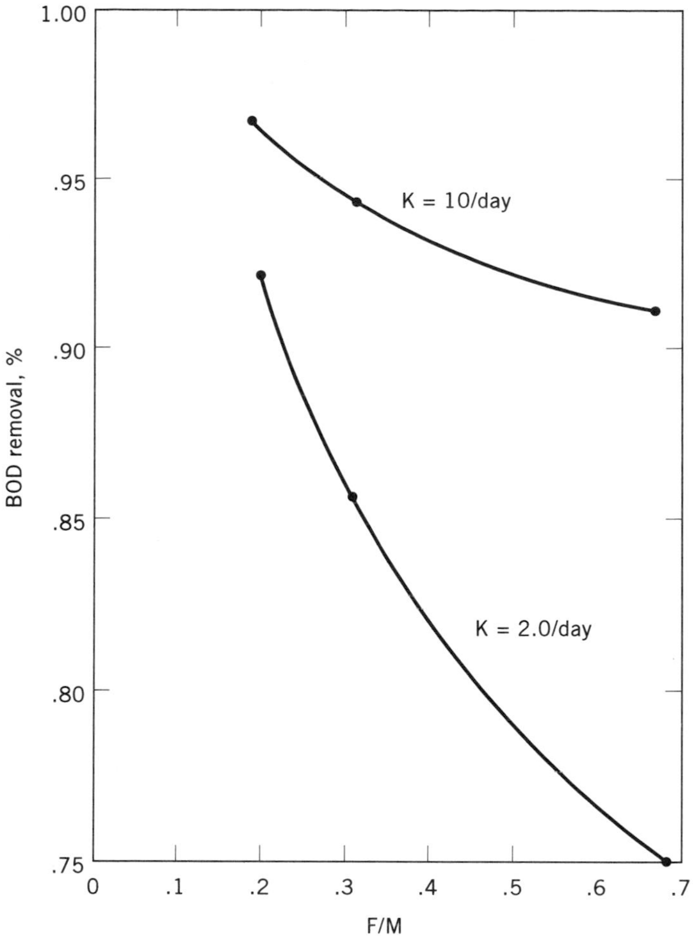

Figure 23. Relationship between F/M and organic removal for wastewaters of different degradability.

- The relationship between age and effluent quality is shown in Figure 24.

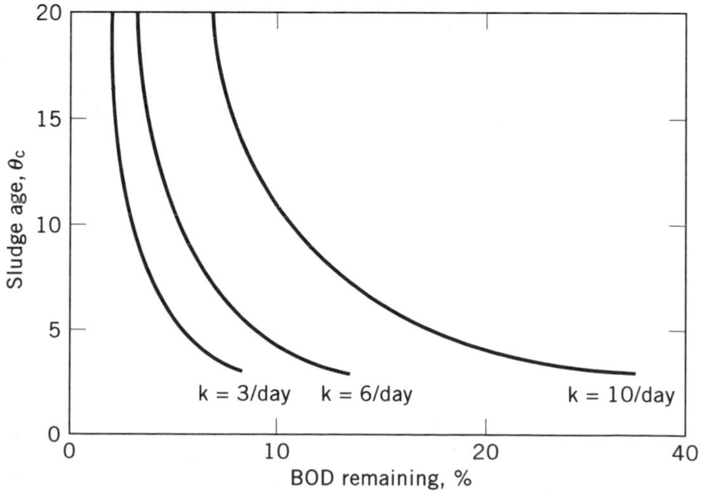

Figure 24. Effect of reaction rate on BOD removal at varying sludge ages.

Table 14. Relative Biodegradability of Certain Organic Compounds

Biodegradable Organic Compounds[a]	Compounds Generally Resistant to Biological Degradation
Acrylic acid	Ethers
Aliphatic acids	Ethylene chlorohydrin
Aliphatic alcohols (normal, iso, secondary)	Isoprene
Aliphatic aldehydes	Methyl vinyl ketone
Aliphatic esters	Morpholine
Alkyl benzene sulfonates with exception of propylene-based benzaldehyde	Oil
Aromatic amines	Polymeric compounds
Dichlorophenols	Polypropylene benzene sulfonates
Ethanolamines	Selected hydrocarbons
Glycols	Aliphatics
Ketones	Aromatics
Methacrylic acid	Alkyl-aryl groups
Methyl methacrylate	Tertiary aliphatic alcohols
Monochlorophenols	Tertiary benzene sulfonates
Nitriles	Trichlorophenols
Phenols	
Primary aliphatic amines	
Styrene	
Vinyl acetate	

[a] Some compounds can be degraded biologically only after extended period of seed acclimation.

- Process performance is affected by temperature. The reaction rate decreases with temperature over a range of 4–31°C. As the temperature decreases, dispersed effluent suspended solids increase; temperatures above 37°C may result in a dispersed floc and poor settling sludge.
- Sludge quality is defined by the sludge volume index (SVI). This is the volume occupied by one dry weight gram after settling for one-half hour and therefore defines the "bulkiness" of the sludge. A bulking sludge is usually caused by an excess of filamentous-type organisms. Filamentous organisms thrive best with readily degradable organics as a food source. If all things are maintained equal, ie, adequate O_2, N and P, and BOD, the floc-forming organisms will dominate. In order to maintain conditions favorable to the floc formers, adequate oxygen nutrients and BOD must diffuse through the floc and reach all the organisms. The filaments with a high surface-area-to-volume ratio can readily obtain these nutrients.
- As the oxygen uptake or F/M increases, the dissolved oxygen must be increased to provide sufficient driving force to penetrate the floc.
- Minimum concentrations of nitrogen and phosphorus are necessary in the effluent.
- High concentrations of BOD are necessary to penetrate the floc requiring a plug flow configuration for readily degradable wastewaters. Alternatively, a selector can be employed to absorb the readily degradable organics so they are not available as a food source for the filaments.

ALTERNATIVE BIOLOGICAL TREATMENT TECHNOLOGIES

Lagoons

Where large land areas are available, lagooning provides a simple and economical treatment for nontoxic or nonhazardous wastewaters. There are several lagoon alternatives:

- The impounding and absorption lagoon has no overflow or there may be an intermittent discharge during periods of high stream flow. These lagoons are particularly suitable to short seasonal operations in arid regions.
- Anaerobic ponds are loaded such that anaerobic conditions prevail throughout the liquid volume.

  One of the big problems with anaerobic ponds is the generation of odors. The odor problem can frequently be eliminated by the addition of sodium nitrate at a dosage equal to 20% of the applied oxygen demand. An alternative is the use of a stratified facultative lagoon in which aerators are suspended 3 m below the liquid surface in order to maintain aerobic surface conditions with anaerobic digestion occurring at the lower depths.
- Aerobic lagoons depend on algae to produce oxygen by photosynthesis. This oxygen in turn is used by the bacteria to oxidize the organics in the wastewater. Since algae are aerobic organisms, the organic loading to the lagoons must be sufficiently low to maintain dissolved oxygen.

Aerated Lagoons

An aerated lagoon system is a two- or three-basin system designed to remove degradable organics (BOD). The first basin is fully mixed, thereby maintaining all solids in suspension. This maximizes the organic removal rate. A second basin operates at a lower power level, thereby permitting solids to deposit on the bottom. The solids undergo anaerobic degradation and stabilization. A third basin is frequently employed for further removal of suspended solids and enhanced clarification. The process is shown in Figure 25.

Aerated lagoons are employed for the treatment of nontoxic or nonhazardous wastewaters such as food processing and pulp and paper.

Retention time varies from 3 to 12 days, so a large land area is usually required.

Activated Sludge

Variants

- *Complete Mix (CMAS).* Applicable to refractory-type wastewaters in which filamentous bulking is not a problem. Has the advantage of dampening fluctuations of influent wastewater quality (Fig. 26).
- *Plug Flow.* Applicable for readily degradable wastewaters subject to filamentous bulking. Requires upstream controls to avoid shock loadings (Fig. 26).
- *Selector.* Applicable for readily degradable wastewaters; also requires upstream controls. In a selector,

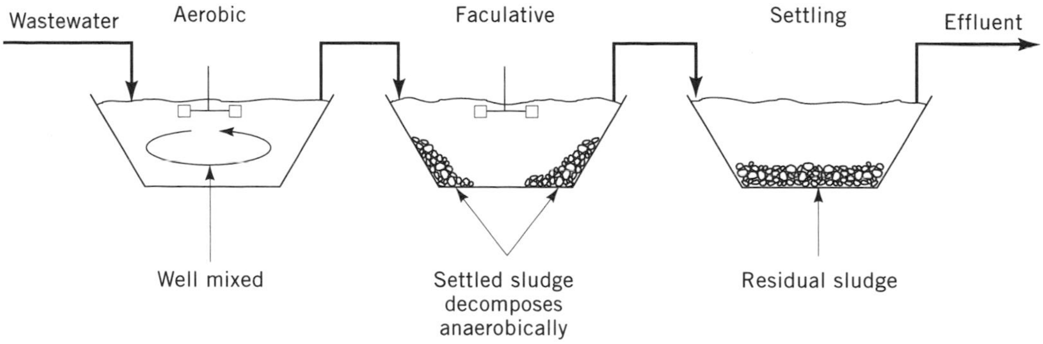

Figure 25. Aerated lagoon types.

degradable organisms are removed by the floc formers by biosorption and therefore not available as a food source for the filaments (Fig. 26).

- *Sequencing Batch Reactor (SBR) or Intermittent Process.* This is a combination of complete mix and plug flow and usually controls filamentous bulking. The nature of the process eliminates the need of an external clarifier (Fig. 27).
- *Oxidation Ditch.* This process is usually considered when nitrogen removal is required.

Performance

- Effluent quality is related to the sludge age; higher sludge ages are required for the more refractory wastewaters.

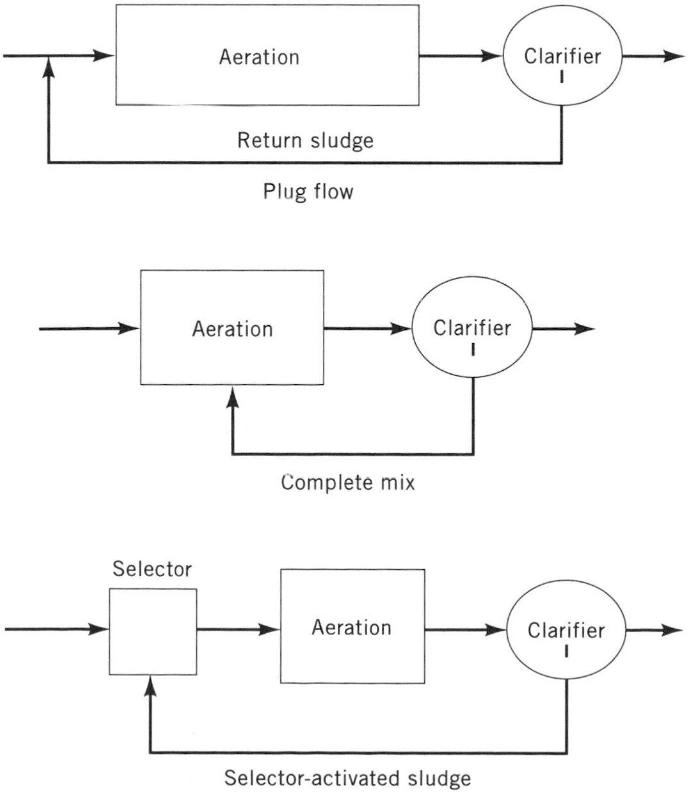

Figure 26. Types of activated sludge processes.

- Degradable priority pollutants can be reduced to microgram-per-liter levels under optimal operating conditions.
- Effluent soluble BOD levels below 10 mg/L are achievable in most cases.
- Nitrification and denitrification can be achieved through process modifications.

FIXED-FILM PROCESSES: TRICKLING FILTER

A trickling filter is a packed bed, usually plastic, on which a biofilm grows. As a wastewater passes over the film, organics and oxygen diffuse into the film where they undergo biodegradation, as shown in Figure 28. The variables affecting performance are the organic loading rate, the hydraulic loading rate, temperature, and the degradability of the wastewater. For the treatment of industrial wastewaters, a trickling filter is considered a pretreatment process usually designed to remove about 50% of the BOD. This is largely due to economic considerations. Trickling filter performance data are shown in Table 15.

ROTATING BIOLOGICAL CONTACTOR (RBC)

An RBC is a fixed-film process in which a biofilm is developed on a rotating plastic cylinder which passes through the wastewater. As the cylinder passes through, the wastewater organics diffuse into the film. As the cylinder passes through the air, oxygen diffuses into the biofilm, causing degradation of the organics. Increased treatment is achieved by increasing the number of stages.

ANAEROBIC TREATMENT (20)

- Complex organics are broken down through a sequence of reactions to end products of methane gas (CH_4) and carbon dioxide (CO_2), as shown in Figure 29.
- Since anaerobic treatment will not reach usual permit discharge levels, it is employed as a pretreatment process prior to discharge to a POTW or to a subsequent aerobic process. Therefore it is most applicable to high-strength wastewaters.

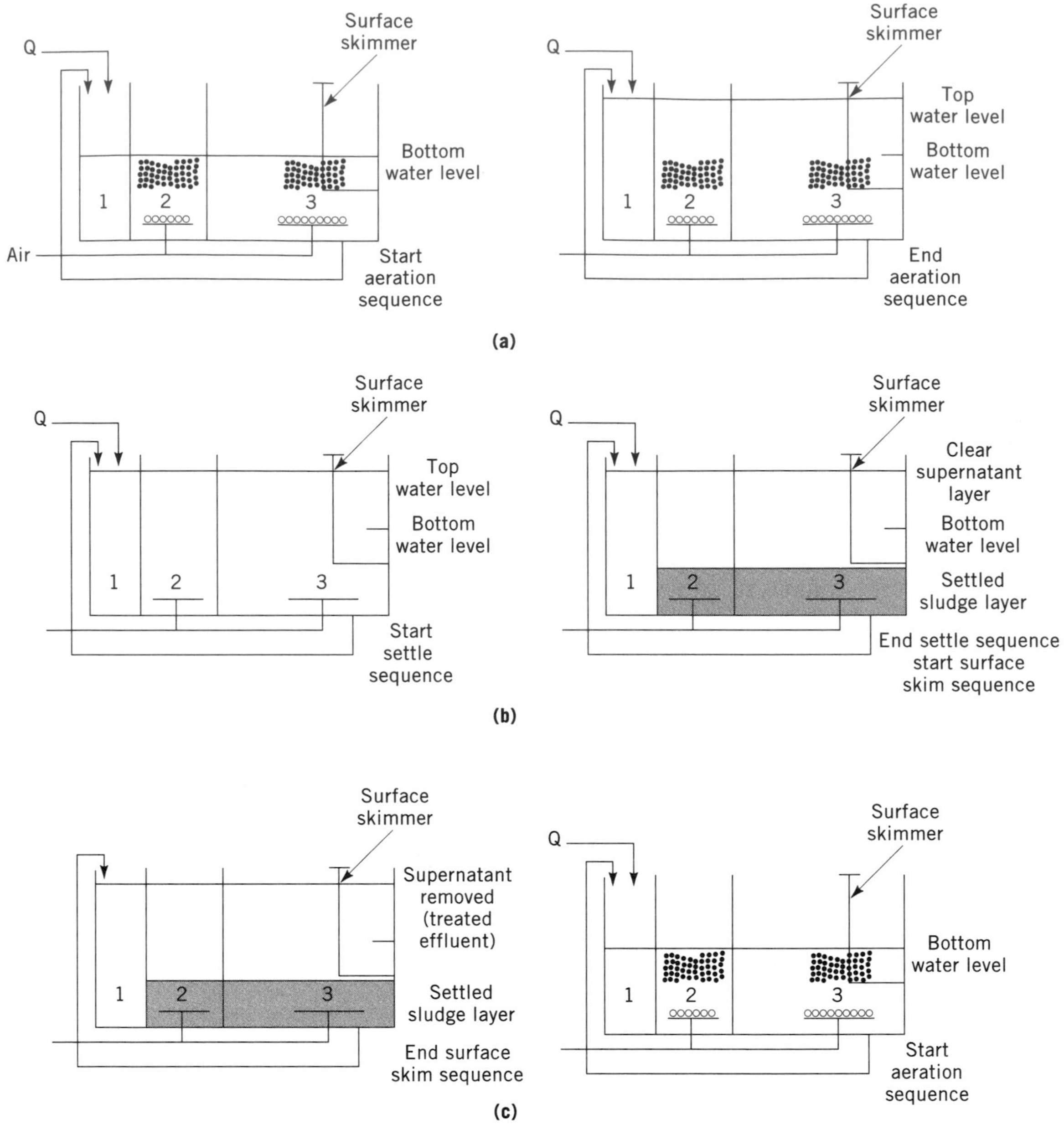

Figure 27. Schematic of cyclic activated sludge sequences: (**a**) aeration sequence; (**b**) settle sequence; (**c**) surface skin sequence.

- While aerobic treatment requires energy to transfer oxygen, anaerobic processes produce energy in the form of methane gas.
- Successful anaerobic process operation depends on maintaining a population of methane organisms. It is therefore critical that the SRT of the anaerobic sludge exceed the growth rate of the methane organisms. At 35°C the common design criteria is an SRT of 10 days or more.
- Anaerobic sludge can be maintained dormant for long periods of time, thereby making the process attractive for seasonal industrial operation such as in the food processing industry.
- A disadvantage to the anaerobic process is that initial startup may take as long as 45–60 days. Should the process be killed by a toxic shock, a long period will be required for a re-startup. Particular care must be taken, therefore, to avoid upset.

TYPES OF ANAEROBIC PROCESSES

There are five principal process variants which are propriety in nature:

- *Anaerobic Filter.* The anaerobic filter is similar to a trickling filter in that a biofilm is generated on a media. The bed is fully submerged and can be operated either upflow or downflow. For very high strength wastewaters, a recycle can be employed.

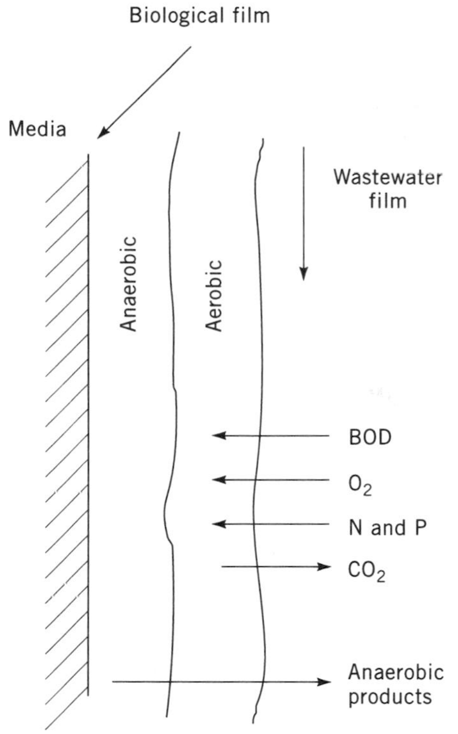

Figure 28. Removal mechanisms in a fixed-film reactor.

- *Anaerobic Contact.* This process can be considered as an aerobic activated sludge since sludge is recycled from a clarifier or separator to the reactor. Since the material leaving the reactor is a gas–liquid–solid mixture, a vacuum degasifier is required to separate the gas and avoid floating sludge in the clarifier.
- *Fluidized Bed.* This reactor consists of a sand bed on which the biomass is grown. Since the sand particles

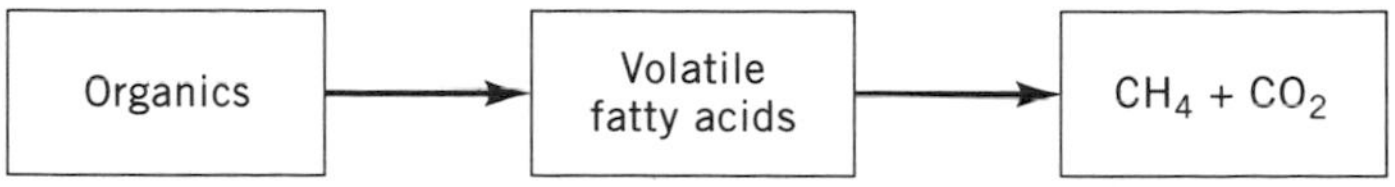

Figure 29. Rotating biological contactors.

are small, a very large biomass can be developed in a small volume of reactor. In order to fluidize the bed, a high recycle is required.
- *Upflow Anaerobic Sludge Blanket (UASB).* Under proper conditions anaerobic sludge will develop as high-density granules. These will form a sludge blanket in the reactor. The wastewater is passed upward through the blanket. Because of its density, a high concentration of biomass can be developed in the blanket.
- *ADI Process.* The ADI is a low-rate anaerobic process which is operated in a reactor resembling a covered football field. Because of the low rate, it is less susceptible to upset compared to the high-rate processes. Its disadvantage is the large land area requirements.

With the exception of the ADI process, anaerobic processes usually operate at a temperature of 35°C. In order to maintain this temperature, the methane gas generated in the process is used to heat the reactor. Anaerobic processes are shown in Figures 30 and 31. Anaerobic treatment performance data are shown in Table 16.

ADVANCED WASTEWATER TREATMENT

New regulations for toxics and priority pollutants frequently cannot be met by conventional technology. Other physical-chemical technologies must therefore be applied.

Table 15. High-Rate Trickling Filtration Performance

Waste	Hydraulic Loading MGAD	Depth (ft)	Raw BOD	Recycle Ratio	BOD Removal, Clarified (%)	Temperature	BOD Loading (lb/thousand ft^3)
Sewage	126	21.6	145	3	88		54
	252		131	3	82		110
	252		175	1	70		250
	63		173	0	78		95
	126		152	0	76		67
	189	10.8	166	0	45		549
	135		165	0	57		390
	95		185	0	51		304
Citrus	72	21.6	542	3	69		199
	189		464	2	42		612
Citrus and sewage	189	21.6	328	2	53		384
Kraft mill	365	18.0	250	0	10	34	
	185	18.0	250	0	24	36	
	200	21.6	250	0	23	40	
	90	21.6	250	0	31	33	
Black liquor	47	18	400	0	73	24	200
	95	18	400	0	58	29	380
	189	18	400	0	58	35	780

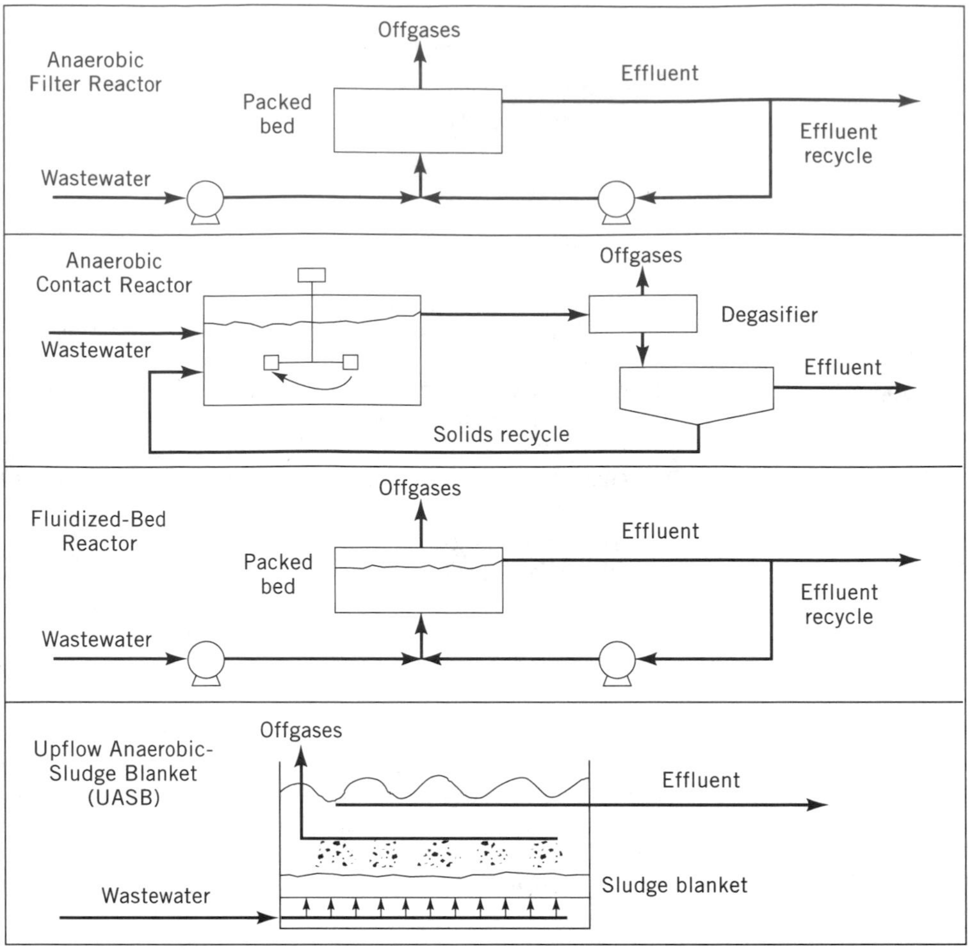

Figure 30. Mechanism of anaerobic digestion.

Chemical Oxidation (21)

Chemical oxidation can be applied in industrial wastewater pretreatment for reduction of toxicity, to oxidize metal complexes to enhance heavy metals removal from wastewaters, or as a posttreatment for toxicity reduction or priority pollutant removal:

- Complex organics and toxics are chemically oxidized to end products of CO_2 and H_2O or to intermediate products which are nontoxic and biodegradable.
- The common oxidants are ozone, hydrogen peroxide (H_2O_2) catalyzed usually with ferrous ion (Fe^{2+}) and in some cases chlorine dioxide and ultraviolet (UV) light. So-called advanced oxidation systems include H_2O_2 + uv, ozone + uv, and H_2O_2, ozone, and uv.

Depending on the application, the oxidation can be complete to end products as in a contaminated groundwater or partial to degradable intermediate products as in a process wastewater. Performance with H_2O_2 is shown in Table 17.

Carbon Adsorption

Carbon can be employed either as granular carbon in columns (GAC) or as powdered carbon added to an activated sludge plant (PACT).

Carbon removes most organics except low-molecular-weight soluble organics such as sugars and alcohols. In general, those organics which adsorb the poorest, biodegrade the best while those which biodegrade poorly adsorb well on carbon.

Design data are available on the specific organics on the priority pollutant list (USEPA) (22). For mixed wastewaters, a laboratory study is needed to determine adsorption characteristics. Wastewater is contacted with a range of concentrations of powdered carbon and adsorption, graphed in the form of a Freundlich isotherm, as shown in Figure 32.

There are several commercial carbons available so that a comparative isotherm study should be made for a wastewater to determine the best carbon.

Granular carbon is regenerated. Regeneration can be accomplished by acid, caustic, solvent, steam, or thermal. For specific organics it is usually feasible to employ one of the first four and operate in situ. For a typical multicomponent wastewater, however, thermal regeneration must be used. This can be accomplished using a multiple hearth furnace or a fluidized bed furnace. Attrition and oxidation losses will range from 5 to 10% by weight. In addition, there will frequently be a capacity loss, particularly for low-molecular-weight organics. A typical GAC system is shown in Figure 33.

GAC can frequently be employed for secondary effluent toxicity reduction, particularly in the case of high-molecu-

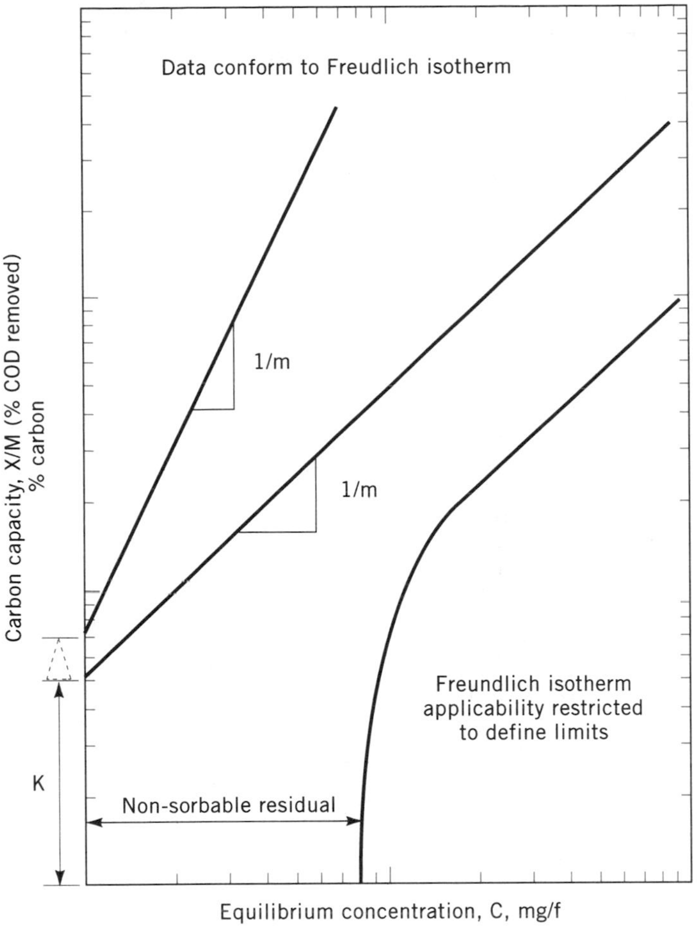

Figure 31. Anaerobic wastewater treatment processes.

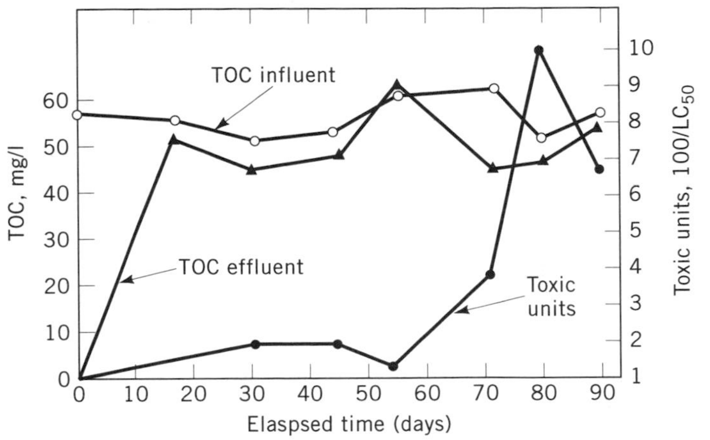

Figure 32. Freundlich adsorption isotherm.

lar-weight oxidation by-products which are strongly adsorbed on carbon, as shown in Figure 34.

PAC is added prior to or directly to the activated sludge aeration basin, as shown in Figure 35. The sludge is therefore a mixture of biomass and carbon. The degradable organics are biodegraded and the nondegradable organics adsorbed on the carbon. Performance data are shown in Table 18.

PAC sludge can be regenerated by wet air oxidation (WAO) or by a multiple hearth furnace. Capacity losses might be high in WAO, particularly with low-molecular-weight organics. Weight loss in a furnace may exceed 20%.

SLUDGE HANDLING AND DISPOSAL

Types of Sludges

Municipal primary sludge consists of organic and inorganic particulates. The sludge must be stabilized before land disposal.

Biological sludge consists of organisms and other particulates not degraded in the biological process.

Chemical sludges consist of chemical precipitates, heavy metals, and other contaminants such as color precipitated from industrial wastewaters.

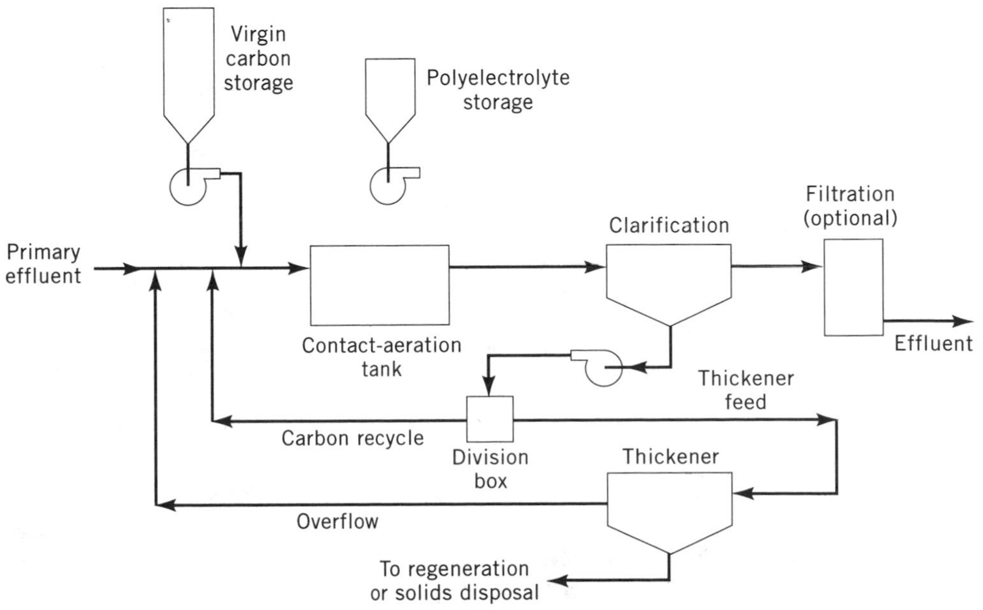

Figure 33. GAC process flow sheet.

Table 16. Performance of Anaerobic Processes

Wastewater	Process	Loading [kg/(m^3·d)]	HRT (h)	Temperature (°C)	Removal (%)
	Anaerobic Contact				
Meat packing		3.2 (BOD)	12	30	95
Meat packing		2.5 (BOD)	13.3	35	95
Keiring		0.085 (BOD)	52.4	30	59
Slaughter house		3.5 (BOD)	12.7	35	95.7
Citrus		3.4 (BOD)	32	34	87
	Upflow Filter				
Synthetic		1.0 (COD)	—	25	90
Pharmaceutical		3.5 (COD)	48	35	98
Pharmaceutical		0.56 (COD)	36	35	80
Guar gum		7.4 (COD)	24	37	60
Rendering		2.0 (COD)	36	35	70
Landfill leachate		7.0 (COD)	—	25	89
Paper-mill foul condensate		10–15 (COD)	24	35	77
	Expanded Bed				
Synthetic		0.8–4.0 (COD)	0.33–6	10–3	80
Paper-mill foul condensate		35–48 (COD)	8.4	35	88
	USAB				
Skimmed milk		71 (COD)	5.3	30	90
Sauerkraut		8–9 (COD)	—	—	90
Potato		25–45 (COD)	4	35	93
Sugar		22.5 (COD)	6	30	94
Champagne		15 (COD)	6.8	30	91
Sugar beet		10 (COD)	4	35	80
Brewery		95 (COD)	—	—	83
Potato		10 (COD)	—	—	90
Paper mill foul condensate					
	ADI-BFV				
Potato		0.2 (COD)	360	25	90
Corn starch		0.45 (COD)	168	35	85
Dairy		0.32 (COD)	240	30	85
Confectionary		0.51 (COD)	336	37	85

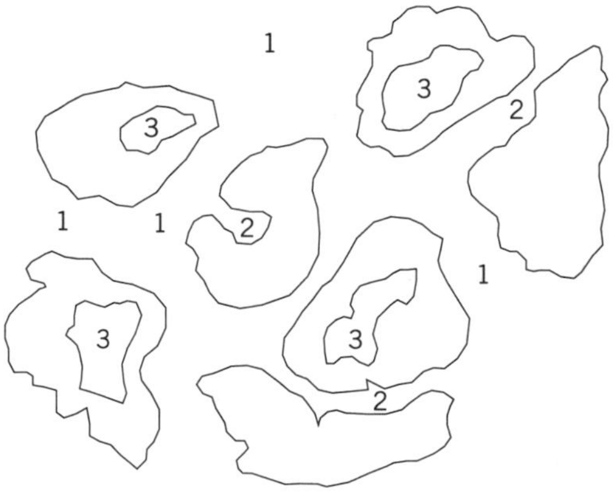

Figure 34. Toxicity removal by GAC.

Most wastewater sludges contain free water, capillary water, and bound water, as shown in Figure 36. Free water can be removed by thickening and capillary water by dewatering. Bound water cannot be removed by conventional dewatering processes, as shown in Figure 37.

Sludge Stabilization

Organic sludges need to be stabilized before ultimate disposal, except in the case of incineration. This is usually achieved by either aerobic or anaerobic digestion. In aerobic digestion, the degradable volatile solids are liquefied and oxidized to CO_2 and H_2O. In anaerobic digestion the solids are liquefied and fermented to CH_4 and CO_2.

Sludge Thickening

Thickening of sludge usually precedes dewatering. Depending on the nature of the sludge, several techniques

Table 17. Hydrogen Peroxide Oxidation of Organics

Compound (mg/L)	Initial Concentration	Percent Reduction COD	Percent Reduction TOC	LC_{50} (%) Before Oxidation	LC_{50} (%) After Oxidation	COD Reduction in 2 Days (%) Before Oxidation	COD Reduction in 2 Days (%) After Oxidation
Nitrobenzene	616	72	38	6.0	76.2	59	31
Aniline	466	77	43	35.7	NT	0	40
o-Cresol	541	75	56	2.5	NT	16	51
m-Cresol	541	73	38	1.3	NT	0	51
p-Cresol	541	72	40	0.4	NT	65	47
o-Chlorophenol	625	75	48	5.1	NT	18	37
m-Chlorophenol	625	75	41	1.8	NT	0	39
p-Chlorophenol	625	76	22	0.3	NT	0	39
2,3-DCP	8.5	70	53	1.0	NT	12	31
2,4-DCP	815	69	50	0.6	NT	9	32
2,5-DCP	815	74	42	1.9	NT	14	38
2,6-DCP	815	61	33	5.7	17.3	0	9
3,5-DCP	815	69	49	0.5	NT	0	9
2,3-DNP	921	80	51	6.3	85.6	0	19
2,4-DNP	921	73	51	2.0	NT	0	49
2,4,6-TCP	800	47	44	2.8	52.2	0	39

Conditions: stoichiometric dosage of H_2O_2, pH 3.5, 50 mg/L Fe^{2+}; NT = not toxic.

Table 18. Wastewater Treatment with Powdered Activated Carbon

	Wastewater Composition (mg/L) BOD	TOC	TSS	Color	Cu	Cr	Ni	Bioassay[a] (LC_{50})
Influent	320	245	70	5,365	0.41	0.09	0.52	—
Biotreatment	3	81	50	3,830	0.36	0.06	0.35	11
+50 mg/L PAC	4	68	41	2,900	0.30	0.05	0.31	25
+100 mg/L PAC	3	53	36	1,650	0.18	0.04	0.27	33
+250 mg/L PAC	2	29	34	323	0.07	0.02	0.24	>75
+500 mg/L PAC	2	17	40	125	0.04	<0.02	0.23	>87

[a] Percentage of wastewater in which 50% of aquatic organisms survive for 48 h.

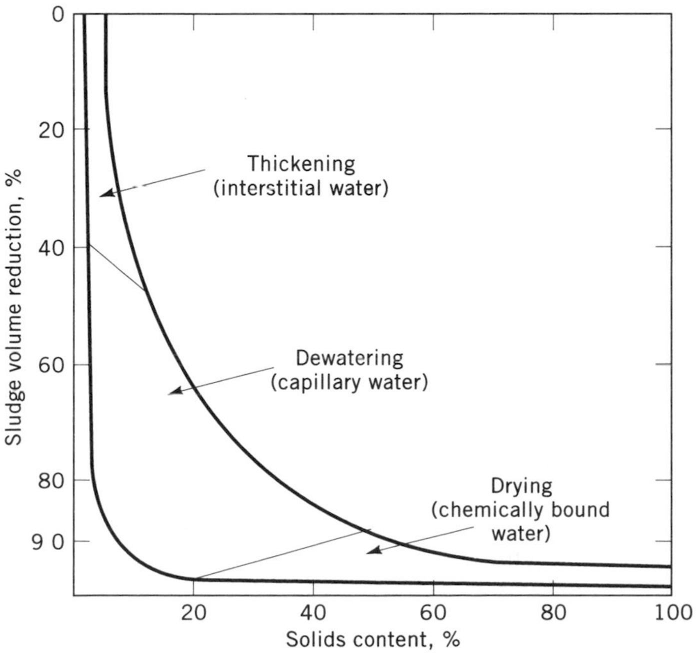

Figure 35. PACT wastewater treatment system general process diagram.

are available for thickening:

- Gravity thickening, applicable to primary municipal sludges and most chemical sludges.
- A technique in which the sludge is passed in a thin sheet over a porous drainage belt. This is particularly applicable to waste-activated sludge.
- Dissolved air flotation (DAF), the same principle as previously described in which air bubbles float the sludge which is then removed by a scraper. Generally applicable to large volumes of waste-activated sludge.
- Centrifuge: Various centrifuge types are used for sludge thickening.

Sludge Dewatering

- Centrifuge: The solid bowl centrifuge concentrates the solids under centrifugal force. Both centrate and cake solids are continuously discharged from the machine. Polymer addition is required for most wastewater sludges. A typical centrifuge is shown in Figure 38.
- Vacuum filter is a cloth-covered drum which operates under an applied vacuum. As the drum passes

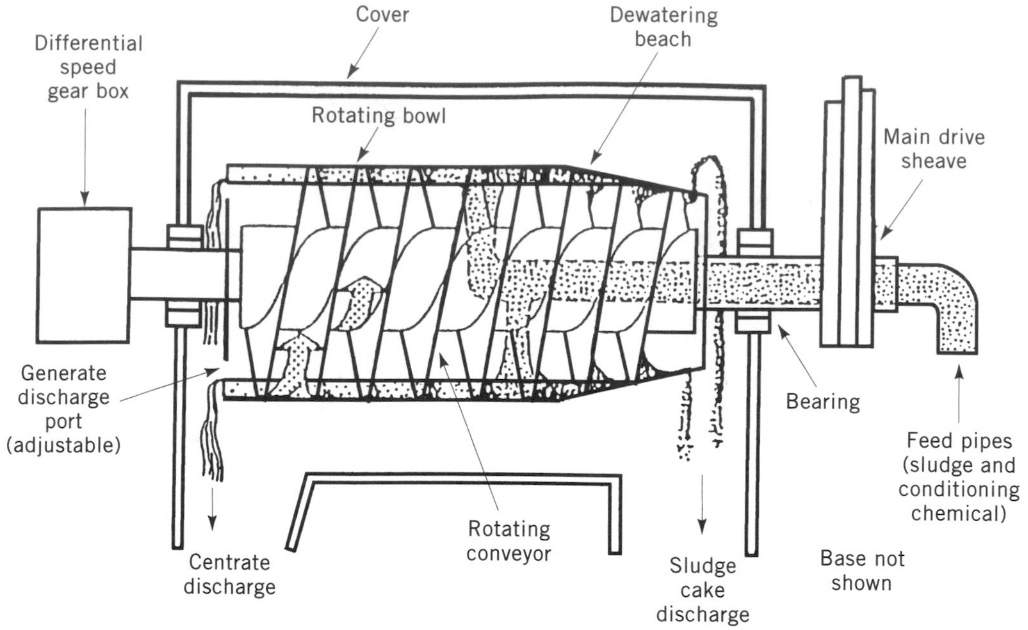

Figure 36. Forms of water in sludges: (1) free water; (2) capillary water; (3) intercellular water.

through the sludge vat, solids are deposited on the filter. As the drum passes through the air, drying of the cake occurs. The cake is continuously discharged to a conveyor belt.

- A belt filter press consists of a gravity drainage belt followed by a series of roller presses which squeeze out water. A belt press is shown in Figure 39.
- A pressure filter is a plate and frame press which operates on an intermittent time cycle. Drier cakes are generally attainable from a filter press.
- Sludge-drying beds are usually used for smaller sludge volumes, which drain and dry rapidly. Their application is usually restricted to the more arid climates.

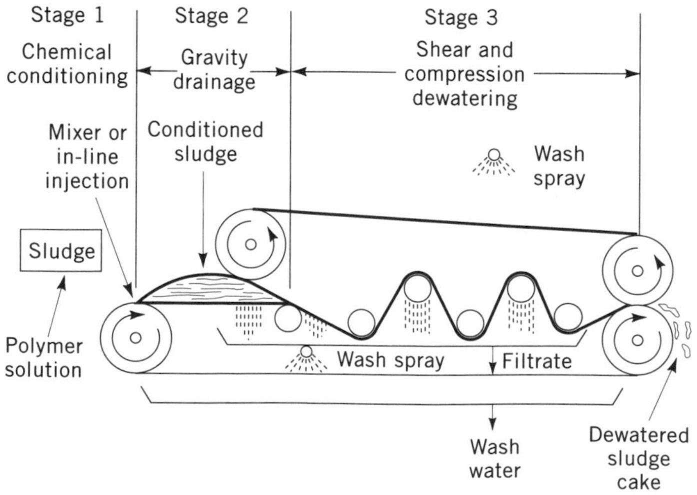

Figure 37. Mechanisms of sludge dewatering.

Thickening and dewatering of wastewater sludges are summarized in Table 19 (9).

Sludge Disposal

Land disposal of wet sludges can be accomplished in a number of ways: by lagooning or the application of liquid sludge to land by truck or a spray system or by pipeline to a remote agricultural or lagoon site.

In lagoons sludge is stored and in the case of organic sludges anaerobically digested. Odor control is achieved either by chemical addition to the overlaying water (Cl_2 or H_2O_2) to oxidize sulfides or by installing aerators in the liquid layer to maintain aerobic conditions.

Biological sludges can be incorporated into the soil. An important consideration is the heavy metal content of the sludge, which will dictate the total years sludge can be applied. The available nitrogen content of the sludge will determine the maximum yearly application.

Dewatered sludges can be employed as a landfill.

Incineration can be accomplished in multiple hearth furnaces in which the sludge passes vertically through a series of hearths, as shown in Figure 40. In a fluidized bed sludge particles are fed into a bed of sand fluidized by upward moving air.

STORMWATER CONTROL

In most industrial plants, it is now necessary to contain and control pollutional discharges from stormwater. Pollutional discharges can be minimized by providing adequate diking around process areas, storage tanks, and liquid transfer points with drainage into the process sewer. Contaminated stormwater is usually collected based on a frequency for the area in question (eg, a 10-year storm) in

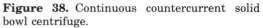

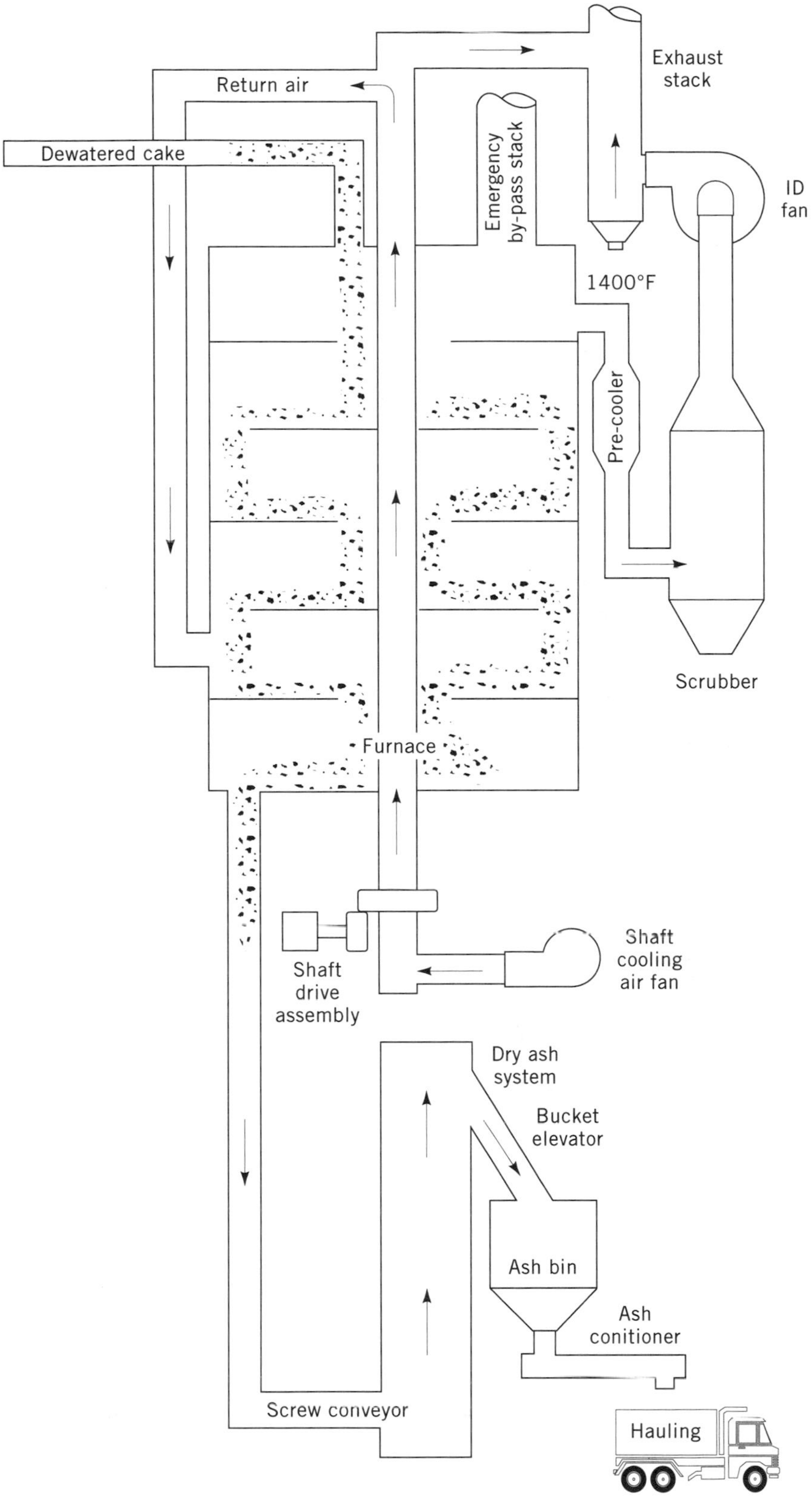

Figure 38. Continuous countercurrent solid bowl centrifuge.

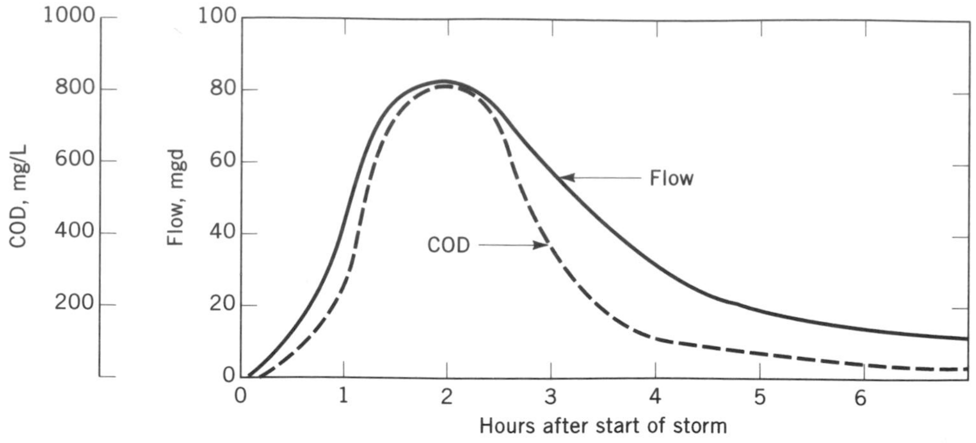

Figure 39. Three basic stages of a belt filter press.

a holding basin. The collected water is then passed through the wastewater treatment plant at a controlled rate.

PROCESS SELECTION

In the treatment of wastewaters it is usually more cost effective to employ biological wastewater treatment where possible. Physical-chemical technologies are employed in those cases where the wastewater is toxic or nonbiodegradable. While these technologies can be applied as a posttreatment (following a biological process), it is usually more effective to employ source treatment before the biological process. A protocol has been developed to select the most cost-effective technology.

The protocol is applied to each individual wastewater stream within the industrial complex.

If the wastewater contains volatile organics (eg, benzene) or heavy metals, these are removed prior to following the protocol. The wastewater is tested for priority pollutants and an aquatic bioassay. If the wastewater proves to be toxic and nonbiodegradable, it is removed for source treatment. If it is biodegradable, priority pollutants and an aquatic bioassay is again run following biodegradation. If the biologically treated effluent is still toxic or if priority pollutant limitations have not been met, the wastewa-

Table 19. Thickening and Dewatering of Wastewater Sludges

Equipment	Type of Sludge	Loading	Cake Solids (%)	Chemicals, Polymer (lb/ton)
Thickening				
Gravity	WAS	5–6 lb/(ft^2·d)	2.5–3.0	None
Gravity	Pulp and paper			
	53% P, 47% WAS	25	4	None
	67% P, 33% WAS	25	6	None
	100% P	25	9	None
Dissolved air flotation	WAS	2.9–4.5 lb/(ft^2·h)	4–5.7	Low dosage
Solid bowl centrifuge	WAS	75–100 gal/min	5–7	None
Basket centrifuge	Citrus, WAS	25–40 gal/min	9–10	10–20
Gravity belt	WAS	315 gal/min	5.5	39
Solid bowl centrifuge	Paper mill, WAS	100 gal/min	11	10
Solid bowl centrifuge	Chemical, WAS	—	7–9	5–10
Dewatering				
Basket centrifuge	Citrus, WAS	25–40 gal/min	9–10	10–20
Basket centrifuge	Paper mill, WAS	60 gal/min	11	5
Belt press	Citrus, WAS	40 gal (min·m)	18	10–20
Belt press	Paper mill, WAS	70 gal/(min·m)	16	6.5
Belt press	Chemical, WAS	—	13–15	10–20
Belt press	Organic chemical, WAS	190 gal/min	15	25
Belt press	Deinking primary	500 L/m	37	4
Belt press	Bleached and unbleached kraft; 67% P, 33% WAS	240 L/m	27	12
Belt press	Kraft linerboard WAS	75 L/m	19	25

Note:
lb/(ft^2·d) = 4.88 kg/(m^2·d)
gal/min = 3.78 × 10^{-3} m^3/min

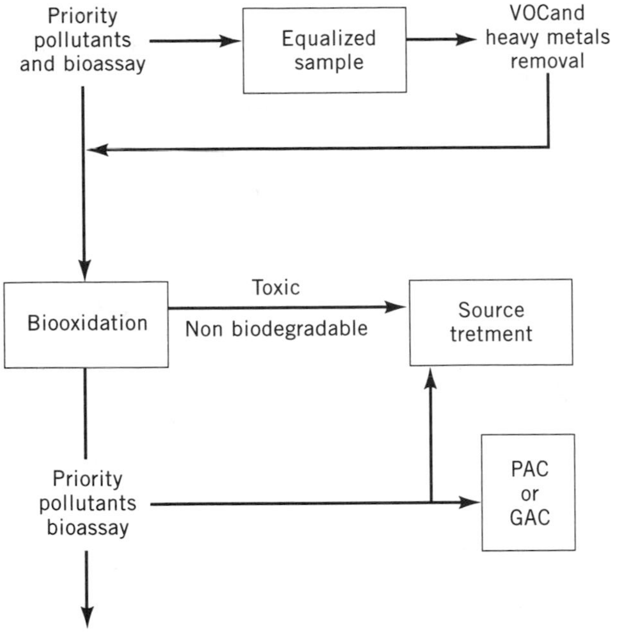

Figure 40. Multiple-hearth system.

ter can either by evaluated for source treatment or tertiary treatment using powdered or granular carbon.

BIBLIOGRAPHY

1. S. R. Weibel et al., "Urban Land Runoff as a Factor in Stream Pollution," *J. Water Pollut. Control Fed.* **36,** 914 (1964).
2. R. O. Sylvester, "Nutrient Content of Drainage Water from Forested, Urban and Agricultural Areas," *Trans. Semin. Algae Metropolitan Waters*, U.S. Public Health Service, Cincinnati, Ohio, 1962.
3. R. V. Thomann and J. A. Mueller, *Principles of Surface Water Quality Modeling and Control*, Harper & Row, New York, 1987.
4. U.S. Environmental Protection Agency (EPA), *USEPA Report of the Committee on Water Quality Criteria*, EPA, Washington, D.C., 1972.
5. G. B. McBride, "Nomographs for Rapid Solutions for the Streeter-Phelps Equation," *J. Water Pollut. Control Fed.* **54**(4), 378 (1982).
6. P. A. Krenkel and V. Novotny, *Water Quality Management*, Academic, New York, 1984.
7. R. V. Thomann, *Systems Analysis and Water Quality Management*, McGraw-Hill, New York, 1972.
8. U.S. Environmental Protection Agency (EPA), *USEPA Manual for Waste Minimization Opportunity Assessments*, EPA 600/2-88/025, EPA, Washington, D.C., 1988.
9. W. W. Eckenfelder, *Industrial Water Pollution Control*, 2nd ed., McGraw-Hill, New York, 1989.
10. W. W. Eckenfelder, *Principles of Water Quality Management*, CBI Publishing, Boston, 1980.
11. R. A. Conway and R. D. Ross, *Handbook of Industrial Waste Disposal*, Van Nostrand Reinhold, New York, 1980.
12. Metcalf & Eddy, *Wastewater Engineering Treatment, Disposal and Reuse*, 3rd ed., McGraw-Hill, New York, 1991.
13. D. L. Ford, "Application of the Total Carbon Analyzer for Industrial Wastewater Evaluation" in *Proceedings 23rd Industrial Waste Conference*, Purdue University, 1968.
14. P. W. Lankford and W. W. Eckenfelder (eds.), *Toxicity Reduction in Industrial Effluents*, Van Nostrand Reinhold, New York, 1990.
15. J. W. Patterson, *Industrial Wastewater Treatment Technology*, Butterworth, Boston, 1985.
16. American Water Works Association (AWWA), *Air Stripping for Volatile Organic Contaminant Removal*, AWWA, Denver, Colo., 1981.
17. C. W. Randall et al., *Design and Retrofit of Wastewater Treatment Plants for Biological Nutrient Removal*, Technomic Publishing, Lancaster, Pa., 1992.
18. R. Sedlak, *Phosphorus and Nitrogen Removal from Municipal Wastewater*, 2nd ed., Lewis Publishers, 1991.
19. W. W. Eckenfelder and P. Grau, *Activated Sludge Process Design and Control-Theory and Practice*, Technomic Publishing, Lancaster, Pa.
20. J. F. Malina and F. G. Pohland, *Design of Anaerobic Processes for the Treatment of Industrial and Municipal Wastes*, Technomic Publishing, Lancaster, Pa., 1992.
21. *Chemical Oxidation Technology for the Nineties*, Technomic Publishing, Lancaster, Pa., 1992.
22. U.S. Environmental Protection Agency (EPA), *EPA Carbon Adsorption Isotherms for Toxic Organics*, EPA-600/8-80-023, EPA, Washington, D.C., April 1980.
23. J. W. Patterson and J. P. Menez, "Equalization Basin Design," *Am. Inst. Chem. Eng. Environ. Progr.* **3,** 2 (1984).

WAVE POWER

GEORGE HAGERMAN
SEASUN Power Systems
Alexandria, Virginia

Considering that three-quarters of the earth's surface is covered by waters that are seldom still, ocean waves have tremendous potential as a renewable source of energy. Yet most people know little of the technology that has been developed to harness this resource, its possible environmental impacts, or the problems of integrating wave power into onshore utility grids.

Wave energy is commonly associated with maritime disasters and coastal destruction, and the prospect of putting it to constructive use seems formidable. Yet should a dolphin ever care to think along such lines, the prospect of designing energy conversion equipment to survive the radiation and temperature extremes of a desert or the dusty, gusty winds of a mountain pass must seem no less daunting. For this perspective, the author is indebted to Stephen Salter, who articulately catalogued the hostilities faced by engineers on dry land, during the IUTAM Symposium on Hydrodynamics of Ocean Wave-Energy Utilization, held in Lisbon, Portugal, July 1985. It is therefore not surprising that civilization has made greater progress in harnessing the sun and winds on land than waves at sea.

Several overviews have been written about wave power. Among these, the most comprehensive are books by McCormick (1) and Shaw (2) and, more recently, reports by Carmichael et al. (3) and Claeson (4). The books by McCormick and Shaw describe wave energy devices that were under development in the late 1970s. They also derive mathematical relationships that describe device response to wave action. Although many of the devices reviewed in these texts are no longer being actively developed, the theoretical concepts are still generally applicable.

Carmichael et al.'s study (3), funded by the Electric Power Research Institute (EPRI), performed economic evaluations of several ocean energy technologies in U.S. waters, including wave, tidal, and ocean thermal energy conversion. For the wave example, the design of a leading British device (the SEA Clam) was adapted to average U.S. East Coast and West Coast wave conditions. Since then, the SEA Clam design has undergone major reconfiguration, resulting in much improved economics. Cost information for the new design was not available at the time the EPRI report was prepared (1985).

The review by Claeson (4) is more recent (1987) and broad, covering wave energy research and development in nine countries (Norway, Sweden, Denmark, the United Kingdom, Ireland, Japan, China, and the United States). In his report, five British devices, two Norwegian devices, and a Swedish device are compared in terms of capital cost, operating and maintenance costs, annual energy production, and the cost of energy. One difficulty with such comparisons is that they are based on plant designs developed for different wave climates. They also reflect international differences in material and labor costs, discount rates, and varying assumptions about plant availability.

This article updates the technology descriptions of the previous works and greatly expands the topics covered, particularly system economics, utility integration issues, and potential environmental impacts. Much of the latter information was derived from regional wave energy studies that the author has conducted for North Carolina (5), northern California (6), and Hawaii (7). Far more information was developed in these studies than is possible to present in this article. Readers interested in one of these geographic regions are encouraged to obtain a copy of the appropriate final report from the sponsoring organization, as cited in the bibliography.

An appreciation of the natural processes that generate, propagate, and dissipate ocean waves is fundamental to understanding the technology that has been developed to convert this energy into useful work. Such an appreciation also leads to a better perspective on the potential applications of wave power and utility integration issues. This article therefore begins by characterizing ocean waves as an energy resource. See also ELECTRIC POWER GENERATION; RENEWABLE RESOURCES.

RESOURCE CHARACTERIZATION

Ocean waves are a derived form of solar energy, with wind being the agent that transfers the sun's energy to the sea surface. Simply stated, unequal solar heating of the earth generates wind, and wind blowing over water generates waves.

Only 0.01% of the earth's solar energy flux is ultimately expressed as wave power. While sunlight falls on the planet at a global rate of 10^{16}–10^{17} W, wave energy is dissipated against the world's coastlines (and renewed by the wind) at a global rate of 10^{12}–10^{13} W (8).

A unique property of ocean waves is their ability to travel vast oceanic distances with negligible loss of energy. Even the longest waves do not begin to "feel the bottom" until they enter water depths of 300 m or less. Consequently, wave energy generated anywhere within an ocean basin ultimately arrives at some island or continental margin of that basin, virtually undiminished.

Thus, while solar radiation and winds are distributed over the planet's entire surface area, wave energy is gathered along its coastlines, which total 336,000 km in length. At a global renewal rate of 2.7×10^{12} W (9), the average wave energy flux worldwide is 8 kW/m of shoreline. By comparison, the annual average flux of solar or wind energy is generally less than 300 W/m^2 of panel area or rotor swept area. Therefore, the energy density of ocean waves is an order of magnitude greater than the natural processes that generate them.

The geographic distribution and temporal variability of wave energy resources are governed by the major wind systems that generate ocean waves: extratropical storms and trade winds. In some areas, notably India, local monsoons can also influence the wave climate.

Geographic Distribution

Extratropical cyclones are born as prevailing westerly winds off continental land masses pick up heat and moisture from western ocean boundary currents such as the Gulf Stream and the Kuroshio Current. These low-pressure systems typically develop sustained wind speeds up to 25 m/s, blowing over a 1000-km fetch for 2–4 days before the storm makes landfall.

In the Northern Hemisphere, extratropical cyclones follow northeasterly tracks, continually building the waves in the storm's southern sector, which are traveling in the same direction as the storm. Consequently, wave energy resources are quite good along the eastern margins of an ocean basin.

As waves grow, their period increases as does their speed. Eventually they travel faster than the storm and leave the region of active wind generation, at which point they are referred to as swell. Such swell arrives ahead of the storm and can give rise to high wave energy levels even when local winds are calm.

On the other hand, waves generated in the northern sector of a Northern Hemisphere cyclone travel opposite the direction of storm advance and have much less exposure to the storm's wind energy. Consequently, swell traveling "backward" from such a storm has much less power than swell leaving the storm's southern sector. As a result, wave resources along the western side of an ocean basin are generally poorer than on the eastern side.

In the North Atlantic Ocean, for example, annual average wave power along the edge of North America's eastern continental shelf ranges from 10 to 20 kW/m. By comparison, shelf-edge wave power off the exposed western coasts of the British Isles averages 60–70 kW/m, decreasing farther south, to about 30 kW/m off central Portugal.

A similar pattern occurs in the North Pacific Ocean. On the western side of the basin, off Taiwan and Japan, wave power averages 5–15 kW/m. On the opposite side, off the coast of northern California, it ranges from 25 to 35 kW/m.

In the Southern Hemisphere, extratropical low-pressure systems develop in the open-ocean expanse that surrounds Antarctica. These storms travel from west to east, uninterrupted by land, and generate high-energy swell that makes landfall along the southwest coasts of South America, Africa, Australia, and New Zealand. This swell also contributes significantly to the wave energy resources of Indonesia and island nations throughout the South Pacific Ocean. As in the Northern Hemisphere, incident wave energy flux decreases with decreasing latitude, averaging more than 100 kW/m just south of New Zealand, dropping to 30–40 kW/m in deep water west of Auckland. Even farther north, in the South Pacific region framed by Vanuatu, Tongatapu, Rarotonga, and Tuvalu, wave power densities at exposed island locations are estimated to be 15–20 kW/m.

As waves leave deep water and travel shoreward across island and continental shelves, they experience refraction and bottom drag. Refraction is the process whereby waves crossing the shelf at an angle bend toward the coast as they enter shoaling water. This spreads adjacent wave rays, which decreases the power per unit width of wave crest. Bottom drag dissipates wave energy through turbulence created at subsurface water motion encounters an unmoving seafloor.

Both processes can lead to a substantial reduction in the wave energy resource. Off northern California, for example, wave power densities in deep water beyond the shelf edge vary from 35 kW/m near the Oregon border to 25 kW/m off Point Conception, 1000 km to the south. By the time waves reach the 10-m-depth contour, their coastwide average power density has dropped to 12 kW/m, and its local distribution over tens of kilometers is quite variable, ranging from 1 kW/m where the shoreline faces south or is sheltered by rocky headlands to 18 kW/m along exposed open coasts (6).

For a given set of wave conditions, refraction is governed primarily by water depth, whereas the amount of energy lost to bottom drag depends not only on depth but also on the horizontal distance that a wave travels. Off North Carolina, for example, average wave power at Cape Hatteras, where the continental shelf is approximately 30 km wide, is more than twice as high as it is in the same water depths farther north, where the shelf is 80 km wide (10). Thus for a given ocean region, the nearshore wave energy resource will be higher along island and continental coasts fronted by narrow shelves.

Temporal Variability

The above estimates represent annual average conditions, yet like the sun and wind, ocean waves are an intermittent energy resource. In order to appreciate some of the differences among the 12 conversion processes described later in this article, it is important to understand wave energy's variability over several time scales:

- Wave to wave (seconds)
- Wave group to wave group (minutes)
- Sea state to sea state (hours to days)
- Season to season (months)
- Power plant service life (years)

Sea states are often described in terms of two statistical parameters: significant wave height and dominant wave period. Significant wave height is defined as the average height of the highest one-third waves. This approximates the height that a shipboard observer will report from visual inspection of the sea state, since such an observer tends to overlook the smaller, less conspicuous waves. Dominant wave period is defined as the period corresponding to the peak in the variance density spectrum of sea surface elevation. It thus represents the harmonic frequency component having the greatest amount of energy in a random wave train. Some wave energy devices can be "tuned" to this frequency in a manner analogous to the tuning of a radio circuit in an electromagnetic wave field.

Although a statistically stationary sea state (ie, waves neither building nor decaying) can be described by its significant wave height and dominant wave period, individual wave height and period vary more or less randomly from one wave to the next. Such variability is more random in seas generated by strong local winds and less random in swell that arrives from far distant storms.

Waves often occur in successive groups of alternately high and low waves, a phenomenon familiar to any surfer. The output of a wave energy device will tend to follow the envelope profile of such wave groups. Incorporating minutes-long storage of mechanical or fluid power (eg, turbine flywheels or hydraulic accumulators) smooths these pulses and enables the use of lower capacity electrical equipment. This saving can more than offset the cost of such storage components, thereby improving overall system economics.

Wave energy levels change over periods of hours to days in response to either local wind conditions or the arrival of swell from distant storms. The random seas built up by strong local winds contain frequency components ranging from 0.04 to 0.5 Hz. To a first approximation, the energy contained in each component travels independently, at a speed that is directly proportional to its period. These different components start out together upon leaving the area of active wind generation, but the long-period waves travel faster and soon outdistance their shorter period counterparts. The energy of the fully developed storm sea is thus dispersed along a corridor emanating from the generation area. The effect is similar to that of a prism, which disperses white light into several bands of color, each one of a different constituent wavelength.

One consequence of wave dispersion is that a wave power plant located in the path of swell from such a storm will experience a rather abrupt increase in dominant wave period as the first, longest waves arrive, followed by a gradual decrease as the slower, shorter period waves arrive. A second consequence of wave dispersion is that it lowers and broadens a storm's energy "pulse," enabling sustained recovery of this energy by a wave power plant located in distant waters, whereas the same plant would be destroyed if located in the path of the storm itself.

Developing long-term statistics associated with the day-to-day variability of wave energy levels is a key step in plant design, since it drives the selection of equipment rating. As illustrated in Figure 1, this involves a trade-off between underutilization of installed capacity and excessive shedding of absorbed power.

Extratropical storms are most frequent and intense during the winter, when monthly average wave energy levels typically are three to five times greater than summer monthly averages. Where utility peak demand is dominated by winter heating and lighting loads (eg, northern Europe), wave energy has a good seasonal load match. Where peak demand is driven by summer air conditioning (eg, California), the seasonal load match is poor.

Although trade winds never approach the intensity of extratropical storm winds, they are much more consistent. Windward island coasts in the tropics are affected not only by trade wind waves but also by swell from extratropical storms at higher latitudes. For example, the

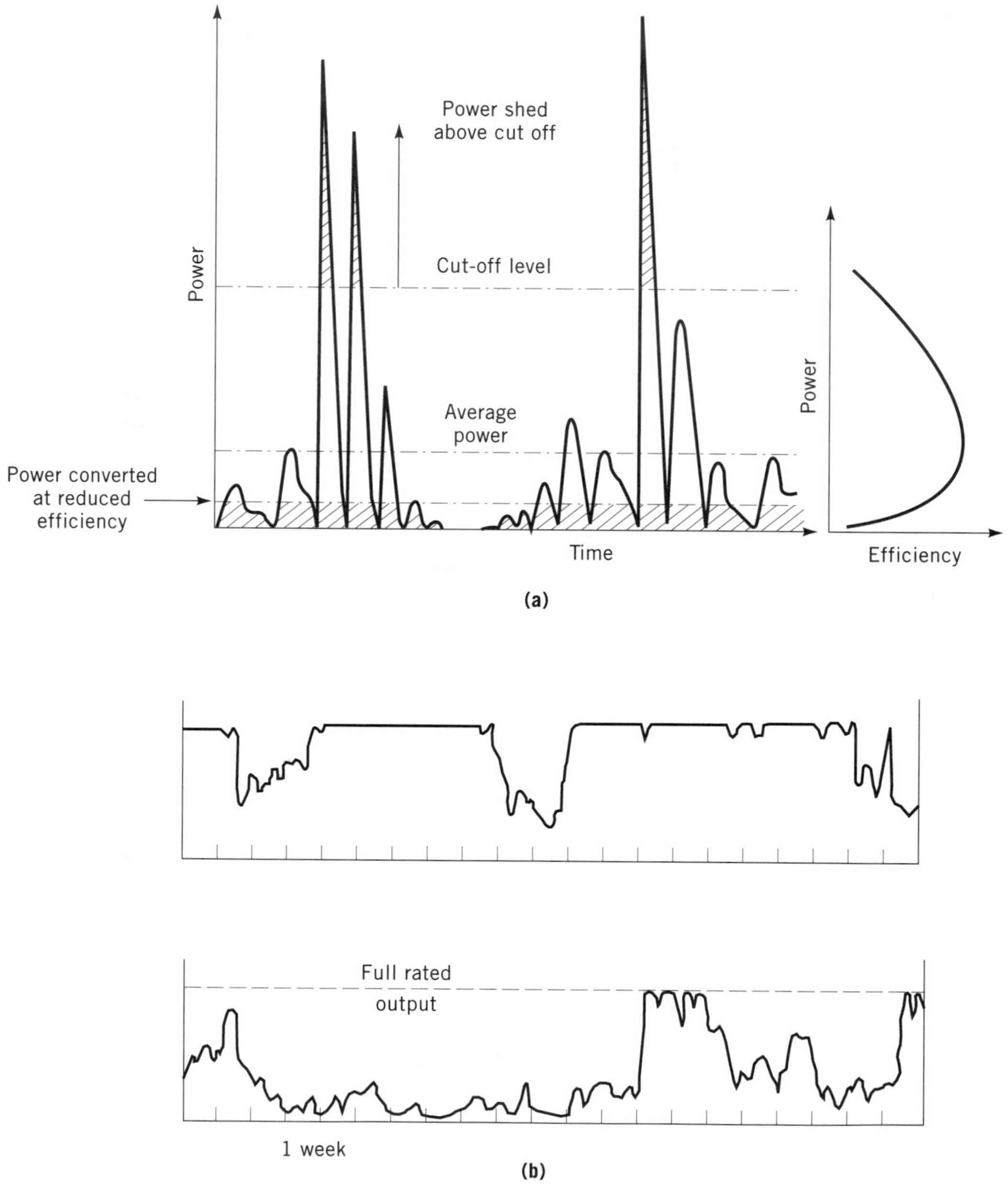

Figure 1. Day-to-day variation of wave power input (**a**) and power plant output (**b**) in response to changing sea states. Top part of (**b**) represents typical output in mid-winter. Lower part represents typical output in mid-summer. Note that in high latitudes, a plant designed for optimal year-round performance will shed much of its wave energy input during the winter yet be significantly underutilized and perform at reduced efficiency during the summer. Although the annual average wave power input would be less in the tropics than in high latitudes, sea states are more consistent from day to day, possibly resulting in improved plant economics. Source: J. K. Dawson, *Wave Energy*, Energy Paper No. 42, Her Majesty's Stationery Office, London, 1979.

wave energy flux at exposed deep-water locations in Hawaii is estimated to be 15 kW/m, divided about equally between northeast trade wind waves and northwest Pacific swell (7).

Finally, year-to-year variability in wave energy levels must be considered, particularly when developing resource data for economic projections. For example, annual average wave energy levels off northern California can vary up to 50% from one year to the next. In many cases, only one or two years of measured wave data may be available from a proposed plant site. Some attempt should be made to determine how well these data represent the long-term wave climate to which a plant would be exposed during its 20–30-year service life.

TECHNOLOGY DESCRIPTION

The basic wave energy conversion process can be stated in very general terms as follows: the force (or torque) of an incident wave causes relative motion between an absorber and a reaction point, which drives a working fluid through a generator prime mover. The periodic nature of ocean waves dictates that this relative motion will be oscillatory and have a frequency range of 3–30 cycles per minute, much less than the hundreds of revolutions per minute required for electric power generation. A variety of working fluids and prime movers are employed to convert these slow-acting, reversing wave forces into high-speed, unidirectional rotation of a generator shaft.

Twelve distinct process variations can be identified, as shown in Figure 2. The main features that distinguish one process from another are mode of energy absorption, type of absorber, and type of reaction point. Energy can be absorbed from heave (vertical motion), surge (horizontal motion in the direction of wave travel), pitch (angular motion about an axis parallel to the wave crests), or some combination of these modes. Absorbers can be fabricated of rigid or flexible material or can be the free surface of the water itself. Reaction points can be inertial masses (suspended plates, buoyant spines, or other absorbers), seafloor anchors (deadweight or pile), or fixed, surface-piercing masses (concrete or land).

Each different wave energy conversion process is described under a separate second-level heading in this section. The descriptions are presented in sequential order according to the circled numbers in Figure 2. Leading system configurations are described for each process, many of which have undergone prototype testing at sea. Alternative configurations are briefly mentioned, particularly where they have unique features. Although work on some of these alternatives has stopped, the research may have application to more advanced configurations of the same process. Indeed, different wave energy conversion processes have many components in common, and much can be learned by consulting references on projects that are no longer active.

Process 1: Reservoir Filled by Wave Surge

This is the simplest conversion process, and onshore versions exist that make use of conventional low-head hydroelectric technology. A leading example is the Norwegian Tapered Channel. Invented by Dr. Even Mehlum, the Tapered Channel consists of a collector, an energy converter, a reservoir, and a power house (Fig. 2). The collector funnels waves into the entrance of the energy converter, which is a vertical-walled channel having a depth of 6–7 m and built up to a height 2–3 m above mean sea level. The channel's width decreases in a shoreward direction, and its end is sealed off. As waves travel along the ever-narrowing channel, they increase in height, spilling water over its sides and into the reservoir. Water then drains back to the sea through a low-head (eg, Kaplan) turbine/generator.

The tapered channel configuration of the converter enables the device to effectively absorb energy from a large range of wave heights. In Stephen Salter's words (11):

> Large waves overtop early and deliver a large volume of water. Small waves must travel further along the channel before they get high enough to reach the top of the wall but nevertheless nearly all waves deliver something. This marks the difference between Tapchan and earlier overtopping schemes with walls parallel to the beach.

A 350-kWe Tapered Channel power plant has been operating since 1986 at Toftestallen, on Norway's North Sea coast. The reservoir for this demonstration project was built by damming two small inlets to the island's interior bay (Fig. 3), and a collector channel was mined into the rock at the head of a natural gully.

The reservoir of a Tapered Channel power plant does not provide long-term storage but smooths the input from one high-energy wave group to the next. For example, the reservoir at Toftestallen is reported to have an area of 8500 m^2, while the turbine is designed for a flow rate of 14–16 m^3/s and an operating head of 3 m (12). Should wave energy levels fall so low that waves no longer overtop the channel walls, the reservoir would drain to mean sea level in about 30 min. It should be noted, however, that the plant is designed to start automatically whenever sufficient head becomes available again.

All rights to the Tapered Channel are held by Norwave A.S., an Oslo-based company incorporated in 1987. The 350-kWe Toftestallen prototype has been operating automatically since 1986 and has survived several extreme storms, including one that destroyed a nearby oscillating water column device. Norwave is now developing a 1.1-MWe commercial plant on the south coast of Java in Indonesia. Cost and performance data for this design are given later in this article.

A different reservoir-filling concept has been proposed for the island nation of Mauritius by Sir A. N. Walton Bott of the Crown Agents in the United Kingdom (13). In this project, waves would surge directly up a ramped seawall built along an existing reef that encloses a natural lagoon, which would act as the reservoir; this plan, however, has not gone forward. Work has also been discontinued on two offshore, fixed reservoir devices: the caisson-based HRS Rectifier, developed by the Hydraulics Research Station in Wallingford during the early years of the British national wave energy program (14), and the DAM-ATOLL device developed by Lockheed Missiles &

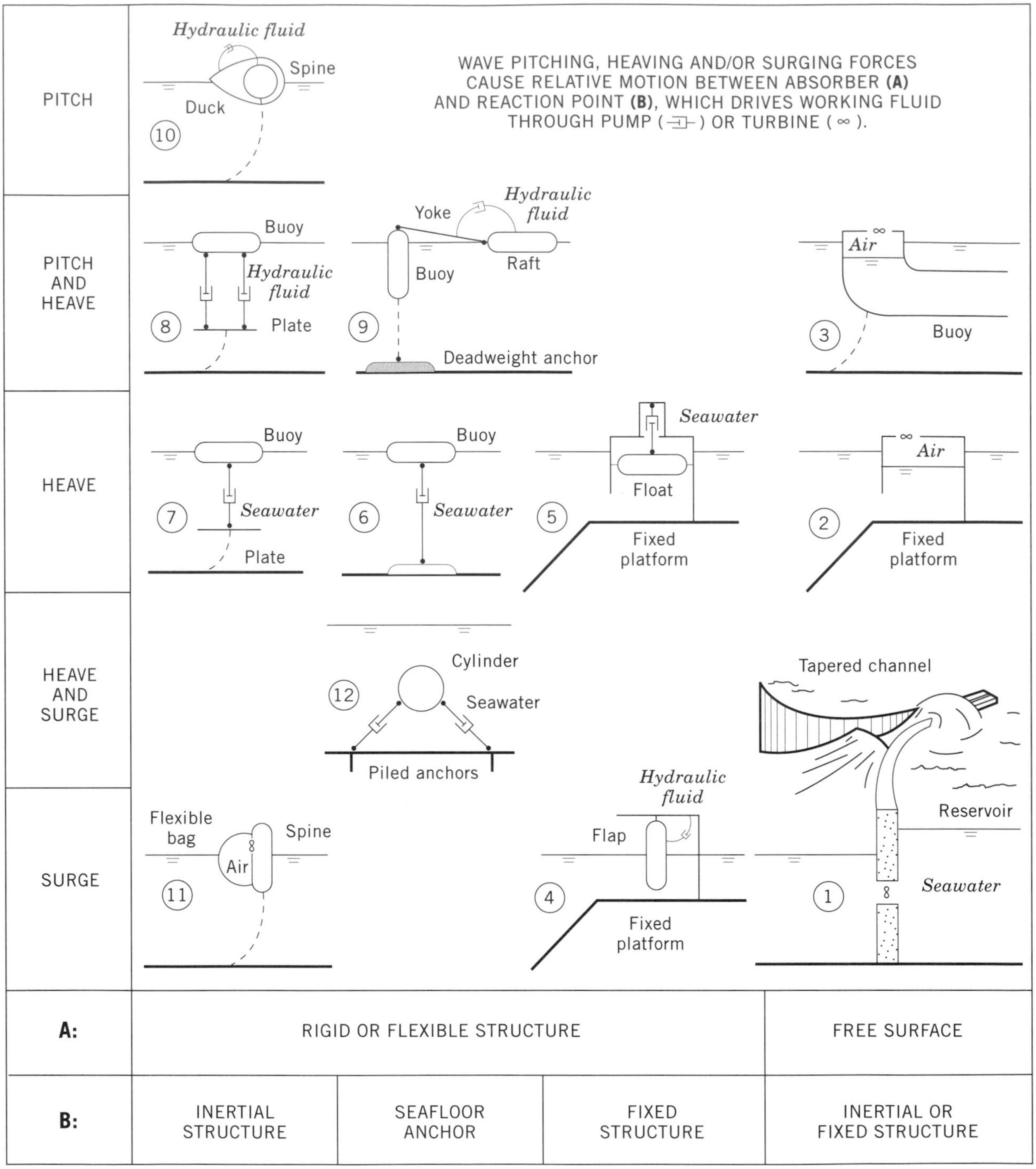

Figure 2. Classification of wave energy conversion processes, based on mode of energy absorption (pitch, heave, surge, or combined modes), type of absorber, and type of reaction point. Circled numbers identify conversion processes described in text. Source: SEASUN Power Systems.

Space Company in the United States, which is a partially submerged dome that causes waves to converge on its central reservoir by refraction (15). Much more recently, Sea Power AB, a Swedish company, has developed a floating device whereby a ramp and reservoir are supported between buoyant pontoons in a catamaran configuration. A 110-kWe prototype was tested at sea off Gothenburg for 2 months in 1991, but data from this project have not been widely published (16).

Process 2: Fixed Oscillating Water Column

Consider a vertical circular cylinder, open at both ends, whose bottom end is submerged. As a wave crest passes, the column of water entrained in the submerged portion of the cylinder rises, pushing air out the top of the cylinder. Likewise, as a wave trough passes, the water column falls, drawing air in. If the natural period of the water column is near resonance with the incident wave period,

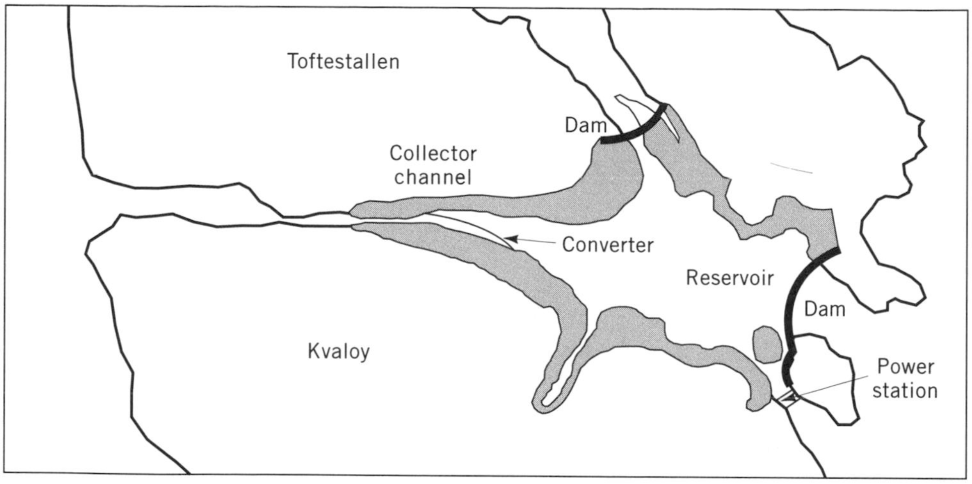

(a)

(b)

Figure 3. Tapered channel concept developed by Norwave A.S. (**a**) Shows the 350-kWe demonstration plant at Toftestallen, on Norway's North Sea coast, which has been operating since 1987. ▦ Indicates maximum flooded area of reservoir (sea level + 3 m). Source: E. Jeffreys, *Veritas* 128–30 (March/May 1985). (**b**) Shows basic concept. Source: A. E. Fredriksen, in J. Twidell, I. Hounam, and C. Lewis, eds., *Energy for Rural and Island Communities IV,* Pergamon Press, Oxford, 1986, pp. 179–182.

then its amplitude of vertical motion can be significantly greater than that of the wave.

The motion of the oscillating water column (OWC) is greatest when the cylinder is completely open, since it is working only against atmospheric pressure. If the cylinder is capped with an orifice plate, the volume of air exchanged with the atmosphere has to flow through an opening that is much smaller in diameter than the cylinder itself. This creates an overpressure ahead of the rising water column and a partial vacuum behind the falling water column, damping the motion in both directions. The volume of air exchanged over a complete wave period is thus reduced in comparison with that of the open cylinder. Even so, since a somewhat reduced volume of air has to flow through a greatly reduced opening, the net effect is to increase the air's flow velocity to the point that it can be converted to electrical energy by a pneumatic turbine. Means of accommodating the reversing air flow so that it spins a generator shaft in only one direction are described below.

The simplest solution is to use one-way valves to rectify the air flow so that it always travels in one direction through the turbine. During the *Kaimei* sea trials (described later in this article), such valves often became stuck in the open or closed position, seriously degrading turbine performance (17). On the other hand, Takenaka

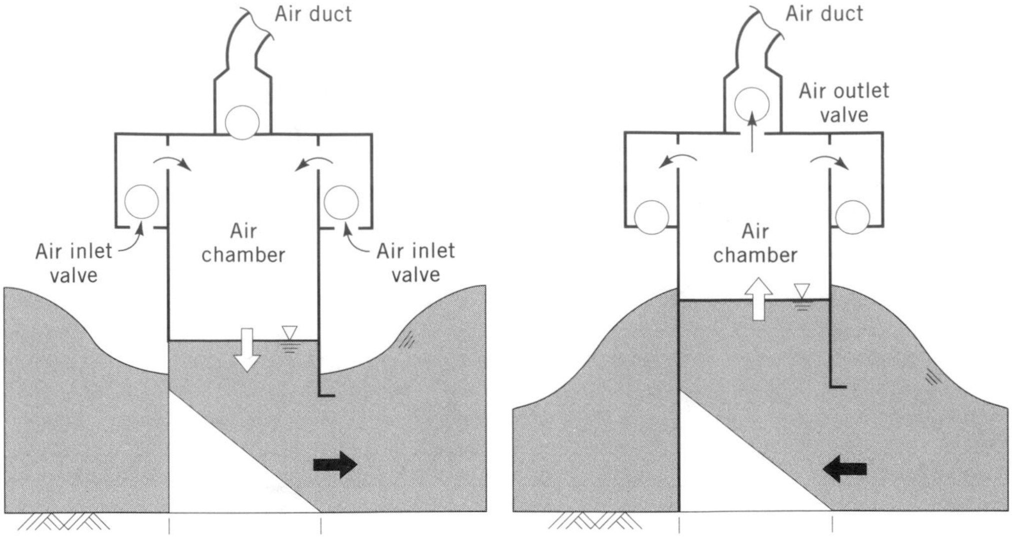

Figure 4. Use of nonreturn valves in the single-acting oscillating water column developed by Takenaka Corporation (18).

Corporation has had good success with its valve arrangement, whereby the rectified air flow from several OWC units supplies an onshore constant-pressure tank (18). This is a single-acting OWC, in that energy is absorbed only from the water column's rise (Fig. 4). There is no sacrifice in efficiency, however, since the column's downward motion is relatively undamped, allowing it to fall further, resulting in a greater buildup of hydrostatic pressure to help drive the column's rise during passage of the next wave crest.

As an alternative to the use of one-way valves, several different self-rectifying turbines have been invented; see reference (19) for a comprehensive review. The one that is in most widespread use is that invented by Dr. Alan Wells of the Queen's University of Belfast in Northern Ireland. The Wells turbine absorbs energy from both the rise and fall of the oscillating water column, which is thus referred to as a double-acting OWC, since the water column's motion is damped in both directions.

In the Wells turbine, the blades have a symmetric aerofoil section and are mounted on the generator shaft in such a way that they have no inclination to the plane of rotation (Fig. 5). The rotation of the blades and the wave-driven axial air flow combine to produce an apparent air flow that generates a lift force, much like that on an airplane wing. Although the axial component periodically reverses, the forward component of the lift force always acts to drive the rotor in one direction, regardless of which way the axial air flow is directed.

Shoreline gullies are naturally tapered channels, and an oscillating water column at the head of such a gully is exposed to higher wave power densities than found at the gully's mouth. Two systems based on this principle have been built, primarily for testing various pneumatic turbine designs. The first of these (20) was a 40-kWe unit that operated for 6 months at Sanzei, on the west coast of Japan (Fig. 6). A second gully-based OWC has been developed by researchers at the Queen's University of Belfast, led by Dr. Trevor Whittaker (21). A 75-kWe prototype is now operating on the Isle of Islay, off Scotland's west coast. Cost and performance data for a 1-MWe commercial plant design based on the Queen's University concept are given later in this article.

Kvaerner-Brug A.S., a large Norwegian hydropower company, has developed a multiresonant OWC system that can be based on land or on free-standing caissons. Kvaerner-Brug's early work on the OWC concentrated on developing means to adjust the natural frequency of the water column in order to tune the device as dominant wave period changes from sea state to sea state. In 1980, they shifted their approach to designing an absorbing structure that would resonate at several frequencies within the range of wave periods expected at a potential plant site. It was thought to be more cost-effective for a device to have several fixed-resonant frequencies rather than a single, continuously variable one.

The structure designed by Kvaerner-Brug to have such multiple resonances consists of a rectangular capture chamber and a harbor formed by extending the side walls of the chamber in a seaward direction. A 500-kWe demonstration plant based on this concept was built at Toftestallen, alongside Norwave's Tapered Channel plant (22). A cleft in the island's cliff wall was enlarged to form the resonant harbor, and the plant operated for 4 years before being destroyed by a severe winter storm.

The Indian Institute of Technology at Madras, under sponsorship from the national government, has developed a free-standing caisson design based on Kvaerner-Brug's resonant harbor concept (Fig. 7). A 150-kWe demonstration plant is now operating in 10 m water depth off India's southwest coast, near the port of Trivandrum (23).

An alternative to free-standing caissons is the use of caissons placed side by side in a breakwater configuration. Such an OWC has been developed by the Japanese Ministry of Transport, under the direction of Dr. Yoshimi Goda, and a 60-kWe prototype has been installed as part

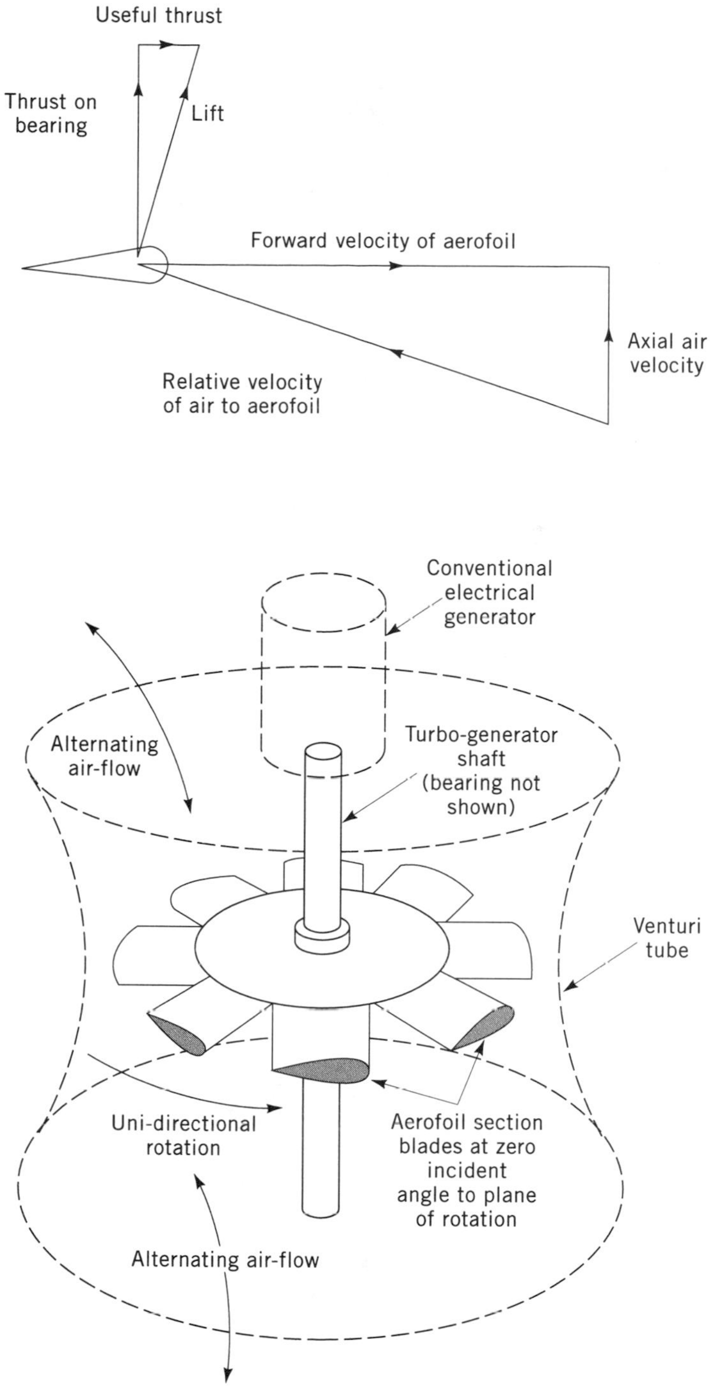

Figure 5. Wells turbine concept, showing the balance of force vectors on symmetrical aerofoil sections as the turbine blades rotate in a plane perpendicular to the direction of air flow (6).

of a new offshore breakwater built at Sakata Port, on the west coast of Japan (24). The new breakwater consists of a row of caissons on a rubble mound foundation. One of the caissons was built with a "curtain wall," which forms the OWC capture chamber (Fig. 8).

The National Engineering Laboratory (NEL) in East Kilbride, Scotland, has a long history of involvement with the British national wave energy program. Like other device teams in the United Kingdom, NEL has developed an evolving series of reference designs for a 2-GWe (nominal) wave power plant off the Outer Hebrides Islands (25). Cost and performance data for the most recent NEL breakwater design are given later in this article.

A noteworthy variant of the caisson-based OWC is a submerged device developed by Vickers Ltd. in the United Kingdom (26). While avoiding the problems of high structural freeboard and wave impact forces, locating the OWC beneath the sea surface entails greater operating and maintenance expense. Although the Vickers resonant duct had energy costs comparable to those of other British devices, active development of this concept has ceased.

Process 3: Floating Oscillating Water Column

This wave energy conversion process was the first to achieve widespread commercial application. As a result of research by Commander Yoshio Masuda in Japan, a battery-charging generator was developed for navigation buoys, driven by the OWC of a central pipe in the buoy's hull. Since Masuda's developmental work in 1964–1965, approximately 700 such generators (rated at 60 W) have been sold by Ryokuseisha Corporation for use in Japan, and another 500 have been exported to other countries (14). Similar units are also being produced by Munster Simms Engineering in Northern Ireland (Fig. 9) and the Guangzhou Institute of Energy Conversion in the People's Republic of China.

Masuda has also pursued the development of OWC technology for larger floating power plants, beginning with his work at the Japan Marine Science and Technology Center (JAMSTEC), on the test ship *Kaimei*, and more recently at Ryokuseisha Corporation, on the Backward Bent Duct Buoy.

JAMSTEC began its wave energy research and development program in 1974 and following 2 years of laboratory testing had the *Kaimei* built as a prototype floating platform for testing relatively large (up to 125-kWe) pneumatic turbine/generators. The *Kaimei* had a length of 80 m, a beam of 12 m, and a design draft of 2.15 m. Thirteen open-bottom capture chambers were built into her hull, each having a waterplane area of 42–50 m^2. The *Kaimei* was deployed twice in 40-m water depth off the west coast of Japan, near the port of Yura.

During her first deployment, from August 1978 to March 1980, eight turbines were tested aboard the *Kaimei*, all using various arrangements of nonreturn valves to rectify the air flow (27). For 4 months during the winter of 1978–1979, wave-generated power from one of the impulse turbines was supplied to the mainland grid. During her second deployment, from July 1985 to July 1986, five turbines were tested, including three impulse turbines (with nonreturn valves), a tandem Wells turbine, and a McCormick counterrotating turbine (28).

Wave tank tests and sea trials indicated that the wave energy absorption efficiency of the *Kaimei* was quite low. In an effort to improve the floating OWC process, Masuda developed the Backward Bent Duct Buoy (BBDB). The BBDB absorbs wave energy from both heave and pitch of a ship-shaped hull, although pitch appears to be the most

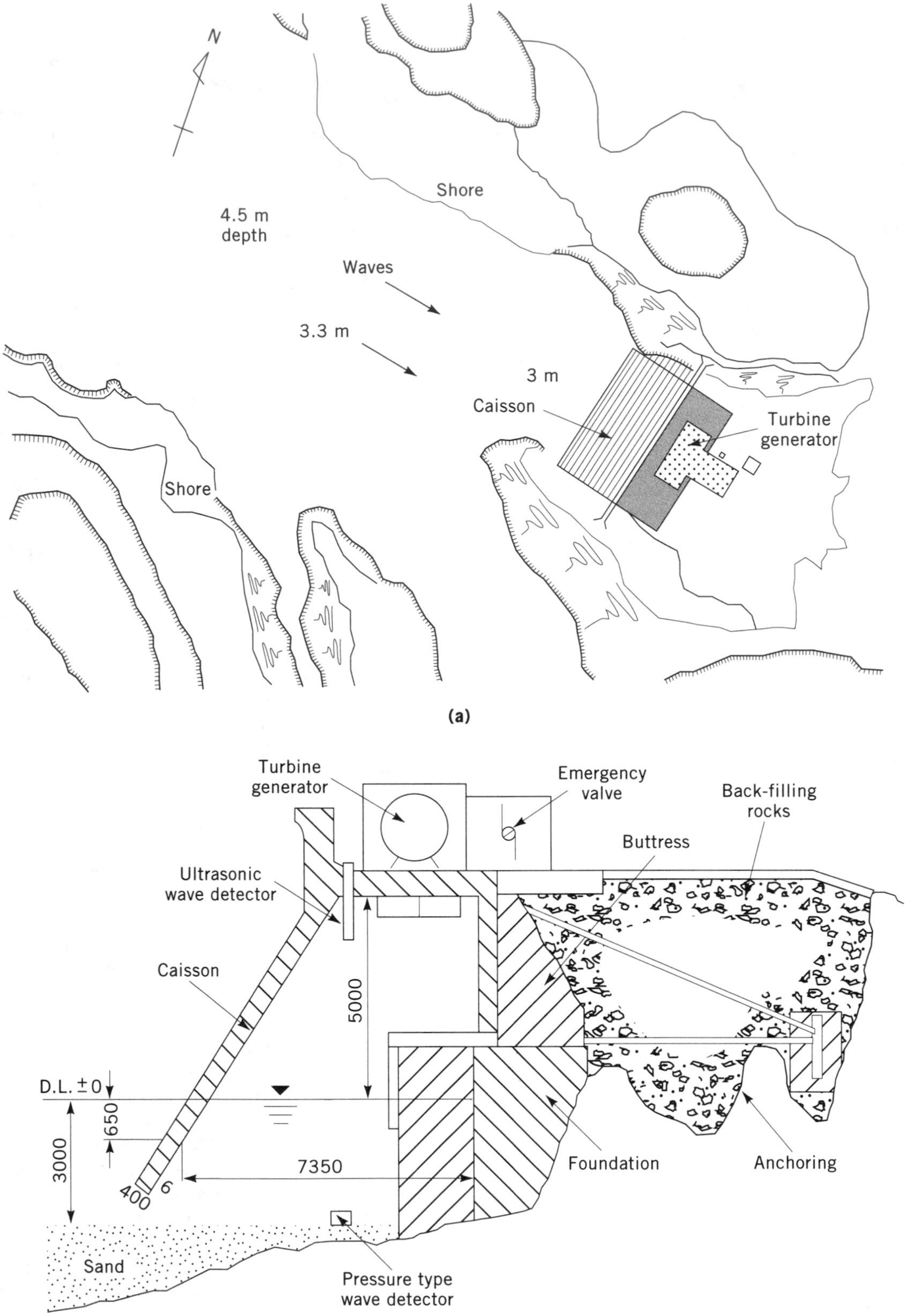

(a)

(b)

Figure 6. (**a**) Plan and (**b**) elevation of the shore-based oscillating water column prototype operated by the Japan Marine Science and Technology Center for 6 months at Sanzei, on the Sea of Japan (20). Elevation dimensions are in mm.

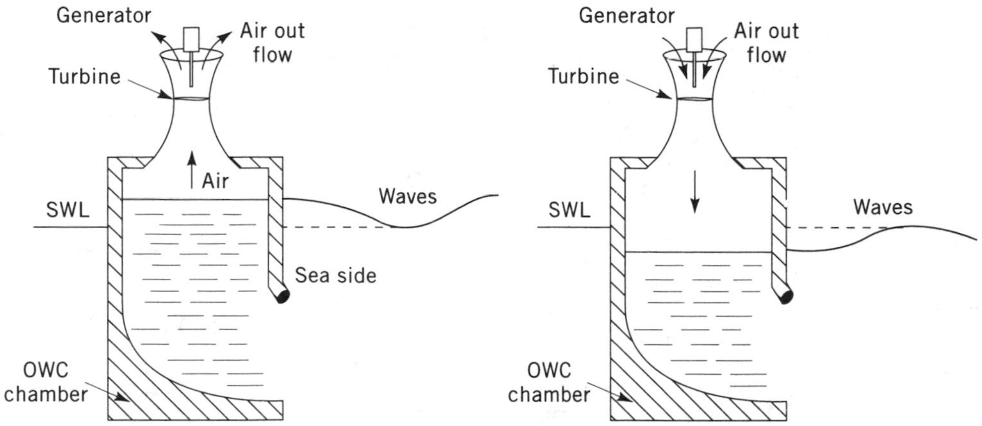

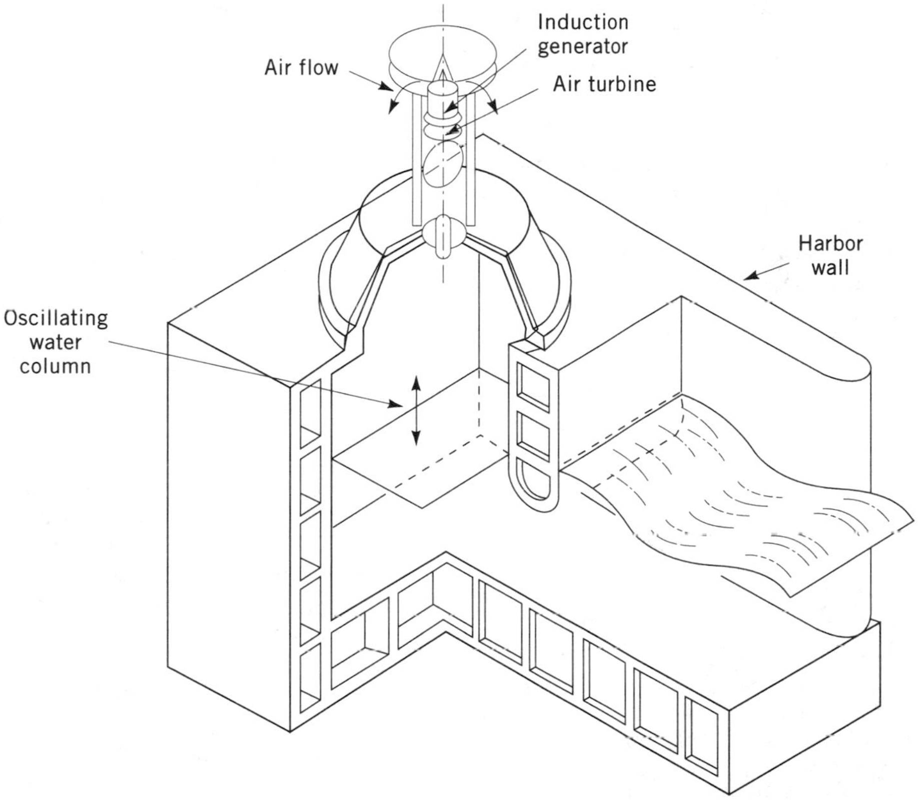

Figure 7. Caisson-based multiresonant oscillating water column concept. The Indian Institute of Technology at Madras has been operating a 150-kWe test plant based on this concept at Trivandrum, on the southwest coast of India, since 1992 (23).

important absorption mode. Wave tank tests have demonstrated the superior performance of the BBDB. For a given capture chamber area and wave height, the BBDB absorbs 3 times the power of a center-pipe navigation buoy and 10 times that of the *Kaimei* (29).

Masuda's most recent experimental work has been carried out in cooperation with researchers at the Guangzhou Institute of Energy Conversion in the People's Republic of China (30). Wave tank test results indicate that addition of a half-cylindrical body just aft of the BBDB's riser section greatly improves the performance of this device, leading to a higher and broader peak in energy absorption efficiency and a shifting of this peak to longer wave periods (meaning that a shorter, less costly hull can be used in a given wave climate).

While the BBDB makes use of hull motions for wave energy absorption, two alternative floating OWC devices have been developed in Japan that are stable in pitch and heave, so that their performance is more like that of a caisson-based OWC. Unlike caissons, however, such floating platforms can be moored in deeper water, taking advantage of the greater wave energy resource there. The first of these stable floating OWC devices is the Mighty Whale, developed at JAMSTEC under the leadership of Dr. Takeaki Miyazaki. OWC capture chambers line the front of the device, with buoyancy chambers behind these.

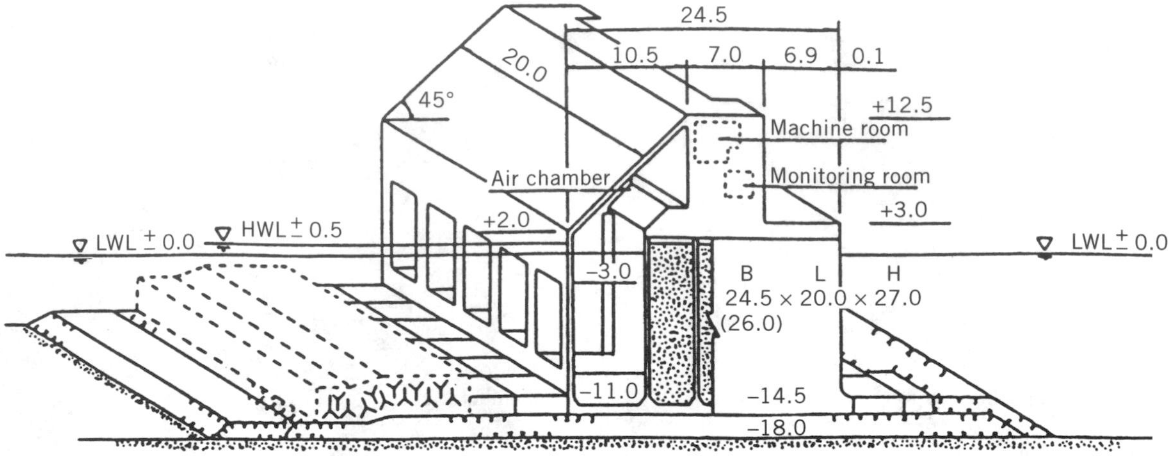

Figure 8. Breakwater-based oscillating water column concept developed by the Japanese Ministry of Transport. A 60-kWe prototype caisson has been installed as part of a new breakwater built at Sakata Port in 1988 (24). Dimensions are in meters.

A flat ramp slopes down and back from the capture chambers into the water, damping the pitching motion of the device. Wave tank tests have shown that mooring forces on the Mighty Whale are low, and under some wave conditions, it actually advances into the waves (31).

Also in Japan, Taisei Corporation has conducted laboratory and mathematical modeling of an OWC tension leg platform (32), which is stable in both pitch and heave. This makes it an ideal candidate device for powering floating airports and other offshore platforms requiring good stability.

Process 4: Pivoting Flap

For over a decade, the Muroran Institute of Technology in Japan has developed a caisson-based pivoting flap device, which it calls the "Pendulor" System (33). The Pendulor itself is a stiffened steel plate that hangs down into a recessed capture chamber. Unlike fixed OWC systems, where the capture chamber extends below the sea surface, the Pendulor System caisson is entirely open to the sea (Fig. 10). Incident waves interact with waves reflected off the back of the chamber to create a standing wave. The Pendulor is positioned at the node of this standing wave, where horizontal forces are at a maximum. As the Pendulor swings in response to these surge forces, it drives a double-acting hydraulic cylinder, which pumps fluid to a hydraulic motor/generator. Cost and performance data for a conceptual design of a 1.5-MWe Pendulor System are reviewed later in this article.

In April 1983, a 5-kW (hydraulic motor rating) prototype was installed at Muroran Port, on the south coast of Hokkaido. The prototype caisson is sited in front of an existing seawall, in a water depth that ranges from 2.5 m at low tide to 4 m at high tide. Two capture chambers have been built into the caisson, but only one has been fitted with a Pendulor.

Twenty months after its installation, the Pendulor was bent during a severe storm, and the shock absorbers for the end stops, which prevent overstroking of the cylinder, had to be redesigned. A new Pendulor was installed in November 1985, and the prototype has survived several severe storms since then without damage.

A small Pendulor System that generates electric power was deployed in 1981. Rated at 20 kWe, this unit is used to heat the public bath of a fishing cooperative at Mashike Harbor, on Hokkaido's west coast (34). Unfortunately, its Pendulor was also damaged by a storm, just 3 months after installation. It was replaced by a shorter Pendulor in 1983, which left a considerable gap at the bottom of the capture chamber. While this has prevented further damage, it has also lowered the system's conversion efficiency. Nevertheless, the plant continues to operate.

Kansai Electric Company has developed an alternative caisson-based pivoting flap device in which the flap is hinged at the bottom rather than top, and a 1-kWe test unit has been installed on Wakasa Bay, northwest of Kyoto (35). In the United States, Q Corporation has developed a tandem-flap device, which consists of two bottom-hinged flaps placed one behind the other in the direction of wave travel. The flaps are mounted in an open-frame platform, which can be installed in deeper water than a caisson. The company has funded a series of scale model tests (36) and with cofunding from the U.S. Department of Energy, tested a 20-kWe prototype in Lake Michigan during the summer and fall of 1987 (37).

Two floating pivoting-flap devices have also been developed, both in the United Kingdom. One is the Triplate Converter developed by Dr. Farley and his co-workers at the Royal Military College of Science (38), and the other is a pitching and surging flap known as PS Frog, invented by Dr. Michael French at Lancaster University (39).

Process 5: Heaving Float in Caisson

Wave forces are large but act very slowly, and some sort of speed increase is necessary in order to drive an electrical generator. The OWC processes described above accomplish this by an area reduction between the capture chamber waterplane and the air turbine inlet.

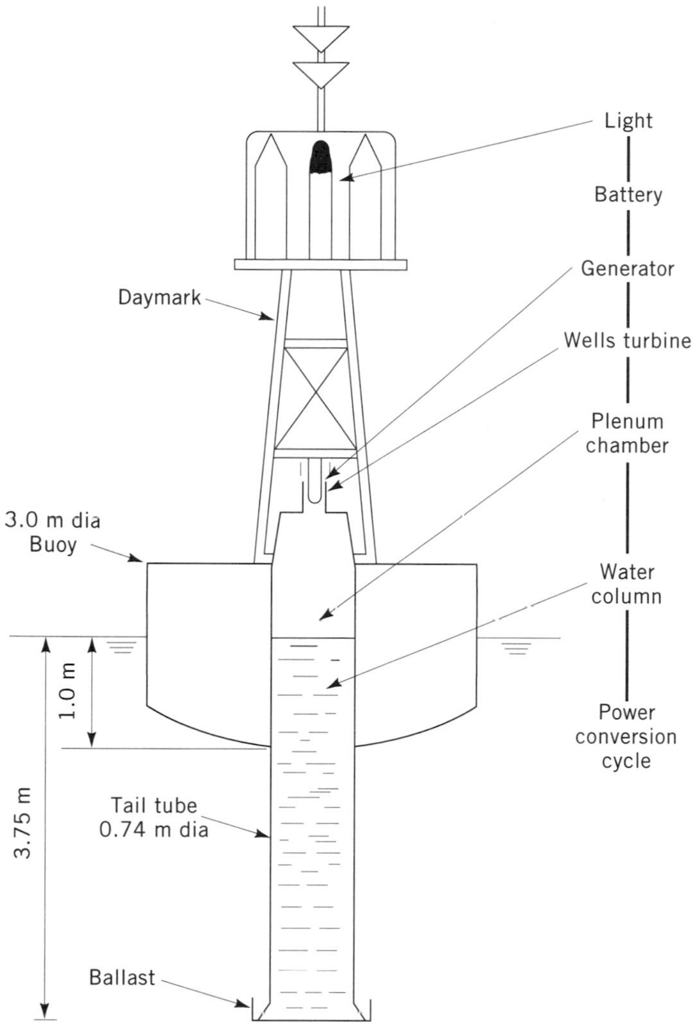

Figure 9. Oscillating water column system for battery charging on heaving navigation buoys. Source: Munster Simms Engineering Ltd.

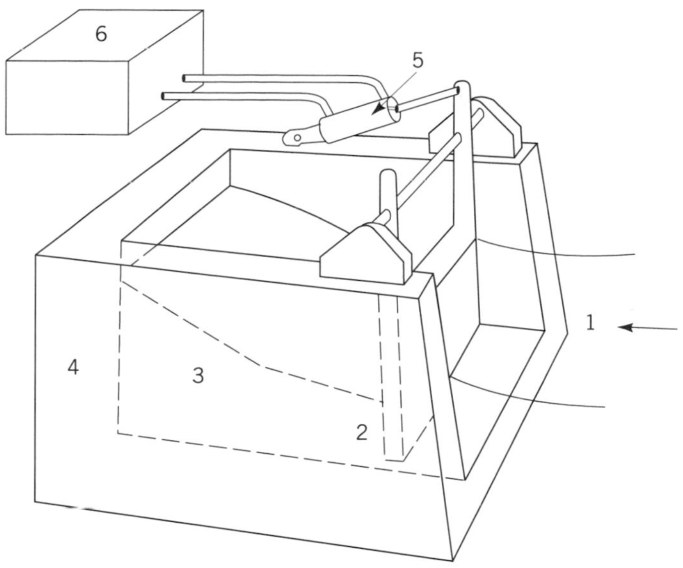

Figure 10. Caisson-based pivoting flap concept developed at Muroran Institute of Technology in Japan. 1, Incident wave; 2, pendulum; 3, water chamber; 4, caisson; 5, hydraulic cylinder; 6, rectifier (33).

A heaving float in a caisson accomplishes much the same thing. Wave forces distributed over the float's waterplane area are transmitted to a much smaller diameter pump, thereby achieving a pressure increase and developing sufficient head to run a high-speed water turbine. Use of an air turbine involves less equipment (no pump or float) but is not as efficient at converting water column motion into electrical energy. A leading example of this process is the Neptune System.

Originally developed by Wave Power Industries of Arcadia, California, patent rights to this process have since been assumed by Ocean Resources Engineering, Inc., also of Arcadia. In 1987, Taiwan Power Company undertook a wave energy resource and technology assessment for its service area, and the Neptune System was selected for further investigation. In April 1988, wave measurements were initiated at a promising site, Lan-Yu Island, off Taiwan's southeast coast, and a conceptual design was prepared, based on the first year of measured wave data (40).

The design concept for Lan-Yu consists of an offshore caisson having a circular capture chamber just over 20 m in diameter, sited in 12 m water depth (Fig. 11). The 19.2-m-diameter float drives a double-acting piston pump having an inner diameter of 0.91 m, which gives the system a pressure amplification ratio of 440:1. Seawater is pumped to an onshore reservoir located 305 m above sea level, which has a capacity of 1170 m^3. At an operating head of 3.07 MPa (445 psi), water from the reservoir drains back through a penstock to a 1-MWe Pelton turbine/generator. Assuming that the turbine/generator operates at rated capacity with 90% efficiency, the reservoir would drain in about 5 min without wave input. As with the Tapered Channel, the reservoir serves to smooth fluctuations due to high and low wave groups and is not intended for long-term energy storage.

A quite similar float-in-caisson pump has been developed by Kajima Corporation in Japan (41). As with the Neptune System, this device has been tested in a laboratory wave tank at $\frac{1}{10}$ to $\frac{1}{15}$ scale; neither device has been tested at sea.

Process 6: Freely Heaving Float and Seafloor Reaction Point

By eliminating the caisson with its large waterplane area, a freely heaving float can be economically moved into much deeper water. Two heaving buoy systems have been developed that utilize deadweight anchors on the seafloor as reaction points but with markedly different applications. One system has been developed in Denmark for central station electric power generation, while another has been developed in the United States for small-scale freshwater production.

Invented by Dr. Kim Nielsen, the Danish heaving buoy device was originally conceived as a heaving and pitching float, and a 1-kWe prototype of this configuration was tested in 1985 (42). Lessons learned from these tests have been incorporated into the present design, which is a circular buoy that absorbs wave energy mainly from heave and to a lesser extent from surge.

In its present configuration, each module consists of a 10-m-diameter buoy tethered to a plastic-lined concrete piston, which is housed in a steel cylinder 22 m high and 3.75 m in diameter mounted on a concrete gravity base. The ratio of buoy waterplane area to pump cross-sectional area is about 7 : 1, and the device operates at a very low head, less than 0.1 MPa (15 psi).

The gravity base is 20–24 m in diameter and 19–25 m high. A 130-kWe submersible Flygt turbine/generator is mounted on the base shell, with an outlet diffuser that extends down into a chamber within the base structure (Fig. 12). This chamber is a seawater storage buffer that operates in the following manner. When the buoy heaves

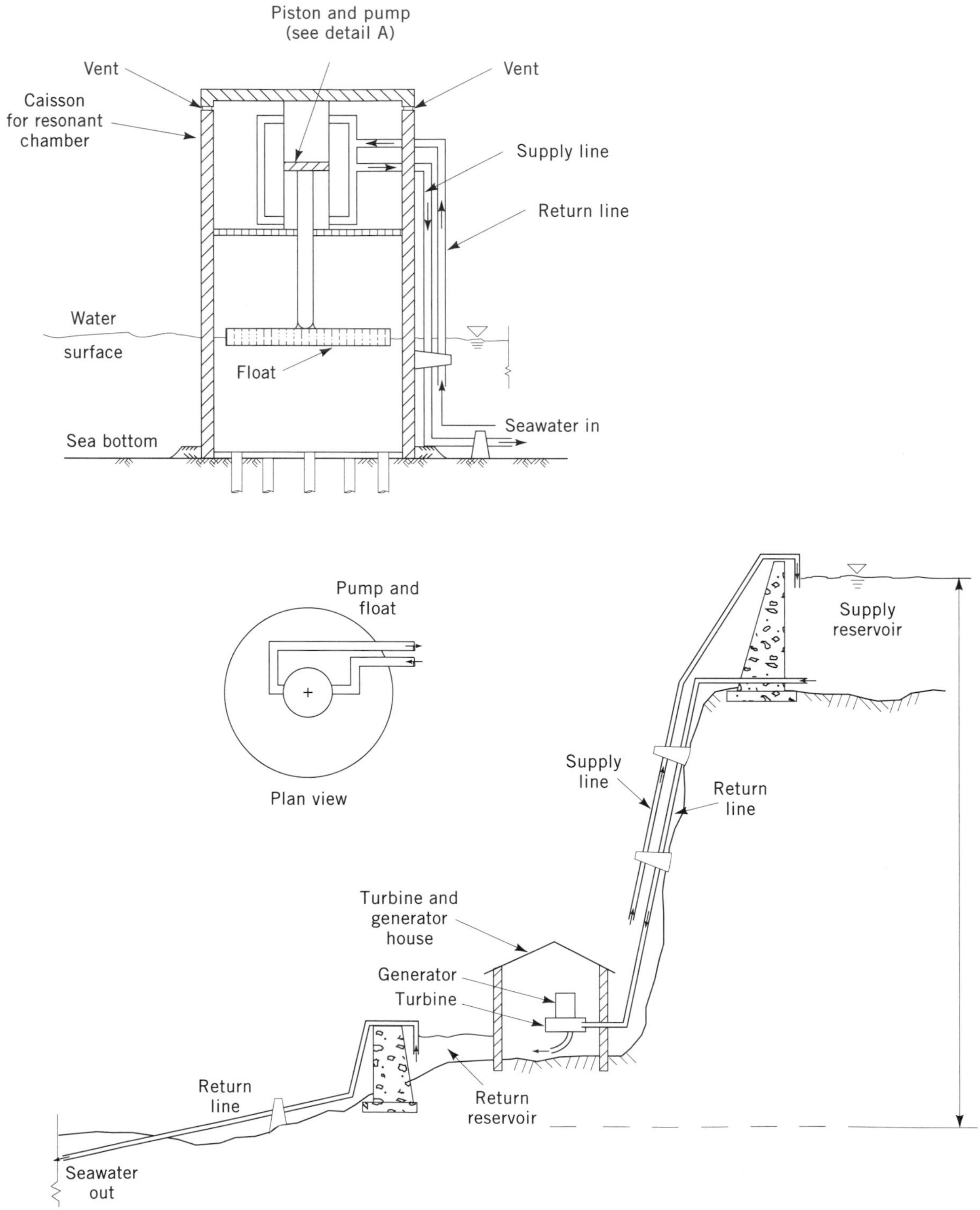

Figure 11. Heaving float in caisson design developed by E. O. Tech, Inc., of Arcadia, Calif., for Lan-Yu Island in Taiwan (40).

up on a wave crest, raising the piston, water is drawn from the chamber into the cylinder through a pair of one-way flap valves. This lowers the chamber pressure, causing water from the surrounding sea to flow in through the turbine. When the buoy drops down in a wave trough, the weight of the piston forces water from the cylinder out to sea through a second pair of flap valves. By partially evacuating the air chamber and creating lower than ambient pressure, the seawater buffer can be "charged" and will maintain steady intake flow through the turbine while the pump alternately withdraws water on the up stroke and discharges it back to sea on the down stroke.

Danish Wave Power Aps, a consortium of four companies, was formed in 1989 to further develop this device. Later that year, a 45-kWe prototype was tested in 30 m water depth off Denmark's northwest coast, near the port of Hanstholm (43). Based on this experience, a conceptual design for a 300-MWe commercial plant was prepared, and cost and performance data were developed for five different sites around the world (44). These data are summarized later in this article.

In the United States, at the University of Delaware, Doug Hicks and Michael Pleass developed the DELBUOY system for the desalination of seawater by direct reverse osmosis (RO). Since 1982, they have deployed an evolving series of prototypes off the southwest coast of Puerto Rico, where they have demonstrated freshwater production at a continuous rate of 950 L (250 gal) per day from a single buoy (45). License to this device is now held by CHPT, Inc., a Lewes-based company specializing in the design and fabrication of high-pressure hydraulic and structural components from composite materials.

DELBUOY employs a 2.1-m-diameter buoy tethered to a seafloor anchor by a single-acting hydraulic cylinder with a bore diameter of 4 cm (Fig. 13). This gives a pressure amplification ratio of nearly 2800 : 1, such that relatively small waves can generate a pump pressure of 5.5 MPa (800 psi), which is adequate to develop reverse osmotic flow. Six buoy/pump moorings supply one RO module that delivers 5680 L (1500 gal) of fresh water per day in waves 1 m high having a period of 3–6 s, which are typical of trade wind wave climates.

Process 7: Freely Heaving Float and Inertial Reaction Point

Use of a seafloor anchor makes the performance of a heaving buoy sensitive to tidal changes in sea level. By employing an inertial reaction point, this sensitivity can be eliminated. A suspended-plate concept developed in Sweden (46) is the most advanced example of this process and makes use of a unique seawater pump that does not have the sliding seals or end stops of a hydraulic cylinder.

Consider the series of buoys illustrated in Figure 14. Each buoy is attached to a collecting line by a length of specially designed elastomeric hose. During the passage of a wave crest, the buoy heaves up, stretching the hose. The helically wound, steel reinforcing wires in the hose wall cause the hose to constrict as it is stretched, thereby reducing its internal volume. This forces seawater out of the hose, through a check valve, and into the collecting line. After the wave crest has passed and the buoy drops down into the succeeding trough, the pump returns to its original length, restoring the hose diameter to its unstretched value. This increase in internal volume draws water into the hose through another check valve, which is open to the sea. The hose is thus primed to pump seawater into the collecting line during passage of the next wave crest.

In order to provide a reaction point for its pumping action, the bottom end of each hose is tethered to a hori-

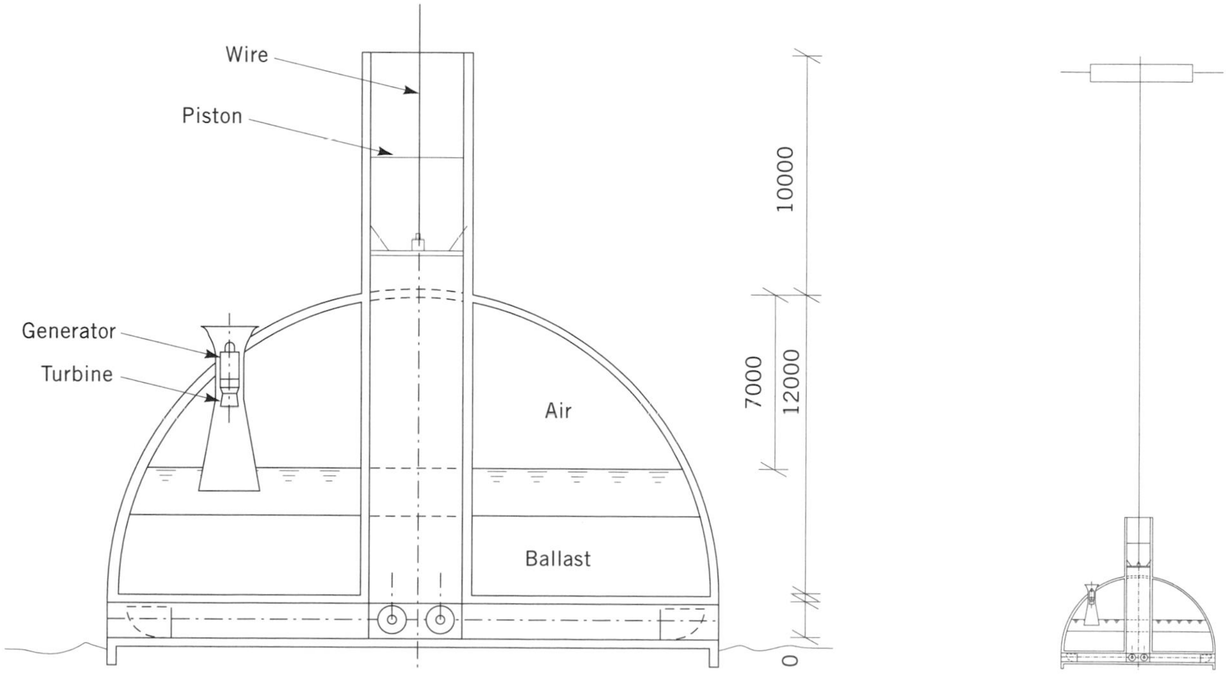

Figure 12. Danish Wave Power's heaving buoy, incorporating a submersible turbine/generator and seawater buffer in a concrete gravity base (44). Dimensions are in mm.

zontal damper plate. When the buoy heaves up, it attempts to move this plate vertically through the water. Due to its large surface area, the plate acts like a sea anchor, and its motion lags behind that of the buoy. Thus, relative motion between each buoy and damper plate alternately constricts and dilates the hose between them.

A power plant would consist of several star-shaped clusters of buoys. Each arm of the star is a collecting line for the output of up to 10 buoy/hose pump modules. These lines radiate symmetrically from a central underwater habitat that contains a vertical-axis Pelton turbine/generator. Pressures as high as 4 MPa (580 psi) can be developed in the collecting line as seawater is pumped through the line-end nozzle. After striking the Pelton wheel, the water falls back into the sea through an air pocket in the habitat. Diverter valves prevent excess flow from reaching the turbine in extreme waves.

Developed by Gotaverken Energy and Technocean AB, a prototype system was tested near Vinga Island, off Sweden's west coast, in 1983–1984. Three buoys were deployed about 600 m from the island and supplied an onshore 30-kWe Pelton turbine/generator. Each buoy had a diameter of 5 m (about one-third that of a commercial plant buoy). The wave energy absorption efficiency of this prototype agreed well with that predicted by a numerical model of the system. Cost and performance data for a 10-MWe design based on this concept are given later in this article.

Another Swedish heaving buoy device uses a piston contained within an open-ended pipe rigidly attached to the buoy hull as the inertial reaction point (47). Developed by Interproject Service AB, a half-scale prototype called *Elskling* was tested at sea at 1979–1981. The prototype buoy had a diameter of 3 m, with a center pipe 1 m in diameter and 20 m long.

Process 8: Contouring Float and Inertial Reaction Point

In the buoy systems described above, heave is the predominant energy absorption mode, due to the relatively small ratio between float diameter and wavelength. Larger floats not only heave but also develop angular momentum as they attempt to follow the contours of the sea surface. A leading example of such a wave-contouring float is the Wave Energy Module (WEM).

The WEM was invented in the late 1970s by Harold Hopfe, and its development was pursued by U.S. Wave Energy, Inc., of Longmeadow, Massachusetts. A series of tests has been conducted on a 1-kWe model in Lake Champlain (48).

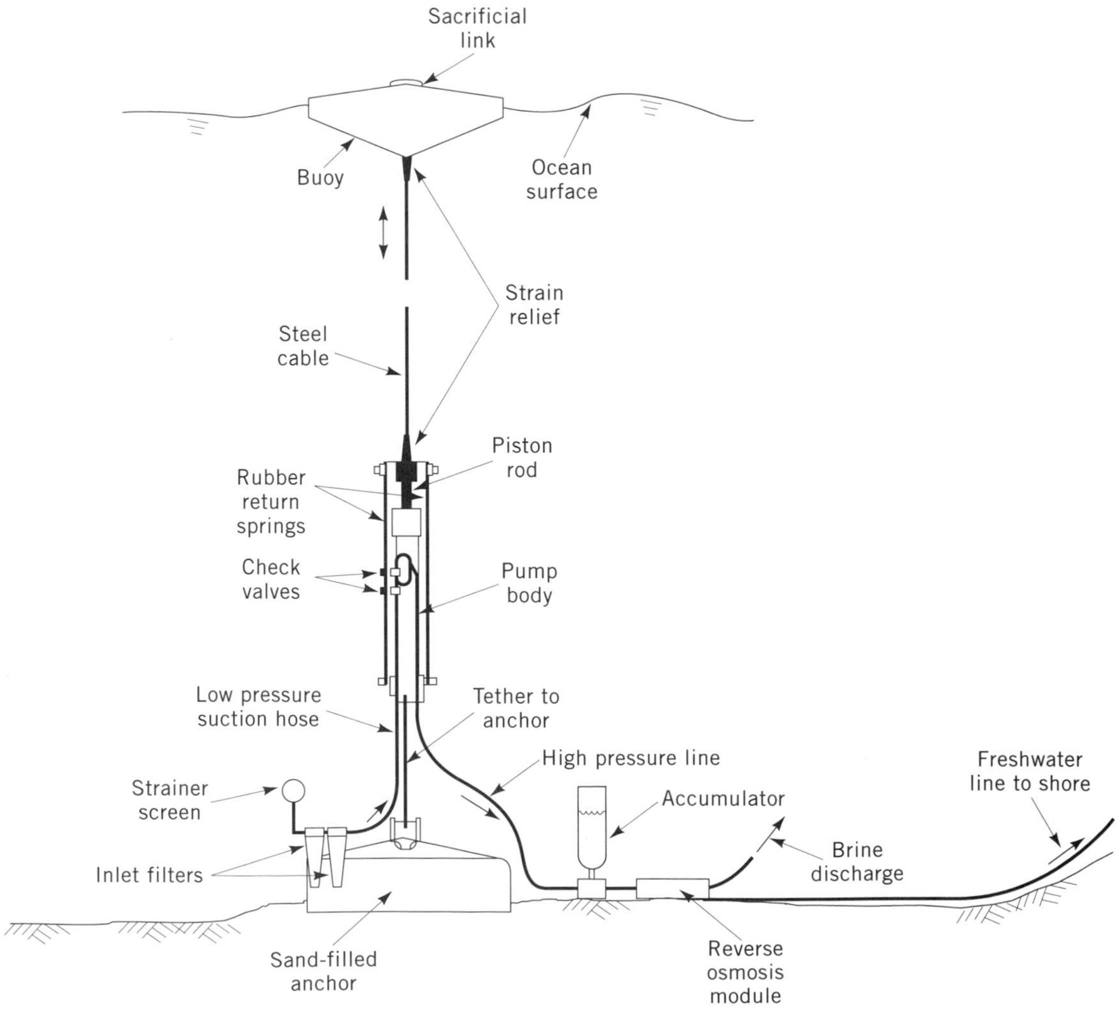

Figure 13. CHPT's heaving buoy for direct reverse osmosis. Source: F. Yeaple, *Design News* (March 27, 1989).

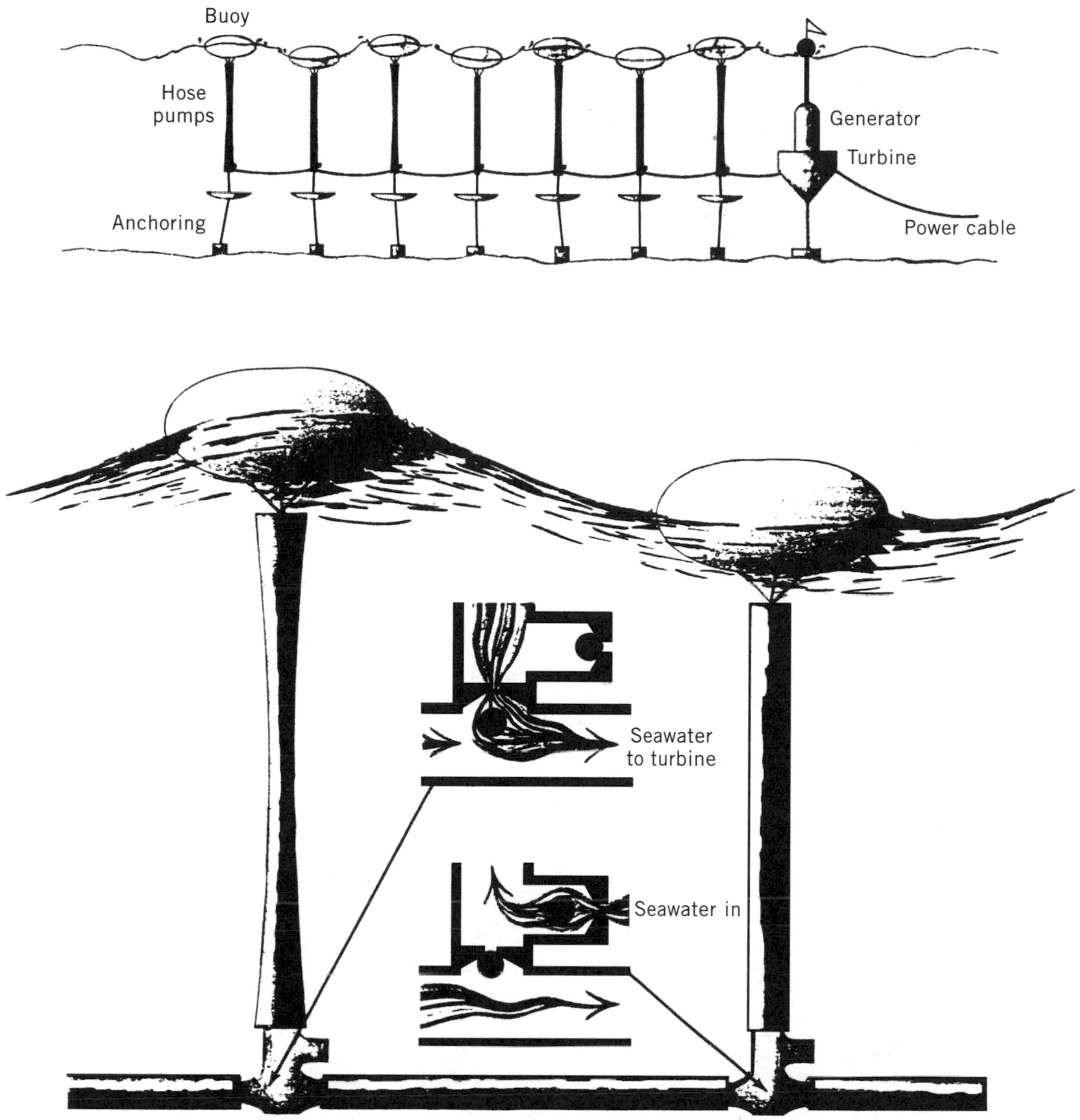

Figure 14. Swedish heaving buoy concept, in which the reaction point is a suspended damper plate rather than a deadweight anchor. Relative motion between each buoy and damper plate alternately constricts and dilates the elastomeric hose between them, pumping seawater into a collecting line that leads to a submerged Pelton turbine/generator. Source: G. Svensson, "Swedish Idea for Wave Power," *Energy Ahead 85,* Vattenfall, Annual R & D Report, Stockholm, Sweden, 1985.

The WEM's operating principle has much in common with that of Gotaverken's heaving buoy device, in that fluid power is generated by wave-induced motion of a circular float relative to a suspended damper plate (Fig. 15). Unlike Gotaverken's concept, however, where each buoy is simply an absorber, the WEM buoy contains the entire power conversion system. During their power stroke, the cylinders pump hydraulic fluid to a high-pressure accumulator. When the buoy heaves downward, or when pitch or roll causes one side of the buoy to tilt toward the damper plate, the cylinders are returned to their neutral position by fluid from a low-pressure accumulator and are thus primed for their next power stroke. The high-pressure accumulator discharges fluid to a hydraulic motor that is coupled to a synchronous electrical generator.

Process 9: Contouring Float and Seafloor Reaction Point

While a circular float such as the WEM follows the sea surface contours by heaving, pitching, and rolling, a long-narrow raft absorbs wave energy primarily from heave and pitch. In the mid-1970s, two contouring raft concepts were invented independently and almost simultaneously: one in the United States by Glenn Hagen and the other in the United Kingdom by Sir Christopher Cockerell, inventor of the Hovercraft. Hagen's patent is assigned to Williams, Inc., a Louisiana-based company, who formed Sea Energy Corporation to further develop Hagen's contouring raft concept.

In the original concepts of Hagen and Cockerell, several floating rafts are hinged together, end to end, and oriented to meet incoming waves head on. Passage of a

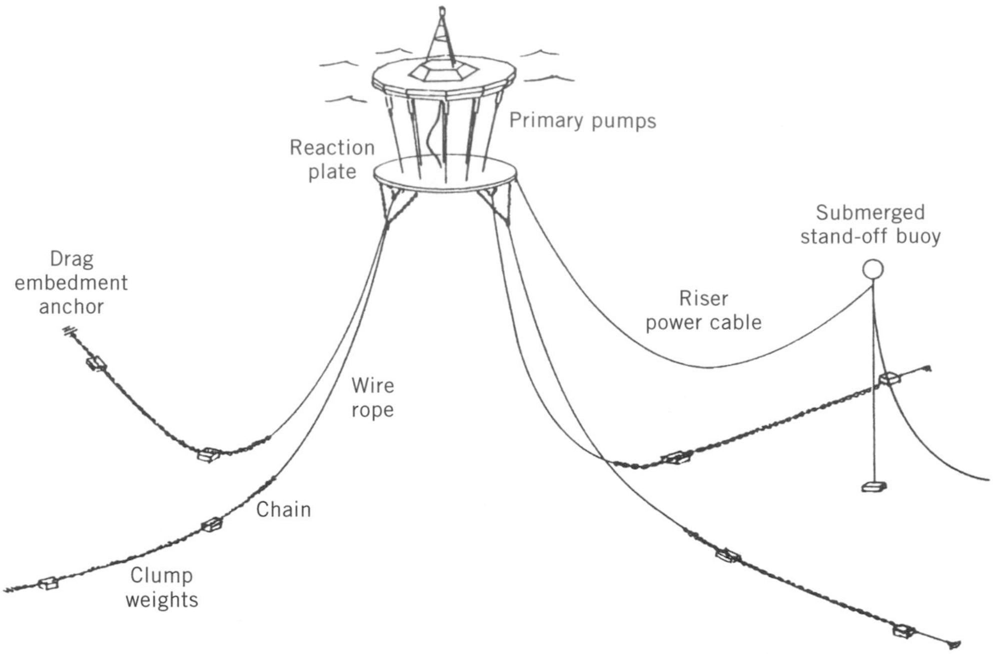

Figure 15. Wave Energy Module concept. Note that the float is much larger than the heaving buoys described earlier and therefore follows the contours of the sea surface, causing energy to be absorbed by the primary pumps from heave, pitch, and roll. Source: U.S. Wave Energy, Inc.

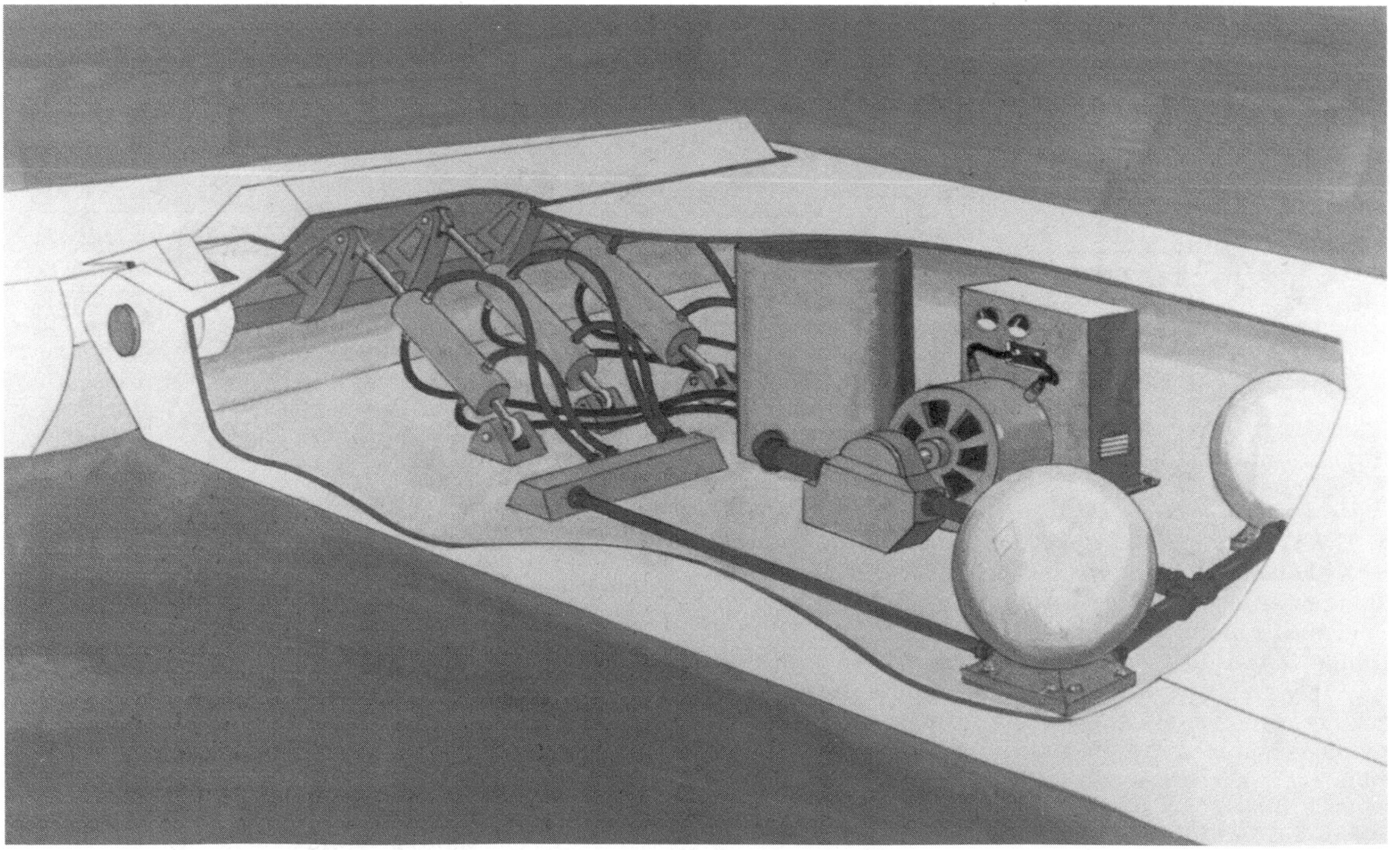

Figure 16. Sea Energy Corporation's contouring raft concept, showing the double-acting hydraulic cylinders driven by rotation of the raft hinge pin. These draw hydraulic fluid from a cylindrical reservoir and pump it to spherical accumulators, which supply a hydraulic motor/generator. Source: Sea Energy Corporation.

wave causes relative pitching between adjacent rafts, and this motion is converted into fluid power by hydraulic cylinders mounted across the hinges. Sea Energy Corporation's early test results suggested that if a practical means could be found for absorbing energy from the wave-following motions of just a single raft, it would absorb energy from both pitch and heave, rather than pitch alone, thereby improving its efficiency. Such a device would require less waterplane area than a multiple-raft unit to capture the same amount of energy in a given sea state. Hull costs would drop by a comparable amount, and the problems associated with hinging several rafts together could be avoided (49).

The single-raft concept is described as follows. A ship- or barge-shaped hull is connected by a rigid yoke to a reference point fixed in heave, such as a taut moored buoy. The yoke is pinned at either end and pierces the raft hull via sealed bearings. Inside the raft, the hinge pin is fitted with lever arms spaced at regular intervals. Each of these is pinned to the rod end of a hydraulic cylinder that is clevis-mounted on the raft (Fig. 16). As the raft pitches or heaves in passing waves, the hinge pin rotates relative to the raft and strokes these cylinders, thereby converting the surface-following motions of the hull into fluid power. Due to the unsteady nature of raft motion in high- and low-wave groups, cylinder pumping action is intermittent. Large-volume accumulators are incorporated into the hydraulic circuit to smooth these pulses and supply fluid at constant pressure to a hydraulic motor/generator.

An alternative hydraulic conversion system has been developed by the Japan Institute for Shipbuilding Advancement (50). In 1984–1985, prototype tests were conducted aboard the specially built jack-up rig *Kaiyo* off Iriomote Island, southwest of Okinawa. Two floats were contained in the open bays of *Kaiyo* and were linked to the rig in such a way that heave, pitch, or surge of the floats stroked hydraulic cylinders. Unlike Sea Energy Corporation's conversion system, where fluid is pumped to accumulators that are simultaneously charging and discharging, the accumulators on *Kaiyo* were arranged in pairs. One member of each pair was charged over a period of time (approximately 10 minutes), while the other member discharged fluid to a hydraulic motor/generator. When the charging accumulator had been brought back up to operating pressure by the pumping action of the floats, the accumulators were switched.

Process 10: Pitching Float and Inertial Reaction Point

Hydraulic cylinders, which are common to many of the processes described above, have an inherent end-stop

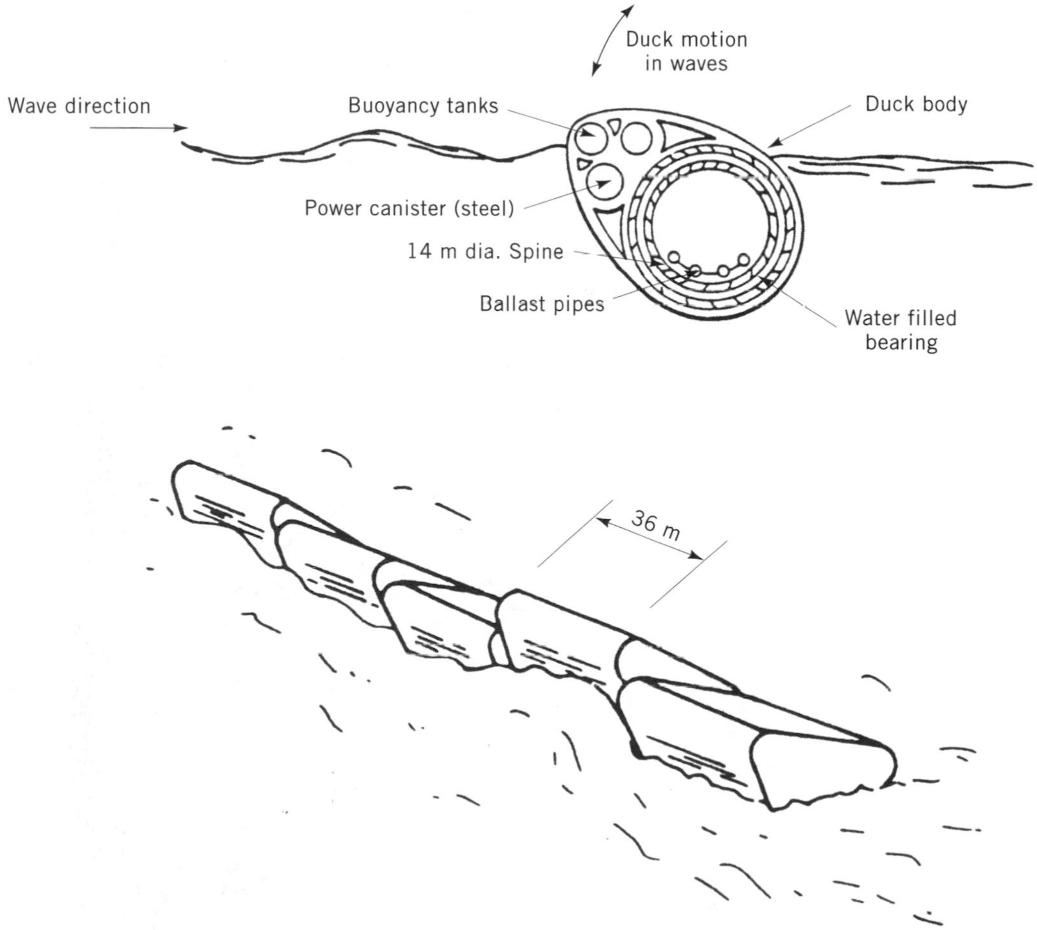

Figure 17. Nodding duck concept developed at the University of Edinburgh. The power conversion system is contained within sealed cannisters and uses a gyroscopic inertial reaction point (6).

problem. If they are designed to have sufficiently long stroke capability to handle the wave power densities of storms (up to 300 kW/m), their volumetric efficiency will be very poor under routine operating conditions (typically less than 30 kW/m). Such long-stroked cylinders also tend to be very costly. Alternatively, short-stroked hydraulic cylinders can be used with absorber end stops to avoid bottoming out the cylinders. Since the shock energies can be quite high, however, there is risk of structural damage to the absorber, as occurred, for example, with the Pendulor test unit at Muroran Port, described earlier. By absorbing energy only from pitch and using a rotary power take-off mechanism, the end-stop problem can be avoided altogether. A leading example of such a device is the Edinburgh Duck.

Wave energy research at the University of Edinburgh dates back to 1973 and has focused on the development of Dr. Stephen Salter's nodding duck concept. This device employs pitching floats that rotate about a long spine oriented parallel to the incident wave crests (Fig. 17). In waves, their action resembles that of ducks upending for food, hence the device's name. Models built at one-tenth scale have been tested on Loch Ness, and the projected cost of energy from this system has steadily decreased (51).

The Edinburgh Duck is widely recognized as one of the most innovative wave energy devices developed to date and is considered by many to be a "second-generation" system, with perhaps 15–20 years further development before a prototype would be ready for deployment (25). The Edinburgh Duck was conceived as a device that would recover the maximum amount from the potential wave energy resource around the United Kingdom. To accomplish this, it would have to operate in deeper water, where the waves are more energetic, but where the engineering demands on the device are also greater. In acknowledging the long time scale for development of the Duck, the Edinburgh design team has not limited itself to current technology but has incorporated a variety of novel features, particularly in structural design and the hydraulic power take-off system.

The 2-GWe reference design for the Outer Hebrides Islands (25) consists of eight spine strings, each 5 km long and supporting 108 Ducks. The Duck bodies are 36 m wide and rotate around a spine section that is 14 m in diameter. Within each Duck body are two independent power canisters, which are completely sealed units housing the power conversion system. The nodding motion of the Duck in waves is reacted against by a gyroscopic inertial frame of reference within each canister. Relative rotation between the gyroscope and the canister wall drives ring cam pumps, which supply hydraulic fluid to a high-pressure main. This ring main feeds several vari-axial hydraulic motor/pumps, which are connected to both the gyroscope and an electrical generator. This arrangement provides short-term energy storage via gyroscopic flywheels, enabling steady input to the synchronous generator.

Process 11: Flexible Bag and Inertial Reaction Point

Having described several seawater and hydraulic fluid systems, it is evident that the use of air as a working fluid has much to recommend it in terms of equipment simplicity (no pumps or valves). Enclosing the air in a flexible bag prevents the turbine from being exposed to corrosive salt air, as happens with oscillating water column systems. Use of a flexible rather than rigid structure also has the benefit of providing a large absorber volume with a relatively modest quantity of fabricated material. A leading example of this process is the SEA Clam.

Sea Energy Associates, Ltd. (SEA), is a British industrial consortium that has collaborated with Coventry Polytechnic under the leadership of Dr. Norman Bellamy to develop the Clam system. This effort began in 1978 and grew out of the group's earlier work on the spine-based Edinburgh Duck.

In its earliest form, the SEA Clam consisted of a straight, hollow spine, with a broad flap hinged along its bottom, seaward-facing edge. A flexible diaphragm was sandwiched between the flap and the spine, much like a clam in its shell, hence the device's name. Wave surge against the flap caused the diaphragm to act in the manner of a bellows, pumping air into and out of the hollow spine. It was soon found that more efficient transfer of wave power to air power could be obtained if the flap was removed, and the bag allowed to "breathe" freely, as shown in Figure 18. Air power is converted to electrical power by a Wells turbine placed in a duct between the bag and spine.

As part of the British national wave program, a 2-GWe reference design based on the straight-spine configuration was prepared for the Outer Hebrides Islands (52). Although this reference design was reported to be one of the most economical, there were several problems with the straight spine. Because the system was enclosed, the energy from any wave whose crest was parallel to the spine could not be absorbed, since the air had no place to go if all the bags were surged against at once. Therefore, an asymmetric mooring bridle was required to hold the spine at an angle to the prevailing wave crests. In a real sea, however, there will always be some wave energy coming from a direction other than the prevailing one, and those components whose crests were parallel to the spine would still not be captured.

A more serious limitation of the straight spine arises from the balance of forces required for its hydrostatic stability. Experimental and theoretical work has shown that the maximum amount of power that can be absorbed by the SEA Clam is directly proportional to the total displacement of the air bags. A larger submerged bag volume allows the waves to develop larger flow rates through the turbines. Examination of Figure 18, however, reveals that such an increase is limited by the pitch-restoring moment that can be provided by ballast in the spine section.

Wrapping the straight spine back on itself in a circle overcomes these problems (53). Efficient bags, having a low spring rate, can be used, and the circle's radius of curvature is such that components of wave energy from all directions can be absorbed. More importantly, the circular spine is pitch and roll stable, achieving a higher ratio of bag capacity to total device displacement. Cost and performance data for the circular SEA Clam are given later in this article.

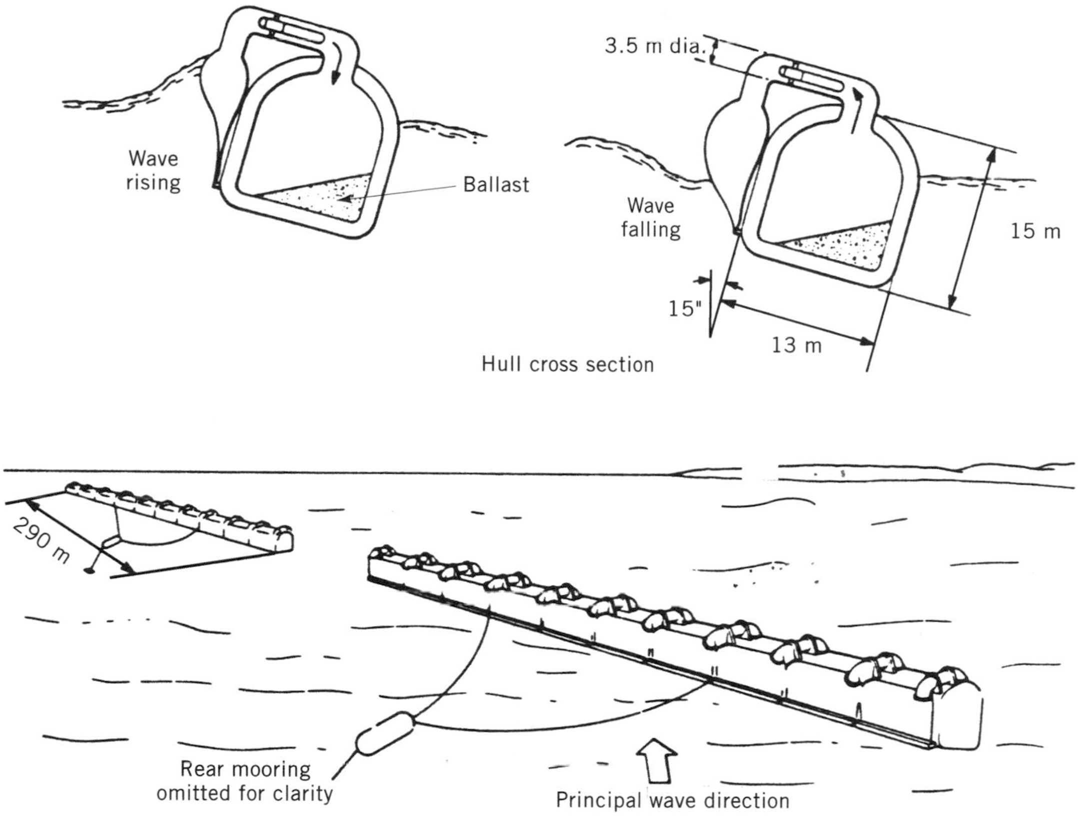

Figure 18. SEA Clam concept and straight-spine design developed during the earlier British wave energy program. Source: Energy Technology Support Unit, *Wave Energy Review Progress Report No. 1,* Harwell Laboratory, Oxfordshire, 1990.

Another important diaphragm device is the Lancaster Flexible Bag, invented by Dr. Michael French and subsequently developed by Wavepower Limited in the United Kingdom (54). Each device consists of a straight, buoyant spine 257 m in length oriented perpendicular to the prevailing wave crests. A series of flexible bags is mounted on each side of the spine, manifolded into high- and low-pressure air ducts running along the spine's interior. The bags are compressed as wave crests pass by, forcing air through nonreturn valves into the high-pressure duct. The bags expand during passage of wave troughs, withdrawing air from the low-pressure duct through another set of nonreturn valves. A conventional air turbine is mounted in the passage between high- and low-pressure ducts at the spine's end. The manifolded duct system thus acts as a short-term storage buffer, delivering a relatively steady flow to the turbine.

Process 12: Submerged Buoyant Absorber and Seafloor Reaction Point

Although heave is the predominant energy absorption mode for floats with a seafloor reaction point, they will also absorb energy in surge, particularly if sited in shallow water. By submerging the float and providing another pump/mooring to seaward, the resulting geometrical configuration will absorb a large amount of surge energy, comparable to or greater than that absorbed in heave. A leading example of such a device is the Bristol Cylinder.

Invented by Dr. David Evans, the Bristol Cylinder was developed in conjunction with Sir Robert McAlpine & Sons, Ltd., in the United Kingdom (55). As with the NEL Breakwater, the Edinburgh Duck, the SEA Clam, and the Lancaster Flexible Bag, its development was pursued largely under the British national program aimed at establishing the feasibility of a 2-GWe central station in the Outer Hebrides Islands.

The absorber for the 2-GWe reference design is a hollow cylinder, fabricated from precast reinforced concrete sections, 100 m long with a diameter of 16 m, submerged 6 m beneath the sea surface in a water depth of 42 m. Each cylinder is anchored by eight mooring legs, with two double-acting pumps splayed off each end (Fig. 19). The mooring legs are polymer tube springs, which are similar to the Swedish hose pump described earlier. The spring rate of the cylinder, and hence its tuning to incident wave frequencies, can be adjusted by changing the internal pressure of the tube springs. It is important to note that although the tube springs are similar to hose pumps, seawater pumping in the Bristol Cylinder is accomplished by stainless steel hydraulic rams.

In the 1982 reference design, high-pressure seawater was pumped to 120-MWe Pelton turbine generators located on fixed platforms. In the most recent (1991) design configuration, a closed-circuit hydraulic conversion system has been adopted, with a complete 1-MWe power plant located in each of eight mooring legs (25). These

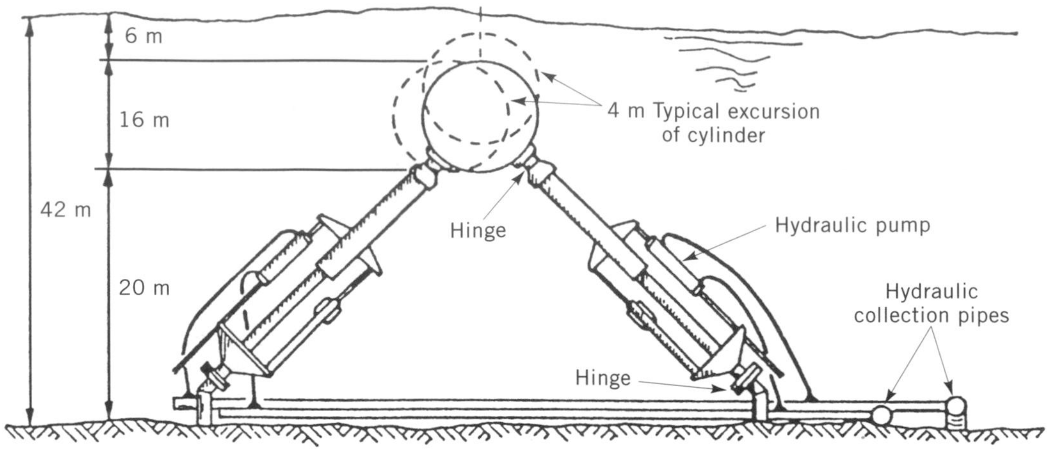

Figure 19. Bristol Cylinder. Source: P. G. Davies, *Wave Energy—The Department of Energy's R & D Programme 1974–83*. Energy Technology Support Unit, Harwell Laboratory, ETSU R26, Oxfordshire, 1985.

also contain linear pumps for tuning the device, and the tube springs have been eliminated. Each mooring leg has a 3.3-kV submarine power cable connecting it to a fixed platform, where the power is transformed to 275 kV for transmission to shore. This new configuration is quite different than the earlier design, and a considerable development program would be required to prove the practicality and reliability of the new mooring leg concept.

UTILITY APPLICATIONS

Electric utility applications of wave energy conversion are distributed generation, to meet local demands on the coast, and central stations designed to export bulk power inland to the main grid. A gas utility application of wave power would be the production of hydrogen as a winter heating fuel. Each of these applications are described separately below.

Distributed Generation

Distributed generation is defined as a wave power plant designed to meet the demand of a coastal load center. A submarine power cable would be landed into an existing onshore substation, and the necessary transformers and switchgear would be provided to connect to the local distribution system. Export of power inland would be incidental only to variations in local demand, and it is assumed that the installed wave-generating capacity would be sufficiently small so as not to require any reinforcement of existing inland transmission lines.

Coastal towns in high latitudes seldom require air conditioning in the summer, and where natural gas service is not available for winter heating, the seasonal variation in electricity demand is well matched to that of the wave energy resource. The remoteness of the coast also makes it attractive for "mini-utilities," whereby winter peak load growth is met by building small, local wave power plants rather than additional transmission lines from the inland grid. A survey of coastal substations in northern California suggests that the useful size of such a wave power plant would range from 1 to 10 MW (6). It should be noted that this range applies to large mainland utility grids; for many islands, plant sizes on the order of tens to hundreds of kilowatts would be more appropriate for distributed generation.

In addition to meeting coastal load growth, a few megawatts of distributed wave-generating capacity may benefit the regional transmission network by providing grid stability and voltage support, particularly in remote coastal areas with load centers at the end of a long transmission circuit. A recent study by Pacific Gas and Electric Company (PG&E) has suggested that small-scale photovoltaic (PV) plants would provide avoided-cost benefits to the local distribution system comparable to the value of the PV plant itself (56). A small wave power plant can have the same effect.

As important as distributed wave generation may be at the local or regional level, its impact on an entire grid may be insignificant. For example, demand growth along the northern California coast in PG&E's service territory is expected to be less than 100 MWe over the next 10 years (6), which represents about 1.5% of PG&E's projected gridwide demand growth over the same time period. If wave power is to make a significant contribution to global energy demand, then large central generating stations must be successfully integrated into the world's industrial utility grids.

Central Station Generation

A major consideration for central station generation is inland transmission of the power once it lands ashore. A second concern is managing the power so it can be most effectively combined with other generating capacity.

In many regions of the world, good wave energy resources are located along coastal stretches that are relatively unpopulated and remote from main-grid load centers. In PG&E's service territory, for example, there are nearly 350 km of coastline with no existing utility grid whatsoever. These include wilderness and park areas, where acquisition of transmission right-of-way and substation land is likely to be difficult. Such isolated coastal stretches represent over one-third of PG&E's wave energy development potential (6).

Even where the grid extends to the coast, it is often not designed to handle large amounts of power. For example, although two-thirds of PG&E's coastline is served by the grid, locations with good wave exposure are typically served by relatively slender 60-kV transmission lines. This is true even in the heavily populated San Francisco Bay area. It has been estimated that inland export of wave power via existing transmission lines would be limited to less than 200 MVA, which is only about 3% of PG&E's projected growth in gridwide demand over the next 10 years.

Another possible inland route for wave power is surplus transmission capacity at thermal power plants located on the coast for access to cooling water; however, these plants may be far removed from the best wave energy resources. Furthermore, much of the surplus may be unavailable, due to other utility requirements for these lines.

Therefore, substantial development of the wave energy resource for bulk power generation will require reinforcement or extension of the onshore utility grid. Recent studies in the United Kingdom suggest that for a 2-GWe wave-generating station off the Outer Hebrides Islands, reinforcement of the onshore grid will be as expensive as sea-to-shore power transmission (25).

Once the power reaches the central grid, there are further difficulties. Although waves are more persistent than the winds that generate them, output from a wave power plant can fall below rated capacity at irregular intervals lasting from a few hours to several days. Modern wave forecasting models exist that can accurately predict wave conditions 24–48 h in advance, which eases the dispatch problem but does not eliminate the need for back-up generation. It is not practical to start up and shut down large coal-fired plants over this time scale. Either small gas- or oil-fired plants (which have higher energy costs than coal) must be available or a large coal-fired plant would have to be kept on part load (spinning reserve), which is relatively inefficient. Alternatively, long-term energy storage, such as onshore pumped hydro stations, could be integrated with wave power development but would be limited by the availability of suitable reservoirs.

Siting two or three smaller wave power plants at widely spaced intervals will provide a greater level of firm capacity than siting the same amount of wave generation in one location. Another way to increase wave energy's baseload contribution is to deploy more absorbers per unit generator rating. This might be economical if absorber fabrication was only a small part of the total plant cost, but as documented later in this article, such is not the case.

Hydrogen Production

A possibly more effective use of central station wave power would be the production of hydrogen as a winter heating fuel. This provides a good match to seasonal demand, eliminates the dispatchability problem, and avoids the environmental and economic costs of reinforcing or extending the onshore utility grid. Furthermore, a recent study by PG&E suggests that hydrogen for heating is a more economical carrier of renewable energy than either hydrogen in fuel cells or electric power transmission with pumped storage (57). Only where no storage is required, as with a naturally dispatchable PV plant (for meeting afternoon peak air conditioning demand), does electric power transmission become more economical than hydrogen for heating.

A recent economic analysis suggests that electricity costs must be less than 2¢/kW·h before hydrogen produced by electrolysis can become competitive with that now produced by steam reforming of methane (58). As natural gas reserves are depleted, however, the cost of methane reformation will rise. Meanwhile, with the commercial deployment of small coastal plants for distributed generation, as described earlier, the cost of wave-generated electricity should fall. As these two trends act together, hydrogen from wave energy could become the lower cost source.

WAVE ENERGY COPRODUCTS

Wave energy conversion processes must be proven in distributed generation before they can be seriously considered for central station applications. The economic viability of such small wave power plants can be enhanced by a variety of coproducts obtained incidental to the generation of electricity, including:

- Seawater renewal for closed-pond aquaculture
- Fresh water for household use, livestock watering, and crop irrigation
- Breakwater protection for ports and harbors

Coproduct potential varies among different wave energy devices, depending on whether they are fixed or floating and whether their working fluid is air, hydraulic oil, or seawater. Some devices could be modified to generate electricity and all three coproducts in a single plant. For example, the closed-circuit hydraulic system of a caisson-based pivoting flap could be replaced with a high-pressure seawater system. Some of the pumped seawater could be diverted to onshore aquaculture ponds, while the rest would enter a RO module. Only part of this flow would pass through the RO membrane as fresh water; the remaining high-pressure brine would be discharged through a Pelton turbine to generate electricity. The caissons themselves would provide harbor protection. Such a combined system, however, is now a matter of speculation. The following text describes single coproduct applications that have been studied for state-of-the-art wave energy devices.

Aquaculture

The reservoir of a Tapered Channel power plant does not provide long-term storage but smooths the input from one high-energy wave group to the next. The reservoir at Toftestallen has an area of 5500 m^2, while the 350-kWe turbine/generator has an operating head of 3 m and a flow rate of 14–16 m/s (12). Assuming a turbine/generator efficiency of 90%, the upper meter of the reservoir is replaced once every 6 min, which is a tremendous rate of seawater renewal. The Crown Agents of London have estimated that the value of fish farmed in a Tapered Channel reservoir "may cover all capital costs of wave collectors, converters and basin" (59).

Deep oceanic water is cold, nutrient rich, and relatively free of pathogens, which makes it an ideal resource for aquaculture, providing ease of temperature regulation, enhanced productivity, and reduced exposure to infection. Its use for open-ocean aquaculture has been proposed in connection with ocean thermal energy conversion (OTEC), but wave-powered pumps also have been proposed for the same purpose (60).

Fresh Water

Although the vast majority of wave energy systems are designed for electric power generation, the previously described DELBUOY was designed primarily for freshwater production. Desalination of seawater is not limited to heaving buoys but can be accomplished by any device that uses seawater as a working fluid and has a ratio of absorber area to pump cross section sufficiently high to develop pressures of 5.5–6.9 MPa (800–1000 psi).

Reverse osmosis is a cross-flow molecular filtration process whereby the pressure differential (in excess of feed stream osmotic pressure) across an RO membrane drives water through the membrane. Salt ions also migrate through the membrane but are driven by concentration gradient rather than by pressure. Therefore, a high seawater flow rate is required to flush away rejected salts that would otherwise build up in a high-concentration layer on the feed side of the membrane surface, resulting in excessive salt migration into the permeate.

Depending on the flow rate and pressure of the seawater feed, only 10–40% of the feed flow passes through an RO membrane to be recovered as low-salinity permeate. The remaining 60–90% is rejected as high-pressure brine. If fresh water is the sole desired product, then turbine-driven pumps can be used to recover energy from the reject brine to boost seawater feed pressure. Alternatively, the reject brine can be discharged through a Pelton turbine/generator for the coproduction of electricity.

Breakwater Protection

Since caissons reflect all wave energy that they do not absorb, there is the possibility of a distinct low-energy "shadow" developing behind the plant, depending on how closely the individual caissons are spaced. This maximizes their potential environmental impact on nearshore sedimentary processes. It is this reflective feature of caissons, however, that makes them ideal candidates for combining wave energy generation with breakwater protection at coastal harbors. In such cases their visual appearance would probably not be considered intrusive, since local scenery has already been altered by existing harbor development.

Even in remote coastal areas, an occasional wave energy breakwater might be acceptable for creating a harbor of refuge for small craft in the event of sudden storms, medical emergencies, or engine problems. The coproduction of fresh water and electricity would have obvious advantages in such an application.

It should be noted, however, that the potential market for wave energy breakwaters is relatively small. For example, PG&E has an open-ocean coast approximately 1000 km in length. Yet a recent survey identified only two existing harbors that needed additional breakwater protection and five coves that could be developed into harbors of refuge (6). The potential wave-generating capacity at these seven sites was estimated to be only 12.5 MWe, which is insignificant in terms of PG&E's gridwide demand. Nevertheless, if even the smallest of these (700 kWe) was developed, it would double the world's present wave-generating capacity. Therefore, breakwaters are an important stepping stone in the commercialization of wave power.

DEMONSTRATION PROJECTS

As of this writing, the total capacity of grid-connected wave power is 685 kWe worldwide. This is dominated by a 350-kWe Tapered Channel plant in Norway and a 150-kWe caisson-based OWC in India. Three smaller plants (a 20-kWe caisson-based pivoting flap, a 30-kWe land-based OWC, and a 60-kWe caisson-based OWC) are now operating in Japan. The most recent addition is a 75-kWe onshore, gully-based OWC in the United Kingdom, which came on line in 1991.

Offshore wave energy conversion processes are considerably less advanced. The specially built ship, *Kaimei*, was deployed twice in the Sea of Japan to test various pneumatic turbines (used in OWC processes and the SEA Clam). Electricity was landed ashore from only a single 125-kWe turbine/generator, however, and this for only 4 months during the winter of 1978–1979. Two heaving buoy systems also have been grid connected for a period of months: a 30-kWe Swedish system in 1983–1984 and a 45-kWe Danish system in the spring of 1990. There have been three offshore projects that did not involve grid connection: a 110-kWe floating reservoir filled by wave surge, deployed off the coast of Sweden in 1991; the jack-up rig *Kaiyo*, which was deployed near Okinawa in 1984–1985; and a platform-based 20-kWe tandem flap deployed in Lake Michigan in 1987. None of these operated continuously for longer than a few months.

SYSTEM ECONOMICS

It is significant that the six demonstration plants now operating are all land- or caisson-based systems. Floating devices have accumulated much less sea time, and as a result, assessments of their economic feasibility are less certain.

On the other hand, the cost and performance of land- and caisson-based systems are more sensitive to local site conditions. The cost of site preparation for a Tapered

Channel power plant or land-based OWC depends greatly on shoreline geology and topography. To the extent that cast-in-place concrete construction is required, the economic feasibility of these schemes also depends on the local availability of aggregate material. Likewise, the cost of a wave energy breakwater is influenced by the need for seafloor foundation leveling as well as the local availability of suitable rock for a rubble mound on which the caissons would rest.

The sheltering effect of coastal features, such as headlands and peninsulas, and the effects of wave refraction are much greater nearshore than offshore. Consequently, the performance of land- or caisson-based systems for a given deep-water wave resource is highly dependent on the exact coastal location of the plant.

Therefore, while cost and performance projections for land- and caisson-based wave power plants draw on greater experience, they are also more site specific. Projections for offshore systems are less certain but are more generally applicable for a given regional wave climate.

Designs Selected for Economic Assessment

Cost and performance data were compiled for 13 wave power plant designs based on 7 different conversion processes. The main features of these designs are listed in Table 1, and additional details are given below.

The Tapered Channel power plant on the south coast of Java is the only design actually intended for construction (61). At its seaward entrance, the channel is 124 m wide, narrowing to a width of just 7 m at its reservoir end over a length of 60 m. The reservoir has an area of 7000 m^2 and drains through a 1.1-MWe, vertical-axis Kaplan turbine/generator. The total cost of the project is estimated at 50 million Norwegian crowns (7.2 million U.S. dollars). Only three-quarters of this is process capital (50% for civil construction and 25% for electrical equipment). The remainder is for engineering, construction management, technology transfer, and project contingency. Annual operating and maintenance expenses are expected to be 2% of the total project cost.

It is interesting to note that the Tapered Channel reservoir makes much more energy-efficient use of coastal land area than would a wind energy project. The annual output of the Java plant is estimated to be 6.1×10^6 kW·h. By comparison, a wind farm operating at 25% capacity factor (typical of a good wind energy resource) would require ten 275-kWe turbines to generate the same amount of electricity. A state-of-the-art wind turbine of this capacity would have a rotor diameter of 26 m (62), and assuming that 10 such turbines were arranged in two rows, with eight rotor diameters between rows and three rotor diameters between individual turbines in a row, the wind project would occupy a land area of 64,900 m^2, about nine times that of the Tapered Channel reservoir. Moreover, just as wind farm land can be used for livestock grazing, Tapered Channel reservoirs can be used for closed-pond aquaculture, as explained earlier.

The 1-MWe gully-based OWC design was developed as part of the United Kingdom's most recent wave energy review (6). During construction of the 75-kWe prototype plant at Islay, a sheet pile cofferdam was built across the mouth of the gully to shelter the construction work from wave action (63). Building the cofferdam was time consuming, and there was more water movement behind it than expected, making construction work difficult, even during calm weather. The design in Table 1 is based on an alternative construction approach, whereby a "designer gully" is excavated by blasting into virgin rock some 20 m in from the natural shoreline (Fig. 20). Construction of the OWC capture chamber and installation of the mechanical and electrical plant occurs in the dry. When construction is complete, the remaining rock in front would be cleared away, opening the plant to the sea. Not only does this facilitate construction, it also enables better shaping of the harbor in front of the capture chamber, eliminating the energy-dissipating irregularities that are characteristic of natural gullies.

The latest NEL breakwater design (6) consists of 660 individual caissons 64 m wide with three OWC capture chambers and a weight of 22,500 t per caisson. The breakwater would be located in 20 m water depth 5 km offshore the island of South Uist. In order to accurately compare this design with Muroran's much smaller Pendular System, cost and performance projections were derived for a single NEL caisson having a generation capacity of 4.1 MWe, assuming that it was shore connected, without the cost and power losses associated with a submarine power cable, and with ready access by a shore-based maintenance crew. Note that unlike the Pendular System, which relies solely on caisson weight for stability against extreme wave forces, the NEL breakwater is further anchored by wire-rope tendons.

The caisson-based Pendulor System is designed for installation in much shallower water. The 1.5-MWe plant consists of 10 caissons, each 25 m wide and weighing 1750 t, placed side by side in a shore-connected breakwater configuration. There are three pivoting flaps per caisson, driving six hydraulic cylinders that pump fluid to a 150-kWe motor/generator (Fig. 21).

The five heaving buoy plant designs by Danish Wave Power involve much greater generation capacity than any of the other designs in this economic analysis. They also employ the smallest individual generators (130 kWe). As a result, their power transmission scheme is considerably more elaborate, making use of extensive subsea cabling as well as four offshore platforms (Fig. 22).

In recent feasibility studies for Pacific Gas and Electric Company (6) and the State of Hawaii (7), SEASUN Power Systems developed reference designs for 10- and 30-MWe wave power plants, based on the Swedish heaving buoy concept, moored in 80 m water depth at two hypothetical deployment sites, one 24 km offshore Half Moon Bay, just south of San Francisco, and the other 10 km offshore Makapuu Point, on the west coast of Oahu. The basic plant design consists of one or more star-shaped clusters of buoys, with 6 collecting lines per star and 10 buoys per collecting line (Fig. 23). The buoys have a diameter of 16–17 m and are spaced on 30-m centers, giving the star a total diameter of 600 m. An underwater habitat at the center of each star houses a vertical-axis, 10-MWe Pelton turbine/generator and a 2.5/35-kV transformer. A three-

Table 1. Cost and Performance Data for Wave Power Plant Designs[a]

Plant Location	Distance Offshore, Water Depth, and Incident Wave Power	No. and Size of Structural Modules	Process Capital (PC) Cost	Fixed O&M Cost	Annual Output
Conversion Process: Tapered Channel (Plant Capacity: 1.1 MWe), 1 module					
South coast of Java, Indonesia	Shore based, 17 kW/m	7000 m^2 reservoir	\$4790/kW (total PC)	2% of total PC/yr	6.1 GWh/yr (63% CF)
Gully-based OWC (1 MWe), 1 module					
South Uist, Outer Hebrides Islands, U.K.	Shore based, 18 kW/m	150 m^2 OWC, 313 m^2 harbor	\$2030/kW (total PC)	\$64/kW/yr	2.2 GWh/yr (25% CF)
Caisson-based Pivoting Flap (1.5 MWe), 10 modules					
Muroran Port, Hokkaido, Japan	Shore connected, 5 m depth, 10.6 kW/m	1750 t caisson	\$2420/kW (BOP)	1% of total PC/yr	6.9 GWh/yr (53% CF)
Caisson-based OWC (4.1 MWe), 1 module					
South Uist, Outer Hebrides Islands, U.K.	5 km offshore, 20 m depth, 30 kW/m	22,500 t caisson 675 m^2 OWC	\$1810/kW (BOP)	\$88/kW/yr	7.5 GWh/yr (21% CF)
Circular SEA Clam (12.5 MWe), 5 modules					
Lewis, Outer Hebrides Islands, U.K.	1 km offshore, 40 m depth, 34 kW/m	5005 t, 60 m dia. hull	\$840/kW (BOP)	\$118/kW/yr	16 GWh/yr (15% CF)
South Uist, Outer Hebrides Islands, U.K.	17 km offshore, 40 m depth, 52 kW/m	5005 t, 60 m dia. hull	\$1120/kW (BOP)	\$118/kW/yr	26 GWh/yr (24% CF)
Danish Heaving Buoy (343 MWe), 2640 modules					
Northwest coast of Jutland, Denmark	80–120 km offshore, 50 m depth, 15 kW/m	1040 t base, 10 m dia. buoy	\$3380/kW (BOP)	1.5% of total PC/yr	575 GWh/yr (19% CF)
North of Utsira, west coast of Norway	10–18 km offshore, 200 m depth, 22 kW/m	1,485 t base, 10 m dia. buoy	\$2820/kW (BOP)	1.5% of total PC/yr	728 GWh/yr (24% CF)
West of Auckland, New Zealand	30 km offsore, 100 m depth, 30 kW/m	1,410 t-base, 10 m dia. buoy	\$3110/kW (BOP)	1.5% of total PC/yr	945 GWh/yr (31% CF)
Galway Bay, west coast of Ireland	35–40 km offshore, 100 m depth, 66 kW/m	1410 t base, 10 m dia. buoy	\$3160/kW (BOP)	1.5% of total PC/yr	1186 GWh/yr (39% CF)
Aveiro, west coast of Portugal	32 km offshore, 100 m depth, 26 kW/m	1410 t base, 10 m dia. buoy	\$2890/kW (BOP)	1.5% of total PC/yr	744 GWh/yr (25% CF)
Swedish Heaving Buoy (10 MWe), 60 modules					
Half Moon Bay, Calif.	24 km offshore, 80 m depth, 25 kW/m	17 m dia. buoy, damper plate[b]	\$1070/kW (BOP)	4% of total PC/yr[c]	25 GWh/yr (29% CF)
Makapuu Point, Oahu, Hawaii	10 km offshore, 80 m depth, 15 kW/m	16 m dia. buoy, damper plate[b]	\$1020/kW (BOP)	4% of total PC/yr[c]	37 GWh/yr (42% CF)

[a] Capacity factor (CF) is indicated in parentheses in last column. Balance of plant (BOP) cost does not include structural fabrication of modules which is accounted for by multiplying number of modules by module weight and by concrete fabrication costs in different countries (see Appendix). Abbreviations: O&M, operation and maintenance; OWC, oscillating water column; PC, process capital.
[b] Buoys and damper plates contain internal steel structural reinforcement. The combined equivalent concrete weight is 327 t for a 17-m-dia. buoy and damper plate, or 289 t for a 16-m-dia.-buoy and damper plate.
[c] Plus periodic hose and mooring hardware replacement at \$25/kW/yr in California, or \$30/kW/yr in Hawaii.

phase AC submarine power cable links adjacent stars and transmits power to shore.

In the recently completed review of wave energy in the United Kingdom (6), the circular SEA Clam emerged as the most economical of the main British offshore devices (Fig. 24). The circular spine configuration was developed after the 2-GWe central station program had ended in the United Kingdom. A conceptual design and cost estimate have been prepared, however, for a 12.5-MWe plant design containing five Clam modules deployed at two in the Outer Hebrides Islands. Each module has a toroidal hull 60 m in diameter assembled as a dodecahedron from 12 concrete sections 4.5 m wide and 8 m deep (Fig. 25). Bags from adjacent sections communicate with each other via metal ductwork. Ten 250-kWe Wells turbine/generators are placed within the main duct, and the entire air system is maintained at a pressure of 15 kPa (2 psi). Projections of electrical output are based on $\frac{1}{15}$-scale model tests in Loch Ness. Although detailed cost and performance data were available only for a 60-m device, significant economic improvement is expected if the diameter of the concrete hull is increased to 80 m (64).

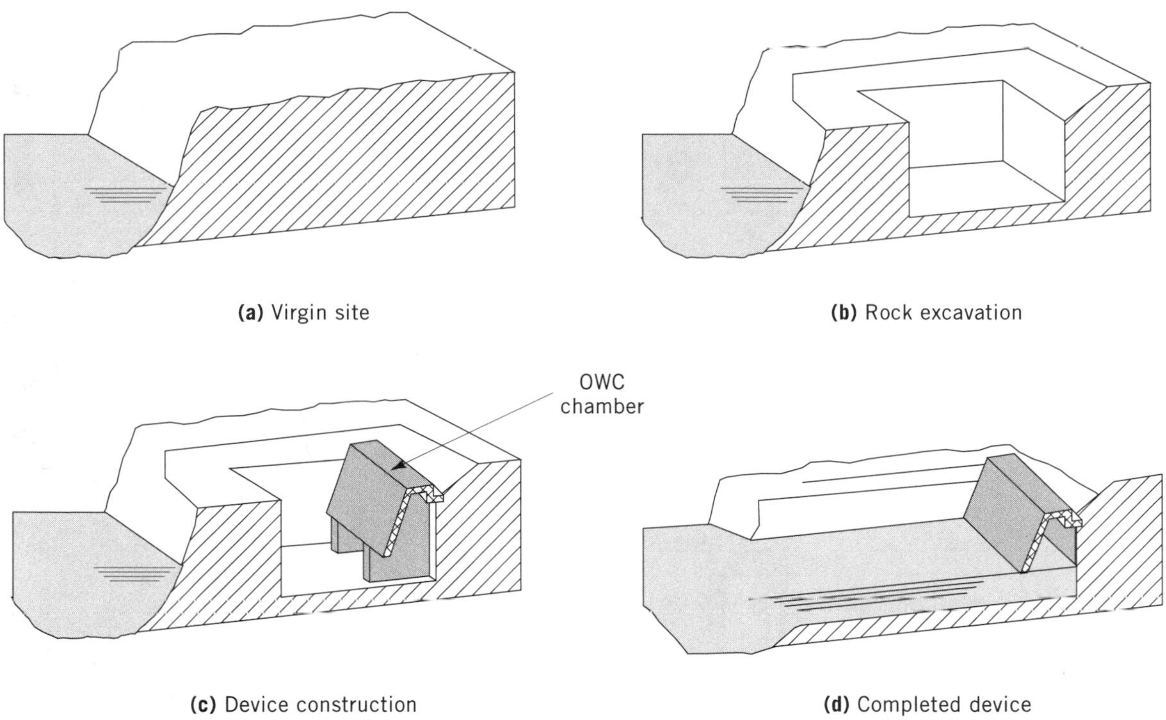

Figure 20. Construction sequence for a "designer gully", as part of the 1-MWe oscillating water column design developed by the Queen's University of Belfast (Shoreline OWC in Fig. 24) (25).

Economic Assessment Methodology

Levelized energy costs were computed using financial parameters specified in the EPRI Technical Assessment Guide (EPRI TAG; parameter values are given in the Appendix at the end of this article). These reflect the life cycle costs that would be experienced by a regulated, investor-owned utility in the United States. It should be noted, however, that the first commercial wave power plants probably will be financed as independent projects, where the owner/operator sells power to a utility. There is no standard formula for computing energy costs in such cases because of the wide variability in project financing schemes. For example, the plant may be sold to the utility after an initial period of successful operation under a power-purchase agreement, but well before the end of its service life. Under these circumstances, a single levelized energy cost would be difficult to define.

Nevertheless, a standard formula is required if economic comparisons are to reflect differences in technology cost-effectiveness rather than differences in accounting methods, financing schemes, or government monetary policies. The EPRI TAG was chosen because it has been developed for just this type of analysis and is well documented (65).

Economic Comparison among Different Wave Energy Devices

The levelized cost of wave energy is plotted as a function of wave energy resource magnitude in Fig. 26. A range of energy costs is given for all but the two land-based designs, reflecting the wide range of unit costs reported for concrete structural fabrication in different countries (see Appendix). As clearly shown by this plot, the economics of a wave power plant in a given part of the world depend much more on local labor and materials costs than they do on the local wave climate. Wave power development looks particularly promising in Portugal, Ireland, and New Zealand, all of which have good wave energy resources and low structural fabrication costs.

As shown in Table 1, devices that use a fixed reaction point (concrete caisson or gravity base) have a structural weight requirement of 5500 t or more per megawatt of generating capacity. On the other hand, devices that use an inertial reaction point (buoyant spine or suspended damper plate) require just 2000 t or less of concrete per megawatt of generating capacity. Consequently, their economic feasibility is markedly less dependent on local structural fabrication costs.

The capital and operating components of energy cost and the distribution of capital costs are given for fixed-reaction-point designs in Figure 27 and for inertial-reaction-point designs in Figure 28. Not surprisingly, mooring and deployment costs are proportional to the mass of the structures that must be positioned and held in place.

Sea-to-shore transmission cost, when normalized against distance offshore, depends greatly on the number of subsea connections. For example, consider three offshore designs that are located at somewhat similar distances offshore (17–30 km). The Danish Wave Power

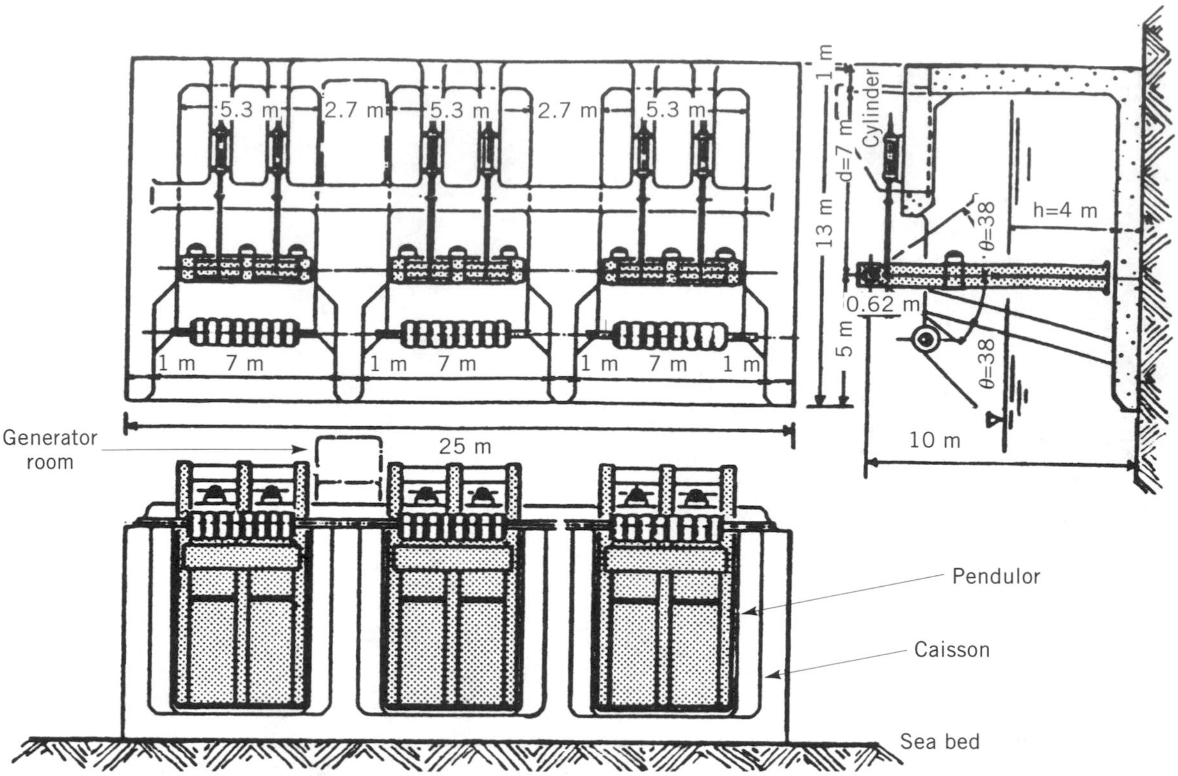

Figure 21. Caisson-based pivoting flap design developed by Muroran Institute of Technology. Each caisson contains three Pendulors, which drive a 150-kWe power conversion system (33).

transmission scheme, which involves 2640 modules to be interconnected, costs $26/kWe km for the New Zealand plant. The SEA Clam design requires that only five modules be interconnected, with a transmission cost of $22/kWe km for the South Uist plant. The Swedish heaving buoy design involves just a single submarine power cable, costing $16/kWe km for the Half Moon Bay plant.

The SEA Clam design at South Uist has a higher capital cost than the same design at Lewis, due to the much longer sea-to-shore transmission distance. The farther offshore site is exposed to much higher wave energy levels, however, resulting in a 60% increase in annual output and an overall lower cost of energy.

Comparison of the two Swedish heaving buoy designs reveals several interesting features. Oahu's island shelf is much narrower than the continental shelf off Half Moon Bay, and only a 10-km submarine power cable is required to reach the 80-m-depth contour. Smaller diameter buoys are optimal in the shorter period waves off Makapuu Point, which represents an even bigger cost saving. On the other hand, offshore deployment equipment will be more costly in Hawaii. The total capital investment is less, however, as higher deployment costs are more than offset by lower transmission and buoy fabrication costs.

The Swedish heaving buoy plant also has a much greater capacity factor in Hawaii than off northern California. This is because the buoys are more efficient absorbers in shorter period waves, which prevail along Hawaiian island coasts exposed to northeast trade winds. On the other hand, the wave climate off northern California is dominated by long-period swell, in which buoy performance is much poorer. Off Makapuu Point, the average wave energy absorption efficiency is 43%, compared with only 17% off Half Moon Bay. Trade wind waves are also more persistent than swell generated by winter storms in the North Pacific, which further contributes to the plant's higher capacity factor in Hawaii.

Economic Comparison of Wave Power with Other Energy Sources

Given the potential market that exists for wave power in the Pacific Ocean, Hawaii represents a good site for comparing wave energy costs with other alternatives. Cost and performance data for competing technologies were derived from Hawaii Electric Light Company's supply-side report that was recently prepared as part of its integrated resource plan (66). These data are summarized in Table 2 and include both coal- and oil-fired generation, together with four renewable energy sources: wind, biomass, geothermal, and OTEC.

In the wave energy costs plotted in Figures 26–28, only process capital and interest paid during plant construction were included in the capital carrying charge. A real project would also include several indirect costs, such as land-based general facilities (offices, shops, and ware-

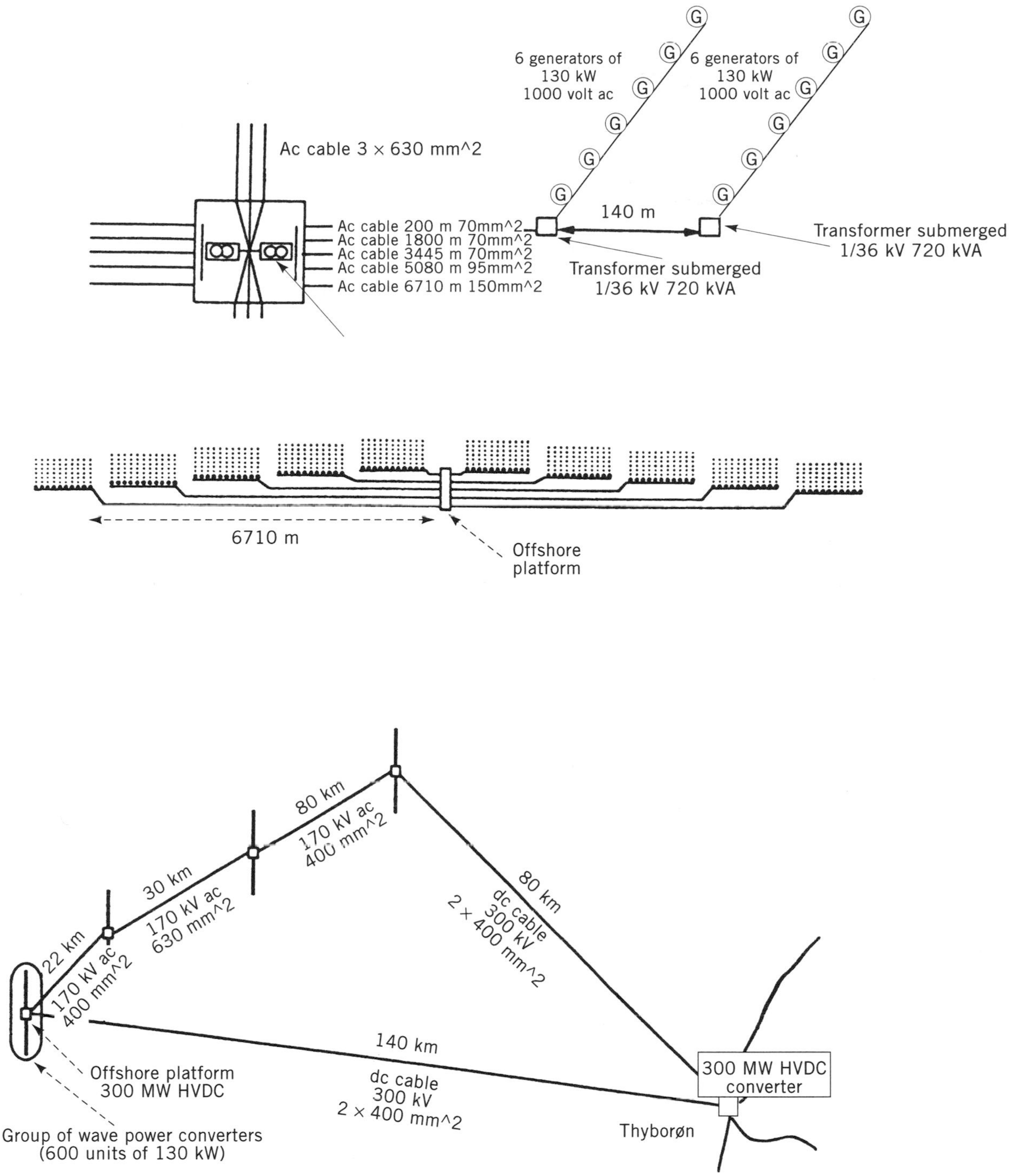

Figure 22. Sea-to-shore power transmission scheme of Danish Wave Power's heaving buoy design for the northwest coast of Jutland (44).

houses), field engineering, and project management. In the Hawaii reference baseline designs for the Swedish heaving buoy power plant, indirect costs were estimated as 20% of process capital, and an overall project contingency of 10% was used to calculate the total capital costs shown in Table 2.

As before, the EPRI TAG (65) was used to compute levelized energy costs, assuming a 30-year service life for all projects (see Appendix for financial parameters used). Results are plotted in Figure 29, which shows that although wave energy is more costly than wind (by about 1.5¢/kW·h for 10-MW plants), it is comparable to the cost of diesel engine generation and geothermal power. The trend lines with increasing capacity suggest that wave energy is also competitive with oil-fired combined cycle generation and is less costly (by at least 2¢/kWh for comparably sized plants) than OTEC, biomass combustion, and oil-fired single-cycle generation.

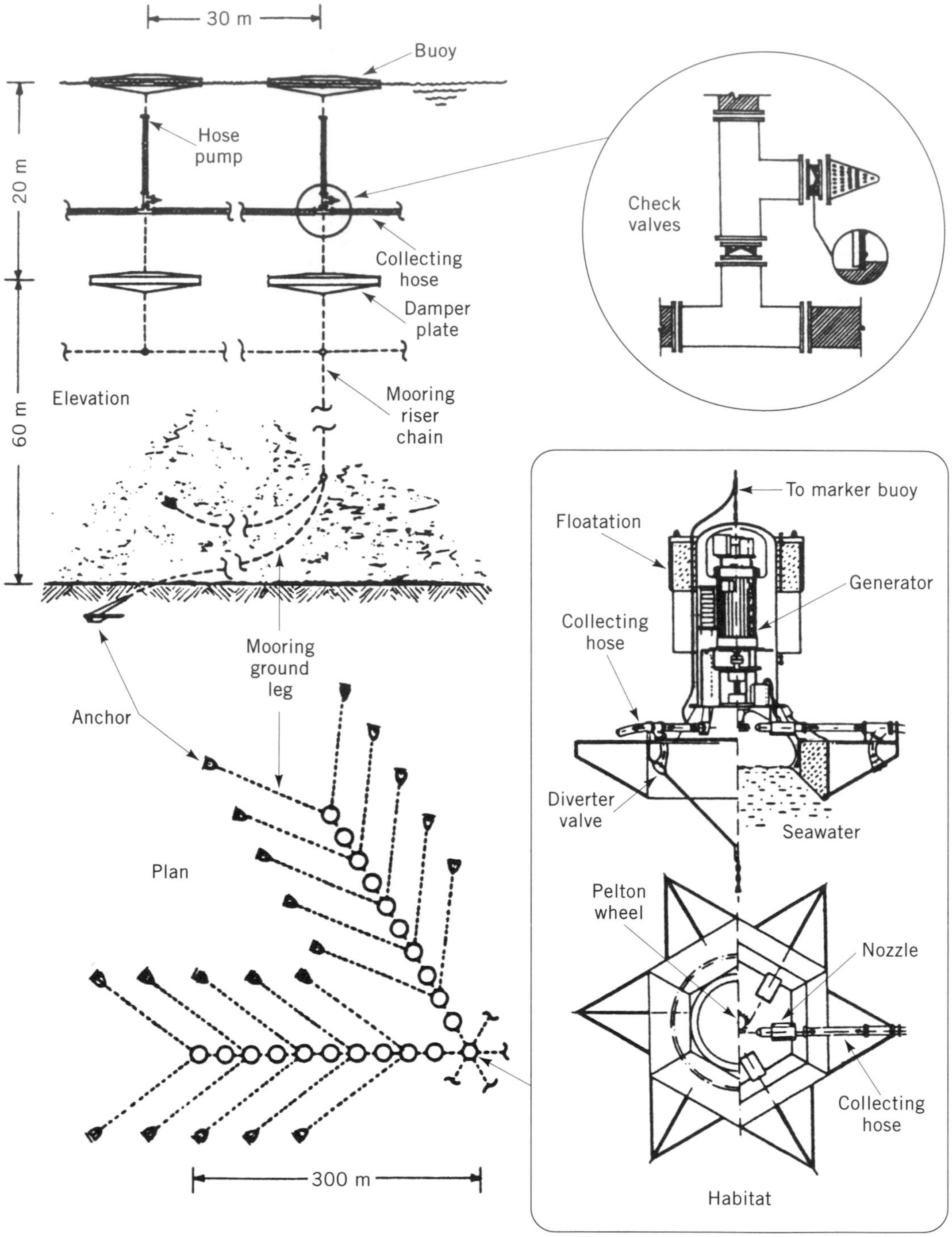

Figure 23. Arrangement details of Swedish heaving buoy design developed by SEASUN Power Systems for 80 m water depth off Half Moon Bay, Calif., and Makapuu Point, Oahu, Hawaii (6).

Freshwater revenues have a large impact on the cost of energy from OTEC power plants. Based on the current rate schedule (1994) of the County of Hawaii Department of Water Supply, \$0.4/m^3 (\$1.50 per thousand gallons) represents the most that an OTEC plant could earn for its freshwater production. At today's water prices, a 10-MWe open-cycle OTEC plant with a single condenser stage is more economical than one with two condenser stages. Likewise, for nominal 50-MWe plants, a closed cycle would be more economical than a hybrid cycle. If water prices double, however, the reverse would be true.

Although no designs have been prepared for a wave power plant producing both fresh water and electricity, this is clearly possible, as explained earlier. Depending on water demand and pricing, it is conceivable that wave energy costs could also be lowered by credit for the production of fresh water.

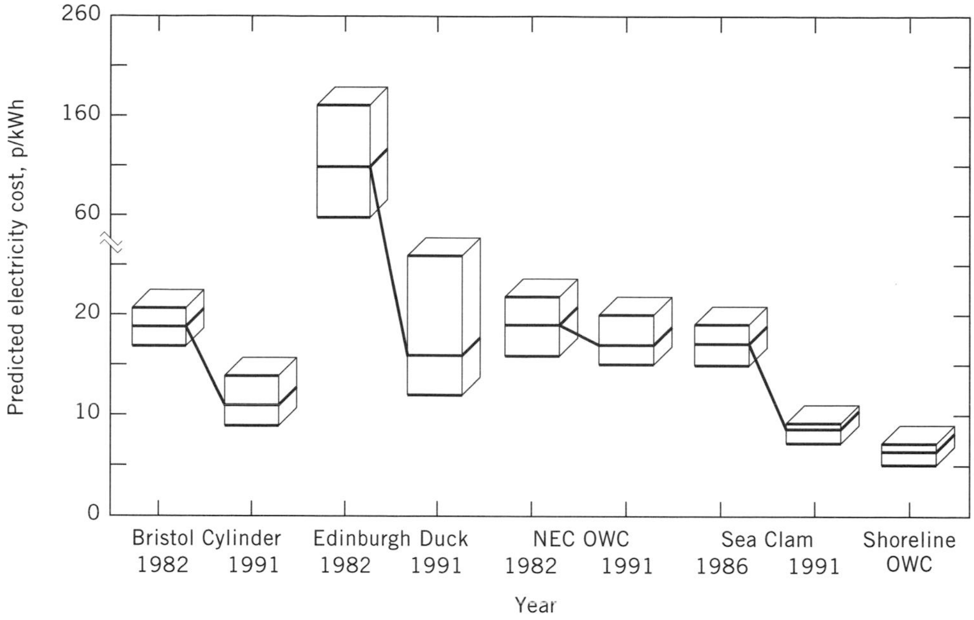

Figure 24. Levelized energy costs for the five main devices studied in the most recent review of wave energy in the United Kingdom. Four of the devices were also studied in the earlier British program. Greatest improvement has occurred with the offshore devices, particularly the highly advanced nodding duck concept. Such dramatic improvements indicate the relative immaturity of offshore technology compared with the better established technology of caisson- and land-based systems (25).

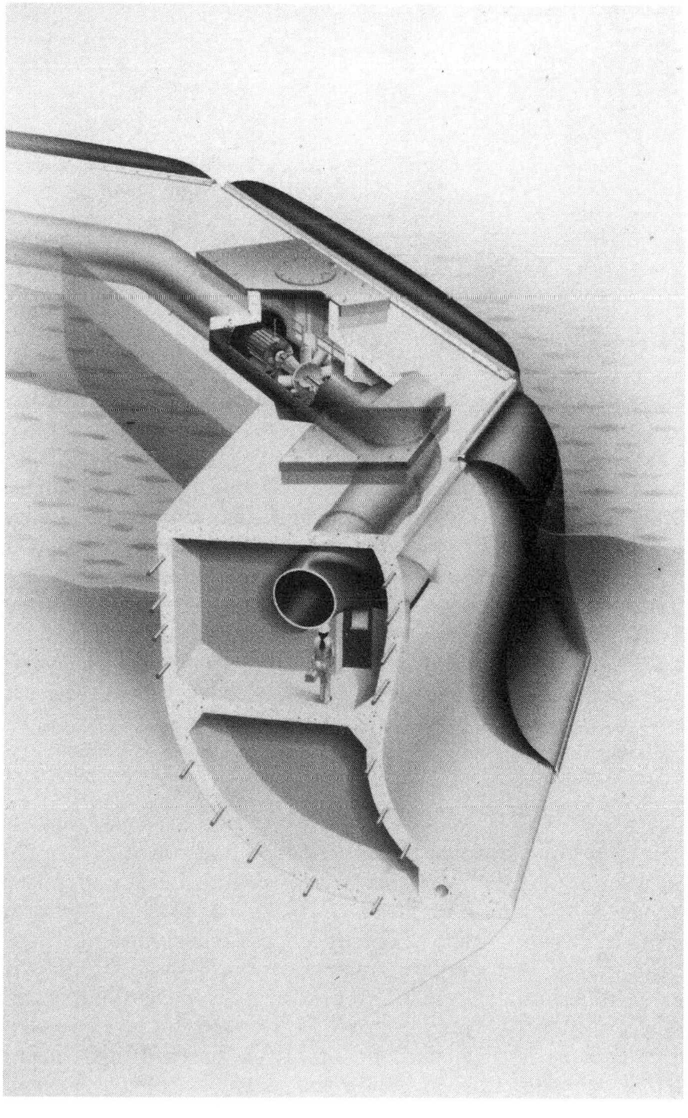

Figure 25. The most recent circular SEA Clam design, based on concrete hull construction. Source: Sea Energy Associates.

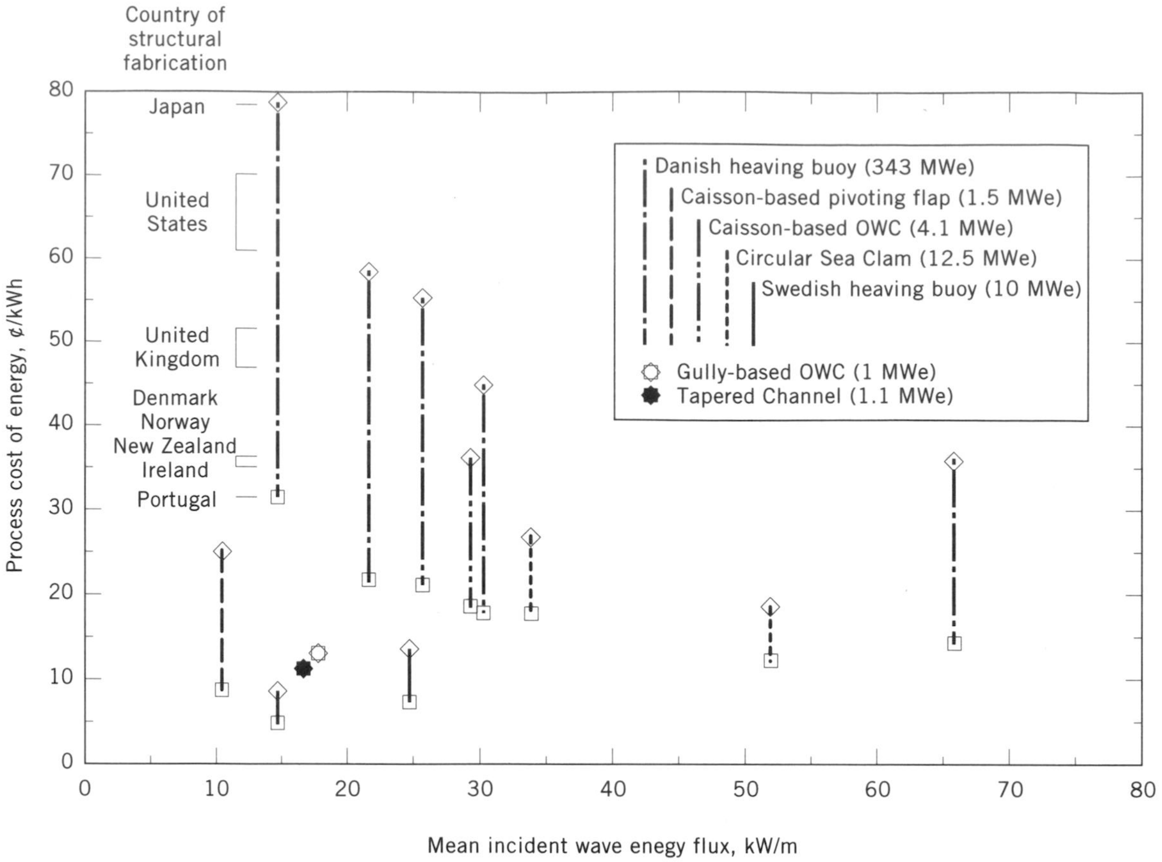

Figure 26. Wave energy cost plotted against annual average incident wave energy flux. Local structural fabrication costs have a much more dramatic effect on plant economics than the local wave energy resource, particularly for those devices utilizing a fixed reaction point. Source: SEASUN Power Systems.

POTENTIAL ENVIRONMENTAL IMPACTS

Various natural processes might be affected if significant amounts of wave energy are removed from the coastal ecosystem, including sediment transport and the functioning of nearshore biological communities. Marine mammal and seabird populations also could be affected by the physical presence of wave energy structures. Depending on the type of conversion process, wave power plants might be a potential source of chemical and noise pollution as well as presenting a visual intrusion on the offshore seascape. Substantial development of the wave energy resource could conflict with other human uses of coastal sea space, and these potential impacts are also reviewed.

Coastal Sedimentary Processes

The shoaling and breaking of waves carry water into the surf zone over a broad stretch of coastline. When waves arrive at an angle of more than 5°–10° to the coast, the longshore component of this mass transport sets up a continuous current that is powerful enough to carry sand in suspension and in sheets along the bottom (bedload transport). Longshore movement of sediment by waves is referred to as littoral drift and is often conceptualized as a "river of sand" flowing along the coast.

Principal sediment sources for littoral drift are coastal rivers and streams as well as sand carried shoreward by the net mass transport of shoaling waves. Once sediment enters the littoral drift, it works its way down the coast until it reaches a barrier. In some cases, this can be a submarine canyon that cuts across the shelf, while in others, it can be a cape or rocky headland. The combination of one or more riverine sediment sources, a coastal zone of active longshore transport, and a downcoast barrier constitutes a littoral cell.

Man-made structures that extend across the surf zone, such as jetties and groins, intercept the littoral drift, causing deposition upcoast of the structure. Beaches downcoast of the structure no longer receive sand, but longshore currents continue to carry sand away, causing erosion. A structure does not have to be attached to the shoreline to have such an effect. For example, in 1934, a detached breakwater 600 m in length was constructed parallel to the coast about 600 m offshore Santa Monica, California. Because the breakwater blocked the wave energy necessary to maintain littoral drift, sediment could no longer be transported and was deposited in the lee of the breakwater. Although no structure was built across the surf zone, longshore sediment transport was interrupted, and erosion occurred downcoast of the breakwater (67).

Wave power plants could have a similar effect, by absorbing energy from waves before they reach the surf

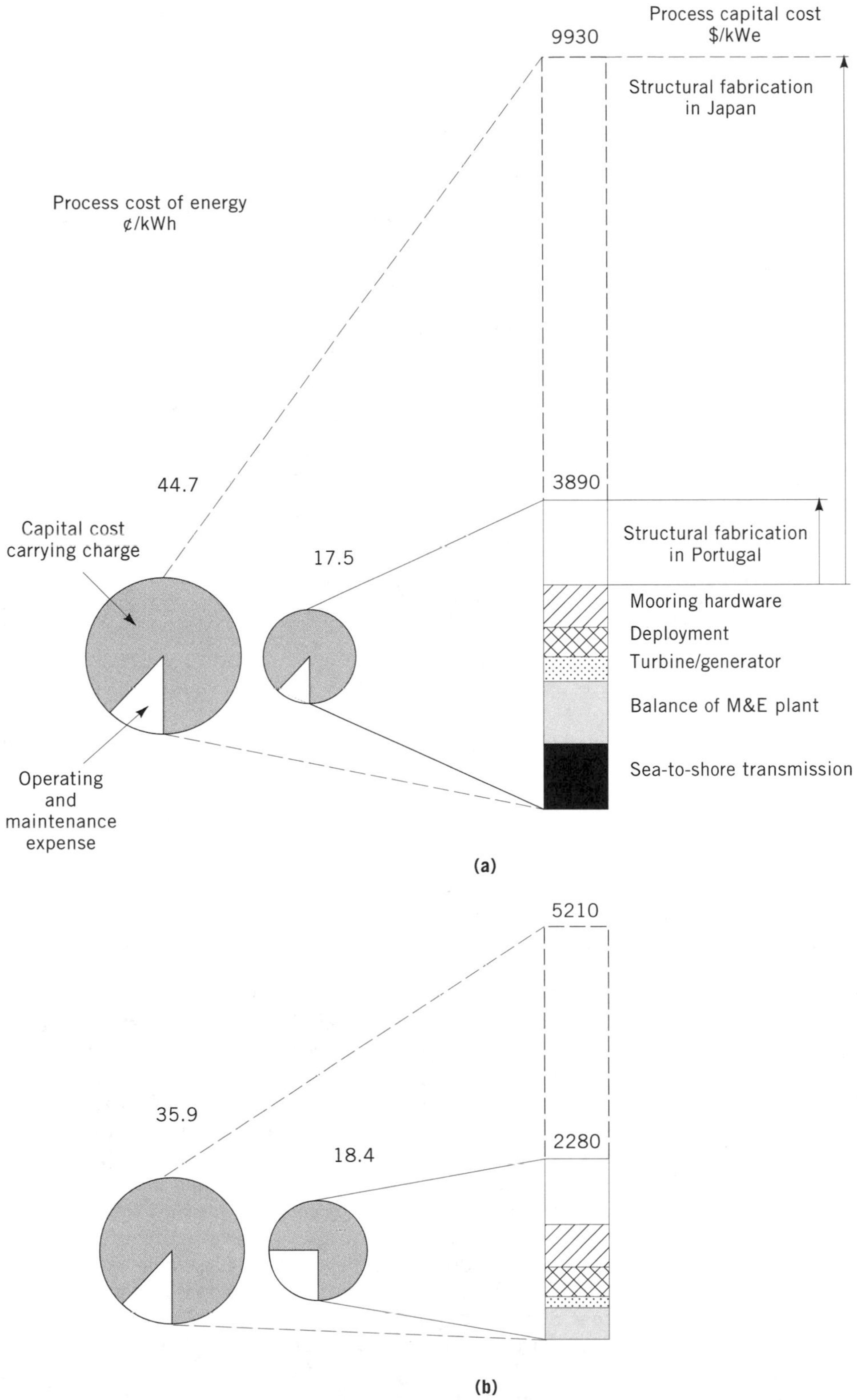

Figure 27. Energy cost breakdown and capital cost distribution for wave power plants utilizing a fixed reaction point. Source: SEASUN Power Systems. (**a**) Danish heavy buoy (343 MWe), west of Auckland, New Zealand. Mean wave energy flux 30 kW/m. Plant capacity factor: 31%. (**b**) Carsson-based OWC (4.1 MWe), South West, Outer Hebrides, UK. Mean wave energy flux 30 kW/m. Plant capacity factor: 21%.

zone. The degree of potential impact depends on the process (floating or fixed structure), how closely individual devices are spaced, and how far offshore the plant is located.

Floating devices have a low potential impact, because any wave energy that is not absorbed will pass through the plant and continue to travel shoreward, where it can power the littoral drift. Furthermore, the undiminished wave energy that passes to either side of the plant will spread by diffraction into the lower energy area immediately behind the plant.

Wave diffraction behind offshore breakwaters and islands is a well-documented phenomenon and can be readily observed from the air. It occurs as wave energy is transferred laterally along wave crests from a region of large wave height to a region of low wave height. Thus

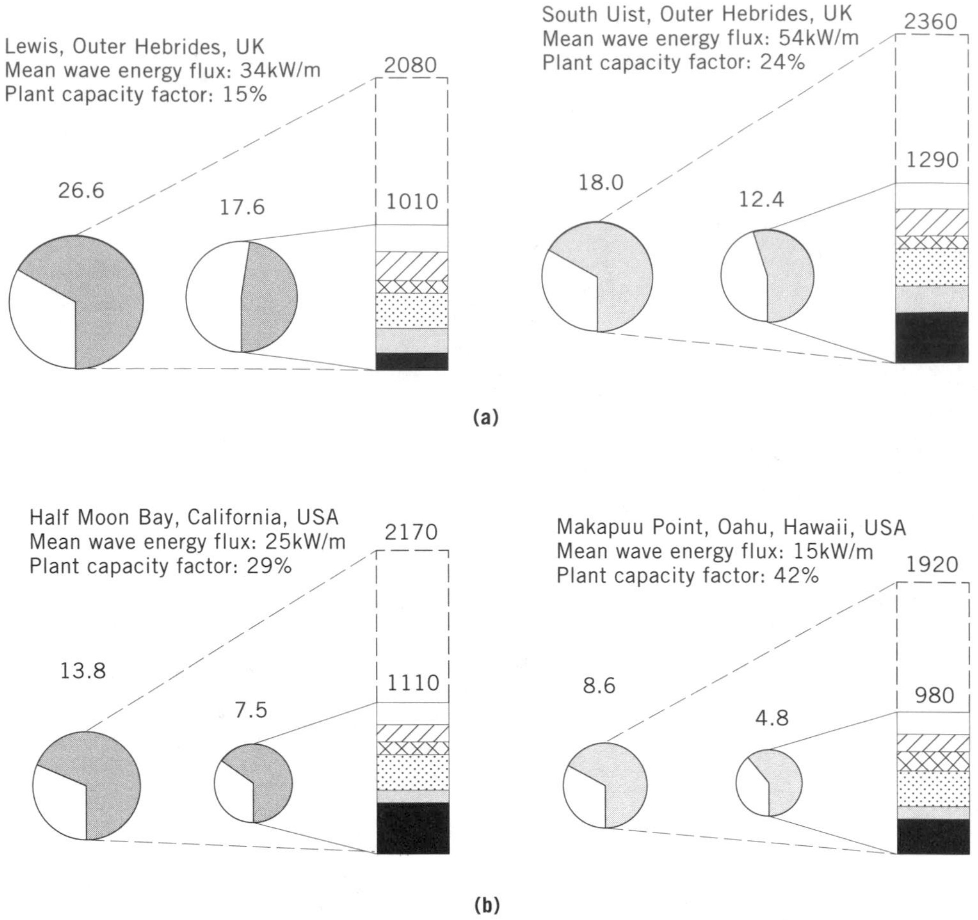

Figure 28. Energy cost breakdown and capital cost distribution for wave power plants utilizing an inertial reaction point. (**a**) Circular Sea Clam (12.5 MWe). (**b**) Swedish Heavy Buoy (10 MWe). Source: SEASUN Power Systems.

while a noticeably calmer area might develop immediately in the lee of an offshore wave power plant, the waves would be substantially reestablished by diffraction 3–4 km shoreward of the plant. The effects at the coast probably would not even be measurable until several plants were deployed.

Caisson-based devices have a higher potential impact, since the caissons will reflect any incident wave energy that is not absorbed and because they would be sited much closer to shore. A distinct low-energy "shadow zone" could be created behind a row of such devices. Therefore, care must be taken to locate caisson-based wave power plants near the downcoast boundary of a littoral cell rather than near its sediment sources.

Nearshore Biological Communities

Kelp beds are the nearshore benthic ecosystem most likely to be affected by wave power development. There are two predominant species of kelp that form beds off the west coast of North America: the giant kelp (*Macrocystis pyrifera*) and the bull kelp (*Nereocystis leutkeana*). The giant kelp is found from Alaska to Baja California but only forms forests south of Point Ano Nuevo. The bull kelp is found from Alaska to Santa Barbara but only forms forests north of Point Conception. The two species differ in both their appearance and life history.

Giant kelp plants consist of numerous fronds emerging from a holdfast that is attached to the seafloor. Each frond consists of a long stemlike stipe, with numerous blades along its entire length. At the base of each blade is a small gas-filled bladder that floats the frond off the bottom. The life history of *Macrocystis* is perennial, with new fronds constantly replacing old ones, which are sloughed off after a few months. The plant itself may live for several years, provided that the holdfast is not destroyed by grazing sea urchins or uprooted by storm waves.

The appearance of bull kelp is markedly different. A single stipe emerges from the holdfast and terminates in a single large float, which in turn gives rise to numerous long, streaming blades. This growth habit is more resistant to uprooting by storm waves than that of giant kelp.

Forests formed by both species of kelp, including mixed stands, provide shelter and food for a variety of marine life. Kelp beds support both commercial and sport fisheries. Some fish graze the kelp directly, while others are carnivores that eat the grazers. Kelp beds are also the favored habitat of the California sea otter and provide temporary refuge for northward migrating gray whales. Both of these marine mammals are endangered species.

The most serious impact of wave power development on kelp would occur if a wave energy device was actually sited within a bed. Not only would the kelp in the "footprint" area of the device be destroyed, but any kelp seaward of the device would have to be removed. This would be particularly important during the summer months, when incident wave energy is low to begin with and the kelp's surface canopy is well developed.

Table 2. Cost and Performance Data for Energy Supply Alternatives on the Big Island of Hawaii[a]

Technology	Plant Size (MW)	Capital Cost ($/kW)	Fixed O&M Cost ($/kW/yr)	Variable O&M Cost (¢/kW·h)	Annual Output (GW·h/yr)	Heat Rate (Btu/kW·h)	Fuel Cost ($/MBtu)
Pulverized coal (PC)	30	3,970	221	0.27	171	12,360	2.23
Atmospheric fluidized bed coal (AFBC)	30	4,197	202	0.45	171	12,780	2.23
Oil-fired combustion turbine (OCT-1)[b]	22.5	1,788	121	0.133	104	10,690	4.91
Oil-fired combustion turbine (OCT-2)[b]	44.9	1,390	103	0.133	104	10,690	4.91
Oil-fired combustion turbine (OCT-3)[b]	62.7	1,421	92.1	0.095	357	7,660	4.91
Oil-fired steam turbine (OST)	30	2,255	106	0.079	171	12,350	4.91
Diesel engine generator (DEG)	15.2	1,981	111	0.26	86.5	8,520	4.91
	26.6	1,643	72.2	0.26	152	8,520	4.91
Biomass combustion (BMC)[c]	30	2,335	47	0.31	171	17,630	3.23
		3,410	101	0.67			6.46
Geothermal flashed steam (GFS)[c]	25	3,200	87	0.70	142	24,000	None
		4,920					
Ocean thermal energy conversion, open cycle with single condenser stage (OC-OTEC [1])	10	10,700	161	2.50 water credit[d]	70	ND	None
Ocean thermal energy conversion, open cycle with second condenser stage (OC-OTEC [2])	9	14,700	221	6.49 water credit[d]	63	ND	None
Ocean thermal energy conversion, hybrid cycle (HC-OTEC)	40	9,400	141	2.59 water credit[d]	280	ND	None
Ocean thermal energy conversion, closed cycle (CC-OTEC)	48	6,000	90	none	336	ND	None
Wind energy conversion (WIND)	10	1,390	18.4	0.91	21.9	NA	None
Wave energy conversion (WAVE)	10	2,300	122	none	37	NA	None
	30	1,970	108	none	111	NA	None

[a] Abbreviations: ND, no data; NA, not applicable; O&M, operation and maintenance.
[b] Industrial frame gas turbine construction. Phases 1 and 2 are simple combustion cycles. Phase 3 is a combined cycle plant, where heat is recovered from the two gas turbines to power a steam turbine.
[c] Where two numbers are given for a single plant size, mid-range values are used to compute levelized cost of energy.
[d] Based on water production capacities of 15,000 m^3/day (OC-OTEC [1]), 35,000 m^3/day (OC-OTEC [2]), or 62,000 m^3/day (HC-OTEC), operating at 80% capacity factor, with water sold for \$0.40/m^3 (\$1.50/kgal).

Siting a wave power plant behind a kelp bed is akin to siting a wind farm behind a grove of trees. Computer simulations of a large storm wave (6.1 m high, 20-s period) passing through a small kelp farm in 15 m water depth indicated an 80% reduction in wave height, to 1.2 m (68). Plant spacing in this simulation was 1.1 m, which is comparable to that measured in natural stands of kelp off the Monterey Peninsula (69).

If a wave power plant can be sited seaward of the zone in which kelp grows, then no kelp would have to be removed. The seaward margin of kelp growth is limited by the amount of light reaching the seafloor, which depends on water clarity. In turbid waters, giant kelp beds are limited to depths of 15–20 m, whereas in clear waters, they may extend to depths of 25–30 m (70).

Siting a caisson-based wave power plant seaward of the kelp is unlikely to be economical, due to the increase in survival wave heights and higher platform overturning moments. Caisson-based wave energy devices thus have a high potential for disturbing kelp beds. Floating devices that can be sited well seaward of the kelp are expected to have much less impact. A comparative study of kelp beds along a natural wave exposure gradient off the Monterey Peninsula (9) suggests that the following changes might occur as a result of extensive offshore wave energy development:

- In regions where the two forest-forming species coexist, *Nereocystis* tends to be excluded because it is shaded out by the large, perennial surface canopy of *Macrocystis*. In areas of high wave exposure, however, *Nereocystis* predominates, since its growth form is more resistant to breakage by waves. To the extent that offshore wave energy development reduces storm wave action, it will increase the competitive advantage of *Macrocystis* in mixed stands, allowing it to grow more abundantly off exposed coasts that were formerly dominated by *Nereocystis*.

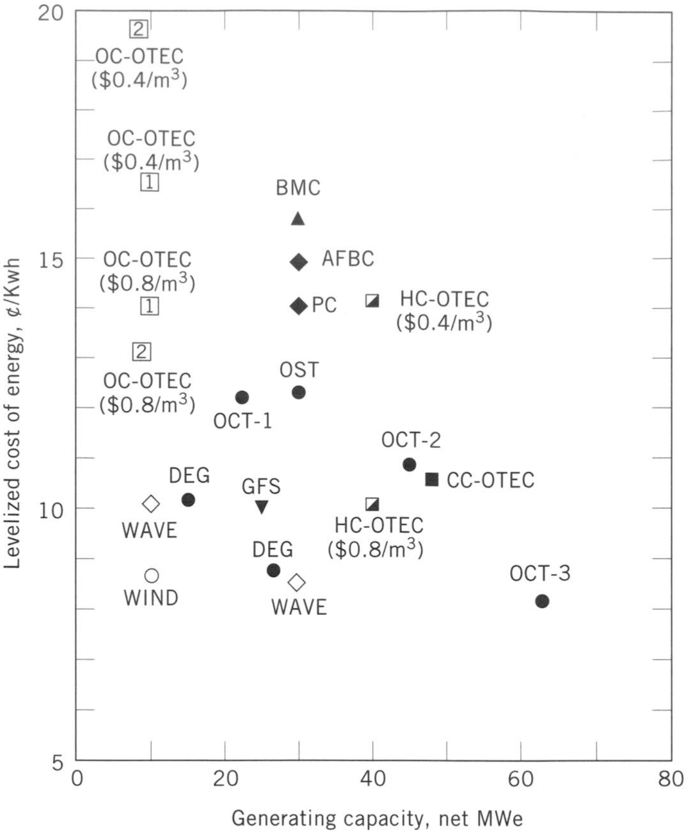

Figure 29. Levelized energy cost comparison among different energy supply alternatives on the Big Island of Hawaii. Source: SEASUN Power Systems.

- The thinning of *Macrocystis* canopies by wave action on exposed coasts allows not only *Nereocystis* to grow more abundantly but also permits the increased growth of understory kelps, such as *Pterygophora* and *Laminaria*. As already indicated, offshore wave power plants could act to increase the abundance of *Macrocystis*, which would then shade out these understory kelps.
- Coralline red algae are more resistant to wave action than fleshy red algae, but because the latter grow more rapidly, they will outcompete coralline forms in protected coastal waters. Therefore, offshore wave energy development might increase the abundance of fleshy red algae in coastal areas that were formerly dominated by coralline algae.

Even if the algal species composition of the kelp beds changes, they will still provide food and shelter for mammals, fish, and invertebrates. In the study cited above, the composition of invertebrate communities in sheltered and exposed kelp beds did not differ significantly.

Marine Mammals and Seabirds

Although marine mammals and seabirds are highly migratory, wave power development could still have an impact on these populations. For example, gray whales will have to swim around wave energy devices located in their coastal migratory path. Pinnipeds (seals and sea lions) may attempt to haul out on floating wave energy devices with low freeboard. Finally, seabirds that commonly nest on offshore rocks and stacks may attempt to colonize caisson-based wave energy devices, since these are intended to be unmanned, with only occasional service visits. These potential impacts are described separately below.

Along the west coast of North America, the gray whale is noted for its annual migrations between feeding grounds in the Arctic seas off Alaska and calving grounds in the coastal lagoons of Baja, California. The following description of gray whale movements was compiled from environmental impact statements prepared for offshore oil and gas leasing by the U.S. Department of the Interior (70,71).

The southern migration takes place in November and December. Southbound migrating whales may travel up to 185 km/day, requiring at least 30 days to complete their entire journey. Aerial surveys report the majority of whales within 2 nautical miles (3.7 km) of the shoreline.

The northern migration occurs in two pulses: the first in March and April, the second in May. Northbound whales tend to move closer inshore as they travel farther north. Off southern California, the majority of whales occur within 1.5 nautical miles (2.8 km) of the shoreline, while in the northern third of the state, they occur "literally in the surf zone" (70). The second pulse consists entirely of mother–calf pairs, which stay extremely close to shore throughout their journey, traveling within kelp beds or just seaward of the breakers.

Any wave power plant that can be sited more than 4 km offshore will have virtually no impact on gray whale migration, provided that construction activities that cross their path (seafloor surveys, laying of submarine power cables) are carried out at times of the year when whales are not migrating. Such activities would almost certainly be carried out in calm-weather months (July through September), so interference should not occur.

As mentioned earlier, caisson-based plants are only economical in water depths less than 20 m, which generally lie within 2 km of shore along PG&E's service area. Installation of such devices may involve destruction of the kelp forest refuge for northbound mother–calf pairs. Furthermore, if the noise emissions from wave energy conversion machinery are perceived as threatening, the whales may give the plant a wide berth. Gray whales have readily acclimated to the noise of offshore oil production platforms, however, which are often used by human observers to watch the migrations (72).

There is no doubt that pinnipeds will attempt to haul out on any floating wave energy device that has a freeboard of less than 1 or 2 m, at least during calm weather. Steller sea lion and harbor seal populations may particularly benefit from such artificial hauling areas, since their use of natural coastal land peaks during the calmer summer months, when floating wave energy devices would be easier to board.

Although such an impact might be positive from a pinniped's point of view, it may create other, negative impacts. For example, it has been reported that sea lions hauling out on a new jetty will follow sport fishing charter boats as they leave harbor and steal the catch of frustrated anglers (73). This could create a problem if a wave energy device is sited close to a popular sport fishing area. Device inspection, maintenance, and repair could be difficult where large numbers of seals or sea lions are hauled out near docking and access points.

If large-scale development of the wave energy resource creates a significant amount of new hauling area, pinniped populations may increase over the long term, with many animals relying on this artificial space. This could cause problems when a plant is decommissioned at the end of its service life, since the natural environment would not have the carrying capacity for these population additions. One solution to this potential problem would be to deploy a new plant as the old one is taken out of service, in a module-by-module fashion. If a wave power plant is not to be replaced, then its decommissioning should be phased over several years, so that the pinniped population can gradually adjust to the loss of carrying capacity.

Caisson-based devices will have sufficient freeboard to discourage even the most determined pinniped, since conversion machinery has to be placed high enough to prevent its inundation by extreme storm waves. In order to be compatible with coastal recreation, airborne noise emissions from such devices will have to be muffled, and unless the residual noise frightens them away, birds may attempt to colonize this high-and-dry space.

The same concern regarding plant decommissioning, described above for pinnipeds, applies to this potential impact as well. In addition, care must be taken to avoid disturbing seabird nesting areas during plant service visits, particularly in breeding and brooding seasons.

Plant Emissions

The possible sources of pollution from a wave power plant depend on the particular conversion process. Devices that incorporate a closed-circuit hydraulic system have the potential for a hydraulic fluid spill, whereas devices that use seawater or air as a working fluid are free of this concern. On the other hand, all devices may have to use toxic chemicals to inhibit marine biofouling. Finally, high-frequency noise may be a problem with devices that utilize a Wells turbine. These three potential sources of pollution are discussed below.

The potential impact of hydraulic fluid spills can be mitigated to some extent by using a water-based fluid, which also reduces the on-board fire hazard. In addition, isolation valves that can be reliably controlled from shore would minimize the volume of any spill once a leak is reported by the plant monitoring system.

Devices with the highest risk of such a spill are those such as the Wave Energy Module and the latest version of the Bristol Cylinder, where the hydraulic cylinders are actually located in the water. The potential for fluid release into the ocean is much less in devices such as the Pendulor System, the Edinburgh Duck, and Sea Energy Corporation's contouring raft, where the hydraulic cylinders are out of the water. In any of these devices, however, the potential for a spill exists during the offshore replacement of hydraulic components or replenishment of hydraulic fluid.

Although fouling by marine organisms could conceivably reduce wave energy absorption efficiency, due to its added mass and drag, this might not be the case. For example, wave tank testing of Cockerell's multiple-raft concept with simulated seaweed growth on the raft bottoms indicated negligible effect (74).

If fouling control is necessary, then it can be accomplished either by periodic cleaning (requires divers) or the use of antifouling coatings (requires drydocking). If the coating option is selected, then the use of an organotin compound, such as tributyl tin (TBT), would almost certainly be considered, since it entails a recoating interval of 6–7 years, compared with 1 or 2 years for copper-based paints. Reference 75 presents a complete review of the environmental problems and legal regulations associated with the use of organotin coatings. The typical legal limit for average TBT release rate is 5 $\mu g/cm^2$ of hull wetted surface area per day. U.S. Navy experience has been that release rates well below this level (on the order of 0.1 $\mu g/cm^2/day$) are fully effective in preventing hard fouling (76). Therefore, even if antifouling coatings are required for a wave energy device, an environmentally acceptable solution to the problem appears to exist.

Flexible reinforced rubber surfaces (the Swedish hose pump, SEA Clam's flexible bag) cannot be coated. This is a potential problem of particular importance to the hose pump, since fouling on the interior hose walls would reduce overall conversion efficiency due to increased fluid friction losses. Based on the fouling of OTEC heat exchanger inlet structures and test loops that occurs when seawater pumping stops for any length of time (77), hose interiors would be especially subject to fouling during periods of summer calm. Ocean test experience suggests, however, that even a small amount of hose flexing is adequate to prevent fouling organisms from taking hold.

If hose fouling does become a problem, there are commercially available rubber formulations that contain TBT, and these could be used to line the interior (and exterior, if necessary) of the hoses during manufacture. Because the TBT is chemically incorporated into the rubber's structure, its release rates are much lower than the problem-causing TBT paints, yet still effective in preventing fouling (78). Again, there appears to be an environmentally acceptable solution to any problem that may arise.

Airborne noise from Wells turbines used in OWC devices has been reported from a variety of different prototype installations. The most complete report comes from a 150-kWe, caisson-based OWC prototype that is now operating near Trivandrum, on the southwest coast of India (79). In the caisson itself, the noise is too loud for normal conversation. Measurements indicate a sound intensity of 70–90 db at the seaward end of the breakwater, where the caisson is located. On shore, 650 m from the caisson, the measured intensity is less than 60 db. The sound in nearby villages has been likened to that of a small single-engine airplane flying overhead.

Even if airborne turbine noise is muffled by a silencer, the sound may also carry into the surrounding water, potentially interfering with military acoustic tracking operations. Underwater noise would also be generated by hydraulic machinery, particularly high-speed Pelton turbines such as used in the Swedish heaving buoy device and an earlier version of the Bristol Cylinder. It should be noted, however, that noise from wave power plant machinery will generally increase in proportion to the ambient background noise associated with surface wave conditions, thus tending to minimize its noticeable effect.

Visual Appearance

Construction of a Tapered Channel wave power plant or a gully-based OWC involves significant blasting and rock removal. Not only the appearance, but also the very fabric of the shoreline, would be permanently altered, and this may not be acceptable in certain locations. Likewise, a wave energy breakwater also might be considered an unacceptable addition to coastal scenery except at those locations where harbor development already exists.

The visual impact of a wave power plant depends on six factors:

- Offshore distance of the plant
- Elevation of the shoreline observer
- Coastal weather conditions
- Size (waterplane area and freeboard) of the individual devices that make up the plant
- Color contrast between the devices and the sea
- Presence of natural or other artificial structures in the offshore seascape

An observer 6 ft (1.8 m) tall standing at the surf's edge (i.e., at mean sea level) has an offshore horizon of 5.2 km, which can be seen even in light haze (International Visibility Code 6, range of 2–10 km). In the United States, this corresponds quite closely with the 3-nautical-mile (5.6-km) geographical limit, which marks the boundary between waters under state control and the Outer Continental Shelf (OCS), which is under federal jurisdiction.

As mentioned previously, caisson-based wave energy devices are only economical in water depths shallower than 20 m. Along northern California and in Hawaii, such depths generally lie within 2 km of the shoreline, where a wave energy caisson would be visible in any weather condition better than thin fog (International Visibility Code 5). On the other hand, along the mid-Atlantic coast of the United States, the 20-m-depth contour is typically more than 10 km offshore, where a wave power plant would be obscured by any natural sea haze that might be present.

Floating wave energy devices can be anchored anywhere on the continental shelf; beyond the shelf break, however, the bottom falls off rapidly, and mooring costs become prohibitive. Along continental coastlines, the shelf break generally lies more than 10 km offshore, but off volcanic islands the shelf is much narrower. Thus in Hawaii, a floating wave power plant would be visible from shore, even to an observer standing at sea level and even if some haze was present.

Heaving buoys have low freeboard and small waterplane area, and their visual impact on the seascape would be low, even for a wave power plant located relatively close to shore. To get an idea of such a plant's appearance, hold a ruler at arm's length and imagine that at the same time you are looking over the ruler's edge at a 10-MWe Swedish heaving buoy plant located 3 km offshore. A buoy having a diameter of 16 m would span 4 mm along the ruler's edge (about $\frac{3}{16}$ in.). With a freeboard of only 1–2 m, however, individual buoys would tend to be obscured by wave action unless viewed from a high elevation. A 600-m-diameter cluster of buoys would span a distance of 14 cm (about $5\frac{1}{2}$ in.) and would certainly be visible to someone who knew where to look. Unless the buoys were painted a highly contrasting color, however, the plant would not immediately draw one's attention, since it would occupy less than 5% of a 180° field of view from the shoreline.

Because of the high level of fishing activity in offshore shelf waters, floating devices will have to be appropriately marked as a navigation hazard. In addition to lights, sound signals, and radar reflectors, highly contrasting day markers will be required. The U.S. Coast Guard specifies that such markers be in the form of a diamond-shaped sign, 3 × 3 ft (0.9 m on a side), with black lettering on a white background and an orange reflective border. While such a sign meets the requirement of being visible within 1 nautical mile (1.8 km), it would be below the perceptual threshold of most observers beyond a distance of 4 nautical miles (7.4 km). Therefore, navigation markers on offshore wave power plants are expected to have negligible visual impact when viewed from shore.

Conflict with Other Uses of Sea Space

All such potential conflicts are site specific and, to a lesser extent, process-specific, depending on a device's conversion efficiency and its clearance requirements relative to adjacent devices, both of which combine to determine the amount of sea space occupied by a wave power plant of given capacity. Most problems can be avoided by early consultation with those involved in any foreseeable conflict.

From an economic standpoint, the most important uses of nearshore and shelf waters where a wave power plant might be deployed are offshore (oil and gas production and commercial fishing (including kelp harvesting). The revenues associated with coastal recreation and tourism are also significant; although visual intrusion is the most obvious conflict, sport fishing and recreational boating might also be adversely affected by large-scale wave power development. Use of coastal sea space by commercial shipping traffic and its use for military exercises or scientific research represent other potential sources of conflict. Finally, the designation of certain ocean areas as marine sanctuaries may preclude wave power development within their boundaries.

In addition to the principal activities described above, coastal sea space is also used by submarine communications cables, municipal wastewater outfalls, and designated dump sites. These are so highly localized and few in number that they are not expected to significantly limit wave energy's development potential. Nevertheless, when actually siting a wave power plant, these should be identified early enough so that they can be avoided.

HISTORICAL PERSPECTIVE AND CURRENT RESEARCH

The 1973 oil embargo by the Organization of Petroleum Exporting Countries ushered in a decade of dramatic price increases for fossil fuels around the world. As with other renewable energy technologies, this led to the establishment of several national government-sponsored programs in wave energy research and development. The greatest efforts by far were in the United Kingdom, Norway, and Japan. Although not as large, significant national programs involving ocean tests of fully operational prototypes have been conducted in Sweden and Denmark. Other national programs that more recently have undertaken significant technology development and ocean testing of hardware include India and China. In 1991, Europe's first multinational wave energy program began and is expected to continue at least through 1998. For each of the programs named above, a brief history and description of current research are given below, in alphabetical order. In describing device development and wave power projects, parenthetical reference is made to the type of conversion process as numbered in the second section (and Fig. 2) of this article. The survey concludes with a review of work in the United States and other less-active countries.

China

Wave energy research in the Peoples' Republic of China has been carried out at the Guangzhou Institute of Energy Conversion. With support from the National Science and Technology Committee of China and additional assistance from Japan, a 20-kWe cliff-based OWC test plant (process 2) is under development on Dawanshan Island, at the mouth of the Pearl River estuary on the South China Sea (80). In addition, at least eighty OWC-powered battery-charging systems for navigation buoys (process 3) have been built and deployed in China's coastal waters (81).

Denmark

Since 1979, Denmark has invested about $3 million in wave energy research and development, 80–90% by the national government and the rest by private industry. All of this has been directed toward the development of a heaving buoy system (process 6) under the direction of Danish Wave Power, Aps, a consortium of four industrial firms. A 45-kWe prototype module consisting of a single 6-m-diameter buoy operated for 2 months during the spring of 1990. In addition to Danish Wave Power's lead role in the European Commission's research program (see below), the national government is supporting the sea trial of a much smaller module, to test the air buffering concept described earlier in this article.

Europe

In 1991, the Commission of the European Communities, through its Directorate General for Science, Research, and Development (DGXII), launched the Commission's first wave energy research program. With funding of $1.5 million under its JOULE I program (Joint Opportunities for Unconventional or Longterm Energy), four Preliminary Actions were carried out over an 18-month period.

The results of this effort were reported at the First European Wave Energy Symposium held July 1993 in Edinburgh, Scotland. That same month, proposals were accepted for the main research and development program, to be carried out under JOULE II. With $3.5 million from the Commission and matching funds from government, academic, and industrial sources in member countries, the overall European effort will be approximately $7 million over the next two years (1994–1995), broken down as follows:

- Produce a European Wave Energy Atlas (10–15% of total program; main centers of data evaluation are in Portugal, Scotland, and Norway)
- Undertake systematic modeling of device hydrodynamics and directed research into primary components (20–30% of total program; coordinated by Danish Wave Power, with wave tank test programs at University College Cork, Ireland, and Edinburgh University, Scotland)
- Begin preliminary pilot plant designs (50–60% of total program; the three device teams selected were the Queen's University of Belfast for a second gully-based OWC at Islay, the Technical University of Lisbon for a gully-based OWC on the island of Pico in the Azores, and Applied Research and Technology for a nearshore, caisson-based OWC off the north coast of Scotland)
- Continue activities and further develop the European Wave Energy Research Network (5–10% of total program; centered at University College Cork, Ireland)

India

The Indian Institute of Technology at Madras, under sponsorship from the national government, has developed a stand-alone wave-absorbing caisson, based on Kvaerner Brug's multi-resonant OWC (process 2). A test unit incorporating a 150-kWe Wells turbine/generator was built for deployment off India's southwest coast, near the port of Trivandrum. Although the first caisson was damaged in deployment and had to be abandoned, a second caisson was successfully deployed in December 1990. The plant was commissioned in October 1991, and based on the lessons learned since then, a wave energy breakwater is being designed for a new fishing harbor at Thangassery, near Quilon, on the southwest coast of India.

Japan

As a result of pioneering research by Commander Yoshio Masuda, a battery-charging generator for navigation buoys was developed in 1964-1965, driven by the OWC of a central pipe in the buoy's hull (process 3). Approximately 700 such generators (rated at 60 W) have been sold by Ryokuseisha Corporation for use in Japan, and another 500 have been exported to other countries. Masuda has also pursued the development of OWC technology for larger floating power plants, beginning with his work at the Japan Marine Science and Technology Center (JAMSTEC) on the test ship *Kaimei* and more recently at Ryokuseisha Corporation on the Backward Bent Duct Buoy (BBDB). The BBDB has undergone ocean testing of $\frac{1}{10}$-scale models in the Sea of Japan and in Mikawa Bay, on Honshu's east coast.

Funded by Japan's Science and Technology Agency, JAMSTEC began its wave energy research and development program in 1974 and, following 2 years of laboratory testing, had the *Kaimei* built as a floating platform for testing relatively large pneumatic turbines. With contributing support from the International Energy Agency (IEA), *Kaimei* was deployed twice in the Sea of Japan: from August 1978 to March 1980 and from July 1985 to July 1986. JAMSTEC is now pursuing the development of a floating OWC device (process 3) known as the "Mighty Whale."

Six other wave energy devices have been ocean tested or built as demonstration plants in Japan:

- A 40-kWe onshore OWC (process 2) at Sanzei, on the Sea of Japan, operated from September 1983 until March 1984. Developed by JAMSTEC and Fuji Electric Company.
- A 30-kWe single-acting OWC (process 2) with an onshore, constant-pressure air tank at Kujukuri, on Honshu's east coast, operating since March 1988. Developed by Takenaka Corporation and Kawasaki Steel Corporation.
- A 60-kWe caisson-based, double-acting OWC (process 2) incorporated into a newly built outer harbor breakwater at Sakata Port, on the Sea of Japan, north of Niigata, operating since November 1989. Developed by the Ministry of Transport's Port and Harbour Research Institute.
- A 20-kWe caisson-based Pendulor System (process 4) at Mashike Harbor, Japan, located on the west coast of Hokkaido, operating since 1983. Developed by the Muroran Institute of Technology and Hitachi Zosen Corporation. A 5-kW (fluid power) test caisson is also located at Muroran Port.
- A 1-kWe caisson-based pivoting flap (process 4), hinged at the bottom rather than from above, such that flap-restoring force is provided by buoyancy rather than weight. Built at Miyazu Energy Research Center on Wakasa Bay, on the Sea of Japan, northwest of Kyoto, operating since 1989. Developed by Kansai Electric Corporation.
- A 20-kWe heaving and pitching float (process 9), based on a specially built jack-up rig named *Kaiyo*, located off Iriomote Island, southwest of Okinawa, operated from August 1984 until the summer of 1987. Developed by the Institute of Ocean Environmental Technology and the Japan Shipbuilding Industry Foundation.

Altogether, approximately $40 million have been spent on wave energy research in Japan. Half of this was for the *Kaimei* program ($17 million from the Japanese government and $3 million from its IEA partners).

Norway

The Norwegian Ministry of Petroleum and Energy has supported wave energy research and development since 1978. In addition to funding an extensive wave measurement and resource assessment program along its North Sea coast, the national government has sponsored device development by three research teams:

- The Central Institute for Industrial Research, whose work on wave focusing has expanded to include development of the Tapered Channel (process 1) and whose key personnel formed Norwave A.S.
- The Norwegian Hydrodynamic Laboratories and Kvaerner Brug A/S, whose work led to the development of the multiresonant oscillating water column (MOWC; process 2)
- The University of Trondheim, where fundamental research has been conducted under the leadership of Dr. Johannes Falnes on the hydrodynamics of point absorbers, particularly heaving buys (process 6)

In the spring of 1984, construction began on two demonstration plants at Toftestallen: a 350-kWe Tapered Channel and a 500-kWe MOWC. The building of these projects was 50% cost shared by their private industry developers and the national government. Norwave's plant began operating in the third quarter of 1986. Kvaerner Brug's plant began intermittent generation of electirc power in November 1985 and was set for fully automatic operation in January 1987. The MOWC plant was destroyed by a severe storm in January 1989, and Kvaerner Brug is no longer actively involved in developing this technology. The Tapered Channel plant continues to operate and is considered sufficiently proven that its developers have received a commercial order for a 1.1-MWe system on the south coast of Java, in Indonesia. By 1986, the Norwegian government had spent $12 million on its national program.

Sweden

Since 1977, Sweden has invested approximately $5 million in wave energy research and development. Half of this funding has come from the national government's energy research program, while the balance represents commercial investment by Swedish industry. Wave energy conversion technology is viewed not only as a potential export commodity for Sweden, but also as a contribution toward meeting that country's future energy needs. Sweden's wave resource is comparatively modest, but a portion of Norway's development potential also is available to Sweden, through the interconnected Nordic electric transmission network.

Three wave energy conversion devices have been ocean tested in Sweden:

- A heaving buoy system that uses a specially reinforced elastomeric hose pump tethered to a submerged damper plate (process 7), developed by Gotaverken Energy Systems AB. A 30-kWe test plant was deployed intermittently off Gothenburg in 1983–1984, consisting of three 5-m-diameter buoys. A battery charging system for navigation buoys, based on the same concept, has been licensed to AB Pharos Marine, Ltd., in the United Kingdom, and is now in limited commercial production.
- Another heaving buoy that also uses an inertial reaction point (process 7) but a different power conversion system was developed by Interproject Services AB, which has conducted sea trials of a 3-m-diameter buoy, *Elskling*.
- A floating ramp and reservoir filled by wave surge (process 1) has been developed by SEA POWER AB.

A 110-kWe pontoon was tested at sea off Sweden's west coast during August and September of 1991.

United Kingdom

This country will be reviewed in some detail, because it represents one of the most massive national programs to date, covering a wide variety of wave energy conversion processes that have gone through several stages of evolution. The British program is also significant because the perceived "failure" of the first 10-year effort has been cited by the U.S. Department of Energy as one reason for its relatively low interest in wave energy, and this has had a decidedly negative impact on the technology's development in that country.

The British national program formally began in February 1974, when the newly formed Department of Energy (DEn) commissioned an introductory assessment by the National Engineering Laboratory (NEL). As a result of this review, the government began a 2-year feasibility study in 1976, based on four devices: the Edinburgh Duck (process 10), Sir Christopher Cockerell's contouring raft (process 9), NEL's floating OWC (process 3), and a caisson-based surge device (process 1) developed at the Hydraulics Research Station in Wallingford and known as the HRS Rectifier.

By 1978, wave data had been collected off South Uist, in the Outer Hebrides Islands off Scotland's northwest coast, and were used to develop reference designs for a 2000-MWe central station at that site, where offshore wave power densities of 40–50 kW/m were measured. Physical models at $\frac{1}{10}$ scale had been tested in natural waves for two devices: the Edinburgh Duck in Loch Ness and Cockerell's contouring raft in the Solent, near the Isle of Wight. A British 125-kW pneumatic turbine was also tested during the first deployment of the *Kaimei*.

The second round of design work began in 1979–1980, with two devices dropped from the program (the Cockerell raft and the HRS Rectifier) and five more added: a submerged OWC known as the Vickers Resonant Duct (process 2), a free-standing caisson-based OWC developed by the Queen's University of Belfast (process 2), the Bristol Cylinder (process 12), and two flexible-diaphragm devices (process 11)—Coventry Polytechnic's SEA Clam and the Lancaster Flexible Bag. In addition, the NEL OWC was now investigated as a fixed breakwater rather than a floating device.

By 1982, eight reference designs had been prepared for the remaining devices (two for the Vickers Resonant Duct, one an attenuator, the other a terminator). The Energy Technology Support Unit (ETSU) at Harwell, who managed the program for DEn, had also hired the consulting firm of Rendel, Palmer, and Tritton (RPT) to review the designs and prepare independent cost and performance estimates. Not surprisingly, RPT's costs were less optimistic than those estimated by the research groups (known as "device teams"). Nevertheless, they were considerably lower than the energy costs of the 1978 reference designs.

Despite this progress, the DEn's Advisory Council on Research and Development announced in March 1982 that the program would be closed down. This decision was based on a strategic review of renewable energy technologies, which concluded that the economic prospects for large-scale offshore wave energy did not look promising in comparison with other technologies, notably wind energy.

Limited support continued for small-scale applications, and DEn cost shared the testing of a $\frac{1}{14}$-scale straight-spine SEA Clam and, later, a $\frac{1}{5}$-scale circular-spine SEA Clam, both on Loch Ness. By 1986, the national government had spent \$20 million on wave energy research and development.

Success of the Norwegian demonstration plants at Toftestallen, coupled with well-documented evidence of bureaucratic failures in the 1980–1982 program, led the British government to reexamine the feasibility of wave energy in the United Kingdom. This change in attitude is reflected by DEn's funding of a 75-kWe gully-based OWC (process 3) on the Scottish Isle of Islay, which became fully operational in May 1991. In addition, ETSU was commissioned to review and update the 1982 reference designs of four main devices from the earlier program (the NEL breakwater, the Edinburgh Duck, the SEA Clam, and the Bristol Cylinder) as well as the Queen's University of Belfast gully-based OWC. Three newer devices were also examined, although in much less detail: Applied Research and Technology's OSPREY, a caisson-based OWC (process 2); Lancaster University's PS Frog, a floating flap that pitches and surges against an internal mass (closest to process 4 but with an inertial reaction point); and Ecovision's Lillypad, in which a buoyant flexible membrane is tethered by hose pumps to a submerged damper membrane (process 7). The results of this study were published in December 1992 (25).

Funding for the most recent British effort has been approximately \$2.5 million, with about 40% supporting the 75-kWe demonstration project at Islay (the capital cost of the plant itself is estimated at just over \$0.4 million). Although the review is now complete, continued support is expected for the Islay project, as well as matching funds toward the two British device teams participating in the European pilot plant project. With the abolishment of the DEn in 1992, the government's wave energy program in the United Kingdom has shifted to the Department of Trade and Industry.

United States

Since 1979, the U.S. Department of Energy has spent approximately \$240 million on its ocean energy program, devoted at almost exclusively to OTEC. Only \$1.5 million was spent on wave-related activities, two-thirds of which went for construction and testing of a prototype 125-kW counterrotating air turbine, invented by Dr. Michael McCormick of the U.S. Naval Academy, as part of its IEA contribution to the *Kaimei* project. The balance supported work on the Tandem Flap (process 4) developed by the Q Corporation, of Troy, Michigan, including temporary deployment of a 20-kWe test unit in Lake Michigan.

In addition to the Q Corporation, eight other American companies have supported development of their own wave energy devices to the extent of conducting model tests in the laboratory or at sea:

- Lockheed Missiles and Space Company, Inc., of Sunnyvale, California, has conducted laboratory wave tank tests of an artificial island surging wave device known as Dam-Atoll (process 1).

- E O Tech, Inc., of Arcadia, California, has conducted laboratory wave tank tests of a heaving-float-in-caisson device, which drives a hydraulic cylinder, known as the Neptune System (process 5).
- Hydropower, Inc., of Laguna Hills, California, has conducted laboratory bench tests of a heaving-float-in-caisson device, with a water turbine suspended from the float, known as the Seamill (process 5).
- Wave Energy, Inc., of Bartlesville, Oklahoma, has tested a 2.4-m heaving buoy (process 6) at Scripps Institution of Oceanography in La Jolla, California.
- E. I. duPont de Nemours & Company, of Wilmington, Delaware, has ocean tested a 2.5-m scrap-tire heaving buoy (process 6) at the U.S. Army Corps of Engineers Field Research Facility in Duck, North Carolina.
- The University of Delaware has deployed a series of 2.1-m heaving buoys (process 6) off the southwest coast of Puerto Rico, designed for desalination rather than electric power generation. The most recent system produced 1100 L (300 gal) of fresh water per day in trade wind waves having a height of 1–1.5 m and periods of 4–6 s. Of all American wave energy devices, this is the closest to commercial production, and the technology has been licensed to Composite High Pressure Technologies, Inc., of Lewes, Delaware.
- U.S. Wave Energy, Inc., of Longmeadow, Massachusetts, has tested a 1-kWe prototype contouring float with damper plate (process 8) on Lake Champlain in Vermont.
- Sea Energy Corporation, of New Orleans, Louisiana, has conducted laboratory wave tank tests of a heaving and pitching raft (process 9).

Despite the national government's lack of interest in wave power, several private utilities and state government organizations have conducted wave energy resource assessments for their service territories, including Virginia Power, the North Carolina Alternative Energy Corporation, Pacific Gas and Electric Company, and the State of Hawaii.

Other Countries

In Canada and New Zealand, wave energy research has focused on assessment of the resource, including the development of new instrumentation and evaluation methods. The South Pacific Applied Geosciences Commission has carried out an extensive regional wave energy resource assessment, encompassing the Cook Islands, Fiji, Tonga, Tuvalu, Vanuatu, and Western Samoa. Wave energy resource measurements have also been made in Argentina, Australia, Indonesia, Mauritius, and Taiwan. Countries that have conducted both resource assessment and device development include Ireland and Portugal (both of which are major participants in the European wave energy program described above) as well as South Africa (at the University of Stellenbosch).

DEVELOPMENT TRENDS AND FUTURE PROSPECTS

Industry Structure

Unlike photovoltaics, where specially designed manufacturing plants must be built to produce solar cells and modules, wave power plants generally make use of off-the-shelf equipment manufactured for other industries. Absorber and reaction structures can be built in existing shipyards or temporary concrete casting yards. Breakwater caissons and offshore devices can be installed using conventional marine construction equipment.

The order of a 1–10-MW plant is not expected to exceed present component suppliers' ability to manufacture the necessary equipment within a 1-year lead time. Likewise, fabrication and deployment of such a small plant would not strain existing construction facilities, and such work would be welcome where local shipbuilding and associated industries are depressed.

The order of a 100–1000-MW central station, however, would certainly entail the construction of specialized fabrication plant and may exceed the manufacturing capabilities of existing component suppliers. This is not considered to be a significant deployment hurdle, since such large stations would not be ordered until several smaller stations were successfully operating. By then, prospective plant owners would have the confidence to invest in new manufacturing capability.

Some process-specific components are not off-the-shelf items but were developed specifically for wave energy applications. The most important of these are the Wells turbine, used in OWC processes and the SEA Clam; the reinforced elastomeric diaphragm used as an air pump in the SEA Clam and Lancaster Flexible Bag; and the specially designed elastomeric hose used as a seawater pump in the Swedish heaving buoy system and as an adjustable compliant mooring for the Bristol Cylinder.

Technology Development

The primary drivers for continued development and improvement of wave energy technology can be divided into two categories. In the first category are drivers that result from market forces and policy making in the outside world. In the second category are drivers that relate directly to the technology itself.

As with other renewable energy technologies, wave energy development will be influenced by the cost of fossil fuels. An increase in the price of oil would enhance the competitiveness of wave energy technology in its most immediate market, which is the supply of electric power and fresh water in island and remote coastal communities now served by diesel plants.

On the other hand, an increase in oil prices will also raise the capital cost of a wave power plant. Marine vessels and construction equipment burn more fuel per unit horsepower than their land-based counterparts, and this will be reflected in higher offshore deployment costs. Even caisson-based devices will require ocean tows from the ports where suitable graving docks exist for their fabrication. Furthermore, lease rates for such equipment tend to go up when offshore oil and gas activity increases (as it would with higher oil prices).

The net effect would be beneficial, but not as much so

as it would be for other renewables such as wind and photovoltaics. This is particularly true for wave energy technologies that make use of materials having petrochemical feedstocks. An example would be the synthetic rubber polymers used in the manufacture of hose pumps and flexible bags.

Likewise, legislation aimed at reducing carbon dioxide emissions is less likely to benefit wave energy than other renewables if a complete accounting is made of CO_2 emissions during plant construction as well as operation. In the long run, however, if coal gasification is used to replace depleted reserves of fossil methane, then wave-generated hydrogen would realize a competitive advantage from any sort of "carbon tax" in those applications where it can be used interchangeably with natural gas.

A far more important outside driver would be the development of an onshore infrastructure for the handling, transportation, and storage of hydrogen as a heating fuel. This does not require a full-blown "hydrogen energy economy," where hydrogen is also used for transportation and electric power generation. It does require, however, sufficient infrastructure that hydrogen can be used as a replacement fuel for major heating applications now served by natural gas. As mentioned previously, this overcomes several of the problems associated with integrating wave power into onshore utility grids, including:

- A winter peak in the seasonal resource profile
- Lack of "natural dispatch" in the daily resource profile
- Limited onshore transmission capacity at the coast

Furthermore, should an onshore infrastructure develop to receive hydrogen, wave energy has several advantages over competing technologies to provide it. Unlike steam reformation of methane or coal gasification, hydrogen production via electrolysis produces no carbon dioxide. Water supply would not be a problem for wave-powered electrolysis, but it could be a limiting factor in those inland areas having good wind or solar resources. While these renewables could gain access to water by moving them offshore, wave power makes much more efficient use of sea space, due to its greater energy density and the fact that wave energy conversion devices tend to be more efficient than their wind or solar counterparts.

The development of a hydrogen infrastructure does not have to depend on wave energy to provide an inexpensive source of hydrogen. A conventional hydroelectric plant has all the advantages of wave energy and is much less costly.

Consequently, the first large-scale electrolytic production of hydrogen will probably come from off-peak hydroelectric generation. Therefore, the development of a hydrogen infrastructure will drive wave energy development, rather than the other way around. Simply stated, wave energy's role would be to meet future hydrogen demand that cannot be met by hydroelectric production.

There are two technology-based drivers of major consequence. The first is the high cost of fixed and floating marine structures, which has led wave energy developers to spend a considerable amount of effort on increasing device absorption efficiency. The second is the lack of long-term ocean test data on which to base projections of operating and maintenance costs, component service lives, and hydrodynamic performance.

Even with the most efficient devices, structure fabrication costs typically account for more than 50% of the initial capital investment in a wave power plant. Thus the development of less expensive materials and fabrication processes would greatly accelerate the commercialization of wave energy technology.

Absorber and reaction structures are usually built of steel or reinforced concrete, and their fabrication makes use of conventional shipbuilding and concrete-forming techniques. Considerable savings may be possible by building these structures out of ferrocement. The quality control problems and high labor costs associated with early ferrocement boat building in the 1960s can be avoided by using the shotcrete laminating process developed in the United States by Martin Iorns of West Sacramento, California (82).

Shotcrete laminating involves the spray application of mortar onto the surface of a female mould. Expanded metal mesh reinforcement is then pressed into the soft mortar, with successive layers of mortar and mesh applied in an overlapping pattern until the laminate reaches design thickness.

This process represents a considerable improvement over earlier ferrocement boat-building methods, where mortar was hand troweled into prepositioned layers of mesh supported at intervals by steel rods. Quality control in the earlier methods was difficult, and poor mortar penetration into the mesh resulted in voids, which weakened the section, allowing seawater corrosion of the reinforcing material. Furthermore, the supporting steel rods gave the finished sections a lower strength-to-weight ratio and poorer impact resistance than sections where mesh alone was used as the reinforcement (83).

The shotcrete laminating process developed by Iorns overcomes these problems and can be largely automated, greatly saving on labor costs. It also eliminates the need for double forms, which are required in more conventional concrete work. Consequently, a floating mould can be used, which greatly facilitates the handling of the finished structure.

Several examples exist of floating docks and boat hulls built by the shotcrete laminating process, which are considerably less expensive than their steel counterparts (84). Neither of these structures, however, is designed for the same service as a wave energy absorber. The potential cost savings can only be demonstrated by the fabrication and endurance testing at sea of a prototype absorber.

This leads naturally to a discussion of the second technology-based driver, which is the lack of long-term ocean test data on which to base cost and performance projections. Until several years sea time can be accumulated on full-scale components and systems, estimates of operating and maintenance costs are highly uncertain.

An even greater risk is associated with uncertainty about the absorption efficiency of full-scale systems in real ocean waves. Again, several years operation of a prototype or demonstration plant would be necessary to convince potential investors that the output of a commercial plant will be sufficiently high to generate projected revenues.

Market Development

Wave power plants can be based on land, on caissons in relatively shallow water (5–15 m depth), or in deeper, offshore waters. Land- and caisson-based systems have achieved the greatest development progress to date and may be the first to find commercial application. Land-based systems, however, involve significant shoreline modification and attendant environmental impacts, which may severely limit their deployment. Likewise, caisson-based wave power plants are likely to be acceptable only at existing ports or where construction of a new small-craft harbor has been approved.

Despite their less-advanced development status, offshore systems have much wider deployment potential, since they do not involve shoreline modification or breakwater construction. Floating devices have been developed that use air or seawater working fluids, eliminating the risk of chemical pollution. High-pressure seawater is particularly attractive as a working fluid, since it can produce both electricity and fresh water.

The near-term market for wave power is clearly in the Pacific Ocean, which has an energetic wave climate (despite its name) and a large number of island nations where the need for small-scale electric power and fresh water is great. If electrified at all, these islands are typically served by old diesel units, some dating back to World War II, operating at the far end of the oil distribution network.

Where coastal land is not at a premium and where a reservoir can be built without extensive blasting or dam construction, Norwave's Tapered Channel is expected to capture a fair share of the Pacific market. Japanese caisson-based systems are also expected to enter this market within the next 5 years, particularly where resort development is tied to construction of a small-craft harbor.

The commercial export of offshore systems is expected to lag behind that of land- and caisson-based systems. Significant improvement in cost and performance will not occur until a floating device is successfully demonstrated at full scale for a period of years. Commercial sales will not occur without a successful demonstration, and without sales, the developers of such systems will not have the financial resources or operating experience necessary to make significant improvements.

Without the long-term component testing afforded by a demonstration plant, it is doubtful whether any sort of offshore wave energy system would become economically feasible in the foreseeable future. Under this scenario, wave power development in industrial countries will be largely limited to caisson-based systems deployed as breakwaters at locations where their environmental impact is consistent with the existing level of onshore harbor development. This would probably amount to only tens of megawatts worldwide.

Without the development of an onshore hydrogen infrastructure, utilization of wave power will be limited by the economic and environmental costs of reinforcing or extending the utility grid and wave energy's lack of dispatchability. Under this scenario, it is expected that wave energy would be used primarily for distributed generation, as described previously. This would probably amount to only hundreds of megawatts worldwide.

With both drivers in place, however, it is conceivable that industrial countries could meet a large part of their energy needs from wave power. Under this scenario, tens of megawatts of caisson-based capacity would be deployed over the next decade. These would be demonstration plants designed primarily for harbor protection, incorporating components used in various offshore wave energy systems for purposes of long-term endurance testing.

By the end of this century, sufficient component sea time would be accumulated to enable confident projections of the cost and performance of offshore wave power plants. These would then be commercially deployed: first in island applications where the competing technology is diesel generation, later in distributed "mini-utilities" along mainland coasts, where the competing technology is extension of the central grid.

Meanwhile, an onshore hydrogen infrastructure would have been developed, fueled initially by surplus hydroelectric generation. The incentive for this development might come from depletion of fossil methane reserves (the feedstock for existing hydrogen production by steam reformation), coupled with legislation designed to reduce carbon dioxide emissions. The stage would then be set for large-scale, offshore, wave-powered production of hydrogen as a sustainable ocean resource.

Offshore Utilization of Ocean Space

Very large floating structures (VLFSs) have been proposed for a variety of offshore applications, including airports, space launch facilities, naval bases, living and recreation complexes, and production of energy-intensive materials and fuels. Bargelike VLFS modules are less costly than semisubmersibles but require more protection from wave action. Fabricating the perimeter of such a VLFS to accommodate wave energy absorbers such as OWCs, pivoting flaps, or flexible bags should not add significantly to its total construction cost.

The large inertia of a VLFS provides the necessary reaction point for these absorbers, which in turn act to remove energy from the wave before it passes under the platform. As previously noted, reaction structures are the single largest capital cost item for caisson-based wave energy devices and the SEA Clam. If the fabrication cost of a VLFS is supported by revenue derived from its main function, then wave energy costs for power conversion equipment and other balance-of-plant may be half or less than that now projected for systems connected to onshore utility grids (refer back to Figs. 26 and 27).

Production of hydrogen as a heating fuel is likely to be the first significant use of VLFS-based offshore wave power for mainland consumption. As onshore hydrogen use expands to include electric power generation and transportation, wave energy has the ultimate potential to supply a large portion of global energy demand, as explained below.

A steady wind blowing over water imparts its energy to waves, which grow in size until they become unstable and break. Shorter period waves reach limiting steepness first, while longer period waves continue to grow. Longer period waves also have a higher group velocity (the speed at which wave energy propagates across the ocean surface), and for a given wind speed, waves longer than a certain period travel faster than the wind. Once waves of this period have reached limiting steepness, the sea can absorb no more wind energy and is said to be fully developed. At this point, energy input from the wind is offset by energy losses due to wave breaking. If energy is absorbed from a fully developed sea by a wave power plant, however, the waves can absorb more energy from the wind than otherwise. Given sufficient fetch downwind of the plant, the waves will again become fully developed, regaining the energy that was absorbed. To use an agricultural analogy, waves regrow as they are "grazed," and the overall productivity of the sea surface (in terms of rate of energy yield, or power) increases. If such cropping can be achieved economically on a worldwide scale, then the global wave energy resource base exceeds that of ocean thermal energy conversion and is comparable to civilization's present power demand (9). The persistent trade winds of the tropics provide a wave climate in which such cropping is a realistic proposition. Mathematical modeling indicates that at a wind speed of 10 m/s, wave regrowth between power plants spaced 50 km apart along a 600-km fetch increases wave energy yield by 40% (85).

The ultimate prospect of large-scale production of wave-generated hydrogen suggests certain criteria for wave energy conversion devices that might be targeted for accelerated research and development during the next 5 years. One criterion is an absorber configuration that can be easily integrated into VLFS perimeter construction. Another is the use of high-pressure seawater as a working fluid. Efficient electrolyzers require fresh water, and DELBUOY has demonstrated that with a sufficiently large ratio of absorber area to pump cross section, seawater can be desalinated directly via reverse osmosis, without an electrical intermediary step. Thus, high-pressure seawater systems have both long-term commercial potential for efficient hydrogen production offshore and near-term commercial potential for cogeneration of fresh water and electricity at remote island and coastal locations.

APPENDIX: ENERGY COSTING METHOD AND FINANCIAL PARAMETERS

The energy costing method used in this article computes the levelized annual revenue requirement over the book life of a power plant and divides this by the plant's annual output. If all electricity generated by the plant were sold for this amount, the total collected revenue would have the same present value as the sum of all fixed charges and expenses paid out during the life of the plant. Levelization thus makes it possible to compare investment alternatives in terms of a single cost.

It should be noted that depreciation of the plant and periodic reinvestment in new equipment will cause the actual revenue requirement to change from year to year, whereas the levelized revenue requirement represents a constant annual payment and does not indicate the effect that a particular investment will have on actual cash flow.

Energy costs can be computed in either constant dollars or nominal dollars. Constant dollars were used, because they have a purchasing power more akin to the reader's recent experience. Nominal dollars represent what the energy would actually cost at the time payments are made.

Revenue requirements accounted for by this method are listed below. A more detailed description, including an explanation of levelization, can be found in Volume 3 of the EPRI Technical Assessment Guide (65).

The annual revenue that must be collected to pay for the construction and operation of a power plant consists of fixed charges and expenses. Fixed charges are long-term financial obligations associated with building the plant and periodically replacing equipment that may wear out during the plant's life. They are "fixed" in the sense that they must be paid, regardless of how much electricity the plant generates. The following fixed charges were applied to the computation of energy costs presented in this article:

- Return on capital investment, in the form of debt return to creditors and equity return to shareholders
- Book depreciation
- Property taxes and insurance
- Income taxes; depreciation, property taxes, insurance, and interest on debt assumed to be tax deductible

Expenses are annual payments associated with operation and maintenance (O&M) and include fixed O&M, variable O&M, and fuel. Note that the sale of coproducts, such as fresh water, is credited toward annual expenses. If coproduct sales exceed expenses, then the difference (which represents profit) is subject to income tax. In comparing renewable energy technologies with fossil fuel technologies, it was assumed that coal and oil prices would not be subject to real price escalation over the life of the plant.

Financial Parameters

The financial parameters used to compute levelized energy costs in this chapter were taken from Volume 1, Appendix A, of the EPRI Technical Assessment Guide. They are as follows:

- Constant-dollar cost of debt financing: 4.6%
- Constant-dollar cost of equity financing: 7.65%
- Debt-to-equity ratio: 50%
- Annual inflation rate: 6%

(The above combination gives a nominal discount rate of 12.5%.)

- Federal and state income tax rate: 38%
- Annual property tax and insurance rate: 2%

In addition, straight-line depreciation was assumed over the entire project book life, with zero net end-of-life salvage value (salvage pays for decommissioning). It was also assumed that there would be no tax preferences or credits.

Concrete Fabrication Costs for Absorber/Reaction Structures

Estimates of reinforced concrete fabrication cost per tonne (t) were gathered from a variety of different sources, as indicated below. Reinforced concrete density was assumed to be 2.5 t/m^3 for U.S. estimates. Estimates originating outside the United States were converted to dollars using the exchange rate in effect at the time the original cost estimate was made. Exchange rates were obtained from the International Monetary Fund in Washington, D.C., and were 138 ¥/US$, 1.78 US$/£, and 6.40 DKR/US$.

Country	Cost in Original Currency	Source	Cost
Japan (caisson)	85,700 ¥/t	33, 86	620 US$/t
United States (at $1000/$yd^3$)	525 US$/t	87	525 US$/t
United States (at $800/$yd^3$)	420 US$/t	87	420 US$/t
United Kingdom (cast in place)	178 £/t	88	315 US$/t
United Kingdom (precast unit)	146 £/t	88	260 US$/t
Denmark	2,250 DKR/t	89	125 US$/t
Norway	2,200 DKR/t	89	
New Zealand	2,050 DKR/t	89	
Ireland	2,000 DKR/t	89	140 US$/t
Portugal	1,360 DKR/t	89	85 US$/t

Buoys and damper plates contain substantial internal structural steel reinforcement. An equivalent total concrete weight of 0.72 t/m^2 of waterplane area was used to estimate fabrication costs for buoys ranging in diameter from 10 to 17 m. This factor is based on a cost estimate of $62,000 for a 14.5-m-diameter buoy using reinforced concrete at $1000/$yd^3$.

BIBLIOGRAPHY

1. M. E. McCormick, *Ocean Wave Energy Conversion*, Wiley, New York, 1981.
2. R. Shaw, *Wave Energy A Design Challenge*. Ellis Horwood, Chichester, 1982.
3. A. D. Carmichael, E. E. Adams, and M. A. Gluckman, *Ocean Energy Technologies: The State of the Art*, Electric Power Research Institute, Palo Alto, Calif., 1986.
4. L. Claeson, *Energi fran havets vagor*, Energiforskningsnamnden, Efn-rapport nr 21 (in Swedish, with English summary), Stockholm, Sweden, 1987.
5. SEASUN Power Systems, *Wave Energy Resource and Technology Assessment for Coastal North Carolina*, North Carolina Alternative Energy Corporation, Contract UY1873, Research Triangle Park, N.C., 1988.
6. SEASUN Power Systems, *Ocean Energy Technology Information Module*. Pacific Gas and Electric Company, Department of Research and Development, 007.691.4, San Ramon, Calif., 1991.
7. SEASUN Power Systems, *Wave Energy Resource and Economic Assessment for the State of Hawaii*, State of Hawaii, Department of Business and Economic Development, Energy Division, Honolulu, Hawaii, 1992.
8. N. N. Panicker, "Power Resource Potential of Ocean Surface Waves," in *Proceedings of the Wave and Salinity Gradient Workshop, Newark, Delaware*, U.S. Energy Research and Development Administration, Washington, D.C., 1976, pp. J1–J48.
9. J. D. Isaacs and W. R. Schmitt, "Ocean Energy: Forms and Prospects," *Science* **207,** 265–273 (1980).
10. SEASUN Power Systems, *Wave Energy Resource and Economics Update for Coastal North Carolina*, North Carolina Alternative Energy Corporation, Contract UY1893, Research Triangle Park, N.C., 1991.
11. S. H. Salter, "World Progress in Wave Energy—1988," *Int. J. Ambient Energy* **10,** 3–24, 1989.
12. Norwegian Royal Ministry of Petroleum and Energy, *Norwegian Wave Power Plants 1987*. Royal Ministry of Petroleum and Energy, Oslo, Norway, 1987.
13. A. N. Walton-Bott, P. D. Hunter, and J. S. M. Hailey, "The Mauritius Wave Energy Scheme—Present Status," in D. V. Evans, ed., *Euromechanics Colloquium 243 Energy from Ocean Waves*, extended abstract, University of Bristol, Bristol, United Kingdom, 1988.
14. P. J. Rance, "The Development of the H. R. S. Rectifier," in *Proceedings of the Wave Energy Conference, London—Heathrow*, 1978, pp. 49–54.
15. C. P. Sherburne and P. H. Davidoff, "DAMATOLL Development Progress," in *Proceedings of the 8th Ocean Energy Conference*, vol. 1, U.S. Department of Energy, DOE/Conf-810622—EXC, Washington, D.C., 1981, pp. 199–204.
16. B. Lundgren, SEA POWER AB, Gothenburg, Sweden, personal communication, May 1993.
17. Y. Masuda, "Experiences in Pneumatic Wave Energy Conversion in Japan," in *Utilization of Ocean Waves—Wave to Energy Conversion*, American Society of Civil Engineers, New York, 1987, pp. 1–33.
18. H. Chino, K. Nishihara, and Y. Nakakuki, "Verification Test of a Wavepower Generating System with a Constant Air Pressure Tank," in *Air Conditioning with Heat Pump*, Department of Business and Economic Development, Energy Division (photocopy on file), Honolulu, Hawaii, 1989, pp. 1–26.
19. K. Kaneko, T. Setoguchi, and S. Raghunathan, "Selfrectifying Turbines for Wave Energy Conversion," in *Proceedings of the First (1991) International Offshore and Polar Engineering Conference*, vol. 1, International Society of Offshore and Polar Engineers, 1991, pp. 385–392.
20. H. Hotta, Y. Washio, S. Ishii, Y. Masuda, T. Miyazaki, and K. Kudo, "The Operational Test on the Shore Fixed OWC Type Wave Power Generator," in *Proceedings of the Fifth International Offshore Mechanics and Arctic Engineering Symposium*, vol. 2, American Society of Mechanical Engineers, New York, 1986, pp. 546–552.
21. J. J. T. Whittaker and S. McIlwaine, "Shoreline Wave Power Experience with the Islay Prototype," in *Proceedings of the First (1991) International and Polar Engineering Conference*, vol. 1, International Society of Offshore and Polar Engineers, 1991, pp. 393–397.
22. K. Bonke and N. Ambli, "Prototype Wave Power Stations in Norway," in M. E. McCormick and Y. C. Kim, eds., *Utilization of Ocean Waves—Wave to Energy Conversion*, American Society of Civil Engineers, New York, 1987, pp. 34–44.
23. P. V. Inderesan and S. S. Murthy, "Generating Electrical Power from Wave Energy the Indian Experiment," in W. D. Jackson and D. A. Hull, eds., *24th Intersociety Energy Conversion Engineering Conference*, vol. 5, Institute of Electrical and Electronics Engineers, 89-CH-27813, New York, 1989, pp. 2121–2126.
24. M. Ohno, H. Funakoshi, T. Saito, K. Oikawa, and S. Takahashi, "Interim Report on the Second Stage of Field Experi-

ments on a Wave Power Extracting Caisson in Sakata Port," in *Proceedings of International Symposium on Ocean Energy Development for Overcoming the Energy & Environmental Crises*, Muroran Institute of Technology. Muroran, Hokkaido, Japan, 1993, pp. 173–182.

25. T. W. Thorpe, *A Review of Wave Energy*, Energy Technology Support Unit, ETSU-R72, Oxfordshire, United Kingdom, 1992.
26. S. D. Drew, "Recent Developments with the Vickers O. W. C. Devices," in H. Berge, ed., *Proceedings of the Second International Conference on Wave Energy Utilization*, Tapir Publishers, Trondheim, Norway, 192, pp. 191–209.
27. S. Ishii, T. Miyazaki, Y. Masuda, and G. Kai, "Reports and Future Plans for the Kaimei Project," in H. Berge, ed., *Proceedings of the Second International Symposium on Wave Energy Utilization*, Tapir Publishers, Trondheim, Norway, 1982, pp. 305–321.
28. H. Hotta, T. Miyazaki, Y. Washio, and S. I. Ishii, "On the Performance of the Wave Power Device Kaimei—The Results on the Open Sea Tests," in *Proceedings of the Seventh International Conference on Offshore Mechanics and Arctic Engineering*, American Society of Mechanical Engineers, New York, 1988, pp. 91–96.
29. Y. Masuda, T. Yamazaki, Y. Outa, and M. E. McCormick, "Study of Backward Bent Duct Buoy," in *Oceans '87 Proceedings*, vol. 2, Marine Technology Society, Washington, D.C., 1987, pp. 384–389.
30. X. Liang, W. Wang, N. Jiang, and X. Gao, "An Experimental Research on Performance of the 5kW BBDB Model," in *Proceedings of International Symposium on Ocean Energy Development for Overcoming the Energy & Environmental Crises*, Muroran Institute of Technology, Muroran, Hokkaido, Japan, 1993, pp. 227–235.
31. T. Miyazaki, Y. Washio, and N. Kato, "Performance of the Floating Wave Energy Converter Mighty Whale," in *Proceedings of International Symposium on Ocean Energy Development for Overcoming Energy & Environmental Crises*, Muroran Institute of Technology, Muroran, Hokkaido, Japan, 1993, pp. 197–204.
32. Y. Tanaka, K. Furukawa, and Y. Motora, "Experimental and Theoretical Study of a Double OWC Floating Wave Power Extractor," in *Proceedings of International Symposium on Ocean Energy Development for Overcoming the Energy & Environmental Crises*, Muroran Institute of Technology, Muroran, Hokkaido, Japan, 1993, pp. 209–220.
33. T. Watabe and H. Kondo, "Hydraulic Technology and Utilization of Ocean Wave Power," in *JHPS International Symposium on Fluid Power Tokyo, March 1989*, 1989, pp. 301–308.
34. M. Kuroi, "On a Flap-Type Wave Energy Converter at the Coastline," in *Conference Record of the 11th Meeting of the United States-Japan Cooperative Program in Natural Resources (UJNR) Panel on Marine Facilities.*
35. T. Miyazaki, "Wave Energy Research and Development in Japan," in *Oceans '91, Honolulu, Hawaii*, SEASUN Power Systems (reprint on file), Alexandria, Va., 1991, pp. 1–11.
36. R. O. Wilke, "Theoretical and Experimental Evaluation of an Engineering Model of the TandemFlap Wave Power Device," in *Proceedings of the Fifth International Offshore Mechanics and Arctic Engineering Symposium*, Vol. 2, American Society of Mechanical Engineers, New York, 1986, pp. 566–573.
37. R. O. Wilke, "First Openwater Test Program for a Fullscale Model of the TandemFlap Wave Energy Conversion Device," in *Proceedings of the Eighth International Conference on Offshore Mechanics and Arctic Engineering*, American Society of Mechanical Engineers, New York, 1989, pp. 435–438.
38. H. Altmann and F. J. M. Farley, "Latest Developments with the Triplate Wave Energy Converter," in *Proceedings of the First Symposium on Wave Energy Utilization*, Chalmers University of Technology, Gothenburg, Sweden, 1979, pp. 525–556.
39. Lancaster University, "*Design Study of the Engineering of a Reactionless Wave Energy Converter Buoy Working in Pitch and Surge (PS Frog)*, Lancaster University, Energy Technology Support Unit Contract E/5A/CON/1681/1852, Lancaster, United Kingdom, 1988.
40. F. H. Y. Wu, and T. T. L. Liao, "Wave Power Development in Taiwan," in H.-J. Krock, ed., *Proceedings of the International Conference on Ocean Energy Recovery*, American Society of Civil Engineers, New York, 1990, pp. 93–100.
41. A. Shiki and K. Iwase, "Research and Development of Wave Pump" in H.-J. Krock, ed., *Proceedings of the International Conference on Ocean Energy Recovery*, American Society of Civil Engineers, New York, 1990, pp. 76–83.
42. K. Nielsen, "On the Performance of a Wave Power Converter," in M. E. McCormick and Y. C. Kim, eds., *Utilization of Ocean Waves—Wave to Energy Conversion*, American Society of Civil Engineers, New York, 1987, pp. 164–183.
43. K. Nielsen and C. Scholten, "Planning a Fullscale Wave Power Conversion Test: 1988–1989," in H.-J. Krock, ed., *Proceedings of the International Conference on Ocean Energy Recovery*, American Society of Civil Engineers, New York, 1990, pp. 111–120.
44. Danish Wave Power Aps, *Feasibility Study of 300MW Wave Power Plants*, Danish Wave Power Aps, Copenhagen, Denmark, 1992.
45. D. C. Hicks, C. M. Pleass, and G. R. Mitcheson, "DELBOUY: Wavepowered Desalination System," in *Oceans '88 Proceedings*, vol. 3, Institute of Electrical and Electronics Engineers, New York, 1988, pp. 1049–1055.
46. G. Svensson, "Swedish Idea for Wave Power," in *Energy Ahead 85*, Vattenfall, Annual Research and Development Report, Stockholm, Sweden, 1985, pp. 40–44.
47. L. Claeson, J. Forsberg, A. Rylander, and B. O. Sjostrom, "Contribution to the Theory and Experience of Energy Production and Transmission from the Buoyconcept," in H. Berge, ed., *Proceedings of the Second International Conference on Wave Energy Utilization*, Tapir Publishers, Trondheim, Norway, 1982, pp. 345–370.
48. H. H. Hopfe and A. D. Grant, "The Wave Energy Module," in J. Twidell, I. Hounam, and C. Lewis, eds., *Energy for Rural and Island Communities IV*, Pergamon, Oxford, 1986, pp. 243–248.
49. U.S. Pat. 4, 781, 023, (Nov. 1988) to C. K. Gordon.
50. A. Yazaki, S. Takezawa, and K. Sugawara, "Experiences on Field Tests of "Kaiyo" Electricity Generating System by Natural Wave Energy," in *Proceedings of the Fifth International Offshore Mechanics and Arctic Engineering Symposium*, Vol. 2, American Society of Mechanical Engineers, New York, 1986, pp. 538–545.
51. S. Salter, "Progress on Edinburgh Ducks," in D. V. Evans and A. F. de O. Falcao, eds., *Hydrodynamics of Ocean Wave-Energy Utilization Proceedings of the IUTAM Symposium, Lisbon, Portugal*, Springer-Verlag, Berlin, 1986, pp. 57–67.
52. N. W. Bellamy, "Development of the SEA Clam Wave Energy Converter," in H. Berge, ed., *Proceedings of the Second International Conference on Wave Energy Utilization*, Tapir Publishers, Trondheim, Norway, 1982, pp. 175–190.
53. N. W. Bellamy, "The Circular SEA Clam Wave Energy Converter," in D. V. Evans and A. F. de O. Falcao, eds., *Hydrodynamics of Ocean Wave-Energy Utilization Proceedings of the IUTAM Symposium, Lisbon, Portugal*, Springer-Verlag, Berlin, 1986, pp. 69–79.

54. M. J. Platts, "Engineering Design of the Lancaster Wave Energy System," in H. Berge, ed., *Proceedings of the Second International Conference on Wave Energy Utilization*, Tapir Publishers, Trondheim, Norway, 1982, pp. 253–274.
55. R. Clare, D. V. Evans, and T. L. Shaw, "Harnessing Sea Wave Energy by a Submerged Cylinder Device," in *Proceedings of the Institution of Civil Engineers, Part 2*, Vol. 73, 1982, pp. 565–585.
56. D. S. Shugar, "Photovoltaics in the Utility Distribution System: The Evaluation of System and Distributed Benefits," in *Proceedings of the 21st IEEE PV Specialists Conference*, in press.
57. G. W. Braun, A. Suchard, and J. Martin, "Hydrogen and Electricity as Carriers of Solar and Wind Energy for the 1990s and Beyond," *Solar Energy Materials* **24,** 62–75 (1991).
58. S. H. Browne, P. K. Takahashi, and J. P. Suyderhoud, "Evaluation of Hydrogen Production Processes," in *Proceedings of the International Renewable Energy Conference*, Department of Business and Economic Development, Energy Division, Honolulu, Hawaii, 1988, pp. 400–412.
59. E. Mehlum, P. Anderssen, T. Hysing, J. J. Stamnes, O. Eriksen, and F. SerckHanssen, "The Status of Wave Energy Projects and Plants in Norway. Part 2: Norwave TAPCHAN—a Commercial Overview," in D. W. Behrens and M. A. Champ, eds., *Proceedings: Oceans '89 Special International Symposium A Global Review of the Development of Wave Energy Technologies*, Pacific Gas and Electric Company, Department of Research and Development, San Ramon, Calif., 1990, pp. 29–36.
60. C. C. K. Liu and H. H. Chen, "Conceptual Design and Analysis of a Wavedriven Artificial Upwelling Device," in *Oceans '91 Proceedings*, vol. 1, Institute of Electrical and Electronics Engineers, New York, 1991, pp. 406–412.
61. K. J. Tjugen, "Tapchan Ocean Wave Energy Project," in *Proceedings of the 1993 European Wave Energy Symposium*, Paper G3, National Engineering Laboratory, East Kilbride, Scotland, in press.
62. K. Connover, R. Lynette & Associates, Redmond, Wash., personal communication, Jan. 1994.
63. T. T. J. Whittaker, "Progress of the Islay Shoremounted Oscillating Water Column Device," in *UK-Ises Conference C57—Wave Energy Devices*, SEASUN Power Systems (preprint on file), Alexandria, Va., 1989, pp. 1–8.
64. N. M. Bellamy, "High Efficiency, Low Cost, Wave Energy Conversion System," in *Proceedings of International Symposium on Ocean Energy Development for Overcoming the Energy & Environmental Crises*, Muroran Institute of Technology, Muroran, Hokkaido, Japan, 1993, pp. 189–195.
65. Electric Power Research Institute, *Technical Assessment Guide*, vol. 3: *Fundamentals and Methods, Supply—1986*, Electric Power Research Institute, P4463-SR, Palo Alto, Calif., 1987.
66. Black & Veatch, *Supply-Side Resource Option Portfolio Development*, Hawaii Electric Light Company, Hilo, Hawaii, 1993.
67. P. D. Komar, *Beach Processes and Sedimentation*, Prentice-Hall, Englewood Cliffs., N.J., 1976.
68. Argonne National Laboratory, *Physical Engineering and Environmental Aspects of Ocean Kelp Farming*. Gas Research Institute, GRI-81/0111, Chicago, Ill., 1982.
69. C. Harrold, J. Watanabe, and S. Lisin, "Spatial Variation in the Structure of Kelp Forest Communities along a Wave Exposure Gradient," *P.S.Z.N.I.: Marine Ecology*, vol. 9, 1988, pp. 131–156.
70. U.S. Department of the Interior, *Final Environmental Impact Statement; Proposed 1981 Outer Continental Shelf Oil and Gas Lease Sale Offshore Central and Southern California; OCS Scale 53*, U.S. Department of the Interior, Bureau of Land Management, Los Angeles, Calif., 1980.
71. U.S. Department of the Interior, *Pacific Outer Continental Shelf; Northern California Proposed Oil and Gas Lease Sale 91; Draft Environmental Impact Statement*, U.S. Department of the Interior, Minerals Management Service, MMS 87-0032, Los Angeles, Calif., 1987.
72. S. Schwartz, Marine Mammal Commission, Washington, D.C., personal communication, Dec. 1988.
73. R. Hoffman, Marine Mammal Commission, Washington, D.C., personal communication, Dec. 1988.
74. C. Cockerell, M. J. Platts, and R. ComynsCarr, "The Development of the Wavecontouring Raft," in *Proceedings of the Wave Energy Conference, LondonHeathrow*, 1978, pp. 7–16.
75. M. A. Champ and W. L. Pugh, "Tributyltin Antifouling Paints: Introduction and Overview," in *Oceans '87 Proceedings*, vol. 4, Marine Technology Society, Washington, D.C., 1987, pp. 1296–1313.
76. P. Schatzburg, "Organotin Antifouling Hull Paints and the U.S. Navy—A Historical Perspective," in *Oceans '87 Proceedings*, vol. 4, Marine Technology Society, Washington, D.C., 1987, pp. 1324–1333.
77. T. O. Morgan, D. S. Sasscer, and C. J. Rivera, "Macrofouling in Simulated OTEC Evaporators at Punta Tuna, Puerto Rico," in *Proceedings of the 8th Ocean Energy Conference*, vol. 1, U.S. Department of Energy, DOE/Conf-810622—EXC, Washington, D.C., 1981, pp. 415–420.
78. M. McCullough, "Contending with Marine Biofouling Evaluation of Antifoulant Materials, June 1978 through November 1982," in *Oceans '83 Proceedings*, vol. 1, Marine Technology Society, Washington, D.C., 1983, pp. 522–526.
79. P. M. Koola, M. Ravindran, and V. S. Raju, "Design Options for a Multipurpose Wave Energy Breakwater," in *Proceedings of International Symposium on Ocean Energy Development for Overcoming the Energy & Environmental Crises*, Muroran Institute of Technology, Muroran, Hokkaido, Japan, 1993, pp. 163–172.
80. K. Gao, "A Developing 20kW OWC Wave Power Plant," in *Proceedings of International Symposium on Ocean Energy Development for Overcoming the Energy & Environmental Crises*, Muroran Institute of Technology, Muroran, Hokkaido, Japan, 1993, pp. 205–208.
81. X. Gao, "Ocean Energy Activities in China," H.-J. Krock, ed., in *Proceedings of the International Conference on Ocean Energy Recovery*, American Society of Civil Engineers, New York, 1990, pp. 18–23.
82. M. E. Iorns, "Offshore Construction of Very Large Floating Platforms," in *Oceans '91 Proceedings*, vol. 2, Institute of Electrical and Electronics Engineers, New York, 1991, pp. 1126–1130.
83. M. E. Iorns, "Cost Reduction and Quality Control in Ferrocement and Marine Concrete," in *Proceedings of the Concrete Ships and Floating Structures Convention*, Thomas Reed Publications, London, 1979, pp. 1–6 (Day 2, Paper 5).
84. M. E. Iorns, "Cost Comparison Ferrocement and Concrete Versus Steel," *Concrete International*, November 1983, pp. 45–50.
85. L. Claeson and B.-O. Sjostrom, "Optimal Wave Power Plant Spacing in the Tropical North Pacific Ocean, in *Oceans '91 Proceedings*, Vol. 1 Institute of Electrical and Electronics Engineers, New York, 1991, pp. 547–549.

86. T. Watabe, H. Kondo, and M. Kobiyama, "A Case Study on the Utilization of Ocean Wave Energy for Fish Farming in Hokkaido," in *Proceedings, 1989 International Symposium on Cold Regions Heat Transfer*, SEASUN Power Systems (reprint on file), Alexandria, Va., 1989, pp. 159–164.
87. M. Lanier, ABAM Engineers Inc., Federal Way, Wash., personal communication April 1988 and Dec. 1993.
88. Atkins Oil and Gas Engineering, *A Parametric Costing Model for Wave Energy Technology*, Energy Technology Support Unit, WV 1685, Oxfordshire, United Kingdom, 1992.
89. Danish Wave Power Aps, *Technical Annex to: Feasibility Study of 300MW Wave Power Plants*, Danish Wave Power Aps, Copenhagen, Denmark, 1992.

WIND POWER

JON G. MCGOWAN
University of Massachusetts
Amherst, Massachusetts

Wind power systems involve the conversion of the earth's winds, a renewable energy source, to rotating shaft power and/or electrical power output. The overall system that accomplishes this energy conversion process is generally called a wind turbine generator (WTG). In wind power systems the actual conversion process uses the basic aerodynamic forces of lift or drag to produce a net positive torque on a rotating shaft, resulting in the production of mechanical power. In addition to the large-scale, megaWatt (MW) sized, production of electrical energy, wind power systems can be applied to smaller sized applications in developing countries as well as for energy supply in remote or specialized applications. These categories can include water pumping, battery charging, and heating end uses as well as hybrid energy (wind/diesel, wind/PV) systems.

As pointed out in recent U.S. studies (1), the scope of the wind resource, both in the United States and globally, is enormous and is less dependent on latitude than other solar based renewable energy technologies. For example, the theoretical wind resource in the United States has been conservatively estimated to be capable of providing values of more than ten times the electrical power currently consumed. If wind machine technology and land accessibility factors are included, the practical size of this resource is estimated to be on the order of 20% of the current U.S. electric consumption (2).

Compared to solar energy, the wind is a very complex resource, possessing a three-dimensional value as compared to solar's two-dimensional qualities. The wind resource is more intermittent and is strongly influenced by terrain or geography factors. Also, due to fluid mechanics considerations, there is a nonlinear (cubic) relationship between wind speed and the power production from a wind turbine. An illustration of the last factor is a comparison of the wind energy productivity of good, excellent, and outstanding wind power sites having average wind speeds of 13, 16, and 19 mph, respectively. The 3 mph difference results in the excellent site having a potential energy production per unit area of 86% more than the good site, while the outstanding site has a potential energy production per unit area of 212% more than the good site. See also ELECTRIC POWER GENERATION; RENEWABLE RESOURCES.

Background

There are comprehensive and illustrated overviews of the history of wind power as well as a complete history of wind turbine design from the ancient Persians to the mid-1950s (3–6). Also, in addition discussion of innovative types of wind turbines, a summary of the historic uses of wind power, of wind-electric generation, and of the U.S. research work conducted during 1970–1985 is available (7). A different perspective on wind power history with a historical perspective of the key design components of a modern wind turbine system is also available (8).

With the exceptions of the 1.25 MW Smith-Putnam turbine program in the United States in 1940 (9), the Electrical Research Association (ERA) work in the U.K. (6), and the Danish Gedser and German Hutter machines (7), the international development of wind power was at a basic standstill following World War II. The energy crisis of the early 1970s, however, caused a worldwide rebirth in the development of wind energy technology. In the United States the Federal Wind Energy Program had its beginnings in 1972 when a joint Solar Energy Panel of the National Science Foundation (NSF) and the National Aeronautics and Space Administration (NASA) recommended that wind energy be developed to broaden the nation's energy options for new energy sources (10). During the 1970s and 1980s, various U.S. organizations and the Department of Energy took on a wide scope of wind power development work from small wind turbine systems (defined by machines up to about 50 kW rated power) to large-scale wind turbines (11–13).

The majority of the initial U.S. wind energy funds went toward a multimegawatt wind turbine development program which produced four series of machines ranging from the 100 kW (38 m diameter) NASA MOD-0 machine to the 3.2 MW Boeing MOD-5B machine with its 98 m diameter and a variable speed generator. Although none of these machines ever reached commercial status, they provided the wind industry with valuable lessons and a technical information base. On an international scale, other large wind turbines are currently being developed by a number of countries. In general, the development of large turbines has been slower, more expensive, and technologically more difficult than that of smaller machines.

In wind turbine technology, the horizontal axis type of wind turbine has emerged as the predominant type. There are still many possible configurations of this type of machine and choices of specific system designs. For example, there are different rotor materials and number of turbine blades, upwind vs downwind machines, fixed vs variable pitch. Also, one of the key questions that has been addressed many times since the 1970s is concerned with the optimum size of wind turbines. System analysis studies conducted during the early part of the U.S. wind energy program indicated optimum wind turbine sizes to be about 1–3 MW. Such studies gave some impetus to other country's, eg, Sweden, U.K., and Germany, national wind programs to fund the development of MW machines.

At present, the economic benefits of volume production of the large machines has not been determined. Despite this, on a worldwide basis in 1993 over ten multiMW wind turbine prototypes are being operated, being tested, or are in the final stages of development or construction.

For large-scale electrical power generation, the development of wind energy conversion systems during the early 1990s progressed significantly. From its problem-ridden beginnings and a heavily subsidized growth era, by the early 1990s the wind industry attained status as an increasingly reliable, competitive, power source. A key factor in this progress was the development of wind farms, large arrays of interconnected wind turbines used for the generation and delivery of large amounts of electricity to a utility grid. These types of energy generation systems have evolved rapidly since the 1980s, and there are now substantial installations in three regions of California, in Hawaii, and in Denmark and other parts of Europe. Through 1988, over three billion dollars worldwide were invested in the purchase and installation of over 20,000 turbines, with the market reaching a peak of about 600 MW installed during 1985 (14). In the United States, California contains the world's largest concentration of wind turbines (15) with approximately 15,500 wind machines located in three windy mountain passes. These machines represent about 1600 MW of generating capacity, and each year they generate 1.2% of the state's electricity. Also, on certain occasions, wind turbines have provided about 8% of Pacific Gas & Electric's load in Northern California.

The rapid growth of wind energy in California did not occur without problems. Several wind turbine manufacturers became involved in the industry without an appreciation of the technical challenges associated with the design and maintenance of wind turbines. During the early 1990s the industry experienced consolidations and major changes, especially with the phaseout of Federal and California tax credits. For example, the total number of U.S. manufacturers of wind turbines dropped from several dozen to less than ten. U.S. Windpower (Kenetech), the remaining large-scale producer of wind turbines in the United States is still the largest developer, operator, and manufacturer of wind turbines. As of 1993 about thirty serious wind turbine manufacturers exist worldwide, with the majority based in Europe (16). The remaining manufacturers are committed to improving reliability, reducing operation and maintenance costs, and increasing productivity. According to the latest analyses of the subject (1,15,17), they appear to be succeeding.

Wind Turbines

Power Output Prediction. It is possible to define the energy conversion performance of a wind turbine system that has been designed to produce electrical power without considering the technical details of its various components. The machine's power curve, which gives the electrical power output as a function of the wind speed must be known; for horizontal axis machines this is usually defined as the wind speed at the hub height. Figure 1 presents an example of a power output curve for an idealized wind turbine. Typically such a curve could be obtained

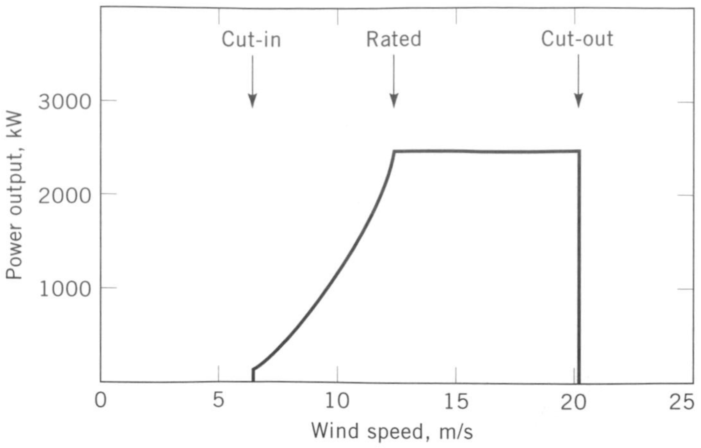

Figure 1. Power output curve for wind turbine.

from a manufacturer's machine specification sheet (18). The performance of a given wind turbine generator can be related to three key points on the velocity scale: cut-in speed, the minimum wind speed at which the machine will deliver useful power; rated wind speed, the wind speed at which the rated power (generally the maximum power output of the electrical generator) is reached; and cut-out speed, the maximum wind speed at which the turbine is allowed to deliver power (usually limited by engineering design and safety factor constraints).

In the overall design of a complete wind turbine system, it is possible to predict the exact shape of the power curve for a given machine. Such a process, however, is by no means a simple engineering task as it involves determination of the power characteristics of the wind turbine rotor, electrical generator characteristics, transmission gear ratios, and component efficiencies. Details of this process are described in numerous references (6–8).

Basic Wind Turbine Aerodynamics. In the performance determination and design of wind turbine systems, it is important to know (from both a resource and engineering design standpoint) the maximum power (per given cross flow area) that can be extracted from a wind stream of a known velocity, V. The kinetic energy in the wind per unit volume is $\frac{1}{2}\rho V^2$, with ρ representing the density of the air. Moving at velocity V, the energy flow rate or power per unit area is given by: $P/A = \frac{1}{2}\rho V^3$. Although this equation shows the cubic relation between the wind speed and the power in the wind, in practice no device can extract all the energy from the wind or the air would have to be completely brought to rest. That would prevent further air from passing through the system and would reduce power extraction efficiency. Slowing the air down too little or too much gives a low efficiency. The optimum power extraction efficiency for a simple rotor does not depend on the particular means of slowing down the air, whether with few or many blades, or with a horizontal or vertical axis design (19).

To determine the maximum turbine rotor extraction possible from the wind, a one-dimensional model, shown in Figure 2, can be used to model the flow through the wind turbine rotor. Following conventional analysis (8), the rotor is replaced with an 'actuator disc.' It can be assumed that there is steady, incompressible, homogeneous

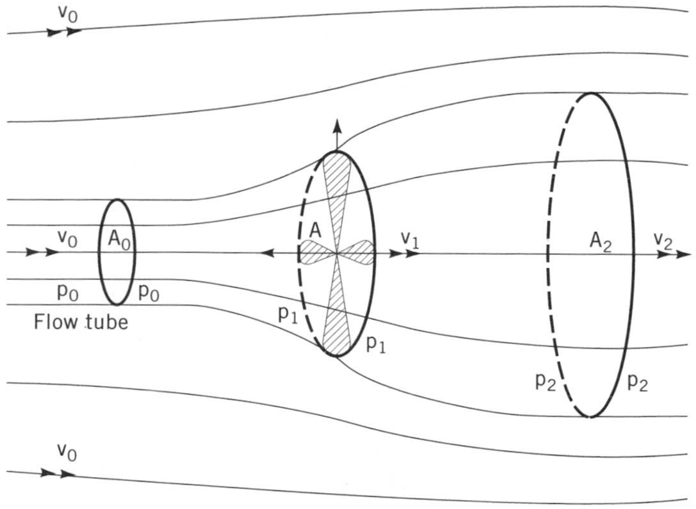

Figure 2. Model of flow through a wind turbine rotor.

wind with uniform flow at disc; that there are no obstructions to upstream or downstream wind flow; that wind flow passing through the disc is separable from remaining flow by a well defined stream tube; and that no rotation of flow is produced by the disc.

By applying both continuity equations between sections 0 and 2 and Bernoulli's theorem upstream and downstream of the disk, the following expression for the power extraction from the air results

$$P = \tfrac{1}{2}\,\rho\, AV_0^3\, 4a(1 - a)^2$$

where a is the axial interference factor, the fractional decrease in wind speed between the free stream and the rotor plane, ie, $a = V_1/V_0$.

By normal convention C_p is defined as the power coefficient, as follows

$$C_p = P/\tfrac{1}{2}\,\rho\, AV_0^3$$

Thus,

$$C_p = 4a(1 - a)^2$$

The maximum value of C_p occurs when $a = \frac{1}{3}$. Therefore

$$C_p,\ \max = \tfrac{16}{27} = 0.593$$

Finally, the maximum power per unit area that can be extracted by a simple horizontal or vertical axis turbine is given by $P_{\max}/A = 0.593\ (\frac{1}{2}\ \rho\ AV_0^3)$. For the design of actual wind turbines, namely the calculation of C_p, both wake rotational effects and two-dimensional airfoil theory must be used in the analytical modeling of the turbine rotor. In general, C_p can be related to the tip speed ratio, λ, of the turbine, defined as the ratio of the outer blade tip speed to the free stream wind speed, V_0. A good summary of the analytical approaches for the calculation of C_p is given (8,20,21).

Type and Classification of Wind Turbine Generators. More U.S. patents for wind systems have been applied for than nearly any other type of energy conversion device (4). These concepts have produced a number of designs, all intended to extract power from the wind, that can be classified in a number of ways. However, there are basic design similarities among most wind turbines (22).

The axis of rotation may be parallel, a horizontal axis wind turbine (HAWT), or perpendicular, usually a vertical axis wind machine (VAWT), to the wind stream.

The predominant force of a machine may be lift or drag. Drag-based machines can have no part moving faster than the wind while lift-type machines rely on aerodynamic lift forces and can have parts that move faster than the wind.

Solidity, the total blade area divided by the swept frontal area of the machine, is also a means of classification. For many turbines the solidity is described by giving the number of blades. In general, high solidity machines start easily with high initial torque, but reach maximum power at low rotational speed. Similarly, low solidity machines may require starting, but reach maximum power at high rotational speeds.

If the purpose of the wind turbine is direct mechanical output, the device is simply a windmill or a wind turbine. If electrical production is desired, the combination of turbine and generator may be called a wind turbine generator (WTG). Sometimes this combination is called a wind energy conversion system (WECS), and small machines use the acronym SWECS.

Whether the frequency of rotation of the wind turbine is maintained at a constant value varies with wind velocity is another consideration. At present, most WECS that are linked to a strong ac electrical grid are controlled by the grid to rotate at nearly constant frequency. However, there are some potential advantages to running a WECS at a variable speed. Some of the new designs of wind turbine systems, based on the use of power electronics, are using this type of design.

Finally, the turbine of the WECS may be directly coupled to the generator, or there may be an intermediate energy store that acts as a smoothing device. Since the wind velocity fluctuates rapidly, the inertia of the wind turbine and the softness of the turbine-generator coupling can be used to prevent fluctuations in the power output. Thus, this type of decoupling could filter out high frequency turbine fluctuations, and allow better matching of wind turbine to wind, and generator to load, than a direct coupled design. Similar effects occur if the blades are independently hinged against a spring, or hinged together.

There are, however, some wind turbine designs that are based on other design considerations. For example, these include the potential orientation of horizontal axis wind machines (downwind or upwind), wind concentrators or augmenters, and other innovative type wind turbine designs. There are comprehensive reviews of the various types of wind machines (4,23), and some of the more innovative designs are documented in past work supported by the U.S. Department of Energy (24,25).

Regardless of the type of wind turbine system, a key design parameter for the performance of a particular wind turbine design is the power coefficient, C_p, which was previously defined as the ratio of the power captured by the rotor to the power in the wind. Typical behavior curves for the rotor power coefficient as a function of tip

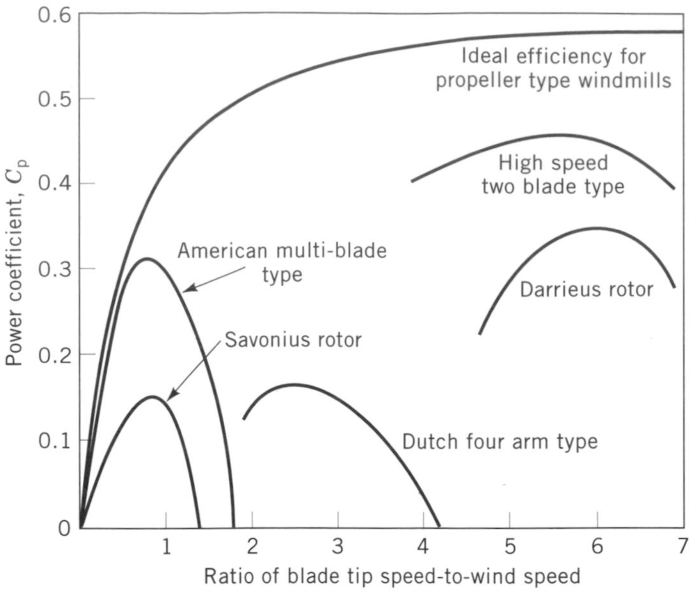

Figure 3. Rotor power coefficient as a function of tip speed ratio.

speed ratio are shown in Figure 3. The differences in efficiencies and tip speeds of the various designs are apparent. For example, the higher starting torque and the lower tip speed ratio of the multiblade design makes it ideal for low speed applications such as water pumping. On the other hand, the potentially high efficiencies of the two (or three) bladed horizontal axis propeller type design are also apparent. This type of design forms the vast majority of all the worldwide turbine systems.

Wind and Site Characteristics

Four areas of wind energy conversion systems require specific wind characteristics (26). For example, wind characteristics needed for performance evaluation include actual mean wind speeds, and seasonal and diurnal distributions at a given location of interest. For wind machine systems design, wind characteristics of interest include representative average conditions, eg, mean wind distribution, and extreme conditions, eg, peak gust speeds. Resource-based wind power siting requirements encompass the prediction or assessment of the desirability of candidate sites for one wind turbine or for a wind farm consisting of a number of interconnected WTGs. Wind characteristics influencing operations influence such subjects as load management planning, operational procedures, eg, start-up, shut-down, directional tracking, etc, and system lifetime and maintenance, eg, wind shear and gust influence on rotor design.

The importance of wind resource and site characterization has been emphasized in numerous texts and technical reports (6,7,9,21,27,28).

Basic Meteorological Data/Measurement. In wind systems resource assessment, the two meteorological variables of most interest are wind speed and wind direction. These variables are readily obtained with a variety of sensors for the measurement of wind speed (anemometers) and wind direction (wind vanes). For wind speed, the most common types of measurement instruments are the cup or propellor anemometers (6,21,28). The most dramatic improvement in resource assessment occurred in the 1980s with advances in data collection, reduction, and analysis made possible by the digital and personal computer.

Most countries have national meteorological services that collect, analyze, and report weather related data, including wind speed and direction (26). Most of this type of data is of questionable use for wind power prediction, however, since measurements are usually taken at a relatively low (10 m) height at airports or in urban locations where wind machines are unlikely to be sited. On the other hand, this type of data can provide a starting point for regional assessments of wind power availability.

National wind energy resource assessments have been in progress in the United States since 1975 (29), and a complete wind energy resource assessment (twelve regional wind energy atlases covering the United States and its territories) was provided by Department of Energy researchers in the early 1980s (30). This work was revised in 1987 (31) and served as the basis for a detailed assessment of the available windy land area and wind energy potential in the contiguous United States (2). In Europe, a similar wind energy atlas was prepared in 1989 (32).

Variation with Height, Location, and Time. An important parameter in the characterization of the wind resource is the variation of wind speed with height above the ground. This variation is strongly dependent on the local terrain (specifically, surface roughness) and has both short-term and long-term time aspects (26). The terrain effects are related to the boundary layer nature of the atmospheric wind flow. A review of the complex nature of terrain effects is given in several wind energy texts (21,26), and the effects of local terrain effects on the wind resource are addressed in a number of siting handbooks that provide guidelines on the siting of small and large wind turbines (27,28,33).

Time for the variation of wind speed with height above the ground can be divided into two categories: short-term intervals (on the order of minutes) and long-term intervals (on the order of months or years). The short-term variation is a complex function of the atmospheric boundary layer and is dependent on surface roughness and the solar radiation input to the atmosphere. The details of this variation have been considered by many researchers using boundary layer similarity theory (26). For wind resource assessments and wind turbine design studies, a power law relationship is usually used as an engineering approximation for long-term time intervals. Its most general form is

$$\frac{U_2}{U_1} = \left[\frac{z_2}{z_1}\right]^{\alpha}$$

where U_2 = wind speed at height z_2, U_1 = wind speed at height z_1, and α = power law exponent.

As documented by numerous researchers (28), the power law exponent is not a simple parameter; it is a complex function of a number of parameters including terrain variables, wind speed and temperature, meterorological factors, and time of day or season. For engineering approximations, a value of $\alpha = \frac{1}{7}$ is often used. It should be noted, however, this average value should only be used if site specific data are not available because of the wide range of values that the power law exponent can assume.

Widely varying time and length constraints on the type of wind data are required for resource assessments and the design and performance evaluation of wind energy conversion systems. For example, any realistic description of the wind flow field can not assume that the incident flow field on a wind turbine is constant in both time and space over the area swept out by the wind turbine rotor. Thus, detailed flow field data should include those characteristics responsible for the fluctuating components of the wind turbine loads, especially the turbulent flow characteristics of the wind. Information on this subject is available (21,22,28).

Distributions/Statistical Representation and Characterization. In general, wind speed data, averaged over an appropriate time interval, can be plotted in a number of forms including velocity-duration curves, histograms, and probability density curves.

For engineering analysis, the use of statistical models to characterize the wind resource via a probability density function is a particularly valuable tool. The use of these models allows a rapid assessment of the energy potential at a particular site with a particular machine. In wind engineering, the most accepted forms for probability density functions are the Weibull and Rayleigh distributions (26).

The Weibull distribution is generally the most used and has the form

$$p(V)_{\text{Weibull}} = \left(\frac{k}{c}\right)\left(\frac{V}{c}\right)^{k-1} \exp\left\{\left(\frac{V}{c}\right)^{k}\right\}$$

where c is the scale factor (units of speed), related to the mean value of V

$$(\overline{V} = \int_{o}^{n} VP(V)dV)$$

by

$$\overline{V} = c\Gamma(1 + 1/k)$$

with Γ the gamma function and k the shape factor (dimensionless) which is related to the variance of the speed.

The Rayleigh distribution is a one parameter (only average velocity is required) probability distribution function, which can be obtained from the Weibull distribution by setting $k = 2$, and is given by

$$p(V)_{\text{Rayleigh}} = \left(\frac{\pi V}{2\overline{V}^2}\right) \exp\left\{\frac{-\pi}{4}\left(\frac{V}{\overline{V}}\right)^{2}\right\}$$

A detailed discussion of the application of the Weibull and Rayleigh distribution statistics as specifically applied to wind resource assessment is available (7).

Siting. A major difference between the siting of wind turbines and conventional power plants is that the performance of a given wind system (measured by total power output and its time varying characteristics) is completely defined by the location of the wind turbine. Thus, the siting of a potential wind turbine or group of turbines is an important part of any wind energy conversion system feasibility study (6,9). Today, a number of siting handbooks exist that provide guidelines on the siting of small (27,31) and large (28,33) wind turbines.

There are basically two approaches to the siting of wind turbine generators: wind prospecting and the evaluation of a predetermined site (28). In the first approach, the region of interest is screened for locations that experience suitably high winds with sufficient frequency such that wind generated electricity may be economically attractive. Once these places are located, estimates or detailed measurements of the wind are needed to determine the economics of a potential site. The major steps in wind prospecting include: (*1*) analyzing the region of interest. Here a large region is screened for candidate resource areas that appear promising. (*2*) Evaluating candidate areas. The candidate resource area is screened for potential candidate sites that have usable winds and that satisfy pertinent land use and accessibility criteria. (*3*) Screening candidate sites. The list of potential candidate sites is narrowed here. (*4*) Evaluating a candidate site. This step involves the collection of actual wind data and a detailed evaluation of the site. (*5*) Developing the site. After a site is chosen, the best locations for individual machines are determined.

The other approach, evaluating a predetermined site, is generally followed when the potential candidate sites are known. This is the case when a developer or utility owns a large plot of land in an area with a known wind resource. For this case, the siting procedure consists of first establishing feasibility, which involves a preliminary evaluation of the wind energy potential; then evaluating the site; and finally developing the site.

In any type of wind siting study, the estimation of the wind resource at an early stage is important. Since detailed wind information is generally not available for individual wind sites, or what data is available is not from the same location, a technique for wind resource assessment is required. Fortunately, there are a number of techniques for this purpose, including numerical or physical modeling of flow over terrain; topographical or biological indicators of wind resource potential; and geomorphological or social and cultural indicators of wind energy resource (28).

The detail and complexity of these siting tools ranges dramatically in complexity, ease of use, and potential accuracy. For example, there are a number of numerical models available that can be used for the prediction of the performance of arrays of wind turbines (2).

Preliminary Estimation of Wind Turbine Power Output. In any wind energy resource assessment, it is important to have an engineering tool that gives a preliminary estimate of the energy production using a minimum of characteristics of a specific wind machine and the wind at a particular site. This method (28,34) assumes a Rayleigh

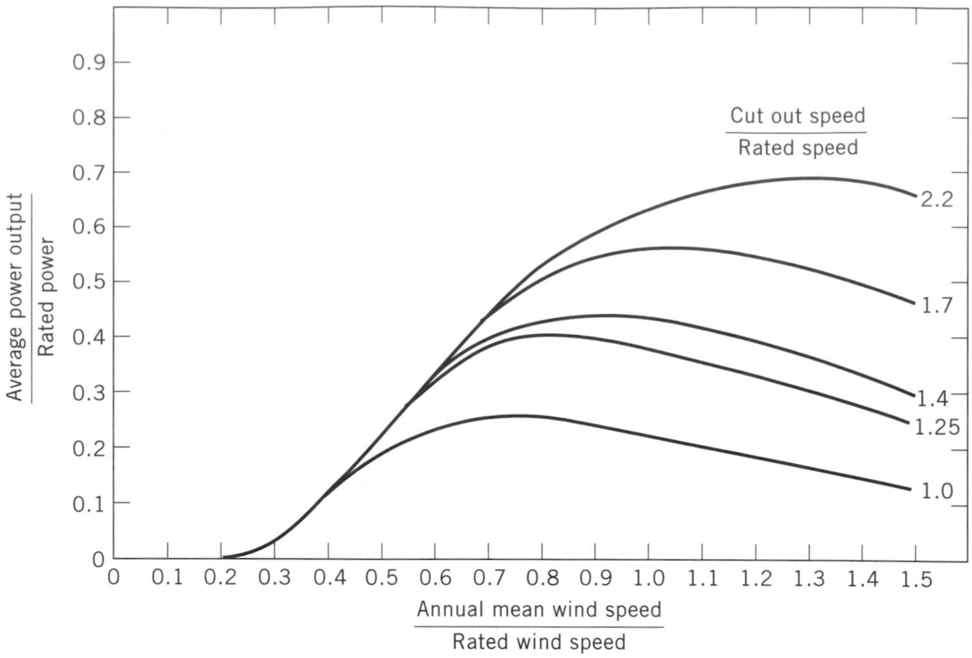

Figure 4. Prediction curve for average power output of a WTG.

distribution for the wind (only an estimate of the average wind is required) and a simplified wind turbine performance curve (see Fig. 1) with a linear fit between cut-in and rated speeds.

Figure 4 illustrates the results of this simplified model giving the average power output of a wind turbine as a function of annual mean wind speed with only a minimum of wind turbine parameters to know (rated power output, rated wind speed, and cut-out speed). For example, suppose a wind turbine has a rated power of 100 kW at a wind speed of 25 mph and a cut-out speed of 50 mph. If the average velocity at a site is 15 mph, the average power output divided by the rated power (capacity factor) is about 0.3. Thus the average power output of this machine is about 30 kW.

There are many more performance variables, eg, percent time running at rated power, that can be determined from an analysis of this time. Furthermore, the prediction can be made more accurate via the use of a Weibull wind distribution.

Wind Power Systems Design

The principal elements of a typical horizontal axis wind turbine are shown in Figure 5. These include the rotor, consisting of the blades and the supporting hub; the drive train, consisting of the low speed shaft (LSS), the transmission, and the high speed shaft (HSS); the generator and required power conditioning equipment (such as power electronics); the tower and the machine bedplate, or supporting frame and the yaw bearing or yaw orientation system; and the machine controls.

The modern wind turbine industry, especially the horizontal axis type design (HAWT), has developed rapidly in the last decade. For the individual components of the system, the engineering principles are generally well understood; they become quite complex when the individual components are combined to form an operating WTG system. Furthermore, even considering only the HAWT design, which is the type predominantly used in the approximately 20,000 wind turbines now in worldwide service, there is currently no consensus as to how a machine should look, or even what the most cost-effective size is (21). For example, for a typical HAWT, some of the main options in machine design and construction include (18)

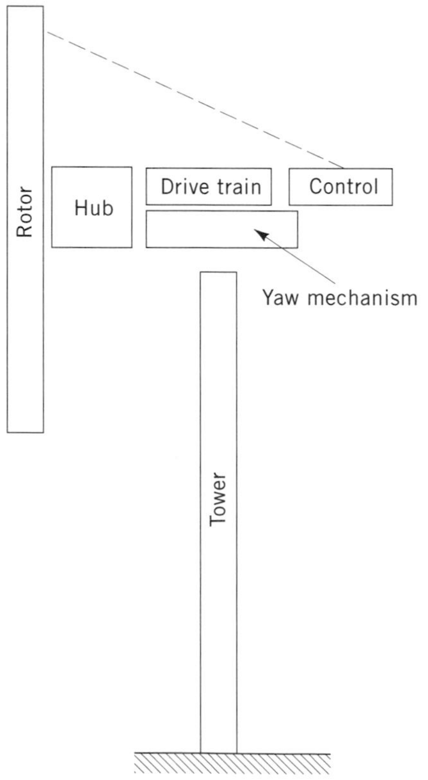

Figure 5. Principal components of a horizontal axis wind turbine.

number of blades (commonly two or three); blade material, construction method, and profile; fixed or variable pitch blades; operation downwind or upwind of tower; fixed or variable turbine speed; orientation by tail vane, fan-tails, self centering action, or servo control; synchronous, induction, dc, or other type of generator; gear, belt, chain, hydraulic transmission, or direct drive generator.

In wind turbine design, many areas, including the overall machine and its cost, require attention. It should be apparent that with this number of design variables, even in the most developed class of wind turbines, opportunities for innovation and development still exist.

Component Design of Wind Turbine Systems. *Rotor.* The hub and blade (denoted as the rotor) assembly of a wind turbine can be considered to be its most important component from both a performance and overall cost standpoint. As pointed out by numerous wind turbine designers, due to the large number of potential wind turbine design variations, the importance of analytically modeling the performance of a wind turbine before its construction cannot be emphasized enough. Several investigators (35) note that two fundamental design tendencies exist in the majority of current wind turbines: (*1*) rotors with medium speed (usually three rotor blades) with hingeless, stiff structure (stiff, heavy rotor blades and hingeless hub), and with passive aerodynamic power limitation (stall); and (*2*) high speed rotors (usually two rotor blades), load balancing rotor hinges (teetering), and active blade pitch control. The first design category (reflective of the majority of the current successful wind farm machines) possesses high construction weight, while the second category reduces it considerably. However, use of the second option leads to higher mechanical and dynamic design sensitivity, and usually more mechanical complexity.

The subject of rotor design issues has been addressed in a comprehensive study by the U.S. National Academy of Sciences (36), with specific summaries on the subjects of materials and manufacturing of wind turbine blades, as well as the modeling of such systems. They note that wind turbine blades are a high technology product that must be produced at moderate cost for the resulting energy to be competitive in price (currently, both fiberglass-reinforced and wood–epoxy composites have been shown to have the combination of strength and low material and fabrication costs required for competitive blade manufacture). In addition, this reference gives important information on the design considerations (such as choice of blade airfoils) that must be applied to this component of a wind turbine system.

Drive Train. The drive train of a WTG consists of the low speed shaft (rotor side), the transmission, and a high speed shaft (generator side). Other components of the drive train include support bearings and brakes. The purpose of the transmission is to speed up the rate of rotation of the blades from a low value (tens of rpm) to a rate suitable for driving a standard generator (hundreds or thousands of rpm). Two types of transmissions are used in wind turbines: parallel shaft and planetary (21). For larger machines (say over 500 kW), the weight and size advantages of planetary transmissions become more pronounced. There is also a very large difference in torque and ratio between large and small turbines, eg, a 3 MW turbine gearbox may require double the step-up ratio and over one hundred times the torque capacity of a 60 kW turbine. The design of wind turbine drive train components usually follows conventional mechanical engineering machine design practice, and numerous drive train design parameters are usually important inputs to dynamic analytical system models. It should also be noted, however, that design constraints on the important transmission component, specific to wind turbine application, may impose the requirement of other than standard components for this application.

Generator/Electrical. Utility scale grid-connected wind turbine generators have generally used synchronous or induction generators. Both these designs require a constant or near-constant rotational speed of the generator, which can affect the performance output of a wind turbine operating under variable wind speed conditions. The majority of wind turbines installed in wind farm applications use induction generators, generally a four pole squirrel cage induction machine. An induction machine operates within a narrow range of speeds centered about the synchronous speed of a machine (for a four pole induction machine designed for use on a 60 Hz grid, the synchronous speed is 1800 rpm). The range of operating speeds is defined by the slip range. Typically, the slip range for a four pole induction machine is about 2%. Thus, used as a generator, this type of machine begins to develop power at about 1800 rpm and reaches full power at about 1835 rpm. The design and performance characterization of these two types of electrical generators is summarized in a number of wind turbine textbooks (7,21). In addition to the generator itself, other standard electrical components are required for overall system control and for electrical power conditioning. For example, the use of an induction type generator may require power factor correction capacitor components.

A potential option for electrical power generation involves the use of a variable speed wind turbine electrical generating system. As summarized by numerous recent technical studies, there are a number of benefits that such a system offers, including the reduction of wear and tear on the wind turbine and potential operation of the wind turbine at near optimum tip speed ratio over a wind range of wind speeds, yielding increased energy capture. Although there are a large number of potential hardware options for variable speed operation of wind turbines (37), the use of power electrics components is the choice of variable speed machines under current design.

Tower Components. Under this component category, one could include the tower structure, the machine bedplate or supporting frame, and the yaw bearing or yaw orientation system. The importance of the design of the bedplate and yaw orientation system should not be understated, as both can markedly influence the capital cost of a wind turbine system, and, for the yaw system component, influence the performance and lifetime of the turbine rotor system. The principal type of tower design in use currently (21,35) are the free standing type using either steel tubes, lattice masts, or concrete towers (for small systems, guyed towers have also been used). Furthermore, the selection of the construction material, the geometrical shape, and the overall component layout are determined by tower height, erection cost considerations, and stiffness requirements. The first two design parameter requirements, tower height and erection cost, are economically related, and also greatly influenced by the characteristics of the site. Also, the stiffness design of the tower is a major factor in wind turbine system dynamics, ie, coupled vibrations of the rotor and tower. The location of the tower's first natural bending frequency relative to the dynamical excitation of the rotor is an important design factor (35). Also, especially for downwind oriented turbines, the effect of tower shadow on the rotor must be considered in a dynamic analysis of the wind turbine.

Controls. The control system for a wind turbine represents a key component with respect to both machine operation and power production. Defining the requirements for wind turbine control systems is an involved subject that generally involves the areas of power limitation, power control, and system control. The design of control systems design for wind turbines application follows traditional engineering control systems theory practices. Wind turbine control involves three primary aspects and the judicious balancing of their requirements (36): (*1*) setting upper bounds on and limiting the torque and power experienced by the drive train, principally the low speed shaft; (*2*) minimizing the fatigue life extraction from the rotor drive train and other structural components due to changes in the wind direction, speed (including gusts), and turbulence, as well as start–stop cycles of the wind turbine; and (*3*) maximizing the energy production.

Modeling of Wind Turbine Systems. For the prediction of performance and the lifetime of a wind turbine, there are three principal engineering tools: theoretical analysis or modeling, subscale or development testing, and full-scale system testing. For wind turbine design, all three tools are in developmental stages, and none can really be subordinated to another in the long run (8). For new machine development, the importance of a disciplined analysis involving detailed analytical modeling, development testing, and full-scale field testing should be emphasized (38). Because of the lack of high quality and generalized data from actual wind turbines, however, a wind turbine designer is usually forced to concentrate on the use of the first tool, analytical modeling. With today's rapid development of analytical modeling techniques supported by the digital computer and resulting programs and codes, the use of analytical modeling has become a most valuable design tool. For simplicity, and following conventional practice, the modeling of a WECS can be divided into separate parts for the mechanical and electrical components.

Mechanical Components. The mechanical components of a wind turbine are shown schematically in Figure 6 and include a number of interacting elements. Since the external forcing is derived from the aerodynamic loads on the rotor, an aerodynamic performance prediction model is very important. In this respect, the use of the two-dimensional strip or annulus theory for rotor modeling and design is common practice. Analytical models here use momentum-blade element theory and simulate the rotor aerodynamics based on the combined use of the conservation of axial and tangential momentum with blade element (strip) theory. A key assumption in such models is that the air flow is composed of an uncoupled summation of annular stream tube flows. Advanced versions of computer codes based on this type of a model are designed to take into account such nonidealities as tip correction, turbulent wake effects, tower influence, and operation in yaw. Major portions of reference books on wind turbine design are devoted to the explanation and use of such models (8,21), and a review of engineering tools developed over the last ten years for blade and rotor design is available (39).

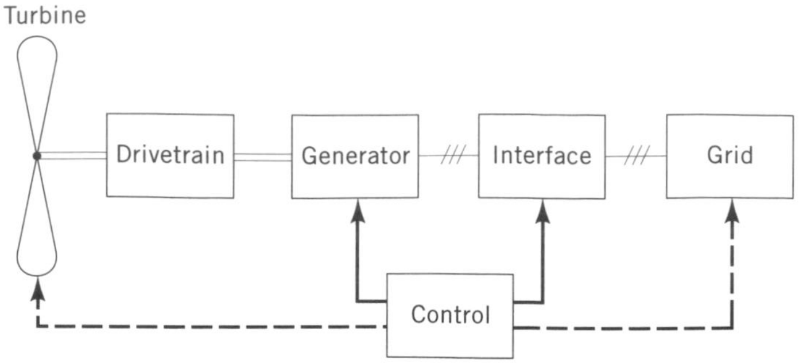

Figure 6. Mechanical components of a WTG system.

The calculation of the aerodynamic loads on a wind turbine are relatively straightforward compared to the calculation of the dynamic response that those loads produce when applied to the structure. This complexity is illustrated by the fact that the dynamic behavior of wind turbines can be divided into three categories: stability, forced response to deterministic loads, and forced response to stochastic loads. In principal, all components of a wind turbine may contribute to the dynamic behavior of the system including the blades, rotor, hub, power train, control system, tower, and foundation. For conventional designs, it may be possible to neglect some of these components, or at least consider them uncoupled from the rest. The dynamic modeling of a wind turbine, usually concentrating on the rotor component, is a complex subject, and can involve a number of complex analytical models (21). In addition to dynamic models for the incoming wind field (including the effects of multiple turbines in wind farms) they include analytical submodels that solve for the deterministic and stochastic loads, as well as their combined contribution to fatigue damage. In addition to numerous technical papers and reports on the subject (36), several books on wind turbine design summarize the fundamentals and use of such models (8,21).

Electrical Component. Since the majority of applications of wind turbine generators is to supply utility grade electrical power, much work has been carried out on the modeling of the electrical aspects of wind turbine systems connected to an electrical grid. In such systems, due to the variability of the wind, large power and frequency fluctuations can occur in the system's output, which can affect the stability of the WTG system and, if the wind turbine power output is relatively large, the grid to which it is connected.

A schematic of a WTG system connected to a grid is shown in Figure 7. It consists of the wind turbine rotor, drive train, and generator as well as an electrical interface and a control system. For the purpose of studying the electrical behavior of the system, it is important to include simple models of the mechanical components as well as the electrical components when constructing dy-

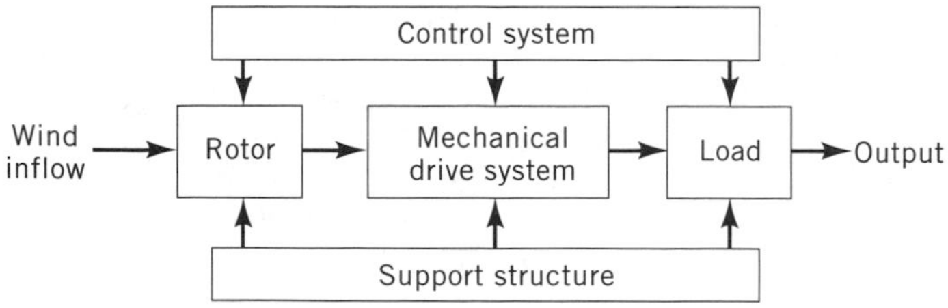

Figure 7. Schematic of a WTG system connected to an electrical grid.

namic models of the system as both the electrical and mechanical systems may exhibit nonlinearities that require detailed analytical modeling. In addition, it is desirable to have a good model of the system's control system, and it is important to take into account interactions between the electrical and mechanical components, since they can exhibit dynamic behavior on the same time scale.

The complexity of analytical models for the electrical systems of a wind turbine generator system can vary significantly. For example, a wind turbine system model may just require the modeling of the electrical performance of the generator as would be the case when the generator has to be matched to the wind turbine rotor, or require the use of conventional control theory modeling to simulate the performance of a control system design. On the other hand, if the design and sizing of power electronics components is required, a complex dynamic model capable of simulating time scales on the order of milliseconds would be required. Excellent summaries of the state-of-the-art of modeling electrical systems for both types of systems are available (7,21,40).

Wind System Economics

The development of the modern wind farm concept (based on the use of a large array of smaller sized wind turbines (about 100 to 300 kW each) connected to the utility grid) during the past ten years has considerably changed the overall economics picture for wind energy. Today, major installations exist in California, Hawaii, and many countries in Europe. California contains the majority of the world's installed wind generating capacity, with approximately 16,000 wind machines in three mountain pass locations representing 1690 MW of installed capacity. In 1992 these machines generated about 2.7 billion kWh of electricity (over 1% of the annual load in California). In Europe, electricity generation by wind turbines exceeded one billion kWh for the first time in 1992.

The development of the California wind farms has contributed to a rapid fall in the capital and operation and maintenance costs of wind turbines, resulting in a lower cost of electricity. In this respect, the major factors that reduced the capital costs of wind turbine systems included improved designs and economies of scale achieved by manufacturing in a common size, purchasing, and production. Also, improvements in machine design and maintenance programs with large numbers of turbines have taken advantage of economies of scale to reduce unit operation and maintenance costs.

The installed costs for installations of new or second generation wind turbines in the United States has decreased from about $4000/kW in 1980 to somewhat less than $1000/kW at present. And, more significantly, the cost of electricity from current wind farm installed wind turbines has decreased from about 30¢/kW to between 7 to 9¢/kW (generally less than the delivered cost of electricity supplied by conventional powerplants). Even lower costs are reported at wind farm sites with better wind resources. That is, the California Energy Commission (15) estimates that current investor owned wind power plants can generate electricity from 4.7 to 7.2¢/kWh. Also, according to U.S. Department of Energy studies (1), there is every indication that, at numerous wind farm sites in the United States, the cost of electricity produced will be on the order of 5¢/kW. In Europe, essentially the same type of dramatic reduction in wind generated electricity costs has occurred.

In order to determine the economics of wind systems, the cost of electricity (COE) is usually found using a life-cycle cost analysis method. In the United States, for utility applications, an established method for computing wind turbine cost of electricity using utility economic assumptions has recently been used (41). This method is based on the Electric Power Research Institute (EPRI) *Technical Assessment Guide*, and requires key inputs such as system capital costs, operation and maintenance costs, wind turbine performance normalized to kWh/m^2, and other standard economic parameters (42). Similar methods are used by European investigators to determine the economic viability of wind turbine systems (21). In addition, a description of various economic analysis methods that can be used for hybrid wind electric systems, such as wind diesel systems, is available (40).

Wind turbines are the principal cost component of a wind farm, representing about 60–75% of the wind farm's total installed cost. As such, it is important to maximize the cost effectiveness of the individual components of a wind turbine, as well as the overall system. While there is no generalized method for determining the costs of the major components of a wind turbine, there are some examples in the current literature from which to make estimates, ie, the cost divided by major component of a utility-based 200 kW machine breaks down as follows: hub and pitch mechanism, 8%; gearbox and brake, 18%; yaw system, 5%; rotor blades, 21%; electrical and controls, 25%; and tower and nacelle, 23% (43).

Future improvements in wind turbine technology are expected to reduce the cost of the major components of wind turbines, the operation and maintenance costs of wind turbine systems, and to increase the relative performance of wind turbine systems (via improved siting techniques or machines).

Applications

In addition to utility electrical power generation there are other nongrid connected (stand alone) applications for wind power systems, including ones that were well established before the recent wind farms development. Some of the more actively investigated systems include water pumping systems, small wind energy systems for specialized applications, and hybrid wind diesel systems.

Water Pumping. Water pumping applications are either drinking water pumping or irrigation. Historically, there has been considerable progress in the development of wind pumps for irrigation, cattle watering, and the supply of domestic water. Most estimates regarding the size of this type of worldwide market conclude that about a million classical wind pumps, generally defined as American farm wind pumps, are now in use. The design of wind turbine systems for water pumping applications follows a somewhat different path than the design of large-scale electrical generating machines especially from the standpoint of matching the performance of the wind turbine and the water pump. In addition, most wind-powered water pumping systems must be highly reliable, able to run unattended for long periods of time, and require a minimum of maintenance. Recent advances in this area have included the development and production of revised traditional farm wind pump designs, including the pumping mechanism, and the development and implementation of wind systems based on wind–electric water-pumping designs. Details of the specific design process for water pumping systems and current advances in this area are summarized in several references on the subject (3,18,20,44,45).

Small Wind Energy Systems for Specialized Applications. Although the recent major emphasis has been on medium to large-scale wind turbine development for utility scale electricity production, there has been some significant progress in the development and worldwide marketing of small (ca 100 W to 2 kW) wind machines for specialized applications. These applications include small radio transmitters, power sources for offshore oil rigs, navigational aids, and battery charging systems for remote or isolated systems. In most of these applications, the power system involves interfacing the wind turbine with a battery storage system and the associated electronic components. Furthermore, there are a number of distinctive aspects of the design of very small wind turbine systems that make them unique. The design aspects of this type of wind energy system are discussed in detail in several references (3,45).

Hybrid or Wind–Diesel Systems. In the remote or non-grid connected areas of the world, the most common way to supply electricity is by the use of a diesel engine driving a generator set. The size of such systems ranges from small systems (on the order of KW) to large systems with multiple diesel engines, which may have MW-sized loads or small grids to supply. Although diesel supplied electric power is generally reliable, it is often expensive, primarily due to high diesel fuel costs.

One possible way to reduce the high fuel costs of such systems is to incorporate a wind turbine into the system, creating a hybrid or wind–diesel system. By conventional definition, a wind–diesel system is any autonomous electricity generating system using wind turbines and diesel generators in which the impact of the wind turbine on the operation of the system may be significant (46). Or, from a wind engineering standpoint, the solution to the problem of using wind power capacity at a remote location is to compensate for the variability of the wind by using a diesel generator to make up any shortfall. The purpose of such systems is to extend the diesel system with a wind turbine (or a number of turbines) probably with some form of storage and other electrical energy convertors, in such a way that the additional costs will be paid by the savings in fuel costs while maintaining an acceptable power quality. The concept is not so simple, however, and a combination of short-term storage, dump load, and load control is generally required for such a system before system stability, performance, and economics improve.

A generalized wind–diesel system consists of the following major components: one or more wind turbines; one or more diesel generator sets; a consumer load; an additional controllable or dump load; a storage system; and a control unit, possibly including load management. There are many possibilities for different design configurations of such systems. In addition to recent hardware and system developments and improvements, considerable technical work has been applied to the analytical modeling of the steady state and dynamic performance of such systems. Significant work on this important wind engineering topic documented both in wind energy texts (21) and in a recent International Energy Agency task force study (40).

Resource Estimates. The area of wind resource assessment has received a great deal of attention, with major wind energy feasibility studies published in the 1990s in the United States and Europe. In addition to expanded data collection and improved analysis techniques, resource assessments in the 1990s are much more realistic since they include assumptions about wind machine characteristics, eg, hub height, machine spacing restrictions, system losses, as well as land exclusion, eg, parks, urban areas, wetlands, and a substantial fraction of forest and agricultural lands, considerations. For example, the Battelle Pacific Northwest Laboratory (PNL) has recently used this approach in conjunction with a detailed wind energy atlas of the United States (2,31). This study, published in 1991, stated that wind turbine technology could supply 20% of the U.S. electrical energy needs. The study also stated that wind could be exploited in locations where the average annual wind resource is class 5 or higher, locations where the average wind speed is 16 mph and higher at a height of 30 m. Specifically, in order to provide this fraction of the nation's total electrical energy demand, about 600 billion kWh per year, 0.6% of the land (ca 18,000 square miles) in the lower 48 states would have to be developed. Also, these investigators noted that less than 5% of this land would be occupied by turbines, equipment, and access roads. Since 80% of this area is located in North Dakota, Wyoming, and Montana, however, the actual use of this resource would involve other siting considerations, such as transmission line access.

On the subject of wind turbine technology, the Battelle study also addressed the point that further advances in wind turbine technology will further enhance wind energy's potential by permitting areas with lower wind resource to be developed. For example, if class 3 areas, locations where the average wind speed is 14 mph and higher at a height of 50 m could be used, approximately 13% of the contiguous U.S. land area could be considered for wind turbine siting. This land area has the potential to supply about four times the current electrical energy consumption of the nation.

Similarly, based on recent European Community studies (32), it has been estimated that at least 10% of Europe's electrical power could be supplied by using current wind turbine technology. The total land required is no larger than the island of Crete; furthermore, most of this land could still be used for agriculture. Even with this resource, the siting of large groups of wind turbines or wind farms in highly populated areas could represent a siting problem. Thus, several European countries and industrial developers have recently turned their attention to the exploitation of Europe's enormous offshore wind resource. The development of offshore wind resources may also be a topic of future research in the northeastern United States where siting problems dominate.

Institutional Constraints and Opportunities. The institutional constraints to the future large-scale deployment of wind energy primarily involve two areas: environmental considerations and electric utility integration (1). On the environmental side, in terms of conventional large-scale electrical generating systems, wind power systems are environmentally benign (no gaseous or particulate emissions, and no hazardous by-products to deal with). In general, wind energy has the same basic environmental impacts associated with the other renewable energy technologies: little or no effect on flora, fauna, climate, materials, and human health, but a potential negative effect on land use. On the other hand, there are some siting considerations that could have negative implications. In addition to land use, these include visual impact, noise, telecommunications interference, and impact on wildlife and natural habitat (especially hazards to birds). Each of these subjects has been addressed in detail in numerous studies in the United States and Europe (47), and none has proven to be a major problem to large-scale wind system implementation if addressed early in the siting process. The California wind farm experience has shown such factors to be minimal as long as wind turbines are not located in proximity to populated areas.

On the environmentally positive side, the three-dimensional nature of the resource provides a distinct advantage over other solar technologies. Specifically, because the general rule for wind turbine siting is "the higher the better" and turbines are usually spaced about two to three diameters apart crosswind and ten diameters apart downwind, only a small fraction of a wind farm land area is actually occupied (1). Thus, only 1 to 5% of the land area is occupied by the wind turbines and their access roads.

Concerning utility integration, issues such as interfacing of wind turbines with the utility grid, transmission capacity, energy storage requirements, and utility capacity credits have had to be addressed. Despite initial reluctance towards integration of wind turbines into the utility grid, utility experience primarily in California has provided a better understanding of wind power production on utility engineering, operations, and planning. For example, the operational hardware required to interconnect wind turbines with the utility grid (such as switchgear, protective relays, and controls) is essentially the same as used on conventional power plants. In the initial wind farm applications, some engineering problems with the wind farm and utility interface appeared. These included harmonic current control, reactive power compensation, and power regulation capacity. However, modern developments in power electronics state-of-the-art and wind turbine control systems have successfully addressed these concerns or have completely eliminated them.

The operations of a utility can be affected by the variability of the power output from a wind turbine or wind farm. In general this variability takes place over three time scales: short-term (seconds to minutes), medium-term (minutes to hours), and long-term (hours to many days). Utilities deal with short-term variations in power input the same way as they deal with normal short-term variations in load. Fortunately, numerous industry studies have shown that the short-term power output of a wind farm varies significantly less than that of a single turbine. The medium-term variations of wind farm output generally have a small effect on utility operations if the wind farm electrical power output represents a small percentage of the total electric load of the utility (defined as grid penetration). For large penetration levels, generally considered by utilities in the United States and Europe to range from 10 to 20%, a utility may have to maintain additional generating capacity on line during periods of rapidly changing wind; thus wind generated electricity will lose some of its economic benefits, reflected in lower utility capacity value. Transmission line capacity has been identified by a number of U.S. utilities as an important issue to be addressed if the penetration into the grid of wind farm systems exceeds 10 to 20%. In the United States many of the largest wind resources in the country are located at a distance from utility load centers and do not have readily available transmission lines. Thus if the full potential of the wind energy resource is to be reached in the U.S., the issue of transmission line access/capacity needs to be addressed.

Technical Constraints and Opportunities. Wind farm machines are cost competitive with regard to delivered energy. Although the machines represent an improvement in performance over initially installed machines, there is still opportunity for technical and cost-effectiveness improvement of utility scale wind machines. Wind turbine designs will have to perform better at good sites than machines in the 1990s do at excellent sites (1). In the United States and Europe, active development programs are in progress to establish the objective of providing a new class of commercially improved wind turbines. The new generation of wind turbines will incorporate technological improvements ranging from blades engineered to enhance energy capture to sophisticated power electronics. It is expected that these machines will be larger, lighter weight, more efficient, and less costly, with projected average energy costs of $0.05/kWh or less. According to the most recent advanced wind turbine program in the United States (48), some of the improvements expected by the end of the 1990s for wind turbines include new blade designs, variable speed operation, rotor hub enhancements, and taller towers.

Furthermore, the same program is expected to yield wind turbine designs that will produce commercial turbines by the year 2000 with advanced new technologies including aerodynamic controls, advanced drive trains, advanced structures, advanced power electronics, and expert control systems.

In light of the past experience and rapid development of the world's wind industry there is every reason to believe that these technical objectives will be accomplished.

Worldwide Opportunities. Starting with the importation of large numbers of European wind turbines to the California wind farms, the center of gravity of the wind industry and expertise has shifted from the United States to Europe. During the 1980s, the development of wind energy in Europe has proceeded at a steady pace, with an ever increasing commitment from several countries for research and development and market incentives. As a result of this program, Europe has over twenty-five wind turbine manufacturers who have commercially available utility scale wind machines, in sizes ranging from medium machines applicable for wind farm operation to very large MW-sized utility machines. Outside Europe the two important manufacturers are U.S. Windpower (California) and Mitsubishi (Japan).

The market stimulation carried out by the European countries via subsidies, production incentives, and tax credits has contributed to the growth of wind energy development, and new European wind installations have exceeded in number those installed in the United States during the early 1990s. The American Wind Energy Association noted that European development is now expected to dwarf that in the United States throughout the 1990s (15). They predict that Denmark, England, Spain, and the Netherlands will each surpass the United States in new installations of wind turbines. In addition to grid connected systems in the developed countries, the worldwide wind power market is expected to grow in grid connected applications in the developing countries, and in standalone power systems. Thus, there is every indication that this renewable energy resource will command a significant portion of the world energy market in the twenty-first century.

BIBLIOGRAPHY

1. U.S. Department of Energy, *The Potential of Renewable Energy*, SERI/TP-260-3674, U.S. Department of Energy, Office of Policy, Planning and Analysis, Washington, D.C., 1990.
2. D. L. Elliott, L. L. Wendell, and G. L. Gower, *An Assessment of the Available Windy Land Area and Wind Energy Potential in the Contiguous United States*, PNL-7789 UC-261, Pacific Northwest Laboratory, Richland, Wash., 1991.
3. J. Park, *The Wind Power Book*, Chesire Books, Palo Alto, Calif., 1981.
4. F. R. Eldridge, *Wind Machines*, 2nd ed., Van Nostrand Reinhold, New York, 1980.
5. D. R. Inglis, *Windpower and Other Energy Options*, University of Michigan Press, Ann Arbor, 1978.
6. E. W. Golding, *The Generation of Electricity by Wind Power*, E. & F. N. Spon, London, 1977.
7. G. L. Johnson, *Wind Energy Systems*, Prentice Hall, Englewood Cliffs, N.J., 1985.
8. D. M. Eggleston and F. S. Stoddard, *Wind Turbine Engineering Design*, Van Nostrand Reinhold, New York, 1987.
9. P. C. Putnam, *Power From the Wind*, Van Nostrand Reinhold, New York, 1948.
10. National Science Foundation Panel, *An Assessment of Solar Energy as a National Energy Resource*, National Science Foundation, Washington, D.C., 1972.
11. J. Naar, *The New Wind Power*, Penguin Books, Middlesex, U.K., 1982.
12. V. Torrey, *Wind-Catchers American Windmills of Yesterday and Tomorrow*, Stephen Greene Press, Brattleboro, Vt., 1976.
13. V. D. Hunt, *Windpower—A Handbook on Wind Energy Conversion Systems*, Van Nostrand Reinhold, New York, 1981.
14. D. Lindley, "The Commercialization of Wind Energy," *Proceedings of the Euroforum—New Energies Congress*, Saarbrucken, Germany, 1988.
15. *1992 Wind Technology Status Report*, American Wind Energy Association, Washington, D.C., 1992.
16. A. Garrad, *Wind Energy In Europe*, European Wind Energy Association Report, 1991.
17. R. Lynette, "Status of Wind Power Industry," *Proceedings 1989 AWEA Annual Meeting*, American Wind Energy Association, Washington, D.C., 1989.
18. D. F. Warne, *Wind Power Equipment*, E. & F. N. Spon, London, 1983.
19. British Wind Energy Association, *Wind Energy for the Eighties*, Peter Peregrinius Ltd., U.K., 1983.
20. E. H. Lysen, *Introduction to Wind Energy*, 2nd ed. publication SWD 82-1, Steering Committee Wind Energy Developing Countries, Amersfoort, the Netherlands, 1983.
21. L. L. Freris, *Wind Energy Conversion Systems*, Prentice Hall, New York, 1990.
22. J. W. Twidell, and A. D. Weir, *Renewable Energy Resources*, E. & F. N. Spon, London, 1986.
23. D. Le Gourieres, *Wind Power Plants*, Pergamon Press, Oxford, 1982.
24. U.S. Department of Energy, *Wind Energy Innovative Systems Conference Proceedings*, Solar Energy Research Institute, Golden, Colo., 1979.
25. U.S. Department of Energy, *SERI Second Wind Energy Innovative Systems Conference*, Solar Energy Research Institute, Golden Colo., 1980.
26. C. G. Justus, *Winds and Wind System Performance*, Franklin Institute Press, Philadelphia, Pa., 1978.
27. H. L. Wegley and co-workers, *A Siting Handbook for Small Wind Energy Conversion Systems*, PNL-2521, Rev. 1, NTIS, Battelle Pacific Northwest Laboratory, Richland, Wash., 1980.
28. T. R. Hiester and W. T. Pennell, *The Meteorological Aspects of Siting Large Wind Turbines*, PNL-2522, NTIS, Battelle Pacific Northwest Laboratory, Richland, Wash., 1983.
29. J. W. Reed, *Wind Power Climatology of the U.S.*, SAND, 74-0348, NTIS, Sandia Laboratories, Alburquerque, N.M., 1975.
30. D. L. Elliott and W. R. Barchet, "National Wind Energy Resource Assessment," *Fifth Wind Energy Conference and Workshop*, Solar Energy Research Institute, Golden, Colo., 1981.
31. D. L. Elliott and co-workers, *Wind Energy Resource Atlas of the United States*, DOE/CH10094-4, Solar Energy Research Institute, Golden, Colo., 1987.
32. I. Troen and E. I. Petersen, *European Wind Atlas*, Riso National Laboratory Publication, Riso, Denmark, 1989.
33. W. T. Pennell, *Siting Guidelines for Utility Application of Wind Urbines*, RP 1520-1, Electric Power Research Institute, Palo Alto, Calif., 1982.
34. W. C. Cliff, *The Effect of Generalized Wind Characteristics on Annual Power Estimated from Wind Turbine Generators*, PNL-2436, NTIS, Pacific Northwest Laboratory, Richland, Wash., 1977.

35. E. Hau and W. Kleinkauf, "Essential Features of Horizontal Axis Wind Turbines," *Proceedings International Solar Energy Society: Advances in Solar Energy Technology*, Pergamon Press, New York, 1987.
36. National Research Council Committee on Assessment of Research Needs for Wind Turbine Rotor Materials Technology, *Assessment of Research Needs for Wind Turbine Rotor Materials Technology*, National Academy Press, Washington, D.C., 1991.
37. J. F. Manwell, J. G. McGowan, and B. H. Bailey, *Solar Energy* **43** (1990).
38. R. Thresher and S. Hock, "An Overview: Challenges in Wind Technology Development," *Proceedings Windpower '91*, Palm Springs, Calif., AWEA, Washington, D.C., 1991.
39. F. S. Stoddard, "Wind Turbine Blade Technology: A Decade of Lessons Learned," *Proceedings of the 1st World Renewable Energy Congress*, Reading, U.K., Pergamon Press, New York, 1990.
40. R. Hunter and G. Elliot, *Wind Diesel Systems*, Cambridge Univ. Press, U.K., 1994.
41. J. M. Cohen and co-workers, "A Methodology for Computing Wind Turbine Cost of Electricity using Utility Economic Assumptions," *Proceedings Windpower '89*, AWEA, Washington, D.C., 1989.
42. *TAG Technical Assessment Guide, Electricity Supply, 1986*, vol. 1, Electric Power Research Institute, Palo Alto, Calif., 1986.
43. D. Lindley and C. Gamble, *J. Wind Engineering and Industrial Aerodynamics* **27** (1988).
44. P. Fraenkel, *Water Pumping Devices*, Intermediate Technology Publications, London, 1986.
45. J. Twidell, *A Guide to Small Wind Energy Conversion Systems*, Cambridge University Press, U.K., 1987.
46. J. G. McGowan, J. F. Manwell, and S. R. Connors, *Solar Energy* **41**(6) (1988).
47. D. T. Swift-Hook, *Wind Energy and the Environment*, Peter Peregrinus Ltd., Exeter, U.K., 1989.
48. *Wind Energy Program Review, Fiscal Years 1990–1991*, U.S. Department of Energy, Washington, D.C., 1991.

WOOD FOR HEATING AND COOKING

JOHN ZERBE
Forest Products Laboratory
Madison, Wisconsin

Wood fuel is a primitive source of heat for cooking and providing warmth. It also has many modern applications including combustion as a solid fuel in state-of-the-art furnaces, conversion to charcoal for refining steel, and conversion to liquid and gaseous fuels for special uses. In the United States and other industrial countries, wood fuel is used mainly to provide warmth for space heating and process heat, but in developing countries wood fuel is used mainly for cooking.

Wood, together with bark, is most widely converted into energy by direct combustion in many types of burners. Black liquor, a wood pulp byproduct, is also used to produce energy at pulp plants. Some wood or black liquor is used to produce electricity in cogeneration or utility plants. Technology is also available, although not always economical, to convert (a) wood to gas by thermochemical gasification and burn it in boilers, driers, and kilns; use it as engine fuel; or make it into synthesis gas for manufacture of fuel or feedstocks; (b) by pyrolysis, wood to solids such as charcoal, gases for fuel gas, and liquids for fuel; (c) wood to other liquid fuels such as ethanol by hydrolysis and fermentation; and (d) natural oils from some woods to diesel fuel. There are also less developed means of converting wood to energy such as anaerobic digestion, fuel cells, and magnetohydrodynamics.

At the other end of the spectrum, in developing countries wood or charcoal is usually burned inefficiently on open hearths. Less frequently, wood or charcoal fuels are burned more effectively in inexpensive stoves. Brazilian and Argentinean steel industries depend upon wood charcoal as a fuel and as a source of carbon. This is also true of cement industries in some countries. In the United States, wood has been used to substitute for natural gas in brick kilns.

Recent and future technology improvements in converting wood to energy will improve wood energy's competitive position relative to alternate fuels and increase wood energy use. Technology improvements will also improve the efficiency of wood conversion and hold down wood demand for energy. See also AIR POLLUTION; AIR POLLUTION CONTROL METHODS; FOREST RESOURCES; EXHAUST CONTROL, INDUSTRIAL.

ENVIRONMENTAL IMPACTS OF WOOD FUEL USAGE

With decreasing fossil fuel supplies and environmental and economic problems in the use of other alternatives such as nuclear energy, the overall tendency in wood use for energy is increased consumption for this purpose. Wood use for energy has both environmental benefits and the potential for causing additional environmental damage.

Wood, in contrast to much coal and some petroleum, has little or no sulfur and appears to have less problems in production of oxides of nitrogen during combustion. It is therefore less threatening to the production of acid rain. Wood is also free from metals, including heavy metals, which is not the case with coal.

Wood is a renewable fuel that can be produced on a sustainable plan. It has advantages over other renewable forms of energy such as solar and wind, in that the energy is self-stored in the fuel itself. The fuel is therefore an energy carrier, and means for storing energy such as heated mass, salts, or batteries are not required. Production of wood on a sustaining basis also means that as wood is consumed and converted to atmospheric carbon dioxide, new wood is grown with the removal of atmospheric carbon dioxide through the process of photosynthesis. The net effect is therefore no change in atmospheric carbon dioxide. This is in contrast to fossil fuels, which increase atmospheric carbon dioxide content and may cause damage because of the greenhouse effect.

However, caution must be used in removal of more of the biomass in forest harvests to avoid adverse impacts from nutrient depletion or the increased potential for soil erosion. Clearing of forest land for purposes other than farming or forestry will displace plants that consume carbon dioxide and increase atmospheric carbon dioxide. This is a severe problem in tropical forests of developing countries, where forest land acreage through shifting agriculture continues to be lost at an alarming rate.

Other concerns about using wood for fuel include the potential for reducing biological diversity in forests. This topic was covered in detail at a Biomass Workshop sponsored by the Audobon Society and Princeton University on May 6, 1991 (1). Discussions at the workshop led to the conclusion that over the long run, the goals of ecological health and biomass productivity are likely to be mutually supportive. It was felt that the biomass energy system will affect biodiversity of plant and animal species most directly through the choice of land for biomass production. The impact of biomass plantations, in which trees are grown through short-rotation intensive culture, depends strongly on the previous character of the land. If natural forests were replaced by monoculture energy plantations (single species of trees), the result could be substantial loss of biodiversity. But if the biomass were grown on lands that are currently degraded or lands, such as modern farmlands, that have been greatly simplified and have become ecologically impoverished, biological diversity could be improved locally.

The impact of humans on the environment, also known as anthropogenic impact, through combustion of all types of organic matter has assumed importance for modeling the carbon cycle and accounting for carbon sinks. The atmosphere is one of these sinks, and atmospheric carbon concentration is increasing. This is leading to some concern about future environmental damage as a result of global warming, because of the increased carbon adding to the greenhouse effect of the atmosphere.

Sources of anthropogenic emissions to the atmosphere each year include fossil fuel combustion (58%), biomass fuel combustion (12%, but renewable), crop residue burning (1%, but renewable), grassland burning (20%, but renewable), shifting agriculture (4.2%), cement production (1.4%), and solid waste production (1.1%). Total excess is estimated at 6.5 ± 1.5 billion tons of carbon. These annual fluxes are viewed against an estimated stock of 560 ± 100 billion tons in the living biomass systems.

Most of the problems associated with deforestation of tropical forests occur because of land-clearing operations and conversion of forests to other uses. Because the wood and its properties and suitable methods for processing are unknown, most trees are felled and burned to clear sites. Research is needed to improve the utilization of such material. This will enhance the value of forests and improve their prospects for sustainable management.

Combustion of wood in inefficient combustors without proper controls adds smoke and particulate emissions to the air. Recently, there has been concern about proper combustion of wood contaminated with other materials such as paint, adhesives, and/or preservatives. Research is underway to determine how these fuels may be combusted or disposed satisfactorily.

HISTORICAL CONTEXT

Since the late nineteenth century, the United States has depended on fossil fuels for energy. Coal was the primary source until the 1920s when oil came into its own.

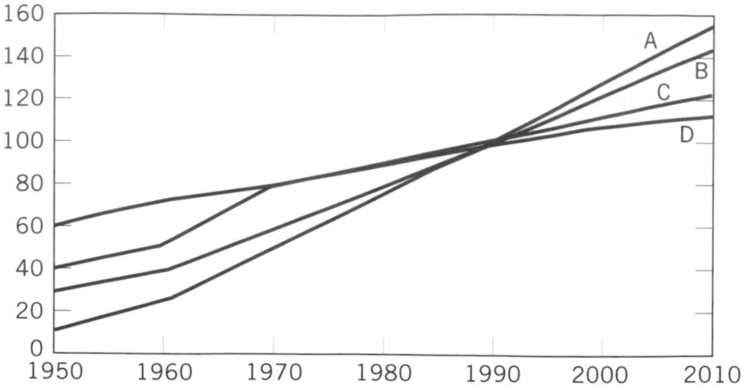

Figure 1. United States indices for energy, population, and economic growth: *A*, population; *B*, energy; *C*, GNP; *D*, electricity.

Throughout this period, U.S. energy consumption rose in almost direct proportion to the rise in gross national product (GNP). However as a result of energy conservation efforts beginning in the early 1970s, U.S. consumption fell slightly as GNP continued to rise (see Fig. 1).

For 100 years before the early 1970s, use of biomass fuels dropped as fossil fuel consumption rose (see Fig. 2). This long-term trend in the diminishing importance of biomass energy was reversed following the OPEC (Office of Petroleum Exporting Countries) oil embargo. Biomass received another boost after the Iranian revolution in 1979. Residential, industrial, and institutional wood energy use rose from about 1.3 quads in 1972 to about 2.7 quads in 1984. There has not been much gain since. Steady gains in industrial, commercial, and institutional use have been counteracted by a decrease in residential use. Since the early 1980s, however, electrical power generation and cogeneration plants in Maine, New Hampshire, Vermont, Michigan, Wisconsin, and California have become significant new users of wood boiler fuel.

Use of wood fuel in residential applications has not been tracked nationwide since 1984, but the level appears to have remained about the same or declined slightly. In some areas, notably Missoula, Montana, and locations in Colorado, burning of wood in stoves is prohibited during atmospheric temperature inversions. In Oregon, newly purchased wood stoves must meet stringent environmental requirements. Such requirements probably have

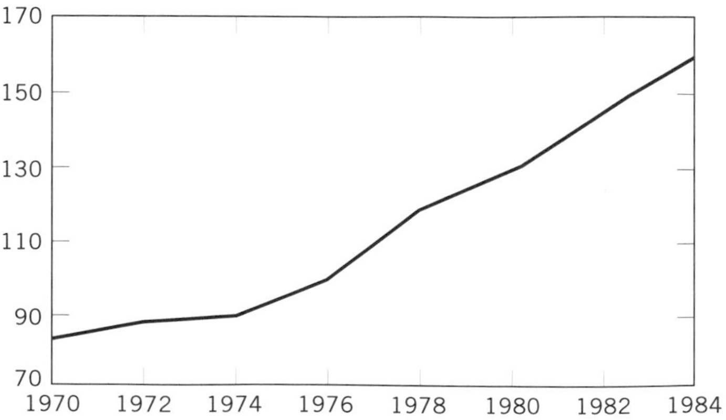

Figure 2. Growth in wood energy consumption.

caused a decline in the use of wood stoves. Other factors are decreased oil and natural gas prices and disenchantment with the effort required to obtain and use wood for residential use.

The Public Utilities Regulatory Policies Act of 1979 provided incentives for cogeneration and small power production facilities. These incentives, which led to the construction of new wood-burning facilities, are designed to avoid the capital costs for building expensive, large, new central power generating units.

In developing countries, the use of wood for cooking fuel continues to be predominant, as it always has been. In 1975, 50% of all wood harvested in the world was fuelwood. In 1986 the percentage was 52. In 1988, in some regions of the world, most notably sub-Saharan Africa, three-quarters of harvested wood was burned for cooking.

The Food and Agriculture Organization (FAO) of the United Nations (2) reported worldwide statistics for fuelwood consumption for 1975 and 1986 and made projections for 1990, 1995, and 2000. These statistics are shown in Table 1.

WOOD FUELS IN FOREST PRODUCTS INDUSTRIES

Until the early 1970s, many U.S. sawmills had "teepee" burners to incinerate nonmarketable leftovers, and most used fossil fuel, not wood, to meet their energy needs. Pulpmills burned some wood, bark, and black liquor for fuel but also relied heavily on coal, gas, and oil for their energy supplies.

The National Environmental Policy Act of 1969 created air quality standards too stringent for continued use of most teepee burners. Moreover, the rapid rise in fossil fuel prices and natural gas shortages in the mid-1970s created additional incentives for forest product firms to use wood fuel.

Many sawmills installed wood-burning facilities. Pulp and paper industries, which were about 38% energy self-sufficient in 1973, started to generate more of their energy with wood fuel, and some new plants were built to operate entirely on wood fuel.

Natural gas use by the U.S. paper industry declined by 34.9% between 1972 and 1986. In 1972, oil represented 21.3% of the total energy consumed by the U.S. paper industry; in 1986, it represented 8.5%. On the other hand, self-generated and residue sources of energy accounted for 56.7% of total energy consumption by 1986.

ECONOMICS

In many applications, use of wood for energy is economical and a contributing factor to prosperity of local, often rural, economies. However, improved technology may make wood for energy more competitive, even with oil prices advancing more slowly than was anticipated. A major factor in using more wood for energy is high cost of forest harvesting. It is prudent to use wood that is less valuable and less suited for use in other consumer products for energy. However, the lower value wood is often from smaller trees that are more expensive to harvest, although this factor does not appear to be as important as stand volume and percentage of hardwoods in comparison to softwoods.

APPLICATIONS FOR WOOD FUELS

As a very rough approximation, in industrial countries use of wood for energy is exemplified by three categories of comparable consumption quantities. These are residential, black liquor, and industrial. Lesser, but growing, amounts of wood are consumed in power generation and commercial and institutional applications. In developing countries the principal use is for domestic cooking.

For residential use of wood for energy in industrial countries the traditional approach has been roundwood consumption in furnaces, simple stoves, or fireplaces.

Table 1. Worldwide Fuelwood Consumption with Projections to Year 2000

FAO Regions and Subregions	Actual		Projections		
	1975	1986	1990	1995	2000
World	1,283,803	1,680,424	1,777,707	1,915,871	2,053,364
Developed economies	70,530	160,885	164,253	168,000	171,675
North America	22,911	108,192	112,071	115,233	118,045
West Europe	37,996	41,687	40,399	39,885	39,589
Oceania	1,618	2,930	2,785	2,855	2,912
Others	8,006	8,077	8,999	10,027	11,129
Developing Economies	950,043	1,209,083	1,299,678	1,423,809	1,549,038
Africa	262,598	367,926	409,009	470,165	540,656
Central America and Caribbean	34,666	47,400	50,771	55,328	59,833
South America	172,156	221,939	228,923	238,322	241,794
Near East—Africa	15,118	20,700	22,787	25,725	28,891
Near East—Asia	41,136	19,133	24,497	26,556	28,385
Far East	419,339	526,111	557,017	600,329	641,360
Oceania	5,031	5,874	6,674	7,384	8,119
Centrally planned economies	263,229	310,457	313,776	324,062	332,651
Asia	166,894	206,398	213,500	226,169	238,460
Europe and former USSR	96,335	104,059	100,276	97,893	94,191

Note: Quantities given in units of 1000 cm^3.

Fireplaces are inherently inefficient and are more esthetic than utilitarian. However, fireplaces are being used more efficiently with newer technology developments in the control of makeup air and hot air distribution and in the use of better designed insert units (stoves) for fireplace spaces. Stoves are being designed to use roundwood more efficiently with better control of air for combustion. Automated controls are being more widely used. Efficiencies are being improved and pollutant emissions reduced with the use of improved technologies such as catalytic converters. Like stoves, furnaces are benefiting from some of the same improvements.

A newer development is the combining of improved fuels with improved combustion units to attain more efficient and more automatic operation. Fuels may be made more effective and easier to handle by control of size and moisture content. Examples are dried chips and pellets. A new product is chunkwood, which comes in larger size particles and may be more efficient to produce, dry, handle, and store. Chunkwood particles are about the size of a man's fist. More sophisticated stoves and furnaces have been designed to take advantage of further improved fuels such as pellets. Sawdust burners have been used for residential, institutional, and some commercial applications since World War II and continue to be used in lumber-producing areas of the country.

In industrial applications, older technologies such as the Dutch oven and traveling grate are still operating satisfactorily, but new technologies including the fluidized bed and gasification are providing advantages in combustion and emission control. Promising developments for industry in the future are a gravel bed combustor, new technology for gas, liquid, and char fuels, and burning wood in combination with coal.

In developing countries there is much potential for using simple, low-cost stoves to significantly improve the efficiency of wood and charcoal fuel use for cooking. However, there is much inertia to overcome in promoting stoves over traditional cooking habits. Traditionally cooking is done indoors around an open fire confined by three strategically placed stones. This is usually done after nightfall, since darkness occurs around 6:00 p.m. year-round in the tropics. Thus the fire provides some light as well as warming heat in addition to heat for cooking. It also generates smoke, which is unhealthful but is sometimes desired to drive off insects that might otherwise be attracted to the thatch in the roofs of the dwellings.

NEW GAS TURBINE TECHNOLOGY

Advanced industrial and utility power systems often use gas or liquid-fueled gas turbine engines. Burning fuels directly in a turbine, without going through an intermediate heat exchanger to heat air for use in the turbine, is an efficient means of generating electricity in combination with process heat. Using coal or wood to directly power a gas turbine has yet to be accomplished commercially, primarily because the ash can cause erosion, deposition, and corrosion of the turbine blades. The size, distribution, concentration, and composition of the ash, as well as the turbine design, determine the lifetime of the turbine blades. New direct combustion turbines for wood to alleviate these problems are under development.

NEW GASIFICATION, LIQUEFACTION, AND CARBONIZATION TECHNOLOGIES

If wood is to become a viable replacement for oil in industrial countries as oil becomes scarcer and more expensive, wood needs to be applied in ways other than as a boiler fuel and residential space heating fuel. Wood may be converted to liquid and gaseous fuels and to improved forms of solid fuel such as charcoal. Technology is available to make ethanol from wood at a cost comparable to making ethanol from corn, but in today's market, this technology is only economical with a large subsidy.

Provision of gaseous fuel from wood can be achieved with known technology, but the cost of gas derived from wood is much higher than the cost of natural gas. Charcoal from wood is not competitive with coal in many applications.

Gasification and pyrolysis research may lead to more economical liquid fuels from wood such as methanol, pyrolysis oils, or conventional gasoline. For the near term, development of a viable methanol from wood process is realistic to expect. Other potential products are gas for operation of internal combustion engines, turbines, and lime kilns and pyrolysis oils for diesel fuel.

LIQUID FUELS FROM WOOD

There are several approaches to produce liquid fuels from wood such as gasification, pyrolysis, and hydrolysis followed by fermentation. The fuels most often considered are ethanol and methanol because they can be used as motor fuels.

Ethanol can be produced from the carbohydrate in wood by acid or enzyme hydrolysis to form monosaccharides that ferment to ethanol. Through acid hydrolysis, about 20% of the energy potential of wood can be recovered in ethanol. While applicable to a wide variety of forest residues, acid hydrolysis destroys all fiber properties and creates significant waste products that must be disposed. The kinetics of acid hydrolysis are well understood, and sufficient data exist to design pilot process equipment. Higher efficiencies of energy recovery can be realized through pyrolysis and methanol production. Neither acid hydrolysis for ethanol production nor wood pyrolysis for methanol production are currently economical. Further research is needed to make liquid fuels from biomass more competitive with gasoline and diesel fuels from petroleum.

Enzyme hydrolysis of wood has been promoted as a process that will result in higher ethanol yields with smaller scale process efficiencies. Economic studies of *cellulose* hydrolysis have not shown this to be competitive in the wood-processing industry. However, enzymatic hydrolysis and fermentation of *hemicellulose* or *recycled fibers* may hold better prospects. In such a process, ethanol would be produced from the readily hydrolyzed hemicellulosic fraction while the cellulosic and lignin fractions of wood would be recovered for their fiber and polymeric properties. For enzyme hydrolysis, additional research is needed to develop enzyme production methods and demonstrate the hydrolysis process on a pilot scale.

For methanol production the next step is the planned demonstration of a large-scale wood gasifier for the production of synthesis gas that can be converted to methanol.

An intensive effort is underway to evaluate a broad range of plant and seed oils and even waste cooking fats as raw materials for conversion to diesel fuel substitutes and blending components. Also under investigation are tall oils from wood and the sap from the *Copaifera* tree of the Amazon rain forest.

BURNING WOOD IN COMBINATION WITH COAL

Federal regulations stipulate that for coal boilers with capacities of 100×10^6 Btu/h or more, the emission limit is either 0.05 lb/10^6 Btu heat input or 0.1 lb/10^6 Btu heat input depending on whether and in what amount other fuels are cofired with coal. Emissions of sulfur dioxide and oxides of nitrogen from combustion of coal and wood are based on total heat input, no matter what the fraction of wood used. These regulations provide a positive advantage for burning wood in combination with coal in large boilers, particularly in the case of high-sulfur coals.

Cofiring coal with biomass can also reduce emissions of CO_2, the main greenhouse gas, to the atmosphere. It has been estimated that each coal-fired power generating plant in the United States could be retrofit to provide an average of about 10 MW of biomass cofiring. This could produce over 17,000 MW of biomass capacity in existing coal-fired utility boilers. Therefore, over 17,000 MW of biomass capacity could be introduced in existing coal-fired utility boilers, displacing 80×10^6 tons of CO_2 to supply 20% of the industry's requested reduction to meet goals of not exceeding 1990 emission levels in the year 2000.

POTENTIAL FOR INCREASED USE OF WOOD FOR ENERGY

Increased use of wood for energy is highly dependent on favorable economics, economics that may be enhanced by technological improvements in harvesting, conversion, and use. Today about 3.7% or 2.7 quads of our energy comes from woody biomass. Our production of energy from woody biomass is comparable to our use of hydropower and nuclear power. This could be increased to about 10 quads or about 13.5% of our current usage (3). Much wood for energy could come from material that is not now used but that would be satisfactory for use as fuel. Such material includes forest residues, plant manufacturing residues, land-clearing residues, urban wood residues, wood from insect-, disease-, and fire-killed trees, and excess of wood growth in comparison to annual cut.

Of the estimated 18×10^9 ft^3 of roundwood timber harvested in the United States in 1986, 3.1×10^9 ft^3, the equivalent of 0.9 quads, was used for fuelwood (U.S. Department of Agriculture, Forest Service, 1988). Of this only 26% came from material classified as growing stock, i.e., live sawtimber trees, poletimber trees, saplings, and seedlings meeting specified standards of quality or vigor and not including cull trees.

Much of municipal solid waste consists of paper, demolition waste, tree trimmings, and other forms of biomass. If more of this material were used for fuel, it could mean a significant savings in constantly increasing costs for landfills. By the same token, in public and private forests under multiple-use management for timber production and other purposes, much of the management costs result from cleanup after logging operations. Often brush from logging operations is concentrated and broadcast-burned to prepare land for new tree growth. This consumes management funds and subjects the atmosphere to more particulate loading as well. Now, however, in some parts of the country broadcast burning is avoided through cleanup credits for harvesting excess wood for energy. This means that wood is burned under controlled conditions, instead of being burned in the open, and emissions to the environment are therefore much reduced. In other instances, dense brush in forests at urban–forest interface areas is being successfully harvested for energy, thereby providing a significantly decreased fire hazard to houses at the forest perimeter.

BIBLIOGRAPHY

1. The Audubon Society and Princeton University, "Toward Ecological Guidelines for Large-Scale Biomass Energy Development," *Proceedings of the Audobon / Princeton Biomass Workshop*, New York, 1991.
2. Food and Agriculture Organization (FAO) of the United Nations, "Forest Products: World Outlook Projections," FAO Forestry Paper 84, Rome, 1988.
3. J. I. Zerbe and K. E. Skog, "Sources and Uses of Wood for Energy in Energy Options for the Year 2000," in J. Heinonen et al., eds., *Contemporary Concepts in Technology and Policy*, volume 1, University of Delaware Press, Newar, 1988.
4. J. I. Zerbe, "Biofuels: Production and Potential," *For. Appl. Res. Public Policy*, Winter 1988.
5. U. S. Department of Energy, *Estimates of U.S. Wood Energy Consumption from 1949 to 1981*, U.S. Government Printing Office, Washington, D.C., 1982.
6. J. W. Koning and K. E. Skog, "Use of Wood for Energy in the United States: A Threat or Challenge?" in D. L. Klass, ed., *Energy from Biomass and Wastes*, Institute of Gas Technology, Chicago, 1987.

Z

ZERO EMISSIONS

See Air pollution: automobile; Air pollution control methods.

INDEX